MATHEMATICAL ELASTICITY

VOLUME III: THEORY OF SHELLS

STUDIES IN MATHEMATICS AND ITS APPLICATIONS

VOLUME 29

ELSEVIER
AMSTERDAM – LAUSANNE – NEW YORK – OXFORD – SHANNON – SINGAPORE – TOKYO

MATHEMATICAL ELASTICITY

VOLUME III: THEORY OF SHELLS

PHILIPPE G. CIARLET

Institut Universitaire de France
Université Pierre et Marie Curie,
Paris, France

With 31 figures by the author

2000

ELSEVIER

AMSTERDAM – LAUSANNE – NEW YORK – OXFORD – SHANNON – SINGAPORE – TOKYO

ELSEVIER SCIENCE B.V.
Sara Burgerhartstraat 25
P.O. Box 211, 1000 AE Amsterdam, The Netherlands

First edition 2000

Library of Congress Cataloging in Publication Data
A catalog record from the Library of Congress has been applied for.

ISBN: 0 444 82891 5

∞ The paper used in this publication meets the requirements of ANSI/NISO Z39.48-1992 (Permanence of Paper).
Printed in The Netherlands.

Shell theory attempts the impossible: to provide a two-dimensional representation of an intrinsically three-dimensional phenomenon.

Warner KOITER and James SIMMONDS.

Foundations of shell theory, in *Proceedings of the Thirteenth International Congress of Theoretical and Applied Mechanics* (Moscow, 1972).

MATHEMATICAL ELASTICITY: GENERAL PLAN

VOLUME I: THREE-DIMENSIONAL ELASTICITY

Part A. Description of three-dimensional elasticity

Chapter 1. Geometrical and other preliminaries
Chapter 2. The equations of equilibrium and the principle of virtual work
Chapter 3. Elastic materials and their constitutive equations
Chapter 4. Hyperelasticity
Chapter 5. The boundary value problems of three-dimensional elasticity

Part B. Mathematical methods in three-dimensional elasticity

Chapter 6. Existence theory based on the implicit function theorem
Chapter 7. Existence theory based on the minimization of the energy

VOLUME II: THEORY OF PLATES

Part A. Linear plate theory

Chapter 1. Linearly elastic plates
Chapter 2. Junctions in linearly elastic multi-structures
Chapter 3. Linearly elastic shallow shells in Cartesian coordinates

Part B. Nonlinear plate theory

Chapter 4. Nonlinearly elastic plates
Chapter 5. The von Kármán equations

VOLUME III: THEORY OF SHELLS

Part A. Linear shell theory

Chapter 1. Three-dimensional linearized elasticity and Korn's inequalities in curvilinear coordinates

Chapter 2. Inequalities of Korn's type on surfaces

Chapter 3. Asymptotic analysis of linearly elastic shells: Preliminaries and outline

Chapter 4. Linearly elastic elliptic membrane shells

Chapter 5. Linearly elastic generalized membrane shells

Chapter 6. Linearly elastic flexural shells

Chapter 7. Koiter's equations and other linear shell theories

Part B. Nonlinear shell theory

Chapter 8. Asymptotic analysis of nonlinearly elastic shells: Preliminaries

Chapter 9. Nonlinearly elastic membrane shells

Chapter 10. Nonlinearly elastic flexural shells

Chapter 11. Koiter's equations and other nonlinear shell theories

MATHEMATICAL ELASTICITY: GENERAL PREFACE[1]

This treatise, which comprises three volumes, is intended to be both a *thorough introduction to contemporary research in elasticity* and a *working textbook at the graduate level for courses in pure or applied mathematics or in continuum mechanics.*

During the past decades, elasticity has become the object of a considerable renewed interest, both in its physical foundations and in its mathematical theory. One reason behind this recent attention is that it has been increasingly acknowledged that the classical *linear equations* of elasticity, whose mathematical theory is now firmly established, have a limited range of applicability, outside of which they should be replaced by the genuine *nonlinear equations* that they in effect approximate.

Another reason, similar in its principle, is that the validity of the classical *lower-dimensional equations*, such as the two-dimensional von Kármán equations for nonlinearly elastic plates or the two-dimensional Koiter equations for linearly elastic shells, is no longer left unquestioned. A need has been felt for a better assessment of their relation to the corresponding three-dimensional equations that they are supposed to "replace".

Thanks to the ever-increasing power of available computers, sophisticated mathematical models that were previously intractable by *approximate methods* are now amenable to numerical simulations. This is one more reason why these models should be established on firm grounds.

This treatise illustrates at length these recent trends, as shown by the *main topics covered*:

- A thorough *description*, with a pervading emphasis on the *nonlinear aspects*, of the two existing mathematical models of *three-dimensional elasticity*, either as a *boundary value problem* consisting of a system of *three quasilinear partial differential equations* of

[1] This "General preface" is an updated excerpt from the "Preface" to the first edition (1988) of Volume I.

the *second order* together with specific *boundary conditions*, or as a *minimization problem* for the associated *energy* over an *ad hoc* set of *admissible deformations* (Vol. I, Part A);

- A *mathematical analysis* of these models, comprising in particular *complete proofs of all the available existence results*, relying either on the *implicit function theorem*, or on the *direct methods of the calculus of variations* (Vol. I, Part B);

- A mathematical justification of the well-known two-dimensional *linear Kirchhoff-Love theory of plates*, by means of *convergence theorems* as the thickness of the plate approaches zero (Vol. II, Part A);

- Similar justifications of mathematical models of *junctions in linearly elastic multi-structures* and of *linearly elastic shallow shells* (Vol. II, Part A);

- A systematic *derivation of two-dimensional plate models from nonlinear three-dimensional elasticity* by means of the *method of formal asymptotic expansions*, which includes a justification of well-known plate models, such as the *nonlinear Kirchhoff-Love theory* and the *von Kármán equations* (Vol. II, Part B);

- A derivation of the *large deformation, frame-indifferent, nonlinear planar membrane and flexural theories* by means of the method of formal asymptotic expansions and a justification of *nonlinear planar membrane equations* by means of a *convergence theorem* (Vol. II, Part B);

- A *mathematical analysis of the two-dimensional, linear and nonlinear, plate equations*, which includes in particular a review of the *existence* and *regularity theorems* in the nonlinear case and an introduction to *bifurcation theory* (Vol. II, Parts A and B);

- A mathematical justification by means of *convergence theorems* of the *two-dimensional membrane, flexural,* and *Koiter equations of a linearly elastic shell* (Vol. III, Part A);

- A systematic derivation of the *two-dimensional membrane* and *flexural equations of a nonlinearly elastic shell* by means of the *method of formal asymptotic expansions* and a justification of *nonlinear membrane shell equations* by means of a *convergence theorem* (Vol. III, Part B).

- A *mathematical analysis of the two-dimensional, linear and nonlinear, shell equations,* with a particular emphasis on the *existence* theory (Vol. III, Parts A and B).

Although the emphasis is definitely on the mathematical side, every effort has been made to keep the prerequisites, whether from mathematics or continuum mechanics, to a minimum, notably by making this treatise as largely *self-contained* as possible. Its reading only presupposes some familiarity with basic topics from analysis and functional analysis.

Naturally, frequent references are made to Vol. I in Vol. II, and to Vols. I and II in Vol. III. However, I have also tried to render *each volume as self-contained as possible.* In particular, all relevant notions from three-dimensional elasticity are (at least briefly) recalled wherever they are needed in Vols. II and III.

References are also made to Vol. I regarding various *mathematical notions* (properties of domains in $\mathbb{R}^n$, differential calculus in normed vector spaces, Sobolev spaces, weak lower semi-continuity, etc.). This is a mere convenience, reflecting that I also regard the three volumes as forming a coherent whole. I am otherwise well aware that Vol. I is neither a text on analysis nor one on functional analysis. Any reader interested in a deeper understanding of such notions should consult the more standard texts referred to in Vol. I.

Each volume is divided into consecutively numbered *chapters.* Chapter m contains an introduction, several sections numbered Sect. $m.1$, Sect. $m.2$, etc., and is concluded by a set of *exercises.* Within Sect. $m.n$, *theorems* are consecutively numbered, as Thm. $m.n$-1, Thm. $m.n$-2, etc., and *figures* are likewise consecutively numbered, as Fig. $m.n$-1, Fig. $m.n$-2, etc. *Remarks* and *formulas* are not numbered. The end of the proof of a theorem, or the end of a remark, is indicated by the symbol ■ in the right margin. In Chapter m, exercises are numbered as Ex. $m.1$, Ex. $m.2$, etc.

All the important results are stated in the form of *theorems* (there are no lemmas, propositions, or corollaries), which therefore represent the core of the text. At the other extreme, the *remarks* are intended to point out some interpretations, extensions, counter-examples, relations with other results, that in principle can be skipped during a first reading; yet, they could be helpful for a better understanding of the material. When a term is defined, it is set in boldface if it is deemed important, or in italics otherwise. Terms that are only given a loose or intuitive meaning are put between quotation marks.

Special attention has been given to the *notation,* which so often has a distractive and depressing effect in a first encounter with elasticity. In particular, each volume begins with special sections, which the reader is urged to consult first, about the notations and the rules that have guided their choice. The same sections also review the main *definitions* and *formulas* that will be used throughout the text.

Complete proofs are generally given. In particular, whenever a mathematical result is of particular significance in elasticity, its proof has been included. More standard mathematical prerequisites are presented (usually without proofs) in special *starred sections*, scattered according to the local needs. The proofs of some *advanced*, or *more specialized, topics*, are sometimes only sketched, in order to keep the length of each volume within reasonable limits; in this case, *ad hoc* references are always provided. These topics are assembled in special sections marked with the symbol $^\flat$, usually at the end of a chapter.

Exercises of varying difficulty are included at the end of each chapter. Some are straightforward applications of, or complements to, the text; others, which are more challenging, are usually provided with hints or references.

This treatise would have never seen the light had I not had the good fortune of having met, and worked with, many exceptional students and colleagues, who helped me over the past three decades decipher the arcane subtleties of mathematical elasticity; their names are listed in the preface to each volume. To all of them, my heartfelt thanks!

I am also particularly indebted to Arjen Sevenster, whose constant interest and understanding were an invaluable help in this seemingly endless enterprise!

Last but not least, this treatise is dedicated to Jacques-Louis Lions, as an expression of my deep appreciation and gratitude.

August, 1986 and October, 1999 Philippe G. Ciarlet

PREFACE TO VOLUME I[1]

A fascinating aspect of three-dimensional elasticity is that, in the course of its study, one *naturally* feels the need for studying *basic mathematical techniques* of *matrix theory, analysis*, and *functional analysis*; how could one find a better motivation? For instance:

- Both common and uncommon results from *matrix theory* are often needed, such as the polar factorization theorem (Thm. 3.2-2), or the celebrated Rivlin-Ericksen representation theorem (Thm. 3.6-1). In the same spirit, who would think that the inequality $|\operatorname{tr} \mathbf{AB}| \leq \sum_i v_i(\mathbf{A}) v_i(\mathbf{B})$, where $v_i(\mathbf{A})$ and $v_i(\mathbf{B})$ denote the singular values, arranged in increasing order, of the matrices $\mathbf{A}$ and $\mathbf{B}$, arises naturally in the analysis of a wide class of actual stored energy functions? Incidentally, this seemingly innocuous inequality is not easy to prove (Thm. 3.2-4)!

- The understanding of the "geometry of deformations" relies on a perhaps elementary, but "applicable", knowledge of *differential geometry.* For instance, my experience is that, among those of my students who had been previously exposed to modern differential geometry, very few could effectively produce the formula $da^{\varphi} = |\mathbf{Cof}\, \boldsymbol{\nabla}\boldsymbol{\varphi}\, \boldsymbol{n}| da$ relating reference and deformed area elements (Thm. 1.7-1).

- The study of geometrical properties (orientation-preserving character, injectivity) of mappings in $\mathbb{R}^3$ naturally leads to using such basic tools as the *invariance of domain theorem* (Thms. 1.2-5 and 1.2-6) or the *topological degree* (Sect. 5.4); yet these are unfortunately all too often left out from standard analysis courses.

- *Differential calculus in Banach* spaces is an indispensable tool which is used throughout this volume, and the unaccustomed reader should quickly become convinced of the many merits of the *Fréchet derivative* and of the *implicit function theorem*, which are the keystones to the existence theory developed in Chap. 6.

- The fundamental *Cauchy-Lipschitz existence theorem for ordinary differential equations in Banach spaces*, as well as the conver-

[1]This "Preface to Volume I" is an updated excerpt from the "Preface" to the first edition (1988) of Volume I.

gence of its *approximation by Euler's method*, are needed in the analysis of incremental methods, often used in the numerical approximation of the equations for nonlinearly elastic structures (Chap. 6).

- Basic topics from *functional analysis* and the *calculus of variations*, such as *Sobolev spaces* (which in elasticity are simply the "spaces of finite energy"), *weak convergence, existence of minimizers for weakly lower semi-continuous functionals*, pervade the treatment of existence results in three-dimensional elasticity (Chaps. 6 and 7).

- Key results about *elliptic linear systems of partial differential equations*, notably sufficient conditions for the $W^{2,p}(\Omega)$-*regularity* of their solutions (Thm. 6.3-6), are needed preliminaries for the existence theory of Chap. 6.

- As a result of John Ball's seminal work in three-dimensional elasticity, *convexity* and the subtler *polyconvexity* play a particularly important rôle throughout this volume. In particular, we shall naturally be led to finding nontrivial examples of *convex hulls*, such as that of the set of all square matrices whose determinant is > 0 (Thm. 4.7-4), and of *convex functions of matrices.* For instance, functions such as $\boldsymbol{F} \to \sum_i \{\lambda_i(\boldsymbol{F}^{\mathrm{T}}\boldsymbol{F})\}^{\alpha/2}$ with $\alpha \geq 1$ naturally arise in the study of *Ogden's materials* in Chap. 4; while proving that such functions are convex is elementary for $\alpha = 2$, it becomes surprisingly difficult for the other values of $\alpha \geq 1$ (Sect. 4.9). Such functions are examples of John Ball's *polyconvex stored energy functions*, a concept of major importance in elasticity (Chaps. 4 and 7).

- In Chap. 7, we shall come across the notion of *compensated compactness.* This technique, discovered and studied by François Murat and Luc Tartar, is now recognized as a powerful tool for studying nonlinear partial differential equations.

Another fascinating aspect of three-dimensional elasticity is that it gives rise to a number of *open problems*, for instance:

- The extension of the "local" analysis of Chap. 6 (existence theory, continuation of the solution as the forces increase, analysis of incremental methods) to genuine mixed displacement-traction problems;

- "Filling the gap" between the existence results based on the implicit function theorem (Chap. 6) and the existence results based on the minimization of the energy (Chap. 7);

- An analysis of the nonuniqueness of solutions (cf. the examples given in Sect. 5.8);

- A mathematical analysis of contact with friction (contact, or self-contact, without friction is studied in Chaps. 5 and 7);

- Finding reasonable conditions under which the minimizers of the energy (Chap. 7) are solutions of the associated Euler-Lagrange equations;

- While substantial progress has been made in the *study of statics* (which is all that I consider here), the analysis of time-dependent elasticity is still at an early stage. Deep results have been recently obtained for one space variable, but formidable difficulties stand in the way of further progress in this area.

This volume will have fulfilled its purposes if the above messages have been conveyed to its readers, that is,

- if it has convinced its more application-minded readers, such as continuum mechanicists, engineers, "applied" mathematicians, that mathematical analysis is an indispensable tool for a genuine understanding of three-dimensional elasticity, whether it be for its modeling or for its analysis, essentially because more and more emphasis is put on the nonlinearities (e.g., injectivity of deformations, polyconvexity, nonuniqueness of solutions, etc.), whose consideration requires, even at the onset, some degree of mathematical sophistication;

- if it has convinced its more mathematically oriented readers that three-dimensional elasticity, far from being a dusty classical field, is on the contrary a prodigious source of challenging open problems.

Although more than 570 items are listed in the bibliography, there has been no attempt to compile an exhaustive list of references. The interested readers should look at the extensive bibliography covering the years 1678-1965 in the treatise of Truesdell & Noll [1965], at the additional references found in the books by Marsden & Hughes [1983], Hanyga [1985], Oden [1986], and especially Antman [1995], and in the papers of Antman [1983] and Truesdell [1983], which give short and illuminating historical perspectives on the interplay between elasticity and analysis.

The readers of this volume are strongly advised to complement the material given here by consulting a few other books, and in this respect, I particularly recommend the following *general references on three-dimensional elasticity* (general references on lower-dimensional theories of plates, shells and rods are given in Vols. II and III):

- In-depth perspectives in continuum mechanics in general, and in elasticity in particular: The treatises of Truesdell & Toupin [1960] and Truesdell & Noll [1965], and the books by Germain [1972], Gurtin [1981b], Eringen [1962], and Truesdell [1991].

- Classical and modern expositions of elasticity: Love [1927], Murnaghan [1951], Timoshenko [1951], Novozhilov [1953], Sokolnikoff [1956], Novozhilov [1961], Landau & Lifchitz [1967], Green & Zerna [1968], Stoker [1968], Green & Adkins [1970], Knops & Payne [1971], Duvaut & Lions [1972], Fichera [1972a, 1972b], Gurtin [1972], Wang & Truesdell [1973], Villaggio [1977], Gurtin [1981a], Nečas & Hlaváček [1981], and Ogden [1984].

- Mathematically oriented treatments in nonlinear elasticity: Marsden & Hughes [1983], Hanyga [1985], Oden [1986], and the landmark book of Antman [1995].

- The comprehensive survey of numerical methods in nonlinear three-dimensional elasticity of Le Tallec [1994].

In my description of continuum mechanics and elasticity, I have only singled out *two axioms*: The stress principle of Euler and Cauchy (Sect. 2.2) and the axiom of material frame-indifference (Sect. 3.3), thus considering that all the other notions are *a priori* given. The reader interested in a more axiomatic treatment of the basic concepts, such as frame of reference, body, reference configuration, mass, forces, material frame-indifference, isotropy, should consult the treatise of Truesdell & Noll [1965], the books of Wang & Truesdell [1973] and Truesdell [1991], and the fundamental contributions of Noll [1959, 1966, 1972, 1973, 1978].

At the risk of raising the eyebrows of some of my readers, and at the expense of various *abus de langage*, I have also ignored in this volume the difference between second-order tensors and matrices. The readers disturbed by this approach should look at the books of Abraham, Marsden & Ratiu [1983] and, especially, of Marsden & Hughes [1983], where they will find all the tensorial and differential geometric aspects of elasticity explained in depth and put in their proper perspective. Likewise, Vol. III should be also helpful in this respect.

This volume is an outgrowth of lectures on elasticity that I have given over the past 15 years at the Tata Institute of Fundamental Research, the University of Stuttgart, the Ecole Normale Supérieure, and the Université Pierre et Marie Curie.

I am particularly indebted to the many students and colleagues I worked with on that subject during the same period; in particular: Michel Bernadou, Dominique Blanchard, Jean-Louis Davet, Philippe Destuynder, Giuseppe Geymonat, Hu Jian-wei, Srinivasan Kesavan, Klaus Kirchgässner, Florian Laurent, Hervé Le Dret, Jindřich Nečas, Robert Nzengwa, Jean-Claude Paumier, Peregrina Quintela-Estevez, Patrick Rabier, and Annie Raoult.

Special thanks are also due to Stuart Antman, Irene Fonseca, Morton Gurtin, Patrick Le Tallec, Bernadette Miara, François Murat, Tinsley Oden, and Gérard Tronel, who were kind enough to read early drafts of this volume and to suggest significant improvements.

For their especially expert and diligent assistance as regards the material realization of this volume, I very sincerely thank Hélène Bugler, Monique Damperat, and Liliane Ruprecht.

August, 1986 and October, 1999 Philippe G. Ciarlet

PREFACE TO VOLUME II [1]

Lower-dimensional plate, shell, and *rod, theories* that rely on *a priori* assumptions of a mechanical or geometrical nature have been proposed by A.-L. Cauchy, Sophie Germain, G. Kirchhoff, T. von Kármán, A.E.H. Love, E. Reissner, Jakob Bernoulli, C.-L.-M.-H. Navier, L. Euler, S.-D. Poisson, E. and F. Cosserat, L.H. Donnell, W. Flügge, S.P. Timoshenko, V.V. Novozhilov, I.N. Vekua, A.E. Green, W.T. Koiter, J.G. Simmonds, P.M. Naghdi, and others.

There are two reasons why these lower-dimensional theories are so often preferred to the three-dimensional theory that they are supposed to "replace" when the thickness, or the diameter of the cross-section, is "small enough".

One reason is their *simpler mathematical structure*, which in turn generates a richer variety of results. For instance, the existence, regularity, or bifurcation, theories, and more generally the "global analysis", are by now on firm mathematical grounds for nonlinearly elastic rods (see Antman [1995] for a scholarly and comprehensive exposition) or for nonlinearly elastic von Kármán plates (see Ciarlet & Rabier [1980]). By contrast, these theories of global analysis are still partly in their infancies for nonlinear three-dimensional elasticity (see Marsden & Hughes [1983] and Vol. I for comprehensive surveys): After the fundamental ideas set forth by Ball [1977], who was able to establish the existence of a minimizer of the energy for a wide class of realistic nonlinearly elastic materials, there indeed remain manifold challenging open problems; for instance, there is no known set of sufficient conditions guaranteeing that such a minimizer satisfies the equilibrium equations even in the weak sense of the principle of virtual work (another existence theory, based on the implicit function theorem, does not share this drawback, but it is restricted to problems with smooth data and to special boundary conditions, unrealistic in practice; see Vol. I and the comprehensive treatment of Valent [1988]).

[1] A substantial portion of this preface is an excerpt from the "Introduction" in Ciarlet & Lods [1996b].

Another virtue of lower-dimensional theories is their far better amenability to *numerical computations*. For instance, directly approximating the three-dimensional displacement field of a cooling tower seems out of reach at the present time, even in the linearly elastic realm: The existing codes use two-dimensional equations, such as those of W.T. Koiter; see Bernadou [1994] for a comprehensive account. Likewise, although substantial progress has recently been achieved for directly approximating the "three-dimensional" displacement field of a linearly elastic rectangular plate, current codes are almost invariably based on two-dimensional equations, such as those of the Kirchhoff-Love or Reissner-Mindlin theories, whose numerical approximation is by now on essentially safe theoretical grounds; see, e.g., Ciarlet [1978, 1991], Glowinski [1984], Hughes [1987], Robert & Thomas [1991], Brezzi & Fortin [1991], Brenner & Scott [1994], Destuynder & Salaun [1996].

Be that as it may, the *locking phenomenon*, and the proper handling of *boundary layers*, in the numerical approximation of two-dimensional plate or shell equations still pose challenging problems; for plates, see notably Arnold [1981], Bathe & Brezzi [1985], Brezzi & Fortin [1986], Hughes & Franca [1988], Pitkäranta [1988], Bathe, Brezzi & Fortin [1989], Arnold & Falk [1989, 1990, 1996], Brezzi, Fortin & Stenberg [1991], Arnold & Brezzi [1993], Lyly, Stenberg & Vihinen [1993], Chenais & Paumier [1994], Schwab [1994, 1996], Schwab & Suri [1994], Schwab & Wright [1995], Suri, Babuška & Schwab [1995], Pitkäranta & Suri [1996].

Lower-dimensional models being thus widely used, two essential, and in fact intimately related, questions arise:

Given a "lower-dimensional" elastic body, together with specific loadings and boundary conditions, how to choose between the manifold lower-dimensional models that are available? For instance, given a linearly elastic shell, which theory should be preferred, among those of Koiter, Naghdi, Novozhilov, Budiansky-Sanders, etc.? This question is of paramount *practical* importance, for *it makes no sense to devise accurate methods for approximating the solution of a "wrong" model*! Consequently, before approximating the exact solution of a given lower-dimensional model, we should first know whether it is "close enough" to the exact solution of the three-dimensional model it is intended to approximate. This observation leads to the second question:

How to mathematically justify in a rational fashion a lower-dimensional model from the three-dimensional model? This question has been answered through three different approaches (only scant references are given here to these approaches, as many additional ones are provided throughout the text).

The *first approach* consists in *directly estimating the difference between the three-dimensional solution and the solution of a given, i.e., "known in advance", lower-dimensional model* (this difference makes sense once the three-dimensional solution is properly averaged or the lower-dimensional one is extended in some fashion to a three-dimensional field). For *linearly elastic plates*, the first such estimate seems to be due to Morgenstern [1959], who cleverly used the Hellinger-Reissner variational principle of the linear theory; see also Morgenstern & Szabó [1961], Goldenveizer [1969], Nordgren [1971, 1972], Simmonds [1971a], Ladevèze [1976, 1980], Shoiket [1976], and Kohn & Vogelius [1985]. This approach was likewise successfully applied to *linearly elastic shells* by Koiter [1970], Simmonds [1971b], and Koiter & Simmonds [1973].

The *second approach*, essentially due to Naghdi [1972] for *plates* and *shells*, consists in using a *hierarchic method*, whose governing principle is an *a priori* assumption that *the admissible displacement fields are restricted to a specific form.* For a plate (to fix ideas), such "test functions" are finite sums of products of unspecified functions of the in-plane variables times given linearly independent functions of the "transverse" variable. The functions of the in-plane variables are then determined by inserting these test functions into the three-dimensional equations or into the three-dimensional energy, a process that leads to the solution of a finite number of two-dimensional boundary value problems. Increasing the number of linearly independent functions of the transverse variable thus yields a "hierarchy" of models, which may be deemed two-dimensional, as they are determined by solving two-dimensional problems.

References to this approach are numerous. For *plates*, see notably Naghdi [1972], Destuynder [1980, Chap. 5], Miara [1989], Babuška & Li [1991, 1992], Schwab [1994, 1995, 1996], Alessandrini, Arnold, Falk & Madureira [1999], and Madureira [1999]; for *rods*, see Antman [1972], Miara & Trabucho [1992], Mascarenhas & Trabucho [1992], Figueiredo & Trabucho [1993], and Antman [1995]; for *shallow shells*, see Figueiredo & Trabucho [1992]; for a general analysis, see Antman [1976] and Antman & Marlow [1991]. See also the related *"con-*

straint method", advocated by Podio-Guidugli [1989, 1990] for modeling *plates* and *shells.*

These two approaches nevertheless rely on some *a priori* assumptions of a mechanical or geometrical nature, intended to account for the "smallness" of a geometrical parameter and intended to be more effective as this parameter approaches zero. Hence the need arises to *mathematically justify these* a priori *assumptions, together with the lower-dimensional theories they engender, directly from three-dimensional elasticity.* Otherwise, these assumptions and theories can be thought of as being "handed down by some higher power (a Hungarian wizard, say)", to quote Truesdell [1978].

This direct justification is achieved by the *third approach*, which consists in applying an *asymptotic method.* It has recently received considerable attention, as exemplified by the books of Destuynder [1986] and Ciarlet [1990] for plates; Le Dret [1991] for plates and beams (straight rods); Trabucho & Viaño [1996] for beams; and Ciarlet [1990], Le Dret [1991], Kozlov, Maz'ya & Movchan [1999] for multi-structures.

In a *formal asymptotic method*, the three-dimensional solution (the displacement field and, in some cases, the stress field) is first *"scaled"* in an appropriate manner so as to be defined on a *fixed* domain, then expanded as a *formal series expansion* in terms of a *"small" parameter* ε, which is the "dimensionless" half-thickness of a plate or a shell, or the "dimensionless" diameter of the cross-section of the rod. "Dimensionless" means that ε measures the *ratio* between the thickness or diameter and some "characteristic" dimension. For a cooling tower, for instance, where common values for the average thickness and height are 0.3m and 150m, the ratio 2ε is thus equal to $1/500$. It is worthwhile to keep in mind this order of magnitude.

The formal series expansion of the scaled three-dimensional solution is then inserted into the three-dimensional problem, and sufficiently many factors of the successive powers of ε found in this fashion are equated to zero until the leading term of the expansion can be computed and, presumably, identified with the scaled solution of a known lower-dimensional problem. Such a method is "formal" in that the series is *not* expected to converge (as an infinite series in powers of ε); in fact, the successive terms of the expansion, except the leading one, cannot usually satisfy the boundary conditions of the three-dimensional problem! This situation is typical of such *singular perturbation problems*; see in this respect the comprehensive treatments given in Lions [1973] and Eckhaus [1979].

The fundamental contributions of Friedrichs & Dressler [1961] and Goldenveizer [1962, 1964] for plates, Rigolot [1972, 1976] for rods, Goldenveizer [1963, 1964] for shells, are among the first successful attempts to apply formal asymptotic methods in linearized elasticity. Some restrictions or *a priori* assumptions were, however, still needed. Another shortcoming is the *lack of convergence theorems* of the scaled three-dimensional solution to the leading term of its formal expansion as $\varepsilon \to 0$, essentially because the asymptotic method is applied in these works to the *partial differential equations* of the three-dimensional problem; in this case, convergence results usually rely on a *maximum principle* (see Eckhaus [1979]), which does *not* hold for the system of linearized three-dimensional elasticity.

Ciarlet & Destuynder [1979a, 1979b] applied instead the formal asymptotic method to the *variational*, or *weak*, formulation of the three-dimensional boundary value problems of *linearly* and *nonlinearly elastic plates.* Without making any *a priori* assumption of a mechanical or geometrical nature, they justified in this fashion the linear and nonlinear *Kirchhoff-Love plate theories* (only the magnitudes of the components of the applied loads and of the Lamé constants must behave as appropriate powers of the thickness, but, as shown in a systematic way by Miara [1994a, 1994b], such asymptotic behaviors are unavoidable).

This approach was extended to *von Kármán plates* by Ciarlet [1980], to *Marguerre-von Kármán shallow shells* by Ciarlet & Paumier [1986] and Busse [1997], to *general nonlinear constitutive equations* by Davet [1986], to nonlinearly elastic plates with varying thickness by Quintela-Estevez [1989], and to *nonlinear elastodynamics* by Raoult [1988] and Karwowski [1993].

By allowing a larger class of behaviors on the applied loads, Fox, Raoult & Simo [1993] were also able to justify in this fashion *two-dimensional nonlinear "planar membrane"* and *"planar flexural" theories* that are valid for "large" deformations and frame-indifferent, in that they share the same invariances as the three-dimensional theory (while Miara [1994b] assumed at the outset that the nonlinear two-dimensional models found by the formal asymptotic method have to reduce to the classical ones once linearized, this assumption was not made by Fox, Raoult & Simo [1993], who were thus able to consider other classes of behaviors).

The one-dimensional equations of a *nonlinearly elastic beam* (a beam is a straight rod) were likewise justified by Cimetière, Geymonat, Le Dret, Raoult & Tutek [1988] and Karwowski [1990]. *Nonlinear beam theory* has also been related to the three-dimensional theory by Mielke [1988, 1990], who justified *St Venant's principle* by a remarkable use of the center manifold theorem. Various classes of lower-dimensional equations modeling *"shallow" or "arbitrarily curved" nonlinearly elastic rods* have been similarly identified by Karwowski [1996a, 1996b].

The most noticeable virtue of the asymptotic method applied to the *weak* formulation of *linear* elasticity problems is its amenability to a rigorous *asymptotic analysis*, which shows that the three-dimensional scaled solution converges in some Hilbert spaces (H^1 or L^2) to the leading term of the formal asymptotic expansion as the "small" parameter approaches zero. Such *convergence theorems* have been established by Destuynder [1980, 1981], Caillerie [1980], Ciarlet & Kesavan [1981], Kohn & Vogelius [1984, 1985, 1986], Raoult [1985], Blanchard & Francfort [1987], Paumier [1991], Cioranescu & Saint Jean Paulin [1995, 1999], Destuynder & Gruais [1995], Aganović, Marušič-Paloka & Tutek [1995], Dauge & Gruais [1996, 1998a, 1998b], Paumier & Raoult [1997], Aganović, Jurak, Marušić-Paloka & Tutek [1998], Dauge, Djurdjevic & Rössle [1998a, 1998b], Andreoiu [1999b], Dauge, Djurdjevic, Faou & Rössle [1999], Dauge, Gruais & Rössle [1999], Djurdjevic [1999], Rössle [1999a, 1999b], and Rössle, Bischoff, Wendland & Ramm [1999] *for linearly elastic plates*; Ciarlet & Miara [1992a] and Busse, Ciarlet & Miara [1996] for *linearly elastic shallow shells*; Bermudez & Viaño [1984], Aganovič & Tutek [1986], Geymonat, Krasucki & Marigo [1987], Trabucho & Viaño [1987], Raoult [1988], Veiga [1995], and Le Dret [1995] for *linearly elastic beams* (see also the comprehensive survey of Trabucho & Viaño [1996] and the works cited therein). Special mention must also be made of the approach of Mielke [1995], who keeps the thickness fixed, but lets the lateral boundary of the plate "go away to infinity".

In these works, the *proofs* essentially rely on the ideas and methods described and developed in Lions [1973] for analyzing "abstract" linear variational problems that contain a small parameter.

Convergence theorems can also be obtained from Γ-*convergence* theory, as in Bourquin, Ciarlet, Geymonat & Raoult [1992] and Anzellotti, Baldo & Percivale [1994] for *linearly elastic plates.* A remarkable feature of Γ-convergence theory is that it also led to the *first* convergence result for *planar nonlinearly elastic bodies*, due to Le Dret & Raoult [1995a] who themselves based their approach on that of Acerbi, Buttazzo & Percivale [1991] for *strings*.

After the earlier formal attempts of A.L. Goldenveizer cited *supra*, a first major step for *linearly elastic shells* was achieved by Destuynder [1980] in his doctoral dissertation (see also Destuynder [1985]), where a convergence theorem for membrane shells was "almost proved"; another major step was achieved by Sanchez-Palencia [1990], who clearly delineated the kinds of geometries of the middle surface and boundary conditions that yield either two-dimensional *membrane*, or two-dimensional *flexural*, equations when the method of formal asymptotic expansions is applied to the variational equations of three-dimensional linearized elasticity (see also Caillerie & Sanchez-Palencia [1995b] and Miara & Sanchez-Palencia [1996]).

Then Ciarlet & Lods [1996b, 1996d] and Ciarlet, Lods & Miara [1996] carried out an *asymptotic analysis of linearly elastic shells that covers all possible cases*: Under three distinct sets of assumptions on the geometry of the middle surface, the boundary conditions, and the order of magnitude of the applied forces, they established *convergence theorems* in H^1, in L^2, or in *ad hoc* completion spaces, that justify either the linear two-dimensional equations of an *"elliptic membrane shell"*, or those of a *"generalized membrane shell"*, or those of a *"flexural shell"*.

Combining these convergences with results of Destuynder [1985] and Sanchez-Palencia [1989a, 1989b, 1992] (see also Sanchez-Hubert & Sanchez-Palencia [1997]), Ciarlet & Lods [1996c] have also justified the well-known two-dimensional *Koiter equations of a linearly elastic shell* (Koiter [1970]), again in all possible cases.

For *nonlinearly elastic shells*, a first noteworthy achievement is due to John [1965, 1971], who showed that, in the absence of surface loads and "away from the edge", the state of stress is "approximately planar" and the stresses "parallel to the middle surface" vary "approximately linearly" across the thickness if the thickness is sufficiently small. These remarkable results laid the ground for the two-dimensional, linear and nonlinear, shell theories of Koiter [1966, 1970] and Koiter & Simmonds [1973]. However, in spite of their elegance and depth, John's results hold only for special cases of loadings; besides, they do not provide information "up to the boundary" (of the middle surface of the shell), let alone about the boundary conditions of the associated two-dimensional problem.

What is especially remarkable in John's analysis (based on exceedingly delicate *a priori* estimates) is that the constitutive equation is of the most general form. His results were therefore the first indication that general stress-strain laws could also be successfully handled by an asymptotic analysis.

Formal asymptotic methods can be applied to *nonlinearly elastic shells* as well. In this direction, the earlier attempts of Green [1962] and Green & Naghdi [1965], also described in Green & Zerna [1968, Chap. 16], are particularly worthy of interest.

A major step was then achieved by Miara [1998] and Lods & Miara [1998], who showed in this fashion that the leading term of the asymptotic expansion of the scaled three-dimensional displacement, again in terms of the thickness as the "small" parameter, can be identified with the solution of *nonlinear two-dimensional "membrane", or "flexural", shell equations*, according to the geometry of the middle surface and the boundary conditions as in the linear case. See also Rao [1994] for spherical shells.

Another approach has been proposed by Ge, Kruse & Marsden [1996] for justifying *time-dependent, nonlinear Cosserat shell theories.* Based on the Hamiltonian structure of the equations of three-dimensional nonlinear elastodynamics, this approach combines the features of both the hierarchic and asymptotic methods; see also Kirchgässner & Djurdjevic [1997].

Another major step is due to Le Dret & Raoult [1996], who established the first *convergence theorem* for *nonlinearly elastic shells.* To this end, they used Γ-*convergence theory* for justifying a *nonlinear "membrane" shell model* (which coincides with that obtained by Miara [1998] only for specific classes of deformations).

Linear and nonlinear shell theories constitute the themes of Volume III.

The *objective of this volume* is to show how *asymptotic methods, with the thickness as the "small" parameter*, indeed *provide a powerful means of justifying two-dimensional plate theories.*

More specifically, without any recourse to any *a priori* assumptions of a geometrical or mechanical nature, it is shown in the *linear case* that the three-dimensional *displacements*, once properly *scaled*, *converge* in H^1 towards a limit that satisfies the well-known two-dimensional equations of the *linear Kirchhoff-Love theory*; the *convergence* of the *stresses* is also established.

In the *nonlinear case*, again after *ad hoc* scalings have been performed, it is shown that the leading term of a formal asymptotic expansion of the three-dimensional solution satisfies well-known two-dimensional equations, such as those of the *nonlinear Kirchhoff-Love theory*, or the *von Kármán equations*. Special attention is also given to the first convergence result obtained by Le Dret & Raoult [1995a] in this case, which leads to two-dimensional *large deformation, frame-indifferent, nonlinear planar membrane theories*.

It is also shown that asymptotic methods can likewise be used for justifying other "lower-dimensional" theories, some known, some new, such as the two-dimensional equations of *elastic shallow shells*, and the coupled "pluri-dimensional" equations of *elastic multi-structures*, i.e., structures with *junctions*.

In each case, the existence, uniqueness or multiplicity, and regularity of solutions to the "limit" equations obtained in this fashion are also studied.

Although I have chosen here the viewpoint of *asymptotic methods*, I have simultaneously tried to provide reasonable introductions, and references, to other approaches, such as the Reissner-Mindlin theory, Naghdi's theory, hierarchic plate theories, theories derived by the constraint method, etc.

Fortunately, there remains an abundance of challenging *open problems*; for instance:

- Finding a rigorous justification of the Reissner-Mindlin equations;

- Justification of the nonlinear Kirchhoff-Love theory by a convergence theorem as the thickness approaches zero;

- Existence of solutions of three-dimensional nonlinear plate problems obtained through a proper extension of the two-dimensional solutions (known to exist);

- Existence theory for two-dimensional plate equations, without any restrictions on the boundary conditions or on the magnitude of the applied forces;

- Numerical comparison (essentially lacking at the present time, even in the linear case) between three-dimensional and two-dimensional solutions of plate equations; etc.

For the reader's convenience, this volume is written in such a way that, to a large extent, *each chapter can be read independently of the others*: For instance, a devotee of the von Kármán's equations may proceed directly to Chap. 5, without having necessarily mastered Chaps. 1 to 4, although I obviously do not wish that this always be the case!

To this end, each chapter begins with a substantial introduction detailing the scalings and assumptions on the data, and expounding the main ideas and results. A reader in a hurry may thus get a quick idea of the content of this volume by reading the introductions of the five chapters; consulting the preliminary sections titled "Plate equations at a glance" and "Shallow shell equations at a glance" should also be helpful in this respect.

As in Vol. I, I have tried to provide a "reasonably complete" bibliography, but in view of the formidable existing literature on plates, I am also well aware that the appended list of references is far from being exhaustive. I apologize for any significant reference that I may have inadvertently overlooked.

This volume is an outgrowth of series of lectures that I have given during the past twenty years at Tel Aviv University, Fudan University, the University of Stuttgart, the University of Bucharest, the Ecole Polytechnique Fédérale de Lausanne, Neresheim (Seminar der Deutschen Mathematiker-Vereinigung), the Eidgenössische Technische Hochschule (Zürich), the Chinese University of Hong Kong, and the Université Pierre et Marie Curie.

Substantial portions of the manuscript were also completed during stays at many other places, notably at New York University (Courant Institute of Mathematical Sciences), Cornell University (Mathematical Sciences Institute), Brown University, the Istituto Mauro Picone (Roma), the University of Texas at Austin, Kyoto University, Stanford University, the Universidade de Santiago de Compostela, and the Istituto di Analisi Numerica (Pavia). I am in this respect particularly indebted to my hosts in all these institutions, as their kind hospitality greatly contributed to the completion of this enterprise! The support of the project "Junctions in Elastic Multi-Structures" of the European Cooperation *"S.C.I.E.N.C.E." Programme* is also gratefully acknowledged.

This volume is also an updated, completely re-organized, and considerably expanded version (about twice as long) of my earlier monograph *"Plates and Junctions in Elastic Multi-Structures: An Asymptotic Analysis"*, co-published in 1990 by Masson, Paris, and Springer-Verlag, Berlin. These lecture notes were based on a course taught over the years at the Université Pierre et Marie Curie, Paris, as part of our Doctoral School "D.E.A. d'Analyse Numérique".

This volume is for the most part the result of joint efforts. I am deeply indebted in this respect to Philippe Destuynder, Hervé Le Dret, Patrick Rabier, and Annie Raoult, whose fundamental contributions and kind cooperations were essential to the success of this enterprise.

I am also very grateful to the many other students and colleagues who either collaborated with me on, or brought their own contributions to, the theory of plates: Martial Aufranc, Michel Bernadou, Dominique Blanchard, Frédéric Bourquin, Monique Dauge, Jean-Louis Davet, Giuseppe Geymonat, Isabelle Gruais, Frédéric d'Hennezel, Srinivasan Kesavan, Christophe Lebeltel, Bernadette Miara, Robert Nzengwa, Paula Oliveira, Jean-Claude Paumier, Peregrina Quintela-Estevez, José M. Rodríguez, Luis Trabucho de Campos, Juan M. Viaño Rey, Xiang Yan.

Special thanks are also due to Daniel Coutand, Karine Genevey, Hervé Le Dret, Cristinel Mardare, Bernadette Miara, Arnaud Montenay, Véronique Lods, and Sebastian Slicaru, who were kind enough to read preliminary versions of the manuscript and to propose many improvements.

I express my particular appreciation to Stuart Antman for the manifold "grammatically elastic" advices he provided me with over the years; he is in particular responsible for suggesting the convenient terminologies "nonlinearly elastic" and "linearly elastic" that I so often use.

I also thank Geneviève Raugel and Alice Traynard, who kindly translated for me the seminal article of von Kármán [1910]. A reproduction of p. 350 of this article, where the celebrated "von Kármán equations" appeared for the first time in print, is shown on page lxiii.

Last but not least, I express my heartfelt gratitude to Mathieu Ciarlet, who greatly helped me through the arduous task of compiling the bibliography.

January, 1997 and October, 1999 Philippe G. Ciarlet

PREFACE TO VOLUME III [1]

The *objective of this volume* is to lay down the proper *mathematical foundations of the two-dimensional theory of shells.* To this end, it provides, without any recourse to any *a priori* assumptions of a geometrical or mechanical nature, a *mathematical justification* of *two-dimensional nonlinear and linear shell theories*, by means of *asymptotic methods, with the thickness as the "small" parameter.*

A major virtue of this approach is that it naturally leads to precise *mathematical definitions* of *"membrane"* and *"flexural" shells*, be they *nonlinearly* or *linearly elastic.* Another noteworthy feature is that it automatically provides in each case the "limit" *two-dimensional energy*, together with the *function space* over which it should be minimized.

This process highlights in particular the rôle played by *two fundamental tensors*, each associated with a displacement field of the middle surface, the *change of metric* and *change of curvature tensors* (either *in extenso* in the nonlinear theories, or in their linearized version in the linear theories).

More specifically, under fundamentally distinct sets of assumptions bearing on the *geometry of the middle surface*, on the *boundary conditions*, and on the *order of magnitude of the applied forces*, it is shown that in the *linear* case, the three-dimensional displacements, once properly *scaled*, converge (in H^1, or in L^2, or in *ad hoc* completions) towards a "two-dimensional" limit that satisfies either the *linear two-dimensional equations of a "membrane" shell* (themselves divided into two subclasses) or the *linear two-dimensional equations of a "flexural" shell.*

Under the same assumptions, the *two-dimensional linear Koiter equations* are also justified for all the above classes of linearly elastic shells.

The *existence* and *uniqueness* of solutions to each of these linear

[1] The preceding "Preface to Volume II" should be read in conjunction with the present one, since most of its content is as relevant to shell theory as it is to plate theory.

equations are also studied in detail, essentially by means of crucial *inequalities of Korn's type on surfaces.*

In the *nonlinear case*, again under fundamentally distinct sets of assumptions on the *geometry of the middle surface*, on the *boundary conditions*, and on the *order of magnitude of the applied forces*, it is shown that the leading term of a formal asymptotic expansion of the three-dimensional solutions, once properly scaled, satisfies either the *nonlinear two-dimensional equations of a "membrane" shell*, or the *nonlinear two-dimensional equations of a "flexural" shell.*

Another class of *two-dimensional equations for a nonlinearly elastic "membrane" shell*, this time justified by means of a *convergence theorem*, is also discussed.

Finally the *existence* of solutions to the "limit" nonlinear equations obtained in the above fashions is also studied.

Thanks to the timely availability of S.S. Antman's admirable book "Nonlinear Problems of Elasticity" (Springer-Verlag, 1995), I do not dwell, however, on the approach that consists in "directly" viewing a shell as a two-dimensional deformable body, in the spirit of the *Cosserat shell theories.*

Fortunately, there remains an abundance of challenging *open problems*; for instance:

- A justification, by a convergence theorem as the thickness approaches zero, of the nonlinear membrane and flexural theories obtained by a formal method in Chaps. 8 to 10;
- A clarification of the respective applicabilities of the two nonlinear membrane shell theories described in Chap. 9;
- A higher-order asymptotic analysis, including error estimates, for linearly elatic shells (such an analysis has been successfully developed for linearly elastic plates; cf. Vol. II, Sect. 1.12);
- A "local" theory of shells, accounting for the possible coexistence of regions with a "membrane-dominated" behavior with ones with a "flexural-dominated" behavior (as the theory presented in this volume is a "limit" one, it gives rise to a single, "global", behavior);
- An analysis of singularities "away from the boundary" (i.e., other than boundary layers) that appear in some "generalized membrane" shells;
- An asymptotic justification of Naghdi's shell equations;
- A refined classification of nonlinearly elastic shells (the definitions proposed in this volume still do not account for some examples

and do not seem to be entirely coherent with the definitions of the linear theory; see the commentaries in Sects. 9.1 and 10.2);

- A numerical comparison between three- and two-dimensional solutions, so as to test the "range of validity" (in terms of data) of the limit two-dimensional equations; etc.

Naturally, this volume relies heavily on elementary notions from *differential geometry in* $\mathbb{R}^3$. However, I do *not* assume any *a priori* knowledge of this subject, whose needed prerequesites are expounded at length in the first two chapters. This volume is thus entirely self-contained in this respect. Be that as it may, this volume is neither a treatise on differential geometry nor one on tensor analysis: The incursions into these fields have been kept to the minimum required here (for instance, the definitions of a "tensor" is nowhere to be found in this volume).

In the same spirit, I have consistently given explicit, but occasionally lenghty, formulas rather than more condensed, intrinsic, ones. This *parti pris* may irritate some differential geometers or shell *aficionados*, but it should render the book more accessible to browsers and non-experts; for the same reason, I have systematically avoided using the (otherwise quite convenient!) terminology "immersion".

At the expense of intentional repetitions, most chapters are written in such a way that they are as *self-contained* as possible. I apologize to those readers, brave enough to read the entire volume, who might be annoyed by such a procedure.

As in Vol. II, each chapter begins with a substantial introduction expounding its main results. A reader in a hurry may thus get a quick idea of the contents of this volume by reading the introductions of the eleven chapters. Consulting the preliminary sections titled *"Two-dimensional linear shell equations at a glance"* and *"Two-dimensional nonlinear shell equations at a glance"* should also be helpful in this respect.

I have tried to provide a "reasonably complete" bibliography about "mathematical" shell theory (only scant incursions are otherwise provided into the engineering or computationally-oriented literature), but I am also well aware that the appended list of references is far from being exhaustive: Shell theory has offered so many challenges during its long history that it has generated an immense literature.

As a complement, I advise the readers of this volume to consult great classics, such as Flügge [1934], Pogorelov [1956, 1966, 1967], Vlasov [1958], Novozhilov [1959, 1970], Goldenveizer [1961], Mushtari & Galimov [1961], Green & Zerna [1968], Timoshenko & Woinowsky-Krieger [1970], Flügge [1973], Rutten [1973], as well as more recent treatments, such as Vinson [1974], Pietraszkiewicz [1977, 1979], Lukasiewicz [1979], Dikmen [1982], Calladine [1983], Basar & Krätzig [1985], Niordson [1985], Vekua [1986], Axelrad [1987], Gould [1987], Pogorelov [1988], Destuynder [1990], Dym [1990], Bernadou [1994], Antman [1995], Stolarski, Belytschko & Lee [1995], Valid [1995], Sanchez-Hubert & Sanchez-Palencia [1997], Libai & Simmonds [1998], Vorovich [1999].

I also recommend the short, but illuminating, accounts by Sanchez-Palencia [1995] and Chapelle & Bathe [1998a] of the often elusive behavior of shells and of the difficulties inherent in their finite element analysis.

Substantial portions of the manuscript were completed during stays at the Romanian Academy (Bucharest), the University of Bucharest, the Istituto di Analisi Numerica (Pavia), Fudan University, the University of Stuttgart, and the Liu Bie-ju Centre for Mathematical Sciences of the City University of Hong Kong. I am deeply indebted to my hosts in these institutions: Viorel Barbu, Marius Iosifescu, George Dincă, Franco Brezzi, Li Ta-tsien, Klaus Kirchgässner, and Roderick Wong.

In particular, the first three chapters are an outgrowth of a series of lectures that I delivered at the University of Stuttgart, thanks to the Alexander von Humboldt-Stiftung, whose support is gratefully acknowledged. These lectures, which greatly beneficiated from the critical comments of Mariana Haragus-Courcelle, Ivica Djurdjevic, and Andreas Rössle, later became a monograph *"Introduction to Linear Shell Theory"*, co-published in 1998 by Gauthier-Villars, Paris and North-Holland, Amsterdam; cf. Ciarlet [1998c].

The support of the project "Shells: Mathematical Modeling and Analysis, Scientific Computing" by the "Human Capital and Mobility" Programme of the Commission of the European Communities is also gratefully acknowledged. This project culminated in an "International Conference on Shells", organized with the highest *maestria* by Juan Manuel Viaño in Santiago de Compostela in 1997.

I also wish to express my deep appreciation to the far-sightedness of those who created the Institut Universitaire de France. Without this remarkable institution, the writing of this volume would have required several additional years!

This volume is for the most part the result of joint efforts. I am deeply indebted in this respect to Michel Bernadou, Philippe Destuynder, Hervé Le Dret, Véronique Lods, Cristinel Mardare, Bernadette Miara, Annie Raoult, and Evariste Sanchez-Palencia, whose fundamental contributions and kind cooperations were essential to the success of this enterprise.

I am very grateful to the many other students or colleagues who also contributed in some way to the content of this volume: Georgiana Andreoiu, Adel Blouza, Stéphane Busse, Denis Caillerie, Maïté Carrive, Dominique Chapelle, Christophe Collard, Daniel Coutand, Monique Dauge, Isabel de Figueiredo, Karine Genevey, Giuseppe Geymonat, Patrick Giroud, Liliana Gratie, Oana Iosifescu, Srinivasan Kesavan, François Larsonneur, Patrick Le Tallec, Arnaud Montenay, Jean-Claude Paumier, Paolo Podio-Guidugli, Olivier Ramos, Rao Bopeng, Anne Roquefort, Sebastian Slicaru, Xiao Li-ming.

Special thanks are also due to Elena Baderko, Adel Blouza, Dominique Chapelle, Daniel Coutand, Liliana Gratie, Hervé Le Dret (in particular, for stimulating discussions about the nonlinear membrane theories!), Cristinel Mardare, Qin Tie-hu, Anne Roquefort, and Karim Trabelsi, who kindly read preliminary versions of the manuscript and suggested many improvements.

I renew my particular appreciation for Stuart Antman, whose kind semantic assistance has been an invaluable help over so many years.

October 1999 Philippe G. Ciarlet

TABLE OF CONTENTS[1]

Mathematical Elasticity: General plan ii

Mathematical Elasticity: General preface v

Preface to Volume I ix

Preface to Volume II xv

Preface to Volume III xxvii

Differential geometry at a glance xxxix

Three-dimensional elasticity in curvilinear coordinates at a glance xlvii

Two-dimensional linear shell equations at a glance . . . li

Two-dimensional nonlinear shell equations at a glance . lvii

PART A. LINEAR SHELL THEORY

Chapter 1. Three-dimensional linearized elasticity and Korn's inequalities in curvilinear coordinates 3

Introduction 3

1.1. Three-dimensional linearized elasticity in Cartesian coordinates 5

1.2. Curvilinear coordinates and metric tensor in a three-dimensional domain 12

1.3. The variational equations of three-dimensional linearized elasticity in curvilinear coordinates 22

1.4. Covariant derivatives and Christoffel symbols in a three-dimensional domain 32

1.5. Linearized change of metric tensor in curvilinear coordinates 35

1.6. The boundary value problem of three-dimensional linearized elasticity in curvilinear coordinates 37

[1]The symbol $^\flat$ indicates a section where most results are stated without proof.

1.7. A lemma of J. L. Lions; three-dimensional Korn's inequalities and infinitesimal rigid displacement lemma in curvilinear coordinates . . . 40
1.8. Existence and uniqueness theorem in curvilinear coordinates . . . 50
$1.9^\flat$. Complement: Recovery of a three-dimensional manifold from its metric tensor field . . . 54
Exercises . . . 56

Chapter 2. Inequalities of Korn's type on surfaces . . . 61
Introduction . . . 61
2.1. Curvilinear coordinates and metric tensor on a surface 63
2.2. Curvature tensor on a surface . . . 73
2.3. Covariant derivatives and Christoffel symbols on a surface . . . 85
2.4. Linearized change of metric tensor on a surface . . . 90
2.5. Linearized change of curvature tensor on a surface . . 93
2.6. Inequalities of Korn's type and infinitesimal rigid displacement lemma on a general surface . . . 100
2.7. Inequality of Korn's type and infinitesimal rigid displacement lemma on an elliptic surface . . . 117
$2.8^\flat$. Complement: Recovery of a surface from its metric and curvature tensor fields . . . 130
Exercises . . . 132

Chapter 3. Asymptotic analysis of linearly elastic shells: Preliminaries and outline . . . 137
Introduction . . . 137
3.1. The three-dimensional equations of a linearly elastic shell . . . 141
3.2. The three-dimensional equations over a domain independent of ε . . . 149
3.3. Geometrical and mechanical preliminaries . . . 154
3.4. The two-dimensional equations of linearly elastic "membrane" and "flexural" shells derived by means of a formal asymptotic analysis . . . 161
3.5. Summary of the convergence theorems . . . 183
Exercises . . . 190

Chapter 4. Linearly elastic elliptic membrane shells . . 193
Introduction 193
4.1. Linearly elastic elliptic membrane shells: Definition, example, and assumptions on the data; the three-dimensional equations over a domain independent of ε 196
4.2. Averages with respect to the transverse variable . . . 201
4.3. A three-dimensional inequality of Korn's type for a family of linearly elastic elliptic membrane shells . . . 205
4.4. Convergence of the scaled displacements as $\varepsilon \to 0$ . . 209
4.5. The two-dimensional equations of a linearly elastic elliptic membrane shell; existence, uniqueness, and regularity of solutions; formulation as a boundary value problem 223
4.6. Justification of the two-dimensional equations of a linearly elastic elliptic membrane shell; commentary and refinements 230
Exercises 235

Chapter 5. Linearly elastic generalized membrane shells 241
Introduction 241
5.1. Linearly elastic generalized membrane shells: Definition and assumptions on the data; the three-dimensional equations over a domain independent of ε 245
5.2. Analytical preliminaries 248
5.3. A three-dimensional inequality of Korn's type for a family of linearly elastic shells 258
5.4. Generalized membrane shells of the first and second kinds 261
5.5. Admissible applied forces 264
5.6. Convergence of the scaled displacements as $\varepsilon \to 0$ . . 266
5.7. The two-dimensional equations of a linearly elastic generalized membrane shell; existence and uniqueness of solutions 287
5.8. Justification of the two-dimensional equations of a linearly elastic generalized membrane shell; examples, commentary, and refinements 291
Exercises 297

Chapter 6. Linearly elastic flexural shells 299
Introduction 299
6.1. Linearly elastic flexural shells: Definition, examples, and assumptions on the data; the three-dimensional equations over a domain independent of ε 302
6.2. Convergence of the scaled displacements as $\varepsilon \to 0$ . . 308
6.3. The two-dimensional equations of a linearly elastic flexural shell; existence and uniqueness of solutions . 317
6.4. Justification of the two-dimensional equations of a linearly elastic flexural shell; commentary and refinements 323
Exercises 328

Chapter 7. Koiter's equations and other two-dimensional linear shell theories 333
Introduction 333
7.1. The two-dimensional Koiter equations for a linearly elastic shell: Existence, uniqueness, and regularity of solutions; formulation as a boundary value problem . 335
7.2. Justification of Koiter's equations for all types of linearly elastic shells 345
7.3. Koiter's equations: Additional commentary and bibliographical notes 360
7.4. The two-dimensional Naghdi equations for a linearly elastic shell; existence and uniqueness of solutions . . 363
7.5. Other linear shell theories 367
7.6. Linear shallow shell theories 369
Exercises 372

PART B. NONLINEAR SHELL THEORY

Chapter 8. Asymptotic analysis of nonlinearly elastic shells: Preliminaries 381
Introduction 381
8.1. Three-dimensional nonlinear elasticity in Cartesian coordinates 386
8.2. Three-dimensional nonlinear elasticity in curvilinear coordinates 392

8.3. The three-dimensional equations of a nonlinearly elastic shell 403
8.4. The three-dimensional equations over a domain independent of ε 407
8.5. Geometrical and mechanical preliminaries 411
8.6. The method of formal asymptotic expansions 413
8.7. The leading term is of order zero 415
8.8. Identification of a two-dimensional variational problem satisfied by the leading term 424
Exercises 430

Chapter 9. Nonlinearly elastic membrane shells 433
Introduction 433
9.1. Nonlinearly elastic membrane shells: Definition, examples, and assumptions on the data 436
9.2. The two-dimensional equations as a variational problem 443
9.3. The two-dimensional equations as a minimization problem 445
9.4. The two-dimensional equations of a nonlinearly elastic membrane shell derived by means of a formal asymptotic analysis; commentary 447
9.5$^\flat$. The two-dimensional equations of a nonlinearly elastic membrane shell derived by means of Γ-convergence theory; commentary 453
Exercises 465

Chapter 10. Nonlinearly elastic flexural shells 469
Introduction 469
10.1. Identification of a two-dimensional variational problem satisfied by the leading term when there are nonzero admissible inextensional displacements 472
10.2. Nonlinearly elastic flexural shells: Definition, examples, and assumptions on the data 502
10.3. The two-dimensional equations as a variational problem 507
10.4. The two-dimensional equations as a minimization problem 519

10.5. The two-dimensional equations of a nonlinearly elastic flexural shell derived by means of a formal asymptotic analysis; commentary 521
10.6. Existence of solutions to the minimization problem . 526
Exercises 539

Chapter 11. Koiter's equations and other two-dimensional nonlinear shell theories 545
Introduction 545
11.1. The two-dimensional Koiter equations for a nonlinearly elastic shell 545
11.2. Other nonlinear shell theories 549
11.3. Nonlinear shallow shell theories 552

References 557

Index 583

DIFFERENTIAL GEOMETRY AT A GLANCE[1]

1. General conventions
2. Differential geometry of three-dimensional domains in $\mathbb{R}^3$
3. Differential geometry of surfaces in $\mathbb{R}^3$
4. Korn's and other inequalities in curvilinear coordinates

1. GENERAL CONVENTIONS

(i) *Latin indices and exponents:* i, j, p, ... , take their values in the set $\{1, 2, 3\}$, unless otherwise indicated as when they are used for indexing sequences.

(ii) *Greek indices and exponents:* α, β, σ, ... , *except* ε and ν in the outer normal derivative operator ∂_ν, take their values in the set $\{1, 2\}$.

(iii) The *repeated index summation convention* is systematically used in conjunction with conventions (i) and (ii).

(iv) The symbol "ε" designates a parameter that is > 0 and approaches zero.

2. DIFFERENTIAL GEOMETRY OF THREE-DIMENSIONAL DOMAINS IN $\mathbb{R}^3$

$\boldsymbol{a} \cdot \boldsymbol{b}$: Euclidean inner product of $\boldsymbol{a} \in \mathbb{R}^3$ and $\boldsymbol{b} \in \mathbb{R}^3$.

$\boldsymbol{a} \wedge \boldsymbol{b}$: exterior product of $\boldsymbol{a} \in \mathbb{R}^3$ and $\boldsymbol{b} \in \mathbb{R}^3$.

$|\boldsymbol{a}|$: Euclidean norm of $\boldsymbol{a} \in \mathbb{R}^3$.

---

[1]Only the notations and definitions that are specific to shell theory are listed in this section and the next ones, which otherwise complement the section "Main notation, definitions, and formulas" in Vol. I.

Ω: domain in $\mathbb{R}^3$ (open, bounded, connected subset of $\mathbb{R}^3$ with a Lipschitz-continuous boundary, the set Ω being locally on one side of its boundary).

$x = (x_i)$: generic point in $\overline{\Omega}$.

dx: volume element in Ω.

$\partial_i = \dfrac{\partial}{\partial x_i}$.

Γ: boundary of Ω.

$d\Gamma$: area element along Γ.

(n_i): unit outer normal vector (defined $d\Gamma$-almost everywhere) along Γ.

$\boldsymbol{\Theta} : \overline{\Omega} \subset \mathbb{R}^3 \to \mathbb{R}^3$: injective and smooth enough mapping such that the three vectors $\partial_i \boldsymbol{\Theta}(x)$ are linearly independent at each point $x \in \overline{\Omega}$.

$\boldsymbol{g}_i = \partial_i \boldsymbol{\Theta}$: vectors of the covariant bases in the set $\boldsymbol{\Theta}(\overline{\Omega})$; this means that, at each $x \in \overline{\Omega}$, the three vectors $\boldsymbol{g}_i(x) = \partial_i \boldsymbol{\Theta}(x)$ form the covariant basis at the point $\boldsymbol{\Theta}(x)$.

$\boldsymbol{g}^i$: vectors of the contravariant bases in the set $\boldsymbol{\Theta}(\overline{\Omega})$; the vectors $\boldsymbol{g}^i(x)$ are defined at each $x \in \overline{\Omega}$ by the relations $\boldsymbol{g}^i(x) \cdot \boldsymbol{g}_j(x) = \delta^i_j$.

$g_{ij} = \boldsymbol{g}_i \cdot \boldsymbol{g}_j$: covariant components of the metric tensor of the set $\boldsymbol{\Theta}(\overline{\Omega})$.

$g = \det(g_{ij})$.

$g^{ij} = \boldsymbol{g}^i \cdot \boldsymbol{g}^j$: contravariant components of the metric tensor of the set $\boldsymbol{\Theta}(\overline{\Omega})$.

$\Gamma^p_{ij} = \boldsymbol{g}^p \cdot \partial_j \boldsymbol{g}_i$: Christoffel symbols.

$v_{i\|j} = \partial_j v_i - \Gamma^p_{ij} v_p$: covariant derivatives of a vector field $v_i \boldsymbol{g}^i$ with covariant components $v_i : \overline{\Omega} \to \mathbb{R}$.

$\sigma^{ij}\|_k = \partial_k \sigma^{ij} + \Gamma^i_{pk} \sigma^{pj} + \Gamma^j_{kq} \sigma^{iq}$: covariant derivatives of a tensor field with contravariant components $\sigma^{ij} : \overline{\Omega} \to \mathbb{R}$.

$g_{ij}(\boldsymbol{v})$: covariant components of the metric tensor of the set $(\boldsymbol{\Theta} + v_k \boldsymbol{g}^k)(\overline{\Omega})$.

$e_{i\|j}(\boldsymbol{v}) = \frac{1}{2}[g_{ij}(\boldsymbol{v}) - g_{ij}]^{lin} = \frac{1}{2}(\partial_j \tilde{\boldsymbol{v}} \cdot \boldsymbol{g}_i + \partial_i \tilde{\boldsymbol{v}} \cdot \boldsymbol{g}_j)$
$= \frac{1}{2}(v_{i\|j} + v_{j\|i}) = \frac{1}{2}(\partial_j v_i + \partial_i v_j) - \Gamma^p_{ij} v_p$: covariant components of the linearized change of metric tensor associated with a

displacement field $\tilde{\boldsymbol{v}} = v_i \boldsymbol{g}^i$ of the set $\boldsymbol{\Theta}(\overline{\Omega})$; the notation $[\cdots]^{lin}$ denotes the linear part with respect to $\boldsymbol{v}$ in $[\cdots]$.

$E_{i\|j}(\boldsymbol{v}) = \frac{1}{2}(g_{ij}(\boldsymbol{v}) - g_{ij})$

$\qquad = \frac{1}{2}(v_{i\|j} + v_{j\|i} + g^{mn} v_{m\|i} v_{n\|j})$: covariant components of the change of metric tensor, also called the Green-St Venant strain tensor, associated with a displacement field $v_i \boldsymbol{g}^i$ of the set $\boldsymbol{\Theta}(\overline{\Omega})$.

3. DIFFERENTIAL GEOMETRY OF SURFACES IN $\mathbb{R}^3$

ω: domain in $\mathbb{R}^2$ (open, bounded, connected subset with a Lipschitz-continuous boundary, the set ω being locally on one side of its boundary).

$y = (y_\alpha)$: generic point in $\overline{\omega}$.

dy: area element in ω.

$$\partial_\alpha = \frac{\partial}{\partial y_\alpha}, \; \partial_{\alpha\beta} = \frac{\partial^2}{\partial y_\alpha \partial y_\beta}.$$

γ: boundary of the set ω.

$d\gamma$: length element along γ.

(ν_α): unit outer normal vector along γ.

(τ_α) with $\tau_1 = -\nu_2$, $\tau_2 = \nu_1$: unit tangent vector along γ.

$\partial_\nu \theta = \nu_\alpha \partial_\alpha \theta$: outer normal derivative of θ along γ.

$\partial_\tau \theta = \tau_\alpha \partial_\alpha \theta$: tangential derivative of θ along γ.

$\boldsymbol{\theta} : \overline{\omega} \subset \mathbb{R}^2 \to \mathbb{R}^3$: injective and smooth enough mapping such that the two vectors $\partial_\alpha \boldsymbol{\theta}(y)$ are linearly independent at each point $y \in \overline{\omega}$.

$S = \boldsymbol{\theta}(\overline{\omega})$: surface.

$\boldsymbol{a}_\alpha = \partial_\alpha \boldsymbol{\theta}$: vectors of the covariant bases of the tangent planes; this means that, at each $y \in \overline{\omega}$, the two vectors $\partial_\alpha \boldsymbol{\theta}(y)$ form the covariant basis of the plane tangent to S at the point $\boldsymbol{\theta}(y) \in S$.

$\boldsymbol{a}^\alpha$: vectors of the contravariant bases of the tangent planes; at each $y \in \overline{\omega}$, the vectors $\boldsymbol{a}^\alpha(y)$ of the plane tangent to S at $\boldsymbol{\theta}(y)$ are defined by the relations $\boldsymbol{a}^\alpha(y) \cdot \boldsymbol{a}_\beta(y) = \delta^\alpha_\beta$.

$$\boldsymbol{a}_3 = \boldsymbol{a}^3 = \frac{\boldsymbol{a}_1 \wedge \boldsymbol{a}_2}{|\boldsymbol{a}_1 \wedge \boldsymbol{a}_2|}.$$

$a_{\alpha\beta} = \boldsymbol{a}_\alpha \cdot \boldsymbol{a}_\beta$: covariant components of the metric tensor of the surface S.

$a = \det(a_{\alpha\beta})$.

$a^{\alpha\beta} = \boldsymbol{a}^\alpha \cdot \boldsymbol{a}^\beta$: contravariant components of the metric tensor of S.

$b_{\alpha\beta} = \boldsymbol{a}^3 \cdot \partial_\beta \boldsymbol{a}_\alpha$: covariant components of the curvature tensor of S.

$b_\alpha^\beta = a^{\beta\sigma} b_{\sigma\alpha}$: mixed components of the curvature tensor of S.

$\Gamma_{\alpha\beta}^\sigma = \boldsymbol{a}^\sigma \cdot \partial_\beta \boldsymbol{a}_\alpha$: Christoffel symbols.

$\eta_{\alpha|\beta} = \partial_\beta \eta_\alpha - \Gamma_{\alpha\beta}^\sigma \eta_\sigma$ and $\eta_{3|\beta} = \partial_\beta \eta_3$: covariant derivatives of a vector field $\eta_i \boldsymbol{a}^i$ with covariant components $\eta_i : \overline{\omega} \to \mathbb{R}$.

$\eta^\alpha|_\beta = \partial_\beta \eta^\alpha + \Gamma_{\beta\sigma}^\alpha \eta^\sigma$: covariant derivatives of a tangential vector field $\eta^\alpha \boldsymbol{a}_\alpha$ with contravariant components $\eta^\alpha : \overline{\omega} \to \mathbb{R}$.

$\eta_{\alpha\|\beta} = \eta_{\alpha|\beta} - b_{\alpha\beta}\eta_3$ and $\eta_{3\|\beta} = \eta_{3|\beta} - b_\beta^\sigma \eta_\sigma$.

$\eta^{\alpha\beta}|_\sigma = \partial_\sigma n^{\alpha\beta} + \Gamma_{\sigma\tau}^\alpha n^{\beta\tau} + \Gamma_{\sigma\tau}^\beta n^{\alpha\tau}$: covariant derivatives of a tensor field with contravariant components $n^{\alpha\beta} : \overline{\omega} \to \mathbb{R}$.

$b_\alpha^\beta|_\sigma = \partial_\sigma b_\alpha^\beta + \Gamma_{\sigma\tau}^\beta b_\alpha^\tau - \Gamma_{\alpha\sigma}^\tau b_\tau^\beta$: covariant derivatives of the curvature tensor, defined here by means of its mixed components $b_\alpha^\beta : \overline{\omega} \to \mathbb{R}$.

$a_{\alpha\beta}(\boldsymbol{\eta})$: covariant components of the metric tensor of the surface $(\boldsymbol{\theta} + \eta_i \boldsymbol{a}^i)(\overline{\omega})$.

$$\begin{aligned}\gamma_{\alpha\beta}(\boldsymbol{\eta}) &= \tfrac{1}{2}\left[a_{\alpha\beta}(\boldsymbol{\eta}) - a_{\alpha\beta}\right]^{lin} = \tfrac{1}{2}\left(\partial_\beta \tilde{\boldsymbol{\eta}} \cdot \boldsymbol{a}_\alpha + \partial_\alpha \tilde{\boldsymbol{\eta}} \cdot \boldsymbol{a}_\beta\right) \\ &= \tfrac{1}{2}(\eta_{\alpha|\beta} + \eta_{\beta|\alpha}) = \tfrac{1}{2}(\partial_\beta \eta_\alpha + \partial_\alpha \eta_\beta) - \Gamma_{\alpha\beta}^\sigma \eta_\sigma - b_{\alpha\beta}\eta_3\end{aligned}$$

: covariant components of the linearized change of metric tensor associated with a displacement field $\tilde{\boldsymbol{\eta}} = \eta_i \boldsymbol{a}^i$ of the surface S; the notation $[\cdots]^{lin}$ denotes the linear part with respect to $\boldsymbol{\eta}$ in $[\cdots]$.

$\tilde{\gamma}_{\alpha\beta}(\tilde{\boldsymbol{\eta}}) = \frac{1}{2}(\partial_\beta \tilde{\boldsymbol{\eta}} \cdot \boldsymbol{a}_\alpha + \partial_\alpha \tilde{\boldsymbol{\eta}} \cdot \boldsymbol{a}_\beta)$.

$b_{\alpha\beta}(\boldsymbol{\eta})$: covariant components of the curvature tensor of the surface $(\boldsymbol{\theta} + \eta_i \boldsymbol{a}^i)(\overline{\omega})$ (only defined at those points in $\overline{\omega}$ where the two vectors $\partial_\alpha(\boldsymbol{\theta} + \eta_i \boldsymbol{a}^i)$ are linearly independent).

$$\begin{aligned}\rho_{\alpha\beta}(\boldsymbol{\eta}) &= [b_{\alpha\beta}(\boldsymbol{\eta}) - b_{\alpha\beta}]^{lin} = (\partial_{\alpha\beta}\tilde{\boldsymbol{\eta}} - \Gamma_{\alpha\beta}^\sigma \partial_\sigma \tilde{\boldsymbol{\eta}}) \cdot \boldsymbol{a}_3 \\ &= \eta_{3|\alpha\beta} - b_\alpha^\sigma b_{\sigma\beta}\eta_3 + b_\alpha^\sigma \eta_\sigma|_\beta + b_\beta^\tau \eta_{\tau|\alpha} + b_\beta^\tau|_\alpha \eta_\tau \\ &= \partial_{\alpha\beta}\eta_3 - \Gamma_{\alpha\beta}^\sigma \partial_\sigma \eta_3 - b_\alpha^\sigma b_{\sigma\beta}\eta_3 \\ &\quad + b_\alpha^\sigma(\partial_\beta \eta_\sigma - \Gamma_{\beta\sigma}^\tau \eta_\tau) + b_\beta^\tau(\partial_\alpha \eta_\tau - \Gamma_{\alpha\tau}^\sigma \eta_\sigma) \\ &\quad + (\partial_\alpha b_\beta^\tau + \Gamma_{\alpha\sigma}^\tau b_\beta^\sigma - \Gamma_{\alpha\beta}^\sigma b_\sigma^\tau)\eta_\tau\end{aligned}$$

: covariant components of

the linearized change of curvature tensor associated with a displacement field $\tilde{\boldsymbol{\eta}} = \eta_i \boldsymbol{a}^i$ of the surface S; the notation $[\cdots]^{lin}$ denotes the linear part with respect to $\boldsymbol{\eta}$ in $[\cdots]$.

$\tilde{\rho}_{\alpha\beta}(\tilde{\boldsymbol{\eta}}) = (\partial_{\alpha\beta}\tilde{\boldsymbol{\eta}} - \Gamma^{\sigma}_{\alpha\beta}\partial_\sigma\tilde{\boldsymbol{\eta}}) \cdot \boldsymbol{a}_3$.

$G_{\alpha\beta}(\boldsymbol{\eta}) = \frac{1}{2}(a_{\alpha\beta}(\boldsymbol{\eta}) - a_{\alpha\beta}) = \frac{1}{2}(\eta_{\alpha\|\beta} + \eta_{\beta\|\alpha} + a^{mn}\eta_{m\|\alpha}\eta_{n\|\beta})$, where $a^{mn} = \boldsymbol{a}^m \cdot \boldsymbol{a}^n$: covariant components of the change of metric tensor associated with a displacement field $\eta_i \boldsymbol{a}^i$ of the surface S.

$R_{\alpha\beta}(\boldsymbol{\eta}) = b_{\alpha\beta}(\boldsymbol{\eta}) - b_{\alpha\beta}$: covariant components of the change of metric tensor associated with a displacement field $\eta_i \boldsymbol{a}^i$ of the surface S (only defined if the two vectors $\partial_\alpha(\boldsymbol{\theta} + \eta_i \boldsymbol{a}^i)$ are linearly independent in ω).

4. KORN'S AND OTHER INEQUALITIES IN CURVILINEAR COORDINATES[1]

Given a domain Ω in $\mathbb{R}^n$, the norm in $L^2(\Omega)$ or $\mathbf{L}^2(\Omega)$ is noted $|\cdot|_{0,\Omega}$ and the norm in $H^m(\Omega)$ or $\mathbf{H}^m(\Omega)$, $m \geq 1$, is noted $\|\cdot\|_{m,\Omega}$.

Three-dimensional Korn's inequality "without boundary conditions" in curvilinear coordinates (Thm. 1.7-2):

$$\|\boldsymbol{v}\|_{1,\Omega} \leq C_0 \Big\{ \sum_i |v_i|^2_{0,\Omega} + \sum_{i,j} |e_{i\|j}(\boldsymbol{v})|^2_{0,\Omega} \Big\}^{1/2}$$

for all $\boldsymbol{v} = (v_i) \in \mathbf{H}^1(\Omega)$.

Three-dimensional Korn's inequality in curvilinear coordinates (Thm. 1.7-4):

$$\|\boldsymbol{v}\|_{1,\Omega} \leq C \Big\{ \sum_{i,j} |e_{i\|j}(\boldsymbol{v})|^2_{0,\Omega} \Big\}^{1/2} \text{ for all } \boldsymbol{v} \in \mathbf{V}(\Omega), \text{ where}$$

$$\mathbf{V}(\Omega) = \{\boldsymbol{v} = (v_i) \in \mathbf{H}^1(\Omega);\ \boldsymbol{v} = \mathbf{0} \text{ on } \Gamma_0\}.$$

[1]The definitions of the sets Ω, ω, etc., and of the functions $e_{i\|j}(\boldsymbol{v})$, $\gamma_{\alpha\beta}(\boldsymbol{\eta})$, etc., used in the subsequent inequalities are given in the preceding subsections.

Uniform positive definiteness of the three-dimensional elasticity tensor (Thm. 1.8-1):

$$\sum_{i,j} |t_{ij}|^2 \leq C_e A^{ijkl}(x) t_{kl} t_{ij}$$
for all $x \in \overline{\Omega}$ and all symmetric matrices (t_{ij}).

Inequality of Korn's type "without boundary conditions" on a general surface (Thm. 2.6-1):

$$\Big\{ \sum_\alpha \|\eta_\alpha\|^2_{1,\omega} + \|\eta_3\|^2_{2,\omega} \Big\}^{1/2}$$
$$\leq c_0 \Big\{ \sum_\alpha |\eta_\alpha|^2_{0,\omega} + \|\eta_3\|^2_{1,\omega} + \sum_{\alpha,\beta} |\gamma_{\alpha\beta}(\boldsymbol{\eta})|^2_{0,\omega} + \sum_{\alpha,\beta} |\rho_{\alpha\beta}(\boldsymbol{\eta})|^2_{0,\omega} \Big\}^{1/2}$$

for all $\boldsymbol{\eta} = (\eta_i) \in H^1(\omega) \times H^1(\omega) \times H^2(\omega)$.

Inequality of Korn's type on a general surface (Thm. 2.6-4):

$$\Big\{ \sum_\alpha \|\eta_\alpha\|^2_{1,\omega} + \|\eta_3\|^2_{2,\omega} \Big\}^{1/2}$$
$$\leq c \Big\{ \sum_{\alpha,\beta} |\gamma_{\alpha\beta}(\boldsymbol{\eta})|^2_{0,\omega} + \sum_{\alpha,\beta} |\rho_{\alpha\beta}(\boldsymbol{\eta})|^2_{0,\omega} \Big\}^{1/2}$$
for all $\boldsymbol{\eta} = (\eta_i) \in \mathbf{V}_K(\omega)$, where
$$\mathbf{V}_K(\omega) = \{\boldsymbol{\eta} = (\eta_i) \in H^1(\omega) \times H^1(\omega) \times H^2(\omega);\ \eta_i = \partial_\nu \eta_3 = 0 \text{ on } \gamma_0\}.$$

Inequality of Korn's type on a general surface with little regularity (Thm. 2.6-6):

$$\Big\{ \|\tilde{\boldsymbol{\eta}}\|^2_{1,\omega} + \sum_{\alpha,\beta} |\partial_{\alpha\beta}\tilde{\boldsymbol{\eta}} \cdot \boldsymbol{a}_3|^2_{0,\omega} \Big\}^{1/2}$$
$$\leq c \Big\{ \sum_{\alpha,\beta} |\tilde{\gamma}_{\alpha\beta}(\tilde{\boldsymbol{\eta}})|^2_{0,\omega} + \sum_{\alpha,\beta} |\tilde{\rho}_{\alpha\beta}(\tilde{\boldsymbol{\eta}})|^2_{0,\omega} \Big\}^{1/2} \text{ for all } \tilde{\boldsymbol{\eta}} \in \tilde{\mathbf{V}}^s_K(\omega),$$
where $\tilde{\mathbf{V}}^s_K(\omega) = \{\tilde{\boldsymbol{\eta}} \in \mathbf{H}^1_0(\omega);\ \partial_{\alpha\beta}\tilde{\boldsymbol{\eta}} \cdot \boldsymbol{a}_3 \in L^2(\omega)\}$.

Second inequality of Korn's type "without boundary conditions" on a general surface (Thm. 2.7-1):

$$\Big\{\sum_\alpha \|\eta_\alpha\|^2_{1,\omega} + |\eta_3|^2_{0,\omega}\Big\}^{1/2} \le c_0\Big\{\sum_i |\eta_i|^2_{0,\omega} + \sum_{\alpha,\beta}|\gamma_{\alpha\beta}(\boldsymbol{\eta})|^2_{0,\omega}\Big\}^{1/2}$$

for all $\boldsymbol{\eta} = (\eta_i) \in H^1(\omega) \times H^1(\omega) \times L^2(\omega)$.

Inequality of Korn's type on an elliptic surface (Thm. 2.7-3):

$$\Big\{\sum_\alpha \|\eta_\alpha\|^2_{1,\omega} + |\eta_3|^2_{0,\omega}\Big\}^{1/2} \le c_M\Big\{\sum_{\alpha,\beta}|\gamma_{\alpha\beta}(\boldsymbol{\eta})|^2_{0,\omega}\Big\}^{1/2}$$

for all $\boldsymbol{\eta} = (\eta_i) \in \mathbf{V}_M(\omega) = H^1_0(\omega) \times H^1_0(\omega) \times L^2(\omega)$.

Uniform positive definiteness of the scaled two-dimensional elasticity tensor of a shell (Thm. 3.3-2):

$$\sum_{\alpha,\beta}|t_{\alpha\beta}|^2 \le c_e a^{\alpha\beta\sigma\tau}(y)t_{\sigma\tau}t_{\alpha\beta}$$

for all $y \in \overline{\omega}$ and all symmetric matrices $(t_{\alpha\beta})$.

Uniform positive definiteness of the scaled three-dimensional elasticity tensor of a shell (Thm. 3.3-2):

$$\sum_{i,j}|t_{ij}|^2 \le C_e A^{ijkl}(\varepsilon)(x)t_{kl}t_{ij}$$

for all $0 < \varepsilon \le \varepsilon_0$, all $x \in \overline{\Omega}$, and all symmetric matrices (t_{ij}).

Three-dimensional inequality of Korn's type for a family of linearly elastic shells (Thm. 5.3-1):

$$\|\boldsymbol{v}\|_{1,\Omega} \le \frac{C}{\varepsilon}\Big\{\sum_{i,j}|e_{i\|j}(\varepsilon;\boldsymbol{v})|^2_{0,\Omega}\Big\}^{1/2}$$

for all $0 < \varepsilon \le \varepsilon_1$ and all $\boldsymbol{v} \in \mathbf{V}(\Omega)$.

Three-dimensional inequality of Korn's type for a family of linearly elastic elliptic membrane shells (Thm. 4.3-1):

$$\Big\{\sum_{\alpha}\|v_\alpha\|^2_{1,\Omega}+|v_3|^2_{0,\Omega}\Big\}^{1/2}\le C_M\Big\{\sum_{i,j}|e_{i\|j}(\varepsilon;\boldsymbol{v})|^2_{0,\Omega}\Big\}^{1/2}$$

for all $0<\varepsilon\le\varepsilon_1$ and all $\boldsymbol{v}=(v_i)\in\mathbf{V}(\Omega)$.

THREE-DIMENSIONAL ELASTICITY IN CURVILINEAR COORDINATES AT A GLANCE

1. Notations common to linear and nonlinear three-dimensional elasticity
2. Three-dimensional linearized elasticity in curvilinear coordinates
3. Three-dimensional nonlinear elasticity in curvilinear coordinates

Frequent use is made in this section of notations whose definitions, not repeated here, are found in the preceding subsection "Differential geometry of three-dimensional domains in $\mathbb{R}^3$".

1. NOTATIONS COMMON TO LINEAR AND NONLINEAR THREE-DIMENSIONAL ELASTICITY

Ω: domain in $\mathbb{R}^3$ with boundary Γ.

$\Gamma = \Gamma_0 \cup \Gamma_1$: partition of the boundary of Ω with *area* $\Gamma_0 > 0$.

$\boldsymbol{\Theta} : \overline{\Omega} \subset \mathbb{R}^3 \to \mathbb{R}^3$: injective and smooth enough mapping such that the three vectors $\partial_i \boldsymbol{\Theta}$ are linearly independent at all points in $\overline{\Omega}$.

$\boldsymbol{\Theta}(\overline{\Omega}) \subset \mathbb{R}^3$: reference configuration of an elastic body.

$A^{ijkl} = \lambda g^{ij} g^{kl} + \mu(g^{ik} g^{jl} + g^{il} g^{jk})$: contravariant components of the three-dimensional elasticity tensor.

$\lambda > 0$ and $\mu > 0$: Lamé constants of a homogeneous, isotropic, elastic material, whose reference configuration is a natural state.

$f^i \in L^2(\Omega)$: contravariant components of the applied body force density.

$h^i \in L^2(\Gamma_1)$: contravariant components of the applied surface force density.

$\boldsymbol{u} = (u_i)$: unknown vector field, whose components $u_i : \overline{\Omega} \to \mathbb{R}$ are the covariant components of the displacement field; this means that $u_i(x) \boldsymbol{g}^i(x)$ is the displacement vector of the point $\boldsymbol{\Theta}(x) \in \boldsymbol{\Theta}(\overline{\Omega})$.

2. THREE DIMENSIONAL LINEARIZED ELASTICITY IN CURVILINEAR COORDINATES

Minimization problem (Thm. 1.8-2):

$$\boldsymbol{u} \in \mathbf{V}(\Omega) := \{\boldsymbol{v} = (v_i) \in \mathbf{H}^1(\Omega);\ \boldsymbol{v} = \mathbf{0} \text{ on } \Gamma_0\},$$
$$J(\boldsymbol{u}) = \inf_{\boldsymbol{v} \in \mathbf{V}(\Omega)} J(\boldsymbol{v}), \text{ where}$$
$$J(\boldsymbol{v}) = \frac{1}{2}\int_\Omega A^{ijkl} e_{k\|l}(\boldsymbol{v}) e_{i\|j}(\boldsymbol{v})\sqrt{g}\,dx - \left\{\int_\Omega f^i v_i \sqrt{g}\,dx + \int_{\Gamma_1} h^i v_i \sqrt{g}\,d\Gamma\right\}.$$

Variational problem (Thm. 1.3-1):

$$\boldsymbol{u} \in \mathbf{V}(\Omega) \text{ and, for all } \boldsymbol{v} \in \mathbf{V}(\Omega),$$
$$\int_\Omega A^{ijkl} e_{k\|l}(\boldsymbol{u}) e_{i\|j}(\boldsymbol{v})\sqrt{g}\,dx = \int_\Omega f^i v_i \sqrt{g}\,dx + \int_{\Gamma_1} h^i v_i \sqrt{g}\,dx.$$

Boundary value problem (Thm. 1.6-1):

$$\begin{aligned} -\sigma^{ij}\|_j &= f^i \quad \text{in } \Omega, \text{ where } \sigma^{ij} = A^{ijkl} e_{k\|l}(\boldsymbol{u}),\\ u_i &= 0 \quad \text{on } \Gamma_0,\\ \sigma^{ij} n_j &= h^i \quad \text{on } \Gamma_1. \end{aligned}$$

3. THREE-DIMENSIONAL NONLINEAR ELASTICITY IN CURVILINEAR COORDINATES

Only *St Venant-Kirchhoff materials* are considered in this subsection.

Minimization problem (Thm. 8.2-1):

$$\begin{aligned}
&\boldsymbol{u} \in \mathbf{W}(\Omega) := \{\boldsymbol{v} = (v_i) \in \mathbf{W}^{1,4}(\Omega);\, \boldsymbol{v} = \mathbf{0} \text{ on } \Gamma_0\},\\
&J(\boldsymbol{u}) = \inf_{\boldsymbol{v} \in \mathbf{W}(\Omega)} J(\boldsymbol{v}),\\
&J(\boldsymbol{v}) = \frac{1}{2}\int_\Omega A^{ijkl} E_{k\|l}(\boldsymbol{v}) E_{i\|j}(\boldsymbol{v})\sqrt{g}\,dx\\
&\qquad\qquad - \left\{\int_\Omega f^i v_i \sqrt{g}\,dx + \int_{\Gamma_1} h^i v_i \sqrt{g}\,d\Gamma\right\}.
\end{aligned}$$

Variational problem (Thm. 8.2-3)

$$\begin{aligned}
&\boldsymbol{u} \in \mathbf{W}(\omega) \text{ and, for all } \boldsymbol{v} \in \mathbf{W}(\Omega),\\
&\int_\Omega A^{ijkl} E_{k\|l}(\boldsymbol{u})(E'_{i\|j}(\boldsymbol{u})\boldsymbol{v})\sqrt{g}\,dx\\
&\qquad\qquad = \int_\Omega f^i v_i \sqrt{g}\,dx + \int_{\Gamma_1} h^i v_i \sqrt{g}\,d\Gamma, \text{ where}\\
&E'_{i\|j}(\boldsymbol{u})\boldsymbol{v} = \frac{1}{2}\left(v_{i\|j} + v_{j\|i} + g^{mn}\{u_{m\|i} v_{n\|j} + u_{n\|j} v_{m\|i}\}\right).
\end{aligned}$$

Boundary value problem (Thm. 8.2-4):

$$\begin{aligned}
-(\sigma^{ij} + \sigma^{kj} g^{il} u_{l\|k})\|_j &= f_i \text{ in } \Omega, \text{ where } \sigma^{ij} = A^{ijkl} E_{k\|l}(\boldsymbol{u}),\\
u_i &= 0 \text{ on } \Gamma_0,\\
(\sigma^{ij} + \sigma^{kj} g^{il} u_{l\|k}) n_j &= h_i \text{ on } \Gamma_1.
\end{aligned}$$

TWO-DIMENSIONAL LINEAR SHELL EQUATIONS AT A GLANCE[1]

1. General notations for shell equations
2. Linearly elastic elliptic membrane shells
3. Linearly elastic generalized membrane shells
4. Linearly elastic flexural shells
5. Koiter's equations for a linearly elastic shell

All the equations listed in this section (save the last minimization problem) are justified by a *convergence theorem.*

In each case (save the last minimization problem), the shell is subjected to a homogeneous boundary condition of place along the portion $\boldsymbol{\Theta}(\gamma_0 \times [-\varepsilon, \varepsilon])$ of its lateral face, i.e., the displacement field vanishes there.

Frequent use is made of notations whose definitions, not repeated here, are found in the preceding subsection "Differential geometry of surfaces in $\mathbb{R}^3$".

1. GENERAL NOTATIONS FOR SHELL EQUATIONS

ω: domain in $\mathbb{R}^2$ with boundary γ.

$\gamma = \gamma_0 \cup \gamma_1$: partition of the boundary of ω with *length* $\gamma_0 > 0$.

$\boldsymbol{\theta} \in \mathcal{C}^3(\overline{\omega}; \mathbb{R}^3)$: injective and smooth enough mapping such that the two vectors $\boldsymbol{a}_\alpha = \partial_\alpha \boldsymbol{\theta}$ are linearly independent at all points of $\overline{\omega}$.

$S = \boldsymbol{\theta}(\overline{\omega})$: middle surface of the shell.

2ε: thickness of the shell.

$\Omega^\varepsilon = \omega \times]-\varepsilon, \varepsilon[, \ \Gamma_0^\varepsilon = \gamma_0 \times [-\varepsilon, \varepsilon], \ \Gamma_\pm^\varepsilon = \omega \times \{\pm\varepsilon\}$.

[1] Linear *shallow shell* equations in curvilinear coordinates are not reproduced here, as their justifications rely on asymptotic analyses similar to those found in Vol. II, Chap. 3; they are nevertheless reviewed in Sect. 7.6.

$\boldsymbol{\Theta}(y, x_3^\varepsilon) = \boldsymbol{\theta}(y) + x_3^\varepsilon \boldsymbol{a}_3(y)$ for all $(y, x_3^\varepsilon) \in \overline{\Omega}^\varepsilon$.

$\boldsymbol{\Theta}(\overline{\Omega}^\varepsilon) \subset \mathbb{R}^3$: reference configuration of the shell.

$a^{\alpha\beta\sigma\tau, \varepsilon} = \dfrac{4\lambda^\varepsilon \mu^\varepsilon}{\lambda^\varepsilon + 2\mu^\varepsilon} a^{\alpha\beta} a^{\sigma\tau} + 2\mu^\varepsilon (a^{\alpha\sigma} a^{\beta\tau} + a^{\alpha\tau} a^{\beta\sigma})$: contravariant components of the two-dimensional elasticity tensor of the shell.

$\lambda^\varepsilon > 0$ and $\mu^\varepsilon > 0$: Lamé constants of the material constituting the shell.

$f^{i, \varepsilon} \in L^2(\Omega^\varepsilon)$: contravariant components of the applied body force density acting in the interior $\boldsymbol{\Theta}(\Omega^\varepsilon)$ of the shell.

$h^{i, \varepsilon}(\cdot, \pm\varepsilon) \in L^2(\omega)$: contravariant components of the applied surface force density acting on the "upper" and "lower" faces $\boldsymbol{\Theta}(\Gamma_+^\varepsilon)$ and $\boldsymbol{\Theta}(\Gamma_-^\varepsilon)$ of the shell.

$$p^{i, \varepsilon} = \int_{-\varepsilon}^{\varepsilon} f^{i, \varepsilon} dx_3^\varepsilon + h^{i, \varepsilon}(\cdot, +\varepsilon) + h^{i, \varepsilon}(\cdot, -\varepsilon).$$

$$\tilde{\boldsymbol{p}}^\varepsilon = p^{i, \varepsilon} \boldsymbol{a}_i.$$

$\boldsymbol{\zeta}^\varepsilon = (\zeta_i^\varepsilon)$: unknown vector field, whose components $\zeta_i^\varepsilon : \overline{\omega} \to \mathbb{R}$ are the covariant components of the displacement field $\zeta_i^\varepsilon \boldsymbol{a}^i$ of the middle surface S. This means that $\zeta_i^\varepsilon(y) \boldsymbol{a}^i(y)$ is the displacement vector of the point $\boldsymbol{\theta}(y) \in S$.

$\tilde{\boldsymbol{\zeta}}^\varepsilon = \zeta_i^\varepsilon \boldsymbol{a}^i$: unknown displacement field of the middle surface S.

$\boldsymbol{\zeta}_K^\varepsilon = (\zeta_{K,i}^\varepsilon)$: unknown vector field for the linear and nonlinear Koiter equations; the components $\zeta_{K,i}^\varepsilon : \overline{\omega} \to \mathbb{R}$ are the covariant components of the displacement field of S.

2. LINEARLY ELASTIC ELLIPTIC MEMBRANE SHELLS

Definition (Sect. 4.1): A shell is a *linearly elastic elliptic membrane shell* if its middle surface S is elliptic and the displacement field vanishes on its entire lateral face $\boldsymbol{\Theta}(\gamma \times [-\varepsilon, \varepsilon])$, i.e., $\gamma_0 = \gamma$.

Minimization problem (Thm. 4.5-2):

$$\boldsymbol{\zeta}^\varepsilon \in \mathbf{V}_M(\omega) = H_0^1(\omega) \times H_0^1(\omega) \times L^2(\omega),$$
$$j_M^\varepsilon(\boldsymbol{\zeta}^\varepsilon) = \inf_{\boldsymbol{\eta} \in \mathbf{V}_M(\omega)} j_M^\varepsilon(\boldsymbol{\eta}), \text{ where}$$
$$j_M^\varepsilon(\boldsymbol{\eta}) = \frac{1}{2} \int_\omega \varepsilon a^{\alpha\beta\sigma\tau,\varepsilon} \gamma_{\sigma\tau}(\boldsymbol{\eta}) \gamma_{\alpha\beta}(\boldsymbol{\eta}) \sqrt{a}\, dy - \int_\omega p^{i,\varepsilon} \eta_i \sqrt{a}\, dy.$$

Variational problem (Thm. 4.5-2):

$$\boldsymbol{\zeta}^\varepsilon \in \mathbf{V}_M(\omega) \text{ and, for all } \boldsymbol{\eta} \in \mathbf{V}_M(\omega),$$
$$\varepsilon \int_\omega a^{\alpha\beta\sigma\tau,\varepsilon} \gamma_{\sigma\tau}(\boldsymbol{\zeta}^\varepsilon) \gamma_{\alpha\beta}(\boldsymbol{\eta}) \sqrt{a}\, dy = \int_\omega p^{i,\varepsilon} \eta_i \sqrt{a}\, dy.$$

Boundary value problem (Thm. 4.5-2):

$$-n^{\alpha\beta,\varepsilon}|_\beta = p^{\alpha,\varepsilon} \text{ in } \omega, \text{ where } n^{\alpha\beta,\varepsilon} = \varepsilon a^{\alpha\beta\sigma\tau,\varepsilon} \gamma_{\sigma\tau}(\boldsymbol{\zeta}^\varepsilon),$$
$$-b_{\alpha\beta} n^{\alpha\beta,\varepsilon} = p^{3,\varepsilon} \text{ in } \omega,$$
$$\zeta_\alpha^\varepsilon = 0 \text{ on } \gamma.$$

3. LINEARLY ELASTIC GENERALIZED MEMBRANE SHELLS

Definition (Sect. 5.1): A shell is a *linearly elastic generalized membrane shell* if the space

$$\mathbf{V}_F(\omega) = \{\boldsymbol{\eta} = (\eta_i) \in H^1(\omega) \times H^1(\omega) \times H^2(\omega);$$
$$\eta_i = \partial_\nu \eta_3 = 0 \text{ on } \gamma_0,\ \gamma_{\alpha\beta}(\boldsymbol{\eta}) = 0 \text{ in } \omega\}$$

reduces to $\{\mathbf{0}\}$, but the shell is not a linearly elastic elliptic membrane shell (in which case the space $\mathbf{V}_F(\omega)$ also reduces to $\{\mathbf{0}\}$).

Note: Only linearly elastic generalized membrane shells "of the first kind" (Sect. 5.4) are considered here.

Preliminary definitions (the functions $\varphi^{\alpha\beta} \in L^2(\omega)$ found in the linear form L_M^ε originate from the assumed "admissibility" of the applied forces; cf. Sect. 5.5 and Thm. 5.6-1):

$$\begin{aligned}
\mathbf{V}(\omega) &= \{\boldsymbol{\eta} = (\eta_i) \in \mathbf{H}^1(\omega);\ \boldsymbol{\eta} = \mathbf{0} \text{ on } \gamma_0\},\\
\mathbf{V}_M^\sharp(\omega) &= \text{completion of } \mathbf{V}(\omega) \text{ with respect to } |\cdot|_\omega^M, \text{ where}\\
|\boldsymbol{\eta}|_\omega^M &= \Big\{\sum_{\alpha,\beta} |\gamma_{\alpha\beta}(\boldsymbol{\eta})|_{0,\omega}^2\Big\}^{1/2},\\
B_M^\varepsilon(\boldsymbol{\zeta}, \boldsymbol{\eta}) &= \varepsilon \int_\omega a^{\alpha\beta\sigma\tau,\varepsilon}\gamma_{\sigma\tau}(\boldsymbol{\zeta})\gamma_{\alpha\beta}(\boldsymbol{\eta})\sqrt{a}\,dy \text{ for } \boldsymbol{\zeta}, \boldsymbol{\eta} \in \mathbf{V}(\omega),\\
L_M^\varepsilon(\boldsymbol{\eta}) &= \varepsilon \int_\omega \varphi^{\alpha\beta}\gamma_{\alpha\beta}(\boldsymbol{\eta})\sqrt{a}\,dy \text{ for } \boldsymbol{\eta} \in \mathbf{V}(\omega),
\end{aligned}$$

$B_M^{\sharp\varepsilon}$: unique continuous extension of the bilinear form B_M^ε from $\mathbf{V}(\omega)$ to $\mathbf{V}_M^\sharp(\omega)$,

$L_M^{\sharp\varepsilon}$: unique continuous extension of the linear form L_M^ε from $\mathbf{V}(\omega)$ to $\mathbf{V}_M^\sharp(\omega)$.

Minimization problem (Thm. 5.7-2):

$$\boldsymbol{\zeta}^\varepsilon \in \mathbf{V}_M^\sharp(\omega) \text{ and } j_M^{\sharp\varepsilon}(\boldsymbol{\zeta}^\varepsilon) = \inf_{\boldsymbol{\eta}\in\mathbf{V}_M^\sharp(\omega)} j_M^{\sharp\varepsilon}(\boldsymbol{\eta}), \text{ where}$$
$$j_M^{\sharp\varepsilon}(\boldsymbol{\eta}) = \frac{1}{2}B_M^{\sharp\varepsilon}(\boldsymbol{\eta}, \boldsymbol{\eta}) - L_M^{\sharp\varepsilon}(\boldsymbol{\eta}).$$

Variational problem (Thm. 5.7-2):

$$\boldsymbol{\zeta}^\varepsilon \in \mathbf{V}_M^\sharp(\omega) \text{ and } B_M^{\sharp\varepsilon}(\boldsymbol{\zeta}^\varepsilon, \boldsymbol{\eta}) = L_M^{\sharp\varepsilon}(\boldsymbol{\eta}) \text{ for all } \boldsymbol{\eta} \in \mathbf{V}_M^\sharp(\omega).$$

4. LINEARLY ELASTIC FLEXURAL SHELLS

Definition (Sect. 6.1): A shell is a *linearly elastic flexural shell* if the space

$$\begin{aligned} \mathbf{V}_F(\omega) = \{&\boldsymbol{\eta} = (\eta_i) \in H^1(\omega) \times H^1(\omega) \times H^2(\omega); \\ &\eta_i = \partial_\nu \eta_3 = 0 \text{ on } \gamma_0,\ \gamma_{\alpha\beta}(\boldsymbol{\eta}) = 0 \text{ in } \omega\} \end{aligned}$$

contains nonzero elements.

Minimization problem (Thm. 6.3-2):

$$\begin{aligned} &\boldsymbol{\zeta}^\varepsilon \in \mathbf{V}_F(\omega) \text{ and } j_F^\varepsilon(\boldsymbol{\zeta}^\varepsilon) = \inf_{\boldsymbol{\eta} \in \mathbf{V}_F(\omega)} j_F^\varepsilon(\boldsymbol{\eta}), \text{ where} \\ &j_F^\varepsilon(\boldsymbol{\eta}) = \frac{1}{6} \int_\omega \varepsilon^3 a^{\alpha\beta\sigma\tau,\varepsilon} \rho_{\sigma\tau}(\boldsymbol{\eta}) \rho_{\alpha\beta}(\boldsymbol{\eta}) \sqrt{a}\, dy - \int_\omega p^{i,\varepsilon} \eta_i \sqrt{a}\, dy. \end{aligned}$$

Variational problem (Thm. 6.3-2):

$$\begin{aligned} &\boldsymbol{\zeta}^\varepsilon \in \mathbf{V}_F(\omega) \text{ and, for all } \boldsymbol{\eta} \in \mathbf{V}_F(\omega), \\ &\frac{\varepsilon^3}{3} \int_\omega a^{\alpha\beta\sigma\tau,\varepsilon} \rho_{\sigma\tau}(\boldsymbol{\zeta}^\varepsilon) \rho_{\alpha\beta}(\boldsymbol{\eta}) \sqrt{a}\, dy = \int_\omega p^{i,\varepsilon} \eta_i \sqrt{a}\, dy. \end{aligned}$$

5. KOITER'S EQUATIONS FOR A LINEARLY ELASTIC SHELL

Minimization problem (Thm. 7.1-1):

$$\begin{aligned} &\boldsymbol{\zeta}_K^\varepsilon \in \mathbf{V}_K(\omega) = \{\boldsymbol{\eta} = (\eta_i) \in H^1(\omega) \times H^1(\omega) \times H^2(\omega); \\ &\qquad\qquad \eta_i = \partial_\nu \eta_3 = 0 \text{ on } \gamma_0\}, \\ &j_K^\varepsilon(\boldsymbol{\zeta}_K^\varepsilon) = \inf_{\boldsymbol{\eta} \in \mathbf{V}_K(\omega)} j_K^\varepsilon(\boldsymbol{\eta}), \text{ where} \\ &j_K^\varepsilon(\boldsymbol{\eta}) = \frac{1}{2} \int_\omega \Big\{ \varepsilon a^{\alpha\beta\sigma\tau,\varepsilon} \gamma_{\sigma\tau}(\boldsymbol{\eta}) \gamma_{\alpha\beta}(\boldsymbol{\eta}) \\ &\qquad + \frac{\varepsilon^3}{3} a^{\alpha\beta\sigma\tau,\varepsilon} \rho_{\sigma\tau}(\boldsymbol{\eta}) \rho_{\alpha\beta}(\boldsymbol{\eta}) \Big\} \sqrt{a}\, dy - \int_\omega p^{i,\varepsilon} \eta_i \sqrt{a}\, dy. \end{aligned}$$

Variational problem (Sect. 7.1):

$$\boldsymbol{\zeta}_K^\varepsilon \in \mathbf{V}_K(\omega) \text{ and, for all } \boldsymbol{\eta} \in \mathbf{V}_K(\omega),$$

$$\int_\omega \Big\{ \varepsilon a^{\alpha\beta\sigma\tau,\varepsilon} \gamma_{\sigma\tau}(\boldsymbol{\zeta}_K^\varepsilon)\gamma_{\alpha\beta}(\boldsymbol{\eta}) + \frac{\varepsilon^3}{3} a^{\alpha\beta\sigma\tau,\varepsilon} \rho_{\sigma\tau}(\boldsymbol{\zeta}_K^\varepsilon)\rho_{\alpha\beta}(\boldsymbol{\eta}) \Big\} \sqrt{a}\, dy$$
$$= \int_\omega p^{i,\varepsilon} \eta_i \sqrt{a}\, dy \text{ for all } \boldsymbol{\eta} = (\eta_i) \in \mathbf{V}_K(\omega),$$

Boundary value problem (Thm. 7.1-2 and Ex. 7.2):

$$\begin{aligned}
m^{\alpha\beta,\varepsilon}|_{\alpha\beta} - b_\alpha^\sigma b_{\sigma\beta} m^{\alpha\beta,\varepsilon} - b_{\alpha\beta} n^{\alpha\beta,\varepsilon} &= p^{3,\varepsilon} \text{ in } \omega,\\
-(n^{\alpha\beta,\varepsilon} + b_\sigma^\alpha m^{\sigma\beta,\varepsilon})|_\beta - b_\sigma^\alpha (m^{\sigma\beta,\varepsilon}|_\beta) &= p^{\alpha,\varepsilon} \text{ in } \omega,\\
\zeta_{i,K}^\varepsilon = \partial_\nu \zeta_{3,K}^\varepsilon &= 0 \text{ on } \gamma_0,\\
m^{\alpha\beta,\varepsilon} \nu_\alpha \nu_\beta &= 0 \text{ on } \gamma_1,\\
(m^{\alpha\beta,\varepsilon}|_\alpha)\nu_\beta + \partial_\tau (m^{\alpha\beta,\varepsilon} \nu_\alpha \tau_\beta) &= 0 \text{ on } \gamma_1,\\
(n^{\alpha\beta,\varepsilon} + 2 b_\sigma^\alpha m^{\sigma\beta,\varepsilon})\nu_\beta &= 0 \text{ on } \gamma_1,
\end{aligned}$$

where

$$m^{\alpha\beta,\varepsilon} = \frac{\varepsilon^3}{3} a^{\alpha\beta\sigma\tau,\varepsilon} \rho_{\sigma\tau}(\boldsymbol{\zeta}_K^\varepsilon) \text{ and } n^{\alpha\beta,\varepsilon} = \varepsilon a^{\alpha\beta\sigma\tau,\varepsilon} \gamma_{\sigma\tau}(\boldsymbol{\zeta}_K^\varepsilon).$$

Minimization problem for a simply supported shell whose middle surface has little regularity (Thm. 7.1-2):

$$\tilde{\boldsymbol{\zeta}}_K^\varepsilon \in \widetilde{\mathbf{V}}_K^s(\omega) = \{\tilde{\boldsymbol{\eta}} \in \mathbf{H}_0^1(\omega);\ \partial_{\alpha\beta}\tilde{\boldsymbol{\eta}} \cdot \boldsymbol{a}_3 \in L^2(\omega)\},$$

$$\tilde{j}_K^\varepsilon(\tilde{\boldsymbol{\zeta}}_K^\varepsilon) = \inf_{\tilde{\boldsymbol{\eta}} \in \widetilde{\mathbf{V}}_K^s(\omega)} \tilde{j}_K^\varepsilon(\tilde{\boldsymbol{\eta}}), \text{ where}$$

$$\tilde{j}_K^\varepsilon(\tilde{\boldsymbol{\eta}}) = \frac{1}{2}\int_\omega \Big\{ \varepsilon a^{\alpha\beta\sigma\tau,\varepsilon} \tilde{\gamma}_{\sigma\tau}(\tilde{\boldsymbol{\eta}})\tilde{\gamma}_{\alpha\beta}(\tilde{\boldsymbol{\eta}})$$
$$+ \frac{\varepsilon^3}{3} a^{\alpha\beta\sigma\tau,\varepsilon} \tilde{\rho}_{\sigma\tau}(\tilde{\boldsymbol{\eta}})\tilde{\rho}_{\alpha\beta}(\tilde{\boldsymbol{\eta}}) \Big\} \sqrt{a}\, dy - \int_\omega \tilde{\boldsymbol{p}}^\varepsilon \cdot \tilde{\boldsymbol{\eta}} \sqrt{a}\, dy.$$

TWO-DIMENSIONAL NONLINEAR SHELL EQUATIONS AT A GLANCE[1]

1. Nonlinearly elastic membrane shell (equations derived by means of a formal asymptotic analysis)
2. Nonlinearly elastic flexural shells (equations derived by means of a formal asymptotic analysis)
3. Nonlinearly elastic membrane shells (equations derived by means of Γ-convergence theory)

The equations listed in Subsects. 1 and 2 are justified by a *formal asymptotic method* for nonlinearly elastic *St Venant-Kirchhoff materials*. Those in Subsect. 3 are justified by a *convergence theorem* for a more general class of hyperelastic materials, which includes St Venant-Kirchhoff ones.

In each case, the shell is subjected to a boundary condition of place along the portion $\boldsymbol{\Theta}(\gamma_0 \times [-\varepsilon, \varepsilon])$ of its lateral face, i.e., the displacement field vanishes there.

Frequent use is made of notations whose definitions, not repeated here, are found in the preceding subsections *"Differential geometry of surfaces in* $\mathbb{R}^3$*"* and *"General notations for shell equations"*.

[1]Nonlinear *shallow shell* equations in curvilinear coordinates are not reproduced here, as their justifications rely on asymptotic analyses similar to those of Vol. II, Sects. 4.14 and 5.12; they are nevertheless reviewed in Sect. 11.3.

1. NONLINEARLY ELASTIC MEMBRANE SHELL (EQUATIONS DERIVED BY MEANS OF A FORMAL ASYMPTOTIC ANALYSIS)

Definition (Sect. 9.1): A shell is a *nonlinearly elastic membrane shell* if the manifold

$$\mathcal{M}_0(\omega) = \{\boldsymbol{\eta} = (\eta_i) \in \mathbf{W}^{1,4}(\omega);\ \boldsymbol{\eta} = \mathbf{0} \text{ on } \gamma_0,\ a_{\alpha\beta}(\boldsymbol{\eta}) - a_{\alpha\beta} = 0 \text{ in } \omega\}$$

reduces to $\{\mathbf{0}\}$.

Minimization problem (Thm. 9.4-1):

$$\boldsymbol{\zeta}^\varepsilon \in \mathbf{W}_M(\omega) = \{\boldsymbol{\eta} \in \mathbf{W}^{1,4}(\omega);\ \boldsymbol{\eta} = \mathbf{0} \text{ on } \gamma_0\},$$

$$j_M^\varepsilon(\boldsymbol{\zeta}^\varepsilon) = \inf_{\boldsymbol{\eta} \in \mathbf{W}_M(\omega)} j_M^\varepsilon(\boldsymbol{\eta}), \text{ where}$$

$$j_M^\varepsilon(\boldsymbol{\eta}) = \frac{\varepsilon}{8} \int_\omega a^{\alpha\beta\sigma\tau,\varepsilon}(a_{\sigma\tau}(\boldsymbol{\eta}) - a_{\sigma\tau})(a_{\alpha\beta}(\boldsymbol{\eta}) - a_{\alpha\beta})\sqrt{a}\,dy - \int_\omega p^{i,\varepsilon}\eta_i\sqrt{a}\,dy.$$

Variational problem (Thm. 9.4-1):

$$\boldsymbol{\zeta}^\varepsilon \in \mathbf{W}_M(\omega) \text{ and, for all } \boldsymbol{\eta} \in \mathbf{W}_M(\omega),$$

$$\varepsilon \int_\omega a^{\alpha\beta\sigma\tau,\varepsilon} G_{\sigma\tau}(\boldsymbol{\zeta}^\varepsilon)(G'_{\alpha\beta}(\boldsymbol{\zeta}^\varepsilon)\boldsymbol{\eta})\sqrt{a}\,dy = \int_\omega p^{i,\varepsilon}\eta_i\sqrt{a}\,dy.$$

Boundary value problem (Thm. 9.4-2):

$$\begin{aligned}
-(n^{\alpha\beta,\varepsilon} + n^{\sigma\beta,\varepsilon}a^{\alpha\tau}\zeta^\varepsilon_{\tau\|\sigma})|_\beta + b^\alpha_\beta n^{\sigma\beta,\varepsilon}\zeta^\varepsilon_{3\|\sigma} &= p^{\alpha,\varepsilon} \text{ in } \omega,\\
-b_{\alpha\beta}(n^{\alpha\beta,\varepsilon} + n^{\sigma\beta,\varepsilon}a^{\alpha\tau}\zeta^\varepsilon_{\tau\|\sigma}) - (n^{\alpha\beta,\varepsilon}\zeta^\varepsilon_{3\|\alpha})|_\beta &= p^{3,\varepsilon} \text{ in } \omega,\\
\zeta^\varepsilon_i &= 0 \text{ on } \gamma_0,\\
(n^{\alpha\beta,\varepsilon} + n^{\sigma\beta,\varepsilon}a^{\alpha\tau}\zeta^\varepsilon_{\tau\|\sigma})\nu_\beta &= 0 \text{ on } \gamma_1,\\
n^{\alpha\beta}\zeta^\varepsilon_{3\|\alpha}\nu_\beta &= 0 \text{ on } \gamma_1,
\end{aligned}$$

where $n^{\alpha\beta,\varepsilon} = \varepsilon a^{\alpha\beta\sigma\tau,\varepsilon}\mathcal{E}_{\alpha\|\beta}(\zeta^\varepsilon)$.

2. NONLINEARLY ELASTIC FLEXURAL SHELLS (EQUATIONS DERIVED BY MEANS OF A FORMAL ASYMPTOTIC ANALYSIS)

Definition (Sect. 10.2): A shell is a *nonlinearly elastic flexural shell* if the manifold

$$\boldsymbol{\mathcal{M}}_F(\omega) = \{\boldsymbol{\eta} \in \mathbf{W}^{2,4}(\omega); \boldsymbol{\eta} = \partial_\nu \boldsymbol{\eta} = \mathbf{0} \text{ on } \gamma_0, \\ a_{\alpha\beta}(\boldsymbol{\eta}) - a_{\alpha\beta} = 0 \text{ in } \omega\}$$

contains nonzero elements and, at each $\boldsymbol{\zeta} \in \boldsymbol{\mathcal{M}}_F(\omega)$, the tangent space to the manifold $\boldsymbol{\mathcal{M}}_F(\omega)$ contains nonzero elements.

Minimization problem (Thm. 10.5-1):

$$\boldsymbol{\zeta}^\varepsilon \in \boldsymbol{\mathcal{M}}_F(\omega) \text{ and } j_F^\varepsilon(\boldsymbol{\zeta}) = \inf_{\boldsymbol{\eta} \in \boldsymbol{\mathcal{M}}_F(\omega)} j_F^\varepsilon(\boldsymbol{\eta}), \text{ where}$$
$$j_F^\varepsilon(\boldsymbol{\eta}) = \frac{\varepsilon^3}{6} \int_\omega a^{\alpha\beta\sigma\tau,\varepsilon} (b_{\sigma\tau}(\boldsymbol{\eta}) - b_{\sigma\tau})(b_{\alpha\beta}(\boldsymbol{\eta}) - b_{\alpha\beta}) \sqrt{a}\, dy \\ - \int_\omega p^{i,\varepsilon} \eta_i \sqrt{a}\, dy.$$

Variational problem (cf. Thm. 10.5-1; the functions $R^\flat_{\alpha\beta}(\boldsymbol{\eta})$ are defined in Thm. 10.3-1):

$$\boldsymbol{\zeta}^\varepsilon \in \boldsymbol{\mathcal{M}}_F(\omega) \text{ and}$$
$$\frac{\varepsilon^3}{3} \int_\omega a^{\alpha\beta\sigma\tau,\varepsilon} R^\flat_{\sigma\tau}(\boldsymbol{\zeta}^\varepsilon)\big((R^\flat_{\alpha\beta})'(\boldsymbol{\zeta}^\varepsilon)\boldsymbol{\eta}\big) \sqrt{a}\, dy = \int_\omega p^{i,\varepsilon} \eta_i \sqrt{a}\, dy$$
$$\text{for all } \boldsymbol{\eta} \in \mathbb{T}_{\boldsymbol{\zeta}^\varepsilon} \boldsymbol{\mathcal{M}}_F(\omega) = \{\boldsymbol{\eta} \in \mathbf{W}^{2,4}(\omega);\ \boldsymbol{\eta} = \partial_\nu \boldsymbol{\eta} = \mathbf{0} \text{ on } \gamma_0, \\ G'_{\alpha\beta}(\boldsymbol{\zeta}^\varepsilon)\boldsymbol{\eta} = 0 \text{ in } \omega\}.$$

3. NONLINEARLY ELASTIC MEMBRANE SHELLS (EQUATIONS DERIVED BY MEANS OF Γ-CONVERGENCE THEORY)

The shell is made of a homogeneous hyperelastic material characterized by a stored energy function $\hat{W} : \mathbb{M}^3 \to \mathbb{R}$ satisfying *ad hoc* growth assumptions "governed" by an exponent $1 < p < \infty$. The shell is subjected to a boundary condition of place along its entire lateral face, i.e., $\gamma_0 = \gamma$.

The minimization problem is expressed in terms of the usual *curvilinear* coordinates (those of the points of ω). However, the unknowns are no longer the covariant components, but instead the *Cartesian components*, of the displacement field $\tilde{\boldsymbol{\zeta}}^\varepsilon$ of the middle surface.

Minimization problem (Thm. 9.5-1):

$$\tilde{\boldsymbol{\zeta}}^\varepsilon = \zeta_i^\varepsilon \boldsymbol{a}^i \in \mathbf{W}_0^{1,p}(\omega) \text{ and } \tilde{j}_M^\varepsilon(\tilde{\boldsymbol{\zeta}}^\varepsilon) = \inf_{\tilde{\boldsymbol{\eta}} \in \mathbf{W}_0^{1,p}(\omega)} \tilde{j}_M^\varepsilon(\tilde{\boldsymbol{\eta}}), \text{ where}$$

$$\tilde{j}_M^\varepsilon(\boldsymbol{\eta}) = 2\varepsilon \int_\omega \mathcal{Q}\hat{W}_0(y, (\boldsymbol{a}_1 + \partial_1 \tilde{\boldsymbol{\eta}}; \boldsymbol{a}_2 + \partial_2 \tilde{\boldsymbol{\eta}}))\sqrt{a}\, dy - \int_\omega \tilde{\boldsymbol{p}}^\varepsilon \cdot \tilde{\boldsymbol{\eta}} \sqrt{a}\, dy,$$

$$\hat{W}_0(y, (\boldsymbol{b}_1; \boldsymbol{b}_2)) = \inf_{\boldsymbol{b}_3 \in \mathbb{R}^3} \hat{W}\big((\boldsymbol{b}_1; \boldsymbol{b}_2; \boldsymbol{b}_3)(\boldsymbol{a}^1(y), \boldsymbol{a}^2(y), \boldsymbol{a}^3(y))\big),$$

$\mathcal{Q}\hat{W}_0(y, \cdot)$: quasiconvex envelope of $\hat{W}_0(y, \cdot)$ at $y \in \overline{\omega}$.

Let $E^{\alpha\beta\lambda\mu}$ denote the tensor of elastic moduli for plane stress at the middle surface

$$E^{\alpha\beta\lambda\mu}=G\left[a^{\alpha\lambda}a^{\beta\mu}+a^{\alpha\mu}a^{\beta\lambda}+\frac{2\nu}{1-\nu}a^{\alpha\beta}a^{\lambda\mu}\right], \tag{8.1}$$

where G is the shear modulus and ν is POISSON's ratio. We have shown elsewhere [14] that the elastic strain energy per unit area of the undeformed middle surface, under the assumptions (a) – (c), is approximated by

$$\tfrac{1}{2}hE^{\alpha\beta\lambda\mu}\gamma_{\alpha\beta}\gamma_{\lambda\mu}+\frac{1}{24}h^3E^{\alpha\beta\lambda\mu}\gamma_{\alpha\beta\|3}\gamma_{\lambda\mu\|3}. \tag{8.2}$$

Let ε denote the maximum principal strain, R the smallest principal radius of curvature of the middle surface, and L the "wave length" of the deformation pattern on the middle surface. The relative error of the approximation (8.2) is then of order ε, h/R or h^2/L^2, whichever of these quantities may be critical. The relative error also remains of the same order, if we replace the normal covariant derivatives of the strain tensor in (8.2) by the corresponding partial derivatives or by any of the measures for the changes of curvature $\bar{\varrho}_{\alpha\beta}$, $\varrho_{\alpha\beta}$ or $\varrho^*_{\alpha\beta}$ of section 4.

It has recently been shown by JOHN [10] that the assumption (c) of approximately plane stress, in the case of vanishing surface loads, is actually a consequence of assumptions (a) and (b). Moreover, it was proved in [10] that the wave length L may be replaced by the distance of the point on the middle surface under consideration to the edge curve. Finally, JOHN's analysis provides a means to replace the qualitative error estimate in [14] by a concrete quantitative estimate. Expression (8.2) for the elastic energy per unit area of the undeformed middle surface, and similar expressions in the alternative measures for the changes of curvature, may now be considered firmly established.

Let $\bar{V}$ denote the approximate expression for the strain energy per unit area of the undeformed middle surface in terms of $\gamma_{\alpha\beta}$ and $\bar{\varrho}_{\alpha\beta}$

$$\bar{V}=\tfrac{1}{2}hE^{\alpha\beta\lambda\mu}\gamma_{\alpha\beta}\gamma_{\lambda\mu}+\frac{1}{24}h^3E^{\alpha\beta\lambda\mu}\bar{\varrho}_{\alpha\beta}\bar{\varrho}_{\lambda\mu}. \tag{8.3}$$

The internal work per unit area of the deformed middle surface in an arbitrary virtual deformation, specified by (6.1), may be equated to the variation of the elastic energy in the same virtual deformation, also evaluated per unit area of the deformed middle surface. Hence we have

$$\bar{n}^{\alpha\beta}=\tfrac{1}{2}\sqrt{a/\bar{a}}\left(\frac{\partial\bar{V}}{\partial\gamma_{\alpha\beta}}+\frac{\partial\bar{V}}{\partial\gamma_{\beta\alpha}}\right),\quad \bar{m}^{\alpha\beta}=\tfrac{1}{2}\sqrt{a/\bar{a}}\left(\frac{\partial\bar{V}}{\partial\bar{\varrho}_{\alpha\beta}}+\frac{\partial\bar{V}}{\partial\bar{\varrho}_{\beta\alpha}}\right). \tag{8.4}$$

In view of our assumption of small strains, we may again put $\sqrt{a/\bar{a}}$ equal to unity, and we obtain the constitutive equations for the strain measures

Reprinted from KOITER, W.T.: On the nonlinear theory of thin elastic shells, in *Proceedings, Koninklijke Nederlandse Akademie van Wetenschappen, Series B*, Vol. 69 (1966), pp. 1-54.

PART A

LINEAR SHELL THEORY

CHAPTER 1

THREE-DIMENSIONAL LINEARIZED ELASTICITY AND KORN'S INEQUALITIES IN CURVILINEAR COORDINATES

INTRODUCTION

The *equations of three-dimensional linearized elasticity* have been extensively studied, but mostly in *Cartesian coordinates.* It is well known for instance that the *existence* and *uniqueness* of their solution depend in a crucial way on *Korn's inequality*, again expressed in *Cartesian coordinates.*

The linear shell theories justified in this volume from three-dimensional elasticity require, however, that these equations be rather expressed in terms of *curvilinear coordinates* that "follow the geometry" of the shell in a most natural way.

Accordingly, the purpose of this preliminary chapter is to provide a thorough derivation and a mathematical treatment of the equations of linearized three-dimensional elasticity in terms of arbitrary *curvilinear coordinates.* This chapter includes in particular a direct proof of the *three-dimensional Korn inequality in curvilinear coordinates.*

The treatment is entirely self-contained, in that no *a priori* knowledge of differential geometry is assumed.

More specifically, we begin by reviewing (Sect. 1.2) the basic definitions and properties arising when a three-dimensional set of the form $\{\hat{\Omega}\}^- = \boldsymbol{\Theta}(\overline{\Omega})$, where Ω is a three-dimensional domain in $\mathbb{R}^3$ and $\boldsymbol{\Theta}$ is a smooth injective mapping with other *ad hoc* properties, is equipped with the coordinates of the points of $\overline{\Omega}$ as its *curvilinear coordinates.* Of particular importance is the *metric tensor* of the domain $\{\hat{\Omega}\}^-$, whose covariant and contravariant components $g_{ij} = g_{ji} : \overline{\Omega} \to \mathbb{R}$ and $g^{ij} = g^{ji} : \overline{\Omega} \to \mathbb{R}$ are given by (Latin indices or exponents take their values in $\{1, 2, 3\}$):

$$g_{ij} = \boldsymbol{g}_i \cdot \boldsymbol{g}_j \text{ and } g^{ij} = \boldsymbol{g}^i \cdot \boldsymbol{g}^j, \text{ where } \boldsymbol{g}_i = \partial_i \boldsymbol{\Theta} \text{ and } \boldsymbol{g}^j \cdot \boldsymbol{g}_i = \delta_i^j.$$

Further properties of the metric tensor are also reviewed in Sect. 1.9.

We then examine how the equations of three-dimensional linearized elasticity expressed in terms of the *Cartesian coordinates* of the points in the set $\{\hat{\Omega}\}^-$ (this formulation is reviewed in Sect. 1.1) are transformed when they are expressed in terms of *curvilinear coordinates*, i.e., in terms of the coordinates of the points in the set $\overline{\Omega} = \boldsymbol{\Theta}^{-1}(\{\hat{\Omega}\}^-)$. We begin by showing, by means of a direct (but not so obvious!) computation (Thm. 1.3-1), that the *variational*, or *weak, formulation* of these equations takes the form:

$$\boldsymbol{u} = (u_i) \in \mathbf{V}(\Omega) := \{\boldsymbol{v} = (v_i) \in \mathbf{H}^1(\Omega);\ \boldsymbol{v} = \mathbf{0} \text{ on } \Gamma_0\},$$

$$\int_\Omega A^{ijkl} e_{k\|l}(\boldsymbol{u}) e_{i\|j}(\boldsymbol{v}) \sqrt{g}\, dx = \int_\Omega f^i v_i \sqrt{g}\, dx + \int_{\Gamma_1} h^i v_i \sqrt{g}\, d\Gamma$$

for all $\boldsymbol{v} = (v_i) \in \mathbf{V}(\Omega)$, where $u_i \boldsymbol{g}^i$ is the unknown displacement vector field inside the set $\{\hat{\Omega}\}^-$, Γ_0 is a subset of $\partial\Omega$ with *area* $\Gamma_0 > 0$,

$$A^{ijkl} = \lambda g^{ij} g^{kl} + \mu(g^{ik} g^{jl} + g^{il} g^{jk}),$$

λ and μ are the Lamé constants of the constituting elastic material, $g = \det(g_{ij})$, the functions $f^i \in L^2(\Omega)$ and $h^i \in L^2(\Gamma_1)$, where $\Gamma_1 = \partial\Omega - \Gamma_0$, account for the applied body and surface forces, and the *linearized strains in curvilinear coordinates* $e_{i\|j}(\boldsymbol{v}) \in L^2(\Omega)$ are defined for each $\boldsymbol{v} = (v_i) \in \mathbf{H}^1(\Omega)$ by

$$e_{i\|j}(\boldsymbol{v}) = \frac{1}{2}(\partial_j v_i + \partial_i v_j) - \Gamma^p_{ij} v_p, \text{ where } \Gamma^p_{ij} = \boldsymbol{g}^p \cdot \partial_i \boldsymbol{g}_j.$$

The interest of this calculation is that it *naturally* introduces essential notions, such as the *covariant derivatives* $v_{i\|j} = \partial_j v_i - \Gamma^p_{ij} v_p$ of a vector field $v_i \boldsymbol{g}^i$ and the *linearized change of metric tensor* associated with this vector field, found here by means of its covariant components

$$e_{i\|j}(\boldsymbol{v}) = \frac{1}{2}(v_{i\|j} + v_{j\|i}).$$

These notions are then studied in greater details for their own sake (Sects. 1.4 and 1.5).

We also describe in detail (Thm. 1.6-1) the equivalent *boundary value problem* when it is likewise expressed in terms of *curvilinear coordinates.*

Finally, we show how a fundamental *lemma of J.L. Lions* (Thm. 1.7-1) can be put to use for directly establishing *Korn's inequality in curvilinear coordinates*: This inequality, which asserts the existence of a constant C such that (Thm. 1.7-4)

$$\|\boldsymbol{v}\|_{1,\Omega} \leq C\Big\{\sum_{i,j} |e_{i\|j}(\boldsymbol{v})|^2_{0,\Omega}\Big\}^{1/2} \text{ for all } \boldsymbol{v} \in \mathbf{V}(\Omega),$$

in turn yields (Thm. 1.8-2) the *existence* and *uniqueness* of a solution to the variational formulation of the equations of three-dimensional linearized elasticity, again directly in *curvilinear coordinates.*

1.1. THREE-DIMENSIONAL LINEARIZED ELASTICITY IN CARTESIAN COORDINATES

Throughout this volume, *Latin* indices and exponents vary in the set $\{1, 2, 3\}$ (except if otherwise indicated, as when they are used for indexing sequences), and the summation convention with respect to repeated indices or exponents is systematically used in conjunction with this rule. Thus for instance,

$$\hat{x}_i \hat{\boldsymbol{e}}^i = \sum_{i=1}^{3} \hat{x}_i \hat{\boldsymbol{e}}^i, \quad f^i v_i = \sum_{i=1}^{3} f^i v_i,$$

$$\hat{e}_{pp}(\hat{\boldsymbol{u}}) = \sum_{p=1}^{3} \hat{e}_{pp}(\hat{\boldsymbol{u}}), \quad \hat{e}_{ij}(\hat{\boldsymbol{u}})\hat{e}_{ij}(\hat{\boldsymbol{v}}) = \sum_{i,j=1}^{3} \hat{e}_{ij}(\hat{\boldsymbol{u}})\hat{e}_{ij}(\hat{\boldsymbol{v}}),$$

"$e_{i\|j}(\boldsymbol{v}) = 0$ in Ω" means "$e_{i\|j}(\boldsymbol{v}) = 0$ in Ω for $i, j = 1, 2, 3$,"

"$\hat{A}^{ijkl}\hat{e}_{kl}(\hat{\boldsymbol{u}})\hat{n}_j = \hat{h}^i$ on $\hat{\Gamma}_1$"

means

"$\sum_{j,k,l=1}^{3} \hat{A}^{ijkl}\hat{e}_{kl}(\hat{\boldsymbol{u}})\hat{n}_j = \hat{h}^i$ on $\hat{\Gamma}_1$ for $i = 1, 2, 3$," etc.

As a model of the three-dimensional "physical" space $\mathcal{E}^3$, we take a **real three-dimensional affine Euclidean space**, i.e., a set in which a point $\hat{O}$ has been chosen as the **origin** and with which is associated a **real three-dimensional Euclidean space**, denoted $\mathbf{E}^3$, endowed with an **orthonormal basis** consisting of three vectors $\hat{\boldsymbol{e}}^i = \hat{\boldsymbol{e}}_i$; "orthonormal" means that $\hat{\boldsymbol{e}}_i \cdot \hat{\boldsymbol{e}}_j = \delta_{ij}$ where $\boldsymbol{a} \cdot \boldsymbol{b} = a_i b_i$ denotes the **Euclidean inner product** of $\boldsymbol{a} = a_i \hat{\boldsymbol{e}}^i \in \mathbf{E}^3$

and $\boldsymbol{b} = b_i \hat{\boldsymbol{e}}^j \in \mathbf{E}^3$, and δ_{ij} designates the Kronecker symbol. The definition of $\mathcal{E}^3$ as an affine Euclidean space means that with any point $x \in \mathcal{E}^3$ is associated a uniquely defined vector in $\mathbf{E}^3$, denoted $\hat{\boldsymbol{O}}\hat{\boldsymbol{x}}$.

The origin $\hat{O} \in \mathcal{E}^3$ and the vectors $\hat{\mathbf{e}}_i \in \mathbf{E}^3$ together constitute a **Cartesian frame** in $\mathcal{E}^3$ and the three components $\hat{x}_i$ of the vector $\hat{\boldsymbol{O}}\hat{\boldsymbol{x}}$ over the basis formed by the vectors $\hat{\mathbf{e}}^i$ are called the **Cartesian coordinates** of $\hat{x} \in \mathcal{E}^3$ or the **Cartesian components** of $\hat{\boldsymbol{O}}\hat{\boldsymbol{x}} \in \mathbf{E}^3$. Once a Cartesian frame has been chosen, any point $\hat{x} \in \mathcal{E}^3$ may thus be *identified* with the vector $\hat{\boldsymbol{O}}\hat{\boldsymbol{x}} = \hat{x}_i \hat{\mathbf{e}}^i \in \mathbf{E}^3$, or with the vector $(\hat{x}_i) \in \mathbb{R}^3$; we then let $\hat{\partial}_i := \partial / \partial \hat{x}_i$.

Remark. The notion of affine space affords the definition of "$\hat{y}+\boldsymbol{a}$" as a point in $\mathcal{E}^3$ whenever $\hat{y}$ is a point in $\mathcal{E}^3$ and $\boldsymbol{a}$ is a vector in $\mathbf{E}^3$ in such a way that $\hat{x} = \hat{O} + \hat{\boldsymbol{O}}\hat{\boldsymbol{x}}$ for all $\hat{x} \in \mathcal{E}^3$; we refer to Schwartz [1992, Chap. 3] for an excellent introduction to affine Euclidean spaces. ■

Finally, we let $|\boldsymbol{a}| := \sqrt{\boldsymbol{a} \cdot \boldsymbol{a}}$ denote the **Euclidean norm** of $\boldsymbol{a} \in \mathbf{E}^3$ and we let $\boldsymbol{a} \wedge \boldsymbol{b} := \hat{\varepsilon}^{ijk} a_j b_k \hat{\mathbf{e}}_i$ denote the **exterior product** of $\boldsymbol{a} = a_i \hat{\mathbf{e}}^i \in \mathbf{E}^3$ and $\boldsymbol{b} = b_i \hat{\mathbf{e}}^i \in \mathbf{E}^3$, where $\hat{\varepsilon}^{ijk} = +1$ if $\{i, j, k\}$ is an even permutation of $\{1, 2, 3\}$, $\hat{\varepsilon}^{ijk} = -1$ if $\{i, j, k\}$ is an odd permutation of $\{1, 2, 3\}$, and $\hat{\varepsilon}^{ijk} = 0$ otherwise.

A **domain** in $\mathcal{E}^3$ is a bounded, open, and connected subset $\hat{\Omega}$ of $\mathcal{E}^3$ with a Lipschitz-continuous boundary $\hat{\Gamma}$, the set $\hat{\Omega}$ being locally on one side of $\hat{\Gamma}$ (this definition holds *verbatim* in $\mathbb{R}^n$; cf. Sect. 1.7).

Let $\hat{\Omega}$ be a domain in $\mathcal{E}^3$, let $d\hat{x}$ denote the *volume element* in $\hat{\Omega}$, let $d\hat{\Gamma}$ denote the *area element* along $\hat{\Gamma}$, and let $\hat{\boldsymbol{n}} = \hat{n}_i \hat{\mathbf{e}}^i$ denote the *unit* ($|\hat{\boldsymbol{n}}| = 1$) *outer normal vector* along $\hat{\Gamma}$ ($d\hat{\Gamma}$ is well defined and $\hat{\boldsymbol{n}}$ is defined $d\hat{\Gamma}$-almost everywhere since $\hat{\Gamma}$ is assumed to be Lipschitz-continuous). Finally, let $\hat{\Gamma} = \hat{\Gamma}_0 \cup \hat{\Gamma}_1$ be a $d\hat{\Gamma}$-measurable partition ($\hat{\Gamma}_0 \cap \hat{\Gamma}_1 = \phi$) of the boundary $\hat{\Gamma}$ that satisfies

$$\boxed{\textit{area}\ \hat{\Gamma}_0 > 0.}$$

The set $\{\hat{\Omega}\}^-$ is the **reference configuration** occupied by an **elastic body** in the absence of applied forces. We assume that $\{\hat{\Omega}\}^-$ is a **natural state**, i.e., that the body is "stress-free" in this configuration. We also assume that the constituting **elastic material** is **isotropic** and **homogeneous** (details about these notions are found in Vol. I[1], Sects. 3.1, 3.4, and 3.6). Under these assumptions, the behavior of the elastic material for "small strains" is entirely governed by *two constants*, called the **Lamé constants** of its constituting material (Vol. I, Sect. 3.8). Experimental evidence shows that the Lamé constants, denoted λ and μ, of usual elastic materials such as steel or aluminum satisfy (Vol. I, Fig. 3.8-4)

$$\boxed{\lambda > 0 \text{ and } \mu > 0.}$$

The body is subjected to **applied body forces** in its interior and to **applied surface forces** on the portion $\hat{\Gamma}_1$ of its boundary, given by their *densities* $(\hat{f}_i) : \hat{\Omega} \to \mathbb{R}^3$ and $(\hat{h}_i) : \hat{\Gamma}_1 \to \mathbb{R}^3$ (Vol. I, Sect. 2.1). For definiteness, we henceforth assume that

$$\boxed{\hat{f}_i \in L^2(\hat{\Omega}) \text{ and } \hat{h}_i \in L^2(\hat{\Gamma}_1).}$$

Remark. In order to avoid any confusion with the notation $\boldsymbol{g}_i$ used later for the vectors of the covariant basis (Sect. 1.2), the components of the applied surface force density are denoted $\hat{h}_i$ in this volume; otherwise the notation $\hat{g}_i$ would have been consistent with that used in Vols. I and II[2]. ■

The *unknown* is the vector field $\hat{\boldsymbol{u}} = (\hat{u}_i) : \{\hat{\Omega}\}^- \to \mathbb{R}^3$, where the three functions $\hat{u}_i : \{\hat{\Omega}\}^- \to \mathbb{R}$ are the **Cartesian components** of the **displacement field** $\hat{u}_i \hat{\boldsymbol{e}}^i : \{\hat{\Omega}\}^- \to \mathbf{E}^3$ that the body undergoes when it is subjected to the applied forces. This means that $\hat{u}_i(\hat{x})\hat{\boldsymbol{e}}^i$ *is the displacement of the point* $\hat{x} \in \{\hat{\Omega}\}^-$ (Fig. 1.1-1). As $\mathbf{E}^3$ may be identified with $\mathbb{R}^3$, we may identify $\hat{u}_i(\hat{x})\hat{\boldsymbol{e}}^i \in \mathbf{E}^3$ with the vector $\hat{\boldsymbol{u}}(\hat{x}) \in \mathbb{R}^3$ and write accordingly the displacement field as

$$\boxed{\hat{\boldsymbol{u}} = \hat{u}_i \hat{\boldsymbol{e}}^i.}$$

Note that this identification has been implicitly made throughout Vols. I and II.

[1] "Vol. I" stands for "Ciarlet, P.G. [1988]: *Mathematical Elasticity, Volume I: Three-Dimensional Elasticity*, North Holland, Amsterdam."

[2] "Vol. II" stands for "Ciarlet, P.G. [1997]: *Mathematical Elasticity, Volume II: Theory of Plates*, North-Holland, Amsterdam."

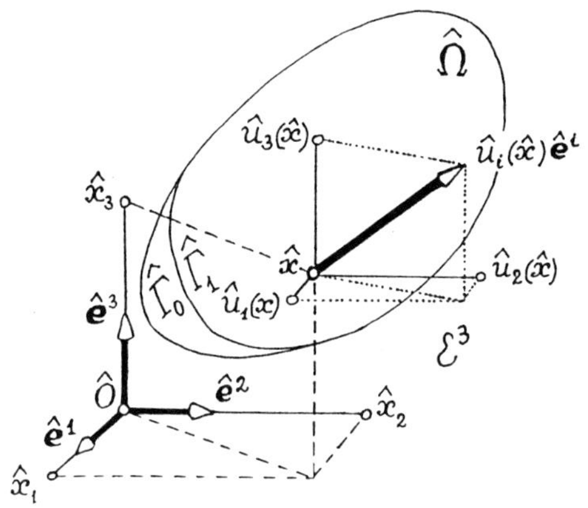

Fig. 1.1-1: *Three-dimensional elasticity in Cartesian coordinates.* Let there be given an elastic body whose reference configuration is the closure of a domain $\hat{\Omega}$ in the "physical space", modeled as a real three-dimensional affine Euclidean space $\mathcal{E}^3$. The displacement vector $\hat{u}_i(\hat{x})\hat{\mathbf{e}}^i$ of the point $\hat{x} \in \{\hat{\Omega}\}^-$ is defined by its Cartesian components $\hat{u}_i(\hat{x})$ in a Cartesian frame of $\mathcal{E}^3$, consisting of an origin $\hat{O}$ and an orthonormal basis formed by three vectors $\hat{\mathbf{e}}^i$.

It is further assumed that *the displacement field vanishes on the set* $\hat{\Gamma}_0$, i.e., that it satisfies the **boundary condition of place**

$$\boxed{\hat{\boldsymbol{u}} = \mathbf{0} \text{ on } \hat{\Gamma}_0.}$$

Note that, unless otherwise specified, a "boundary condition of place" is *always* meant in this volume to be *homogeneous*.

Then in *linearized elasticity, the unknown* $\hat{\boldsymbol{u}} = (\hat{u}_i)$ *satisfies the following variational problem*, which constitutes the **variational**, or **weak, formulation**, of the **equations of three-dimensional linearized elasticity in Cartesian coordinates** (Vol. I, Sects. 6.2 and 6.3):

$$\boxed{\begin{aligned}&\hat{\boldsymbol{u}} \in \mathbf{V}(\hat{\Omega}) := \{\hat{\boldsymbol{v}} = (\hat{v}_i) \in \mathbf{H}^1(\hat{\Omega});\ \hat{\boldsymbol{v}} = \mathbf{0} \text{ on } \hat{\Gamma}_0\},\\ &\int_{\hat{\Omega}} \{\lambda \hat{e}_{pp}(\hat{\boldsymbol{u}})\hat{e}_{qq}(\hat{\boldsymbol{v}}) + 2\mu \hat{e}_{ij}(\hat{\boldsymbol{u}})\hat{e}_{ij}(\hat{\boldsymbol{v}})\}\, d\hat{x}\\ &\qquad = \int_{\hat{\Omega}} \hat{f}_i \hat{v}_i\, d\hat{x} + \int_{\hat{\Gamma}_1} \hat{h}_i \hat{v}_i\, d\hat{\Gamma} \text{ for all } \hat{\boldsymbol{v}} \in \mathbf{V}(\hat{\Omega}),\end{aligned}}$$

where $\mathbf{H}^1(\hat{\Omega})$ denotes the space of vector fields $\hat{\boldsymbol{v}} = (\hat{v}_i)$ with components $\hat{v}_i$ in the Sobolev space $H^1(\hat{\Omega})$, and the functions

$$\boxed{\hat{e}_{ij}(\hat{\boldsymbol{v}}) := \frac{1}{2}(\hat{\partial}_j \hat{v}_i + \hat{\partial}_i \hat{v}_j) \in L^2(\hat{\Omega})}$$

denote the components of the **linearized change of metric tensor**

$$\boxed{\hat{\boldsymbol{e}}(\hat{\boldsymbol{v}}) := (\hat{e}_{ij}(\hat{\boldsymbol{v}}))}$$

in Cartesian coordinates associated with an arbitrary displacement field $\hat{\boldsymbol{v}} = \hat{v}_i \hat{\boldsymbol{e}}^i = (\hat{v}_i) \in \mathbf{V}(\hat{\Omega})$. The functions $\hat{e}_{ij}(\hat{\boldsymbol{v}})$ are also called the **linearized strains in Cartesian coordinates** and the tensor $\hat{\boldsymbol{e}}(\hat{\boldsymbol{v}})$ is also called the **linearized strain tensor**.

In view of its importance and of its subsequent generalization to curvilinear coordinates (Sect. 1.5), we briefly recall the significance of this tensor. Let $\hat{\boldsymbol{v}} : \{\hat{\Omega}\}^- \to \mathbb{R}^3$ be an arbitrary displacement field of the reference configuration $\{\hat{\Omega}\}^-$ and let

$$\hat{\boldsymbol{\psi}} = (\hat{\psi}_i) := \boldsymbol{id}_{\mathbb{R}^3} + \hat{\boldsymbol{v}}$$

be the associated **deformation**. Then the "new" metric in the associated **deformed configuration** $\hat{\boldsymbol{\psi}}(\{\hat{\Omega}\}^-)$ is given by the associated (right) **Cauchy-Green strain tensor** defined by (Vol. I, Sect. 1.8)

$$\hat{\boldsymbol{\nabla}}\hat{\boldsymbol{\psi}}^T \hat{\boldsymbol{\nabla}}\hat{\boldsymbol{\psi}} = \mathbf{I} + \hat{\boldsymbol{\nabla}}\hat{\boldsymbol{v}}^T + \hat{\boldsymbol{\nabla}}\hat{\boldsymbol{v}} + \hat{\boldsymbol{\nabla}}\hat{\boldsymbol{v}}^T \hat{\boldsymbol{\nabla}}\hat{\boldsymbol{v}},$$

where the matrices $\hat{\boldsymbol{\nabla}}\hat{\boldsymbol{v}} = (\hat{\partial}_j \hat{v}_i)$ and $\hat{\boldsymbol{\nabla}}\hat{\boldsymbol{\psi}} = (\hat{\partial}_j \hat{\psi}_i)$ (i is the row index and j is the column index) respectively denote the corresponding *displacement* and *deformation gradients.* Hence

$$\hat{e}(\hat{\boldsymbol{v}}) = \frac{1}{2}[\hat{\boldsymbol{\nabla}}\hat{\boldsymbol{\psi}}^T \hat{\boldsymbol{\nabla}}\hat{\boldsymbol{\psi}} - \mathbf{I}]^{lin},$$

where $[\cdots]^{lin}$ means that only the linear part with respect to $\hat{\boldsymbol{v}}$ is retained in the expression $[\cdots]$. In other words, *the tensor* $\hat{\mathbf{e}}(\hat{\boldsymbol{v}})$ *measures* (half of) *the linearized difference between the "new" metric in* $\hat{\varphi}(\{\hat{\Omega}\}^-)$ *and the "old" metric in* $\{\hat{\Omega}\}^-$ (whose right Cauchy-Green strain tensor is $\mathbf{I}$), both expressed in terms of the Cartesian coordinates $\hat{x}_i$. This is why $\hat{\mathbf{e}}(\hat{\boldsymbol{v}})$ is called the *linearized "change of metric" tensor in Cartesian coordinates.*

It is well known that the proof of the *existence* and *uniqueness* of a solution to this *variational problem* relies on *Korn's inequality "in Cartesian coordinates"* (see, e.g., Vol. I, Sect. 6.3, or Vol. II, Sects. 1.1 and 1.2). In fact, we shall provide in this volume a direct proof of a more general *Korn's inequality "in curvilinear coordinates"* (Thm. 1.7-4), which contains Korn's inequality in Cartesian coordinates as a special case.

As is easily seen by means of Green's formula (Vol. I, Thm. 6.3-1), the above variational problem is, at least *formally*, equivalent to the following *boundary value problem* (δ_{ij} denotes the Kronecker symbol):

$$\begin{aligned} -\hat{\partial}_j\{\lambda\hat{e}_{pp}(\hat{\boldsymbol{u}})\delta_{ij} + 2\mu\hat{e}_{ij}(\hat{\boldsymbol{u}})\} &= \hat{f}_i \quad \text{in} \quad \hat{\Omega}, \\ \hat{u}_i &= 0 \quad \text{on} \quad \hat{\Gamma}_0, \\ \{\lambda\hat{e}_{pp}(\hat{\boldsymbol{u}})\delta_{ij} + 2\mu\hat{e}_{ij}(\hat{\boldsymbol{u}})\}\,\hat{n}_j &= \hat{h}_i \quad \text{on} \quad \hat{\Gamma}_1. \end{aligned}$$

Either formulation constitutes the **equations of three-dimensional linearized elasticity in Cartesian coordinates** for a *linearized displacement-traction problem* (cf. Vol. I, Sect. 6.2; "displacement-traction" refers to the kind of boundary conditions considered here).

A solution of the *variational problem*, which as such may be only in $\mathbf{H}^1(\hat{\Omega})$, is called a *weak solution*, while a solution of the *boundary value problem* (e.g., in $\mathcal{C}^2(\{\hat{\Omega}\}^-)$) is called a *classical solution.* Except in some *very* special cases, such as when the boundary $\hat{\Gamma}$ is smooth and $\hat{\Gamma}_1 = \phi$ or $\hat{\Gamma}_0$ and $\hat{\Gamma}_1$ are two connected components of $\hat{\Gamma}$, "weak solutions are seldom classical" (Vol. I, Sect. 6.3).

Remark. This boundary value problem is the linearization about $\hat{\boldsymbol{u}} = \mathbf{0}$ of the corresponding boundary value problem of *nonlinear* elasticity (considered later; cf. Sect. 8.1). This linearization is detailed in Vol. I, Sect. 6.2. ■

Let us also denote by δ^{ij} the Kronecker symbol and by $\hat{f}^i := \hat{f}_i$ and $\hat{h}^i := \hat{h}_i$ the components of the applied forces, and let the components $\hat{A}^{ijkl}$ of the **three-dimensional elasticity tensor in Cartesian coordinates** be defined by

$$\hat{A}^{ijkl} := \lambda\delta^{ij}\delta^{kl} + \mu(\delta^{ik}\delta^{jl} + \delta^{il}\delta^{jk}).$$

Then the above variational equations may also be written as

$$\int_{\hat{\Omega}} \hat{A}^{ijkl}\hat{e}_{kl}(\hat{\boldsymbol{u}})\hat{e}_{ij}(\hat{\boldsymbol{v}})\,d\hat{x} = \int_{\hat{\Omega}} \hat{f}^i\hat{v}_i\,d\hat{x} + \int_{\hat{\Gamma}_1} \hat{h}^i\hat{v}_i\,d\hat{\Gamma} \quad \text{for all } \hat{\boldsymbol{v}} \in \mathbf{V}(\hat{\Omega}),$$

and the above boundary value problem may also be written as

$$\begin{aligned} -\hat{\partial}_j\left\{\hat{A}^{ijkl}\hat{e}_{kl}(\hat{\boldsymbol{u}})\right\} &= \hat{f}^i \quad \text{in} \quad \hat{\Omega}, \\ \hat{u}_i &= 0 \quad \text{on} \quad \hat{\Gamma}_0, \\ \hat{A}^{ijkl}\hat{e}_{kl}(\hat{\boldsymbol{u}})\hat{n}_j &= \hat{h}^i \quad \text{on} \quad \hat{\Gamma}_1. \end{aligned}$$

If we introduce the **linearized stresses:**

$$\hat{\sigma}^{ij} := \hat{A}^{ijkl}\hat{e}_{kl}(\hat{\boldsymbol{u}}),$$

the variational equations take the even shorter form

$$\int_{\hat{\Omega}} \hat{\sigma}^{ij}\hat{e}_{ij}(\hat{\boldsymbol{v}})\,d\hat{x} = \int_{\hat{\Omega}} \hat{f}^i\hat{v}_i\,d\hat{x} + \int_{\hat{\Gamma}_1} \hat{h}^i\hat{v}_i\,d\hat{\Gamma} \text{ for all } \hat{\boldsymbol{v}} \in \mathbf{V}(\hat{\Omega}),$$

which constitutes the **linearized principle of virtual work in Cartesian coordinates**. The equations

$$\begin{aligned} -\hat{\partial}_j\hat{\sigma}^{ij} &= \hat{f}^i \quad \text{in} \quad \hat{\Omega}, \\ \hat{\sigma}^{ij}\hat{n}_j &= \hat{h}^i \quad \text{on} \quad \hat{\Gamma}_1, \end{aligned}$$

constitute the **linearized equations of equilibrium in Cartesian coordinates.**

The above principle of virtual work and equations of equilibrium are special cases of more general relations *"in curvilinear coordinates"* (Sects. 1.3 and 1.6). Thanks to the new notations δ^{ij}, $\hat{f}^i$, and $\hat{h}^i$, they now obey the proper rules concerning "covariant indices" and "contravariant exponents".

We also note that *the three-dimensional elasticity tensor satisfies the inequality*

$$\boxed{2\mu \sum_{i,j} |t_{ij}|^2 \leq \hat{A}^{ijkl} t_{kl} t_{ij}}$$

for all symmetric matrices (t_{ij}) and that its components satisfy the *symmetries*

$$\boxed{\hat{A}^{ijkl} = \hat{A}^{jikl} = \hat{A}^{klij}.}$$

These properties are again special cases of more general properties "in curvilinear coordinates" (Thms. 1.3-1 and 1.8-1).

Remark. Should we let instead $\hat{A}^{ijkl} = \lambda\delta^{ij}\delta^{kl} + 2\mu\delta^{ik}\delta^{jl}$, the left-hand side of the variational equations would be unaltered, but the symmetries $\hat{A}^{ijkl} = \hat{A}^{jikl}$ would be lost. Yet these are crucially needed for establishing the relation

$$\hat{A}^{ijkl}\hat{e}_{kl}(\hat{\boldsymbol{u}})\hat{e}_{ij}(\hat{\boldsymbol{v}}) = \hat{A}^{ijkl}\hat{e}_{kl}(\hat{\boldsymbol{u}})\hat{\partial}_j\hat{v}_i,$$

used in turn for deriving the partial differential equations $-\hat{\partial}_j\{\hat{A}^{ijkl}\hat{e}_{kl}(\hat{\boldsymbol{u}})\} = \hat{f}^i$ in $\hat{\Omega}$. ■

1.2. CURVILINEAR COORDINATES AND METRIC TENSOR IN A THREE-DIMENSIONAL DOMAIN

The theme of this volume is the study of **elastic shells**, i.e., of elastic bodies whose reference configuration $\{\hat{\Omega}^\varepsilon\}^- \subset \mathcal{E}^3$ consists of all points within a distance $\leq \varepsilon$ from a given surface $S \subset \mathcal{E}^3$ and $\varepsilon > 0$ is thought of as being "small" (but this last property is irrelevant at this stage); cf. Fig. 1.2-1.

Assume that the surface S is defined as the image $\boldsymbol{\theta}(\overline{\omega})$ of the closure of a domain ω in $\mathbb{R}^2$, where $\boldsymbol{\theta} : \overline{\omega} \to \mathcal{E}^3$ is a smooth injective

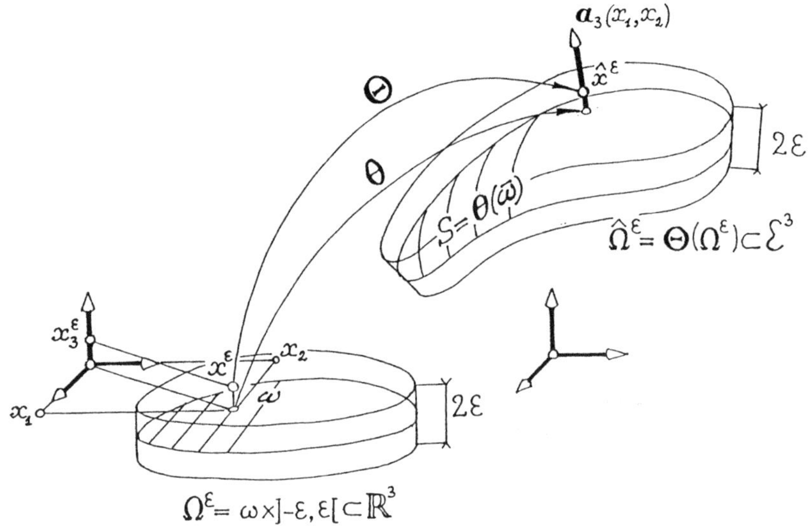

Fig. 1.2-1: *A shell and its "natural" curvilinear coordinates.* A shell is a three-dimensional elastic body occupying a reference configuration $\{\hat{\Omega}^\varepsilon\}^-$ of a specific shape, consisting of all points within a distance $\leq \varepsilon$ from a surface S, where $\varepsilon > 0$ is given. The surface S is defined as $S = \boldsymbol{\theta}(\bar{\omega})$, where $\omega \subset \mathbb{R}^2$ and $\boldsymbol{\theta} : \bar{\omega} \to \mathcal{E}^3$ is a smooth injective mapping. Each point $\hat{\boldsymbol{x}}^\varepsilon$ of $\{\hat{\Omega}^\varepsilon\}^-$ is thus of the form $\hat{\boldsymbol{x}}^\varepsilon = \boldsymbol{\Theta}(x_1, x_2, x_3^\varepsilon) = \boldsymbol{\theta}(x_1, x_2) + x_3^\varepsilon \boldsymbol{a}_3(x_1, x_2)$, where $\boldsymbol{a}_3(x_1, x_2)$ is a unit vector normal to S at the point $\boldsymbol{\theta}(x_1, x_2)$ and $(x_1, x_2, x_3^\varepsilon) \in \overline{\Omega}^\varepsilon = \bar{\omega} \times [-\varepsilon, \varepsilon]$. For $\varepsilon > 0$ small enough, the mapping $\boldsymbol{\Theta} : \overline{\Omega}^\varepsilon = \bar{\omega} \times [-\varepsilon, \varepsilon]$ is also injective and the curvilinear coordinates of any point $\hat{\boldsymbol{x}}^\varepsilon \in \{\hat{\Omega}^\varepsilon\}^-$ are then defined as the coordinates of the unique point $\boldsymbol{x}^\varepsilon = (x_1, x_2, x_3^\varepsilon) \in \overline{\Omega}^\varepsilon$ such that $\hat{\boldsymbol{x}}^\varepsilon = \boldsymbol{\Theta}(\boldsymbol{x}^\varepsilon)$.

mapping. Let $\boldsymbol{a}_3$ denote a continuously varying unit normal vector along S, and let

$$\boxed{\Omega^\varepsilon := \omega \times]-\varepsilon, \varepsilon[.}$$

Hence the set $\{\hat{\Omega}^\varepsilon\}^-$ is given by

$$\boxed{\{\hat{\Omega}^\varepsilon\}^- = \boldsymbol{\Theta}\left(\overline{\Omega}^\varepsilon\right),}$$

where the mapping $\boldsymbol{\Theta} : \overline{\Omega}^\varepsilon \subset \mathbb{R}^3 \to \mathcal{E}^3$ is defined by

$$\boxed{\begin{aligned}\boldsymbol{\Theta}\left(x_1, x_2, x_3^\varepsilon\right) := \boldsymbol{\theta}(x_1, x_2) + x_3^\varepsilon \boldsymbol{a}_3(x_1, x_2)\\ \text{for all } (x_1, x_2, x_3^\varepsilon) \in \overline{\Omega}^\varepsilon.\end{aligned}}$$

If the mapping $\boldsymbol{\Theta} : \overline{\Omega}^\varepsilon \to \mathcal{E}^3$ is injective (as is the case if $\varepsilon > 0$ is small enough; cf. Thm. 3.1-1), each point $\hat{x}^\varepsilon = (\hat{x}_1^\varepsilon, \hat{x}_2^\varepsilon, \hat{x}_3^\varepsilon) \in \{\hat{\Omega}^\varepsilon\}^-$ is the image $\hat{x}^\varepsilon = \boldsymbol{\Theta}(x^\varepsilon)$ of a *unique* point $x^\varepsilon = (x_1, x_2, x_3^\varepsilon) \in \overline{\Omega}^\varepsilon$, the three coordinates x_1, x_2, x_3^ε of which are called the **"natural" curvilinear coordinates** of $\hat{x}^\varepsilon$.

Remark. Later on, the curvilinear coordinates x_α will be denoted y_α, so as to afford the short notation y for a generic point in $\overline{\omega}$. ■

If a linearly elastic shell is subjected to applied forces and to a boundary condition of place, the displacement field inside $\{\hat{\Omega}^\varepsilon\}^-$ satisfies the *equations of three-dimensional linearized elasticity in Cartesian coordinates* described in Sect. 1.1. As we shall see throughout this volume, it turns out that an essential preliminary to the asymptotic analysis of such a shell problem when ε approaches zero consists in *rewriting these equations in terms of the curvilinear coordinates* x_1, x_2, x_3^ε (the "new" variables) *instead of the Cartesian coordinates* $\hat{x}_1^\varepsilon$, $\hat{x}_2^\varepsilon$, $\hat{x}_3^\varepsilon$ (the "old" variables).

As a preparation to this rewriting, we review in this section basic definitions and properties of *curvilinear coordinates in a "general" three-dimensional domain* $\hat{\Omega} \subset \mathcal{E}^3$; thus the exponents ε have no longer any *raison d'être*.

Let there be given a three-dimensional affine Euclidean space $\mathcal{E}^3$, a Cartesian frame in $\mathcal{E}^3$, and a domain $\hat{\Omega} \subset \mathcal{E}^3$ as in Sect. 1.1. In addition, let there be given a *three-dimensional vector space* in which three vectors $\boldsymbol{e}^i = \boldsymbol{e}_i$ form a basis; *this space will accordingly be identified with* $\mathbb{R}^3$. We let x_i denote the coordinates of a point x in this space and we let $\partial_i := \partial/\partial x_i$ and $\partial_{ij} := \partial^2/\partial x_i \partial x_j$.

Assume that there exist a *domain* Ω in $\mathbb{R}^3$ and an *injective* mapping $\boldsymbol{\Theta} : \overline{\Omega} \to \mathcal{E}^3$ such that $\boldsymbol{\Theta}(\overline{\Omega}) = \{\hat{\Omega}\}^-$. Hence each point

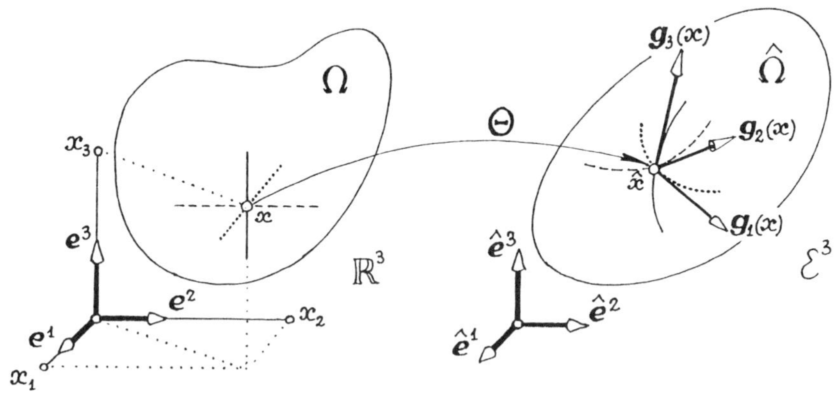

Fig. 1.2-2: *Curvilinear coordinates and covariant bases in a three-dimensional domain.* The three coordinates x_1, x_2, x_3 of $x \in \overline{\Omega}$ are the curvilinear coordinates of $\hat{x} = \boldsymbol{\Theta}(x) \in \{\hat{\Omega}\}^-$. Whenever they are linearly independent, the three vectors $\boldsymbol{g}_i(x) = \partial_i \boldsymbol{\Theta}(x)$ form the covariant basis at $\hat{x} = \boldsymbol{\Theta}(x)$; they are then tangent to the coordinate lines passing through $\hat{x}$.

$\hat{x} \in \{\hat{\Omega}\}^-$ can be unambiguously written as

$$\boxed{\hat{x} = \boldsymbol{\Theta}(x),\ x \in \overline{\Omega},}$$

and the three coordinates x_i of x are called the **curvilinear coordinates** of $\hat{x}$ (Fig. 1.2-2). Naturally, *there are infinitely many ways of defining curvilinear coordinates in a given domain* $\hat{\Omega}$, depending on how the domain Ω and the mapping $\boldsymbol{\Theta}$ are chosen!

Examples of curvilinear coordinates include the well-known *cylindrical* and *spherical coordinates* (see Fig. 1.2-3 and also Ex. 1.1). Another instance is provided by the curvilinear coordinates x_1, x_2, x_3^ε used for defining a shell (Fig. 1.2-1); note that, in this instance, each set of curvilinear coordinates x_1, x_2 on the surface S gives rise to a set of curvilinear coordinates in $\{\hat{\Omega}^\varepsilon\}^-$.

Assume that the mapping $\boldsymbol{\Theta} = \Theta_i \hat{e}^i : \overline{\Omega} \subset \mathbb{R}^3 \to \{\hat{\Omega}\}^- \subset \mathcal{E}^3$ is differentiable at a point $x \in \overline{\Omega}$. If $\boldsymbol{\delta x}$ is such that $(x + \boldsymbol{\delta x}) \in \overline{\Omega}$, we

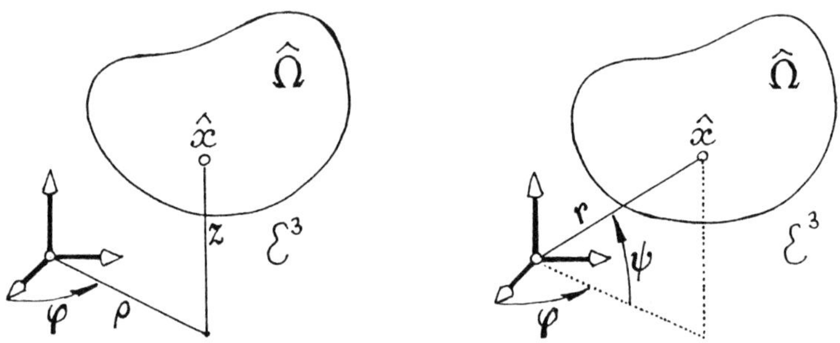

Fig. 1.2-3: *Two familiar examples of curvilinear coordinates.* Let $\hat{\Omega}$ be a three-dimensional domain. The cylindrical coordinates of $\hat{x} \in \hat{\Omega}$ are φ, ρ, z. The spherical coordinates of $\hat{x} \in \hat{\Omega}$ are φ, ψ, r. Special care must of course be exercised in order that such curvilinear coordinates be unambiguously defined, since the same point is defined by (φ, ρ, z), or $(\varphi + 2k\pi, \rho, z)$, or $(\varphi + k\pi, -\rho, z)$, $k \in \mathbb{Z}$; in addition, neither φ nor ψ are defined at the origin $\hat{O}$.

As is customary, these coordinates are appended directly to the set $\hat{\Omega}$, but they are in fact coordinates in a different domain Ω, not represented here!

thus have

$$\boldsymbol{\Theta}(x + \boldsymbol{\delta x}) = \boldsymbol{\Theta}(x) + \boldsymbol{\nabla}\boldsymbol{\Theta}(x)\boldsymbol{\delta x} + o(\boldsymbol{\delta x}),$$

where the matrix $\boldsymbol{\nabla}\boldsymbol{\Theta}(x)$ is given by

$$\boldsymbol{\nabla}\boldsymbol{\Theta}(x) := \begin{pmatrix} \partial_1\Theta_1 & \partial_2\Theta_1 & \partial_3\Theta_1 \\ \partial_1\Theta_2 & \partial_2\Theta_2 & \partial_3\Theta_2 \\ \partial_1\Theta_3 & \partial_2\Theta_3 & \partial_3\Theta_3 \end{pmatrix}(x).$$

Let the three vectors $\boldsymbol{g}_i(x) \in \mathbb{R}^3$ be defined by

$$\boldsymbol{g}_i(x) := \partial_i\boldsymbol{\Theta}(x) = \begin{pmatrix} \partial_i\Theta_1 \\ \partial_i\Theta_2 \\ \partial_i\Theta_3 \end{pmatrix}(x),$$

so that $\boldsymbol{g}_i(x)$ *is the* i-th *column vector of the matrix* $\boldsymbol{\nabla\Theta}(x)$. Let $\boldsymbol{\delta x} = \delta x^i \boldsymbol{e}_i$; then the expansion of $\boldsymbol{\Theta}$ about x may also be written as

$$\boxed{\boldsymbol{\Theta}(x + \boldsymbol{\delta x}) = \boldsymbol{\Theta}(x) + \delta x^i \boldsymbol{g}_i(x) + o(\boldsymbol{\delta x}).}$$

If in particular $\boldsymbol{\delta x}$ is of the form $\boldsymbol{\delta x} = \delta t \boldsymbol{e}_i$, where $\delta t \in \mathbb{R}$ and $\boldsymbol{e}_i$ is one of the basis vectors in $\mathbb{R}^3$, this relation reduces to

$$\boxed{\boldsymbol{\Theta}(x + \delta t \boldsymbol{e}_i) = \boldsymbol{\Theta}(x) + \delta t \boldsymbol{g}_i(x) + o(\delta t).}$$

We henceforth assume that *the three vectors* $\boldsymbol{g}_i(x)$ *are linearly independent*, in which case they constitute the **covariant basis** at the point $\hat{x} = \boldsymbol{\Theta}(x)$. The last relation thus shows that, in this case, *each vector* $\boldsymbol{g}_i(x)$ *is tangent to the* i-th **coordinate line** *passing through* $\hat{x} = \boldsymbol{\Theta}(x)$, defined as the image by $\boldsymbol{\Theta}$ of the points of $\hat{\Omega}$ that lie on the line parallel to $\boldsymbol{e}_i$ passing through x; cf. Fig. 1.2-2 (there exist t_0 and t_1 with $t_0 < t_1$ and $0 \in [t_0, t_1]$ such that the equation of the i-th coordinate line is $t \in [t_0, t_1] \to \boldsymbol{f}_i(t) := \boldsymbol{\Theta}(x + t\boldsymbol{e}_i)$ in a sufficiently small neighborhood of $\hat{x}$; hence $\boldsymbol{f}_i'(0) = \partial_i \boldsymbol{\Theta}(x) = \boldsymbol{g}_i(x)$).

Naturally, we are committing here a convenient *abus de langage*: The vector tangent to the i-th coordinate line at $\hat{x} = \boldsymbol{\Theta}(x)$ is in fact that *vector in the affine space* $\mathcal{E}^3$ that is parallel to $\boldsymbol{g}_i(x)$ and has $\hat{x}$ as its origin.

Returning to a general increment $\boldsymbol{\delta x} = \delta x^i \boldsymbol{e}_i$, we also infer from the expansion of $\boldsymbol{\Theta}$ about x that (recall that we use the summation convention):

$$\boxed{\begin{aligned} |\boldsymbol{\Theta}(x + \boldsymbol{\delta x}) - \boldsymbol{\Theta}(x)|^2 &= \boldsymbol{\delta x}^T \boldsymbol{\nabla\Theta}(x)^T \boldsymbol{\nabla\Theta}(x) \boldsymbol{\delta x} + o\left(|\boldsymbol{\delta x}|^2\right) \\ &= \delta x^i \boldsymbol{g}_i(x) \cdot \boldsymbol{g}_j(x) \delta x^j + o\left(|\boldsymbol{\delta x}|^2\right). \end{aligned}}$$

In other words, the principal part of the length between the points $\boldsymbol{\Theta}(x + \boldsymbol{\delta x})$ and $\boldsymbol{\Theta}(x)$ is $\{\delta x^i \boldsymbol{g}_i(x) \cdot \boldsymbol{g}_j(x) \delta x^j\}^{1/2}$. This observation suggests the introduction of the *symmetric matrix* $(g_{ij}(x))$ *of order three*, whose elements

$$\boxed{g_{ij}(x) := \boldsymbol{g}_i(x) \cdot \boldsymbol{g}_j(x) = (\boldsymbol{\nabla\Theta}(x)^T \boldsymbol{\nabla\Theta}(x))_{ij}}$$

are the **covariant components** of the **metric tensor** at $\hat{x} = \boldsymbol{\Theta}(x)$. Note that *the matrix* $\boldsymbol{\nabla}\boldsymbol{\Theta}(x)$ *is invertible* and that *the matrix* $(g_{ij}(x))$ *is positive definite*, since the vectors $\boldsymbol{g}_i(x)$ are assumed to be linearly independent.

A word of caution. We refrain here from following the common, but *improper*, usage of calling *"metric tensor"* the matrix $(g_{ij}(x))$ itself. In spite of its convenience, it is a confusing and flagrant *abus de langage*, for the "genuine" metric tensor also has "contravariant components" $g^{ij}(x)$ (see the next theorem); it even has "mixed components" $g^i_j(x)$ (which are simply the Kronecker symbols δ^i_j). See, e.g., Antman [1996, Chap. 11]. ■

In the next theorem, we review fundamental formulas showing how *volume, area,* and *length elements at a point* $\hat{x} = \boldsymbol{\Theta}(x)$ *in the set* $\{\hat{\Omega}\}^-$ can be expressed either in terms of the matrix $\boldsymbol{\nabla}\boldsymbol{\Theta}(x)$ or in terms of the matrix $(g_{ij}(x))$ or of its inverse matrix $(g^{ij}(x))$; see also Fig. 1.2-4. Parts (a) and (b) will be immediately put to use in the next section; part (c) provides the essence of the metric tensor. Otherwise, we refer to Vol. I, Sects. 1.5 to 1.8, for comments and references.

If $\mathbf{A}$ is a square matrix, $\mathbf{Cof\,A}$ denotes the *cofactor matrix* of $\mathbf{A}$. If $\mathbf{A}$ is invertible, we thus have $\mathbf{Cof\,A} = (\det \mathbf{A})\mathbf{A}^{-T}$ (details about the cofactor matrix may be found in Vol. I, Sect. 1.1). Also note that the assumptions made on the mapping $\boldsymbol{\Theta}$ guarantee that $\{\hat{\Omega}\}^- = \boldsymbol{\Theta}(\overline{\Omega})$ and that the boundaries $\hat{\Gamma}$ of $\hat{\Omega}$ and Γ of Ω are related by $\hat{\Gamma} = \boldsymbol{\Theta}(\Gamma)$ (Vol. I, Thm. 1.2-8 and Ex. 1.7).

Theorem 1.2-1. *Let Ω be a domain in $\mathbb{R}^3$, let $\boldsymbol{\Theta} : \overline{\Omega} \to \mathcal{E}^3$ be an injective and smooth enough mapping such that the three vectors $\boldsymbol{g}_i(x) := \partial_i\boldsymbol{\Theta}(x)$ are linearly independent at all points $x \in \overline{\Omega}$, and let $\hat{\Omega} = \boldsymbol{\Theta}(\Omega)$.*

(a) *The volume element $d\hat{x}$ at $\hat{x} = \boldsymbol{\Theta}(x) \in \hat{\Omega}$ is given in terms of the volume element dx at $x \in \Omega$ by*

$$\boxed{d\hat{x} = |\det \boldsymbol{\nabla}\boldsymbol{\Theta}(x)|\, dx = \sqrt{g(x)}\, dx,}$$

where

$$\boxed{g(x) := \det(g_{ij}(x)) \text{ and } g_{ij}(x) := \boldsymbol{g}_i(x) \cdot \boldsymbol{g}_j(x).}$$

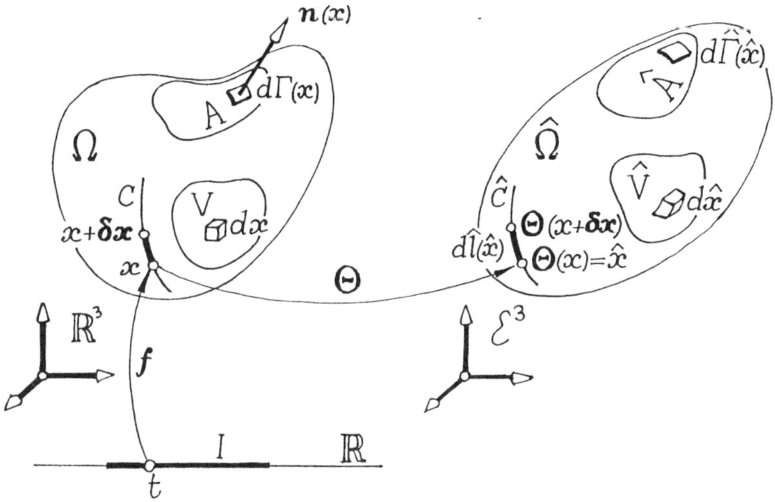

Fig. 1.2-4: *Volume, area, and length elements in curvilinear coordinates.* The elements $d\hat{x}$, $d\hat{\Gamma}(\hat{x})$, and $d\hat{l}(\hat{x})$ at $\hat{x} = \boldsymbol{\Theta}(x) \in \{\hat{\Omega}\}^-$ are expressed in terms of dx, $d\Gamma(x)$, and $\boldsymbol{\delta x}$ at $x \in \overline{\Omega}$ by means of the covariant and contravariant components of the metric tensor; cf. Thm. 1.2-1. The corresponding relations are used for computing the volume of a subdomain $\hat{V} = \boldsymbol{\Theta}(V) \subset \{\hat{\Omega}\}^-$, the area of a surface $\hat{A} = \boldsymbol{\Theta}(A) \subset \partial\hat{\Omega}$, and the length of a curve $\hat{C} = \Theta(C) \subset \{\hat{\Omega}\}^-$, where $C = \boldsymbol{f}(I)$ and $I \subset \mathbb{R}$.

(b) *The area element $d\hat{\Gamma}(\hat{x})$ at $\hat{x} = \boldsymbol{\Theta}(x) \in \partial\hat{\Omega}$ is given in terms of the area element $d\Gamma(x)$ at $x \in \partial\Omega$ by*

$$\boxed{\begin{aligned} d\hat{\Gamma}(\hat{x}) &= |\mathbf{Cof}\,\boldsymbol{\nabla\Theta}(x)\boldsymbol{n}(x)|\,d\Gamma(x) \\ &= \sqrt{g(x)}\sqrt{n_i(x)g^{ij}(x)n_j(x)}\,d\Gamma(x), \end{aligned}}$$

where $\boldsymbol{n}(x) := n_i(x)\boldsymbol{e}^i$ denotes the unit outer normal vector at $x \in \partial\Omega$, *and the matrix* $(g^{ij}(x))$ *is the inverse of the matrix* $(g_{ij}(x))$:

$$(g^{ij}(x)) := (g_{ij}(x))^{-1}.$$

The components $g^{ij}(x)$ *of this matrix are called the* **contravariant components** *of the* **metric tensor** *at* x.

(c) *The length element* $d\hat{l}(\hat{x})$ *at* $\hat{x} = \boldsymbol{\Theta}(x) \in \{\hat{\Omega}\}^-$ *is given by*

$$d\hat{l}(\hat{x}) = \left\{\boldsymbol{\delta x}^T \boldsymbol{\nabla\Theta}(x)^T \boldsymbol{\nabla\Theta}(x) \boldsymbol{\delta x}\right\}^{1/2} = \left\{\delta x^i g_{ij}(x) \delta x^j\right\}^{1/2},$$

where $\boldsymbol{\delta x} = \delta x^i \boldsymbol{e}_i$.

Proof. The relation $d\hat{x} = |\det \boldsymbol{\nabla\Theta}(x)|\, dx$ between the volume elements is well known. The relation $g(x) = |\det \boldsymbol{\nabla\Theta}(x)|^2$ follows from the relation $(g_{ij}(x)) = \boldsymbol{\nabla\Theta}(x)^T \boldsymbol{\nabla\Theta}(x)$.

Indications about the proof of the relation between the area elements $d\hat{\Gamma}(\hat{x})$ and $d\Gamma(x)$ given in (b) are found in Vol. I, Thm. 1.7-1 (naturally, $\boldsymbol{n}(x) = n_i(x)\boldsymbol{e}^i$ is identified here with the column vector in $\mathbb{R}^3$ with $n_i(x)$ as its components). Using the relations $\mathbf{Cof}(\mathbf{A}^T) = (\mathbf{Cof}\,\mathbf{A})^T$ and $\mathbf{Cof}(\mathbf{AB}) = (\mathbf{Cof}\,\mathbf{A})(\mathbf{Cof}\,\mathbf{B})$, we next have:

$$\begin{aligned} |\mathbf{Cof}\,\boldsymbol{\nabla\Theta}(x)\boldsymbol{n}(x)|^2 &= \boldsymbol{n}(x)^T \mathbf{Cof}\left(\boldsymbol{\nabla\Theta}(x)^T \boldsymbol{\nabla\Theta}(x)\right) \boldsymbol{n}(x) \\ &= g(x) n_i(x) g^{ij}(x) n_j(x). \end{aligned}$$

Either expression of the length element given in (c) simply recalls that $d\hat{l}(\hat{x})$ is by definition the "principal part" with respect to $\boldsymbol{\delta x}$ of the length $|\boldsymbol{\Theta}(x + \boldsymbol{\delta x}) - \boldsymbol{\Theta}(x)|$, whose expression precisely led to the introduction of the matrix $(g_{ij}(x))$. ∎

The relations found in Thm. 1.2-1 are used for computing *volume integrals, areas,* and *lengths* inside $\{\hat{\Omega}\}^-$: Let V be a subdomain of Ω, let $\hat{V} := \boldsymbol{\Theta}(V)$, and let $\hat{f} : \hat{V} \to \mathbb{R}$ be a $d\hat{x}$-measurable function. Then

$$\int_{\hat{V}} \hat{f}(\hat{x})\, d\hat{x} = \int_V (\hat{f} \circ \boldsymbol{\Theta})(x) \sqrt{g(x)}\, dx.$$

In particular, the *volume* of $\hat{V}$ is given by

$$\boxed{vol\ \hat{V} := \int_{\hat{V}} d\hat{x} = \int_V \sqrt{g(x)}\, dx.}$$

Next, let $A \subset \partial\Omega$ be a surface, let $\hat{A} := \boldsymbol{\Theta}(A) \subset \partial\hat{\Omega}$, and let $\hat{h} : \hat{A} \to \mathbb{R}$ be a $d\hat{\Gamma}$-measurable function. Then

$$\boxed{\int_{\hat{A}} \hat{h}(\hat{x})\, d\hat{\Gamma}(\hat{x}) = \int_A (\hat{h} \circ \boldsymbol{\Theta})(x)\sqrt{g(x)}\sqrt{n_i(x)g^{ij}(x)n_j(x)}\, d\Gamma(x).}$$

In particular, the *area* of $\hat{A}$ is given by

$$\boxed{area\ \hat{A} := \int_{\hat{A}} d\hat{\Gamma}(\hat{x}) = \int_A \sqrt{g(x)}\sqrt{n_i(x)g^{ij}(x)n_j(x)}\, d\Gamma(x).}$$

Finally, consider a *curve* $C = \boldsymbol{f}(I)$ in $\overline{\Omega}$, where I is a compact interval of $\mathbb{R}$ and $\boldsymbol{f} = f^i\boldsymbol{e}_i : I \to \overline{\Omega}$ is a smooth enough injective mapping. Then the *length* of the *curve* $\hat{C} := \boldsymbol{\Theta}(C) \subset \{\hat{\Omega}\}^-$ is given by

$$\boxed{length\ \hat{C} := \int_I \left|\frac{d}{dt}(\boldsymbol{\Theta} \circ \boldsymbol{f})(t)\right| dt = \int_I \sqrt{g_{ij}(\boldsymbol{f}(t))\frac{df^i}{dt}(t)\frac{df^j}{dt}(t)}\, dt.}$$

Let δ^i_j be yet another way of designating the Kronecker symbol. Given a point $x \in \overline{\Omega}$, *the nine relations*

$$\boxed{\boldsymbol{g}^i(x) \cdot \boldsymbol{g}_j(x) = \delta^i_j}$$

unambiguously define three linearly independent vectors $\boldsymbol{g}^i(x)$, which form the **contravariant basis** at the point $\hat{x} = \boldsymbol{\Theta}(x)$. To see this, let *a priori* $\boldsymbol{g}^i(x) = X^{ik}(x)\boldsymbol{g}_k(x)$ in the relations $\boldsymbol{g}^i(x) \cdot \boldsymbol{g}_j(x) = \delta^i_j$. This gives $X^{ik}(x)g_{kj}(x) = \delta^i_j$; consequently, $X^{ik}(x) = g^{ik}(x)$, where $(g^{ij}(x)) := (g_{ij}(x))^{-1}$ (Thm. 1.2-1 (b)). This also shows that

$$\begin{aligned}\boldsymbol{g}^i(x) \cdot \boldsymbol{g}^j(x) &= \left(g^{ik}(x)\boldsymbol{g}_k(x)\right) \cdot \left(g^{jl}(x)\boldsymbol{g}_l(x)\right)\\ &= g^{ik}(x)g^{jl}(x)g_{kl}(x) = g^{ik}(x)\delta^j_k = g^{ij}(x),\end{aligned}$$

and thus the vectors $\boldsymbol{g}^i(x)$ are linearly independent since the matrix $(g^{ij}(x))$ is positive definite. We would likewise establish that $\boldsymbol{g}_i(x) = g_{ij}(x)\boldsymbol{g}^j(x)$.

Let us record for convenience these fundamental relations between the vectors of the covariant and contravariant bases and the covariant and contravariant components of the metric tensor:

$$g_{ij}(x) = \boldsymbol{g}_i(x) \cdot \boldsymbol{g}_j(x) \text{ and } g^{ij}(x) = \boldsymbol{g}^i(x) \cdot \boldsymbol{g}^j(x),$$
$$\boldsymbol{g}_i(x) = g_{ij}(x)\boldsymbol{g}^j(x) \text{ and } \boldsymbol{g}^i(x) = g^{ij}(x)\boldsymbol{g}_j(x).$$

In Sect. 1.9, we shall briefly discuss the reciprocal question of *recovering a three-dimensional manifold* $\boldsymbol{\Theta}(\Omega)$ *from its metric tensor field:* Given a positive definite symmetric matrix field (g_{ij}) on Ω, find conditions under which there exists a mapping $\boldsymbol{\Theta} : \Omega \to \mathcal{E}^3$ such that

$$\partial_i \boldsymbol{\Theta} \cdot \partial_j \boldsymbol{\Theta} = g_{ij} \text{ in } \Omega.$$

Indications about the meaning of the adjectives "covariant" and "contravariant" used in the definitions of the components of the metric tensor are provided in Ex. 1.3. Otherwise they can be simply seen as particularly effective (though possibly mysterious!) *definitions.*

1.3. THE VARIATIONAL EQUATIONS OF THREE-DIMENSIONAL LINEARIZED ELASTICITY IN CURVILINEAR COORDINATES

Our *point of departure* is the variational formulation of the equations of three-dimensional linearized elasticity described in Sect. 1.1, which consist in finding $\hat{\boldsymbol{u}} = (\hat{u}_i)$ such that

$$\hat{\boldsymbol{u}} \in \mathbf{V}(\hat{\Omega}) = \{\hat{\boldsymbol{v}} = (\hat{v}_i) \in \mathbf{H}^1(\hat{\Omega}); \hat{\boldsymbol{v}} = \mathbf{0} \text{ on } \hat{\Gamma}_0\},$$
$$\int_{\hat{\Omega}} \hat{A}^{ijkl} \hat{e}_{kl}(\hat{\boldsymbol{u}}) \hat{e}_{ij}(\hat{\boldsymbol{v}})\, d\hat{x} = \int_{\hat{\Omega}} \hat{f}^i \hat{v}_i \, d\hat{x} + \int_{\hat{\Gamma}_1} \hat{h}^i \hat{v}_i \, d\hat{\Gamma}$$
$$\text{for all } \hat{\boldsymbol{v}} = (\hat{v}_i) \in \mathbf{V}(\hat{\Omega}),$$

where $\hat{\Omega}$ is a domain, with boundary $\hat{\Gamma}$ partitioned as $\hat{\Gamma} = \hat{\Gamma}_0 \cup \hat{\Gamma}_1$, in the affine Euclidean space $\mathcal{E}^3$,

$$\boxed{\begin{aligned} \hat{A}^{ijkl} &= \lambda\delta^{ij}\delta^{kl} + \mu(\delta^{ik}\delta^{jl} + \delta^{il}\delta^{jk}), \\ \hat{e}_{ij}(\hat{\boldsymbol{v}}) &= \frac{1}{2}(\hat{\partial}_j \hat{v}_i + \hat{\partial}_i \hat{v}_j), \end{aligned}}$$

and $\hat{f}^i \in L^2(\hat{\Omega})$ and $\hat{h}^i \in L^2(\hat{\Gamma}_1)$ are given functions.

These equations are expressed in terms of the *Cartesian coordinates* $\hat{x}_i$ of the points $\hat{x} = (\hat{x}_i) \in \{\hat{\Omega}\}^-$ and of the *Cartesian components* $\hat{u}_i$, $\hat{f}^i$, $\hat{h}^i$ of the displacement and force densities. As in Sect. 1.2, we assume that we are also given a domain Ω, with boundary Γ, in $\mathbb{R}^3$ and a smooth enough injective mapping $\boldsymbol{\Theta} : \overline{\Omega} \to \mathcal{E}^3$ such that $\boldsymbol{\Theta}(\overline{\Omega}) = \{\hat{\Omega}\}^-$ and the three vectors $\boldsymbol{g}_i(x) = \partial_i \boldsymbol{\Theta}(x)$ are linearly independent at all points $x \in \overline{\Omega}$.

Our *objective* consists in expressing these equations in terms of the *curvilinear coordinates* x_i of the points $\hat{x} = \boldsymbol{\Theta}(x) \in \{\hat{\Omega}\}^-$, where $x = (x_i) \in \overline{\Omega}$. In other words, we wish to carry out a *change of variables*, from the "old" variables $\hat{x}_i$ to the "new" variables x_i, in each one of the integrals appearing in the above variational equations, which we thus wish to write as

$$\int_{\hat{\Omega}} \cdots d\hat{x} = \int_{\Omega} \cdots dx \quad \text{and} \quad \int_{\hat{\Gamma}_1} \cdots d\hat{\Gamma} = \int_{\Gamma_1} \cdots d\Gamma,$$

where $\Gamma_1 \subset \Gamma$ and $\boldsymbol{\Theta}(\Gamma_1) = \hat{\Gamma}_1$.

Because the "old" unknowns $\hat{u}_i : \{\hat{\Omega}\}^- \to \mathbb{R}$ are the components of a *vector field*, some care must evidently be exercised in the definition of the "new" unknowns, which must be related to the old ones by means of an "intrinsic" quantity, i.e., having "physical invariance". Observing that the *displacement vector* $\hat{u}_i(\hat{x})\hat{\boldsymbol{e}}^i$ at each point $\hat{x} \in \{\hat{\Omega}\}^-$ possesses this property, we define three *new unknowns* $u_i : \overline{\Omega} \to \mathbb{R}$ by requiring that (Fig. 1.3-1)

$$\boxed{\hat{u}_i(\hat{x})\hat{\boldsymbol{e}}^i =: u_i(x)\boldsymbol{g}^i(x) \text{ for all } \hat{x} = \boldsymbol{\Theta}(x),\ x \in \overline{\Omega},}$$

where the three vectors $\boldsymbol{g}^i(x)$ form the *contravariant basis* at $\hat{x} = \boldsymbol{\Theta}(x)$ (Sect. 1.2). Using the relations $\boldsymbol{g}^i(x) \cdot \boldsymbol{g}_j(x) = \delta^i_j$ and $\hat{\boldsymbol{e}}^i \cdot \hat{\boldsymbol{e}}_j = \delta^i_j$, we immediately find how the old and new unknowns are related, viz.,

$$u_j(x) = u_i(x)\boldsymbol{g}^i(x) \cdot \boldsymbol{g}_j(x) = \hat{u}_i(\hat{x})\hat{\boldsymbol{e}}^i \cdot \boldsymbol{g}_j(x),$$
$$\hat{u}_i(\hat{x}) = \hat{u}_j(\hat{x})\hat{\boldsymbol{e}}^j \cdot \hat{\boldsymbol{e}}_i = u_j(x)\boldsymbol{g}^j(x) \cdot \hat{\boldsymbol{e}}_i.$$

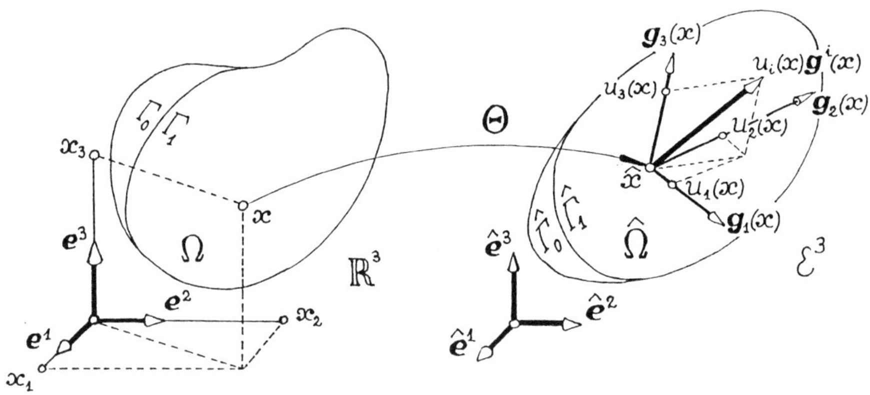

Fig. 1.3-1: *Three-dimensional elasticity in curvilinear coordinates.* In curvilinear coordinates, the displacement vector $u_i(x)\boldsymbol{g}^i(x)$ of the point $\hat{x} = \boldsymbol{\Theta}(x) \in \hat{\Omega}$ is defined by its covariant components $u_i(x)$ over the contravariant basis vectors $\boldsymbol{g}^i(x)$ defined by $\boldsymbol{g}^i(x) \cdot \boldsymbol{g}_j(x) = \delta^i_j$, where $\boldsymbol{g}_j(x) = \partial_j \boldsymbol{\Theta}(x)$. In Cartesian coordinates, the same displacement vector $\hat{u}_i(\hat{x})\hat{\mathbf{e}}^i$ is defined by its Cartesian components $\hat{u}_i(\hat{x})$ over the vectors $\hat{\mathbf{e}}^i$ of the Cartesian frame chosen in $\mathcal{E}^3$; cf. Fig. 1.1-1.

Let

$$[\boldsymbol{g}_j(x)]^i := \boldsymbol{g}_j(x) \cdot \hat{\mathbf{e}}^i \text{ and } [\boldsymbol{g}^j(x)]_i := \boldsymbol{g}^j(x) \cdot \hat{\mathbf{e}}_i,$$

i.e., $[\boldsymbol{g}_j(x)]^i$ denotes the i-th component of the vector $\boldsymbol{g}_j(x)$ and $[\boldsymbol{g}^j(x)]_i$ denotes the i-th component of the vector $\boldsymbol{g}^j(x)$ over the basis $\{\hat{\mathbf{e}}^1, \hat{\mathbf{e}}^2, \hat{\mathbf{e}}^3\} = \{\hat{\mathbf{e}}_1, \hat{\mathbf{e}}_2, \hat{\mathbf{e}}_3\}$ of the Euclidean space $\mathbf{E}^3$ associated with the affine space $\mathcal{E}^3$; cf. Sect. 1.1. In terms of these notations, the preceding relations thus become

$$u_j(x) = \hat{u}_i(\hat{x})[\boldsymbol{g}_j(x)]^i \text{ and } \hat{u}_i(\hat{x}) = u_j(x)[\boldsymbol{g}^j(x)]_i, \ \hat{x} = \boldsymbol{\Theta}(x).$$

The three components $u_i(x)$ are called the **covariant components** *of the* **displacement vector at** $\hat{x}$, and the three functions

$u_i : \overline{\Omega} \to \mathbb{R}$ defined in this fashion are called the **covariant components** *of the* **displacement field** $u_i \boldsymbol{g}^i : \overline{\Omega} \to \mathbf{E}^3$.

A word of caution. The "old" unknown $\hat{\boldsymbol{u}}(\hat{x}) = (\hat{u}_i(\hat{x})) \in \mathbb{R}^3$ could be justifiably identified with the *displacement vector* $\hat{u}_i(\hat{x})\hat{\boldsymbol{e}}^i$ itself since the space $\mathbf{E}^3$ may be identified with the space $\mathbb{R}^3$ once the basis $\{\hat{\boldsymbol{e}}^1, \hat{\boldsymbol{e}}^2, \hat{\boldsymbol{e}}^3\}$ is given in $\mathbf{E}^3$. However, *this is no longer true for the "new" unknown* $\boldsymbol{u}(x) = (u_i(x)) \in \mathbb{R}^3$, since its components $u_i(x)$ now represent the components of the displacement field over the basis $\{\boldsymbol{g}^1(x), \boldsymbol{g}^2(x), \boldsymbol{g}^3(x)\}$, *which varies with* $x \in \overline{\Omega}$. ∎

We likewise associate "new" functions $v_i : \overline{\Omega} \to \mathbb{R}$ with the "old" functions $\hat{v}_i : \{\hat{\Omega}\}^- \to \mathbb{R}$ appearing in the variational equations by letting

$$\boxed{\hat{v}_i(\hat{x})\hat{\boldsymbol{e}}^i =: v_i(x)\boldsymbol{g}^i(x) \text{ for all } \hat{x} = \boldsymbol{\Theta}(x),\ x \in \overline{\Omega}.}$$

Remark. As the functions $\hat{u}_i$ belong to the Sobolev space $H^1(\hat{\Omega})$, they are in fact *equivalence classes* of functions, and any function in a class is in fact only defined almost everywhere. Accordingly, the relations defining the (equivalence classes of) functions u_i (which likewise belong to the Sobolev space $H^1(\Omega)$ under *ad hoc* assumptions on the mapping $\boldsymbol{\Theta}$; cf. Thm. 1.3-1) need hold only *for almost all* $\hat{x} \in \boldsymbol{\Theta}(\overline{\Omega})$. But in order to avoid cumbersome statements, we blithely omit such mentions. ∎

We begin the change of variables by considering the integrals found in the *right-hand side* of the variational equations, i.e., those corresponding to the applied forces. With the Cartesian components $\hat{f}^i : \hat{\Omega} = \boldsymbol{\Theta}(\Omega) \to \mathbb{R}$ of the applied body force density, let there be associated its **contravariant components** $f^i : \Omega \to \mathbb{R}$, defined by

$$\boxed{\hat{f}^i(\hat{x})\hat{\boldsymbol{e}}_i =: f^i(x)\boldsymbol{g}_i(x) \text{ for all } \hat{x} = \boldsymbol{\Theta}(x),\ x \in \Omega.}$$

This definition shows that

$$f^i(x) = \hat{f}^j(\hat{x})[\boldsymbol{g}^i(x)]_j,$$

and consequently that $f^i \in L^2(\Omega)$ if $\hat{f}^j \in L^2(\hat{\Omega})$. It also implies that

$$\hat{f}^i(\hat{x})\hat{v}_i(\hat{x}) = (\hat{f}^i(\hat{x})\hat{\boldsymbol{e}}_i) \cdot (\hat{v}_j(\hat{x})\hat{\boldsymbol{e}}^j) = (f^i(x)\boldsymbol{g}_i(x)) \cdot (v_j(x)\boldsymbol{g}^j(x)) = f^i(x)v_i(x)$$

for all $\hat{x} = \boldsymbol{\Theta}(x)$, $x \in \Omega$. Hence

$$\boxed{\hat{f}^i(\hat{x})\hat{v}_i(\hat{x})\,d\hat{x} = f^i(x)v_i(x)\sqrt{g(x)}\,dx \text{ for all } \hat{x} = \boldsymbol{\Theta}(x), x \in \Omega,}$$

since $d\hat{x} = \sqrt{g(x)}\,dx$ (Thm. 1.2-1 (a)), and thus

$$\boxed{\int_{\hat{\Omega}} \hat{f}^i\hat{v}_i\,d\hat{x} = \int_{\Omega} f^i v_i \sqrt{g}\,dx.}$$

Remarks. (1) What has just been proved is in effect the *invariance of the number $f^i(x)v_i(x)$ with respect to changes of curvilinear coordinates*, provided one vector appears by means of its "contravariant" components (i.e., on the "covariant" basis) and the other by means of its "covariant" components (i.e., on the "contravariant" basis). Naturally, this number is nothing but the *Euclidean inner product of the two vectors*!

(2) The adjectives "covariant" and "contravariant" used in the definitions of the components of the displacement and force vector fields are given a proper interpretation in Ex. 1.2. ∎

With the Cartesian components $\hat{h}^i : \hat{\Gamma}_1 = \boldsymbol{\Theta}(\Gamma_1) \to \mathbb{R}$ of the applied surface force density, let there be likewise associated its **contravariant components** $h^i : \Gamma_1 \to \mathbb{R}$, defined by

$$\boxed{\hat{h}^i(\hat{x})\hat{\boldsymbol{e}}_i d\hat{\Gamma}(\hat{x}) =: h^i(x)\boldsymbol{g}_i(x)\sqrt{g(x)}d\Gamma(x) \text{ for all } \hat{x} = \boldsymbol{\Theta}(x),\ x \in \Gamma_1,}$$

where the area elements $d\hat{\Gamma}(\hat{x})$ at $\hat{x} = \boldsymbol{\Theta}(x) \in \hat{\Gamma}_1$ and $d\Gamma(x)$ at $x \in \Gamma_1$ are related by (Thm. 1.2-1 (b))

$$d\hat{\Gamma}(\hat{x}) = \sqrt{g(x)}\sqrt{n_k(x)g^{kl}(x)n_l(x)}\,d\Gamma(x).$$

This definition shows that

$$\boxed{\begin{aligned}\hat{h}^i(\hat{x})\hat{\boldsymbol{e}}_i = \left\{n_k(x)g^{kl}(x)n_l(x)\right\}^{-1/2} h^i(x)\boldsymbol{g}_i(x) \\ \text{for all } \hat{x} = \boldsymbol{\Theta}(x),\ x \in \Gamma_1,\end{aligned}}$$

hence that $h^i \in L^2(\Gamma_1)$ if $\hat{h}^j \in L^2(\hat{\Gamma}_1)$, since

$$h^i(x) = \sqrt{n_k(x)g^{kl}(x)n_l(x)}\, \hat{h}^j(\hat{x})[\boldsymbol{g}^i(x)]_j.$$

The factor $\{n_k g^{kl} n_l\}^{-1/2}$ is introduced in the definition of the functions h^i in order that

$$\boxed{\int_{\hat{\Gamma}_1} \hat{h}^i \hat{v}_i \, d\hat{\Gamma} = \int_{\Gamma_1} h^i v_i \sqrt{g}\, d\Gamma,}$$

i.e., in order that the *same* factor $\sqrt{g}$ appears in both integrals $\int_\Omega f^i v_i \sqrt{g}\, dx$ and $\int_{\Gamma_1} h^i v_i \sqrt{g}\, d\Gamma$. It in turn gives rise to a more "natural" boundary condition on Γ_1 when the variational equations are transformed into a boundary value problem (Thm. 1.6-1).

Transforming the integrals appearing in the *left-hand side* of the variational equations seems to be a similarly innocuous enterprise, simply requiring in addition applications of the chain rule, since first-order derivatives occur in the integrands. In fact, carrying out this transformation in a finite time is a subtle task, which relies in particular on the notion of *covariant differentiation of a vector field*; cf. part (iii) of the next proof.

Theorem 1.3-1. *Let $\hat{\Omega}$ be a domain in $\mathcal{E}^3$, let $\hat{f}_i \in L^2(\hat{\Omega})$ and $\hat{h}_i \in L^2(\hat{\Gamma}_1)$ be given functions, and let $\hat{\boldsymbol{u}} = (\hat{u}_i) \in \mathbf{V}(\hat{\Omega})$ denote the weak solution of the associated linearized elasticity problem in Cartesian coordinates* (Sect. 1.1).

Let Ω be a domain in $\mathbb{R}^3$ and let $\boldsymbol{\Theta}$ be a $\mathcal{C}^2$-diffeomorphism of $\overline{\Omega}$ onto $\{\hat{\Omega}\}^- = \boldsymbol{\Theta}(\overline{\Omega})$, so that the vectors $\boldsymbol{g}_i(x) = \partial_i \boldsymbol{\Theta}(x)$ are linearly independent at all points $x \in \overline{\Omega}$. Let the vectors $\boldsymbol{g}^i(x)$ be defined by the relations $\boldsymbol{g}^i(x) \cdot \boldsymbol{g}_j(x) = \delta^i_j$ and let $g(x) = \det(\boldsymbol{g}_i(x) \cdot \boldsymbol{g}_j(x))$ and $g^{ij}(x) = \boldsymbol{g}^i(x) \cdot \boldsymbol{g}^j(x)$, $x \in \overline{\Omega}$.

Then the vector field $\boldsymbol{u} = (u_i) : \overline{\Omega} \to \mathbb{R}^3$ defined by

$$\boxed{\hat{u}_i(\hat{x})\hat{\boldsymbol{e}}^i =: u_i(x)\boldsymbol{g}^i(x) \text{ for all } \hat{x} = \boldsymbol{\Theta}(x),\ x \in \overline{\Omega},}$$

satisfies the following variational problem:

$$\boldsymbol{u} \in \mathbf{V}(\Omega) := \{\boldsymbol{v} = (v_i) \in \mathbf{H}^1(\Omega);\ \boldsymbol{v} = \mathbf{0} \text{ on } \Gamma_0\},$$

$$\int_\Omega A^{ijkl} e_{k\|l}(\boldsymbol{u}) e_{i\|j}(\boldsymbol{v}) \sqrt{g}\, dx = \int_\Omega f^i v_i \sqrt{g}\, dx + \int_{\Gamma_1} h^i v_i \sqrt{g}\, d\Gamma$$
$$\text{for all } \boldsymbol{v} = (v_i) \in \mathbf{V}(\Omega),$$

where $\Gamma_0 := \boldsymbol{\Theta}^{-1}(\hat{\Gamma}_0)$, $\Gamma_1 := \boldsymbol{\Theta}^{-1}(\hat{\Gamma}_1)$, *the functions* $f^i \in L^2(\Omega)$ *and* $h^i \in L^2(\Gamma_1)$ *are defined by:*

$$\hat{f}^i(\hat{x})\hat{\boldsymbol{e}}_i\, d\hat{x} =: \sqrt{g(x)} f^i(x) \boldsymbol{g}_i(x)\, dx,\ \hat{x} = \boldsymbol{\Theta}(x),\ x \in \Omega,$$
$$\hat{h}^i(\hat{x})\hat{\boldsymbol{e}}_i\, d\hat{\Gamma}(\hat{x}) =: \sqrt{g(x)} h^i(x) \boldsymbol{g}_i(x)\, d\Gamma(x),\ \hat{x} = \boldsymbol{\Theta}(x),\ x \in \Gamma_1,$$

the functions $A^{ijkl} = A^{jikl} = A^{klij} \in \mathcal{C}^1(\overline{\Omega})$ *are defined by:*

$$A^{ijkl} := \lambda g^{ij} g^{kl} + \mu\big(g^{ik} g^{jl} + g^{il} g^{jk}\big),$$

and finally, the functions $e_{i\|j}(\boldsymbol{v}) = e_{j\|i}(\boldsymbol{v}) \in L^2(\Omega)$ *are defined for all* $\boldsymbol{v} \in \mathbf{H}^1(\Omega)$ *by:*

$$e_{i\|j}(\boldsymbol{v}) := \frac{1}{2}(\partial_j v_i + \partial_i v_j) - \Gamma^p_{ij} v_p, \text{ where}$$
$$\Gamma^p_{ij} := \boldsymbol{g}^p \cdot \partial_i \boldsymbol{g}_j = \Gamma^p_{ji} \in \mathcal{C}^0(\overline{\Omega}).$$

Proof. The following *convention* holds throughout this proof: The simultaneous appearance of $\hat{x}$ and x in an equality means that they are related by $\hat{x} = \boldsymbol{\Theta}(x)$ and that the equality in question holds either for all $x \in \overline{\Omega}$ or for almost all $x \in \Omega$.

(i) *Another expression of* $[\boldsymbol{g}^i(x)]_k := \boldsymbol{g}^i(x) \cdot \hat{\boldsymbol{e}}_k$.

Let $\boldsymbol{\Theta}(x) = \Theta^k(x)\hat{\boldsymbol{e}}_k$ and $\hat{\boldsymbol{\Theta}}(\hat{x}) = \hat{\Theta}^i(\hat{x})\boldsymbol{e}_i$, where $\hat{\boldsymbol{\Theta}} : \{\hat{\Omega}\}^- \to \overline{\Omega}$ denotes the inverse mapping of $\boldsymbol{\Theta}$. Since $\hat{\boldsymbol{\Theta}}(\boldsymbol{\Theta}(x)) = x$ for all $x \in \overline{\Omega}$, the chain rule (see, e.g., Vol. I, Thm. 1.2-1) shows that the matrices

$\boldsymbol{\nabla\Theta}(x) := (\partial_j \Theta^k(x))$ (the row index is k) and $\hat{\boldsymbol{\nabla}}\hat{\boldsymbol{\Theta}}(\hat{x}) := (\hat{\partial}_k \hat{\Theta}^i(\hat{x}))$ (the row index is i) satisfy

$$\hat{\boldsymbol{\nabla}}\hat{\boldsymbol{\Theta}}(\hat{x})\boldsymbol{\nabla\Theta}(x) = \mathbf{I},$$

or equivalently,

$$\begin{aligned}\hat{\partial}_k \hat{\Theta}^i(\hat{x})\partial_j \Theta^k(x) &= \begin{pmatrix}\hat{\partial}_1 \hat{\Theta}^i(\hat{x}) & \partial_2 \hat{\Theta}^i(\hat{x}) & \partial_3 \hat{\Theta}^i(\hat{x})\end{pmatrix} \begin{pmatrix}\partial_j \Theta^1(x) \\ \partial_j \Theta^2(x) \\ \partial_j \Theta^3(x)\end{pmatrix} \\ &= \delta^i_j.\end{aligned}$$

The components of the above column vector being precisely those of the vector $\boldsymbol{g}_j(x)$, the components of the above row vector must be those of the vector $\boldsymbol{g}^i(x)$ since $\boldsymbol{g}^i(x)$ is uniquely defined for each exponent i by the three relations $\boldsymbol{g}^i(x) \cdot \boldsymbol{g}_j(x) = \delta^i_j$, $j = 1, 2, 3$. Hence the k-th component of $\boldsymbol{g}^i(x)$ over the basis $\{\hat{\boldsymbol{e}}_1, \hat{\boldsymbol{e}}_2, \hat{\boldsymbol{e}}_3\}$ has the following expression in terms of the inverse mapping $\hat{\boldsymbol{\Theta}}$:

$$[\boldsymbol{g}^i(x)]_k = \hat{\partial}_k \hat{\Theta}^i(\hat{x}).$$

(ii) *Definition of the Christoffel symbols.*

We next compute the derivatives $\partial_l \boldsymbol{g}^q(x)$ (the fields $\boldsymbol{g}^q = g^{qr}\boldsymbol{g}_r$ are of class $\mathcal{C}^1$ on $\overline{\Omega}$ since $\boldsymbol{\Theta}$ is assumed to be of class $\mathcal{C}^2$) as they will clearly be needed (see (iii)) for expressing the derivatives $\hat{\partial}_j \hat{u}_i(\hat{x})$ as functions of x (recall that $\hat{u}_i(\hat{x}) = u_k(x)[\boldsymbol{g}^k(x)]_i$). The vectors $\boldsymbol{g}^k(x)$ forming a basis, we may write *a priori*

$$\partial_l \boldsymbol{g}^q(x) = -\Gamma^q_{lk}(x)\boldsymbol{g}^k(x),$$

thereby unambiguously defining functions $\Gamma^q_{lk} : \overline{\Omega} \to \mathbb{R}$, which are called the *Christoffel symbols*. To find their expressions in terms of the mappings $\boldsymbol{\Theta}$ and $\hat{\boldsymbol{\Theta}}$, we observe that

$$\Gamma^q_{lk}(x) = \Gamma^q_{lm}(x)\delta^m_k = \Gamma^q_{lm}(x)\boldsymbol{g}^m(x) \cdot \boldsymbol{g}_k(x) = -\partial_l \boldsymbol{g}^q(x) \cdot \boldsymbol{g}_k(x),$$

and, noting that $\partial_l(\boldsymbol{g}^q(x) \cdot \boldsymbol{g}_k(x)) = 0$ and $[\boldsymbol{g}^q(x)]_p = \hat{\partial}_p \hat{\Theta}^q(\hat{x})$, we obtain

$$\Gamma^q_{lk}(x) = \boldsymbol{g}^q(x) \cdot \partial_l \boldsymbol{g}_k(x) = \hat{\partial}_p \hat{\Theta}^q(\hat{x})\partial_{lk}\Theta^p(x) = \Gamma^q_{kl}(x).$$

Since $\boldsymbol{\Theta} \in \mathcal{C}^2(\overline{\Omega}; \mathcal{E}^3)$ and $\hat{\boldsymbol{\Theta}} \in \mathcal{C}^1(\{\hat{\Omega}\}^-; \mathbb{R}^3)$ by assumption, the last relations show that $\Gamma^q_{kl} \in \mathcal{C}^0(\overline{\Omega})$.

(iii) *Let $\hat{\boldsymbol{v}} = (\hat{v}_i)$ be given in the space $\mathbf{V}(\hat{\Omega})$. Then the vector field $\boldsymbol{v} = (v_i)$ defined by $\hat{v}_i(\hat{x})\hat{\boldsymbol{e}}^i = v_i(x)\boldsymbol{g}^i(x)$ is in the space $\mathbf{V}(\Omega)$; moreover,*

$$\hat{\partial}_j \hat{v}_i(\hat{x}) = v_{k\|l}(x)[\boldsymbol{g}^k(x)]_i[\boldsymbol{g}^l(x)]_j,$$

where

$$v_{k\|l}(x) := \partial_l v_k(x) - \Gamma^q_{lk}(x) v_q(x)$$

denotes a "covariant derivative" of the vector field $v_k \boldsymbol{g}^k$ at x, $[\boldsymbol{g}^k(x)]_i$ denotes the i-th component of $\boldsymbol{g}^k(x)$ over the basis $\{\hat{\mathbf{e}}_1, \hat{\mathbf{e}}_2, \hat{\mathbf{e}}_3\}$, and $\Gamma^q_{lk}(x) = \boldsymbol{g}^q(x) \cdot \partial_l \boldsymbol{g}_k(x)$ are the Christoffel symbols introduced in (ii).

By a classical result about composite mappings (see, e.g., Nečas [1967, Chap. 2, Lemma 3.2] or Adams [1975, Thm. 3.35]), a function $\hat{w} \circ \boldsymbol{\Theta}$ is in $H^1(\Omega)$ if $\hat{w} \in H^1(\hat{\Omega})$ and $\boldsymbol{\Theta} : \overline{\Omega} \to \{\hat{\Omega}\}^- = \boldsymbol{\Theta}(\overline{\Omega})$ is a bijection such that both $\boldsymbol{\Theta}$ and its inverse mapping $\hat{\boldsymbol{\Theta}}$ are Lipschitz-continuous; consequently, the functions $\hat{v}_j \circ \boldsymbol{\Theta}$ are in $H^1(\Omega)$. Since the functions $[\boldsymbol{g}_i]^j$ are in $\mathcal{C}^1(\overline{\Omega})$, the functions $v_i = (\hat{v}_j \circ \boldsymbol{\Theta})[\boldsymbol{g}_i]^j$ are thus in $H^1(\Omega)$; besides, they satisfy $v_i = 0$ on Γ_0 since $\hat{v}_j = 0$ on $\hat{\Gamma}_0$.

We next compute the partial derivatives $\hat{\partial}_j \hat{v}_i(\hat{x})$ as functions of x by means of the relation $\hat{v}_i(\hat{x}) = v_k(x)[\boldsymbol{g}^k(x)]_i$. To this end, we first note that a differentiable function $w : \overline{\Omega} \to \mathbb{R}$ satisfies

$$\hat{\partial}_j w(\hat{\boldsymbol{\Theta}}(\hat{x})) = \partial_l w(x) \hat{\partial}_j \hat{\Theta}^l(\hat{x}) = \partial_l w(x)[\boldsymbol{g}^l(x)]_j,$$

by the chain rule and by (i). In particular then,

$$\begin{aligned} \hat{\partial}_j \hat{v}_i(\hat{x}) &= \hat{\partial}_j v_k(\hat{\boldsymbol{\Theta}}(\hat{x}))[\boldsymbol{g}^k(x)]_i + v_q(x)\hat{\partial}_j[\boldsymbol{g}^q(\hat{\boldsymbol{\Theta}}(\hat{x}))]_i \\ &= \partial_l v_k(x)[\boldsymbol{g}^l(x)]_j[\boldsymbol{g}^k(x)]_i + v_q(x)\big(\partial_l[\boldsymbol{g}^q(x)]_i\big)[\boldsymbol{g}^l(x)]_j \\ &= \big(\partial_l v_k(x) - \Gamma^q_{lk}(x) v_q(x)\big)[\boldsymbol{g}^k(x)]_i[\boldsymbol{g}^l(x)]_j, \end{aligned}$$

since $\partial_l \boldsymbol{g}^q(x) = -\Gamma^q_{lk}(x)\boldsymbol{g}^k(x)$ by (ii).

(iv) *The integrand $\hat{A}^{ijkl}\hat{e}_{kl}(\hat{\boldsymbol{u}})\hat{e}_{ij}(\hat{\boldsymbol{v}}) : \hat{\Omega} \to \mathbb{R}$ appearing in the left-hand side of the variational equations over $\hat{\Omega}$ satisfies*

$$\left(\hat{A}^{ijkl}\hat{e}_{kl}(\hat{\boldsymbol{u}})\hat{e}_{ij}(\hat{\boldsymbol{v}})\right)(\hat{x}) = \left(A^{ijkl}e_{k\|l}(\boldsymbol{u})e_{i\|j}(\boldsymbol{v})\right)(x),$$

where (covariant derivatives such as $v_{i\|j}$ have been introduced in (iii)):

$$A^{ijkl} := \lambda g^{ij} g^{kl} + \mu(g^{ik} g^{jl} + g^{il} g^{jk}) = A^{jikl} = A^{klij} \in \mathcal{C}^1(\overline{\Omega}),$$

$$e_{i\|j}(\boldsymbol{v}) := \frac{1}{2}(v_{i\|j} + v_{j\|i}) = \frac{1}{2}(\partial_j v_i + \partial_i v_j) - \Gamma^p_{ij} v_p = e_{j\|i}(\boldsymbol{v}) \in L^2(\Omega).$$

We first have, by (iii),

$$\begin{aligned}\hat{e}_{ij}(\hat{\boldsymbol{v}})(\hat{x}) &= \frac{1}{2}\left(\hat{\partial}_j \hat{v}_i + \hat{\partial}_i \hat{v}_j\right)(\hat{x}) \\ &= \frac{1}{2}\left(v_{k\|l}[\boldsymbol{g}^k]_i[\boldsymbol{g}^l]_j + v_{k\|l}[\boldsymbol{g}^k]_j[\boldsymbol{g}^l]_i\right)(x) \\ &= \left(e_{k\|l}(\boldsymbol{v})[\boldsymbol{g}^k]_i[\boldsymbol{g}^l]_j\right)(x).\end{aligned}$$

Since the Christoffel symbols satisfy $\Gamma^q_{kl} = \Gamma^q_{lk}$ (cf. (ii)), the functions $e_{k\|l}(\boldsymbol{v})$ likewise satisfy

$$e_{k\|l}(\boldsymbol{v}) = e_{l\|k}(\boldsymbol{v}).$$

Recalling that $\boldsymbol{g}^i(x) \cdot \boldsymbol{g}^j(x) = g^{ij}(x)$, we next have

$$\delta^{ij}\hat{e}_{ij}(\hat{\boldsymbol{v}})(\hat{x}) = \hat{e}_{pp}(\hat{\boldsymbol{v}})(\hat{x}) = \left(e_{k\|l}(\boldsymbol{v})[\boldsymbol{g}^k]_p[\boldsymbol{g}^l]_p\right)(x) = \left(e_{k\|l}(\boldsymbol{v})g^{kl}\right)(x),$$

and thus

$$\begin{aligned}\delta^{ij}\delta^{kl}\hat{e}_{kl}(\hat{\boldsymbol{u}})(\hat{x})\hat{e}_{ij}(\hat{\boldsymbol{v}})(\hat{x}) &= \hat{e}_{pp}(\hat{\boldsymbol{u}})(\hat{x})\hat{e}_{qq}(\hat{\boldsymbol{v}})(\hat{x}) \\ &= \left(g^{ij} g^{kl} e_{k\|l}(\boldsymbol{u}) e_{i\|j}(\boldsymbol{v})\right)(x).\end{aligned}$$

Likewise,

$$\begin{aligned}\frac{1}{2}\left(\delta^{ik}\delta^{jl} + \delta^{il}\delta^{jk}\right)\hat{e}_{kl}(\hat{\boldsymbol{u}})(\hat{x})\hat{e}_{ij}(\hat{\boldsymbol{v}})(\hat{x}) &= \hat{e}_{ij}(\hat{\boldsymbol{u}})(\hat{x})\hat{e}_{ij}(\hat{\boldsymbol{v}})(\hat{x}) \\ &= \left(e_{k\|l}(\boldsymbol{u})e_{m\|n}(\boldsymbol{v})[\boldsymbol{g}^k]_i[\boldsymbol{g}^l]_j[\boldsymbol{g}^m]_i[\boldsymbol{g}^n]_j\right)(x) \\ &= \left(g^{km} g^{ln} e_{k\|l}(\boldsymbol{u}) e_{m\|n}(\boldsymbol{v})\right)(x) \\ &= \frac{1}{2}\left((g^{ik} g^{jl} + g^{il} g^{jk}) e_{k\|l}(\boldsymbol{u}) e_{i\|j}(\boldsymbol{v})\right)(x),\end{aligned}$$

since $e_{k\|l}(\boldsymbol{u}) = e_{l\|k}(\boldsymbol{u})$.

(v) *Conclusions*: Since $d\hat{x} = \sqrt{g(x)}\, dx$ (Thm. 1.2-1 (a)), part (iv) shows that

$$\int_{\hat{\Omega}} \hat{A}^{ijkl}\hat{e}_{kl}(\hat{\boldsymbol{u}})\hat{e}_{ij}(\hat{\boldsymbol{v}})\, d\hat{x} = \int_\Omega A^{ijkl} e_{k\|l}(\boldsymbol{u}) e_{i\|j}(\boldsymbol{v})\sqrt{g}\, dx,$$

on the one hand. At the beginning of this section, it was also shown that the definition of the functions f^i and h^i implies that

$$\int_{\hat{\Omega}} \hat{f}^i \hat{v}_i \, d\hat{x} = \int_{\Omega} f^i v_i \sqrt{g} \, dx,$$
$$\int_{\hat{\Gamma}_1} \hat{h}^i \hat{v}_i \, d\hat{\Gamma} = \int_{\Gamma_1} h^i v_i \sqrt{g} \, d\Gamma,$$

on the other, and thus the proof is complete. ■

Naturally, if $\mathcal{E}^3$ is identified with $\mathbb{R}^3$ and $\boldsymbol{\Theta} = \boldsymbol{id}_{\mathbb{R}^3}$, each vector $\boldsymbol{g}_i(x)$ is equal to $\hat{\boldsymbol{e}}_i$ and thus $\boldsymbol{g}^i(x) = \hat{\boldsymbol{e}}^i$, $g(x) = 1$, $g^{ij}(x) = \delta^{ij}$, and $\Gamma^p_{ij}(x) = 0$ for all $x \in \overline{\Omega}$; in addition, the fields (u_i) and $(\hat{u}_i)$, (f^i) and $(\hat{f}^i)$, (h^i) and $(\hat{h}^i)$, and (A^{ijkl}) and $(\hat{A}^{ijkl})$ coincide.

The variational problem found in Thm. 1.3-1 constitutes the **variational,** or **weak, formulation** of the **equations of three-dimensional linearized elasticity in curvilinear coordinates.**

The functions $A^{ijkl} : \overline{\Omega} \to \mathbb{R}$ introduced in Thm. 1.3-1 are called the **contravariant components** *of the* **three-dimensional elasticity tensor in curvilinear coordinates**; they thus generalize to arbitrary curvilinear coordinates the components $\hat{A}^{ijkl} : \{\hat{\Omega}\}^- \to \mathbb{R}$ of the *elasticity tensor in Cartesian coordinates* (Sect. 1.1).

The boundary condition $\boldsymbol{u} = (u_i) = \boldsymbol{0}$ on Γ_0, or the equivalent relation $u_i \boldsymbol{g}^i = 0$ on $\boldsymbol{\Theta}(\Gamma_0)$, constitutes a (homogeneous) **boundary condition of place.**

Remark. The interpretation of the adjective "contravariant" attached to the components of the elasticity tensor is given in Ex. 1.4. ■

1.4. COVARIANT DERIVATIVES AND CHRISTOFFEL SYMBOLS IN A THREE-DIMENSIONAL DOMAIN

We now record as a theorem several important definitions, relations, or properties all originating from the proof of Thm. 1.3-1. Together with those of Sect. 1.2, these constitute our first encounter with *differential geometry "in a three-dimensional manifold in* $\mathbb{R}^3$*"*. Other related notions are treated later, such as the *covariant derivatives of a tensor field* (Thm. 1.6-1) and the *recovery of a three-dimensional manifold from its metric tensor field* (Sect. 1.9).

For further details and complements, see classical texts such as Malliavin [1972], Choquet-Bruhat, Dewitt-Morette & Dillard-Bleick [1977], or Abraham, Marsden & Ratiu [1983]. The books by Green & Zerna [1968], Marsden & Hughes [1983], and Simmonds [1994] provide treatments of differential geometry and tensor analysis that are essentially motivated by, and thus well adapted to, three-dimensional elasticity.

Theorem 1.4-1. *Let the assumptions on the mapping* $\boldsymbol{\Theta}$ *and the definitions of the vector fields* $\boldsymbol{g}_i$ *and* $\boldsymbol{g}^j$ *be as in* Thm. 1.3-1, *and let there be given a vector field* $v_i \boldsymbol{g}^i$ *on* $\overline{\Omega}$ *with smooth enough covariant components* $v_i : \overline{\Omega} \to \mathbb{R}$.

(a) *The components of the vectors* $\boldsymbol{g}_i(x)$ *and* $\boldsymbol{g}^j(x)$ *are given by*

$$\begin{aligned} [\boldsymbol{g}_i(x)]^k &= \boldsymbol{g}_i(x) \cdot \hat{\boldsymbol{e}}^k = \partial_i \Theta^k(x), \\ [\boldsymbol{g}^j(x)]_l &= \boldsymbol{g}^j(x) \cdot \hat{\boldsymbol{e}}_l = \hat{\partial}_l \hat{\Theta}^j(\hat{x}), \quad \hat{x} = \boldsymbol{\Theta}(x). \end{aligned}$$

In other words, the components of $\boldsymbol{g}_i(x)$ *are those of the i-th column of the matrix* $\boldsymbol{\nabla}\boldsymbol{\Theta}(x)$ *while those of* $\boldsymbol{g}^j(x)$ *are those of the j-th row of the matrix* $\hat{\boldsymbol{\nabla}}\hat{\boldsymbol{\Theta}}(\hat{x})$.

(b) *The* **first-order covariant derivatives** $v_{i\|j} : \overline{\Omega} \to \mathbb{R}$ *of the vector field* $v_i \boldsymbol{g}^i$, *which are defined by*

$$v_{i\|j} := \partial_j v_i - \Gamma^p_{ij} v_p, \text{ where } \Gamma^p_{ij} := \boldsymbol{g}^p \cdot \partial_i \boldsymbol{g}_j,$$

can be also defined by the relations

$$\partial_j(v_i \boldsymbol{g}^i) = v_{i\|j} \boldsymbol{g}^i \iff v_{i\|j} = \left\{ \partial_j (v_k \boldsymbol{g}^k) \right\} \cdot \boldsymbol{g}_i.$$

(c) *The* **Christoffel symbols** $\Gamma^p_{ij} := \boldsymbol{g}^p \cdot \partial_i \boldsymbol{g}_j = \Gamma^p_{ji} : \overline{\Omega} \to \mathbb{R}$ *satisfy the relations*

$$\partial_i \boldsymbol{g}^p = -\Gamma^p_{ij} \boldsymbol{g}^j \text{ and } \partial_j \boldsymbol{g}_q = \Gamma^i_{jq} \boldsymbol{g}_i.$$

(d) *Let* $\hat{v}_i\hat{e}^i$ *be the vector field defined on* $\{\hat{\Omega}\}^- = \boldsymbol{\Theta}(\overline{\Omega})$ *by*

$$\boxed{\hat{v}_i(\hat{x})\hat{e}^i := v_i(x)\boldsymbol{g}^i(x) \text{ for all } \hat{x} = \boldsymbol{\Theta}(x),\ x \in \overline{\Omega},}$$

and let $[\boldsymbol{g}^k(x)]_i$ *denotes the i-th component of* $\boldsymbol{g}^k(x)$ *over the basis* $\{\hat{\boldsymbol{e}}_1, \hat{\boldsymbol{e}}_2, \hat{\boldsymbol{e}}_3\}$. *Then, for all* $x \in \overline{\Omega}$,

$$\boxed{\hat{\partial}_j\hat{v}_i(\hat{x}) = \left(v_{k\|l}[\boldsymbol{g}^k]_i[\boldsymbol{g}^l]_j\right)(x),\ \hat{x} = \boldsymbol{\Theta}(x).}$$

■

Proof. It remains to verify that the covariant derivatives $v_{i\|j}$, defined in part (iii) of the proof of Thm. 1.3-1 by

$$v_{i\|j} = \partial_j v_i - \Gamma^p_{ij}v_p,$$

may be equivalently defined by the relations

$$\partial_j(v_i\boldsymbol{g}^i) = v_{i\|j}\boldsymbol{g}^i,$$

which unambiguously define the functions $v_{i\|j} = \{\partial_j(v_k\boldsymbol{g}^k)\}\cdot\boldsymbol{g}_i$ since the vectors $\boldsymbol{g}^i$ are linearly independent at all points of $\overline{\Omega}$ by assumption. To this end, simply note that, by definition, the Christoffel symbols satisfy $\partial_i\boldsymbol{g}^p = -\Gamma^p_{ij}\boldsymbol{g}^j$ (cf. part (ii) of the proof of Thm. 1.3-1); hence

$$\partial_j(v_i\boldsymbol{g}^i) = (\partial_j v_i)\boldsymbol{g}^i + v_i\partial_j\boldsymbol{g}^i = (\partial_j v_i)\boldsymbol{g}^i - v_i\Gamma^i_{jk}\boldsymbol{g}^k = v_{i\|j}\boldsymbol{g}^i.$$

To establish the other relations $\partial_j\boldsymbol{g}_q = \Gamma^i_{jq}\boldsymbol{g}_i$, note that

$$0 = \partial_j(\boldsymbol{g}^p\cdot\boldsymbol{g}_q) = -\Gamma^p_{ji}\boldsymbol{g}^i\cdot\boldsymbol{g}_q + \boldsymbol{g}^p\cdot\partial_j\boldsymbol{g}_q = -\Gamma^p_{qj} + \boldsymbol{g}^p\cdot\partial_j\boldsymbol{g}_q;$$

hence

$$\partial_j\boldsymbol{g}_q = (\partial_j\boldsymbol{g}_q\cdot\boldsymbol{g}^p)\boldsymbol{g}_p = \Gamma^p_{qj}\boldsymbol{g}_p.$$

■

If the affine space $\mathcal{E}^3$ is identified with $\mathbb{R}^3$ and $\boldsymbol{\Theta} = \boldsymbol{id}$, the relation $\partial_j(v_i\boldsymbol{g}^i)(x) = (v_{i\|j}\boldsymbol{g}^i)(x)$ found in Thm. 1.4-1 (b) reduces to $\hat{\partial}_j(\hat{v}_i(\hat{x})\hat{\boldsymbol{e}}^i) = (\hat{\partial}_j\hat{v}_i(\hat{x}))\hat{\boldsymbol{e}}^i$. In this sense, a covariant derivative of the first order constitutes a generalization of a partial derivative of the first order in Cartesian coordinates.

Remarks. (1) The relation between the functions $\hat{\partial}_j \hat{v}_i$ and $v_{k\|l}$ (Thm. 1.4-1 (d)) can be inverted; cf. Ex. 1.5.

(2) The formula

$$\hat{\partial}_p \hat{v}_p(\hat{x}) = (v_{k\|l}[\boldsymbol{g}^k]_p[\boldsymbol{g}^l]_p)(x) = (v_{k\|l}g^{kl})(x),$$

established in part (iv) of the proof of Thm. 1.3-1, is the expression of the *divergence of a vector field* in *curvilinear coordinates.* Formulas likewise expressing the *gradient, curl,* and *Laplacian operators* in *curvilinear coordinates* are given in Ex. 1.6. ■

1.5. LINEARIZED CHANGE OF METRIC TENSOR IN CURVILINEAR COORDINATES

Let there be given an arbitrary **displacement field** of the set $\boldsymbol{\Theta}(\overline{\Omega})$, i.e., an arbitrary vector field $v_i \boldsymbol{g}^i$ defined by means of its **covariant components** $v_i : \overline{\Omega} \to \mathbb{R}$; this means that $v_i(x)\boldsymbol{g}^i(x)$ is the displacement of the point $\hat{x} = \boldsymbol{\Theta}(x)$.

We then show that the matrix

$$(e_{i\|j}(\boldsymbol{v})) = \left(\frac{1}{2}(\partial_j v_i + \partial_i v_j) - \Gamma^p_{ij} v_p\right),$$

introduced in Thm. 1.3-1, generalizes to arbitrary *curvilinear coordinates* the *linearized strain tensor*

$$\hat{\boldsymbol{e}}(\hat{\boldsymbol{v}}) := (\hat{e}_{ij}(\hat{\boldsymbol{v}})) = \left(\frac{1}{2}(\hat{\partial}_j \hat{v}_i + \hat{\partial}_i \hat{v}_j)\right)$$

in Cartesian coordinates (Sect. 1.1).

More specifically, we show that *the matrix* $(e_{i\|j}(\boldsymbol{v}))$ *likewise measures* (half of) *the linearized difference between the "new" metric in the deformed configuration* $(\boldsymbol{\Theta} + v_i \boldsymbol{g}^i)(\overline{\Omega})$ *and the "old" metric in the reference configuration* $\boldsymbol{\Theta}(\overline{\Omega})$, but now expressed by means of their covariant components in terms of the *curvilinear coordinates* x_i. We also record for future reference the relation between the functions $\hat{e}_{ij}(\hat{\boldsymbol{v}})$ and $e_{k\|l}(\boldsymbol{v})$ that was established in part (iv) of the proof of Thm. 1.3-1 (this relation can be inverted; cf. Ex. 1.5).

Theorem 1.5-1. *Let Ω be a domain in $\mathbb{R}^3$, let $\boldsymbol{\Theta} : \overline{\Omega} \to \mathcal{E}^3$ be a smooth enough injective mapping such that the three vectors $\boldsymbol{g}_i = \partial_i \boldsymbol{\Theta}$ are linearly independent at all points of $\overline{\Omega}$, and let the vectors $\boldsymbol{g}^i$ be defined by $\boldsymbol{g}^i \cdot \boldsymbol{g}_j = \delta^i_j$. Given an arbitrary displacement field*

$$\boxed{\tilde{\boldsymbol{v}} := v_i \boldsymbol{g}^i}$$

of the set $\boldsymbol{\Theta}(\overline{\Omega})$ with smooth enough covariant components $v_i : \overline{\Omega} \to \mathbb{R}$, let the **covariant components** *of the* **linearized strain**, *or* **linearized change of metric, tensor** *associated with this vector field be defined by*

$$\boxed{e_{i\|j}(\boldsymbol{v}) := \frac{1}{2}[g_{ij}(\boldsymbol{v}) - g_{ij}]^{lin},}$$

where g_{ij} and $g_{ij}(\boldsymbol{v})$ denote the covariant components of the metric tensors respectively attached to the sets $\boldsymbol{\Theta}(\overline{\Omega})$ and $(\boldsymbol{\Theta} + v_i \boldsymbol{g}^i)(\overline{\Omega})$, and $[\cdots]^{lin}$ denotes the linear part with respect to $\boldsymbol{v} := (v_i)$ in the expression $[\cdots]$. Then

$$\boxed{\begin{aligned} e_{i\|j}(\boldsymbol{v}) &= \frac{1}{2}(\partial_j \tilde{\boldsymbol{v}} \cdot \boldsymbol{g}_i + \partial_i \tilde{\boldsymbol{v}} \cdot \boldsymbol{g}_j) = e_{j\|i}(\tilde{\boldsymbol{v}}) \\ &= \frac{1}{2}(v_{i\|j} + v_{j\|i}) \\ &= \frac{1}{2}(\partial_j v_i + \partial_i v_j) - \Gamma^p_{ij} v_p, \end{aligned}}$$

where the functions $v_{i\|j} = \partial_j v_i - \Gamma^p_{ij} v_p : \overline{\Omega} \to \mathbb{R}$ and $\Gamma^p_{ij} : \overline{\Omega} \to \mathbb{R}$ are respectively the first-order covariant derivatives of the vector field $v_i \boldsymbol{g}^i$ and the Christoffel symbols of the surface S (Thm. 1.4-1). *The functions $e_{i\|j}(\boldsymbol{v})$ are also called the* **linearized strains in curvilinear coordinates.** *The linearized strains in Cartesian and curvilinear coordinates are related by:*

$$\boxed{\begin{aligned} \hat{e}_{ij}(\hat{\boldsymbol{v}})(\hat{x}) &:= \frac{1}{2}\big(\hat{\partial}_j \hat{v}_i + \hat{\partial}_i \hat{v}_j\big)(\hat{x}) \\ &= \big(e_{k\|l}(\boldsymbol{v})[\boldsymbol{g}^k]_i [\boldsymbol{g}^l]_j\big)(x),\ \hat{x} = \boldsymbol{\Theta}(x). \end{aligned}}$$

Proof. The covariant components $g_{ij}(\boldsymbol{v})$ are defined in $\overline{\Omega}$ by (Sect. 1.2)

$$g_{ij}(\boldsymbol{v}) = \partial_i(\boldsymbol{\Theta} + \tilde{\boldsymbol{v}}) \cdot \partial_j(\boldsymbol{\Theta} + \tilde{\boldsymbol{v}}).$$

Note that *both* the reference configuration $\boldsymbol{\Theta}(\overline{\Omega})$ and the deformed configuration $(\boldsymbol{\Theta} + \tilde{\boldsymbol{v}})(\overline{\Omega})$ are equipped with the *same* curvilinear coordinates x_i. The relations

$$\partial_i(\boldsymbol{\Theta} + \tilde{\boldsymbol{v}}) = \boldsymbol{g}_i + \partial_i \tilde{\boldsymbol{v}}$$

then show that

$$\begin{aligned} g_{ij}(\boldsymbol{v}) &= (\boldsymbol{g}_i + \partial_i \tilde{\boldsymbol{v}})(\boldsymbol{g}_j + \partial_j \tilde{\boldsymbol{v}}) \\ &= g_{ij} + \partial_j \tilde{\boldsymbol{v}} \cdot \boldsymbol{g}_i + \partial_i \tilde{\boldsymbol{v}} \cdot \boldsymbol{g}_j + \partial_i \tilde{\boldsymbol{v}} \cdot \partial_j \tilde{\boldsymbol{v}}, \end{aligned}$$

hence that

$$e_{i\|j}(\boldsymbol{v}) = \frac{1}{2}[g_{ij}(\boldsymbol{v}) - g_{ij}]^{lin} = \frac{1}{2}(\partial_j \tilde{\boldsymbol{v}} \cdot \boldsymbol{g}_i + \partial_i \tilde{\boldsymbol{v}} \cdot \boldsymbol{g}_j).$$

The other expressions of the functions $e_{i\|j}(\boldsymbol{v})$ follow from the relations $\partial_j \tilde{\boldsymbol{v}} = v_{k\|j}\boldsymbol{g}^k$ and $v_{k\|j} = \partial_j v_k - \Gamma^p_{jk} v_p$ (Thm. 1.4-1). ∎

1.6. THE BOUNDARY VALUE PROBLEM OF THREE-DIMENSIONAL LINEARIZED ELASTICITY IN CURVILINEAR COORDINATES

While deriving the *boundary value problem* that is (at least formally) equivalent to the variational equations of three-dimensional linearized elasticity in *Cartesian coordinates* simply amounts to applying the fundamental Green formula, doing so in *curvilinear coordinates* is more subtle. As we next show, it relies in particular on the notion of *covariant differentiation of a tensor field.* We recall that $n_i(x)\boldsymbol{e}^i$ denotes the unit outer normal vector at $x \in \Gamma$ and that $f^i \in L^2(\Omega)$ and $h^i \in L^2(\Gamma_1)$ by assumption.

Theorem 1.6-1. *Let the notations and assumptions be as in* Thm. 1.3-1. *If the solution* $\boldsymbol{u} = (u_i)$ *to the variational problem:*

$$\boxed{\begin{aligned} &\boldsymbol{u} \in \mathbf{V}(\Omega) := \{\boldsymbol{v} = (v_i) \in \mathbf{H}^1(\Omega);\ \boldsymbol{v} = \mathbf{0} \text{ on } \Gamma_0\},\\ &\int_\Omega A^{ijkl} e_{k\|l}(\boldsymbol{u}) e_{i\|j}(\boldsymbol{v})\sqrt{g}\,dx = \int_\Omega f^i v_i \sqrt{g}\,dx + \int_{\Gamma_1} h^i v_i \sqrt{g}\,d\Gamma\\ &\qquad\qquad\qquad\qquad \text{for all } \boldsymbol{v} = (v_i) \in \mathbf{V}(\Omega), \end{aligned}}$$

is smooth enough, it also satisfies the following **boundary value problem of three-dimensional linearized elasticity in curvilinear coordinates:**

$$\boxed{\begin{aligned} -\sigma^{ij}\|_j &= f^i \quad \text{in} \quad \Omega,\\ u_i &= 0 \quad \text{on} \quad \Gamma_0,\\ \sigma^{ij} n_j &= h^i \quad \text{on} \quad \Gamma_1, \end{aligned}}$$

where the functions

$$\boxed{\sigma^{ij} := A^{ijkl} e_{k\|l}(\boldsymbol{u})}$$

are the **contravariant components** *of the* **linearized stress tensor** *field, and the functions*

$$\boxed{\sigma^{ij}\|_k := \partial_k \sigma^{ij} + \Gamma^i_{pk}\sigma^{pj} + \Gamma^j_{kq}\sigma^{iq}}$$

are **first-order covariant derivatives** *of this tensor field* (naturally, $\Gamma^j_{jq} = \sum_{j=1}^3 \Gamma^j_{jq}$ when $k = j$ according to the summation convention).

Proof. (i) We first establish the relations (needed in part (ii)):

$$\partial_j \sqrt{g} = \sqrt{g}\,\Gamma^q_{qj}.$$

To this end, we recall that $\sqrt{g} = |\det \boldsymbol{\nabla\Theta}|$, that the column vectors of the matrix $\boldsymbol{\nabla\Theta}$ are $\boldsymbol{g}_1, \boldsymbol{g}_2, \boldsymbol{g}_3$ (in this order), and that the vectors $\boldsymbol{g}_i$ are linearly independent at all points in $\overline{\Omega}$. Assume for instance that $\det \boldsymbol{\nabla\Theta} > 0$, so that

$$\sqrt{g} = \det \boldsymbol{\nabla\Theta} = \det(\boldsymbol{g}_1, \boldsymbol{g}_2, \boldsymbol{g}_3) \text{ in } \overline{\Omega}.$$

Then

$$\begin{aligned}\partial_j\sqrt{g} &= \det(\partial_j \boldsymbol{g}_1, \boldsymbol{g}_2, \boldsymbol{g}_3) + \det(\boldsymbol{g}_1, \partial_j\boldsymbol{g}_2, \boldsymbol{g}_3) + \det(\boldsymbol{g}_1, \boldsymbol{g}_2, \partial_j\boldsymbol{g}_3)\\ &= \Gamma^p_{1j}\det(\boldsymbol{g}_p, \boldsymbol{g}_2, \boldsymbol{g}_3) + \Gamma^p_{2j}\det(\boldsymbol{g}_1, \boldsymbol{g}_p, \boldsymbol{g}_3) + \Gamma^p_{3j}\det(\boldsymbol{g}_1, \boldsymbol{g}_2, \boldsymbol{g}_p)\end{aligned}$$

since $\partial_j \boldsymbol{g}_q = \Gamma^p_{qj}\boldsymbol{g}_p$ (Thm. 1.4-1 (c)); hence

$$\partial_j\sqrt{g} = \left(\Gamma^1_{1j} + \Gamma^2_{2j} + \Gamma^3_{3j}\right)\det(\boldsymbol{g}_1, \boldsymbol{g}_2, \boldsymbol{g}_3) = \Gamma^q_{qj}\sqrt{g}.$$

The proof is similar if $\sqrt{g} = -\det \boldsymbol{\nabla\Theta}$ in $\overline{\Omega}$.

(ii) Assume that the functions σ^{ij} are in $H^1(\Omega)$ (as is the case if $\boldsymbol{u} \in \mathbf{H}^2(\Omega)$ and the assumptions on the mapping $\boldsymbol{\Theta}$ are those of Thm. 1.3-1), so that we can apply the following *Green formula in Sobolev spaces* (see, e.g., Vol. I, Thm. 6.1-9):

$$\int_\Omega (\partial_j v)w\,dx = -\int_\Omega v\partial_j w\,dx + \int_\Gamma vwn_j\,d\Gamma \text{ for all } v,\, w \in H^1(\Omega).$$

Taking also into account the symmetries $\sigma^{ij} = \sigma^{ji}$ (which themselves follow from the symmetries $A^{ijkl} = A^{jikl}$; cf. Thm. 1.3-1) and the relations $\partial_j\sqrt{g} = \sqrt{g}\,\Gamma^q_{qj}$ (part (i)), we obtain:

$$\begin{aligned}&\int_\Omega A^{ijkl}e_{k\|l}(\boldsymbol{u})e_{i\|j}(\boldsymbol{v})\sqrt{g}\,dx = \int_\Omega \sigma^{ij}e_{i\|j}(\boldsymbol{v})\sqrt{g}\,dx\\ &\quad= \int_\Omega \sqrt{g}\sigma^{ij}\left(\frac{1}{2}(\partial_j v_i + \partial_j v_i) - \Gamma^p_{ij}v_p\right)dx\\ &\quad= \int_\Omega \sqrt{g}\sigma^{ij}\partial_j v_i\,dx - \int_\Omega \sqrt{g}\sigma^{ij}\Gamma^p_{ij}v_p\,dx\\ &\quad= -\int_\Omega \partial_j\left(\sqrt{g}\sigma^{ij}\right)v_i\,dx + \int_\Gamma \sqrt{g}\sigma^{ij}n_jv_i\,d\Gamma - \int_\Omega \sqrt{g}\sigma^{pj}\Gamma^i_{pj}v_i\,dx\\ &\quad= -\int_\Omega \sqrt{g}\left(\partial_j\sigma^{ij} + \Gamma^i_{pj}\sigma^{pj} + \Gamma^j_{jq}\sigma^{iq}\right)v_i\,dx + \int_\Gamma \sqrt{g}\sigma^{ij}n_jv_i\,d\Gamma\\ &\quad= -\int_\Omega \sqrt{g}(\sigma^{ij}\|_j)v_i\,dx + \int_\Gamma \sqrt{g}\sigma^{ij}n_jv_i\,d\Gamma\end{aligned}$$

for all $\boldsymbol{v} = (v_i) \in \mathbf{H}^1(\Omega)$. Hence the variational equations imply that

$$\int_\Omega \sqrt{g}\left(\sigma^{ij}\|_j + f^i\right)v_i\,dx = \int_{\Gamma_1}\sqrt{g}\left(\sigma^{ij}n_j - h^i\right)v_i\,d\Gamma$$

for all $(v_i) \in \mathbf{V}(\Omega)$. Letting the functions v_i vary in $\mathcal{D}(\Omega)$ shows that $\sigma^{ij}\|_j + f^i = 0$ in Ω and letting the functions (v_i) vary in $\mathbf{V}(\Omega)$ in the remaining equations shows that $\sigma^{ij}n_j - h^i = 0$ on Γ_1. ■

Naturally, the same boundary value problem in curvilinear coordinates can be directly obtained from its Cartesian counterpart:

$$\begin{aligned} -\hat{\partial}_j \hat{\sigma}^{ij} &= \hat{f}^i \quad \text{in} \quad \hat{\Omega}, \\ \hat{u}_i &= 0 \quad \text{on} \quad \hat{\Gamma}_0, \\ \hat{\sigma}^{ij}\hat{n}_j &= \hat{h}^i \quad \text{on} \quad \hat{\Gamma}_1, \end{aligned}$$

where $\hat{\sigma}^{ij} = \hat{A}^{ijkl}\hat{e}_{kl}(\hat{\boldsymbol{u}})$ (Sect. 1.1); but again this requires some care; cf. Ex. 1.7.

The variational equations

$$\int_\Omega \sigma^{ij} e_{i\|j}(\boldsymbol{v})\sqrt{g}\,dx = \int_\Omega f^i v_i \sqrt{g}\,dx + \int_{\Gamma_1} h^i v_i \sqrt{g}\,d\Gamma \quad \text{for all } \boldsymbol{v} = (v_i) \in \mathbf{V}(\Omega)$$

constitute the **linearized principle of virtual work in curvilinear coordinates.** The equations

$$\begin{aligned} -\sigma^{ij}\|_j &= f^i \quad \text{in} \quad \Omega, \\ \sigma^{ij}n_j &= h^i \quad \text{on} \quad \Gamma_1, \end{aligned}$$

constitute the **linearized equations of equilibrium in curvilinear coordinates.**

1.7. A LEMMA OF J. L. LIONS; THREE-DIMENSIONAL KORN'S INEQUALITIES AND INFINITESIMAL RIGID DISPLACEMENT LEMMA IN CURVILINEAR COORDINATES

We first review some essential definitions and notations, together with a fundamental *lemma of J.L. Lions* (Thm. 1.7-1); we borrow here the beginning of Sect. 1.1 from Vol. II.

A **domain** Ω in $\mathbb{R}^n$ is an *open, bounded, connected* subset of $\mathbb{R}^n$ with a *Lipschitz-continuous boundary* $\Gamma = \partial\Omega$, the set Ω being *locally* *on one side of* Γ. As Γ is Lipschitz-continuous, a *measure* $d\Gamma$ can be defined along Γ and a **unit outer normal vector** $\boldsymbol{\nu} = (\nu_i)_{i=1}^n$ ("unit" means that its Euclidean norm is one) exists $d\Gamma$-almost everywhere along Γ (see, e.g., Vol. I, Sect. 1.6). For the same reason, the classical spaces $\mathcal{C}^m(\overline{\Omega})$ or $\mathcal{C}^{m,\alpha}(\overline{\Omega})$, $0 < \alpha \leq 1$, can be unambiguously defined for any integer $m \geq 0$; see, e.g., Nečas [1967, Chap. 2], Adams [1975, Chap. 1], Stein [1970].

Let Ω be a domain in $\mathbb{R}^n$. For each integer $m \geq 1$, $H^m(\Omega)$ and $H_0^m(\Omega)$ denote the usual **Sobolev spaces**; in particular,

$$\begin{aligned} H^1(\Omega) &:= \{v \in L^2(\Omega);\ \partial_i v \in L^2(\Omega),\ 1 \leq i \leq n\}, \\ H^2(\Omega) &:= \{v \in H^1(\Omega);\ \partial_{ij} v \in L^2(\Omega),\ 1 \leq i,\ j \leq n\}, \end{aligned}$$

where $\partial_i v$ and $\partial_{ij} v$ denote the partial derivatives of the first and second order in the sense of distributions, and

$$\begin{aligned} H_0^1(\Omega) &:= \{v \in H^1(\Omega);\ v = 0 \text{ on } \Gamma\}, \\ H_0^2(\Omega) &:= \{v \in H^2(\Omega);\ v = \partial_\nu v = 0 \text{ on } \Gamma\}, \end{aligned}$$

where the relations on Γ are to be understood in the sense of traces, and $\partial_\nu v := \sum_{i=1}^n \nu_i \partial_i v$ denotes the outer normal derivative of v along Γ. Boldface letters denote vector-valued or matrix-valued functions, also called *vector fields* or *matrix fields*, and their associated function spaces. The norm in $L^2(\Omega)$ or $\mathbf{L}^2(\Omega)$ is noted $|\cdot|_{0,\Omega}$ and the norm in $H^m(\Omega)$ or $\mathbf{H}^m(\Omega)$, $m \geq 1$, is noted $\|\cdot\|_{m,\Omega}$. In particular then,

$$|v|_{0,\Omega} := \left\{ \int_\Omega |v|^2\, dx \right\}^{1/2} \text{ if } v \in L^2(\Omega),$$

$$|\boldsymbol{v}|_{0,\Omega} := \left\{ \sum_{i=1}^n |v_i|_{0,\Omega}^2 \right\}^{1/2} \text{ if } \boldsymbol{v} = (v_i)_{i=1}^n \in \mathbf{L}^2(\Omega),$$

$$\|v\|_{1,\Omega} := \left\{ |v|_{0,\Omega}^2 + \sum_{i=1}^n |\partial_i v|_{0,\Omega}^2 \right\}^{1/2} \text{ if } v \in H^1(\Omega),$$

$$\|\boldsymbol{v}\|_{1,\Omega} := \left\{ \sum_{i=1}^n \|v_i\|_{1,\Omega}^2 \right\}^{1/2} \text{ if } \boldsymbol{v} = (v_i)_{i=1}^n \in \mathbf{H}^1(\Omega),$$

$$\|v\|_{2,\Omega} := \left\{ |v|_{0,\Omega}^2 + \sum_{i=1}^n |\partial_i v|_{0,\Omega}^2 + \sum_{i,j=1}^n |\partial_{ij} v|_{0,\Omega}^2 \right\}^{1/2} \text{ if } v \in H^2(\Omega).$$

Most properties needed in this volume about the Sobolev spaces $H^m(\Omega)$, and also about the Sobolev spaces $W^{m,p}(\Omega)$, $1 \leq p \leq \infty$, are recalled in Vol. I, Sect. 6.1. Detailed treatments are found in Nečas [1967], Lions & Magenes [1968], Dautray & Lions [1984, Chaps. 1-3], Adams [1975]. An excellent introduction is given in Brezis [1983].

In this section, we also consider the Sobolev space

$$H^{-1}(\Omega) := \text{ dual space of } H_0^1(\Omega).$$

Another possible definition of the space $H_0^1(\Omega)$ being

$$H_0^1(\Omega) = \text{ closure of } \mathcal{D}(\Omega) \text{ with respect to } \|\cdot\|_{1,\Omega},$$

where $\mathcal{D}(\Omega)$ denotes the space of infinitely differentiable real-valued functions defined over Ω whose support is a compact subset of Ω, it is clear that

$$v \in L^2(\Omega) \Rightarrow v \in H^{-1}(\Omega) \text{ and } \partial_i v \in H^{-1}(\Omega),\ 1 \le i \le n,$$

since (the duality between the spaces $\mathcal{D}(\Omega)$ and $\mathcal{D}'(\Omega)$ is denoted $<\cdot,\cdot>$):

$$|<v, \varphi>| = \left|\int_\Omega v\varphi\, dx\right| \le |v|_{0,\Omega}\|\varphi\|_{1,\Omega},$$

$$|<\partial_i v, \varphi>| = |-<v, \partial_i\varphi>| = \left|-\int_\Omega v\partial_i\varphi\, dx\right| \le |v|_{0,\Omega}\|\varphi\|_{1,\Omega}$$

for all $\varphi \in \mathcal{D}(\Omega)$. It is *remarkable*, but also *remarkably difficult to prove*, that the converse implication holds:

Theorem 1.7-1 (lemma of J.L. Lions). *Let Ω be a domain in $\mathbb{R}^n$ and let v be a distribution on Ω. Then*

$$\left\{v \in H^{-1}(\Omega) \text{ and } \partial_i v \in H^{-1}(\Omega),\ 1 \le i \le n\right\} \Rightarrow v \in L^2(\Omega).$$

∎

This implication was first proved by J.L. Lions, as stated in Magenes & Stampacchia [1958, p. 320, Note ([27])]. Its first published proof for domains with smooth boundaries appeared in Duvaut & Lions [1972, p. 111]; another proof was also given by Tartar [1978]. Various extensions to "genuine" domains, i.e., with Lipschitz-continuous boundaries, are given in Bolley & Camus [1976], Geymonat & Suquet

[1986], and Borchers & Sohr [1990]; Amrouche & Girault [1994, Prop. 2.10] even proved that the more general implication

$$\{v \in \mathcal{D}'(\Omega) \text{ and } \partial_i v \in W^{m,p}(\Omega),\ 1 \leq i \leq n\} \Rightarrow v \in W^{m+1,p}(\Omega)$$

holds for *arbitrary* integers $m \in \mathbb{Z}$ and real numbers $1 < p < \infty$.

Some minimal regularity of the boundary is anyway required: Geymonat & Gilardi [1998] have shown that the lemma of J.L. Lions does *not* hold if the open set Ω satisfies only the "segment property".

Remark. Although Thm. 1.7-1 shall be referred to as *"the" lemma of J.L. Lions* in this volume, there are other results of his that bear the same name in the literature, such as his "compactness lemmas" (Lions [1961, Prop. 4.1, p. 59] or Lions [1969, Sect. 5.2, p. 57]) or his "singular perturbation lemma" (Lions [1973, Lemma 5.1, p. 126]). ∎

Proving the existence and uniqueness of a solution to the variational problem of three-dimensional linearized elasticity found in Thm. 1.3-1 amounts to showing that *the associated bilinear form is* $\mathbf{V}(\Omega)$*-elliptic*, as all the other assumptions of the *Lax-Milgram lemma* are immediately verified.

To this end, an essential step consists in establishing a *three-dimensional Korn inequality in curvilinear coordinates* (Thm. 1.7-4) as in Ciarlet [1993]; see also Ciarlet [1998a, 1998b]. In essence, this inequality asserts that, given a domain $\Omega \subset \mathbb{R}^3$ with boundary Γ and $\Gamma_0 \subset \Gamma$ with *area* $\Gamma_0 > 0$, *the* $\mathbf{L}^2(\Omega)$*-norm of the matrix field* $(e_{i\|j}(\boldsymbol{v}))$ *is equivalent to the* $\mathbf{H}^1(\Omega)$*-norm of all vector fields* $\boldsymbol{v} \in \mathbf{H}^1(\Omega)$ *that vanish on* Γ_0 (the $\mathbf{V}(\Omega)$-ellipticity of the bilinear form also relies on the positive definiteness of the three-dimensional elasticity tensor introduced in Sect. 1.3, a property that will be established in Thm. 1.8-1). We recall that the functions $e_{i\|j}(\boldsymbol{v})$ are the covariant components of the linearized strain tensor associated with a displacement field $v_i \boldsymbol{g}^i$ (Thm. 1.5-1).

Such a Korn inequality is obtained in *three stages* (Thm. 1.7-2, 1.7-3, and 1.7-4). The first one (Thm. 1.7-2) consists in establishing, as a consequence of the *lemma of J.L. Lions*, a *Korn inequality valid for all vector fields* $\boldsymbol{v} = (v_i) \in \mathbf{H}^1(\Omega)$, i.e., that need not satisfy any boundary condition on Γ.

As its Cartesian special case, this inequality is truly remarkable, since only *six* different combinations of first-order partial derivatives, viz., $\frac{1}{2}(\partial_j v_i + \partial_i v_j)$, occur in its right-hand side, while all *nine* partial derivatives $\partial_i v_j$ occur in its left-hand side! A similarly striking observation applies to part (ii) of the next proof.

Note that we again adhere to our rule governing the usage of Latin indices.

Theorem 1.7-2 (Korn's inequality "without boundary conditions" in curvilinear coordinates). *Let Ω be a domain in $\mathbb{R}^3$ and let $\boldsymbol{\Theta}$ be a $\mathcal{C}^2$-diffeomorphism of $\overline{\Omega}$ onto $\{\hat{\Omega}\}^- = \boldsymbol{\Theta}(\overline{\Omega})$, so that the three vectors $\boldsymbol{g}_i = \partial_i \boldsymbol{\Theta}$ are linearly independent at all points of $\overline{\Omega}$. Given $\boldsymbol{v} = (v_i) \in \mathbf{H}^1(\Omega)$, let*

$$e_{i\|j}(\boldsymbol{v}) := \left\{\frac{1}{2}(\partial_j v_i + \partial_i v_j) - \Gamma^p_{ij} v_p\right\} \in L^2(\Omega)$$

denote the covariant components of the linearized change of metric tensor associated with the displacement field $v_i \boldsymbol{g}^i$. Then there exists a constant $C_0 = C_0(\Omega, \boldsymbol{\Theta})$ such that

$$\|\boldsymbol{v}\|_{1,\Omega} \leq C_0 \Big\{ \sum_i |v_i|^2_{0,\Omega} + \sum_{i,j} |e_{i\|j}(\boldsymbol{v})|^2_{0,\Omega} \Big\}^{1/2} \quad \text{for all } \boldsymbol{v} = (v_i) \in \mathbf{H}^1(\Omega).$$

Proof. The proof given here is essentially an extension of that given in Duvaut & Lions [1972, p. 110] for proving Korn's inequality without boundary conditions in *Cartesian coordinates.*

(i) *Define the space*

$$\mathbf{W}(\Omega) := \left\{ \boldsymbol{v} = (v_i) \in \mathbf{L}^2(\Omega);\ e_{i\|j}(\boldsymbol{v}) \in L^2(\Omega) \right\}.$$

Then, equipped with the norm $\|\cdot\|_\Omega$ defined by

$$\|\boldsymbol{v}\|_\Omega := \Big\{ \sum_i |v_i|^2_{0,\Omega} + \sum_{i,j} |e_{i\|j}(\boldsymbol{v})|^2_{0,\Omega} \Big\}^{1/2},$$

the space $\mathbf{W}(\Omega)$ is a Hilbert space.

Note that the relations "$e_{i\|j}(\boldsymbol{v}) \in L^2(\Omega)$" are understood in the sense of distributions, i.e., they mean that there exist functions in

$L^2(\Omega)$, *denoted* $e_{i\|j}(\boldsymbol{v})$, such that

$$\int_\Omega e_{i\|j}(\boldsymbol{v})\varphi\,dx = -\int_\Omega \left\{\frac{1}{2}(v_i\partial_j\varphi + v_j\partial_i\varphi) + \Gamma_{ij}^p v_p\varphi\right\} dx$$

$$\text{for all } \varphi \in \mathcal{D}(\Omega).$$

Let there be given a Cauchy sequence $(\boldsymbol{v}^k)_{k=1}^\infty$ with elements $\boldsymbol{v}^k = (v_i^k) \in \mathbf{W}(\Omega)$. The definition of the norm $\|\cdot\|_\Omega$ shows that there exist functions $v_i \in L^2(\Omega)$ and $e_{i\|j} \in L^2(\Omega)$ such that

$$v_i^k \to v_i \text{ in } L^2(\Omega) \text{ and } e_{i\|j}(\boldsymbol{v}^k) \to e_{i\|j} \text{ in } L^2(\Omega) \text{ as } k \to \infty,$$

since the space $L^2(\Omega)$ is complete. Given a function $\varphi \in \mathcal{D}(\Omega)$, letting $k \to \infty$ in the relations

$$\int_\Omega e_{i\|j}(\boldsymbol{v}^k)\varphi\,dx = -\int_\Omega \left\{\frac{1}{2}(v_i^k\partial_j\varphi + v_j^k\partial_i\varphi) + \Gamma_{ij}^p v_p^k\varphi\right\} dx,\ k \ge 1,$$

shows that $e_{i\|j} = e_{i\|j}(\boldsymbol{v})$.

(ii) *The spaces* $\mathbf{W}(\Omega)$ *and* $\mathbf{H}^1(\Omega)$ *coincide.*

Clearly, $\mathbf{H}^1(\Omega) \subset \mathbf{W}(\Omega)$. To establish the other inclusion, let $\boldsymbol{v} = (v_i) \in \mathbf{W}(\Omega)$. Then

$$e_{ij}(\boldsymbol{v}) := \frac{1}{2}(\partial_j v_i + \partial_i v_j) = \left\{e_{i\|j}(\boldsymbol{v}) + \Gamma_{ij}^p v_p\right\} \in L^2(\Omega),$$

since $e_{i\|j}(\boldsymbol{v}) \in L^2(\Omega)$, $\Gamma_{ij}^p \in \mathcal{C}^0(\overline{\Omega})$, and $v_p \in L^2(\Omega)$. We thus have

$$\partial_k v_i \in H^{-1}(\Omega),$$

$$\partial_j(\partial_k v_i) = \{\partial_j e_{ik}(\boldsymbol{v}) + \partial_k e_{ij}(\boldsymbol{v}) - \partial_i e_{jk}(\boldsymbol{v})\} \in H^{-1}(\Omega),$$

since $w \in L^2(\Omega)$ implies $\partial_k w \in H^{-1}(\Omega)$. Hence $\partial_k v_i \in L^2(\Omega)$ by the *lemma of J.L. Lions* (Thm. 1.7-1) and thus $\boldsymbol{v} \in \mathbf{H}^1(\Omega)$.

(iii) *Korn's inequality without boundary conditions.*

The identity mapping ι from the space $\mathbf{H}^1(\Omega)$ equipped with $\|\cdot\|_{1,\Omega}$ into the space $\mathbf{W}(\Omega)$ equipped with $\|\cdot\|_\Omega$ is injective, continuous (there clearly exists a constant c such that $\|\boldsymbol{v}\|_\Omega \le c\|\boldsymbol{v}\|_{1,\Omega}$ for all $\boldsymbol{v} \in \mathbf{H}^1(\Omega)$), and surjective by (ii). Since both spaces are complete (cf. (i)), the *closed graph theorem* (see, e.g., Brezis [1983, p. 19] for a proof) then shows that the inverse mapping ι^{-1} is also continuous; this continuity is exactly what is expressed by *Korn's inequality without boundary conditions.* ■

Our next objective is to "get rid" of the norms $|v_i|_{0,\Omega}$ in the right-hand side of the Korn inequality established in Thm. 1.7-2 when the fields $\boldsymbol{v} = (v_i) \in \mathbf{H}^1(\Omega)$ are subjected to the boundary condition $\boldsymbol{v} = \mathbf{0}$ on $\Gamma_0 \subset \Gamma$ and *area* $\Gamma_0 > 0$. As a preliminary, we establish the weaker property that the *semi-norm*

$$\boldsymbol{v} \to \Big\{ \sum_{i,j} |e_{i\|j}(\boldsymbol{v})|^2_{0,\Omega} \Big\}^{1/2}$$

becomes a *norm* for such fields, by generalizing to curvilinear coordinates the well-known *infinitesimal rigid displacement lemma in Cartesian coordinates* (see, e.g., part (ii) of the proof of Thm. 6.3-4 in Vol. I); "infinitesimal" reminds that if $e_{i\|j}(\boldsymbol{v}) = 0$ in Ω, i.e., if only the *linearized* part of the change of metric tensor vanishes, the corresponding displacement field $v_i \boldsymbol{g}^i$ is likewise only the *linearized* part of a genuine rigid displacement (for more details in Cartesian coordinates, see Vol. I, Ex. 6.2).

Part (a) in the next theorem is an *infinitesimal rigid displacement lemma "without boundary conditions"*, while part (b) is an *infinitesimal rigid displacement lemma "with boundary conditions"*.

Theorem 1.7-3 (infinitesimal rigid displacement lemma in curvilinear coordinates). *Let the assumptions be as in* Thm. 1.7-2.

(a) *Let* $\boldsymbol{v} = (v_i) \in \mathbf{H}^1(\Omega)$ *be such that*

$$\boxed{e_{i\|j}(\boldsymbol{v}) = 0 \text{ in } \Omega.}$$

Then the vector field $v_i \boldsymbol{g}^i$ *is an* **infinitesimal rigid displacement**, *in the sense that there exist two vectors* $\hat{\mathbf{c}}$, $\hat{\boldsymbol{d}} \in \mathbb{R}^3$ *such that*

$$\boxed{v_i(x)\boldsymbol{g}^i(x) = \hat{\mathbf{c}} + \hat{\boldsymbol{d}} \wedge \boldsymbol{\Theta}(x) \text{ for all } x \in \overline{\Omega}.}$$

(b) *Let* Γ_0 *be a* $d\Gamma$*-measurable subset of* $\Gamma = \partial\Omega$ *that satisfies*

$$\boxed{\textit{area } \Gamma_0 > 0.}$$

Then

$$\left.\begin{aligned} \boldsymbol{v} = (v_i) \in \mathbf{H}^1(\Omega),\ \boldsymbol{v} = \mathbf{0} \text{ on } \Gamma_0, \\ e_{i\|j}(\boldsymbol{v}) = 0 \text{ in } \Omega \end{aligned}\right\} \Rightarrow \boldsymbol{v} = \mathbf{0} \text{ in } \Omega.$$

Proof. In part (iv) of the proof of Thm. 1.3-1, we established the relation

$$\hat{e}_{ij}(\hat{\boldsymbol{v}})(\hat{x}) = \left(e_{k\|l}(\boldsymbol{v})[\boldsymbol{g}^k]_i[\boldsymbol{g}^l]_j\right)(x) \text{ for all } \hat{x} = \boldsymbol{\Theta}(x),\ x \in \Omega,$$

where $\hat{e}_{ij}(\hat{\boldsymbol{v}}) = \frac{1}{2}(\hat{\partial}_j\hat{v}_i + \hat{\partial}_i\hat{v}_j)$ and the vector fields $\hat{\boldsymbol{v}} = (\hat{v}_i) \in \mathbf{H}^1(\hat{\Omega})$ and $\boldsymbol{v} = (v_i) \in \mathbf{H}^1(\Omega)$ are related by

$$\hat{v}_i(\hat{x})\hat{\boldsymbol{e}}^i = v_i(x)\boldsymbol{g}^i(x) \text{ for all } \hat{x} = \boldsymbol{\Theta}(x),\ x \in \Omega.$$

Hence

$$e_{i\|j}(\boldsymbol{v}) = 0 \text{ in } \Omega \Rightarrow \hat{e}_{ij}(\hat{\boldsymbol{v}}) = 0 \text{ in } \hat{\Omega},$$

and the identity (the same as in the proof of Thm. 1.7-2)

$$\hat{\partial}_j(\hat{\partial}_k\hat{v}_i) = \hat{\partial}_j\hat{e}_{ik}(\hat{\boldsymbol{v}}) + \hat{\partial}_k\hat{e}_{ij}(\hat{\boldsymbol{v}}) - \hat{\partial}_i\hat{e}_{jk}(\hat{\boldsymbol{v}}) \text{ in } \mathcal{D}'(\hat{\Omega})$$

further shows that

$$\hat{e}_{ij}(\hat{\boldsymbol{v}}) = 0 \text{ in } \hat{\Omega} \Rightarrow \hat{\partial}_j(\hat{\partial}_k\hat{v}_i) = 0 \text{ in } \mathcal{D}'(\hat{\Omega}).$$

By a classical result from distribution theory (Schwartz [1966, p. 60]), each function $\hat{v}_i$ is therefore a polynomial of degree ≤ 1 (recall that the set $\hat{\Omega}$ is connected). In other words, there exist constants $\hat{c}_i$ and $\hat{d}_{ij}$ such that

$$\hat{v}_i(\hat{x}) = \hat{c}_i + \hat{d}_{ij}\hat{x}_j \text{ for all } \hat{x} = (\hat{x}_i) \in \hat{\Omega}.$$

But $\hat{e}_{ij}(\hat{\boldsymbol{v}}) = 0$ also implies that $\hat{d}_{ij} = -\hat{d}_{ji}$; hence there exist two vectors $\hat{\boldsymbol{c}}$, $\hat{\boldsymbol{d}} \in \mathbb{R}^3$ such that

$$\hat{\boldsymbol{v}}_i(\hat{x})\hat{\boldsymbol{e}}^i = \hat{\boldsymbol{c}} + \hat{\boldsymbol{d}} \wedge \hat{\boldsymbol{O}}\hat{\boldsymbol{x}} \text{ for all } \hat{x} \in \hat{\Omega},$$

hence such that

$$v_i(x)\boldsymbol{g}^i(x) = \hat{\boldsymbol{c}} + \hat{\boldsymbol{d}} \wedge \boldsymbol{\Theta}(x) \text{ for all } x \in \Omega.$$

Since the set where such a vector field $\hat{v}_i \hat{\boldsymbol{e}}^i$ vanishes is always of zero area unless $\hat{\boldsymbol{c}} = \hat{\boldsymbol{d}} = \mathbf{0}$ (as is easily proved; see, e.g., Vol. I, Thm. 6.3-4), it follows that $\hat{\boldsymbol{v}} = \mathbf{0}$ when *area* $\Gamma_0 > 0$. ∎

Remarks. (1) Since the fields $\boldsymbol{g}_i$ are of class $\mathcal{C}^1$ on $\overline{\Omega}$ by assumption, the components v_i of a field $\boldsymbol{v} = (v_i) \in \mathbf{H}^1(\Omega)$ satisfying $e_{i\|j}(\boldsymbol{v}) = 0$ in Ω are thus automatically in $\mathcal{C}^1(\overline{\Omega})$ since $v_i = (v_j \boldsymbol{g}^j) \cdot \boldsymbol{g}_i$.

(2) Remarkably, the field $v_i \boldsymbol{g}^i = \hat{\boldsymbol{c}} + \hat{\boldsymbol{d}} \wedge \boldsymbol{\Theta}$ inherits in this case even more regularity, as it is of class $\mathcal{C}^2$ on $\overline{\Omega}$ (a similar phenomenon will be encountered in Thm. 2.6-3)!

(3) The assertion in part (a) illustrates a key idea from differential geometry: An *"intrinsic"* property (the vector field $v_i \boldsymbol{g}^i$ is an infinitesimal rigid displacement) is derived from relations (the assumptions $e_{i\|j}(\boldsymbol{v}) = 0$ in Ω) written in a *particular* system of curvilinear coordinates. ∎

We are now in a position to prove a *Korn inequality "with boundary conditions"*, which plays a fundamental rôle in three-dimensional linearized elasticity in curvilinear coordinates (Thm. 1.8-2).

Theorem 1.7-4 (Korn's inequality in curvilinear coordinates). *Let Ω be a domain in $\mathbb{R}^3$ and let $\boldsymbol{\Theta}$ be a $\mathcal{C}^2$-diffeomorphism of $\overline{\Omega}$ onto $\{\hat{\Omega}\}^- = \boldsymbol{\Theta}(\overline{\Omega})$, so that the three vectors $\boldsymbol{g}_i = \partial_i \boldsymbol{\Theta}$ are linearly independent at all points of $\overline{\Omega}$, let Γ_0 be a $d\Gamma$-measurable subset of $\Gamma = \partial\Omega$ that satisfies*

$$\boxed{\textit{area}\ \Gamma_0 > 0,}$$

and let the space $\mathbf{V}(\Omega)$ be defined by

$$\boxed{\mathbf{V}(\Omega) := \left\{\boldsymbol{v} = (v_i) \in \mathbf{H}^1(\Omega);\ \boldsymbol{v} = \mathbf{0} \text{ on } \Gamma_0\right\}.}$$

Then there exists a constant $C = C(\Omega, \Gamma_0, \boldsymbol{\Theta})$ such that

$$\boxed{\|\boldsymbol{v}\|_{1,\Omega} \le C\Big\{\sum_{i,j} |e_{i\|j}(\boldsymbol{v})|_{0,\Omega}^2\Big\}^{1/2} \text{ for all } \boldsymbol{v} \in \mathbf{V}(\Omega).}$$

Proof. Given $\boldsymbol{v} = (v_i) \in \mathbf{H}^1(\Omega)$, let

$$|\boldsymbol{v}|_\Omega := \Big\{ \sum_{i,j} |e_{i\|j}(\boldsymbol{v})|_{0,\Omega}^2 \Big\}^{1/2}.$$

If the announced inequality is false, there exists a sequence $(\boldsymbol{v}^k)_{k=1}^\infty$ of elements $\boldsymbol{v}^k \in \mathbf{V}(\Omega)$ such that

$$\|\boldsymbol{v}^k\|_{1,\Omega} = 1 \text{ for all } k \text{ and } \lim_{k\to\infty} |\boldsymbol{v}^k|_\Omega = 0.$$

Since the sequence $(\boldsymbol{v}^k)_{k=1}^\infty$ is bounded in $\mathbf{H}^1(\Omega)$, a subsequence $(\boldsymbol{v}^l)_{l=1}^\infty$ converges in $\mathbf{L}^2(\Omega)$ by the *Rellich-Kondrašov theorem* (see, e.g., Vol. I, Thm. 6.1-5); furthermore, since $\lim_{l\to\infty} |\boldsymbol{v}^l|_\Omega = 0$, each sequence $(e_{i\|j}(\boldsymbol{v}^l))_{l=1}^\infty$ also converges in $L^2(\Omega)$ (to 0, but this information is not used at this stage). The subsequence $(\boldsymbol{v}^l)_{l=1}^\infty$ is thus a Cauchy sequence with respect to the norm

$$\boldsymbol{v} = (v_i) \to \Big\{ \sum_i |v_i|_{0,\Omega}^2 + \sum_{i,j} |e_{i\|j}(\boldsymbol{v})|_{0,\Omega}^2 \Big\}^{1/2},$$

hence with respect to the norm $\|\cdot\|_{1,\Omega}$ by *Korn's inequality without boundary conditions* (Thm. 1.7-2).

The space $\mathbf{V}(\Omega)$ being complete as a closed subspace of $\mathbf{H}^1(\Omega)$, there exists $\boldsymbol{v} \in \mathbf{V}(\Omega)$ such that

$$\boldsymbol{v}^l \to \boldsymbol{v} \text{ in } \mathbf{H}^1(\Omega),$$

and the limit $\boldsymbol{v}$ satisfies $|e_{i\|j}(\boldsymbol{v})|_{0,\Omega} = \lim_{l\to\infty} |e_{i\|j}(\boldsymbol{v}^l)|_{0,\Omega} = 0$; hence $\boldsymbol{v} = \mathbf{0}$ by Thm. 1.7-3. But this contradicts the relations $\|\boldsymbol{v}^l\|_{1,\Omega} = 1$ for all $l \geq 1$, and the proof is complete. ∎

Identifying $\mathcal{E}^3$ with $\mathbb{R}^3$ and letting $\boldsymbol{\Theta} = \boldsymbol{id}_{\mathbb{R}^3}$ shows that Thms. 1.7-2, 1.7-3, and 1.7-4 contain as special cases the Korn inequalities and the infinitesimal rigid displacement lemma in *Cartesian coordinates* (see, e.g., Duvaut & Lions [1972, p. 110], Vol. I, Sect. 6.3, or Vol. II, Sect. 1.1).

1.8. EXISTENCE AND UNIQUENESS THEOREM IN CURVILINEAR COORDINATES

As a preliminary to showing that the bilinear form found in Thm. 1.3-1 is $\mathbf{V}(\Omega)$-elliptic, we first need to establish the *uniform positive definiteness of the three-dimensional elasticity tensor* ("uniform" means with respect to points in $\overline{\Omega}$ and to symmetric matrices of order three). Incidentally, the proof of this property has a flavor typical of differential geometry: Although the proof is seemingly innocuous, it may not be that innocuous to find (the reader should verify this assertion)!

Theorem 1.8-1. *Let the assumptions on the mapping* $\boldsymbol{\Theta}$ *be as in* Thm. 1.3-1, *let the contravariant components* $A^{ijkl} : \overline{\Omega} \to \mathbb{R}$ *of the elasticity tensor be defined by*

$$A^{ijkl} = \lambda g^{ij} g^{kl} + \mu \big(g^{ik} g^{jl} + g^{il} g^{jk} \big),$$

and assume that

$$\lambda \geq 0 \text{ and } \mu > 0.$$

Then there exists a constant $C_e = C_e(\Omega, \boldsymbol{\Theta}, \mu) > 0$ *such that*

$$\sum_{i,j} |t_{ij}|^2 \leq C_e A^{ijkl}(x) t_{kl} t_{ij}$$

for all $x \in \overline{\Omega}$ *and all symmetric matrices* (t_{ij}).

Proof. We first note that

$$g^{ij}(x) g^{kl}(x) t_{kl} t_{ij} = \Big(g^{ij}(x) t_{ij} \Big)^2 \geq 0$$

for all $x \in \overline{\Omega}$ and all matrices (t_{ij}). Using the symmetries $t_{ij} = t_{ji}$ and the relations

$$g^{ik}(x) = \boldsymbol{g}^i(x) \cdot \boldsymbol{g}^k(x) = [\boldsymbol{g}^i(x)]_m [\boldsymbol{g}^k(x)]_m,$$

we next have

$$\frac{1}{2}\left(g^{ik}(x)g^{jl}(x)+g^{il}(x)g^{jk}(x)\right)t_{kl}t_{ij}=g^{ik}(x)g^{jl}(x)t_{kl}t_{ij}$$
$$=[\boldsymbol{g}^i(x)]_m[\boldsymbol{g}^j(x)]_n t_{ij}[\boldsymbol{g}^k(x)]_m[\boldsymbol{g}^l(x)]_n t_{kl}$$
$$=\sum_{m,n}\left([\boldsymbol{g}^i(x)]_m[\boldsymbol{g}^j(x)]_n t_{ij}\right)^2\geq 0$$

for all $x\in\overline{\Omega}$ and all symmetric matrices (t_{ij}).

We recall that the vectors $\boldsymbol{g}^i$ and the mappings $\boldsymbol{\Theta}=\Theta^m\hat{\boldsymbol{e}}_m$ and $\hat{\boldsymbol{\Theta}}=\hat{\Theta}^i\boldsymbol{e}_i$ satisfy the relations (both established in part (i) of the proof of Thm. 1.3-1):

$$[\boldsymbol{g}^i(x)]_m=\hat{\partial}_m\hat{\Theta}^i(\hat{x})\text{ and }\partial_p\Theta^m(x)\hat{\partial}_m\hat{\Theta}^i(\hat{x})=\delta^i_p$$

for all $\hat{x}=\boldsymbol{\Theta}(x)$, $x\in\overline{\Omega}$. These relations therefore imply that

$$g^{ik}(x)g^{jl}(x)t_{kl}t_{ij}=\sum_{m,n}|\hat{t}_{mn}(\hat{x})|^2\geq 0,$$

where

$$\hat{t}_{mn}(\hat{x}):=\hat{\partial}_m\hat{\Theta}^i(\hat{x})\hat{\partial}_n\hat{\Theta}^j(\hat{x})t_{ij}.$$

We have thus proved that $g^{ik}(x)g^{jl}(x)t_{kl}t_{ij}$ is always ≥ 0 and vanishes if and only if $\hat{t}_{mn}(\hat{x})=0$ for all m, n. But the relation between the components $\hat{t}_{mn}(\hat{x})$ and t_{ij} can be inverted as

$$\begin{aligned}\partial_p\Theta^m(x)\partial_q\Theta^n(x)\hat{t}_{mn}(\hat{x}) &= \partial_p\Theta^m(x)\hat{\partial}_m\hat{\Theta}^i(\hat{x})\partial_q\Theta^n(x)\hat{\partial}_n\hat{\Theta}^j(\hat{x})t_{ij}\\ &= \delta^i_p\delta^j_q t_{ij} = t_{pq},\end{aligned}$$

and thus

$$\hat{t}_{mn}(\hat{x})=\hat{\partial}_m\hat{\Theta}^i(\hat{x})\hat{\partial}_n\hat{\Theta}^j(\hat{x})t_{ij}=0\text{ for all }m,\,n\Rightarrow t_{ij}=0\text{ for all }i,\,j.$$

To sum up, we have shown that for all $x\in\overline{\Omega}$ and all symmetric matrices (t_{ij}),

$$g^{ik}(x)g^{jl}(x)t_{kl}t_{ij}>0\text{ unless }t_{ij}=0\text{ for all }i,\,j.$$

Let $\mathbb{S}^3$ denote the space of all symmetric matrices. Since the mapping

$$(x,\,(t_{ij}))\in\mathbf{K}:=\overline{\Omega}\times\left\{(t_{ij})\in\mathbb{S}^3;\,\sum_{i,j}|t_{ij}|^2=1\right\}\to g^{ik}(x)g^{jl}(x)t_{kl}t_{ij}$$

is continuous and its domain of definition is compact, we infer that

$$c := \inf_{(x,(t_{ij}))\in \mathbf{K}} g^{ik}(x)g^{jl}(x)t_{kl}t_{ij} > 0.$$

Hence

$$\sum_{i,j} |t_{ij}|^2 \leq c^{-1} g^{ik}(x)g^{jl}(x)t_{kl}t_{ij},$$

and thus

$$\sum_{i,j} |t_{ij}|^2 \leq C_e A^{ijkl}(x)t_{kl}t_{ij}$$

for all $x \in \overline{\Omega}$ and all symmetric matrices (t_{ij}) with $C_e := (2\mu c)^{-1}$ since $\lambda \geq 0$ and $\mu > 0$ by assumption. ∎

Remarks. (1) Another proof is suggested in Ex. 1.8.

(2) The elasticity tensor retains its uniform positive definiteness "even if λ is slightly negative"; cf. Ex. 1.9. ∎

Combined with Korn's inequality, the positive definiteness of the elasticity tensor leads to the existence and uniqueness of a **weak solution**, i.e., a solution to the *variational* equations of three-dimensional linearized elasticity in curvilinear coordinates.

Theorem 1.8-2. *Let Ω be a domain in $\mathbb{R}^3$ and let $\boldsymbol{\Theta}$ be a $\mathcal{C}^2$-diffeomorphism of $\overline{\Omega}$ onto its image $\boldsymbol{\Theta}(\overline{\Omega})$, so that the three vectors $\boldsymbol{g}_i(x) = \partial_i \boldsymbol{\Theta}(x)$ are linearly independent at all points $x \in \overline{\Omega}$. Let Γ_0 be a $d\Gamma$-measurable subset of $\Gamma = \partial\Omega$ that satisfies* area $\Gamma_0 > 0$. *Finally, let there be given constants $\lambda \geq 0$ and $\mu > 0$ and functions $f^i \in L^{6/5}(\Omega)$ and $h^i \in L^{4/3}(\Gamma_1)$, where $\Gamma_1 := \Gamma - \Gamma_0$.*

Then there is one and only one solution $\boldsymbol{u} = (u_i)$ to the variational problem:

$$\boldsymbol{u} \in \mathbf{V}(\Omega) := \{\boldsymbol{v} = (v_i) \in \mathbf{H}^1(\Omega);\ \boldsymbol{v} = \mathbf{0} \text{ on } \Gamma_0\},$$

$$\int_\Omega A^{ijkl} e_{k\|l}(\boldsymbol{u}) e_{i\|j}(\boldsymbol{v}) \sqrt{g}\, dx = \int_\Omega f^i v_i \sqrt{g}\, dx + \int_{\Gamma_1} h^i v_i \sqrt{g}\, d\Gamma$$

$$\text{for all } \boldsymbol{v} = (v_i) \in \mathbf{V}(\Omega),$$

where

$$\begin{aligned} A^{ijkl} &= \lambda g^{ij} g^{kl} + \mu \left(g^{ik} g^{jl} + g^{il} g^{jk} \right), \\ e_{i\|j}(\boldsymbol{v}) &= \frac{1}{2}(\partial_j v_i + \partial_i v_j) - \Gamma^p_{ij} v_p, \quad \Gamma^p_{ij} = \boldsymbol{g}^p \cdot \partial_i \boldsymbol{g}_j. \end{aligned}$$

The field $\boldsymbol{u}$ *is also a solution to the minimization problem:*

$$\boldsymbol{u} \in \mathbf{V}(\Omega) \text{ and } J(\boldsymbol{u}) = \inf_{\boldsymbol{v} \in \mathbf{V}(\Omega)} J(\boldsymbol{v}), \text{ where}$$
$$J(\boldsymbol{v}) := \frac{1}{2} \int_\Omega A^{ijkl} e_{k\|l}(\boldsymbol{v}) e_{i\|j}(\boldsymbol{v}) \sqrt{g}\, dx - \left\{ \int_\Omega f^i v_i \sqrt{g}\, dx + \int_{\Gamma_1} h^i v_i \sqrt{g}\, d\Gamma \right\}.$$

Proof. As a closed subspace of $\mathbf{H}^1(\Omega)$, the space $\mathbf{V}(\Omega)$ is a Hilbert space. The assumptions made on the mapping $\boldsymbol{\Theta}$ ensure in particular that the functions A^{ijkl}, Γ^p_{ij}, and g are continuous on $\overline{\Omega}$. Hence *the bilinear form*

$$(\boldsymbol{u}, \boldsymbol{v}) \longrightarrow \int_\Omega A^{ijkl} e_{k\|l}(\boldsymbol{u}) e_{i\|j}(\boldsymbol{v}) \sqrt{g}\, dx$$

is continuous over $\mathbf{H}^1(\Omega)$.

The continuous imbedding $H^1(\Omega) \hookrightarrow L^6(\Omega)$ and the continuity of the trace operator $tr : H^1(\Omega) \to L^4(\Gamma)$ (see, e.g., Vol. I, Thms. 6.1-3 and 6.1-7) imply that *the linear form*

$$\boldsymbol{v} \longrightarrow \left\{ \int_\Omega f^i v_i \sqrt{g}\, dx + \int_{\Gamma_1} h^i v_i \sqrt{g}\, d\Gamma \right\}$$

is continuous over $\mathbf{H}^1(\Omega)$.

Since the symmetric matrix $(g_{ij}(x))$ is positive definite for all $x \in \overline{\Omega}$ and $\overline{\Omega}$ is compact by assumption, there exists a constant g_0 such that

$$g(x) = \det(g_{ij}(x)) \geq g_0 > 0 \text{ for all } x \in \overline{\Omega}.$$

Then *Korn's inequality in curvilinear coordinates* (Thm. 1.7-4) and the *uniform positive definiteness of the elasticity tensor* (Thm. 1.8-1) together imply that

$$\int_\Omega A^{ijkl} e_{k\|l}(\boldsymbol{v}) e_{i\|j}(\boldsymbol{v}) \sqrt{g}\, dx \geq C_e^{-1} C^{-2} \sqrt{g_0} \|\boldsymbol{v}\|^2_{1,\Omega} \text{ for all } \boldsymbol{v} \in \mathbf{V}(\Omega).$$

Hence *the bilinear form is* $\mathbf{V}(\Omega)$*-elliptic.*

The bilinear form being also *symmetric* since $A^{ijkl} = A^{klij}$, all the assumptions of the *Lax-Milgram lemma* in its "symmetric" version (see, e.g., Vol. I, Thm. 6.3-2) are satisfied: The variational equations have one and only one solution, which may be equivalently characterized as the solution of the minimization problem stated in the theorem. ∎

An immediate corollary with a more "intrinsic" flavor to the above result is the existence and uniqueness of a *displacement field* $u_i \boldsymbol{g}^i$, whose covariant components $u_i \in H^1(\Omega)$ are thus obtained by finding the solution $\boldsymbol{u} = (u_i)$ to the variational problem of Thm. 1.8-2. Since the vector fields $\boldsymbol{g}^i$ of the contravariant bases belong to the space $\mathcal{C}^1(\overline{\Omega})$ by assumption, the displacement field $u_i \boldsymbol{g}^i$ also has its Cartesian components in $H^1(\Omega)$.

Naturally, the existence and uniqueness result of Thm. 1.8-2 implies the same result in *Cartesian coordinates* (to see this, identify $\mathcal{E}^3$ with $\mathbb{R}^3$ and let $\boldsymbol{\Theta} = \boldsymbol{id}_{\mathbb{R}^3}$). The *converse* implication also holds; cf. Ex. 1.10.

Remarks. (1) When $\Gamma_0 = \Gamma$ and the boundary Γ is smooth enough, *regularity results* can be obtained, showing that the weak solution obtained in Thm. 1.8-2 is also a *"classical solution"*, i.e., a solution of the corresponding *boundary value problem* (Thm. 1.6-1); cf. Ex. 1.11.

(2) The assumptions $\hat{f}^i \in L^2(\hat{\Omega})$ and $\hat{h}^i \in L^2(\hat{\Gamma}_1)$ made for definiteness in Sect. 1.1 can be slightly weakened as in Thm. 1.8-2, to $\hat{f}^i \in L^{6/5}(\hat{\Omega})$ and $\hat{h}^i \in L^{4/3}(\hat{\Gamma}_1)$. ∎

1.9♭. COMPLEMENT: RECOVERY OF A THREE-DIMENSIONAL MANIFOLD FROM ITS METRIC TENSOR FIELD

Let Ω be an open subset in $\mathbb{R}^3$ and let $\boldsymbol{\Theta} : \Omega \to \mathcal{E}^3$ be a thrice continuously differentiable mapping such that the three vectors $\boldsymbol{g}_i = \partial_i \boldsymbol{\Theta}$ are linearly independent everywhere in Ω. As before, let $g_{ij} = \boldsymbol{g}_i \cdot \boldsymbol{g}_j$, let the three vectors $\boldsymbol{g}^j$ be defined by the relations $\boldsymbol{g}^j \cdot \boldsymbol{g}_i = \delta_i^j$, let $g^{ij} = \boldsymbol{g}^i \cdot \boldsymbol{g}^j$, and finally, let the Christoffel symbols be defined by $\Gamma_{ij}^p = \boldsymbol{g}^p \cdot \partial_j \boldsymbol{g}_i$. It is easily found that the Christoffel symbols satisfy the following *compatibility conditions* (Ex. 1.12):

$$\partial_l \Gamma_{ik}^j - \partial_k \Gamma_{il}^j + \Gamma_{ik}^m \Gamma_{ml}^j - \Gamma_{il}^m \Gamma_{mk}^j = 0 \text{ in } \Omega.$$

These conditions are in effect relations between partial derivatives of the first, second, and third order of the mapping $\boldsymbol{\Theta}$. They are also relations between the functions g_{ij} and their derivatives, since the Christoffel symbols Γ_{ij}^p themselves may be directly defined in terms of the functions g_{ij} as follows (Ex. 1.12):

$$\Gamma_{ij}^p = g^{pq}\Gamma_{ijq}, \text{ where } \Gamma_{ijq} := \frac{1}{2}(\partial_j g_{iq} + \partial_i g_{jq} - \partial_q g_{ij}).$$

The functions Γ_{ijq} are called the *Christoffel symbols of the first kind* and the functions Γ_{ij}^p are called *in this context* the *Christoffel symbols of the second kind.* The above compatibility conditions may also be expressed in terms of the Christoffel symbols of the first kind as (Ex. 1.12):

$$\partial_l \Gamma_{ikj} - \partial_k \Gamma_{ilj} + g^{mn}(\Gamma_{iln}\Gamma_{jkm} - \Gamma_{ikn}\Gamma_{jlm}) = 0 \text{ in } \Omega.$$

Remarkably, these *necessary* conditions (in either one of their equivalent forms) are also *sufficient* for the *existence* of a mapping $\boldsymbol{\Theta} : \Omega \subset \mathbb{R}^3 \to \mathcal{E}^3$ whose metric tensor field is given on Ω:

Theorem 1.9-1. *Let Ω be a simply connected open subset of $\mathbb{R}^3$, and let there be given a twice continuously differentiable, symmetric, and positive definite matrix field (g_{ij}) on Ω that satisfies*

$$\boxed{\begin{aligned}&\partial_l \Gamma_{ikj} - \partial_k \Gamma_{ilj} + g^{mn}(\Gamma_{iln}\Gamma_{jkm} - \Gamma_{ikn}\Gamma_{jlm}) = 0 \text{ in } \Omega, \text{ where}\\ &\Gamma_{ijq} := \frac{1}{2}(\partial_j g_{iq} + \partial_i g_{jq} - \partial_q g_{ij}), \quad (g^{pq}) := (g_{ij})^{-1}.\end{aligned}}$$

Then there exists a mapping $\boldsymbol{\Theta} \in \mathcal{C}^3(\Omega; \mathbb{R}^3)$ such that

$$\boxed{\partial_i \boldsymbol{\Theta} \cdot \partial_j \boldsymbol{\Theta} = g_{ij} \text{ in } \Omega.}$$

Furthermore, this mapping is unique "up to rigid deformations in $\mathbb{R}^3$": This means that any other solution is necessarily of the form

$x \in \Omega \to \mathbf{c} + \mathbf{Q}\boldsymbol{\Theta}(x)$, where $\mathbf{c}$ is a vector in $\mathbb{R}^3$ and $\mathbf{Q}$ is an orthogonal matrix of order three. ∎

This result is a consequence of the compatibility relations satisfied by the Christoffel symbols, of the simple connexity of Ω, and of a deep existence result of Thomas [1934]; see Blume [1989].

EXERCISES

1.1. Compute the vectors of the covariant and contravariant bases, the volume, area, and length elements, the Christoffel symbols, and the covariant and contravariant components of the metric tensors (Sect. 1.2) corresponding to cylindrical and spherical coordinates (Fig. 1.2-3).

1.2. This exercise explains in what sense the components of a *vector* may be *"covariant"* or *"contravariant"*.

Let Ω and $\tilde{\Omega}$ be two domains in $\mathbb{R}^3$ and let $\boldsymbol{\Theta} : \overline{\Omega} \to \mathcal{E}^3$ and $\tilde{\boldsymbol{\Theta}} : \{\tilde{\Omega}\}^- \to \mathcal{E}^3$ be two $\mathcal{C}^1$-diffeomorphisms such that $\boldsymbol{\Theta}(\Omega) = \tilde{\boldsymbol{\Theta}}(\tilde{\Omega}) = \hat{\Omega}$ and such that the vectors $\boldsymbol{g}_i(x) := \partial_i \boldsymbol{\Theta}(x)$ and $\tilde{\boldsymbol{g}}_i(\tilde{x}) = \tilde{\partial}_i \boldsymbol{\Theta}(\tilde{x})$ of the covariant bases at the *same* point $\boldsymbol{\Theta}(x) = \tilde{\boldsymbol{\Theta}}(\tilde{x}) = \hat{x} \in \{\hat{\Omega}\}^-$ are linearly independent. Let $\boldsymbol{g}^i(x)$ and $\tilde{\boldsymbol{g}}^i(\tilde{x})$ be the vectors of the corresponding contravariant bases at the same point $\hat{x}$.

(1) Show that

$$\boldsymbol{g}_i(x) = \frac{\partial \chi^j}{\partial x_i}(x)\tilde{\boldsymbol{g}}_j(\tilde{x}) \text{ and } \boldsymbol{g}^i(x) = \frac{\partial \tilde{\chi}^i}{\partial \tilde{x}_j}(\tilde{x})\tilde{\boldsymbol{g}}^j(\tilde{x}),$$

where $\boldsymbol{\chi} = (\chi^j) := \tilde{\boldsymbol{\Theta}}^{-1} \circ \boldsymbol{\Theta} : \overline{\Omega} \to \{\tilde{\Omega}\}^-$ (hence $\tilde{x} = \boldsymbol{\chi}(x)$) and $\tilde{\boldsymbol{\chi}} = (\tilde{\chi}^i) := \boldsymbol{\chi}^{-1} : \{\tilde{\Omega}\}^- \to \overline{\Omega}$.

(2) Let $v_i(x)$ and $\tilde{v}_i(\tilde{x})$ be the covariant components and let $v^i(x)$ and $\tilde{v}^i(\tilde{x})$ be the contravariant components of the *same* vector at $\hat{x}$, i.e.,

$$v_i(x)\boldsymbol{g}^i(x) = \tilde{v}_i(\tilde{x})\tilde{\boldsymbol{g}}^i(\tilde{x}) = v^i(x)\boldsymbol{g}_i(x) = \tilde{v}^i(\tilde{x})\tilde{\boldsymbol{g}}_i(\tilde{x}).$$

Show that

$$v_i(x) = \frac{\partial \chi^j}{\partial x_i}(x)\tilde{v}_j(\tilde{x}) \text{ and } v^i(x) = \frac{\partial \tilde{\chi}^i}{\partial \tilde{x}_j}(\tilde{x})\tilde{v}^j(\tilde{x}).$$

In other words, the components $v_i(x)$ "vary like" the vectors $\boldsymbol{g}_i(x)$ of the *covariant* basis under a change of curvilinear coordinates, while the components $v^i(x)$ of the same vector "vary like" the vectors $\boldsymbol{g}^i(x)$ of the *contravariant* basis: This is why they are respectively called *"covariant"* and *"contravariant"*.

1.3. This exercise shows why the *"covariant"* and *"contravariant"* components of the *metric tensor* are so named. The notations and assumptions are those of Ex. 1.2.

Let $g_{ij}(x)$ and $\tilde{g}_{ij}(\tilde{x})$ be the covariant components and let $g^{ij}(x)$ and $\tilde{g}^{ij}(\tilde{x})$ be the contravariant components of the metric tensor (Sect. 1.2) at the *same* point $\boldsymbol{\Theta}(x) = \tilde{\boldsymbol{\Theta}}(\tilde{x}) = \hat{x} \in \{\hat{\Omega}\}^-$. Show that

$$g_{ij}(x) = \frac{\partial \chi^k}{\partial x_i}(x)\frac{\partial \chi^l}{\partial x_j}(x)\tilde{g}_{kl}(\tilde{x}) \text{ and } g^{ij}(x) = \frac{\partial \tilde{\chi}^i}{\partial \tilde{x}_k}(\tilde{x})\frac{\partial \tilde{\chi}^j}{\partial \tilde{x}_l}(\tilde{x})\tilde{g}^{kl}(\tilde{x}).$$

These formulas explain why the components $g_{ij}(x)$ and $g^{ij}(x)$ are respectively called "covariant" and "contravariant": *Each* index in $g_{ij}(x)$ "varies like" that of the corresponding vector of the *covariant* basis under a change of curvilinear coordinates (see Ex. 1.2 (1)), while *each* exponent in $g^{ij}(x)$ "varies like" that of the corresponding vector of the *contravariant* basis (see *ibid.*).

Remark. What is exactly the "second-order tensor" hidden behind its covariant components $g_{ij}(x)$ or its contravariant exponents $g^{ij}(x)$ is beautifully explained in the gentle introduction to tensors given by Antman [1995, Chap. 11, Sects. 1 to 3]; it is also shown in *ibid.* that the same "tensor" also has "mixed" components $g^i_j(x)$, which turn out to be simply the Kronecker symbols δ^i_j!

Exhaustive treatments of tensor analysis, particularly as regards its relevance to elasticity, may be found in Boothby [1975], Marsden & Hughes [1983, Chap. 1], or Simmonds [1994].

1.4. The assumptions are those of Ex. 1.2.

(1) Let $\varepsilon^{ijk}(x) = \dfrac{1}{\sqrt{g(x)}}$ if $\{i, j, k\}$ is an even permutation of $\{1, 2, 3\}$, let $\varepsilon^{ijk}(x) = -\dfrac{1}{\sqrt{g(x)}}$ if $\{i, j, k\}$ is an odd permutation of $\{1, 2, 3\}$, and let $\varepsilon^{ijk}(x) = 0$ otherwise. Show that each exponent in the functions $\varepsilon^{ijk}(x)$ "varies like" that of the corresponding vector of the *contravariant* basis under a change of curvilinear coordinates.

(2) Show that each exponent in the "contravariant" components $A^{ijkl}(x)$ of the three-dimensional elasticity tensor in curvilinear co-

ordinates introduced in Sect. 1.3 again "varies like" that of the corresponding vector of the *contravariant* basis under a change of curvilinear coordinates.

Remark. See again Antman [1995, Chap. 11, Sects. 1 to 3] to decipher the *"third-order tensor"* and *"fourth-order tensor"* hidden respectively behind their contravariant components $\varepsilon^{ijk}(x)$ and $A^{ijkl}(x)$.

1.5. In part (iii) of the proof of Thm. 1.3-1, it is shown that

$$\hat{\partial}_j \hat{v}_i(\hat{x}) = (v_{k\|l}[\boldsymbol{g}^k]_i[\boldsymbol{g}^l]_j)(x) \text{ for all } \hat{x} = \boldsymbol{\Theta}(x) \in \hat{\Omega}.$$

Show that, conversely, each covariant derivative $v_{i\|j}(x)$ can be expressed as a linear combination of the partial derivatives $\hat{\partial}_l \hat{v}_k(\hat{x})$.

1.6. This exercise provides the expression of the *gradient, Laplacian,* and *curl operators* in *curvilinear coordinates* (that of the divergence operator is given at the end of Sect. 1.4). The notations and assumptions are those of Thm. 1.3-1.

(1) Given a function $\hat{v} : \hat{\Omega} \to \mathbb{R}$, let $\widehat{\mathbf{grad}}\, \hat{v}$ be the vector field with components $\hat{\partial}_i \hat{v}$, let $\hat{\Delta} \hat{v} := \sum_i \frac{\partial^2 \hat{v}}{\partial \hat{x}_i^2}$, and let the function $v : \Omega \to \mathbb{R}$ be defined by $v(x) = \hat{v}(\hat{x})$ for all $\hat{x} = \boldsymbol{\Theta}(x) \in \hat{\Omega}$.

Show that

$$(\widehat{\mathbf{grad}}\, \hat{v})(\hat{x}) = \left((\partial_i v)\boldsymbol{g}^i\right)(x),$$

$$\hat{\Delta} \hat{v}(\hat{x}) = \left(\frac{1}{\sqrt{g}} \partial_i(\sqrt{g}\, g^{ij} \partial_j v)\right)(x).$$

(2) Given a vector field $\hat{\boldsymbol{v}} = (\hat{v}_i) : \hat{\Omega} \to \mathbb{R}^3$, let $\widehat{\mathbf{curl}}\, \hat{\boldsymbol{v}}$ be the vector field with components $\hat{\varepsilon}^{ijk} \hat{\partial}_j \hat{v}_i$, where $\hat{\varepsilon}^{ijk} = +1$ if $\{i, j, k\}$ is an even permutation of $\{1, 2, 3\}$, $\hat{\varepsilon}^{ijk} = -1$ if $\{i, j, k\}$ is an odd permutation of $\{1, 2, 3\}$, and $\hat{\varepsilon}^{ijk} = 0$ otherwise. Show that

$$\widehat{\mathbf{curl}}\, \hat{\boldsymbol{v}}(\hat{x}) = (\varepsilon^{ijk} v_{j\|i} \boldsymbol{g}_k)(x),$$

where $\varepsilon^{ijk}(x) := \frac{1}{\sqrt{g(x)}} \hat{\varepsilon}^{ijk}$.

1.7. Show that the boundary value problem of three-dimensional linearized elasticity in curvilinear coordinates (Thm. 1.6-1), viz.,

$$-\sigma^{ij}\|_j = f^i \text{ in } \Omega, \quad u_i = 0 \text{ on } \Gamma_0, \quad \sigma^{ij} n_j = h^i \text{ on } \Gamma_1,$$

where $\sigma^{ij} = A^{ijkl} e_{k\|l}(\boldsymbol{u})$, can be directly derived from its Cartesian counterpart (Sect. 1.1), viz.,

$$-\hat{\partial}_j \hat{\sigma}^{ij} = \hat{f}^i \text{ in } \hat{\Omega},\ \hat{u}_i = 0 \text{ on } \hat{\Gamma}_0,\ \hat{\sigma}^{ij}\hat{n}_j = \hat{h}^i \text{ on } \hat{\Gamma}_1,$$

where $\hat{\sigma}^{ij} = \hat{A}^{ijkl}\hat{e}_{kl}(\hat{\boldsymbol{u}})$ ("directly" means without recourse to the variational equations as in Thm. 1.6-1).

1.8. The notations and assumptions are those of Thm. 1.8-1. Letting $\boldsymbol{t}_i(x) := t_{ij}\boldsymbol{g}^j(x)$ and $[\boldsymbol{t}_i(x)]^j := \boldsymbol{t}_i(x) \cdot \hat{\boldsymbol{e}}^j$, show that

$$g^{ik}(x)g^{jl}(x)t_{kl}t_{ij} = \sum_j |[\boldsymbol{t}_i(x)]^j \boldsymbol{g}^i(x)|^2.$$

Then infer from this equality another proof of this theorem.

1.9. (1) Show that the three-dimensional elasticity tensor in Cartesian coordinates, defined in Sect. 1.1 by its components

$$\hat{A}^{ijkl} = \lambda\delta^{ij}\delta^{kl} + \mu(\delta^{ik}\delta^{jl} + \delta^{il}\delta^{jk}),$$

remains positive definite "even if λ is slightly negative", in the following sense: There exists a constant $\hat{C} > 0$ such that

$$\hat{A}^{ijkl}\hat{t}_{kl}\hat{t}_{ij} \geq \hat{C}\sum_{i,j} |\hat{t}_{ij}|^2 \text{ for all symmetric matrices } (\hat{t}_{ij})$$

if and only if $\mu > 0$ and $2\lambda + 3\mu > 0$.

(2) Show that a similar property holds for the elasticity tensor defined in curvilinear coordinates by its contravariant components

$$A^{ijkl} = \lambda g^{ij}g^{kl} + \mu(g^{ik}g^{jl} + g^{il}g^{jk}).$$

1.10. Combining the relation (Thm. 1.3-1)

$$\int_{\hat{\Omega}} \hat{A}^{ijkl}\hat{e}_{kl}(\hat{\boldsymbol{u}})e_{ij}(\hat{\boldsymbol{v}})\,d\hat{x} = \int_{\Omega} A^{ijkl}e_{k\|l}(\boldsymbol{u})e_{i\|j}(\boldsymbol{v})\sqrt{g}\,dx$$

with the three-dimensional Korn's inequality in Cartesian coordinates (see, e.g., Vol. I, Sect. 6.3 or Vol. II, Sect. 1.1), show directly that the bilinear form

$$(\boldsymbol{u},\boldsymbol{v}) \in \mathbf{V}(\Omega) \times \mathbf{V}(\Omega) \to \int_{\Omega} A^{ijkl}e_{k\|l}(\boldsymbol{u})e_{i\|j}(\boldsymbol{v})\sqrt{g}\,dx$$

is $\mathbf{V}(\Omega)$-elliptic, thus providing another proof to Thm. 1.8-2.

Hint: Use a classical result about composite mappings in Sobolev spaces (see, e.g., Nečas [1967, Chap. 2, Lemma 3.2] or Adams [1975, Thm. 3.35]).

1.11. (1) Show that, if the boundary Γ and the mapping $\boldsymbol{\Theta}$ are sufficiently smooth, if $\Gamma_0 = \Gamma$, and if $f^i \in L^p(\Omega)$, $p \geq \frac{6}{5}$, the weak solution $\boldsymbol{u} \in \mathbf{V}(\Omega) = \mathbf{H}_0^1(\Omega)$ found in Thm. 1.8-2 is in the space $\mathbf{W}^{2,p}(\Omega)$.

(2) Show that $\boldsymbol{u}$ satisfies the equations (Thm. 1.6-1)

$$-A^{ijkl}e_{k\|l}(\boldsymbol{u})\|_j = f^i \text{ in } L^p(\Omega).$$

Hint: Use the regularity of the weak solution of the corresponding boundary value problem of three-dimensional linearized elasticity in Cartesian coordinates (Vol. I, Thm. 6.3-6).

1.12. (1) Show that the Christoffel symbols of the second kind satisfy the relations

$$\partial_l\Gamma_{ik}^j - \partial_k\Gamma_{il}^j + \Gamma_{ik}^m\Gamma_{ml}^j - \Gamma_{il}^m\Gamma_{mk}^j = 0 \text{ in } \Omega.$$

(2) In the text, the Christoffel symbols of the second kind are defined as $\Gamma_{ij}^p = \boldsymbol{g}^p \cdot \partial_j\boldsymbol{g}_i$ (Thm. 1.4-1), i.e., by means of the vectors of the covariant and contravariant bases. It is remarkable that they can also be defined in terms of the covariant and contravariant components of the metric tensor. More precisely, show that

$$\Gamma_{ij}^p = g^{pq}\Gamma_{ijq} \text{ where } \Gamma_{ijq} := \frac{1}{2}(\partial_j g_{iq} + \partial_i g_{jq} - \partial_q g_{ij}).$$

(3) Show that the relations satisfied by the Christoffel symbols of the second kind (cf. (1)) may be equivalently expressed in terms of the Christoffel symbols of the first kind as

$$\partial_l\Gamma_{ikj} - \partial_k\Gamma_{ilj} + g^{mn}\left(\Gamma_{iln}\Gamma_{jkm} - \Gamma_{ikn}\Gamma_{jlm}\right) = 0 \text{ in } \Omega.$$

CHAPTER 2

INEQUALITIES OF KORN'S TYPE ON SURFACES

INTRODUCTION

We shall see in the next chapters that the theory of linearly elastic shells leads to two-dimensional equations that are "posed on the middle surface S of the shell", i.e., that are expressed in terms of *curvilinear coordinates* of the surface S.

The purpose of this chapter is to study such equations *per* themselves. To this end, we first provide all the necessary preliminaries from the *differential geometry of surfaces in* $\mathbb{R}^3$ (Sects. 2.1 to 2.3). We then provide complete proofs of the *existence* and *uniqueness* of the solutions to three fundamental classes of linear shell equations, the *elliptic membrane, flexural*, and *Koiter* equations (Sects. 2.6 and 2.7).

As in Chap. 1, *the treatment is entirely self-contained*, in that no *a priori* knowledge of differential geometry is assumed.

More specifically, recall that a *"three-dimensional manifold"* $\boldsymbol{\Theta}(\overline{\Omega})$ *in* $\mathbb{R}^3$, where Ω is a three-dimensional domain in $\mathbb{R}^3$ and $\boldsymbol{\Theta} : \overline{\Omega} \to \mathbb{R}^3$ is an *ad hoc* injective mapping, is unambiguously defined (up to rigid deformations) by a *single* tensor field, the *metric tensor* field, whose covariant components $g_{ij} = g_{ji} : \overline{\Omega} \to \mathbb{R}$ are given by $g_{ij} = \partial_i \boldsymbol{\Theta} \cdot \partial_j \boldsymbol{\Theta}$ (cf. Sects. 1.2 and 1.9). Consider instead a *surface* $S = \boldsymbol{\theta}(\overline{\omega})$ *in* $\mathbb{R}^3$, where ω is a two-dimensional domain in $\mathbb{R}^2$ and $\boldsymbol{\theta} : \overline{\omega} \to \mathbb{R}^3$ is a smooth injective mapping with other *ad hoc* properties. Then by contrast, such a "two-dimensional manifold" requires *two* tensor fields for its definition (again up to rigid deformations), the *metric tensor* field *and* the *curvature tensor* field, whose covariant components $a_{\alpha\beta} = a_{\beta\alpha} : \overline{\omega} \to \mathbb{R}$ and $b_{\alpha\beta} = b_{\beta\alpha} : \overline{\omega} \to \mathbb{R}$ are respectively given by (Greek indices or exponents take their values in $\{1, 2\}$):

$$a_{\alpha\beta} = \boldsymbol{a}_\alpha \cdot \boldsymbol{a}_\beta \text{ and } b_{\alpha\beta} = \boldsymbol{a}_3 \cdot \partial_\alpha \boldsymbol{a}_\beta,$$

where $\boldsymbol{a}_\alpha = \partial_\alpha \boldsymbol{\theta}$ and $\boldsymbol{a}_3 = \dfrac{\boldsymbol{a}_1 \wedge \boldsymbol{a}_2}{|\boldsymbol{a}_1 \wedge \boldsymbol{a}_2|}$. These two tensors, which are also called the *first* and *second fundamental forms* of the surface, are studied in Sects. 2.1 and 2.2; see also Sect. 2.8.

In Sects. 2.4 and 2.5, we introduce two other *fundamental tensors*, which play a key rôle in the two-dimensional theory of linearly elastic shells, the *linearized change of metric tensor* and the *linearized change of curvature tensor*, each one being associated with a displacement vector field $\eta_i \boldsymbol{a}^i$ of the surface $S = \boldsymbol{\theta}(\overline{\omega})$, where the vectors $\boldsymbol{a}^i$ are defined by the relations $\boldsymbol{a}^i \cdot \boldsymbol{a}_j = \delta^i_j$.

The covariant components of these tensors are given by

$$\begin{aligned}\gamma_{\alpha\beta}(\boldsymbol{\eta}) &= \frac{1}{2}(\partial_\alpha \eta_\beta + \partial_\beta \eta_\alpha) - \Gamma^\sigma_{\alpha\beta}\eta_\sigma - b_{\alpha\beta}\eta_3,\\ \rho_{\alpha\beta}(\boldsymbol{\eta}) &= \partial_{\alpha\beta}\eta_3 - \Gamma^\sigma_{\alpha\beta}\partial_\sigma\eta_3 - b^\sigma_\alpha b_{\sigma\beta}\eta_3 + b^\sigma_\alpha(\partial_\beta\eta_\sigma - \Gamma^\tau_{\beta\sigma}\eta_\tau)\\ &\quad + b^\tau_\beta(\partial_\alpha\eta_\tau - \Gamma^\sigma_{\alpha\tau}\eta_\sigma) + (\partial_\alpha b^\tau_\beta + \Gamma^\tau_{\alpha\sigma}b^\sigma_\beta - \Gamma^\sigma_{\alpha\beta}b^\tau_\sigma)\eta_\tau,\end{aligned}$$

where $\Gamma^\sigma_{\alpha\beta} = \boldsymbol{a}^\sigma \cdot \partial_\alpha \boldsymbol{a}_\beta$ are the *Christoffel symbols* of S (Sect. 2.3) and b^β_α are the mixed components of the curvature tensor of S (Sect. 2.2). The functions $\gamma_{\alpha\beta}(\boldsymbol{\eta})$ and $\rho_{\alpha\beta}(\boldsymbol{\eta})$ may be also written in more condensed manners, either by means of *covariant derivatives* or through the introduction of the vector field $\eta_i \boldsymbol{a}^i$ (Thms. 2.4-1 and 2.5-1). Note that

$$\begin{aligned}\gamma_{\alpha\beta}(\boldsymbol{\eta}) \in L^2(\omega) \;\text{ if }\; \boldsymbol{\eta} = (\eta_i) \in H^1(\omega) \times H^1(\omega) \times L^2(\omega),\\ \rho_{\alpha\beta}(\boldsymbol{\eta}) \in L^2(\omega) \;\text{ if }\; \boldsymbol{\eta} = (\eta_i) \in H^1(\omega) \times H^1(\omega) \times H^2(\omega).\end{aligned}$$

An *inequality of Korn's type "on a general surface"* is then established (Thm. 2.6-4): Given any subset γ_0 of $\gamma = \partial\omega$ satisfying $0 <$ *length* $\gamma_0 \leq$ *length* γ, define the space

$$\mathbf{V}_K(\omega) = \{\boldsymbol{\eta} = (\eta_i) \in H^1(\omega) \times H^1(\omega) \times H^2(\omega); \eta_i = \partial_\nu \eta_3 = 0 \text{ on } \gamma_0\}.$$

Then this inequality asserts the existence of a constant c such that

$$\Big\{\sum_\alpha \|\eta_\alpha\|^2_{1,\omega} + \|\eta_3\|^2_{2,\omega}\Big\}^{1/2} \leq c\Big\{\sum_{\alpha,\beta} |\gamma_{\alpha\beta}(\boldsymbol{\eta})|^2_{0,\omega} + \sum_{\alpha,\beta} |\rho_{\alpha\beta}(\boldsymbol{\eta})|^2_{0,\omega}\Big\}^{1/2}$$

for all $\boldsymbol{\eta} = (\eta_i) \in \mathbf{V}_K(\omega)$.

The *existence* and *uniqueness* of a solution to the *two-dimensional Koiter equations for a linearly elastic shell* and to the *two-dimensional equations of a linearly elastic flexural shell* follow from this inequality.

We also show how this inequality of Korn's type has been recently extended to *surfaces "with little regularity"* (Thm. 2.6-6).

If $\gamma_0 = \gamma$ and the surface S is *elliptic*, i.e., both principal radii of curvature are everywhere of the same sign, it is remarkable that an *inequality of Korn's type "on an elliptic surface"* that only involves the linearized change of metric tensor in its right-hand side can be established (Thm. 2.7-3). This inequality asserts the existence of a constant c_M such that

$$\Big\{\sum_\alpha \|\eta_\alpha\|^2_{1,\omega} + |\eta_3|^2_{0,\omega}\Big\}^{1/2} \le c_M \Big\{\sum_{\alpha,\beta} |\gamma_{\alpha\beta}(\boldsymbol{\eta})|^2_{0,\omega}\Big\}^{1/2}$$

for all $\boldsymbol{\eta} = \mathbf{V}_M(\omega) = H^1_0(\omega) \times H^1_0(\omega) \times L^2(\omega)$.

The *existence* and *uniqueness* of a solution to the *two-dimensional equations of a linearly elastic elliptic membrane shell* then follow from this inequality.

Note that Chaps. 1 and 2 together provide a *unified presentation of inequalities of Korn's type "in curvilinear coordinates", whether in a three-dimensional domain in* $\mathbb{R}^3$ *or on a surface in* $\mathbb{R}^3$, notably by showing that all these inequalities of Korn's type hinge crucially on the *lemma of J.L. Lions* recalled in Thm. 1.7-1 and on *ad hoc infinitesimal rigid displacement lemmas.*

2.1. CURVILINEAR COORDINATES AND METRIC TENSOR ON A SURFACE

In addition to the rules governing Latin indices that we set in Sect. 1.1, we henceforth require that *Greek* indices and exponents, *except* ε and ν and τ in the notations ∂_ν and ∂_τ, vary in the set $\{1, 2\}$ and that the summation convention be systematically used in conjunction with these rules. For instance,

"$\gamma_{\alpha\beta}(\boldsymbol{\eta}) = 0$ in ω" means "$\gamma_{\alpha\beta}(\boldsymbol{\eta}) = 0$ in ω for α, $\beta = 1, 2$",

the relation

$$\partial_\alpha(\eta_i \boldsymbol{a}^i) = (\eta_{\beta|\alpha} - b_{\alpha\beta}\eta_3)\boldsymbol{a}^\beta + (\eta_{3|\alpha} + b^\beta_\alpha \eta_\beta)\boldsymbol{a}^3$$

means that

$$\partial_\alpha\Big(\sum_{i=1}^3 \eta_i \boldsymbol{a}^i\Big) = \sum_{\beta=1}^2 (\eta_{\beta|\alpha} - b_{\alpha\beta}\eta_3)\boldsymbol{a}^\beta + \Big(\eta_{3|\alpha} + \sum_{\beta=1}^2 b^\beta_\alpha \eta_\beta\Big)\boldsymbol{a}^3$$

for $\alpha = 1, 2$, etc.

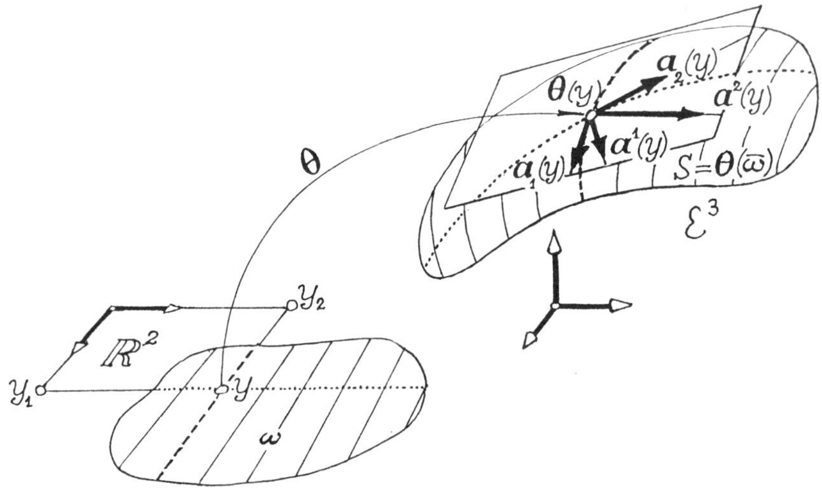

Fig. 2.1-1: *Curvilinear coordinates on a surface and covariant and contravariant bases of the tangent plane.* Let $S = \boldsymbol{\theta}(\overline{\omega})$ be a surface in $\mathcal{E}^3$. The two coordinates y_1, y_2 of $y \in \overline{\omega}$ are the curvilinear coordinates of $\hat{y} = \boldsymbol{\theta}(y) \in S$. If they are linearly independent, the two vectors $\boldsymbol{a}_\alpha(y) = \partial_\alpha \boldsymbol{\theta}(y)$, which are tangent to the coordinate lines passing through $\hat{y}$, form the covariant basis of the tangent plane to S at $\hat{y} = \boldsymbol{\theta}(y)$; the two vectors $\boldsymbol{a}^\alpha(y)$ defined by $\boldsymbol{a}^\alpha(y) \cdot \boldsymbol{a}_\beta(y) = \delta^\alpha_\beta$ form its contravariant basis.

Let there be given as in Sect. 1.1 a *real three-dimensional affine Euclidean space* $\mathcal{E}^3$, equipped with a *Cartesian frame* consisting of an origin $\hat{O}$ together with three vectors $\hat{\mathbf{e}}^i = \hat{\mathbf{e}}_i$ forming an orthonormal basis, and let $\boldsymbol{a} \cdot \boldsymbol{b}$, $|\boldsymbol{a}|$, and $\boldsymbol{a} \wedge \boldsymbol{b}$ denote the Euclidean inner product, the Euclidean norm, and the vector product of vectors $\boldsymbol{a}$, $\boldsymbol{b}$ in the associated space $\mathbf{E}^3$.

In addition, let there be given a *two-dimensional vector space*, in which two vectors $\boldsymbol{e}^\alpha = \boldsymbol{e}_\alpha$ form a basis; *this space will accordingly be identified with* $\mathbb{R}^2$. We let y_α denote the coordinates of a point y in this space and we let $\partial_\alpha := \partial/\partial y_\alpha$ and $\partial_{\alpha\beta} := \partial^2/\partial y_\alpha \partial y_\beta$.

Finally, let there be given a *domain* ω in $\mathbb{R}^2$ (according to the definition given at the beginning of Sect. 1.7) and an *injective* mapping

$\boldsymbol{\theta} : \overline{\omega} \to \mathcal{E}^3$. The set

$$\boxed{S := \boldsymbol{\theta}(\overline{\omega})}$$

is called a **surface** in $\mathcal{E}^3$. Each point $\hat{y} \in S$ may thus be unambiguously written as

$$\boxed{\hat{y} = \boldsymbol{\theta}(y), \quad y \in \overline{\omega},}$$

and the two coordinates y_α of y are called the **curvilinear coordinates** of $\hat{y}$ (Fig. 2.1-1). Well-known *examples* of surfaces and of curvilinear coordinates and their corresponding coordinate lines (defined below) are given in Figs. 2.1-2 and 2.1-3.

Remark. According to the usual terminology, the set $\boldsymbol{\theta}(\omega)$ is a *two-dimensional manifold*, while the surface $\boldsymbol{\theta}(\overline{\omega})$ is a *two-dimensional manifold "with boundary", in* $\mathcal{E}^3$. By contrast, *a two-dimensional manifold "without boundary" in* $\mathcal{E}^3$, such as a sphere or a torus, requires for its description *several* mappings $\boldsymbol{\theta}_p : \omega_p \subset \mathbb{R}^2 \to \mathcal{E}^3$, $p = 1, 2, \ldots, P$, satisfying *ad hoc* compatibility conditions on the "overlapping submanifolds" $\boldsymbol{\theta}_p(\omega_p) \cap \boldsymbol{\theta}_q(\omega_q)$, $p \neq q$, of S.

Excellent (and particularly readable!) introductions to finite-dimensional manifolds are given in Schwartz [1992, Chap. 3, Sect. 9], Schwartz [1993, Chap. VI, Sects. 8-10], and do Carmo [1994, Chaps. 3 and 4]. ■

Naturally, once a surface S is defined as $S = \boldsymbol{\theta}(\overline{\omega})$, *there are infinitely many other ways of defining curvilinear coordinates on* S, depending on how the domain ω and the mapping $\boldsymbol{\theta}$ are chosen. For instance, a portion S of a sphere may be represented by means of *Cartesian coordinates, spherical coordinates*, or *stereographic coordinates* (Fig. 2.1-3). Incidentally, this example illustrates the variety of restrictions that have to be imposed on S according to which kind of curvilinear coordinates it is equipped with!

Assume that the mapping

$$\boldsymbol{\theta} = \theta_i \hat{\boldsymbol{e}}^i : \overline{\omega} \subset \mathbb{R}^2 \to \boldsymbol{\theta}(\overline{\omega}) = S \subset \mathcal{E}^3$$

is differentiable at a point $y \in \overline{\omega}$. If $\boldsymbol{\delta y}$ is such that $(y + \boldsymbol{\delta y}) \in \overline{\omega}$, we thus have

$$\boxed{\boldsymbol{\theta}(y + \boldsymbol{\delta y}) = \boldsymbol{\theta}(y) + \boldsymbol{\nabla\theta}(y)\boldsymbol{\delta y} + o(\boldsymbol{\delta y}),}$$

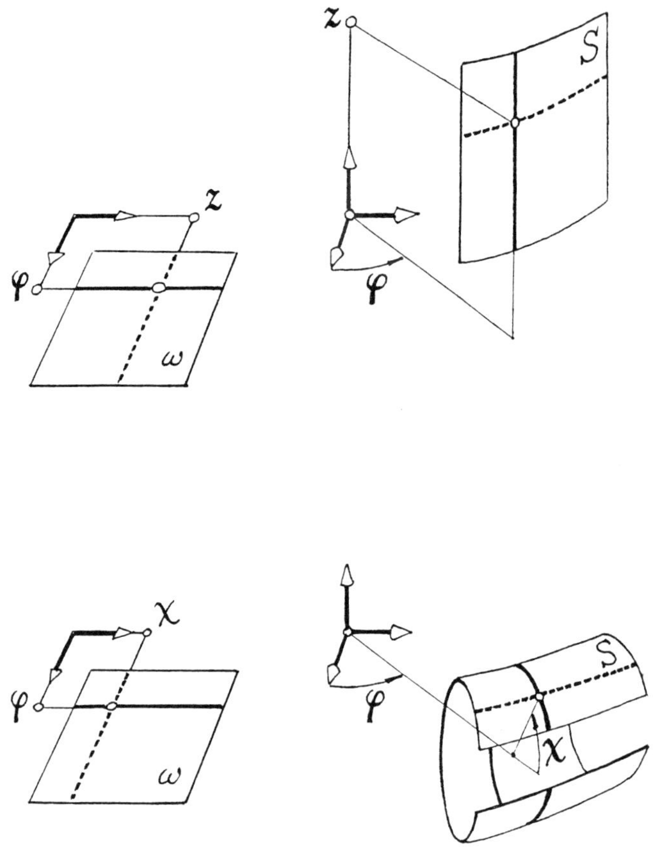

Fig. 2.1-2: *Two familiar examples of surfaces, curvilinear coordinates, and coordinate lines.* A portion S of a circular cylinder of radius R corresponds to a mapping $\boldsymbol{\theta}$ of the form

$$(\varphi,\, z) \to (R\cos\varphi,\, R\sin\varphi,\, z).$$

A portion S of a torus corresponds to a mapping $\boldsymbol{\theta}$ of the form

$$(\varphi,\, \chi) \to ((R + r\cos\chi)\cos\varphi,\, (R + r\cos\chi)\sin\varphi,\, R\sin\chi),$$

with $R > r$. In each case, the corresponding coordinate lines are represented on S with self-explanatory graphical conventions; see also Ex. 2.1.

where the matrix $\boldsymbol{\nabla\theta}(y)$ is given by

$$\boldsymbol{\nabla\theta}(y) := \begin{pmatrix} \partial_1\theta_1 & \partial_2\theta_1 \\ \partial_1\theta_2 & \partial_2\theta_2 \\ \partial_1\theta_3 & \partial_2\theta_3 \end{pmatrix}(y).$$

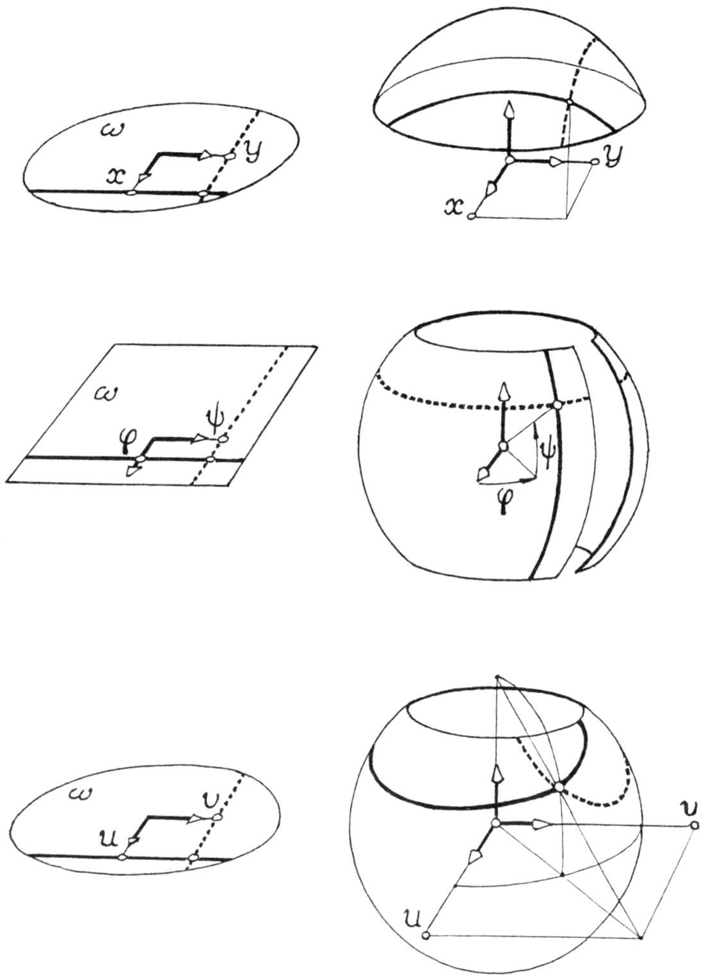

Fig. 2.1-3: *Several systems of curvilinear coordinates on a sphere.* Let Σ be a sphere of radius R. A portion $S \subset \Sigma$ contained in the "northern hemisphere" can be represented by means of Cartesian coordinates:

$$(x, y) \to (x, y, \{R^2 - (x^2 + y^2)\}^{1/2}).$$

A portion $S \subset \Sigma$ that excludes a neighborhood of both "poles" and of a "meridian" (to fix ideas) can be represented by means of spherical coordinates:

$$(\varphi, \psi) \to (R\cos\psi\cos\varphi,\ R\cos\psi\sin\varphi,\ R\sin\psi).$$

A portion $S \subset \Sigma$ that excludes a neighborhood of the "North pole" can be represented by means of stereographic coordinates:

$$(u, v) \to \left(\frac{2R^2u}{u^2+v^2+R^2},\ \frac{2R^2v}{u^2+v^2+R^2},\ R\frac{u^2+v^2-R^2}{u^2+v^2+R^2}\right).$$

As in Fig. 2.1-2, the corresponding coordinate lines are represented in each case with self-explanatory graphical conventions (see also Ex. 2.1).

A word of caution. The presentation in this section follows that of Sect. 1.2, "the mapping $\boldsymbol{\theta} : \overline{\omega} \subset \mathbb{R}^2 \to \mathcal{E}^3$ replacing the mapping $\boldsymbol{\Theta} : \overline{\Omega} \subset \mathbb{R}^3 \to \mathcal{E}^3$". As such, there are strong *similarities* between the two presentations, such as the way the metric tensor is defined in each case (see below), but there are also sharp *differences*. For instance, the matrix $\boldsymbol{\nabla}\boldsymbol{\theta}(y)$ is *not* a square matrix, while the matrix $\boldsymbol{\nabla}\boldsymbol{\Theta}(x)$ *is* one! ■

Let the two vectors $\boldsymbol{a}_\alpha(y) \in \mathbb{R}^3$ be defined by

$$\boxed{\boldsymbol{a}_\alpha(y) := \partial_\alpha \boldsymbol{\theta}(y) = \begin{pmatrix} \partial_\alpha \theta_1 \\ \partial_\alpha \theta_2 \\ \partial_\alpha \theta_3 \end{pmatrix} (y),}$$

so that $\boldsymbol{a}_\alpha(y)$ *is the* α*-th column vector of the matrix* $\boldsymbol{\nabla}\boldsymbol{\theta}(y)$. Let $\boldsymbol{\delta y} = \delta y^\alpha \boldsymbol{e}_\alpha$; then the expansion of $\boldsymbol{\theta}$ about y may be also written as

$$\boxed{\boldsymbol{\theta}(y + \boldsymbol{\delta y}) = \boldsymbol{\theta}(y) + \delta y^\alpha \boldsymbol{a}_\alpha(y) + o(\boldsymbol{\delta y}).}$$

If in particular $\boldsymbol{\delta y}$ is of the form $\boldsymbol{\delta y} = \delta t \boldsymbol{e}_\alpha$, where $\delta t \in \mathbb{R}$ and $\boldsymbol{e}_\alpha$ is one of the basis vectors in $\mathbb{R}^2$, this relation reduces to

$$\boxed{\boldsymbol{\theta}(y + \delta t \boldsymbol{e}_\alpha) = \boldsymbol{\theta}(y) + \delta t \boldsymbol{a}_\alpha(y) + o(\delta t).}$$

We henceforth assume that the two vectors $\boldsymbol{a}_\alpha(y)$ *are linearly independent.* The last relation thus shows that in this case *each vector* $\boldsymbol{a}_\alpha(y)$ *is tangent to the* α-th **coordinate line** *passing through* $\hat{y} = \boldsymbol{\theta}(y)$, defined as the image by $\boldsymbol{\theta}$ of the points of $\overline{\omega}$ that lie on a line parallel to $\boldsymbol{e}_\alpha$ passing through y; cf. Fig. 2.1-1 (there exist t_0 and t_1 with $t_0 < t_1$ and $0 \in [t_0, t_1]$ such that the equation of the α-th coordinate line is $t \in [t_0, t_1] \to \boldsymbol{f}_\alpha(t) := \boldsymbol{\theta}(y + t\boldsymbol{e}_\alpha)$ in a sufficiently small neighborhood of $\hat{y}$; hence $\boldsymbol{f}'_\alpha(0) = \partial_\alpha \boldsymbol{\theta}(y) = \boldsymbol{a}_\alpha(y)$). The vectors $\boldsymbol{a}_\alpha(y)$, which thus span the *tangent plane* to the surface S at $\hat{y} = \boldsymbol{\theta}(y)$, form the **covariant basis of the tangent plane** to S at $\hat{y}$; cf. Fig. 2.1-1.

Remark. A differentiable mapping $\boldsymbol{\theta} : \overline{\omega} \subset \mathbb{R}^2 \to \mathbb{R}^3$ such that the two vectors $\partial_\alpha \boldsymbol{\theta}(y)$ are linearly independent at all points $y \in \overline{\omega}$

is called an *immersion*. ■

Naturally, we are committing here two convenient *abus de langage*: The vector tangent to the α-th coordinate line at $\hat{y} = \boldsymbol{\theta}(y)$ is in fact the *vector in the affine space* $\mathcal{E}^3$ that is parallel to $\boldsymbol{a}_\alpha(y)$ and has $\hat{y}$ as its origin. Likewise, the tangent plane to S at $\hat{y}$ is in fact the *affine plane in the affine space* $\mathcal{E}^3$ that passes through $\hat{y}$ and is parallel to the vectors $\boldsymbol{a}_\alpha(y)$.

Returning to a general increment $\boldsymbol{\delta y} = \delta y^\alpha \boldsymbol{e}_\alpha$, we also infer from the expansion of $\boldsymbol{\theta}$ about y that

$$\boxed{\begin{aligned} |\boldsymbol{\theta}(y + \boldsymbol{\delta y}) - \boldsymbol{\theta}(y)|^2 &= \boldsymbol{\delta y}^T \boldsymbol{\nabla\theta}(y)^T \boldsymbol{\nabla\theta}(y) \boldsymbol{\delta y} + o(|\boldsymbol{\delta y}|^2) \\ &= \delta y^\alpha \boldsymbol{a}_\alpha(y) \cdot \boldsymbol{a}_\beta(y) \delta y^\beta + o(|\boldsymbol{\delta y}|^2). \end{aligned}}$$

In other words, the principal part of the length between the points $\boldsymbol{\theta}(y + \boldsymbol{\delta y})$ and $\boldsymbol{\theta}(y)$ is $\{\delta y^\alpha \boldsymbol{a}_\alpha(y) \cdot \boldsymbol{a}_\beta(y) \delta y^\beta\}^{1/2}$. This observation suggests the introduction of the *symmetric matrix* $(a_{\alpha\beta}(y))$ *of order two*, whose elements

$$\boxed{a_{\alpha\beta}(y) := \boldsymbol{a}_\alpha(y) \cdot \boldsymbol{a}_\beta(y) = \left(\boldsymbol{\nabla\theta}(y)^T \boldsymbol{\nabla\theta}(y)\right)_{\alpha\beta}}$$

are called the **covariant components** of the **metric tensor**, also called the **first fundamental form**, *of the surface* S at $\hat{y} = \boldsymbol{\theta}(y)$. Note that *the matrix* $(a_{\alpha\beta}(y))$ *is positive definite* since the vectors $\boldsymbol{a}_\alpha(y)$ are assumed to be linearly independent.

In the next theorem, we review the fundamental formulas expressing the *area* and *length elements* at a point $\hat{y} = \boldsymbol{\theta}(y)$ of the surface S in terms of the matrix $(a_{\alpha\beta}(y))$; see also Fig. 2.1-4.

Theorem 2.1-1. *Let ω be a domain in $\mathbb{R}^2$, let $\boldsymbol{\theta} : \overline{\omega} \to \mathcal{E}^3$ be a smooth enough and injective mapping such that the two vectors $\boldsymbol{a}_\alpha(y) = \partial_\alpha \boldsymbol{\theta}(y)$ are linearly independent at all points $y \in \overline{\omega}$, and let $S = \boldsymbol{\theta}(\overline{\omega})$.*

(a) *The area element $dS(\hat{y})$ at $\hat{y} = \boldsymbol{\theta}(y) \in S$ is given in terms of the area element dy at $y \in \overline{\omega}$ by*

$$\boxed{dS(\hat{y}) = \sqrt{a(y)}\, dy,}$$

where

$$\boxed{a(y) := \det(a_{\alpha\beta}(y)) \text{ and } a_{\alpha\beta}(y) = \boldsymbol{a}_\alpha(y) \cdot \boldsymbol{a}_\beta(y).}$$

(b) *The length element $d\hat{l}(\hat{y})$ at $\hat{y} = \boldsymbol{\theta}(y) \in S$ is given by*

$$\boxed{d\hat{l}(\hat{y}) = \left\{\delta y^\alpha a_{\alpha\beta}(y) \delta y^\beta\right\}^{1/2}.}$$

Proof. The relation (a) between the area elements is well known. In particular, it can also be deduced from the relation between the area elements $d\hat{\Gamma}(\hat{x})$ and $d\Gamma(x)$ given in Thm. 1.2-1 (b) by means of an *ad hoc* "three-dimensional extension" of the mapping $\boldsymbol{\theta}$; cf. Ex. 2.2.

The expression of the length element in (b) simply recalls that $d\hat{l}(\hat{y})$ is by definition the principal part with respect to $\boldsymbol{\delta y} = \delta y^\alpha \boldsymbol{e}_\alpha$ of the length $|\boldsymbol{\theta}(y + \boldsymbol{\delta y}) - \boldsymbol{\theta}(y)|$, whose expression precisely leads to the introduction of the matrix $(a_{\alpha\beta}(y))$. ■

Remark. The (otherwise natural) notation $d\hat{a}(\hat{y})$ for the area element at $\hat{y} = \boldsymbol{\theta}(y)$ is avoided as it bears too much resemblance with the customarily used notation $a(y)$ for $\det(a_{\alpha\beta}(y))$. ■

The relations in Thm. 2.1-1 are used for computing *surface integrals* and *lengths* on the surface S (Fig. 2.1-4): Let A be a dy-measurable subset of $\overline{\omega}$, let $\hat{A} := \boldsymbol{\theta}(A)$, and let $\hat{f} : \hat{A} \to \mathbb{R}$ be a dS-measurable function. Then

$$\boxed{\int_{\hat{A}} \hat{f}(\hat{y})\, d\hat{S}(\hat{y}) = \int_A (\hat{f} \circ \boldsymbol{\theta})(y) \sqrt{a(y)}\, dy.}$$

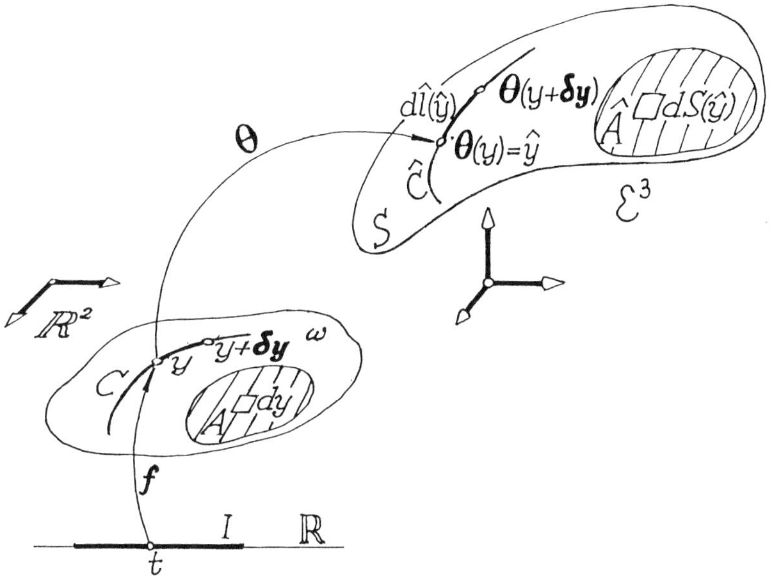

Fig. 2.1-4: *Area and length elements on a surface.* The elements $dS(\hat{y})$ and $d\hat{l}(\hat{y})$ at $\hat{y} = \boldsymbol{\theta}(y) \in S$ are related to dy and $\boldsymbol{\delta y}$ by means of the covariant components of the metric tensor of the surface S; cf. Thm. 2.1-1. The corresponding relations are used for computing the area of a surface $\hat{A} = \boldsymbol{\theta}(A) \subset S$ and the length of a curve $\hat{C} = \boldsymbol{\theta}(C) \subset S$, where $C = \boldsymbol{f}(I)$ and $I \subset \mathbb{R}$.

In particular, the *area* of $\hat{A}$ is given by

$$\boxed{\mathit{area}\ \hat{A} := \int_{\hat{A}} dS(\hat{y}) = \int_A \sqrt{a(y)}\, dy.}$$

Consider next a *curve* $C = \boldsymbol{f}(I)$ in $\overline{\omega}$, where I is a compact interval in $\mathbb{R}$ and $\boldsymbol{f} = f^\alpha \boldsymbol{e}_\alpha : I \to \overline{\omega}$ is a smooth enough injective mapping. Then the *length* of the *curve* $\hat{C} := \boldsymbol{\theta}(C)$ is given by

$$\textit{length } \hat{C} := \int_I \left| \frac{d}{dt}(\boldsymbol{\theta} \circ \boldsymbol{f})(t) \right| dt = \int_I \sqrt{a_{\alpha\beta}(\boldsymbol{f}(t)) \frac{df^\alpha}{dt}(t) \frac{df^\beta}{dt}(t)}\, dt.$$

Remark. The first fundamental form is also used for computing *angles* between coordinate lines; cf. Ex. 2.3. ■

Let δ^α_β denote the Kronecker symbol. Given a point $y \in \overline{\omega}$, *the four relations*

$$\boldsymbol{a}^\alpha(y) \cdot \boldsymbol{a}_\beta(y) = \delta^\alpha_\beta$$

unambiguously define two linearly independent vectors $\boldsymbol{a}^\alpha(y)$ *in the tangent plane*, which form the **contravariant basis of the tangent plane** (Fig. 2.1-1). To see this, let *a priori* $\boldsymbol{a}^\alpha(y) = Y^{\alpha\sigma}(y)\boldsymbol{a}_\sigma(y)$ in the relations $\boldsymbol{a}^\alpha(y) \cdot \boldsymbol{a}_\beta(y) = \delta^\alpha_\beta$. This gives $Y^{\alpha\sigma}(y)a_{\sigma\beta}(y) = \delta^\alpha_\beta$; hence $Y^{\alpha\sigma}(y) = a^{\alpha\sigma}(y)$, where (recall that the symmetric matrix $(a_{\alpha\beta}(y))$ is positive definite by assumption):

$$(a^{\alpha\beta}(y)) := (a_{\alpha\beta}(y))^{-1}.$$

The elements $a^{\alpha\beta}(y)$ of the symmetric matrix $(a^{\alpha\beta}(y))$ are called the **contravariant components** of the **metric tensor**, or **first fundamental form**, of the surface S at $\hat{y} = \boldsymbol{\theta}(y)$. The relations $\boldsymbol{a}^\alpha(y) = a^{\alpha\sigma}(y)\boldsymbol{a}_\sigma(y)$ also show that

$$\begin{aligned} \boldsymbol{a}^\alpha(y) \cdot \boldsymbol{a}^\beta(y) &= a^{\alpha\sigma}(y)a^{\beta\tau}(y)\boldsymbol{a}_\sigma(y) \cdot \boldsymbol{a}_\tau(y) \\ &= a^{\alpha\sigma}(y)a^{\beta\tau}(y)a_{\sigma\tau}(y) = a^{\alpha\sigma}(y)\delta^\beta_\sigma = a^{\alpha\beta}(y), \end{aligned}$$

and thus the vectors $\boldsymbol{a}^\alpha(y)$ are linearly independent since the matrix $(a^{\alpha\beta}(y))$ is positive definite.

Let us assemble for convenience these fundamental relations between the vectors of the covariant and contravariant bases of the tangent plane and the covariant and contravariant components of the metric tensor:

$$\begin{aligned} a_{\alpha\beta}(y) = \boldsymbol{a}_\alpha(y) \cdot \boldsymbol{a}_\beta(y) \quad &\text{and} \quad a^{\alpha\beta}(y) = \boldsymbol{a}^\alpha(y) \cdot \boldsymbol{a}^\beta(y), \\ \boldsymbol{a}_\alpha(y) = a_{\alpha\beta}(y)\boldsymbol{a}^\beta(y) \quad &\text{and} \quad \boldsymbol{a}^\alpha(y) = a^{\alpha\beta}(y)\boldsymbol{a}_\beta(y). \end{aligned}$$

2.2. CURVATURE TENSOR ON A SURFACE

While a "three-dimensional manifold" in $\mathcal{E}^3$, i.e., a set $\boldsymbol{\Theta}(\Omega)$ associated with a mapping $\boldsymbol{\Theta} : \Omega \subset \mathbb{R}^3 \to \mathcal{E}^3$, is well defined by its "metric" (uniquely up to rigid deformations in $\mathbb{R}^3$) provided *ad hoc* compatibility conditions are satisfied by the covariant components $g_{ij} : \Omega \to \mathbb{R}$ of its *metric tensor* (Sect. 1.9), a "two-dimensional manifold" $\boldsymbol{\theta}(\omega)$ in $\mathcal{E}^3$, i.e., a set $\boldsymbol{\theta}(\omega)$ associated with a mapping $\boldsymbol{\theta} : \omega \subset \mathbb{R}^2 \to \mathcal{E}^3$, *cannot* be defined by its "metric" alone. We shall not dwell into this fascinating question in this section, referring instead to Fig. 2.2-1 for an intuitive example and to Thm. 2.8-1 for a precise statement. The "missing information" is provided by the "curvature" of the surface, introduced in this section by means of the covariant components of its *curvature tensor*.

Consider as in Sect. 2.1 a *surface* $S = \boldsymbol{\theta}(\overline{\omega})$ in $\mathcal{E}^3$, where the injective mapping $\boldsymbol{\theta} : \overline{\omega} \subset \mathbb{R}^2 \to \mathcal{E}^3$ is smooth enough and such that, for each $y \in \overline{\omega}$, the two vectors $\boldsymbol{a}_\alpha(y) = \partial_\alpha \boldsymbol{\theta}(y)$ are linearly independent. For each $y \in \overline{\omega}$, the vector

$$\boxed{\boldsymbol{a}_3(y) := \frac{\boldsymbol{a}_1(y) \wedge \boldsymbol{a}_2(y)}{|\boldsymbol{a}_1(y) \wedge \boldsymbol{a}_2(y)|}}$$

is thus well defined, has Euclidean norm one, and is normal to the surface S at the point $\hat{y} = \boldsymbol{\theta}(y)$. This is another *abus de langage* as the vector normal to S at $\hat{y}$ is in fact the *vector in the affine space* $\mathcal{E}^3$ that has $\hat{y}$ as its origin and is parallel to $\boldsymbol{a}_3(y)$. Note that the denominator in the definition of $\boldsymbol{a}_3(y)$ may be also written as

$$\boxed{|\boldsymbol{a}_1(y) \wedge \boldsymbol{a}_2(y)| = \sqrt{a(y)},}$$

where $a(y) = \det(a_{\alpha\beta}(y))$ (Sect. 2.1). This relation, which holds in fact even if $a(y) = 0$, will be established in the proof of Thm. 2.5-1.

Fix $y \in \overline{\omega}$ and consider a *plane P normal to S at $\hat{y} = \boldsymbol{\theta}(y)$*, i.e., a plane that "contains the vector $\boldsymbol{a}_3(y)$"; by the same *abus de langage* as above, "a plane containing $\boldsymbol{a}_3(y)$" means in fact any *plane in the affine space* $\mathcal{E}^3$ that contains the point $\hat{y}$ and is parallel to the vector $\boldsymbol{a}_3(y)$. The intersection $\hat{C} = P \cap S$ is thus a *planar curve* on S.

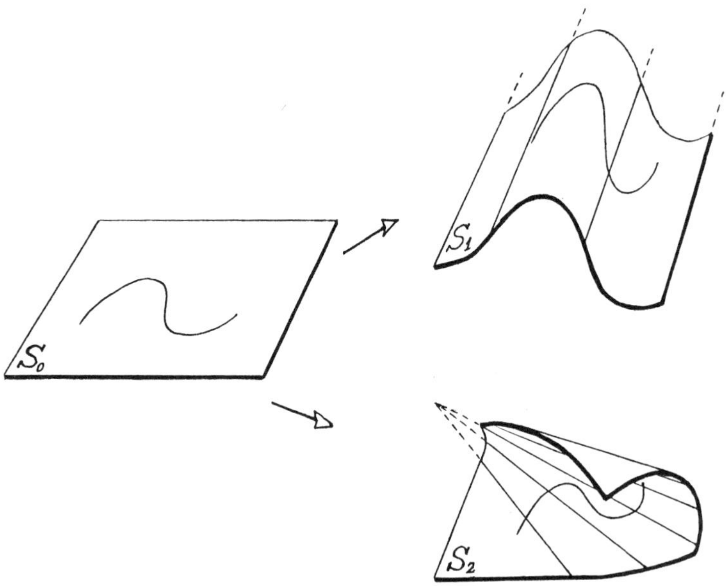

Fig. 2.2-1: *A metric does not define a surface in $\mathcal{E}^3$.* A flat surface S_0 may be deformed into a portion S_1 of a cylinder or a portion S_2 of a cone without altering the length of any curve drawn on it (cylinders and cones are instances of "developable surfaces"). Yet it should be clear that in general S_0 and S_1, or S_0 and S_2, or S_1 and S_2, are not identical surfaces *modulo* a rigid deformation of $\mathbb{R}^3$!

As shown in Thm. 2.2-1, it is remarkable that the *curvature* of $\hat{C}$ at $\hat{y}$ can be computed by means of the covariant components $a_{\alpha\beta}(y)$ of the first fundamental form of S introduced in Sect. 2.1, *together with* the covariant components $b_{\alpha\beta}(y)$ of the *"second" fundamental form* of S; the definition of the **curvature** *of a planar curve* is recalled in Fig. 2.2-2.

If the algebraic curvature of $\hat{C}$ at $\hat{y}$ is $\neq 0$, it can be written as $\dfrac{1}{R}$, and R is then called the *algebraic* **radius of curvature** of the curve $\hat{C}$ at $\hat{y}$; this means that the **center of curvature** of the curve $\hat{C}$ at $\hat{y}$ is the point $(\hat{y} + R\boldsymbol{a}_3(y))$; cf. Fig. 2.2-3. While R is

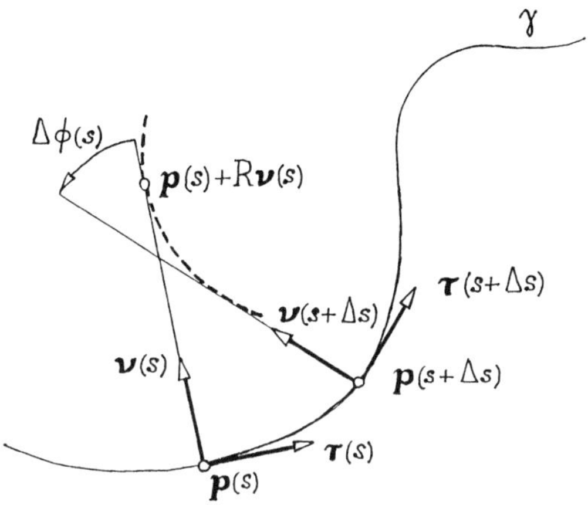

Fig. 2.2-2: *Curvature of a planar curve.* Let γ be a smooth enough planar curve, parametrized by its curvilinear abscissa s. Consider two points $\boldsymbol{p}(s)$ and $\boldsymbol{p}(s+\Delta s)$ with curvilinear abscissae s and $s+\Delta s$ and let $\Delta\phi(s)$ be the algebraic angle between the two normals $\boldsymbol{\nu}(s)$ and $\boldsymbol{\nu}(s+\Delta s)$ (oriented in the usual way) to γ at those points. When $\Delta s \to 0$, the ratio $\dfrac{\Delta\phi(s)}{\Delta s}$ has a limit, called the "curvature" of γ at $\boldsymbol{p}(s)$. If this limit is non-zero, its inverse R is called the (algebraic) "radius of curvature" of γ at $\boldsymbol{p}(s)$ (the sign of R depends on the orientation chosen on γ). The point $\boldsymbol{p}(s) + R\boldsymbol{\nu}(s)$, which is intrinsically defined, is called the "center of curvature" of γ at $\boldsymbol{p}(s)$: It is the center of the "osculating circle" at $\boldsymbol{p}(s)$, i.e., the limit as $\Delta s \to 0$ of the circle tangent to γ at $\boldsymbol{p}(s)$ that passes through the point $\boldsymbol{p}(s+\Delta s)$. The center of curvature is also the limit as $\Delta s \to 0$ of the intersection of the normals $\boldsymbol{\nu}(s)$ and $\boldsymbol{\nu}(s+\Delta s)$; consequently, the centers of curvature of γ lie on a curve (dashed on the figure), called "la développée" in French, that is tangent to the normals to γ.

not intrinsically defined (its sign changes in any system of curvilinear coordinates where "$\boldsymbol{a}_3(y)$ is replaced by $-\boldsymbol{a}_3(y)$"), the center of curvature *is* intrinsically defined. If the curvature of $\hat{C}$ at $\hat{y}$ is 0, the radius of curvature of the curve $\hat{C}$ at $\hat{y}$ is said to be *infinite*; for this reason, it is customary to still write the curvature as $\dfrac{1}{R}$ in this case.

Note that *the real number* $\dfrac{1}{R}$ *is always well defined* by the formula given in the next theorem, since the symmetric matrix $(a_{\alpha\beta}(y))$ is positive definite. This implies in particular that *the radius of curvature never vanishes along a surface* $S = \boldsymbol{\theta}(\overline{\omega})$ *defined by a mapping* $\boldsymbol{\theta}$

satisfying the assumptions of the next theorem, hence in particular of class $\mathcal{C}^2$ on $\overline{\omega}$. It is intuitively clear that the vanishing of R implies that $\boldsymbol{\theta}$ cannot be "too" smooth: Think of a surface made of two portions of planes intersecting along a segment, which thus constitutes a *fold* on the surface; or think of a surface $\boldsymbol{\theta}(\overline{\omega})$ with $0 \in \omega$ and $\boldsymbol{\theta}(y_1, y_2) = |y_1|^{1+\alpha}$ for some $0 < \alpha < 1$, so that $\boldsymbol{\theta} \in \mathcal{C}^1(\overline{\omega}; \mathcal{E}^3)$ but $\boldsymbol{\theta} \notin \mathcal{C}^2(\overline{\omega}; \mathcal{E}^3)$: The radius of curvature of a curve corresponding to a constant y_2 vanishes at $y_1 = 0$.

Theorem 2.2-1. *Let ω be a domain in $\mathbb{R}^2$ and let $\boldsymbol{\theta} \in \mathcal{C}^2(\overline{\omega}; \mathcal{E}^3)$ be an injective mapping such that the two vectors $\boldsymbol{a}_\alpha(y) = \partial_\alpha \boldsymbol{\theta}(y)$ are linearly independent at all points $y \in \overline{\omega}$.*

Fix $y \in \overline{\omega}$ and consider a plane P normal to $S = \boldsymbol{\theta}(\overline{\omega})$ at the point $\hat{y} = \boldsymbol{\theta}(y)$. The intersection $P \cap S$ is a curve $\hat{C}$ on S, which is the image $\hat{C} = \boldsymbol{\theta}(C)$ of a curve C in the set $\overline{\omega}$. Assume that the curve C is represented in a sufficiently small neighborhood of y as the image $\boldsymbol{f}(I)$ of a closed interval $I \subset \mathbb{R}$ with a non-empty interior, where $\boldsymbol{f} = f^\alpha \boldsymbol{e}_\alpha : I \to \mathbb{R}$ is a smooth enough injective mapping that satisfies $\dfrac{df^\alpha}{dt}(t)\, \boldsymbol{e}_\alpha \neq \boldsymbol{0}$ where $t \in I$ is such that $y = \boldsymbol{f}(t)$ (Fig. 2.2-3). *Then the curvature $\dfrac{1}{R}$ of the planar curve $\hat{C}$ at $\hat{y}$ is given by the ratio*

$$\boxed{\frac{1}{R} = \frac{b_{\alpha\beta}(\boldsymbol{f}(t)) \dfrac{df^\alpha}{dt}(t) \dfrac{df^\beta}{dt}(t)}{a_{\alpha\beta}(\boldsymbol{f}(t)) \dfrac{df^\alpha}{dt}(t) \dfrac{df^\beta}{dt}(t)},}$$

where $a_{\alpha\beta}(y)$ are the covariant components of the metric tensor of S at y (Sect. 2.1) *and*

$$\boxed{b_{\alpha\beta}(y) := \boldsymbol{a}_3(y) \cdot \partial_\alpha \boldsymbol{a}_\beta(y) = -\partial_\alpha \boldsymbol{a}_3(y) \cdot \boldsymbol{a}_\beta(y) = b_{\beta\alpha}(y).}$$

Proof. (i) We first establish a well-known formula for the curvature $\dfrac{1}{R}$ of a planar curve. Using the notations of Fig. 2.2-2, we note that

$$\sin \Delta\phi(s) = \boldsymbol{\nu}(s) \cdot \boldsymbol{\tau}(s + \Delta s) = -\{\boldsymbol{\nu}(s + \Delta s) - \boldsymbol{\nu}(s)\} \cdot \boldsymbol{\tau}(s + \Delta s),$$

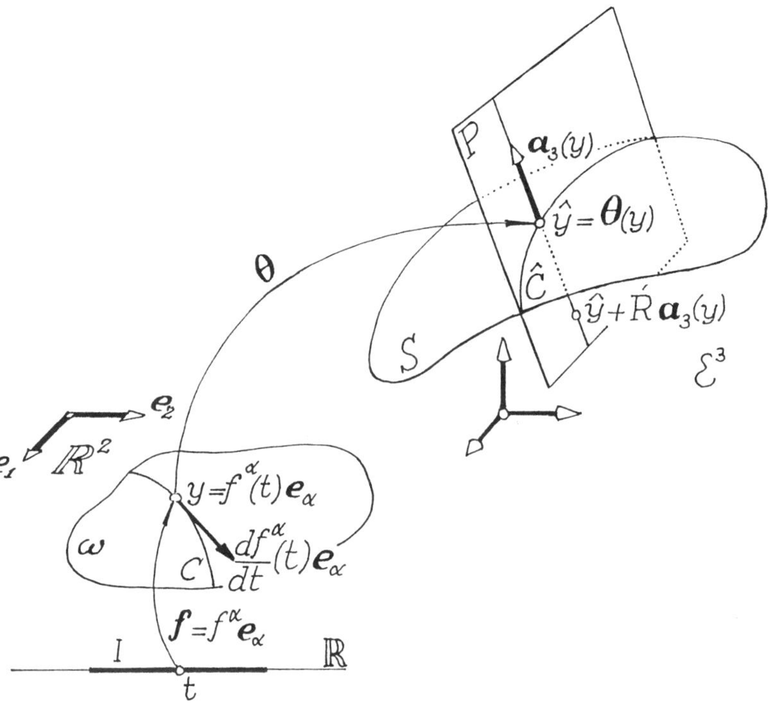

Fig. 2.2-3: *Curvature on a surface.* Let P be a plane containing the vector $\boldsymbol{a}_3(y) = \dfrac{\boldsymbol{a}_1(y) \wedge \boldsymbol{a}_2(y)}{|\boldsymbol{a}_1(y) \wedge \boldsymbol{a}_2(y)|}$, which is normal to S. The algebraic curvature $\dfrac{1}{R}$ of the planar curve $\hat{C} = P \cap S = \boldsymbol{\theta}(C)$ at $\hat{y} = \boldsymbol{\theta}(y)$ is given by the ratio

$$\frac{1}{R} = \frac{b_{\alpha\beta}(\boldsymbol{f}(t))\dfrac{df^\alpha}{dt}(t)\,\dfrac{df^\beta}{dt}(t)}{a_{\alpha\beta}(\boldsymbol{f}(t))\dfrac{df^\alpha}{dt}(t)\,\dfrac{df^\beta}{dt}(t)},$$

where $a_{\alpha\beta}(y)$ and $b_{\alpha\beta}(y)$ are the covariant components of the metric and curvature tensors (also called the first and second fundamental forms) of the surface S at $\hat{y}$ and $\dfrac{df^\alpha}{dt}(t)$ are the components of the vector tangent to the curve $C = \boldsymbol{f}(I)$ at $y = \boldsymbol{f}(t) = f^\alpha(t)\boldsymbol{e}_\alpha$. If $\dfrac{1}{R} \neq 0$, the center of curvature of the curve $\hat{C}$ at $\hat{y}$ is the point $(\hat{y} + R\boldsymbol{a}_3(y))$, which is intrinsically defined in the affine Euclidean space $\mathcal{E}^3$.

so that

$$\frac{1}{R} := \lim_{\Delta s \to 0} \frac{\Delta\phi(s)}{\Delta s} = \lim_{\Delta s \to 0} \frac{\sin \Delta\phi(s)}{\Delta s} = -\frac{d\boldsymbol{\nu}}{ds}(s) \cdot \boldsymbol{\tau}(s).$$

(ii) The curve $(\boldsymbol{\theta} \circ \boldsymbol{f})(I)$, which is *a priori* parametrized by $t \in I$, can be also parametrized by its curvilinear abscissa s in a neighborhood of the point $\hat{y}$. There thus exist an interval $\tilde{I} \subset I$ and a mapping $\boldsymbol{p} : J \to P$, where $J \subset \mathbb{R}$ is an interval, such that

$$(\boldsymbol{\theta} \circ \boldsymbol{f})(t) = \boldsymbol{p}(s) \text{ and } (\boldsymbol{a}_3 \circ \boldsymbol{f})(t) = \boldsymbol{\nu}(s) \text{ for all } t \in \tilde{I},\ s \in J.$$

By (i), the curvature $\dfrac{1}{R}$ of $\hat{C}$ is given by

$$\frac{1}{R} = -\frac{d\boldsymbol{\nu}}{ds}(s) \cdot \boldsymbol{\tau}(s),$$

where

$$\begin{aligned}
\frac{d\boldsymbol{\nu}}{ds}(s) &= \frac{d(\boldsymbol{a}_3 \circ \boldsymbol{f})}{dt}(t)\frac{dt}{ds} = \partial_\alpha \boldsymbol{a}_3(\boldsymbol{f}(t))\frac{df^\alpha}{dt}(t)\,\frac{dt}{ds},\\
\boldsymbol{\tau}(s) &= \frac{d\boldsymbol{p}}{ds}(s) = \frac{d(\boldsymbol{\theta} \circ \boldsymbol{f})}{dt}(t)\frac{dt}{ds}\\
&= \partial_\beta \boldsymbol{\theta}(\boldsymbol{f}(t))\frac{df^\beta}{dt}(t)\,\frac{dt}{ds} = \boldsymbol{a}_\beta(\boldsymbol{f}(t))\frac{df^\beta}{dt}(t)\,\frac{dt}{ds}.
\end{aligned}$$

Hence

$$\frac{1}{R} = -\partial_\alpha \boldsymbol{a}_3(\boldsymbol{f}(t)) \cdot \boldsymbol{a}_\beta(\boldsymbol{f}(t))\frac{df^\alpha}{dt}(t)\,\frac{df^\beta}{dt}(t)\left(\frac{dt}{ds}\right)^2.$$

To obtain the announced expression for $\dfrac{1}{R}$, it suffices to note that

$$-\partial_\alpha \boldsymbol{a}_3(\boldsymbol{f}(t)) \cdot \boldsymbol{a}_\beta(\boldsymbol{f}(t)) = b_{\alpha\beta}(\boldsymbol{f}(t)),$$

by definition of the functions $b_{\alpha\beta}$ and that (Thm. 2.1-1)

$$ds = \left\{\delta y^\alpha a_{\alpha\beta}(y)\delta y^\beta\right\}^{1/2} = \left\{a_{\alpha\beta}(\boldsymbol{f}(t))\frac{df^\alpha}{dt}(t)\,\frac{df^\beta}{dt}(t)\right\}^{1/2} dt.$$

■

Remark. The knowledge of the curvatures of curves contained in *planes normal to* S suffices for computing the curvature of *any* curve on S. More specifically, the radius of curvature $\tilde{R}$ at $\hat{y}$ of any smooth enough curve $\tilde{C}$ (planar or not) on the surface S is given by $\frac{\cos\varphi}{\tilde{R}} = \frac{1}{R}$, where φ is the angle between the "principal normal" to $\tilde{C}$ at $\hat{y}$ and $\boldsymbol{a}_3(y)$ and $\frac{1}{R}$ is given in Thm. 2.2-1; see, e.g., Stoker [1969, Chap. 4, Sect. 12]. ∎

Precise information about the shape of the surface in a neighborhood of its point $\hat{y} = \boldsymbol{\theta}(y)$ can thus be gathered by "letting the plane P turn around the normal vector $\boldsymbol{a}_3(y)$" and by following in this process the variations of the curvatures at $\hat{y}$ of the corresponding planar curves $\hat{C} = P \cap C$, as given in Thm. 2.2-1. As we now show, *these curvatures span a compact interval of* $\mathbb{R}$, i.e., they "stay away from infinity".

Theorem 2.2-2. (a) *Let the assumptions and notations be as in* Thm. 2.2-1. *For a fixed $y \in \overline{\omega}$, consider the set $\mathcal{P}$ of all planes P normal to S at $\hat{y} = \boldsymbol{\theta}(y)$. Then the set of curvatures of the associated planar curves $P \cap S$, $P \in \mathcal{P}$, is a compact interval of $\mathbb{R}$, denoted* $\left[\frac{1}{R_1(y)}, \frac{1}{R_2(y)}\right]$ (either one of its end-points, or both, can be 0).

(b) *Let the matrix $(b_\alpha^\beta(y))$, α being the row index, be defined by*

$$\boxed{b_\alpha^\beta(y) := a^{\beta\sigma}(y)b_{\alpha\sigma}(y),}$$

where $(a^{\alpha\beta}(y)) = (a_{\alpha\beta}(y))^{-1}$ (Sect. 2.1) *and the matrix $(b_{\alpha\beta}(y))$ is defined as in* Thm. 2.2-1. *Then*

$$\boxed{\begin{aligned}\frac{1}{R_1(y)} + \frac{1}{R_2(y)} &= b_1^1(y) + b_2^2(y),\\ \frac{1}{R_1(y)R_2(y)} &= b_1^1(y)b_2^2(y) - b_1^2(y)b_2^1(y) = \frac{\det(b_{\alpha\beta}(y))}{\det(a_{\alpha\beta}(y))}.\end{aligned}}$$

(c) *If $\frac{1}{R_1(y)} \neq \frac{1}{R_2(y)}$, there is a unique pair of orthogonal planes $P_1 \in \mathcal{P}$ and $P_2 \in \mathcal{P}$ such that the curvatures of the associated planar curves $P_1 \cap S$ and $P_2 \cap S$ are precisely $\frac{1}{R_1(y)}$ and $\frac{1}{R_2(y)}$.*

Proof. (i) Let $\Delta(P)$ denote the intersection of $P \in \mathcal{P}$ with the tangent plane to S at $\hat{y}$, and let $\hat{C}(P)$ denote the intersection of P with the surface S; hence $\Delta(P)$ is tangent to $\hat{C}(P)$ at $\hat{y} \in S$. Since we assume as in Thm. 2.2-1 that $\hat{C}(P) = (\boldsymbol{\theta} \circ \boldsymbol{f}(P))(I)$ for some mapping $\boldsymbol{f}(P) : I \subset \mathbb{R} \to \overline{\omega}$, the line $\Delta(P)$ is given by

$$\Delta(P) = \left\{ \hat{y} + \lambda \frac{d(\boldsymbol{\theta} \circ \boldsymbol{f}(P))}{dt}(t);\ \lambda \in \mathbb{R} \right\} = \{\hat{y} + \lambda \xi^\alpha \boldsymbol{a}_\alpha(y);\ \lambda \in \mathbb{R}\},$$

where $\xi^\alpha := \dfrac{df^\alpha(P)}{dt}(t)$ and $\xi^\alpha \boldsymbol{e}_\alpha \neq \boldsymbol{0}$ by assumption. Since the line $\{y + \mu \xi^\alpha \boldsymbol{e}_\alpha;\ \mu \in \mathbb{R}\}$ is tangent to the curve $C(P) := \boldsymbol{\theta}^{-1}(\hat{C}(P))$ at $y \in \overline{\omega}$ (the mapping $\boldsymbol{\theta} : \overline{\omega} \to \mathbb{R}^3$ is injective by assumption) for each such parametrizing function $\boldsymbol{f}(P) : I \to C$ and the vectors $\boldsymbol{a}_\alpha(y)$ are linearly independent, there exists a bijection between the set of all lines $\Delta(P) \subset T$, $P \in \mathcal{P}$, and the set of all lines supporting the nonzero tangent vectors to the curve $C(P)$.

Hence Thm. 2.2-1 shows that when P varies in $\mathcal{P}$, the curvature of the corresponding curves $\hat{C} = \hat{C}(P)$ at $\hat{y}$ takes the same values as does the ratio $\dfrac{b_{\alpha\beta}(y)\xi^\alpha\xi^\beta}{a_{\alpha\beta}(y)\xi^\alpha\xi^\beta}$ when $\boldsymbol{\xi} := (\zeta_\alpha)$ varies in $\mathbb{R}^2 - \{\boldsymbol{0}\}$.

(ii) Let the symmetric matrices $\mathbf{A}$ and $\mathbf{B}$ of order two be defined by

$$\mathbf{A} := (a_{\alpha\beta}(y)) \text{ and } \mathbf{B} := (b_{\alpha\beta}(y)).$$

Since $\mathbf{A}$ is positive definite, it has a (unique) *square root* $\mathbf{C}$, i.e., a symmetric positive definite matrix $\mathbf{C}$ such that $\mathbf{A} = \mathbf{C}^2$ (for a proof, see, e.g., Vol. I, Thm. 3.2-1). Hence the ratio

$$\frac{b_{\alpha\beta}(y)\xi^\alpha\xi^\beta}{a_{\alpha\beta}(y)\xi^\alpha\xi^\beta} = \frac{\boldsymbol{\xi}^T\mathbf{B}\boldsymbol{\xi}}{\boldsymbol{\xi}^T\mathbf{A}\boldsymbol{\xi}} = \frac{\boldsymbol{\eta}^T\mathbf{C}^{-1}\mathbf{B}\mathbf{C}^{-1}\boldsymbol{\eta}}{\boldsymbol{\eta}^T\boldsymbol{\eta}}, \text{ where } \boldsymbol{\eta} = \mathbf{C}\boldsymbol{\xi},$$

is nothing but the *Rayleigh quotient* associated with the symmetric matrix $\mathbf{C}^{-1}\mathbf{B}\mathbf{C}^{-1}$. When $\boldsymbol{\eta}$ varies in $\mathbb{R}^2 - \{\boldsymbol{0}\}$, this Rayleigh quotient thus spans the compact interval of $\mathbb{R}$ whose end-points are the smallest and largest eigenvalue, respectively denoted $\dfrac{1}{R_1(y)}$ and $\dfrac{1}{R_2(y)}$, of the matrix $\mathbf{C}^{-1}\mathbf{B}\mathbf{C}^{-1}$ (for a proof, see, e.g., Ciarlet [1982, Thm. 1.3-1]). This proves (a).

Furthermore, the relation

$$\mathbf{C}(\mathbf{C}^{-1}\mathbf{B}\mathbf{C}^{-1})\mathbf{C}^{-1} = \mathbf{B}\mathbf{C}^{-2} = \mathbf{B}\mathbf{A}^{-1}$$

shows that the eigenvalues of the symmetric matrix $\mathbf{C}^{-1}\mathbf{B}\mathbf{C}^{-1}$ coincide with those of the (non-symmetric in general) matrix $\mathbf{B}\mathbf{A}^{-1}$. Note that $\mathbf{B}\mathbf{A}^{-1} = (b_\alpha^\beta(y))$ with $b_\alpha^\beta(y) := a^{\beta\sigma}(y)b_{\alpha\sigma}(y)$, α being the row index, since $\mathbf{A}^{-1} = (a^{\alpha\beta}(y))$; hence the relations in (b) simply express that the sum and the product of the eigenvalues of the matrix $\mathbf{B}\mathbf{A}^{-1}$ are respectively equal to its trace and to its determinant, which may be also written as $\dfrac{\det(b_{\alpha\beta}(y))}{\det(a_{\alpha\beta}(y))}$ since $\mathbf{B}\mathbf{A}^{-1} = (b_\alpha^\beta(y))$. This proves (b).

(iii) Let $\boldsymbol{\eta}_1 = (\eta_1^\alpha) = \mathbf{C}\boldsymbol{\xi}_1$ and $\boldsymbol{\eta}_2 = (\eta_2^\beta) = \mathbf{C}\boldsymbol{\xi}_2$, with $\boldsymbol{\xi}_1 = (\xi_1^\alpha)$ and $\boldsymbol{\xi}_2 = (\xi_2^\beta)$, be two orthogonal ($\boldsymbol{\eta}_1^T\boldsymbol{\eta}_2 = 0$) eigenvectors of the symmetric matrix $\mathbf{C}^{-1}\mathbf{B}\mathbf{C}^{-1}$, corresponding to the eigenvalues $\dfrac{1}{R_1(y)}$ and $\dfrac{1}{R_2(y)}$, respectively. Hence

$$0 = \boldsymbol{\eta}_1^T\boldsymbol{\eta}_2 = \boldsymbol{\xi}_1^T\mathbf{C}^T\mathbf{C}\boldsymbol{\xi}_2 = \boldsymbol{\xi}_1^T\mathbf{A}\boldsymbol{\xi}_2 = 0,$$

since $\mathbf{C}^T = \mathbf{C}$. By (i), the corresponding lines $\Delta(P_1)$ and $\Delta(P_2)$ of the tangent plane are parallel to the vectors $\xi_1^\alpha \boldsymbol{a}_\alpha(y)$ and $\xi_2^\beta \boldsymbol{a}_\beta(y)$, which are orthogonal since

$$\{\xi_1^\alpha \boldsymbol{a}_\alpha(y)\} \cdot \{\xi_2^\beta \boldsymbol{a}_\beta(y)\} = a_{\alpha\beta}(y)\xi_1^\alpha\xi_2^\beta = \boldsymbol{\xi}_1^T\mathbf{A}\boldsymbol{\xi}_2.$$

If $\dfrac{1}{R_1(y)} \neq \dfrac{1}{R_2(y)}$, the directions of the vectors $\boldsymbol{\eta}_1$ and $\boldsymbol{\eta}_2$ are uniquely determined and the lines $\Delta(P_1)$ and $\Delta(P_2)$ are likewise uniquely determined. This proves (c). ∎

We are now in a position to state several *fundamental definitions*: The elements $b_{\alpha\beta}(y)$ of the (symmetric) matrix $(b_{\alpha\beta}(y))$ defined in Thm. 2.2-1 and the element $b_\alpha^\beta(y)$ of the (non-symmetric in general) matrix $(b_\alpha^\beta(y))$ defined in Thm. 2.2-2 are respectively called the **covariant components** and the **mixed components** of the **curvature tensor**, also called the **second fundamental form**, of the surface S at $\hat{y} = \boldsymbol{\theta}(y)$.

A word of caution. With this definition of "curvature tensor", I depart from other texts, where "Curvature tensor" of S often means the "Riemann curvature tensor" of S (the covariant components of this other tensor are defined in the last exercise of this chapter). ∎

The real numbers $\frac{1}{R_1(y)}$ and $\frac{1}{R_2(y)}$ (one or both of which being possibly 0) found in Thm. 2.2-2 are called the **principal curvatures** of S at $\hat{y}$.

If $\frac{1}{R_1(y)} = \frac{1}{R_2(y)}$, the curvatures of the planar curves $P \cap S$ are the same in all directions, i.e., for all $P \in \mathcal{P}$. If $\frac{1}{R_1(y)} = \frac{1}{R_2(y)} = 0$, the point $\hat{y} = \boldsymbol{\theta}(y)$ is called a **planar point**; if $\frac{1}{R_1(y)} = \frac{1}{R_2(y)} \neq 0$, $\hat{y}$ is called an **umbilical point**; cf. Exs. 2.4 and 2.5.

Let $\hat{y} = \boldsymbol{\theta}(y) \in S$ be a point that is neither planar nor umbilical; in other words, the principal curvatures at $\hat{y}$ are not equal. Then the two orthogonal lines tangent to the planar curves $P_1 \cap S$ and $P_2 \cap S$ (Thm. 2.2-2 (c)) are called the **principal directions** at $\hat{y}$.

A **line of curvature** on S is a curve on S that is tangent to a principal direction at each one of its points. It can be shown that *a point that is neither planar nor umbilical possesses a neighborhood where two orthogonal families of lines of curvature can be chosen as coordinate lines*; see, e.g., Klingenberg [1973, Lemma 3.6.6].

If both $\frac{1}{R_1(y)}$ and $\frac{1}{R_2(y)}$ are $\neq 0$, the real numbers $R_1(y)$ and $R_2(y)$ are called the (algebraic) **principal radii of curvature** of S at $\hat{y}$. If, e.g., $\frac{1}{R_1(y)} = 0$, the corresponding principal radius of curvature $R_1(y)$ is said to be *infinite*. While the principal radii of curvature may simultaneously change their signs in another system of curvilinear coordinates (as already noted), the associated *centers of curvature* are intrinsically defined.

The numbers $\left(\frac{1}{R_1(y)} + \frac{1}{R_2(y)}\right)$ and $\frac{1}{R_1(y)R_2(y)}$, identified as the principal invariants of the matrix $(b_\alpha^\beta(y))$ in Thm. 2.2-1, are respectively called the **mean curvature** and the **Gaussian, or total, curvature** of S at $\hat{y}$.

As shown in Fig. 2.2-4, the points on a surface can be classified as **elliptic, parabolic**, or **hyperbolic**, according as their *Gaussian curvature* is > 0, $= 0$ but they are not planar, or < 0.

An **asymptotic line** is a curve on a surface that is everywhere tangent to a direction along which the radius of curvature is infinite; any point along an asymptotic line is thus either parabolic or hyperbolic. It can be shown that, *if all the points of a surface are hyperbolic, any point possesses a neighborhood where two intersecting families of asymptotic lines can be chosen as coordinate lines*; see, e.g., Klingenberg [1973, Lemma 3.6.12].

As intuitively suggested at the beginning of this section, a surface in $\mathbb{R}^3$ cannot be defined by its *metric* alone, i.e., by its first fundamental form alone, since its *curvature* must be in addition specified through its second fundamental form; see Sect. 2.8. But quite surprisingly, the *Gaussian curvature* can also be expressed solely in terms of the functions $a_{\alpha\beta}$ and their derivatives! This is the celebrated *Theorema egregium* ("outstanding theorem") of Gauß [1828]; see Sect. 2.8.

Another striking result where the *Gaussian curvature* is used is the equally celebrated *Gauß-Bonnet theorem*, so named after Gauß [1828] and Bonnet [1848] (for a "modern" proof, see, e.g., Klingenberg [1973, Thm. 6.3-5] or do Carmo [1994, Chap. 6, Thm. 1]): *Let S be a smooth enough, "closed", "orientable", and compact surface in $\mathbb{R}^3$* (a "closed" surface is one "without boundary", such as a sphere or a torus; "orientable" surfaces, which exclude for instance Klein bottles, are defined, e.g., in Klingenberg [1973, Sect. 5.5]) *and let $K : S \to \mathbb{R}$ denote its Gaussian curvature. Then*

$$\int_S K(\hat{y})\, dS(\hat{y}) = 2\pi(2 - 2g(S)),$$

where the **genus** *$g(S)$ is the number of "holes" of S* (for instance, a sphere has genus zero, while a torus has genus one; cf. Ex. 2.6). The integer $\chi(S)$ defined by $\chi(S) := (2 - 2g(S))$ is the **Euler characteristic** of S.

A **developable surface** is one whose *Gaussian curvature* vanishes everywhere (this is the definition of Stoker [1969, Chap. 5, Sect. 2]; developable surfaces are otherwise often defined as "ruled" surfaces whose Gaussian curvature vanishes everywhere, as in, e.g., Klingenberg [1973, Sect. 3.7]). A portion of a plane provides a first example, in fact the only one of a developable surface *all* points of

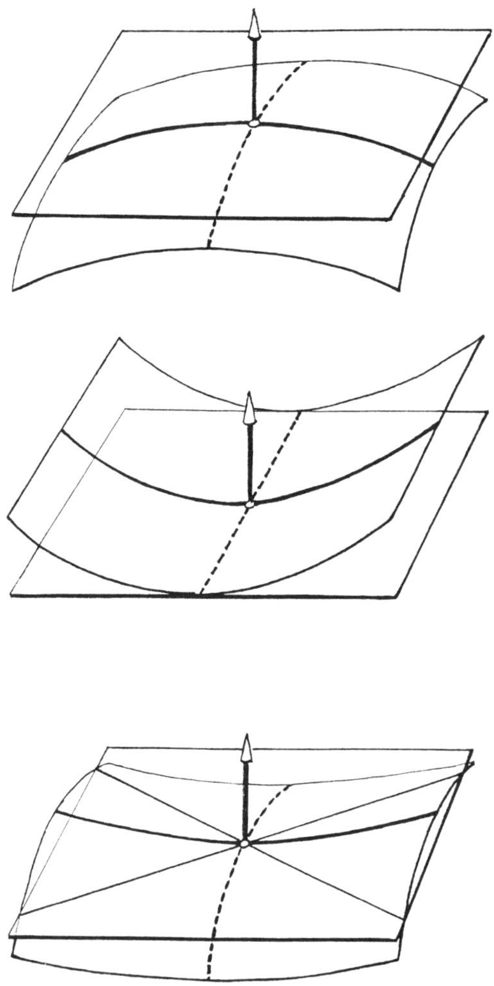

Fig. 2.2-4: *Different kinds of points on a surface.* A point is elliptic if the Gaussian curvature is > 0 or equivalently, if the two principal radii of curvature are of the same sign; the surface is then locally on one side of its tangent plane. A point is parabolic if exactly one of the two principal radii of curvature is infinite; the surface is again locally on one side of its tangent plane. A point is hyperbolic if the Gaussian curvature is < 0 or equivalently, if the two principal radii of curvature are of different signs; the surface then intersects its tangent plane along two curves.

which are planar (Ex. 2.4). Any developable surface *all* points of which are *parabolic* can be likewise fully described: It is *either* a portion of a cylinder, *or* a portion of a cone, *or* a portion of a surface spanned by the tangents to a skewed curve. The description of a developable surface comprising both planar and parabolic points is more subtle (although the above examples are in a sense the only ones possible, at least locally; see Stoker [1969, Chap. 5, Sects. 2 to 6]).

The interest of developable surfaces is that they can be, at least locally, continuously "rolled out", or "developed" (hence their name), onto a plane *without changing the metric* of the intermediary surfaces in the process.

Remarks. (1) That such *"continuously varying isometric surfaces"* exist plays a key rôle in the theory of "flexural shells" (Chaps. 6 and 10).

(2) Interesting connections between the theory of developable surfaces and the appearance of *folds* on plates and shells subjected to applied forces have been exhibited and studied by Sanchez-Palencia [1995], Pomeau [1995], Ben Amar & Pomeau [1997]. ■

In Sect. 2.8, we shall briefly discuss the reciprocal question of *recovering a surface $\boldsymbol{\theta}(\omega)$ from its metric and curvature tensor fields*: Given a positive definite symmetric matrix field $(a_{\alpha\beta})$ on ω and a symmetric matrix field $(b_{\alpha\beta})$ on ω, find conditions under which there exists a mapping $\boldsymbol{\theta} : \omega \to \mathcal{E}^3$ such that

$$\begin{aligned} \partial_\alpha \boldsymbol{\theta} \cdot \partial_\beta \boldsymbol{\theta} &= a_{\alpha\beta} \text{ in } \omega, \\ \frac{\partial_1 \boldsymbol{\theta} \wedge \partial_2 \boldsymbol{\theta}}{|\partial_1 \boldsymbol{\theta} \wedge \partial_2 \boldsymbol{\theta}|} \cdot \partial_{\alpha\beta} \boldsymbol{\theta} &= b_{\alpha\beta} \text{ in } \omega. \end{aligned}$$

2.3. COVARIANT DERIVATIVES AND CHRISTOFFEL SYMBOLS ON A SURFACE

The content of Sects. 2.1 and 2.2 and of the present section constitute our first encounter with *differential geometry "on a surface"*. Other related notions are treated later, such as *covariant derivatives of a tensor field on a surface* (Thm. 2.5-1) and the *recovery of a surface from its metric and curvature tensor fields* (Sect. 2.8).

For further details and complements, also for the notion of *"tensors on a surface"*, see classical texts such as Valiron [1950, Chap. 13], Struik [1961], Stoker [1969], Klingenberg [1973], Spivak [1975], do Carmo [1976], or Berger & Gostiaux [1992, Chaps. 10 and 11] and the more recent books of do Carmo [1994] and Hsiung [1997]; Sanchez-Hubert & Sanchez-Palencia [1997] provide a wealth of complements particularly well adapted to shell theory.

More "intrinsic" approaches to the differential geometry of surfaces have also been advocated in shell theory, notably by Destuynder [1990, Chap. 2], Podio-Guidugli [1991], Carrive [1995], Carrive, Le Tallec & Mouro [1995], Valid [1995], Delfour & Zolesio [1997].

We begin by some definitions. As in Sects. 2.1 and 2.2, consider a surface $S = \boldsymbol{\theta}(\overline{\omega})$ in $\mathcal{E}^3$, where $\boldsymbol{\theta} : \overline{\omega} \subset \mathbb{R}^2 \to \mathbb{R}^3$ is a smooth enough injective mapping such that the two vectors $\boldsymbol{a}_\alpha(y) = \partial_\alpha \boldsymbol{\theta}(y)$ are linearly independent at all $y \in \overline{\omega}$ and let

$$\boxed{\boldsymbol{a}_3(y) = \boldsymbol{a}^3(y) := \frac{\boldsymbol{a}_1(y) \wedge \boldsymbol{a}_2(y)}{|\boldsymbol{a}_1(y) \wedge \boldsymbol{a}_2(y)|}, \quad y \in \overline{\omega}.}$$

Then the vectors $\boldsymbol{a}_\alpha(y)$ (which form the covariant basis of the tangent plane to S at $\hat{y} = \boldsymbol{\theta}(y)$; see Fig. 2.1-1) together with the vector $\boldsymbol{a}_3(y)$ (which is normal to S and has Euclidean norm one) form the **covariant basis** *at* $\hat{y}$.

Let the vectors $\boldsymbol{a}^\alpha(y)$ of the tangent plane to S at $\hat{y}$ be defined by the relations $\boldsymbol{a}^\alpha(y) \cdot \boldsymbol{a}_\beta(y) = \delta^\alpha_\beta$. Then the vectors $\boldsymbol{a}^\alpha(y)$ (which form the contravariant basis of the tangent plane at $\hat{y}$; see again Fig. 2.1-1) together with the vector $\boldsymbol{a}^3(y)$ form the **contravariant basis** *at* $\hat{y}$; cf. Fig. 2.3-1. Note that the vectors of the covariant and contravariant bases at $\hat{y}$ satisfy

$$\boxed{\boldsymbol{a}^i(y) \cdot \boldsymbol{a}_j(y) = \delta^i_j.}$$

Consider an arbitrary **displacement field** of the middle surface S, i.e., an arbitrary vector field $\eta_i \boldsymbol{a}^i$ defined by means of its **covariant components** $\eta_i : \overline{\omega} \to \mathbb{R}$; this means that $\eta_i(y)\boldsymbol{a}^i(y)$ is the displacement of the point $\hat{y} = \boldsymbol{\theta}(y) \in S$ (Fig. 2.3-1).

As we shall see, finding the explicit forms of the associated linearized change of metric tensor (Thm. 2.4-1) and linearized change of curvature tensor (Thm. 2.5-1) requires the explicit forms of the

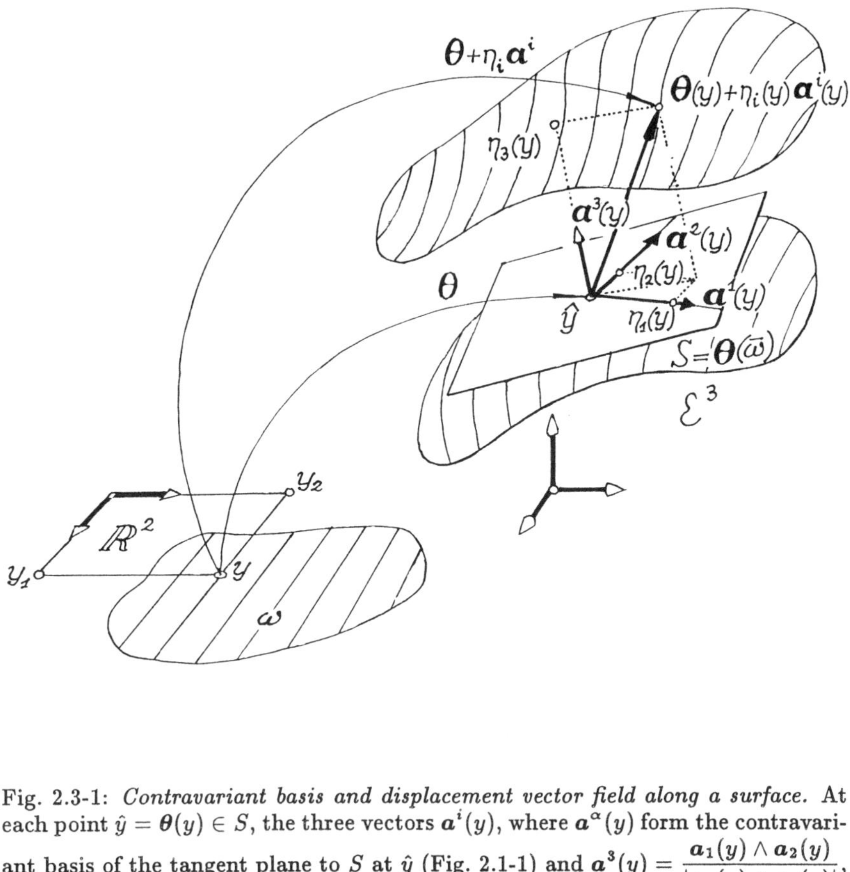

Fig. 2.3-1: *Contravariant basis and displacement vector field along a surface.* At each point $\hat{y} = \boldsymbol{\theta}(y) \in S$, the three vectors $\boldsymbol{a}^i(y)$, where $\boldsymbol{a}^\alpha(y)$ form the contravariant basis of the tangent plane to S at $\hat{y}$ (Fig. 2.1-1) and $\boldsymbol{a}^3(y) = \dfrac{\boldsymbol{a}_1(y) \wedge \boldsymbol{a}_2(y)}{|\boldsymbol{a}_1(y) \wedge \boldsymbol{a}_2(y)|}$, form the contravariant basis at $\hat{y}$. An arbitrary displacement field of S may then be defined by its covariant components $\eta_i : \overline{\omega} \to \mathbb{R}$: This means that $\eta_i(y)\boldsymbol{a}^i(y)$ is the displacement of the point $\hat{y}$. The mapping $(\boldsymbol{\theta} + \eta_i\boldsymbol{a}^i) : \overline{\omega} \to \mathcal{E}^3$ defines a surface $(\boldsymbol{\theta} + \eta_i\boldsymbol{a}^i)(\overline{\omega})$ in $\mathcal{E}^3$, which is thus equipped with the same curvilinear coordinates (the coordinates y_1, y_2 of $y \in \overline{\omega}$) as those of the surface $S = \boldsymbol{\theta}(\overline{\omega})$.

partial derivatives $\partial_\alpha(\eta_i\boldsymbol{a}^i)$ of this vector field. These are found in the next theorem, as immediate consequences of two basic formulas, those of *Gauß* and *Weingarten*. The *Christoffel symbols "on a surface"* and the *covariant derivatives "on a surface"* are also naturally introduced in this process.

Whenever no confusion should arise, we henceforth drop the explicit dependence on a particular point. For instance, the relation "$\partial_\alpha \boldsymbol{a}^3 = -b_{\alpha\beta}\boldsymbol{a}^\beta$" means that "$\partial_\alpha \boldsymbol{a}^3(y) = -b_{\alpha\beta}(y)\boldsymbol{a}^\beta(y)$ for all $y \in \overline{\omega}$"; "$\partial_\alpha \boldsymbol{a}^3$ is in the tangent plane" means "$\partial_\alpha \boldsymbol{a}^3(y)$ is in the tangent plane to S at $\hat{y} = \boldsymbol{\theta}(y)$ for all $y \in \overline{\omega}$"; etc.

Theorem 2.3-1. *Let ω be a domain in $\mathbb{R}^2$ and let $\boldsymbol{\theta} \in \mathcal{C}^2(\overline{\omega}; \mathcal{E}^3)$ be an injective mapping such that the two vectors $\boldsymbol{a}_\alpha = \partial_\alpha \boldsymbol{\theta}$ are linearly independent at all points of $\overline{\omega}$.*

(a) *The derivatives of the vectors of the covariant and contravariant bases are given by the* **formulas of Gauß:**

$$\partial_\alpha \boldsymbol{a}_\beta = \Gamma^\sigma_{\alpha\beta}\boldsymbol{a}_\sigma + b_{\alpha\beta}\boldsymbol{a}_3 \quad \text{and} \quad \partial_\alpha \boldsymbol{a}^\beta = -\Gamma^\beta_{\alpha\sigma}\boldsymbol{a}^\sigma + b^\beta_\alpha \boldsymbol{a}^3,$$

and **Weingarten:**

$$\partial_\alpha \boldsymbol{a}_3 = \partial_\alpha \boldsymbol{a}^3 = -b_{\alpha\beta}\boldsymbol{a}^\beta = -b^\sigma_\alpha \boldsymbol{a}_\sigma,$$

where the covariant and mixed components $b_{\alpha\beta}$ and b^β_α of the curvature tensor of S are defined in Thms. 2.2-1 *and* 2.2-2 *and*

$$\Gamma^\sigma_{\alpha\beta} := \boldsymbol{a}^\sigma \cdot \partial_\alpha \boldsymbol{a}_\beta = -\partial_\alpha \boldsymbol{a}^\sigma \cdot \boldsymbol{a}_\beta = \Gamma^\sigma_{\beta\alpha}$$

are the **Christoffel symbols of the surface** S.

(b) *Let there be given a vector field $\eta_i \boldsymbol{a}^i$ with covariant components $\eta_i \in H^1(\omega)$. Then $\eta_i \boldsymbol{a}^i \in \mathbf{H}^1(\omega)$ and the partial derivatives $\partial_\alpha(\eta_i \boldsymbol{a}^i) \in \mathbf{L}^2(\omega)$ are given by*

$$\begin{aligned}\partial_\alpha(\eta_i \boldsymbol{a}^i) &= (\partial_\alpha \eta_\beta - \Gamma^\sigma_{\alpha\beta}\eta_\sigma - b_{\alpha\beta}\eta_3)\boldsymbol{a}^\beta + (\partial_\alpha \eta_3 + b^\beta_\alpha \eta_\beta)\boldsymbol{a}^3 \\ &= (\eta_{\beta|\alpha} - b_{\alpha\beta}\eta_3)\boldsymbol{a}^\beta + (\eta_{3|\alpha} + b^\beta_\alpha \eta_\beta)\boldsymbol{a}^3,\end{aligned}$$

where

$$\eta_{\beta|\alpha} := \partial_\alpha \eta_\beta - \Gamma^\sigma_{\alpha\beta}\eta_\sigma \text{ and } \eta_{3|\alpha} := \partial_\alpha \eta_3$$

denote the **first-order covariant derivatives** *of the vector field $\eta_i \boldsymbol{a}^i$.*

Proof. Fix $\alpha \in \{1, 2\}$ throughout this proof. Since any vector $\boldsymbol{c}$ in the tangent plane can be expanded as $\boldsymbol{c} = (\boldsymbol{c} \cdot \boldsymbol{a}_\beta)\boldsymbol{a}^\beta = (\boldsymbol{c} \cdot \boldsymbol{a}^\sigma)\boldsymbol{a}_\sigma$, since $\partial_\alpha \boldsymbol{a}^3$ is in the tangent plane ($\partial_\alpha \boldsymbol{a}^3 \cdot \boldsymbol{a}^3 = \frac{1}{2}\partial_\alpha(\boldsymbol{a}^3 \cdot \boldsymbol{a}^3) = 0$), and since $\partial_\alpha \boldsymbol{a}^3 \cdot \boldsymbol{a}_\beta = -b_{\alpha\beta}$ (Thm. 2.2-1), it follows that

$$\partial_\alpha \boldsymbol{a}^3 = (\partial_\alpha \boldsymbol{a}^3 \cdot \boldsymbol{a}_\beta)\boldsymbol{a}^\beta = -b_{\alpha\beta}\boldsymbol{a}^\beta,$$

and thus the first formula of Weingarten is established. This formula, together with the definition of the functions b_α^β (Thm. 2.2-2), implies that

$$\partial_\alpha \boldsymbol{a}_3 = (\partial_\alpha \boldsymbol{a}_3 \cdot \boldsymbol{a}^\sigma)\boldsymbol{a}_\sigma = -b_{\alpha\beta}(\boldsymbol{a}^\beta \cdot \boldsymbol{a}^\sigma)\boldsymbol{a}_\sigma = -b_{\alpha\beta}a^{\beta\sigma}\boldsymbol{a}_\sigma = -b_\alpha^\sigma \boldsymbol{a}_\sigma,$$

and thus the second formula of Weingarten is established.

Any vector $\boldsymbol{c}$ can be expanded as $\boldsymbol{c} = (\boldsymbol{c} \cdot \boldsymbol{a}^i)\boldsymbol{a}_i = (\boldsymbol{c} \cdot \boldsymbol{a}_j)\boldsymbol{a}^j$. In particular,

$$\partial_\alpha \boldsymbol{a}_\beta = (\partial_\alpha \boldsymbol{a}_\beta \cdot \boldsymbol{a}^\sigma)\boldsymbol{a}_\sigma + (\partial_\alpha \boldsymbol{a}_\beta \cdot \boldsymbol{a}^3)\boldsymbol{a}_3 = \Gamma_{\alpha\beta}^\sigma \boldsymbol{a}_\sigma + b_{\alpha\beta}\boldsymbol{a}_3,$$

by definition of $\Gamma_{\alpha\beta}^\sigma$ and $b_{\alpha\beta}$. Finally,

$$\partial_\alpha \boldsymbol{a}^\beta = (\partial_\alpha \boldsymbol{a}^\beta \cdot \boldsymbol{a}_\sigma)\boldsymbol{a}^\sigma + (\partial_\alpha \boldsymbol{a}^\beta \cdot \boldsymbol{a}_3)\boldsymbol{a}^3 = -\Gamma_{\alpha\sigma}^\beta \boldsymbol{a}^\sigma + b_\alpha^\beta \boldsymbol{a}^3,$$

since

$$\partial_\alpha \boldsymbol{a}^\beta \cdot \boldsymbol{a}_3 = -\boldsymbol{a}^\beta \cdot \partial_\alpha \boldsymbol{a}_3 = b_\alpha^\sigma \boldsymbol{a}_\sigma \cdot \boldsymbol{a}^\beta = b_\alpha^\beta,$$

by the second formula of Weingarten; the formulas of Gauß are thus established.

That $\eta_i \boldsymbol{a}^i \in \mathbf{H}^1(\omega)$ if $\eta_i \in H^1(\omega)$ is clear since $\boldsymbol{a}^i \in \boldsymbol{C}^1(\overline{\omega})$ if $\boldsymbol{\theta} \in \mathcal{C}^2(\overline{\omega}; \mathcal{E}^3)$. The formulas of Gauß and Weingarten immediately lead to the announced expression of $\partial_\alpha(\eta_i \boldsymbol{a}^i)$. ∎

The definition of the covariant derivatives $\eta_{\alpha|\beta} = \partial_\beta \eta_\alpha - \Gamma_{\alpha\beta}^\sigma \eta_\sigma$ "on a surface in $\mathcal{E}^3$" given in Thm. 2.3-1 is highly reminiscent of the definition of the covariant derivatives $v_{i\|j} = \partial_j v_i - \Gamma_{ij}^p v_p$ "in a three-dimensional manifold in $\mathcal{E}^3$" given in Chap. 1. However, the former are more subtle to apprehend than the latter. To see this, recall that the covariant derivatives $v_{i\|j} = \partial_j v_i - \Gamma_{ij}^p v_p$ may be also defined by the relations (Thm. 1.4-1)

$$v_{i\|j}\boldsymbol{g}^j = \partial_j(v_i \boldsymbol{g}^i).$$

By contrast, *even if only tangential vector fields* $\eta_\alpha \boldsymbol{a}^\alpha$ *on* S *are considered* (i.e., vector fields $\eta_i \boldsymbol{a}^i$ for which $\eta_3 = 0$), *their covariant derivatives* $\eta_{\alpha|\beta} = \partial_\beta \eta_\alpha - \Gamma^\sigma_{\alpha\beta}\eta_\sigma$ *only satisfy the more general relation*

$$\eta_{\alpha|\beta}\boldsymbol{a}^\alpha = \mathbf{P}\left\{\partial_\beta(\eta_\alpha \boldsymbol{a}^\alpha)\right\},$$

where $\mathbf{P}$ denotes the projection operator on the tangent plane in the direction of the normal vector (i.e., $\mathbf{P}(c_i \boldsymbol{a}^i) := c_\alpha \boldsymbol{a}^\alpha$), since

$$\partial_\beta(\eta_\alpha \boldsymbol{a}^\alpha) = \eta_{\alpha|\beta}\boldsymbol{a}^\alpha + b^\alpha_\beta \eta_\alpha \boldsymbol{a}^3$$

for such tangential fields by Thm. 2.3-1. The reason behind this discrepancy is simple: While a "three-dimensional manifold in $\mathbb{R}^3$" has *zero curvature* (Sect. 1.9), a surface has in general a *nonzero curvature*, manifesting itself here by the "extra term" $b^\alpha_\beta \eta_\alpha \boldsymbol{a}^3$. This term vanishes in $\overline{\omega}$ if S is a *portion of a plane*, since in this case $b^\alpha_\beta = b_{\alpha\beta} = 0$. Note that, *again in this case*, the formula giving the partial derivatives in Thm. 2.3-1 (b) reduces to

$$\partial_\alpha(\eta_i \boldsymbol{a}^i) = (\eta_{i|\alpha})\boldsymbol{a}^i.$$

We shall not dwell any further in this section into these aspects, referring instead to the references listed at the beginning of this section for more details. Meanwhile, the covariant derivatives $\eta_{i|\alpha}$ may be simply viewed as convenient *notations*.

Remark. The Christoffel symbols $\Gamma^\sigma_{\alpha\beta} = \boldsymbol{a}^\sigma \cdot \partial_\alpha \boldsymbol{a}_\beta$ "on a surface in $\mathcal{E}^3$" are to be carefully distinguished from the Christoffel symbols $\Gamma^p_{ij} = \boldsymbol{g}^p \cdot \partial_i \boldsymbol{g}_j$ "in a three-dimensional manifold in $\mathcal{E}^3$" introduced in Chap. 1. To avoid any confusion due to the identical notation when $i, j, p \in \{1, 2\}$, specific notations shall always be introduced, should the two kinds of Christoffel symbols coexist in a given argument. ■

2.4. LINEARIZED CHANGE OF METRIC TENSOR ON A SURFACE

We recall that, given an arbitrary displacement field $v_i \boldsymbol{g}^i$ of a three-dimensional manifold $\boldsymbol{\Theta}(\overline{\Omega})$ in $\mathcal{E}^3$ defined by its covariant components $v_i : \overline{\Omega} \to \mathbb{R}$, the covariant components $e_{i\|j}(\boldsymbol{v})$ of the associated linearized change of metric tensor are defined by (see Thm.

1.5-1; also for the notations, not recalled here)

$$e_{i\|j}(\boldsymbol{v}) = \frac{1}{2}\left[g_{ij}(\boldsymbol{v}) - g_{ij}\right]^{lin}.$$

It was then found that

$$e_{i\|j}(\boldsymbol{v}) = \frac{1}{2}(v_{i\|j} + v_{j\|i}) = \frac{1}{2}(\partial_j v_i + \partial_i v_j) - \Gamma_{ij}^p v_p.$$

Since a surface, i.e., a two-dimensional manifold in $\mathcal{E}^3$, also has a metric tensor (Sect. 2.1), it is natural to likewise define the covariant components of a *"linearized change of metric tensor"* associated with any displacement field defined on it; this is the object of this section. Notice the strong analogy between the next theorem and Thm. 1.5-1.

Theorem 2.4-1. *Let* ω *be a domain in* $\mathbb{R}^2$ *and let* $\boldsymbol{\theta} \in \mathcal{C}^2(\overline{\omega}; \mathcal{E}^3)$ *be an injective mapping such that the two vectors* $\boldsymbol{a}_\alpha = \partial_\alpha \boldsymbol{\theta}$ *are linearly independent at all points of* $\overline{\omega}$*, let* $\boldsymbol{a}_3 = \dfrac{\boldsymbol{a}_1 \wedge \boldsymbol{a}_2}{|\boldsymbol{a}_1 \wedge \boldsymbol{a}_2|}$*, and let the vectors* $\boldsymbol{a}^i$ *be defined by* $\boldsymbol{a}^i \cdot \boldsymbol{a}_j = \delta^i_j$*. Given an arbitrary displacement field*

$$\boxed{\tilde{\boldsymbol{\eta}} := \eta_i \boldsymbol{a}^i}$$

of the surface $S = \boldsymbol{\theta}(\overline{\omega})$ *with smooth enough covariant components* $\eta_i : \overline{\omega} \to \mathbb{R}$*, let the* **covariant components** *of the* **linearized change of metric tensor** *associated with this vector field be defined by*

$$\boxed{\gamma_{\alpha\beta}(\boldsymbol{\eta}) := \frac{1}{2}\left[a_{\alpha\beta}(\boldsymbol{\eta}) - a_{\alpha\beta}\right]^{lin},}$$

where $a_{\alpha\beta}$ *and* $a_{\alpha\beta}(\boldsymbol{\eta})$ *are the covariant components of the metric tensor of the surfaces* $\boldsymbol{\theta}(\overline{\omega})$ *and* $(\boldsymbol{\theta} + \eta_i \boldsymbol{a}^i)(\overline{\omega})$ (Fig. 2.3-1), *and* $[\cdots]^{lin}$ *denotes the linear part with respect to* $\boldsymbol{\eta} = (\eta_i)$ *in the expression* $[\cdots]$*. Then*

$$\boxed{\begin{aligned}\gamma_{\alpha\beta}(\boldsymbol{\eta}) &= \frac{1}{2}\left(\partial_\beta \tilde{\boldsymbol{\eta}} \cdot \boldsymbol{a}_\alpha + \partial_\alpha \tilde{\boldsymbol{\eta}} \cdot \boldsymbol{a}_\beta\right) = \gamma_{\beta\alpha}(\boldsymbol{\eta}) \\ &= \frac{1}{2}(\eta_{\alpha|\beta} + \eta_{\beta|\alpha}) - b_{\alpha\beta}\eta_3 \\ &= \frac{1}{2}(\partial_\beta \eta_\alpha + \partial_\alpha \eta_\beta) - \Gamma^\sigma_{\alpha\beta}\eta_\sigma - b_{\alpha\beta}\eta_3,\end{aligned}}$$

where the Christoffel symbols $\Gamma^{\sigma}_{\alpha\beta}$ *and the covariant derivatives* $\eta_{\alpha|\beta}$ *are defined in* Thm. 2.3-1 *and the covariant components* $b_{\alpha\beta}$ *of the curvature tensor of* S *are defined in* Thm. 2.2-1. *Consequently,*

$$\boxed{\eta_\alpha \in H^1(\omega) \text{ and } \eta_3 \in L^2(\omega) \Rightarrow \gamma_{\alpha\beta}(\boldsymbol{\eta}) \in L^2(\omega).}$$

Proof. The covariant components $a_{\alpha\beta}(\boldsymbol{\eta})$ of the metric tensor of the surface $(\boldsymbol{\theta} + \eta_i \boldsymbol{a}^i)(\overline{\omega})$ are by definition (Sect. 2.1) given by

$$a_{\alpha\beta}(\boldsymbol{\eta}) = \partial_\alpha(\boldsymbol{\theta} + \tilde{\boldsymbol{\eta}}) \cdot \partial_\beta(\boldsymbol{\theta} + \tilde{\boldsymbol{\eta}}).$$

Note that *both* surfaces $\boldsymbol{\theta}(\overline{\omega})$ and $(\boldsymbol{\theta} + \eta_i \boldsymbol{a}^i)(\overline{\omega})$ are thus equipped with the *same* curvilinear coordinates y_α. The relations

$$\partial_\alpha(\boldsymbol{\theta} + \tilde{\boldsymbol{\eta}}) = \boldsymbol{a}_\alpha + \partial_\alpha \tilde{\boldsymbol{\eta}}$$

then show that

$$\begin{aligned} a_{\alpha\beta}(\boldsymbol{\eta}) &= (\boldsymbol{a}_\alpha + \partial_\alpha \tilde{\boldsymbol{\eta}}) \cdot (\boldsymbol{a}_\beta + \partial_\beta \tilde{\boldsymbol{\eta}}) \\ &= a_{\alpha\beta} + \partial_\beta \tilde{\boldsymbol{\eta}} \cdot \boldsymbol{a}_\alpha + \partial_\alpha \tilde{\boldsymbol{\eta}} \cdot \boldsymbol{a}_\beta + \partial_\alpha \tilde{\boldsymbol{\eta}} \cdot \partial_\beta \tilde{\boldsymbol{\eta}}, \end{aligned}$$

hence that

$$\gamma_{\alpha\beta}(\boldsymbol{\eta}) = \frac{1}{2}[a_{\alpha\beta}(\boldsymbol{\eta}) - a_{\alpha\beta}]^{lin} = \frac{1}{2}(\partial_\beta \tilde{\boldsymbol{\eta}} \cdot \boldsymbol{a}_\alpha + \partial_\alpha \tilde{\boldsymbol{\eta}} \cdot \boldsymbol{a}_\beta).$$

The other expressions of $\gamma_{\alpha\beta}(\boldsymbol{\eta})$ immediately follow from the expression of $\partial_\alpha \tilde{\boldsymbol{\eta}} = \partial_\alpha(\eta_i \boldsymbol{a}^i)$ given in Thm. 2.3-1 (b). ■

While the expression of $\gamma_{\alpha\beta}(\boldsymbol{\eta})$ in terms of the covariant components η_i of the displacement field is well known, that in terms of $\tilde{\boldsymbol{\eta}} = \eta_i \boldsymbol{a}^i$ seems to be less known; it was recently put to efficient use by Blouza & Le Dret [1994a, 1994b, 1999], who noticed that it has the advantage of still making sense under substantially weaker smoothness assumptions on the mapping $\boldsymbol{\theta}$; see Thms. 2.6-5 and 2.6-6.

A word of caution. *The vector fields*

$$\boldsymbol{\eta} = (\eta_i) \text{ and } \tilde{\boldsymbol{\eta}} = \eta_i \boldsymbol{a}^i,$$

which are both defined on $\overline{\omega}$, *must be carefully distinguished*! While the latter has an *intrinsic character*, the former has not; it only provides a means of recovering the field $\tilde{\boldsymbol{\eta}}$ *via* its *covariant components* η_i.

■

2.5. LINEARIZED CHANGE OF CURVATURE TENSOR ON A SURFACE

As a surface S possesses a curvature tensor in addition to its metric tensor (Sect. 2.2), it is likewise natural to associate a *"linearized change of curvature tensor"* with any displacement field defined on S, through a process akin to that followed in Thm. 2.4-1 for defining the *linearized change of metric tensor.* Together, *these two tensors play a major rôle in the theory of linearly elastic shells,* as it shall be amply demonstrated in the next chapters.

Theorem 2.5-1. *Let ω be a domain in $\mathbb{R}^2$ and let $\boldsymbol{\theta} \in \mathcal{C}^3(\overline{\omega}; \mathcal{E}^3)$ be an injective mapping such that the two vectors $\boldsymbol{a}_\alpha = \partial_\alpha \boldsymbol{\theta}$ are linearly independent at all points of $\overline{\omega}$. Given a displacement field*

$$\boxed{\tilde{\boldsymbol{\eta}} := \eta_i \boldsymbol{a}^i}$$

of the surface $S = \boldsymbol{\theta}(\overline{\omega})$ with smooth enough and "small enough" covariant components $\eta_i : \overline{\omega} \to \mathbb{R}$, let the **covariant components** *of the* **linearized change of curvature tensor** *associated with such a vector field be defined by*

$$\boxed{\rho_{\alpha\beta}(\boldsymbol{\eta}) := [b_{\alpha\beta}(\boldsymbol{\eta}) - b_{\alpha\beta}]^{lin},}$$

where $b_{\alpha\beta}$ and $b_{\alpha\beta}(\boldsymbol{\eta})$ are the covariant components of the curvature tensors of the surfaces $\boldsymbol{\theta}(\overline{\omega})$ and $(\boldsymbol{\theta}+\eta_i\boldsymbol{a}^i)(\overline{\omega})$ (Fig. 2.3-1), *and $[\cdots]^{lin}$ denotes the linear part with respect to $\boldsymbol{\eta} = (\eta_i)$ in the expression $[\cdots]$. Then*

$$\boxed{\begin{aligned}
\rho_{\alpha\beta}(\boldsymbol{\eta}) &= (\partial_{\alpha\beta}\tilde{\boldsymbol{\eta}} - \Gamma^\sigma_{\alpha\beta}\partial_\sigma\tilde{\boldsymbol{\eta}}) \cdot \boldsymbol{a}_3 = \rho_{\beta\alpha}(\boldsymbol{\eta}) \\
&= \eta_{3|\alpha\beta} - b^\sigma_\alpha b_{\sigma\beta}\eta_3 + b^\sigma_\alpha \eta_{\sigma|\beta} + b^\tau_\beta \eta_{\tau|\alpha} + b^\tau_\beta|_\alpha \eta_\tau \\
&= \partial_{\alpha\beta}\eta_3 - \Gamma^\sigma_{\alpha\beta}\partial_\sigma\eta_3 - b^\sigma_\alpha b_{\sigma\beta}\eta_3 \\
&\quad + b^\sigma_\alpha(\partial_\beta\eta_\sigma - \Gamma^\tau_{\beta\sigma}\eta_\tau) + b^\tau_\beta(\partial_\alpha\eta_\tau - \Gamma^\sigma_{\alpha\tau}\eta_\sigma) \\
&\quad + (\partial_\alpha b^\tau_\beta + \Gamma^\tau_{\alpha\sigma} b^\sigma_\beta - \Gamma^\sigma_{\alpha\beta} b^\tau_\sigma)\eta_\tau,
\end{aligned}}$$

where the Christoffel symbols $\Gamma^\sigma_{\alpha\beta}$ *and the covariant derivatives* $\eta_{\alpha|\beta}$ *are defined as in* Thm. 2.3-1,

$$\eta_{3|\alpha\beta} := \partial_{\alpha\beta}\eta_3 - \Gamma^\sigma_{\alpha\beta}\partial_\sigma\eta_3 \text{ and } b^\tau_\beta|_\alpha := \partial_\alpha b^\tau_\beta + \Gamma^\tau_{\alpha\sigma}b^\sigma_\beta - \Gamma^\sigma_{\alpha\beta}b^\tau_\sigma$$

denote respectively a **second-order covariant derivative** *of the vector field* $\eta_i \boldsymbol{a}^i$ *and a* **first-order covariant derivative** *of the curvature tensor of* S*, defined here by means of its mixed components. Consequently,*

$$\eta_\alpha \in H^1(\omega) \text{ and } \eta_3 \in H^2(\omega) \Rightarrow \rho_{\alpha\beta}(\boldsymbol{\eta}) \in L^2(\omega).$$

The covariant derivatives $b^\tau_\beta|_\alpha$ *satisfy the symmetry relations*

$$b^\tau_\beta|_\alpha = b^\tau_\alpha|_\beta.$$

Proof. For convenience, the proof is divided into five parts. In parts (i) and (ii), we establish elementary relations satisfied by the vectors $\boldsymbol{a}_i$ and $\boldsymbol{a}^i$ of the covariant and contravariant bases along S.

(i) *The two vectors* $\boldsymbol{a}_\alpha = \partial_\alpha \boldsymbol{\theta}$ *satisfy* $|\boldsymbol{a}_1 \wedge \boldsymbol{a}_2| = \sqrt{a}$*, where* $a = \det(a_{\alpha\beta})$.

Let $\boldsymbol{a}_3(y) := \dfrac{\boldsymbol{a}_1(y) \wedge \boldsymbol{a}_2(y)}{|\boldsymbol{a}_1(y) \wedge \boldsymbol{a}_2(y)|}$ and $\boldsymbol{\Theta}(y, x_3) := \boldsymbol{\theta}(y) + x_3\boldsymbol{a}_3(y)$ for all $y \in \overline{\omega}$ and $x_3 \in \mathbb{R}$. Using the notations of Sect. 1.2 (see notably Thm. 1.2-1), we have $\det \boldsymbol{\nabla\Theta} = (\boldsymbol{g}_1 \wedge \boldsymbol{g}_2) \cdot \boldsymbol{g}_3$, where

$$\boldsymbol{g}_\alpha := \partial_\alpha \boldsymbol{\Theta} = \boldsymbol{a}_\alpha + x_3 \partial_\alpha \boldsymbol{a}_3 \text{ and } \boldsymbol{g}_3 := \partial_3 \boldsymbol{\Theta} = \boldsymbol{a}_3.$$

Assume for instance that $\det \boldsymbol{\nabla\Theta} > 0$ for $|x_3|$ small enough, so that $\sqrt{g} = \det \boldsymbol{\nabla\Theta}$ for $|x_3|$ small enough, where $g := \det(g_{ij})$ and $g_{ij} := \boldsymbol{g}_i \cdot \boldsymbol{g}_j$. Hence

$$\sqrt{g}\big|_{x_3=0} = \det \boldsymbol{\nabla\Theta}\big|_{x_3=0} = (\boldsymbol{g}_1 \wedge \boldsymbol{g}_2)\cdot\boldsymbol{g}_3|_{x_3=0} = (\boldsymbol{a}_1 \wedge \boldsymbol{a}_2)\cdot\boldsymbol{a}_3 = |\boldsymbol{a}_1 \wedge \boldsymbol{a}_2|$$

on the one hand. Since

$$g|_{x_3=0} = \det \begin{pmatrix} \boldsymbol{g}_1\cdot\boldsymbol{g}_1 & \boldsymbol{g}_1\cdot\boldsymbol{g}_2 & \boldsymbol{g}_1\cdot\boldsymbol{g}_3 \\ \boldsymbol{g}_2\cdot\boldsymbol{g}_1 & \boldsymbol{g}_2\cdot\boldsymbol{g}_2 & \boldsymbol{g}_2\cdot\boldsymbol{g}_3 \\ \boldsymbol{g}_3\cdot\boldsymbol{g}_1 & \boldsymbol{g}_3\cdot\boldsymbol{g}_2 & \boldsymbol{g}_3\cdot\boldsymbol{g}_3 \end{pmatrix} \Big|_{x_3=0}$$

$$= \det \begin{pmatrix} \boldsymbol{a}_1\cdot\boldsymbol{a}_1 & \boldsymbol{a}_1\cdot\boldsymbol{a}_2 & 0 \\ \boldsymbol{a}_2\cdot\boldsymbol{a}_1 & \boldsymbol{a}_2\cdot\boldsymbol{a}_2 & 0 \\ 0 & 0 & 1 \end{pmatrix}$$

$$= \det (a_{\alpha\beta}) = a$$

on the other, the assertion is proved. The case where $\det \boldsymbol{\nabla\Theta} < 0$ for $|x_3|$ small enough is treated analogously.

(ii) *The vectors $\boldsymbol{a}_i$ and $\boldsymbol{a}^\alpha$ are related by $\boldsymbol{a}_1 \wedge \boldsymbol{a}_3 = -\sqrt{a}\boldsymbol{a}^2$ and $\boldsymbol{a}_3 \wedge \boldsymbol{a}_2 = -\sqrt{a}\boldsymbol{a}^1$.*

To prove that two vectors $\boldsymbol{c}$ and $\boldsymbol{d}$ coincide, it suffices to prove that $\boldsymbol{c} \cdot \boldsymbol{a}_i = \boldsymbol{d} \cdot \boldsymbol{a}_i$ for $i \in \{1, 2, 3\}$. In the present case,

$$(\boldsymbol{a}_1 \wedge \boldsymbol{a}_3) \cdot \boldsymbol{a}_1 = 0 \text{ and } (\boldsymbol{a}_1 \wedge \boldsymbol{a}_3) \cdot \boldsymbol{a}_3 = 0,$$

$$(\boldsymbol{a}_1 \wedge \boldsymbol{a}_3) \cdot \boldsymbol{a}_2 = -(\boldsymbol{a}_1 \wedge \boldsymbol{a}_2) \cdot \boldsymbol{a}_3 = -\sqrt{a},$$

since $\sqrt{a}\boldsymbol{a}_3 = \boldsymbol{a}_1 \wedge \boldsymbol{a}_2$ by (i), on the one hand; on the other,

$$-\sqrt{a}\boldsymbol{a}^2 \cdot \boldsymbol{a}_1 = -\sqrt{a}\boldsymbol{a}^2 \cdot \boldsymbol{a}_3 = 0 \text{ and } -\sqrt{a}\boldsymbol{a}^2 \cdot \boldsymbol{a}_2 = -\sqrt{a},$$

since $\boldsymbol{a}^i \cdot \boldsymbol{a}_j = \delta^i_j$. Hence $\boldsymbol{a}_1 \wedge \boldsymbol{a}_3 = -\sqrt{a}\boldsymbol{a}^2$. The other relation is similarly established.

(iii) *The covariant components $b_{\alpha\beta}(\boldsymbol{\eta})$ and $b_{\alpha\beta}$ satisfy*

$$b_{\alpha\beta}(\boldsymbol{\eta}) = b_{\alpha\beta} + (\partial_{\alpha\beta}\tilde{\boldsymbol{\eta}} - \Gamma^\sigma_{\alpha\beta}\partial_\sigma\tilde{\boldsymbol{\eta}}) \cdot \boldsymbol{a}_3 + \text{h.o.t.},$$

where, here or in any other similar expansion found subsequently in this part of the proof, "h.o.t." *stands for "terms of order higher than linear with respect to $\boldsymbol{\eta} = (\eta_i)$". Consequently,*

$$\rho_{\alpha\beta}(\boldsymbol{\eta}) := [b_{\alpha\beta}(\boldsymbol{\eta}) - b_{\alpha\beta}]^{lin} = \left(\partial_{\alpha\beta}\tilde{\boldsymbol{\eta}} - \Gamma^\sigma_{\alpha\beta}\partial_\sigma\tilde{\boldsymbol{\eta}}\right) \cdot \boldsymbol{a}_3 = \rho_{\beta\alpha}(\boldsymbol{\eta}).$$

Since the vectors $\boldsymbol{a}_\alpha = \partial_\alpha\boldsymbol{\theta}$ are linearly independent in $\overline{\omega}$ and the fields $\boldsymbol{\eta} = (\eta_i)$ are smooth enough by assumption, *the vectors*

$\partial_\alpha(\boldsymbol{\theta} + \eta_i \boldsymbol{a}^i)$ *are also linearly independent in* $\overline{\omega}$ *provided the fields* $\boldsymbol{\eta}$ *are "small enough"*, e.g., with respect to the norm of the space $\mathcal{C}^1(\overline{\omega}; \mathbb{R}^3)$. The following computations are therefore licit as they apply to a linearization around $\boldsymbol{\eta} = \mathbf{0}$.

Let

$$\boldsymbol{a}_\alpha(\boldsymbol{\eta}) := \partial_\alpha(\boldsymbol{\theta} + \tilde{\boldsymbol{\eta}}) = \boldsymbol{a}_\alpha + \partial_\alpha \tilde{\boldsymbol{\eta}} \text{ and } \boldsymbol{a}_3(\boldsymbol{\eta}) := \frac{\boldsymbol{a}_1(\boldsymbol{\eta}) \wedge \boldsymbol{a}_2(\boldsymbol{\eta})}{\sqrt{a(\boldsymbol{\eta})}},$$

where (cf. (i))

$$a(\boldsymbol{\eta}) := \det(a_{\alpha\beta}(\boldsymbol{\eta})) \text{ and } a_{\alpha\beta}(\boldsymbol{\eta}) := a_\alpha(\boldsymbol{\eta}) \cdot \boldsymbol{a}_\beta(\eta).$$

Then

$$\begin{aligned} b_{\alpha\beta}(\boldsymbol{\eta}) &= \partial_\alpha \boldsymbol{a}_\beta(\boldsymbol{\eta}) \cdot \boldsymbol{a}_3(\boldsymbol{\eta}) \\ &= \frac{1}{\sqrt{a(\boldsymbol{\eta})}}(\partial_\alpha \boldsymbol{a}_\beta + \partial_{\alpha\beta}\tilde{\boldsymbol{\eta}}) \cdot (\boldsymbol{a}_1 \wedge \boldsymbol{a}_2 + \boldsymbol{a}_1 \wedge \partial_2 \tilde{\boldsymbol{\eta}} + \partial_1 \tilde{\boldsymbol{\eta}} \wedge \boldsymbol{a}_2 + \text{h.o.t.}) \\ &= \frac{1}{\sqrt{a(\boldsymbol{\eta})}} \left\{ \sqrt{a}(b_{\alpha\beta} + \partial_{\alpha\beta}\tilde{\boldsymbol{\eta}} \cdot \boldsymbol{a}_3) \right\} \\ &\quad + \frac{1}{\sqrt{a(\boldsymbol{\eta})}} \left\{ (\Gamma^\sigma_{\alpha\beta} \boldsymbol{a}_\sigma + b_{\alpha\beta}\boldsymbol{a}_3) \cdot (\boldsymbol{a}_1 \wedge \partial_2 \tilde{\boldsymbol{\eta}} + \partial_1 \tilde{\boldsymbol{\eta}} \wedge \boldsymbol{a}_2) + \text{h.o.t.} \right\}, \end{aligned}$$

since $b_{\alpha\beta} = \partial_\alpha \boldsymbol{a}_\beta \cdot \boldsymbol{a}_3$ and $\partial_\alpha \boldsymbol{a}_\beta = \Gamma^\sigma_{\alpha\beta}\boldsymbol{a}_\sigma + b_{\alpha\beta}\boldsymbol{a}_3$ by the formula of Gauß (Thm. 2.3-1 (a)). Next,

$$\begin{aligned} &(\Gamma^\sigma_{\alpha\beta}\boldsymbol{a}_\sigma + b_{\alpha\beta}\boldsymbol{a}_3) \cdot (\boldsymbol{a}_1 \wedge \partial_2 \tilde{\boldsymbol{\eta}}) \\ &\qquad = \Gamma^2_{\alpha\beta}\boldsymbol{a}_2 \cdot (\boldsymbol{a}_1 \wedge \partial_2 \tilde{\boldsymbol{\eta}}) - b_{\alpha\beta}\partial_2 \tilde{\boldsymbol{\eta}} \cdot (\boldsymbol{a}_1 \wedge \boldsymbol{a}_3) \\ &\qquad = \sqrt{a}\left(-\Gamma^2_{\alpha\beta}\partial_2 \tilde{\boldsymbol{\eta}} \cdot \boldsymbol{a}_3 + b_{\alpha\beta}\partial_2 \tilde{\boldsymbol{\eta}} \cdot \boldsymbol{a}^2\right), \end{aligned}$$

since, by (ii), $\boldsymbol{a}_2 \cdot (\boldsymbol{a}_1 \wedge \partial_2 \tilde{\boldsymbol{\eta}}) = -\partial_2 \tilde{\boldsymbol{\eta}} \cdot (\boldsymbol{a}_1 \wedge \boldsymbol{a}_2) = -\sqrt{a}\partial_2 \tilde{\boldsymbol{\eta}} \cdot \boldsymbol{a}_3$ and $\boldsymbol{a}_1 \wedge \boldsymbol{a}_3 = -\sqrt{a}\boldsymbol{a}^2$; likewise,

$$(\Gamma^\sigma_{\alpha\beta}\boldsymbol{a}_\sigma + b_{\alpha\beta}\boldsymbol{a}_3) \cdot (\partial_1 \tilde{\boldsymbol{\eta}} \wedge \boldsymbol{a}_2) = \sqrt{a}(-\Gamma^1_{\alpha\beta}\partial_1 \tilde{\boldsymbol{\eta}} \cdot \boldsymbol{a}_3 + b_{\alpha\beta}\partial_1 \boldsymbol{\eta} \cdot \boldsymbol{a}^1).$$

Consequently,

$$b_{\alpha\beta}(\boldsymbol{\eta}) = \sqrt{\frac{a}{a(\boldsymbol{\eta})}} \left\{ b_{\alpha\beta}(1 + \partial_\sigma \tilde{\boldsymbol{\eta}} \cdot \boldsymbol{a}^\sigma) + (\partial_{\alpha\beta}\tilde{\boldsymbol{\eta}} - \Gamma^\sigma_{\alpha\beta}\partial_\sigma \tilde{\boldsymbol{\eta}}) \cdot \boldsymbol{a}_3 + \text{h.o.t.} \right\}.$$

There remains to find the linear term with respect to $\boldsymbol{\eta} = (\eta_i)$ in the expansion $\frac{1}{\sqrt{a(\boldsymbol{\eta})}} = \frac{1}{\sqrt{a}}(1 + \cdots)$. To this end, we use the formula (see, e.g., Vol. I, Sect. 1.2)

$$\det(\mathbf{A} + \mathbf{H}) = (\det \mathbf{A})(1 + \operatorname{tr} \mathbf{A}^{-1}\mathbf{H} + o(\mathbf{H})),$$

with $\mathbf{A} := (a_{\alpha\beta})$ and $\mathbf{A} + \mathbf{H} := (a_{\alpha\beta}(\boldsymbol{\eta}))$. In this case,

$$\mathbf{H} = (\partial_\beta \tilde{\boldsymbol{\eta}} \cdot \boldsymbol{a}_\alpha + \partial_\alpha \tilde{\boldsymbol{\eta}} \cdot \boldsymbol{a}_\beta + \text{h.o.t.}),$$

since $[a_{\alpha\beta}(\boldsymbol{\eta}) - a_{\alpha\beta}]^{lin} = \partial_\beta \tilde{\boldsymbol{\eta}} \cdot \boldsymbol{a}_\alpha + \partial_\alpha \tilde{\boldsymbol{\eta}} \cdot \boldsymbol{a}_\beta$ (Thm. 2.4-1). Therefore,

$$a(\boldsymbol{\eta}) = \det(a_{\alpha\beta}(\boldsymbol{\eta})) = \det(a_{\alpha\beta})(1 + 2\partial_\alpha \tilde{\boldsymbol{\eta}} \cdot \boldsymbol{a}^\alpha + \text{h.o.t.}),$$

since $\mathbf{A}^{-1} = (a^{\alpha\beta})$; consequently,

$$\frac{1}{\sqrt{a(\boldsymbol{\eta})}} = \frac{1}{\sqrt{a}}(1 - \partial_\alpha \tilde{\boldsymbol{\eta}} \cdot \boldsymbol{a}^\alpha + \text{h.o.t.}).$$

Noting that there are no linear terms with respect to $\boldsymbol{\eta} = (\eta_i)$ in the product $(1 - \partial_\alpha \tilde{\boldsymbol{\eta}} \cdot \boldsymbol{a}^\alpha)(1 + \partial_\sigma \tilde{\boldsymbol{\eta}} \cdot \boldsymbol{a}^\sigma)$, we find the announced expansion, viz.,

$$b_{\alpha\beta}(\boldsymbol{\eta}) = b_{\alpha\beta} + (\partial_{\alpha\beta} \tilde{\boldsymbol{\eta}} - \Gamma^\sigma_{\alpha\beta} \partial_\sigma \tilde{\boldsymbol{\eta}}) \cdot \boldsymbol{a}_3 + \text{h.o.t.}$$

(iv) *The components $\rho_{\alpha\beta}(\boldsymbol{\eta})$ can be also written as*

$$\rho_{\alpha\beta}(\boldsymbol{\eta}) = \eta_{3|\alpha\beta} - b^\sigma_\alpha b_{\sigma\beta} \eta_3 + b^\sigma_\alpha \eta_{\sigma|\beta} + b^\tau_\beta \eta_{\tau|\alpha} + b^\tau_\beta|_\alpha \eta_\tau,$$

where the covariant derivatives $\eta_{3|\alpha\beta}$ and $b^\tau_\beta|_\alpha$ are defined in the statement of the theorem.

By Thm. 2.3-1 (b),

$$\partial_\sigma \tilde{\boldsymbol{\eta}} = (\partial_\sigma \eta_\beta - \Gamma^\tau_{\sigma\beta} \eta_\tau - b_{\sigma\beta} \eta_3) \boldsymbol{a}^\beta + (\partial_\sigma \eta_3 + b^\tau_\sigma \eta_\tau) \boldsymbol{a}^3.$$

Hence

$$-\Gamma^\sigma_{\alpha\beta} \partial_\sigma \tilde{\boldsymbol{\eta}} \cdot \boldsymbol{a}_3 = -\Gamma^\sigma_{\alpha\beta}(\partial_\sigma \eta_3 + b^\tau_\sigma \eta_\tau),$$

since $\boldsymbol{a}^i \cdot \boldsymbol{a}_3 = \delta^i_3$. Again by Thm. 2.3-1 (b),

$$\begin{aligned}
\partial_{\alpha\beta}\tilde{\boldsymbol{\eta}} \cdot \boldsymbol{a}_3 &= \partial_\alpha \left\{ (\partial_\beta \eta_\sigma - \Gamma^\tau_{\beta\sigma}\eta_\tau - b_{\beta\sigma}\eta_3)\boldsymbol{a}^\sigma \right. \\
&\quad \left. +(\partial_\beta \eta_3 + b^\tau_\beta \eta_\tau)\boldsymbol{a}^3 \right\} \cdot \boldsymbol{a}_3 \\
&= (\partial_\beta \eta_\sigma - \Gamma^\tau_{\beta\sigma}\eta_\tau - b_{\beta\sigma}\eta_3)\partial_\alpha \boldsymbol{a}^\sigma \cdot \boldsymbol{a}_3 \\
&\quad +(\partial_{\alpha\beta}\eta_3 + (\partial_\alpha b^\tau_\beta)\eta_\tau + b^\tau_\beta \partial_\alpha \eta_\tau)\boldsymbol{a}^3 \cdot \boldsymbol{a}_3 \\
&\quad +(\partial_\beta \eta_3 + b^\tau_\beta \eta_\tau)\partial_\alpha \boldsymbol{a}^3 \cdot \boldsymbol{a}_3 \\
&= b^\sigma_\alpha(\partial_\beta \eta_\sigma - \Gamma^\tau_{\beta\sigma}\eta_\tau) - b^\sigma_\alpha b_{\sigma\beta}\eta_3 + \partial_{\alpha\beta}\eta_3 \\
&\quad +(\partial_\alpha b^\tau_\beta)\eta_\tau + b^\tau_\beta \partial_\alpha \eta_\tau,
\end{aligned}$$

since

$$\partial_\alpha \boldsymbol{a}^\sigma \cdot \boldsymbol{a}_3 = (-\Gamma^\sigma_{\alpha\tau}\boldsymbol{a}^\tau + b^\sigma_\alpha \boldsymbol{a}^3) \cdot \boldsymbol{a}_3 = b^\sigma_\alpha,$$
$$\partial_\alpha \boldsymbol{a}^3 \cdot \boldsymbol{a}_3 = -b_{\alpha\sigma}\boldsymbol{a}^\sigma \cdot \boldsymbol{a}_3 = 0,$$

by the formulas of Gauß and Weingarten (Thm. 2.3-1 (a)). We thus obtain

$$\begin{aligned}
\rho_{\alpha\beta}(\boldsymbol{\eta}) &= (\partial_{\alpha\beta}\tilde{\boldsymbol{\eta}} - \Gamma^\sigma_{\alpha\beta}\partial_\sigma \tilde{\boldsymbol{\eta}}) \cdot \boldsymbol{a}_3 \\
&= b^\sigma_\alpha(\partial_\beta \eta_\sigma - \Gamma^\tau_{\beta\sigma}\eta_\tau) - b^\sigma_\alpha b_{\sigma\beta}\eta_3 + \partial_{\alpha\beta}\eta_3 + (\partial_\alpha b^\tau_\beta)\eta_\tau + b^\tau_\beta \partial_\alpha \eta_\tau \\
&\quad -\Gamma^\sigma_{\alpha\beta}(\partial_\sigma \eta_3 + b^\tau_\sigma \eta_\tau).
\end{aligned}$$

While this relation seemingly involves only the covariant derivatives $\eta_{3|\alpha\beta}$ and $\eta_{\sigma|\beta}$, it may be easily rewritten so as to involve in addition the covariant derivatives $\eta_{\tau|\alpha}$ and $b^\tau_{\beta|\alpha}$. The stratagem simply consists in using the relation $\Gamma^\tau_{\alpha\sigma}b^\sigma_\beta \eta_\tau - \Gamma^\sigma_{\alpha\tau}b^\tau_\beta \eta_\sigma = 0$! This gives

$$\begin{aligned}
\rho_{\alpha\beta}(\boldsymbol{\eta}) &= (\partial_{\alpha\beta}\eta_3 - \Gamma^\sigma_{\alpha\beta}\partial_\tau \eta_3) - b^\sigma_\alpha b_{\sigma\beta}\eta_3 \\
&\quad +b^\sigma_\alpha(\partial_\beta \eta_\sigma - \Gamma^\tau_{\beta\sigma}\eta_\tau) + b^\tau_\beta(\partial_\alpha \eta_\tau - \Gamma^\sigma_{\alpha\tau}\eta_\tau) \\
&\quad +(\partial_\alpha b^\tau_\beta + \Gamma^\tau_{\alpha\sigma}b^\sigma_\beta b^\sigma_\beta - \Gamma^\sigma_{\alpha\beta}b^\tau_\sigma)\eta_\tau.
\end{aligned}$$

(v) *The covariant derivatives* $b^\tau_\beta|_\alpha$ *are symmetric with respect to the indices* α *and* β.

Thanks to the formulas of Gauß and Weingarten (Thm. 2.3-1 (a)), we can write

$$\begin{aligned}
\mathbf{0} &= \partial_{\alpha\beta}\boldsymbol{a}^\tau - \partial_{\beta\alpha}\boldsymbol{a}^\tau = \partial_\alpha\left(-\Gamma^\tau_{\beta\sigma}\boldsymbol{a}^\sigma + b^\tau_\beta \boldsymbol{a}^3\right) - \partial_\beta\left(-\Gamma^\tau_{\alpha\sigma}\boldsymbol{a}^\sigma + b^\tau_\alpha \boldsymbol{a}^3\right) \\
&= -(\partial_\alpha \Gamma^\tau_{\beta\sigma})\boldsymbol{a}^\sigma + \Gamma^\tau_{\beta\sigma}\Gamma^\sigma_{\alpha\nu}\boldsymbol{a}^\nu - \Gamma^\tau_{\beta\sigma}b^\sigma_\alpha \boldsymbol{a}^3 + (\partial_\alpha b^\tau_\beta)\boldsymbol{a}^3 - b^\tau_\beta b_{\alpha\sigma}\boldsymbol{a}^\sigma
\end{aligned}$$

$$+(\partial_\beta \Gamma^\tau_{\alpha\sigma})\boldsymbol{a}^\sigma - \Gamma^\tau_{\alpha\sigma}\Gamma^\sigma_{\beta\mu}\boldsymbol{a}^\mu + \Gamma^\tau_{\alpha\sigma} b^\sigma_\beta \boldsymbol{a}^3 - (\partial_\beta b^\tau_\alpha)\boldsymbol{a}^3 + b^\tau_\alpha b_{\beta\sigma}\boldsymbol{a}^\sigma.$$

Consequently,

$$0 = (\partial_{\alpha\beta}\boldsymbol{a}^\tau - \partial_{\beta\alpha}\boldsymbol{a}^\tau) \cdot \boldsymbol{a}^3 = \partial_\alpha b^\tau_\beta - \partial_\beta b^\tau_\alpha + \Gamma^\tau_{\alpha\sigma} b^\sigma_\beta - \Gamma^\tau_{\beta\sigma} b^\sigma_\alpha,$$

on the one hand. On the other, we immediately infer from the definition of the covariant derivatives $b^\tau_\beta|_\alpha$ that we also have

$$b^\tau_\beta|_\alpha - b^\tau_\alpha|_\beta = \partial_\alpha b^\tau_\beta - \partial_\beta b^\tau_\alpha + \Gamma^\tau_{\alpha\sigma} b^\sigma_\beta - \Gamma^\tau_{\beta\sigma} b^\sigma_\alpha,$$

and thus the proof is complete. ∎

Remarks. (1) Covariant derivatives $b_{\alpha\beta|\sigma}$ can be likewise defined when the curvature tensor is defined by means of its covariant components $b_{\alpha\beta}$; see Ex. 2.7.

(2) The functions $c_{\alpha\beta} := b^\sigma_\alpha b_{\sigma\beta} = c_{\beta\alpha}$ appearing in the expression of $\rho_{\alpha\beta}(\boldsymbol{\eta})$ are the covariant components of the *third fundamental form* of S. For details, see, e.g., Stoker [1969, p. 98] or Klingenberg [1973, p. 48].

(3) While the functions $b_{\alpha\beta}(\boldsymbol{\eta})$ are not always well defined (the vectors $a_\alpha(\boldsymbol{\eta})$ must be linearly independent in $\overline{\omega}$), the functions $\rho_{\alpha\beta}(\boldsymbol{\eta})$ are *always* well defined. ∎

While the expression of the components $\rho_{\alpha\beta}(\boldsymbol{\eta})$ in terms of the covariant components η_i of the displacement field is fairly complicated but well known (see, e.g., Koiter [1970]), that in terms of $\tilde{\boldsymbol{\eta}} = \eta_i \boldsymbol{a}^i$ is *remarkably simple* but seems to be new. As its "linearized change of metric counterpart" (Thm. 2.4-1), it was efficiently put to use by Blouza & Le Dret [1994a, 1994b, 1999], who showed that its principal merit is to afford the definition of the components $\rho_{\alpha\beta}(\boldsymbol{\eta})$ under substantially weaker regularity assumptions on the mapping $\boldsymbol{\theta}$; see Thms. 2.6-5 and 2.6-6.

A word of caution. As already observed after Thm. 2.4-1, *the vector fields*

$$\boldsymbol{\eta} = (\eta_i) \text{ and } \tilde{\boldsymbol{\eta}} = \eta_i \boldsymbol{a}^i,$$

which are both defined on $\overline{\omega}$, must be carefully distinguished! ∎

Remarks. (1) The symmetry $\rho_{\alpha\beta}(\boldsymbol{\eta}) = \rho_{\beta\alpha}(\boldsymbol{\eta})$ is an immediate consequence of the expression $\rho_{\alpha\beta}(\boldsymbol{\eta}) = (\partial_{\alpha\beta}\tilde{\boldsymbol{\eta}} - \Gamma^{\sigma}_{\alpha\beta}\partial_\tau\tilde{\boldsymbol{\eta}}) \cdot \boldsymbol{a}_3$ found there. By contrast, deriving the same symmetry from the other expression of $\rho_{\alpha\beta}(\boldsymbol{\eta})$ requires proving first that the covariant derivatives $b^{\sigma}_{\beta}|_{\alpha}$ are themselves symmetric with respect to the indices α and β (cf. part (v) of the proof of Thm. 2.5-1). While a proof of the symmetry $b^{\tau}_{\beta}|_{\alpha} = b^{\tau}_{\alpha}|_{\beta}$ is thus not essential here, it shall definitely be later in the proof of Thm. 2.6-2.

(2) We assume that $\boldsymbol{\theta} \in \mathcal{C}^3(\overline{\omega}; \mathcal{E}^3)$ in Thm. 2.5-1 (while we only assumed $\boldsymbol{\theta} \in \mathcal{C}^2(\overline{\omega}; \mathcal{E}^3)$ in Thm. 2.4-1) in order that $\rho_{\alpha\beta}(\boldsymbol{\eta}) \in L^2(\omega)$ if $\boldsymbol{\eta} \in H^1(\omega) \times H^1(\omega) \times H^2(\omega)$: The culprits are the functions $b^{\tau}_{\beta}|_{\alpha}$ appearing in the functions $\rho_{\alpha\beta}(\boldsymbol{\eta})$. As already noted, Blouza & Le Dret [1994a, 1994b, 1999] have shown that such regularity assumptions on $\boldsymbol{\theta}$ can be weakened if only the expressions of $\gamma_{\alpha\beta}(\boldsymbol{\eta})$ and $\rho_{\alpha\beta}(\boldsymbol{\eta})$ in terms of the field $\tilde{\boldsymbol{\eta}}$ are considered. ■

2.6. INEQUALITIES OF KORN'S TYPE AND INFINITESIMAL RIGID DISPLACEMENT LEMMA ON A GENERAL SURFACE

The *two-dimensional Koiter equations for a linearly elastic shell*, so named after Koiter [1970], take the following form: The unknowns are the covariant components $\zeta^{\varepsilon}_{i,K} : \overline{\omega} \to \mathbb{R}$ of the displacement field $\zeta^{\varepsilon}_{i,K}\boldsymbol{a}^i$ of the middle surface $S = \boldsymbol{\theta}(\overline{\omega})$ of the shell and $\boldsymbol{\zeta}^{\varepsilon}_K := (\zeta^{\varepsilon}_{i,K})$ satisfies (∂_ν denotes the outer normal derivative operator along the boundary of ω):

$$\boldsymbol{\zeta}^{\varepsilon}_K \in \mathbf{V}_K(\omega) := \big\{\boldsymbol{\eta} = (\eta_i) \in H^1(\omega) \times H^1(\omega) \times H^2(\omega); \\ \eta_i = \partial_\nu \eta_3 = 0 \text{ on } \gamma_0\big\},$$

$$\int_\omega \Big\{\varepsilon a^{\alpha\beta\sigma\tau,\varepsilon}\gamma_{\sigma\tau}(\boldsymbol{\zeta}^{\varepsilon}_K)\gamma_{\alpha\beta}(\boldsymbol{\eta}) + \frac{\varepsilon^3}{3} a^{\alpha\beta\sigma\tau,\varepsilon}\rho_{\sigma\tau}(\boldsymbol{\zeta}^{\varepsilon}_K)\rho_{\alpha\beta}(\boldsymbol{\eta})\Big\}\sqrt{a}\, dy \\ = \int_\omega p^{i,\varepsilon}\eta_i\sqrt{a}\, dy \text{ for all } \boldsymbol{\eta} = (\eta_i) \in \mathbf{V}_K(\omega),$$

where γ_0 is a subset of $\partial\omega$ with *length* $\gamma_0 > 0$, $2\varepsilon > 0$ is the *thickness* of the shell,

$$a^{\alpha\beta\sigma\tau,\,\varepsilon} := \frac{4\lambda^\varepsilon \mu^\varepsilon}{\lambda^\varepsilon + 2\mu^\varepsilon} a^{\alpha\beta} a^{\sigma\tau} + 2\mu^\varepsilon (a^{\alpha\sigma} a^{\beta\tau} + a^{\alpha\tau} a^{\beta\sigma})$$

denote the contravariant components of the *two-dimensional elasticity tensor of the shell* (λ^ε and μ^ε are the Lamé constants of the elastic material constituting the shell), $\gamma_{\alpha\beta}(\boldsymbol{\eta})$ and $\rho_{\alpha\beta}(\boldsymbol{\eta})$ denote the covariant components of the *linearized change of metric* and *change of curvature tensors* associated with a displacement field $\eta_i \boldsymbol{a}^i$ of S (Sects. 2.4 and 2.5; their definitions are recalled in Thm. 2.6-1 below), and the given functions $p^{i,\,\varepsilon} \in L^2(\omega)$ account for the applied forces; finally, the boundary conditions $\eta_i = \partial_\nu \eta_3 = 0$ on γ_0 express that the shell is *clamped* along the portion $\boldsymbol{\theta}(\gamma_0)$ of its middle surface (other boundary conditions are possible). These equations will be fully justified from three-dimensional linearized elasticity in Chap. 7, where various refinements and comments are also to be found.

We shall later establish (Thm. 3.3-2) that there exists a constant $c_e = c_e(\omega, \boldsymbol{\theta}, \mu^\varepsilon) > 0$ such that

$$\sum_{\alpha,\beta} |t_{\alpha\beta}|^2 \le c_e a^{\alpha\beta\sigma\tau,\,\varepsilon}(y) t_{\sigma\tau} t_{\alpha\beta}$$

for all $y \in \overline{\omega}$ and all symmetric matrices $(t_{\alpha\beta})$. Establishing the existence and uniqueness of a solution to this variational problem by means of the *Lax-Milgram lemma* thus amounts to establishing the existence of a constant c such that

$$\Big\{\sum_\alpha \|\eta_\alpha\|^2_{1,\,\omega} + \|\eta_3\|^2_{2,\,\omega}\Big\}^{1/2}$$
$$\le c\Big\{\sum_{\alpha,\beta} |\gamma_{\alpha\beta}(\boldsymbol{\eta})|^2_{0,\,\omega} + \sum_{\alpha,\beta} |\rho_{\alpha\beta}(\boldsymbol{\eta})|^2_{0,\,\omega}\Big\}^{1/2} \text{ for all } \boldsymbol{\eta} \in \mathbf{V}_K(\omega).$$

The objective of this section consists in showing that such *an "inequality of Korn's type" indeed holds on a general surface* (Thm. 2.6-4). As is readily checked, the *same* inequality of Korn's type, together with the *Lax-Milgram lemma*, also provides an existence and uniqueness theorem for the *two-dimensional equations of a linearly elastic flexural shell.* These equations will be fully justified in Chap. 6

through an asymptotic analysis of the three-dimensional solutions, under the assumption that the space

$$\boxed{\mathbf{V}_F(\omega) := \{\boldsymbol{\eta} \in \mathbf{V}_K(\omega);\ \gamma_{\alpha\beta}(\boldsymbol{\eta}) = 0 \text{ in } \omega\}}$$

does not reduce to $\{\mathbf{0}\}$. The unknowns are again the covariant components, now denoted $\zeta_i^\varepsilon : \overline{\omega} \to \mathbb{R}$, of the displacement field $\zeta_i^\varepsilon \boldsymbol{a}^i$ of the middle surface S of the shell and $\boldsymbol{\zeta}^\varepsilon := (\zeta_i^\varepsilon)$ satisfies:

$$\boldsymbol{\zeta}^\varepsilon \in \mathbf{V}_F(\omega),$$

$$\frac{\varepsilon^3}{3}\int_\omega a^{\alpha\beta\sigma\tau,\varepsilon}\rho_{\sigma\tau}(\boldsymbol{\zeta}^\varepsilon)\rho_{\alpha\beta}(\boldsymbol{\eta})\sqrt{a}\,dy = \int_\omega p^{i,\varepsilon}\eta_i\sqrt{a}\,dy$$

$$\text{for all } \boldsymbol{\eta} = (\eta_i) \in \mathbf{V}_F(\omega).$$

In addition, such an inequality of Korn's type presents a *mathematical* interest *per se*: In Sect. 1.7, we established "three-dimensional" Korn inequalities, first without (Thm. 1.7-2), then with (Thm. 1.7-4), boundary conditions (the second one depending on a "three-dimensional" infinitesimal rigid displacement lemma; cf. Thm. 1.7-3): Both inequalities involved the covariant components $e_{i\|j}(\boldsymbol{v})$ of the *"three-dimensional" linearized change of metric tensor.*

But while only *one* tensor, the metric tensor, is attached to a *three-dimensional manifold in* $\mathcal{E}^3$, *two* tensors, the metric and curvature tensors, are attached to a *surface in* $\mathcal{E}^3$ (Sects. 2.1 and 2.2). It is thus natural to likewise establish *inequalities of Korn's type "on a surface"*, first without (Thm. 2.6-1), then with (Thm. 2.6-4) boundary conditions (the second one again depending on an *infinitesimal rigid displacement lemma "on a surface"*; cf. Thm. 2.6-3), such inequalities now involving the covariant components $\gamma_{\alpha\beta}(\boldsymbol{\eta})$ and $\rho_{\alpha\beta}(\boldsymbol{\eta})$ of its *linearized change of metric tensor* and *linearized change of curvature tensor*, respectively defined in Sects. 2.4 and 2.5.

We demonstrate that these inequalities are valid for a *"general" surface* $S = \boldsymbol{\theta}(\overline{\omega})$, i.e., corresponding to a "general" injective mapping $\boldsymbol{\theta} : \overline{\omega} \to \mathcal{E}^3$ (save that $\boldsymbol{\theta}$ should be smooth enough) and for *any subset* γ_0 *of* γ *with* length $\gamma_0 > 0$. In other words, *no restriction is imposed on the "geometry" of* S *nor on the set* γ_0 (by contrast, specific restrictions must be imposed on $\boldsymbol{\theta}$ and γ_0 in order that the inequality of Korn's type established in the next section hold).

The infinitesimal rigid displacement lemma (Thm. 2.6-3) and the inequality of Korn's type "with boundary conditions" (Thm. 2.6-4) were first established by Bernadou & Ciarlet [1976]. A simpler presentation, which we follow here, was then proposed by Ciarlet & Miara [1992b] (see also Bernadou, Ciarlet & Miara [1994]). Its first stage consists (as in dimension three; cf. Thm. 1.7-2) in establishing an inequality of Korn's type "without boundary condition", again as a consequence of the *lemma of J.L. Lions*:

Theorem 2.6-1 (inequality of Korn's type "without boundary conditions" on a general surface). *Let ω be a domain in $\mathbb{R}^2$ and let $\boldsymbol{\theta} \in \mathcal{C}^3(\overline{\omega}; \mathcal{E}^3)$ be an injective mapping such that the two vectors $\boldsymbol{a}_\alpha = \partial_\alpha \boldsymbol{\theta}$ are linearly independent at all points of $\overline{\omega}$. Given $\boldsymbol{\eta} = (\eta_i) \in H^1(\omega) \times H^1(\omega) \times H^2(\omega)$, let*

$$\begin{aligned}
\gamma_{\alpha\beta}(\boldsymbol{\eta}) &:= \Big\{\frac{1}{2}(\partial_\beta \eta_\alpha + \partial_\alpha \eta_\beta) - \Gamma^\sigma_{\alpha\beta}\eta_\sigma - b_{\alpha\beta}\eta_3\Big\} \in L^2(\omega),\\
\rho_{\alpha\beta}(\boldsymbol{\eta}) &:= \Big\{\partial_{\alpha\beta}\eta_3 - \Gamma^\sigma_{\alpha\beta}\partial_\sigma\eta_3 - b^\sigma_\alpha b_{\sigma\beta}\eta_3\\
&\qquad + b^\sigma_\alpha(\partial_\beta\eta_\sigma - \Gamma^\tau_{\beta\sigma}\eta_\tau) + b^\tau_\beta(\partial_\alpha\eta_\tau - \Gamma^\sigma_{\alpha\tau}\eta_\sigma)\\
&\qquad + (\partial_\alpha b^\tau_\beta + \Gamma^\tau_{\alpha\sigma}b^\sigma_\beta - \Gamma^\sigma_{\alpha\beta}b^\tau_\sigma)\eta_\tau\Big\} \in L^2(\omega)
\end{aligned}$$

denote the covariant components of the linearized change of metric and change of curvature tensors associated with the displacement field $\eta_i \boldsymbol{a}^i$ of the surface $S = \boldsymbol{\theta}(\overline{\omega})$. Then there exists a constant $c_0 = c_0(\omega, \boldsymbol{\theta})$ such that (the notations $|\cdot|_{0,\omega}$ and $\|\cdot\|_{m,\omega}$ respectively designate the norms in $L^2(\omega)$ and $H^m(\omega)$, $m \geq 1$; cf. Sect. 1.7):

$$\begin{aligned}
&\Big\{\sum_\alpha \|\eta_\alpha\|^2_{1,\omega} + \|\eta_3\|^2_{2,\omega}\Big\}^{1/2}\\
&\leq c_0\Big\{\sum_\alpha |\eta_\alpha|^2_{0,\omega} + \|\eta_3\|^2_{1,\omega} + \sum_{\alpha,\beta}|\gamma_{\alpha\beta}(\boldsymbol{\eta})|^2_{0,\omega} + \sum_{\alpha,\beta}|\rho_{\alpha\beta}(\boldsymbol{\eta})|^2_{0,\omega}\Big\}^{1/2}\\
&\qquad\qquad \text{for all } \boldsymbol{\eta} = (\eta_i) \in H^1(\omega) \times H^1(\omega) \times H^2(\omega).
\end{aligned}$$

Proof. (i) *Define the space*

$$\begin{aligned}
\mathbf{W}_K(\omega) := \{\boldsymbol{\eta} = (\eta_i) \in L^2(\omega) \times L^2(\omega) \times H^1(\omega);\\
\gamma_{\alpha\beta}(\boldsymbol{\eta}) \in L^2(\omega),\ \rho_{\alpha\beta}(\boldsymbol{\eta}) \in L^2(\omega)\}.
\end{aligned}$$

Then, equipped with the norm $\|\cdot\|_\omega^K$ defined by

$$\|\boldsymbol{\eta}\|_\omega^K := \Big\{\sum_\alpha |\eta_\alpha|_{0,\omega}^2 + \|\eta_3\|_{1,\omega}^2 + \sum_{\alpha,\beta} |\gamma_{\alpha\beta}(\boldsymbol{\eta})|_{0,\omega}^2 + \sum_{\alpha,\beta} |\rho_{\alpha\beta}(\boldsymbol{\eta})|_{0,\omega}^2\Big\}^{1/2},$$

the space $\mathbf{W}_K(\omega)$ is a Hilbert space.

The relations "$\gamma_{\alpha\beta}(\boldsymbol{\eta}) \in L^2(\omega)$" and "$\rho_{\alpha\beta}(\boldsymbol{\eta}) \in L^2(\omega)$" appearing in the definition of the space $\mathbf{W}_K(\omega)$ are to be understood in the sense of distributions. They mean that $\boldsymbol{\eta} \in L^2(\omega) \times L^2(\omega) \times H^1(\omega)$ belongs to $\mathbf{W}_K(\omega)$ if there exist functions in $L^2(\omega)$, *denoted* $\gamma_{\alpha\beta}(\boldsymbol{\eta})$ and $\rho_{\alpha\beta}(\boldsymbol{\eta})$, such that for all $\varphi \in \mathcal{D}(\omega)$,

$$\int_\omega \gamma_{\alpha\beta}(\boldsymbol{\eta})\varphi\, dy = -\int_\omega \Big\{\frac{1}{2}(\eta_\beta \partial_\alpha \varphi + \eta_\alpha \partial_\beta \varphi) + \Gamma_{\alpha\beta}^\sigma \eta_\sigma \varphi + b_{\alpha\beta}\eta_3\varphi\Big\} dy,$$

$$\begin{aligned}\int_\omega \rho_{\alpha\beta}(\boldsymbol{\eta})\varphi\, dy = -\int_\omega \{&\partial_\alpha\eta_3\partial_\beta\varphi + \Gamma_{\alpha\beta}^\sigma \partial_\sigma\eta_3\varphi + b_\alpha^\sigma b_{\sigma\beta}\eta_3\varphi \\ &+ \eta_\sigma\partial_\beta(b_\alpha^\sigma\varphi) + b_\alpha^\sigma\Gamma_{\beta\sigma}^\tau\eta_\tau\varphi \\ &+ \eta_\tau\partial_\alpha(b_\beta^\tau\varphi) + b_\beta^\tau\Gamma_{\alpha\tau}^\sigma\eta_\sigma\varphi \\ &- \big(\partial_\alpha b_\beta^\tau + \Gamma_{\alpha\sigma}^\tau b_\beta^\sigma - \Gamma_{\alpha\beta}^\sigma b_\sigma^\tau\big)\eta_\tau\varphi\} dy.\end{aligned}$$

Let there be given a Cauchy sequence $(\boldsymbol{\eta}^k)_{k=1}^\infty$ with elements $\boldsymbol{\eta}^k = (\eta_i^k) \in \mathbf{W}_K(\omega)$. The definition of the norm $\|\cdot\|_\omega^K$ shows that there exist $\eta_\alpha \in L^2(\omega)$, $\eta_3 \in H^1(\omega)$, $\gamma_{\alpha\beta} \in L^2(\omega)$, and $\rho_{\alpha\beta} \in L^2(\omega)$ such that

$$\eta_\alpha^k \to \eta_\alpha \text{ in } L^2(\omega), \quad \eta_3^k \to \eta_3 \text{ in } H^1(\omega),$$
$$\gamma_{\alpha\beta}(\boldsymbol{\eta}^k) \to \gamma_{\alpha\beta} \text{ in } L^2(\omega), \quad \rho_{\alpha\beta}(\boldsymbol{\eta}^k) \to \rho_{\alpha\beta} \text{ in } L^2(\omega)$$

as $k \to \infty$. Given a function $\varphi \in \mathcal{D}(\omega)$, letting $k \to \infty$ in the relations $\int_\omega \gamma_{\alpha\beta}(\boldsymbol{\eta}^k)\varphi\, d\omega = \dots$ and $\int_\omega \rho_{\alpha\beta}(\boldsymbol{\eta}^k)\varphi\, d\omega = \dots$ then shows that $\gamma_{\alpha\beta} = \gamma_{\alpha\beta}(\boldsymbol{\eta})$ and $\rho_{\alpha\beta} = \rho_{\alpha\beta}(\boldsymbol{\eta})$.

(ii) *The spaces $\mathbf{W}_K(\omega)$ and $H^1(\omega) \times H^1(\omega) \times H^2(\omega)$ coincide.*

Clearly, $H^1(\omega) \times H^1(\omega) \times H^2(\omega) \subset \mathbf{W}_K(\omega)$. To prove the other inclusion, let $\boldsymbol{\eta} = (\eta_i) \in \mathbf{W}_K(\omega)$. The relations

$$e_{\alpha\beta}(\boldsymbol{\eta}) := \frac{1}{2}(\partial_\alpha\eta_\beta + \partial_\beta\eta_\alpha) = \gamma_{\alpha\beta}(\boldsymbol{\eta}) + \Gamma_{\alpha\beta}^\sigma\eta_\sigma + b_{\alpha\beta}\eta_3$$

then imply that $e_{\alpha\beta}(\boldsymbol{\eta}) \in L^2(\omega)$ since the functions $\Gamma^\sigma_{\alpha\beta}$ and $b_{\alpha\beta}$ are continuous on $\overline{\omega}$ (even continuously differentiable, since we assume $\boldsymbol{\theta} \in \mathcal{C}^3(\overline{\omega}; \mathcal{E}^3)$). Therefore,

$$\partial_\sigma \eta_\alpha \in H^{-1}(\omega),$$

$$\partial_\beta(\partial_\sigma \eta_\alpha) = \{\partial_\beta e_{\alpha\sigma}(\boldsymbol{\eta}) + \partial_\sigma e_{\alpha\beta}(\boldsymbol{\eta}) - \partial_\alpha e_{\beta\sigma}(\boldsymbol{\eta})\} \in H^{-1}(\omega),$$

since $\chi \in L^2(\omega)$ implies $\partial_\sigma \chi \in H^{-1}(\omega)$. Hence $\partial_\sigma \eta_\alpha \in L^2(\omega)$ by the *lemma of J. L. Lions* (Thm. 1.7-1) and thus $\eta_\alpha \in H^1(\omega)$.

The definition of the functions $\rho_{\alpha\beta}(\boldsymbol{\eta})$, the continuity over $\overline{\omega}$ of the functions $\Gamma^\sigma_{\alpha\beta}$, $b_{\sigma\beta}$, b^σ_α, and $\partial_\alpha b^\tau_\beta$, and the relations $\rho_{\alpha\beta}(\boldsymbol{\eta}) \in L^2(\omega)$ then imply that $\partial_{\alpha\beta}\eta_3 \in L^2(\omega)$, hence that $\eta_3 \in H^2(\omega)$.

(iii) *Inequality of Korn's type without boundary conditions.*

The identity mapping ι from the space $H^1(\omega) \times H^1(\omega) \times H^2(\omega)$ equipped with its product norm $\boldsymbol{\eta} = (\eta_i) \to \{\sum_\alpha \|\eta_\alpha\|^2_{1,\omega} + \|\eta_3\|^2_{2,\omega}\}^{1/2}$ into the space $\mathbf{W}_K(\omega)$ equipped with $\|\cdot\|^K_\omega$ is injective, continuous, and surjective by (ii). Since both spaces are complete (cf. (i)), the *closed graph theorem* then shows that the inverse mapping ι^{-1} is also continuous or equivalently, that the inequality of Korn's type without boundary conditions holds. ■

Remark. An earlier "inequality of Korn's type on a surface" is due to Rougée [1969, Lemma 2.4], who showed the existence of a constant c such that

$$\Big\{\sum_\alpha \|\eta_\alpha\|^2_{1,\omega}\Big\}^{1/2} \leq c\Big\{\sum_\alpha |\eta_\alpha|^2_{0,\omega} + \sum_{\alpha,\beta} \Big|\frac{1}{2}(\partial_\alpha \eta_\beta + \partial_\beta \eta_\alpha) - \Gamma^\sigma_{\alpha\beta}\eta_\sigma\Big|^2_{0,\omega}\Big\}$$

for all $(\eta_\alpha) \in H^1(\omega) \times H^1(\omega)$. ■

In order to establish an inequality of Korn's type "with boundary conditions", we have to identify classes of boundary conditions to be imposed on the fields $\boldsymbol{\eta} = (\eta_i) \in H^1(\omega) \times H^1(\omega) \times H^2(\omega)$ in order that we can "get rid" of the norms $|\eta_\alpha|_{0,\omega}$ and $\|\eta_3\|_{1,\omega}$ in the right-hand side of the above inequality, i.e., situations where the *semi-norm*

$$\boldsymbol{\eta} = (\eta_i) \to \Big\{\sum_{\alpha,\beta} |\gamma_{\alpha\beta}(\boldsymbol{\eta})|^2_{0,\omega} + \sum_{\alpha,\beta} |\rho_{\alpha\beta}(\boldsymbol{\eta})|^2_{0,\omega}\Big\}^{1/2}$$

becomes a *norm,* which should be in addition *equivalent* to the product norm.

To this end, we begin by establishing in Thm. 2.6-3 (as in dimension three; cf. Thm. 1.7-3) an *infinitesimal rigid displacement lemma*, which provides in particular one instance of boundary conditions implying that this semi-norm becomes a norm.

The elegant proof given here was suggested to me by Chapelle [1994]. It relies on the preliminary observation, quite worthwhile *per se*, that a vector field $\eta_i \boldsymbol{a}^i$ on a surface may be "canonically" extended to a three-dimensional vector field $v_i \boldsymbol{g}^i$ in such a way that all the components $e_{i\|j}(\boldsymbol{v})$ of the associated "three-dimensional" linearized change of metric tensor have remarkable expressions in terms of the components $\gamma_{\alpha\beta}(\boldsymbol{\eta})$ and $\rho_{\alpha\beta}(\boldsymbol{\eta})$ of the linearized change of metric and curvature tensors of the surface vector field:

Theorem 2.6-2. *Let the assumptions on the mapping* $\boldsymbol{\theta} : \overline{\omega} \to \mathcal{E}^3$ *be as in* Thm. 2.6-1. *By* Thm. 3.1-1, *there exists* $\varepsilon_0 > 0$ *such that the canonical extension* $\boldsymbol{\Theta}$ *of the mapping* $\boldsymbol{\theta}$ *defined by*

$$\boxed{\boldsymbol{\Theta}(y, x_3) := \boldsymbol{\theta}(y) + x_3 \boldsymbol{a}_3(y) \text{ for all } (y, x_3) \in \overline{\Omega}_0,}$$

where $\Omega_0 := \omega \times]-\varepsilon_0, \varepsilon_0[$ *and* $\boldsymbol{a}_3 := \dfrac{\boldsymbol{a}_1 \wedge \boldsymbol{a}_2}{|\boldsymbol{a}_1 \wedge \boldsymbol{a}_2|}$, *is a* $\mathcal{C}^1$*-diffeomorphism from* $\overline{\Omega}_0$ *onto* $\boldsymbol{\Theta}(\overline{\Omega}_0)$ *and the three vectors* $\boldsymbol{g}_i := \partial_i \boldsymbol{\Theta}$ *are linearly independent at all points of* $\overline{\Omega}_0$.

With any vector field $\eta_i \boldsymbol{a}^i$ *with covariant components* η_α *in* $H^1(\omega)$ *and* η_3 *in* $H^2(\omega)$, *let there be associated the vector field* $v_i \boldsymbol{g}^i$ *defined on* $\overline{\Omega}_0$ *by*

$$\boxed{v_i(y, x_3)\boldsymbol{g}^i(y, x_3) = \eta_i(y)\boldsymbol{a}^i(y) + x_3 \mathcal{X}_\alpha(y)\boldsymbol{a}^\alpha(y)}$$

for all $(y, x_3) \in \overline{\Omega}_0$, *where the vectors* $\boldsymbol{g}^i$ *form the contravariant basis associated with the vectors* $\boldsymbol{g}_i$ (i.e., $\boldsymbol{g}^i \cdot \boldsymbol{g}_j = \delta^i_j$; cf. Sect. 1.2) *and*

$$\boxed{\mathcal{X}_\alpha := -(\partial_\alpha \eta_3 + b^\sigma_\alpha \eta_\sigma).}$$

Then the covariant components v_i *of the vector field* $v_i \boldsymbol{g}^i$ *are in* $H^1(\Omega_0)$ *and the covariant components* $e_{i\|j}(\boldsymbol{v}) \in L^2(\Omega_0)$ *of the asso-*

ciated linearized change of metric tensor (Sect. 1.5) *are given by*

$$\begin{aligned} e_{\alpha\|\beta}(\boldsymbol{v}) &= \gamma_{\alpha\beta}(\boldsymbol{\eta}) - x_3\rho_{\alpha\beta}(\boldsymbol{\eta}) \\ &\quad + \frac{x_3^2}{2}\{b_\alpha^\sigma\rho_{\beta\sigma}(\boldsymbol{\eta}) + b_\beta^\tau\rho_{\alpha\tau}(\boldsymbol{\eta}) - 2b_\alpha^\sigma b_\beta^\tau\gamma_{\sigma\tau}(\boldsymbol{\eta})\}, \\ e_{i\|3}(\boldsymbol{v}) &= 0. \end{aligned}$$

Proof. As in the above expressions of the functions $e_{\alpha\|\beta}(\boldsymbol{v})$, the dependence on x_3 is explicit, but the dependence with respect to $y \in \overline{\omega}$ is omitted, throughout the proof.

(i) *Given functions* η_α, $\mathcal{X}_\alpha \in H^1(\omega)$ *and* $\eta_3 \in H^2(\omega)$, *let the vector field* $v_i\boldsymbol{g}^i$ *be defined on* $\overline{\Omega}_0$ *by*

$$v_i\boldsymbol{g}^i = \eta_i\boldsymbol{a}^i + x_3\mathcal{X}_\alpha\boldsymbol{a}^\alpha$$

(in other words, we momentarily ignore that the functions $\mathcal{X}_\alpha$ have the specific forms indicated in the theorem). *Then the functions* v_i *are in* $H^1(\Omega_0)$ *and the covariant components* $e_{i\|j}(\boldsymbol{v})$ *of the linearized change of metric tensor associated with the field* $v_i\boldsymbol{g}^i$ *are given by*

$$\begin{aligned} e_{\alpha\|\beta}(\boldsymbol{v}) &= \left\{\frac{1}{2}(\eta_{\alpha|\beta} + \eta_{\beta|\alpha}) - b_{\alpha\beta}\eta_3\right\} \\ &+ \frac{x_3}{2}\left\{\mathcal{X}_{\alpha|\beta} + \mathcal{X}_{\beta|\alpha} - b_\alpha^\sigma(\eta_{\sigma|\beta} - b_{\beta\sigma}\eta_3) - b_\beta^\tau(\eta_{\tau|\alpha} - b_{\alpha\tau}\eta_3)\right\} \\ &+ \frac{x_3^2}{2}\left\{ - b_\alpha^\sigma\mathcal{X}_{\sigma|\beta} - b_\beta^\tau\mathcal{X}_{\tau|\alpha}\right\}, \\ e_{\alpha\|3}(\boldsymbol{v}) &= \frac{1}{2}(\mathcal{X}_\alpha + \partial_\alpha\eta_3 + b_\alpha^\sigma\eta_\sigma), \\ e_{3\|3}(\boldsymbol{v}) &= 0, \end{aligned}$$

where $\eta_{\alpha|\beta} = \partial_\beta\eta_\alpha - \Gamma_{\alpha\beta}^\sigma\eta_\sigma$ *and* $\mathcal{X}_{\alpha|\beta} = \partial_\beta\mathcal{X}_\alpha - \Gamma_{\alpha\beta}^\sigma\mathcal{X}_\sigma$ *designate the covariant derivatives of the fields* $\eta_i\boldsymbol{a}^i$ *and* $\mathcal{X}_i\boldsymbol{a}^i$ *with* $\mathcal{X}_3 = 0$ (Thm. 2.3-1).

Since

$$\partial_\alpha\boldsymbol{a}_3 = -b_\alpha^\sigma\boldsymbol{a}_\sigma$$

by the second formula of Weingarten (Thm. 2.3-1), *the vectors of the covariant basis associated with the mapping* $\boldsymbol{\Theta} = \boldsymbol{\theta} + x_3\boldsymbol{a}_3$ *are given by*

$$\boldsymbol{g}_\alpha = \boldsymbol{a}_\alpha - x_3 b_\alpha^\sigma\boldsymbol{a}_\sigma \text{ and } \boldsymbol{g}_3 = \boldsymbol{a}_3.$$

The assumed regularities of the functions η_i and $\mathcal{X}_\alpha$ imply that

$$v_i = (v_j \boldsymbol{g}^j) \cdot \boldsymbol{g}_i = (\eta_j \boldsymbol{a}^j + x_3 \mathcal{X}_\alpha \boldsymbol{a}^\alpha) \cdot \boldsymbol{g}_i \in H^1(\Omega_0)$$

since $\boldsymbol{g}_i \in \boldsymbol{C}^1(\overline{\Omega}_0)$. The announced expressions for the functions $e_{i\|j}(\boldsymbol{v})$ are obtained by simple computations, based on the relations (Thms. 1.4-1 and 1.5-1):

$$v_{i\|j} = \{\partial_j(v_k \boldsymbol{g}^k)\} \cdot \boldsymbol{g}_i \text{ and } e_{i\|j}(\boldsymbol{v}) = \frac{1}{2}(v_{i\|j} + v_{j\|i}),$$

the latter advantageously avoiding the computation of the Christoffel symbols Γ^p_{ij}.

(ii) *When* $\mathcal{X}_\alpha = -(\partial_\alpha \eta_3 + b^\sigma_\alpha \eta_\sigma)$, *the functions* $e_{i\|j}(\boldsymbol{v})$ *in* (i) *take the expressions announced in the statement of the theorem.*

We first note that $\mathcal{X}_\alpha \in H^1(\omega)$ (since $b^\sigma_\alpha \in \mathcal{C}^1(\overline{\omega})$) and that $e_{\alpha\|3}(\boldsymbol{v}) = 0$ in this case. It thus remains to find the explicit forms of the functions $e_{\alpha\|\beta}(\boldsymbol{v})$. Replacing the functions $\mathcal{X}_\alpha$ by their expressions and using the symmetry relations $b^\sigma_\alpha|_\beta = b^\sigma_\beta|_\alpha$ (Thm. 2.5-1), we find that

$$\frac{1}{2}\left\{\mathcal{X}_{\alpha|\beta} + \mathcal{X}_{\beta|\alpha} - b^\sigma_\alpha(\eta_{\sigma|\beta} - b_{\beta\sigma}\eta_3) - b^\tau_\beta(\eta_{\tau|\alpha} - b_{\alpha\tau}\eta_3)\right\}$$
$$= -\eta_{3|\alpha\beta} - b^\sigma_\alpha \eta_{\sigma|\beta} - b^\tau_\beta \eta_{\tau|\alpha} - b^\tau_{\beta|\alpha}\eta_\tau + b^\sigma_\alpha b_{\sigma\beta}\eta_3,$$

i.e., the factor of x_3 in $e_{\alpha\|\beta}(\boldsymbol{v})$ is equal to $-\rho_{\alpha\beta}(\boldsymbol{\eta})$. Finally,

$$\begin{aligned}
&-b^\sigma_\alpha \mathcal{X}_{\sigma|\beta} - b^\tau_\beta \mathcal{X}_{\tau|\alpha} \\
&\quad = b^\sigma_\alpha(\eta_{3|\beta\sigma} + b^\tau_{\sigma|\beta}\eta_\tau + b^\tau_\sigma \eta_{\tau|\beta}) + b^\tau_\beta(\eta_{3|\alpha\tau} + b^\sigma_{\tau|\alpha}\eta_\sigma + b^\sigma_\tau \eta_{\sigma|\alpha}) \\
&\quad = b^\sigma_\alpha(\rho_{\beta\sigma}(\boldsymbol{\eta}) - b^\tau_\beta \eta_{\tau|\sigma} + b^\tau_\beta b_{\tau\sigma}\eta_3) + b^\tau_\beta(\rho_{\alpha\tau}(\boldsymbol{\eta}) - b^\sigma_\alpha \eta_{\sigma|\tau} + b^\sigma_\alpha b_{\sigma\tau}\eta_3) \\
&\quad = b^\sigma_\alpha \rho_{\beta\sigma}(\boldsymbol{\eta}) + b^\tau_\beta \rho_{\alpha\tau}(\boldsymbol{\eta}) - 2 b^\sigma_\alpha b^\tau_\beta \gamma_{\sigma\tau}(\boldsymbol{\eta}),
\end{aligned}$$

i.e., the factor of $\dfrac{x_3^2}{2}$ in $e_{\alpha\|\beta}(\boldsymbol{v})$ is that announced in the theorem. ∎

Remark. The otherwise mysterious definition of the functions $\mathcal{X}_\alpha$ is thus motivated by the requirement that $e_{\alpha\|3}(\boldsymbol{v}) = 0$; cf. part (ii). ∎

We now establish an *infinitesimal rigid displacement lemma "on a surface"*; "infinitesimal" reminds that only the *linearized* parts of the change of metric and curvature tensors are required to vanish. Thanks to Thm. 2.6-2, this lemma becomes a simple consequence of the "three-dimensional" infinitesimal rigid displacement lemma in curvilinear coordinates (Thm. 1.7-3), to which it should be profitably compared.

This lemma is due to Bernadou & Ciarlet [1976, Thms. 5.1-1 and 5.2-1], who gave a more direct, but less "transparent", proof (see also Bernadou [1994, Part 1, Lemma 5.1.4]). As shown by Blouza & Le Dret [1999], it still holds under weaker smoothness assumptions on the mapping $\boldsymbol{\theta}$ (see Thm. 2.6-5).

Part (a) in the next theorem is an *infinitesimal rigid displacement lemma "without boundary conditions"*, while part (b) is an *infinitesimal rigid displacement lemma "with boundary conditions"*.

Theorem 2.6-3 (infinitesimal rigid displacement lemma on a general surface). *Let there be given a domain ω in $\mathbb{R}^2$ and an injective mapping $\boldsymbol{\theta} \in \mathcal{C}^3(\overline{\omega}; \mathcal{E}^3)$ such that the two vectors $\boldsymbol{a}_\alpha = \partial_\alpha \boldsymbol{\theta}$ are linearly independent at all points of $\overline{\omega}$.*

(a) *Let $\boldsymbol{\eta} = (\eta_i) \in H^1(\omega) \times H^1(\omega) \times H^2(\omega)$ be such that*

$$\boxed{\gamma_{\alpha\beta}(\boldsymbol{\eta}) = \rho_{\alpha\beta}(\boldsymbol{\eta}) = 0 \text{ in } \omega.}$$

Then the vector field $\eta_i \boldsymbol{a}^i$ is an **infinitesimal rigid displacement** *of the surface $S = \boldsymbol{\theta}(\overline{\omega})$, in the sense that there exist two vectors $\hat{\mathbf{c}}$, $\hat{\boldsymbol{d}} \in \mathbb{R}^3$ such that*

$$\boxed{\eta_i(y)\boldsymbol{a}^i(y) = \hat{\mathbf{c}} + \hat{\boldsymbol{d}} \wedge \boldsymbol{\theta}(y) \text{ for all } y \in \overline{\omega}.}$$

(b) *Let γ_0 be a $d\gamma$-measurable subset of $\gamma = \partial\omega$ that satisfies*

$$\boxed{\textit{length } \gamma_0 > 0.}$$

Then (∂_ν denotes the outer normal derivative operator along γ):

$$\boxed{\left.\begin{aligned} \boldsymbol{\eta} = (\eta_i) \in H^1(\omega) \times H^1(\omega) \times H^2(\omega), \\ \eta_i = \partial_\nu \eta_3 = 0 \text{ on } \gamma_0, \\ \gamma_{\alpha\beta}(\boldsymbol{\eta}) = \rho_{\alpha\beta}(\boldsymbol{\eta}) = 0 \text{ in } \omega \end{aligned}\right\} \Rightarrow \boldsymbol{\eta} = \mathbf{0} \text{ in } \omega.}$$

Proof. Let the set $\Omega_0 = \omega \times]-\varepsilon_0, \varepsilon_0[$ and the vector field $\boldsymbol{v} = (v_i) \in \mathbf{H}^1(\Omega_0)$ be defined as in Thm. 2.6-2. By the same theorem,

$$\gamma_{\alpha\beta}(\boldsymbol{\eta}) = \rho_{\alpha\beta}(\boldsymbol{\eta}) = 0 \text{ in } \omega \Rightarrow e_{i\|j}(\boldsymbol{v}) = 0 \text{ in } \Omega_0,$$

and thus, by Thm. 1.7-3 (a), there exist two vectors $\hat{\mathbf{c}}$, $\hat{\boldsymbol{d}} \in \mathbb{R}^3$ such that

$$v_i(y, x_3)\boldsymbol{g}^i(y, x_3) = \hat{\mathbf{c}} + \hat{\boldsymbol{d}} \wedge \{\boldsymbol{\theta}(y) + x_3\boldsymbol{a}_3(y)\} \text{ for all } (y, x_3) \in \overline{\Omega}_0.$$

Hence

$$\eta_i(y)\boldsymbol{a}^i(y) = v_i(y, x_3)\boldsymbol{g}^i(y, x_3)|_{x_3=0} = \hat{\mathbf{c}} + \hat{\boldsymbol{d}} \wedge \boldsymbol{\theta}(y) \text{ for all } y \in \overline{\omega},$$

and part (a) is established.

If $\eta_i = \partial_\nu\eta_3 = 0$ on γ_0, the functions $\mathcal{X}_\alpha = -(\partial_\alpha\eta_3 + b_\alpha^\sigma\eta_\sigma)$ vanish on γ_0, since $\eta_3 = \partial_\nu\eta_3 = 0$ on γ_0 implies $\partial_\alpha\eta_3 = 0$ on γ_0; consequently (Thm. 2.6-2),

$$v_i = (v_j\boldsymbol{g}^j) \cdot \boldsymbol{g}_i = (\eta_j\boldsymbol{a}^j + x_3\mathcal{X}_\alpha\boldsymbol{a}^\alpha) \cdot \boldsymbol{g}_i = 0 \text{ on } \Gamma_0 := \gamma_0 \times [-\varepsilon_0, \varepsilon_0].$$

Since *area* $\Gamma_0 > 0$, Thm. 1.7-3 (b) implies that $\boldsymbol{v} = \mathbf{0}$ in $\overline{\Omega}_0$, hence that $\boldsymbol{\eta} = \mathbf{0}$ on $\overline{\omega}$. ∎

Remarks. (1) If a field $\boldsymbol{\eta} = (\eta_i) \in H^1(\omega) \times H^1(\omega) \times H^2(\omega)$ satisfies $\gamma_{\alpha\beta}(\boldsymbol{\eta}) = \rho_{\alpha\beta}(\boldsymbol{\eta}) = 0$ in ω, its three components η_i are automatically in $\mathcal{C}^2(\overline{\omega})$ since $\eta_i = (\eta_j\boldsymbol{a}^j) \cdot \boldsymbol{a}_i$ and the fields $\boldsymbol{a}_i$ are of class $\mathcal{C}^2$ on $\overline{\omega}$. Remarkably, the field $\eta_i\boldsymbol{a}^i = \hat{\mathbf{c}} + \hat{\boldsymbol{d}} \wedge \boldsymbol{\theta}$ inherits in this case even *more* regularity, as it is of class $\mathcal{C}^3$ on $\overline{\omega}$!

(2) The constant vector $\hat{\boldsymbol{d}}$ can be expressed in terms of the functions η_i; cf. Ex. 2.8.

(3) If a field $\boldsymbol{\eta} = (\eta_i)$ satisfies the assumptions of part (a), the boundary conditions $\eta_i = \partial_\nu\eta_3 = 0$ on γ_0 of part (b) are not the only ones that lead to $\boldsymbol{\eta} = \mathbf{0}$ in ω; cf. Ex. 2.9 for another instance.

(4) Crucial information can still be gathered about the displacement field $\eta_i\boldsymbol{a}^i$ if only the components $\gamma_{\alpha\beta}(\boldsymbol{\eta})$ vanish in ω; cf. Ex. 2.8. ∎

We are now in a position to prove an *inequality of Korn's type "with boundary conditions"* that plays a *fundamental rôle in the analysis of linearly elastic shells*, in particular for establishing the existence and uniqueness of the solution to the *two-dimensional Koiter equations for a linearly elastic "clamped" shell* (another class of boundary conditions is considered in Ex. 2.9) and of the solution to the *two-dimensional equations of a linearly elastic "flexural shell"*, as explained at the beginning of this section.

This inequality was first proved by Bernadou & Ciarlet [1976]. It was later given other proofs by Bernadou, Ciarlet & Miara [1994], then by Blouza & Le Dret [1999], who showed that it still holds under a less stringent smoothness assumption on the mapping $\boldsymbol{\theta}$ (see Thm. 2.6-6).

Theorem 2.6-4 (inequality of Korn's type on a general surface). *Let ω be a domain in $\mathbb{R}^2$, let $\boldsymbol{\theta} \in \mathcal{C}^3(\overline{\omega}; \mathcal{E}^3)$ be an injective mapping such that the two vectors $\boldsymbol{a}_\alpha = \partial_\alpha \boldsymbol{\theta}$ are linearly independent at all points of $\overline{\omega}$, let γ_0 be a $d\gamma$-measurable subset of $\gamma = \partial\omega$ that satisfies*

$$\boxed{\textit{length } \gamma_0 > 0,}$$

and let the space $\mathbf{V}_K(\omega)$ be defined as (∂_ν denotes the outer normal derivative operator along γ):

$$\boxed{\begin{aligned}\mathbf{V}_K(\omega) := \{\boldsymbol{\eta} = (\eta_i) \in H^1(\omega) \times H^1(\omega) \times H^2(\omega);\\ \eta_i = \partial_\nu \eta_3 = 0 \text{ on } \gamma_0\}.\end{aligned}}$$

Given $\boldsymbol{\eta} = (\eta_i) \in H^1(\omega) \times H^1(\omega) \times H^2(\omega)$, let

$$\boxed{\begin{aligned}\gamma_{\alpha\beta}(\boldsymbol{\eta}) &:= \Big\{\frac{1}{2}(\partial_\beta \eta_\alpha + \partial_\alpha \eta_\beta) - \Gamma^\sigma_{\alpha\beta}\eta_\sigma - b_{\alpha\beta}\eta_3\Big\} \in L^2(\omega),\\ \rho_{\alpha\beta}(\boldsymbol{\eta}) &:= \Big\{\partial_{\alpha\beta}\eta_3 - \Gamma^\sigma_{\alpha\beta}\partial_\sigma\eta_3 - b^\sigma_\alpha b_{\sigma\beta}\eta_3\\ &\quad + b^\sigma_\alpha(\partial_\beta\eta_\sigma - \Gamma^\tau_{\beta\sigma}\eta_\tau) + b^\tau_\beta(\partial_\alpha\eta_\tau - \Gamma^\sigma_{\alpha\tau}\eta_\sigma)\\ &\quad + (\partial_\alpha b^\tau_\beta + \Gamma^\tau_{\alpha\sigma}b^\sigma_\beta - \Gamma^\sigma_{\alpha\beta}b^\tau_\sigma)\eta_\tau\Big\} \in L^2(\omega)\end{aligned}}$$

denote the covariant components of the linearized change of metric and linearized change of curvature tensors associated with the dis-

placement field $\eta_i \boldsymbol{a}^i$ of the surface $S = \boldsymbol{\theta}(\overline{\omega})$. Then there exists a constant $c = c(\omega, \gamma_0, \boldsymbol{\theta})$ such that

$$\Big\{ \sum_\alpha \|\eta_\alpha\|^2_{1,\omega} + \|\eta_3\|^2_{2,\omega} \Big\}^{1/2} \leq c \Big\{ \sum_{\alpha,\beta} |\gamma_{\alpha\beta}(\boldsymbol{\eta})|^2_{0,\omega} + \sum_{\alpha,\beta} |\rho_{\alpha\beta}(\boldsymbol{\eta})|^2_{0,\omega} \Big\}^{1/2} \quad \text{for all } \boldsymbol{\eta} = (\eta_i) \in \mathbf{V}_K(\omega).$$

Proof. Let

$$\|\boldsymbol{\eta}\|_{H^1(\omega)\times H^1(\omega)\times H^2(\omega)} := \Big\{ \sum_\alpha \|\eta_\alpha\|^2_{1,\omega} + \|\eta_3\|^2_{2,\omega} \Big\}^{1/2},$$

and

$$|\boldsymbol{\eta}|^K_\omega := \Big\{ \sum_{\alpha,\beta} |\gamma_{\alpha\beta}(\boldsymbol{\eta})|^2_{0,\omega} + \sum_{\alpha,\beta} |\rho_{\alpha\beta}(\boldsymbol{\eta})|^2_{0,\omega} \Big\}^{1/2}.$$

If the announced inequality is false, there exists a sequence $(\boldsymbol{\eta}^k)_{k=1}^\infty$ of functions $\boldsymbol{\eta}^k \in \mathbf{V}_K(\omega)$ such that

$$\|\boldsymbol{\eta}^k\|_{H^1(\omega)\times H^1(\omega)\times H^2(\omega)} = 1 \text{ for all } k \text{ and } \lim_{k\to\infty} |\boldsymbol{\eta}^k|^K_\omega = 0.$$

Since the sequence $(\boldsymbol{\eta}^k)_{k=1}^\infty$ is bounded in $H^1(\omega) \times H^1(\omega) \times H^2(\omega)$, a subsequence $(\boldsymbol{\eta}^l)_{l=1}^\infty$ converges in $L^2(\omega) \times L^2(\omega) \times H^1(\omega)$ by the *Rellich-Kondrašov theorem* (see, e.g., Vol. I, Thm. 6.1-5); furthermore, each sequence $(\gamma_{\alpha\beta}(\boldsymbol{\eta}^l))_{l=1}^\infty$ and $(\rho_{\alpha\beta}(\boldsymbol{\eta}^l))_{l=1}^\infty$ also converges in $L^2(\omega)$ (to 0, but this information is not used at this stage) since $\lim_{l\to\infty} |\boldsymbol{\eta}^l|^K_\omega = 0$. The subsequence $(\boldsymbol{\eta}^l)_{l=1}^\infty$ is thus a Cauchy sequence with respect to the norm

$$\boldsymbol{\eta} \to \Big\{ \sum_\alpha |\eta_\alpha|^2_{0,\omega} + \|\eta_3\|^2_{1,\omega} + \sum_{\alpha,\beta} |\gamma_{\alpha\beta}(\boldsymbol{\eta})|^2_{0,\omega} + \sum_{\alpha,\beta} |\rho_{\alpha\beta}(\boldsymbol{\eta})|^2_{0,\omega} \Big\}^{1/2},$$

hence with respect to the norm $\|\cdot\|_{H^1(\omega)\times H^1(\omega)\times H^2(\omega)}$ by *Korn's inequality without boundary conditions* (Thm. 2.6-1).

The space $\mathbf{V}_K(\omega)$ being complete as a closed subspace of the space $H^1(\omega) \times H^1(\omega) \times H^2(\omega)$, there exists $\boldsymbol{\eta} \in \mathbf{V}_K(\omega)$ such that

$$\boldsymbol{\eta}^l \to \boldsymbol{\eta} \text{ in } H^1(\omega) \times H^1(\omega) \times H^2(\omega),$$

and the limit $\boldsymbol{\eta}$ satisfies

$$|\gamma_{\alpha\beta}(\boldsymbol{\eta})|_{0,\omega} = \lim_{l\to\infty} |\gamma_{\alpha\beta}(\boldsymbol{\eta}^l)|_{0,\omega} = 0,$$

$$|\rho_{\alpha\beta}(\boldsymbol{\eta})|_{0,\omega} = \lim_{l\to\infty} |\rho_{\alpha\beta}(\boldsymbol{\eta}^l)|_{0,\omega} = 0.$$

Hence $\boldsymbol{\eta} = \mathbf{0}$ by Thm. 2.6-3. But this contradicts the relations $\|\boldsymbol{\eta}^l\|_{H^1(\omega)\times H^1(\omega)\times H^2(\omega)} = 1$ for all $l \geq 1$, and the proof is complete. ■

If the mapping $\boldsymbol{\theta}$ is of the form $\boldsymbol{\theta}(y_1, y_2) = (y_1, y_2, 0)$ for all $(y_1, y_2) \in \overline{\omega}$, the inequality of Thm. 2.6-4 reduces to two distinct inequalities (obtained by letting first $\eta_\alpha = 0$, then $\eta_3 = 0$):

$$\|\eta_3\|_{2,\omega} \leq c\left\{\sum_{\alpha,\beta} |\partial_{\alpha\beta}\eta_3|^2_{0,\omega}\right\}^{1/2}$$

for all $\eta_3 \in H^2(\omega)$ satisfying $\eta_3 = \partial_\nu \eta_3 = 0$ on γ_0, and

$$\left\{\sum_\alpha \|\eta_\alpha\|^2_{1,\omega}\right\}^{1/2} \leq c\left\{\sum_\alpha \left|\frac{1}{2}(\partial_\beta\eta_\alpha + \partial_\alpha\eta_\beta)\right|^2_{0,\omega}\right\}^{1/2}$$

for all $\eta_\alpha \in \mathbf{H}^1(\omega)$ satisfying $\eta_\alpha = 0$ on γ_0. The first inequality is a well-known property of Sobolev spaces (see, e.g., Vol II, Thm. 1.5-1 and its proof); the second is the *two-dimensional Korn inequality in Cartesian coordinates*. Both play a central rôle in the *existence theory* for *linear two-dimensional plate equations* (Vol. II, Thms. 1.5-1 and 1.5-2).

As shown by Blouza & Le Dret [1994a, 1994b, 1999], *the regularity assumptions made on the mapping $\boldsymbol{\theta}$ and on the field $\boldsymbol{\eta} = (\eta_i)$ in both the infinitesimal rigid displacement lemma and the inequality of Korn's type* (Thms. 2.6-3 and 2.6-4) *can be substantially weakened.*

This improvement relies on the observation that the covariant components of the linearized change of metric and change of curvature tensors, viz.,

$$\gamma_{\alpha\beta}(\boldsymbol{\eta}) = \frac{1}{2}(\partial_\beta\eta_\alpha + \partial_\alpha\eta_\beta) - \Gamma^\sigma_{\alpha\beta}\eta_\sigma - b_{\alpha\beta}\eta_3$$

and

$$\begin{aligned}\rho_{\alpha\beta}(\boldsymbol{\eta}) = {} & \partial_{\alpha\beta}\eta_3 - \Gamma^\sigma_{\alpha\beta}\partial_\sigma\eta_3 - b^\sigma_\alpha b_{\sigma\beta}\eta_3 \\ & + b^\sigma_\alpha(\partial_\beta\eta_\sigma - \Gamma^\tau_{\beta\sigma}\eta_\tau) + b^\tau_\beta(\partial_\alpha\eta_\tau - \Gamma^\sigma_{\alpha\tau}\eta_\sigma) \\ & + (\partial_\alpha b^\tau_\beta + \Gamma^\tau_{\alpha\sigma}b^\sigma_\beta - \Gamma^\sigma_{\alpha\beta}b^\tau_\sigma)\eta_\tau,\end{aligned}$$

are also given by

$$\gamma_{\alpha\beta}(\boldsymbol{\eta}) = \frac{1}{2}(\partial_\beta \tilde{\boldsymbol{\eta}} \cdot \boldsymbol{a}_\alpha + \partial_\alpha \tilde{\boldsymbol{\eta}} \cdot \boldsymbol{a}_\beta) =: \tilde{\gamma}_{\alpha\beta}(\tilde{\boldsymbol{\eta}})$$

and

$$\rho_{\alpha\beta}(\boldsymbol{\eta}) = (\partial_{\alpha\beta} \tilde{\boldsymbol{\eta}} - \Gamma^\sigma_{\alpha\beta} \partial_\sigma \tilde{\boldsymbol{\eta}}) \cdot \boldsymbol{a}_3 =: \tilde{\rho}_{\alpha\beta}(\tilde{\boldsymbol{\eta}}),$$

in terms of the field

$$\tilde{\boldsymbol{\eta}} := \eta_i \boldsymbol{a}^i.$$

The interest of the new expressions $\tilde{\gamma}_{\alpha\beta}(\tilde{\boldsymbol{\eta}})$ and $\tilde{\rho}_{\alpha\beta}(\tilde{\boldsymbol{\eta}})$ is that they still define *bona fide* distributions under significantly weaker smoothness assumptions than those of Thm. 2.6-4, viz., $\boldsymbol{\theta} \in \mathcal{C}^3(\overline{\omega}; \mathcal{E}^3)$ and $\boldsymbol{\eta} = (\eta_i) \in H^1(\omega) \times H^1(\omega) \times H^2(\omega)$. More specifically, it is easily verified that $\tilde{\gamma}_{\alpha\beta}(\tilde{\boldsymbol{\eta}})) \in L^2(\omega)$ and $\tilde{\rho}_{\alpha\beta}(\tilde{\boldsymbol{\eta}}) \in H^{-1}(\omega)$ if $\boldsymbol{\theta} \in W^{2,\infty}(\omega; \mathcal{E}^3)$ and $\tilde{\boldsymbol{\eta}} \in \mathbf{H}^1(\omega)$.

Note that, to avoid any confusion, we intentionally employ the *new* notations $\tilde{\gamma}_{\alpha\beta}(\tilde{\boldsymbol{\eta}})$ and $\tilde{\rho}_{\alpha\beta}(\tilde{\boldsymbol{\eta}})$.

Using this observation, Blouza & Le Dret [1999, Thm. 6] first establish the following extension of Thm. 2.6-3:

Theorem 2.6-5 (infinitesimal rigid displacement lemma on a general surface with little regularity). *Let ω be a domain in $\mathbb{R}^2$ and let $\boldsymbol{\theta} \in W^{2,\infty}(\omega; \mathcal{E}^3)$ be an injective mapping such that the two vectors $\boldsymbol{a}_\alpha = \partial_\alpha \boldsymbol{\theta}$ are linearly independent at all points of $\overline{\omega}$.*

Given $\tilde{\boldsymbol{\eta}} \in \mathbf{H}^1(\omega)$, let the distributions $\tilde{\gamma}_{\alpha\beta}(\tilde{\boldsymbol{\eta}}) \in L^2(\omega)$ and $\tilde{\rho}_{\alpha\beta}(\tilde{\boldsymbol{\eta}}) \in H^{-1}(\omega)$ be defined by

$$\boxed{\begin{aligned} \tilde{\gamma}_{\alpha\beta}(\tilde{\boldsymbol{\eta}}) &:= \frac{1}{2}(\partial_\beta \tilde{\boldsymbol{\eta}} \cdot \boldsymbol{a}_\alpha + \partial_\alpha \tilde{\boldsymbol{\eta}} \cdot \boldsymbol{a}_\beta), \\ \tilde{\rho}_{\alpha\beta}(\tilde{\boldsymbol{\eta}}) &:= (\partial_{\alpha\beta} \tilde{\boldsymbol{\eta}} - \Gamma^\sigma_{\alpha\beta} \partial_\sigma \tilde{\boldsymbol{\eta}}) \cdot \boldsymbol{a}_3. \end{aligned}}$$

Let $\tilde{\boldsymbol{\eta}} \in \mathbf{H}^1(\omega)$ be such that

$$\boxed{\tilde{\gamma}_{\alpha\beta}(\tilde{\boldsymbol{\eta}}) = \tilde{\rho}_{\alpha\beta}(\tilde{\boldsymbol{\eta}}) = 0 \text{ in } \omega.}$$

Then $\tilde{\boldsymbol{\eta}}$ *is an* **infinitesimal rigid displacement** *of the surface* $S = \boldsymbol{\theta}(\overline{\omega})$, *in the sense that there exist two vectors* $\hat{\mathbf{c}}, \hat{\boldsymbol{d}} \in \mathbb{R}^3$ *such that*

$$\boxed{\tilde{\boldsymbol{\eta}}(y) = \hat{\mathbf{c}} + \hat{\boldsymbol{d}} \wedge \boldsymbol{\theta}(y) \text{ for all } y \in \overline{\omega}.}$$

■

Blouza & Le Dret [1999, Lemma 11] then proceed to establish the following variant of Thm. 2.6-4, which is stated here with the boundary conditions $\eta_i = 0$ on γ corresponding to a shell that is *simply supported along its entire boundary.*

Theorem 2.6-6 (inequality of Korn's type on a general surface with little regularity). *Let the assumptions on the mapping* $\boldsymbol{\theta}$ *be as in* Thm. 2.6-5 *and let the space* $\widetilde{\mathbf{V}}_K^s(\omega)$ *be defined as*

$$\boxed{\widetilde{\mathbf{V}}_K^s(\omega) = \{\tilde{\boldsymbol{\eta}} \in \mathbf{H}_0^1(\omega);\ \partial_{\alpha\beta}\tilde{\boldsymbol{\eta}} \cdot \boldsymbol{a}_3 \in L^2(\omega)\}.}$$

Then there exists a constant c such that

$$\boxed{\begin{aligned}&\Big\{\|\tilde{\boldsymbol{\eta}}\|_{1,\omega}^2 + \sum_{\alpha,\beta} |\partial_{\alpha\beta}\tilde{\boldsymbol{\eta}} \cdot \boldsymbol{a}_3|_{0,\omega}^2\Big\}^{1/2} \\ &\qquad \leq c\Big\{\sum_{\alpha,\beta} |\tilde{\gamma}_{\alpha\beta}(\tilde{\boldsymbol{\eta}})|_{0,\omega}^2 + \sum_{\alpha,\beta} |\tilde{\rho}_{\alpha\beta}(\tilde{\boldsymbol{\eta}})|_{0,\omega}^2\Big\}^{1/2} \text{ for all } \tilde{\boldsymbol{\eta}} \in \widetilde{\mathbf{V}}_K^s(\omega),\end{aligned}}$$

where the distributions $\tilde{\gamma}_{\alpha\beta}(\tilde{\boldsymbol{\eta}})$ *and* $\tilde{\rho}_{\alpha\beta}(\tilde{\boldsymbol{\eta}})$ *are defined as in* Thm. 2.6-5 (note that $\tilde{\rho}_{\alpha\beta}(\tilde{\boldsymbol{\eta}}) \in L^2(\omega)$ if $\tilde{\boldsymbol{\eta}} \in \widetilde{\mathbf{V}}_K^s(\omega)$). ■

Remark. As expected, the space $\widetilde{\mathbf{V}}_K^s(\omega)$ becomes isomorphic to the space $H_0^1(\omega) \times H_0^1(\omega) \times \{H^2(\omega) \times H_0^1(\omega)\}$ when $\boldsymbol{\theta}$ is smooth enough; cf. Ex. 2.10. ■

This theorem thus establishes as a corollary the *existence* and *uniqueness* of the solution to *the two-dimensional Koiter equations for a simply supported shell whose middle surface has little regularity,* once these equations are rewritten as follows: Find $\tilde{\boldsymbol{\zeta}}_K^\varepsilon$ such that

$$\tilde{\boldsymbol{\zeta}}_K^\varepsilon \in \tilde{\mathbf{V}}_K^s(\omega),$$

$$\int_\omega \left\{ \varepsilon a^{\alpha\beta\sigma\tau,\varepsilon} \tilde{\gamma}_{\sigma\tau}(\tilde{\boldsymbol{\zeta}}_K^\varepsilon) \tilde{\gamma}_{\alpha\beta}(\tilde{\boldsymbol{\eta}}) + \frac{\varepsilon^3}{3} a^{\alpha\beta\sigma\tau,\varepsilon} \tilde{\rho}_{\sigma\tau}(\tilde{\boldsymbol{\zeta}}_K^\varepsilon) \tilde{\rho}_{\alpha\beta}(\tilde{\boldsymbol{\eta}}) \right\} \sqrt{a}\, dy$$
$$= \int_\omega \tilde{\boldsymbol{p}}^\varepsilon \cdot \tilde{\boldsymbol{\eta}} \sqrt{a}\, dy \text{ for all } \tilde{\boldsymbol{\eta}} \in \mathbf{V}_K^s(\omega),$$

where the given field $\tilde{\boldsymbol{p}}^\varepsilon \in \mathbf{L}^2(\omega)$ accounts for the applied forces.

A word of caution. In this approach, the unknown is the displacement field $\tilde{\boldsymbol{\zeta}}_K^\varepsilon$ of the middle surface $S = \boldsymbol{\theta}(\omega)$, viewed as a *vector field with Cartesian components in* $H^1(\omega)$. The displacement field is thus no longer recovered by means of its covariant components on the contravariant bases. ■

Remark. Boundary conditions of *clamping*, viz., $\eta_i = \partial_\nu \eta_3 = 0$ on $\gamma_0 \subset \gamma$, as considered in Thm. 2.6-4, can be also handled *via* the present approach, provided they are first re-interpreted so as to make sense for vector fields $\tilde{\boldsymbol{\eta}}$ that only satisfy "$\tilde{\boldsymbol{\eta}} \in \mathbf{H}^1(\omega)$ and $\partial_{\alpha\beta}\tilde{\boldsymbol{\eta}} \cdot \boldsymbol{a}_3 \in L^2(\omega)$"; cf. Blouza & Le Dret [1999, Sect. 6]. ■

On the one hand, this approach clearly widens the class of shells that can be modeled by Koiter's equations, since discontinuities in the second derivatives of the mapping $\boldsymbol{\theta}$ are allowed, provided these derivatives stay in $L^\infty(\omega)$. For instance, it affords the consideration of a shell whose middle surface is composed of a portion of a plane and a portion of a circular cylinder meeting along a segment and having a common tangent plane along this segment.

On the other, the asymptotic analysis that will ultimately justify the two-dimensional Koiter equations in the form described at the beginning of this section (i.e., expressed in terms of the functions $\gamma_{\alpha\beta}(\boldsymbol{\eta})$ and $\rho_{\alpha\beta}(\boldsymbol{\eta})$) does seem, however, to require that $\boldsymbol{\theta} \in \mathcal{C}^3(\overline{\omega}; \mathcal{E}^3)$; cf. Sect. 7.2.

2.7. INEQUALITY OF KORN'S TYPE AND INFINITESIMAL RIGID DISPLACEMENT LEMMA ON AN ELLIPTIC SURFACE

The *two-dimensional equations of a linearly elastic elliptic membrane shell*, which will be fully justified in Chap. 4 through an asymptotic analysis of the three-dimensional solutions *under specific restrictions on the boundary conditions and on the "geometry" of the middle surface*, take the following form: The unknowns are the covariant components $\zeta_i^\varepsilon : \overline{\omega} \to \mathbb{R}$ of the displacement $\zeta_i^\varepsilon \boldsymbol{a}^i$ of the middle surface $S = \boldsymbol{\theta}(\overline{\omega})$ of the shell and $\boldsymbol{\zeta}^\varepsilon := (\zeta_i^\varepsilon)$ satisfies:

$$\boxed{\begin{aligned} &\boldsymbol{\zeta}^\varepsilon \in \mathbf{V}_M(\omega) := H_0^1(\omega) \times H_0^1(\omega) \times L^2(\omega), \\ &\int_\omega \varepsilon a^{\alpha\beta\sigma\tau,\,\varepsilon} \gamma_{\sigma\tau}(\boldsymbol{\zeta}^\varepsilon) \gamma_{\alpha\beta}(\boldsymbol{\eta}) \sqrt{a}\, dy \\ &\qquad = \int_\omega p^{i,\,\varepsilon} \eta_i \sqrt{a}\, dy \text{ for all } \boldsymbol{\eta} = (\eta_i) \in \mathbf{V}_M(\omega), \end{aligned}}$$

where $2\varepsilon > 0$ is the thickness of the shell,

$$a^{\alpha\beta\sigma\tau,\,\varepsilon} := \frac{4\lambda^\varepsilon \mu^\varepsilon}{\lambda^\varepsilon + 2\mu^\varepsilon} a^{\alpha\beta} a^{\sigma\tau} + 2\mu^\varepsilon (a^{\alpha\sigma} a^{\beta\tau} + a^{\alpha\tau} a^{\beta\sigma})$$

denote the contravariant components of the *two-dimensional elasticity tensor of the shell* (the same as in Sect. 2.6), $\gamma_{\alpha\beta}(\boldsymbol{\eta})$ denote the covariant components of the linearized change of metric tensor associated with a displacement field $\eta_i \boldsymbol{a}^i$ of S, and the given functions $p^{i,\,\varepsilon} \in L^2(\omega)$ account for the applied forces. As already noted in Sect. 2.6, there exists a constant $c_e = c_e(\omega, \boldsymbol{\theta}, \mu^\varepsilon) > 0$ such that

$$\sum_{\alpha,\beta} |t_{\alpha\beta}|^2 \leq c_e a^{\alpha\beta\sigma\tau,\,\varepsilon}(y) t_{\sigma\tau} t_{\alpha\beta}$$

for all $y \in \overline{\omega}$ and all symmetric matrices $(t_{\alpha\beta})$ (Thm. 3.3-2). Establishing the existence and uniqueness of a solution to the above variational problem by means of the *Lax-Milgram lemma* thus amounts to proving the existence of a constant c_M such that

$$\Big\{ \sum_\alpha \|\eta_\alpha\|_{1,\,\omega}^2 + |\eta_3|_{0,\,\omega}^2 \Big\}^{1/2} \leq c_M \Big\{ \sum_{\alpha,\beta} |\gamma_{\alpha\beta}(\boldsymbol{\eta})|_{0,\,\omega}^2 \Big\}^{1/2} \quad \text{for all } \boldsymbol{\eta} = (\eta_i) \in \mathbf{V}_M(\omega).$$

The objective of this section, based on Ciarlet & Lods [1996a] and Ciarlet & Sanchez-Palencia [1996], is to find *sufficient conditions* (which, remarkably, turn out to be also necessary; cf. Thm. 2.7-4), *essentially bearing on the "geometry" of the surface S, guaranteeing that such an "inequality of Korn's type" holds.* It is also worth noticing that the justification alluded to above of these two-dimensional membrane shell equations from three-dimensional elasticity will be carried out under precisely the *same* assumptions on the geometry of S (Chap. 4).

We follow the usual pattern, i.e., we begin by proving an *inequality of Korn's type "without boundary condition"*, which in fact holds for *"arbitrary" geometries*, even though *it only involves the linearized change of metric tensor.*

Everything has its price, however: The norm $\|\eta_3\|_{2,\omega}$ appearing in the left-hand side of the "first" inequality of Korn's type on a general surface (Thm. 2.6-1) is now replaced by the norm $|\eta_3|_{0,\omega}$.

Theorem 2.7-1 (second inequality of Korn's type "without boundary conditions" on a general surface). *Let ω be a domain in $\mathbb{R}^2$ and let $\boldsymbol{\theta} \in \mathcal{C}^2(\overline{\omega}; \mathcal{E}^3)$ be an injective mapping such that the two vectors $\boldsymbol{a}_\alpha = \partial_\alpha \boldsymbol{\theta}$ are linearly independent at all points of $\overline{\omega}$.*

Given $\boldsymbol{\eta} = (\eta_i) \in H^1(\omega) \times H^1(\omega) \times L^2(\omega)$, let

$$\boxed{\gamma_{\alpha\beta}(\boldsymbol{\eta}) = \left\{\frac{1}{2}(\partial_\beta \eta_\alpha + \partial_\alpha \eta_\beta) - \Gamma^\sigma_{\alpha\beta}\eta_\sigma - b_{\alpha\beta}\eta_3\right\} \in L^2(\omega)}$$

denote the covariant components of the linearized change of metric tensor associated with the displacement field $\eta_i \boldsymbol{a}^i$ of the surface $S = \boldsymbol{\theta}(\overline{\omega})$. Then there exists a constant $c_0 = c_0(\omega, \boldsymbol{\theta})$ such that

$$\boxed{\begin{aligned}&\Big\{\sum_\alpha \|\eta_\alpha\|^2_{1,\omega} + |\eta_3|^2_{0,\omega}\Big\}^{1/2} \leq c_0 \Big\{\sum_i |\eta_i|^2_{0,\omega} + \sum_{\alpha,\beta} |\gamma_{\alpha\beta}(\boldsymbol{\eta})|^2_{0,\omega}\Big\}^{1/2} \\ &\qquad\qquad \text{for all } \boldsymbol{\eta} = (\eta_i) \in H^1(\omega) \times H^1(\omega) \times L^2(\omega).\end{aligned}}$$

Proof. The proof is analogous to that of Thm. 2.6-1 and, for this reason, is only sketched; it relies on the following steps: First, the space

$$\mathbf{W}_M(\omega) := \{\boldsymbol{\eta} = (\eta_i) \in \mathbf{L}^2(\omega); \gamma_{\alpha\beta}(\boldsymbol{\eta}) \in L^2(\omega)\}$$

becomes a Hilbert space when it is equipped with the norm $\|\cdot\|_\omega^M$ defined by

$$\|\boldsymbol{\eta}\|_\omega^M := \Big\{ \sum_i |\eta_i|_{0,\omega}^2 + \sum_{\alpha,\beta} |\gamma_{\alpha\beta}(\boldsymbol{\eta})|_{0,\omega}^2 \Big\}^{1/2}.$$

Next, *the two spaces* $\mathbf{W}_M(\omega)$ *and* $H^1(\omega)\times H^1(\omega)\times L^2(\omega)$ *coincide*, thanks again to the identities

$$\partial_{\alpha\beta}\eta_\sigma = \partial_\alpha e_{\beta\sigma}(\boldsymbol{\eta}) + \partial_\beta e_{\alpha\sigma}(\boldsymbol{\eta}) - \partial_\sigma e_{\alpha\beta}(\boldsymbol{\eta})$$

and to the *lemma of J.L. Lions* (Thm. 1.7-1).

Finally, the *closed graph theorem* shows that the identity mapping from the space $H^1(\omega) \times H^1(\omega) \times L^2(\omega)$ equipped with the product norm $\boldsymbol{\eta} = (\eta_i) \to \{\sum_\alpha \|\eta_\alpha\|_{1,\omega}^2 + |\eta_3|_{0,\omega}^2\}^{1/2}$ onto the space $\mathbf{W}_M(\omega)$ equipped with $\|\cdot\|_\omega^M$ has a continuous inverse. Hence the announced inequality holds. ∎

The next step consists in identifying sufficient conditions affording the "elimination" of the norms $|\eta_i|_{0,\omega}$ in the right-hand side of the above inequality of Korn's type. Whether it be for the three-dimensional Korn inequality in curvilinear coordinates (Thm. 1.7-4) or for the inequality of Korn's type on a general surface (Thm. 2.6-4), the corresponding eliminations simply resulted from imposing *ad hoc* boundary conditions on the displacement fields.

This idyllic scheme needs to be drastically amended in the present case!

If we were following the same pattern, we would first prove an *"infinitesimal rigid displacement lemma without boundary conditions"*, characterizing those displacement fields $\eta_i \boldsymbol{a}^i$ with covariant components $\eta_\alpha \in H^1(\omega)$ and $\eta_3 \in L^2(\omega)$ that satisfy $\gamma_{\alpha\beta}(\boldsymbol{\eta}) = 0$ in ω. The most powerful result in this direction is due to Blouza & Le Dret [1999, Thm. 6]: *If the three covariant components η_i of the displacement field are in $L^2(\omega)$ and $\gamma_{\alpha\beta}(\boldsymbol{\eta}) = 0$ in ω, then there exists $\boldsymbol{\psi} \in \mathbf{H}^{-1}(\omega)$ such that*

$$\partial_\alpha(\eta_i \boldsymbol{a}^i) = \boldsymbol{\psi} \wedge \partial_\alpha \boldsymbol{\theta}.$$

However, this characterization does not provide an *explicit* form of the corresponding displacement field (compare with Thm. 2.6-3, or with Thm. 1.7-3 in the three-dimensional case); in fact it does not even allow to conclude that the space formed by such fields is finite-dimensional!

We are thus left to *directly* establishing an *"infinitesimal rigid displacement lemma with boundary conditions"*, i.e., to finding in particular *boundary conditions* guaranteeing that the *semi-norm*

$$\boldsymbol{\eta} = (\eta_i) \to \Big\{ \sum_{\alpha,\beta} |\gamma_{\alpha\beta}(\boldsymbol{\eta})|^2_{0,\omega} \Big\}^{1/2}$$

becomes a *norm* for the displacement fields $\eta_i \boldsymbol{a}^i$ that satisfy them. Since η_3 is only in $L^2(\omega)$, the only choice consists in "trying" the boundary conditions

$$\eta_\alpha = 0 \text{ on } \gamma_0 \subset \gamma, \text{ with } \textit{length } \gamma_0 > 0.$$

It then turns out that such an infinitesimal rigid displacement lemma does hold, but only for *special geometries of the surface S* and *special subsets $\boldsymbol{\theta}(\gamma_0)$ of the "boundary" of S*. We refer to Lods & Mardare [1998a], Mardare [1998c], and Slicaru [1998] who have identified various situations of interest where this lemma holds; such situations are described in Chaps. 4 and 5.

Note in passing that *such geometrical restrictions occur "for the first time" in our derivations of inequalities of Korn's type.*

But the worst is yet to come! For, *even though this infinitesimal rigid displacement lemma "often" holds, it "seldom" implies that the norm $\boldsymbol{\eta} \to \{\sum_{\alpha,\beta} |\gamma_{\alpha\beta}(\boldsymbol{\eta})|^2_{0,\omega}\}^{1/2}$ is equivalent to the product norm* $\boldsymbol{\eta} = (\eta_i) \to \{\sum_\alpha \|\eta_\alpha\|^2_{1,\omega} + |\eta_3|^2_{0,\omega}\}^{1/2}$.

More precisely, we shall see (Thm. 2.7-3) that, under *ad hoc* regularity assumptions on the mapping $\boldsymbol{\theta}$, *these two norms are equivalent if $\gamma = \gamma_0$ and the surface S is "elliptic"* according to the definition given below. In addition, Slicaru [1997] has shown the remarkable result that, even under "minimal" regularity assumptions, *the same sufficient conditions are also necessary for the equivalence of these norms* (see Thm. 2.7-4 for a more precise statement), which thus "seldom" occurs indeed!

We begin by a fundamental definition: Let a surface $S = \boldsymbol{\theta}(\overline{\omega})$ be given, where $\boldsymbol{\theta} \in \mathcal{C}^2(\overline{\omega}; \mathcal{E}^3)$ is an injective mapping such that the two vectors $\boldsymbol{a}_\alpha = \partial_\alpha \boldsymbol{\theta}$ are linearly independent at all points of $\overline{\omega}$. Then S is **elliptic** if the symmetric matrix $(b_{\alpha\beta}(y))$ formed by the covariant components of the curvature tensor of S is positive, *or* negative, definite at all points $y \in \overline{\omega}$; or equivalently if there exists a constant c such that

$$\boxed{\sum_{\alpha} |\xi^{\alpha}|^2 \leq c|b_{\alpha\beta}(y)\xi^{\alpha}\xi^{\beta}| \text{ for all } y \in \overline{\omega} \text{ and all } (\xi^{\alpha}) \in \mathbb{R}^2;}$$

or equivalently if *the Gaussian curvature of* S (Sect. 2.2) *is everywhere strictly positive, i.e., if*

$$\boxed{\frac{1}{R_1(y)R_2(y)} > 0 \text{ for all } y \in \overline{\omega},}$$

where $R_1(y)$ and $R_2(y)$ are the principal radii of curvature of S at $\boldsymbol{\theta}(y)$. A portion of an *ellipsoid* provides an instance of elliptic surface.

In the next theorem, *analytic* functions of two real variables in an open subset of $\mathbb{R}^2$ are considered; we simply recall here their definition, referring to Dieudonné [1968] for a particularly elegant treatment of analytic functions of any finite number of real or complex variables: Let ω be an open subset of $\mathbb{R}^2$; a function $f : \omega \to \mathbb{R}$ is analytic if, given any $y = (y_1, y_2) \in \omega$, there exists $r > 0$ and $a_{mn} \in \mathbb{R}$, $m \geq 0$, $n \geq 0$, such that the open ball of radius r centered at y is in ω and

$$f(y_1', y_2') = \sum_{m,n=0}^{\infty} a_{mn}(y_1' - y_1)^m (y_2' - y_2)^n$$

for all $y' = (y_1', y_2')$ such that $|y' - y| < r$, these series being absolutely convergent.

We also recall that a function belongs to the space $\mathcal{C}^{2,1}(\overline{\omega})$ if it is in $\mathcal{C}^2(\overline{\omega})$ and its second-order partial derivatives are Lipschitz-continuous on $\overline{\omega}$.

We now prove the announced "infinitesimal rigid displacement lemma", directly under the assumptions ($\gamma_0 = \gamma$ and S is elliptic) that will eventually lead to the equivalence of norms.

For pedagogical purposes, we assume in our proof more smoothness on the boundary γ and on the mapping $\boldsymbol{\theta}$ than is necessary; in so doing, we essentially combine the proofs given by Ciarlet & Lods [1996a] and Ciarlet & Sanchez-Palencia [1996]. We refer to

Lods & Mardare [1998a] for a proof under the more general assumptions stated in the theorem below. An earlier version of this lemma is due to Vekua [1962], who proved it under the assumptions that γ is of class $\mathcal{C}^3$ and $\boldsymbol{\theta} \in W^{3,p}(\omega; \mathcal{E}^3)$, $p > 1$, using the theory of "generalized analytic functions".

Theorem 2.7-2 (infinitesimal rigid displacement lemma on an elliptic surface). *Let there be given a domain ω in $\mathbb{R}^2$ and an injective mapping $\boldsymbol{\theta} \in \mathcal{C}^{2,1}(\overline{\omega}; \mathcal{E}^3)$ such that the two vectors $\boldsymbol{a}_\alpha = \partial_\alpha \boldsymbol{\theta}$ are linearly independent at all points of $\overline{\omega}$ and such that the surface $S = \boldsymbol{\theta}(\overline{\omega})$ is elliptic. Then*

$$\left.\begin{array}{r} \boldsymbol{\eta} = (\eta_i) \in H_0^1(\omega) \times H_0^1(\omega) \times L^2(\omega), \\ \gamma_{\alpha\beta}(\boldsymbol{\eta}) = 0 \text{ in } \omega \end{array}\right\} \Rightarrow \boldsymbol{\eta} = \mathbf{0} \text{ in } \omega.$$

Proof. We give the proof under the additional assumptions that *the boundary γ is of class $\mathcal{C}^3$* and that *the components of the mapping $\boldsymbol{\theta}$ are restrictions to $\overline{\omega}$ of analytic functions in an open set $\omega' \subset \mathbb{R}^2$ containing $\overline{\omega}$.*

(i) We first note that establishing this implication is equivalent to proving a *uniqueness theorem*, viz., $\boldsymbol{\eta} = (\eta_i) = \mathbf{0}$ is the only solution in the space $H^1(\omega) \times H^1(\omega) \times L^2(\omega)$ of the linear system formed by the three partial differential equations $\gamma_{\alpha\beta}(\boldsymbol{\eta}) = 0$ in ω together with the two boundary conditions (understood in the sense of traces) $\eta_\alpha = 0$ on γ, or *in extenso*,

$$\begin{aligned} \partial_1\eta_1 \qquad\qquad\qquad - \Gamma_{11}^\sigma \eta_\sigma - b_{11}\eta_3 &= 0 \text{ in } \omega, \\ \frac{1}{2}\partial_2\eta_1 + \frac{1}{2}\partial_1\eta_2 - \Gamma_{12}^\sigma \eta_\sigma - b_{12}\eta_3 &= 0 \text{ in } \omega, \\ \partial_2\eta_2 - \Gamma_{22}^\sigma \eta_\sigma - b_{22}\eta_3 &= 0 \text{ in } \omega, \\ \eta_1 &= 0 \text{ on } \gamma, \\ \eta_2 &= 0 \text{ on } \gamma. \end{aligned}$$

(ii) *Any field $\boldsymbol{\eta} = (\eta_i) \in H_0^1(\omega) \times H_0^1(\omega) \times L^2(\omega)$ satisfying $\gamma_{\alpha\beta}(\boldsymbol{\eta}) = 0$ in ω and $\eta_\alpha = 0$ on γ is in the space $\mathcal{C}^1(\overline{\omega}) \times \mathcal{C}^1(\overline{\omega}) \times \mathcal{C}^0(\overline{\omega})$.*

This regularity result relies on a crucial observation, due to Geymonat & Sanchez-Palencia [1991, 1995]: The partial differential equations $\gamma_{\alpha\beta}(\boldsymbol{\eta}) = 0$ in ω constitute a system, of the first order with respect to the functions η_α and of order zero with respect to the function η_3, that is *"uniformly elliptic"* in the sense of Agmon, Douglis & Nirenberg [1964]. This means that there exists a constant $A > 0$ such that (here and subsequently in this part of the proof, we use the notations of Agmon, Douglis & Nirenberg [1964]):

$$A^{-1} \sum_\alpha |\xi_\alpha|^2 \leq |L(y, \boldsymbol{\xi})| \leq A \sum_\alpha |\xi_\alpha|^2$$

for all $y \in \overline{\omega}$ and $\boldsymbol{\xi} = (\xi_\alpha) \in \mathbb{R}^2$, where

$$L(y, \boldsymbol{\xi}) := \det \begin{pmatrix} \xi_1 & 0 & -b_{11}(y) \\ \frac{1}{2}\xi_2 & \frac{1}{2}\xi_1 & -b_{12}(y) \\ 0 & \xi_2 & -b_{22}(y) \end{pmatrix}.$$

The way the above matrix of order three is constructed from the equations $\gamma_{\alpha\beta}(\boldsymbol{\eta}) = 0$ should be clear; suffice it to specify that only the coefficients of the partial derivatives of the highest order for each unknown (one for η_α and zero for η_3) are taken into account. The uniform ellipticity of the system "$\gamma_{\alpha\beta}(\boldsymbol{\eta}) = 0$ in ω" thus holds since

$$L(y, \boldsymbol{\xi}) = -\frac{1}{2}(\xi_2 \ \ -\xi_1) \begin{pmatrix} b_{11}(y) & b_{12}(y) \\ b_{21}(y) & b_{22}(y) \end{pmatrix} \begin{pmatrix} \xi_2 \\ -\xi_1 \end{pmatrix}$$

in the present case, and the symmetric matrix $(b_{\alpha\beta}(y))$ is either positive, or negative, definite at all points $y \in \overline{\omega}$ by the assumed *ellipticity* of S.

In addition, the *"supplementary condition on L"* (which needs to be verified only in two dimensions, as here) is also satisfied: The degree m of the polynomial L with respect to ξ_1 and ξ_2 being two, the polynomial

$$\tau \in \mathbb{C} \to L(y, \boldsymbol{\xi} + \tau\boldsymbol{\eta}) \in \mathbb{C}$$

has exactly $\frac{m}{2} =$ *one* root τ^+ with $\text{Im}\tau^+ > 0$ for all $y \in \overline{\omega}$ and all linearly independent vectors $\boldsymbol{\xi} = (\xi_\alpha)$ and $\boldsymbol{\eta} = (\eta_\alpha)$ in $\mathbb{R}^2$.

Finally, when $\frac{m}{2} =$ *one* boundary condition, e.g., $\eta_1 = 0$ on γ, is appended to the equations "$\gamma_{\alpha\beta}(\boldsymbol{\eta}) = 0$ in ω", the *"complementing boundary condition"* is also satisfied: This means in this case that the polynomial $\tau \in \mathbb{C} \to (\tau - \tau^+)$ divides the polynomials $\tau \to c(\xi_1 + \tau\eta_1)$ and $\tau \to c(\xi_2 + \tau\eta_2)$ only if the constant c vanishes.

It then follows from Agmon, Douglis & Nirenberg [1964, Thm. 10.5] that, if γ is of class $\mathcal{C}^3$ and the coefficients of the uniformly elliptic system $\gamma_{\alpha\beta}(\boldsymbol{\eta}) = 0$ are in the space $\mathcal{C}^2(\overline{\omega})$, any solution $\boldsymbol{\eta} \in H^1(\omega) \times H^1(\omega) \times L^2(\omega)$ of $\gamma_{\alpha\beta}(\boldsymbol{\eta}) = 0$ in ω together with, e.g., $\eta_1 = 0$ on γ is in the space $H^3(\omega) \times H^3(\omega) \times H^2(\omega)$. The assertion then follows from the continuous imbeddings $H^m(\omega) \hookrightarrow \mathcal{C}^{m-2}(\overline{\omega})$, $m \geq 2$.

(iii) *Local uniqueness of the solution of "$\gamma_{\alpha\beta}(\boldsymbol{\eta}) = 0$ in ω and $\eta_\alpha = 0$ on γ".*

The assumed ellipticity of the surface S shows that there exists a constant $c > 0$ such that $|b_{11}(y)| \geq c$ for all $y \in \overline{\omega}$. Hence the unknown η_3 may be eliminated by means of the equation $\gamma_{11}(\boldsymbol{\eta}) = 0$ (naturally, η_3 could be likewise eliminated by means of the equation $\gamma_{22}(\boldsymbol{\eta}) = 0$). This elimination shows that

$$\eta_3 = \frac{1}{b_{11}}(\partial_1 \eta_1 - \Gamma_{11}^\sigma \eta_\sigma)$$

and that η_1 and η_2 are solutions of the *"reduced" system*

$$\begin{aligned}
-2\frac{b_{12}}{b_{11}}\partial_1\eta_1 + \partial_2\eta_1 + \partial_1\eta_2 - 2\Big(\Gamma_{12}^\sigma - \frac{b_{12}}{b_{11}}\Gamma_{11}^\sigma\Big)\eta_\sigma &= 0 \text{ in } \omega,\\
-\frac{b_{22}}{b_{11}}\partial_1\eta_1 + \partial_2\eta_2 - \Big(\Gamma_{22}^\sigma - \frac{b_{22}}{b_{11}}\Gamma_{11}^\sigma\Big)\eta_\sigma &= 0 \text{ in } \omega,\\
\eta_1 &= 0 \text{ on } \gamma,\\
\eta_2 &= 0 \text{ on } \gamma.
\end{aligned}$$

Since *the coefficients of this reduced system are analytic in ω'* and since *the boundary γ is of class $\mathcal{C}^3$ and is not a characteristic curve for this system*, as is easily verified by using again the assumed *ellipticity* of the surface S, *Holmgren's uniqueness theorem* (see, e.g., Courant & Hilbert [1962, p. 238], Bers, John & Schechter [1964, p. 47], or Dautray & Lions [1984, Chap. 5, Sect. 1]) shows that "locally", i.e., *in a small enough open neighborhood $\tilde{\omega} \subset \omega'$ of any point of γ, $(\eta_1, \eta_2) = (0, 0)$ is the unique solution to this reduced system in $\mathcal{C}^1(\tilde{\omega}) \times \mathcal{C}^1(\tilde{\omega})$*. Recalling that any solution $\boldsymbol{\eta} = (\eta_i)$ of the "full" system is such that $\eta_\alpha \in \mathcal{C}^1(\overline{\omega})$ by (ii), we have thus shown: *Any point of γ possesses an open neighborhood $\tilde{\omega} \subset \omega'$ such that the only solution $\boldsymbol{\eta} = (\eta_i) \in H^1(\omega) \times H^1(\omega) \times L^2(\omega)$ of the "full" system "$\gamma_{\alpha\beta}(\boldsymbol{\eta}) = 0$ in ω and $\eta_\alpha = 0$ on γ" is $\boldsymbol{\eta} = \mathbf{0}$ in $\tilde{\omega} \cap \overline{\omega}$.*

(iv) *Global uniqueness of the solution of "$\gamma_{\alpha\beta}(\boldsymbol{\eta}) = 0$ in ω and $\eta_\alpha = 0$ on γ".*

By a theorem of Morrey & Nirenberg [1957], *any solution of a uniformly elliptic system whose coefficients are analytic in ω is analytic in ω.* Since any solution $\boldsymbol{\eta} = (\eta_i)$ of "$\gamma_{\alpha\beta}(\boldsymbol{\eta}) = 0$ in ω and $\eta_\alpha = 0$ on γ" vanishes in a neighborhood of γ by (iii) and $\boldsymbol{\eta} = \mathbf{0}$ is an analytic solution, the *analytic continuation theorem* for analytic functions of several variables (see, e.g., Dieudonné [1968, Thm. 9.4.2]) thus shows that $\boldsymbol{\eta} = \mathbf{0}$ is the only solution. ∎

Several comments are in order about this theorem and its proof:

As expected, the infinitesimal rigid displacement lemma is an "intrinsic" property of an elliptic surface: It also holds if the same surface is equipped with another system of curvilinear coordinates; cf. Ex. 2.11.

The regularity theorem of Agmon, Douglis & Nirenberg [1964] used in part (ii) requires that only *one* boundary condition be added to the partial differential equations $\gamma_{\alpha\beta}(\boldsymbol{\eta}) = 0$ in ω, but this boundary condition must be imposed on the *entire* boundary γ. Hence this analysis precludes the consideration of boundary conditions such as $\eta_\alpha = 0$ on γ_0 with *length* $\gamma_0 <$ *length* γ.

A particularly interesting discussion about Holmgren's theorem and uniformly elliptic systems, especially adapted to linear shell theory, is given in Sanchez-Hubert & Sanchez-Palencia [1997, Chaps. 2 and 3].

We already mentioned that the regularity assumptions "$\boldsymbol{\theta}$ analytic in $\omega' \supset \overline{\omega}$ and γ of class $\mathcal{C}^3$" can be substantially weakened. For instance, Ciarlet & Lods [1996a] assumed instead "$\boldsymbol{\theta} \in \mathcal{C}^5(\overline{\omega}; \mathcal{E}^3)$ and γ of class $\mathcal{C}^4$". The analog of part (ii) in this case shows that the functions η_1 and η_2 are in $\mathcal{C}^2(\overline{\omega})$, again by resorting to the regularity theorem of Agmon, Douglis & Nirenberg [1964]; the "local uniqueness" of the solution η_1 and η_2 of the reduced system (part (iii)) is then obtained by combining results of Carleman [1938] and Calderón [1958] (by contrast with Holmgren's theorem, these results do not require the assumption of analyticity of the coefficients); finally, the "global uniqueness" (part (iv)) is obtained through the unique continuation theorem of Aronszajn [1956] applied to an elliptic equation satisfied by a single unknown, the auxiliary function $(\partial_1\eta_2 - \partial_2\eta_1)$ (also introduced in a different context by Bernadou, Ciarlet & Miara [1994]).

The final word in this direction seems to have been achieved by Lods & Mardare [1998a] who were able to weaken the assumptions to those stated in Thm. 2.7-2, i.e., to "$\boldsymbol{\theta} \in \mathcal{C}^{2,1}(\overline{\omega}; \mathcal{E}^3)$ and γ is Lipschitz-continuous". To this end, they notably made use of the unique continuation theorem of Hörmander [1983].

In some special cases, Thm. 2.7-2 is easier to establish. For instance, an elegant and short proof, due to Brezzi [1994], applies to the special case where $\boldsymbol{\theta}(y_1, y_2) = (y_1, y_2, f(y_1, y_2))$ (Ex. 2.12); as noted by Ciarlet & Lods [1996a], a portion of a sphere is likewise amenable to a simple proof (Ex. 2.13).

It seems, however, that under the minimal regularity assumption "$\boldsymbol{\theta} \in \mathcal{C}^2(\overline{\omega}; \mathcal{E}^3)$", all that can be proved is the *finite-dimensionality* of the space

$$\left\{\boldsymbol{\eta} = (\eta_i) \in H_0^1(\omega) \times H_0^1(\omega) \times L^2(\omega);\ \gamma_{\alpha\beta}(\boldsymbol{\eta}) = 0 \text{ in } \omega\right\}.$$

This finite-dimensionality was established by Ciarlet & Lods [1996a], who reduced this issue to showing that the pair (η_1, η_2) belongs to the kernel of an operator of the form $(\mathbf{I} - \mathbf{T})$, with $\mathbf{T}$ compact. As noted by Geymonat & Sanchez-Palencia [1991], this issue can also be resolved by resorting to Agmon, Douglis & Nirenberg [1964] or Geymonat [1965], under the assumptions "$\boldsymbol{\theta} \in \mathcal{C}^2(\overline{\omega}; \mathcal{E}^3)$ and γ of class $\mathcal{C}^1$".

We are now in a position to prove the main result of this section, due to Ciarlet & Lods [1996a] and Ciarlet & Sanchez-Palencia [1996]; special mention must also be made of the early existence and uniqueness theorem for elliptic surfaces of Destuynder [1985, Thms. 6.1 and 6.5], obtained under the additional assumptions that the surface S can be covered by a single system of lines of curvature (Sect. 2.2) and that the $\mathcal{C}^0(\overline{\omega})$-norms of the corresponding Christoffel symbols of S are small enough.

In particular, Thm. 2.7-3 yields the *existence* and *uniqueness* of the solution to the *two-dimensional equations of a linearly elastic membrane shell*, as explained at the beginning of this section. The subscript "M" is appended to the constant as a reminder of this property.

Theorem 2.7-3 (inequality of Korn's type on an elliptic surface). *Let ω be a domain in $\mathbb{R}^2$ and let $\boldsymbol{\theta} \in \mathcal{C}^{2,1}(\overline{\omega}; \mathcal{E}^3)$ be an injective mapping such that the two vectors $\boldsymbol{a}_\alpha = \partial_\alpha \boldsymbol{\theta}$ are linearly* *independent at all points of* $\overline{\omega}$ *and such that the surface* $S = \boldsymbol{\theta}(\overline{\omega})$ *is elliptic.*

Given $\boldsymbol{\eta} = (\eta_i) \in H^1(\omega) \times H^1(\omega) \times L^2(\omega)$, *let*

$$\gamma_{\alpha\beta}(\boldsymbol{\eta}) = \Big\{\frac{1}{2}(\partial_\beta \eta_\alpha + \partial_\alpha \eta_\beta) - \Gamma^\sigma_{\alpha\beta}\eta_\sigma - b_{\alpha\beta}\eta_3\Big\} \in L^2(\omega)$$

denote the covariant components of the linearized change of metric tensor associated with the displacement field $\eta_i \boldsymbol{a}^i$ *of the surface* S. *Then there exists a constant* $c_M = c_M(\omega, \boldsymbol{\theta})$ *such that*

$$\Big\{\sum_\alpha \|\eta_\alpha\|^2_{1,\omega} + |\eta_3|^2_{0,\omega}\Big\}^{1/2} \le c_M\Big\{\sum_{\alpha,\beta} |\gamma_{\alpha\beta}(\boldsymbol{\eta})|^2_{0,\omega}\Big\}^{1/2}$$
$$\text{for all } \boldsymbol{\eta} = (\eta_i) \in \mathbf{V}_M(\omega) := H^1_0(\omega) \times H^1_0(\omega) \times L^2(\omega).$$

Proof. (i) By the *second inequality of Korn's type "without boundary conditions" on a general surface* (Thm. 2.7-1), there exists a constant c_0 such that

$$\|\boldsymbol{\eta}\|_{H^1(\omega)\times H^1(\omega)\times L^2(\omega)} := \Big\{\sum_\alpha \|\eta_\alpha\|^2_{1,\omega} + |\eta_3|^2_{0,\omega}\Big\}^{1/2}$$
$$\le c_0\Big\{\sum_i |\eta_i|^2_{0,\omega} + \sum_{\alpha,\beta} |\gamma_{\alpha\beta}(\boldsymbol{\eta})|^2_{0,\omega}\Big\}^{1/2}$$

for all $\boldsymbol{\eta} \in H^1(\omega) \times H^1(\omega) \times L^2(\omega) \supset \mathbf{V}_M(\omega)$. Hence *it suffices to show that there exists a constant c such that*

$$\Big\{\sum_i |\eta_i|^2_{0,\omega}\Big\}^{1/2} \le c\Big\{\sum_{\alpha,\beta} |\gamma_{\alpha\beta}(\boldsymbol{\eta})|^2_{0,\omega}\Big\}^{1/2} \text{ for all } \boldsymbol{\eta} \in \mathbf{V}_M(\omega).$$

(ii) If the last inequality is false, there exists a sequence $(\boldsymbol{\eta}^k)_{k=1}^\infty$ of functions $\boldsymbol{\eta}^k = (\eta_i^k) \in \mathbf{V}_M(\omega)$ such that

$$\Big\{\sum_i |\eta_i^k|^2_{0,\omega}\Big\}^{1/2} = 1 \text{ for all } k \text{ and } \lim_{k\to\infty}\Big\{\sum_{\alpha,\beta} |\gamma_{\alpha\beta}(\boldsymbol{\eta}^k)|^2_{0,\omega}\Big\}^{1/2} = 0.$$

In particular then, the sequence $(\boldsymbol{\eta}^k)_{k=1}^\infty$ is bounded with respect to the norm $\|\cdot\|_{H^1(\omega)\times H^1(\omega)\times L^2(\omega)}$, thanks again to the *second inequality of Korn's type* of Thm. 2.7-1. Since any bounded sequence in a Hilbert space contains a weakly convergent subsequence (see, e.g., Vol. I, Thm. 7.1-4), *there exists a subsequence* $(\boldsymbol{\eta}^l)_{l=1}^\infty$ *and an element* $\boldsymbol{\eta} = (\eta_i) \in \mathbf{V}_M(\omega)$ *such that*

$$\eta_\alpha^l \rightharpoonup \eta_\alpha \text{ in } H^1(\omega) \text{ and } \eta_\alpha^l \to \eta_\alpha \text{ in } L^2(\omega),$$

$$\eta_3^l \rightharpoonup \eta_3 \text{ in } L^2(\omega),$$

where $\rightharpoonup$ and $\to$ denote weak and strong convergences (the compact imbedding $H^1(\omega) \Subset L^2(\omega)$ is also used here; see, e.g., Vol. I, Thm. 6.1-5).

(iii) Naturally, the difficulty rests with the subsequence $(\eta_3^l)_{l=1}^\infty$, which converges only *weakly* in $L^2(\omega)$. Our recourse for showing that it in fact *strongly* converges in $L^2(\omega)$ will be the assumed ellipticity of the surface S (see (iv)), but *first, we prove that* $\boldsymbol{\eta} = (\eta_i) = \mathbf{0}$. To this end, we simply note that $\eta_\alpha^l \rightharpoonup \eta_\alpha$ in $H^1(\omega)$ and $\eta_3^l \rightharpoonup \eta_3$ in $L^2(\omega)$ imply that

$$\gamma_{\alpha\beta}(\boldsymbol{\eta}^l) \rightharpoonup \gamma_{\alpha\beta}(\boldsymbol{\eta}) \text{ in } L^2(\omega)$$

on the one hand; since

$$\gamma_{\alpha\beta}(\boldsymbol{\eta}^l) \to 0 \text{ in } L^2(\omega)$$

on the other, we conclude that $\gamma_{\alpha\beta}(\boldsymbol{\eta}) = 0$. Hence $\boldsymbol{\eta} = \mathbf{0}$ by Thm. 2.7-2.

(iv) *We next show that* $\eta_3^l \to 0$ in $L^2(\omega)$. The strong convergences

$$\gamma_{\alpha\beta}(\boldsymbol{\eta}^l) \to 0 \text{ in } L^2(\omega) \text{ and } \eta_\alpha^l \to 0 \text{ in } L^2(\omega)$$

combined with the definition of the functions $\gamma_{\alpha\beta}(\boldsymbol{\eta})$ imply the following strong convergences:

$$\begin{aligned}
\partial_1\eta_1^l - b_{11}\eta_3^l &= \{\gamma_{11}(\boldsymbol{\eta}^l) + \Gamma_{11}^\sigma \eta_\sigma^l\} \quad \to 0 \text{ in } L^2(\omega),\\
\partial_2\eta_1^l + \partial_1\eta_2^l - 2b_{12}\eta_3^l &= \{2\gamma_{12}(\boldsymbol{\eta}^l) + 2\Gamma_{12}^\sigma \eta_\sigma^l\} \to 0 \text{ in } L^2(\omega),\\
\partial_2\eta_2^l - b_{22}\eta_3^l &= \{\gamma_{22}(\boldsymbol{\eta}^l) + \Gamma_{22}^\sigma \eta_\sigma^l\} \quad \to 0 \text{ in } L^2(\omega).
\end{aligned}$$

Since the function $b_{11} \in \mathcal{C}^0(\overline{\omega})$ does not vanish in $\overline{\omega}$ by the assumed *ellipticity* of the surface S, we can eliminate η_3^l between the first and second, and between the first and third, relations; this elimination yields:

$$\left\{\partial_2\eta_1^l + \partial_1\eta_2^l - 2\frac{b_{12}}{b_{11}}\partial_1\eta_1^l\right\} \to 0 \text{ in } L^2(\omega),$$

$$\left\{\partial_2\eta_2^l - \frac{b_{22}}{b_{11}}\partial_1\eta_1^l\right\} \to 0 \text{ in } L^2(\omega).$$

Multiplying the first relation by $\partial_2\eta_1^l$ and the second by $\partial_1\eta_1^l$, then integrating over ω, we get

$$\int_\omega \left\{(\partial_2\eta_1^l)^2 + \partial_2\eta_1^l\partial_1\eta_2^l - 2\frac{b_{12}}{b_{11}}\partial_1\eta_1^l\partial_2\eta_1^l\right\}dy \to 0,$$

$$\int_\omega \left\{\partial_1\eta_1^l\partial_2\eta_2^l - \frac{b_{22}}{b_{11}}(\partial_1\eta_1^l)^2\right\}dy \to 0,$$

since each sequence $(\partial_\alpha\eta_1^l)_{l=1}^\infty$ is bounded in $L^2(\omega)$ (each sequence even weakly converges to 0 in $L^2(\omega)$). Subtracting the last two relations and using the relation $\int_\omega \partial_2\eta_1\partial_1\eta_2 dy = \int_\omega \partial_1\eta_1\partial_2\eta_2 dy$ satisfied by all $(\eta_1, \eta_2) \in H_0^1(\omega) \times H_0^1(\omega)$, we thus obtain

$$\int_\omega \left\{\left(\partial_2\eta_1^l - \frac{b_{12}}{b_{11}}\partial_1\eta_1^l\right)^2 + \frac{1}{(b_{11})^2}\left(b_{11}b_{22} - (b_{12})^2\right)\left(\partial_1\eta_1^l\right)^2\right\}dy \to 0.$$

Consequently,

$$\partial_1\eta_1^l \to 0 \text{ in } L^2(\omega),$$

since $b_{11}b_{22} - (b_{12})^2 = \det(b_{\alpha\beta}) \in \mathcal{C}^0(\overline{\omega})$ does not vanish in $\overline{\omega}$ by the assumed *ellipticity* of S. Hence

$$\eta_3^l = \left\{\frac{1}{b_{11}}\partial_1\eta_1^l - \frac{1}{b_{11}}\left(\partial_1\eta_1^l - b_{11}\eta_3^l\right)\right\} \to 0 \text{ in } L^2(\omega).$$

(v) The relations $\eta_i^l \to 0$ in $L^2(\omega)$ established in parts (iii) and (iv) thus contradict the relations $\{\sum_i |\eta_i^l|_{0,\omega}^2\}^{1/2} = 1$ for all l, and the proof is complete. ∎

Finally, we state without proof a noteworthy result of Slicaru [1997, 1998], who showed that the *sufficient* conditions of Thm. 2.7-3 for obtaining an inequality of Korn's type (of the form given in this theorem) are also *necessary*!

Theorem 2.7-4. *Let ω be a domain in $\mathbb{R}^2$, let γ_0 be a $d\gamma$-measurable subset of $\gamma = \partial\omega$, and let $\boldsymbol{\theta} \in \mathcal{C}^2(\overline{\omega}; \mathcal{E}^3)$ be an injective mapping such that the two vectors $\boldsymbol{a}_\alpha = \partial_\alpha\boldsymbol{\theta}$ are linearly independent at all points of $\overline{\omega}$. Assume that there exists a constant c such that*

$$\left\{\sum_\alpha \|\eta_\alpha\|_{1,\omega}^2 + |\eta_3|_{0,\omega}^2\right\}^{1/2} \le c\left\{\sum_{\alpha,\beta} |\gamma_{\alpha\beta}(\boldsymbol{\eta})|_{0,\omega}^2\right\}^{1/2}$$

for all $\boldsymbol{\eta} = (\eta_i) \in \mathbf{V}(\omega; \gamma_0)$, *where*

$$\mathbf{V}(\omega; \gamma_0) := \{\boldsymbol{\eta} = (\eta_i) \in H^1(\omega) \times H^1(\omega) \times L^2(\omega);\ \eta_\alpha = 0 \text{ on } \gamma_0\}.$$

Then $\gamma_0 = \gamma$ *and the surface* $S = \boldsymbol{\theta}(\overline{\omega})$ *is elliptic.* ■

2.8♭. COMPLEMENT: RECOVERY OF A SURFACE FROM ITS METRIC AND CURVATURE TENSOR FIELDS

The content of this section should be advantageously compared to that of Sect. 1.9. Let ω be an open subset in $\mathbb{R}^2$ and let $\boldsymbol{\theta} : \omega \to \mathcal{E}^3$ be a thrice continuously differentiable mapping such that the two vectors $\boldsymbol{a}_\alpha = \partial_\alpha \boldsymbol{\theta}$ are linearly independent everywhere in ω. As before, let $\boldsymbol{a}_3 = \dfrac{\boldsymbol{a}_1 \wedge \boldsymbol{a}_2}{|\boldsymbol{a}_1 \wedge \boldsymbol{a}_2|}$, let the three vectors $\boldsymbol{a}^i$ be defined by the relations $\boldsymbol{a}^j \cdot \boldsymbol{a}_i = \delta_i^j$, let $a_{\alpha\beta} = \boldsymbol{a}_\alpha \cdot \boldsymbol{a}_\beta$ and $b_{\alpha\beta} = \boldsymbol{a}_3 \cdot \partial_\beta \boldsymbol{a}_\alpha$, and let the Christoffel symbols be defined by $\Gamma_{\alpha\beta}^\sigma = \boldsymbol{a}^\sigma \cdot \partial_\beta \boldsymbol{a}_\alpha$. In addition, let the *Christoffel symbols of the first kind* $\Gamma_{\alpha\beta\tau}$ be defined by

$$\Gamma_{\alpha\beta\sigma} := a_{\sigma\tau}\Gamma_{\alpha\beta}^\tau.$$

In this context, the functions $\Gamma_{\alpha\beta}^\sigma$ are also called the *Christoffel symbols of the second kind.*

Then it can be shown that the Christoffel symbols and the covariant components of the curvature tensor satisfy the following *compatibility conditions* (Ex. 2.14):

$$\begin{aligned}
\partial_\tau \Gamma_{\alpha\sigma\beta} - \partial_\sigma \Gamma_{\alpha\tau\beta} + \Gamma_{\alpha\tau}^\mu \Gamma_{\beta\sigma\mu} - \Gamma_{\alpha\sigma}^\lambda \Gamma_{\beta\tau\lambda} &= b_{\alpha\sigma} b_{\beta\tau} - b_{\alpha\tau} b_{\beta\sigma} \text{ in } \omega, \\
\partial_\sigma b_{\alpha\beta} - \partial_\beta b_{\alpha\sigma} + \Gamma_{\alpha\beta}^\tau b_{\tau\sigma} - \Gamma_{\alpha\sigma}^\tau b_{\tau\beta} &= 0 \text{ in } \omega.
\end{aligned}$$

These conditions are in effect relations between partial derivatives of the first, second, and third order of the mapping $\boldsymbol{\theta}$. They are also relations between the covariant components $a_{\alpha\beta}$ and $b_{\alpha\beta}$ of the metric and curvature tensors and their derivatives, since the Christoffel symbols $\Gamma_{\alpha\beta\tau}$ may be directly defined in terms of the functions $a_{\alpha\beta}$ as (Ex. 2.14):

$$\Gamma_{\alpha\beta\tau} = \frac{1}{2}(\partial_\beta a_{\alpha\tau} + \partial_\alpha a_{\beta\tau} - \partial_\tau a_{\alpha\beta}).$$

Remarkably, these *necessary* conditions are also *sufficient* for the existence of a mapping $\boldsymbol{\theta} : \omega \subset \mathbb{R}^2 \to \mathcal{E}^3$ whose metric and curvature tensor fields are given on ω:

Theorem 2.8-1. *Let ω be a simply connected open subset of $\mathbb{R}^2$, and let there be given a twice continuously differentiable, symmetric, and positive definite matrix field $(a_{\alpha\beta})$ on ω and a continuously differentiable symmetric matrix field $(b_{\alpha\beta})$ on ω that together satisfy*

$$\partial_\tau \Gamma_{\alpha\sigma\beta} - \partial_\sigma \Gamma_{\alpha\tau\beta} + \Gamma^\mu_{\alpha\tau}\Gamma_{\beta\sigma\mu} - \Gamma^\lambda_{\alpha\sigma}\Gamma_{\beta\tau\lambda} = b_{\alpha\sigma}b_{\beta\tau} - b_{\alpha\tau}b_{\beta\sigma} \text{ in } \omega,$$

$$\partial_\sigma b_{\alpha\beta} - \partial_\beta b_{\alpha\sigma} + \Gamma^\tau_{\alpha\beta}b_{\tau\sigma} - \Gamma^\tau_{\alpha\sigma}b_{\tau\beta} = 0 \text{ in } \omega, \text{ where}$$

$$\Gamma_{\alpha\beta\tau} := \frac{1}{2}(\partial_\beta a_{\alpha\tau} + \partial_\alpha a_{\beta\tau} - \partial_\tau a_{\alpha\beta}),$$

$$\Gamma^\sigma_{\alpha\beta} := a^{\sigma\tau}\Gamma_{\alpha\beta\tau}, \quad (a^{\sigma\tau}) := (a_{\alpha\beta})^{-1}.$$

Then there exists a mapping $\boldsymbol{\theta} \in \mathcal{C}^3(\omega; \mathbb{R}^3)$ such that

$$\partial_\alpha\boldsymbol{\theta} \cdot \partial_\beta\boldsymbol{\theta} = a_{\alpha\beta} \text{ in } \omega,$$

$$\frac{\partial_1\boldsymbol{\theta} \wedge \partial_2\boldsymbol{\theta}}{|\partial_1\boldsymbol{\theta} \wedge \partial_2\boldsymbol{\theta}|} \cdot \partial_{\alpha\beta}\boldsymbol{\theta} = b_{\alpha\beta} \text{ in } \omega.$$

Furthermore, this mapping is unique "up to rigid deformations in $\mathbb{R}^3$": This means that any other solution is necessarily of the form $y \in \omega \to \mathbf{c} + \mathbf{Q}\boldsymbol{\theta}(y)$, where $\mathbf{c}$ is a vector in $\mathbb{R}^3$ and $\mathbf{Q}$ is an orthogonal matrix of order three. ■

For a direct proof of this delicate result, see Klingenberg [1973, Thm. 3.8.8]. Otherwise, Thm. 2.8-1 can be also obtained as a corollary to Thm. 1.9-1; see Ciarlet & Larsonneur [2000].

The first relations constitute one version of the *"Theorema egregium"* of Gauß [1828] (Ex. 2.14), while the second ones are equivalent to the "*Codazzi-Mainardi identities*" (Ex. 2.7).

EXERCISES

2.1. (1) Compute the vectors of the covariant and contravariant bases, the area and length elements, the Christoffel symbols, the covariant and contravariant components of the metric tensor, and the covariant and mixed components of the curvature tensor corresponding to a portion of a circular cylinder and to a portion of a torus equipped with the curvilinear coordinates shown in Fig. 2.1-2.

(2) Carry out the same computations for a portion of a sphere equipped with the three kinds of curvilinear coordinates shown in Fig. 2.1-3.

(3) In each case considered in (2), verify that the radius of the sphere indeed satisfies the relation established in Thm. 2.2-1.

(4) Carry out the same computations as in (1) for a portion of a hyperbolic paraboloid represented by a mapping of the form

$$(x, y) \to (x, y, \frac{h}{ab} xy),$$

where a, b, and h are three given lengths.

2.2. Show that the relation $dS(\hat{y}) = \sqrt{a(y)}\, dy$ providing the area element at $\hat{y} = \boldsymbol{\theta}(y) \in S = \boldsymbol{\theta}(\overline{\omega})$ (Thm. 2.1-1) can be recovered from the relation (Thm. 1.2-1)

$$d\hat{\Gamma}(\hat{x}) = |\mathbf{Cof}\,\boldsymbol{\nabla\Theta}(x)\boldsymbol{n}(x)|\, d\Gamma(x) \text{ at } \hat{x} = \boldsymbol{\Theta}(x) \in \hat{\Gamma} = \boldsymbol{\Theta}(\Gamma),$$

by specializing the mapping $\boldsymbol{\Theta}$ to be the canonical extension of the mapping $\boldsymbol{\theta}$ (Thm. 3.1-1).

2.3. The notations and assumptions are as in Sect. 2.1. Let $y \in \overline{\omega}$ be given. Show that the angle $\alpha(y)$ between the coordinate lines passing through the point $\boldsymbol{\theta}(y)$ satisfies

$$\cos \alpha(y) = \frac{a_{12}(y)}{\{a_{11}(y)a_{22}(y)\}^{1/2}}.$$

2.4. Let the assumptions and notations be as in Thm. 2.2-1. A point $\boldsymbol{\theta}(y) \in S$ is called a *planar point* if $\frac{1}{R_1(y)} = \frac{1}{R_2(y)} = 0$. Show that, if all the points of S are planar, S is a portion of a plane.

Hint: See, e.g., Stoker [1969, p. 87].

2.5. Let the assumptions and notations be as in Thm. 2.2-1 and assume in addition that $\boldsymbol{\theta} \in \mathcal{C}^3(\overline{\omega}; \mathcal{E}^3)$. A point $\boldsymbol{\theta}(y) \in S$ is called an *umbilical point* if $\dfrac{1}{R_1(y)} = \dfrac{1}{R_2(y)} \neq 0$. Show that, if all the points of S are umbilical, S is a portion of a sphere.

Hint: See, e.g., Stoker [1969, p. 99].

2.6. Let $K : S \to \mathbb{R}$ denote the Gaussian curvature of a (smooth enough) surface S.

(1) Let S be a sphere; show that $\int_S K(\hat{y})\, dS(\hat{y}) = 4\pi$.

(2) Let S be a torus; show that $\int_S K(\hat{y})\, dS(\hat{y}) = 0$.

Remark. These are instances of applications of the *Gauß-Bonnet theorem* (Sect. 2.2).

2.7. Let $b_{\alpha\beta|\sigma} := \partial_\sigma b_{\alpha\beta} - \Gamma^\tau_{\alpha\sigma} b_{\tau\beta} - \Gamma^\tau_{\beta\sigma} b_{\alpha\tau}$ denote the *first-order covariant derivatives of the curvature tensor*, defined here by means of its *covariant* components. Show that these covariant derivatives satisfy the *Codazzi-Mainardi identities*

$$b_{\alpha\beta|\sigma} = b_{\alpha\sigma|\beta},$$

which are themselves equivalent to the relations (Thm. 2.8-1)

$$\partial_\sigma b_{\alpha\beta} - \partial_\beta b_{\alpha\sigma} + \Gamma^\tau_{\alpha\beta} b_{\tau\sigma} - \Gamma^\tau_{\alpha\sigma} b_{\tau\beta} = 0.$$

Hint: The proof is analogous to that given in Thm. 2.5-1 for establishing the relations $b^\tau_\beta|_\alpha = b^\tau_\alpha|_\beta$.

2.8. Let the assumptions on the mapping $\boldsymbol{\theta}$ be as in Thm. 2.6-3.

(1) Let there be given $\boldsymbol{\eta} = (\eta_i) \in \mathbf{H}^1(\omega)$ such that $\gamma_{\alpha\beta}(\boldsymbol{\eta}) = 0$ in ω. Show that

$$\partial_\alpha(\eta_i \boldsymbol{a}^i) = \boldsymbol{\psi} \wedge \partial_\alpha \boldsymbol{\theta},$$

where

$$\boldsymbol{\psi} := \varepsilon^{\alpha\beta}(\partial_\beta \eta_3 + b^\sigma_\beta \eta_\sigma)\boldsymbol{a}_\alpha + \frac{1}{2}\varepsilon^{\alpha\beta}\eta_{\beta|\alpha}\boldsymbol{a}_3,$$

with $\varepsilon^{11} = \varepsilon^{22} = 0$, $\varepsilon^{12} = \dfrac{1}{\sqrt{a}}$, and $\varepsilon^{21} = -\dfrac{1}{\sqrt{a}}$.

Remark. The vector field $\boldsymbol{\psi}$ is the *infinitesimal rotation field*, introduced by Vekua [1962], then put to various uses by Bernadou & Ciarlet [1976, Lemma 2.5], Choi [1993], Choi & Sanchez-Palencia [1993]. Its existence under minimal regularity assumptions is established by Blouza & Le Dret [1999, Thm. 6].

(2) Let there be given $\boldsymbol{\eta} = (\eta_i) \in H^1(\omega) \times H^1(\omega) \times H^2(\omega)$ such that $\gamma_{\alpha\beta}(\boldsymbol{\eta}) = \rho_{\alpha\beta}(\boldsymbol{\eta}) = 0$ in ω. Show that the vector field $\boldsymbol{\psi}$ found in (1) is constant in this case and equal to the vector $\hat{\boldsymbol{d}}$ found in Thm. 2.6-3 (a).

2.9. Let the assumptions on the mapping $\boldsymbol{\theta}$ be as in Thm. 2.6-3 and let γ_0 be a relatively open subset of γ that satisfies *length* $\gamma_0 > 0$.

(1) Show that the implication (the analog of Thm. 2.6-3 (b) for another class of boundary conditions):

$$\left.\begin{aligned} \boldsymbol{\eta} = (\eta_i) \in H^1(\omega) \times H^1(\omega) \times H^2(\omega), \\ \eta_i = 0 \text{ on } \gamma_0, \\ \gamma_{\alpha\beta}(\boldsymbol{\eta}) = \rho_{\alpha\beta}(\boldsymbol{\eta}) = 0 \text{ in } \omega \end{aligned}\right\} \Rightarrow \boldsymbol{\eta} = \boldsymbol{0} \text{ in } \omega$$

holds if and only if $\boldsymbol{\theta}(\gamma_0)$ is not a subset of a straight line.

Hint: One way to solve this problem is to combine Thm. 2.6-3 (a) with Ex. 2.8.

(2) Assuming that the implication of (1) holds, show that an *inequality of Korn's type on a general surface* analogous to that of Thm. 2.6-4 holds, where the space

$$\mathbf{V}_K^s(\omega) := \{\boldsymbol{\eta} = (\eta_i) \in H^1(\omega) \times H^1(\omega) \times H^2(\omega);\ \eta_i = 0 \text{ on } \gamma_0\}$$

replaces the space $\mathbf{V}_K(\omega)$ found in *ibid.* The exponent "s" reminds that the space $\mathbf{V}_K^s(\omega)$ corresponds to a shell that is *simply supported* along the curve $\boldsymbol{\theta}(\gamma_0)$, i.e., whose displacement field $\zeta_i \boldsymbol{a}^i$ satisfies the *two-dimensional boundary conditions of simple support* $\zeta_i = 0$ on γ_0.

2.10. (1) The notations and assumptions are the same as in Thm. 2.6-6. Show that the space $\widetilde{\mathbf{V}}_K^s(\omega)$ becomes a Hilbert space when it is equipped with the norm

$$\tilde{\boldsymbol{\eta}} \to \Big\{ \|\tilde{\boldsymbol{\eta}}\|_{1,\omega}^2 + \sum_{\alpha,\beta} |\partial_{\alpha\beta}\tilde{\boldsymbol{\eta}} \cdot \boldsymbol{a}_3|_{0,\omega}^2 \Big\}^{1/2}.$$

(2) Assume in addition that $\boldsymbol{\theta} \in W^{3,\infty}(\omega; \mathcal{E}^3)$. Show that the mapping $\boldsymbol{\eta} = (\eta_i) \to \tilde{\boldsymbol{\eta}} = \eta_i \boldsymbol{a}^i$ is an isomorphism between the space

$$\mathbf{V}_K^s(\omega) := H_0^1(\omega) \times H_0^1(\omega) \times \{H^2(\omega) \cap H_0^1(\omega)\},$$

equipped with the product norm $\boldsymbol{\eta} \to \{\sum_\alpha \|\eta_\alpha\|_{1,\omega}^2 + \|\eta_3\|_{2,\omega}\}^{1/2}$, and the space $\widetilde{\mathbf{V}}_K^s(\omega)$ equipped with the norm of (1).

Remark. This observation is due to Blouza & Le Dret [1999, Lemma 4].

2.11. Let an elliptic surface be equipped with two different systems of curvilinear coordinates. Show that, if the infinitesimal rigid displacement lemma on an elliptic surface (Thm. 2.7-2) holds for one system, it holds for the other.

2.12. Let $\omega \subset \mathbb{R}^2$ be a domain with a boundary of class $\mathcal{C}^2$.

(1) Let there be given functions $A_{\alpha\beta} \in \mathcal{C}^1(\overline{\omega})$ satisfying the following "ellipticity condition": There exists a constant $c > 0$ such that

$$A_{\alpha\beta}(y)\xi^\alpha\xi^\beta \geq c\sum_\alpha |\xi^\alpha|^2 \text{ for all } y \in \overline{\omega} \text{ and all } (\xi^\alpha) \in \mathbb{R}^2.$$

Show that, if $\chi \in H_0^1(\omega)$ satisfies $A_{\alpha\beta}\partial_{\alpha\beta}\chi \in L^2(\omega)$, then $\chi \in H^2(\omega)$.

Hint: Indications may be found, e.g., in Brezis [1983, Sect. 9.6].

(2) Let $S = \boldsymbol{\theta}(\overline{\omega})$ be an elliptic surface with a mapping $\boldsymbol{\theta}$ of the form $\boldsymbol{\theta} : (y_1, y_2) \in \overline{\omega} \to (y_1, y_2, f(y_1, y_2))$, where $f \in \mathcal{C}^3(\overline{\omega})$. Show that a field $\boldsymbol{\eta} = (\eta_i) \in H^1(\omega) \times H^1(\omega) \times L^2(\omega)$ satisfies $\gamma_{\alpha\beta}(\boldsymbol{\eta}) = 0$ in ω if and only if there exists $g \in L^2(\omega)$ such that

$$e_{\alpha\beta}(\boldsymbol{\eta}) := \frac{1}{2}(\partial_\beta\eta_\alpha + \partial_\alpha\eta_\beta) = g\,\partial_{\alpha\beta}f.$$

(3) Show that, for all $\eta_\alpha \in H_0^1(\omega)$ and all $\chi \in H_0^1(\omega) \cap H^2(\omega)$,

$$\int_\omega \{e_{11}(\boldsymbol{\eta})\partial_{22}\chi - 2e_{12}(\boldsymbol{\eta})\partial_{12}\chi + e_{22}(\boldsymbol{\eta})\partial_{11}\chi\}\,dy = 0.$$

(4) Show that, if $\boldsymbol{\eta} \in H_0^1(\omega) \times H_0^1(\omega) \times L^2(\omega)$ satisfies $\gamma_{\alpha\beta}(\boldsymbol{\eta}) = 0$ in ω, then $\boldsymbol{\eta} = \mathbf{0}$ in ω.

Remark. This result, which is due to Brezzi [1994], thus provides a considerable simplification of the proof of Thm. 2.7-2 for the special class of elliptic surfaces considered in (2).

2.13. Let S be a portion of a sphere with radius R, equipped with stereographic coordinates u and v (Fig. 2.1-3). Given any element $\boldsymbol{\eta} = (\eta_i) \in H_0^1(\omega) \times H_0^1(\omega) \times L^2(\omega)$, let

$$\chi_\alpha := \eta_\alpha(u^2 + v^2 + R^2)^2.$$

Show that the relations $\gamma_{\alpha\beta}(\boldsymbol{\eta}) = 0$ in ω imply that

$$\partial_2\chi_1 + \partial_1\chi_2 = 0 \text{ and } \partial_2\chi_2 - \partial_1\chi_1 = 0 \text{ in } \omega,$$

and conclude in this fashion that $\eta_\alpha = 0$ in ω, thus providing in this case a quick proof of Thm. 2.7-2 (this observation is found in Ciarlet & Lods [1996a, Sect. 3]).

2.14. (1) The Christoffel symbols are defined as $\Gamma^\sigma_{\alpha\beta} = \boldsymbol{a}^\sigma \cdot \partial_\beta \boldsymbol{a}_\alpha$ in the text (Thm. 2.3-1), i.e., by means of the vectors of the covariant and contravariant bases of the tangent plane. It is remarkable that they can be also defined solely in terms of the metric tensor of the surface. More precisely, show that

$$\Gamma^\sigma_{\alpha\beta} = a^{\sigma\tau}\Gamma_{\alpha\beta\tau}, \text{ where } \Gamma_{\alpha\beta\tau} := \frac{1}{2}(\partial_\beta a_{\alpha\tau} + \partial_\alpha a_{\beta\tau} - \partial_\tau a_{\alpha\beta}).$$

(2) Show that the Christoffel symbols satisfy the compatibility conditions

$$\begin{aligned} \partial_\tau \Gamma_{\alpha\sigma\beta} - \partial_\sigma \Gamma_{\alpha\tau\beta} + \Gamma^\mu_{\alpha\tau}\Gamma_{\beta\sigma\mu} - \Gamma^\lambda_{\alpha\sigma}\Gamma_{\beta\tau\lambda} = b_{\alpha\sigma}b_{\beta\tau} - b_{\alpha\tau}b_{\beta\sigma} \text{ in } \omega, \\ \partial_\sigma b_{\alpha\beta} - \partial_\beta b_{\alpha\sigma} + \Gamma^\tau_{\alpha\beta}b_{\tau\sigma} - \Gamma^\tau_{\alpha\sigma}b_{\tau\beta} = 0 \text{ in } \omega. \end{aligned}$$

(3) Let

$$R_{\alpha\beta\sigma\tau} := \partial_\tau \Gamma_{\alpha\sigma\beta} - \partial_\sigma \Gamma_{\alpha\tau\beta} + \Gamma^\mu_{\alpha\tau}\Gamma_{\beta\sigma\mu} - \Gamma^\lambda_{\alpha\sigma}\Gamma_{\beta\tau\lambda}.$$

Show that

$$\begin{aligned} R_{\alpha\beta\sigma\tau} &= -R_{\alpha\beta\tau\sigma} = -R_{\beta\alpha\sigma\tau} = R_{\sigma\tau\alpha\beta}, \\ R_{\alpha\beta 11} &= R_{\alpha\beta 22} = R_{11\sigma\tau} = R_{22\sigma\tau} = 0, \\ R_{1212} &= R_{2121} = -R_{1221} = -R_{2112}, \end{aligned}$$

and that, consequently, the compatibility conditions found in (2) in fact reduce to

$$R_{1212} = b_{11}b_{22} - (b_{12})^2.$$

(4) Deduce from (3) that the Gaussian curvature K (Sect. 2.2) satisfies

$$K = \left(a_{11}a_{22} - (a_{12})^2\right)^{-1}(\partial_2\Gamma_{112} - \partial_1\Gamma_{122} + \Gamma^\mu_{12}\Gamma_{21\mu} - \Gamma^\lambda_{11}\Gamma_{22\lambda}).$$

This is the astonishing *Theorema egregium* of Gauß [1828]: The Gaussian curvature depends only on the knowledge of the *first* fundamental form of a surface, i.e., only on its metric!

Remark. The functions $R_{\alpha\beta\sigma\tau}$ are the covariant components of the *Riemann curvature tensor* of the surface.

CHAPTER 3

ASYMPTOTIC ANALYSIS OF LINEARLY ELASTIC SHELLS: PRELIMINARIES AND OUTLINE

INTRODUCTION

The purpose of this chapter is twofold: *First* and foremost, it gathers the fundamental *preliminaries* needed in the remainder of this volume for carrying out the asymptotic analysis of *all* the kinds of linearly elastic shells that we shall encounter: After *ad hoc "scalings" of the unknowns* (the covariant components of the displacement field) and *ad hoc "asymptotic" assumptions on the data* (the Lamé constants and applied force densities) have been made, we transform the problem of a linearly elastic clamped shell with thickness $2\varepsilon > 0$ into a *"scaled problem"*, defined over a domain that is *independent of* ε.

Second, it justifies the *two fundamentally distinct classes of two-dimensional equations* that mathematically model *linearly elastic shells*, through a *formal asymptotic analysis* of the solution of the scaled problems, where ε is considered as the "small" parameter. Since this formal analysis will be later replaced by the more satisfactory (but also considerably more delicate) convergence theorems of the next three chapters, this chapter thus essentially serves a pedagogical purpose.

More specifically, let ω be a domain in $\mathbb{R}^2$ with boundary γ and let $\boldsymbol{\theta} : \overline{\omega} \to \mathbb{R}^3$ be a smooth enough injective mapping such that the two vectors $\boldsymbol{a}_\alpha = \partial_\alpha \boldsymbol{\theta}$ are linearly independent at all points in $\overline{\omega}$. Consider a *shell* with middle surface $S = \boldsymbol{\theta}(\overline{\omega})$ and thickness $2\varepsilon > 0$, i.e., a body whose reference configuration is the set $\boldsymbol{\Theta}(\overline{\Omega}^\varepsilon)$, where $\Omega^\varepsilon = \omega \times]-\varepsilon, \varepsilon[$,

$$\boldsymbol{\Theta}(x^\varepsilon) = \boldsymbol{\theta}(y) + x_3^\varepsilon \boldsymbol{a}_3(y) \text{ for all } x^\varepsilon = (y, x_3^\varepsilon) = (x_i^\varepsilon) \in \overline{\Omega}^\varepsilon,$$

and $\boldsymbol{a}_3(y)$ is the unit outer vector normal to S at $\boldsymbol{\theta}(y)$ constructed as in Sect. 2.2.

The shell is subjected to *applied body forces* with contravariant components $f^{i,\varepsilon} : \Omega^\varepsilon \to \mathbb{R}$ and is subjected to a *boundary condition of place* along a portion $\boldsymbol{\Theta}(\gamma_0 \times [-\varepsilon, \varepsilon])$ of its lateral face $\boldsymbol{\Theta}(\gamma \times [-\varepsilon, \varepsilon])$, where $\gamma_0 \subset \gamma$ and *length* $\gamma_0 > 0$. *Applied surface forces* acting on the upper and lower faces $\boldsymbol{\Theta}(\omega \times \{\varepsilon\})$ and $\boldsymbol{\Theta}(\omega \times \{-\varepsilon\})$ may be also considered; for simplicity, we assume in this introduction that they vanish. Let λ^ε and μ^ε denote the Lamé constants of the elastic material constituting the shell.

Then in *linearized elasticity*, the unknown $\boldsymbol{u}^\varepsilon = (u_i^\varepsilon)$, where $u_i^\varepsilon : \overline{\Omega}^\varepsilon \to \mathbb{R}$ are the covariant components of the displacement field of the points of the shell, satisfies the following three-dimensional equations in terms of the "natural" curvilinear coordinates x_i^ε of the shell (we quickly review in Sect. 3.1 the main features of this variational problem, otherwise extensively studied in Chap. 1):

$$\boldsymbol{u}^\varepsilon \in \mathbf{V}(\Omega^\varepsilon) = \{\boldsymbol{v}^\varepsilon = (v_i^\varepsilon) \in \mathbf{H}^1(\Omega^\varepsilon);\ \boldsymbol{v}^\varepsilon = \mathbf{0} \text{ on } \gamma_0 \times [-\varepsilon,\, \varepsilon]\},$$

$$\int_{\Omega^\varepsilon} A^{ijkl,\varepsilon} e^\varepsilon_{k\|l}(\boldsymbol{u}^\varepsilon) e^\varepsilon_{i\|j}(\boldsymbol{v}^\varepsilon)\sqrt{g^\varepsilon}\, dx^\varepsilon = \int_{\Omega^\varepsilon} f^{i,\varepsilon} v_i^\varepsilon \sqrt{g^\varepsilon}\, dx^\varepsilon \text{ for all } \boldsymbol{v}^\varepsilon \in \mathbf{V}(\Omega^\varepsilon),$$

where

$$A^{ijkl,\varepsilon} = \lambda^\varepsilon g^{ij,\varepsilon} g^{kl,\varepsilon} + \mu^\varepsilon \left(g^{ik,\varepsilon} g^{jl,\varepsilon} + g^{il,\varepsilon} g^{jk,\varepsilon}\right),$$

$$e^\varepsilon_{i\|j}(\boldsymbol{v}^\varepsilon) = \frac{1}{2}(v^\varepsilon_{i\|j} + v^\varepsilon_{j\|i}),$$

$$v^\varepsilon_{i\|j} = \partial^\varepsilon_j v^\varepsilon_i - \Gamma^{p,\varepsilon}_{ij} v^\varepsilon_p.$$

In Sect. 3.2, we transform this problem into an equivalent problem, but now posed over the set $\Omega = \omega \times]-1, 1[$, which is *independent of* ε. This transformation relies in a crucial way on appropriate *scalings of the unknowns* u_i^ε and *assumptions on the data* λ^ε, μ^ε *and* $f^{i,\varepsilon}$. More specifically, we define the *scaled unknown* $\boldsymbol{u}(\varepsilon) = (u_i(\varepsilon))$ by letting

$$u_i^\varepsilon(x^\varepsilon) = u_i(\varepsilon)(x) \text{ for all } x^\varepsilon = \pi^\varepsilon x \in \overline{\Omega}^\varepsilon,$$

where $\pi^\varepsilon(x_1, x_2, x_3) = (x_1, x_2, \varepsilon x_3)$. We then *assume* that there exist constants $\lambda > 0$, $\mu > 0$ and functions f^i *independent of* ε such that

$$\lambda^\varepsilon = \lambda \text{ and } \mu^\varepsilon = \mu,$$

$$f^{i,\varepsilon}(x^\varepsilon) = \varepsilon^p f^i(x) \text{ for all } x^\varepsilon = \pi^\varepsilon x \in \Omega^\varepsilon,$$

where the exponent p is *unspecified at this stage.*

It is found in this fashion that the scaled unknown $\boldsymbol{u}(\varepsilon)$ satisfies a variational problem of the form (Thm. 3.2-1):

$$\boldsymbol{u}(\varepsilon) \in \mathbf{V}(\Omega) = \left\{\boldsymbol{v} = (v_i) \in \mathbf{H}^1(\Omega);\ \boldsymbol{v} = \mathbf{0} \text{ on } \gamma_0 \times [-1, 1]\right\},$$

$$\int_\Omega A^{ijkl}(\varepsilon) e_{k\|l}(\varepsilon;\ \boldsymbol{u}(\varepsilon)) e_{i\|j}(\varepsilon;\ \boldsymbol{v}) \sqrt{g(\varepsilon)}\, dx$$

$$= \varepsilon^p \int_\Omega f^i v_i \sqrt{g(\varepsilon)}\, dx \text{ for all } \boldsymbol{v} \in \mathbf{V}(\Omega),$$

where, for any vector field $\boldsymbol{v} = (v_i) \in \mathbf{H}^1(\Omega)$, the *scaled linearized strains* $e_{i\|j}(\varepsilon; \boldsymbol{v}) = e_{j\|i}(\varepsilon;\ \boldsymbol{v}) \in L^2(\Omega)$ are defined by:

$$\begin{aligned}
e_{\alpha\|\beta}(\varepsilon;\ \boldsymbol{v}) &= \frac{1}{2}(\partial_\beta v_\alpha + \partial_\alpha v_\beta) - \Gamma^p_{\alpha\beta}(\varepsilon) v_p,\\
e_{\alpha\|3}(\varepsilon;\ \boldsymbol{v}) &= \frac{1}{2}\left(\frac{1}{\varepsilon}\partial_3 v_\alpha + \partial_\alpha v_3\right) - \Gamma^\sigma_{\alpha 3}(\varepsilon) v_\sigma,\\
e_{3\|3}(\varepsilon;\ \boldsymbol{v}) &= \frac{1}{\varepsilon}\partial_3 v_3.
\end{aligned}$$

The specific form of this variational problem suggests that we use the *method of formal asymptotic expansions*, i.e., we let

$$\boldsymbol{u}(\varepsilon) = \boldsymbol{u}^0 + \varepsilon \boldsymbol{u}^1 + \varepsilon^2 \boldsymbol{u}^2 + \varepsilon^3 \boldsymbol{u}^3 + \varepsilon^4 \boldsymbol{u}^4 + \cdots, \text{ with } \boldsymbol{u}^0 \neq \mathbf{0},$$

in the variational equations, and then we equate to zero the factors of the successive powers of ε found in the resulting equations until the leading term $\boldsymbol{u}^0$ can be fully identified, without imposing any restriction on the applied force densities. In so doing, we are led after a series of delicate computations (which themselves rely on various "geometrical" and "mechanical" preliminaries established in Sect. 3.3) to the conclusion that the leading term $\boldsymbol{u}^0$ satisfies a *two-dimensional problem* that falls in *one* of the following *two* categories:

Assume first that *the space*

$$\mathbf{V}_0(\omega) := \{\boldsymbol{\eta} \in \mathbf{H}^1(\omega);\ \boldsymbol{\eta} = \mathbf{0} \text{ on } \gamma_0,\ \gamma_{\alpha\beta}(\boldsymbol{\eta}) = 0 \text{ in } \omega\}$$

contains only the function $\boldsymbol{\eta} = \mathbf{0}$. Then the functions $f^{i,\varepsilon}$ must be of the form

$$f^{i,\varepsilon}(x^\varepsilon) = f^{i,0}(x) \text{ for all } x^\varepsilon = \pi^\varepsilon x \in \Omega^\varepsilon,$$

where the functions $f^{i,0}$ are *independent* of ε (i.e., the exponent p must be set equal to 0 in this case). Furthermore, *the leading term* $\boldsymbol{u}^0 : \overline{\Omega} \to \mathbb{R}^3$ *is independent of the transverse variable* x_3 *and* $\boldsymbol{\zeta}^0 = \frac{1}{2}\int_{-1}^{1} \boldsymbol{u}^0 \, dx_3$ *should satisfy the following* (scaled) *two-dimensional variational problem of a linearly elastic "membrane" shell* (see Thm. 3.4-2; this problem is in fact provisional, since the function spaces and boundary conditions appearing in the definition of the space $\mathbf{V}(\omega)$ definitely need to be modified; see Chaps. 4 and 5, or Sect. 3.5):

$$\boldsymbol{\zeta}^0 \in \mathbf{V}(\omega) = \{\boldsymbol{\eta} = (\eta_i) \in \mathbf{H}^1(\omega); \ \boldsymbol{\eta} = \mathbf{0} \text{ on } \gamma_0\},$$

$$\int_\omega a^{\alpha\beta\sigma\tau}\gamma_{\sigma\tau}(\boldsymbol{\zeta}^0)\gamma_{\alpha\beta}(\boldsymbol{\eta})\sqrt{a}\,dy = \int_\omega \left\{\int_{-1}^{1} f^{i,0}\,dx_3\right\}\eta_i\sqrt{a}\,dy$$

for all $\boldsymbol{\eta} = (\eta_i) \in \mathbf{V}(\omega)$, where

$$\begin{aligned}\gamma_{\alpha\beta}(\boldsymbol{\eta}) &= \frac{1}{2}(\partial_\beta\eta_\alpha + \partial_\alpha\eta_\beta) - \Gamma^\sigma_{\alpha\beta}\eta_\sigma - b_{\alpha\beta}\eta_3,\\ a^{\alpha\beta\sigma\tau} &= \frac{4\lambda\mu}{\lambda + 2\mu}a^{\alpha\beta}a^{\sigma\tau} + 2\mu(a^{\alpha\sigma}a^{\beta\tau} + a^{\alpha\tau}a^{\beta\sigma}),\end{aligned}$$

$a = \det(a_{\alpha\beta})$, and $a^{\alpha\beta} = \boldsymbol{a}^\alpha \cdot \boldsymbol{a}^\beta$ are the contravariant components of the metric tensor of S.

Assume next that *the space* $\mathbf{V}_0(\omega)$ *contains nonzero functions* $\boldsymbol{\eta}$. Then the functions $f^{i,\varepsilon}$ must be of the form

$$f^{i,\varepsilon}(x^\varepsilon) = \varepsilon^2 f^{i,2}(x) \text{ for all } x^\varepsilon = \pi^\varepsilon x \in \Omega^\varepsilon,$$

where the functions $f^{i,2}$ are *independent* of ε (i.e., the exponent p must be set equal to 2 in this case). Furthermore, *the leading term* $\boldsymbol{u}^0 : \overline{\Omega} \to \mathbb{R}^3$ *is independent of the transverse variable* x_3 *and the field* $\boldsymbol{\zeta}^0 = \frac{1}{2}\int_{-1}^{1} \boldsymbol{u}^0\,dx_3$ *satisfies the following* (scaled) *two-dimensional variational problem of a linearly elastic "flexural" shell* (Thm. 3.4-3):

$$\begin{aligned}\boldsymbol{\zeta}^0 \in \mathbf{V}_F(\omega) := \{&\boldsymbol{\eta} = (\eta_i) \in H^1(\omega) \times H^1(\omega) \times H^2(\omega);\\ &\eta_i = \partial_\nu\eta_3 = 0 \text{ on } \gamma_0, \ \gamma_{\alpha\beta}(\boldsymbol{\eta}) = 0 \text{ in } \omega\},\end{aligned}$$

$$\frac{1}{3}\int_\omega a^{\alpha\beta\sigma\tau}\rho_{\sigma\tau}(\boldsymbol{\zeta}^0)\rho_{\alpha\beta}(\boldsymbol{\eta})\sqrt{a}\,dy = \int_\omega \left\{\int_{-1}^{1} f^{i,2}\,dx_3\right\}\eta_i\sqrt{a}\,dy$$

for all $\boldsymbol{\eta} = (\eta_i) \in \mathbf{V}_F(\omega)$, *where*

$$\begin{aligned}\rho_{\alpha\beta}(\boldsymbol{\eta}) = {} & \partial_{\alpha\beta}\eta_3 - \Gamma^\sigma_{\alpha\beta}\partial_\sigma\eta_3 - b^\sigma_\alpha b_{\sigma\beta}\eta_3 + b^\sigma_\alpha(\partial_\beta\eta_\sigma - \Gamma^\tau_{\beta\sigma}\eta_\tau)\\ & + b^\tau_\beta(\partial_\alpha\eta_\tau - \Gamma^\sigma_{\alpha\tau}\eta_\sigma) + (\partial_\alpha b^\tau_\beta + \Gamma^\tau_{\alpha\sigma}b^\sigma_\beta - \Gamma^\sigma_{\alpha\beta}b^\tau_\sigma)\eta_\tau,\end{aligned}$$

and the functions $\gamma_{\alpha\beta}(\boldsymbol{\eta})$ and $a^{\alpha\beta\sigma\tau}$ are the same as above.

Although the method of formal asymptotic expansions thus admirably serves the purpose of clearly identifying two fundamental classes of possible limit equations, it needs to be substantiated by a *convergence analysis.*

This is why we conclude this chapter by a quick review (Sect. 3.5) of the *convergence theorems* that will be established in the next three chapters, showing that the scaled unknown $\boldsymbol{u}(\varepsilon)$ has indeed a *limit as* $\varepsilon \to 0$ in an *ad hoc* functional space, which in each case depends on the *geometry of the middle surface* S and on the *subset* γ_0 *of* γ. This "refined" analysis also illustrates the limits of the otherwise quite efficient formal approach, as it shows that the "membrane" shells found above need themselves to be subdivided into two subcategories.

3.1. THE THREE-DIMENSIONAL EQUATIONS OF A LINEARLY ELASTIC SHELL

To begin with, we briefly review the main notations, definitions, and results, mostly from three-dimensional linearized elasticity in curvilinear coordinates and from the differential geometry of surfaces, that will be needed in the sequel, and we provide *ad hoc* cross-references to Chaps. 1 and 2, where these notions have been expounded in detail. Note that the real three-dimensional affine Euclidean space formerly denoted $\mathcal{E}^3$ in Chaps. 1 and 2 will be henceforth denoted $\mathbb{R}^3$.

Greek indices and exponents (except ε) belong to the set $\{1, 2\}$, *Latin* indices and exponents (except when otherwise indicated, as when they are used to index sequences) belong to the set $\{1, 2, 3\}$, and the *summation convention* with respect to repeated indices and exponents is systematically used. Symbols such as δ^α_β or δ^i_j designate the Kronecker's symbol. The Euclidean scalar product and the exterior product of $\boldsymbol{a}, \boldsymbol{b} \in \mathbb{R}^3$ are noted $\boldsymbol{a} \cdot \boldsymbol{b}$ and $\boldsymbol{a} \wedge \boldsymbol{b}$; the Euclidean norm of $\boldsymbol{a} \in \mathbb{R}^3$ is noted $|\boldsymbol{a}|$.

A *domain* Ω *in* $\mathbb{R}^n$ is a bounded, open, connected subset of $\mathbb{R}^n$ with a Lipschitz-continuous boundary $\partial\Omega$, the set Ω being locally on one side of $\partial\Omega$. For each integer $m \geq 1$, $H^m(\Omega)$ and $H^m_0(\Omega)$ denote the usual Sobolev spaces of real-valued functions. Boldface letters denote vector-valued functions and their associated function spaces. The norm in $L^2(\Omega)$ or $\mathbf{L}^2(\Omega)$ is noted $|\cdot|_{0,\Omega}$ and the norm in $H^m(\Omega)$

or $\mathbf{H}^m(\Omega)$, $m \geq 1$, is noted $\|\cdot\|_{m,\Omega}$ (Sect. 1.7).

Let ω be a domain in $\mathbb{R}^2$ with boundary γ. Let $y = (y_\alpha)$ denote a generic point in the set $\overline{\omega}$ and let $\partial_\alpha := \partial/\partial y_\alpha$. Let $\boldsymbol{\theta} \in \mathcal{C}^2(\overline{\omega}; \mathbb{R}^3)$ be an *injective* mapping such that the two vectors

$$\boldsymbol{a}_\alpha(y) := \partial_\alpha \boldsymbol{\theta}(y)$$

are *linearly independent at all points* $y \in \overline{\omega}$. These two vectors form the *covariant basis* of the tangent plane to the *surface*

$$S := \boldsymbol{\theta}(\overline{\omega})$$

at the point $\boldsymbol{\theta}(y)$ and the two vectors $\boldsymbol{a}^\alpha(y)$ of the tangent plane at $\boldsymbol{\theta}(y)$ defined by the relations

$$\boldsymbol{a}^\alpha(y) \cdot \boldsymbol{a}_\beta(y) = \delta^\alpha_\beta$$

form its *contravariant basis*. Also let

$$\boldsymbol{a}_3(y) = \boldsymbol{a}^3(y) := \frac{\boldsymbol{a}_1(y) \wedge \boldsymbol{a}_2(y)}{|\boldsymbol{a}_1(y) \wedge \boldsymbol{a}_2(y)|}.$$

Then $|\boldsymbol{a}_3(y)| = 1$, the vector $\boldsymbol{a}_3(y)$ is normal to S at the point $\boldsymbol{\theta}(y)$, and the three vectors $\boldsymbol{a}^i(y)$ form the *contravariant basis at* $\boldsymbol{\theta}(y)$; cf. Fig. 2.3-1. Note that (y_1, y_2) constitutes a system of *curvilinear coordinates* (Sect. 2.1) for describing the surface S.

The *covariant* and *contravariant components* $a_{\alpha\beta}$ and $a^{\alpha\beta}$ of the *metric tensor*, also called the *first fundamental form* (Sect. 2.1), the *covariant* and *mixed components* $b_{\alpha\beta}$ and b^β_α of the *curvature tensor*, also called the *second fundamental form* (Sect. 2.2), and the *Christoffel symbols* $\Gamma^\sigma_{\alpha\beta}$ (Sect. 2.3), *of the surface* S are then defined by letting (whenever no confusion should arise, we henceforth drop the explicit dependence on the variable $y \in \overline{\omega}$):

$$a_{\alpha\beta} := \boldsymbol{a}_\alpha \cdot \boldsymbol{a}_\beta, \qquad a^{\alpha\beta} := \boldsymbol{a}^\alpha \cdot \boldsymbol{a}^\beta,$$
$$b_{\alpha\beta} := \boldsymbol{a}^3 \cdot \partial_\beta \boldsymbol{a}_\alpha, \qquad b^\beta_\alpha := a^{\beta\sigma} b_{\sigma\alpha},$$
$$\Gamma^\sigma_{\alpha\beta} := \boldsymbol{a}^\sigma \cdot \partial_\beta \boldsymbol{a}_\alpha.$$

Note the symmetries:

$$a_{\alpha\beta} = a_{\beta\alpha}, \quad a^{\alpha\beta} = a^{\beta\alpha}, \quad b_{\alpha\beta} = b_{\beta\alpha}, \quad \Gamma^\sigma_{\alpha\beta} = \Gamma^\sigma_{\beta\alpha}.$$

The *area element* along S is $\sqrt{a}\,dy$, where

$$a := \det(a_{\alpha\beta}).$$

All the functions $a_{\alpha\beta}$, $a^{\alpha\beta}$, $b_{\alpha\beta}$, b_α^β, $\Gamma^\sigma_{\alpha\beta}$, and a are thus at least continuous over the set $\overline{\omega}$.

Let γ_0 denote a $d\gamma$-measurable subset of the boundary γ of ω satisfying

$$\boxed{\textit{length}\ \gamma_0 > 0.}$$

For each $\varepsilon > 0$, we define the sets

$$\boxed{\begin{gathered}\Omega^\varepsilon := \omega \times]-\varepsilon, \varepsilon[, \\ \Gamma^\varepsilon_+ := \omega \times \{\varepsilon\}, \quad \Gamma^\varepsilon_- := \omega \times \{-\varepsilon\}, \quad \Gamma^\varepsilon_0 := \gamma_0 \times [-\varepsilon, \varepsilon].\end{gathered}}$$

Let $x^\varepsilon = (x_i^\varepsilon)$ denote a generic point in the set $\overline{\Omega}^\varepsilon$ and let $\partial_i^\varepsilon := \partial/\partial x_i^\varepsilon$; hence $x_\alpha^\varepsilon = y_\alpha$ and $\partial_\alpha^\varepsilon = \partial_\alpha$.

Consider an **elastic shell** with **middle surface** $S = \boldsymbol{\theta}(\overline{\omega})$ and **thickness** $2\varepsilon > 0$, i.e., an elastic body whose *reference configuration* consists of all points within a distance $\leq \varepsilon$ from S (Fig. 1.2-1). In other words, *the reference configuration of the shell is the image* $\boldsymbol{\Theta}(\overline{\Omega}^\varepsilon) \subset \mathbb{R}^3$ *of the set* $\overline{\Omega}^\varepsilon \subset \mathbb{R}^3$ *through a mapping* $\boldsymbol{\Theta} : \overline{\Omega}^\varepsilon \to \mathbb{R}^3$ *given by*

$$\boxed{\boldsymbol{\Theta}(x^\varepsilon) := \boldsymbol{\theta}(y) + x_3^\varepsilon \boldsymbol{a}_3(y) \text{ for all } x^\varepsilon = (y, x_3^\varepsilon) = (y_1, y_2, x_3^\varepsilon) \in \overline{\Omega}^\varepsilon.}$$

Remark. Shells whose middle surface S has "no boundary", such as an ellipsoid or a torus, are thus not covered by the present analysis, which assumes that $S = \boldsymbol{\theta}(\overline{\omega})$ can be described by a *single* injective mapping $\boldsymbol{\theta} : \overline{\omega} \to \mathbb{R}^3$. References for such shells are provided in Sects. 4.6 and 5.8. ■

The next theorem is essential for describing the geometry of such a shell: It shows that, *if the injective mapping* $\boldsymbol{\theta} : \overline{\omega} \to \mathbb{R}^3$ *is smooth enough, the mapping* $\boldsymbol{\Theta} : \overline{\Omega}^\varepsilon \to \mathbb{R}^3$ *is also injective for* $\varepsilon > 0$ *small enough.*

Theorem 3.1-1. *Let ω be a domain in $\mathbb{R}^2$, let $\boldsymbol{\theta} \in \mathcal{C}^2(\overline{\omega}; \mathbb{R}^3)$ be an injective mapping such that the two vectors $\boldsymbol{a}_\alpha = \partial_\alpha \boldsymbol{\theta}$ are linearly independent at all points of $\overline{\omega}$, and let $\boldsymbol{a}_3 = \dfrac{\boldsymbol{a}_1 \wedge \boldsymbol{a}_2}{|\boldsymbol{a}_1 \wedge \boldsymbol{a}_2|}$.*

Then there exists $\varepsilon_0 > 0$ such that the mapping $\boldsymbol{\Theta} : \overline{\Omega}_0 \to \mathbb{R}^3$ defined by

$$\boldsymbol{\Theta}(y, x_3) := \boldsymbol{\theta}(y) + x_3 \boldsymbol{a}_3(y) \text{ for all } (y, x_3) \in \overline{\Omega}_0, \text{ where } \Omega_0 := \omega \times]-\varepsilon_0, \varepsilon_0[,$$

is a $\mathcal{C}^1$-diffeomorphism from $\overline{\Omega}_0$ onto $\boldsymbol{\Theta}(\overline{\Omega}_0)$ and $\det(\boldsymbol{g}_1, \boldsymbol{g}_2, \boldsymbol{g}_3) > 0$ *in $\overline{\Omega}_0$, where $\boldsymbol{g}_i := \partial_i \boldsymbol{\Theta}$. The mapping $\boldsymbol{\Theta} : \overline{\Omega}_0 \to \mathbb{R}^3$ is called the* **canonical extension** *of the mapping $\boldsymbol{\theta} : \overline{\omega} \to \mathbb{R}^3$.*

Proof. The assumed regularity on the mapping $\boldsymbol{\theta}$ implies that $\boldsymbol{\Theta} \in \mathcal{C}^1(\overline{\omega} \times [-\varepsilon, \varepsilon]; \mathbb{R}^3)$ for any $\varepsilon > 0$. The relations

$$\boldsymbol{g}_\alpha = \partial_\alpha \boldsymbol{\Theta} = \boldsymbol{a}_\alpha + x_3 \partial_\alpha \boldsymbol{a}_3 \text{ and } \boldsymbol{g}_3 = \partial_3 \boldsymbol{\Theta} = \boldsymbol{a}_3$$

imply that

$$\det(\boldsymbol{g}_1, \boldsymbol{g}_2, \boldsymbol{g}_3)|_{x_3=0} = \det(\boldsymbol{a}_1, \boldsymbol{a}_2, \boldsymbol{a}_3) = \sqrt{a}$$

(the last equality is proved in part (i) of the proof of Thm. 2.5-1). Hence it is clear that $\det(\boldsymbol{g}_1, \boldsymbol{g}_2, \boldsymbol{g}_3) > 0$ on $\overline{\omega} \times [-\varepsilon, \varepsilon]$ if $\varepsilon > 0$ is small enough.

Hence the *implicit function theorem* can be applied: It shows that, locally, the mapping $\boldsymbol{\Theta}$ is a $\mathcal{C}^1$-diffeomorphism: Given any $y \in \overline{\omega}$, there exist a neighborhood $U(y)$ of y in $\overline{\omega}$ and $\varepsilon(y) > 0$ such that $\boldsymbol{\Theta}$ is a $\mathcal{C}^1$-diffeomorphism from the set $U(y) \times [-\varepsilon(y), \varepsilon(y)]$ onto $\boldsymbol{\Theta}(U(y) \times [-\varepsilon(y), \varepsilon(y)])$. See, e.g., Schwartz [1992, Chap. 3, Sect. 8]; the proof of the implicit function theorem, which is almost invariably given for functions defined over *open* sets, can be easily extended to functions defined over *closures of domains*, such as the sets $\overline{\omega} \times [-\varepsilon, \varepsilon]$ (in particular, the definition of differentiable functions over a domain poses no difficulty; see Stein [1970]).

To establish that the mapping $\boldsymbol{\Theta} : \overline{\omega} \times [-\varepsilon, \varepsilon] \to \mathbb{R}^3$ is injective provided $\varepsilon > 0$ is small enough, we proceed by contradiction: If this property is false, there exist $\varepsilon_n > 0$, (y^n, x_3^n), and $(\tilde{y}^n, \tilde{x}_3^n)$ such that

$$\varepsilon_n \to 0, \quad y^n \in \overline{\omega}, \quad \tilde{y}^n \in \overline{\omega}, \quad |x_3^n| \leq \varepsilon_n, \quad |\tilde{x}_3^n| \leq \varepsilon_n,$$
$$(y^n, x_3^n) \neq (\tilde{y}^n, \tilde{x}_3^n) \text{ and } \boldsymbol{\Theta}(y^n, x_3^n) = \boldsymbol{\Theta}(\tilde{y}^n, \tilde{x}_3^n).$$

Since the set $\overline{\omega}$ is compact, there exist $y \in \overline{\omega}$, $\tilde{y} \in \overline{\omega}$, and a subsequence, still indexed by n for convenience, such that

$$y^n \to y, \quad \tilde{y}^n \to \tilde{y}, \quad x_3^n \to 0, \quad \tilde{x}_3^n \to 0$$

as $n \to \infty$. Hence

$$\boldsymbol{\theta}(y) = \lim_{n\to\infty} \boldsymbol{\Theta}(y^n, x_3^n) = \lim_{n\to\infty} \boldsymbol{\Theta}(\tilde{y}^n, \tilde{x}_3^n) = \boldsymbol{\theta}(\tilde{y}),$$

by the continuity of the mapping $\boldsymbol{\Theta}$ and thus $y = \tilde{y}$ since the mapping $\boldsymbol{\theta}$ is injective by assumption. But these properties contradict the local injectivity (established *supra*) of the mapping $\boldsymbol{\Theta}$. Hence there exists $\varepsilon_0 > 0$ such that $\boldsymbol{\Theta}$ is injective on the set $\overline{\Omega}_0 = \overline{\omega} \times [-\varepsilon_0, \varepsilon_0]$.

Showing that $\boldsymbol{\Theta}^{-1} : \boldsymbol{\Theta}(\overline{\Omega}_0) \to \overline{\Omega}_0$ is also of class $\mathcal{C}^1$ is akin to proving the differentiability of an implicit function (see, e.g., Schwartz [1992, Chap. 3, Sect. 8]). Again, such a proof is usually given for functions defined over *open* sets, but it can be easily extended to functions defined over *closures of domains.* ■

Thm. 3.1-1 thus shows that, if $\varepsilon_0 > 0$ is small enough, $(y_1, y_2, x_3^\varepsilon)$ constitutes a *bona fide* system of *"natural" curvilinear coordinates* (Sect. 1.2) for describing the reference configuration $\boldsymbol{\Theta}(\overline{\Omega}^\varepsilon)$ of the shell and the physical problem is meaningful as *the set* $\boldsymbol{\Theta}(\overline{\Omega}^\varepsilon)$ *does not "interpenetrate itself"*. The curvilinear coordinate $x_3^\varepsilon \in [-\varepsilon, \varepsilon]$ is called the **transverse variable.**

The same theorem also shows that, again for $\varepsilon > 0$ small enough, the three vectors

$$\boldsymbol{g}_i^\varepsilon(x^\varepsilon) := \partial_i^\varepsilon \boldsymbol{\Theta}(x^\varepsilon)$$

form the *covariant basis* (of the tangent space, here $\mathbb{R}^3$, to the manifold $\boldsymbol{\Theta}(\overline{\Omega}^\varepsilon)$) at the point $\boldsymbol{\Theta}(x^\varepsilon)$, and the three vectors $\boldsymbol{g}^{i,\varepsilon}(x^\varepsilon)$ defined by

$$\boldsymbol{g}^{i,\varepsilon}(x^\varepsilon) \cdot \boldsymbol{g}_j^\varepsilon(x^\varepsilon) = \delta_j^i$$

form the *contravariant basis* at $\boldsymbol{\Theta}(x^\varepsilon)$ (Sect. 1.2).

We also define the *covariant* and *contravariant* components g_{ij}^ε and $g^{ij,\varepsilon}$ of the *metric tensor* (Sect. 1.2) and the *Christoffel symbols* $\Gamma_{ij}^{p,\varepsilon}$ (Sect. 1.4) of the set $\boldsymbol{\Theta}(\overline{\Omega}^\varepsilon)$ by letting (we omit the explicit dependence on x^ε):

$$g_{ij}^\varepsilon := \boldsymbol{g}_i^\varepsilon \cdot \boldsymbol{g}_j^\varepsilon, \quad g^{ij,\varepsilon} := \boldsymbol{g}^{i,\varepsilon} \cdot \boldsymbol{g}^{j,\varepsilon},$$

$$\Gamma_{ij}^{p,\varepsilon} := \boldsymbol{g}^{p,\varepsilon} \cdot \partial_i^\varepsilon \boldsymbol{g}_j^\varepsilon.$$

Note the symmetries:

$$g_{ij}^{\varepsilon} = g_{ji}^{\varepsilon}, \quad g^{ij,\varepsilon} = g^{ji,\varepsilon}, \quad \Gamma_{ij}^{p,\varepsilon} = \Gamma_{ji}^{p,\varepsilon}.$$

The *volume element* in the set $\boldsymbol{\Theta}(\overline{\Omega}^{\varepsilon})$ is $\sqrt{g^{\varepsilon}}\, dx^{\varepsilon}$, where

$$g^{\varepsilon} := \det(g_{ij}^{\varepsilon}).$$

We assume that the material constituting the shell is *homogeneous* and *isotropic* and that $\boldsymbol{\Theta}(\overline{\Omega}^{\varepsilon})$ is a *natural state,* so that the material is characterized by its two *Lamé constants* $\lambda^{\varepsilon} > 0$ and $\mu^{\varepsilon} > 0$. The *unknown* of the problem is the vector field $\boldsymbol{u}^{\varepsilon} = (u_i^{\varepsilon}) : \overline{\Omega}^{\varepsilon} \to \mathbb{R}^3$, where the three functions $u_i^{\varepsilon} : \overline{\Omega}^{\varepsilon} \to \mathbb{R}$ are the *covariant components* of the displacement field $u_i^{\varepsilon}\boldsymbol{g}^{i,\varepsilon} : \overline{\Omega}^{\varepsilon} \to \mathbb{R}^3$ that the shell undergoes under the influence of *applied forces*. This means that $u_i^{\varepsilon}(x^{\varepsilon})\boldsymbol{g}^{i,\varepsilon}(x^{\varepsilon})$ is the displacement of the point $\boldsymbol{\Theta}(x^{\varepsilon})$; see Fig. 3.1-1.

Finally, we assume that the shell is subjected to the *boundary condition of place* $\boldsymbol{u}^{\varepsilon} = \boldsymbol{0}$ on Γ_0^{ε}, i.e., that the displacement field vanishes along the portion $\boldsymbol{\Theta}(\Gamma_0^{\varepsilon})$ of its **lateral face** $\boldsymbol{\Theta}(\gamma \times [-\varepsilon, \varepsilon])$.

A word of caution. We refrain from calling "boundary conditions of clamping" the boundary conditions $u_i^{\varepsilon} = 0$ on Γ_0^{ε}, in order to avoid any confusion with the *"two-dimensional boundary conditions of clamping"*, which apply only to some *specific* classes of *two-dimensional* shell equations (Chaps. 6 and 10). ■

In *linearized elasticity*, the unknown $\boldsymbol{u}^{\varepsilon} = (u_i^{\varepsilon})$ then satisfies the following **three-dimensional variational problem $\mathcal{P}(\Omega^{\varepsilon})$ of a linearly elastic shell in curvilinear coordinates**, i.e., expressed in terms of the "natural" curvilinear coordinates x_i^{ε} of the shell (Thm. 1.3-1):

$$\boldsymbol{u}^{\varepsilon} \in \mathbf{V}(\Omega^{\varepsilon}) := \{\boldsymbol{v}^{\varepsilon} = (v_i^{\varepsilon}) \in \mathbf{H}^1(\Omega^{\varepsilon});\ \boldsymbol{v}^{\varepsilon} = \boldsymbol{0} \text{ on } \Gamma_0^{\varepsilon}\},$$

$$\int_{\Omega^{\varepsilon}} A^{ijkl,\varepsilon} e_{k\|l}^{\varepsilon}(\boldsymbol{u}^{\varepsilon}) e_{i\|j}^{\varepsilon}(\boldsymbol{v}^{\varepsilon}) \sqrt{g^{\varepsilon}}\, dx^{\varepsilon}$$

$$= \int_{\Omega^{\varepsilon}} f^{i,\varepsilon} v_i^{\varepsilon} \sqrt{g^{\varepsilon}}\, dx^{\varepsilon} + \int_{\Gamma_+^{\varepsilon} \cup \Gamma_-^{\varepsilon}} h^{i,\varepsilon} v_i^{\varepsilon} \sqrt{g^{\varepsilon}}\, d\Gamma^{\varepsilon}$$

$$\text{for all } \boldsymbol{v}^{\varepsilon} \in \mathbf{V}(\Omega^{\varepsilon}),$$

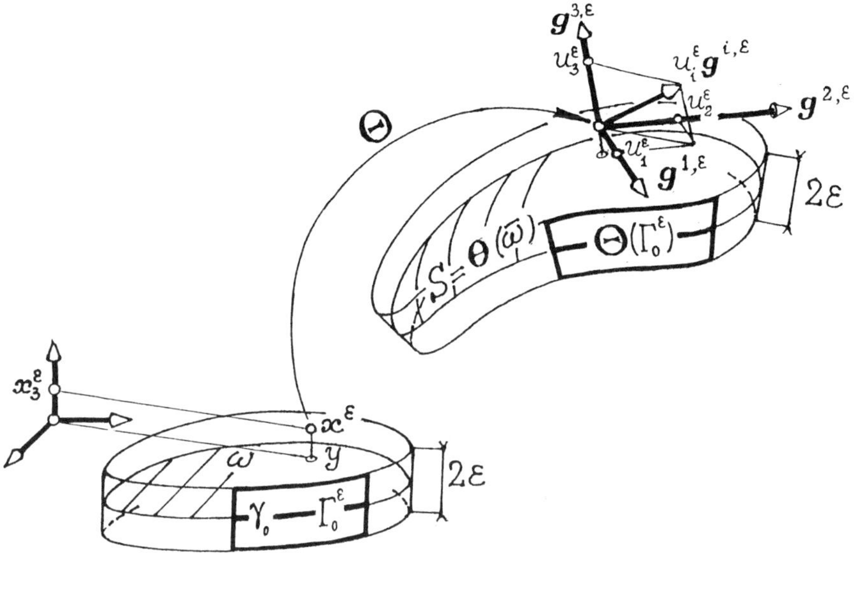

Fig. 3.1-1: *The three-dimensional shell problem.* Let $\Omega^\varepsilon = \omega\times]-\varepsilon, \varepsilon[$. The set $\Theta(\overline{\Omega}^\varepsilon)$, where $\Theta(y, x_3^\varepsilon) = \boldsymbol{\theta}(y) + x_3^\varepsilon \boldsymbol{a}_3(y)$ for all $\boldsymbol{x}^\varepsilon = (y, x_3^\varepsilon) \in \overline{\Omega}^\varepsilon$, is the reference configuration of a shell, with thickness 2ε and middle surface $S = \boldsymbol{\theta}(\overline{\omega})$, which is subjected to a boundary condition of place along the portion $\Theta(\Gamma_0^\varepsilon)$ of its lateral face (i.e., the displacement vanishes on $\Theta(\Gamma_0^\varepsilon)$), where $\Gamma_0^\varepsilon = \gamma_0 \times [-\varepsilon, \varepsilon]$ and $\gamma_0 \subset \gamma = \partial\omega$. The shell is subjected to applied body forces in its interior $\Theta(\Omega^\varepsilon)$ and to applied surface forces on its upper and lower faces $\Theta(\Gamma_+^\varepsilon)$ and $\Theta(\Gamma_-^\varepsilon)$ where $\Gamma_\pm^\varepsilon = \overline{\omega} \times \{\pm\varepsilon\}$. Under the influence of these forces, a point $\Theta(\boldsymbol{x}^\varepsilon)$ undergoes a displacement $u_i^\varepsilon(\boldsymbol{x}^\varepsilon)\boldsymbol{g}^{i,\varepsilon}(\boldsymbol{x}^\varepsilon)$, where the three vectors $\boldsymbol{g}^{i,\varepsilon}(\boldsymbol{x}^\varepsilon)$ form the contravariant basis at $\Theta(\boldsymbol{x}^\varepsilon)$. The unknowns of the problem are the three covariant components $u_i^\varepsilon : \overline{\Omega}^\varepsilon \to \mathbb{R}$ of the displacement field $u_i^\varepsilon \boldsymbol{g}^{i,\varepsilon} : \overline{\Omega}^\varepsilon \to \mathbb{R}^3$ of the points of $\Theta(\overline{\Omega}^\varepsilon)$, which thus satisfy the boundary conditions $u_i^\varepsilon = 0$ on Γ_0^ε. The objective consists in finding *ad hoc* conditions affording the replacement of this three-dimensional problem by a "two-dimensional problem posed over the middle surface S" if ε is "small enough"; see Fig. 3.1-2.

Note that, for the sake of visual clarity, the thickness is overly exaggerated.

where

$$\boxed{A^{ijkl,\varepsilon} := \lambda^\varepsilon g^{ij,\varepsilon} g^{kl,\varepsilon} + \mu^\varepsilon(g^{ik,\varepsilon} g^{jl,\varepsilon} + g^{il,\varepsilon} g^{jk,\varepsilon})}$$

designate the *contravariant* components of the *three-dimensional elas-*

ticity tensor,

$$e^{\varepsilon}_{i\|j}(\boldsymbol{v}^{\varepsilon}) := \frac{1}{2}\left(\partial^{\varepsilon}_j v^{\varepsilon}_i + \partial^{\varepsilon}_i v^{\varepsilon}_j\right) - \Gamma^{p,\varepsilon}_{ij} v^{\varepsilon}_p$$

designate the *linearized strains in curvilinear coordinates* associated with an arbitrary displacement field $v^{\varepsilon}_i \boldsymbol{g}^{i,\varepsilon}$ of the set $\boldsymbol{\Theta}(\overline{\Omega}^{\varepsilon})$ (Thm. 1.5-1), $f^{i,\varepsilon} \in L^2(\Omega^{\varepsilon})$ and $h^{i,\varepsilon} \in L^2(\Gamma^{\varepsilon}_{+} \cup \Gamma^{\varepsilon}_{-})$ denote the contravariant components of the *applied body* and *surface force densities*, respectively applied to the interior $\boldsymbol{\Theta}(\Omega^{\varepsilon})$ of the shell and to its "**upper**" and "**lower**" **faces** $\boldsymbol{\Theta}(\Gamma^{\varepsilon}_{+})$ and $\boldsymbol{\Theta}(\Gamma^{\varepsilon}_{-})$, and $d\Gamma^{\varepsilon}$ designates the area element along $\partial\Omega^{\varepsilon}$. We thus assume that there are *no* surface forces applied to the portion $\boldsymbol{\Theta}((\gamma - \gamma_0) \times [-\varepsilon, \varepsilon])$ of the lateral face of the shell.

We record in passing the symmetries

$$A^{ijkl,\varepsilon} = A^{jikl,\varepsilon} = A^{klij,\varepsilon},$$

and the relations (satisfied because the mapping $\boldsymbol{\Theta}$ is of the special form given earlier)

$$\Gamma^{3,\varepsilon}_{\alpha 3} = \Gamma^{p,\varepsilon}_{33} = 0 \text{ and } A^{\alpha\beta\sigma 3,\varepsilon} = A^{\alpha 333,\varepsilon} = 0 \text{ in } \overline{\Omega}^{\varepsilon}.$$

The above three-dimensional equations of a linearly elastic shell have one and only one solution for $\varepsilon > 0$ *small enough.* As shown in Thm. 1.8-2, this existence and uniqueness result relies on *Korn's inequality in curvilinear coordinates* (Thm. 1.7-4), combined with the *uniform positive definiteness of the elasticity tensor* (Thm. 1.8-1).

Our objective in the rest of this chapter consists in showing that, if $\varepsilon > 0$ is small enough and the data are of an appropriate order with respect to ε, the above three-dimensional problems are "asymptotically equivalent" to a *"two-dimensional problem posed over the middle surface of the shell"*. This means that *the new unknown should be* $\boldsymbol{\zeta}^{\varepsilon} = (\zeta^{\varepsilon}_i)$, *where* ζ^{ε}_i *are the covariant components of the displacement* $\zeta^{\varepsilon}_i \boldsymbol{a}^i : \overline{\omega} \to \mathbb{R}^3$ *of the points of the middle surface* $S = \boldsymbol{\theta}(\overline{\omega})$. In other words, $\zeta^{\varepsilon}_i(y)\boldsymbol{a}^i(y)$ is the displacement of the point $\boldsymbol{\theta}(y) \in S$; see Fig. 3.1-2.

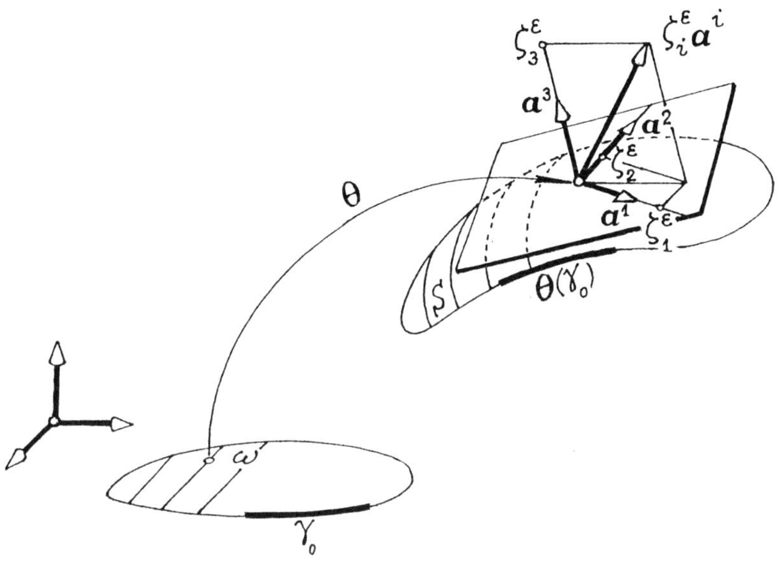

Fig. 3.1-2: *A two-dimensional shell problem.* For $\varepsilon > 0$ "small enough" and data of *ad hoc* orders of magnitude, the three-dimensional shell problem (Fig. 3.1-1) is replaced by a "two-dimensional shell problem". This means that the new unknowns are the three covariant components $\zeta_i^\varepsilon : \overline{\omega} \to \mathbb{R}$ of the displacement field $\zeta_i^\varepsilon \boldsymbol{a}^i : \overline{\omega} \to \mathbb{R}^3$ of the points of the middle surface $S = \boldsymbol{\theta}(\overline{\omega})$. In this process, the "three-dimensional" boundary conditions on Γ_0^ε are replaced by *ad hoc* "two-dimensional" boundary conditions on γ_0.

3.2. THE THREE-DIMENSIONAL EQUATIONS OVER A DOMAIN INDEPENDENT OF ε

We describe in this section the *basic preliminaries of the asymptotic analysis of a linearly elastic shell*, as set forth by Sanchez-Palencia [1990, 1992] in the slightly different, but in fact equivalent, framework of a "multi-scale" asymptotic analysis, then by Miara & Sanchez-Palencia [1996], Ciarlet & Lods [1996b, 1996c, 1996d], and Ciarlet, Lods & Miara [1996].

"Asymptotic analysis" means that *our objective is to study the behavior of the displacement field* $u_i^\varepsilon \boldsymbol{g}^{i,\varepsilon} : \overline{\Omega}^\varepsilon \to \mathbb{R}^3$ *as* $\varepsilon \to 0$, an endeavor that will be achieved by studying the *behavior as* $\varepsilon \to 0$ *of the covariant components* $u_i^\varepsilon : \overline{\Omega}^\varepsilon \to \mathbb{R}$ *of the displacement field*, i.e, *the behavior of the unknown* $\boldsymbol{u}^\varepsilon = (u_i^\varepsilon) : \overline{\Omega}^\varepsilon \to \mathbb{R}^3$ of the three-dimensional variational problem $\mathcal{P}(\Omega^\varepsilon)$ described in Sect. 3.1.

Since these fields are defined on sets $\overline{\Omega}^\varepsilon$ that themselves vary with ε, our first task naturally consists in *transforming the three-dimensional problems $\mathcal{P}(\Omega^\varepsilon)$ into problems posed over a set that does not depend on ε.* The underlying *principle* is thus *identical* to that followed for plates (Vol. II, Sect. 1.3). As we shall see, there are, however, *striking differences*!

Let

$$\Omega := \omega\times]-1, 1[,$$
$$\Gamma_0 := \gamma_0 \times [-1, 1], \quad \Gamma_+ := \omega \times \{1\}, \quad \Gamma_- := \omega \times \{-1\}.$$

Let $x = (x_1, x_2, x_3)$ denote a generic point in the set $\overline{\Omega}$ and let $\partial_i := \partial/\partial x_i$; hence $x_\alpha = y_\alpha$ (a generic point in the set $\overline{\omega}$ is denoted $y = (y_1, y_2)$; cf. Sect. 3.1). With each point $x \in \overline{\Omega}$, we associate the point $x^\varepsilon \in \overline{\Omega}^\varepsilon$ through the bijection (Fig. 3.2-1)

$$\pi^\varepsilon : x = (x_1, x_2, x_3) \in \overline{\Omega} \longrightarrow x^\varepsilon = (x_i^\varepsilon) = (x_1, x_2, \varepsilon x_3) \in \overline{\Omega}^\varepsilon.$$

Consequently,

$$\partial_\alpha^\varepsilon = \partial_\alpha \text{ and } \partial_3^\varepsilon = \frac{1}{\varepsilon}\partial_3.$$

The coordinate $x_3 \in [-1, 1]$ will be also called **transverse variable**, like $x_3^\varepsilon \in [-\varepsilon, \varepsilon]$ in Sect. 3.1 ("scaled" transverse variable could be preferred; however, no confusion should arise).

In order to carry out our asymptotic treatment of the solutions $\boldsymbol{u}^\varepsilon = (u_i^\varepsilon)$ of problems $\mathcal{P}(\Omega^\varepsilon)$ by considering ε as a small parameter, we must:

(i) *specify the way the unknown $\boldsymbol{u}^\varepsilon = (u_i^\varepsilon)$ and more generally the vector fields $\boldsymbol{v} = (v_i^\varepsilon)$ appearing in the formulation of problems $\mathcal{P}(\Omega^\varepsilon)$ are mapped into vector fields over the set $\overline{\Omega}$;*

(ii) *control the way the Lamé constants and the applied forces depend on the parameter ε.*

With the unknown $\boldsymbol{u}^\varepsilon = (u_i^\varepsilon) : \overline{\Omega}^\varepsilon \to \mathbb{R}^3$ and the vector fields $\boldsymbol{v}^\varepsilon = (v_i^\varepsilon) : \overline{\Omega}^\varepsilon \to \mathbb{R}^3$ appearing in the three-dimensional problem $\mathcal{P}(\Omega^\varepsilon)$, we associate the **scaled unknown** $\boldsymbol{u}(\varepsilon) = (u_i(\varepsilon)) : \overline{\Omega} \to \mathbb{R}^3$

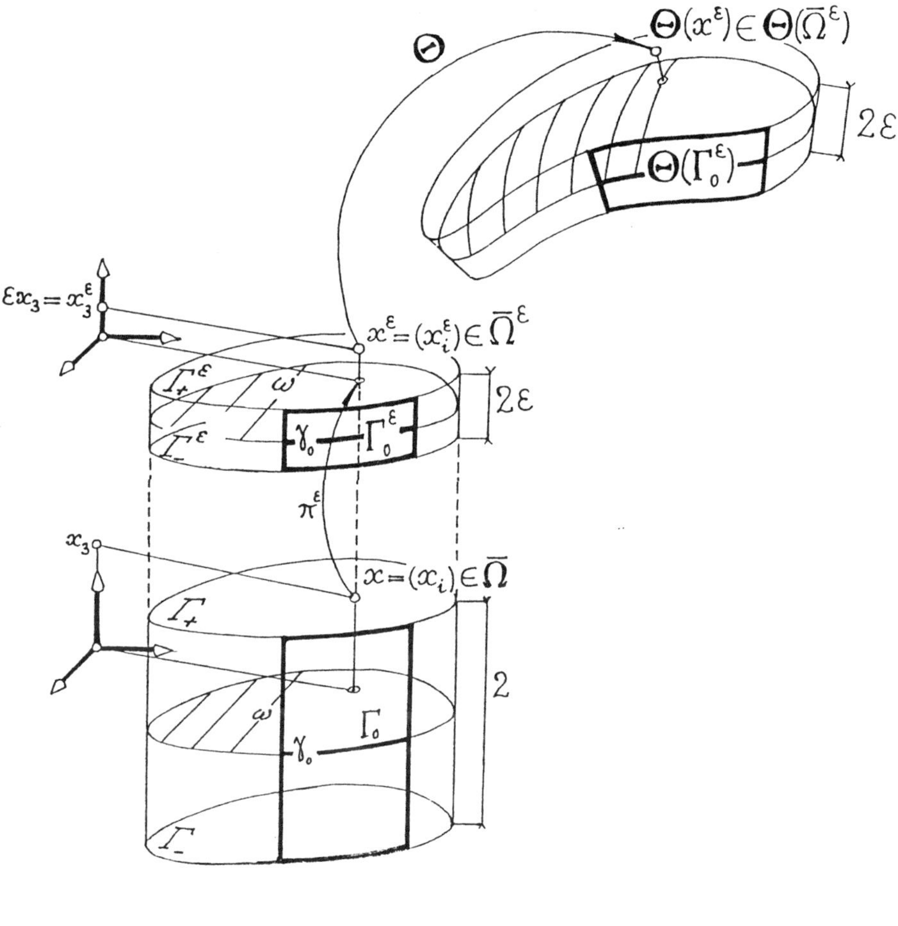

Fig. 3.2-1: *Transformation of the shell problem into a scaled problem posed over a set* $\overline{\Omega}$ *that is independent of* ε. Each point of the reference configuration of the shell is the image $\Theta(x^\varepsilon)$ of a point $x^\varepsilon = (x_i^\varepsilon)$ of the set $\overline{\Omega}^\varepsilon = \overline{\omega} \times [-\varepsilon, \varepsilon]$, over which the shell problem is posed in terms of the "natural" curvilinear coordinates x_i^ε. Each point $x^\varepsilon = (x_i^\varepsilon) \in \overline{\Omega}^\varepsilon$ is itself the image $\pi^\varepsilon x$ of the point $x = (x_i)$ of the set $\overline{\Omega} = \overline{\omega} \times [-1, 1]$ with $x_\alpha^\varepsilon = x_\alpha$ and $x_3^\varepsilon = \varepsilon x_3$. Thanks to these changes of variables, the shell problem is transformed into a problem posed over the set $\overline{\Omega}$.

and the **scaled vector fields** $\boldsymbol{v} = (v_i) : \overline{\Omega} \to \mathbb{R}^3$ defined by the **scalings:**

$$u_i^\varepsilon(x^\varepsilon) =: u_i(\varepsilon)(x) \text{ and } v_i^\varepsilon(x^\varepsilon) =: v_i(x) \text{ for all } x^\varepsilon = \pi^\varepsilon x \in \overline{\Omega}^\varepsilon.$$

The three components $u_i(\varepsilon)$ of the scaled unknown $\boldsymbol{u}(\varepsilon)$ are called the **scaled displacements.**

We next make the following **assumptions on the data**, i.e., on the *Lamé constants* and on the *applied body and surface forces*: There exist constants $\lambda > 0$ and $\mu > 0$ *independent of* ε and there exist functions $f^i \in L^2(\Omega)$ and $h^i \in L^2(\Gamma_+ \cup \Gamma_-)$ *independent of* ε such that

$$\lambda^\varepsilon = \lambda \text{ and } \mu^\varepsilon = \mu,$$

$$\begin{aligned} f^{i,\varepsilon}(x^\varepsilon) &= \varepsilon^p f^i(x) && \text{for all } x^\varepsilon = \pi^\varepsilon x \in \Omega^\varepsilon, \\ h^{i,\varepsilon}(x^\varepsilon) &= \varepsilon^{p+1} h^i(x) && \text{for all } x^\varepsilon = \pi^\varepsilon x \in \Gamma_+^\varepsilon \cup \Gamma_-^\varepsilon, \end{aligned}$$

where *the exponent p is for the time being left unspecified*; needless to say, p is *not* subjected to the usual rule governing Latin exponents!

Since the problem is linear, we assume without loss of generality that the scaled unknown $\boldsymbol{u}(\varepsilon)$ *is "of order* 0 *with respect to* ε*".* This means that the leading term of a formal asymptotic expansion of $\boldsymbol{u}(\varepsilon)$ is *a priori* assumed to be of order 0 (Sect. 3.4), an assumption later justified by the proof that the scaled unknown $\boldsymbol{u}(\varepsilon)$ converges in *ad hoc* space to a *nonzero limit* (Chaps. 4, 5, and 6), when the applied forces are of the right orders.

A complete justification of the scalings of the unknowns and the assumptions on the data, including the determination of the exponent p, will be given in Sect. 3.4. At this preliminary stage, we simply record three observations:

(i) For *plates*, the "passage from Ω^ε to Ω" is identical, but *different* scalings can be made on the *"horizontal"* components $u^{\alpha,\varepsilon}$ and *"vertical"* components $u^{3,\varepsilon}$ of the unknown, and *different* assumptions can be made on the "horizontal" components $f^{\alpha,\varepsilon}$ or $h^{\alpha,\varepsilon}$ and "vertical" components $f^{3,\varepsilon}$ or $h^{3,\varepsilon}$ of the applied forces. This was the basis of the asymptotic analysis set forth for plates by Ciarlet & Destuynder [1979a], also described at length in Vol. II, Sect. 1.3.

(ii) As the problem is *linear*, there is no loss of generality in choosing $\varepsilon^0 = 1$ as the (same) scaling factor for all three covariant components of the displacement field.

(iii) For definiteness, we assume that *the Lamé constants are independent of* ε. However, this assumption is merely a *special case* among a whole *class of assumptions*, which permit in particular the Lamé constants to vary with ε as $\varepsilon \to 0$ if one so wishes. More precisely, a multiplication of both Lamé constants by a factor ε^t, $t \in \mathbb{R}$, is always possible, as we shall explain with more details in the next chapters. The choice $t = 0$ is merely made here for definiteness.

A simple computation then yields the variational problem that the scaled unknown $\boldsymbol{u}(\varepsilon)$ satisfies over the set Ω, thus over a domain that is *independent of* ε (as noted in Sect. 3.1, the Christoffel symbols $\Gamma^{3,\varepsilon}_{\alpha 3}$ and $\Gamma^{p,\varepsilon}_{33}$ vanish in Ω^ε for the special class of mappings $\boldsymbol{\Theta}$ considered here; consequently, the functions $\Gamma^3_{\alpha 3}(\varepsilon)$ and $\Gamma^p_{33}(\varepsilon)$ defined below likewise vanish in Ω):

Theorem 3.2-1. *Let ω be a domain in $\mathbb{R}^2$, let $\boldsymbol{\theta} \in \mathcal{C}^2(\overline{\omega}; \mathbb{R}^3)$ be an injective mapping such that the two vectors $\boldsymbol{a}_\alpha = \partial_\alpha \boldsymbol{\theta}$ are linearly independent at all points of $\overline{\omega}$, and let $\varepsilon_0 > 0$ be as in* Thm. 3.1-1. *With the functions $\Gamma^{p,\varepsilon}_{ij}, g^\varepsilon, A^{ijkl,\varepsilon} : \overline{\Omega}^\varepsilon \to \mathbb{R}$ appearing in problem $\mathcal{P}(\Omega^\varepsilon)$* (Sect. 3.1) *we associate for each $0 < \varepsilon \leq \varepsilon_0$ the "scaled" functions $\Gamma^p_{ij}(\varepsilon)$, $g(\varepsilon)$, $A^{ijkl}(\varepsilon) : \overline{\Omega} \to \mathbb{R}$ defined by*

$$
\boxed{
\begin{aligned}
\Gamma^p_{ij}(\varepsilon)(x) &:= \Gamma^{p,\varepsilon}_{ij}(x^\varepsilon) && \text{for all } x^\varepsilon = \pi^\varepsilon x \in \overline{\Omega}^\varepsilon,\\
g(\varepsilon)(x) &:= g^\varepsilon(x^\varepsilon) && \text{for all } x^\varepsilon = \pi^\varepsilon x \in \overline{\Omega}^\varepsilon,\\
A^{ijkl}(\varepsilon)(x) &:= A^{ijkl,\varepsilon}(x^\varepsilon) && \text{for all } x^\varepsilon = \pi^\varepsilon x \in \overline{\Omega}^\varepsilon.
\end{aligned}
}
$$

With any vector field $\boldsymbol{v} = (v_i) \in \mathbf{H}^1(\Omega)$, we associate the **scaled linearized strains** $e_{i\|j}(\varepsilon; \boldsymbol{v}) = e_{j\|i}(\varepsilon; \boldsymbol{v}) \in L^2(\Omega)$ *defined by:*

$$
\boxed{
\begin{aligned}
e_{\alpha\|\beta}(\varepsilon; \boldsymbol{v}) &:= \frac{1}{2}(\partial_\beta v_\alpha + \partial_\alpha v_\beta) - \Gamma^p_{\alpha\beta}(\varepsilon) v_p,\\
e_{\alpha\|3}(\varepsilon; \boldsymbol{v}) &:= \frac{1}{2}\Big(\frac{1}{\varepsilon}\partial_3 v_\alpha + \partial_\alpha v_3\Big) - \Gamma^\sigma_{\alpha 3}(\varepsilon) v_\sigma,\\
e_{3\|3}(\varepsilon; \boldsymbol{v}) &:= \frac{1}{\varepsilon}\partial_3 v_3.
\end{aligned}
}
$$

Let the assumptions on the data be as above. Then, for each $0 < \varepsilon \leq \varepsilon_0$, *the scaled unknown* $\boldsymbol{u}(\varepsilon)$ *satisfies the following* **scaled three-dimensional variational problem $\mathcal{P}(\varepsilon; \Omega)$ of a linearly elastic shell:**

$$\begin{aligned}
&\boldsymbol{u}(\varepsilon) \in \mathbf{V}(\Omega) := \left\{\boldsymbol{v} = (v_i) \in \mathbf{H}^1(\Omega);\ \boldsymbol{v} = \mathbf{0} \text{ on } \Gamma_0\right\},\\
&\int_\Omega A^{ijkl}(\varepsilon) e_{k\|l}(\varepsilon; \boldsymbol{u}(\varepsilon)) e_{i\|j}(\varepsilon; \boldsymbol{v}) \sqrt{g(\varepsilon)}\, dx\\
&\quad = \varepsilon^p \left\{ \int_\Omega f^i v_i \sqrt{g(\varepsilon)}\, dx + \int_{\Gamma_+ \cup \Gamma_-} h^i v_i \sqrt{g(\varepsilon)}\, d\Gamma \right\}\\
&\qquad \text{for all } \boldsymbol{v} \in \mathbf{V}(\Omega).
\end{aligned}$$

■

The functions $A^{ijkl}(\varepsilon)$ are called the **contravariant components** of the **scaled three-dimensional elasticity tensor of the shell.** The functions $e_{i\|j}(\varepsilon; \boldsymbol{v})$ are called *"scaled" linearized strains* because they satisfy

$$e_{i\|j}(\varepsilon; \boldsymbol{v})(x) = e^\varepsilon_{i\|j}(\boldsymbol{v}^\varepsilon)(x^\varepsilon) \text{ for all } x^\varepsilon = \pi^\varepsilon x \in \overline{\Omega}^\varepsilon.$$

Note that the scaled strains $e_{i\|3}(\varepsilon; \boldsymbol{v})$ are *not* defined for $\varepsilon = 0$. Hence problems $\mathcal{P}(\varepsilon; \Omega)$ provide instances of *singular perturbation problems* in variational form, as considered and extensively studied by Lions [1973].

Remark. By contrast, the functions $\Gamma^p_{ij}(\varepsilon)$, $g(\varepsilon)$, $A^{ijkl}(\varepsilon)$ converge in the space $\mathcal{C}^0(\overline{\Omega})$ as $\varepsilon \to 0$ (Thm. 3.3-1). ■

3.3. GEOMETRICAL AND MECHANICAL PRELIMINARIES

Our first result, based on Ciarlet & Lods [1996b, Lemma 3.1], Ciarlet & Lods [1996d, Lemma 3.1], and Ciarlet, Lods & Miara [1996, Lemma 3.1], gathers all the *"geometrical" preliminaries* needed in the sequel for the asymptotic analysis of linearly elastic shells, regarding *the behavior of the functions* $\Gamma^p_{ij}(\varepsilon)$ *and* $g(\varepsilon)$ (defined in Thm. 3.2-1) *as* $\varepsilon \to 0$. A noteworthy conclusion is that, while these are functions of $x = (y, x_3) \in \overline{\Omega} = \overline{\omega} \times [-1, 1]$, *their limits for* $\varepsilon = 0$ *are functions of* $y \in \overline{\omega}$ *only, i.e., the limits are independent of the transverse variable* x_3; note in this respect that the functions $\Gamma^\sigma_{\alpha\beta}$, $b_{\alpha\beta}$, b^σ_α, and a are identified in the next theorem with functions defined over the set $\overline{\Omega}$ by letting these be constant with respect to x_3. Observe also that the notational distinction between the "three-dimensional" and "two-dimensional" Christoffel symbols $\Gamma^\sigma_{\alpha\beta}(\varepsilon)$ and $\Gamma^\sigma_{\alpha\beta}$ is automatic since the symbol ε appears only in the former.

Theorem 3.3-1. *Let* ω *be a domain in* $\mathbb{R}^2$, *let* $\boldsymbol{\theta} \in \mathcal{C}^3(\overline{\omega}; \mathbb{R}^3)$ *be an injective mapping such that the two vectors* $\boldsymbol{a}_\alpha = \partial_\alpha \boldsymbol{\theta}$ *are linearly independent at all points of* $\overline{\omega}$ *and let* $\varepsilon_0 > 0$ *be as in* Thm. 3.1-1. *The functions* $\Gamma^p_{ij}(\varepsilon) = \Gamma^p_{ji}(\varepsilon)$ *and* $g(\varepsilon)$ *are defined in* Thm. 3.2-1, *the functions* $b_{\alpha\beta}$, b^σ_α, $\Gamma^\sigma_{\alpha\beta}$, a *are defined in* Sect. 3.1, *the covariant derivatives* $b^\sigma_{\beta|\alpha}$ *are defined by* (Thm. 2.5-1)

$$b^\sigma_\beta|_\alpha = \partial_\alpha b^\sigma_\beta + \Gamma^\sigma_{\alpha\tau} b^\tau_\beta - \Gamma^\tau_{\alpha\beta} b^\sigma_\tau,$$

and the functions $b_{\alpha\beta}$, b^σ_α, $\Gamma^\sigma_{\alpha\beta}$, $b^\tau_{\beta|\alpha}$, *and* a *are identified with functions in* $\mathcal{C}^0(\overline{\Omega})$. *Then*

$$\begin{aligned}
\Gamma^\sigma_{\alpha\beta}(\varepsilon) &= \Gamma^\sigma_{\alpha\beta} - \varepsilon x_3 b^\sigma_\beta|_\alpha + O(\varepsilon^2),\\
\Gamma^3_{\alpha\beta}(\varepsilon) &= b_{\alpha\beta} - \varepsilon x_3 b^\sigma_\alpha b_{\sigma\beta},\\
\partial_3 \Gamma^p_{\alpha\beta}(\varepsilon) &= O(\varepsilon),\\
\Gamma^\sigma_{\alpha 3}(\varepsilon) &= -b^\sigma_\alpha - \varepsilon x_3 b^\tau_\alpha b^\sigma_\tau + O(\varepsilon^2),\\
\Gamma^3_{\alpha 3}(\varepsilon) = \Gamma^p_{33}(\varepsilon) &= 0,\\
g(\varepsilon) &= a + O(\varepsilon),
\end{aligned}$$

for all $0 < \varepsilon \le \varepsilon_0$, *where the order symbols* $O(\varepsilon)$ *and* $O(\varepsilon^2)$ *are meant with respect to the norm* $\|\cdot\|_{0,\infty,\overline{\Omega}}$ *defined by*

$$\|w\|_{0,\infty,\overline{\Omega}} = \sup\left\{|w(x)|;\, x \in \overline{\Omega}\right\}.$$

Finally, there exist constants a_0, g_0, *and* g_1 *such that*

$$\begin{aligned}
&0 < a_0 \le a(y) \text{ for all } y \in \overline{\omega},\\
&0 < g_0 \le g(\varepsilon)(x) \le g_1 \text{ for all } x \in \overline{\Omega} \text{ and all } 0 < \varepsilon \le \varepsilon_0.
\end{aligned}$$

Proof. The vectors $\boldsymbol{a}_i$, $\boldsymbol{a}^i$, $\boldsymbol{g}_i^\varepsilon$, $\boldsymbol{g}^{i,\varepsilon}$ and the scalars $a_{\alpha\beta}$, g_{ij}^ε, $g^{ij,\varepsilon}$ are defined in Sect. 3.1. Let

$$\boldsymbol{g}_i(\varepsilon)(x) := \boldsymbol{g}_i^\varepsilon(x^\varepsilon) \text{ and } \boldsymbol{g}^i(\varepsilon)(x) := \boldsymbol{g}^{i,\varepsilon}(x^\varepsilon) \text{ for all } x^\varepsilon = \pi^\varepsilon x \in \overline{\Omega}^\varepsilon,$$
$$g_{ij}(\varepsilon)(x) := g_{ij}^\varepsilon(x^\varepsilon) \text{ and } g^{ij}(\varepsilon)(x) := g^{ij,\varepsilon}(x^\varepsilon) \text{ for all } x^\varepsilon = \pi^\varepsilon x \in \overline{\Omega}^\varepsilon.$$

By definition,

$$\boldsymbol{g}_\alpha(\varepsilon) = \boldsymbol{a}_\alpha + \varepsilon x_3 \partial_\alpha \boldsymbol{a}_3 = \boldsymbol{a}_\alpha - \varepsilon x_3 b_{\alpha\sigma} \boldsymbol{a}^\sigma,$$

since $\partial_\alpha \boldsymbol{a}_3 = -b_{\alpha\sigma}\boldsymbol{a}^\sigma$ by the formula of Weingarten (Thm. 2.3-1). Hence

$$g_{\alpha\beta}(\varepsilon) = a_{\alpha\beta} - 2\varepsilon x_3 b_{\alpha\beta} + O(\varepsilon^2) \text{ and } g_{i3}(\varepsilon) = \delta_{i3},$$

since $\boldsymbol{g}_3(\varepsilon) = \boldsymbol{a}_3$. The matrix $(g_{ij}(\varepsilon))$ being thus of the form

$$(g_{ij}(\varepsilon)) = \mathbf{G} + \varepsilon\mathbf{H} + O(\varepsilon^2),$$

its inverse satisfies

$$(g_{ij}(\varepsilon))^{-1} = (g^{ij}(\varepsilon)) = \mathbf{G}^{-1} - \varepsilon\mathbf{G}^{-1}\mathbf{H}\mathbf{G}^{-1} + O(\varepsilon^2).$$

Combined with the expressions found *supra* for the elements of the matrices $\mathbf{G}$ and $\mathbf{H}$ and the relations $b_\tau^\alpha = a^{\alpha\sigma} b_{\sigma\tau}$, the last equality gives

$$g^{\alpha\beta}(\varepsilon) = a^{\alpha\beta} + 2\varepsilon x_3 a^{\alpha\sigma} b_\sigma^\beta + O(\varepsilon^2) \text{ and } g^{i3}(\varepsilon) = \delta^{i3},$$

so that

$$\boldsymbol{g}^\alpha(\varepsilon) = g^{\alpha i}(\varepsilon)\boldsymbol{g}_i(\varepsilon) = \boldsymbol{a}^\alpha + \varepsilon x_3 b_\sigma^\alpha \boldsymbol{a}^\sigma + O(\varepsilon^2).$$

By definition, $\Gamma_{\alpha j}^\sigma(\varepsilon) = \boldsymbol{g}^\sigma(\varepsilon) \cdot \partial_\alpha \boldsymbol{g}_j(\varepsilon)$; consequently,

$$\Gamma_{\alpha\beta}^\sigma(\varepsilon) = \big(\boldsymbol{a}^\sigma + \varepsilon x_3 b_\tau^\sigma \boldsymbol{a}^\tau + O(\varepsilon^2)\big) \cdot (\partial_\alpha \boldsymbol{a}_\beta + \varepsilon x_3 \partial_{\alpha\beta} \boldsymbol{a}_3),$$
$$\Gamma_{\alpha 3}^\sigma(\varepsilon) = \big(\boldsymbol{a}^\sigma + \varepsilon x_3 b_\tau^\sigma \boldsymbol{a}^\tau + O(\varepsilon^2)\big) \cdot \partial_\alpha \boldsymbol{a}_3.$$

Combining these expressions with the formulas of Gauß and Weingarten (Thm. 2.3-1), viz.,

$$\partial_\alpha \boldsymbol{a}_\beta = \Gamma_{\alpha\beta}^\sigma \boldsymbol{a}_\sigma + b_{\alpha\beta}\boldsymbol{a}_3 \quad\text{and}\quad \partial_\alpha \boldsymbol{a}_3 = -b_\alpha^\sigma \boldsymbol{a}_\sigma,$$

together with the definition of the covariant derivatives $b_{\beta|\alpha}^\sigma$, we find that

$$\Gamma_{\alpha\beta}^\sigma(\varepsilon) = \Gamma_{\alpha\beta}^\sigma - \varepsilon x_3 b_\beta^\sigma|_\alpha + O(\varepsilon^2),$$
$$\Gamma_{\alpha 3}^\sigma(\varepsilon) = -b_\alpha^\sigma - \varepsilon x_3 b_\alpha^\tau b_\tau^\sigma + O(\varepsilon^2).$$

Since $\boldsymbol{g}^3(\varepsilon) = g^{3i}(\varepsilon)\boldsymbol{g}_i(\varepsilon) = \boldsymbol{a}^3$, we likewise have, thanks again to the formulas of Gauß and Weingarten,

$$\Gamma^3_{\alpha\beta}(\varepsilon) = \boldsymbol{g}^3(\varepsilon) \cdot \partial_\alpha \boldsymbol{g}_\beta(\varepsilon) = \boldsymbol{a}^3 \cdot (\partial_\alpha \boldsymbol{a}_\beta + \varepsilon x_3 \partial_{\alpha\beta}\boldsymbol{a}_3) = b_{\alpha\beta} - \varepsilon x_3 b^\sigma_\alpha b_{\sigma\beta}.$$

The relations

$$\Gamma^p_{\alpha\beta}(\varepsilon) = \boldsymbol{g}^p(\varepsilon) \cdot \partial_\beta \boldsymbol{g}_\alpha(\varepsilon) = \boldsymbol{g}^p(\varepsilon) \cdot (\partial_\beta \boldsymbol{a}_\alpha + \varepsilon x_3 \partial_{\alpha\beta}\boldsymbol{a}_3)$$

show that

$$\partial_3 \Gamma^p_{\alpha\beta}(\varepsilon) = \partial_3 \boldsymbol{g}^p(\varepsilon) \cdot \partial_\beta \boldsymbol{a}_\alpha + O(\varepsilon).$$

Let

$$\|w\|_{1,\infty,\overline{\Omega}} = \|w\|_{0,\infty,\overline{\Omega}} + \sum_i \|\partial_i w\|_{0,\infty,\overline{\Omega}}.$$

Since only first-order and second-order derivatives of $\boldsymbol{\theta}$ occur in the functions

$$g_{\alpha\beta}(\varepsilon) = (\boldsymbol{a}_\alpha + \varepsilon x_3 \partial_\alpha \boldsymbol{a}_3) \cdot (\boldsymbol{a}_\beta + \varepsilon x_3 \partial_\beta \boldsymbol{a}_3),$$

and since $\boldsymbol{\theta} \in \mathcal{C}^3(\overline{\omega}; \mathbb{R}^3)$, we conclude that there exists a constant c_1 such that

$$g_{\alpha\beta}(\varepsilon) = a_{\alpha\beta} + \varepsilon g^1_{\alpha\beta}(\varepsilon) \text{ with } \|g^1_{\alpha\beta}(\varepsilon)\|_{1,\infty,\overline{\Omega}} \leq c_1.$$

Consequently,

$$g(\varepsilon) - a = O(\varepsilon),$$

and there likewise exists a constant c_2 such that

$$g^{\alpha\beta}(\varepsilon) = a^{\alpha\beta} + \varepsilon g^{\alpha\beta,1}(\varepsilon) \text{ with } \|g^{\alpha\beta,1}(\varepsilon)\|_{1,\infty,\overline{\Omega}} \leq c_2.$$

The relations $\partial_3 \Gamma^p_{\alpha\beta}(\varepsilon) = O(\varepsilon)$ then follow by noting that

$$\boldsymbol{g}^\alpha(\varepsilon) = g^{\alpha\beta}(\varepsilon)\boldsymbol{g}_\beta(\varepsilon) \text{ and } \boldsymbol{g}^3(\varepsilon) = \boldsymbol{a}^3.$$

Since the vectors $\boldsymbol{a}_\alpha$ are linearly independent at all points of $\overline{\omega}$, there exist constants a_0 and a_1 such that

$$0 < a_0 \leq a(y) \leq a_1 \text{ for all } y \in \overline{\omega}.$$

Hence the above asymptotic behavior of the function $g(\varepsilon)$ shows that, if $\varepsilon_0 > 0$ is chosen as in Thm. 3.1-1, there exist constants g_0 and g_1 such that

$$0 < g_0 \leq g(\varepsilon)(x) \leq g_1 \text{ for all } x \in \overline{\Omega} \text{ and all } 0 < \varepsilon \leq \varepsilon_0.$$

Finally, it is clear that the relations $\Gamma^3_{\alpha 3}(\varepsilon) = \Gamma^p_{33}(\varepsilon) = 0$ are immediate consequences of the relations $\Gamma^{3,\varepsilon}_{\alpha 3} = \Gamma^{p,\varepsilon}_{33} = 0$ (Sect. 3.1). ■

Our second result, based on Bernadou, Ciarlet & Miara [1994, Lemma 2.1] and Ciarlet & Lods [1996b, Lemma 3.1], gathers all the *"mechanical" preliminaries* needed in the sequel.

It comprises two parts: In part (a), the *uniform positive definiteness of the scaled two-dimensional elasticity tensor of the shell*, given here in terms of its *contravariant components* $a^{\alpha\beta\sigma\tau}$, is established; this tensor will appear naturally in the formulation of all the "limit" scaled two-dimensional shell equations found at the outcome of the asymptotic analysis.

In part (b), the *behavior of the scaled contravariant components* $A^{ijkl}(\varepsilon)$ *of the three-dimensional elasticity tensor* (defined in Thm. 3.2-1) *as* $\varepsilon \to 0$ is analyzed; it is shown that their limits are functions of $y \in \overline{\omega}$ only and that *the positive definiteness of the associated scaled three-dimensional elasticity tensor is uniform*, not only with respect to $x \in \overline{\Omega}$, but also with respect to $0 < \varepsilon \leq \varepsilon_0$.

Theorem 3.3-2. (a) *Let* ω *be a domain in* $\mathbb{R}^2$, *let* $\boldsymbol{\theta} \in \mathcal{C}^1(\overline{\omega}; \mathbb{R}^3)$ *be an injective mapping such that the two vectors* $\boldsymbol{a}_\alpha = \partial_\alpha \boldsymbol{\theta}$ *are linearly independent at all points of* $\overline{\omega}$, *let* $a^{\alpha\beta}$ *denote the contravariant components of the metric tensor of* $S = \boldsymbol{\theta}(\overline{\omega})$, *let the* **contravariant components** *of the* **scaled two-dimensional elasticity tensor of the shell** *be defined by*

$$\boxed{a^{\alpha\beta\sigma\tau} := \frac{4\lambda\mu}{\lambda + 2\mu} a^{\alpha\beta} a^{\sigma\tau} + 2\mu(a^{\alpha\sigma} a^{\beta\tau} + a^{\alpha\tau} a^{\beta\sigma}),}$$

and assume that

$$\boxed{\lambda \geq 0 \text{ and } \mu > 0.}$$

Then there exists a constant $c_e = c_e(\omega, \boldsymbol{\theta}, \mu) > 0$ *such that*

$$\boxed{\sum_{\alpha,\beta} |t_{\alpha\beta}|^2 \leq c_e a^{\alpha\beta\sigma\tau}(y) t_{\sigma\tau} t_{\alpha\beta}}$$

for all $y \in \overline{\omega}$ *and all symmetric matrices* $(t_{\alpha\beta})$.

(b) *Let* $\boldsymbol{\theta} \in \mathcal{C}^2(\overline{\omega}; \mathbb{R}^3)$ *and let the other assumptions on the mapping* $\boldsymbol{\theta}$ *and the definition of* ε_0 *be as in* Thm. 3.1-1. *The contravariant components* $A^{ijkl}(\varepsilon) = A^{jikl}(\varepsilon) = A^{klij}(\varepsilon)$ *of the scaled three-dimensional elasticity tensor* (Thm. 3.2-1) *satisfy*

$$\boxed{A^{ijkl}(\varepsilon) = A^{ijkl}(0) + O(\varepsilon) \text{ and } A^{\alpha\beta\sigma 3}(\varepsilon) = A^{\alpha 333}(\varepsilon) = 0}$$

for all $0 < \varepsilon \le \varepsilon_0$, *where the order symbol* $O(\varepsilon)$ *is meant with respect to the norm* $\|\cdot\|_{0,\infty,\overline{\Omega}}$ (defined in Thm. 3.3-1) *and*

$$\boxed{\begin{aligned} A^{\alpha\beta\sigma\tau}(0) &:= \lambda a^{\alpha\beta} a^{\sigma\tau} + \mu(a^{\alpha\sigma} a^{\beta\tau} + a^{\alpha\tau} a^{\beta\sigma}), \\ A^{\alpha\beta 33}(0) &:= \lambda a^{\alpha\beta}, \ \ A^{\alpha 3\sigma 3}(0) := \mu a^{\alpha\sigma}, \ \ A^{3333}(0) := \lambda + 2\mu, \\ A^{\alpha\beta\sigma 3}(0) &= A^{\alpha 333}(0) := 0. \end{aligned}}$$

Finally, there exists a constant $C_e > 0$ *such that*

$$\boxed{\sum_{i,j} |t_{ij}|^2 \le C_e A^{ijkl}(\varepsilon)(x) t_{kl} t_{ij}}$$

for all $0 < \varepsilon \le \varepsilon_0$, *all* $x \in \overline{\Omega}$, *and all symmetric matrices* (t_{ij}).

Proof. We note, first that

$$a^{\alpha\beta}(y) a^{\sigma\tau}(y) t_{\sigma\tau} t_{\alpha\beta} = \Big(a^{\alpha\beta}(y) t_{\alpha\beta}\Big)^2 \ge 0$$

for all $y \in \overline{\omega}$ and all matrices $(t_{\alpha\beta})$, next that

$$\Big(a^{\alpha\sigma}(y) a^{\beta\tau}(y) + a^{\alpha\tau}(y) a^{\beta\sigma}(y)\Big) t_{\sigma\tau} t_{\alpha\beta} = 2\boldsymbol{t}^T \mathbf{A}(y) \boldsymbol{t}$$

for all symmetric matrices $(t_{\alpha\beta})$, where

$$\begin{aligned} \mathbf{A}(y) &:= \begin{pmatrix} a^{11}a^{11} & 2a^{11}a^{12} & a^{12}a^{12} \\ 2a^{11}a^{12} & 2(a^{12}a^{12} + a^{11}a^{22}) & 2a^{12}a^{22} \\ a^{12}a^{12} & 2a^{12}a^{22} & a^{22}a^{22} \end{pmatrix}(y), \\ \boldsymbol{t} &:= \begin{pmatrix} t_{11} \\ t_{12} \\ t_{22} \end{pmatrix}. \end{aligned}$$

Since

$$a^{11}a^{11} > 0 \text{ in } \overline{\omega},$$

$$\det \begin{pmatrix} a^{11}a^{11} & 2a^{11}a^{12} \\ 2a^{11}a^{12} & 2(a^{12}a^{12} + a^{11}a^{22}) \end{pmatrix} = 2\frac{a^{11}a^{11}}{a} > 0 \text{ in } \overline{\omega},$$

$$\det \mathbf{A} = \frac{2}{a^3} > 0 \text{ in } \overline{\omega},$$

where $a = \det(a_{\alpha\beta})$, we infer from a well-known characterization that the symmetric matrix $\mathbf{A}(y)$ is positive definite at all $y \in \overline{\omega}$, and the existence of the constant $c_e > 0$ follows from the compactness of the set $\overline{\omega}$; this proves (a).

An argument similar to that used in the proof of Thm. 3.3-1 shows that

$$g^{\alpha\beta}(\varepsilon) = a^{\alpha\beta} + O(\varepsilon) \text{ and } g^{i3}(\varepsilon) = \delta^{i3},$$

where $g^{ij}(\varepsilon)(x) := g^{ij,\varepsilon}(x^\varepsilon)$ for all $x^\varepsilon = \pi^\varepsilon x \in \overline{\Omega}^\varepsilon$. These relations, combined with the definitions (Thm. 3.2-1)

$$A^{ijkl}(\varepsilon) = \lambda g^{ij}(\varepsilon)g^{kl}(\varepsilon) + \mu(g^{ik}(\varepsilon)g^{jl}(\varepsilon) + g^{il}(\varepsilon)g^{jk}(\varepsilon)),$$

imply that

$$A^{ijkl}(\varepsilon) = A^{ijkl}(0) + O(\varepsilon) \text{ and } A^{\alpha\beta\sigma 3}(\varepsilon) = A^{\alpha 333}(\varepsilon) = 0.$$

For each $0 < \varepsilon \leq \varepsilon_0$, the three-dimensional elasticity tensor defined by means of its contravariant components $A^{ijkl,\varepsilon}$ is positive definite, uniformly with respect to $x^\varepsilon \in \overline{\Omega}^\varepsilon$ (Thm. 1.8-1). This implies that there exists $c(\varepsilon) > 0$ such that

$$\sum_{i,j} |t_{ij}|^2 \leq c(\varepsilon)A^{ijkl}(\varepsilon)(x)t_{kl}t_{ij}$$

for all $x \in \overline{\Omega}$ and all $(t_{ij}) \in \mathbb{S}$, where $\mathbb{S}$ denotes the set of all symmetric matrices of order three, on the one hand. The definitions of the functions $A^{ijkl}(0)$ imply that

$$\begin{aligned} A^{ijkl}(0)t_{kl}t_{ij} &= \lambda(a^{\alpha\beta}t_{\alpha\beta} + t_{33})^2 + \mu(a^{\alpha\sigma}a^{\beta\tau} + a^{\alpha\tau}a^{\beta\sigma})t_{\sigma\tau}t_{\alpha\beta} \\ &\quad + 4\mu a^{\alpha\beta}t_{\alpha 3}t_{\beta 3} + (\lambda + 2\mu)t_{33}t_{33}, \end{aligned}$$

on the other. Since there exists c_1 such that (see the first part of this proof):

$$\sum_{\alpha,\beta} |t_{\alpha\beta}|^2 \leq c_1 \left(a^{\alpha\sigma}(y)a^{\beta\tau}(y) + a^{\alpha\tau}(y)a^{\beta\sigma}(y)\right) t_{\sigma\tau}t_{\alpha\beta}$$

for all $y \in \overline{\omega}$ and all symmetric matrices $(t_{\alpha\beta})$, it follows that there exists c_2 such that

$$\sum_{i,j} |t_{ij}|^2 \le c_2 A^{ijkl}(0)(x) t_{kl} t_{ij}$$

for all $x \in \overline{\Omega}$ and all $(t_{ij}) \in \mathbb{S}$. The continuity of the mapping

$$(\varepsilon, x, (t_{ij})) \in [0, \varepsilon_0] \times \overline{\Omega} \times \mathbb{S}_1 \to A^{ijkl}(\varepsilon)(x) t_{kl} t_{ij},$$

where $\mathbb{S}_1 := \{(t_{ij}) \in \mathbb{S}; \sum_{i,j} |t_{ij}|^2 = 1\}$, then yields the existence of a constant $C_e > 0$ such that the inequality

$$\sum_{i,j} |t_{ij}|^2 \le C_e A^{ijkl}(\varepsilon)(x) t_{kl} t_{ij}$$

holds for all $(t_{ij}) \in \mathbb{S}$, $x \in \overline{\Omega}$, and $0 \le \varepsilon \le \varepsilon_0$; this proves (b). ■

Remarks. (1) The uniform positive definiteness of the scaled two-dimensional elasticity tensor of the shell (part (a)) can be established in a different manner; cf. Ex. 3.1.

(2) Although $\lambda < 0$ is not realistic from a mechanical standpoint (the Lamé constants of actual elastic materials are strictly positive), the elasticity tensor remains positive definite even if λ is "slightly negative"; cf. Ex. 3.2. ■

Other "technical", but of a more specialized nature, preliminaries will be also proved in the next chapters; see notably Sects. 4.2, 4.3, 5.2, and 5.3.

3.4. THE TWO-DIMENSIONAL EQUATIONS OF LINEARLY ELASTIC "MEMBRANE" AND "FLEXURAL" SHELLS DERIVED BY MEANS OF A FORMAL ASYMPTOTIC ANALYSIS

The *formal* asymptotic analysis of linearly elastic "thin" shells is a relatively recent subject. After the landmark contributions of Goldenveizer [1963, 1964], a major step was achieved by Destuynder [1980] in his Doctoral Dissertation, where a convergence theorem for elliptic membrane shells (as they will be defined in the next chapter) was in particular "almost proved". Another major step was achieved in the pioneering *Note aux Comptes Rendus* of Sanchez-Palencia [1990] (see also Sanchez-Palencia [1992]), who for the first time clearly demonstrated that the asymptotic behavior of a shell as its thickness goes to zero is governed, *either* by *"membrane"*, *or* by *"flexural"*, *two-dimensional equations* and that the distinction rests on whether a certain *space of linearized inextensional displacements* reduces to $\{\mathbf{0}\}$ or not. His analysis was further refined by Miara & Sanchez-Palencia [1996], whom we faithfully follow in this section. Finally, Caillerie & Sanchez-Palencia [1995b] have shown that, *by contrast*, the "membrane" and "flexural" terms remain *coupled* in the limit equations (again obtained *via* a formal asymptotic analysis) when the linearly elastic material is *anisotropic* and *nonhomogeneous*.

Within the linear theory, this section has an optional character: Stricto sensu, only the content of Sects. 3.1, 3.2, and 3.3 is needed for the asymptotic analysis of *linearly elastic shells* carried out in the next three chapters; besides, its conclusions, which are based on the method of *formal* asymptotic expansions, might not have the same appeal as those derived from the *convergence theorems* proved in the next chapters. *Nevertheless, its content admirably serves four major purposes:*

First and foremost, it clearly leads through a perhaps formal, but certainly subtle, analysis to the somewhat unexpected conclusion that *there are two fundamentally distinct kinds of linearly elastic shells*, the *"membrane" shells* and the *"flexural" shells* (the "membrane" shells being themselves later subdivided into two subclasses; cf. Chaps. 4 and 5). This conclusion is of course in sharp contrast with that drawn for *linearly elastic plates* in Vol. II, Chap. 1.

Secondly, *it justifies the scalings on the unknowns* u_i^ε *and the assumptions on the data* $f^{i,\varepsilon}$ *and* $h^{i,\varepsilon}$ made in Sect. 3.2, notably by specifying how the exponent p must be chosen in each case.

Thirdly, it explains why the *logical order of presentation* consists in treating the "membrane" shells *before* the "flexural" ones; otherwise it makes as much sense, from a *mathematical* standpoint at least, to first consider the "flexural" shells.

Fourthly, it plays an indispensable rôle in the justification of the two-dimensional equations of the *nonlinear theory* (Chaps. 8 to 10) by the method of formal asymptotic expansions.

Consider the *three-dimensional shell problems* $\mathcal{P}(\Omega^\varepsilon)$, as they are defined in Sect. 3.1 for each $\varepsilon > 0$. We assume that the mapping $\boldsymbol{\theta}$

that defines the middle surface of the shell satisfies

$$\boxed{\boldsymbol{\theta} \in \mathcal{C}^3(\overline{\omega}; \mathbb{R}^3).}$$

We also assume *ab initio* that *the Lamé constants* λ^ε *and* μ^ε *are independent of* ε, i.e., that there exist constants $\lambda > 0$ and $\mu > 0$ such that

$$\boxed{\lambda^\varepsilon = \lambda \text{ and } \mu^\varepsilon = \mu \text{ for all } \varepsilon > 0.}$$

First, vector fields $\boldsymbol{u}(\varepsilon) = (u_i(\varepsilon)) \in \mathbf{H}^1(\Omega)$ and $\boldsymbol{v} = (v_i) \in \mathbf{H}^1(\Omega)$ and functions $f^i(\varepsilon) \in L^2(\Omega)$ and $h^i(\varepsilon) \in L^2(\Gamma_+ \cup \Gamma_-)$ are defined by letting

$$\boxed{\begin{aligned} u_i^\varepsilon(x^\varepsilon) = u_i(\varepsilon)(x) \text{ and } v_i^\varepsilon(x^\varepsilon) &= v_i(x) \text{ for all } x^\varepsilon = \pi^\varepsilon x \in \overline{\Omega}^\varepsilon, \\ f^{i,\varepsilon}(x^\varepsilon) &= f^i(\varepsilon)(x) \text{ for all } x^\varepsilon = \pi^\varepsilon x \in \Omega^\varepsilon, \\ h^{i,\varepsilon}(x^\varepsilon) &= h^i(\varepsilon)(x) \text{ for all } x^\varepsilon = \pi^\varepsilon x \in \Gamma_+^\varepsilon \cup \Gamma_-^\varepsilon. \end{aligned}}$$

In this fashion, each variational problem $\mathcal{P}(\Omega^\varepsilon)$, $\varepsilon > 0$, is transformed into the following variational problem $\mathcal{P}^*(\varepsilon; \Omega)$:

$$\boxed{\begin{aligned} &\boldsymbol{u}(\varepsilon) \in \mathbf{V}(\Omega) = \{\boldsymbol{v} = (v_i) \in \mathbf{H}^1(\Omega); \boldsymbol{v} = \mathbf{0} \text{ on } \Gamma_0\}, \\ &\int_\Omega A^{ijkl}(\varepsilon) e_{k\|l}(\varepsilon; \boldsymbol{u}(\varepsilon)) e_{i\|j}(\varepsilon; \boldsymbol{v}) \sqrt{g(\varepsilon)}\, dx \\ &\qquad = \int_\Omega f^i(\varepsilon) v_i \sqrt{g(\varepsilon)}\, dx + \frac{1}{\varepsilon} \int_{\Gamma_+ \cup \Gamma_-} h^i(\varepsilon) v_i \sqrt{g(\varepsilon)}\, d\Gamma \\ &\qquad\qquad\qquad\qquad \text{for all } \boldsymbol{v} \in \mathbf{V}(\Omega), \end{aligned}}$$

the "scaled" functions $A^{ijkl}(\varepsilon)$ and $g(\varepsilon)$ and the scaled linearized strains $e_{i\|j}(\varepsilon; \boldsymbol{v})$ being defined as in Thm. 3.2-1, where they appeared in the problems then denoted $\mathcal{P}(\varepsilon; \Omega)$. As expected, the *left-hand sides* of the variational equations of problems $\mathcal{P}^*(\varepsilon; \Omega)$ and $\mathcal{P}(\varepsilon; \Omega)$ coincide since the same transformation was already performed in Thm. 3.2-1 on the covariant components of the unknown displacement. The difference lies in the *right-hand sides, where the "right" scalings are yet to be found* on the contravariant components of the applied forces. The notation $\mathcal{P}^*(\varepsilon; \Omega)$ is precisely intended to remind of this difference.

Second, it is assumed that *there exists a formal asymptotic expansion of the scaled unknown* of the form:

$$\boxed{\boldsymbol{u}(\varepsilon) = \boldsymbol{u}^0 + \varepsilon \boldsymbol{u}^1 + \varepsilon^2 \boldsymbol{u}^2 + \varepsilon^3 \boldsymbol{u}^3 + \varepsilon^4 \boldsymbol{u}^4 + \cdots, \quad \text{with } \boldsymbol{u}^0 \neq \boldsymbol{0}.}$$

As already observed in Sect. 3.2, there is no loss of generality in assuming that the leading term $\boldsymbol{u}^0$ is of order 0 with respect to ε since the problem is *linear.* Note that, *by contrast, this freedom is lost in the nonlinear case*; cf. Sect. 8.6.

According to the basic *Ansatz* of the *method of formal asymptotic expansions* (more details and references about this method are found in Sect. 8.6, where it is put to use in an even more general setting), we wish to identify the leading term $\boldsymbol{u}^0$ by equating to 0 the factors of the successive powers of ε found in problem $\mathcal{P}^*(\varepsilon; \Omega)$ when $\boldsymbol{u}(\varepsilon)$ is replaced by its formal expansion, then by solving the resulting variational equations. In this process, each *needed* term $\boldsymbol{u}^q$, $q \geq 0$, is required to be in the space $\mathbf{H}^1(\Omega)$, but the *boundary condition* on Γ_0 is imposed on it only if it is needed for determining $\boldsymbol{u}^0$.

Finally, it is required that *no restriction should be imposed on the applied force densities entering the right-hand side of the equations used for determining the leading term eventually found,* since no such restriction is imposed on the original problems $\mathcal{P}(\Omega^\varepsilon)$. It is this natural requirement that allows to successively "eliminate" unwanted models.

The above formal expansion of the scaled unknown combined with the asymptotic behavior of the functions $\Gamma^p_{ij}(\varepsilon)$ (Thm. 3.3-1) induces formal asymptotic expansions of the scaled linearized strains of the form:

$$\boxed{e_{i\|j}(\varepsilon; \boldsymbol{u}(\varepsilon)) = \frac{1}{\varepsilon} e^{-1}_{i\|j} + e^0_{i\|j} + \varepsilon e^1_{i\|j} + \varepsilon^2 e^2_{i\|j} + \varepsilon^3 e^3_{i\|j} + \cdots,}$$

where

$$\left\{\begin{aligned} e^{-1}_{\alpha\|\beta} &= 0, \\ e^{-1}_{\alpha\|3} &= \frac{1}{2}\partial_3 u^0_\alpha, \\ e^{-1}_{3\|3} &= \partial_3 u^0_3, \end{aligned}\right. \qquad \left\{\begin{aligned} e^0_{\alpha\|\beta} &= \frac{1}{2}(\partial_\beta u^0_\alpha + \partial_\alpha u^0_\beta) - \Gamma^\sigma_{\alpha\beta} u^0_\sigma - b_{\alpha\beta} u^0_3, \\ e^0_{\alpha\|3} &= \frac{1}{2}(\partial_3 u^1_\alpha + \partial_\alpha u^0_3) + b^\sigma_\alpha u^0_\sigma, \\ e^0_{3\|3} &= \partial_3 u^1_3, \end{aligned}\right.$$

$$\begin{cases} e^1_{\alpha\|\beta} = \dfrac{1}{2}(\partial_\beta u^1_\alpha + \partial_\alpha u^1_\beta) - \Gamma^\sigma_{\alpha\beta} u^1_\sigma - b_{\alpha\beta} u^1_3 + x_3\{b^\sigma_\beta|_\alpha u^0_\sigma + b^\sigma_\alpha b_{\sigma\beta} u^0_3\}, \\ e^1_{\alpha\|3} = \dfrac{1}{2}(\partial_3 u^2_\alpha + \partial_\alpha u^1_3) + b^\sigma_\alpha u^1_\sigma + x_3 b^\tau_\alpha b^\sigma_\tau u^0_\sigma, \\ e^1_{3\|3} = \partial_3 u^2_3, \end{cases}$$

and each function $e^q_{i\|j}$, $q \geq 2$, is a linear combination of $\boldsymbol{u}^{q-1}$, $\boldsymbol{u}^q$, $\boldsymbol{u}^{q+1}$ and of their first-order partial derivatives (only the values $q = 2$ and $q = 3$ are needed in the sequel).

The functions $e_{i\|j}(\varepsilon; \boldsymbol{v})$ likewise admit formal asymptotic expansions (again resulting from the asymptotic behavior of the functions $\Gamma^p_{ij}(\varepsilon)$) of the form:

$$\boxed{e_{i\|j}(\varepsilon; \boldsymbol{v}) = \frac{1}{\varepsilon} e^{-1}_{i\|j}(\boldsymbol{v}) + e^0_{i\|j}(\boldsymbol{v}) + \varepsilon e^1_{i\|j}(\boldsymbol{v}) + \varepsilon^2 e^2_{i\|j}(\boldsymbol{v}) + \cdots,}$$

where

$$\begin{cases} e^{-1}_{\alpha\|\beta}(\boldsymbol{v}) = 0, \\ e^{-1}_{\alpha\|3}(\boldsymbol{v}) = \dfrac{1}{2}\partial_3 v_\alpha, \\ e^{-1}_{3\|3}(\boldsymbol{v}) = \partial_3 v_3, \end{cases} \quad \begin{cases} e^0_{\alpha\|\beta}(\boldsymbol{v}) = \dfrac{1}{2}(\partial_\beta v_\alpha + \partial_\alpha v_\beta) - \Gamma^\sigma_{\alpha\beta} v_\sigma - b_{\alpha\beta} v_3, \\ e^0_{\alpha\|3}(\boldsymbol{v}) = \dfrac{1}{2}\partial_\alpha v_3 + b^\sigma_\alpha v_\sigma, \\ e^0_{3\|3}(\boldsymbol{v}) = 0, \end{cases}$$

$$\begin{cases} e^1_{\alpha\|\beta}(\boldsymbol{v}) = x_3 b^\sigma_\beta|_\alpha v_\sigma + x_3 b^\sigma_\alpha b_{\sigma\beta} v_3, \\ e^1_{\alpha\|3}(\boldsymbol{v}) = x_3 b^\tau_\alpha b^\sigma_\tau v_\sigma, \\ e^1_{3\|3}(\boldsymbol{v}) = 0. \end{cases}$$

The asymptotic behavior of the functions $g(\varepsilon)$ and $A^{ijkl}(\varepsilon)$ (Thms. 3.3-1 and 3.3-2) also implies that (recall that $\boldsymbol{\theta} \in \mathcal{C}^3(\overline{\omega}; \mathbb{R}^3)$ by assumption)

$$\boxed{\begin{aligned} g(\varepsilon) &= a + O(\varepsilon), \\ A^{ijkl}(\varepsilon)\sqrt{g(\varepsilon)} &= A^{ijkl}(0)\sqrt{a} + \varepsilon B^{ijkl,1} + \varepsilon^2 B^{ijkl,2} + o(\varepsilon^2), \end{aligned}}$$

where the functions $A^{ijkl}(0)$ are defined in Thm. 3.3-2.

Note that the above expansions comprise all the terms that will actually *appear* in an equation at some stage of the subsequent analysis. It does not mean, however, that each one of their *specific expressions* will be actually needed.

As a last preliminary, we also record a simple result frequently used in this volume (as in Vol. II):

Theorem 3.4-1. *Let ω be a domain in $\mathbb{R}^2$ with boundary γ, let $\Omega = \omega \times]-1, 1[$, and let $w \in L^p(\Omega)$, $p > 1$, be a function such that*

$$\int_\Omega w \partial_3 v \, dx = 0 \text{ for all } v \in \mathcal{C}^\infty(\overline{\Omega}) \text{ that satisfy } v = 0 \text{ on } \gamma \times [-1, 1].$$

Then $w = 0$.

Proof. Let φ be an arbitrary function in the space $\mathcal{D}(\Omega)$ and let the function $v : \overline{\Omega} \to \mathbb{R}$ be defined by

$$v(x_1, x_2, x_3) = \int_{-1}^{x_3} \varphi(x_1, x_2, t) \, dt \text{ for all } (x_1, x_2, x_3) \in \overline{\Omega}.$$

Then $v \in \mathcal{C}^\infty(\overline{\Omega})$ and $v = 0$ on $\gamma \times [-1, 1]$; hence

$$\int_\Omega w\varphi \, dx = \int_\Omega w \partial_3 v \, dx = 0,$$

and thus $w = 0$ by a classical property of the space $L^p(\Omega)$, $p > 1$. ■

The above implication *a fortiori* holds if $\int_\Omega w \partial_3 v \, dx = 0$ for all $v \in H^1(\Omega)$ that vanish on Γ_0; this is how it is used in the remainder of this section.

We are now in a position to *start the cancellation of the factors of the successive powers of ε in the variational equations of problem $\mathcal{P}^*(\varepsilon; \Omega)$, until the leading term $\boldsymbol{u}^0$ of the formal asymptotic expansion can be fully identified.* It will be found in this fashion that $\boldsymbol{u}^0$ *satisfies a two-dimensional variational problem*, which may be either that of a *"membrane" shell* or that of a *"flexural" shell* and that in each case *the "right" orders on the components of the applied forces* (which we are seeking) *appearing in the right-hand sides are simultaneously determined.*

Note that, by contrast with our usual practice, these major conclusions will be formally stated (Thms. 3.4-2 and 3.4-3) only *after* they have been established, in a *series of eight steps* numbered (i) to (viii), as this procedure seems more natural in the present case.

(i) Since the lowest power of ε appearing in $\mathcal{P}^*(\varepsilon;\Omega)$ is ε^{-2}, we are naturally led to *first try*

$$f^i(\varepsilon) = \frac{1}{\varepsilon^2} f^{i,-2} \text{ and } h^i(\varepsilon) = \frac{1}{\varepsilon} h^{i,-1},$$

where, *here and subsequently, any function* $f^{i,r} \in L^2(\Omega)$ *and any function* $h^{i,r+1} \in L^2(\Gamma_+ \cup \Gamma_-)$, $r \geq -2$, *is meant to be independent of* ε. The cancellation of the coefficient of ε^{-2} in the variational equations of problem $\mathcal{P}^*(\varepsilon;\Omega)$ then leads to the equations

$$\int_\Omega A^{ijkl}(0) e^{-1}_{k\|l} e^{-1}_{i\|j}(\boldsymbol{v}) \sqrt{a}\, dx = \int_\Omega f^{i,-2} v_i \sqrt{a}\, dx + \int_{\Gamma_+\cup\Gamma_-} h^{i,-1} v_i \sqrt{a}\, d\Gamma$$

for all $\boldsymbol{v} \in \mathbf{V}(\Omega)$. These equations imply that *the functions* $f^{i,-2}$ *and* $h^{i,-1}$ *cannot be chosen arbitrarily*, since we should have in particular

$$\int_\Omega f^{i,-2} v_i \sqrt{a}\, dx + \int_{\Gamma_+\cup\Gamma_-} h^{i,-1} v_i \sqrt{a}\, d\Gamma = 0$$

for all $\boldsymbol{v} \in \mathbf{V}(\Omega)$ that are *independent of* x_3; to see this, recall that $e^{-1}_{\alpha\|\beta}(\boldsymbol{v}) = 0$, $e^{-1}_{\alpha\|3}(\boldsymbol{v}) = \frac{1}{2}\partial_3 v_\alpha$, and $e^{-1}_{3\|3}(\boldsymbol{v}) = \partial_3 v_3$. Consequently, *we must let* $f^{i,-2} = 0$ *and* $h^{i,-1} = 0$. The expressions of the functions $A^{ijkl}(0)$ and $e^{-1}_{k\|l}$ then show that we must have

$$\begin{aligned}
&\int_\Omega A^{ijkl}(0) e^{-1}_{k\|l} e^{-1}_{i\|j}(\boldsymbol{v}) \sqrt{a}\, dx \\
&= \int_\Omega \left\{ 4A^{\alpha 3\sigma 3}(0) e^{-1}_{\sigma\|3} e^{-1}_{\alpha\|3}(\boldsymbol{v}) + A^{3333}(0) e^{-1}_{3\|3} e^{-1}_{3\|3}(\boldsymbol{v}) \right\} \sqrt{a}\, dx \\
&= \int_\Omega \{ \mu a^{\alpha\sigma} \partial_3 u^0_\sigma \partial_3 v_\alpha + (\lambda + 2\mu) \partial_3 u^0_3 \partial_3 v_3 \} \sqrt{a}\, dx = 0
\end{aligned}$$

for all $\boldsymbol{v} \in \mathbf{V}(\Omega)$. Combined with Thm. 3.4-1, these equations in turn imply that

$$\partial_3 u^0_i = 0 \text{ in } \Omega,$$

since $a^{\alpha\sigma}\partial_3 u^0_\sigma = 0$ in Ω implies that $\partial_3 u^0_\alpha = 0$ in Ω (the matrix $(a^{\alpha\beta})$ is positive definite); hence *the leading term* $\boldsymbol{u}^0 \in \mathbf{V}(\Omega)$ *is independent of the transverse variable* x_3 and it can therefore be identified with a vector field $\zeta^0 \in \mathbf{H}^1(\omega)$ satisfying $\zeta^0 = \mathbf{0}$ on γ_0. Canceling the coefficient of ε^{-2} thus yields the relations:

$$\boxed{\begin{aligned}&\zeta^0 \in \mathbf{V}(\omega) := \{\boldsymbol{\eta} = (\eta_i) \in \mathbf{H}^1(\omega);\ \boldsymbol{\eta} = \mathbf{0} \text{ on } \gamma_0\},\\ &e^{-1}_{i\|j} = 0 \text{ in } \Omega.\end{aligned}}$$

(ii) Our *second try* is thus

$$f^i(\varepsilon) = \frac{1}{\varepsilon} f^{i,-1} \text{ and } h^i(\varepsilon) = h^{i,0}.$$

Combined with the relations $e^{-1}_{i\|j} = 0$ obtained in step (i), the cancellation of the coefficient of ε^{-1} in the variational equations of problem $\mathcal{P}^*(\varepsilon;\Omega)$ then leads to the equations

$$\int_\Omega A^{ijkl}(0) e^0_{k\|l} e^{-1}_{i\|j}(\boldsymbol{v}) \sqrt{a}\, dx = \int_\Omega f^{i,-1} v_i \sqrt{a}\, dx + \int_{\Gamma_+ \cup \Gamma_-} h^{i,0} v_i \sqrt{a}\, d\Gamma$$

for all $\boldsymbol{v} \in \mathbf{V}(\Omega)$. Letting $\boldsymbol{v} \in \mathbf{V}(\Omega)$ be independent of x_3 then implies as in step (i) that *the functions* $f^{i,-1}$ *and* $h^{i,0}$ *cannot be chosen arbitrarily*, since for such functions $\boldsymbol{v}$,

$$\int_\Omega f^{i,-1} v_i \sqrt{a}\, dx + \int_{\Gamma_+ \cup \Gamma_-} h^{i,0} v_i \sqrt{a}\, d\Gamma = 0.$$

Consequently, *we must let* $f^{i,-1} = 0$ *and* $h^{i,0} = 0$ and, accordingly, we must then have

$$\begin{aligned}
&\int_\Omega A^{ijkl}(0) e^0_{k\|l} e^{-1}_{i\|j}(\boldsymbol{v}) \sqrt{a}\, dx \\
&= \int_\Omega 4A^{\alpha 3\sigma 3}(0) e^0_{\alpha\|3} e^{-1}_{\sigma\|3}(\boldsymbol{v}) \sqrt{a}\, dx \\
&\quad + \int_\Omega \left(A^{\alpha\beta 33}(0) e^0_{\alpha\|\beta} + A^{3333}(0) e^0_{3\|3} \right) e^{-1}_{3\|3}(\boldsymbol{v}) \sqrt{a}\, dx \\
&= \int_\Omega \left\{ 2\mu a^{\alpha\sigma} e^0_{\alpha\|3} \partial_3 v_\sigma + \left(\lambda a^{\alpha\beta} e^0_{\alpha\|\beta} + (\lambda + 2\mu) e^0_{3\|3} \right) \partial_3 v_3 \right\} \sqrt{a}\, dx \\
&= 0
\end{aligned}$$

for all $\boldsymbol{v} \in \mathbf{V}(\Omega)$. Combined with Thm. 3.4-1, these equations in turn imply that

$$\boxed{e^0_{\alpha\|3} = 0 \text{ and } e^0_{3\|3} = -\frac{\lambda}{\lambda + 2\mu} a^{\alpha\beta} e^0_{\alpha\|\beta} \text{ in } \Omega.}$$

(iii) Our *third try* is thus

$$f^i(\varepsilon) = f^{i,0} \text{ and } h^i(\varepsilon) = \varepsilon h^{i,1}.$$

The cancellation of the coefficient of ε^0 in the variational equations of problem $\mathcal{P}^*(\varepsilon;\Omega)$ then leads to the equations (recall that $e^{-1}_{i\|j} = 0$; cf. step (i)):

$$\int_\Omega A^{ijkl}(0)\{e^0_{k\|l}e^0_{i\|j}(\boldsymbol{v}) + e^1_{k\|l}e^{-1}_{i\|j}(\boldsymbol{v})\}\sqrt{a}\,dx$$
$$+ \int_\Omega B^{ijkl,1}e^0_{k\|l}e^{-1}_{i\|j}(\boldsymbol{v})\,dx$$
$$= \int_\Omega f^{i,0}v_i\sqrt{a}\,dx + \int_{\Gamma_+\cup\Gamma_-} h^{i,1}v_i\sqrt{a}\,d\Gamma$$

for all $\boldsymbol{v} \in \mathbf{V}(\Omega)$. Observe in passing that the occurrence of the functions $e^1_{k\|l}$ in these equations implies that the formal asymptotic expansion of $\boldsymbol{u}(\varepsilon)$ should be "at least" of the form

$$\boldsymbol{u}(\varepsilon) = \boldsymbol{u}^0 + \varepsilon\boldsymbol{u}^1 + \varepsilon^2\boldsymbol{u}^2 + \cdots.$$

Let $\boldsymbol{v} = \boldsymbol{\eta} \in \mathbf{V}(\omega)$, i.e., $\boldsymbol{v} \in \mathbf{V}(\Omega)$ and $\boldsymbol{v}$ is independent of x_3. Noting that $e^{-1}_{i\|j}(\boldsymbol{v}) = 0$ for such functions $\boldsymbol{v}$ and using the expressions of the functions $A^{ijkl}(0)$, the relations satisfied by the functions $e^0_{i\|3}$ (step (ii)), and the definitions of the functions $e^0_{\sigma\|\tau}$ and $e^0_{\alpha\|\beta}(\boldsymbol{\eta})$, we find that

$$\int_\Omega A^{ijkl}(0)e^0_{k\|l}e^0_{i\|j}(\boldsymbol{\eta})\sqrt{a}\,dx$$
$$= \int_\Omega \left(A^{\alpha\beta\sigma\tau}(0)e^0_{\sigma\|\tau} + A^{\alpha\beta 33}(0)e^0_{3\|3}\right)e^0_{\alpha\|\beta}(\boldsymbol{\eta})\sqrt{a}\,dx$$
$$= \int_\Omega \left\{\lambda a^{\alpha\beta}a^{\sigma\tau} + \mu(a^{\alpha\sigma}a^{\beta\tau} + a^{\alpha\tau}a^{\beta\sigma})\right\}e^0_{\sigma\|\tau}e^0_{\alpha\|\beta}(\boldsymbol{\eta})\sqrt{a}\,dx$$
$$+ \int_\Omega \lambda a^{\alpha\beta}e^0_{3\|3}e^0_{\alpha\|\beta}(\boldsymbol{\eta})\sqrt{a}\,dx$$
$$= \frac{1}{2}\int_\Omega a^{\alpha\beta\sigma\tau}e^0_{\sigma\|\tau}e^0_{\alpha\|\beta}(\boldsymbol{\eta})\sqrt{a}\,dx$$
$$= \int_\Omega f^{i,0}\eta_i\sqrt{a}\,dx + \int_{\Gamma_+\cup\Gamma_-} h^{i,1}\eta_i\sqrt{a}\,d\Gamma$$

for all $\boldsymbol{\eta} \in \mathbf{V}(\omega)$, where the functions

$$a^{\alpha\beta\sigma\tau} := \frac{4\lambda\mu}{\lambda + 2\mu} a^{\alpha\beta} a^{\sigma\tau} + 2\mu(a^{\alpha\sigma} a^{\beta\tau} + a^{\alpha\tau} a^{\beta\sigma})$$

are the *contravariant components* of the **scaled two-dimensional elasticity tensor of the shell** (already introduced, but without any justification, in Thm. 3.3-2).

Let

$$\gamma_{\alpha\beta}(\boldsymbol{\eta}) := \frac{1}{2}(\partial_\beta \eta_\alpha + \partial_\alpha \eta_\beta) - \Gamma^\sigma_{\alpha\beta}\eta_\sigma - b_{\alpha\beta}\eta_3$$

denote the *covariant components* of the **linearized change of metric tensor** associated with a displacement field $\eta_i \boldsymbol{a}^i$ of surface S (Sect. 2.4). Note that

$$\gamma_{\alpha\beta}(\boldsymbol{\eta}) \in L^2(\omega) \text{ if } \boldsymbol{\eta} = (\eta_i) \in H^1(\omega) \times H^1(\omega) \times L^2(\omega).$$

Then remarkably, we precisely have

$$e^0_{\alpha\|\beta} = \gamma_{\alpha\beta}(\boldsymbol{\zeta}^0) \text{ and } e^0_{\alpha\|\beta}(\boldsymbol{\eta}) = \gamma_{\alpha\beta}(\boldsymbol{\eta}) \text{ for all } \boldsymbol{\eta} \in \mathbf{V}(\omega).$$

(iv) Define the space

$$\mathbf{V}_0(\omega) := \{\boldsymbol{\eta} \in \mathbf{V}(\omega);\ \gamma_{\alpha\beta}(\boldsymbol{\eta}) = 0 \text{ in } \omega\},$$

and *assume first that*

$$\mathbf{V}_0(\omega) = \{\mathbf{0}\},$$

in which case the shell will be called a *linearly elastic "membrane" shell.*

A word of caution. This definition is essentially *provisional*, as it will have to be subsequently amended; cf. Sects. 4.1 and 5.1. The quotation marks emphasize this temporary character. ■

Then by step (iii), *the vector field* $\boldsymbol{\zeta}^0$ *should satisfy the following two-dimensional variational problem "*$\mathcal{P}_M(\omega)$*"*:

$$\boldsymbol{\zeta}^0 \in \mathbf{V}(\omega) = \{\boldsymbol{\eta} = (\eta_i) \in \mathbf{H}^1(\omega);\ \boldsymbol{\eta} = \mathbf{0} \text{ on } \gamma_0\},$$
$$\int_\omega a^{\alpha\beta\sigma\tau}\gamma_{\sigma\tau}(\boldsymbol{\zeta}^0)\gamma_{\alpha\beta}(\boldsymbol{\eta})\sqrt{a}\,dy = \int_\omega p^{i,0}\eta_i\sqrt{a}\,dy \text{ for all } \boldsymbol{\eta} \in \mathbf{V}(\omega),$$
$$\text{where } p^{i,0} := \int_{-1}^{1} f^{i,0}\,dx_3 + h^{i,1}(\cdot,\,1) + h^{i,1}(\cdot,\,-1).$$

Besides, the induction stops here since "$\mathcal{P}_M(\omega)$" is a *bona fide* variational problem, which can be studied for its own sake:

In this direction, a first observation is that the issue of *uniqueness* of a solution to problem "$\mathcal{P}_M(\omega)$" is immediately resolved *when* $\mathbf{V}_0(\omega) = \{\mathbf{0}\}$. For, if $p^{i,0} = 0$, we must have in particular

$$\int_\omega a^{\alpha\beta\sigma\tau}\gamma_{\sigma\tau}(\boldsymbol{\zeta}^0)\gamma_{\alpha\beta}(\boldsymbol{\zeta}^0)\sqrt{a}\,dy = 0,$$

and thus $\gamma_{\alpha\beta}(\boldsymbol{\zeta}^0) = 0$ in ω (the scaled two-dimensional elasticity tensor is uniformly positive definite; cf. Thm. 3.3-2 (a)). Therefore, $\boldsymbol{\zeta}^0 \in \mathbf{V}_0(\omega)$ and thus $\boldsymbol{\zeta}^0 = \mathbf{0}$.

By contrast, the issue of *existence* of a solution is a delicate one. *If* $\gamma_0 = \gamma$ *and the surface* $S = \boldsymbol{\theta}(\overline{\omega})$ *is elliptic*, we have already shown through a delicate analysis that problem "$\mathcal{P}_M(\omega)$" has a solution, albeit in the *larger* space $\mathbf{V}_M(\omega) = H_0^1(\omega) \times H_0^1(\omega) \times L^2(\omega)$ (Thms. 2.7-3 and 3.3-2). In the other cases where $\mathbf{V}_0(\omega) = \{\mathbf{0}\}$, the conclusions are even more subtle since the "right" spaces where to seek a solution turn out to be "abstract" *completions* (see in particular Thm. 5.6-1)!

Such observations already indicate why the definition of a "membrane" shell and that of problem "$\mathcal{P}_M(\omega)$" (note the quotation marks) will have both to be modified later.

(v) *Assume next that the space* $\mathbf{V}_0(\omega)$ *contains nonzero functions*, i.e., that

$$\mathbf{V}_0(\omega) \neq \{\mathbf{0}\},$$

in which case the shell will be called a *linearly elastic "flexural" shell.*

A word of caution. Like that of a "membrane" shell, this definition is essentially *provisional*, as it will have to be subsequently amended; cf. Sect. 6.1. The quotation marks again emphasize this temporary character. ■

If $\mathbf{V}_0(\omega) \neq \{\mathbf{0}\}$, *the functions* $f^{i,0}$ *and* $h^{i,1}$ *cannot be chosen arbitrarily* since we must have

$$\int_\omega p^{i,0}\eta_i\sqrt{a}\,dy = \int_\Omega f^{i,0}\eta_i\sqrt{a}\,dx + \int_{\Gamma_+\cup\Gamma_-} h^{i,1}\eta_i\sqrt{a}\,d\Gamma = 0$$

for all $\boldsymbol{\eta} \in (\mathbf{V}_0(\omega) - \{\mathbf{0}\})$ by definition of problem "$\mathcal{P}_M(\omega)$". Hence *we must let* $f^{i,0} = 0$ *and* $h^{i,1} = 0$ and accordingly pursue the induction.

Before continuing the induction proper (step (vi)), let us have a second look at the cancellation of the coefficients of ε^{-1} and ε^0 (steps (ii) and (iii)). Since $p^{i,0} = 0$, letting $\boldsymbol{\eta} = \boldsymbol{\zeta}_0$ in the variational equations of problem "$\mathcal{P}_M(\omega)$" gives

$$\int_\omega a^{\alpha\beta\sigma\tau}\gamma_{\sigma\tau}(\boldsymbol{\zeta}^0)\gamma_{\alpha\beta}(\boldsymbol{\zeta}^0)\sqrt{a}\,dy = 0.$$

Hence $\gamma_{\alpha\beta}(\boldsymbol{\zeta}^0) = 0$ and thus $\boldsymbol{\zeta}^0 \in \mathbf{V}_0(\omega)$.

Since then $e^0_{\alpha\|\beta} = \gamma_{\alpha\beta}(\boldsymbol{\zeta}^0) = 0$, the relations

$$e^0_{\alpha\|3} = 0 \text{ and } (\lambda + 2\mu)e^0_{3\|3} + \lambda a^{\alpha\beta}e^0_{\alpha\|\beta} = 0 \text{ in } \Omega$$

established in step (ii) imply that

$$\partial_3 u^1_3 = e^0_{3\|3} = 0$$

and

$$e^0_{\alpha\|3} = \frac{1}{2}(\partial_\alpha u^0_3 + \partial_3 u^1_\alpha) + b^\sigma_\alpha u^0_\sigma = \frac{1}{2}(\partial_\alpha \zeta^0_3 + \partial_3 u^1_\alpha) + b^\sigma_\alpha \zeta^0_\sigma = 0.$$

Hence

$$\partial_3 u^1_\alpha = -(\partial_\alpha\zeta^0_3 + 2b^\sigma_\alpha\zeta^0_\sigma).$$

Assume that $\boldsymbol{u}^1 \in \mathbf{V}(\Omega)$. Since each function $(\partial_\alpha\zeta^0_3 + 2b^\sigma_\alpha\zeta^0_\sigma)$ is independent of x_3, there exists $\boldsymbol{\zeta}^1 = (\zeta^1_i) \in \mathbf{V}(\omega)$ such that

$$u^1_\alpha = \zeta^1_\alpha - x_3(\partial_\alpha\zeta^0_3 + 2b^\sigma_\alpha\zeta^0_\sigma) \text{ and } u^1_3 = \zeta^1_3,$$

the first relations forcing in addition the function ζ^0_3 to be in the space $H^2(\omega)$ and to satisfy the boundary condition $\partial_\nu\zeta^0_3 = 0$ on γ_0

(∂_ν denotes the outer normal derivative along γ) since $\zeta^0_\alpha = 0$ on γ_0 by step (i) (Ex. 1.1 in Vol. II can be handy here). To sum up, we have shown that

$$\begin{aligned}
e^0_{i\|j} &= 0 \text{ in } \Omega,\\
\boldsymbol{\zeta}^0 &\in \mathbf{V}_F(\omega) := \{\boldsymbol{\eta} = (\eta_i) \in H^1(\omega)\times H^1(\omega)\times H^2(\omega);\\
&\qquad \eta_i = \partial_\nu \eta_3 = 0 \text{ on } \gamma_0,\ \gamma_{\alpha\beta}(\boldsymbol{\eta}) = 0 \text{ in } \omega\},\\
u^1_\alpha &= \zeta^1_\alpha - x_3(\partial_\alpha \zeta^0_3 + 2b^\sigma_\alpha \zeta^0_\sigma) \text{ and } u^1_3 = \zeta^1_3,\\
&\qquad \text{where } \boldsymbol{\zeta}^1 = (\zeta^1_i) \in \mathbf{V}(\omega).
\end{aligned}$$

Since $e^0_{i\|j} = 0$, the cancellation of the coefficient of ε^0 in $\mathcal{P}^*(\varepsilon;\Omega)$ reduces to

$$\int_\Omega A^{ijkl}(0) e^1_{k\|l} e^{-1}_{i\|j}(\boldsymbol{v})\sqrt{a}\,dx = 0 \text{ for all } \boldsymbol{v} \in \mathbf{V}(\Omega).$$

By a computation similar to that in step (ii), we then conclude that

$$e^1_{\alpha\|3} = 0 \text{ and } e^1_{3\|3} = -\frac{\lambda}{\lambda+2\mu} a^{\alpha\beta} e^1_{\alpha\|\beta} \text{ in } \Omega.$$

It remains to compute the functions $e^1_{\alpha\|\beta}$, as their expressions play a key rôle later. Combining their definitions with the expressions found *supra* for the components u^1_i in terms of ζ^0_i and ζ^1_i and using the symmetry relations (Thm. 2.5-1)

$$b^\sigma_\alpha|_\beta := \partial_\beta b^\sigma_\alpha + \Gamma^\sigma_{\beta\tau} b^\tau_\alpha - \Gamma^\tau_{\alpha\beta} b^\sigma_\tau = b^\sigma_\beta|_\alpha,$$

we obtain, after some manipulations,

$$\begin{aligned}
e^1_{\alpha\|\beta} &= \frac{1}{2}(\partial_\beta u^1_\alpha + \partial_\alpha u^1_\beta) - \Gamma^\sigma_{\alpha\beta} u^1_\sigma - b_{\alpha\beta} u^1_3 + x_3\{b^\sigma_\beta|_\alpha u^0_\sigma + b^\sigma_\alpha b_{\sigma\beta} u^0_3\}\\
&= \frac{1}{2}(\partial_\beta \zeta^1_\alpha + \partial_\alpha \zeta^1_\beta) - \Gamma^\sigma_{\alpha\beta}\zeta^1_\sigma - b_{\alpha\beta}\zeta^1_3\\
&\quad - x_3\{\partial_{\alpha\beta}\zeta^0_3 - \Gamma^\sigma_{\alpha\beta}\partial_\sigma \zeta^0_3 - b^\sigma_\alpha b_{\sigma\beta}\zeta^0_3\\
&\quad + b^\sigma_\alpha(\partial_\beta \zeta^0_\sigma - \Gamma^\tau_{\beta\sigma}\zeta^0_\tau) + b^\tau_\beta(\partial_\alpha \zeta^0_\tau - \Gamma^\sigma_{\alpha\tau}\zeta^0_\sigma) + b^\tau_\beta|_\alpha \zeta^0_\tau\}.
\end{aligned}$$

Let

$$\begin{aligned}
\rho_{\alpha\beta}(\boldsymbol{\eta}) &:= \partial_{\alpha\beta}\eta_3 - \Gamma^\sigma_{\alpha\beta}\partial_\sigma \eta_3 - b^\sigma_\alpha b_{\sigma\beta}\eta_3\\
&\quad + b^\sigma_\alpha(\partial_\beta \eta_\sigma - \Gamma^\tau_{\beta\sigma}\eta_\tau) + b^\tau_\beta(\partial_\alpha \eta_\tau - \Gamma^\sigma_{\alpha\tau}\eta_\sigma) + b^\tau_\beta|_\alpha \eta_\tau
\end{aligned}$$

denote the *covariant components* of the **linearized change of curvature tensor** associated with a displacement field $\eta_i \boldsymbol{a}^i$ of the surface S (Sect. 2.5). Note that

$$\rho_{\alpha\beta}(\boldsymbol{\eta}) \in L^2(\omega) \text{ if } \boldsymbol{\eta} = (\eta_i) \in H^1(\omega) \times H^1(\omega) \times H^2(\omega).$$

Then remarkably, the functions $e^1_{\alpha\|\beta}$ have the following simple expressions in terms of the functions $\gamma_{\alpha\beta}(\boldsymbol{\zeta}^1)$ (the functions $\gamma_{\alpha\beta}(\boldsymbol{\eta})$ are defined in step (iii) for any $\boldsymbol{\eta} \in \mathbf{V}(\omega)$) and $\rho_{\alpha\beta}(\boldsymbol{\zeta}^0)$:

$$\boxed{e^1_{\alpha\|\beta} = \gamma_{\alpha\beta}(\boldsymbol{\zeta}^1) - x_3 \rho_{\alpha\beta}(\boldsymbol{\zeta}^0).}$$

(vi) We are now in a position to continue the induction *when* $\mathbf{V}_0(\omega) \neq \{\mathbf{0}\}$, with

$$f^i(\varepsilon) = \varepsilon f^{i,1} \text{ and } h^i(\varepsilon) = \varepsilon^2 h^{i,2}$$

as our *fourth try*. The cancellation of the coefficient of ε in $\mathcal{P}^*(\varepsilon;\Omega)$ then shows that (recall that $e^{-1}_{i\|j} = e^0_{i\|j} = 0$; cf. steps (i) and (v)):

$$\int_\Omega A^{ijkl}(0)\{e^1_{k\|l} e^0_{i\|j}(\boldsymbol{v}) + e^2_{k\|l} e^{-1}_{i\|j}(\boldsymbol{v})\}\sqrt{a}\,dx$$
$$+ \int_\Omega B^{ijkl,1} e^1_{k\|l} e^{-1}_{i\|j}(\boldsymbol{v})\,dx = \int_\Omega f^{i,1} v_i \sqrt{a}\,dx + \int_{\Gamma_+\cup\Gamma_-} h^{i,2} v_i \sqrt{a}\,d\Gamma$$

for all $\boldsymbol{v} \in \mathbf{V}(\Omega)$. Let $\boldsymbol{v} = \boldsymbol{\eta} \in \mathbf{V}(\omega)$. Noting that $e^{-1}_{i\|j}(\boldsymbol{v}) = 0$ for such functions $\boldsymbol{v}$ and using the expressions of the functions $A^{ijkl}(0)$ and the relations (found in step (v)) satisfied by the functions $e^1_{i\|j}$, we are led to the equations

$$\int_\Omega A^{ijkl}(0) e^1_{k\|l} e^0_{i\|j}(\boldsymbol{\eta})\sqrt{a}\,dx$$
$$= \frac{1}{2}\int_\Omega a^{\alpha\beta\sigma\tau} e^0_{\sigma\|\tau} e^0_{\alpha\|\beta}(\boldsymbol{\eta})\sqrt{a}\,dx$$
$$= \int_\omega a^{\alpha\beta\sigma\tau}\gamma_{\sigma\tau}(\boldsymbol{\zeta}^1)\gamma_{\alpha\beta}(\boldsymbol{\eta})\sqrt{a}\,dy$$
$$= \int_\Omega f^{i,1}\eta_i\sqrt{a}\,dx + \int_{\Gamma_+\cup\Gamma_-} h^{i,2}\eta_i\sqrt{a}\,d\Gamma \text{ for all } \boldsymbol{\eta} \in \mathbf{V}(\omega).$$

Hence *the functions* $f^{i,1}$ and $h^{i,2}$ *cannot be chosen arbitrarily*, since we must have

$$\int_\omega f^{i,1}\eta_i\sqrt{a}\,dy + \int_{\Gamma_+\cup\Gamma_-} h^{i,2}\eta_i\sqrt{a}\,d\Gamma = 0$$

for all $\boldsymbol{\eta} \in (\mathbf{V}_0(\omega) - \{\mathbf{0}\})$; consequently, *we must let* $f^{i,1} = 0$ *and* $h^{i,2} = 0$. But then, letting $\boldsymbol{\eta} = \boldsymbol{\zeta}^1$ in the above variational problem shows that

$$\int_\omega a^{\alpha\beta\sigma\tau}\gamma_{\sigma\tau}(\boldsymbol{\zeta}^1)\gamma_{\alpha\beta}(\boldsymbol{\zeta}^1)\sqrt{a}\,dy = 0.$$

Hence $\gamma_{\alpha\beta}(\boldsymbol{\zeta}^1) = 0$ and consequently,

$$\boxed{\boldsymbol{\zeta}^1 \in \mathbf{V}_0(\omega),}$$

since on the other hand, $\boldsymbol{\zeta}^1 \in \mathbf{V}(\omega)$ by step (v).

(vii) As a preparation to step (viii), let us further exploit the cancellation of the coefficient of ε in $\mathcal{P}^*(\varepsilon;\Omega)$ *when* $\mathbf{V}_0(\omega) \neq \{\mathbf{0}\}$, which now leads to the equations

$$\int_\Omega A^{ijkl}(0)\{e^1_{k\|l}e^0_{i\|j}(\boldsymbol{v}) + e^2_{k\|l}e^{-1}_{i\|j}(\boldsymbol{v})\}\sqrt{a}\,dx + \int_\Omega B^{ijkl,1}e^1_{k\|l}e^{-1}_{i\|j}(\boldsymbol{v})\,dx = 0 \text{ for all } \boldsymbol{v} \in \mathbf{V}(\Omega),$$

since we are now assuming that $f^{i,1} = 0$ and $h^{i,2} = 0$.

Given an arbitrary element $\boldsymbol{\eta}$ in the space $\mathbf{V}_F(\omega)$ defined in step (v), let $\boldsymbol{v}(\boldsymbol{\eta}) = (v_i(\boldsymbol{\eta}))$ be defined by

$$v_\alpha(\boldsymbol{\eta}) := x_3\{2b^\sigma_\alpha\eta_\sigma + \partial_\alpha\eta_3\} \text{ and } v_3(\boldsymbol{\eta}) := 0.$$

Since $\boldsymbol{v}(\boldsymbol{\eta}) \in \mathbf{V}(\Omega)$, we may let $\boldsymbol{v} = \boldsymbol{v}(\boldsymbol{\eta})$ in the above equations; this leads to the equations:

$$\boxed{\begin{aligned}&\int_\Omega A^{ijkl}(0)e^1_{k\|l}e^0_{i\|j}(\boldsymbol{v}(\boldsymbol{\eta}))\sqrt{a}\,dx\\ &+4\int_\Omega A^{\alpha3\sigma3}(0)e^2_{\sigma\|3}\Big(b^\tau_\alpha\eta_\tau + \frac{1}{2}\partial_\alpha\eta_3\Big)\sqrt{a}\,dx\\ &+4\int_\Omega B^{\alpha3\sigma3,1}e^1_{\sigma\|3}\Big(b^\tau_\alpha\eta_\tau + \frac{1}{2}\partial_\alpha\eta_3\Big)\sqrt{a}\,dx = 0 \text{ for all } \boldsymbol{\eta} \in \mathbf{V}_F(\omega).\end{aligned}}$$

Remark. The relations $\gamma_{\alpha\beta}(\boldsymbol{\eta}) = 0$ in ω satisfied by the functions $\boldsymbol{\eta} \in \mathbf{V}_F(\omega)$ are not needed at this stage; in other words, the above equations are in fact valid for all $\boldsymbol{\eta} = (\eta_i) \in H^1(\omega) \times H^1(\omega) \times H^2(\omega)$ that satisfy the boundary conditions $\eta_i = \partial_\nu \eta_3 = 0$ on γ_0. ■

(viii) We conclude the induction *when* $\mathbf{V}_0(\omega) \neq \{\mathbf{0}\}$, with

$$f^i(\varepsilon) = \varepsilon^2 f^{i,2} \text{ and } h^i(\varepsilon) = \varepsilon^3 h^{i,3}$$

as our *fifth*, and *final, try*. The cancellation of the coefficient of ε^2 in $\mathcal{P}^*(\varepsilon;\Omega)$ then shows that (recall that $e^{-1}_{i\|j} = e^0_{i\|j} = 0$):

$$\begin{aligned}
&\int_\Omega A^{ijkl}(0)\{e^1_{k\|l}e^1_{i\|j}(\boldsymbol{v}) + e^2_{k\|l}e^0_{i\|j}(\boldsymbol{v}) + e^3_{k\|l}e^{-1}_{i\|j}(\boldsymbol{v})\}\sqrt{a}\,dx \\
&+ \int_\Omega B^{ijkl,1}\{e^1_{k\|l}e^0_{i\|j}(\boldsymbol{v}) + e^2_{k\|l}e^{-1}_{i\|j}(\boldsymbol{v})\}\,dx \\
&+ \int_\Omega B^{ijkl,2}e^1_{k\|l}e^{-1}_{i\|j}(\boldsymbol{v})\,dx = \int_\Omega f^{i,2}v_i\sqrt{a}\,dx + \int_{\Gamma_+\cup\Gamma_-} h^{i,3}v_i\sqrt{a}\,d\Gamma
\end{aligned}$$

for all $\boldsymbol{v} \in \mathbf{V}(\Omega)$. Observe in passing that the occurrence of the functions $e^3_{k\|l}$ in these equations implies that the formal asymptotic expansion of $\boldsymbol{u}(\varepsilon)$ be "at least" of the form

$$\boldsymbol{u}(\varepsilon) = \boldsymbol{u}^0 + \varepsilon\boldsymbol{u}^1 + \varepsilon^2\boldsymbol{u}^2 + \varepsilon^3\boldsymbol{u}^3 + \varepsilon^4\boldsymbol{u}^4 + \cdots,$$

as $e^3_{k\|l}$ is a function of $\boldsymbol{u}^2, \boldsymbol{u}^3$, and $\boldsymbol{u}^4$.

Let $\boldsymbol{v} = \boldsymbol{\eta} \in \mathbf{V}_F(\omega)$. Then for such functions $\boldsymbol{v}$,

$$e^{-1}_{i\|j}(\boldsymbol{v}) = 0,$$

$$e^0_{\alpha\|\beta}(\boldsymbol{v}) = \gamma_{\alpha\beta}(\boldsymbol{\eta}) = 0, \quad e^0_{\alpha\|3}(\boldsymbol{v}) = \frac{1}{2}\partial_\alpha\eta_3 + b^\sigma_\alpha\eta_\sigma, \quad e^0_{3\|3}(\boldsymbol{v}) = 0,$$

so that we are left with the equations

$$\begin{aligned}
&\int_\Omega A^{ijkl}(0)e^1_{k\|l}e^1_{i\|j}(\boldsymbol{\eta})\sqrt{a}dx \\
&+4\int_\Omega A^{\alpha3\sigma3}(0)e^2_{\sigma\|3}\{b^\tau_\alpha\eta_\tau + \frac{1}{2}\partial_\alpha\eta_3\}\sqrt{a}\,dx \\
&+4\int_\Omega B^{\alpha3\sigma3,1}e^1_{\sigma\|3}\{b^\tau_\alpha\eta_\tau + \frac{1}{2}\partial_\alpha\eta_3\}\sqrt{a}\,dx \\
&\qquad = \int_\omega p^{i,2}\eta_i\sqrt{a}\,dy \text{ for all } \boldsymbol{\eta} \in \mathbf{V}_F(\omega),
\end{aligned}$$

where

$$p^{i,2} := \int_{-1}^{1} f^{i,2}\, dx_3 + h^{i,3}(\cdot, 1) + h^{i,3}(\cdot, -1).$$

Subtracting the equations found in step (vii) from these equations, we find that

$$\int_\Omega A^{ijkl}(0) e^1_{k\|l} \{ e^1_{i\|j}(\boldsymbol{\eta}) - e^0_{i\|j}(\boldsymbol{v}(\boldsymbol{\eta})) \} \sqrt{a}\, dx = \int_\omega p^{i,2} \eta_i \sqrt{a}\, dy$$

for all $\boldsymbol{\eta} \in \mathbf{V}_F(\omega)$. First, the relations (established in step (v))

$$e^1_{\alpha\|3} = 0 \text{ and } e^1_{3\|3} = -\frac{\lambda}{\lambda + 2\mu} a^{\alpha\beta} e^1_{\alpha\|\beta} \text{ in } \Omega,$$

and the relations $e^1_{3\|3}(\boldsymbol{\eta}) = e^0_{3\|3}(\boldsymbol{v}(\boldsymbol{\eta})) = 0$ together imply that

$$\begin{aligned} & A^{ijkl}(0) e^1_{k\|l} \{ e^1_{i\|j}(\boldsymbol{\eta}) - e^0_{i\|j}(\boldsymbol{v}(\boldsymbol{\eta})) \} \\ &\quad = A^{\alpha\beta\sigma\tau}(0) e^1_{\sigma\|\tau} \{ e^1_{\alpha\|\beta}(\boldsymbol{\eta}) - e^0_{\alpha\|\beta}(\boldsymbol{v}(\boldsymbol{\eta})) \} \\ &\quad + A^{\alpha\beta 33}(0) e^1_{3\|3} \{ e^1_{\alpha\|\beta}(\boldsymbol{\eta}) - e^0_{\alpha\|\beta}(\boldsymbol{v}(\boldsymbol{\eta})) \} \\ &\quad = \frac{1}{2} a^{\alpha\beta\sigma\tau} e^1_{\sigma\|\tau} \{ e^1_{\alpha\|\beta}(\boldsymbol{\eta}) - e^0_{\alpha\|\beta}(\boldsymbol{v}(\boldsymbol{\eta})) \}. \end{aligned}$$

Second,

$$e^1_{\sigma\|\tau} = -x_3 \rho_{\sigma\tau}(\boldsymbol{\zeta}^0)$$

by steps (v) and (vi). Finally,

$$\begin{aligned} e^1_{\alpha\|\beta}(\boldsymbol{\eta}) - e^0_{\alpha\|\beta}(\boldsymbol{v}(\boldsymbol{\eta})) &= x_3 \{ b^\sigma_\beta |_\alpha \eta_\sigma + b^\sigma_\alpha b_{\sigma\beta} \eta_3 \} \\ & -x_3 \{ \partial_\alpha (b^\tau_\beta \eta_\tau) + \partial_\beta (b^\sigma_\alpha \eta_\sigma) + \partial_{\alpha\beta} \eta_3 - \Gamma^\sigma_{\alpha\beta} \partial_\sigma \eta_3 - 2\Gamma^\sigma_{\alpha\beta} b^\tau_\sigma \eta_\tau \}, \end{aligned}$$

and a computation analogous to that performed in step (v) then shows that

$$\boxed{e^1_{\alpha\|\beta}(\boldsymbol{\eta}) - e^0_{\alpha\|\beta}(\boldsymbol{v}(\boldsymbol{\eta})) = -x_3 \rho_{\alpha\beta}(\boldsymbol{\eta}) \text{ for all } \boldsymbol{\eta} \in \mathbf{V}_F(\omega).}$$

We have thus derived the equations

$$\begin{aligned} & \int_\Omega A^{ijkl}(0) e^1_{k\|l} \{ e^1_{i\|j}(\boldsymbol{\eta}) - e^0_{i\|j}(\boldsymbol{v}(\boldsymbol{\eta})) \} \sqrt{a}\, dx \\ &\quad = \frac{1}{2} \int_\Omega x_3^2 a^{\alpha\beta\sigma\tau} \rho_{\sigma\tau}(\boldsymbol{\zeta}^0) \rho_{\alpha\beta}(\boldsymbol{\eta}) \sqrt{a}\, dx \\ &\quad = \frac{1}{3} \int_\omega a^{\alpha\beta\sigma\tau} \rho_{\sigma\tau}(\boldsymbol{\zeta}^0) \rho_{\alpha\beta}(\boldsymbol{\eta}) \sqrt{a}\, dy \text{ for all } \boldsymbol{\eta} \in \mathbf{V}_F(\omega). \end{aligned}$$

To sum up, we have established that, *when* $\mathbf{V}_0(\omega) \neq \{\mathbf{0}\}$, *the vector field* $\boldsymbol{\zeta}^0$ *must satisfy the following variational problem* $\mathcal{P}_F(\omega)$:

$$\boldsymbol{\zeta}^0 \in \mathbf{V}_F(\omega) = \{\boldsymbol{\eta} = (\eta_i) \in H^1(\omega) \times H^1(\omega) \times H^2(\omega);\ \eta_i = \partial_\nu \eta_3 = 0 \text{ on } \gamma_0, \gamma_{\alpha\beta}(\boldsymbol{\eta}) = 0 \text{ in } \omega\},$$

$$\frac{1}{3}\int_\omega a^{\alpha\beta\sigma\tau}\rho_{\sigma\tau}(\boldsymbol{\zeta}^0)\rho_{\alpha\beta}(\boldsymbol{\eta})\sqrt{a}\,d\omega = \int_\omega p^{i,2}\eta_i\sqrt{a}\,dy \text{ for all } \boldsymbol{\eta} \in \mathbf{V}_F(\omega),$$

$$\text{where } p^{i,2} := \int_{-1}^{1} f^{i,2}\,dx_3 + h^{i,3}(\cdot, 1) + h^{i,3}(\cdot, -1).$$

The issues of *existence* and *uniqueness* of a solution to problem $\mathcal{P}_F(\omega)$ are already resolved, as corollaries to Thms. 2.6-4 and 3.3-2.

Let us now formally recapitulate the main conclusions of the above formal asymptotic analysis. We recall that the following theorems, which are found in Miara & Sanchez-Palencia [1996], were themselves based on a prior work of Sanchez-Palencia [1990]. Another noteworthy reference is Caillerie & Sanchez-Palencia [1995b], who analogously carried out the formal asymptotic analysis of linearly elastic shells made of *nonhomogeneous* and *anisotropic* materials.

We logically begin with "membrane" shells (an example is proposed in Ex. 3.3):

Theorem 3.4-2. *Assume that the space*

$$\mathbf{V}_0(\omega) := \{\boldsymbol{\eta} \in \mathbf{H}^1(\omega);\ \boldsymbol{\eta} = \mathbf{0} \text{ on } \gamma_0,\ \gamma_{\alpha\beta}(\boldsymbol{\eta}) = 0 \text{ in } \omega\}$$

contains only $\boldsymbol{\eta} = \mathbf{0}$ *and that the scaled unknown* $\boldsymbol{u}(\varepsilon) = (u_i(\varepsilon))$, *where* $u_i(\varepsilon)(x) := u_i^\varepsilon(x^\varepsilon)$ *for all* $x^\varepsilon = \pi^\varepsilon x \in \overline{\Omega}^\varepsilon$, *admits a formal asymptotic expansion of the form*

$$\boldsymbol{u}(\varepsilon) = \boldsymbol{u}^0 + \varepsilon\boldsymbol{u}^1 + \varepsilon^2\boldsymbol{u}^2 + \cdots, \text{ with } \boldsymbol{u}^0 \in \mathbf{V}(\Omega),\ \boldsymbol{u}^0 \neq \mathbf{0}, \text{ and } \boldsymbol{u}^q \in \mathbf{H}^1(\Omega),\ q = 1, 2.$$

Assume for definiteness that the Lamé constants are independent of ε. *Then in order that no restriction be put on the applied forces* *entering the equations satisfied by the leading term $\boldsymbol{u}^0$, their components must be of the form*

$$\boxed{\begin{aligned} f^{i,\varepsilon}(x^\varepsilon) &= f^{i,0}(x) \text{ for all } x^\varepsilon = \pi^\varepsilon x \in \Omega^\varepsilon, \\ h^{i,\varepsilon}(x^\varepsilon) &= \varepsilon h^{i,1}(x) \text{ for all } x^\varepsilon = \pi^\varepsilon x \in \Gamma_+^\varepsilon \cup \Gamma_-^\varepsilon, \end{aligned}}$$

where the functions $f^{i,0} \in L^2(\Omega)$ and $h^{i,1} \in L^2(\Gamma_+ \cup \Gamma_-)$ are independent of ε. This being the case, the leading term $\boldsymbol{u}^0 : \overline{\Omega} \to \mathbb{R}^3$ is independent of the transverse variable and, once identified with a vector field $\boldsymbol{\zeta}^0 : \overline{\omega} \to \mathbb{R}^3$, it should satisfy the following scaled two-dimensional variational problem "$\mathcal{P}_M(\omega)$" of a linearly elastic "membrane" shell:

$$\boxed{\begin{aligned} &\boldsymbol{\zeta}^0 \in \mathbf{V}(\omega) := \{\boldsymbol{\eta} = (\eta_i) \in \mathbf{H}^1(\omega);\ \boldsymbol{\eta} = \mathbf{0} \text{ on } \gamma_0\}, \\ &\int_\omega a^{\alpha\beta\sigma\tau}\gamma_{\sigma\tau}(\boldsymbol{\zeta}^0)\gamma_{\alpha\beta}(\boldsymbol{\eta})\sqrt{a}\,dy = \int_\omega p^{i,0}\eta_i\sqrt{a}\,dy \\ &\qquad\qquad\qquad \text{for all } \boldsymbol{\eta} = (\eta_i) \in \mathbf{V}(\omega), \end{aligned}}$$

where

$$\boxed{\begin{aligned} \gamma_{\alpha\beta}(\boldsymbol{\eta}) &:= \frac{1}{2}(\partial_\alpha\eta_\beta + \partial_\beta\eta_\alpha) - \Gamma_{\alpha\beta}^\sigma\eta_\sigma - b_{\alpha\beta}\eta_3, \\ a^{\alpha\beta\sigma\tau} &:= \frac{4\lambda\mu}{\lambda + 2\mu}a^{\alpha\beta}a^{\sigma\tau} + 2\mu(a^{\alpha\sigma}a^{\beta\tau} + a^{\alpha\tau}a^{\beta\sigma}), \\ p^{i,0} &:= \int_{-1}^1 f^{i,0}\,dx_3 + h_+^{i,1} + h_-^{i,1} \text{ and } h_\pm^{i,1} = h^{i,1}(\cdot, \pm 1). \end{aligned}}$$

■

A word of caution. The variational problem "$\mathcal{P}_M(\omega)$" is *not* well posed in general. More precisely, the space $\mathbf{V}(\omega)$ has to be replaced, either with the space

$$\mathbf{V}_M(\omega) := H_0^1(\omega) \times H_0^1(\omega) \times L^2(\omega)$$

if the shell is an "elliptic membrane" one (Thm. 4.4-1), or with the space

$$\mathbf{V}_M(\omega) := \text{completion of } \mathbf{V}(\omega) \text{ with respect to } |\cdot|_\omega^M,$$

where

$$|\boldsymbol{\eta}|_{\omega}^{M} := \left\{ \sum_{\alpha,\beta} |\gamma_{\alpha\beta}(\boldsymbol{\eta})|_{0,\omega}^{2} \right\}^{1/2}$$

if the shell is a "generalized membrane" one (Thm. 5.6-1); see also Sect. 3.5. ■

We likewise summarize the analysis of "flexural" shells (an example is proposed in Ex. 3.4):

Theorem 3.4-3. *Assume that the space*

$$\boxed{\mathbf{V}_0(\omega) := \{\boldsymbol{\eta} \in \mathbf{H}^1(\omega);\ \boldsymbol{\eta} = \mathbf{0} \text{ on } \gamma_0,\ \gamma_{\alpha\beta}(\boldsymbol{\eta}) = 0 \text{ in } \omega\}}$$

contains nonzero fields $\boldsymbol{\eta}$ and that the scaled unknown $\boldsymbol{u}(\varepsilon) = (u_i(\varepsilon))$, where $u_i(\varepsilon)(x) := u_i^\varepsilon(x^\varepsilon)$ for all $x^\varepsilon = \pi^\varepsilon x \in \overline{\Omega}^\varepsilon$, admits a formal asymptotic expansion of the form

$$\boxed{\begin{aligned}&\boldsymbol{u}(\varepsilon) = \boldsymbol{u}^0 + \varepsilon \boldsymbol{u}^1 + \varepsilon^2 \boldsymbol{u}^2 + \varepsilon^3 \boldsymbol{u}^3 + \varepsilon^4 \boldsymbol{u}^4 + \cdots, \text{ with}\\ &\boldsymbol{u}^0,\ \boldsymbol{u}^1 \in \mathbf{V}(\Omega),\ \boldsymbol{u}^0 \neq \mathbf{0}, \text{ and } \boldsymbol{u}^q \in \mathbf{H}^1(\Omega),\ q = 2, 3, 4.\end{aligned}}$$

Assume for definiteness that the Lamé constants are independent of ε. Then in order that no restriction be put on the applied forces entering the equations satisfied by the leading term $\boldsymbol{u}^0$, their components must be of the form

$$\boxed{\begin{aligned} f^{i,\varepsilon}(x^\varepsilon) &= \varepsilon^2 f^{i,2}(x) \text{ for all } x^\varepsilon = \pi^\varepsilon x \in \Omega^\varepsilon,\\ h^{i,\varepsilon}(x^\varepsilon) &= \varepsilon^3 h^{i,3}(x) \text{ for all } x^\varepsilon = \pi^\varepsilon x \in \Gamma_+^\varepsilon \cup \Gamma_-^\varepsilon,\end{aligned}}$$

where the functions $f^{i,2} \in L^2(\Omega)$ and $h^{i,3} \in L^2(\Gamma_+ \cup \Gamma_-)$ are independent of ε. This being the case, the leading term $\boldsymbol{u}^0 : \overline{\Omega} \to \mathbb{R}^3$ is independent of the transverse variable and, once identified with a vector field $\boldsymbol{\zeta}^0 : \overline{\omega} \to \mathbb{R}^3$, it should satisfy the following scaled two-dimensional variational problem $\mathcal{P}_F(\omega)$ of a linearly elastic "flexural" *shell:*

$$\zeta^0 \in \mathbf{V}_F(\omega) := \{\boldsymbol{\eta} = (\eta_i) \in H^1(\omega) \times H^1(\omega) \times H^2(\omega);\\ \eta_i = \partial_\nu \eta_3 = 0 \text{ on } \gamma_0,\ \gamma_{\alpha\beta}(\boldsymbol{\eta}) = 0 \text{ in } \omega\},$$

$$\frac{1}{3}\int_\omega a^{\alpha\beta\sigma\tau} \rho_{\sigma\tau}(\boldsymbol{\zeta}^0)\rho_{\alpha\beta}(\boldsymbol{\eta})\sqrt{a}\,dy = \int_\omega p^{i,2}\eta_i\sqrt{a}\,dy$$

$$\text{for all } \boldsymbol{\eta} = (\eta_i) \in \mathbf{V}_F(\omega),$$

where

$$\gamma_{\alpha\beta}(\boldsymbol{\eta}) := \frac{1}{2}(\partial_\alpha\eta_\beta + \partial_\beta\eta_\alpha) - \Gamma^\sigma_{\alpha\beta}\eta_\sigma - b_{\alpha\beta}\eta_3,$$

$$\rho_{\alpha\beta}(\boldsymbol{\eta}) := \partial_{\alpha\beta}\eta_3 - \Gamma^\sigma_{\alpha\beta}\partial_\sigma\eta_3 - b^\sigma_\alpha b_{\sigma\beta}\eta_3 + b^\sigma_\alpha(\partial_\beta\eta_\sigma - \Gamma^\tau_{\beta\sigma}\eta_\tau) \\ + b^\tau_\beta(\partial_\alpha\eta_\tau - \Gamma^\sigma_{\alpha\tau}\eta_\sigma) + (\partial_\alpha b^\tau_\beta + \Gamma^\tau_{\alpha\sigma}b^\sigma_\beta - \Gamma^\sigma_{\alpha\beta}b^\tau_\sigma)\eta_\tau,$$

$$a^{\alpha\beta\sigma\tau} := \frac{4\lambda\mu}{\lambda+2\mu}a^{\alpha\beta}a^{\sigma\tau} + 2\mu(a^{\alpha\sigma}a^{\beta\tau} + a^{\alpha\tau}a^{\beta\sigma}),$$

$$p^{i,2} := \int_{-1}^{1} f^{i,2}\,dx_3 + h^{i,3}_+ + h^{i,3}_- \text{ and } h^{i,3}_\pm = h^{i,3}(\cdot, \pm 1).$$

■

Several important comments are in order about these two theorems:

First, it is remarkable that the *linearized change of metric and change of curvature tensors* have been exactly recovered (by means of their covariant components $\gamma_{\alpha\beta}(\boldsymbol{\eta})$ and $\rho_{\alpha\beta}(\boldsymbol{\eta})$) as the asymptotic analysis proceeded, *without any explicit reference to their geometrical definitions* (Thms. 2.4-1 and 2.5-1)!

No less remarkable is the *increase of regularity* (from $H^1(\omega)$ to $H^2(\omega)$) that is automatically induced on the normal component ζ_3 by the assumption $\mathbf{V}_0(\omega) \neq \{\mathbf{0}\}$ satisfied by a "flexural" shell; see step (v) of the recursion.

Another virtue of this procedure is to likewise automatically provide the expression of the *scaled two-dimensional elasticity tensor of the shell,* by means of its contravariant components

$$a^{\alpha\beta\sigma\tau} = \frac{4\lambda\mu}{\lambda+2\mu}a^{\alpha\beta}a^{\sigma\tau} + 2\mu(a^{\alpha\sigma}a^{\sigma\tau} + a^{\alpha\tau}a^{\beta\sigma}).$$

A word of caution. Although reminiscent of those of the "original" *three-dimensional elasticity tensor*, viz.,

$$A^{ijkl,\varepsilon} = \lambda g^{ij,\varepsilon} g^{kl,\varepsilon} + \mu(g^{ik,\varepsilon} g^{jl,\varepsilon} + g^{il,\varepsilon} g^{jk,\varepsilon})$$

(recall that the Lamé constants are assumed to be independent of ε; hence $\lambda^\varepsilon = \lambda$ and $\mu^\varepsilon = \mu$), these components *differ* in that "the coefficient λ is replaced by $\dfrac{2\lambda\mu}{\lambda + 2\mu}$". *The asymptotic analysis thus spares the arduous task of justifying such an intriguing "replacement"*! ■

A guiding principle for carrying to its term the asymptotic analysis was that, whenever applied forces of a given order had to satisfy some compatibility conditions, they were replaced by higher order ones. It should be emphasized, however, that "unwanted" models discarded for this reason likewise deserve to be studied for their own sake, as they have been in the case of plates by Dauge, Djurdjevic & Rössle [1998a].

We conclude this section by linking the present formal asymptotic analysis to the content of the next chapters, where what was defined in this section as the leading term of the formal asymptotic expansion of the scaled unknown $\boldsymbol{u}(\varepsilon)$ will be rigorously identified as the *limit of* $\boldsymbol{u}(\varepsilon)$ *when* ε *approaches* 0 in *ad hoc* functional spaces; see also Sect. 3.5 for a brief review of the corresponding *convergence theorems*.

The present distinction between "membrane" shells and "flexural" shells depends on whether $\mathbf{V}_0(\omega) = \{\mathbf{0}\}$ or $\mathbf{V}_0(\omega) \neq \{\mathbf{0}\}$, where the space $\mathbf{V}_0(\omega)$ is defined by

$$\mathbf{V}_0(\omega) = \{\boldsymbol{\eta} = (\eta_i) \in \mathbf{H}^1(\omega);\ \boldsymbol{\eta} = \mathbf{0} \text{ on } \gamma_0,\ \gamma_{\alpha\beta}(\boldsymbol{\eta}) = 0 \text{ in } \omega\}.$$

The convergence analyses of the next chapters will, however, entail a crucial amendment of the present classification: *The decisive criterion will be instead whether* $\mathbf{V}_F(\omega) = \{\mathbf{0}\}$ *or* $\mathbf{V}_F(\omega) \neq \{\mathbf{0}\}$, *where the space* $\mathbf{V}_F(\omega)$ (already encountered in Thm. 3.4-3) *is defined by*

$$\begin{aligned}\mathbf{V}_F(\omega) = \{\boldsymbol{\eta} = (\eta_i) \in H^1(\omega) \times H^1(\omega) \times H^2(\omega);\\ \eta_i = \partial_\nu \eta_3 = 0 \text{ on } \gamma_0,\ \gamma_{\alpha\beta}(\boldsymbol{\eta}) = 0 \text{ in } \omega\},\end{aligned}$$

Besides, the linearly elastic *membrane shells*, i.e., those corresponding to $\mathbf{V}_F(\omega) = \{\mathbf{0}\}$, will have to be themselves separated into two distinct categories, formed either by the *elliptic membrane shells*

(Chap. 4) or by the *generalized membrane shells* (Chap. 5). Those corresponding to $\mathbf{V}_F(\omega) \neq \{\mathbf{0}\}$ are the *flexural shells* (Chap. 6). See also Sect. 3.5.

Note that *we are, and will henceforth be, no longer using quotation marks in these "final" definitions.* By contrast, the quotation marks appended *in this section* to the definitions of "membrane" and "flexural" shells served the purpose of emphasizing their provisional character. As already noted, problem "$\mathcal{P}_M(\omega)$" has likewise a temporary character as it will be later replaced by *one* among *two* closely related, but not identical, variational problems, respectively denoted $\mathcal{P}_M(\omega)$ and $\mathcal{P}_M^\sharp(\omega)$ in Chaps. 4 and 5. By chance, problem $\mathcal{P}_F(\omega)$ is already in its "final" form; it will thus be re-stated *verbatim* in Chap. 6. See also Sect. 3.5.

3.5. SUMMARY OF THE CONVERGENCE THEOREMS

As Sect. 3.4, *this section has an optional character*: Its purpose is simply to briefly review the way the *formal* asymptotic analysis of Sect. 3.4 will be superseded in Chaps. 4 to 6 by rigorous, but substantially more delicate, *convergence* theorems, whose lengthy proofs may thus be spared to the reader in haste... This section also provides references to the main results.

Inspired by the *formal* analysis of Sanchez-Palencia [1990] and by its later refinement by Miara & Sanchez-Palencia [1996] (the content of Sect. 3.4), Ciarlet & Lods [1996b, 1996d] and Ciarlet, Lods & Miara [1996] have carried out a *no longer formal* asymptotic analysis of linearly elastic shells. Under *three* distinct and carefully circumscribed sets of assumptions that cover *all possible cases*, they established the *convergence of the scaled unknown in* ad hoc *functional spaces as* $\varepsilon \to 0$, justifying in this fashion the linear two-dimensional equations of *linearly elastic elliptic membrane*, or *generalized membrane*, or *flexural, shells*. The *main conclusions* of their analysis, which should be advantageously compared to those drawn in Thms. 3.4-2 and 3.4-3, are summarized below. Note that the definition of each category of shell is governed by the *geometry of the middle surface*, the *boundary conditions*, and the *order of magnitude of the applied forces*.

Consider a family of linearly elastic shells with *thickness* 2ε *approaching zero*, with each formed by the *same* homogeneous and isotropic material, with each having the *same middle surface* $S = \boldsymbol{\theta}(\overline{\omega})$ and with each subjected to a boundary condition of place along a portion of its lateral face having the *same* set $\boldsymbol{\theta}(\gamma_0)$ as its middle line, where $\gamma_0 \subset \gamma = \partial\omega$ and *length* $\gamma_0 > 0$. For each $\varepsilon > 0$, the shells are subjected to applied body and surface forces.

For each $\varepsilon > 0$, let $u_i(\varepsilon)$ denote as in Sect. 3.2 the three *scaled displacements,* i.e., the scaled covariant components of the displacement field of the points of the shell; each scaled displacement $u_i(\varepsilon)$ is thus defined over the fixed domain $\Omega = \omega\times]-1, 1[$.

Assume first that $\gamma_0 = \gamma$ *and* that the surface S is *elliptic*, in the sense that its Gaussian curvature is > 0 everywhere. Then Ciarlet & Lods [1996b] have shown that, if the applied body force density is $O(1)$ and the applied surface force density is $O(\varepsilon)$ with respect to ε in the sense of Thm. 3.4-2, *the scaled unknown* $\boldsymbol{u}(\varepsilon) = (u_i(\varepsilon))$ *converges in* $H^1(\Omega) \times H^1(\Omega) \times L^2(\Omega)$ *as* $\varepsilon \to 0$ *to a limit* $\boldsymbol{u}$, *which is independent of the transverse variable* (this convergence result has been recently improved by Mardare [1998a], who obtained an $O(\varepsilon^{1/6})$ error estimate). Furthermore, *the average* $\boldsymbol{\zeta} := \frac{1}{2}\int_{-1}^{1} \boldsymbol{u}\, dx_3$ *satisfies the* **scaled two-dimensional variational problem $\mathcal{P}_M(\omega)$ of a linearly elastic elliptic membrane shell:**

$$\boxed{\begin{aligned}
&\boldsymbol{\zeta} \in \mathbf{V}_M(\omega) := H_0^1(\omega) \times H_0^1(\omega) \times L^2(\omega),\\
&\int_\omega a^{\alpha\beta\sigma\tau}\gamma_{\sigma\tau}(\boldsymbol{\zeta})\gamma_{\alpha\beta}(\boldsymbol{\eta})\sqrt{a}\, dy\\
&\qquad = \int_\omega p^{i,0}\eta_i\sqrt{a}\, dy \text{ for all } \boldsymbol{\eta} = (\eta_i) \in \mathbf{V}_M(\omega),
\end{aligned}}$$

where the functions $a^{\alpha\beta\sigma\tau}$, $\gamma_{\alpha\beta}(\boldsymbol{\eta})$, and $p^{i,0}$ are defined as in Thm. 3.4-2. Notice, however, that the space $\mathbf{V}_M(\omega)$ does *not* coincide with the space $\mathbf{V}(\omega)$ found in Thm. 3.4-2!

The detailed proof of this convergence result, which occupies most of Chap. 4, is concluded with Thm. 4.4-1.

Naturally, these equations have to be "de-scaled", so as to be expressed in terms of *"physical"* unknowns and data. To this end, we let $\boldsymbol{\zeta}^\varepsilon(y) = \boldsymbol{\zeta}(y)$ for $y \in \overline{\omega}$ (in view of the scalings $\boldsymbol{u}^\varepsilon(x^\varepsilon) = \boldsymbol{u}(\varepsilon)(x)$ for $x^\varepsilon = \pi^\varepsilon x \in \overline{\Omega}$); it is found in this fashion that the de-scaled unknown $\boldsymbol{\zeta}^\varepsilon$ satisfies the two-dimensional variational problem:

$$\boldsymbol{\zeta}^\varepsilon \in \mathbf{V}_M(\omega),$$
$$\varepsilon \int_\omega a^{\alpha\beta\sigma\tau,\varepsilon} \gamma_{\sigma\tau}(\boldsymbol{\zeta}^\varepsilon) \gamma_{\alpha\beta}(\boldsymbol{\eta}) \sqrt{a}\, dy$$
$$= \int_\omega p^{i,\varepsilon} \eta_i \sqrt{a}\, dy \text{ for all } \boldsymbol{\eta} = (\eta_i) \in \mathbf{V}_M(\omega),$$

where

$$a^{\alpha\beta\sigma\tau,\varepsilon} := \frac{4\lambda^\varepsilon \mu^\varepsilon}{\lambda^\varepsilon + 2\mu^\varepsilon} a^{\alpha\beta} a^{\sigma\tau} + 2\mu^\varepsilon (a^{\alpha\sigma} a^{\beta\tau} + a^{\alpha\tau} a^{\beta\sigma}),$$
$$p^{i,\varepsilon} := \int_{-\varepsilon}^{\varepsilon} f^{i,\varepsilon}\, dx_3^\varepsilon + h_+^{i,\varepsilon} + h_-^{i,\varepsilon} \text{ and } h_\pm^{i,\varepsilon} := h^{i,\varepsilon}(\cdot, \pm\varepsilon).$$

Note that these equations are of the form announced in Sect. 2.7.

As shown by Ciarlet & Lods [1996a] and Ciarlet & Sanchez-Palencia [1996] (under rather stringent regularity assumptions, later substantially relaxed by Lods & Mardare [1998a]; cf. Thm. 2.7-2), the conjunction of the two assumptions "$\gamma_0 = \gamma$" and "S is elliptic" provides a *first* instance where the space

$$\mathbf{V}_F(\omega) := \{\boldsymbol{\eta} = (\eta_i) \in H^1(\omega) \times H^1(\omega) \times H^2(\omega);$$
$$\eta_i = \partial_\nu \eta_3 = 0 \text{ on } \gamma_0,\ \gamma_{\alpha\beta}(\boldsymbol{\eta}) = 0 \text{ in } \omega\}$$

contains only the function $\boldsymbol{\eta} = \mathbf{0}$. The space $\mathbf{V}_F(\omega)$ was introduced by Sanchez-Palencia [1989a].

Then Ciarlet and Lods [1996d] studied all the *"remaining" cases where* $\mathbf{V}_F(\omega) = \{\mathbf{0}\}$, for instance, when the surface S is elliptic but $0 <$ *length* $\gamma_0 <$ *length* γ (as shown by Vekua [1962] and Lods & Mardare [1998a]), or when S is a portion of a hyperboloid of revolution and γ_0 is "large enough" (as shown by Sanchez-Palencia [1993] and Mardare [1998c]), etc. To give a flavor of their results, consider the *most common* case, where the space

$$\mathbf{V}_0(\omega) := \{\boldsymbol{\eta} \in \mathbf{H}^1(\omega);\ \boldsymbol{\eta} = \mathbf{0} \text{ on } \gamma_0,\ \gamma_{\alpha\beta}(\boldsymbol{\eta}) = 0 \text{ in } \omega\},$$

which contains $\mathbf{V}_F(\omega)$, *"already" reduces to* $\{\mathbf{0}\}$, *or equivalently, when the semi-norm*

$$|\cdot|_\omega^M : \boldsymbol{\eta} = (\eta_i) \to |\boldsymbol{\eta}|_\omega^M := \left\{ \sum_{\alpha,\beta} |\gamma_{\alpha\beta}(\boldsymbol{\eta})|_{0,\omega}^2 \right\}^{1/2}$$

becomes a norm over the space

$$\mathbf{V}(\omega) := \{\boldsymbol{\eta} \in \mathbf{H}^1(\omega);\ \boldsymbol{\eta} = \mathbf{0} \text{ on } \gamma_0\}.$$

In this case, if the applied forces are *"admissible"* in a specific sense (too technical to be reproduced here; cf. Sect. 5.5), *the average* $\frac{1}{2}\int_{-1}^{1} \boldsymbol{u}(\varepsilon)\,dx_3$ *converges as* $\varepsilon \to 0$ *in the space*

$$\mathbf{V}_M^\sharp(\omega) := \text{ completion of } \mathbf{V}(\omega) \text{ with respect to } |\cdot|_\omega^M.$$

A convergence result also holds for the scaled unknown $\boldsymbol{u}(\varepsilon)$ itself (again too technical to be reproduced here). *Furthermore, the limit* $\boldsymbol{\zeta}$ *satisfies the* **scaled two-dimensional variational problem** $\mathcal{P}_M^\sharp(\omega)$ **of a linearly elastic generalized membrane shell:**

$$\boldsymbol{\zeta} \in \mathbf{V}_M^\sharp(\omega) \text{ and } B_M^\sharp(\boldsymbol{\zeta}, \boldsymbol{\eta}) = L_M^\sharp(\boldsymbol{\eta}) \text{ for all } \boldsymbol{\eta} \in \mathbf{V}_M^\sharp(\omega),$$

where $B_M^\sharp$ is the unique extension to $\mathbf{V}_M^\sharp(\omega)$ of the bilinear form

$$(\boldsymbol{\zeta}, \boldsymbol{\eta}) \to \int_\omega a^{\alpha\beta\sigma\tau} \gamma_{\sigma\tau}(\boldsymbol{\zeta}) \gamma_{\alpha\beta}(\boldsymbol{\eta}) \sqrt{a}\,dy$$

already found in the variational equations of an elliptic membrane shell (see *supra*) and $L_M^\sharp : \mathbf{V}_M^\sharp(\omega) \to \mathbb{R}$ is an *ad hoc* linear form, determined by the behavior as $\varepsilon \to 0$ of the admissible forces. Notice that the space $\mathbf{V}_M^\sharp(\omega)$ neither coincides with the space $\mathbf{V}(\omega)$ found in Thm. 3.4-2!

The detailed proof of this convergence result, which occupies most of Chap. 5, is concluded with Thm. 5.6-1.

A particularly noteworthy feature of such variational problems is that they provide instances of "*sensitive problems*", introduced and studied by Lions & Sanchez-Palencia [1994, 1996, 1997].

Remark. In the "last", but seemingly uncommon, case where $\mathbf{V}_F(\omega) = \{\mathbf{0}\}$ but $|\cdot|_\omega^M$ is a "genuine" semi-norm over the space $\mathbf{V}(\omega)$, a similar convergence result can be established, but instead in the *completion* $\dot{\mathbf{V}}_M^\sharp(\omega)$ with respect to $|\cdot|_\omega^M$ of the *quotient* space $\dot{\mathbf{V}}_M(\omega) := \mathbf{V}(\omega)/\mathbf{V}_0(\omega)$.

The limit found in this fashion in the space $\dot{\mathbf{V}}_M^\sharp(\omega)$ then satisfies the *scaled two-dimensional variational problem of a linearly elastic generalized membrane shell "of the second kind"*; see Thm. 5.6-2. ■

Finally, Ciarlet, Lods & Miara [1996] have considered the case where *the space* $\mathbf{V}_F(\omega)$ *contains non-zero functions.* This assumption is satisfied in particular if S is a portion of a cylinder and $\boldsymbol{\theta}(\gamma_0)$ is contained in one or two generatrices of S (as shown by Lods & Mardare [1998a]; see also Ex. 3.4) or if S is contained in a plane, i.e., the shell is a *plate.*

They then showed that, if the applied body force density is $O(\varepsilon^2)$ and the applied surface force density is $O(\varepsilon^3)$ with respect to ε in the sense of Thm. 3.4-3, *the scaled unknown* $\boldsymbol{u}(\varepsilon) = (u_i(\varepsilon))$ *converges in* $\mathbf{H}^1(\Omega)$ *to a limit* $\boldsymbol{u}$, *which is independent of the transverse variable.* Furthermore, *the average* $\boldsymbol{\zeta} := \frac{1}{2}\int_{-1}^{1} \boldsymbol{u}\, dx_3$ *satisfies the* **scaled two-dimensional variational problem $\mathcal{P}_F(\omega)$ of a linearly elastic flexural shell:**

$$\boldsymbol{\zeta} \in \mathbf{V}_F(\omega) := \{\boldsymbol{\eta} = (\eta_i) \in H^1(\omega) \times H^1(\omega) \times H^2(\omega);$$
$$\eta_i = \partial_\nu \eta_3 = 0 \text{ on } \gamma_0, \gamma_{\alpha\beta}(\boldsymbol{\eta}) = 0 \text{ in } \omega\},$$
$$\frac{1}{3}\int_\omega a^{\alpha\beta\sigma\tau} \rho_{\sigma\tau}(\boldsymbol{\zeta}) \rho_{\alpha\beta}(\boldsymbol{\eta}) \sqrt{a}\, dy = \int_\omega p^{i,2} \eta_i \sqrt{a}\, dy$$
$$\text{for all } \boldsymbol{\eta} = (\eta_i) \in \mathbf{V}_F(\omega),$$

where the functions $a^{\alpha\beta\sigma\tau}$, $\rho_{\alpha\beta}(\boldsymbol{\eta})$, and $p^{i,2}$ are defined as in Thm. 3.4-3. Notice that, for once, the "right" space $\mathbf{V}_F(\omega)$ was already found in this theorem!

The detailed proof of this convergence result is the object of Thm. 6.2-1.

In order to acquire physical significance, these equations must be *de-scaled*, like those of a membrane shell. The de-scaling then shows that the de-scaled unknown $\boldsymbol{\zeta}^\varepsilon$ satisfies the two-dimensional variational problem:

$$\boxed{\begin{aligned}&\boldsymbol{\zeta}^\varepsilon \in \mathbf{V}_F(\omega),\\ &\frac{\varepsilon^3}{3}\int_\omega a^{\alpha\beta\sigma\tau,\varepsilon}\rho_{\sigma\tau}(\boldsymbol{\zeta}^\varepsilon)\rho_{\alpha\beta}(\boldsymbol{\eta})\sqrt{a}\,dy = \int_\omega p^{i,\varepsilon}\eta_i\sqrt{a}\,dy\\ &\qquad\qquad\qquad\qquad\qquad\text{for all } \boldsymbol{\eta} = (\eta_i) \in \mathbf{V}_F(\omega),\end{aligned}}$$

where

$$\boxed{\begin{aligned}a^{\alpha\beta\sigma\tau,\varepsilon} &:= \frac{4\lambda^\varepsilon\mu^\varepsilon}{\lambda^\varepsilon + 2\mu^\varepsilon}a^{\alpha\beta}a^{\sigma\tau} + 2\mu^\varepsilon(a^{\alpha\sigma}a^{\beta\tau} + a^{\alpha\tau}a^{\beta\sigma}),\\ p^{i,\varepsilon} &:= \int_{-\varepsilon}^{\varepsilon} f^{i,\varepsilon}\,dx_3^\varepsilon + h_+^{i,\varepsilon} + h_-^{i,\varepsilon} \text{ and } h_\pm^{i,\varepsilon} := h^{i,\varepsilon}(\cdot, \pm\varepsilon).\end{aligned}}$$

Note that these equations are of the form announced in Sect. 2.6.

All these convergence results heavily rely on the *inequalities of Korn's type on a surface* established in Chap. 2, first for a general surface (Thm. 2.6-4), then for an elliptic surface (Thm. 2.7-3). Not only do these inequalities play a crucial rôle for establishing the existence and uniqueness of the solution to the "limit" two-dimensional equations (as already shown in Sects. 2.6 and 2.7), but they also constitute the basis for proving *"three-dimensional" inequalities of Korn's type*, either for a *family of general shells* or for a family of *elliptic membrane shells*, that are themselves the keystones of the convergence proofs.

These convergences for the displacements have been recently complemented by Collard & Miara [1999] who have established the convergence of the corresponding *linearized stresses*, by Giroud [1998] who has considered linearly elastic shells made of *nonhomogeneous* and *anisotropic* materials, and by Xiao Li-ming [1998, 1999a] who has considered *time-dependent* linearly elastic membrane and flexural shells.

Ciarlet & Lods [1996c, 1996d] have also fully justified the *two-dimensional Koiter equations for a linearly elastic shell*, described at the beginning of Sect. 2.6, by combining the above convergences with former asymptotic analyses of Koiter's equations due to Destuynder [1985] and Sanchez-Palencia [1989a, 1989b]. More specifically, Ciarlet & Lods [1996c, 1996d] have shown that *the average* $\frac{1}{2\varepsilon}\int_{-\varepsilon}^{\varepsilon} \boldsymbol{u}^{\varepsilon}\, dx_3^{\varepsilon}$ *and the solution* $\boldsymbol{\zeta}_K^{\varepsilon}$ *of Koiter's equation have for each kind of shell the same asymptotic behavior as* $\varepsilon \to 0$.

The detailed proof of this result is the object of Thms. 7.2-1 to 7.2-3.

A somewhat unexpected phenomenon, identified by Ciarlet [1992a, 1992b], arises when *"a shell becomes a plate"*: Suppose *first* that the middle surface S "converges" in a natural sense to a planar domain, while the thickness 2ε is held fixed: Then the solution of the three-dimensional shell equations converges toward that of the three-dimensional plate equations (Ex. 3.5). If *then* ε approaches zero, the solution of the three-dimensional plate equations converges toward the solution of the two-dimensional plate equations, i.e., those of the linear Kirchhoff-Love theory, which *simultaneously* incorporate *membrane* and *flexural* terms (Vol. II, Thm. 1.4-1), regardless of the geometry of S and of the boundary conditions.

By contrast, suppose *first* that ε approaches zero while the middle surface is held fixed. Then the above convergence theorems show that the three-dimensional solutions converges towards a two-dimensional limit that satisfies *either* the elliptic membrane, *or* the generalized membrane, *or* the flexural equations, according to the geometry of S and to the boundary conditions. If *then* the surface S becomes planar, the limit equations obtained in this process are thus *either* the membrane, *or* the flexural, plate equations. In other words, *"the two passages to the limits do not commute"*! Additional considerations about this puzzling question are found in Sanchez-Palencia [1994] and Lods [1996].

As shown by Busse, Ciarlet & Miara [1997], the two-dimensional equations of a *linearly elastic "shallow" shell in curvilinear coordinates* (these equations are given in Sect. 7.6) can be also justified by means of a convergence theorem, which is reminiscent of that used for a linearly elastic shallow shell in *Cartesian coordinates* (Vol. II, Chap. 3). We recall that, according to the definition justified by Ciarlet & Paumier [1986] in the nonlinear case and by Ciarlet & Miara [1992a] in the linear case, a shell is "shallow" if, in its reference configuration, the deviation of its middle surface from a plane is of the order of its thickness (see again Vol. II, Chap. 3 for details).

EXERCISES

3.1. Let the assumptions be as in Thm. 3.3-2.

(1) Show that the uniform positive definiteness of the *two-dimensional elasticity tensor on* S can be also established in a manner reminiscent of that used in Thm. 1.8-1 for establishing the uniform positive definiteness of the three-dimensional elasticity tensor.

(2) Show that yet another proof consists in adapting that suggested in Ex. 1.8 in the three-dimensional case.

3.2. Let $a^{\alpha\beta\sigma\tau}$ denote the contravariant components of the *two-dimensional elasticity tensor* of a shell (Thm. 3.3-2), and let $\mu > 0$ be given. Show that there exists a constant $\lambda_0(\mu) < 0$ such that, if $\lambda > \lambda_0(\mu)$, there exists a constant $c > 0$ such that

$$\sum_{\alpha,\beta} |t_{\alpha\beta}|^2 \leq c a^{\alpha\beta\sigma\tau}(y) t_{\sigma\tau} t_{\alpha\beta}$$

for all $y \in \overline{\omega}$ and all symmetric matrices $(t_{\alpha\beta})$.

Remark. Such a uniform positive definiteness of the two-dimensional elasticity tensor is established in Thm. 3.3-2 under the assumptions $\lambda \geq 0$ and $\mu > 0$.

3.3. Consider a linearly elastic shell whose middle surface is a portion $S = \boldsymbol{\theta}(\overline{\omega})$ of a sphere, with (Fig. 2.1-3)

$$\omega := \{(u, v) \in \mathbb{R}^2;\ 0 < u^2 + v^2 < r^2\},$$
$$\boldsymbol{\theta}(u, v) := \left(\frac{2R^2 u}{u^2 + v^2 + R^2}, \frac{2R^2 v}{u^2 + v^2 + R^2}, R\frac{u^2 + v^2 - R^2}{u^2 + v^2 + R^2}\right),$$

i.e., stereographic coordinates are used for representing the surface S. This *spherical shell* is subjected to a boundary condition of place along its entire lateral face, i.e., $\gamma_0 = \gamma$.

(1) Show that such a shell is a *"membrane" shell*, according to the definition given in Thm. 3.4-2.

Hint: Use Ex. 2.13.

(2) Compute explicitly the functions $a^{\alpha\beta\sigma\tau}$, $\gamma_{\alpha\beta}(\boldsymbol{\eta})$, and a appearing in the formulation of the associated variational problem "$\mathcal{P}_M(\omega)$" (Thm. 3.4-2).

(3) Carry out the same computations when the surface S is represented by means of spherical coordinates (Fig. 2.1-3).

Remark. The expressions sought in (2) and (3) immediately follow from Ex. 2.1 (2).

3.4. Consider a linearly elastic shell whose middle surface is a portion $S = \boldsymbol{\theta}(\overline{\omega})$ of a circular cylinder, with (Fig. 2.1-2)

$$\begin{aligned} \omega &:= \{(\varphi, z) \in \mathbb{R}^2;\, 0 < \varphi < \pi,\, 0 < z < h\}, \\ \boldsymbol{\theta}(\varphi, z) &:= (R\cos\varphi,\, R\sin\varphi,\, z). \end{aligned}$$

This *cylindrical shell* is subjected to a boundary condition of place along both "vertical" portions of its lateral face, i.e.,

$$\gamma_0 = \{(0, z) \in \mathbb{R}^2;\, 0 \le z \le h\} \cup \{(\pi, z) \in \mathbb{R}^2;\, 0 \le z \le h\}.$$

(1) Show that such a shell is a *"flexural" shell*, according to the definition given in Thm. 3.4-3.

(2) Compute explicitly the functions $a^{\alpha\beta\sigma\tau}$, $\gamma_{\alpha\beta}(\boldsymbol{\eta})$, $\rho_{\alpha\beta}(\boldsymbol{\eta})$, and a appearing in the formulation of the associated variational problem $\mathcal{P}_F(\omega)$ (Thm. 3.4-3).

Remark. The sought expressions follow from Ex. 2.1 (1).

(3) Assume now that this shell is subjected to a boundary condition of place along its entire lateral face, i.e., that $\gamma_0 = \gamma$. Is it still a "flexural" shell?

3.5. Let there be given an injective mapping $\boldsymbol{\theta} = (\theta_i) \in \mathcal{C}^2(\overline{\omega}; \mathbb{R}^3)$ such that the two vectors $\partial_\alpha \boldsymbol{\theta}$ are linearly independent at all points of $\overline{\omega}$. Following Ciarlet [1992a], consider a one-parameter family of mappings $\boldsymbol{\theta}(t) = (\theta_i(t)) : \overline{\omega} \to \mathbb{R}^3$ defined for $0 \le t \le 1$ and $(y_1, y_2) \in \overline{\omega}$ by

$$\begin{aligned} \theta_\alpha(t)(y_1, y_2) &= t\theta_\alpha(y_1, y_2) + (1 - t)y_\alpha, \\ \theta_3(t)(y_1, y_2) &= t\theta_3(y_1, y_2), \end{aligned}$$

and assume for simplicity that the contravariant components of the applied forces are independent of t. Let $\boldsymbol{u}^\varepsilon(t) \in \mathbf{V}(\Omega^\varepsilon)$ denote for $0 \le t \le 1$ the solutions of the corresponding variational problems $\mathcal{P}(\Omega^\varepsilon)$, as described in Sect. 3.1 but now parametrized by t.

Show that the mapping $t \in [0, 1] \to \boldsymbol{u}^\varepsilon(t) \in \mathbf{V}(\Omega^\varepsilon)$ is continuous. In particular, $\boldsymbol{u}^\varepsilon(t) \to \boldsymbol{u}^\varepsilon(0)$ in $\mathbf{H}^1(\Omega^\varepsilon)$ when $t \to 0^+$, i.e., when *"the shell becomes a plate"*, where $\boldsymbol{u}^\varepsilon(0) \in \mathbf{V}(\Omega^\varepsilon)$ satisfies the three-dimensional equations of a clamped plate:

$$\int_{\Omega^\varepsilon} B^{ijkl,\varepsilon} e^\varepsilon_{ij}(\boldsymbol{u}^\varepsilon(0)) e^\varepsilon_{kl}(\boldsymbol{v}^\varepsilon)\, dx^\varepsilon = \int_{\Omega^\varepsilon} f^{i,\varepsilon} v^\varepsilon_i \, dx^\varepsilon + \int_{\Gamma^\varepsilon_+ \cup \Gamma^\varepsilon_-} h^{i,\varepsilon} v^\varepsilon_i \, d\Gamma^\varepsilon,$$

where

$$B^{ijkl,\varepsilon} = \lambda^\varepsilon \delta^{ij}\delta^{jk} + \mu^\varepsilon(\delta^{ik}\delta^{jl} + \delta^{il}\delta^{jk}).$$

Hint: Observing that the Lax-Milgram lemma can be proved by means of the contraction mapping theorem (Lions & Stampacchia [1967]), use the local uniform ellipticity of the associated bilinear forms to show that the contractions are also locally uniformly contracting.

CHAPTER 4

LINEARLY ELASTIC ELLIPTIC MEMBRANE SHELLS

INTRODUCTION

A *linearly elastic elliptic membrane shell* is one whose middle surface $S = \boldsymbol{\theta}(\overline{\omega})$ is *elliptic*, i.e., that has a Gaussian curvature that is everywhere > 0, and which is subjected to a boundary condition of place along its *entire* lateral face (Sect. 4.1).

The *infinitesimal rigid displacement lemma on an elliptic surface* (Thm. 2.7-2) thus shows that an elliptic membrane shell provides *one* instance where the space

$$\mathbf{V}_F(\omega) = \{\boldsymbol{\eta} = (\eta_i) \in H^1(\omega) \times H^1(\omega) \times H^2(\omega);\ \eta_i = \partial_\nu \eta_3 = 0 \text{ on } \gamma_0, \\ \gamma_{\alpha\beta}(\boldsymbol{\eta}) = 0 \text{ in } \omega\}$$

reduces to $\{\mathbf{0}\}$. The *other* instances where $\mathbf{V}_F(\omega) = \{\mathbf{0}\}$ are provided by the *"generalized membrane" shells*, which will be studied in the next chapter.

As we shall see, whether $\mathbf{V}_F(\omega) = \{\mathbf{0}\}$ or $\mathbf{V}_F(\omega) \neq \{\mathbf{0}\}$ is the fundamental criterion for classifying linearly elastic shells: Those corresponding to $\mathbf{V}_F(\omega) = \{\mathbf{0}\}$ constitute the *membrane shells*, while those corresponding to $\mathbf{V}_F(\omega) \neq \{\mathbf{0}\}$ constitute the *flexural shells* (Chap. 6).

The purpose of this chapter is to *identify* and to *mathematically justify* the *two-dimensional equations of a linearly elastic elliptic membrane shell*, by establishing the *convergence in* ad hoc *functional spaces of the three-dimensional displacements as the thickness of such a shell approaches zero.*

More specifically, consider a *family of linearly elastic elliptic membrane shells* with thickness 2ε approaching zero and with each having the *same* middle surface $S = \boldsymbol{\theta}(\overline{\omega})$. As in Chap. 3, the associated three-dimensional problems, posed in curvilinear coordinates over the sets $\Omega^\varepsilon = \omega \times]-\varepsilon, \varepsilon[$, are transformed for each $\varepsilon > 0$ into equivalent problems, but now posed over the set $\Omega = \omega \times]-1, 1[$, which is *independent of* ε. This transformation relies in a crucial way on appropriate *scalings of the unknowns* u_i^ε (the covariant components of the displacement field) and *assumptions on the Lamé constants* λ^ε *and* μ^ε *and on the contravariant components* $f^{i,\varepsilon}$ *of the applied body forces* (for simplicity, we assume in this introduction that there are no applied surface forces). More specifically, we define the *scaled unknown* $\boldsymbol{u}(\varepsilon) = (u_i(\varepsilon))$ by letting

$$u_i^\varepsilon(x^\varepsilon) = u_i(\varepsilon)(x) \text{ for all } x^\varepsilon = \pi^\varepsilon x \in \overline{\Omega}^\varepsilon,$$

where $\pi^\varepsilon(x_1, x_2, x_3) = (x_1, x_2, \varepsilon x_3)$. Guided by the formal asymptotic analysis of Sect. 3.4, we then *assume* that there exist constants $\lambda > 0$ and $\mu > 0$ and functions $f^i \in L^2(\Omega)$ *independent of* ε such that

$$\lambda^\varepsilon = \lambda \quad \text{and} \quad \mu^\varepsilon = \mu,$$
$$f^{i,\varepsilon}(x^\varepsilon) = f^i(x) \text{ for all } x^\varepsilon = \pi^\varepsilon x \in \Omega^\varepsilon.$$

It is found in this fashion that the scaled unknown $\boldsymbol{u}(\varepsilon)$ satisfies a variational problem of the form (Thm. 4.1-1):

$$\boldsymbol{u}(\varepsilon) = (u_i(\varepsilon)) \in \mathbf{V}(\Omega) = \left\{\boldsymbol{v} = (v_i) \in \mathbf{H}^1(\Omega);\ \boldsymbol{v} = \mathbf{0} \text{ on } \gamma \times [-1, 1]\right\},$$

$$\int_\Omega A^{ijkl}(\varepsilon)e_{k\|l}(\varepsilon; \boldsymbol{u}(\varepsilon))e_{i\|j}(\varepsilon; \boldsymbol{v})\sqrt{g(\varepsilon)}\,dx$$
$$= \int_\Omega f^i v_i \sqrt{g(\varepsilon)}\,dx \text{ for all } \boldsymbol{v} \in \mathbf{V}(\Omega),$$

where, for any vector field $\boldsymbol{v} = (v_i) \in \mathbf{H}^1(\Omega)$, the *scaled linearized strains* $e_{i\|j}(\varepsilon; \boldsymbol{v}) = e_{j\|i}(\varepsilon; \boldsymbol{v}) \in L^2(\Omega)$ are defined by:

$$\begin{aligned}
e_{\alpha\|\beta}(\varepsilon; \boldsymbol{v}) &= \frac{1}{2}(\partial_\beta v_\alpha + \partial_\alpha v_\beta) - \Gamma^p_{\alpha\beta}(\varepsilon)v_p,\\
e_{\alpha\|3}(\varepsilon; \boldsymbol{v}) &= \frac{1}{2}\left(\frac{1}{\varepsilon}\partial_3 v_\alpha + \partial_\alpha v_3\right) - \Gamma^\sigma_{\alpha 3}(\varepsilon)v_\sigma,\\
e_{3\|3}(\varepsilon; \boldsymbol{v}) &= \frac{1}{\varepsilon}\partial_3 v_3.
\end{aligned}$$

To begin with, we establish several properties of "averages with respect to the transverse variable" that will be of frequent use (Thm. 4.2-1). Then we prove a crucial *three-dimensional inequality of Korn's type* (Thm. 4.3-1): Given a family of *linearly elastic elliptic membrane shells* with each having the same (elliptic) middle surface $S = \boldsymbol{\theta}(\overline{\omega})$, there exists a constant C such that, for ε small enough and for all $\boldsymbol{v} = (v_i) \in \mathbf{V}(\Omega)$,

$$\Big\{\sum_\alpha \|v_\alpha\|_{1,\Omega}^2 + |v_3|_{0,\Omega}^2\Big\}^{1/2} \le C\Big\{\sum_{i,j} |e_{i\|j}(\varepsilon; \boldsymbol{v})|_{0,\Omega}^2\Big\}^{1/2}.$$

The restriction that the shells be *elliptic membrane* ones is essential here!

Equipped with these preliminaries, we then establish the *main result of this chapter* (Thm. 4.4-1), by showing that *the family* $(\boldsymbol{u}(\varepsilon))_{\varepsilon>0}$ *strongly converges in the space* $H^1(\Omega) \times H^1(\Omega) \times L^2(\Omega)$ *as* $\varepsilon \to 0$ and that $\boldsymbol{u} = \lim_{\varepsilon\to 0} \boldsymbol{u}(\varepsilon)$ *is obtained by solving a two-dimensional problem.*

More specifically, we show that *the limit* $\boldsymbol{u}$ *is independent of the transverse variable* and that $\boldsymbol{\zeta} = \frac{1}{2}\int_{-1}^{1} \boldsymbol{u}\, dx_3$ satisfies the following (scaled) *two-dimensional variational problem of a linearly elastic membrane shell:*

$$\boldsymbol{\zeta} = (\zeta_i) \in \mathbf{V}_M(\omega) = H_0^1(\omega) \times H_0^1(\omega) \times L^2(\omega),$$

$$\int_\omega a^{\alpha\beta\sigma\tau}\gamma_{\sigma\tau}(\boldsymbol{\zeta})\gamma_{\alpha\beta}(\boldsymbol{\eta})\sqrt{a}\, dy = \int_\omega \Big\{\int_{-1}^{1} f^i\, dx_3\Big\}\eta_i\sqrt{a}\, dy$$

for all $\boldsymbol{\eta} = (\eta_i) \in \mathbf{V}_M(\omega)$, *where*

$$\gamma_{\alpha\beta}(\boldsymbol{\eta}) = \frac{1}{2}(\partial_\beta\eta_\alpha + \partial_\alpha\eta_\beta) - \Gamma_{\alpha\beta}^\sigma\eta_\sigma - b_{\alpha\beta}\eta_3,$$

$$a^{\alpha\beta\sigma\tau} = \frac{4\lambda\mu}{\lambda + 2\mu}a^{\alpha\beta}a^{\sigma\tau} + 2\mu(a^{\alpha\sigma}a^{\beta\tau} + a^{\alpha\tau}a^{\beta\sigma}),$$

the functions $\gamma_{\alpha\beta}(\boldsymbol{\eta})$ and $a^{\alpha\beta\sigma\tau}$ being respectively the covariant components of the *linearized change of metric tensor* associated with a displacement field $\eta_i\boldsymbol{a}^i$ of the middle surface S (Sect. 2.4) and the contravariant components of the (scaled) *two-dimensional elasticity tensor of the shell.*

To conclude this chapter, we review the *existence, uniqueness*, and *regularity* properties of the solution to the above variational problem and we describe the associated *minimization problem* and *boundary value problem* (Thm. 4.5-1). We also rewrite these two-dimensional equations and the fundamental convergence theorem in terms of *de-scaled unknowns and data*, thus providing a justification in terms of "physical" quantities of the *two-dimensional equations of a linearly elastic membrane shell* (Thms. 4.5-2 and 4.6-1).

In particular, the *limit displacement field* $\zeta_i^\varepsilon \boldsymbol{a}^i$ of the middle surface is such that $\boldsymbol{\zeta}^\varepsilon = (\zeta_i^\varepsilon)$ satisfies the following *minimization problem:*

$$\boldsymbol{\zeta}^\varepsilon \in \mathbf{V}_M(\omega) \text{ and } j_M^\varepsilon(\boldsymbol{\zeta}^\varepsilon) = \inf_{\boldsymbol{\eta} \in \mathbf{V}_M(\omega)} j_M^\varepsilon(\boldsymbol{\eta}),$$

where the *two-dimensional energy* $j_M^\varepsilon : \mathbf{V}_M(\omega) \to \mathbb{R}$ is defined by

$$j_M^\varepsilon(\boldsymbol{\eta}) = \frac{1}{2}\int_\omega \varepsilon a^{\alpha\beta\sigma\tau,\varepsilon}\gamma_{\sigma\tau}(\boldsymbol{\eta})\gamma_{\alpha\beta}(\boldsymbol{\eta})\sqrt{a}\,dy - \int_\omega \Big\{\int_{-\varepsilon}^{\varepsilon} f^{i,\varepsilon}\,dx_3^\varepsilon\Big\}\eta_i\sqrt{a}\,dy,$$

and

$$a^{\alpha\beta\sigma\tau,\varepsilon} = \frac{4\lambda^\varepsilon\mu^\varepsilon}{\lambda^\varepsilon + 2\mu^\varepsilon}a^{\alpha\beta}a^{\sigma\tau} + 2\mu^\varepsilon(a^{\alpha\sigma}a^{\beta\tau} + a^{\alpha\tau}a^{\beta\sigma}).$$

4.1. LINEARLY ELASTIC ELLIPTIC MEMBRANE SHELLS: DEFINITION, EXAMPLE, AND ASSUMPTIONS ON THE DATA; THE THREE-DIMENSIONAL EQUATIONS OVER A DOMAIN INDEPENDENT OF ε

Let ω be a domain in $\mathbb{R}^2$ with boundary γ and let $\boldsymbol{\theta} \in \mathcal{C}^2(\overline{\omega}; \mathbb{R}^3)$ be an injective mapping such that the two vectors $\partial_\alpha\boldsymbol{\theta}(y)$ are linearly independent at all points $y \in \overline{\omega}$. A *linearly elastic shell* with middle surface $S = \boldsymbol{\theta}(\overline{\omega})$ is called a **linearly elastic elliptic membrane shell** if the following two conditions are simultaneously satisfied (the definitions and notations are those of Sect. 3.1):

(i) The shell is subjected to a (homogeneous) **boundary condition of place** along its *entire* lateral face $\boldsymbol{\Theta}(\gamma \times [-\varepsilon, \varepsilon])$, i.e., *the displacement field vanishes there*; equivalently,

$$\boxed{\gamma_0 = \gamma.}$$

(ii) *The middle surface S is* **elliptic**, in the sense that there exists a constant c such that

$$\boxed{\sum_\alpha |\xi^\alpha|^2 \le c|b_{\alpha\beta}(y)\xi^\alpha\xi^\beta| \text{ for all } y \in \overline{\omega} \text{ and all } (\xi^\alpha) \in \mathbb{R}^2,}$$

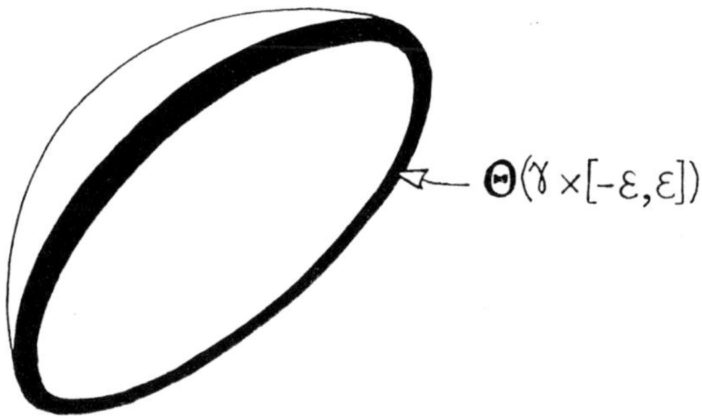

Fig. 4.1-1: *A linearly elastic elliptic membrane shell.* A linearly elastic shell whose middle surface $S = \boldsymbol{\theta}(\overline{\omega})$ is a portion of an ellipsoid E, and which is subjected to a boundary condition of place (i.e., of vanishing displacement field) along its entire lateral face $\boldsymbol{\Theta}(\gamma \times [-\varepsilon, \varepsilon])$ (darkened on the figure), provides an instance of an elliptic membrane shell. Note that stereographic coordinates (Fig. 2.1-3) afford the representation of such a surface S.

Elliptic membrane shells provide a first instance ($\gamma_0 = \gamma$ and S is elliptic) where the space

$$\mathbf{V}_F(\omega) = \{\boldsymbol{\eta} = (\eta_i) \in H^1(\omega) \times H^1(\omega) \times H^2(\omega);\ \eta_i = \partial_\nu \eta_3 = 0 \text{ on } \gamma_0,\\ \gamma_{\alpha\beta}(\boldsymbol{\eta}) = 0 \text{ in } \omega\},$$

which is the key to the classification of linearly elastic shells, reduces to $\{\mathbf{0}\}$ (in fact, a larger space "already" reduces to $\{\mathbf{0}\}$ in this case; cf. Thm. 2.7-2). The "generalized membrane shells" (Chap. 5) exhaust all the remaining cases where $\mathbf{V}_F(\omega) = \{\mathbf{0}\}$.

where the functions $b_{\alpha\beta} : \overline{\omega} \to \mathbb{R}$ are the covariant components of the curvature tensor of S (this definition was already given in Sect. 2.7). This assumption means that *the Gaussian curvature of S is everywhere* > 0; equivalently, *the two principal radii of curvature are either both > 0 at all points of S, or both < 0 at all points of S* (see Sect. 2.2 for a detailed exposition of these notions).

A shell subjected to a boundary condition of place along its entire lateral face and whose middle surface is a portion of an ellipsoid provides an example of a linearly elastic membrane shell (Fig. 4.1-1).

Let there be given a linearly elastic elliptic membrane shell such that $\boldsymbol{\theta} \in \mathcal{C}^{2,1}(\overline{\omega}; \mathbb{R}^3)$. Then the following inequality of Korn's type on the elliptic surface $S = \boldsymbol{\theta}(\overline{\omega})$ holds (Thm. 2.7-3): *There exists a* *constant c_M such that*

$$\Big\{\sum_\alpha \|\eta_\alpha\|^2_{1,\omega} + |\eta_3|^2_{0,\omega}\Big\}^{1/2} \le c_M \Big\{\sum_{\alpha,\beta} |\gamma_{\alpha\beta}(\boldsymbol{\eta})|^2_{0,\omega}\Big\}^{1/2}$$
$$\text{for all } \boldsymbol{\eta} = (\eta_i) \in \mathbf{V}_M(\omega) := H^1_0(\omega) \times H^1_0(\omega) \times L^2(\omega),$$

where the functions

$$\gamma_{\alpha\beta}(\boldsymbol{\eta}) := \frac{1}{2}(\partial_\alpha \eta_\beta + \partial_\beta \eta_\alpha) - \Gamma^\sigma_{\alpha\beta}\eta_\sigma - b_{\alpha\beta}\eta_3$$

are the covariant components of the linearized change of metric tensor associated with a displacement field $\eta_i \boldsymbol{a}^i$ of the surface S; the subscript "M" announces that $\mathbf{V}_M(\omega)$ is the functional space over which the limit two-dimensional equations of such a membrane shell are posed (Thm. 4.4-1). *This inequality, consequence of the definition of an elliptic membrane shell, will be the key to the ensuing analysis of this chapter.*

A word of caution. *The definition of a linearly elastic elliptic membrane shell thus depends only on the subset of the lateral face where the shell is subjected to a boundary condition of place* (this subset should be the entire lateral face) *and on the "geometry" of its middle surface* (its Gaussian curvature should be >0 everywhere). This definition is thus independent of the particular system of curvilinear coordinates employed for representing the surface S. ■

If assumptions (i) and (ii) are satisfied and $\boldsymbol{\theta} \in \mathcal{C}^{2,1}(\overline{\omega}; \mathbb{R}^3)$, we thus have

$$\{\boldsymbol{\eta} = (\eta_i) \in H^1_0(\omega) \times H^1_0(\omega) \times L^2(\omega);\ \gamma_{\alpha\beta}(\boldsymbol{\eta}) = 0 \text{ in } \omega\} = \{\mathbf{0}\}.$$

Hence *linearly elastic elliptic membrane shells provide a first instance* ($\gamma_0 = \gamma$ and S is elliptic) *where the space* (already introduced in Thm. 3.4-3)

$$\mathbf{V}_F(\omega) := \{\boldsymbol{\eta} = (\eta_i) \in H^1(\omega) \times H^1(\omega) \times H^2(\omega);$$
$$\eta_i = \partial_\nu \eta_3 = 0 \text{ on } \gamma_0,\ \gamma_{\alpha\beta}(\boldsymbol{\eta}) = 0 \text{ in } \omega\}$$

a fortiori *reduces to* $\{\mathbf{0}\}$ (we recall that ∂_ν denotes the outer normal derivative operator along γ; the subscript "F" reminds that this space is central to the study of *flexural shells*, undertaken in Chap. 6).

The importance of this observation lies in that *the issue of whether* $\mathbf{V}_F(\omega) = \{\mathbf{0}\}$ *or* $\mathbf{V}_F(\omega) \neq \{\mathbf{0}\}$ *is the basis of the classification of linearly elastic shells*: A shell is called a **linearly elastic membrane shell** if $\mathbf{V}_F(\omega) = \{\mathbf{0}\}$ or a **linearly elastic flexural shell** if $\mathbf{V}_F(\omega) \neq \{\mathbf{0}\}$.

In this direction, note that *elliptic membrane shells far from exhaust all the instances where* $\mathbf{V}_F(\omega) = \{\mathbf{0}\}$, the remaining instances corresponding to the *generalized membrane shells* studied in the next chapter.

The formal analysis of Sect. 3.4 then naturally leads us to make the following **scalings of the unknowns** and **assumptions on the data** *for a family of linearly elastic elliptic membrane shells* with each having the *same* (elliptic) middle surface $S = \boldsymbol{\theta}(\overline{\omega})$ as their thickness 2ε approaches zero.

First, we define the **scaled unknown** $\boldsymbol{u}(\varepsilon) = (u_i(\varepsilon))$ by letting

$$\boxed{u_i^\varepsilon(x^\varepsilon) := u_i(\varepsilon)(x) \text{ for all } x^\varepsilon = \pi^\varepsilon x \in \overline{\Omega}^\varepsilon.}$$

Next, we *require* that the *Lamé constants* and the *applied body and surface force densities* be such that

$$\boxed{\begin{array}{rll} \lambda^\varepsilon = \lambda & \text{and} & \mu^\varepsilon = \mu, \\ f^{i,\varepsilon}(x^\varepsilon) = f^i(x) & \text{for all} & x^\varepsilon = \pi^\varepsilon x \in \Omega^\varepsilon, \\ h^{i,\varepsilon}(x^\varepsilon) = \varepsilon h^i(x) & \text{for all} & x^\varepsilon = \pi^\varepsilon x \in \Gamma_+^\varepsilon \cup \Gamma_-^\varepsilon, \end{array}}$$

where the constants $\lambda > 0$ *and* $\mu > 0$ *and the functions* $f^i \in L^2(\Omega)$ *and* $h^i \in L^2(\Gamma_+ \cup \Gamma_-)$ *are independent of* ε (Fig. 3.2-1 recapitulates the definitions of the sets Ω^ε, Ω, Γ_+^ε, Γ_+, Γ_-^ε, and Γ_-).

Remark. For notational brevity, the functions f^i and h^i stand for the functions that were respectively denoted $f^{i,0}$ and $h^{i,1}$ in Sect. 3.4. ■

As an immediate corollary to Thm. 3.2-1 (simply corresponding to $p = 0$), we obtain the *problems satisfied by the scaled unknown over the set* Ω, thus over a domain that is *independent of* ε:

Theorem 4.1-1. *Let ω be a domain in $\mathbb{R}^2$, let $\boldsymbol{\theta} \in \mathcal{C}^2(\overline{\omega}; \mathbb{R}^3)$ be an injective mapping such that the two vectors $\boldsymbol{a}_\alpha = \partial_\alpha \boldsymbol{\theta}$ are linearly independent at all points of $\overline{\omega}$, and let $\varepsilon_0 > 0$ be as in* Thm. 3.1-1. *Consider a family of linearly elastic elliptic membrane shells with thickness 2ε with each having the same elliptic middle surface $S = \boldsymbol{\theta}(\overline{\omega})$.*

Let the assumptions on the data be as above. Then, for each $0 < \varepsilon \leq \varepsilon_0$, the scaled unknown $\boldsymbol{u}(\varepsilon) = (u_i(\varepsilon))$ satisfies the following **scaled three-dimensional variational problem $\mathcal{P}(\varepsilon; \Omega)$ of a linearly elastic elliptic membrane shell:**

$$\boldsymbol{u}(\varepsilon) \in \mathbf{V}(\Omega) := \{\boldsymbol{v} = (v_i) \in \mathbf{H}^1(\Omega);\ \boldsymbol{v} = \mathbf{0} \text{ on } \gamma \times [-1, 1]\},$$

$$\int_\Omega A^{ijkl}(\varepsilon) e_{k\|l}(\varepsilon; \boldsymbol{u}(\varepsilon)) e_{i\|j}(\varepsilon; \boldsymbol{v}) \sqrt{g(\varepsilon)}\, dx$$

$$= \int_\Omega f^i v_i \sqrt{g(\varepsilon)}\, dx + \int_{\Gamma_+ \cup \Gamma_-} h^i v_i \sqrt{g(\varepsilon)}\, d\Gamma \text{ for all } \boldsymbol{v} \in \mathbf{V}(\Omega),$$

where the scaled linearized strains $e_{i\|j}(\varepsilon; \boldsymbol{v})$ are given by

$$e_{\alpha\|\beta}(\varepsilon; \boldsymbol{v}) = \frac{1}{2}(\partial_\beta v_\alpha + \partial_\alpha v_\beta) - \Gamma^p_{\alpha\beta}(\varepsilon) v_p,$$

$$e_{\alpha\|3}(\varepsilon; \boldsymbol{v}) = \frac{1}{2}\left(\frac{1}{\varepsilon}\partial_3 v_\alpha + \partial_\alpha v_3\right) - \Gamma^\sigma_{\alpha 3}(\varepsilon) v_\sigma,$$

$$e_{3\|3}(\varepsilon; \boldsymbol{v}) = \frac{1}{\varepsilon}\partial_3 v_3,$$

and the "scaled" functions $A^{ijkl}(\varepsilon)$, $\Gamma^p_{ij}(\varepsilon)$, $g(\varepsilon)$ are defined as in Thm. 3.2-1. ∎

Our main objective in this chapter consists in analyzing the *behavior of the solutions $\boldsymbol{u}(\varepsilon) \in \mathbf{H}^1(\Omega)$ of problems $\mathcal{P}(\varepsilon; \Omega)$ as $\varepsilon \to 0$.*

To this end, we begin by proving various "analytical" preliminaries, which complement the "geometrical" and "mechanical" preliminaries proved in Sect. 3.3. Those of Sect. 4.2 are common to *all* types of linearly elastic shells while the inequality of Sect. 4.3 is special to elliptic membrane shells.

4.2. AVERAGES WITH RESPECT TO THE TRANSVERSE VARIABLE

Averages with respect to the transverse variable x_3 play a fundamental rôle in this chapter as well as in the next chapters. If v, or $\boldsymbol{v}$, are real-valued, or vector-valued, functions defined a.e. over $\Omega = \omega \times]-1, 1[$ (the abbreviation "a.e." stands for "almost everywhere"), these **averages** $\bar{v}$, or $\overline{\boldsymbol{v}}$, are the real-valued, or vector-valued, functions defined a.e. over ω by letting

$$\bar{v}(y) := \frac{1}{2}\int_{-1}^{1} v(y, x_3)\, dx_3, \quad \overline{\boldsymbol{v}}(y) := \frac{1}{2}\int_{-1}^{1} \boldsymbol{v}(y, x_3)\, dx_3$$

for almost all $y \in \omega$ whenever these definitions make sense (Thm. 4.2-1 (a) provides such an instance).

The next theorem gathers the properties of averages that are relevant in the ensuing analysis. Note that the functions $\gamma_{\alpha\beta}(\boldsymbol{\eta})$ introduced in part (d) are the covariant components of the *linearized change of metric*, or *strain, tensor*, associated with an arbitrary displacement field $\eta_i \boldsymbol{a}^i$ of the surface S (Sect. 2.4). We follow here Ciarlet & Lods [1996b, Lemma 3.2].

Observe that the convergence established in part (d) holds *whether the surface* $S = \boldsymbol{\theta}(\overline{\omega})$ *is elliptic or not.*

Theorem 4.2-1. *Let* ω *be a domain in* $\mathbb{R}^2$ *and let* $\Omega = \omega \times]-1, 1[$.

(a) *Let* $v \in L^2(\Omega)$. *Then* $\overline{v}(y)$ *as defined* supra *is finite for almost all* $y \in \omega$, *the function* $\overline{v}$ *defined in this fashion belongs to* $L^2(\omega)$, *and*

$$|\overline{v}|_{0,\omega} \le \frac{1}{\sqrt{2}} |v|_{0,\Omega}.$$

If $\partial_3 v = 0$ *in the sense of distributions, i.e., if* $\int_\Omega v \partial_3 \varphi \, dx = 0$ *for all* $\varphi \in \mathcal{D}(\Omega)$, *then* v *does not depend on* x_3, *and*

$$v(y, x_3) = \overline{v}(y) \text{ for almost all } (y, x_3) \in \Omega.$$

(b) *Let* $v \in H^1(\Omega)$. *Then* $\overline{v} \in H^1(\omega)$, $\partial_\alpha \overline{v} = \overline{\partial_\alpha v}$, *and*

$$\|\overline{v}\|_{1,\omega} \le \frac{1}{\sqrt{2}} \|v\|_{1,\Omega}.$$

Let γ_0 denote a $d\gamma$-measurable subset of γ. If $v = 0$ on $\gamma_0 \times [-1, 1]$, then $\overline{v} = 0$ on γ_0; in particular, $\overline{v} \in H_0^1(\omega)$ if $v = 0$ on $\gamma \times [-1, 1]$.

(c) *Let $(v(\varepsilon))_{\varepsilon>0}$ be a sequence of functions $v(\varepsilon) \in H^1(\Omega)$ and let $\overline{v} \in L^2(\omega)$ be such that*

$$\partial_3 v(\varepsilon) \to 0 \text{ in } L^2(\Omega) \text{ and } \overline{v}(\varepsilon) \to \overline{v} \text{ in } L^2(\omega) \text{ as } \varepsilon \to 0.$$

Then

$$v(\varepsilon) \to \overline{v} \text{ in } L^2(\Omega) \text{ as } \varepsilon \to 0,$$

where the function $\overline{v} \in L^2(\omega)$ is identified with a function in $L^2(\Omega)$ by letting $\overline{v}(y, x_3) := \overline{v}(y)$ for almost all $(y, x_3) \in \Omega$.

(d) *Let the assumptions on the mapping $\boldsymbol{\theta}$ and the definitions be as in* Thm. 3.3-1. *Let there be given a sequence $(\boldsymbol{v}(\varepsilon))_{\varepsilon>0}$ of vector fields $\boldsymbol{v}(\varepsilon) = (v_i(\varepsilon)) \in \mathbf{H}^1(\Omega)$ that is bounded in $\mathbf{L}^2(\Omega)$, let the scaled linearized strains $e_{\alpha\|\beta}(\varepsilon; \boldsymbol{v}(\varepsilon))$ be defined as in* Thm. 3.2-1, *and let*

$$\gamma_{\alpha\beta}(\boldsymbol{\eta}) := \frac{1}{2}(\partial_\beta \eta_\alpha + \partial_\alpha \eta_\beta) - \Gamma_{\alpha\beta}^\sigma \eta_\sigma - b_{\alpha\beta}\eta_3$$

for any $\boldsymbol{\eta} = (\eta_i) \in H^1(\omega) \times H^1(\omega) \times L^2(\omega)$. Then

$$\{\overline{e_{\alpha\|\beta}(\varepsilon; \boldsymbol{v}(\varepsilon))} - \gamma_{\alpha\beta}(\overline{\boldsymbol{v}(\varepsilon)})\} \to 0 \text{ in } L^2(\omega) \text{ as } \varepsilon \to 0.$$

Proof. (i) Let $v \in L^2(\Omega)$. For almost all $y \in \omega$, the function $v(y, \cdot)$ belongs to the space $L^2(]-1, 1[)$ by Fubini's theorem. For such points y, Cauchy-Schwarz inequality gives

$$\left| \int_{-1}^1 v(y, x_3)\, dx_3 \right|^2 \le 2 \int_{-1}^1 |v(y, x_3)|^2\, dx_3 < +\infty.$$

Hence

$$\int_\omega |\overline{v}(y)|^2\, dy \le \frac{1}{2} \int_\omega \left\{ \int_{-1}^1 |v(y, x_3)|^2\, dx_3 \right\} dy = \frac{1}{2} \int_\Omega |v|^2\, dx,$$

and thus $|\overline{v}|_{0,\omega} \le \frac{1}{\sqrt{2}} |v|_{0,\Omega}$.

If $\partial_3 v = 0$ in the sense of distributions, there exists $\eta \in L^2(\omega)$ such that $v(y, x_3) = \eta(y)$ for almost all $(y, x_3) \in \Omega$ (Ex 4.1). But $\overline{v} = \eta$ in this case, and thus (a) is proved.

(ii) Let $v \in H^1(\Omega)$. Given an arbitrary function $\varphi \in \mathcal{D}(\omega)$, let $\Phi : \Omega \to \mathbb{R}$ be defined by $\Phi(y, x_3) = \varphi(y)$ for all $(y, x_3) \in \Omega$. Since Φ vanishes on $\gamma \times [-1, 1]$ and the "horizontal" components of the unit outer normal vector vanish on $\Gamma_+ \cup \Gamma_-$, Green's formula in $H^1(\Omega)$ gives

$$\int_\Omega v \partial_\alpha \Phi \, dx = - \int_\Omega \partial_\alpha v \, \Phi \, dx.$$

Since $v \in L^2(\Omega)$, $\partial_\alpha v \in L^2(\Omega)$, and Φ and $\partial_\alpha \Phi$ are independent of x_3, Fubini's theorem yields

$$\int_\omega \overline{v} \partial_\alpha \varphi \, dy = - \int_\omega \overline{\partial_\alpha v} \, \varphi \, dy;$$

hence $\overline{v} \in H^1(\omega)$ and $\partial_\alpha \overline{v} = \overline{\partial_\alpha v}$. These relations, combined with the inequality proved in (i), imply that $\|\overline{v}\|_{1,\omega} \leq \frac{1}{\sqrt{2}} \|v\|_{1,\Omega}$.

Let γ_0 be a $d\gamma$-measurable subset of γ, and assume in addition that $v = 0$ on $\Gamma_0 := \gamma_0 \times [-1, 1]$. There exists a sequence $(\Phi^k)_{k=1}^\infty$ of functions $\Phi^k \in \mathcal{C}^\infty(\overline{\Omega})$ such that $\Phi^k \to v$ in $H^1(\Omega)$ as $k \to \infty$. Hence, ${\Phi^k}_{|\Gamma_0} \to 0 = v_{|\Gamma_0}$ in $L^2(\Gamma_0)$ as $k \to \infty$; consequently, ${\overline{\Phi}^k}_{|\gamma_0} \to 0$ in $L^2(\gamma_0)$ since $\|\overline{\Phi}^k\|_{L^2(\gamma_0)} \leq \frac{1}{\sqrt{2}} \|\Phi^k\|_{L^2(\Gamma_0)}$, and thus ${\overline{\Phi}^k}_{|\gamma_0} \to 0 = \overline{v}_{|\gamma_0}$ in $L^2(\gamma_0)$. This proves (b).

(iii) Let $v \in L^2(\Omega)$ be such that its derivative $\partial_3 v$ in the sense of distributions belongs to $L^2(\Omega)$. Then we may write, for almost all $(y, s) \in \omega \times \,]-1, 1[$ (Ex. 4.1),

$$v(y, s) = v(y, -1) + \int_{-1}^s \partial_3 v(y, x_3) \, dx_3,$$

and thus

$$\overline{v}(y) = \frac{1}{2} \int_{-1}^1 v(y, s) ds = v(y, -1) + \frac{1}{2} \int_{-1}^1 \left(\int_{-1}^t \partial_3 v(y, x_3) dx_3 \right) dt.$$

Hence the following identity holds:

$$v(y, s) = \overline{v}(y) + \int_{-1}^s \partial_3 v(y, x_3) dx_3 - \frac{1}{2} \int_{-1}^1 \left(\int_{-1}^t \partial_3 v(y, x_3) dx_3 \right) dt.$$

This identity, combined with the triangular inequality and the relations

$$\int_\omega \left\{ \int_{-1}^{1} |\overline{v}(y)|^2 \, ds \right\} dy = 2|\overline{v}|_{0,\omega}^2,$$

$$\int_\omega \left\{ \int_{-1}^{1} \left| \int_{-1}^{s} \partial_3 v(y, x_3) \, dx_3 \right|^2 ds \right\} dy \leq 4|\partial_3 v|_{0,\Omega}^2,$$

$$\int_\omega \left\{ \int_{-1}^{1} \left| \int_{-1}^{1} \left(\int_{-1}^{t} \partial_3 v(y, x_3) \, dx_3 \right) dt \right|^2 ds \right\} dy \leq 8|\partial_3 v|_{0,\Omega}^2,$$

shows that

$$|v|_{0,\Omega} \leq \sqrt{2}|\overline{v}|_{0,\omega} + \left(2 + \sqrt{2}\right)|\partial_3 v|_{0,\Omega}.$$

The desired convergence $v(\varepsilon) \to \overline{v}$ in $L^2(\Omega)$ is then proved by letting $v = v(\varepsilon) - \overline{v}$ in the last inequality, and thus (c) is proved.

(iv) From (ii), we infer that

$$\frac{1}{2}\overline{(\partial_\alpha v_\beta(\varepsilon) + \partial_\beta v_\alpha(\varepsilon))} = \frac{1}{2}(\partial_\alpha \overline{v_\beta(\varepsilon)} + \partial_\beta \overline{v_\alpha(\varepsilon)}).$$

Hence in order to establish the convergence announced in (d), it remains to establish that

$$\{\overline{\Gamma^p_{\alpha\beta}(\varepsilon) v_p(\varepsilon)} - \Gamma^p_{\alpha\beta}(0)\overline{v_p(\varepsilon)}\} \to 0 \text{ in } L^2(\omega) \text{ as } \varepsilon \to 0,$$

where the functions $\Gamma^\sigma_{\alpha\beta}(0) := \Gamma^\sigma_{\alpha\beta}$ and $\Gamma^3_{\alpha\beta}(0) := b_{\alpha\beta}$ are independent of x_3.

But the behavior of the functions $\Gamma^p_{\alpha\beta}(\varepsilon)$ as $\varepsilon \to 0$ (Thm. 3.3-1) and the inequality $|\overline{v}|_{0,\omega} \leq \frac{1}{\sqrt{2}}|v|_{0,\Omega}$ of part (i) together imply that

$$|\overline{\Gamma^p_{\alpha\beta}(\varepsilon) v_p(\varepsilon)} - \Gamma^p_{\alpha\beta}(0)\overline{v_p(\varepsilon)}|_{0,\omega} = |\overline{\Gamma^p_{\alpha\beta}(\varepsilon) v_p(\varepsilon) - \Gamma^p_{\alpha\beta}(0) v_p(\varepsilon)}|_{0,\omega}$$

$$\leq \frac{1}{\sqrt{2}}|\Gamma^p_{\alpha\beta}(\varepsilon) v_p(\varepsilon) - \Gamma^p_{\alpha\beta}(0) v_p(\varepsilon)|_{0,\Omega} \leq C_1 \sqrt{\frac{3}{2}}\varepsilon |\boldsymbol{v}(\varepsilon)|_{0,\Omega},$$

and the conclusion follows from the assumed boundedness of the sequence $(\boldsymbol{v}(\varepsilon))_{\varepsilon>0}$ in the space $\mathbf{L}^2(\Omega)$. This proves (d). ∎

Remark. It is easily verified that the result of part (d) still holds if $\boldsymbol{\theta} \in \mathcal{C}^2(\overline{\omega}; \mathbb{R}^3)$ (it was assumed that $\boldsymbol{\theta} \in \mathcal{C}^3(\overline{\omega}; \mathbb{R}^3)$ in Thm. 3.3-1, but then sharper asymptotic behaviors of the scaled Christoffel symbols $\Gamma^p_{\alpha\beta}(\varepsilon)$ were sought). ∎

4.3. A THREE-DIMENSIONAL INEQUALITY OF KORN'S TYPE FOR A FAMILY OF LINEARLY ELASTIC ELLIPTIC MEMBRANE SHELLS

Let γ_0 be a subset of γ with *length* $\gamma_0 > 0$. If no particular assumption is made on the geometry of the surface S, it will be shown in the next chapter (Thm. 5.3-1) that there exist constants $\varepsilon_1 > 0$ and $C > 0$ such that, for all $0 < \varepsilon \leq \varepsilon_1$ and all fields $\boldsymbol{v} = (v_i) \in \mathbf{H}^1(\Omega)$ vanishing on the set $\gamma_0 \times [-1, 1]$,

$$\left\{ \sum_i \|v_i\|_{1,\Omega}^2 \right\}^{1/2} \leq \frac{C}{\varepsilon} \left\{ \sum_{i,j} |e_{i\|j}(\varepsilon; \boldsymbol{v})|_{0,\Omega}^2 \right\}^{1/2}.$$

This relation is a *three-dimensional inequality of Korn's type for a family of "general " shells*, the scaled strains $e_{i\|j}(\varepsilon; \boldsymbol{v})$ "in curvilinear coordinates" replacing the classical strains $\frac{1}{2}(\partial_j v_i + \partial_i v_j)$ "in Cartesian coordinates".

It is remarkable that, *for a family of elliptic membrane shells*, the "constant" C/ε may be replaced by a constant that is *independent of* ε, at the expense, however, of replacing $\|v_3\|_{1,\Omega}$ by $|v_3|_{0,\Omega}$ in the left-hand side. More specifically, another *three-dimensional inequality of Korn's type* holds for such a family, which plays a key rôle in the proof of the *convergence theorem* (Thm. 4.4-1): It is used there to establish the fundamental *a priori* bounds that the family $(\boldsymbol{u}(\varepsilon))_{\varepsilon>0}$ satisfies.

The following result is due to Ciarlet & Lods [1996b, Thm. 4.1].

Theorem 4.3-1. *Assume that* $\boldsymbol{\theta} \in \mathcal{C}^3(\overline{\omega}; \mathbb{R}^3)$ *and let* $\varepsilon_0 > 0$ *be defined as in* Thm. 3.1-1. *Consider a family of linearly elastic elliptic membrane shells with thickness* 2ε *with each having the same elliptic middle surface* $S = \boldsymbol{\theta}(\overline{\omega})$. *Define the space*

$$\mathbf{V}(\Omega) := \{\boldsymbol{v} = (v_i) \in \mathbf{H}^1(\Omega);\ \boldsymbol{v} = \mathbf{0} \text{ on } \gamma \times [-1, 1]\}.$$

Then there exist a constant ε_1 *satisfying* $0 < \varepsilon_1 \leq \varepsilon_0$ *and a constant* C *such that, for all* $0 < \varepsilon \leq \varepsilon_1$, *the following* **three-dimensional inequality of Korn's type for a family of linearly**

elastic elliptic membrane shells *holds*:

$$\Big\{\sum_\alpha \|v_\alpha\|^2_{1,\Omega} + |v_3|^2_{0,\Omega}\Big\}^{1/2} \le C\Big\{\sum_{i,j} |e_{i\|j}(\varepsilon;\, \boldsymbol{v})|^2_{0,\Omega}\Big\}^{1/2}$$
$$\text{for all } \boldsymbol{v} = (v_i) \in \mathbf{V}(\Omega),$$

where the scaled linearized strains $e_{i\|j}(\varepsilon;\, \boldsymbol{v})$ *are defined by*

$$\begin{aligned}
e_{\alpha\|\beta}(\varepsilon;\, \boldsymbol{v}) &:= \frac{1}{2}(\partial_\beta v_\alpha + \partial_\alpha v_\beta) - \Gamma^p_{\alpha\beta}(\varepsilon) v_p,\\
e_{\alpha\|3}(\varepsilon;\, \boldsymbol{v}) &:= \frac{1}{2}\Big(\frac{1}{\varepsilon}\partial_3 v_\alpha + \partial_\alpha v_3\Big) - \Gamma^\sigma_{\alpha 3}(\varepsilon) v_\sigma,\\
e_{3\|3}(\varepsilon;\, \boldsymbol{v}) &:= \frac{1}{\varepsilon}\partial_3 v_3.
\end{aligned}$$

Proof. For the sake of brevity, we set the following rule governing the usage of *constants* in this proof; Whenever c_1, c_2, etc., appears in an inequality, it means that *there exists a constant, denoted by this symbol, that is* > 0 *and independent of all the "variables" that enter the inequality such that this inequality holds.*

(i) *We first establish that, for all* $0 < \varepsilon \le \varepsilon_0$,

$$\Big\{\sum_\alpha \|v_\alpha\|^2_{1,\Omega} + \|\varepsilon v_3\|^2_{1,\Omega}\Big\}^{1/2} \le c_1\Big\{\sum_{i,j} |e_{i\|j}(\varepsilon;\, \boldsymbol{v})|^2_{0,\Omega} + \sum_i |v_i|^2_{0,\Omega}\Big\}^{1/2}$$

for all $\boldsymbol{v} = (v_i) \in \mathbf{H}^1(\Omega)$.

Given $\boldsymbol{v} = (v_i) \in \mathbf{H}^1(\Omega)$, let

$$e_{ij}(\boldsymbol{v}) := \frac{1}{2}(\partial_j v_i + \partial_i v_j) \text{ and } \boldsymbol{v}(\varepsilon) := (v_1,\, v_2,\, \varepsilon v_3) \text{ for } \varepsilon \le \varepsilon_0.$$

Then

$$\begin{aligned}
e_{\alpha\beta}(\boldsymbol{v}(\varepsilon)) &= e_{\alpha\|\beta}(\varepsilon;\, \boldsymbol{v}) + \Gamma^p_{\alpha\beta}(\varepsilon) v_p,\\
e_{\alpha 3}(\boldsymbol{v}(\varepsilon)) &= \varepsilon e_{\alpha\|3}(\varepsilon;\, \boldsymbol{v}) + \varepsilon\Gamma^\sigma_{\alpha 3}(\varepsilon) v_\sigma,\\
e_{33}(\boldsymbol{v}(\varepsilon)) &= \varepsilon^2 e_{3\|3}(\varepsilon;\, \boldsymbol{v}),
\end{aligned}$$

and consequently, the estimates on the functions $\Gamma^p_{\alpha i}(\varepsilon)$ established in Thm. 3.3-1 show that

$$\left\{\sum_{i,j} |e_{ij}(\boldsymbol{v}(\varepsilon))|^2_{0,\Omega}\right\}^{1/2} \leq c_2\left\{\sum_{i,j} |e_{i\|j}(\varepsilon;\, \boldsymbol{v})|^2_{0,\Omega} + \sum_i |v_i|^2_{0,\Omega}\right\}^{1/2},$$

since we may assume without loss of generality that $\varepsilon_0 \leq 1$. By the "classical" three-dimensional Korn inequality in Cartesian coordinates (see, e.g., Vol. II, Thm. 1.1-2 or let $\boldsymbol{\Theta} = \boldsymbol{id}_{\mathbb{R}^3}$ in Thm. 1.7-4 in this volume),

$$\begin{aligned} \|\boldsymbol{v}(\varepsilon)\|_{1,\Omega} &= \left\{\sum_\alpha \|v_\alpha\|^2_{1,\Omega} + \|\varepsilon v_3\|^2_{1,\Omega}\right\}^{1/2} \\ &\leq c_3\left\{\sum_{i,j} |e_{ij}(\boldsymbol{v}(\varepsilon))|^2_{0,\Omega} + |\boldsymbol{v}(\varepsilon)|^2_{0,\Omega}\right\}^{1/2}, \end{aligned}$$

and the assertion follows from the last two inequalities.

(ii) *We next establish that there exists* $0 < \varepsilon_1 \leq \varepsilon_0$ *such that, for all* $0 < \varepsilon \leq \varepsilon_1$,

$$\left\{\sum_i |v_i|^2_{0,\Omega}\right\}^{1/2} \leq c_4\left\{\sum_{i,j} |e_{i\|j}(\varepsilon;\, \boldsymbol{v})|^2_{0,\Omega}\right\}^{1/2} \text{ for all } \boldsymbol{v} \in \mathbf{V}(\Omega).$$

The proof will then be complete, since this inequality and that of (i) together imply that

$$c_1^{-2}\sum_\alpha \|v_\alpha\|^2_{1,\Omega} + |v_3|^2_{0,\Omega} \leq (1 + 2c_4^2)\sum_{i,j} |e_{i\|j}(\varepsilon;\, \boldsymbol{v})|^2_{0,\Omega}$$

for all $\boldsymbol{v} \in \mathbf{V}(\Omega)$.

If the announced inequality is false, there exist $\varepsilon_m > 0$ and a sequence $(\boldsymbol{v}^m)_{m=1}^\infty$ of vector fields $\boldsymbol{v}^m = (v_i^m) \in \mathbf{V}(\Omega)$ such that (the Latin letters m and n are used here for indexing sequences):

$$\begin{aligned} \varepsilon_m &\to 0 \text{ as } m \to \infty, \\ e_{i\|j}(\varepsilon_m;\, \boldsymbol{v}^m) &\to 0 \text{ in } L^2(\Omega) \text{ as } m \to \infty, \\ \sum_i |v_i^m|^2_{0,\Omega} &= 1 \text{ for all } m. \end{aligned}$$

Both sequences $(v_\alpha^m)_{m=0}^\infty$ are then bounded in $H^1(\Omega)$ by (i). Hence there exist subsequences $(v_\alpha^n)_{n=0}^\infty$ and there exist functions $v_\alpha \in H^1(\Omega)$ satisfying $v_\alpha = 0$ on $\gamma \times [-1, 1]$ and a function $v_3 \in L^2(\Omega)$ such that

$$v_\alpha^n \rightharpoonup v_\alpha \text{ in } H^1(\Omega) \text{ and } v_\alpha^n \to v_\alpha \text{ in } L^2(\Omega),$$
$$v_3^n \rightharpoonup v_3 \text{ in } L^2(\Omega) \text{ as } n \to \infty,$$

where $\to$ and $\rightharpoonup$ denote strong and weak convergences, respectively. The remainder of the proof consists in showing that the sequence $(v_3^n)_{n=0}^\infty$ converges *strongly* in $L^2(\Omega)$ and that the three limit functions v_i vanish, in contradiction with $\sum_i |v_i^m|_{0,\Omega}^2 = 1$ for all m. To these ends, we proceed in *three stages:*

First, we show that (averages such as $\overline{\boldsymbol{v}}^n$ are defined in Sect. 4.2)

$$\overline{\boldsymbol{v}}^n \to \mathbf{0} \text{ in the space } \mathbf{V}_M(\omega) \text{ as } n \to \infty.$$

To see this, observe that, by Thm. 4.2-1 (a),

$$e_{\alpha\|\beta}(\varepsilon_n; \boldsymbol{v}^n) \to 0 \text{ in } L^2(\Omega) \Longrightarrow \overline{e_{\alpha\|\beta}(\varepsilon_n; \boldsymbol{v}^n)} \to 0 \text{ in } L^2(\omega),$$

and that, by Thm. 4.2-1 (d),

$$\overline{e_{\alpha\|\beta}(\varepsilon_n; \boldsymbol{v}^n)} \to 0 \text{ in } L^2(\omega) \Longrightarrow \gamma_{\alpha\beta}(\overline{\boldsymbol{v}}^n) \to 0 \text{ in } L^2(\omega),$$

as $n \to \infty$. Hence the convergence $\overline{\boldsymbol{v}}^n \to \mathbf{0}$ in $\mathbf{V}_M(\omega)$ follows from the *inequality of Korn's type on an elliptic surface* (Thm. 2.7-3).

Second, we show that

$$v_\alpha^n \to 0 \text{ in } L^2(\Omega).$$

The estimates on the functions $\Gamma_{\alpha 3}^\sigma(\varepsilon)$ established in Thm. 3.3-1 and the assumed properties of the sequences $(\varepsilon_n)_{n=1}^\infty$ and $(\boldsymbol{v}^n)_{n=1}^\infty$ imply that

$$\partial_3 v_\alpha^n + \varepsilon_n \partial_\alpha v_3^n = 2\varepsilon_n e_{\alpha\|3}(\varepsilon_n; \boldsymbol{v}^n) + 2\varepsilon_n \Gamma_{\alpha 3}^\sigma(\varepsilon_n) v_\sigma^n \to 0 \text{ in } L^2(\Omega).$$

Let $\varphi \in \mathcal{D}(\Omega)$; since the sequence $(v_3^n)_{n=0}^\infty$ is bounded in $L^2(\Omega)$, we have (recall that $v_\alpha := \lim_{n\to\infty} v_\alpha^n$ in $L^2(\Omega)$)

$$\begin{aligned}\int_\Omega \partial_3 v_\alpha \, \varphi \, dx &= -\int_\Omega v_\alpha \partial_3 \varphi \, dx \\ &= -\lim_{n\to\infty} \left\{ \int_\Omega v_\alpha^n \partial_3 \varphi \, dx + \varepsilon_n \int_\Omega v_3^n \partial_\alpha \varphi \, dx \right\} \\ &= \lim_{n\to\infty} \left\{ \int_\Omega (\partial_3 v_\alpha^n + \varepsilon_n \partial_\alpha v_3^n) \, \varphi \, dx \right\} = 0,\end{aligned}$$

and thus $\partial_3 v_\alpha = 0$ in $L^2(\Omega)$. Hence v_α can be identified with $\overline{v}_\alpha$ by Thm. 4.2-1 (a); but

$$v_\alpha^n \to v_\alpha \text{ in } L^2(\Omega) \Longrightarrow \overline{v}_\alpha^n \to \overline{v}_\alpha \text{ in } L^2(\omega)$$

on the one hand and $\overline{v}_\alpha = 0$ on the other. Hence $v_\alpha^n \to 0$ in $L^2(\Omega)$.

Finally, we show that

$$v_3^n \to 0 \text{ in } L^2(\Omega).$$

By the assumed convergence $e_{3\|3}(\varepsilon_n; \boldsymbol{v}^n) \to 0$ in $L^2(\Omega)$,

$$\partial_3 v_3^n = \varepsilon_n e_{3\|3}(\varepsilon_n; \boldsymbol{v}^n) \to 0 \text{ in } L^2(\Omega)$$

on the one hand and $\overline{v}_3^n \to 0$ in $L^2(\omega)$ on the other. The convergence $v_3^n \to 0$ in $L^2(\Omega)$ then follows from Thm. 4.2-1 (c). We have therefore reached a contradiction, and the proof is complete. ■

4.4. CONVERGENCE OF THE SCALED DISPLACEMENTS AS $\varepsilon \to 0$

We now establish the main results of this chapter: Consider a *family of linearly elastic elliptic membrane shells with thickness* $2\varepsilon > 0$ with each having the *same* middle surface $S = \boldsymbol{\theta}(\overline{\omega})$. Then *the solutions* $\boldsymbol{u}(\varepsilon)$ *of the associated scaled three-dimensional problems* $\mathcal{P}(\varepsilon; \Omega)$ (Thm. 4.1-1) *converge in* $H^1(\Omega) \times H^1(\Omega) \times L^2(\Omega)$ *as* $\varepsilon \to 0$ *toward a limit* $\boldsymbol{u}$ *and this limit, which is independent of the transverse variable* x_3, *can be identified with the solution* $\overline{\boldsymbol{u}}$ *of a two-dimensional variational problem posed over the set* ω. This limit problem will be later identified (Thm. 4.5-1) as the scaled *two-dimensional variational problem of an elliptic membrane shell.*

The functions $\gamma_{\alpha\beta}(\boldsymbol{\eta})$ and $a^{\alpha\beta\sigma\tau}$ used in the next theorem respectively represent the covariant components of the **linearized change of metric tensor** associated with a displacement field $\eta_i \boldsymbol{a}^i$ of the middle surface S (Sect. 2.4) and the contravariant components of the **scaled two-dimensional elasticity tensor of the shell**; we recall that $\varepsilon_0 > 0$ is defined in Thm. 3.1-1.

The following result is due to Ciarlet & Lods [1996b, Thm. 5.1].

Theorem 4.4-1. *Assume that $\boldsymbol{\theta} \in \mathcal{C}^3(\overline{\omega}; \mathbb{R}^3)$. Consider a family of linearly elastic elliptic membrane shells with thickness 2ε approaching zero and with each having the same elliptic middle surface $S = \boldsymbol{\theta}(\overline{\omega})$, and let the assumptions on the data be as in* Sect. 4.1.

Let $\boldsymbol{u}(\varepsilon)$ denote for $0 < \varepsilon \le \varepsilon_0$ the solution of the associated scaled three-dimensional problems $\mathcal{P}(\varepsilon; \Omega)$ (Thm. 4.1-1). *Then there exist functions $u_\alpha \in H^1(\Omega)$ satisfying $u_\alpha = 0$ on $\gamma \times [-1, 1]$ and a function $u_3 \in L^2(\Omega)$ such that:*

$$\begin{gathered} u_\alpha(\varepsilon) \to u_\alpha \text{ in } H^1(\Omega) \text{ and } u_3(\varepsilon) \to u_3 \text{ in } L^2(\Omega) \text{ as } \varepsilon \to 0, \\ \boldsymbol{u} := (u_i) \text{ is independent of the transverse variable } x_3. \end{gathered}$$

Furthermore, the average

$$\overline{\boldsymbol{u}} := \frac{1}{2}\int_{-1}^{1} \boldsymbol{u}\, dx_3$$

satisfies the following **scaled variational two-dimensional problem $\mathcal{P}_M(\omega)$ of a linearly elastic elliptic membrane shell:**

$$\begin{aligned} &\overline{\boldsymbol{u}} = (\overline{u}_i) \in \mathbf{V}_M(\omega) := H_0^1(\omega) \times H_0^1(\omega) \times L^2(\omega), \\ &\int_\omega a^{\alpha\beta\sigma\tau}\gamma_{\sigma\tau}(\overline{\boldsymbol{u}})\gamma_{\alpha\beta}(\boldsymbol{\eta})\sqrt{a}\, dy = \int_\omega p^i \eta_i \sqrt{a}\, dy \\ &\qquad\qquad\qquad\qquad \text{for all } \boldsymbol{\eta} = (\eta_i) \in \mathbf{V}_M(\omega), \end{aligned}$$

where (the definitions of the functions $\Gamma^\sigma_{\alpha\beta}$, $b_{\alpha\beta}$, $a^{\alpha\beta}$, and a are recalled in Sect. 3.1):

$$\gamma_{\alpha\beta}(\boldsymbol{\eta}) := \frac{1}{2}(\partial_\beta \eta_\alpha + \partial_\alpha \eta_\beta) - \Gamma^\sigma_{\alpha\beta}\eta_\sigma - b_{\alpha\beta}\eta_3,$$

$$a^{\alpha\beta\sigma\tau} := \frac{4\lambda\mu}{\lambda+2\mu}a^{\alpha\beta}a^{\sigma\tau} + 2\mu(a^{\alpha\sigma}a^{\beta\tau} + a^{\alpha\tau}a^{\beta\sigma}),$$

$$p^i := \int_{-1}^{1} f^i\,dx_3 + h^i_+ + h^i_- \text{ and } h^i_\pm := h^i(\cdot, \pm 1).$$

Proof. For the sake of clarity, the proof is divided into ten parts, numbered (i) to (x). As it essentially involves technicalities, the consideration of surface forces is postponed until part (ix); in other words, *we assume in parts* (i) *to* (viii) *that the scaled unknown* $\boldsymbol{u}(\varepsilon)$ *satisfies the following variational problem* $\mathcal{P}(\varepsilon; \Omega)$:

$$\boldsymbol{u}(\varepsilon) \in \mathbf{V}(\Omega) := \{\boldsymbol{v} = (v_i) \in \mathbf{H}^1(\Omega);\ \boldsymbol{v} = \mathbf{0} \text{ on } \gamma \times [-1, 1]\},$$

$$\int_\Omega A^{ijkl}(\varepsilon)e_{k\|l}(\varepsilon; \boldsymbol{u}(\varepsilon))e_{i\|j}(\varepsilon; \boldsymbol{v})\sqrt{g(\varepsilon)}\,dx = \int_\Omega f^i v_i \sqrt{g(\varepsilon)}\,dx$$

for all $\boldsymbol{v} \in \mathbf{V}(\Omega)$. For notational brevity, we let

$$e_{i\|j}(\varepsilon) := e_{i\|j}(\varepsilon; \boldsymbol{u}(\varepsilon))$$

throughout the proof.

(i) A priori *bounds and extraction of weakly convergent sequences: The norms* $|e_{i\|j}(\varepsilon)|_{0,\Omega}$, $\|u_\alpha(\varepsilon)\|_{1,\Omega}$, *and* $|u_3(\varepsilon)|_{0,\Omega}$ *are bounded independently of* $0 < \varepsilon \le \varepsilon_1$, *where* $\varepsilon_1 > 0$ *is given by* Thm. 4.3-1. *Consequently, there exists a subsequence, still denoted* $(\boldsymbol{u}(\varepsilon))_{\varepsilon>0}$ *for convenience, and there exist functions* $e_{i\|j} \in L^2(\Omega)$, $u_\alpha \in H^1(\Omega)$ *satisfying* $u_\alpha = 0$ *on* $\gamma \times [-1, 1]$, *and* $u_3 \in L^2(\Omega)$ *such that* (we recall that $\to$ and $\rightharpoonup$ respectively denote strong and weak convergences):

$$e_{i\|j}(\varepsilon) \rightharpoonup e_{i\|j} \text{ in } L^2(\Omega),$$

$$u_\alpha(\varepsilon) \rightharpoonup u_\alpha \text{ in } H^1(\Omega) \text{ and thus } u_\alpha(\varepsilon) \to u_\alpha \text{ in } L^2(\Omega),$$

$$u_3(\varepsilon) \rightharpoonup u_3 \text{ in } L^2(\Omega).$$

From the variational equations satisfied by the scaled unknown $\boldsymbol{u}(\varepsilon)$ (Thm. 4.1-1), the asymptotic behavior of the function $g(\varepsilon)$

(Thm. 3.3-1), the uniform positive definiteness of the scaled three-dimensional elasticity tensor (Thm. 3.3-2), and the *three-dimensional inequality of Korn's type for an elliptic membrane shell* (Thm. 4.3-1), we infer that

$$\begin{aligned} C_M^{-2} \sum_i |u_i(\varepsilon)|_{0,\Omega}^2 &\leq C_M^{-2} \Big(\sum_\alpha \|u_\alpha(\varepsilon)\|_{1,\Omega}^2 + |u_3(\varepsilon)|_{0,\Omega}^2 \Big) \\ &\leq \sum_{i,j} |e_{i\|j}(\varepsilon)|_{0,\Omega}^2 \leq C_e g_0^{-1/2} \int_\Omega A^{ijkl}(\varepsilon) e_{k\|l}(\varepsilon) e_{i\|j}(\varepsilon) \sqrt{g(\varepsilon)}\, dx \\ &= C_e g_0^{-1/2} \int_\Omega f^i u_i(\varepsilon) \sqrt{g(\varepsilon)}\, dx \\ &\leq C_e g_0^{-1/2} g_1^{1/2} \Big\{ \sum_i |f^i|_{0,\Omega}^2 \Big\}^{1/2} \Big\{ \sum_i |u_i(\varepsilon)|_{0,\Omega}^2 \Big\}^{1/2}; \end{aligned}$$

hence all the assertions of (i) follow.

(ii) *The limits u_i found in* (i) *are independent of x_3.*

By (i),

$$\partial_3 u_\alpha(\varepsilon) + \varepsilon \partial_\alpha u_3(\varepsilon) = 2\varepsilon \left\{ e_{\alpha\|3}(\varepsilon) + \Gamma_{\alpha 3}^\sigma(\varepsilon) u_\sigma(\varepsilon) \right\} \to 0 \text{ in } L^2(\Omega),$$

since the functions $\Gamma_{\alpha 3}^\sigma(\varepsilon)$ converge in $\mathcal{C}^0(\overline{\Omega})$ (Thm. 3.3-1). Let there be given $\varphi \in \mathcal{D}(\Omega)$; since $u_\alpha(\varepsilon) \rightharpoonup u_\alpha$ in $H^1(\Omega)$ and $(u_3(\varepsilon))_{\varepsilon>0}$ is bounded in $L^2(\Omega)$ by (i),

$$\int_\Omega \partial_3 u_\alpha \, \varphi \, dx = \lim_{\varepsilon \to 0} \int_\Omega \partial_3 u_\alpha(\varepsilon)\, \varphi \, dx,$$

$$\lim_{\varepsilon \to 0} \left\{ \int_\Omega \varepsilon \partial_\alpha u_3(\varepsilon)\, \varphi \, dx \right\} = - \lim_{\varepsilon \to 0} \left\{ \int_\Omega \varepsilon u_3(\varepsilon) \partial_\alpha \varphi \, dx \right\} = 0;$$

hence

$$0 = \lim_{\varepsilon \to 0} \left\{ \int_\Omega \big(\partial_3 u_\alpha(\varepsilon) + \varepsilon \partial_\alpha u_3(\varepsilon) \big) \varphi \, dx \right\} = \int_\Omega \partial_3 u_\alpha \, \varphi \, dx,$$

which means that $\partial_3 u_\alpha = 0$ in $L^2(\Omega)$. Part (i) likewise implies that

$$\partial_3 u_3(\varepsilon) = \varepsilon e_{3\|3}(\varepsilon) \to 0 \text{ in } L^2(\Omega).$$

Let $\varphi \in \mathcal{D}(\Omega)$; since $u_3(\varepsilon) \rightharpoonup u_3$ in $L^2(\Omega)$ by (i),

$$\int_\Omega u_3 \partial_3 \varphi \, dx = \lim_{\varepsilon \to 0} \int_\Omega u_3(\varepsilon) \partial_3 \varphi \, dx = - \lim_{\varepsilon \to 0} \int_\Omega \partial_3 u_3(\varepsilon)\, \varphi \, dx = 0,$$

which means that $\partial_3 u_3 = 0$ in the sense of distributions; it then suffices to apply Thm. 4.2-1 (a).

(iii) *The limits* $e_{i\|j}$ *found in* (i) *are independent of* x_3*; they are moreover related to the limit* $\boldsymbol{u} := (u_i)$ *by*

$$\begin{aligned} e_{\alpha\|\beta} &= \gamma_{\alpha\beta}(\boldsymbol{u}) := \frac{1}{2}(\partial_\alpha u_\beta + \partial_\beta u_\alpha) - \Gamma^\sigma_{\alpha\beta} u_\sigma - b_{\alpha\beta} u_3, \\ e_{\alpha\|3} &= 0, \\ e_{3\|3} &= -\frac{\lambda}{\lambda + 2\mu} a^{\alpha\beta} e_{\alpha\|\beta}. \end{aligned}$$

The convergences $e_{\alpha\|\beta}(\varepsilon) \rightharpoonup e_{\alpha\|\beta}$ in $L^2(\Omega)$, $u_\alpha(\varepsilon) \rightharpoonup u_\alpha$ in $H^1(\Omega)$, $u_3(\varepsilon) \rightharpoonup u_3$ in $L^2(\Omega)$ (part (i)), $\Gamma^\sigma_{\alpha\beta}(\varepsilon) \to \Gamma^\sigma_{\alpha\beta}$ and $\Gamma^3_{\alpha\beta}(\varepsilon) \to b_{\alpha\beta}$ in $\mathcal{C}^0(\overline{\Omega})$ (Thm. 3.3-1) together imply that

$$e_{\alpha\|\beta}(\varepsilon) = \frac{1}{2}(\partial_\alpha u_\beta(\varepsilon) + \partial_\beta u_\alpha(\varepsilon)) - \Gamma^p_{\alpha\beta}(\varepsilon) u_p(\varepsilon) \rightharpoonup \gamma_{\alpha\beta}(\boldsymbol{u}) = e_{\alpha\|\beta}$$

in $L^2(\Omega)$. Besides, the functions $e_{\alpha\|\beta}$ are independent of x_3 since the functions u_i are independent of x_3 by (ii).

Let $\boldsymbol{v} = (v_i)$ be an arbitrary function in the space

$$\mathbf{V}(\Omega) = \left\{\boldsymbol{v} = (v_i) \in \mathbf{H}^1(\Omega);\ \boldsymbol{v} = \mathbf{0} \text{ on } \gamma \times [-1, 1]\right\}.$$

The following relations are immediate consequences of the definitions of the scaled strains $e_{i\|j}(\varepsilon; \boldsymbol{v})$:

$$\begin{aligned} \varepsilon e_{\alpha\|\beta}(\varepsilon; \boldsymbol{v}) &\to 0 \text{ in } L^2(\Omega), \\ \varepsilon e_{\alpha\|3}(\varepsilon; \boldsymbol{v}) &\to \frac{1}{2}\partial_3 v_\alpha \text{ in } L^2(\Omega), \\ \varepsilon e_{3\|3}(\varepsilon; \boldsymbol{v}) &= \partial_3 v_3 \text{ for all } \varepsilon > 0. \end{aligned}$$

Using the variational equations of the three-dimensional problem $\mathcal{P}(\varepsilon; \Omega)$ (Thm. 4.1-1) and the relations $A^{\alpha\beta\sigma 3}(\varepsilon) = A^{\alpha 333}(\varepsilon) = 0$ (Thm. 3.3-1), we obtain

$$\begin{aligned} &\int_\Omega A^{ijkl}(\varepsilon)\left\{\varepsilon e_{k\|l}(\varepsilon) e_{i\|j}(\varepsilon; \boldsymbol{v})\right\}\sqrt{g(\varepsilon)}\,dx \\ &= \int_\Omega \left\{A^{\alpha\beta\sigma\tau}(\varepsilon) e_{\sigma\|\tau}(\varepsilon) + A^{\alpha\beta 33}(\varepsilon) e_{3\|3}(\varepsilon)\right\}\left\{\varepsilon e_{\alpha\|\beta}(\varepsilon; \boldsymbol{v})\right\}\sqrt{g(\varepsilon)}\,dx \end{aligned}$$

$$+\int_\Omega \{4A^{\alpha 3\sigma 3}(\varepsilon)e_{\sigma\|3}(\varepsilon)\}\{\varepsilon e_{\alpha\|3}(\varepsilon;\, \boldsymbol{v})\}\sqrt{g(\varepsilon)}\, dx$$

$$+\int_\Omega \{A^{33\sigma\tau}(\varepsilon)e_{\sigma\|\tau}(\varepsilon)+A^{3333}(\varepsilon)e_{3\|3}(\varepsilon)\}\{\varepsilon e_{3\|3}(\varepsilon;\, \boldsymbol{v})\}\sqrt{g(\varepsilon)}\, dx$$

$$=\varepsilon\int_\Omega f^i v_i\sqrt{g(\varepsilon)}\, dx.$$

Keep $\boldsymbol{v}\in\mathbf{V}(\Omega)$ *fixed* and let $\varepsilon\to 0$. The asymptotic behavior of the functions $\varepsilon e_{i\|j}(\varepsilon;\, \boldsymbol{v})$, the asymptotic behavior of the functions $A^{ijkl}(\varepsilon)$ and $g(\varepsilon)$ (Thms. 3.3-1 and 3.3-2), and the weak convergences $e_{i\|j}(\varepsilon)\rightharpoonup e_{i\|j}$ in $L^2(\Omega)$ (part (i)) together imply that:

$$\int_\Omega \Big\{2\mu a^{\alpha\sigma}e_{\sigma\|3}\partial_3 v_\alpha+\big(\lambda a^{\sigma\tau}e_{\sigma\|\tau}+(\lambda+2\mu)e_{3\|3}\big)\partial_3 v_3\Big\}\sqrt{a}\, dx=0.$$

Letting $\boldsymbol{v}$ *vary* in $\mathbf{V}(\Omega)$ then yields the relations satisfied by the limits $e_{i\|3}$ (if $w\in L^2(\Omega)$ satisfies $\int_\Omega w\partial_3 v\, dx=0$ for all $v\in H^1(\Omega)$ that vanish on $\gamma\times[-1,\, 1]$, then $w=0$; cf. Thm. 3.4-1).

(iv) *The function* $\overline{\boldsymbol{u}}:=(\overline{u}_i)$ *satisfies the two-dimensional variational problem* $\mathcal{P}_M(\omega)$ *described in the statement of the theorem, with* $p^i:=\int_{-1}^1 f^i\, dx_3$ (recall that $h^i=0$ for the time being). *Since the solution to this problem is unique* (Thms. 2.7-2 and 3.3-2), *the convergences of* part (i) *hold in fact for the whole family* $(\boldsymbol{u}(\varepsilon))_{\varepsilon>0}$ (if the functions $\overline{u}_i$ are unique, so are the functions u_i and $e_{i\|j}$ by parts (ii) and (iii)).

That $\overline{\boldsymbol{u}}\in\mathbf{V}_M(\omega)$ follows from Thm. 4.2-1 (parts (a) and (b)). Let $\boldsymbol{v}=(v_i)\in\mathbf{V}(\Omega)$ be *independent of the variable* x_3; then the asymptotic behavior of the functions $\Gamma^p_{\alpha\beta}(\varepsilon)$ and $\Gamma^\sigma_{\alpha 3}(\varepsilon)$ (Thm. 3.3-1) shows that

$$e_{\alpha\|\beta}(\varepsilon;\, \boldsymbol{v})\to\gamma_{\alpha\beta}(\boldsymbol{v}):=\frac{1}{2}(\partial_\alpha v_\beta+\partial_\beta v_\alpha)-\Gamma^\sigma_{\alpha\beta}v_\sigma-b_{\alpha\beta}v_3 \text{ in } L^2(\Omega),$$

$$e_{\alpha\|3}(\varepsilon;\, \boldsymbol{v})\to\left\{\frac{1}{2}\partial_\alpha v_3+b^\sigma_\alpha v_\sigma\right\} \text{ in } L^2(\Omega),$$

$$e_{3\|3}(\varepsilon;\, \boldsymbol{v})=0,$$

as $\varepsilon\to 0$. Keep such a function $\boldsymbol{v}\in\mathbf{V}(\Omega)$ *fixed* in the variational equations of problem $\mathcal{P}(\varepsilon;\, \Omega)$ and let $\varepsilon\to 0$. The asymptotic behavior of the functions $e_{i\|j}(\varepsilon;\, \boldsymbol{v})$, the asymptotic behavior of the functions

$A^{ijkl}(\varepsilon)$ and $g(\varepsilon)$ (Thms. 3.3-1 and 3.3-2), the weak convergences $e_{i\|j}(\varepsilon) \rightharpoonup e_{i\|j}$ in $L^2(\Omega)$ (part (i)), and the relations satisfied by the limits $e_{i\|j}$ (part (iii)) together yield the equations

$$\int_\Omega \left\{ \frac{2\lambda\mu}{\lambda+2\mu} a^{\alpha\beta} a^{\sigma\tau} + \mu\left(a^{\alpha\sigma}a^{\beta\tau} + a^{\alpha\tau}a^{\beta\sigma}\right) \right\} e_{\sigma\|\tau} \gamma_{\alpha\beta}(\boldsymbol{v})\sqrt{a}\,dx$$
$$= \int_\Omega f^i v_i \sqrt{a}\,dx,$$

which we may also write as (both functions $\boldsymbol{u}$ and $\boldsymbol{v}$ are independent of x_3):

$$\int_\omega a^{\alpha\beta\sigma\tau}\gamma_{\sigma\tau}(\overline{\boldsymbol{u}})\gamma_{\alpha\beta}(\overline{\boldsymbol{v}})\sqrt{a}\,dy = \int_\omega \left\{ \int_{-1}^{1} f^i\,dx_3 \right\} \overline{v}_i \sqrt{a}\,dy,$$

where the functions $\gamma_{\alpha\beta}(\boldsymbol{\eta})$ and $a^{\alpha\beta\sigma\tau}$ are defined in the theorem.

Given $\boldsymbol{\eta} = (\eta_i) \in \mathbf{H}_0^1(\omega)$, let $\boldsymbol{v} = (v_i)$ be defined by

$$\boldsymbol{v}(y, x_3) = \boldsymbol{\eta}(y) \text{ for } (y, x_3) \in \Omega.$$

Then $\boldsymbol{v} \in \mathbf{V}(\Omega)$ and $\boldsymbol{v}$ is independent of x_3; hence by Thm. 4.2-1 (b), the above variational equations are satisfied with $\overline{\boldsymbol{v}} = \boldsymbol{\eta}$. Since both sides of these equations are continuous linear forms with respect to $\overline{v}_3 = \eta_3 \in L^2(\omega)$ for fixed $\overline{v}_\alpha \in H_0^1(\omega)$, these equations are valid for all $\boldsymbol{\eta} \in \mathbf{V}_M(\omega) = H_0^1(\omega) \times H_0^1(\omega) \times L^2(\omega)$, since $H_0^1(\omega)$ is dense in $L^2(\omega)$.

(v) *The weak convergences* $e_{i\|j}(\varepsilon) \rightharpoonup e_{i\|j}$ *in* $L^2(\Omega)$ *established in* part (i) *are in fact strong:*

$$e_{i\|j}(\varepsilon) \to e_{i\|j} \text{ in } L^2(\Omega).$$

Combining the uniform positive definiteness of the scaled elasticity tensor and the inequalities $0 < g_0 \le g(\varepsilon)(x)$, $x \in \overline{\Omega}$ (Thms. 3.3-1 and 3.3-2) with the variational equations of problem $\mathcal{P}(\varepsilon; \Omega)$ (again, recall that $h^i = 0$ for the time being) where we let $\boldsymbol{v} = \boldsymbol{u}(\varepsilon)$, we first infer that

$$C_e^{-1} g_0^{1/2} \sum_{i,j} |e_{i\|j}(\varepsilon) - e_{i\|j}|_{0,\Omega}^2 \le \Lambda(\varepsilon),$$

where

$$\Lambda(\varepsilon) := \int_\Omega A^{ijkl}(\varepsilon)\left(e_{k\|l}(\varepsilon) - e_{k\|l}\right)\left(e_{i\|j}(\varepsilon) - e_{i\|j}\right)\sqrt{g(\varepsilon)}\,dx$$

$$= \int_\Omega f^i u_i(\varepsilon)\sqrt{g(\varepsilon)}\,dx - \int_\Omega A^{ijkl}(\varepsilon)\big(2e_{k\|l}(\varepsilon) - e_{k\|l}\big)e_{i\|j}\sqrt{g(\varepsilon)}\,dx.$$

Using the weak convergences established in part (i) and the asymptotic behavior of the functions $A^{ijkl}(\varepsilon)$ and $g(\varepsilon)$, we next have

$$\Lambda := \lim_{\varepsilon\to 0}\Lambda(\varepsilon) = \int_\Omega f^i u_i\sqrt{a}\,dx - \int_\Omega A^{ijkl}(0)e_{k\|l}e_{i\|j}\sqrt{a}\,dx.$$

Using the expressions of the functions $A^{ijkl}(0)$ (Thm. 3.3-2) and $e_{i\|3}$ (part (iii)), we finally obtain

$$\begin{aligned}
&\int_\Omega A^{ijkl}(0)e_{k\|l}e_{i\|j}\sqrt{a}\,dx \\
&= \int_\Omega \big\{\lambda a^{\alpha\beta}a^{\sigma\tau} + \mu\big(a^{\alpha\sigma}a^{\beta\tau} + a^{\alpha\tau}a^{\beta\sigma}\big)\big\}e_{\sigma\|\tau}e_{\alpha\|\beta}\sqrt{a}\,dx \\
&\quad + \int_\Omega \lambda a^{\alpha\beta}e_{3\|3}e_{\alpha\|\beta}\sqrt{a}\,dx + \int_\Omega 4\mu a^{\alpha\sigma}e_{\sigma\|3}e_{\alpha\|3}\sqrt{a}\,dx \\
&\quad + \int_\Omega \big\{\lambda a^{\sigma\tau}e_{\sigma\|\tau} + (\lambda+2\mu)e_{3\|3}\big\}e_{3\|3}\sqrt{a}\,dx \\
&= \int_\Omega \Big\{\frac{2\lambda\mu}{\lambda+2\mu}a^{\alpha\beta}a^{\sigma\tau} + \mu\big(a^{\alpha\sigma}a^{\beta\tau} + a^{\alpha\tau}a^{\beta\sigma}\big)\Big\}e_{\sigma\|\tau}e_{\alpha\|\beta}\sqrt{a}\,dx.
\end{aligned}$$

Letting $\boldsymbol{\eta} = \overline{\boldsymbol{u}}$ in the variational equations satisfied by $\overline{\boldsymbol{u}}$ (part (iv)) and using the expression of the function $e_{\alpha\|\beta}$ found in part (iii), we conclude that

$$\Lambda = 0.$$

Consequently, the strong convergences $e_{i\|j}(\varepsilon) \to e_{i\|j}$ in $L^2(\Omega)$ are established.

(vi) *The family* $(\overline{\boldsymbol{u}}(\varepsilon))_{\varepsilon>0}$ *converges strongly to* $\overline{\boldsymbol{u}}$ *in the space* $\mathbf{V}_M(\omega)$, *i.e.*,

$$\overline{u}_\alpha(\varepsilon) \to \overline{u}_\alpha \text{ in } H^1(\omega) \text{ and } \overline{u}_3(\varepsilon) \to \overline{u}_3 \text{ in } L^2(\omega).$$

By virtue of the *inequality of Korn's type on an elliptic surface* (Thm. 2.7-3), proving these strong convergences is equivalent to proving that

$$\gamma_{\alpha\beta}(\overline{\boldsymbol{u}}(\varepsilon)) \to \gamma_{\alpha\beta}(\overline{\boldsymbol{u}}) = \overline{e_{\alpha\|\beta}} \text{ in } L^2(\omega),$$

since $\gamma_{\alpha\beta}(\boldsymbol{u}) = e_{\alpha\|\beta}$ by (iii). But, since $e_{\alpha\|\beta}(\varepsilon) \to e_{\alpha\|\beta}$ in $L^2(\Omega)$ by part (v), we infer from Thm. 4.2-1 (a) that

$$\overline{e_{\alpha\|\beta}(\varepsilon)} \to \overline{e_{\alpha\|\beta}} \text{ in } L^2(\omega),$$

on the one hand, and we infer from Thm. 4.2-1 (d) that

$$\left(\overline{e_{\alpha\|\beta}(\varepsilon)} - \gamma_{\alpha\beta}(\overline{\boldsymbol{u}}(\varepsilon))\right) \to 0 \text{ in } L^2(\omega),$$

on the other (recall that $e_{\alpha\|\beta}(\varepsilon) := e_{\alpha\|\beta}(\varepsilon; \boldsymbol{u}(\varepsilon))$). Hence the announced strong convergences hold.

(vii) *The weak convergence* $u_3(\varepsilon) \rightharpoonup u_3$ *in* $L^2(\Omega)$ *established in* part (i) *is in fact strong*:

$$u_3(\varepsilon) \to u_3 \text{ in } L^2(\Omega).$$

First, we have $\partial_3 u_3(\varepsilon) = \varepsilon e_{3\|3}(\varepsilon) \to 0$ in $L^2(\Omega)$; second, we have already shown in (vi) that $\overline{u}_3(\varepsilon) \to \overline{u}_3$ in $L^2(\omega)$. Hence the conclusion follows from Thm. 4.2-1 (c) and from the independence of the function u_3 with respect to the transverse variable x_3 (cf. (ii)).

(viii) It remains to show that *the weak convergences* $u_\alpha(\varepsilon) \rightharpoonup u_\alpha$ *in* $H^1(\Omega)$ *established in* part (i) *are strong*, i.e., that

$$u_\alpha(\varepsilon) \to u_\alpha \text{ in } H^1(\Omega).$$

To this end, we observe that, by the classical *three-dimensional Korn inequality in Cartesian coordinates*, proving these strong convergences is equivalent to proving that

$$e_{ij}(\boldsymbol{u}'(\varepsilon)) \to e_{ij}(\boldsymbol{u}') \text{ in } L^2(\Omega),$$

where $e_{ij}(\boldsymbol{v}) := \frac{1}{2}(\partial_j v_i + \partial_i v_j)$ and

$$\boldsymbol{u}'(\varepsilon) := (u_1(\varepsilon),\, u_2(\varepsilon),\, 0), \quad \boldsymbol{u}' := (u_1,\, u_2,\, 0).$$

Since $e_{\alpha\|\beta}(\varepsilon) = e_{\alpha\beta}(\boldsymbol{u}'(\varepsilon)) - \Gamma^p_{\alpha\beta}(\varepsilon) u_p(\varepsilon)$, parts (iii) and (v) imply that

$$e_{\alpha\|\beta}(\varepsilon) \to \{e_{\alpha\beta}(\boldsymbol{u}') - \Gamma^\sigma_{\alpha\beta} u_\sigma - b_{\alpha\beta} u_3\} = e_{\alpha\|\beta} \text{ in } L^2(\Omega).$$

The asymptotic behavior of the functions $\Gamma^p_{\alpha\beta}(\varepsilon)$ and the strong convergences $u_i(\varepsilon) \to u_i$ in $L^2(\Omega)$ established in parts (i) and (vii) together imply that

$$e_{\alpha\beta}(\boldsymbol{u}'(\varepsilon)) \to e_{\alpha\beta}(\boldsymbol{u}') \text{ in } L^2(\Omega).$$

Notice in passing that we do not need here the strong convergences $\overline{u}_\alpha(\varepsilon) \to \overline{u}_\alpha$ in $H^1(\omega)$, but that we definitely need the strong convergence $\overline{u}_3(\varepsilon) \to \overline{u}_3$ (all these convergences have been established in part (vi)), which in turn implied in (vii) the strong convergence $u_3(\varepsilon) \to u_3$ in $L^2(\Omega)$.

Since $e_{33}(\boldsymbol{u}'(\varepsilon)) = e_{33}(\boldsymbol{u}') = 0$, it suffices to show that ($\partial_3 u_\alpha = 0$ by (ii))

$$\partial_3 u_\alpha(\varepsilon) = 2e_{\alpha 3}(\boldsymbol{u}'(\varepsilon)) \to 0 = 2e_{\alpha 3}(\boldsymbol{u}') \text{ in } L^2(\Omega),$$

or equivalently, that

$$\partial_3 u_\alpha(\varepsilon) \to 0 \text{ in } H^{-1}(\Omega) \text{ and } \partial_i \partial_3 u_\alpha(\varepsilon) \to 0 \text{ in } H^{-1}(\Omega).$$

The equivalence between the last two relations is yet another consequence of the *lemma of J.L. Lions* (Thm. 1.7-1), which, together with the closed graph theorem, implies that *the mapping*

$$v \in L^2(\Omega) \to (v, \partial_1 v, \partial_2 v, \partial_3 v) \in \mathbf{H}^{-1}(\Omega)$$

is an isomorphism.

Since $\partial_3 u_\alpha(\varepsilon) = 2\varepsilon(e_{\alpha\|3}(\varepsilon) + \Gamma^\sigma_{\alpha 3}(\varepsilon)u_\sigma(\varepsilon)) - \varepsilon\partial_\alpha u_3(\varepsilon)$, we first have, for all $\varphi \in \mathcal{D}(\Omega)$,

$$\int_\Omega \partial_3 u_\alpha(\varepsilon)\varphi \, dx = \varepsilon \int_\Omega \left\{2\left(e_{\alpha\|3}(\varepsilon) + \Gamma^\sigma_{\alpha 3}(\varepsilon)u_\sigma(\varepsilon)\right)\varphi + u_3(\varepsilon)\partial_\alpha\varphi\right\} dx,$$

and consequently, there exists by (i) a constant c_1 independent of ε such that

$$\|\partial_3 u_\alpha(\varepsilon)\|_{-1,\Omega} \leq c_1\varepsilon,$$

where $\|\cdot\|_{-1,\Omega}$ denotes the norm in $H^{-1}(\Omega)$. Hence $\partial_3 u_\alpha(\varepsilon) \to 0$ in $H^{-1}(\Omega)$.

We next have the identity

$$\begin{aligned}\partial_\beta\partial_3 u_\alpha(\varepsilon) &= \partial_3 e_{\alpha\beta}(\boldsymbol{u}'(\varepsilon)) + \partial_\beta\left(\varepsilon e_{\alpha\|3}(\varepsilon) + \varepsilon\Gamma^\sigma_{\alpha 3}(\varepsilon)u_\sigma(\varepsilon)\right) \\ &\quad -\partial_\alpha\left(\varepsilon e_{\beta\|3}(\varepsilon) + \varepsilon\Gamma^\tau_{\beta 3}(\varepsilon)u_\tau(\varepsilon)\right)\end{aligned}$$

in $\mathcal{D}'(\Omega)$. The convergences $e_{\alpha\beta}(\boldsymbol{u}'(\varepsilon)) \to e_{\alpha\beta}(\boldsymbol{u})$ in $L^2(\Omega)$ therefore imply that

$$\partial_3 e_{\alpha\beta}(\boldsymbol{u}'(\varepsilon)) \to \partial_3 e_{\alpha\beta}(\boldsymbol{u}') = 0 \text{ in } H^{-1}(\Omega),$$

since $\partial_3\partial_\beta u_\alpha = 0$ in $\mathcal{D}'(\Omega)$. Denoting by $< \cdot, \cdot >$ the duality between $\mathcal{D}'(\Omega)$ and $\mathcal{D}(\Omega)$, we thus have, for all $\varphi \in \mathcal{D}(\Omega)$,

$$< \partial_\beta \left(\varepsilon e_{\alpha\|3}(\varepsilon) + \varepsilon\Gamma^\sigma_{\alpha 3}(\varepsilon)u_\sigma(\varepsilon)\right), \varphi >$$
$$= -\varepsilon \int_\Omega \left\{e_{\alpha\|3}(\varepsilon) + \Gamma^\sigma_{\alpha 3}(\varepsilon)u_\sigma(\varepsilon)\right\} \partial_\beta\varphi \, dx,$$

and consequently, there exists c_2 independent of ε such that

$$\|\partial_\beta \left(\varepsilon e_{\alpha\|3}(\varepsilon) + \varepsilon\Gamma^\sigma_{\alpha 3}(\varepsilon)u_\sigma(\varepsilon)\right)\|_{-1,\Omega} \le c_2\varepsilon,$$

by (i). The last term in the expression of $\partial_\beta\partial_3 u_\alpha(\varepsilon)$ is treated analogously. Hence $\partial_\beta\partial_3 u_\alpha(\varepsilon) \to 0$ in $H^{-1}(\Omega)$.

Finally, we have, for all $\varphi \in \mathcal{D}(\Omega)$,

$$< \partial_3\partial_3 u_\alpha(\varepsilon), \varphi > = -\int_\Omega \partial_3 u_\alpha(\varepsilon)\partial_3\varphi \, dx$$
$$= -2\varepsilon \int_\Omega \left\{e_{\alpha\|3}(\varepsilon) + \Gamma^\sigma_{\alpha 3}(\varepsilon)u_\sigma(\varepsilon)\right\} \partial_3\varphi \, dx + \varepsilon^2 \int_\Omega e_{3\|3}(\varepsilon)\partial_\alpha\varphi \, dx,$$

and consequently, there exists by (i) a constant c_3 independent of ε such that

$$\|\partial_3\partial_3 u_\alpha(\varepsilon)\|_{-1,\Omega} \le c_3\varepsilon.$$

Hence $\partial_3\partial_3 u_\alpha(\varepsilon) \to 0$ in $H^{-1}(\Omega)$, and the proof is complete when only body forces are considered.

(ix) *Let* $X(\Omega) := \{v \in L^2(\Omega);\ \partial_3 v \in L^2(\Omega)\}$ ($\partial_3 v$ is a derivative in the sense of distributions). *Then the trace $v(\cdot, s)$ of any function $v \in X(\Omega)$ is well defined as a function in $L^2(\omega)$ for all $s \in [-1, 1]$ and the trace operator defined in this fashion is continuous. In particular, there exists a constant c_4 such that*

$$\|v\|_{L^2(\Gamma_+\cup\Gamma_-)} \le c_4 \left\{|v|^2_{0,\Omega} + |\partial_3 v|^2_{0,\Omega}\right\}^{1/2}$$

for all $v \in X(\Omega)$; as a consequence, there exists a constant c_5 such that

$$\|v_3\|_{L^2(\Gamma_+\cup\Gamma_-)} \le c_5 \left\{ \sum_{i,j} |e_{i\|j}(\varepsilon;\, \boldsymbol{v})|^2_{0,\Omega} \right\}^{1/2}$$

for all $\boldsymbol{v} \in \mathbf{V}(\Omega)$ (these inequalities will afford the consideration of surface forces in part (x)):

Let $v \in X(\Omega)$. For almost all $y \in \omega$ and all $s \in [-1, 1]$, we can write (Ex. 4.1)

$$v(y, -1) = v(y, s) - \int_{-1}^{s} \partial_3 v(y, x_3)\, dx_3.$$

Consequently,

$$\begin{aligned}|v(\cdot, -1)|^2_{0,\omega} &\leq 2\int_\omega |v(y, s)|^2\, dy + 2\int_\omega \left|\int_{-1}^{s} \partial_3 v(y, x_3)\, dx_3\right|^2 dy \\ &\leq 2\int_\omega |v(y, s)|^2\, dy + 2\int_\omega (1+s)\left\{\int_{-1}^{s} |\partial_3 v(y, x_3)|^2\, dx_3\right\} dy \\ &\leq 2\int_\omega |v(y, s)|^2\, dy + 4|\partial_3 v|^2_{0,\Omega}.\end{aligned}$$

Integrating the last inequality with respect to $s \in [-1, 1]$ yields

$$\|v\|_{L^2(\Gamma_-)} \leq \left\{|v|^2_{0,\Omega} + 4|\partial_3 v|^2_{0,\Omega}\right\}^{1/2}.$$

Since $\|v\|_{L^2(\Gamma_+)}$ is likewise bounded, the first inequality is proved.

Let next $\boldsymbol{v} = (v_i) \in \mathbf{V}(\Omega)$. Then

$$\begin{aligned}\|v_3\|_{L^2(\Gamma_+ \cup \Gamma_-)} &\leq c_4 \left\{|v_3|^2_{0,\Omega} + |\partial_3 v_3|^2_{0,\Omega}\right\}^{1/2} \\ &\leq c_4 \left\{|v_3|^2_{0,\Omega} + |e_{3\|3}(\varepsilon; \boldsymbol{v})|^2_{0,\Omega}\right\}^{1/2} \\ &\leq c_4(C+1)\left\{\sum_{i,j} |e_{i\|j}(\varepsilon; \boldsymbol{v})|^2_{0,\Omega}\right\}^{1/2},\end{aligned}$$

by the *three-dimensional inequality of Korn's type for an elliptic membrane shell* (Thm. 4.3-1), and the second inequality is proved.

(x) *Consideration of surface forces*:

We henceforth assume that only surface forces act on the shells, i.e., the variational problem $\mathcal{P}(\varepsilon, \Omega)$ satisfied by the scaled unknown $\boldsymbol{u}(\varepsilon)$ now takes the form (cf. Thm. 4.1-1 where $f^i = 0$; recall that $e_{k\|l}(\varepsilon) := e_{k\|l}(\varepsilon; \boldsymbol{u}(\varepsilon))$):

$$\boldsymbol{u}(\varepsilon) \in \mathbf{V}(\Omega) = \left\{\boldsymbol{v} = (v_i) \in \mathbf{H}^1(\Omega);\ \boldsymbol{v} = \mathbf{0} \text{ on } \gamma \times [-1, 1]\right\},$$

$$\int_\Omega A^{ijkl}(\varepsilon) e_{k\|l}(\varepsilon) e_{i\|j}(\varepsilon; \boldsymbol{v}) \sqrt{g(\varepsilon)}\, dx = \int_{\Gamma_+ \cup \Gamma_-} h^i v_i \sqrt{g(\varepsilon)}\, d\Gamma \quad \text{for all } \boldsymbol{v} \in \mathbf{V}(\Omega).$$

Carrying out an asymptotic analysis of $\boldsymbol{u}(\varepsilon)$ as $\varepsilon \to 0$ involves the same eight steps as those corresponding to body forces (cf. parts (i) to (viii)). We only indicate the *modifications* needed for handling the new right-hand side in the variational equations of $\mathcal{P}(\varepsilon; \Omega)$.

The chain of inequalities that led in part (i) to the *a priori* bounds now reads, thanks to the inequality proved in (ix) and to the *three-dimensional inequality of Korn's type for an elliptic membrane shell:*

$$
\begin{aligned}
C_M^{-2}\Big\{ \sum_\alpha \|u_\alpha(\varepsilon)\|^2_{1,\Omega} + |u_3(\varepsilon)|^2_{0,\Omega} \Big\} &\le \sum_{i,j} |e_{i\|j}(\varepsilon)|^2_{0,\Omega} \\
&\le C_e g_0^{-1/2} \Big(\int_{\Gamma_+\cup\Gamma_-} h^\alpha u_\alpha(\varepsilon)\sqrt{g(\varepsilon)}\, d\Gamma \\
&\qquad + \int_{\Gamma_+\cup\Gamma_-} h^3 u_3(\varepsilon)\sqrt{g(\varepsilon)}\, d\Gamma \Big) \\
&\le C_e g_0^{-1/2} g_1^{1/2} \Big(\|h^\alpha\|_{L^2(\Gamma_+\cup\Gamma_-)} \|u_\alpha(\varepsilon)\|_{L^2(\Gamma_+\cup\Gamma_-)} \\
&\qquad + \|h^3\|_{L^2(\Gamma_+\cup\Gamma_-)} \|u_3(\varepsilon)\|_{L^2(\Gamma_+\cup\Gamma_-)} \Big) \\
&\le C_e g_0^{-1/2} g_1^{1/2} \Big(c_6 \Big\{ \sum_\alpha \|u_\alpha(\varepsilon)\|^2_{1,\Omega} \Big\}^{1/2} \\
&\qquad + c_5 \|h^3\|_{L^2(\Gamma_+\cup\Gamma_-)} \Big\{ \sum_{i,j} |e_{i\|j}(\varepsilon)|^2_{0,\Omega} \Big\}^{1/2} \Big) \\
&\le C_e g_0^{-1/2} g_1^{1/2} (c_6 C_M + c_7) \Big\{ \sum_{i,j} \|e_{i\|j}(\varepsilon)\|^2_{0,\Omega} \Big\}^{1/2},
\end{aligned}
$$

and thus the conclusions are the same. The constant c_6 depends on the norms $\|h^\alpha\|_{L^2(\Gamma_+\cup\Gamma_-)}$ and on the norm of the trace operator acting from $H^1(\Omega)$ into $L^2(\Gamma_+ \cup \Gamma_-)$; the constant c_7 depends on $\|h^3\|_{L^2(\Gamma_+\cup\Gamma_-)}$ and on the constant c_5 (cf. part (ix)).

Part (ii) is the same. In (iii), the right-hand side of the relation

$$
\int_\Omega A^{ijkl}(\varepsilon) e_{k\|l}(\varepsilon)\{\varepsilon e_{i\|j}(\varepsilon; \boldsymbol{v})\}\sqrt{g(\varepsilon)}\, dx = \varepsilon \int_{\Gamma_+\cup\Gamma_-} h^i v_i \sqrt{g(\varepsilon)}\, d\Gamma
$$

again converges to 0 as $\varepsilon \to 0$ if $\boldsymbol{v} \in \mathbf{V}(\Omega)$ is fixed, since $g(\varepsilon) \to a$ in $\mathcal{C}^0(\overline{\Omega})$ (Thm. 3.3-1). In part (iv), let again $\boldsymbol{v} = (v_i) \in \mathbf{V}(\Omega)$ be independent of x_3; then, thanks again to the convergence $g(\varepsilon) \to a$

in $\mathcal{C}^0(\overline{\Omega})$ and to Lebesgue's dominated convergence theorem,

$$\lim_{\varepsilon\to 0}\int_{\Gamma_+\cup\Gamma_-} h^i v_i\sqrt{g(\varepsilon)}\,d\Gamma = \int_{\Gamma_+\cup\Gamma_-} h^i v_i\sqrt{a}\,dy = \int_\omega \left(h^i_+ + h^i_-\right)\overline{v}_i\sqrt{a}\,dy.$$

The same denseness argument as in (iv) then shows that $\overline{\boldsymbol{u}}$ satisfies

$$\int_\omega a^{\alpha\beta\sigma\tau}\gamma_{\sigma\tau}(\overline{\boldsymbol{u}})\gamma_{\alpha\beta}(\boldsymbol{\eta})\sqrt{a}\,dy = \int_\omega \left(h^i_+ + h^i_-\right)\eta_i\sqrt{a}\,dy$$

$$\text{for all } \boldsymbol{\eta} = (\eta_i) \in \mathbf{V}_M(\omega).$$

In part (v), we now have

$$\Lambda(\varepsilon) = \int_{\Gamma_+\cup\Gamma_-} h^i u_i(\varepsilon)\sqrt{g(\varepsilon)}\,d\Gamma - \int_\Omega A^{ijkl}(\varepsilon)(2e_{k\|l}(\varepsilon) - e_{k\|l})e_{i\|j}\sqrt{g(\varepsilon)}\,dx.$$

Since a linear mapping that is strongly continuous is also continuous with respect to the weak topologies (see, e.g., Brezis [1983, p. 39]), $u_\alpha(\varepsilon) \rightharpoonup u_\alpha$ in $H^1(\Omega)$ implies $u_\alpha(\varepsilon) \rightharpoonup u_\alpha$ in $L^2(\Gamma_+\cup\Gamma_-)$, and thus

$$\begin{aligned}\lim_{\varepsilon\to 0}&\int_{\Gamma_+\cup\Gamma_-} h^\alpha u_\alpha(\varepsilon)\sqrt{g(\varepsilon)}\,d\Gamma\\ &= \lim_{\varepsilon\to 0}\left\{\int_{\Gamma_+\cup\Gamma_-} h^\alpha u_\alpha(\varepsilon)\left(\sqrt{g(\varepsilon)} - \sqrt{a}\right)d\Gamma + \int_{\Gamma_+\cup\Gamma_-} h^\alpha u_\alpha(\varepsilon)\sqrt{a}\,d\Gamma\right\}\\ &= \int_{\Gamma_+\cup\Gamma_-} h^\alpha u_\alpha\sqrt{a}\,d\Gamma = \int_\omega \left(h^\alpha_+ + h^\alpha_-\right)\overline{u}_\alpha\sqrt{a}\,dy,\end{aligned}$$

on the one hand.

For the same reason, the weak convergences $u_3(\varepsilon) \rightharpoonup u_3$ in $L^2(\Omega)$ and $\partial_3 u_3(\varepsilon) = \varepsilon e_{3\|3}(\varepsilon) \rightharpoonup 0 = \partial_3 u_3$ in $L^2(\Omega)$ (cf. (i) and (ii)) and the first inequality in (ix) together imply that $u_3(\varepsilon)(\cdot,\pm 1) \rightharpoonup u_3(\cdot,\pm 1)$ in $L^2(\omega)$. Therefore,

$$\begin{aligned}\lim_{\varepsilon\to 0}&\int_{\Gamma_+\cup\Gamma_-} h^3 u_3(\varepsilon)\sqrt{g(\varepsilon)}\,d\Gamma\\ &= \int_{\Gamma_+\cup\Gamma_-} h^3 u_3\sqrt{a}\,d\Gamma = \int_\omega \left(h^3_+ + h^3_-\right)\overline{u}_3\sqrt{a}\,dy,\end{aligned}$$

on the other.

In this fashion, we again conclude as in (v) that $\Lambda = 0$. The remaining steps (vi), (vii), and (viii) are unaltered. The proof is thus complete. ∎

Remark. The treatment of surface forces given here (cf. parts (ix) and (x)) is based on an idea of Lods [1995]. It affords a substantially simpler proof than the one originally given in Ciarlet & Lods [1996b, Sect. 6]. ∎

4.5. THE TWO-DIMENSIONAL EQUATIONS OF A LINEARLY ELASTIC ELLIPTIC MEMBRANE SHELL; EXISTENCE, UNIQUENESS, AND REGULARITY OF SOLUTIONS; FORMULATION AS A BOUNDARY VALUE PROBLEM

The next theorem recapitulates the definition and assembles the main properties of the limit two-dimensional problem $\mathcal{P}_M(\omega)$ found at the outcome of the asymptotic analysis carried out in Thm. 4.4-1. Note that $\mathcal{P}_M(\omega)$ is an *atypical variational problem* in that one of the unknowns, viz., ζ_3 "only" lies in the space $L^2(\omega)$.

The *existence* and *uniqueness* theory, which is quickly reviewed in this theorem, is expounded in detail in Sect. 2.7, where *ad hoc* references are also provided.

Theorem 4.5-1. *Let ω be a domain in $\mathbb{R}^2$ and let $\boldsymbol{\theta} \in \mathcal{C}^{2,1}(\overline{\omega}; \mathbb{R}^3)$ be an injective mapping such that the two vectors $\boldsymbol{a}_\alpha = \partial_\alpha \boldsymbol{\theta}$ are linearly independent at all points of $\overline{\omega}$ and such that the surface $S = \boldsymbol{\theta}(\overline{\omega})$ is elliptic.*

(a) *The associated* **scaled two-dimensional variational problem $\mathcal{P}_M(\omega)$ of a linearly elastic elliptic membrane shell:** *Given functions $p^i \in L^2(\omega)$, find $\boldsymbol{\zeta} = (\zeta_i)$ that satisfies*

$$\boxed{\begin{aligned} &\boldsymbol{\zeta} \in \mathbf{V}_M(\omega) := H_0^1(\omega) \times H_0^1(\omega) \times L^2(\omega), \\ &\int_\omega a^{\alpha\beta\sigma\tau} \gamma_{\sigma\tau}(\boldsymbol{\zeta}) \gamma_{\alpha\beta}(\boldsymbol{\eta}) \sqrt{a}\, dy = \int_\omega p^i \eta_i \sqrt{a}\, dy \\ &\qquad\qquad\qquad\qquad \text{for all } \boldsymbol{\eta} = (\eta_i) \in \mathbf{V}_M(\omega), \end{aligned}}$$

where

$$
\begin{aligned}
\gamma_{\alpha\beta}(\boldsymbol{\eta}) &:= \frac{1}{2}(\partial_\beta \eta_\alpha + \partial_\alpha \eta_\beta) - \Gamma^\sigma_{\alpha\beta}\eta_\sigma - b_{\alpha\beta}\eta_3, \\
a^{\alpha\beta\sigma\tau} &:= \frac{4\lambda\mu}{\lambda + 2\mu} a^{\alpha\beta} a^{\sigma\tau} + 2\mu\left(a^{\alpha\sigma} a^{\beta\tau} + a^{\alpha\tau} a^{\beta\sigma}\right),
\end{aligned}
$$

has one and only one solution, which is also the unique solution of the minimization problem: Find $\boldsymbol{\zeta}$ *such that*

$$
\begin{gathered}
\boldsymbol{\zeta} \in \mathbf{V}_M(\omega) \ \text{ and } \ j_M(\boldsymbol{\zeta}) = \inf_{\boldsymbol{\eta} \in \mathbf{V}_M(\omega)} j_M(\boldsymbol{\eta}), \text{ where} \\
j_M(\boldsymbol{\eta}) := \frac{1}{2}\int_\omega a^{\alpha\beta\sigma\tau}\gamma_{\sigma\tau}(\boldsymbol{\eta})\gamma_{\alpha\beta}(\boldsymbol{\eta})\sqrt{a}\,dy - \int_\omega p^i \eta_i \sqrt{a}\,dy,
\end{gathered}
$$

where the functional $j_M : \mathbf{V}_M(\omega)$ *is called the* **scaled two-dimensional energy of a linearly elastic elliptic membrane shell.**

(b) *If the solution* $\boldsymbol{\zeta} = (\zeta_i)$ *of* $\mathcal{P}_M(\omega)$ *is smooth enough, it also satisfies the boundary value problem:*

$$
\begin{aligned}
-n^{\alpha\beta}|_\beta &= p^\alpha \text{ in } \omega, \\
-b_{\alpha\beta} n^{\alpha\beta} &= p^3 \text{ in } \omega, \\
\zeta_\alpha &= 0 \text{ on } \gamma,
\end{aligned}
$$

where the functions

$$
n^{\alpha\beta} := a^{\alpha\beta\sigma\tau}\gamma_{\sigma\tau}(\boldsymbol{\zeta})
$$

are the **contravariant components** *of the* **scaled stress resultant tensor** *field, and the functions*

$$
n^{\alpha\beta}|_\sigma := \partial_\sigma n^{\alpha\beta} + \Gamma^\alpha_{\sigma\tau} n^{\beta\tau} + \Gamma^\beta_{\sigma\tau} n^{\alpha\tau}
$$

are **first-order covariant derivatives** *of this tensor field.*

(c) *Assume that there exist an integer* $m \geq 0$ *and a real number* $q > 1$ *such that* γ *is of class* $\mathcal{C}^{m+3}$, $\boldsymbol{\theta} \in \mathcal{C}^{m+3}(\overline{\omega}; \mathbb{R}^3)$, $p^\alpha \in W^{m,q}(\omega)$, *and* $p^3 \in W^{m+1,q}(\omega)$. *Then*

$$
\boldsymbol{\zeta} = (\zeta_i) \in W^{m+2,q}(\omega) \times W^{m+2,q}(\omega) \times W^{m+1,q}(\omega).
$$

Proof. The existence and uniqueness of a solution to the variational problem $\mathcal{P}_M(\omega)$, or to its equivalent minimization problem, is a consequence of the *inequality of Korn's type on an elliptic surface* (Thm. 2.7-3), of the *uniform positive definiteness of the scaled two-dimensional elasticity tensor of the shell* (Thm. 3.3-2 (a)), of the inequality $a(y) \geq a_0 > 0$, $y \in \overline{\omega}$ (Thm. 3.3-1), and of the *Lax-Milgram lemma.*

In view of finding the associated boundary value problem announced in part (b), we first prove that

$$\partial_\alpha \sqrt{a} = \sqrt{a}\,\Gamma^\sigma_{\sigma\alpha}.$$

To this end, we note that the function $\sqrt{g} = \det \boldsymbol{\nabla\Theta}$ and the Christoffel symbols Γ^p_{ij} associated with the *canonical extension* $\boldsymbol{\Theta}$ *of the mapping* $\boldsymbol{\theta}$, defined for $\varepsilon_0 > 0$ small enough by (Thm. 3.1-1):

$$\boldsymbol{\Theta}(y, x_3) = \boldsymbol{\theta}(y) + x_3 \boldsymbol{a}_3(y) \text{ for } y \in \overline{\omega} \text{ and } |x_3| \leq \varepsilon_0,$$

satisfy

$$(\partial_j \sqrt{g})(y, x_3) = \sqrt{g(y, x_3)}\,\Gamma^q_{qj}(y, x_3)$$

for all $y \in \overline{\omega}$ and $|x_3| \leq \varepsilon_0$ (see part (i) of the proof of Thm. 1.6-1). Letting $x_3 = 0$ and noting that

$$\sqrt{g(y, 0)} = \sqrt{a(y)},\ \Gamma^\sigma_{\sigma\alpha}(y, 0) = \Gamma^\sigma_{\sigma\alpha}(y),\ \Gamma^3_{3\alpha}(y, 0) = 0$$

for all $y \in \overline{\omega}$, we obtain

$$(\partial_\alpha \sqrt{a})(y) = \sqrt{a(y)}\,\Gamma^\sigma_{\sigma\alpha}(y) \text{ for all } y \in \overline{\omega}.$$

Using the *Green formula in Sobolev spaces* (see, e.g., Vol. I, Thm. 6.1-9) and assuming that the functions $n^{\alpha\beta} = n^{\beta\alpha}$ are in $H^1(\omega)$, we next obtain:

$$\begin{aligned}
&\int_\omega a^{\alpha\beta\sigma\tau} \gamma_{\sigma\tau}(\boldsymbol{\zeta}) \gamma_{\alpha\beta}(\boldsymbol{\eta}) \sqrt{a}\, dy = \int_\omega n^{\alpha\beta} \gamma_{\alpha\beta}(\boldsymbol{\eta}) \sqrt{a}\, dy \\
&= \int_\omega \sqrt{a}\, n^{\alpha\beta} \left(\frac{1}{2}(\partial_\beta \eta_\alpha + \partial_\alpha \eta_\beta) - \Gamma^\sigma_{\alpha\beta} \eta_\sigma - b_{\alpha\beta} \eta_3 \right) dy \\
&= \int_\omega \sqrt{a}\, n^{\alpha\beta} \partial_\beta \eta_\alpha\, dy - \int_\omega \sqrt{a} n^{\alpha\beta} \Gamma^\sigma_{\alpha\beta} \eta_\sigma\, dy - \int_\omega \sqrt{a} n^{\alpha\beta} b_{\alpha\beta} \eta_3\, dy \\
&= -\int_\omega \partial_\beta (\sqrt{a}\, n^{\alpha\beta}) \eta_\alpha\, dy - \int_\omega \sqrt{a}\, n^{\alpha\beta} \Gamma^\sigma_{\alpha\beta} \eta_\sigma\, dy - \int_\omega \sqrt{a}\, n^{\alpha\beta} b_{\alpha\beta} \eta_3\, dy
\end{aligned}$$

$$= -\int_\omega \sqrt{a}(\partial_\beta n^{\alpha\beta} + \Gamma^\alpha_{\tau\beta} n^{\tau\beta} + \Gamma^\beta_{\beta\tau} n^{\alpha\tau})\eta_\alpha\, dy - \int_\omega \sqrt{a}\, n^{\alpha\beta} b_{\alpha\beta}\eta_3\, dy$$
$$= -\int_\omega \sqrt{a}\Big\{ (n^{\alpha\beta}|_\beta)\eta_\alpha + b_{\alpha\beta} n^{\alpha\beta}\eta_3 \Big\} dy$$

for all $\boldsymbol{\eta} = (\eta_i) \in \mathbf{V}_M(\omega)$. Hence the variational equations imply that

$$\int_\omega \sqrt{a}\left\{ (n^{\alpha\beta}|_\beta + p^\alpha)\eta_\alpha + (b_{\alpha\beta} n^{\alpha\beta} + p^3)\eta_3 \right\} dy = 0$$

for all $(\eta_i) \in \mathbf{V}_M(\omega)$, and thus $n^{\alpha\beta}|_\beta = p^\alpha$ and $b_{\alpha\beta} n^{\alpha\beta} = p^3$ in ω.

The regularity result of part (c), which is due to Genevey [1996], is left as a problem (Ex. 4.2). ■

Remarks. (1) As shown by Ciarlet & Sanchez-Palencia [1996], the existence and uniqueness of a solution to the variational problem $\mathcal{P}_M(\omega)$ (cf. part (a)) may be also obtained after an *ad hoc* elimination of the unknown ζ_3 has been performed (this elimination is described in Ex. 4.2 (1)).

(2) The proof of part (b) is highly reminiscent of, and should thus be profitably compared to, that of Thm. 1.6-1.

(3) The boundary value problem found in (b) is another instance of a *uniformly elliptic system* in the sense of Agmon, Douglis & Nirenberg [1964]; See Ex. 4.3.

(4) First-order covariant derivatives $b^\alpha_\beta|_\sigma$ of a tensor (the curvature tensor) defined by means of its *mixed* components b^α_β have been defined in Thm. 2.5-1. ■

In order to get physically meaningful formulas, it remains to "de-scale" the unknowns ζ_i that satisfy the limit "scaled" problem $\mathcal{P}_M(\omega)$ found in Thm. 4.4-1. In view of the scalings

$$u_i(\varepsilon)(x) = u_i^\varepsilon(x^\varepsilon) \text{ for all } x^\varepsilon = \pi^\varepsilon x \in \overline{\Omega}^\varepsilon$$

made on the covariant components of the displacement field (Sect. 4.1), we are led to defining for each $\varepsilon > 0$ the **covariant components** $\zeta_i^\varepsilon : \overline{\omega} \to \mathbb{R}$ of the **limit displacement field** $\tilde{\boldsymbol{\zeta}}^\varepsilon : \overline{\omega} \to \mathbb{R}^3$ *of the middle surface S of the shell* by letting (the vectors $\boldsymbol{a}^i$ form the contravariant basis at each point of S):

$$\boxed{\zeta_i^\varepsilon := \zeta_i \text{ and } \tilde{\boldsymbol{\zeta}}^\varepsilon := \zeta_i^\varepsilon \boldsymbol{a}^i.}$$

A word of caution. Naturally, *the field* $\boldsymbol{\zeta}^\varepsilon := (\zeta_i^\varepsilon)$ *and the field* $\tilde{\boldsymbol{\zeta}}^\varepsilon = \zeta_i^\varepsilon \boldsymbol{a}^i$ *must be carefully distinguished* ! The former is essentially a convenient *mathematical "intermediary"*, but only the latter has *physical significance.* ■

Remark. Conceivably, the limit scaled displacement field *across the thickness of the shell* could be also de-scaled, resulting into the *limit displacement field* $\tilde{\boldsymbol{u}}^\varepsilon(0) : \overline{\Omega}^\varepsilon \to \mathbb{R}^3$ *inside the shell* defined by (the vectors $\boldsymbol{g}^{i,\varepsilon}$ form the contravariant basis at each point of the reference configuration $\boldsymbol{\Theta}(\overline{\Omega}^\varepsilon)$; cf. Sect. 3.1):

$$\tilde{\boldsymbol{u}}^\varepsilon(0) := \zeta_i^\varepsilon \boldsymbol{g}^{i,\varepsilon},$$

since the scaled limit $\boldsymbol{u} = (u_i)$ is independent of the transverse variable x_3 (Thm. 4.4-1). For the same reason, however, the de-scaled field does not inherit any remarkable structure as x_3^ε varies across the thickness of the shell.

By contrast, the limit displacement field across a *plate* is a *Kirchhoff-Love* one (Vol. II, Thm. 1.7-1): It does inherit this richer structure because *different* scalings can be made in this case on the "horizontal" and "vertical" components of the displacement field. ■

Recall that $f^{i,\varepsilon} \in L^2(\Omega^\varepsilon)$ and $h^{i,\varepsilon} \in L^2(\Gamma_+^\varepsilon \cup \Gamma_-^\varepsilon)$ represent the contravariant components of the applied body and surface forces *actually* acting on the shell and that λ^ε and μ^ε denote the *actual* Lamé constants of its constituting material. We then have the following immediate corollary to Thms. 4.4-1 and 4.5-1; naturally, the *existence, uniqueness* and *regularity* results of Thm. 4.5-1 apply *verbatim* to the solution of the "de-scaled" problem $\mathcal{P}_M^\varepsilon(\omega)$ (for this reason, they are not reproduced here):

Theorem 4.5-2. *Let the assumptions on the data be as in* Thm. 4.4-1. *Then the vector field* $\boldsymbol{\zeta}^\varepsilon := (\zeta_i^\varepsilon)$ *formed by the covariant components of the limit displacement field* $\zeta_i^\varepsilon \boldsymbol{a}^i$ *of the middle surface* S *satisfies the following* **two-dimensional variational problem** $\mathcal{P}_M^\varepsilon(\omega)$ **of a linearly elastic elliptic membrane shell:**

$$\boldsymbol{\zeta}^\varepsilon \in \mathbf{V}_M(\omega) := H_0^1(\omega) \times H_0^1(\omega) \times L^2(\omega),$$
$$\varepsilon \int_\omega a^{\alpha\beta\sigma\tau,\varepsilon} \gamma_{\sigma\tau}(\boldsymbol{\zeta}^\varepsilon) \gamma_{\alpha\beta}(\boldsymbol{\eta}) \sqrt{a}\, dy = \int_\omega p^{i,\varepsilon} \eta_i \sqrt{a}\, dy$$
$$\text{for all } \boldsymbol{\eta} = (\eta_i) \in \mathbf{V}_M(\omega),$$

where

$$
\begin{aligned}
\gamma_{\alpha\beta}(\boldsymbol{\eta}) &:= \frac{1}{2}(\partial_\beta \eta_\alpha + \partial_\alpha \eta_\beta) - \Gamma^\sigma_{\alpha\beta}\eta_\sigma - b_{\alpha\beta}\eta_3, \\
a^{\alpha\beta\sigma\tau,\varepsilon} &:= \frac{4\lambda^\varepsilon \mu^\varepsilon}{\lambda^\varepsilon + 2\mu^\varepsilon} a^{\alpha\beta} a^{\sigma\tau} + 2\mu^\varepsilon (a^{\alpha\sigma} a^{\beta\tau} + a^{\alpha\tau} a^{\beta\sigma}), \\
p^{i,\varepsilon} &:= \int_{-\varepsilon}^{\varepsilon} f^{i,\varepsilon}\, dx_3^\varepsilon + h_+^{i,\varepsilon} + h_-^{i,\varepsilon} \text{ and } h_\pm^{i,\varepsilon} := h^{i,\varepsilon}(\cdot, \pm\varepsilon).
\end{aligned}
$$

Equivalently, the field $\boldsymbol{\zeta}^\varepsilon$ satisfies the following minimization problem:

$$
\begin{gathered}
\boldsymbol{\zeta}^\varepsilon \in \mathbf{V}_M(\omega) \text{ and } j_M^\varepsilon(\boldsymbol{\zeta}^\varepsilon) = \inf_{\boldsymbol{\eta} \in \mathbf{V}_M(\omega)} j_M^\varepsilon(\boldsymbol{\eta}), \text{ where} \\
j_M^\varepsilon(\boldsymbol{\eta}) := \frac{1}{2}\int_\omega \varepsilon a^{\alpha\beta\sigma\tau,\varepsilon} \gamma_{\sigma\tau}(\boldsymbol{\eta}) \gamma_{\alpha\beta}(\boldsymbol{\eta}) \sqrt{a}\, dy - \int_\omega p^{i,\varepsilon} \eta_i \sqrt{a}\, dy.
\end{gathered}
$$

If the field $\boldsymbol{\zeta}^\varepsilon = (\zeta_i^\varepsilon)$ is smooth enough, it also satisfies the following boundary value problem:

$$
\begin{aligned}
-n^{\alpha\beta,\varepsilon}|_\beta &= p^{\alpha,\varepsilon} \text{ in } \omega, \\
-b_{\alpha\beta} n^{\alpha\beta,\varepsilon} &= p^{3,\varepsilon} \text{ in } \omega, \\
\zeta_\alpha^\varepsilon &= 0 \text{ on } \gamma,
\end{aligned}
$$

where

$$
\begin{aligned}
n^{\alpha\beta,\varepsilon} &:= \varepsilon a^{\alpha\beta\sigma\tau,\varepsilon} \gamma_{\sigma\tau}(\boldsymbol{\zeta}^\varepsilon), \\
n^{\alpha\beta,\varepsilon}|_\sigma &:= \partial_\sigma n^{\alpha\beta,\varepsilon} + \Gamma^\alpha_{\tau\sigma} n^{\tau\beta,\varepsilon} + \Gamma^\beta_{\sigma\tau} n^{\alpha\tau,\varepsilon}.
\end{aligned}
$$

■

Each one of the three formulations found in Thm. 4.5-2 constitutes one version of the **two-dimensional equations of a linearly elastic elliptic membrane shell.** The functions $\gamma_{\alpha\beta}(\boldsymbol{\eta})$ are the *covariant components* of the **linearized change of metric tensor** associated with a displacement field $\eta_i \boldsymbol{a}^i$ of the middle surface S, the functions $a^{\alpha\beta\sigma\tau,\varepsilon}$ are the *contravariant components* of the **two-dimensional elasticity tensor of the shell**, and the functions

$n^{\alpha\beta,\varepsilon}$ are the contravariant components of the **stress resultant tensor** field ("resultant" means "after integration of the limit stresses across the thickness of the shell"; cf. Ex. 4.4).

The functional $j_M^\varepsilon : \mathbf{V}_M(\omega) \to \mathbb{R}$ is the **two-dimensional energy**, and the functional

$$\boxed{\boldsymbol{\eta} \in \mathbf{V}_M(\omega) \to \frac{1}{2}\int_\omega \varepsilon a^{\alpha\beta\sigma\tau,\varepsilon}\gamma_{\sigma\tau}(\boldsymbol{\eta})\gamma_{\alpha\beta}(\boldsymbol{\eta})\sqrt{a}\,dy}$$

is the **two-dimensional strain energy, of a linearly elastic elliptic membrane shell.**

Finally, the equations $-n^{\alpha\beta,\varepsilon}|_\beta = p^{\alpha,\varepsilon}$ and $-b_{\alpha\beta}n^{\alpha\beta,\varepsilon} = p^{3,\varepsilon}$ in ω constitute the **two-dimensional equations of equilibrium,** and the relations $n^{\alpha\beta,\varepsilon} = \varepsilon a^{\alpha\beta\sigma\tau,\varepsilon}\gamma_{\sigma\tau}(\boldsymbol{\zeta}^\varepsilon)$ constitute the **two-dimensional constitutive equation, of a linearly elastic elliptic membrane shell.**

Note that the *same* two-dimensional problem $\mathcal{P}_M^\varepsilon(\omega)$ is evidently obtained if the scalings on the unknowns are the same as before, i.e.,

$$\boxed{u_i^\varepsilon(x^\varepsilon) = u_i(\varepsilon)(x) \text{ for all } x^\varepsilon = \pi^\varepsilon x \in \overline{\Omega}^\varepsilon,}$$

but the following *more general* **assumptions on the data** are made:

$$\boxed{\begin{aligned} \lambda^\varepsilon = \varepsilon^t\lambda \quad &\text{and} \quad \mu^\varepsilon = \varepsilon^t\mu, \\ f^{i,\varepsilon}(x^\varepsilon) = \varepsilon^t f^i(x) \quad &\text{for all} \quad x^\varepsilon = \pi^\varepsilon x \in \Omega^\varepsilon, \\ h^{i,\varepsilon}(x^\varepsilon) = \varepsilon^{1+t}h^i(x) \quad &\text{for all} \quad x^\varepsilon = \pi^\varepsilon x \in \Gamma_+^\varepsilon \cup \Gamma_-^\varepsilon, \end{aligned}}$$

where the constants $\lambda > 0$ and $\mu > 0$ and the functions $f^i \in L^2(\Omega)$ and $h^i \in L^2(\Gamma_+ \cup \Gamma_-)$ are *independent of* ε and t *is an arbitrary real number.* Besides, the analysis of Sect. 3.4 shows that *these assumptions on the data are the only ones possible for elliptic membrane shells.*

Remark. A *different* kind of generalization is possible. For definiteness, assume that the Lamé constants are independent of ε. Then it is easily verified that the same scaled limit problem $\mathcal{P}_M(\omega)$ is obtained under the assumptions that

$$f^{i,\varepsilon}(x^\varepsilon) = f^i(\varepsilon; x) \text{ for all } x^\varepsilon = \pi^\varepsilon x \in \Omega^\varepsilon,$$

$$h^{i,\varepsilon}(x^\varepsilon) = \varepsilon h^i(\varepsilon; x) \text{ for all } x^\varepsilon = \pi^\varepsilon x \in \Gamma_+^\varepsilon \cup \Gamma_-^\varepsilon,$$

and

$$f^i(\varepsilon; \cdot) \to f^i \text{ in } L^2(\Omega) \text{ as } \varepsilon \to 0,$$
$$h^i(\varepsilon; \cdot) \to h^i \text{ in } L^2(\Gamma_+ \cup \Gamma_-) \text{ as } \varepsilon \to 0.$$

The functions p_i^ε appearing in the right-hand sides of problem $\mathcal{P}_M^\varepsilon(\omega)$ are then defined by

$$p_i^\varepsilon := \varepsilon \left\{ \int_{-1}^{1} f^i \, dx_3 + h_+^i + h_-^i \right\}.$$

■

4.6. JUSTIFICATION OF THE TWO-DIMENSIONAL EQUATIONS OF A LINEARLY ELASTIC ELLIPTIC MEMBRANE SHELL; COMMENTARY AND REFINEMENTS

It remains to convert in terms of de-scaled unknowns the fundamental convergence theorem established in Sect. 4.4. As the "original" unknowns u_i^ε are defined over a domain that *varies with* ε (the set Ω^ε), their *averages* $\frac{1}{2\varepsilon}\int_{-\varepsilon}^{\varepsilon} u_i^\varepsilon \, dx_3^\varepsilon$ are more appropriate for this purpose, since they are defined over a *fixed* domain (the set ω).

The convergences $u_\alpha(\varepsilon) \to u_\alpha$ in $H^1(\Omega)$ and $u_3(\varepsilon) \to u_3$ in $L^2(\Omega)$ (Thm. 4.4-1), the scalings $u_i(\varepsilon)(x) = u_i^\varepsilon(x^\varepsilon)$ for all $x^\varepsilon = \pi^\varepsilon x \in \overline{\Omega}^\varepsilon$ (Sect. 4.1), the de-scalings $\zeta_i^\varepsilon = \zeta_i = \frac{1}{2}\int_{-1}^{1} u_i \, dx_3$ (Sect. 4.5) and Thm. 4.2-1 together yield the following convergences for the averages $\frac{1}{2\varepsilon}\int_{-\varepsilon}^{\varepsilon} u_i^\varepsilon \, dx_3^\varepsilon$ of the covariant components of the original three-dimensional displacement:

$$\frac{1}{2\varepsilon}\int_{-\varepsilon}^{\varepsilon} u_\alpha^\varepsilon \, dx_3^\varepsilon \to \zeta_\alpha^\varepsilon \text{ in } H^1(\omega) \text{ and } \frac{1}{2\varepsilon}\int_{-\varepsilon}^{\varepsilon} u_3^\varepsilon \, dx_3^\varepsilon \to \zeta_3^\varepsilon \text{ in } L^2(\omega).$$

However, these convergences can be further improved and given a more intrinsic character by considering instead the *averages* of the *tangential component* $u_\alpha^\varepsilon \boldsymbol{g}^{\alpha,\varepsilon}$ and of the *normal component* $u_3^\varepsilon \boldsymbol{g}^{3,\varepsilon}$ of the *three-dimensional displacement vector* itself; note in this respect that, along a given normal direction to the surface S, the vectors $\boldsymbol{g}^{\alpha,\varepsilon}$ and $\boldsymbol{g}^{3,\varepsilon}$ remain respectively parallel to the tangent plane and normal to S since $\boldsymbol{g}_\alpha^\varepsilon = \boldsymbol{a}_\alpha - x_3^\varepsilon b_\alpha^\sigma \boldsymbol{a}_\sigma$, $\boldsymbol{g}^{3,\varepsilon} = \boldsymbol{a}_3$, and $\boldsymbol{g}_i^\varepsilon \cdot \boldsymbol{g}^{j,\varepsilon} = \delta_i^j$. More specifically, the above convergences combined with the behavior as $\varepsilon \to 0$ of the vectors $\boldsymbol{g}^{i,\varepsilon}$ lead to the following result:

Theorem 4.6-1. *Assume that* $\boldsymbol{\theta} \in \mathcal{C}^3(\overline{\omega}; \mathbb{R}^3)$. *Consider a family of linearly elastic elliptic membrane shells with thickness approaching zero and with each having the same elliptic middle surface* $S = \boldsymbol{\theta}(\overline{\omega})$, *and let the assumptions on the data be as in* Sect. 4.1.

Let $(u_i^\varepsilon) \in \mathbf{H}^1(\Omega^\varepsilon)$ *and* $(\zeta_i^\varepsilon) \in H^1(\omega) \times H^1(\omega) \times L^2(\omega)$ *respectively denote for each* $\varepsilon > 0$ *the solutions to the three-dimensional and two-dimensional problems* $\mathcal{P}(\Omega^\varepsilon)$ *and* $\mathcal{P}_M^\varepsilon(\omega)$ (Sect. 3.1 and Thm. 4.5-2).

Finally, let $(\zeta_i) \in H^1(\omega) \times H^1(\omega) \times L^2(\omega)$ *denote the solution to problem* $\mathcal{P}_M(\omega)$ (Thm. 4.5-1), *which is thus independent of* ε. *Then*

$$\boxed{\begin{aligned} \zeta_\alpha^\varepsilon = \zeta_\alpha \text{ and thus } \zeta_\alpha^\varepsilon \boldsymbol{a}^\alpha &= \zeta_\alpha \boldsymbol{a}^\alpha \text{ in } \mathbf{H}^1(\omega) \text{ for all } \varepsilon > 0, \\ \frac{1}{2\varepsilon}\int_{-\varepsilon}^{\varepsilon} u_\alpha^\varepsilon \boldsymbol{g}^{\alpha,\varepsilon}\, dx_3^\varepsilon &\to \zeta_\alpha \boldsymbol{a}^\alpha \text{ in } \mathbf{H}^1(\omega) \text{ as } \varepsilon \to 0, \end{aligned}}$$

and

$$\boxed{\begin{aligned} \zeta_3^\varepsilon = \zeta_3 \text{ and thus } \zeta_3^\varepsilon \boldsymbol{a}^3 &= \zeta_3 \boldsymbol{a}^3 \text{ in } \mathbf{L}^2(\omega) \text{ for all } \varepsilon > 0, \\ \frac{1}{2\varepsilon}\int_{-\varepsilon}^{\varepsilon} u_3^\varepsilon \boldsymbol{g}^{3,\varepsilon}\, dx_3^\varepsilon &\to \zeta_3 \boldsymbol{a}^3 \text{ in } \mathbf{L}^2(\omega) \text{ as } \varepsilon \to 0. \end{aligned}}$$

Proof. As is easily seen (a similar argument was used in the proof of Thm. 3.3-1), the assumption $\boldsymbol{\theta} \in \mathcal{C}^3(\overline{\omega}; \mathbb{R}^3)$ implies that the vector fields $\boldsymbol{g}^\alpha(\varepsilon) : \overline{\Omega} \to \mathbb{R}^3$ defined by $\boldsymbol{g}^\alpha(\varepsilon)(x) := \boldsymbol{g}^{\alpha,\varepsilon}(x^\varepsilon)$ for all $x^\varepsilon = \pi^\varepsilon x \in \overline{\Omega}^\varepsilon$ are such that (the vector fields $\boldsymbol{a}^\alpha : \overline{\omega} \to \mathbb{R}^3$ are identified here with vector fields defined over the set $\overline{\Omega}$):

$$\boldsymbol{g}^\alpha(\varepsilon) - \boldsymbol{a}^\alpha = O(\varepsilon) \text{ in } \mathcal{C}^1(\overline{\Omega}).$$

Since

$$\begin{aligned} \frac{1}{2\varepsilon}\int_{-\varepsilon}^{\varepsilon} u_\alpha^\varepsilon \boldsymbol{g}^{\alpha,\varepsilon}\, dx_3^\varepsilon - \zeta_\alpha^\varepsilon \boldsymbol{a}^\alpha &= \frac{1}{2}\int_{-1}^{1} u_\alpha(\varepsilon)\boldsymbol{g}^\alpha(\varepsilon)\, dx_3 - \zeta_\alpha \boldsymbol{a}^\alpha \\ &= \frac{1}{2}\int_{-1}^{1} u_\alpha(\varepsilon)\,(\boldsymbol{g}^\alpha(\varepsilon) - \boldsymbol{a}^\alpha)\, dx_3 - (\overline{u_\alpha(\varepsilon)} - \zeta_\alpha)\boldsymbol{a}^\alpha, \end{aligned}$$

the convergences $u_\alpha(\varepsilon) \to u_\alpha$ in $H^1(\Omega)$ and $\boldsymbol{g}^\alpha(\varepsilon) \to \boldsymbol{a}^\alpha$ in $\mathcal{C}^1(\overline{\Omega})$ imply that

$$u_\alpha(\varepsilon)\,(\boldsymbol{g}^\alpha(\varepsilon) - \boldsymbol{a}^\alpha) \to 0 \text{ in } \mathbf{H}^1(\Omega);$$

hence $\frac{1}{2}\int_{-1}^{1} u_\alpha(\varepsilon)(\boldsymbol{g}^\alpha(\varepsilon) - \boldsymbol{a}^\alpha)\,dx_3 \to 0$ in $\mathbf{H}^1(\omega)$ by Thm. 4.2-1 (b). The same theorem also shows that $(\overline{u_\alpha(\varepsilon)} - \zeta_\alpha)\boldsymbol{a}^\alpha \to 0$ in $\mathbf{H}^1(\omega)$. The proof is even simpler for the normal components: Since $\boldsymbol{g}^{3,\varepsilon} = \boldsymbol{a}^3$,

$$\frac{1}{2\varepsilon}\int_{-\varepsilon}^{\varepsilon} u_3^\varepsilon \boldsymbol{g}^{3,\varepsilon}\,dx_3^\varepsilon - \zeta_3^\varepsilon \boldsymbol{a}^3 = (\overline{u_3(\varepsilon)} - \zeta_3)\boldsymbol{a}^3,$$

and Thm. 4.2-1 (a) then shows that $(\overline{u_3(\varepsilon)} - \zeta_3)\boldsymbol{a}^3 \to 0$ in $\mathbf{L}^2(\omega)$. ∎

The fields $\tilde{\boldsymbol{\zeta}}_T^\varepsilon : \overline{\omega} \to \mathbb{R}^3$ and $\tilde{\boldsymbol{\zeta}}_N^\varepsilon : \overline{\omega} \to \mathbb{R}^3$ defined by

$$\boxed{\tilde{\boldsymbol{\zeta}}_T^\varepsilon := \zeta_\alpha^\varepsilon \boldsymbol{a}^\alpha \text{ and } \tilde{\boldsymbol{\zeta}}_N^\varepsilon := \zeta_3^\varepsilon \boldsymbol{a}^3,}$$

which appear in Thm. 4.6-1, are respectively called the **limit tangential displacement field**, and the **limit normal displacement field**, *of the middle surface S of the shell.* Naturally, they are related to the *limit displacement field* $\tilde{\boldsymbol{\zeta}}^\varepsilon = \zeta_i^\varepsilon \boldsymbol{a}^i$ *of S* (Sect. 4.5) by

$$\boxed{\tilde{\boldsymbol{\zeta}}^\varepsilon = \tilde{\boldsymbol{\zeta}}_T^\varepsilon + \tilde{\boldsymbol{\zeta}}_N^\varepsilon.}$$

Under the essential assumptions that $\gamma_0 = \gamma$ and that the surface S is elliptic, we have therefore justified by a convergence result (Thm. 4.6-1) *two-dimensional equations that are called those of a linearly elastic "membrane" shell in the literature* (which, however, usually ignores the distinction between "elliptic" and "generalized" membrane shells); see, e.g., Koiter [1966, eqs. (9.14) and (9.15)], Green & Zerna [1968, Sect. 11.1], Dikmen [1982, eqs. (7.10)], or Niordson [1985, eq. (10.3)]. In so doing, we have also justified the formal asymptotic approach of Sanchez-Palencia [1990] (see also Miara & Sanchez-Palencia [1996] and Caillerie & Sanchez-Palencia [1995b]) when *"bending is well-inhibited"*, according to the terminology of E. Sanchez-Palencia.

A word of caution. In an *elliptic membrane* shell, body forces of order $O(1)$ with respect to ε thus produce a limit displacement field that is also $O(1)$. By contrast, body forces must be of order $O(\varepsilon^2)$ in order to produce an $O(1)$ limit displacement field in a *flexural* shell. See Chap. 6. ■

The *first convergence results* for linearly elastic membrane shells have been obtained by Destuynder [1980] in his Doctoral Dissertation. In particular, the convergences established there in Thm. 7.9 (p. 305) under the assumption that the surface S is elliptic are almost identical to those established in Thm. 4.4-1 for the components $u_\alpha(\varepsilon)$, but "weaker" for the component $u_3(\varepsilon)$, since P. Destuynder only established that $\varepsilon u_3(\varepsilon) \to 0$ in $L^2(\Omega)$. Besides, his justification of the membrane shell equations remained partially formal as it still relied on an assumed asymptotic expansion of $u_3(\varepsilon)$.

Using Γ-*convergence theory* (see, e.g., Vol. II, Sect. 1.11, and the references therein), Acerbi, Buttazzo & Percivale [1988] were able to obtain convergence theorems for linearly elastic shells viewed as "thin inclusions" in a larger, surrounding elastic body. As a consequence, the distinction between membrane shells and flexural shells (Chap. 6) is no longer related to the geometry of the middle surface and to the boundary conditions as here, but instead to the ratio (as a power of ε) between the Lamé constants of the two elastic materials in presence. This asymptotic analysis is thus more reminiscent of that of Ciarlet, Le Dret & Nzengwa [1989] (described at length in Vol. II, Chap. 2), who considered an *elastic multi-structure* consisting of a plate partly inserted in an elastic body; if the shell were a plate, the approach of Acerbi, Buttazzo & Percivale [1988] would only apply to the inserted portion, however.

After the original work of Ciarlet & Lods [1996b] described in this chapter, the asymptotic analysis of linearly elastic membrane shells underwent several refinements and generalizations:

First, Genevey [1999] has shown that the convergence result of Thm. 4.4-1 can be also obtained by resorting to Γ-*convergence* theory.

Using the techniques of Lions [1973], Mardare [1998a] was then able to compute a *corrector*, so as to obtain in this fashion the following remarkable *error estimate: In addition to the hypotheses made in* Thm. 4.4-1, *assume that the boundary of the domain ω is of class C^2, that $\partial_\alpha f^\alpha \in L^2(\Omega)$ and $h^i = 0$, and that*

$$\zeta = \frac{1}{2}\int_{-1}^{1} \boldsymbol{u}\, dx_3 \in \mathbf{H}^2(\omega) \cap \mathbf{V}_M(\omega).$$

Then there exists a constant $C = C(\omega, \boldsymbol{\theta}, f^i, \boldsymbol{\zeta})$ independent of ε such that

$$\|\boldsymbol{u}(\varepsilon) - \boldsymbol{u}\|_{H^1(\Omega)\times H^1(\Omega)\times L^2(\Omega)} \leq C\varepsilon^{1/6},$$

and moreover, the exponent 1/6 is the best possible.

Other useful extensions include the justification by an asymptotic analysis of *linearly elastic membrane shells with variable thickness* (Busse [1998]; see also Ex. 4.5) or made with a *nonhomogeneous* and *anisotropic material* (Giroud [1998]), the *convergence of the* (scaled) *stresses* and the *explicit forms of the limit stresses* (Collard & Miara [1999]; see also Ex. 4.4), an asymptotic analysis of the associated *time-dependent problem* (Xiao Li-ming [1998]), and the extension of the present analysis to "membrane shells" whose *middle surface S is elliptic but has "no boundary"*, such as an entire ellipsoid (Ramos [1995] and Slicaru [1997]).

The surprising phenomena appearing when a linearly elastic elliptic membrane shell *"becomes a plate"* are investigated in Ciarlet [1992a, 1992b], Sanchez-Palencia [1994], and Lods [1995].

The variational formulation of the limit two-dimensional problem of a linearly elastic membrane shell (Thm. 4.5-2) possesses the unusual feature that its third unknown ζ_3^ε "only" belongs to the space $L^2(\omega)$. This explains why the averaged three-dimensional boundary conditions

$$\overline{u}_3^\varepsilon := \frac{1}{2\varepsilon}\int_{-\varepsilon}^{\varepsilon} u_3^\varepsilon \, dx_3^\varepsilon = 0 \text{ on } \gamma$$

are "lost" as $\varepsilon \to 0$, since $\zeta_3^\varepsilon = 0$ on γ does not make sense. As expected, this loss is compensated by the appearance of a *boundary layer* in the unknown ζ_3^ε. The remaining boundary conditions $\zeta_\alpha^\varepsilon = 0$ on γ constitute **two-dimensional boundary conditions of weak simple support.**

Again because the third unknown ζ_3^ε is only in $L^2(\omega)$, the linear operator associated with the variational problem of a linearly elastic elliptic membrane shell is *not* compact (Ex. 4.6) and thus the analysis of the corresponding *eigenvalue problem* requires special care; see Sanchez-Hubert & Sanchez-Palencia [1997, Chap. 10] and, for the finite element approximations, Rappaz, Sanchez-Hubert, Sanchez-Palencia & Vassiliev [1997]. The *controllability* and *homogenization* of elliptic membrane shells likewise pose special difficulties; see Valente [1997] and Hamdache & Sanchez-Palencia [1997].

EXERCISES

4.1. The results of this exercise are needed in the proofs of Thms. 4.2-1 and 4.4-1. Let ω be a domain in $\mathbb{R}^2$, and let $\Omega = \omega \times]-1, 1[$.

(1) Let $v \in L^2(\Omega)$ be such that $\partial_3 v = 0$ in Ω in the sense of distributions. Show that there exists $\eta \in L^2(\omega)$ such that $v(y, x_3) = \eta(y)$ for almost all $(y, x_3) \in \Omega$.

(2) Let $v \in L^2(\Omega)$ be such that its derivative $\partial_3 v$ in the sense of distributions belongs to $L^2(\Omega)$. Show that its trace $v(\cdot, s)$ is well defined as a function in $L^2(\omega)$ for all $s \in [-1, 1]$ and that

$$v(y, s) = v(y, -1) + \int_{-1}^{s} \partial_3 v(y, x_3)\, dx_3$$

for almost all $y \in \omega$ and all $s \in [-1, 1]$.

Remarks. This exercise is solved in, e.g., Le Dret [1991, Lemmas 1.3 and 4.1]. The result of (1) was also needed in the asymptotic analysis of plates; cf. Vol. II, Ex. 1.1.

4.2. The following problem is based on Genevey [1996]; it shows how to establish the *regularity* (announced in Thm. 4.5-1 (c)) of the solution to the two-dimensional equations of a linearly elastic elliptic membrane shell. Unless otherwise specified, the notations and assumptions are those of Thm. 4.5-1.

(1) Assume that $\boldsymbol{\theta} \in \mathcal{C}^2(\overline{\omega}; \mathbb{R}^3)$. Show that $\boldsymbol{\zeta} = (\zeta_i)$ satisfies the *two-dimensional problem $\mathcal{P}_M(\omega)$ of a linearly elastic membrane shell* if and only if $\hat{\boldsymbol{\zeta}} := (\zeta_\alpha)$ satisfies the following variational problem $\hat{\mathcal{P}}_M(\omega)$:

$$\hat{\boldsymbol{\zeta}} \in \hat{\mathbf{V}}_M(\omega) := H_0^1(\omega) \times H_0^1(\omega),$$

$$\int_\omega \hat{a}^{\alpha\beta\sigma\tau} \hat{\gamma}_{\sigma\tau}(\hat{\boldsymbol{\zeta}}) \hat{\gamma}_{\alpha\beta}(\hat{\boldsymbol{\eta}}) \sqrt{a}\, dy = \int_\omega \left\{ p^\alpha \eta_\alpha + d a^{\alpha\beta\sigma\tau} b_{\sigma\tau} \hat{\gamma}_{\alpha\beta}(\hat{\boldsymbol{\eta}}) \right\} \sqrt{a}\, dy$$

for all $\hat{\boldsymbol{\eta}} = (\eta_\alpha) \in \widehat{\mathbf{V}}_M(\omega)$, and

$$\zeta_3 = d a^{\alpha\beta\sigma\tau} b_{\sigma\tau} \hat{\gamma}_{\alpha\beta}(\hat{\boldsymbol{\zeta}}) + d p^3,$$

where (note that the assumption of ellipticity of S ensures that the function d is well defined),

$$\hat{\gamma}_{\alpha\beta}(\hat{\boldsymbol{\eta}}) := \frac{1}{2}(\partial_\alpha \eta_\beta + \partial_\beta \eta_\alpha) - \Gamma^\sigma_{\alpha\beta}\eta_\sigma,$$
$$\hat{a}^{\alpha\beta\sigma\tau} := a^{\alpha\beta\sigma\tau} - c b_{\gamma\delta} a^{\gamma\delta\alpha\beta} a^{\lambda\mu\sigma\tau} b_{\lambda\mu},$$
$$d := \left(a^{\alpha\beta\sigma\tau} b_{\sigma\tau} b_{\alpha\beta}\right)^{-1}.$$

Remark. As shown by Ciarlet & Sanchez-Palencia [1996], this elimination of the unknown ζ_3 is the basis of another proof of the existence and uniqueness of a solution to the two-dimensional problem $\mathcal{P}_M(\omega)$ of an elliptic membrane shell (Thm. 4.5-1).

(2) Assuming that the data and the solution $\hat{\boldsymbol{\zeta}}$ to $\hat{\mathcal{P}}_M(\omega)$ are smooth enough, write the boundary value problem that $\hat{\boldsymbol{\zeta}} = (\zeta_\alpha)$ satisfies, as a system of two second-order partial differential equations in ω, together with the two boundary conditions $\zeta_\alpha = 0$ on γ.

(3) Assume from now on that the boundary γ is of class $\mathcal{C}^3$ and that $\boldsymbol{\theta} \in \mathcal{C}^3(\overline{\omega}; \mathbb{R}^3)$. Show that the second order system found in (2) is "*uniformly elliptic*" and satisfies the "*supplementary condition on L*" and the "*complementing boundary condition*", in the sense of Agmon, Douglis & Nirenberg [1964].

Remark. Similar verifications were made in the proof of Thm. 2.7-2 for the system $\gamma_{\alpha\beta}(\boldsymbol{\eta}) = 0$ in ω.

(4) Show that the second-order system found in (2) is "*strongly elliptic*" in the sense of Nečas [1967, p. 185]. Using Nečas [1967, Lemma 3.2, p. 260]), infer from this property that $\zeta_\alpha \in H^2(\omega) \cap H^1_0(\omega)$ if $p^\alpha \in L^2(\omega)$ and $p^3 \in H^1(\omega)$.

(5) Using Geymonat [1965, Thm. 3.5], deduce from (4) that $\zeta_\alpha \in W^{2,q}(\omega) \cap H^1_0(\omega)$ if $p^\alpha \in L^q(\omega)$ and $p^3 \in W^{1,q}(\omega)$ for some $q > 1$.

(6) Assume that there exist an integer $m \geq 1$ and a real number $q > 1$ such that γ is of class $\mathcal{C}^{m+3}$, $\boldsymbol{\theta} \in \mathcal{C}^{m+3}(\overline{\omega}; \mathbb{R}^3)$, $p^\alpha \in W^{m,q}(\omega)$, and $p^3 \in W^{m+1,q}(\omega)$. Using Agmon, Douglis & Nirenberg [1964, Thm. 10.5], show that $\zeta_\alpha \in W^{m+2,q}(\omega) \cap H^1_0(\omega)$.

Remark. An analogous proof of regularity holds for the *equations of three-dimensional linearized elasticity*; cf. Vol. I, Thm. 6.3-6 and Ex. 1.11.

4.3. Let the notations and assumptions be as in Thm. 4.5-1.

(1) Show that the three partial differential equations

$$-n^{\alpha\beta}|_\beta = p^\alpha \text{ and } -n^{\alpha\beta} b_{\alpha\beta} = p^3 \text{ in } \omega$$

found in Thm. 4.5-1 (b) constitute a system of partial differential equations, of the second order with respect to the unknowns ζ_α and of the first order with respect to the unknowns ζ_3, that is *"uniformly elliptic"* in the sense of Agmon, Douglis & Nirenberg [1964].

(2) Show that, when the two boundary conditions $\zeta_\alpha = 0$ on γ are appended as in Thm. 4.5-1 (b) to the three partial differential equations of (1), the *"complementing boundary condition"* of *ibid.* holds.

Remarks. Another uniformly elliptic system (then of the first order with respect to the first and second unknowns and of order zero with respect to the third unknown) satisfying the complementing boundary condition has already been encountered in Thm. 2.7-2. Incidentally, checking whether the complementing boundary condition is satisfied may require substantial algebraic manipulations; algorithms of *formal computation* overcome this difficulty in some cases; see Connétable [1995].

(3) Show that the boundary value problem in Thm. 4.5-1 (b) is an instance of the elliptic systems studied by Grubb & Geymonat [1977], whose approach thus provides another proof of the *existence* result of Thm. 4.5-1 (a).

Hint: See Sanchez-Hubert & Sanchez-Palencia [1997, Chap. 7, Props. 3.8 and 3.9].

4.4. We established in Thm. 4.4-1 that the scaled displacements converge and we also saw there how their limits can be identified. The present problem shows that, under the same assumptions, the *convergence of the* (scaled) *stresses* can be likewise established and that the corresponding *limit stresses* can be similarly explicitly computed. These results are due to Collard & Miara [1999, Thm. 7].

Given a family of linearly elastic membrane shells that satisfies the assumptions of Thm. 4.4-1, let

$$\sigma^{ij,\varepsilon} = A^{ijkl,\varepsilon} e^\varepsilon_{k\|l}(\boldsymbol{u}^\varepsilon)$$

denote for each $\varepsilon > 0$ the contravariant components of the *linearized stress tensor* field (Thm. 1.6-1) inside each shell and define the *scaled stresses* $\sigma^{ij}(\varepsilon) : \overline{\Omega} \to \mathbb{R}$ by letting

$$\sigma^{ij,\varepsilon}(x^\varepsilon) =: \sigma^{ij}(\varepsilon)(x) \text{ for all } x^\varepsilon = \pi^\varepsilon x \in \overline{\Omega}.$$

Note that the scaled stresses then satisfy

$$\sigma^{ij}(\varepsilon) = A^{ijkl}(\varepsilon) e_{k\|l}(\varepsilon; \boldsymbol{u}(\varepsilon)).$$

(1) Show that

$$\sigma^{\alpha\beta}(\varepsilon) \to \sigma^{\alpha\beta}(0) \text{ in } L^2(\Omega),$$
$$\frac{1}{\varepsilon}\sigma^{i3}(\varepsilon) \to \sigma^{i3,1} \text{ in } H^1(-1, 1; H^{-1}(\omega))$$

as $\varepsilon \to 0$, where the limits are given by

$$\begin{aligned}\sigma^{\alpha\beta}(0) &= a^{\alpha\beta\sigma\tau}\gamma_{\sigma\tau}(\overline{\boldsymbol{u}}),\\ \sigma^{\alpha3,1} &= -h^{\alpha}_{-} - \int_{-1}^{x_3} f^{\alpha}\, dy_3\\ &\quad - (1+x_3)(\partial_\beta \sigma^{\alpha\beta}(0) + \Gamma^{\tau}_{\beta\tau}\sigma^{\alpha\beta}(0) + \Gamma^{\alpha}_{\beta\tau}\sigma^{\beta\tau}(0)),\\ \sigma^{33,1} &= -h^{3}_{-} - \int_{-1}^{x_3} f^{3}\, dy_3 - (1+x_3)\sigma^{\alpha\beta}(0)b_{\alpha\beta}.\end{aligned}$$

(2) Let the *limit stresses* $\sigma^{\alpha\beta,\varepsilon}(0)$ be defined by the de-scalings

$$\sigma^{\alpha\beta,\varepsilon}(0)(x^{\varepsilon}) := \sigma^{\alpha\beta}(0)(x) \text{ for all } x^{\varepsilon} = \pi^{\varepsilon}x \in \overline{\Omega}.$$

Show that the functions $n^{\alpha\beta,\varepsilon}$ appearing in the boundary value problem found in Thm. 4.5-2 satisfy the relations

$$n^{\alpha\beta,\varepsilon} = \int_{-\varepsilon}^{\varepsilon} \sigma^{\alpha\beta,\varepsilon}(0)\, dx_3^{\varepsilon},$$

thus justifying their definition as components of the *stress resultant tensor* field.

4.5. Consider a family of *linearly elastic elliptic membrane shells* ($\gamma_0 = \gamma$ and S is elliptic) *with variable thickness*, i.e., whose reference configurations are the sets $\boldsymbol{\Theta}^e(\overline{\Omega}^{\varepsilon})$, where the mapping $\boldsymbol{\Theta}^e : \overline{\Omega}^{\varepsilon} \to \mathbb{R}^3$ is defined for each $\varepsilon > 0$ by

$$\boldsymbol{\Theta}^e(y, x_3^{\varepsilon}) := \boldsymbol{\theta}(y) + x_3^{\varepsilon} e(y)\boldsymbol{a}_3(y)$$

for all $(y, x_3^{\varepsilon}) \in \overline{\Omega}^{\varepsilon} = \overline{\omega} \times [-\varepsilon, \varepsilon]$, the given function $e \in W^{2,\infty}(\omega)$ satisfying $0 < e_0 \le e(y)$ for all $y \in \overline{\omega}$.

For such a family of elliptic membrane shells, carry out an asymptotic analysis analogous to that of Thm. 4.4-1. Assuming for simplicity that there are no surface forces, show in particular that the limit scaled two-dimensional problem takes the form

$$\int_\omega a^{\alpha\beta\sigma\tau}\gamma^e_{\sigma\tau}(\boldsymbol{\zeta})\gamma^e_{\alpha\beta}(\boldsymbol{\eta}) e\sqrt{a}\, dy = \int_\omega \left\{\int_{-1}^{1} f^i\, dx_3\right\}\eta_i e\sqrt{a}\, dy$$

for all $\boldsymbol{\eta} \in \mathbf{V}_M(\omega) = H_0^1(\omega) \times H_0^1(\omega) \times L^2(\omega)$, where the covariant components of the linearized change of metric tensor are now replaced by the more general expressions:

$$\gamma_{\alpha\beta}^e(\boldsymbol{\eta}) := \frac{1}{2}(\partial_\beta \eta_\alpha + \partial_\alpha \eta_\beta) - \Gamma_{\alpha\beta}^\sigma \eta_\sigma - \frac{1}{e} b_{\alpha\beta} \eta_3.$$

Remark. This asymptotic analysis is carried out in Busse [1998, Thm. 4.2].

4.6. Let the notations and assumptions be as in Thm. 4.5-1. Show that the linear operator

$$\mathbf{A} : \boldsymbol{p} \in \{\mathbf{V}_M(\omega)\}' \to \mathbf{A}\boldsymbol{p} \in \mathbf{V}_M(\omega)$$

defined by

$$\int_\omega a^{\alpha\beta\sigma\tau} \gamma_{\sigma\tau}(\mathbf{A}\boldsymbol{p}) \gamma_{\alpha\beta}(\boldsymbol{\eta}) \sqrt{a}\, dy = \boldsymbol{p}(\boldsymbol{\eta}) \text{ for all } \boldsymbol{\eta} \in \mathbf{V}_M(\omega)$$

is not compact.

Hint (based on an idea of D. Coutand): Assume that $\mathbf{A}$ is compact and consider a sequence $(\boldsymbol{p}^k)_{k=1}^\infty$ of elements of $\{\mathbf{V}_M(\omega)\}'$ of the form $\boldsymbol{p}^k = (0, 0, p^k)$, where $(p^k)_{k=1}^\infty$ is a bounded sequence in $L^2(\omega)$ that admits no strongly convergent subsequence (such a sequence exists since $\dim L^2(\omega) = +\infty$). Then use the relation

$$\zeta_3 = d a^{\alpha\beta\sigma\tau} b_{\sigma\tau} \hat{\gamma}_{\alpha\beta}(\hat{\boldsymbol{\zeta}}) + d p^3$$

established in Ex. 4.2 (1) to reach a contraction.

CHAPTER 5

LINEARLY ELASTIC GENERALIZED MEMBRANE SHELLS

INTRODUCTION

Consider a linearly elastic shell with middle surface $S = \boldsymbol{\theta}(\overline{\omega})$, subjected to a boundary condition of place along a portion of its lateral face with $\boldsymbol{\theta}(\gamma_0)$ as its middle curve, where $\gamma_0 \subset \gamma$. Such a shell is a *linearly elastic generalized membrane shell* if it is *not* an elliptic membrane shell according to the definition given in Chap. 4, yet its associated space

$$\mathbf{V}_F(\omega) = \{\boldsymbol{\eta} = (\eta_i) \in H^1(\omega) \times H^1(\omega) \times H^2(\omega);\\ \eta_i = \partial_\nu \eta_3 = 0 \text{ on } \gamma_0, \gamma_{\alpha\beta}(\boldsymbol{\eta}) = 0 \text{ in } \omega\}.$$

still reduces to $\{\mathbf{0}\}$ (Sect. 5.1). As shown later in this chapter (Sect. 5.8), *examples* of linearly elastic generalized membrane shells abound.

The elliptic membrane shells studied in Chap. 4 provide a first instance of *linearly elastic membrane shells*, i.e., those for which $\mathbf{V}_F(\omega) = \{\mathbf{0}\}$. The generalized membrane shells studied in the present chapter thus exhaust all the remaining linearly elastic membrane shells.

The purpose of this chapter is to *identify* and to *mathematically justify* the *two-dimensional equations of a linearly elastic generalized membrane shell*, by establishing the *convergence in* ad hoc *functional spaces of the three-dimensional displacements as the thickness of such a shell approaches zero.*

More specifically, consider a *family of linearly elastic generalized membrane shells* with thickness 2ε approaching zero, with each having the *same* middle surface $S = \boldsymbol{\theta}(\overline{\omega})$, and with each subjected to a boundary condition of place along a portion of its lateral face having the *same* set $\boldsymbol{\theta}(\gamma_0)$ as its middle curve. The associated three-dimensional problems, posed in curvilinear coordinates over the sets $\Omega^\varepsilon = \omega \times]-\varepsilon, \varepsilon[$, are first transformed for each $\varepsilon > 0$ into equivalent problems, but now posed over the set $\Omega = \omega \times]-1, 1[$, which is *independent of* ε. This transformation relies in a crucial way on appropriate *scalings of the unknowns* u_i^ε (the covariant components of the displacement field) and *assumptions on the Lamé constants* λ^ε *and* μ^ε *and on the contravariant components* $f^{i,\varepsilon}$ *of the applied body forces* (for simplicity, we assume in this introduction that there are no applied surface forces). More specifically, we define the *scaled unknown* $\boldsymbol{u}(\varepsilon) = (u_i(\varepsilon))$ by letting

$$u_i^\varepsilon(x^\varepsilon) = u_i(\varepsilon)(x) \text{ for all } x^\varepsilon = \pi^\varepsilon x \in \overline{\Omega}^\varepsilon,$$

where $\pi^\varepsilon(x_1, x_2, x_3) = (x_1, x_2, \varepsilon x_3)$. Guided by the formal asymptotic analysis of Sect. 3.4, we then *assume* that there exist constants $\lambda > 0$ and $\mu > 0$ and functions $f^i \in L^2(\Omega)$ *independent of* ε such that

$$\lambda^\varepsilon = \lambda \text{ and } \mu^\varepsilon = \mu,$$
$$f^{i,\varepsilon}(x^\varepsilon) = f^i(x) \text{ for all } x^\varepsilon = \pi^\varepsilon x \in \Omega^\varepsilon.$$

It is found in this fashion that the scaled unknown $\boldsymbol{u}(\varepsilon) = (u_i(\varepsilon))$ satisfies a variational problem of the form (Thm. 5.1-1):

$$\boldsymbol{u}(\varepsilon) \in \mathbf{V}(\Omega) = \left\{\boldsymbol{v} = (v_i) \in \mathbf{H}^1(\Omega); \boldsymbol{v} = \mathbf{0} \text{ on } \gamma_0 \times [-1, 1]\right\},$$
$$\int_\Omega A^{ijkl}(\varepsilon) e_{k\|l}(\varepsilon; \boldsymbol{u}(\varepsilon)) e_{i\|j}(\varepsilon; \boldsymbol{v}) \sqrt{g(\varepsilon)}\, dx$$
$$= \int_\Omega f^i v_i \sqrt{g(\varepsilon)}\, dx \text{ for all } \boldsymbol{v} \in \mathbf{V}(\Omega),$$

where, for any vector field $\boldsymbol{v} = (v_i) \in \mathbf{H}^1(\Omega)$, the *scaled linearized strains* $e_{i\|j}(\varepsilon; \boldsymbol{v}) = e_{j\|i}(\varepsilon; \boldsymbol{v}) \in L^2(\Omega)$ are defined by:

$$e_{\alpha\|\beta}(\varepsilon; \boldsymbol{v}) = \frac{1}{2}(\partial_\beta v_\alpha + \partial_\alpha v_\beta) - \Gamma^p_{\alpha\beta}(\varepsilon) v_p,$$
$$e_{\alpha\|3}(\varepsilon; \boldsymbol{v}) = \frac{1}{2}\left(\frac{1}{\varepsilon}\partial_3 v_\alpha + \partial_\alpha v_3\right) - \Gamma^\sigma_{\alpha 3}(\varepsilon) v_\sigma,$$
$$e_{3\|3}(\varepsilon; \boldsymbol{v}) = \frac{1}{\varepsilon}\partial_3 v_3.$$

After establishing several analytical preliminaries (Sect. 5.2), we prove a crucial *three-dimensional inequality of Korn's type* (Thm. 5.3-1): Given a family of *linearly elastic shells*, with each having the *same* middle surface $S = \boldsymbol{\theta}(\overline{\omega})$ and with each subjected to a boundary condition of place along a portion of its lateral face having the *same* set $\boldsymbol{\theta}(\gamma_0)$ as its middle curve, there exists a constant C such that

$$\|\boldsymbol{v}\|_{1,\Omega} \le \frac{C}{\varepsilon}\Big\{\sum_{i,j} |e_{i\|j}(\varepsilon;\boldsymbol{v})|^2_{0,\Omega}\Big\}^{1/2}$$

for $\varepsilon > 0$ small enough and for all $\boldsymbol{v} \in \mathbf{V}(\Omega)$.

Note that this inequality holds whether the space $\mathbf{V}_F(\omega)$ reduces to $\{\mathbf{0}\}$ (as in this chapter or in Chap. 4) or not (as in the next chapter). Recall that *another* three-dimensional inequality of Korn's type holds, this time with a constant independent of ε, but only for a family of elliptic membrane shells; cf. Sect. 4.3.

A last preliminary is needed, in the form of the following definition (Sect. 5.5): The applied forces are said to be *admissible* if there exists a constant κ_0 such that

$$\Big|\int_\Omega f^i v_i \sqrt{g(\varepsilon)}\,dx\Big| \le \kappa_0\Big\{\sum_{i,j} |e_{i\|j}(\varepsilon;\boldsymbol{v})|^2_{0,\Omega}\Big\}^{1/2}$$

for $\varepsilon > 0$ small enough and for all $\boldsymbol{v} = (v_i) \in \mathbf{V}(\Omega)$.

Equipped with these preliminaries, we then establish the main result of this chapter: Consider a generalized membrane shell *"of the first kind"*, i.e., one for which *the space*

$$\mathbf{V}_0(\omega) = \{\boldsymbol{\eta} = (\eta_i) \in \mathbf{H}^1(\omega);\ \boldsymbol{\eta} = \mathbf{0} \text{ on } \gamma_0,\ \gamma_{\alpha\beta}(\boldsymbol{\eta}) = 0 \text{ in } \omega\}$$

"already" reduces to $\{\mathbf{0}\}$ (Sect. 5.4). Equivalently, the semi-norm $|\cdot|^M_\omega$ defined by

$$|\boldsymbol{\eta}|^M_\omega = \Big\{\sum_{\alpha,\beta} |\gamma_{\alpha\beta}(\boldsymbol{\eta})|^2_{0,\omega}\Big\}^{1/2}$$

becomes a norm over the space

$$\mathbf{V}(\omega) = \{\boldsymbol{\eta} = (\eta_i) \in \mathbf{H}^1(\omega);\ \boldsymbol{\eta} = \mathbf{0} \text{ on } \gamma_0\}.$$

This assumption is satisfied for instances if $\gamma_0 \ne \partial\omega$ and S is elliptic, or if S is a portion of a hyperboloid of revolution and γ_0 is "large enough" (these and other examples are reviewed in Sect. 5.8). We also assume that *the applied forces are admissible.*

We then show (Thm. 5.6-1) that *the averages* $\frac{1}{2}\int_{-1}^{1} \boldsymbol{u}(\varepsilon)\,dx_3$ *converge as* $\varepsilon \to 0$ *in the space*

$$\mathbf{V}^\sharp_M(\Omega) = \text{completion of } \mathbf{V}(\omega) \text{ with respect to } |\cdot|^M_\omega,$$

and that *the limit* $\boldsymbol{\zeta} \in \mathbf{V}^{\sharp}_M(\omega)$ *satisfies the following* (scaled) *two-dimensional equations of a linearly elastic generalized membrane shell* of the first kind:

$$B^{\sharp}_M(\boldsymbol{\zeta}, \boldsymbol{\eta}) = L^{\sharp}_M(\boldsymbol{\eta}) \text{ for all } \boldsymbol{\eta} \in \mathbf{V}^{\sharp}_M(\omega),$$

where $B^{\sharp}_M$ is the unique continuous extension to $\mathbf{V}^{\sharp}_M(\omega) \times \mathbf{V}^{\sharp}_M(\omega)$ of the bilinear form already encountered in the (scaled) two-dimensional variational problem of a membrane shell (Thm. 4.5-1), viz.,

$$(\boldsymbol{\zeta}, \boldsymbol{\eta}) \in \mathbf{V}(\omega) \times \mathbf{V}(\omega) \to \int_\omega a^{\alpha\beta\sigma\tau} \gamma_{\sigma\tau}(\boldsymbol{\zeta}) \gamma_{\alpha\beta}(\boldsymbol{\eta}) \sqrt{a}\, dy,$$

and the continuous linear form $L^{\sharp}_M : \mathbf{V}^{\sharp}_M(\omega) \to \mathbb{R}$ reflects the limit behavior of the admissible applied forces as $\varepsilon \to 0$ (a convergence result also holds for the fields $\boldsymbol{u}(\varepsilon)$ themselves; cf. again Thm. 5.6-1).

We recall that

$$\begin{aligned} \gamma_{\alpha\beta}(\boldsymbol{\eta}) &= \frac{1}{2}(\partial_\beta \eta_\alpha + \partial_\alpha \eta_\beta) - \Gamma^\sigma_{\alpha\beta} \eta_\sigma - b_{\alpha\beta} \eta_3, \\ a^{\alpha\beta\sigma\tau} &= \frac{4\lambda\mu}{\lambda + 2\mu} a^{\alpha\beta} a^{\sigma\tau} + 2\mu (a^{\alpha\sigma} a^{\beta\tau} + a^{\alpha\tau} a^{\beta\sigma}), \end{aligned}$$

the functions $\gamma_{\alpha\beta}(\boldsymbol{\eta})$ and $a^{\alpha\beta\sigma\tau}$ being respectively the covariant components of the *linearized change of metric tensor* associated with a displacement field $\eta_i \boldsymbol{a}^i$ of the middle surface S and the contravariant components of the (scaled) *two-dimensional elasticity tensor of the shell.*

We also treat generalized membrane shells *"of the second kind"*, i.e., those for which $|\cdot|^M_\omega$ is no longer a norm over the space $\mathbf{V}(\omega)$ (this is simply another way of saying that $\mathbf{V}_0(\omega) \neq \{\mathbf{0}\}$), but is a norm over the smaller space

$$\begin{aligned} \mathbf{V}_K(\omega) = \{\boldsymbol{\eta} = (\eta_i) \in H^1(\omega) \times H^1(\omega) \times H^2(\omega); \\ \eta_i = \partial_\nu \eta_3 = 0 \text{ on } \gamma_0\} \end{aligned}$$

A convergence theorem can still be established in this case, but only in the completion of the *quotient space* $\mathbf{V}(\omega)/\mathbf{V}_0(\omega)$ with respect to $|\cdot|^M_\omega$ (Thm. 5.6-2).

We conclude this chapter by reviewing the *existence* and *uniqueness* properties of the solution to the above variational problem. We also rewrite the fundamental convergence theorem in terms of *de-scaled unknowns and data*, thus providing a justification in terms of "physical" quantities of the *two-dimensional equations of a linearly elastic generalized membrane shell* (Thm. 5.8-1).

These two-dimensional equations often provide intriguing examples of *"sensitive" variational problems*, according to the terminology recently introduced by J.L. Lions and E. Sanchez-Palencia, in the sense that they may possess two unusual features: They are posed in spaces that are not necessarily contained in spaces of distributions and their solutions may be "highly sensitive" to arbitrarily small smooth perturbations on the data (Sect. 5.8).

In Sect. 5.8, we also review several *examples* of linearly elastic generalized membrane shells.

5.1. LINEARLY ELASTIC GENERALIZED MEMBRANE SHELLS: DEFINITION AND ASSUMPTIONS ON THE DATA; THE THREE-DIMENSIONAL EQUATIONS OVER A DOMAIN INDEPENDENT OF ε

Let ω be a domain in $\mathbb{R}^2$ with boundary γ and let $\boldsymbol{\theta} \in \mathcal{C}^2(\overline{\omega}; \mathbb{R}^3)$ be an injective mapping such that the two vectors $\partial_\alpha \boldsymbol{\theta}(y)$ are linearly independent at all points $y \in \overline{\omega}$. A *linearly elastic shell* with middle surface $S = \boldsymbol{\theta}(\overline{\omega})$ is called a **linearly elastic generalized membrane shell** if the following three conditions are simultaneously satisfied (the definitions and notations are those of Sect. 3.1):

(i) The shell is subjected to a *boundary condition of place* along a portion of its lateral face with $\boldsymbol{\theta}(\gamma_0)$ as its middle curve, where the subset $\gamma_0 \subset \gamma$ satisfies

$$\boxed{\textit{length}\ \gamma_0 > 0.}$$

(ii) Define the space (∂_ν denotes the outer normal derivative operator along γ):

$$\boxed{\begin{aligned}\mathbf{V}_F(\omega) := \{&\boldsymbol{\eta} = (\eta_i) \in H^1(\omega) \times H^1(\omega) \times H^2(\omega);\\ &\eta_i = \partial_\nu \eta_3 = 0 \text{ on } \gamma_0,\ \gamma_{\alpha\beta}(\boldsymbol{\eta}) = 0 \text{ in } \omega\}.\end{aligned}}$$

Then

$$\boxed{\mathbf{V}_F(\omega) = \{\mathbf{0}\}.}$$

We recall that

$$\gamma_{\alpha\beta}(\boldsymbol{\eta}) = \frac{1}{2}(\partial_\beta \eta_\alpha + \partial_\alpha \eta_\beta) - \Gamma^\sigma_{\alpha\beta}\eta_\sigma - b_{\alpha\beta}\eta_3$$

denote the covariant components of the linearized change of metric tensor associated with a displacement field $\eta_i \boldsymbol{a}^i$ of the surface S. The subscript "F" announces the central rôle that the space $\mathbf{V}_F(\omega)$ plays in the definition of flexural shells; cf. Sect. 6.1.

(iii) *The shell is not an elliptic membrane shell.* We recall that a linearly elastic shell is an elliptic membrane shell if $\gamma_0 = \gamma$ and S is elliptic (Sect. 4.1) and that an elliptic membrane shell also provides an instance where the space $\mathbf{V}_F(\omega)$, which in this case is the space $H_0^1(\omega) \times H_0^1(\omega) \times H_0^2(\omega)$, reduces to $\{\mathbf{0}\}$ (if $\boldsymbol{\theta} \in \mathcal{C}^{2,1}(\omega; \mathbb{R}^3)$; cf. Thm. 2.7-2). *Generalized membrane shells thus exhaust all the remaining cases* of **linearly elastic membrane shells**, i.e., those for which $\mathbf{V}_F(\omega) = \{\mathbf{0}\}$.

Examples of linearly elastic generalized membrane shells are given in Sect. 5.8

A word of caution. Like the definition of a linearly elastic elliptic membrane shell, that of a linearly elastic generalized membrane shell depends only on the subset of the lateral face where the shell is subjected to a boundary condition of place (*via* the set γ_0) and on the "geometry" of the middle curve of the shell; cf. Ex. 5.1. ∎

The formal analysis of Sect. 3.4 then naturally leads us to make the following **scalings of the unknowns** and **assumptions on the data** *for a family of linearly elastic generalized membrane shells*, with each having the *same* middle surface $S = \boldsymbol{\theta}(\overline{\omega})$ and with each subjected to a boundary condition of place along a portion of its lateral face having the *same* set $\boldsymbol{\theta}(\gamma_0)$ as its middle curve, as their thickness 2ε approaches zero.

First, we define the **scaled unknown** $\boldsymbol{u}(\varepsilon) = (u_i(\varepsilon))$ by letting

$$\boxed{u_i^\varepsilon(x^\varepsilon) =: u_i(\varepsilon)(x) \text{ for all } x^\varepsilon = \pi^\varepsilon x \in \overline{\Omega}^\varepsilon.}$$

Next, we *require* that the *Lamé constants* and the *applied body and surface force densities* be such that

$$\boxed{\begin{array}{rcl} \lambda^\varepsilon = \lambda & \text{and} & \mu^\varepsilon = \mu, \\ f^{i,\varepsilon}(x^\varepsilon) = f^i(x) & \text{for all} & x^\varepsilon = \pi^\varepsilon x \in \Omega^\varepsilon, \\ h^{i,\varepsilon}(x^\varepsilon) = \varepsilon h^i(x) & \text{for all} & x^\varepsilon = \pi^\varepsilon x \in \Gamma_+^\varepsilon \cup \Gamma_-^\varepsilon, \end{array}}$$

where the constants $\lambda > 0$ *and* $\mu > 0$ *and the functions* $f^i \in L^2(\Omega)$ *and* $h^i \in L^2(\Gamma_+ \cup \Gamma_-)$ *are independent of* ε (Fig. 3.2-1 recapitulates the definitions of the sets Ω^ε, Ω, Γ_+^ε, Γ_+, Γ_-^ε, and Γ_-).

Remark. For notational brevity, the functions f^i and h^i stand (as in the preceding chapter) for the functions that were respectively denoted $f^{i,0}$ and $h^{i,1}$ in Sect. 3.4. ■

A word of caution. In order to carry out our asymptotic analysis of generalized membrane shells, we will have to make in addition a rather stringent assumption on the *applied forces*, which supersedes in fact the present ones, in such a way that the linear form appearing in the variational problem $\mathcal{P}(\varepsilon; \Omega)$ described in Thm. 5.1-1 becomes continuous with respect to an *ad hoc* norm, and uniformly so with respect to ε; cf. Sect. 5.5. ■

As an immediate corollary to Thm. 3.2-1 (simply corresponding to $p = 0$), we obtain the *problems satisfied by the scaled unknown over the set* Ω, thus over a domain that is *independent of* ε:

Theorem 5.1-1. *Let* ω *be a domain in* $\mathbb{R}^2$, *let* $\boldsymbol{\theta} \in \mathcal{C}^2(\overline{\omega}; \mathbb{R}^3)$ *be an injective mapping such that the two vectors* $\boldsymbol{a}_\alpha = \partial_\alpha \boldsymbol{\theta}$ *are linearly independent at all points of* $\overline{\omega}$, *and let* $\varepsilon_0 > 0$ *be as in* Thm. 3.1-1. *Consider a family of linearly elastic generalized membrane shells with thickness* 2ε, *with each having the same middle surface* $S = \boldsymbol{\theta}(\overline{\omega})$ *and with each subjected to a boundary condition of place along a portion of its lateral face having the same set* $\boldsymbol{\theta}(\gamma_0)$ *as its middle curve.*

Let the assumptions on the data be as above. Then, for each $0 < \varepsilon \le \varepsilon_0$, *the scaled unknown* $\boldsymbol{u}(\varepsilon) = (u_i(\varepsilon))$ *satisfies the following* **scaled three-dimensional variational problem** $\mathcal{P}(\varepsilon; \Omega)$ **of a**

linearly elastic generalized membrane shell:

$$\boxed{\begin{aligned}&\boldsymbol{u}(\varepsilon) \in \mathbf{V}(\Omega) := \{\boldsymbol{v} = (v_i) \in \mathbf{H}^1(\Omega);\ \boldsymbol{v} = \mathbf{0} \text{ on } \Gamma_0\},\\ &\int_\Omega A^{ijkl}(\varepsilon) e_{k\|l}(\varepsilon; \boldsymbol{u}(\varepsilon)) e_{i\|j}(\varepsilon; \boldsymbol{v}) \sqrt{g(\varepsilon)}\, dx\\ &= \int_\Omega f^i v_i \sqrt{g(\varepsilon)}\, dx + \int_{\Gamma_+ \cup \Gamma_-} h^i v_i \sqrt{g(\varepsilon)}\, d\Gamma \text{ for all } \boldsymbol{v} \in \mathbf{V}(\Omega),\end{aligned}}$$

where $\Gamma_0 := \gamma_0 \times [-1, 1]$, *the scaled linearized strains* $e_{i\|j}(\varepsilon; \boldsymbol{v})$ *are given by*

$$\boxed{\begin{aligned} e_{\alpha\|\beta}(\varepsilon; \boldsymbol{v}) &= \frac{1}{2}(\partial_\beta v_\alpha + \partial_\alpha v_\beta) - \Gamma^p_{\alpha\beta}(\varepsilon) v_p,\\ e_{\alpha\|3}(\varepsilon; \boldsymbol{v}) &= \frac{1}{2}\left(\frac{1}{\varepsilon}\partial_3 v_\alpha + \partial_\alpha v_3\right) - \Gamma^\sigma_{\alpha 3}(\varepsilon) v_\sigma,\\ e_{3\|3}(\varepsilon; \boldsymbol{v}) &= \frac{1}{\varepsilon}\partial_3 v_3,\end{aligned}}$$

and the functions $A^{ijkl}(\varepsilon)$, $g(\varepsilon)$, $\Gamma^p_{ij}(\varepsilon)$ *are defined as in* Thm. 3.2-1. ■

Our main objective in this chapter consists in analyzing the *behavior of the solutions* $\boldsymbol{u}(\varepsilon) \in \mathbf{H}^1(\Omega)$ *of problems* $\mathcal{P}(\varepsilon; \Omega)$ *as* $\varepsilon \to 0$.

To this end, we begin by proving in Sects. 5.2 and 5.3 various "analytical" preliminaries, which complement the "geometrical" and "mechanical" preliminaries proved in Sect. 3.3. Note that these preliminaries are common to *all types of linearly elastic shells.* In particular, they likewise play an essential rôle in the asymptotic analysis of linearly elastic flexural shells carried out in the next chapter.

5.2. ANALYTICAL PRELIMINARIES

To begin with, we analyze the asymptotic behavior as $\varepsilon \to 0$ of the scaled linearized strains

$$e_{\alpha\|\beta}(\varepsilon; \boldsymbol{v}) = \frac{1}{2}(\partial_\beta v_\alpha + \partial_\alpha v_\beta) - \Gamma^p_{\alpha\beta}(\varepsilon) v_p$$

appearing in the definition of the scaled three-dimensional problem $\mathcal{P}(\varepsilon;\,\Omega)$ (Thm. 5.1-1). To this end, we are naturally led to introduce the "three-dimensional analogs" $\gamma_{\alpha\beta}(\boldsymbol{v})$ and $\rho_{\alpha\beta}(\boldsymbol{v})$ of the covariant components $\gamma_{\alpha\beta}(\boldsymbol{\eta})$ and $\rho_{\alpha\beta}(\boldsymbol{\eta})$ of the "two-dimensional" linearized change of metric and change of curvature tensors (Sects. 2.4 and 2.5). The following result is due to Ciarlet, Lods & Miara [1996, Lemma 3.2].

Theorem 5.2-1. *Let the functions $\Gamma^\sigma_{\alpha\beta}$, $b_{\alpha\beta}$, $b^\beta_\alpha \in \mathcal{C}^0(\overline{\omega})$ be identified with functions in $\mathcal{C}^0(\overline{\Omega})$ and let for any $\boldsymbol{v} = (v_i) \in \mathbf{H}^1(\Omega)$ the functions $\gamma_{\alpha\beta}(\boldsymbol{v}) \in L^2(\Omega)$ and $\rho_{\alpha\beta}(\boldsymbol{v}) \in H^{-1}(\Omega)$ be defined by:*

$$\begin{aligned}
\gamma_{\alpha\beta}(\boldsymbol{v}) &:= \frac{1}{2}(\partial_\beta v_\alpha + \partial_\alpha v_\beta) - \Gamma^\sigma_{\alpha\beta} v_\sigma - b_{\alpha\beta} v_3,\\
\rho_{\alpha\beta}(\boldsymbol{v}) &:= \partial_{\alpha\beta} v_3 - \Gamma^\sigma_{\alpha\beta}\partial_\sigma v_3 + b^\sigma_\beta(\partial_\alpha v_\sigma - \Gamma^\tau_{\alpha\sigma} v_\tau)\\
&\quad + b^\sigma_\alpha(\partial_\beta v_\sigma - \Gamma^\tau_{\beta\sigma} v_\tau) + b^\sigma_\alpha|_\beta v_\sigma - b^\sigma_\alpha b_{\sigma\beta} v_3,
\end{aligned}$$

where $b^\tau_\beta|_\alpha = \partial_\alpha b^\tau_\beta + \Gamma^\tau_{\alpha\sigma} b^\sigma_\beta - \Gamma^\sigma_{\alpha\beta} b^\tau_\sigma = b^\tau_\alpha|_\beta$ (Thm. 2.5-1). *Then there exists a constant C_1 such that, for all $0 < \varepsilon \le \varepsilon_0$ and all $\boldsymbol{v} \in \mathbf{H}^1(\Omega)$, the scaled linearized strains $e_{\alpha\|\beta}(\varepsilon;\,\boldsymbol{v})$ satisfy*

$$\left|\frac{1}{\varepsilon} e_{\alpha\|\beta}(\varepsilon;\,\boldsymbol{v}) - e^1_{\alpha\|\beta}(\varepsilon;\,\boldsymbol{v})\right|_{0,\Omega} \le C_1 \varepsilon \sum_\alpha |v_\alpha|_{0,\Omega},$$

where

$$e^1_{\alpha\|\beta}(\varepsilon;\,\boldsymbol{v}) := \frac{1}{\varepsilon}\gamma_{\alpha\beta}(\boldsymbol{v}) + x_3 b^\sigma_\beta|_\alpha v_\sigma + x_3 b^\sigma_\alpha b_{\sigma\beta} v_3$$

and ($\|\cdot\|_{-1,\Omega}$ denotes the norm in $H^{-1}(\Omega)$):

$$\begin{aligned}
&\left\|\frac{1}{\varepsilon}\partial_3 e_{\alpha\|\beta}(\varepsilon;\,\boldsymbol{v}) + \rho_{\alpha\beta}(\boldsymbol{v})\right\|_{-1,\Omega}\\
&\quad \le C_1\Big\{\sum_i |e_{i\|3}(\varepsilon;\,\boldsymbol{v})|_{0,\Omega} + \varepsilon\sum_\alpha |v_\alpha|_{0,\Omega} + \varepsilon\|v_3\|_{1,\Omega}\Big\}.
\end{aligned}$$

Proof. The constants c_1, c_2 and c_3 appearing in this proof are meant to be independent of ε and of $\boldsymbol{v} \in \mathbf{H}^1(\Omega)$.

The first inequality is a consequence of the definitions of the scaled strains $e_{\alpha\|\beta}(\varepsilon;\,\boldsymbol{v})$ and of the asymptotic behavior of the functions $\Gamma^i_{\alpha\beta}(\varepsilon)$ (Thm. 3.3-1). Next, let the functions $e^0_{\alpha\|3}(\varepsilon;\,\boldsymbol{v}) \in L^2(\Omega)$ and

$\tilde{\rho}_{\alpha\beta}(\boldsymbol{v}) \in H^{-1}(\Omega)$ be defined by

$$\begin{aligned} e^0_{\alpha\|3}(\varepsilon;\boldsymbol{v}) &:= \frac{1}{2}\Big(\partial_\alpha v_3 + \frac{1}{\varepsilon}\partial_3 v_\alpha\Big) + b^\sigma_\alpha v_\sigma, \\ \tilde{\rho}_{\alpha\beta}(\boldsymbol{v}) &:= \partial_{\alpha\beta} v_3 + \partial_\alpha(b^\tau_\beta v_\tau) + \partial_\beta(b^\sigma_\alpha v_\sigma) - \Gamma^\sigma_{\alpha\beta}(\partial_\sigma v_3 + 2b^\tau_\sigma v_\tau) \\ &\quad - b^\sigma_\beta|_\alpha v_\sigma - b^\sigma_\alpha b_{\sigma\beta} v_3. \end{aligned}$$

Then a simple computation shows that

$$\begin{aligned} &\partial_3 e^1_{\alpha\|\beta}(\varepsilon;\boldsymbol{v}) + \tilde{\rho}_{\alpha\beta}(\boldsymbol{v}) = \partial_\alpha e^0_{\beta\|3}(\varepsilon;\boldsymbol{v}) + \partial_\beta e^0_{\alpha\|3}(\varepsilon;\boldsymbol{v}) \\ &\quad -2\Gamma^\sigma_{\alpha\beta} e^0_{\sigma\|3}(\varepsilon;\boldsymbol{v}) - b_{\alpha\beta} e_{3\|3}(\varepsilon;\boldsymbol{v}) + \varepsilon x_3 b^\sigma_\alpha b_{\sigma\beta} e_{3\|3}(\varepsilon;\boldsymbol{v}) \\ &\quad +\varepsilon x_3 b^\sigma_\beta|_\alpha \big(2e^0_{\sigma\|3}(\varepsilon;\boldsymbol{v}) - \partial_\sigma v_3 - 2b^\tau_\sigma v_\tau\big). \end{aligned}$$

Consequently,

$$\begin{aligned} &\|\partial_3 e^1_{\alpha\|\beta}(\varepsilon;\boldsymbol{v}) + \tilde{\rho}_{\alpha\beta}(\boldsymbol{v})\|_{-1,\Omega} \\ &\leq c_1\Big\{\sum_\alpha |e^0_{\alpha\|3}(\varepsilon;\boldsymbol{v})|_{0,\Omega} + |e_{3\|3}(\varepsilon;\boldsymbol{v})|_{0,\Omega} + \varepsilon\sum_\alpha |v_\alpha|_{0,\Omega} + \varepsilon\|v_3\|_{1,\Omega}\Big\} \end{aligned}$$

on the one hand. On the other, the first inequality implies that

$$\|\frac{1}{\varepsilon}\partial_3 e_{\alpha\|\beta}(\varepsilon;\boldsymbol{v}) - \partial_3 e^1_{\alpha\|\beta}(\varepsilon;\boldsymbol{v})\|_{-1,\Omega} \leq c_2\varepsilon\sum_\alpha |v_\alpha|_{0,\Omega},$$

the definitions of the scaled strains $e_{\alpha\|3}(\varepsilon;\boldsymbol{v})$ and of the functions $e^0_{\alpha\|3}(\varepsilon;\boldsymbol{v})$ together with the asymptotic behavior of the functions $\Gamma^\sigma_{\alpha 3}(\varepsilon)$ imply that

$$|e_{\alpha\|3}(\varepsilon;\boldsymbol{v}) - e^0_{\alpha\|3}(\varepsilon;\boldsymbol{v})|_{0,\Omega} \leq c_3\varepsilon\sum_\alpha |v_\alpha|_{0,\Omega},$$

and, finally, an easy computation shows that, for $\boldsymbol{v} \in \mathbf{H}^1(\Omega)$,

$$\rho_{\alpha\beta}(\boldsymbol{v}) = \tilde{\rho}_{\alpha\beta}(\boldsymbol{v}).$$

Hence the second inequality follows from the above relations. ∎

The next theorem, which is due to Ciarlet, Lods & Miara [1996, Lemma 3.3], is *crucial*, as it plays an essential rôle in the proofs of a three-dimensional inequality of Korn's type for a family of linearly elastic shells (Thm. 5.3-1) and of the convergence of the scaled unknown as $\varepsilon \to 0$ (Thms. 5.6-1 and 5.6-2): Consider a sequence $(\boldsymbol{u}(\varepsilon))_{\varepsilon>0}$ of functions in the space $\mathbf{V}(\Omega)$ that converges weakly in $\mathbf{H}^1(\Omega)$, thus also strongly in $\mathbf{L}^2(\Omega)$, and let $\boldsymbol{u}$ be its limit. We first show that, if it so happens that the corresponding sequences $(e_{i\|j}(\varepsilon; \boldsymbol{u}(\varepsilon)))_{\varepsilon>0}$ weakly converge in $L^2(\Omega)$, considerable information can then be gathered about the limit $\boldsymbol{u}$ and the functions $\gamma_{\alpha\beta}(\boldsymbol{u})$ and $\rho_{\alpha\beta}(\boldsymbol{u})$. We also show that, if in addition the corresponding sequences $(\rho_{\alpha\beta}(\boldsymbol{u}(\varepsilon)))_{\varepsilon>0}$ strongly converge in $H^{-1}(\Omega)$, then in fact the sequence $(\boldsymbol{u}(\varepsilon))_{\varepsilon>0}$ strongly converges in $\mathbf{H}^1(\Omega)$.

In the following statement, the symbols $\to$ and $\rightharpoonup$ respectively denote strong and weak convergences and ∂_ν denotes the outer normal derivative operator along the boundary of ω.

Theorem 5.2-2. *Let*

$$\mathbf{V}(\Omega) := \{\boldsymbol{v} = (v_i) \in \mathbf{H}^1(\Omega);\ \boldsymbol{v} = \mathbf{0} \text{ on } \Gamma_0\},$$

and let $e_{i\|j}(\varepsilon; \boldsymbol{v}) \in L^2(\Omega)$, $\gamma_{\alpha\beta}(\boldsymbol{v}) \in L^2(\Omega)$, *and* $\rho_{\alpha\beta}(\boldsymbol{v}) \in H^{-1}(\Omega)$ *be defined for any function* $\boldsymbol{v} \in \mathbf{V}(\Omega)$ *as in* Thms. 5.1-1 *and* 5.2-1. *Let* $(\boldsymbol{u}(\varepsilon))_{\varepsilon>0}$ *be a sequence of functions* $\boldsymbol{u}(\varepsilon) \in \mathbf{V}(\Omega)$ *that satisfies*

$$\boldsymbol{u}(\varepsilon) \rightharpoonup \boldsymbol{u} \text{ in } \mathbf{H}^1(\Omega),$$

$$\frac{1}{\varepsilon} e_{i\|j}(\varepsilon; \boldsymbol{u}(\varepsilon)) \rightharpoonup e^1_{i\|j} \text{ in } L^2(\Omega),$$

as $\varepsilon \to 0$. *Then*

$$\boldsymbol{u} = (u_i) \text{ is independent of the transverse variable } x_3,$$

$$\overline{\boldsymbol{u}} = (\overline{u}_i) := \frac{1}{2}\int_{-1}^{1} \boldsymbol{u}\, dx_3 \in H^1(\omega) \times H^1(\omega) \times H^2(\omega),$$

$$\overline{u}_i = \partial_\nu \overline{u}_3 = 0 \text{ on } \gamma_0,$$

$$\gamma_{\alpha\beta}(\boldsymbol{u}) = 0,$$

$$\rho_{\alpha\beta}(\boldsymbol{u}) \in L^2(\Omega) \text{ and } \rho_{\alpha\beta}(\boldsymbol{u}) = -\partial_3 e^1_{\alpha\|\beta}.$$

If in addition there exist functions $\chi_{\alpha\beta} \in H^{-1}(\Omega)$ *such that*

$$\rho_{\alpha\beta}(\boldsymbol{u}(\varepsilon)) \to \chi_{\alpha\beta} \text{ in } H^{-1}(\Omega) \text{ as } \varepsilon \to 0,$$

then

$$\boldsymbol{u}(\varepsilon) \to \boldsymbol{u} \text{ in } \mathbf{H}^1(\Omega) \text{ as } \varepsilon \to 0,$$

$$\rho_{\alpha\beta}(\boldsymbol{u}) = \chi_{\alpha\beta} \text{ and thus } \chi_{\alpha\beta} \in L^2(\Omega).$$

Proof. For the sake of clarity, the proof is divided into six parts, numbered (i) to (vi). We first recall that (Thm. 4.2-1 (a)), given $v \in L^2(\Omega)$, $\overline{v}(y) := \frac{1}{2}\int_{-1}^{1} v(y, x_3)\, dx_3$ is finite for almost all $y \in \omega$ and the *average*

$$\overline{v} := \frac{1}{2}\int_{-1}^{1} v\, dx_3,$$

defined in this fashion is in $L^2(\omega)$, or in $H^1(\omega)$ if $v \in H^1(\Omega)$; in particular, the functions $\overline{u}_i$ are in $H^1(\omega)$ and they vanish on γ_0 (Thm. 4.2-1 (b)). For notational brevity, we let

$$e_{i\|j}(\varepsilon) := e_{i\|j}(\varepsilon; \boldsymbol{u}(\varepsilon))$$

throughout the proof. Finally, observe that the assumption $\boldsymbol{u}(\varepsilon) \rightharpoonup \boldsymbol{u}$ in $\mathbf{H}^1(\Omega)$ implies that $\boldsymbol{u}(\varepsilon) \to \boldsymbol{u}$ in $\mathbf{L}^2(\Omega)$ as the imbedding from $H^1(\Omega)$ into $L^2(\Omega)$ is compact.

(i) *We first show that* $\boldsymbol{u} = (u_i)$ *is independent of* x_3 *and that* $\overline{\boldsymbol{u}} = \mathbf{0}$ *on* γ_0. Since the sequence $(\boldsymbol{u}(\varepsilon))_{\varepsilon>0}$ is bounded in $\mathbf{H}^1(\Omega)$ and $e_{i\|j}(\varepsilon) \to 0$ in $L^2(\Omega)$ (in a Hilbert space, a weakly convergent sequence is bounded),

$$\partial_3 u_\alpha(\varepsilon) = \varepsilon\{2e_{\alpha\|3}(\varepsilon) - \partial_\alpha u_3(\varepsilon) + 2\Gamma^\sigma_{\alpha 3}(\varepsilon)u_\sigma(\varepsilon)\} \to 0 \text{ in } L^2(\Omega),$$
$$\partial_3 u_3(\varepsilon) = \varepsilon e_{3\|3}(\varepsilon) \to 0 \text{ in } L^2(\Omega).$$

Hence $\partial_3 u_i = 0$, and consequently, by Thm. 4.2-1 (a),

$$\boldsymbol{u}(y, x_3) = \overline{\boldsymbol{u}}(y) \text{ for almost all } (y, x_3) \in \Omega,$$

where the function $\overline{\boldsymbol{u}} = (\overline{u}_i) = \frac{1}{2}\int_{-1}^{1} \boldsymbol{u}\, dx_3$ is in the space $\mathbf{H}^1(\omega)$ and satisfies $\overline{\boldsymbol{u}} = \mathbf{0}$ on γ_0. Thus $\boldsymbol{u} = (u_i)$ is independent of the transverse variable x_3.

(ii) *We next show that* $\overline{u}_3 \in H^2(\omega)$ *and* $\partial_\nu \overline{u}_3 = 0$ *on* γ_0. If $v \in \mathcal{C}^1(\overline{\Omega})$, an integration by parts shows that

$$\int_{-1}^{1} x_3 v(y, x_3)\, dx_3 = \frac{1}{2}\int_{-1}^{1} (1 - x_3^2)\partial_3 v(y, x_3)\, dx_3 \text{ for all } y \in \overline{\omega}.$$

By Thm. 4.2-1, the mappings

$$w \in L^2(\Omega) \to \overline{w} \in L^2(\omega) \text{ and } w \in H^1(\Omega) \to \overline{w} \in H^1(\omega)$$

are both continuous, so that the above relation remains valid (for almost all $y \in \omega$) if $v \in H^1(\Omega)$.

Let

$$\overline{\boldsymbol{u}}^1(\varepsilon) = (\overline{u}_i^1(\varepsilon)) := \frac{1}{2\varepsilon} \int_{-1}^{1} (1 - x_3^2) \partial_3 \boldsymbol{u}(\varepsilon) \, dx_3.$$

Hence we may also write

$$\overline{\boldsymbol{u}}^1(\varepsilon) = \frac{1}{\varepsilon} \int_{-1}^{1} x_3 \boldsymbol{u}(\varepsilon) \, dx_3,$$

and, by Thm. 4.2-1 (b),

$$\overline{\boldsymbol{u}}^1(\varepsilon) \in \mathbf{H}^1(\omega) \text{ and } \overline{\boldsymbol{u}}^1(\varepsilon) = \mathbf{0} \text{ on } \gamma_0.$$

By assumption, $e_{3\|3}(\varepsilon) = \frac{1}{\varepsilon}\partial_3 u_3(\varepsilon) \to 0$ in $L^2(\Omega)$, and thus, by definition of $\overline{u}_3^1(\varepsilon)$,

$$\overline{u}_3^1(\varepsilon) \to 0 \text{ in } L^2(\omega).$$

The assumptions made on the sequence $(\boldsymbol{u}(\varepsilon))_{\varepsilon>0}$ combined with the asymptotic behavior of the functions $\Gamma_{\alpha 3}^{\sigma}(\varepsilon)$ (Thm. 3.3-1) imply that

$$\frac{1}{\varepsilon}\partial_3 u_\alpha(\varepsilon) = 2e_{\alpha\|3}(\varepsilon) - \partial_\alpha u_3(\varepsilon) + 2\Gamma_{\alpha 3}^{\sigma}(\varepsilon) u_\sigma(\varepsilon) \rightharpoonup \{-\partial_\alpha u_3 - 2b_\alpha^\sigma u_\sigma\}$$

in $L^2(\Omega)$, and thus (a strongly continuous linear mapping is also continuous for the weak topologies; see, e.g., Brezis [1983, Thm. III.9], by definition of $\overline{u}_\alpha^1(\varepsilon)$,

$$\overline{u}_\alpha^1(\varepsilon) \rightharpoonup \frac{1}{2} \int_{-1}^{1} (x_3^2 - 1)(\partial_\alpha u_3 + 2b_\alpha^\sigma u_\sigma) \, dx_3 \text{ in } L^2(\omega).$$

Since the functions u_i are independent of x_3, since $\overline{\partial_\alpha u_3} = \partial_\alpha \overline{u}_3$, again by Thm. 4.2-1 (b), and since $\frac{1}{2}\int_{-1}^{1}(x_3^2 - 1)\, dx_3 = -\frac{2}{3}$, it follows that

$$\overline{u}_\alpha^1(\varepsilon) \rightharpoonup \overline{u}_\alpha^1 := -\frac{2}{3}\left(\partial_\alpha \overline{u}_3 + 2b_\alpha^\beta \overline{u}_\beta\right) \text{ in } L^2(\omega).$$

Our next objective consists in showing that, *in fact, the sequences* $(\overline{u}_\alpha^1(\varepsilon))_{\varepsilon>0}$ *weakly converge in* $H^1(\omega)$. To this end, it suffices to establish that the sequences $(\frac{1}{2}(\partial_\alpha \overline{u}_\beta^1(\varepsilon) + \partial_\beta \overline{u}_\alpha^1(\varepsilon)))_{\varepsilon>0}$ weakly converge in $L^2(\omega)$ (by the two-dimensional Korn inequality, the norm $\|\cdot\|_{1,\omega}$ is equivalent to the norm $(\eta_\alpha) \to \{\sum_{\alpha,\beta} |\frac{1}{2}(\partial_\alpha \eta_\beta + \partial_\beta \eta_\alpha)|_{0,\omega}^2\}^{1/2}$ over the space $\{(\eta_\alpha) \in \mathbf{H}^1(\omega);\, \eta_\alpha = 0 \text{ on } \gamma_0\})$. A simple computation, based in particular on the equality $\overline{\boldsymbol{u}}^1(\varepsilon) = \frac{1}{\varepsilon}\int_{-1}^{1} x_3 \boldsymbol{u}(\varepsilon)\, dx_3$, shows that

$$\frac{1}{2}\big(\partial_\alpha \overline{u}^1_\beta(\varepsilon) + \partial_\beta \overline{u}^1_\alpha(\varepsilon)\big) = 2\overline{x_3 e^1_{\alpha\|\beta}(\varepsilon)}$$
$$-2b^\sigma_\beta|_\alpha \overline{x_3^2 u_\sigma(\varepsilon)} - 2b^\sigma_\alpha b_{\sigma\beta}\overline{x_3^2 u_3(\varepsilon)} + b_{\alpha\beta}\overline{u}^1_3(\varepsilon) + \Gamma^\sigma_{\alpha\beta}\overline{u}^1_\sigma(\varepsilon),$$

where

$$e^1_{\alpha\|\beta}(\varepsilon) := e^1_{\alpha\|\beta}(\varepsilon;\, \boldsymbol{u}(\varepsilon)),$$

and the functions $e^1_{\alpha\|\beta}(\varepsilon;\, \boldsymbol{v})$ are defined as in Thm. 5.2-1. The first inequality established in this theorem together with the assumptions $\boldsymbol{u}_\alpha(\varepsilon) \to \boldsymbol{u}_\alpha$ in $\mathbf{L}^2(\Omega)$ then imply that

$$\left\{\frac{1}{\varepsilon} e_{\alpha\|\beta}(\varepsilon) - e^1_{\alpha\|\beta}(\varepsilon)\right\} \to 0 \text{ in } L^2(\Omega),$$

and the assumptions $\frac{1}{\varepsilon} e_{\alpha\|\beta}(\varepsilon) \rightharpoonup e^1_{\alpha\|\beta}$ in $L^2(\Omega)$ in turn imply that

$$e^1_{\alpha\|\beta}(\varepsilon) \rightharpoonup e^1_{\alpha\|\beta} \text{ in } L^2(\Omega).$$

Consequently,

$$\overline{x_3 e^1_{\alpha\|\beta}(\varepsilon)} \rightharpoonup \overline{x_3 e^1_{\alpha\|\beta}} \text{ in } L^2(\omega).$$

Since $\boldsymbol{u}(\varepsilon) \to \boldsymbol{u}$ in $L^2(\Omega)$ as $\boldsymbol{u}(\varepsilon) \rightharpoonup \boldsymbol{u}$ in $\mathbf{H}^1(\Omega)$ by assumption, both sequences $(b^\sigma_\beta|_\alpha \overline{x_3^2 u_\sigma(\varepsilon)})_{\varepsilon>0}$ and $(b^\sigma_\alpha b_{\sigma\beta}\overline{x_3^2 u_3(\varepsilon)})_{\varepsilon>0}$ strongly converge in $L^2(\omega)$; since $\overline{u}^1_3(\varepsilon) \to 0$ in $L^2(\omega)$ (part (ii)), the sequence $(b_{\alpha\beta}\overline{u}^1_3(\varepsilon))_{\varepsilon>0}$ strongly converges (to 0) in $L^2(\omega)$ and since $(\overline{u}^1_\alpha(\varepsilon))_{\varepsilon>0}$ weakly converges in $L^2(\omega)$ (part (ii)), the sequence $(\Gamma^\sigma_{\alpha\beta}\overline{u}^1_\sigma(\varepsilon))_{\varepsilon>0}$ weakly converges in $L^2(\omega)$.

Hence the sequence $(\overline{u}^1_\alpha(\varepsilon))_{\varepsilon>0}$ weakly converges in $H^1(\omega)$; thus $\overline{u}^1_\alpha(\varepsilon) \rightharpoonup \overline{u}^1_\alpha$ in $L^2(\omega)$ implies that $\overline{u}^1_\alpha \in H^1(\omega)$, and therefore that $\partial_\alpha \overline{u}_3 \in H^1(\omega)$ since $b^\beta_\alpha \overline{u}_\beta \in H^1(\omega)$; we have shown in this fashion that

$$\overline{u}_3 \in H^2(\omega).$$

We know from (ii) that $\overline{u}^1_\alpha(\varepsilon) = 0$ on γ_0; therefore, $\overline{u}^1_\alpha(\varepsilon) \rightharpoonup \overline{u}^1_\alpha$ in $H^1(\omega)$ implies that $\overline{u}^1_\alpha = 0$ on γ_0. Since $\overline{u}_i = 0$ on γ_0, the equality $\overline{u}^1_\alpha = -\frac{2}{3}(\partial_\alpha \overline{u}_3 + 2b^\beta_\alpha \overline{u}_\beta)$ shows that $\partial_\alpha \overline{u}_3 = 0$ on γ_0; hence

$$\partial_\nu \overline{u}_3 = 0 \text{ on } \gamma_0.$$

We have thus established that $(\overline{u}_i) \in H^1(\omega) \times H^1(\omega) \times H^2(\omega)$ and that $\overline{u}_i = \partial_\nu \overline{u}_3 = 0$ on γ_0.

(iii) *We next prove that $\gamma_{\alpha\beta}(\boldsymbol{u}(\varepsilon)) \to 0$ in $L^2(\Omega)$ as $\varepsilon \to 0$ and that $\gamma_{\alpha\beta}(\boldsymbol{u}) = 0$.* By definition, the functions $e^1_{\alpha\|\beta}(\varepsilon)$ are given by (Thm. 5.2-1)

$$e^1_{\alpha\|\beta}(\varepsilon) = \frac{1}{\varepsilon}\gamma_{\alpha\beta}(\boldsymbol{u}(\varepsilon)) + x_3 b^\sigma_\beta|_\alpha u_\sigma(\varepsilon) + x_3 b^\sigma_\alpha b_{\sigma\beta} u_3(\varepsilon).$$

Hence

$$|\gamma_{\alpha\beta}(\boldsymbol{u}(\varepsilon)) - e_{\alpha\|\beta}(\varepsilon)|_{0,\Omega} \le \varepsilon |e^1_{\alpha\|\beta}(\varepsilon) - \frac{1}{\varepsilon} e_{\alpha\|\beta}(\varepsilon)|_{0,\Omega}$$
$$+\varepsilon|x_3 b^\sigma_\beta|_\alpha u_\sigma(\varepsilon)|_{0,\Omega} + \varepsilon|x_3 b^\sigma_\alpha b_{\sigma\beta} u_3(\varepsilon)|_{0,\Omega},$$

and thus, by the first inequality in Thm. 5.2-1 and the assumptions made on the sequence $(\boldsymbol{u}(\varepsilon))_{\varepsilon>0}$,

$$\gamma_{\alpha\beta}(\boldsymbol{u}(\varepsilon)) \to 0 \text{ in } L^2(\Omega).$$

By the same assumptions,

$$\gamma_{\alpha\beta}(\boldsymbol{u}(\varepsilon)) \rightharpoonup \gamma_{\alpha\beta}(\boldsymbol{u}) \text{ in } L^2(\Omega).$$

Hence $\gamma_{\alpha\beta}(\boldsymbol{u}) = 0$, as was to be proved.

(iv) *We next prove that $\rho_{\alpha\beta}(\boldsymbol{u}) = -\partial_3 e^1_{\alpha\|\beta}$ in $L^2(\Omega)$.* From the same assumptions and from the second inequality in Thm. 5.2-1, we infer that

$$\left\{\frac{1}{\varepsilon}\partial_3 e_{\alpha\|\beta}(\varepsilon) + \rho_{\alpha\beta}(\boldsymbol{u}(\varepsilon))\right\} \to 0 \text{ in } H^{-1}(\Omega).$$

The operator $\partial_3 : L^2(\Omega) \to H^{-1}(\Omega)$ being also continuous for the weak topologies, we deduce from the assumption $\boldsymbol{u}(\varepsilon) \rightharpoonup \boldsymbol{u}$ in $\mathbf{H}^1(\Omega)$ that

$$\frac{1}{\varepsilon}\partial_3 e_{\alpha\|\beta}(\varepsilon) \rightharpoonup \partial_3 e^1_{\alpha\|\beta} \text{ in } H^{-1}(\Omega),$$

hence that

$$\left\{\rho_{\alpha\beta}(\boldsymbol{u}(\varepsilon)) + \partial_3 e^1_{\alpha\|\beta}\right\} \rightharpoonup 0 \text{ in } H^{-1}(\Omega).$$

Since, again by the assumptions on the sequence $(\boldsymbol{u}(\varepsilon))_{\varepsilon>0}$,

$$\rho_{\alpha\beta}(\boldsymbol{u}(\varepsilon)) \rightharpoonup \rho_{\alpha\beta}(\boldsymbol{u}) \text{ in } H^{-1}(\Omega),$$

the desired equality $\rho_{\alpha\beta}(\boldsymbol{u}) = -\partial_3 e^1_{\alpha\|\beta}$ holds in $H^{-1}(\Omega)$; that it is in fact an equality in $L^2(\Omega)$ follows from the relation $\overline{u}_3 \in H^2(\omega)$ established in part (ii) which, together with the independence of u_3 with respect of x_3 established in part (i), implies that $u_3 \in H^2(\Omega)$.

(v) *We next prove that, under the additional assumption that* $\rho_{\alpha\beta}(\boldsymbol{u}(\varepsilon)) \to \chi_{\alpha\beta}$ *in* $H^{-1}(\Omega)$ *as* $\varepsilon \to 0$, *the sequence* $(u_3(\varepsilon))_{\varepsilon>0}$ *strongly converges in* $H^1(\Omega)$. We first note that

$$\partial_3 u_3(\varepsilon) = \varepsilon e_{3\|3}(\varepsilon) \to 0 = \partial_3 u_3 \text{ in } L^2(\Omega).$$

By the *lemma of J.L. Lions* (Thm. 1.7-1), the mapping

$$v \in L^2(\Omega) \to (v, (\partial_i v)) \in H^{-1}(\Omega) \times \mathbf{H}^{-1}(\Omega)$$

is an isomorphism. In order to prove that

$$\partial_\alpha u_3(\varepsilon) \to \partial_\alpha u_3 \text{ in } L^2(\Omega),$$

it therefore suffices to prove that

$$\partial_\alpha u_3(\varepsilon) \to \partial_\alpha u_3 \text{ in } H^{-1}(\Omega),$$
$$\partial_{\alpha i} u_3(\varepsilon) \to \partial_{\alpha i} u_3 \text{ in } H^{-1}(\Omega).$$

The first convergence is a consequence of the assumed convergence $u_3(\varepsilon) \to u_3$ in $L^2(\Omega)$. The relation $\partial_3 u_3(\varepsilon) \to 0 = \partial_3 u_3$ in $L^2(\Omega)$ likewise implies that

$$\partial_{\alpha 3} u_3(\varepsilon) \to 0 = \partial_{\alpha 3} u_3 \text{ in } H^{-1}(\Omega).$$

From the definition of the functions $\rho_{\alpha\beta}(\boldsymbol{v})$ (Thm. 5.2-1) and from the assumption $\boldsymbol{u}(\varepsilon) \rightharpoonup \boldsymbol{u}$ in $\mathbf{H}^1(\Omega)$, we infer that

$$\rho_{\alpha\beta}(\boldsymbol{u}(\varepsilon)) \rightharpoonup \rho_{\alpha\beta}(\boldsymbol{u}) \text{ in } H^{-1}(\Omega),$$

and thus,

$$\rho_{\alpha\beta}(\boldsymbol{u}(\varepsilon)) \to \chi_{\alpha\beta} = \rho_{\alpha\beta}(\boldsymbol{u}) \text{ in } H^{-1}(\Omega).$$

This relation also shows that $\chi_{\alpha\beta} \in L^2(\Omega)$ since $\rho_{\alpha\beta}(\boldsymbol{u}) \in L^2(\Omega)$ (part (iv)).

Since $u_i(\varepsilon) \to u_i$ in $H^{-1}(\Omega)$ and $\partial_\alpha u_i(\varepsilon) \to \partial_\alpha u_i$ in $H^{-1}(\Omega)$ by assumption, and since

$$\begin{aligned}\partial_{\alpha\beta} u_3(\varepsilon) &= \rho_{\alpha\beta}(\boldsymbol{u}(\varepsilon)) + \Gamma^\sigma_{\alpha\beta}\partial_\sigma u_3(\varepsilon) - b^\sigma_\beta(\partial_\alpha u_\sigma(\varepsilon) - \Gamma^\tau_{\alpha\sigma} u_\tau(\varepsilon)) \\ &\quad - b^\sigma_\alpha(\partial_\beta u_\sigma(\varepsilon) - \Gamma^\tau_{\beta\sigma} u_\tau(\varepsilon)) - b^\sigma_\alpha|_\beta u_\sigma(\varepsilon) + b^\sigma_\alpha b_{\sigma\beta} u_3(\varepsilon),\end{aligned}$$

the convergences $\partial_\alpha u_3(\varepsilon) \to \partial_\alpha u_3$ in $H^{-1}(\Omega)$ imply that

$$\partial_{\alpha\beta} u_3(\varepsilon) \to \partial_{\alpha\beta} u_3 \text{ in } H^{-1}(\Omega).$$

Hence $\partial_{\alpha i} u_3(\varepsilon) \to \partial_{\alpha i} u_3$ in $H^{-1}(\Omega)$ and consequently,

$$\partial_\alpha u_3(\varepsilon) \to \partial_\alpha u_3 \in L^2(\Omega).$$

(vi) *Finally, we prove that the sequences* $(u_\alpha(\varepsilon))_{\varepsilon>0}$ *strongly converge in* $H^1(\Omega)$. By virtue of Korn's inequality applied to functions in the space $\mathbf{V}(\Omega)$, this is equivalent to proving that

$$e_{ij}(\boldsymbol{u}'(\varepsilon)) \to e_{ij}(\boldsymbol{u}') \text{ in } L^2(\Omega),$$

where

$$\begin{aligned} e_{ij}(\boldsymbol{v}) &:= \frac{1}{2}(\partial_j v_i + \partial_i v_j), \\ \boldsymbol{u}'(\varepsilon) &:= (u_1(\varepsilon),\, u_2(\varepsilon),\, 0), \quad \boldsymbol{u}' := (u_1,\, u_2,\, 0). \end{aligned}$$

In part (iii), we have shown that

$$\gamma_{\alpha\beta}(\boldsymbol{u}(\varepsilon)) \to 0 \text{ in } L^2(\Omega).$$

Since

$$e_{\alpha\beta}(\boldsymbol{u}'(\varepsilon)) = \gamma_{\alpha\beta}(\boldsymbol{u}(\varepsilon)) + \Gamma^\sigma_{\alpha\beta} u_\sigma(\varepsilon) + b_{\alpha\beta} u_3(\varepsilon),$$

we conclude that

$$e_{\alpha\beta}(\boldsymbol{u}'(\varepsilon)) \to \{\Gamma^\sigma_{\alpha\beta} u_\sigma + b_{\alpha\beta} u_3\} = e_{\alpha\beta}(\boldsymbol{u}') \text{ in } L^2(\Omega),$$

by assumption. Next,

$$\begin{aligned} e_{\alpha 3}(\boldsymbol{u}'(\varepsilon)) &= \frac{1}{2}\partial_3 u_\alpha(\varepsilon) \\ &= \frac{\varepsilon}{2}\{2e_{\alpha\|3}(\varepsilon) - \partial_\alpha u_3(\varepsilon) + 2\Gamma^\sigma_{\alpha 3}(\varepsilon) u_\sigma(\varepsilon)\} \to 0 = \frac{1}{2}\partial_3 u_\alpha = e_{\alpha 3}(\boldsymbol{u}'), \end{aligned}$$

by part (i); finally,

$$e_{33}(\boldsymbol{u}'(\varepsilon)) = 0 = e_{33}(\boldsymbol{u}'),$$

and the proof is complete. ■

5.3. A THREE-DIMENSIONAL INEQUALITY OF KORN'S TYPE FOR A FAMILY OF LINEARLY ELASTIC SHELLS

The key to the convergence theorems of Sect. 5.6 is a *three-dimensional inequality of Korn's type* (Thm. 5.3-1), which may be viewed as a *"scaled" Korn's inequality in curvilinear coordinates* (Thm. 1.7-4) *for linearly elastic shells.* The "constant" C/ε appearing in this inequality, together with *ad hoc* assumptions on the applied forces (Sect. 5.5), will yield the fundamental *a priori bounds* that the family $(\boldsymbol{u}(\varepsilon))_{\varepsilon>0}$ satisfies (cf. parts (i) and (ii) in the proofs of Thms. 5.6-1 and 5.6-2).

We emphasize that this inequality holds for an *arbitrary "geometry" of the surface* $S = \boldsymbol{\theta}(\overline{\omega})$ and for an *arbitrary subset* γ_0 *of* γ with *length* $\gamma_0 > 0$, whether the space $\mathbf{V}_F(\omega)$ introduced in Sect. 5.1 reduces to $\{\mathbf{0}\}$, as in this chapter or the previous one, or not. Indeed, this inequality is likewise put to intensive use in the convergence theorem of the next chapter (Thm. 6.2-1), where we consider flexural shells, i.e., those for which $\mathbf{V}_F(\omega) \neq \{\mathbf{0}\}$.

Remarks. (1) We have seen in Thm. 4.3-1 that, if $\gamma_0 = \gamma$ and the surface S is elliptic, the "constant" C/ε can be replaced by a "genuine" constant, i.e., that is *independent of* ε; however, the norm $\|v_3\|_{1,\Omega}$ appearing in the left-hand side must then be replaced by the norm $|v_3|_{0,\Omega}$.

(2) Another Korn inequality with a "constant" also of the form C/ε was established in Kohn & Vogelius [1985] (see also Acerbi, Buttazzo & Percivale [1988]). It holds, however, over the "variable" domain Ω^ε and besides, it involves the "usual" functions $\frac{1}{2}(\partial_i^\varepsilon v_j^\varepsilon + \partial_j^\varepsilon v_i^\varepsilon)$. ∎

The following inequality of Korn's type is due to Ciarlet, Lods & Miara [1996, Thm. 4.1]; we recall that $\varepsilon_0 > 0$ is defined in Thm. 3.1-1.

Theorem 5.3-1. *Assume that* $\boldsymbol{\theta} \in \mathcal{C}^3(\overline{\omega}; \mathbb{R}^3)$. *Consider a family of linearly elastic shells with thickness* 2ε, *with each having the same middle surface* $S = \boldsymbol{\theta}(\overline{\omega})$ *and with each subjected to a boundary condition of place along a portion of its lateral face having the same* *set $\boldsymbol{\theta}(\gamma_0)$ as its middle curve, where the subset $\gamma_0 \subset \gamma$ satisfies*

$$\boxed{\textit{length } \gamma_0 > 0.}$$

Define the space

$$\boxed{\begin{aligned}\mathbf{V}(\Omega) := \{\boldsymbol{v} = (v_i) \in \mathbf{H}^1(\Omega);\ \boldsymbol{v} = \mathbf{0} \text{ on } \Gamma_0\},\\ \text{where } \Gamma_0 = \gamma_0 \times [-1, 1].\end{aligned}}$$

Then there exist a constant ε_1 satisfying $0 < \varepsilon_1 \leq \varepsilon_0$ and a constant C such that, for all $0 < \varepsilon \leq \varepsilon_1$, the following **three-dimensional inequality of Korn's type for a family of linearly elastic shells** *holds:*

$$\boxed{\|\boldsymbol{v}\|_{1,\Omega} \leq \frac{C}{\varepsilon}\Big\{\sum_{i,j} |e_{i\|j}(\varepsilon; \boldsymbol{v})|_{0,\Omega}^2\Big\}^{1/2} \text{ for all } \boldsymbol{v} \in \mathbf{V}(\Omega),}$$

where the scaled linearized strains $e_{i\|j}(\varepsilon; \boldsymbol{v})$ are defined by

$$\boxed{\begin{aligned}e_{\alpha\|\beta}(\varepsilon; \boldsymbol{v}) &:= \frac{1}{2}(\partial_\beta v_\alpha + \partial_\alpha v_\beta) - \Gamma^p_{\alpha\beta}(\varepsilon) v_p,\\ e_{\alpha\|3}(\varepsilon; \boldsymbol{v}) &:= \frac{1}{2}\Big(\frac{1}{\varepsilon}\partial_3 v_\alpha + \partial_\alpha v_3\Big) - \Gamma^\sigma_{\alpha 3}(\varepsilon) v_\sigma,\\ e_{3\|3}(\varepsilon; \boldsymbol{v}) &:= \frac{1}{\varepsilon}\partial_3 v_3.\end{aligned}}$$

■

Proof. Assume that this inequality is false. Then there exist $\varepsilon_m > 0$ and $\boldsymbol{v}^m = (v_i^m) \in \mathbf{V}(\Omega)$, $m = 0, 1, \ldots$ (the Latin letters m and n are used here for indexing sequences), such that

$$\begin{aligned}\varepsilon_m &\to 0 \text{ as } m \to \infty,\\ \|\boldsymbol{v}^m\|_{1,\Omega} &= 1 \text{ for all } m,\\ \frac{1}{\varepsilon_m} e_{i\|j}(\varepsilon_m; \boldsymbol{v}^m) &\to 0 \text{ in } L^2(\Omega) \text{ as } m \to \infty.\end{aligned}$$

Since the sequence $(\boldsymbol{v}^m)_{m=0}^{\infty}$ is bounded in $\mathbf{H}^1(\Omega)$, there exist a subsequence $(\boldsymbol{v}^n)_{n=0}^{\infty}$ and a function $\boldsymbol{v} \in \mathbf{V}(\Omega)$ such that

$$\boldsymbol{v}^n \rightharpoonup \boldsymbol{v} \text{ in } \mathbf{H}^1(\Omega) \text{ and } \boldsymbol{v}^n \to \boldsymbol{v} \text{ in } \mathbf{L}^2(\Omega) \text{ as } n \to \infty.$$

Since $\frac{1}{\varepsilon_m} e_{\alpha\|\beta}(\varepsilon_m; \boldsymbol{v}^m) \to 0$ in $L^2(\Omega)$,

$$\frac{1}{\varepsilon_n} \partial_3 e_{\alpha\|\beta}(\varepsilon_n; \boldsymbol{v}^n) \to 0 \text{ in } H^{-1}(\Omega),$$

and, by the second inequality in Thm. 5.2-1,

$$\|\rho_{\alpha\beta}(\boldsymbol{v}^n) + \frac{1}{\varepsilon_n} \partial_3 e_{\alpha\|\beta}(\varepsilon_n; \boldsymbol{v}^n)\|_{-1,\Omega}$$
$$\leq C_1 \Big\{ \sum_i |e_{i\|3}(\varepsilon_n; \boldsymbol{v}^n)|_{0,\Omega} + \varepsilon_n \sum_\alpha |v_\alpha^n|_{0,\Omega} + \varepsilon_n \|v_3^n\|_{1,\Omega} \Big\}.$$

Hence the assumptions on the sequences $(\varepsilon_m)_{m=0}^{\infty}$ and $(\boldsymbol{v}^m)_{m=0}^{\infty}$ imply that

$$\rho_{\alpha\eta}(\boldsymbol{v}^n) \to 0 \text{ in } H^{-1}(\Omega) \text{ as } n \to \infty.$$

We may thus apply Thm. 5.2-2. This gives:

$$\boldsymbol{v} = (v_i) \text{ is independent of } x_3,$$

$$\overline{\boldsymbol{v}} = (\overline{v}_i) := \frac{1}{2} \int_{-1}^{1} \boldsymbol{v} \, dx_3 \in H^1(\omega) \times H^1(\omega) \times H^2(\omega),$$

$$\overline{v}_i = \partial_\nu \overline{v}_3 = 0 \text{ on } \gamma_0,$$

$$\gamma_{\alpha\beta}(\boldsymbol{v}) = \rho_{\alpha\beta}(\boldsymbol{v}) = 0 \text{ in } \Omega.$$

Consequently,

$$\gamma_{\alpha\beta}(\overline{\boldsymbol{v}}) = \rho_{\alpha\beta}(\overline{\boldsymbol{v}}) = 0 \text{ in } \omega.$$

The *infinitesimal rigid displacement lemma on a general surface* (Thm. 2.6-3 (b)) then implies that $\overline{\boldsymbol{v}} = \mathbf{0}$ in ω; hence $\boldsymbol{v} = \mathbf{0}$ in Ω since $\boldsymbol{v}$ is independent of x_3. Thm. 5.2-2 also shows that

$$\boldsymbol{v}^n \to \boldsymbol{v} \text{ in } \mathbf{H}^1(\Omega) \text{ as } n \to \infty,$$

so that $\boldsymbol{v}^n \to \mathbf{0}$ in $\mathbf{H}^1(\Omega)$. But this contradicts $\|\boldsymbol{v}^n\|_{1,\Omega} = 1$ for all n, and the proof is complete. ■

Remark. If the mapping $\boldsymbol{\theta}$ is of the form $\boldsymbol{\theta}(y_1, y_2) = (y_1, y_2, 0)$ for all $(y_1, y_2) \in \overline{\omega}$, the above inequality of Korn's type reduces for $\varepsilon = 1$ to the *three-dimensional Korn inequality in Cartesian coordinates* over the set $\Omega = \omega \times]-1, 1[$. ■

5.4. GENERALIZED MEMBRANE SHELLS OF THE FIRST AND SECOND KINDS

According to the definition given in Sect. 5.1, if a linearly elastic shell is a generalized membrane shell, the space

$$\mathbf{V}_F(\omega) = \{\boldsymbol{\eta} = (\eta_i) \in H^1(\omega) \times H^1(\omega) \times H^2(\omega); \\ \eta_i = \partial_\nu \eta_3 = 0 \text{ on } \gamma_0,\ \gamma_{\alpha\beta}(\boldsymbol{\eta}) = 0 \text{ in } \omega\}$$

reduces to $\{\mathbf{0}\}$. This condition is clearly equivalent to stating that *the semi-norm* $|\cdot|_\omega^M$ *defined by*

$$|\boldsymbol{\eta}|_\omega^M := \Big\{\sum_{\alpha,\beta} |\gamma_{\alpha\beta}(\boldsymbol{\eta})|_{0,\omega}^2\Big\}^{1/2}$$

for all $\boldsymbol{\eta} = (\eta_i) \in H^1(\omega) \times H^1(\omega) \times L^2(\omega)$, *becomes a norm over the space*

$$\mathbf{V}_K(\omega) := \{\boldsymbol{\eta} = (\eta_i) \in H^1(\omega) \times H^1(\omega) \times H^2(\omega); \\ \eta_i = \partial_\nu \eta_3 = 0 \text{ on } \gamma_0\}.$$

The subscript "K" reminds that this space is central to the study of the two-dimensional Koiter equations for a linearly elastic shell; cf. Sect. 2.6 and Chap. 7.

As already noted in Sect. 5.1, a linearly elastic elliptic membrane shell also provides an instance of such an occurrence since, by the *inequality of Korn's type on an elliptic surface* (Thm. 2.7-3), there exists a constant c such that

$$\Big\{\sum_\alpha \|\eta_\alpha\|_{1,\omega}^2 + |\eta_3|_{0,\omega}^2\Big\}^{1/2} \le c|\boldsymbol{\eta}|_\omega^M \text{ for all } \boldsymbol{\eta} \in \mathbf{V}_M(\omega),$$

where

$$\mathbf{V}_M(\omega) := H_0^1(\omega) \times H_0^1(\omega) \times L^2(\omega).$$

Hence if the shell is an elliptic membrane shell, the semi-norm $|\cdot|_\omega^M$ is already a norm over the space $\mathbf{V}_M(\omega)$, thus *a fortiori* over the space $\mathbf{V}_K(\omega)$, which in this case is equal to $H_0^1(\omega) \times H_0^1(\omega) \times H_0^2(\omega)$.

But this inequality of Korn's type shows much more, namely that *the space* $\mathbf{V}_M(\omega)$ *is complete when it is endowed with the norm* $|\cdot|_\omega^M$ (the "other" inequality clearly holds), *a fact that sets elliptic membrane shells apart from generalized membrane shells.* More specifically, the functional spaces in which the limit two-dimensional equations of a generalized membrane shell are well posed will turn out to be *abstract completions*, which are no longer immediately apprehensible!

It seems in this respect that Destuynder [1985, p. 37] was the first to realize the need of considering such abstract completions in linearized membrane shell theory.

As shown in the proof of the convergence theorem of Sect. 5.5, an additional precaution must be exercized, as *generalized membrane shells need themselves to be subdivided into two different categories.* To describe these, we introduce the two spaces

$$\boxed{\begin{aligned} \mathbf{V}(\omega) &:= \{\boldsymbol{\eta} = (\eta_i) \in \mathbf{H}^1(\omega);\ \boldsymbol{\eta} = \mathbf{0} \text{ on } \gamma_0\}, \\ \mathbf{V}_0(\omega) &:= \{\boldsymbol{\eta} \in \mathbf{V}(\omega);\ \gamma_{\alpha\beta}(\boldsymbol{\eta}) = 0 \text{ in } \omega\}. \end{aligned}}$$

Remark. The space $\mathbf{V}_0(\omega)$ thus plays the same rôle with respect to the space $\mathbf{V}(\omega)$ as does $\mathbf{V}_F(\omega)$ with respect to $\mathbf{V}_K(\omega)$. In both cases, they constitute instances of spaces of "linearized inextensional displacements"; cf. Sect. 6.1. ■

Then a *linearly elastic generalized membrane shell is* **of the first kind** if

$$\boxed{\mathbf{V}_0(\omega) = \{\mathbf{0}\},}$$

or equivalently, *if the semi-norm* $|\cdot|_\omega^M$ *is "already" a norm over the space* $\mathbf{V}(\omega)$ (hence *a fortiori* over the space $\mathbf{V}_K(\omega) \subset \mathbf{V}(\omega)$). Otherwise, i.e., if

$$\boxed{\mathbf{V}_F(\omega) = \{\mathbf{0}\} \text{ but } \mathbf{V}_0(\omega) \neq \{\mathbf{0}\},}$$

or equivalently, if $|\cdot|_\omega^M$ is a norm over $\mathbf{V}_K(\omega)$ but *not* over $\mathbf{V}(\omega)$, the shell is a *generalized membrane shell* **of the second kind.**

Examples of linearly elastic generalized membrane shells of the first kind abound; cf. Sect. 5.8.

A word of caution. Whether a generalized membrane shell is of the first or second kind again depends only on the set γ_0 and on the "geometry" of the middle surface S; cf. Ex. 5.1. ■

An appraisal of the difficulties inherent to the analysis of generalized membrane shells lies in the definition of the *ad hoc* completions alluded to above: For a generalized membrane shell of the first kind, the limit two-dimensional equations are well posed in the space (Thm. 5.6-1)

$$\mathbf{V}_M^\sharp(\omega) := \text{completion of } \mathbf{V}(\omega) \text{ with respect to } |\cdot|_\omega^M,$$

while for a generalized membrane shell of the second kind, these equations are well posed in the space (Thm. 5.6-2)

$$\dot{\mathbf{V}}_M^\sharp(\omega) := \text{completion of } \mathbf{V}(\omega)/\mathbf{V}_0(\omega) \text{ with respect to } |\cdot|_\omega^M.$$

Identifying such completions requires in each instance a careful analysis, which leads at times to surprising conclusions; the examples described in Sect. 5.8 are eloquent in this respect.

Observe in this context how easier was the asymptotic analysis of a family of elliptic membrane shells, since $\mathbf{V}(\omega) = \mathbf{H}_0^1(\omega)$ in this case, so that the corresponding completion $\mathbf{V}_M^\sharp(\omega)$ is simply the space $\mathbf{V}_M(\omega) = H_0^1(\omega) \times H_0^1(\omega) \times L^2(\omega)$.

To carry out our asymptotic analysis of a family of generalized membrane shells, we shall also need to consider the space (as usual, $\overline{\boldsymbol{v}}$ denotes the average of $\boldsymbol{v}$ with respect to the transverse variable)

$$\mathbf{V}_0(\Omega) := \{\boldsymbol{v} \in \mathbf{H}^1(\Omega);\ \boldsymbol{v} = \mathbf{0} \text{ on } \Gamma_0,\ \partial_3\boldsymbol{v} = \mathbf{0} \text{ in } \Omega,\ \gamma_{\alpha\beta}(\overline{\boldsymbol{v}}) = 0 \text{ in } \omega\},$$

which is the "three-dimensional analog" of the space $\mathbf{V}_0(\omega)$. We shall likewise need the semi-norm $|\cdot|_\Omega^M$ defined by

$$|\boldsymbol{v}|_\Omega^M := \left\{|\partial_3\boldsymbol{v}|_{0,\Omega}^2 + \left(|\overline{\boldsymbol{v}}|_\omega^M\right)^2\right\}^{1/2} \text{ for all } \boldsymbol{v} \in \mathbf{V}(\Omega),$$

which is the "three-dimensional analog" of the semi-norm $|\cdot|_\omega^M$. Observe in passing that

$$\mathbf{V}_0(\omega) = \{\mathbf{0}\} \Leftrightarrow \mathbf{V}_0(\Omega) = \{\mathbf{0}\},$$

or equivalently, that $|\cdot|_\omega^M$ *is a norm on the space* $\mathbf{V}(\omega)$ *if and only if* $|\cdot|_\Omega^M$ *is a norm over the space* $\mathbf{V}(\Omega)$. In other words, *a linearly elastic generalized membrane shell is of the first kind if and only if* $\mathbf{V}_0(\Omega) = \{\mathbf{0}\}$.

5.5. ADMISSIBLE APPLIED FORCES

In order to derive the fundamental *a priori* estimates that the family $(\boldsymbol{u}(\varepsilon))_{\varepsilon>0}$ of scaled unknowns satisfies, we need to assume that *the applied forces contribute in a special way to the variational problem* $\mathcal{P}(\varepsilon;\,\Omega)$, according to the following considerations.

Consider a *family of linearly elastic generalized membrane shells of the first or second kind*, with thickness 2ε, with each having the same middle surface $S = \boldsymbol{\theta}(\overline{\omega})$ and with each subjected to a boundary condition of place along a portion of its lateral face having the same set $\boldsymbol{\theta}(\gamma_0)$ as its middle curve, and let the assumptions on the data be as in Sect. 5.1. For each $\varepsilon > 0$, let the linear form $L(\varepsilon) : \mathbf{V}(\Omega) \to \mathbb{R}$ be defined by

$$L(\varepsilon)(\boldsymbol{v}) := \int_\Omega f^i v_i \sqrt{g(\varepsilon)}\, dx + \int_{\Gamma_+ \cup \Gamma_-} h^i v_i \sqrt{g(\varepsilon)}\, d\Gamma \quad \text{for all } \boldsymbol{v} \in \mathbf{V}(\Omega).$$

In other words, $L(\varepsilon)(\boldsymbol{v})$ is the right-hand side in problem $\mathcal{P}(\varepsilon;\,\Omega)$ (Thm. 5.1-1), which takes into account the *applied forces* through the functions $f^i \in L^2(\Omega)$ and $h^i \in L^2(\Gamma_+ \cup \Gamma_-)$. Then each linear form $L(\varepsilon) : \mathbf{V}(\Omega) \to \mathbb{R}$ is clearly *continuous with respect to the norm* $\|\cdot\|_{1,\Omega}$ *and uniformly so with respect to* $0 < \varepsilon \leq \varepsilon_0$ (recall that $g(\varepsilon)(x) \leq g_1$ for all $x \in \overline{\Omega}$ and all $0 < \varepsilon \leq \varepsilon_0$; cf. Thm. 3.3-1). By the three-dimensional inequality of Korn's type for a family of linearly elastic shells (Thm. 5.3-1), there thus exists a constant κ_{-1} such that

$$|L(\varepsilon)(\boldsymbol{v})| \leq \frac{\kappa_{-1}}{\varepsilon} \Big\{ \sum_{i,j} |e_{i\|j}(\varepsilon;\, \boldsymbol{v})|_{0,\Omega} \Big\}^{1/2} \text{ for all } \boldsymbol{v} \in \mathbf{V}(\Omega).$$

This inequality shows that the linear forms $L(\varepsilon) : \mathbf{V}(\Omega) \to \mathbb{R}$ are also *continuous with respect to the norm* (itself dependent on ε)

$$\boldsymbol{v} \to \Big\{ \sum_{i,j} |e_{i\|j}(\varepsilon; \boldsymbol{v})|^2_{0,\Omega} \Big\}^{1/2},$$

but no longer uniformly so with respect to ε if no specific assumption is made on the applied forces.

As we shall see, carrying out the asymptotic analysis of a family of generalized membrane shells of either type crucially hinges on the *uniform continuity with respect to* ε of the linear forms with respect to this norm (cf. part (ii) in the proofs of both Thms. 5.6-1 and 5.6-2).

In view of fulfilling this requirement in a concrete manner, we observe that, by the Hahn-Banach extension theorem (see, e.g., Yosida [1966, Chap. 4]), if a linear form $L(\varepsilon) : \mathbf{V}(\Omega) \to \mathbb{R}$ is continuous with respect to this norm, there exist functions $F^{ij}(\varepsilon) = F^{ji}(\varepsilon) \in L^2(\Omega)$ (which are not necessarily uniquely determined) such that

$$L(\varepsilon)(\boldsymbol{v}) = \int_\Omega F^{ij}(\varepsilon) e_{i\|j}(\varepsilon; \boldsymbol{v}) \sqrt{g(\varepsilon)}\, dx \text{ for all } \boldsymbol{v} \in \mathbf{V}(\Omega).$$

The required uniform continuity is thus guaranteed if we impose that the norms $|F^{ij}(\varepsilon)|_{0,\Omega}$ be uniformly bounded with respect to ε.

But we shall also see that the identification of the limit two-dimensional problems found at the outcome of the asymptotic analysis further requires that *the functions* $F^{ij}(\varepsilon)$ *have limits in* $L^2(\Omega)$ *as* $\varepsilon \to 0$ (cf. part (v) in the proofs of both Thms. 5.6-1 and 5.6-2).

Combining these two requirements, we are therefore naturally led to the following definition: *Applied forces acting on a family of linearly elastic generalized membrane shells are said to be* **admissible** if there exist for each $\varepsilon > 0$ functions $F^{ij}(\varepsilon) = F^{ji}(\varepsilon) \in L^2(\Omega)$ and there exist functions $F^{ij} = F^{ji} \in L^2(\Omega)$ such that

$$L(\varepsilon)(\boldsymbol{v}) = \int_\Omega F^{ij}(\varepsilon) e_{i\|j}(\varepsilon; \boldsymbol{v}) \sqrt{g(\varepsilon)}\, dx$$

$$\text{for all } 0 < \varepsilon \leq \varepsilon_0 \text{ and for all } \boldsymbol{v} \in \mathbf{V}(\Omega),$$

$$F^{ij}(\varepsilon) \to F^{ij} \text{ in } L^2(\Omega) \text{ as } \varepsilon \to 0.$$

If the applied forces are admissible, there thus exists a constant κ_0 such that (recall that $g(\varepsilon)(x) \le g_1$, $x \in \overline{\Omega}$; cf. Thm. 3.3-1):

$$|L(\varepsilon)(\boldsymbol{v})| \le \kappa_0 \Big\{ \sum_{i,j} |e_{i\|j}(\varepsilon; \boldsymbol{v})|^2_{0,\Omega} \Big\}^{1/2}$$
$$\text{for all } 0 < \varepsilon \le \varepsilon_0 \text{ and for all } \boldsymbol{v} \in \mathbf{V}(\Omega).$$

As announced *supra*, this inequality will be put to an essential use in part (ii) of both Thms. 5.6-1 and 5.6-2.

Remark. The more convenient definition of admissible forces given here, which is due to Mardare [1998c], slightly differs from that originally given by Ciarlet & Lods [1996d]; cf. Ex. 5.2. ■

This definition is in effect an assumption about the contravariant components $f^{i,\varepsilon} \in L^2(\Omega^\varepsilon)$ and $h^{i,\varepsilon} \in L^2(\Gamma_+^\varepsilon \cup \Gamma_-^\varepsilon)$ of the applied body and surface forces actually acting on the shells of the family under consideration: It means that *the right-hand side in the variational equations of problem* $\mathcal{P}(\Omega^\varepsilon)$ (Sect. 3.1) *can also be written for each* $\varepsilon > 0$ *as*

$$\int_{\Omega^\varepsilon} f^{i,\varepsilon} v_i^\varepsilon \sqrt{g^\varepsilon}\, dx^\varepsilon + \int_{\Gamma_+^\varepsilon \cup \Gamma_-^\varepsilon} h^{i,\varepsilon} v_i^\varepsilon \sqrt{g^\varepsilon}\, d\Gamma^\varepsilon = \varepsilon L(\varepsilon)(\boldsymbol{v}),$$

where $v_i(x) := v_i^\varepsilon(x^\varepsilon)$ for all $x^\varepsilon = \pi^\varepsilon x \in \overline{\Omega}^\varepsilon$ and $L(\varepsilon)(\boldsymbol{v})$ is of the above form. As such, it may be again understood as an assumption on the *orders* of the applied forces as $\varepsilon \to 0$.

Naturally, *admissible forces have to be identified for each type of generalized membrane shells*; see in this respect the examples described in Sect. 5.8.

5.6. CONVERGENCE OF THE SCALED DISPLACEMENTS AS $\varepsilon \to 0$

We now establish the main results of this chapter: To begin with, we consider a *family of linearly elastic generalized membrane shells of the first kind*, with thickness $2\varepsilon > 0$, with each having the *same* middle surface $S = \boldsymbol{\theta}(\overline{\omega})$ and with each subjected to a boundary condition of place along a portion of its lateral face having the *same* set $\boldsymbol{\theta}(\gamma_0)$ as its middle curve. Then the solutions $\boldsymbol{u}(\varepsilon)$ of the associated scaled three-dimensional problems $\mathcal{P}(\varepsilon;\,\Omega)$ (Thm. 5.1-1) converge in an "abstract" *completion* $\mathbf{V}^{\sharp}_{M}(\Omega)$ as $\varepsilon \to 0$. In addition, the *averages*

$$\overline{\boldsymbol{u}(\varepsilon)} = \frac{1}{2}\int_{-1}^{1} \boldsymbol{u}(\varepsilon)\, dx_3$$

of the scaled unknowns likewise converge in an "abstract" *completion* $\mathbf{V}^{\sharp}_{M}(\omega)$ as $\varepsilon \to 0$ and their limit satisfies an "abstract" variational problem posed over the same space $\mathbf{V}^{\sharp}_{M}(\omega)$.

The functions $\gamma_{\alpha\beta}(\boldsymbol{\eta})$ and $a^{\alpha\beta\sigma\tau}$ used in the next theorem respectively represent the covariant components of the *change of metric tensor* associated with a displacement field $\eta_i \boldsymbol{a}^i$ of the surface S and the contravariant components of the (scaled) *two-dimensional elasticity tensor* of *the "limit" shell.* Hence the bilinear form B_M defined *infra* is precisely that found in the scaled variational problem of a *linearly elastic elliptic membrane shell* (Thm. 4.5-1). We recall that $\varepsilon_0 > 0$ is defined in Thm. 3.1-1.

For convenience, we recapitulate the definitions of some spaces and semi-norms appearing in Thm. 5.6-1 and its proof:

$$\begin{aligned}
\mathbf{V}(\Omega) &= \{\boldsymbol{v} = (v_i) \in \mathbf{H}^1(\Omega);\ \boldsymbol{v} = \mathbf{0} \text{ on } \Gamma_0 = \gamma_0 \times [-1,\,1]\},\\
|\boldsymbol{v}|^M_\Omega &= \left\{|\partial_3 \boldsymbol{v}|^2_{0,\Omega} + \left(|\overline{\boldsymbol{v}}|^M_\omega\right)^2\right\}^{1/2}, \text{ where } \overline{\boldsymbol{v}} = \frac{1}{2}\int_{-1}^{1} \boldsymbol{v}\, dx_3,\\
\mathbf{V}(\omega) &= \{\boldsymbol{\eta} = (\eta_i) \in \mathbf{H}^1(\omega);\ \boldsymbol{\eta} = 0 \text{ on } \gamma_0\},\\
|\boldsymbol{\eta}|^M_\omega &= \left\{\sum_{\alpha,\beta} |\gamma_{\alpha\beta}(\boldsymbol{\eta})|^2_{0,\omega}\right\}^{1/2}.
\end{aligned}$$

Notice that the assumption that is central to this chapter, namely that $\mathbf{V}_F(\omega) = \{\mathbf{0}\}$ (Sect. 5.1), is not needed until part (vi) of the proof.

The following result is due to Ciarlet & Lods [1996d, Thm. 5.1]; only the proof had to be slightly modified to take into account the new definition of admissible applied forces given in Sect. 5.5.

Theorem 5.6-1. *Assume that $\boldsymbol{\theta} \in \mathcal{C}^3(\overline{\omega};\,\mathbb{R}^3)$. Consider a family of linearly elastic generalized membrane shells of the first kind* (Sect. 5.4), *with thickness 2ε approaching zero, with each having the same middle surface $S = \boldsymbol{\theta}(\overline{\omega})$, with each subjected to a boundary condition* *of place along a portion of its lateral face having the same set* $\boldsymbol{\theta}(\gamma_0)$ *as its middle curve, and subjected to applied forces that are admissible* (Sect. 5.5). *Let* $\boldsymbol{u}(\varepsilon)$ *denote for* $0 < \varepsilon \le \varepsilon_0$ *the solution of the associated scaled three-dimensional problems* $\mathcal{P}(\varepsilon;\,\Omega)$ (Thm. 5.1-1).

Define the spaces

$$\mathbf{V}_M^\sharp(\Omega) := \text{completion of } \mathbf{V}(\Omega) \text{ with respect to } |\cdot|_\Omega^M,$$
$$\mathbf{V}_M^\sharp(\omega) := \text{completion of } \mathbf{V}(\omega) \text{ with respect to } |\cdot|_\omega^M.$$

Then there exist $\boldsymbol{u} \in \mathbf{V}_M^\sharp(\Omega)$ *and* $\boldsymbol{\zeta} \in \mathbf{V}_M^\sharp(\omega)$ *such that*

$$\boldsymbol{u}(\varepsilon) \to \boldsymbol{u} \text{ in } \mathbf{V}_M^\sharp(\Omega) \text{ as } \varepsilon \to 0,$$
$$\overline{\boldsymbol{u}(\varepsilon)} := \frac{1}{2}\int_{-1}^{1} \boldsymbol{u}(\varepsilon)\,dx_3 \to \boldsymbol{\zeta} \text{ in } \mathbf{V}_M^\sharp(\omega) \text{ as } \varepsilon \to 0.$$

Let

$$a^{\alpha\beta\sigma\tau} := \frac{4\lambda\mu}{\lambda+2\mu}a^{\alpha\beta}a^{\sigma\tau} + 2\mu(a^{\alpha\sigma}a^{\beta\tau} + a^{\alpha\tau}a^{\beta\sigma}),$$

$$\gamma_{\alpha\beta}(\boldsymbol{\eta}) := \frac{1}{2}(\partial_\beta\eta_\alpha + \partial_\alpha\eta_\beta) - \Gamma_{\alpha\beta}^\sigma\eta_\sigma - b_{\alpha\beta}\eta_3,$$

$$B_M(\boldsymbol{\zeta},\boldsymbol{\eta}) := \int_\omega a^{\alpha\beta\sigma\tau}\gamma_{\sigma\tau}(\boldsymbol{\zeta})\gamma_{\alpha\beta}(\boldsymbol{\eta})\sqrt{a}\,dy \text{ for } \boldsymbol{\zeta},\,\boldsymbol{\eta} \in \mathbf{V}(\omega),$$

$$L_M(\boldsymbol{\eta}) := \int_\omega \varphi^{\alpha\beta}\gamma_{\alpha\beta}(\boldsymbol{\eta})\sqrt{a}\,dy \text{ for } \boldsymbol{\eta} \in \mathbf{V}(\omega),$$

$$\varphi^{\alpha\beta} := \int_{-1}^{1}\left\{F^{\alpha\beta} - \frac{\lambda}{\lambda+2\mu}a^{\alpha\beta}F^{33}\right\}dx_3 \in L^2(\omega),$$

where the functions $F^{ij} \in L^2(\Omega)$ *are those used in the definition of admissible forces* (Sect. 5.5), *and let* $B_M^\sharp$ *and* $L_M^\sharp$ *denote the unique continuous extensions from* $\mathbf{V}(\omega)$ *to* $\mathbf{V}_M^\sharp(\omega)$ *of the bilinear form* B_M *and linear form* L_M. *Then the limit* $\boldsymbol{\zeta}$ *satisfies the following* **scaled two-dimensional variational problem** $\mathcal{P}_M^\sharp(\omega)$ **of a linearly elastic generalized membrane shell of the first kind:**

$$\boldsymbol{\zeta} \in \mathbf{V}_M^\sharp(\omega) \text{ and } B_M^\sharp(\boldsymbol{\zeta},\boldsymbol{\eta}) = L_M^\sharp(\boldsymbol{\eta}) \text{ for all } \boldsymbol{\eta} \in \mathbf{V}_M^\sharp(\omega).$$

Proof. The proof is divided into eleven parts. Throughout the proof, we let

$$e_{i\|j}(\varepsilon) := e_{i\|j}(\varepsilon; \boldsymbol{u}(\varepsilon)),$$

and we let $\to$ and $\rightharpoonup$ respectively denote strong and weak convergences.

Whenever c_1, c_2, etc., appears in an inequality, it means that there exists a constant, denoted by this symbol, that is > 0 and independent of all the "variables" entering the inequality such that this inequality holds.

(i) *There exist* $0 < \varepsilon_1 \le \varepsilon_0$ *and* c_0 *such that, for all* $0 < \varepsilon \le \varepsilon_1$,

$$|\boldsymbol{v}|_\Omega^M \le c_0 \Big\{ \sum_{i,j} |e_{i\|j}(\varepsilon; \boldsymbol{v})|_{0,\Omega}^2 \Big\}^{1/2} \quad \text{for all } \boldsymbol{v} \in \mathbf{V}(\Omega).$$

Let

$$\gamma_{\alpha\beta}(\boldsymbol{v}) := \frac{1}{2}(\partial_\beta v_\alpha + \partial_\alpha v_\beta) - \Gamma_{\alpha\beta}^\sigma v_\sigma - b_{\alpha\beta} v_3$$

for all $\boldsymbol{v} = (v_i) \in \mathbf{V}(\Omega)$. The relations

$$\gamma_{\alpha\beta}(\boldsymbol{v}) = e_{\alpha\|\beta}(\varepsilon; \boldsymbol{v}) + (\Gamma_{\alpha\beta}^\sigma(\varepsilon) - \Gamma_{\alpha\beta}^\sigma) v_\sigma - \varepsilon x_3 b_\alpha^\sigma b_{\sigma\beta} v_3,$$

$$\partial_3 v_\alpha = 2\varepsilon e_{\alpha\|3}(\varepsilon; \boldsymbol{v}) - \varepsilon \partial_\alpha v_3 + 2\varepsilon \Gamma_{\alpha 3}^\sigma(\varepsilon) v_\sigma,$$

$$\partial_3 v_3 = \varepsilon e_{3\|3}(\varepsilon; \boldsymbol{v}),$$

$$\|\Gamma_{\alpha\beta}^\sigma(\varepsilon) - \Gamma_{\alpha\beta}^\sigma\|_{0,\infty,\overline{\Omega}} = O(\varepsilon), \quad \|\Gamma_{\alpha 3}^\sigma(\varepsilon)\|_{0,\infty,\overline{\Omega}} = O(1),$$

the last ones being consequences of Thm. 3.3-1, imply that, for all $0 < \varepsilon \le \varepsilon_0$ (without loss of generality, we may assume that $\varepsilon_0 \le 1$) and for all $\boldsymbol{v} \in \mathbf{V}(\Omega)$,

$$\sum_{\alpha,\beta} |\gamma_{\alpha\beta}(\boldsymbol{v})|_{0,\Omega} + \sum_i |\partial_3 v_i|_{0,\Omega} \le c_1 \Big\{ \sum_{i,j} |e_{i\|j}(\varepsilon; \boldsymbol{v})|_{0,\Omega} + \varepsilon \|\boldsymbol{v}\|_{1,\Omega} \Big\}.$$

Hence the desired inequality is obtained by combining the inequalities

$$|\gamma_{\alpha\beta}(\overline{\boldsymbol{v}})|_{0,\omega} \le \frac{1}{\sqrt{2}} |\gamma_{\alpha\beta}(\boldsymbol{v})|_{0,\Omega},$$

which follow from Thm. 4.2-1 (a) and the relations $\gamma_{\alpha\beta}(\overline{\boldsymbol{v}}) = \overline{\gamma_{\alpha\beta}(\boldsymbol{v})}$, with the *three-dimensional inequality of Korn's type for a family of linearly elastic shells* (Thm. 5.3-1).

(ii) A priori *bounds and extraction of weakly convergent sequences: The semi-norms* $|\boldsymbol{u}(\varepsilon)|_{\Omega}^{M}$ *and* $|\overline{\boldsymbol{u}(\varepsilon)}|_{\omega}^{M}$ *and the norms* $\|\varepsilon\boldsymbol{u}(\varepsilon)\|_{1,\Omega}$ *and* $|e_{i\|j}(\varepsilon)|_{0,\Omega}$ *are bounded independently of* $0 < \varepsilon \leq \varepsilon_1$.

Define the spaces

$$\begin{aligned}
\mathbf{V}_M^{\sharp}(\Omega) &:= \text{completion of } \mathbf{V}(\Omega) \text{ with respect to } |\cdot|_{\Omega}^{M}, \\
\mathbf{V}_M^{\sharp}(\omega) &:= \text{completion of } \mathbf{V}(\omega) \text{ with respect to } |\cdot|_{\omega}^{M}.
\end{aligned}$$

Then there is a subsequence, still denoted $(\boldsymbol{u}(\varepsilon))_{\varepsilon>0}$ *for convenience, and there exist* $\boldsymbol{u} \in \mathbf{V}_M^{\sharp}(\Omega)$, $\boldsymbol{u}^{-1} = (u_i^{-1}) \in \mathbf{V}(\Omega)$, $e_{i\|j} \in L^2(\Omega)$, *and* $\boldsymbol{\zeta} \in \mathbf{V}_M^{\sharp}(\omega)$ *such that*

$$\begin{aligned}
\boldsymbol{u}(\varepsilon) &\rightharpoonup \boldsymbol{u} \text{ in } \mathbf{V}_M^{\sharp}(\Omega), \\
\varepsilon\boldsymbol{u}(\varepsilon) &\rightharpoonup \boldsymbol{u}^{-1} \text{ in } \mathbf{H}^1(\Omega) \text{ and thus } \varepsilon\boldsymbol{u}(\varepsilon) \to \boldsymbol{u}^{-1} \text{ in } \mathbf{L}^2(\Omega), \\
e_{i\|j}(\varepsilon) &\rightharpoonup e_{i\|j} \text{ in } L^2(\Omega), \\
\partial_3 u_3(\varepsilon) &= \varepsilon e_{3\|3}(\varepsilon) \to 0 \text{ in } L^2(\Omega), \\
\overline{\boldsymbol{u}(\varepsilon)} &\rightharpoonup \boldsymbol{\zeta} \text{ in } \mathbf{V}_M^{\sharp}(\omega) \text{ as } \varepsilon \to 0.
\end{aligned}$$

Since the applied forces are *admissible*, there exists a constant κ_0 such that (Sect. 5.5)

$$|L(\varepsilon)(\boldsymbol{v})| \leq \kappa_0 \Big\{ \sum_{i,j} |e_{i\|j}(\varepsilon; \boldsymbol{v})|_{0,\Omega}^2 \Big\}^{1/2}$$

for all $\boldsymbol{v} \in \mathbf{V}(\Omega)$ and all $0 < \varepsilon \leq \varepsilon_0$. Combining the variational equations of problem $\mathcal{P}(\varepsilon; \Omega)$ (Thm. 5.1-1) with the asymptotic behavior of the function $g(\varepsilon)$ (Thm. 3.3-1), the uniform positive definiteness of the scaled three-dimensional elasticity tensor (Thm. 3.3-2), the assumption that the applied forces are admissible, and the inequality obtained in part (i), we thus obtain

$$\begin{aligned}
c_0^{-2}\left(|\boldsymbol{u}(\varepsilon)|_{\Omega}^{M}\right)^2 &\leq \sum_{i,j} |e_{i\|j}(\varepsilon)|_{0,\Omega}^2 \\
&\leq C_e g_0^{-1/2} \int_{\Omega} A^{ijkl}(\varepsilon) e_{k\|l}(\varepsilon) e_{i\|j}(\varepsilon) \sqrt{g(\varepsilon)}\, dx \\
&= C_e g_0^{-1/2} L(\varepsilon)(\boldsymbol{u}(\varepsilon)) \\
&\leq \kappa_0 C_e g_0^{-1/2} \Big\{ \sum_{i,j} |e_{i\|j}(\varepsilon)|_{0,\Omega}^2 \Big\}^{1/2}
\end{aligned}$$

for all $0 < \varepsilon \le \varepsilon_1$. The boundedness of $|e_{i\|j}(\varepsilon)|_{0,\Omega}$, hence that of $|\boldsymbol{u}(\varepsilon)|_\Omega^M$ and of $|\overline{\boldsymbol{u}(\varepsilon)}|_\omega^M \le |\boldsymbol{u}(\varepsilon)|_\Omega^M$, follow from this inequality; the boundedness of $\|\varepsilon \boldsymbol{u}(\varepsilon)\|_{1,\Omega}$ follows from the *three-dimensional inequality of Korn's type for a family of linearly elastic shells* (Thm. 5.3-1). Note that the assumption that the applied forces are "*admissible*" is crucially needed here.

(iii) *The limits $e_{i\|j}$ found in* part (ii) *satisfy*

$$e_{\alpha\|3} = \frac{1}{2\mu} a_{\alpha\beta} F^{\beta 3},$$

$$e_{3\|3} = -\frac{\lambda}{\lambda + 2\mu} a^{\alpha\beta} e_{\alpha\|\beta} + \frac{F^{33}}{\lambda + 2\mu},$$

the functions $F^{ij} \in L^2(\Omega)$ being those used in the definition of admissible forces (Sect. 5.5).

Let $\boldsymbol{v} = (v_i)$ be an arbitrary function in the space $\mathbf{V}(\Omega)$. Then

$$\varepsilon e_{\alpha\|\beta}(\varepsilon; \boldsymbol{v}) \to 0 \text{ in } L^2(\Omega),$$
$$\varepsilon e_{\alpha\|3}(\varepsilon; \boldsymbol{v}) \to \frac{1}{2}\partial_3 v_\alpha \text{ in } L^2(\Omega),$$
$$\varepsilon e_{3\|3}(\varepsilon; \boldsymbol{v}) = \partial_3 v_3 \text{ for all } \varepsilon > 0,$$

by definition of the functions $e_{i\|j}(\varepsilon; \boldsymbol{v})$. The variational equations of problem $\mathcal{P}(\varepsilon; \Omega)$ (Thm. 5.1-1) may also be written as (note that $A^{\alpha\beta\sigma 3}(\varepsilon) = A^{\alpha 333}(\varepsilon) = 0$; cf. Thm. 3.3-2 (b)):

$$\begin{aligned}
&\int_\Omega \left\{ A^{\alpha\beta\sigma\tau}(\varepsilon) e_{\sigma\|\tau}(\varepsilon) + A^{\alpha\beta 33}(\varepsilon) e_{3\|3}(\varepsilon) \right\} \{\varepsilon e_{\alpha\|\beta}(\varepsilon; \boldsymbol{v})\} \sqrt{g(\varepsilon)}\, dx \\
&\quad + \int_\Omega \left\{ 4A^{\alpha 3\sigma 3}(\varepsilon) e_{\sigma\|3}(\varepsilon) \right\} \{\varepsilon e_{\alpha\|3}(\varepsilon; \boldsymbol{v})\} \sqrt{g(\varepsilon)}\, dx \\
&\quad + \int_\Omega \left\{ A^{33\sigma\tau}(\varepsilon) e_{\sigma\|\tau}(\varepsilon) + A^{3333}(\varepsilon) e_{3\|3}(\varepsilon) \right\} \{\varepsilon e_{3\|3}(\varepsilon; \boldsymbol{v})\} \sqrt{g(\varepsilon)}\, dx \\
&= \int_\Omega F^{\alpha\beta}(\varepsilon) \{\varepsilon e_{\alpha\|\beta}(\varepsilon; \boldsymbol{v})\} \sqrt{g(\varepsilon)}\, dx \\
&\quad + \int_\Omega 2F^{\alpha 3}(\varepsilon) \{\varepsilon e_{\alpha\|3}(\varepsilon; \boldsymbol{v})\} \sqrt{g(\varepsilon)}\, dx \\
&\quad + \int_\Omega F^{33} \{\varepsilon e_{3\|3}(\varepsilon; \boldsymbol{v})\} \sqrt{g(\varepsilon)}\, dx.
\end{aligned}$$

Keep $\boldsymbol{v} \in \mathbf{V}(\Omega)$ *fixed* and let $\varepsilon \to 0$. The asymptotic behavior of the functions $\varepsilon e_{i\|j}(\varepsilon; \boldsymbol{v})$, the asymptotic behavior of the functions $A^{ijkl}(\varepsilon)$ and $g(\varepsilon)$ (Thms. 3.3-1 and 3.3-2), and the weak convergences $e_{i\|j}(\varepsilon) \rightharpoonup e_{i\|j}$ in $L^2(\Omega)$ (part (ii)) together imply that

$$\int_\Omega \left\{2\mu a^{\alpha\sigma} e_{\sigma\|3}\partial_3 v_\alpha + (\lambda a^{\sigma\tau} e_{\sigma\|\tau} + (\lambda + 2\mu)e_{3\|3})\partial_3 v_3\right\} \sqrt{a}\, dx$$
$$= \int_\Omega \left\{F^{\alpha 3}\partial_3 v_\alpha + F^{33}\partial_3 v_3\right\} \sqrt{a}\, dx.$$

Letting $\boldsymbol{v}$ *vary* in $\mathbf{V}(\Omega)$ then yields the relations satisfied by the limits $e_{i\|3}$ (if $w \in L^2(\Omega)$ satisfies $\int_\Omega w\partial_3 v\, dx = 0$ for all $v \in H^1(\Omega)$ that vanish on Γ_0, then $w = 0$; cf. Thm. 3.4-1).

(iv) *The whole family* $(\boldsymbol{u}(\varepsilon))_{\varepsilon>0}$ *satisfies*:

$$\{\overline{e_{\alpha\|\beta}(\varepsilon)} - \gamma_{\alpha\beta}(\overline{\boldsymbol{u}(\varepsilon)})\} \to 0 \text{ in } L^2(\Omega) \text{ as } \varepsilon \to 0;$$

consequently, the subsequence considered in part (ii) *satisfies*:

$$\gamma_{\alpha\beta}(\overline{\boldsymbol{u}(\varepsilon)}) \rightharpoonup \overline{e_{\alpha\|\beta}} \text{ in } L^2(\omega).$$

The definitions of the functions $e_{\alpha\|\beta}(\varepsilon)$ and $\gamma_{\alpha\beta}(\boldsymbol{u}(\varepsilon))$ and the asymptotic behavior of the functions $\Gamma^p_{\alpha\beta}(\varepsilon)$ (Thm. 3.3-1) together imply that

$$\overline{e_{\alpha\|\beta}(\varepsilon)} - \gamma_{\alpha\beta}(\overline{\boldsymbol{u}(\varepsilon)}) = -\overline{\Gamma^{\sigma,2}_{\alpha\beta}(\varepsilon)u_\sigma(\varepsilon)} + \overline{\varepsilon x_3 b^\sigma_\beta|_\alpha u_\sigma(\varepsilon)} + \overline{\varepsilon x_3 b^\sigma_\alpha b_{\sigma\beta} u_3(\varepsilon)},$$

where the functions

$$\Gamma^{\sigma,2}_{\alpha\beta}(\varepsilon) := \Gamma^\sigma_{\alpha\beta}(\varepsilon) - \left(\Gamma^\sigma_{\alpha\beta} - \varepsilon x_3 b^\sigma_\beta|_\alpha\right)$$

satisfy

$$\|\Gamma^{\sigma,2}_{\alpha\beta}(\varepsilon)\|_{0,\infty,\overline{\Omega}} \le c_2\varepsilon^2$$

for all $0 < \varepsilon \le \varepsilon_0$. Since $|\overline{v}|_{0,\omega} \le \frac{1}{\sqrt{2}}|v|_{0,\Omega}$ (Thm. 4.2-1 (a)) and $\overline{x_3 v} = \overline{\frac{1}{2}(1 - x_3^2)\partial_3 v}$ (this relation, which is easily seen to hold if $v \in \mathcal{C}^1(\overline{\Omega})$, is then extended to $v \in H^1(\Omega)$ by Thm. 4.2-1 (b)),

$$|\overline{x_3 v}|_{0,\omega} = \frac{1}{2}|\overline{(1 - x_3^2)\partial_3 v}|_{0,\omega} \le c_3|\partial_3 v|_{0,\Omega},$$

and thus

$$|\overline{e_{\alpha\|\beta}(\varepsilon)} - \gamma_{\alpha\beta}(\overline{\boldsymbol{u}(\varepsilon)})|_{0,\omega} \leq c_4 \varepsilon \Big\{ \sum_\alpha |\varepsilon u_\alpha(\varepsilon)|_{0,\Omega} + \sum_i |\partial_3 u_i(\varepsilon)|_{0,\Omega} \Big\}.$$

The announced convergence for the whole family thus follows from the boundedness of the norms $\|\varepsilon \boldsymbol{u}(\varepsilon)\|_{1,\Omega}$ and semi-norms $|\boldsymbol{u}(\varepsilon)|_\Omega^M$ established in part (ii).

(v) *The limits $e_{\alpha\|\beta}$ found in part* (ii) *satisfy*

$$\int_\omega a^{\alpha\beta\sigma\tau} \overline{e_{\sigma\|\tau}} \gamma_{\alpha\beta}(\boldsymbol{\eta}) \sqrt{a}\, dy = \int_\omega \varphi^{\alpha\beta} \gamma_{\alpha\beta}(\boldsymbol{\eta}) \sqrt{a}\, dy \text{ for all } \boldsymbol{\eta} \in \mathbf{V}(\omega),$$

where

$$\varphi^{\alpha\beta} := \int_{-1}^{1} \Big\{ F^{\alpha\beta} - \frac{\lambda}{\lambda + 2\mu} a^{\alpha\beta} F^{33} \Big\}\, dx_3 \in L^2(\omega),$$

the functions $F^{ij} \in L^2(\Omega)$ being those used in the definition of admissible forces (Sect. 5.5).

Let $\boldsymbol{v} = (v_i) \in \mathbf{V}(\Omega)$ be *independent of the transverse variable x_3*. For such a function $\boldsymbol{v}$,

$$\begin{aligned} e_{\alpha\|\beta}(\varepsilon; \boldsymbol{v}) &\to \gamma_{\alpha\beta}(\boldsymbol{v}) \text{ in } L^2(\Omega), \\ e_{\alpha\|3}(\varepsilon; \boldsymbol{v}) &\to \Big\{ \frac{1}{2} \partial_\alpha v_3 + b_\alpha^\sigma v_\sigma \Big\} \text{ in } L^2(\Omega), \\ e_{3\|3}(\varepsilon; \boldsymbol{v}) &= 0. \end{aligned}$$

Keep such a function $\boldsymbol{v} \in \mathbf{V}(\Omega)$ *fixed* in the variational equations of problem $\mathcal{P}(\varepsilon; \Omega)$ and let $\varepsilon \to 0$. The asymptotic behavior of the functions $A^{ijkl}(\varepsilon)$ and $g(\varepsilon)$ (Thms. 3.3-1 and 3.3-2), the above asymptotic behavior of the functions $e_{i\|j}(\varepsilon; \boldsymbol{v})$, the weak convergences $e_{i\|j}(\varepsilon) \rightharpoonup e_{i\|j}$ in $L^2(\Omega)$ (part (ii)), and the relations satisfied by the limits $e_{i\|j}$ (part (iii)) together imply that

$$\begin{aligned} &\lim_{\varepsilon \to 0} \int_\Omega A^{ijkl}(\varepsilon) e_{k\|l}(\varepsilon) e_{i\|j}(\varepsilon; \boldsymbol{v}) \sqrt{g(\varepsilon)}\, dx \\ &= \int_\Omega \Big\{ \frac{2\lambda\mu}{\lambda + 2\mu} a^{\alpha\beta} a^{\sigma\tau} + \mu (a^{\alpha\sigma} a^{\beta\tau} + a^{\alpha\tau} a^{\beta\sigma}) \Big\} e_{\sigma\|\tau} \gamma_{\alpha\beta}(\boldsymbol{v}) \sqrt{a}\, dx \\ &\quad + \int_\Omega \Big(\frac{\lambda}{\lambda + 2\mu} a^{\alpha\beta} F^{33} \gamma_{\alpha\beta}(\boldsymbol{v}) + 2F^{\alpha 3} \Big\{ \frac{1}{2} \partial_\alpha v_3 + b_\alpha^\sigma v_\sigma \Big\} \Big) \sqrt{a}\, dx. \end{aligned}$$

The assumption that the applied forces are admissible implies that, for such a function $\boldsymbol{v}$,

$$\begin{aligned}
&\lim_{\varepsilon\to 0} L(\varepsilon)(\boldsymbol{v}) \\
&= \lim_{\varepsilon\to 0}\int_\Omega \Big(F^{\alpha\beta}(\varepsilon)e_{\alpha\|\beta}(\varepsilon;\,\boldsymbol{v}) + 2F^{\alpha 3}(\varepsilon)e_{\alpha\|3}(\varepsilon;\,\boldsymbol{v})\Big)\sqrt{g(\varepsilon)}\,dx \\
&= \int_\Omega \Big(F^{\alpha\beta}\gamma_{\alpha\beta}(\boldsymbol{v}) + 2F^{\alpha 3}\Big\{\frac{1}{2}\partial_\alpha v_3 + b^\sigma_\alpha v_\sigma\Big\}\Big)\sqrt{a}\,dx.
\end{aligned}$$

The conjunction of the above relations then implies that

$$\begin{aligned}
&\int_\Omega \Big\{\frac{2\lambda\mu}{\lambda+2\mu}a^{\alpha\beta}a^{\sigma\tau} + \mu(a^{\alpha\sigma}a^{\beta\tau} + a^{\alpha\tau}a^{\beta\sigma})\Big\}e_{\sigma\|\tau}\gamma_{\alpha\beta}(\boldsymbol{v})\sqrt{a}\,dx \\
&= \int_\omega a^{\alpha\beta\sigma\tau}\overline{e_{\sigma\|\tau}}\gamma_{\alpha\beta}(\overline{\boldsymbol{v}})\sqrt{a}\,dy \\
&= \int_\omega \Big(\int_{-1}^{1}\Big\{F^{\alpha\beta} - \frac{\lambda}{\lambda+2\mu}a^{\alpha\beta}F^{33}\Big\}\,dx_3\Big)\gamma_{\alpha\beta}(\overline{\boldsymbol{v}})\sqrt{a}\,dy,
\end{aligned}$$

where $\overline{\boldsymbol{v}} := \frac{1}{2}\int_{-1}^{1}\boldsymbol{v}\,dx_3$.

Given any function $\boldsymbol{\eta}\in\mathbf{V}(\omega)$, there exists a function $\boldsymbol{v}\in\mathbf{V}(\Omega)$ independent of x_3 such that $\overline{\boldsymbol{v}} = \boldsymbol{\eta}$. Hence the announced relations hold for all $\boldsymbol{\eta}\in\mathbf{V}(\omega)$.

(vi) *The subsequence* $(\boldsymbol{u}(\varepsilon))_{\varepsilon>0}$ *found in* part (ii) *is such that*

$$\begin{aligned}
\varepsilon\boldsymbol{u}(\varepsilon) &\rightharpoonup \mathbf{0} \text{ in } \mathbf{H}^1(\Omega), \\
\partial_3 u_\alpha(\varepsilon) &\rightharpoonup 0 \text{ in } L^2(\Omega),
\end{aligned}$$

as $\varepsilon\to 0$. *Furthermore,*

$$e_{\alpha\|\beta} \text{ is independent of the transverse variable } x_3.$$

The functions

$$\boldsymbol{u}^{-1}(\varepsilon) := \varepsilon\boldsymbol{u}(\varepsilon)\in\mathbf{V}(\Omega)$$

associated with the subsequence found in part (ii) satisfy

$$\begin{aligned}
\boldsymbol{u}^{-1}(\varepsilon) \rightharpoonup \boldsymbol{u}^{-1} \text{ in } \mathbf{H}^1(\Omega) &\text{ and } \boldsymbol{u}^{-1}(\varepsilon)\to\boldsymbol{u}^{-1} \text{ in } \mathbf{L}^2(\Omega), \\
\frac{1}{\varepsilon}e_{i\|j}\left(\varepsilon;\,\boldsymbol{u}^{-1}(\varepsilon)\right) &\rightharpoonup e_{i\|j} \text{ in } L^2(\Omega).
\end{aligned}$$

Hence $\overline{\boldsymbol{u}^{-1}} \in \mathbf{V}_F(\omega)$ by Thm. 5.2-2 and consequently $\overline{\boldsymbol{u}^{-1}} = \mathbf{0}$, *since* $\mathbf{V}_F(\omega) = \{\mathbf{0}\}$ *by assumption.* By the same theorem, $\boldsymbol{u}^{-1}$ is independent of x_3; hence

$$\boldsymbol{u}^{-1} = \mathbf{0},$$

and the convergence $\varepsilon \boldsymbol{u}(\varepsilon) \rightharpoonup \mathbf{0}$ in $\mathbf{H}^1(\Omega)$, hence also the convergence $\varepsilon \boldsymbol{u}(\varepsilon) \to \mathbf{0}$ in $\mathbf{L}^2(\Omega)$, are established. These, combined with the relations (which follow from the definition of the scaled strains $e_{\alpha\|3}(\varepsilon)$):

$$\partial_3 u_\alpha(\varepsilon) = 2\varepsilon e_{\alpha\|3}(\varepsilon) - \varepsilon \partial_\alpha u_3(\varepsilon) + 2\varepsilon \Gamma^\sigma_{\alpha 3}(\varepsilon) u_\sigma(\varepsilon)$$

and the boundedness of the sequence $(\Gamma^\sigma_{\alpha 3}(\varepsilon))_{\varepsilon > 0}$ in $\mathcal{C}^0(\overline{\Omega})$ (Thm. 3.3-1), in turn yield the weak convergences $\partial_3 u_\alpha(\varepsilon) \rightharpoonup 0$ in $L^2(\Omega)$. Again by Thm. 5.2-2,

$$\partial_3 e_{\alpha\|\beta} = -\rho_{\alpha\beta}\left(\boldsymbol{u}^{-1}\right) = 0,$$

since $\boldsymbol{u}^{-1} = \mathbf{0}$. The functions $e_{\alpha\|\beta}$ are thus independent of x_3.

(vii) *The following strong convergences hold as $\varepsilon \to 0$* (so far, they were only known to be weak; cf. parts (ii), (iv), and (vi)):

$$\begin{aligned}
e_{i\|j}(\varepsilon) &\to e_{i\|j} \text{ in } L^2(\Omega),\\
\varepsilon \boldsymbol{u}(\varepsilon) &\to \mathbf{0} \text{ in } \mathbf{H}^1(\Omega),\\
\gamma_{\alpha\beta}(\overline{\boldsymbol{u}(\varepsilon)}) &\to \overline{e_{\alpha\|\beta}} \text{ in } L^2(\omega),\\
\overline{\boldsymbol{u}(\varepsilon)} &\to \boldsymbol{\zeta} \text{ in } \mathbf{V}^\sharp_M(\omega).
\end{aligned}$$

Letting $\boldsymbol{v} = \boldsymbol{u}(\varepsilon)$ in the variational equations of problem $\mathcal{P}(\varepsilon; \Omega)$ and using the asymptotic behavior of the function $g(\varepsilon)$ (Thm. 3.3-1) and the uniform positive definiteness of the scaled elasticity tensor (Thm. 3.3-2), we obtain

$$\sum_{i,j} |e_{i\|j}(\varepsilon) - e_{i\|j}|^2_{0,\Omega} \le c_4 \Lambda(\varepsilon),$$

where

$$\Lambda(\varepsilon) := L(\varepsilon)(\boldsymbol{u}(\varepsilon)) - \int_\Omega A^{ijkl}(\varepsilon)(2e_{k\|l}(\varepsilon) - e_{k\|l})e_{i\|j}\sqrt{g(\varepsilon)}\,dx.$$

The strong convergences $F^{ij}(\varepsilon) \to F^{ij}$ in $L^2(\Omega)$, the weak convergences $e_{i\|j}(\varepsilon) \rightharpoonup e_{i\|j}$ in $L^2(\Omega)$ (part (ii)), and the relations satisfied by the limits $e_{i\|j}$ (parts (iii) and (vi)) together imply that

$$\begin{aligned}
&\lim_{\varepsilon\to 0} L(\varepsilon)(\boldsymbol{u}(\varepsilon)) \\
&= \lim_{\varepsilon\to 0}\left\{ \int_\Omega F^{\alpha\beta}(\varepsilon)e_{\alpha\|\beta}(\varepsilon)\sqrt{g(\varepsilon)}\,dx \right.\\
&\qquad \left. + \int_\Omega \left(2F^{\alpha 3}(\varepsilon)e_{\alpha\|3}(\varepsilon) + F^{33}(\varepsilon)e_{3\|3}(\varepsilon)\right)\sqrt{g(\varepsilon)}\,dx \right\}\\
&= \int_\omega \varphi^{\alpha\beta}\overline{e_{\alpha\|\beta}}\sqrt{a}\,dy + \int_\Omega \left\{\frac{1}{\mu}a_{\alpha\beta}F^{\alpha 3}F^{\beta 3} + \frac{1}{\lambda+2\mu}F^{33}F^{33}\right\}\sqrt{a}\,dx
\end{aligned}$$

as $\varepsilon \to 0$, on the one hand. The asymptotic behavior of the functions $A^{ijkl}(\varepsilon)$ and $g(\varepsilon)$ (Thms. 3.3-1 and 3.3-2), the weak convergences $e_{i\|j}(\varepsilon) \rightharpoonup e_{i\|j}$ in $L^2(\Omega)$ (part (ii)), and the relations satisfied by the limits $e_{i\|j}$ (parts (iii) and (vi)) together imply that

$$\begin{aligned}
&\lim_{\varepsilon\to 0}\int_\Omega A^{ijkl}(\varepsilon)(2e_{k\|l}(\varepsilon) - e_{k\|l})e_{i\|j}\sqrt{g(\varepsilon)}\,dx \\
&= \int_\Omega A^{ijkl}(0)e_{k\|l}e_{i\|j}\sqrt{a}\,dx \\
&= \int_\Omega \left(\frac{1}{2}a^{\alpha\beta\sigma\tau}e_{\sigma\|\tau} + \frac{\lambda}{\lambda+2\mu}a^{\alpha\beta}F^{33}\right)e_{\alpha\|\beta}\sqrt{a}\,dx \\
&\quad + \int_\Omega \left\{4\mu a^{\alpha\sigma}e_{\sigma\|3}e_{\alpha\|3} + F^{33}e_{3\|3}\right\}\sqrt{a}\,dx \\
&= \int_\omega a^{\alpha\beta\sigma\tau}\overline{e_{\sigma\|\tau}}\ \overline{e_{\alpha\|\beta}}\sqrt{a}\,dy \\
&\quad + \int_\Omega \left\{\frac{1}{\mu}a_{\alpha\beta}F^{\alpha 3}F^{\beta 3} + \frac{1}{\lambda+2\mu}F^{33}F^{33}\right\}\sqrt{a}\,dx,
\end{aligned}$$

on the other. Hence

$$\lim_{\varepsilon\to 0}\Lambda(\varepsilon) = \Lambda := \left\{\int_\omega \varphi^{\alpha\beta}\overline{e_{\alpha\|\beta}}\sqrt{a}\,dy - \int_\omega a^{\alpha\beta\sigma\tau}\overline{e_{\sigma\|\tau}}\ \overline{e_{\alpha\|\beta}}\sqrt{a}\,dy\right\}.$$

Since $\overline{\boldsymbol{u}(\varepsilon)} \in \mathbf{V}(\omega)$ for each $\varepsilon > 0$, we have in particular

$$\int_\omega a^{\alpha\beta\sigma\tau}\overline{e_{\sigma\|\tau}}\gamma_{\alpha\beta}(\overline{\boldsymbol{u}(\varepsilon)})\sqrt{a}\,dy = \int_\omega \varphi^{\alpha\beta}\gamma_{\alpha\beta}(\overline{\boldsymbol{u}(\varepsilon)})\sqrt{a}\,dy,$$

by part (v). Using the weak convergences $\gamma_{\alpha\beta}(\overline{\boldsymbol{u}(\varepsilon)}) \rightharpoonup \overline{e_{\alpha\|\beta}}$ in $L^2(\omega)$ (part (iv)), we thus obtain

$$\Lambda = 0.$$

In this fashion, the strong convergences $e_{i\|j}(\varepsilon) \to e_{i\|j}$ in $L^2(\Omega)$ are established.

In Thm. 5.2-1, we showed that

$$\begin{aligned} &\|\frac{1}{\varepsilon}\partial_3 e_{\alpha\|\beta}(\varepsilon; \boldsymbol{v}) + \rho_{\alpha\beta}(\boldsymbol{v})\|_{-1,\Omega} \\ &\qquad \leq C_1\Big\{\sum_i |e_{i\|3}(\varepsilon; \boldsymbol{v})|_{0,\Omega} + \varepsilon\sum_\alpha |v_\alpha|_{0,\Omega} + \varepsilon\|v_3\|_{1,\Omega}\Big\} \end{aligned}$$

for all $\boldsymbol{v} \in \mathbf{H}^1(\Omega)$, where $\|\cdot\|_{-1,\Omega}$ denotes the norm in $H^{-1}(\Omega)$. Letting $\boldsymbol{v} = \boldsymbol{u}^{-1}(\varepsilon) := \varepsilon\boldsymbol{u}(\varepsilon)$ in this inequality thus yields

$$\begin{aligned} &\|\partial_3 e_{\alpha\|\beta}(\varepsilon) + \rho_{\alpha\beta}(\boldsymbol{u}^{-1}(\varepsilon))\|_{-1,\Omega} \\ &\quad \leq C_1\Big\{\sum_i \varepsilon|e_{i\|3}(\varepsilon)|_{0,\Omega} + \varepsilon\sum_\alpha |\varepsilon u_\alpha(\varepsilon)|_{0,\Omega} + \varepsilon\|\varepsilon u_3(\varepsilon)\|_{1,\Omega}\Big\} \leq c_5\varepsilon, \end{aligned}$$

by part (ii). Since $e_{\alpha\|\beta}(\varepsilon) \to e_{\alpha\|\beta}$ in $L^2(\Omega)$ as shown above and since the limits $e_{\alpha\|\beta}$ are independent of the transverse variable (part (vi)),

$$\partial_3 e_{\alpha\|\beta}(\varepsilon) \to \partial_3 e_{\alpha\|\beta} = 0 \text{ in } H^{-1}(\Omega);$$

therefore,

$$\rho_{\alpha\beta}(\boldsymbol{u}^{-1}(\varepsilon)) \to 0 \text{ in } H^{-1}(\Omega).$$

An application of Thm. 5.2-2 thus shows that

$$\boldsymbol{u}^{-1}(\varepsilon) = \varepsilon\boldsymbol{u}(\varepsilon) \to \mathbf{0} \text{ in } \mathbf{H}^1(\Omega).$$

By Thm. 4.2-1 (a), the strong convergences $e_{i\|j}(\varepsilon) \to e_{i\|j}$ in $L^2(\Omega)$ imply in particular the strong convergences

$$\overline{e_{\alpha\|\beta}(\varepsilon)} \to \overline{e_{\alpha\|\beta}} \text{ in } L^2(\omega),$$

which, combined with those of part (iv), yield the strong convergences $\gamma_{\alpha\beta}(\overline{\boldsymbol{u}(\varepsilon)}) \to \overline{e_{\alpha\|\beta}}$ in $L^2(\Omega)$. Hence $(\gamma_{\alpha\beta}(\overline{\boldsymbol{u}(\varepsilon)}))_{\varepsilon>0}$ is a Cauchy sequence in $L^2(\omega)$. Since

$$|\overline{\boldsymbol{u}(\varepsilon)} - \overline{\boldsymbol{u}(\varepsilon')}|_\omega^M = \Big\{\sum_{\alpha,\beta} |\gamma_{\alpha\beta}(\overline{\boldsymbol{u}(\varepsilon)}) - \gamma_{\alpha\beta}(\overline{\boldsymbol{u}(\varepsilon')})|_{0,\omega}^2\Big\}^{1/2},$$

the strong convergence $\overline{\boldsymbol{u}(\varepsilon)} \to \boldsymbol{\zeta}$ in $\mathbf{V}_M^\sharp(\omega)$ is established.

(viii) *The limit* $\boldsymbol{\zeta} \in \mathbf{V}_M^\sharp(\omega)$ *found in part* (vii) *satisfies the equations*

$$B_M^\sharp(\boldsymbol{\zeta}, \boldsymbol{\eta}) = L_M^\sharp(\boldsymbol{\eta}) \text{ for all } \boldsymbol{\eta} \in \mathbf{V}_M^\sharp(\omega),$$

which have a unique solution. Consequently, the convergence

$$\overline{\boldsymbol{u}(\varepsilon)} \to \boldsymbol{\zeta} \textit{ in } \mathbf{V}_M^\sharp(\omega)$$

established in part (vii) *holds for the whole family* $(\overline{\boldsymbol{u}(\varepsilon)})_{\varepsilon>0}$.

Let $\boldsymbol{\eta} \in \mathbf{V}(\omega)$. Since $\overline{\boldsymbol{u}(\varepsilon)} \in \mathbf{V}(\omega)$,

$$\begin{aligned}
\lim_{\varepsilon\to 0} B_M(\overline{\boldsymbol{u}(\varepsilon)}, \boldsymbol{\eta}) &= \lim_{\varepsilon\to 0} \int_\omega a^{\alpha\beta\sigma\tau} \gamma_{\sigma\tau}(\overline{\boldsymbol{u}(\varepsilon)}) \gamma_{\alpha\beta}(\boldsymbol{\eta}) \sqrt{a}\, dy \\
&= \int_\omega a^{\alpha\beta\sigma\tau} \overline{e_{\alpha\|\beta}} \gamma_{\alpha\beta}(\boldsymbol{\eta}) \sqrt{a}\, dy \\
&= L(\boldsymbol{\eta}),
\end{aligned}$$

by part (v) and the convergences $\gamma_{\alpha\beta}(\overline{\boldsymbol{u}(\varepsilon)}) \to \overline{e_{\alpha\|\beta}}$ in $L^2(\omega)$ established in part (vii). But we also have

$$\lim_{\varepsilon\to 0} B_M(\overline{\boldsymbol{u}(\varepsilon)}, \boldsymbol{\eta}) = B_M^\sharp(\boldsymbol{\zeta}, \boldsymbol{\eta}),$$

since $\overline{\boldsymbol{u}(\varepsilon)} \to \boldsymbol{\zeta}$ in $\mathbf{V}_M^\sharp(\omega)$ (part (vii)). Hence

$$B_M^\sharp(\boldsymbol{\zeta}, \boldsymbol{\eta}) = L(\boldsymbol{\eta}) \text{ for all } \boldsymbol{\eta} \in \mathbf{V}(\omega),$$

and thus

$$B_M^\sharp(\boldsymbol{\zeta}, \boldsymbol{\eta}) = L_M^\sharp(\boldsymbol{\eta}) \text{ for all } \boldsymbol{\eta} \in \mathbf{V}_M^\sharp(\omega),$$

by definition of the continuous extensions $B_M^\sharp$ and $L_M^\sharp$.

These variational equations have one and only one solution, for the space $\mathbf{V}_M^\sharp(\omega)$, the bilinear form $B_M^\sharp$, and the linear form $L_M^\sharp$ satisfy all the assumptions of the *Lax-Milgram lemma* (the ellipticity of the bilinear form relies in particular on the uniform positive definiteness of the scaled two-dimensional elasticity tensor of the shell; cf. Thm. 3.3-2).

We next prove a simple corollary of the lemma of J.L. Lions; this corollary will be put to use in part (x).

(ix) *Let Ω be a domain in $\mathbb{R}^3$. Given $\boldsymbol{v} = (v_i) \in \mathbf{L}^2(\Omega)$, define the distributions*

$$e_{ij}(\boldsymbol{v}) := \frac{1}{2}(\partial_i v_j + \partial_j v_i) \in H^{-1}(\Omega).$$

Let there be given a sequence of functions $\boldsymbol{v}^k = (v_i^k) \in \mathbf{L}^2(\Omega)$ such that

$$\boldsymbol{v}^k \to \mathbf{0} \text{ in } \mathbf{H}^{-1}(\Omega) \text{ and } e_{ij}(\boldsymbol{v}^k) \to 0 \text{ in } \mathbf{H}^{-1}(\Omega) \text{ as } k \to \infty.$$

Then

$$\boldsymbol{v}^k \to \mathbf{0} \text{ in } \mathbf{L}^2(\Omega).$$

As shown by Amrouche & Girault [1994], the implication

$$\{w \in H^{-2}(\Omega) \text{ and } \partial_i w \in H^{-2}(\Omega),\ 1 \le i \le n\} \Rightarrow w \in H^{-1}(\Omega)$$

holds if Ω is a domain in $\mathbb{R}^n$, where

$$H^{-2}(\Omega) := \text{ dual space of } H_0^2(\Omega).$$

Let then $\boldsymbol{v} = (v_i)$ be such that

$$\boldsymbol{v} \in \mathbf{H}^{-1}(\Omega) \text{ and } e_{ij}(\boldsymbol{v}) \in \mathbf{H}^{-1}(\Omega).$$

Since $w \in H^{-1}(\Omega)$ implies $\partial_j w \in H^{-2}(\Omega)$, we have

$$\begin{aligned} \partial_j v_i &\in H^{-2}(\Omega), \\ \partial_k(\partial_j v_i) = \{\partial_k e_{ij}(\boldsymbol{v}) + \partial_j e_{ik}(\boldsymbol{v}) - \partial_i e_{jk}(\boldsymbol{v})\} &\in H^{-2}(\Omega), \end{aligned}$$

so that $\partial_j v_i \in H^{-1}(\Omega)$. Using the *lemma of J.L. Lions* (Thm. 1.7-1), we therefore conclude that $\boldsymbol{v} \in \mathbf{L}^2(\Omega)$. In other words,

$$\{\boldsymbol{v} = (v_i) \in \mathbf{H}^{-1}(\Omega);\ e_{ij}(\boldsymbol{v}) \in \mathbf{H}^{-1}(\Omega)\} = \mathbf{L}^2(\Omega).$$

Then the closed graph theorem shows that

$$|\boldsymbol{v}|_{0,\Omega} \le c_6 \Big\{ \|\boldsymbol{v}\|_{-1,\Omega} + \sum_{i,j} \|e_{ij}(\boldsymbol{v})\|_{-1,\Omega} \Big\}$$

for all $\boldsymbol{v} \in \mathbf{L}^2(\Omega)$ and the asserted property follows from this inequality.

(x) *The following strong convergences hold* (so far, they were only known to be weak; cf. parts (ii) and (vi)):

$$\begin{aligned} \boldsymbol{u}(\varepsilon) &\to \boldsymbol{u} \text{ in } \mathbf{V}^\sharp_M(\Omega), \\ \partial_3 u_\alpha(\varepsilon) &\to 0 \text{ in } L^2(\Omega). \end{aligned}$$

To establish the convergence $\boldsymbol{u}(\varepsilon) \to \boldsymbol{u}$ in $\mathbf{V}^\sharp_M(\Omega)$, it suffices to show that $(\boldsymbol{u}(\varepsilon))_{\varepsilon>0}$ is a Cauchy sequence with respect to the norm $|\cdot|^M_\Omega$. By definition,

$$\begin{aligned} &\left(|\boldsymbol{u}(\varepsilon) - \boldsymbol{u}(\varepsilon')|^M_\Omega\right)^2 \\ &\quad = \sum_{\alpha,\beta} |\gamma_{\alpha\beta}(\overline{\boldsymbol{u}(\varepsilon)}) - \gamma_{\alpha\beta}(\overline{\boldsymbol{u}(\varepsilon')})|^2_{0,\omega} + \sum_i |\partial_3 u_i(\varepsilon) - \partial_3 u_i(\varepsilon')|^2_{0,\Omega}. \end{aligned}$$

Consequently, in view of the convergences $\partial_3 u_3(\varepsilon) \to 0$ in $L^2(\Omega)$ and $\gamma_{\alpha\beta}(\overline{\boldsymbol{u}(\varepsilon)}) \to \overline{e_{\alpha\|\beta}}$ in $L^2(\omega)$ (parts (ii) and (vii)), it remains in fact to establish the convergences $\partial_3 u_\alpha(\varepsilon) \to 0$ in $L^2(\Omega)$.

Proving these is equivalent to showing that

$$\partial_3 \boldsymbol{u}'(\varepsilon) \to \mathbf{0} \text{ in } \mathbf{L}^2(\Omega), \text{ where } \boldsymbol{u}'(\varepsilon) := (u_1(\varepsilon), u_2(\varepsilon), 0).$$

By part (ix), this is in turn equivalent to proving that

$$\partial_3 \boldsymbol{u}'(\varepsilon) \to \mathbf{0} \text{ in } \mathbf{H}^{-1}(\Omega),$$

together with

$$e_{ij}(\partial_3 \boldsymbol{u}'(\varepsilon)) \to 0 \text{ in } H^{-1}(\Omega),$$

where

$$e_{ij}(\boldsymbol{v}) := \frac{1}{2}(\partial_j v_i + \partial_i v_j).$$

By definition of the functions $e_{\alpha\|3}(\varepsilon)$,

$$\partial_3 u_\alpha(\varepsilon) = 2\varepsilon e_{\alpha\|3}(\varepsilon) - \varepsilon \partial_\alpha u_3(\varepsilon) + 2\varepsilon \Gamma^\sigma_{\alpha 3}(\varepsilon) u_\sigma(\varepsilon).$$

Hence the boundedness of the sequences $(\Gamma^\sigma_{\alpha 3}(\varepsilon))_{\varepsilon>0}$ in $\mathcal{C}^0(\overline{\Omega})$ (Thm. 3.3-1) and the convergences (parts (vi) and (vii))

$$\begin{aligned} \varepsilon e_{\alpha\|3}(\varepsilon) &\to 0 \text{ in } L^2(\Omega), \\ \varepsilon u_i(\varepsilon) &\to 0 \text{ in } L^2(\Omega), \end{aligned}$$

imply that

$$\partial_3 u_\alpha(\varepsilon) \to 0 \text{ in } H^{-1}(\Omega);$$

this shows that $\partial_3 \boldsymbol{u}'(\varepsilon) \to \mathbf{0}$ in $\mathbf{H}^{-1}(\Omega)$. In view of establishing that $e_{ij}(\partial_3 \boldsymbol{u}'(\varepsilon)) \to 0$ in $H^{-1}(\Omega)$, we first notice that

$$e_{33}(\partial_3 \boldsymbol{u}'(\varepsilon)) = \partial_3 e_{33}(\boldsymbol{u}'(\varepsilon)) = 0.$$

The asymptotic behavior of the functions $\Gamma^{\sigma}_{\alpha 3}(\varepsilon)$ (Thm. 3.3-1) and the convergences $e_{i\|j}(\varepsilon) \to e_{i\|j}$ in $L^2(\Omega)$ (part (vii)) next imply that

$$\{\partial_3 u_\alpha(\varepsilon) + \varepsilon \partial_\alpha u_3(\varepsilon) + 2\varepsilon b^{\sigma}_{\alpha} u_\sigma(\varepsilon)\} \to 0 \text{ in } L^2(\Omega);$$

consequently

$$\partial_{33} u_\alpha(\varepsilon) + \varepsilon \partial_{\alpha 3} u_3(\varepsilon) + 2\varepsilon b^{\sigma}_{\alpha} \partial_3 u_\sigma(\varepsilon) \to 0 \text{ in } H^{-1}(\Omega).$$

Since $\partial_3 u_3(\varepsilon) \to 0$ in $L^2(\Omega)$ (part (ii)) and $\varepsilon u_\sigma(\varepsilon) \to 0$ in $L^2(\Omega)$ (part (vi)), we have

$$2e_{\alpha 3}(\partial_3 \boldsymbol{u}'(\varepsilon)) = \partial_{33} u_\alpha(\varepsilon) \to 0 \text{ in } H^{-1}(\Omega).$$

From the relations $e_{\alpha\|\beta}(\varepsilon) \to e_{\alpha\|\beta}$ in $L^2(\Omega)$ and $\partial_3 e_{\alpha\|\beta} = 0$ (part (vi)), we next infer that $\partial_3 e_{\alpha\|\beta}(\varepsilon) \to 0$ in $H^{-1}(\Omega)$, hence that

$$\left\{\partial_3 e_{\alpha\beta}(\boldsymbol{u}(\varepsilon)) - \partial_3\left(\Gamma^{p}_{\alpha\beta}(\varepsilon) u_p(\varepsilon)\right)\right\} \to 0 \text{ in } H^{-1}(\Omega),$$

by definition of the functions $e_{\alpha\|\beta}(\varepsilon)$. Because the functions $\Gamma^{p}_{\alpha\beta}(\varepsilon)$ are in $\mathcal{C}^1(\overline{\Omega})$, the functions $\Gamma^{p}_{\alpha\beta}(\varepsilon) u_p(\varepsilon)$ are in $H^1(\Omega)$; besides,

$$\partial_3(\Gamma^{p}_{\alpha\beta}(\varepsilon) u_p(\varepsilon)) = (\partial_3 \Gamma^{p}_{\alpha\beta}(\varepsilon)) u_p(\varepsilon) + \Gamma^{p}_{\alpha\beta}(\varepsilon) \partial_3 u_p(\varepsilon),$$

and

$$\Gamma^{p}_{\alpha\beta}(\varepsilon) \partial_3 u_p(\varepsilon) \to 0 \text{ in } H^{-1}(\Omega),$$

since $\partial_3 u_p(\varepsilon) \to 0$ in $H^{-1}(\Omega)$ (we have established that $\partial_3 u_3(\varepsilon) \to 0$ in $L^2(\Omega)$ in part (ii) and we just established that $\partial_3 u_\alpha(\varepsilon) \to 0$ in $H^{-1}(\Omega)$). Finally, the estimates $\|\partial_3 \Gamma^{p}_{\alpha\beta}(\varepsilon)\|_{0,\infty,\overline{\Omega}} \leq C_1 \varepsilon$ (Thm. 3.3-1) and the convergences $\varepsilon \boldsymbol{u}(\varepsilon) \to 0$ in $\mathbf{L}^2(\Omega)$ (part (vi)) imply that

$$\left(\partial_3 \Gamma^{p}_{\alpha\beta}(\varepsilon)\right) u_p(\varepsilon) \to 0 \text{ in } L^2(\Omega).$$

We thus infer from these relations that

$$e_{\alpha\beta}(\partial_3 \boldsymbol{u}'(\varepsilon)) = \partial_3 e_{\alpha\beta}(\boldsymbol{u}(\varepsilon)) \to 0 \text{ in } H^{-1}(\Omega),$$

and all the required convergences $e_{ij}(\partial_3 \boldsymbol{u}'(\varepsilon)) \to 0$ in $H^{-1}(\Omega)$ are thus established.

(xi) *The whole family* $(\boldsymbol{u}(\varepsilon))_{\varepsilon>0}$ *converges strongly to* $\boldsymbol{u}$ *in the space* $\mathbf{V}_M^\sharp(\Omega)$.

We have shown that the whole family $(\overline{\boldsymbol{u}(\varepsilon)})_{\varepsilon>0}$ converges in the space $\mathbf{V}_M^\sharp(\omega)$ (part (viii)) and that $\partial_3 \boldsymbol{u}(\varepsilon) \to \mathbf{0}$ in $\mathbf{L}^2(\Omega)$ for a subsequence (parts (ii) and (x)). Since the limit of this subsequence is unique, the whole family $(\partial_3 \boldsymbol{u}(\varepsilon))_{\varepsilon>0}$ converges in $\mathbf{L}^2(\Omega)$. Hence the family $(\boldsymbol{u}(\varepsilon))_{\varepsilon>0}$ is a Cauchy sequence in the Hilbert space $\mathbf{V}_M^\sharp(\Omega)$, and the proof is complete. ■

Remark. Thm. 5.6-1 applies *a fortiori* to a family of *linearly elastic membrane shells*, in which case the space $\mathbf{V}_M^\sharp(\omega)$ is simply $H_0^1(\omega) \times H_0^1(\omega) \times L^2(\omega)$. The convergence obtained in this fashion "over the set ω" is identical to, but the convergence "over the set Ω" is weaker than, that obtained in Thm. 4.4-1, where we were able to establish the convergence of the family $(\boldsymbol{u}(\varepsilon))_{\varepsilon>0}$ in the space $H^1(\Omega) \times H^1(\Omega) \times L^2(\Omega)$. In addition, we did not have to assume in Thm. 4.4-1 that the forces were "admissible". ■

We next consider a family of *linearly elastic generalized membrane shells of the second kind*, in which case similar convergences hold, again in "abstract" *completions*, but these are now completions of *quotient spaces.*

We use the following notations: The equivalence class of $\boldsymbol{v} \in \mathbf{V}(\Omega)$ in the quotient space $\mathbf{V}(\Omega)/\mathbf{V}_0(\Omega)$ is noted $\dot{\boldsymbol{v}}$; likewise, the equivalence class of $\boldsymbol{\eta} \in \mathbf{V}(\omega)$ in the quotient space $\mathbf{V}(\omega)/\mathbf{V}_0(\omega)$ is noted $\dot{\boldsymbol{\eta}}$. Similar notations designate equivalence classes in the completions $\dot{\mathbf{V}}_M^\sharp(\Omega)$ and $\dot{\mathbf{V}}_M^\sharp(\omega)$.

The following result is due to Ciarlet & Lods [1996d, Thm. 5.1].

Theorem 5.6-2. *Assume that* $\boldsymbol{\theta} \in \mathcal{C}^3(\overline{\omega}; \mathbb{R}^3)$. *Consider a family of linearly elastic generalized membrane shells of the second kind* (Sect. 5.4), *with thickness* 2ε *approaching zero, with each having the same middle surface* $S = \boldsymbol{\theta}(\overline{\omega})$, *and with each subjected to a boundary condition of place along a portion of its lateral face having the same set* $\boldsymbol{\theta}(\gamma_0)$ *as its middle curve, and subjected to applied forces that are admissible* (Sect. 5.5). *Let* $\boldsymbol{u}(\varepsilon)$ *denote for* $0 < \varepsilon \le \varepsilon_0$ *the solution of the associated scaled three-dimensional problems* $\mathcal{P}(\varepsilon; \Omega)$ (Thm. 5.1-1).

Define the spaces

$$\dot{\mathbf{V}}^{\sharp}_M(\Omega) := \text{completion of } \mathbf{V}(\Omega)/\mathbf{V}_0(\Omega) \text{ with respect to } |\cdot|^M_\Omega,$$
$$\dot{\mathbf{V}}^{\sharp}_M(\omega) := \text{completion of } \mathbf{V}(\omega)/\mathbf{V}_0(\omega) \text{ with respect to } |\cdot|^M_\omega,$$

where

$$\mathbf{V}_0(\Omega) := \{\boldsymbol{v} \in \mathbf{V}(\Omega);\, \partial_3 \boldsymbol{v} = \mathbf{0} \text{ in } \Omega,\, \gamma_{\alpha\beta}(\overline{\boldsymbol{v}}) = 0 \text{ in } \omega\} \neq \{\mathbf{0}\},$$
$$\mathbf{V}_0(\omega) := \{\boldsymbol{\eta} \in \mathbf{V}(\omega);\, \gamma_{\alpha\beta}(\boldsymbol{\eta}) = 0 \text{ in } \omega\} \neq \{\mathbf{0}\}.$$

Then there exist $\dot{\boldsymbol{u}} \in \dot{\mathbf{V}}^{\sharp}_M(\Omega)$ *and* $\dot{\boldsymbol{\zeta}} \in \dot{\mathbf{V}}^{\sharp}_M(\omega)$ *such that*

$$\dot{\boldsymbol{u}}(\varepsilon) \to \dot{\boldsymbol{u}} \text{ in } \dot{\mathbf{V}}^{\sharp}_M(\Omega) \text{ as } \varepsilon \to 0,$$
$$\overline{\dot{\boldsymbol{u}}(\varepsilon)} := \overline{\frac{1}{2}\int_{-1}^{1} \boldsymbol{u}(\varepsilon)\, dx_3}^{\,\cdot} \to \dot{\boldsymbol{\zeta}} \text{ in } \dot{\mathbf{V}}^{\sharp}_M(\omega) \text{ as } \varepsilon \to 0.$$

Let the forms B_M and L_M be defined as in Thm. 5.6-1. *For* $\dot{\boldsymbol{\zeta}}, \dot{\boldsymbol{\eta}} \in \{\mathbf{V}(\omega)/\mathbf{V}_0(\omega)\}$, *let*

$$\dot{B}_M(\dot{\boldsymbol{\zeta}}, \dot{\boldsymbol{\eta}}) := B_M(\boldsymbol{\zeta}', \boldsymbol{\eta}') \text{ for any } \boldsymbol{\zeta}' \in \dot{\boldsymbol{\zeta}} \text{ and any } \boldsymbol{\eta}' \in \dot{\boldsymbol{\eta}},$$
$$\dot{L}_M(\dot{\boldsymbol{\eta}}) := L_M(\boldsymbol{\eta}') \text{ for any } \boldsymbol{\eta}' \in \dot{\boldsymbol{\eta}},$$

and let $\dot{B}^{\sharp}_M$ and $\dot{L}^{\sharp}_M$ denote the unique continuous extensions of the bilinear form $\dot{B}_M$ and linear form $\dot{L}_M$ from $\mathbf{V}(\omega)/\mathbf{V}_0(\omega)$ to $\dot{\mathbf{V}}^{\sharp}_M(\omega)$. Then the limit $\dot{\boldsymbol{\zeta}}$ satisfies the following **scaled two-dimensional variational problem $\dot{\mathcal{P}}^{\sharp}_M(\omega)$ of a generalized membrane shell of the second kind:**

$$\dot{\boldsymbol{\zeta}} \in \dot{\mathbf{V}}^{\sharp}_M(\omega) \text{ and } \dot{B}^{\sharp}_M(\dot{\boldsymbol{\zeta}}, \dot{\boldsymbol{\eta}}) = \dot{L}^{\sharp}_M(\dot{\boldsymbol{\eta}}) \text{ for all } \dot{\boldsymbol{\eta}} \in \dot{\mathbf{V}}^{\sharp}_M(\omega).$$

Proof. The proof closely follows that of Thm. 5.6-1. We simply indicate which parts need to be *modified.*

Part (ii) becomes (otherwise its proof is identical): *The semi-norms* $|\boldsymbol{u}(\varepsilon)|_\Omega^M$ *and* $|\overline{\boldsymbol{u}(\varepsilon)}|_\omega^M$, *hence the norms* $|\dot{\boldsymbol{u}}(\varepsilon)|_\Omega^M$ *and* $|\dot{\overline{\boldsymbol{u}(\varepsilon)}}|_\Omega^M$ *and the norms* $\|\varepsilon\boldsymbol{u}(\varepsilon)\|_{1,\Omega}$ *and* $|e_{i\|j}(\varepsilon)|_{0,\Omega}$, *are bounded independently of* $0 < \varepsilon \leq \varepsilon_1$.

Define the spaces

$$\begin{aligned}\dot{\mathbf{V}}_M^\sharp(\Omega) &:= \text{completion of } \mathbf{V}(\Omega)/\mathbf{V}_0(\Omega) \text{ with respect to } |\cdot|_\Omega^M,\\ \dot{\mathbf{V}}_M^\sharp(\omega) &:= \text{completion of } \mathbf{V}(\omega)/\mathbf{V}_0(\omega) \text{ with respect to } |\cdot|_\omega^M.\end{aligned}$$

Then there is a subsequence, still denoted $(\boldsymbol{u}(\varepsilon))_{\varepsilon>0}$ *for convenience, and there exist* $\dot{\boldsymbol{u}} \in \dot{\mathbf{V}}_M^\sharp(\Omega)$, $\boldsymbol{u}^{-1} = (u_i^{-1}) \in \mathbf{V}(\Omega)$, $e_{i\|j} \in L^2(\Omega)$, *and* $\dot{\boldsymbol{\zeta}} \in \dot{\mathbf{V}}_M^\sharp(\omega)$ *such that*

$$\begin{aligned}\dot{\boldsymbol{u}}(\varepsilon) &\rightharpoonup \dot{\boldsymbol{u}} \text{ in } \dot{\mathbf{V}}_M^\sharp(\Omega),\\ \varepsilon\boldsymbol{u}(\varepsilon) &\rightharpoonup \boldsymbol{u}^{-1} \text{ in } \mathbf{H}^1(\Omega) \text{ and thus } \varepsilon\boldsymbol{u}(\varepsilon) \to \boldsymbol{u}^{-1} \text{ in } \mathbf{L}^2(\Omega),\\ e_{i\|j}(\varepsilon) &\rightharpoonup e_{i\|j} \text{ in } L^2(\Omega),\\ \partial_3 u_3(\varepsilon) &= \varepsilon e_{3\|3}(\varepsilon) \to 0 \text{ in } L^2(\Omega),\\ \dot{\overline{\boldsymbol{u}(\varepsilon)}} &\rightharpoonup \dot{\boldsymbol{\zeta}} \text{ in } \dot{\mathbf{V}}_M^\sharp(\omega) \text{ as } \varepsilon \to 0.\end{aligned}$$

Part (vii) becomes: *The following strong convergences hold* as $\varepsilon \to 0$:

$$\begin{aligned}e_{i\|j}(\varepsilon) &\to e_{i\|j} \text{ in } L^2(\Omega),\\ \varepsilon\boldsymbol{u}(\varepsilon) &\to \mathbf{0} \text{ in } \mathbf{H}^1(\Omega),\\ \gamma_{\alpha\beta}(\overline{\boldsymbol{u}(\varepsilon)}) &\to \overline{e_{\alpha\|\beta}} \text{ in } L^2(\omega),\\ \dot{\overline{\boldsymbol{u}(\varepsilon)}} &\to \dot{\boldsymbol{\zeta}} \text{ in } \dot{\mathbf{V}}_M^\sharp(\omega),\end{aligned}$$

The only difference in the proof of part (vii) consists in noting that

$$\begin{aligned}|\dot{\overline{\boldsymbol{u}(\varepsilon)}} - \dot{\overline{\boldsymbol{u}(\varepsilon')}}|_\omega^M &= |\overline{\boldsymbol{u}(\varepsilon)} - \overline{\boldsymbol{u}(\varepsilon')}|_\omega^M\\ &= \left\{\sum_{\alpha,\beta} |\gamma_{\alpha\beta}(\overline{\boldsymbol{u}(\varepsilon)}) - \gamma_{\alpha\beta}(\overline{\boldsymbol{u}(\varepsilon')})|_{0,\omega}^2\right\}^{1/2},\end{aligned}$$

and thus that the strong convergence $\dot{\overline{\boldsymbol{u}(\varepsilon)}} \to \dot{\boldsymbol{\zeta}}$ in $\dot{\mathbf{V}}_M^\sharp(\omega)$ holds.

Part (viii) and its proof become: *The limit* $\dot{\boldsymbol{\zeta}} \in \dot{\mathbf{V}}_M^\sharp(\omega)$ *found in* part (vii) *satisfies the equations*

$$\dot{B}_M^\sharp(\dot{\boldsymbol{\zeta}}, \dot{\boldsymbol{\eta}}) = \dot{L}_M^\sharp(\dot{\boldsymbol{\eta}}) \text{ for all } \dot{\boldsymbol{\eta}} \in \dot{\mathbf{V}}_M^\sharp(\omega),$$

which have a unique solution. Consequently, the convergence

$$\dot{\overline{\boldsymbol{u}(\varepsilon)}} \to \dot{\boldsymbol{\zeta}} \text{ in } \dot{\mathbf{V}}_M^\sharp(\omega)$$

established in part (vii) *holds for the whole family* $(\dot{\overline{\boldsymbol{u}(\varepsilon)}})_{\varepsilon>0}$.

Let $\boldsymbol{\eta} \in \mathbf{V}(\omega)$. Since $\overline{\boldsymbol{u}(\varepsilon)} \in \mathbf{V}(\omega)$,

$$\begin{aligned}
\lim_{\varepsilon\to 0} B_M(\overline{\boldsymbol{u}(\varepsilon)}, \boldsymbol{\eta}) &= \lim_{\varepsilon\to 0} \int_\omega a^{\alpha\beta\sigma\tau} \gamma_{\sigma\tau}(\overline{\boldsymbol{u}(\varepsilon)})\gamma_{\alpha\beta}(\boldsymbol{\eta})\sqrt{a}\, dy \\
&= \int_\omega a^{\alpha\beta\sigma\tau} \overline{e_{\alpha\|\beta}} \gamma_{\alpha\beta}(\boldsymbol{\eta})\sqrt{a}\, dy \\
&= L(\boldsymbol{\eta}),
\end{aligned}$$

by part (v) and by the convergences $\gamma_{\alpha\beta}(\overline{\boldsymbol{u}(\varepsilon)}) \to \overline{e_{\alpha\|\beta}}$ in $L^2(\omega)$ established in part (vii). But we also have $\dot{B}_M(\dot{\overline{\boldsymbol{u}(\varepsilon)}}, \dot{\boldsymbol{\eta}}) = B_M(\overline{\boldsymbol{u}(\varepsilon)}, \boldsymbol{\eta})$ and $\dot{L}(\dot{\boldsymbol{\eta}}) = L(\boldsymbol{\eta})$ by definition of the forms $\dot{B}$ and $\dot{L}$; therefore,

$$\lim_{\varepsilon\to 0} \dot{B}_M(\dot{\overline{\boldsymbol{u}(\varepsilon)}}, \dot{\boldsymbol{\eta}}) = \dot{L}(\dot{\boldsymbol{\eta}}),$$

on the one hand. On the other,

$$\lim_{\varepsilon\to 0} \dot{B}_M(\dot{\overline{\boldsymbol{u}(\varepsilon)}}, \dot{\boldsymbol{\eta}}) = \dot{B}_M^\sharp(\dot{\boldsymbol{\zeta}}, \dot{\boldsymbol{\eta}}),$$

since $\dot{\overline{\boldsymbol{u}(\varepsilon)}} \to \dot{\boldsymbol{\zeta}}$ in $\dot{\mathbf{V}}_M^\sharp(\omega)$ (part (vii)). Hence

$$\dot{B}_M^\sharp(\dot{\boldsymbol{\zeta}}, \dot{\boldsymbol{\eta}}) = \dot{L}(\dot{\boldsymbol{\eta}}) \text{ for all } \dot{\boldsymbol{\eta}} \in \dot{\mathbf{V}}(\omega),$$

and thus

$$\dot{B}_M^\sharp(\dot{\boldsymbol{\zeta}}, \dot{\boldsymbol{\eta}}) = \dot{L}_M^\sharp(\dot{\boldsymbol{\eta}}) \text{ for all } \dot{\boldsymbol{\eta}} \in \dot{\mathbf{V}}_M^\sharp(\omega),$$

by definition of the continuous extensions $\dot{B}_M^\sharp$ and $\dot{L}_M^\sharp$.

These variational equations have one and only one solution, for the space $\dot{\mathbf{V}}_M^\sharp(\omega)$, the bilinear form $\dot{B}_M^\sharp$, and the linear form $\dot{L}_M^\sharp$, satisfy all the assumptions of the *Lax-Milgram lemma* (the ellipticity of the bilinear form relies in particular on the uniform positive definiteness of the scaled two-dimensional elasticity tensor of the shell; cf. Thm. 3.3-2).

Part (x) becomes: *The following strong convergences hold:*

$$\dot{\boldsymbol{u}}(\varepsilon) \to \dot{\boldsymbol{u}} \text{ in } \dot{\mathbf{V}}^{\sharp}_M(\Omega),$$
$$\partial_3 u_\alpha(\varepsilon) \to 0 \text{ in } L^2(\Omega).$$

To establish the convergence $\dot{\boldsymbol{u}}(\varepsilon) \to \dot{\boldsymbol{u}}$ in $\dot{\mathbf{V}}^{\sharp}_M(\Omega)$, it suffices to show that $(\dot{\boldsymbol{u}}(\varepsilon))_{\varepsilon>0}$ is a Cauchy sequence with respect to the norm $|\cdot|^M_\Omega$. By definition,

$$\left(|\dot{\boldsymbol{u}}(\varepsilon) - \dot{\boldsymbol{u}}(\varepsilon')|^M_\Omega\right)^2 = \left(|\boldsymbol{u}(\varepsilon) - \boldsymbol{u}(\varepsilon')|^M_\Omega\right)^2$$
$$= \sum_{\alpha,\beta} |\gamma_{\alpha\beta}(\overline{\boldsymbol{u}(\varepsilon)}) - \gamma_{\alpha\beta}(\overline{\boldsymbol{u}(\varepsilon')})|^2_{0,\omega} + \sum_i |\partial_3 u_i(\varepsilon) - \partial_3 u_i(\varepsilon')|^2_{0,\Omega}.$$

The remainder of the proof of part (x) is the same.

Finally, part (xi) and its proof become: *The whole family* $(\dot{\boldsymbol{u}}(\varepsilon))_{\varepsilon>0}$ *converges strongly to* $\dot{\boldsymbol{u}}$ *in the space* $\dot{\mathbf{V}}^{\sharp}_M(\Omega)$.

We have shown that the whole family $(\dot{\overline{\boldsymbol{u}(\varepsilon)}})_{\varepsilon>0}$ converges in the space $\dot{\mathbf{V}}^{\sharp}_M(\omega)$ (part (viii)) and that $\partial_3\boldsymbol{u}(\varepsilon) \to \mathbf{0}$ in $\mathbf{L}^2(\Omega)$ for a subsequence (parts (ii) and (x)). Since the limit of this subsequence is unique, the whole family $(\partial_3\boldsymbol{u}(\varepsilon))_{\varepsilon>0}$ converges in $\mathbf{L}^2(\Omega)$. Hence the family $(\dot{\boldsymbol{u}}(\varepsilon))_{\varepsilon>0}$ is a Cauchy sequence in the Hilbert space $\dot{\mathbf{V}}^{\sharp}_M(\Omega)$, and the proof is complete. ∎

A scrutiny of the proofs of Thms. 5.6-1 and 5.6-2 shows that they provide worthy information about the asymptotic behavior of the scaled three-dimensional solutions $\boldsymbol{u}(\varepsilon)$ and of their averages $\overline{\boldsymbol{u}(\varepsilon)}$, viz.,

$$e_{i\|j}(\varepsilon; \boldsymbol{u}(\varepsilon)) \to e_{i\|j} \text{ in } L^2(\Omega),$$
$$\gamma_{\alpha\beta}(\overline{\boldsymbol{u}(\varepsilon)}) \to \overline{e_{\alpha\|\beta}} \text{ in } L^2(\omega),$$
$$\partial_3\boldsymbol{u}(\varepsilon) \to \mathbf{0} \text{ in } \mathbf{L}(\Omega),$$
$$\varepsilon\boldsymbol{u}(\varepsilon) \to \mathbf{0} \text{ in } \mathbf{H}^1(\Omega).$$

A noticeable feature of these convergences is that they occur in spaces that are more "decent" that the somewhat "exotic" spaces $\mathbf{V}^{\sharp}_{M}(\omega)$ or $\dot{\mathbf{V}}^{\sharp}_{M}(\omega)$.

For a family of linearly elastic *elliptic membrane*, or *flexural*, shells, the three-dimensional limit $\boldsymbol{u}$ was, or will be, found to be independent of x_3 (Thms. 4.4-1 or 6.2-1). The convergence $\partial_3 \boldsymbol{u}(\varepsilon) \to \mathbf{0}$ in $\mathbf{L}^2(\Omega)$ is an attenuated form of this property for a family of *generalized membrane* shells.

5.7. THE TWO-DIMENSIONAL EQUATIONS OF A LINEARLY ELASTIC GENERALIZED MEMBRANE SHELL; EXISTENCE AND UNIQUENESS OF SOLUTIONS

For brevity, we only consider in this section generalized membrane shells *of the first kind*, as those of the second kind are treated in a similar fashion.

The next theorem recapitulates the definition and assembles the main properties of the limit two-dimensional problem found at the outcome of the asymptotic analysis carried out in Thm. 5.6-1.

Theorem 5.7-1. *Let ω be a domain in $\mathbb{R}^2$, let γ_0 be a subset of the boundary of ω with* length $\gamma_0 > 0$, *and let $\boldsymbol{\theta} \in \mathcal{C}^3(\overline{\omega}; \mathbb{R}^3)$ be an injective mapping such that the two vectors $\boldsymbol{a}_\alpha = \partial_\alpha \boldsymbol{\theta}$ are linearly independent at all points of $\overline{\omega}$. Assume that*

$$\begin{aligned}
\mathbf{V}_0(\omega) &= \{\mathbf{0}\}, \text{ where} \\
\mathbf{V}_0(\omega) &:= \{\boldsymbol{\eta} = (\eta_i) \in \mathbf{H}^1(\omega);\ \boldsymbol{\eta} = \mathbf{0} \text{ on } \gamma_0,\ \gamma_{\alpha\beta}(\boldsymbol{\eta}) = 0 \text{ in } \omega\}, \\
\gamma_{\alpha\beta}(\boldsymbol{\eta}) &:= \frac{1}{2}(\partial_\beta \eta_\alpha + \partial_\alpha \eta_\beta) - \Gamma^\sigma_{\alpha\beta}\eta_\sigma - b_{\alpha\beta}\eta_3,
\end{aligned}$$

and define the spaces

$$\begin{aligned}
\mathbf{V}(\omega) &:= \{\boldsymbol{\eta} = (\eta_i) \in \mathbf{H}^1(\omega);\ \boldsymbol{\eta} = \mathbf{0} \text{ on } \gamma_0\}, \\
\mathbf{V}^{\sharp}_{M}(\omega) &:= \text{completion of } \mathbf{V}(\omega) \text{ with respect to } |\cdot|^M_\omega, \text{ where} \\
|\boldsymbol{\eta}|^M_\omega &:= \Big\{\sum_{\alpha,\beta} |\gamma_{\alpha\beta}(\boldsymbol{\eta})|^2_{0,\omega}\Big\}^{1/2}.
\end{aligned}$$

Let

$$a^{\alpha\beta\sigma\tau} := \frac{4\lambda\mu}{\lambda+2\mu}a^{\alpha\beta}a^{\sigma\tau} + 2\mu(a^{\alpha\sigma}a^{\beta\tau} + a^{\alpha\tau}a^{\beta\sigma}),$$

$$B_M(\boldsymbol{\zeta}, \boldsymbol{\eta}) := \int_\omega a^{\alpha\beta\sigma\tau}\gamma_{\sigma\tau}(\boldsymbol{\zeta})\gamma_{\alpha\beta}(\boldsymbol{\eta})\sqrt{a}\,dy \text{ for } \boldsymbol{\zeta}, \boldsymbol{\eta} \in \mathbf{V}(\omega),$$

$$L(\boldsymbol{\eta}) := \int_\omega \varphi^{\alpha\beta}\gamma_{\alpha\beta}(\boldsymbol{\eta})\sqrt{a}\,dy \text{ for } \boldsymbol{\eta} \in \mathbf{V}(\omega),$$

where the functions $\varphi^{\alpha\beta} \in L^2(\omega)$ *are given, and let* $B_M^\sharp$ *and* $L_M^\sharp$ *denote the unique continuous extensions from* $\mathbf{V}(\omega)$ *to* $\mathbf{V}_M^\sharp(\omega)$ *of the bilinear form* B_M *and linear form* L.

Then the scaled two-dimensional variational problem $\mathcal{P}_M^\sharp(\omega)$ *of a linearly elastic generalized membrane shell of the first kind: Find* $\boldsymbol{\zeta}$ *that satisfies*

$$\boldsymbol{\zeta} \in \mathbf{V}_M^\sharp(\omega) \text{ and } B_M^\sharp(\boldsymbol{\zeta}, \boldsymbol{\eta}) = L_M^\sharp(\boldsymbol{\eta}) \text{ for all } \boldsymbol{\eta} \in \mathbf{V}_M^\sharp(\omega),$$

has one and only one solution.

Proof. The assumption $\mathbf{V}_0(\omega) = \{\mathbf{0}\}$ means that the semi-norm $|\cdot|_\omega^M$ is a norm over the space $\mathbf{V}(\omega)$. The linear form $L : \mathbf{V}(\omega) \to \mathbb{R}$ and the bilinear form $B_M : \mathbf{V}(\omega) \times \mathbf{V}(\omega) \to \mathbb{R}$ are clearly continuous with respect to this norm. Besides,

$$B_M(\boldsymbol{\eta}, \boldsymbol{\eta}) \geq c_e^{-1}\sqrt{a_0}(|\boldsymbol{\eta}|_\omega^M)^2 \text{ for all } \boldsymbol{\eta} \in \mathbf{V}(\omega),$$

since the scaled two-dimensional elasticity tensor of the shell is uniformly positive definite (Thm. 3.3-2 (a)) and $a(y) \geq a_0 > 0$ for all $y \in \overline{\omega}$ (Thm. 3.3-1). These properties remain valid on the space $\mathbf{V}_M^\sharp(\omega)$ since $\mathbf{V}(\omega)$ is by construction dense in $\mathbf{V}_M^\sharp(\omega)$, again with respect to $|\cdot|_\omega^M$. The conclusion thus follows from the *Lax-Milgram lemma*. ∎

Thm. 5.7-1 is established under the assumption that $\mathbf{V}_0(\omega) = \{\mathbf{0}\}$. Therefore it applies not only to *generalized membrane shells* but also to *elliptic membrane shells*, since these satisfy the even stronger assumption (established in Thm. 2.7-2; recall that $\gamma_0 = \gamma$ in this case):

$$\{\boldsymbol{\eta} = (\eta_i) \in H_0^1(\omega) \times H_0^1(\omega) \times L^2(\omega);\ \gamma_{\alpha\beta}(\boldsymbol{\eta}) = 0 \text{ in } \omega\} = \{\mathbf{0}\}.$$

However, Thm. 5.7-1 alone does *not* say anything about the corresponding space $\mathbf{V}_M^\sharp(\omega)$, which in this case turns out to be the space $\mathbf{V}_M(\omega) = H_0^1(\omega) \times H_0^1(\omega) \times L^2(\omega)$. To prove this constitutes the "hard" part of the analysis, which rests on the *inequality of Korn's type on an elliptic surface* (Thm. 2.7-3). The same inequality is also crucially needed for establishing that, again in the case of an elliptic membrane shell, any continuous linear form L on the space $\mathbf{V}_M(\omega)$ may be indeed expressed as in Thm. 5.7-1 by means of functions $\varphi^{\alpha\beta} \in L^2(\omega)$.

In order to get physically meaningful formulas, it remains to "de-scale" the unknown $\boldsymbol{\zeta}$ that satisfies the limit "scaled" problem $\mathcal{P}_M^\sharp(\omega)$ found in Thm. 5.7-1. In view of the scaling

$$\boldsymbol{u}(\varepsilon)(x) = \boldsymbol{u}^\varepsilon(x^\varepsilon) = (u_i^\varepsilon(x^\varepsilon)) \text{ for all } x^\varepsilon = \pi^\varepsilon x \in \overline{\Omega}^\varepsilon$$

made on the covariant components u_i^ε of the displacement field (Sect. 5.1), we are naturally led to defining for each $\varepsilon > 0$ the **limit vector field** $\boldsymbol{\zeta}^\varepsilon$ by letting

$$\boxed{\boldsymbol{\zeta}^\varepsilon := \boldsymbol{\zeta}.}$$

A word of caution. For the *elliptic membrane shells* considered in the previous chapter, or for the *flexural shells* considered in the next chapter, *each* component of the field $\boldsymbol{\zeta} = (\zeta_i)$ can be separately de-scaled, affording in turn the consideration of the *limit tangential displacement field* $\zeta_\alpha^\varepsilon \boldsymbol{a}^\alpha$ and of the *limit normal displacement field* $\zeta_3^\varepsilon \boldsymbol{a}^3$ of the middle surface of the shell, where $\zeta_i^\varepsilon := \zeta_i$ (cf. Sect. 4.6 or 6.4). By contrast, such a componentwise de-scaling is impossible for a *generalized membrane shell*, unless specific knowledge on the components of the "functions" in the *completion* $\mathbf{V}_M^\sharp(\omega)$ allows to do so (the examples of Sect. 5.8 are particularly illuminating in this respect). In other words, the above de-scaling $\boldsymbol{\zeta}^\varepsilon = \boldsymbol{\zeta}$ is to be understood as an *equality in the "abstract" completion* $\mathbf{V}_M^\sharp(\omega)$ and nothing else! ■

Recall that λ^ε and μ^ε denote for each $\varepsilon > 0$ the *actual* Lamé constants of the elastic material constituting the shell. We then have the following immediate corollary to Thm. 5.7-1; naturally, the *existence* and *uniqueness* results of Thm. 5.7-1 apply *verbatim* to the de-scaled problem $\mathcal{P}_M^{\sharp\varepsilon}(\omega)$ (for this reason, they are not reproduced here):

Theorem 5.7-2. *Let the assumptions and definitions not repeated here be as in* Thm. 5.7-1, *let*

$$a^{\alpha\beta\sigma\tau,\varepsilon} := \frac{4\lambda^\varepsilon\mu^\varepsilon}{\lambda^\varepsilon + 2\mu^\varepsilon} a^{\alpha\beta}a^{\sigma\tau} + 2\mu^\varepsilon(a^{\alpha\sigma}a^{\beta\tau} + a^{\alpha\tau}a^{\beta\sigma}),$$

$$B_M^\varepsilon(\boldsymbol{\zeta}, \boldsymbol{\eta}) := \varepsilon \int_\omega a^{\alpha\beta\sigma\tau,\varepsilon}\gamma_{\sigma\tau}(\boldsymbol{\zeta})\gamma_{\alpha\beta}(\boldsymbol{\eta})\sqrt{a}\, dy \text{ for } \boldsymbol{\zeta}, \boldsymbol{\eta} \in \mathbf{V}(\omega),$$

$$L_M^\varepsilon(\boldsymbol{\eta}) := \int_\omega \varphi^{\alpha\beta,\varepsilon}\gamma_{\alpha\beta}(\boldsymbol{\eta})\sqrt{a}\, dy \text{ for } \boldsymbol{\eta} \in \mathbf{V}(\omega),$$

$$\varphi^{\alpha\beta,\varepsilon} := \varepsilon\varphi^{\alpha\beta},$$

and let $B_M^{\sharp\varepsilon}$ *and* $L_M^{\sharp\varepsilon}$ *denote the unique continuous extensions from* $\mathbf{V}(\omega)$ *to* $\mathbf{V}_M^\sharp(\omega)$ *of the bilinear form* B_M^ε *and linear form* L_M^ε. *Then the limit vector field* $\boldsymbol{\zeta}^\varepsilon$ *satisfies the following* **two-dimensional variational problem** $\mathcal{P}_M^{\sharp\varepsilon}(\omega)$ **of a linearly elastic generalized membrane shell of the first kind:**

$$\boxed{\boldsymbol{\zeta}^\varepsilon \in \mathbf{V}_M^\sharp(\omega) \text{ and } B_M^{\sharp\varepsilon}(\boldsymbol{\zeta}^\varepsilon, \boldsymbol{\eta}) = L_M^{\sharp\varepsilon}(\boldsymbol{\eta}) \text{ for all } \boldsymbol{\eta} \in \mathbf{V}_M^\sharp(\omega).}$$

Equivalently, the field $\boldsymbol{\zeta}^\varepsilon$ *satisfies the following minimization problem:*

$$\boxed{\begin{aligned} &\boldsymbol{\zeta}^\varepsilon \in \mathbf{V}_M^\sharp(\omega) \text{ and } j_M^{\sharp\varepsilon}(\boldsymbol{\zeta}^\varepsilon) = \inf_{\boldsymbol{\eta}\in\mathbf{V}_M^\sharp(\omega)} j_M^{\sharp\varepsilon}(\boldsymbol{\eta}), \text{ where} \\ &j_M^{\sharp\varepsilon}(\boldsymbol{\eta}) := \frac{1}{2}B_M^{\sharp\varepsilon}(\boldsymbol{\eta}, \boldsymbol{\eta}) - L_M^{\sharp\varepsilon}(\boldsymbol{\eta}).\end{aligned}}$$

■

Each one of the two formulations found in Thm. 5.7-2 constitutes the **two-dimensional equations of a linearly elastic generalized membrane shell of the first kind.** The functional $j_M^{\sharp\varepsilon} : \mathbf{V}_M^\sharp(\omega) \to \mathbb{R}$ is the **two-dimensional energy**, and the functional

$$\boxed{\boldsymbol{\eta} \in \mathbf{V}_M^\sharp(\omega) \to \frac{1}{2}B_M^{\sharp\varepsilon}(\boldsymbol{\eta}, \boldsymbol{\eta})}$$

is the **two-dimensional strain energy, of a linearly elastic generalized membrane shell of the first kind.** The functions $a^{\alpha\beta\sigma\tau,\varepsilon}$

are the *contravariant components* of the **two-dimensional elasticity tensor of the shell**, already encountered in the two-dimensional equations of a linearly elastic elliptic membrane shell (Thm. 4.5-2).

Observe that the bilinear form $B_M^{\sharp\varepsilon}$ found in the variational equations of a linearly elastic *generalized membrane* shell is an extension of the bilinear form B_M^ε already found in the variational equations of a linearly elastic *elliptic membrane* shell (Thm. 4.5-2). This is why both kinds constitute together the *linearly elastic "membrane" shells.*

5.8. JUSTIFICATION OF THE TWO-DIMENSIONAL EQUATIONS OF A LINEARLY ELASTIC GENERALIZED MEMBRANE SHELL; EXAMPLES, COMMENTARY, AND REFINEMENTS

It remains to convert in terms of *de-scaled unknowns* the fundamental convergence result established in Sect. 5.6. As the "original" unknowns $\boldsymbol{u}^\varepsilon = (u_i^\varepsilon)$ are defined over a domain that *varies with* ε (the set Ω^ε), their *averages* $\frac{1}{2\varepsilon}\int_{-\varepsilon}^{\varepsilon} \boldsymbol{u}^\varepsilon \, dx_3^\varepsilon$ are more appropriate for this purpose, since they are defined over a *fixed* domain (the set ω). For simplicity, we only consider generalized membrane shells *of the first kind.*

Theorem 5.8-1. *Assume that* $\boldsymbol{\theta} \in \mathcal{C}^3(\overline{\omega}; \mathbb{R}^3)$. *Consider a family of linearly elastic generalized membrane shells of the first kind* (Sect. 5.4), *with thickness* 2ε *approaching zero, with each having the same middle surface* $S = \boldsymbol{\theta}(\overline{\omega})$, *with each subjected to a boundary condition of place along a portion of its lateral face having the same set* $\boldsymbol{\theta}(\gamma_0)$ *as its middle curve, and subjected to applied forces that are admissible* (Sect. 5.5).

For each $\varepsilon > 0$, *let* $\boldsymbol{u}^\varepsilon \in \mathbf{H}^1(\Omega^\varepsilon)$ *and* $\boldsymbol{\zeta}^\varepsilon \in \mathbf{V}_M^\sharp(\omega)$ *respectively denote the solutions to the three-dimensional and two-dimensional problems* $\mathcal{P}(\Omega^\varepsilon)$ *and* $\mathcal{P}_M^{\sharp\varepsilon}(\omega)$ (Sect. 3.1 and Thm. 5.7-2).

Finally, let $\boldsymbol{\zeta} \in \mathbf{V}_M^\sharp(\omega)$ *denote the solution to problem* $\mathcal{P}_M^\sharp(\omega)$ (Thm. 5.7-1), *which is thus independent of* ε. *Then*

$$\boldsymbol{\zeta}^\varepsilon = \boldsymbol{\zeta} \text{ and } \frac{1}{2\varepsilon}\int_{-\varepsilon}^{\varepsilon} \boldsymbol{u}^\varepsilon \, dx_3^\varepsilon \to \boldsymbol{\zeta} \text{ in } \mathbf{V}_M^\sharp(\omega) \text{ as } \varepsilon \to 0.$$

Proof. It suffices to combine the convergence

$$\frac{1}{2}\int_{-1}^{1} \boldsymbol{u}(\varepsilon)\, dx_3 \to \boldsymbol{\zeta} \text{ in } \mathbf{V}^{\sharp}_M(\omega)$$

established in Thm. 5.6-1, the relation

$$\frac{1}{2\varepsilon}\int_{-\varepsilon}^{\varepsilon} \boldsymbol{u}^\varepsilon\, dx_3^\varepsilon = \frac{1}{2}\int_{-1}^{1} \boldsymbol{u}(\varepsilon)\, dx_3,$$

which follows from the scaling $\boldsymbol{u}(\varepsilon)(x) := \boldsymbol{u}^\varepsilon(x^\varepsilon)$ for all $x^\varepsilon = \pi^\varepsilon x \in \overline{\Omega}^\varepsilon$ (Sect. 5.1), and the de-scaling $\boldsymbol{\zeta}^\varepsilon := \boldsymbol{\zeta}$ (Sect. 5.7). ■

Remark. Without any specific information about the structure of the vector fields in the space $\mathbf{V}^{\sharp}_M(\omega)$, these relations can no longer be refined (as in Sect. 4.6 for elliptic membrane shells) into ones involving separately the tangential and normal components of the three-dimensional displacement vector fields. ■

Under the essential assumptions that the space $\mathbf{V}_F(\omega)$ *reduces to* $\{\mathbf{0}\}$ *and that the forces are admissible, we have therefore justified by a convergence result* (Thm. 5.8-1) *the two-dimensional equations of a linearly elastic generalized membrane shell.* In so doing, we have also justified the formal asymptotic approach of Caillerie & Sanchez-Palencia [1995b] when *"bending is badly-inhibited"*, according to the terminology of E. Sanchez-Palencia.

The asymptotic analysis of Ciarlet & Lods [1996d] described in this chapter has been extended by Slicaru [1998] to *linearly elastic shells whose middle surface "has no boundary"*, such as a torus.

Among linearly elastic shells, generalized membrane shells possess distinctive characteristics that set them apart:

While *forces* applied to a family of elliptic membrane or flexural shells can be arbitrary (Sects. 4.1 and 6.1), body forces (for instance) applied to a family of generalized membrane shells can no longer be accounted for by an arbitrary linear form of the form $\boldsymbol{v}^\varepsilon = (v_i^\varepsilon) \to \int_{\Omega^\varepsilon} f^{i,\varepsilon} v_i^\varepsilon \sqrt{g^\varepsilon}\, dx^\varepsilon$, i.e., with *arbitrary* contravariant components $f^{i,\varepsilon} \in L^2(\Omega^\varepsilon)$. They must be *admissible "for the three-dimensional equations"*, in order that the associated scaled linear forms be continuous with respect to the norm $\boldsymbol{v} \to \{\sum_{i,j} |e_{i\|j}(\varepsilon; \boldsymbol{v})|^2_{0,\Omega}\}^{1/2}$ and uniformly so with respect to $\varepsilon > 0$ (Sect. 5.5).

The linear form found in the variational equations of the limit two-dimensional problem for such a shell is likewise subjected to a restriction: On the dense subspace $\mathbf{V}(\omega)$ of the space $\mathbf{V}_M^\sharp(\omega)$, it must be of the form $\boldsymbol{\eta} \to \int_\omega \varphi^{\alpha\beta}\gamma_{\alpha\beta}(\boldsymbol{\eta})\sqrt{a}\,dy$ (Thm. 5.6-1). In other words, the applied forces must be also *admissible "for the two-dimensional equations"*, in such a way that the linear form appearing therein must be an element of the *dual space* of $\mathbf{V}_M^\sharp(\omega)$.

As this dual space may be quite "small", the limit variational problem, which otherwise satisfies all the assumptions of the Lax-Milgram lemma (Thm. 5.7-1), possesses the unusual feature that its solution may no longer exist if the data undergo arbitrarily small, yet arbitrarily smooth, perturbations! Another unusual feature of this problem is that the space $\mathbf{V}_M^\sharp(\omega)$ in which its solution is sought may not necessarily be a space of distributions!

Such variational problems fall in the category of *"sensitive problems"* introduced by Lions & Sanchez-Palencia [1994]. Since then, such problems have been extensively studied, either for their own sake or as regards their relations to the theory of generalized membrane shells, by Lions & Sanchez-Palencia [1996, 1997a, 1997b, 1998], Pitkäranta & Sanchez-Palencia [1997], Sanchez-Palencia [1999], Leguillon, Sanchez-Hubert & Sanchez-Palencia [1999].

Examples of linearly elastic generalized membrane shells are numerous. In each case, however, the proof that the space $\mathbf{V}_F(\omega)$ reduces to $\{\mathbf{0}\}$ and the identification of the corresponding space $\mathbf{V}_M^\sharp(\omega)$ and of the admissible applied forces usually require delicate analyses. For this reason, these are not given here; only *ad hoc* references are provided.

A measure of the difficulty inherent to showing that $\mathbf{V}_F(\omega) = \{\mathbf{0}\}$ is given by the proof of the infinitesimal rigid displacement lemma when the surface $S = \boldsymbol{\theta}(\overline{\omega})$ is elliptic and $\gamma_0 = \gamma$ (Thm. 2.7-2).

All the following examples correspond to $\mathbf{V}_0(\omega) = \{\mathbf{0}\}$, i.e., to generalized membrane shells *of the first kind*, according to the definition of Sect. 5.4.

Remark. It seems that there are no known examples of generalized membrane shells of the *second* kind. According to C. Mardare, they should correspond to surfaces S "with little regularity". ■

A *first example* is provided by a shell whose middle surface is a portion of an *elliptic surface* $S = \boldsymbol{\theta}(\overline{\omega})$, i.e., one whose Gaussian

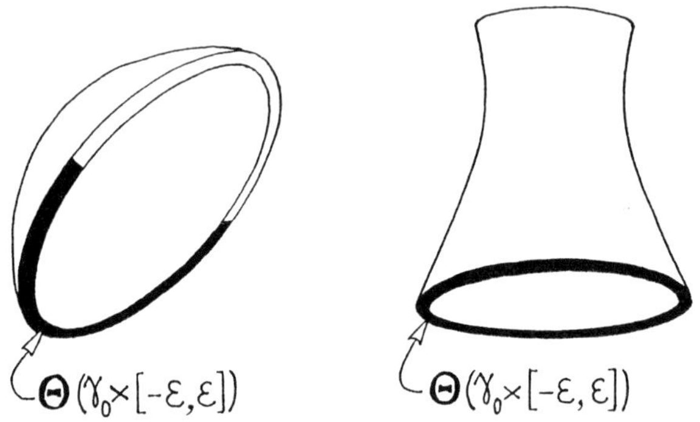

Fig. 5.8-1: *Two linearly elastic generalized membrane shells.* A shell whose middle surface $S = \boldsymbol{\theta}(\overline{\omega})$ is a portion of an ellipsoid and which is subjected to a boundary condition of place (i.e., of vanishing displacement field) along a portion (darkened on the figure) of its lateral face whose middle curve $\boldsymbol{\theta}(\gamma_0)$ is such that $0 < \textit{length}\, \gamma_0 < \textit{length}\, \gamma$, provides an instance of a linearly elastic generalized membrane shell, i.e., one which is not an elliptic membrane shell (Chap. 4), yet whose associated space

$$\mathbf{V}_F(\omega) = \{\boldsymbol{\eta} = (\eta_i) \in H^1(\omega) \times H^1(\omega) \times H^2(\omega);\, \eta_i = \partial_\nu \eta_3 = 0 \text{ on } \gamma_0\}$$

contains only $\boldsymbol{\eta} = \mathbf{0}$ (like that associated with an elliptic membrane shell). A comparison with Fig. 4.1-1 illustrates the crucial rôle played by the set $\boldsymbol{\theta}(\gamma_0)$ for determining the type of a shell!

A shell whose middle surface $S = \boldsymbol{\theta}(\overline{\omega})$ is a portion of a hyperboloid of revolution and which is subjected to a boundary condition of place along its entire "lower" lateral face provides another instance of a linearly elastic generalized membrane shell.

curvature is everywhere > 0 (Sect. 4.1), and which is subjected to a boundary condition of place along a portion of its lateral face whose middle curve $\boldsymbol{\theta}(\gamma_0)$ is such that $0 < \textit{length}\, \gamma_0 < \textit{length}\, \gamma$ (Fig. 5.8-1). Using the unique continuation theorem of Hörmander [1983], Lods & Mardare [1998a] have shown that, under the assumption that $\boldsymbol{\theta} \in \mathcal{C}^{2,1}(\overline{\omega}; \mathbb{R}^3)$, the space

$$\mathbf{R}(\omega) := \{\boldsymbol{\eta} = (\eta_i) \in H^1(\omega) \times H^1(\omega) \times L^2(\omega);\, \eta_\alpha = 0 \text{ on } \gamma_0,\\ \gamma_{\alpha\beta}(\boldsymbol{\eta}) = 0 \text{ in } \omega\}$$

"already" reduces to $\{\mathbf{0}\}$; hence *a fortiori* $\mathbf{V}_0(\omega) = \{\mathbf{0}\}$!

This result extends an earlier result of Vekua [1962] who had shown that $\mathbf{R}(\omega) = \{\mathbf{0}\}$ under the assumptions that $\boldsymbol{\theta} \in W^{3,p}(\omega; \mathbb{R}^3)$, $p > 2$, and that γ is of class $\mathcal{C}^3$. It also contains in the special case where $\gamma_0 = \gamma$ the *infinitesimal rigid displacement lemma on an elliptic surface* (Thm. 2.7-2) due to Ciarlet & Lods [1996a] and Ciarlet & Sanchez-Palencia [1996].

The two-dimensional equations for such a shell constitutes an example of a *sensitive problem:* As shown by Lions & Sanchez-Palencia [1996], the space $\mathbf{V}_M^\sharp(\omega)$ is *not* a space of distributions and there exist functions $p^i \in \mathcal{D}(\omega)$ such that applied forces corresponding to the linear form $\boldsymbol{\eta} \to \int_\omega p^i \eta_i \sqrt{a}\, dy$ are *not* admissible for the two-dimensional equations!

Lods & Mardare [1998a] have also shown that the regularity of the mapping $\boldsymbol{\theta}$ and the assumption of ellipticity of the surface S can be both substantially weakened when S is a portion of a *surface of revolution.*

Remark. While the space $\mathbf{V}_M^\sharp(\omega)$ corresponding to an *elliptic membrane* shell, i.e., one whose middle surface S is smooth and elliptic and which is subjected to a boundary condition of place along its entire lateral face ($\gamma_0 = \gamma$), is a space of distributions (since $\mathbf{V}_M^\sharp(\omega) = H_0^1(\omega) \times H_0^1(\omega) \times L^2(\omega)$ in this case; cf. Chap. 4), it ceases to be so when there is a *fold*, also called an *edge*, on S, in which case the two-dimensional problem becomes *sensitive*; see Lions & Sanchez-Palencia [1998], Gérard & Sanchez-Palencia [1999]. ■

A *second example* is that of a shell whose middle surface is a portion of a *hyperbolic surface* S, i.e., one whose Gaussian curvature is everywhere < 0, provided *the curve intersects all the asymptotic lines* (Sect. 2.2) *of* S; for instance, a shell whose middle surface is a portion of a hyperboloid of revolution and which is subjected to a boundary condition of place along its entire "lower" lateral face (Fig. 5.8-1) fulfills these requirements. Under the assumption that $\boldsymbol{\theta} \in \mathcal{C}^{2,1}(\overline{\omega}; \mathbb{R}^3)$, Mardare [1998c] has shown that $\mathbf{R}(\omega) = \{\mathbf{0}\}$ and that the space $\mathbf{V}_M^\sharp(\omega)$ is a closed subspace of $L^2(\omega) \times L^2(\omega) \times H^{-1}(\omega)$, extending in this fashion earlier results of Vekua [1962] and Sanchez-Palencia [1993]. Interesting complements about shells with hyperbolic middle surfaces are found in Sanchez-Hubert & Sanchez-Palencia [1997, Chap. 7, Sect. 2], Karamian [1998b].

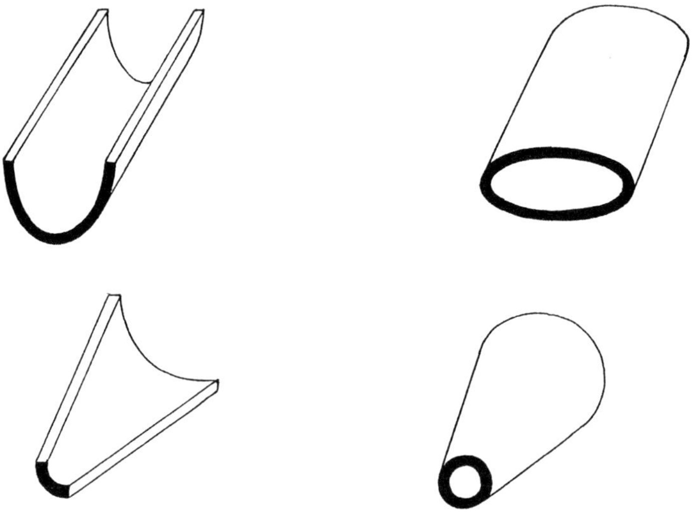

Fig. 5.8-2: *Other examples of linearly elastic generalized membrane shells.* Consider a shell whose middle surface $S = \boldsymbol{\theta}(\overline{\omega})$ is a portion of a cone or a cylinder and which is subjected to a boundary condition of place (i.e., of vanishing displacement field) along a portion (darkened on the figure) of its lateral face with $\boldsymbol{\theta}(\gamma_0)$ as its middle curve. If $\boldsymbol{\theta}(\gamma_0)$ intersects all the generatrices of S, such a shell is a linearly elastic generalized membrane shell.

Remark. If the hyperbolic surface S has a *fold*, the space $\mathbf{V}_M^\sharp(\omega)$ may no longer be a space of distributions, thus yielding another example of *sensitive problem*; see Lions & Sanchez-Palencia [1997b]. ■

A *third example* is that of a shell whose middle surface S is a portion of a *parabolic surface*, i.e., one consisting entirely of parabolic points (Sect. 2.2), again if the set $\boldsymbol{\theta}(\gamma_0)$ is "large enough". Such is the case when S is a portion of a *cylinder* or a portion of a *cone* and *the curve* $\boldsymbol{\theta}(\gamma_0)$ *intersects all the generatrices of* S (which are also the asymptotic lines of S in this case; cf. Sect. 2.2); cf. Fig. 5.8-2. Under the assumption that $\boldsymbol{\theta} \in W^{2,\infty}(\omega; \mathbb{R}^3)$, Mardare [1998c] has shown that $\mathbf{R}(\omega) = \{\mathbf{0}\}$ and that the space $\mathbf{V}_M^\sharp(\omega)$ is a closed subspace of $H^{-1}(\omega) \times H^{-1}(\omega) \times H^{-2}(\omega)$; the regularity assumptions can again be weakened when S is a portion of a surface of revolution. Interesting complements about shells with parabolic middle surfaces are found in Sanchez-Hubert & Sanchez-Palencia [1997, Chap. 7, Sect. 4].

A compact surface "without boundary" is *infinitesimally rigid* if any displacement field satisfying $\gamma_{\alpha\beta}(\boldsymbol{\eta}) = 0$ everywhere on the surface (these relations must be first re-interpreted so as to make sense on such a surface) are infinitesimal rigid displacements in the sense of Thm. 2.6-5. The identification of such infinitesimally rigid surfaces is a classical problem in the differential geometry of surfaces; see, e.g., Nirenberg [1962]. A shell whose middle surface S has no boundary is a linearly elastic generalized membrane shell if S is infinitesimally rigid. Such shells have been extensively studied by Slicaru [1998], who notably extended the analysis of Metskhovrishvili [1957] when S is a torus.

A related phenomenon is the *infinitesimal rigidification* induced by *folds* on a surface and its relations with linear shell theory. In particular, the asymptotic behavior as the thickness approaches zero of a shell with a smooth middle surface and that of one with an arbitrarily close middle surface, but with *folds* (such as a polyhedral surface, i.e., an assembly of planar facets), may be startlingly different! In this direction, see Akian & Sanchez-Palencia [1992], Choi [1993], Choi & Sanchez-Palencia [1993], Geymonat & Sanchez-Palencia [1995], Sanchez-Palencia [1995], Gérard & Sanchez-Palencia [1999].

EXERCISES

5.1. The notations used in (1) should be self-explanatory.

(1) Let a surface $S = \boldsymbol{\theta}(\overline{\omega}) = \tilde{\boldsymbol{\theta}}(\{\tilde{\omega}\}^-)$ be equipped with two systems of curvilinear coordinates $(y_\alpha) \in \overline{\omega}$ and $(\tilde{y}_\alpha) \in \{\tilde{\omega}\}^-$ and assume that $\boldsymbol{\theta}(\gamma_0) = \tilde{\boldsymbol{\theta}}(\tilde{\gamma}_0)$. Show that $\mathbf{V}_F(\omega) = \{\mathbf{0}\}$ implies that $\widetilde{\mathbf{V}}_F(\tilde{\omega}) = \{\mathbf{0}\}$. This means that the definition of a linearly elastic generalized membrane shell (Sect. 5.1) is independent of the system of curvilinear coordinates employed for representing the surface S.

(2) Show that whether such a shell is of the first or of the second kind (Sect. 5.4) is likewise independent of the particular system of curvilinear coordinates employed.

5.2. According to the definition originally given by Ciarlet & Lods [1996d], the applied forces are *admissible* if, for each $\varepsilon > 0$, there exist functions $\varphi^{\alpha\beta}(\varepsilon) = \varphi^{\beta\alpha}(\varepsilon) \in L^2(\omega)$ and $F^i(\varepsilon) \in L^2(\Omega)$ such that $L(\varepsilon)(\boldsymbol{v})$ may be also expressed as

$$L(\varepsilon)(\boldsymbol{v}) = \int_\omega \varphi^{\alpha\beta}(\varepsilon)\gamma_{\alpha\beta}(\overline{\boldsymbol{v}})\sqrt{a}\,dy + \int_\Omega F^i(\varepsilon)\partial_3 v_i \sqrt{g(\varepsilon)}\,dx \quad \text{for all } \boldsymbol{v} \in \mathbf{V}(\Omega),$$

where $\overline{\boldsymbol{v}} = \frac{1}{2}\int_{-1}^{1} \boldsymbol{v}\,dx_3$, and if there exist functions $\varphi^{\alpha\beta} \in L^2(\omega)$ and $F^i \in L^2(\Omega)$ such that

$$\varphi^{\alpha\beta}(\varepsilon) \to \varphi^{\alpha\beta} \text{ in } L^2(\omega) \text{ and } F^i(\varepsilon) \to F^i \text{ in } L^2(\Omega) \text{ as } \varepsilon \to 0.$$

Using this definition (which differs from that of Sect. 5.5), carry out an asymptotic analysis analogous to that of Thm. 5.6-1.

Hint: If the applied forces are admissible, there exists a constant C such that (recall that $g(\varepsilon) = a + O(\varepsilon)$ in $\mathcal{C}^0(\overline{\Omega})$; cf. Thm. 3.3-1):

$$|L(\varepsilon)(\boldsymbol{v})| \le C_3 |\boldsymbol{v}|_\Omega^M \text{ for all } \boldsymbol{v} \in \mathbf{V}(\Omega) \text{ and all } 0 < \varepsilon \le \varepsilon_0.$$

For further details, see Ciarlet & Lods [1996d, Thm. 5.1].

CHAPTER 6

LINEARLY ELASTIC FLEXURAL SHELLS

INTRODUCTION

Consider a linearly elastic shell with middle surface $S = \boldsymbol{\theta}(\overline{\omega})$, subjected to a boundary condition of place along a portion of its lateral face with $\boldsymbol{\theta}(\gamma_0)$ as its middle curve, where $\gamma_0 \subset \gamma$. Such a shell is a *linearly elastic flexural shell* if its associated space

$$\mathbf{V}_F(\omega) = \{\boldsymbol{\eta} = (\eta_i) \in H^1(\omega) \times H^1(\omega) \times H^2(\omega);\\ \eta_i = \partial_\nu \eta_3 = 0 \text{ on } \gamma_0, \gamma_{\alpha\beta}(\boldsymbol{\eta}) = 0 \text{ in } \omega\}.$$

contains *nonzero* functions (Sect. 6.1). *Examples*, where S is either a portion of a cylinder (in which case $\boldsymbol{\theta}(\gamma_0)$ is included in one or two generatrices) or a portion of a cone (in which case $\boldsymbol{\theta}(\gamma_0)$ is included in one generatrix) are presented in Sect. 6.1.

The purpose of this chapter is to *identify* and to *mathematically justify* the *two-dimensional equations of a linearly elastic flexural shell*, by establishing the *convergence in* ad hoc *functional spaces of the three-dimensional displacements as the thickness of such a shell approaches zero.*

More specifically, consider a *family of linearly elastic flexural shells* with thickness 2ε approaching zero, with each having the *same* middle surface $S = \boldsymbol{\theta}(\overline{\omega})$, and with each subjected to a boundary condition of place along a portion of its lateral face having the *same* set $\boldsymbol{\theta}(\gamma_0)$ as its middle curve. The associated three-dimensional problems, posed in curvilinear coordinates over the sets $\Omega^\varepsilon = \omega \times]-\varepsilon, \varepsilon[$, are first transformed for each $\varepsilon > 0$ into equivalent problems, but now posed over the set $\Omega = \omega \times]-1, 1[$, which is *independent of* ε. This transformation relies in a crucial way on appropriate *scalings of the unknowns* u_i^ε (the covariant components of the displacement field) and *assumptions on the Lamé constants* λ^ε *and* μ^ε *and on the contravariant components* $f^{i,\varepsilon}$ *of the applied body forces* (for simplicity, we assume in this introduction that there are no applied surface forces). More specifically, we define the *scaled unknown* $\boldsymbol{u}(\varepsilon) = (u_i(\varepsilon))$ by letting

$$u_i^\varepsilon(x^\varepsilon) = u_i(\varepsilon)(x) \text{ for all } x^\varepsilon = \pi^\varepsilon x \in \overline{\Omega}^\varepsilon,$$

where $\pi^\varepsilon(x_1, x_2, x_3) = (x_1, x_2, \varepsilon x_3)$. We then *assume* that there exist constants $\lambda > 0$ and $\mu > 0$ and functions f^i *independent of* ε such that

$$\lambda^\varepsilon = \lambda \text{ and } \mu^\varepsilon = \mu,$$
$$f^{i,\varepsilon}(x^\varepsilon) = \varepsilon^2 f^i(x) \text{ for all } x^\varepsilon = \pi^\varepsilon x \in \Omega^\varepsilon.$$

It is found in this fashion that the scaled unknown $\boldsymbol{u}(\varepsilon)$ satisfies a variational problem of the form (Thm. 6.1-1):

$$\boldsymbol{u}(\varepsilon) = (u_i(\varepsilon)) \in \mathbf{V}(\Omega) = \{\boldsymbol{v} = (v_i) \in \mathbf{H}^1(\Omega); \\ \boldsymbol{v} = \mathbf{0} \text{ on } \gamma_0 \times [-1, 1]\},$$

$$\int_\Omega A^{ijkl}(\varepsilon) e_{k\|l}(\varepsilon; \boldsymbol{u}(\varepsilon)) e_{i\|j}(\varepsilon; \boldsymbol{v}) \sqrt{g(\varepsilon)}\, dx \\ = \varepsilon^2 \int_\Omega f^i v_i \sqrt{g(\varepsilon)}\, dx \text{ for all } \boldsymbol{v} \in \mathbf{V}(\Omega),$$

where, for any vector field $\boldsymbol{v} = (v_i) \in \mathbf{H}^1(\Omega)$, the *scaled linearized strains* $e_{i\|j}(\varepsilon; \boldsymbol{v}) = e_{j\|i}(\varepsilon; \boldsymbol{v}) \in L^2(\Omega)$ are defined by:

$$e_{\alpha\|\beta}(\varepsilon; \boldsymbol{v}) = \frac{1}{2}(\partial_\beta v_\alpha + \partial_\alpha v_\beta) - \Gamma^p_{\alpha\beta}(\varepsilon) v_p,$$
$$e_{\alpha\|3}(\varepsilon; \boldsymbol{v}) = \frac{1}{2}\left(\frac{1}{\varepsilon}\partial_3 v_\alpha + \partial_\alpha v_3\right) - \Gamma^\sigma_{\alpha 3}(\varepsilon) v_\sigma,$$
$$e_{3\|3}(\varepsilon; \boldsymbol{v}) = \frac{1}{\varepsilon}\partial_3 v_3.$$

Using various analytical preliminaries established in the previous chapter, most notably the *three-dimensional inequality of Korn's type for a family of linearly elastic shells*, we then establish the *main result of this chapter* (Thm. 6.2-1), by showing that *the family* $(\boldsymbol{u}(\varepsilon))_{\varepsilon>0}$ *strongly converges in the space* $\mathbf{H}^1(\Omega)$ *as* $\varepsilon \to 0$ *and that* $\boldsymbol{u} = \lim_{\varepsilon\to 0} \boldsymbol{u}(\varepsilon)$ *is obtained by solving a two-dimensional problem.*

More specifically, we show that *the limit $\boldsymbol{u}$ is independent of the transverse variable* and that $\boldsymbol{\zeta} = \frac{1}{2}\int_{-1}^{1} \boldsymbol{u}\, dx_3$ satisfies the following (scaled) *two dimensional variational problem of a linearly elastic flexural shell:*

$$\boldsymbol{\zeta} \in \mathbf{V}_F(\omega) = \big\{\boldsymbol{\eta} = (\eta_i) \in H^1(\omega) \times H^1(\omega) \times H^2(\omega);$$
$$\eta_i = \partial_\nu \eta_3 = 0 \text{ on } \gamma_0,\ \gamma_{\alpha\beta}(\boldsymbol{\eta}) = 0 \text{ in } \omega\big\},$$

$$\frac{1}{3}\int_\omega a^{\alpha\beta\sigma\tau} \rho_{\sigma\tau}(\boldsymbol{\zeta})\rho_{\alpha\beta}(\boldsymbol{\eta})\sqrt{a}\, dy = \int_\omega \left\{\int_{-1}^{1} f^i\, dx_3\right\}\eta_i \sqrt{a}\, dy$$

for all $\boldsymbol{\eta} = (\eta_i) \in \mathbf{V}_F(\omega)$, where

$$\begin{aligned}
\gamma_{\alpha\beta}(\boldsymbol{\eta}) &= \frac{1}{2}(\partial_\beta \eta_\alpha + \partial_\alpha \eta_\beta) - \Gamma^\sigma_{\alpha\beta}\eta_\sigma - b_{\alpha\beta}\eta_3,\\
\rho_{\alpha\beta}(\boldsymbol{\eta}) &= \partial_{\alpha\beta}\eta_3 - \Gamma^\sigma_{\alpha\beta}\partial_\sigma\eta_3 - b^\sigma_\alpha b_{\sigma\beta}\eta_3 + b^\sigma_\beta(\partial_\alpha\eta_\sigma - \Gamma^\tau_{\alpha\sigma}\eta_\tau)\\
&\quad + b^\sigma_\alpha(\partial_\beta\eta_\sigma - \Gamma^\tau_{\beta\sigma}\eta_\tau) + (\partial_\alpha b^\tau_\beta + \Gamma^\tau_{\alpha\sigma}b^\sigma_\beta - \Gamma^\sigma_{\alpha\beta}b^\tau_\sigma)\eta_\tau,\\
a^{\alpha\beta\sigma\tau} &= \frac{4\lambda\mu}{\lambda + 2\mu}a^{\alpha\beta}a^{\sigma\tau} + 2\mu\big(a^{\alpha\sigma}a^{\beta\tau} + a^{\alpha\tau}a^{\beta\sigma}\big),
\end{aligned}$$

the functions $\gamma_{\alpha\beta}(\boldsymbol{\eta})$, $\rho_{\alpha\beta}(\boldsymbol{\eta})$, and $a^{\alpha\beta\sigma\tau}$ being respectively the covariant components of the *linearized change of metric* and *change of curvature tensors* associated with a displacement field $\eta_i \boldsymbol{a}^i$ of the middle surface S (first encountered in Sects. 2.4 and 2.5) and the contravariant components of the (scaled) *two-dimensional elasticity tensor of the shell.*

We conclude this chapter by reviewing the *existence* and *uniqueness* properties of the solution to the above variational problem and to the associated *minimization problem* (Thm. 6.3-1). We also rewrite the fundamental convergence theorem in terms of *de-scaled unknowns and data*, thus providing a justification in terms of "physical" quantities of the *two-dimensional equations of a linearly elastic flexural shell* (Thm. 6.4-1).

In particular, the *limit displacement field* $\zeta_i^\varepsilon \boldsymbol{a}^i$ of the middle surface S is such that $\boldsymbol{\zeta}^\varepsilon = (\zeta_i^\varepsilon)$ satisfies the following *minimization problem:*

$$\boldsymbol{\zeta}^\varepsilon \in \mathbf{V}_F(\omega) \text{ and } j_F^\varepsilon(\boldsymbol{\zeta}^\varepsilon) = \inf_{\boldsymbol{\eta}\in\mathbf{V}_F(\omega)} j_F^\varepsilon(\boldsymbol{\eta}),$$

where the *two-dimensional energy* $j_F^\varepsilon : \mathbf{V}_F(\omega) \to \mathbb{R}$ is defined by

$$j_F^\varepsilon(\boldsymbol{\eta}) = \frac{1}{6}\int_\omega \varepsilon^3 a^{\alpha\beta\sigma\tau,\varepsilon}\rho_{\sigma\tau}(\boldsymbol{\eta})\rho_{\alpha\beta}(\boldsymbol{\eta})\sqrt{a}\,dy - \int_\omega \Big\{\int_{-\varepsilon}^{\varepsilon} f^{i,\varepsilon}\,dx_3^\varepsilon\Big\}\eta_i\sqrt{a}\,dy,$$

and

$$a^{\alpha\beta\sigma\tau,\varepsilon} = \frac{4\lambda^\varepsilon\mu^\varepsilon}{\lambda^\varepsilon + 2\mu^\varepsilon}a^{\alpha\beta}a^{\sigma\tau} + 2\mu^\varepsilon(a^{\alpha\sigma}a^{\beta\tau} + a^{\alpha\tau}a^{\beta\sigma}).$$

6.1. LINEARLY ELASTIC FLEXURAL SHELLS: DEFINITION, EXAMPLES, AND ASSUMPTIONS ON THE DATA; THE THREE-DIMENSIONAL EQUATIONS OVER A DOMAIN INDEPENDENT OF ε

Let ω be a domain in $\mathbb{R}^2$ with boundary γ and let $\boldsymbol{\theta} \in \mathcal{C}^2(\overline{\omega};\mathbb{R}^3)$ be an injective mapping such that the two vectors $\partial_\alpha\boldsymbol{\theta}(y)$ are linearly independent at all points $y \in \overline{\omega}$. A *linearly elastic shell* with middle surface $S = \boldsymbol{\theta}(\overline{\omega})$ is called a **linearly elastic flexural shell** if the following two conditions are simultaneously satisfied (the definitions and notations are those of Sect. 3.1):

(i) The shell is subjected to a *boundary condition of place* along a portion of its lateral face with $\boldsymbol{\theta}(\gamma_0)$ as its middle curve, where the subset $\gamma_0 \subset \gamma$ satisfies

$$\boxed{\textit{length}\ \gamma_0 > 0.}$$

(ii) Define the space (∂_ν denotes the outer normal derivative operator along γ):

$$\boxed{\begin{aligned}\mathbf{V}_F(\omega) := \big\{\boldsymbol{\eta} = (\eta_i) \in H^1(\omega)\times H^1(\omega)\times H^2(\omega);\\ \eta_i = \partial_\nu\eta_3 = 0 \text{ on } \gamma_0,\ \gamma_{\alpha\beta}(\boldsymbol{\eta}) = 0 \text{ in } \omega\big\}.\end{aligned}}$$

Then *the space* $\mathbf{V}_F(\omega)$ *contains nonzero functions*; equivalently,

$$\boxed{\mathbf{V}_F(\omega) \neq \{\mathbf{0}\}.}$$

We recall that the functions

$$\gamma_{\alpha\beta}(\boldsymbol{\eta}) = \frac{1}{2}(\partial_\beta \eta_\alpha + \partial_\alpha \eta_\beta) - \Gamma^\sigma_{\alpha\beta}\eta_\sigma - b_{\alpha\beta}\eta_3$$

denote the covariant components of the linearized change of metric tensor associated with a displacement field $\eta_i \boldsymbol{a}^i$ of the surface S. The subscript "F" announces that $\mathbf{V}_F(\omega)$ is the functional space over which the limit two-dimensional equations of a linearly elastic flexural shell are posed (Thm. 6.2-1).

In other words, there exist nonzero **admissible linearized inextensional displacements** $\eta_i \boldsymbol{a}^i$ of the middle surface S: *"Admissible"* means that they satisfy *boundary conditions "of clamping"* along the curve $\boldsymbol{\theta}(\gamma_0)$, expressed here by means of the boundary conditions $\eta_i = \partial_\nu \eta_3$ on γ_0 on the associated field $\boldsymbol{\eta} = (\eta_i)$ (these boundary conditions are interpreted later; cf. Sect. 6.4); *"linearized inextensional"* reflects that the functions $\gamma_{\alpha\beta}(\boldsymbol{\eta})$ are the *linearizations* with respect to $\boldsymbol{\eta} = (\eta_i)$ of the covariant components of the *exact* change of metric tensor associated with a displacement field $\eta_i \boldsymbol{a}^i$ of the surface S; cf. Sect. 2.4.

Examples of linearly elastic flexural shells are given in Figs. 6.1-1 to 6.1-3; see also Exs. 6.1 to 6.5.

A word of caution. Like those of linearly elastic elliptic membrane or generalized membrane shells, the definition of a linearly elastic flexural shell depends only on the subset of the lateral face where the shell is subjected to a boundary condition of place (*via* the set γ_0) and on the geometry of the middle surface of the shell; in this respect, see also Ex. 6.6. ■

The formal analysis of Sect. 3.4 then naturally leads us to make the following **scalings of the unknowns** and **assumptions on the data** *for a family of linearly elastic flexural shells*, with each having the *same* middle surface $S = \boldsymbol{\theta}(\overline{\omega})$ and with each subjected to a boundary condition of place along a portion of its lateral face having the *same* set $\boldsymbol{\theta}(\gamma_0)$ as its middle curve, as their thickness 2ε approaches zero.

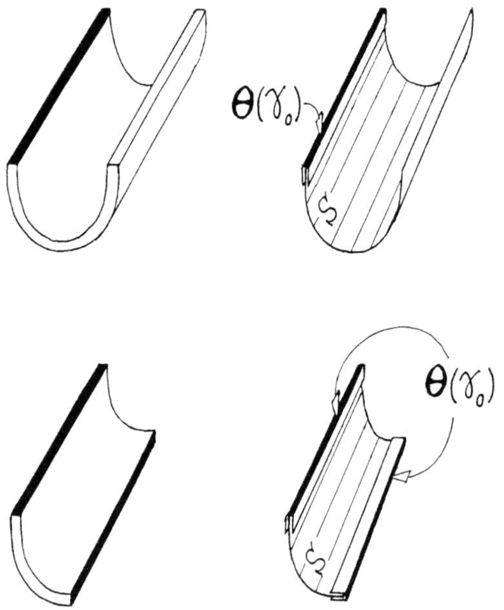

Fig. 6.1-1: *A linearly elastic flexural shell.* A shell whose middle surface $S = \boldsymbol{\theta}(\overline{\omega})$ is a portion of a cylinder and which is subjected to a boundary condition of place (i.e., of vanishing displacement field) along a portion (darkened on the figure) of its lateral face whose middle curve $\boldsymbol{\theta}(\gamma_0)$ is contained in one or two generatrices of S provides an instance of a linearly elastic flexural shell, i.e., one for which the space
$\mathbf{V}_F(\omega) = \{\boldsymbol{\eta} \in H^1(\omega) \times H^1(\omega) \times H^2(\omega);\ \eta_i = \partial_\nu \eta_3 = 0 \text{ on } \gamma_0,\ \gamma_{\alpha\beta}(\boldsymbol{\eta}) = 0 \text{ in } \omega\}$
contains nonzero functions $\boldsymbol{\eta}$; cf. Ex. 6.1.

The "two-dimensional boundary conditions of clamping" $\eta_i = \partial_\nu \eta_3 = 0$ on γ_0 inherited by the limit two-dimensional equations are so named because they mean that the points of, and the tangent spaces to, the deformed and undeformed middle surfaces coincide along the set $\boldsymbol{\theta}(\gamma_0)$ (as suggested in the "two-dimensional" figure); cf. Sect. 6.3.

This example and that of the cylindrical shell from Fig. 5.8-2 illustrate the crucial rôle played by the set $\boldsymbol{\theta}(\gamma_0)$ for determining the type of a shell.

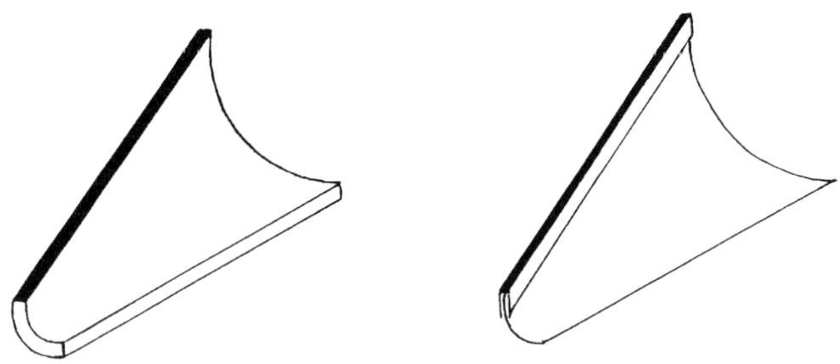

Fig. 6.1-2: *Another example of a linearly elastic flexural shell.* A shell whose middle surface $S = \boldsymbol{\theta}(\overline{\omega})$ is a portion of a cone excluding its vertex and which is subjected to a boundary condition of place along a portion (darkened on the figure) of its lateral face whose middle curve $\boldsymbol{\theta}(\gamma_0)$ is contained in one generatrix of S provides another example where $\mathbf{V}_F(\omega) \neq \{\mathbf{0}\}$; cf. Ex. 6.2.

The two-dimensional boundary conditions of clamping inherited by the limit two-dimensional equations are suggested on the "two-dimensional" figure.

A comparison with Fig. 5.8-2 again illustrates the rôle played by the set $\boldsymbol{\theta}(\gamma_0)$ for determining the type of shell.

First, we define the **scaled unknown** $\boldsymbol{u}(\varepsilon) = (u_i(\varepsilon))$ by letting

$$\boxed{u_i^\varepsilon(x^\varepsilon) =: u_i(\varepsilon)(x) \text{ for all } x^\varepsilon = \pi^\varepsilon x \in \overline{\Omega}^\varepsilon.}$$

Next, we *require* that the *Lamé constants* and the *applied body and surface force densities* be such that

$$\boxed{\begin{aligned} \lambda^\varepsilon = \lambda \quad &\text{and} \quad \mu^\varepsilon = \mu, \\ f^{i,\varepsilon}(x^\varepsilon) = \varepsilon^2 f^i(x) \quad &\text{for all} \quad x^\varepsilon = \pi^\varepsilon x \in \Omega^\varepsilon, \\ h^{i,\varepsilon}(x^\varepsilon) = \varepsilon^3 h^i(x) \quad &\text{for all} \quad x^\varepsilon = \pi^\varepsilon x \in \Gamma_+^\varepsilon \cup \Gamma_-^\varepsilon, \end{aligned}}$$

where the constants $\lambda > 0$ and $\mu > 0$ and the functions $f^i \in L^2(\Omega)$ and $h^i \in L^2(\Gamma_+ \cup \Gamma_-)$ are independent of ε (Fig. 3.2-1 recapitulates the definitions of the sets Ω^ε, Ω, Γ_+^ε, Γ_+, Γ_-^ε, and Γ_-).

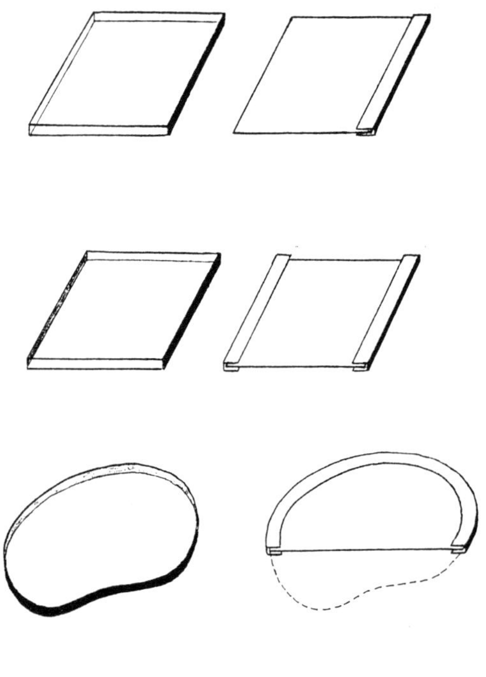

Fig. 6.1-3: *Another example of a linearly elastic flexural shell.* A plate subjected to a boundary condition of place along any portion (darkened on the figure) of its lateral face whose middle line γ_0 satisfies *length* $\gamma_0 > 0$ provides an instance of a linearly elastic flexural shell since $\mathbf{V}_F(\omega) \supset \{\boldsymbol{\eta} = (0, 0, \eta_3);\ \eta_3 \in H_0^2(\omega)\} \neq \{\mathbf{0}\}$ in each case; cf. Ex. 6.3.

The two-dimensional boundary conditions of clamping inherited by the limit two-dimensional equations are suggested on the "two-dimensional" figures (for visual clarity, only a portion of the plate is represented in the last case).

Surprisingly, if γ_0 is "large enough", the same plate is no longer modeled as a flexural shell when it is viewed as a nonlinearly elastic body! See Fig. 9.1-2.

Remark. For notational brevity, the functions f^i and h^i stand for the functions that were respectively denoted $f^{i,2}$ and $h^{i,3}$ in Sect. 3.4. ■

As an immediate corollary to Thm. 3.2-1 (simply corresponding to $p = 2$), we obtain the *problems satisfied by the scaled unknown over the set* Ω, thus over a domain that is *independent of* ε:

Theorem 6.1-1. *Let ω be a domain in $\mathbb{R}^2$, let $\boldsymbol{\theta} \in \mathcal{C}^2(\overline{\omega}; \mathbb{R}^3)$ be an injective mapping such that the two vectors $\boldsymbol{a}_\alpha = \partial_\alpha \boldsymbol{\theta}$ are linearly independent at all points of $\overline{\omega}$, and let $\varepsilon_0 > 0$ be as in* Thm. 3.1-1. *Consider a family of linearly elastic flexural shells with thickness 2ε, with each having the same middle surface $S = \boldsymbol{\theta}(\overline{\omega})$ and with each subjected to a boundary condition of place along a portion of its lateral face having the same set $\boldsymbol{\theta}(\gamma_0)$ as its middle curve.*

Let the assumptions on the data be as above. Then, for each $0 < \varepsilon \le \varepsilon_0$, the scaled unknown $\boldsymbol{u}(\varepsilon) = (u_i(\varepsilon))$ satisfies the following **scaled three-dimensional variational problem $\mathcal{P}(\varepsilon; \Omega)$ of a linearly elastic flexural shell:**

$$\begin{aligned}
&\boldsymbol{u}(\varepsilon) \in \mathbf{V}(\Omega) := \{\boldsymbol{v} = (v_i) \in \mathbf{H}^1(\Omega);\ \boldsymbol{v} = \mathbf{0} \text{ on } \Gamma_0\},\\
&\int_\Omega A^{ijkl}(\varepsilon) e_{k\|l}(\varepsilon; \boldsymbol{u}(\varepsilon)) e_{i\|j}(\varepsilon; \boldsymbol{v}) \sqrt{g(\varepsilon)}\, dx\\
&\qquad = \varepsilon^2 \left\{ \int_\Omega f^i v_i \sqrt{g(\varepsilon)}\, dx + \int_{\Gamma_+ \cup \Gamma_-} h^i v_i \sqrt{g(\varepsilon)}\, d\Gamma \right\}\\
&\qquad\qquad\qquad \text{for all } \boldsymbol{v} \in \mathbf{V}(\Omega),
\end{aligned}$$

where $\Gamma_0 := \gamma_0 \times [-1, 1]$, the scaled strains $e_{i\|j}(\varepsilon; \boldsymbol{v})$ are given by

$$\begin{aligned}
e_{\alpha\|\beta}(\varepsilon; \boldsymbol{v}) &= \frac{1}{2}(\partial_\beta v_\alpha + \partial_\alpha v_\beta) - \Gamma^p_{\alpha\beta}(\varepsilon) v_p,\\
e_{\alpha\|3}(\varepsilon; \boldsymbol{v}) &= \frac{1}{2}\left(\frac{1}{\varepsilon}\partial_3 v_\alpha + \partial_\alpha v_3\right) - \Gamma^\sigma_{\alpha 3}(\varepsilon) v_\sigma,\\
e_{3\|3}(\varepsilon; \boldsymbol{v}) &= \frac{1}{\varepsilon}\partial_3 v_3,
\end{aligned}$$

and the functions $A^{ijkl}(\varepsilon)$, $g(\varepsilon)$, and $\Gamma^p_{ij}(\varepsilon)$ are defined as in Thm. 3.2-1. ■

Our main objective in this chapter consists in analyzing the behavior of the solutions $\boldsymbol{u}(\varepsilon) \in \mathbf{H}^1(\Omega)$ of problems $\mathcal{P}(\varepsilon;\, \Omega)$ as $\varepsilon \to 0$.

To this end, essential uses will be made of the *analytical preliminaries* and of the *three-dimensional inequality of Korn's type for a family of linearly elastic shells* (Sects. 5.2 and 5.3), already needed for the asymptotic analysis of linearly elastic generalized membrane shells carried out in the last chapter.

6.2. CONVERGENCE OF THE SCALED DISPLACEMENTS AS $\varepsilon \to 0$

We now establish the main results of this chapter: Consider a *family of linearly elastic flexural shells* with thickness $2\varepsilon > 0$, with each having the *same* middle surface $S = \boldsymbol{\theta}(\omega)$ and with each subjected to a boundary condition of place along a portion of its lateral face having the *same* set $\boldsymbol{\theta}(\gamma_0)$ as its middle curve. Then *the solutions $\boldsymbol{u}(\varepsilon)$ of the associated scaled three-dimensional problems $\mathcal{P}(\varepsilon;\, \Omega)$* (Thm. 6.1-1) *converge in $\mathbf{H}^1(\Omega)$ as $\varepsilon \to 0$ toward a limit $\boldsymbol{u}$ and this limit, which is independent of the transverse variable x_3, can be identified with the solution $\overline{\boldsymbol{u}}$ of a two-dimensional variational problem posed over the set ω.* This limit problem will be later identified (Thm. 6.3-1) as the scaled *two-dimensional variational problem of a linearly elastic flexural shell.*

The functions $\gamma_{\alpha\beta}(\boldsymbol{\eta})$, $\rho_{\alpha\beta}(\boldsymbol{\eta})$, and $a^{\alpha\beta\sigma\tau}$ used in the next theorem respectively represent the covariant components of the **linearized change of metric**, and **change of curvature, tensors** associated with a displacement field $\eta_i \boldsymbol{a}^i$ of the surface S and the contravariant components of the **scaled two-dimensional elasticity tensor of the shell**; we recall that $\varepsilon_0 > 0$ is defined in Thm. 3.1-1.

The next result is due to Ciarlet, Lods & Miara [1996, Thm. 5.1].

Theorem 6.2-1. *Assume that $\boldsymbol{\theta} \in \mathcal{C}^3(\overline{\omega};\, \mathbb{R}^3)$. Consider a family of linearly elastic flexural shells with thickness 2ε approaching zero, with each having the same middle surface $S = \boldsymbol{\theta}(\overline{\omega})$ and with each subjected to a boundary condition of place along a portion of its lateral face having the same set $\boldsymbol{\theta}(\gamma_0)$ as its middle curve, and let the assumptions on the data be as in* Sect. 6.1.

Let $\boldsymbol{u}(\varepsilon)$ denote for $0 < \varepsilon \le \varepsilon_0$ the solution of the associated scaled three-dimensional problems $\mathcal{P}(\varepsilon;\,\Omega)$ (Thm. 6.1-1). *Then there exists $\boldsymbol{u} = (u_i) \in \mathbf{H}^1(\Omega)$ satisfying $\boldsymbol{u} = \mathbf{0}$ on $\Gamma_0 = \gamma_0 \times [-1,\,1]$ such that*

$$\boxed{\begin{gathered}\boldsymbol{u}(\varepsilon) \to \boldsymbol{u} \text{ in } \mathbf{H}^1(\Omega) \text{ as } \varepsilon \to 0,\\ \boldsymbol{u} := (u_i) \text{ is independent of the transverse variable } x_3.\end{gathered}}$$

Furthermore,

$$\boxed{\overline{\boldsymbol{u}} := \frac{1}{2}\int_{-1}^{1} \boldsymbol{u}\, dx_3}$$

satisfies the following **scaled two-dimensional variational problem $\mathcal{P}_F(\omega)$ of a linearly elastic flexural shell:**

$$\boxed{\begin{aligned}&\overline{\boldsymbol{u}} = (\overline{u}_i) \in \mathbf{V}_F(\omega) := \big\{\boldsymbol{\eta} = (\eta_i) \in H^1(\omega) \times H^1(\omega) \times H^2(\omega);\\ &\qquad\qquad \eta_i = \partial_\nu \eta_3 = 0 \text{ on } \gamma_0,\ \gamma_{\alpha\beta}(\boldsymbol{\eta}) = 0 \text{ in } \omega\big\},\\ &\frac{1}{3}\int_\omega a^{\alpha\beta\sigma\tau}\rho_{\sigma\tau}(\overline{\boldsymbol{u}})\rho_{\alpha\beta}(\boldsymbol{\eta})\sqrt{a}\,dy = \int_\omega p^i \eta_i \sqrt{a}\,dy\\ &\qquad\qquad \text{for all } \boldsymbol{\eta} = (\eta_i) \in \mathbf{V}_F(\omega),\end{aligned}}$$

where (the definitions of the functions $\Gamma^\sigma_{\alpha\beta}$, $b_{\alpha\beta}$, b^β_α, $a^{\alpha\beta}$, and a are recalled in Sect. 3.1)

$$\boxed{\begin{aligned}\gamma_{\alpha\beta}(\boldsymbol{\eta}) &:= \frac{1}{2}(\partial_\beta\eta_\alpha + \partial_\alpha\eta_\beta) - \Gamma^\sigma_{\alpha\beta}\eta_\sigma - b_{\alpha\beta}\eta_3,\\ \rho_{\alpha\beta}(\boldsymbol{\eta}) &:= \partial_{\alpha\beta}\eta_3 - \Gamma^\sigma_{\alpha\beta}\partial_\sigma\eta_3 - b^\sigma_\alpha b_{\sigma\beta}\eta_3 + b^\sigma_\alpha(\partial_\beta\eta_\sigma - \Gamma^\tau_{\beta\sigma}\eta_\tau)\\ &\quad + b^\tau_\beta(\partial_\alpha\eta_\tau - \Gamma^\sigma_{\alpha\tau}\eta_\sigma) + (\partial_\alpha b^\tau_\beta + \Gamma^\tau_{\alpha\sigma}b^\sigma_\beta - \Gamma^\sigma_{\alpha\beta}b^\tau_\sigma)\eta_\tau,\\ a^{\alpha\beta\sigma\tau} &:= \frac{4\lambda\mu}{\lambda + 2\mu}a^{\alpha\beta}a^{\sigma\tau} + 2\mu(a^{\alpha\sigma}a^{\beta\tau} + a^{\alpha\tau}a^{\beta\sigma}),\\ p^i &:= \int_{-1}^{1} f^i\,dx_3 + h^i_+ + h^i_- \text{ and } h^i_\pm := h^i(\cdot,\,\pm 1).\end{aligned}}$$

Proof. For the sake of clarity, the proof is divided into six parts, numbered (i) to (vi). For notational brevity, we let

$$e_{i\|j}(\varepsilon) := e_{i\|j}(\varepsilon; \boldsymbol{u}(\varepsilon))$$

throughout the proof.

(i) A priori *bounds and extraction of weakly convergent sequences: The norms* $|\frac{1}{\varepsilon} e_{i\|j}(\varepsilon)|_{0,\Omega}$ *and* $\|\boldsymbol{u}(\varepsilon)\|_{1,\Omega}$ *are bounded independently of* $0 < \varepsilon \leq \varepsilon_1$, *where* $\varepsilon_1 > 0$ *is given by* Thm. 5.3-1. *Consequently, there exists a subsequence, still denoted* $(\boldsymbol{u}(\varepsilon))_{\varepsilon>0}$ *for convenience, and there exist functions* $e^1_{i\|j} \in L^2(\Omega)$ *and* $\boldsymbol{u} \in \mathbf{H}^1(\Omega)$ *satisfying* $\boldsymbol{u} = \mathbf{0}$ *on* $\Gamma_0 = \gamma_0 \times [-1, 1]$ *such that* (recall that $\to$ and $\rightharpoonup$ respectively denote strong and weak convergences):

$$\boldsymbol{u}(\varepsilon) \rightharpoonup \boldsymbol{u} \text{ in } \mathbf{H}^1(\Omega) \text{ and thus } \boldsymbol{u}(\varepsilon) \to \boldsymbol{u} \text{ in } \mathbf{L}^2(\Omega),$$

$$\frac{1}{\varepsilon} e_{i\|j}(\varepsilon) \rightharpoonup e^1_{i\|j} \text{ in } L^2(\Omega).$$

From the variational equations satisfied by the scaled unknown $\boldsymbol{u}(\varepsilon)$ (Thm. 6.1-1), the asymptotic behavior of the function $g(\varepsilon)$ (Thm. 3.3-1), the uniform positive definiteness of the scaled three-dimensional elasticity tensor (Thm. 3.3-2), and the *three-dimensional inequality of Korn's type for a family of linearly elastic shells* (Thm. 5.3-1), we infer that

$$\varepsilon^2 C^{-2} \|\boldsymbol{u}(\varepsilon)\|^2_{1,\Omega} \leq \sum_{i,j} |e_{i\|j}(\varepsilon)|^2_{0,\Omega}$$

$$\leq C_e g_0^{-1/2} \int_\Omega A^{ijkl}(\varepsilon) e_{k\|l}(\varepsilon) e_{i\|j}(\varepsilon) \sqrt{g(\varepsilon)}\, dx$$

$$= \varepsilon^2 C_e g_0^{-1/2} \left\{ \int_\Omega f^i u_i(\varepsilon) \sqrt{g(\varepsilon)}\, dx + \int_{\Gamma_+ \cup \Gamma_-} h^i u_i(\varepsilon) \sqrt{g(\varepsilon)}\, d\Gamma \right\}.$$

Since there exists a constant c_1 depending on g_1, on the norms $|f^i|_{0,\Omega}$ and $\|h^i\|_{L^2(\Gamma_+ \cup \Gamma_-)}$, and on the norm of the trace operator from $H^1(\Omega)$ into $L^2(\Gamma_+ \cup \Gamma_-)$ such that

$$\left| \int_\Omega f^i u_i(\varepsilon) \sqrt{g(\varepsilon)}\, dx + \int_{\Gamma_+ \cup \Gamma_-} h^i u_i(\varepsilon) \sqrt{g(\varepsilon)}\, d\Gamma \right| \leq c_1 \|\boldsymbol{u}(\varepsilon)\|_{1,\Omega}$$

for all $0 < \varepsilon \leq \varepsilon_0$, the assertions follow.

(ii) *The limit* $\boldsymbol{u}$ *and its average* $\overline{\boldsymbol{u}} = \frac{1}{2}\int_{-1}^{1} \boldsymbol{u}\, dx_3$ *satisfy the following properties:*

$$\boldsymbol{u} = (u_i) \text{ is independent of the transverse variable } x_3,$$
$$\overline{\boldsymbol{u}} = (\overline{u}_i) \in H^1(\omega) \times H^1(\omega) \times H^2(\omega) \text{ and } \overline{u}_i = \partial_\nu \overline{u}_3 = 0 \text{ on } \gamma_0,$$
$$\gamma_{\alpha\beta}(\overline{\boldsymbol{u}}) = 0 \text{ in } \omega.$$

These properties are consequences of the convergences established in part (i) and of Thm. 5.2-2.

(iii) *The limits* $e^1_{i\|j}$ *and* $\boldsymbol{u}$ *found in part* (i) *are related by:*

$$\begin{aligned} -\partial_3 e^1_{\alpha\|\beta} &= \rho_{\alpha\beta}(\boldsymbol{u}) \text{ in } L^2(\Omega), \\ e^1_{\alpha\|3} &= 0, \\ e^1_{3\|3} &= -\frac{\lambda}{\lambda + 2\mu} a^{\alpha\beta} e^1_{\alpha\|\beta}. \end{aligned}$$

The equality $-\partial_3 e^1_{\alpha\|\beta} = \rho_{\alpha\beta}(\boldsymbol{u})$ in $L^2(\Omega)$ again follows from Thm. 5.2-2. Let $\boldsymbol{v} = (v_i)$ be an arbitrary function in the space

$$\mathbf{V}(\Omega) = \{\boldsymbol{v} \in \mathbf{H}^1(\Omega);\ \boldsymbol{v} = \mathbf{0} \text{ on } \Gamma_0\}.$$

Then

$$\begin{aligned} \varepsilon e_{\alpha\|\beta}(\varepsilon; \boldsymbol{v}) &\to 0 \text{ in } L^2(\Omega), \\ \varepsilon e_{\alpha\|3}(\varepsilon; \boldsymbol{v}) &\to \frac{1}{2}\partial_3 v_\alpha \text{ in } L^2(\Omega), \\ \varepsilon e_{3\|3}(\varepsilon; \boldsymbol{v}) &= \partial_3 v_3 \text{ for all } \varepsilon > 0. \end{aligned}$$

Using the variational equations of the scaled variational problem $\mathcal{P}(\varepsilon; \Omega)$ (Thm. 6.1-1) and the relations $A^{\alpha\beta\sigma 3}(\varepsilon) = A^{\alpha 333}(\varepsilon) = 0$ (Thm. 3.3-2), we may write

$$\begin{aligned} &\int_\Omega A^{ijkl}(\varepsilon) e_{k\|l}(\varepsilon) e_{i\|j}(\varepsilon; \boldsymbol{v}) \sqrt{g(\varepsilon)}\, dx \\ &= \frac{1}{\varepsilon}\int_\Omega \{A^{\alpha\beta\sigma\tau}(\varepsilon) e_{\sigma\|\tau}(\varepsilon) + A^{\alpha\beta 33}(\varepsilon) e_{3\|3}(\varepsilon)\}\{\varepsilon e_{\alpha\|\beta}(\varepsilon; \boldsymbol{v})\} \sqrt{g(\varepsilon)}\, dx \\ &+ \frac{1}{\varepsilon}\int_\Omega \{4A^{\alpha 3\sigma 3}(\varepsilon) e_{\sigma\|3}(\varepsilon)\}\{\varepsilon e_{\alpha\|3}(\varepsilon; \boldsymbol{v})\} \sqrt{g(\varepsilon)}\, dx \end{aligned}$$

$$+ \frac{1}{\varepsilon} \int_\Omega \{A^{33\sigma\tau}(\varepsilon) e_{\sigma\|\tau}(\varepsilon) + A^{3333}(\varepsilon) e_{3\|3}(\varepsilon)\}\{\varepsilon e_{3\|3}(\varepsilon; \boldsymbol{v})\}\sqrt{g(\varepsilon)}\, dx$$

$$= \varepsilon^2 \Big\{ \int_\Omega f^i v_i \sqrt{g(\varepsilon)}\, dx + \int_{\Gamma_+ \cup \Gamma_-} h^i v_i \sqrt{g(\varepsilon)}\, d\Gamma \Big\}.$$

Keep $\boldsymbol{v} \in \mathbf{V}(\Omega)$ *fixed* and let $\varepsilon \to 0$. The asymptotic behavior of the functions $\varepsilon e_{i\|j}(\varepsilon; \boldsymbol{v})$, the asymptotic behavior of the functions $A^{ijkl}(\varepsilon)$ and $g(\varepsilon)$ (Thms. 3.3-1 and 3.3-2), and the weak convergences $\frac{1}{\varepsilon} e_{i\|j}(\varepsilon) \rightharpoonup e^1_{i\|j}$ (part (i)) together imply that:

$$\int_\Omega \Big\{ 2\mu a^{\alpha\sigma} e^1_{\sigma\|3} \partial_3 v_\alpha + \Big(\lambda a^{\sigma\tau} e^1_{\sigma\|\tau} + (\lambda + 2\mu) e^1_{3\|3} \Big) \partial_3 v_3 \Big\} \sqrt{a}\, dx = 0.$$

Letting $\boldsymbol{v}$ *vary* in $\mathbf{V}(\Omega)$ then yields the other relations satisfied by the limits $e^1_{i\|3}$ (if $w \in L^2(\Omega)$ satisfies $\int_\Omega w \partial_3 v\, dx = 0$ for all $v \in H^1(\Omega)$ that vanish on $\gamma_0 \times [-1, 1]$, then $w = 0$; cf. Thm. 3.4-1).

(iv) *The function* $\overline{\boldsymbol{u}} := (\overline{u}_i)$ *satisfies the two-dimensional variational problem* $\mathcal{P}_F(\omega)$ *described in the statement of the theorem. Since this problem has a unique solution* (Thms. 2.6-4 and 3.3-2), *the convergences* $\boldsymbol{u}(\varepsilon) \rightharpoonup \boldsymbol{u}$ *in* $\mathbf{H}^1(\Omega)$ *and* $\boldsymbol{u}(\varepsilon) \to \boldsymbol{u}$ *in* $\mathbf{L}^2(\Omega)$ (part (i)) *hold in fact for the whole family* $(\boldsymbol{u}(\varepsilon))_{\varepsilon>0}$ (if the function $\overline{\boldsymbol{u}}$ is unique, so is the function $\boldsymbol{u}$ as it is independent of x_3; cf. part (ii)). Note in passing that reaching the same conclusion for the families $\left(\frac{1}{\varepsilon} e_{i\|j}(\varepsilon)\right)_{\varepsilon>0}$ has to be postponed until part (v).

That $\overline{\boldsymbol{u}} \in \mathbf{V}_F(\omega)$ has been proved in part (ii). Given an arbitrary function $\boldsymbol{\eta} = (\eta_i)$ in the space $\mathbf{V}_F(\omega)$, let the function $\boldsymbol{v}(\varepsilon) = (v_i(\varepsilon))$ be defined almost everywhere in Ω for all $\varepsilon > 0$ by (the idea of constructing such a function is borrowed from Miara & Sanchez-Palencia [1996]):

$$v_\alpha(\varepsilon) := \eta_\alpha - \varepsilon x_3 \theta_\alpha \text{ with } \theta_\alpha := \partial_\alpha \eta_3 + 2 b^\sigma_\alpha \eta_\sigma,$$
$$v_3(\varepsilon) := \eta_3.$$

First, we clearly have

$$\boldsymbol{v}(\varepsilon) \in \mathbf{V}(\Omega) \text{ for all } \varepsilon > 0,$$
$$e_{3\|3}(\varepsilon; \boldsymbol{v}(\varepsilon)) = 0 \text{ for all } \varepsilon > 0.$$

Next, we show that, for a *fixed* element $\boldsymbol{\eta} \in \mathbf{V}_F(\omega)$, identified wherever needed with a function in the space $H^1(\Omega) \times H^1(\Omega) \times H^2(\Omega)$,

$$\boldsymbol{v}(\varepsilon) \to \boldsymbol{\eta} \text{ in } \mathbf{H}^1(\Omega),$$

$$\frac{1}{\varepsilon} e_{\alpha\|\beta}(\varepsilon; \boldsymbol{v}(\varepsilon)) \to \{-x_3 \rho_{\alpha\beta}(\boldsymbol{\eta})\} \text{ in } L^2(\Omega) \text{ as } \varepsilon \to 0,$$

$$\left(\frac{1}{\varepsilon} e_{\alpha\|3}(\varepsilon; \boldsymbol{v}(\varepsilon))\right)_{\varepsilon>0} \text{ converges in } L^2(\Omega).$$

The first relation clearly holds. Using the relations $\gamma_{\alpha\beta}(\boldsymbol{\eta}) = 0$, we find after an easy computation that

$$\begin{aligned} e^1_{\alpha\|\beta}(\varepsilon; \boldsymbol{v}(\varepsilon)) &:= \frac{1}{\varepsilon}\gamma_{\alpha\beta}(\boldsymbol{v}(\varepsilon)) + x_3 b^\sigma_\beta|_\alpha v_\sigma(\varepsilon) + x_3 b^\sigma_\alpha b_{\sigma\beta} v_3(\varepsilon) \\ &= -x_3\left\{\frac{1}{2}(\partial_\beta\theta_\alpha + \partial_\alpha\theta_\beta) - \Gamma^\sigma_{\alpha\beta}\theta_\sigma - b^\sigma_\beta|_\alpha\eta_\sigma - b^\sigma_\alpha b_{\sigma\beta}\eta_3\right\} \\ &\quad -\varepsilon x_3^2 b^\sigma_\beta|_\alpha\theta_\sigma \\ &= -x_3\rho_{\alpha\beta}(\boldsymbol{\eta}) - \varepsilon x_3^2 b^\sigma_\beta|_\alpha\theta_\sigma. \end{aligned}$$

Then, by Thm. 5.2-1,

$$\left|\frac{1}{\varepsilon} e_{\alpha\|\beta}(\varepsilon; \boldsymbol{v}(\varepsilon)) - e^1_{\alpha\|\beta}(\varepsilon; \boldsymbol{v}(\varepsilon))\right|_{0,\Omega} \le C_1\varepsilon \sum_\alpha |v_\alpha(\varepsilon)|_{0,\Omega},$$

and thus

$$\frac{1}{\varepsilon} e_{\alpha\|\beta}(\varepsilon; \boldsymbol{v}(\varepsilon)) \to \{-x_3\rho_{\alpha\beta}(\boldsymbol{\eta})\} \text{ in } L^2(\Omega).$$

Another computation shows that

$$\frac{1}{\varepsilon} e_{\alpha\|3}(\varepsilon; \boldsymbol{v}(\varepsilon)) = -\frac{1}{\varepsilon}(\Gamma^\sigma_{\alpha3}(\varepsilon) + b^\sigma_\alpha)\eta_\sigma + x_3\Gamma^\sigma_{\alpha3}(\varepsilon)\theta_\sigma.$$

This equality, combined with the asymptotic behavior of the functions $\Gamma^\sigma_{\alpha3}(\varepsilon)$ (Thm. 3.3-1), proves that $\left(\frac{1}{\varepsilon} e_{\alpha\|3}(\varepsilon; \boldsymbol{v}(\varepsilon))\right)_{\varepsilon>0}$ converges in $L^2(\Omega)$.

Keep the function $\boldsymbol{\eta} \in \mathbf{V}_F(\omega)$ *fixed*, let $\boldsymbol{v} = \boldsymbol{v}(\varepsilon)$ in the variational equations of problem $\mathcal{P}(\varepsilon; \Omega)$ (Thm. 6.1-1), where $\boldsymbol{v}(\varepsilon) = (v_i(\varepsilon))$ is defined as *supra*, and let $\varepsilon \to 0$. The asymptotic behavior of the functions $\boldsymbol{v}(\varepsilon)$ and $\frac{1}{\varepsilon} e_{i\|j}(\varepsilon; \boldsymbol{v}(\varepsilon))$, the asymptotic behavior of the functions $A^{ijkl}(\varepsilon)$ and $g(\varepsilon)$ (Thms. 3.3-1 and 3.3-2), the weak convergences $\frac{1}{\varepsilon}e_{i\|j}(\varepsilon) \rightharpoonup e^1_{i\|j}$ (part (i)), and the relations satisfied by the limits $e^1_{i\|3}$ (part (iii)) together yield

$$\begin{aligned}
&\lim_{\varepsilon\to 0}\frac{1}{\varepsilon^2}\int_\Omega A^{ijkl}(\varepsilon)e_{k\|l}(\varepsilon)e_{i\|j}(\varepsilon;\boldsymbol{v})\sqrt{g(\varepsilon)}\,dx \\
&= \lim_{\varepsilon\to 0}\Big\{\int_\Omega A^{\alpha\beta\sigma\tau}(\varepsilon)\{\frac{1}{\varepsilon}e_{\sigma\|\tau}(\varepsilon)\}\sqrt{g(\varepsilon)}\,dx \\
&\qquad + \int_\Omega A^{\alpha\beta 33}(\varepsilon)\{\frac{1}{\varepsilon}e_{3\|3}(\varepsilon)\}\{\frac{1}{\varepsilon}e_{\alpha\|\beta}(\varepsilon;\boldsymbol{v})\}\sqrt{g(\varepsilon)}\,dx \\
&\qquad + \int_\Omega 4A^{\alpha 3\sigma 3}(\varepsilon)\{\frac{1}{\varepsilon}e_{\sigma\|3}(\varepsilon)\}\{\frac{1}{\varepsilon}e_{\alpha\|3}(\varepsilon;\boldsymbol{v})\}\sqrt{g(\varepsilon)}\,dx \\
&\qquad + \int_\Omega A^{33\sigma\tau}(\varepsilon)\{\frac{1}{\varepsilon}e_{\sigma\|\tau}(\varepsilon)\}\sqrt{g(\varepsilon)}\,dx \\
&\qquad + \int_\Omega A^{3333}(\varepsilon)\{\frac{1}{\varepsilon}e_{3\|3}(\varepsilon)\}\{\frac{1}{\varepsilon}e_{3\|3}(\varepsilon;\boldsymbol{v})\}\sqrt{g(\varepsilon)}\,dx\Big\} \\
&= -\frac{1}{2}\int_\Omega x_3 a^{\alpha\beta\sigma\tau}e^1_{\sigma\|\tau}\rho_{\alpha\beta}(\boldsymbol{\eta})\sqrt{a}\,dx \\
&= -\int_\Omega x_3\Big\{\frac{2\lambda\mu}{\lambda+2\mu}a^{\alpha\beta}a^{\sigma\tau}+\mu(a^{\alpha\sigma}a^{\beta\tau}+a^{\alpha\tau}a^{\beta\sigma})\Big\}e^1_{\sigma\|\tau}\rho_{\alpha\beta}(\boldsymbol{\eta})\sqrt{a}\,dx \\
&= \lim_{\varepsilon\to 0}\Big\{\int_\Omega f^i v_i(\varepsilon)\sqrt{g(\varepsilon)}\,dx + \int_{\Gamma_+\cup\Gamma_-} h^i v_i(\varepsilon)\sqrt{g(\varepsilon)}\,d\Gamma\Big\} \\
&= \int_\Omega f^i\eta_i\sqrt{a}\,dx + \int_{\Gamma_+\cup\Gamma_-} h^i\eta_i\sqrt{a}\,dy \\
&= \int_\omega\Big\{\int_{-1}^{1} f^i\,dx_3 + h^i_+ + h^i_-\Big\}\eta_i\sqrt{a}\,dy.
\end{aligned}$$

We have yet to take into account the relations $\rho_{\alpha\beta}(\boldsymbol{u}) = -\partial_3 e^1_{\alpha\|\beta}$ in $L^2(\Omega)$ (part (iii)). Since the function $\boldsymbol{u}$ is independent of x_3 (see part (ii)), these relations show that the functions $e^1_{\alpha\|\beta}$ are of the form

$$e^1_{\alpha\|\beta} = \Upsilon_{\alpha\beta} - x_3\rho_{\alpha\beta}(\overline{\boldsymbol{u}}) \text{ with } \Upsilon_{\alpha\beta}\in L^2(\omega).$$

Therefore

$$-\frac{1}{2}\int_\Omega x_3 a^{\alpha\beta\sigma\tau} e^1_{\sigma\|\tau}\rho_{\alpha\beta}(\boldsymbol{\eta})\sqrt{a}\,dx = \frac{1}{2}\int_\Omega x_3^2 a^{\alpha\beta\sigma\tau}\rho_{\sigma\tau}(\overline{\boldsymbol{u}})\rho_{\alpha\beta}(\boldsymbol{\eta})\sqrt{a}\,dx$$
$$= \frac{1}{3}\int_\omega a^{\alpha\beta\sigma\tau}\rho_{\sigma\tau}(\overline{\boldsymbol{u}})\rho_{\alpha\beta}(\boldsymbol{\eta})\sqrt{a}\,dy,$$

and thus we have established that the variational equations of problem $\mathcal{P}_F(\omega)$ are satisfied by $\overline{\boldsymbol{u}}$.

(v) *The weak convergences* $\frac{1}{\varepsilon}e_{i\|j}(\varepsilon) \rightharpoonup e^1_{i\|j}$ *in* $L^2(\Omega)$ *established in part* (i) *are in fact strong:*

$$\frac{1}{\varepsilon}e_{i\|j}(\varepsilon) \to e^1_{i\|j} \text{ in } L^2(\Omega).$$

Besides, the limits $e^1_{i\|j}$ *are unique; hence these convergences hold for the whole family* $\left(\frac{1}{\varepsilon}e_{i\|j}(\varepsilon)\right)_{\varepsilon>0}$.

Combining the uniform positive definiteness of the scaled elasticity tensor and the asymptotic behavior of the function $g(\varepsilon)$ (Thms. 3.3-1 and 3.3-2) with the variational equations of problem $\mathcal{P}(\varepsilon;\Omega)$ where we let $\boldsymbol{v} = \boldsymbol{u}(\varepsilon)$, we obtain

$$C_e^{-1} g_0^{1/2}\sum_{i,j}\left|\frac{1}{\varepsilon}e_{i\|j}(\varepsilon) - e^1_{i\|j}\right|^2_{0,\Omega} \le \Lambda(\varepsilon),$$

where

$$\Lambda(\varepsilon) := \int_\Omega A^{ijkl}(\varepsilon)\left\{\frac{1}{\varepsilon}e_{k\|l}(\varepsilon) - e^1_{k\|l}\right\}\left\{\frac{1}{\varepsilon}e_{i\|j}(\varepsilon) - e^1_{i\|j}\right\}\sqrt{g(\varepsilon)}\,dx$$
$$= \int_\Omega f^i u_i(\varepsilon)\sqrt{g(\varepsilon)}\,dx + \int_{\Gamma_+\cup\Gamma_-} h^i u_i(\varepsilon)\sqrt{g(\varepsilon)}\,d\Gamma$$
$$- \int_\Omega A^{ijkl}(\varepsilon)\left\{\frac{2}{\varepsilon}e_{k\|l}(\varepsilon) - e^1_{k\|l}\right\}e^1_{i\|j}\sqrt{g(\varepsilon)}\,dx.$$

The asymptotic behavior of the functions $A^{ijkl}(\varepsilon)$ and $g(\varepsilon)$ together with the convergences $u_i(\varepsilon) \to u_i$ in $L^2(\Omega)$ and $u_i(\varepsilon) \rightharpoonup u_i$ in $H^1(\Omega)$ (the latter implying that tr $u_i(\varepsilon) \rightharpoonup$ tr u_i in $L^2(\Gamma_+\cup\Gamma_-)$) and the weak convergences $\frac{1}{\varepsilon}e_{i\|j}(\varepsilon) \rightharpoonup e^1_{i\|j}$ in $L^2(\Omega)$ (part (i)) imply that

$$\Lambda := \lim_{\varepsilon\to 0}\Lambda(\varepsilon) = \int_\Omega f^i u_i\sqrt{a}\,dx + \int_{\Gamma_+\cup\Gamma_-} h^i u_i\sqrt{a}\,d\Gamma - \int_\Omega A^{ijkl}(0)e^1_{k\|l}e^1_{i\|j}\sqrt{a}\,dx.$$

Using the expressions of the functions $A^{ijkl}(0)$ (Thm. 3.3-2) and $e^1_{i\|3}$ (part (iii)) and recalling that $\boldsymbol{u} = (u_i)$ is independent of x_3 (part (ii)), we next infer that

$$\Lambda = \int_\omega \left\{\int_{-1}^1 f^i\,dx_3 + h^i_+ + h^i_-\right\}\overline{u}_i\sqrt{a}\,dy - \frac{1}{2}\int_\Omega a^{\alpha\beta\sigma\tau}e^1_{\sigma\|\tau}e^1_{\alpha\|\beta}\sqrt{a}\,dx.$$

The relations $\rho_{\alpha\beta}(\boldsymbol{u}) = -\partial_3 e^1_{\alpha\|\beta}$ in $L^2(\Omega)$ (part (iii)) imply that $e^1_{\alpha\|\beta} = \Upsilon_{\alpha\beta} - x_3\rho_{\alpha\beta}(\overline{\boldsymbol{u}})$ with $\Upsilon_{\alpha\beta} \in L^2(\omega)$; hence

$$\int_\Omega a^{\alpha\beta\sigma\tau}e^1_{\sigma\|\tau}e^1_{\alpha\|\beta}\sqrt{a}\,dx = 2\int_\omega a^{\alpha\beta\sigma\tau}\Upsilon_{\sigma\tau}\Upsilon_{\alpha\beta}\sqrt{a}\,dy + \frac{2}{3}\int_\omega a^{\alpha\beta\sigma\tau}\rho_{\sigma\tau}(\overline{\boldsymbol{u}})\rho_{\alpha\beta}(\overline{\boldsymbol{u}})\sqrt{a}\,dy.$$

Since $\overline{\boldsymbol{u}}$ satisfies the variational equations of problem $\mathcal{P}_F(\omega)$ (see part (iv)), we thus have

$$\Lambda = -\int_\omega a^{\alpha\beta\sigma\tau}\Upsilon_{\sigma\tau}\Upsilon_{\alpha\beta}\sqrt{a}\,dy.$$

Noting that $\Lambda \geq 0$ (the numbers $\Lambda(\varepsilon)$ are ≥ 0 by their definitions) and that the scaled two-dimensional elasticity tensor of the shell is uniformly positive definite in $\overline{\omega}$ (Thm. 3.3-2), we conclude that $\Upsilon_{\alpha\beta} = 0$. These relations have two consequences: First,

$$\Lambda = 0,$$

and thus the strong convergences $\frac{1}{\varepsilon}e_{i\|j}(\varepsilon) \to e^1_{i\|j}$ hold in $L^2(\Omega)$.

Second, the functions $e^1_{\alpha\|\beta}$ are uniquely determined, since they are given by

$$e^1_{\alpha\|\beta} = -x_3\rho_{\alpha\beta}(\overline{\boldsymbol{u}}),$$

and the function $\overline{\boldsymbol{u}}$ is unique (part (iv)). That the functions $e^1_{i\|3}$ are uniquely determined then follows from the relations established in part (iii).

(vi) *The weak convergence* $\boldsymbol{u}(\varepsilon) \rightharpoonup \boldsymbol{u}$ *in* $\mathbf{H}^1(\Omega)$ *established in part* (i) *is in fact strong*:

$$\boldsymbol{u}(\varepsilon) \to \boldsymbol{u} \text{ in } H^1(\Omega).$$

By Thm. 5.2-2, it remains to show that each family $(\rho_{\alpha\beta}(\boldsymbol{u}(\varepsilon)))_{\varepsilon>0}$ strongly converges in $H^{-1}(\Omega)$. Since $\frac{1}{\varepsilon}e_{\alpha\|\beta}(\varepsilon) \to e^1_{\alpha\|\beta}$ in $L^2(\Omega)$ by part (v), we first have

$$\frac{1}{\varepsilon}\partial_3 e_{\alpha\|\beta}(\varepsilon) \to \partial_3 e^1_{\alpha\|\beta} \text{ in } H^{-1}(\Omega).$$

By Thm. 5.2-1, we next have

$$\left\{\rho_{\alpha\beta}(\boldsymbol{u}(\varepsilon)) + \frac{1}{\varepsilon}\partial_3 e_{\alpha\|\beta}(\varepsilon)\right\} \to 0 \text{ in } H^{-1}(\Omega).$$

Hence

$$\rho_{\alpha\beta}(\boldsymbol{u}(\varepsilon)) \to \left\{-\partial_3 e^1_{\alpha\|\beta}\right\} \text{ in } H^{-1}(\Omega),$$

and the proof is complete. ∎

6.3. THE TWO-DIMENSIONAL EQUATIONS OF A LINEARLY ELASTIC FLEXURAL SHELL; EXISTENCE AND UNIQUENESS OF SOLUTIONS

The next theorem recapitulates the definition and assembles the main features of the limit two-dimensional problem $\mathcal{P}_F(\omega)$ found at the outcome of the asymptotic analysis carried out in Thm. 6.2-1.

The *existence* and *uniqueness* theory, which is quickly reviewed in this theorem, is expounded in detail in Sect. 2.6, where *ad hoc* references are also provided.

Theorem 6.3-1. *Let* ω *be a domain in* $\mathbb{R}^2$, *let* γ_0 *be a subset of the boundary of* ω *with* length $\gamma_0 > 0$, *and let* $\boldsymbol{\theta} \in \mathcal{C}^3(\overline{\omega}; \mathbb{R}^3)$ *be an injective mapping such that the two vectors* $\boldsymbol{a}_\alpha = \partial_\alpha \boldsymbol{\theta}$ *are linearly* *independent at all points of $\overline{\omega}$ and such that*

$$\begin{aligned}
&\mathbf{V}_F(\omega) \neq \{\mathbf{0}\}, \text{ where}\\
&\mathbf{V}_F(\omega) := \{\boldsymbol{\eta} = (\eta_i) \in H^1(\omega) \times H^1(\omega) \times H^2(\omega);\\
&\qquad\qquad \eta_i = \partial_\nu \eta_3 = 0 \text{ on } \gamma_0,\ \gamma_{\alpha\beta}(\boldsymbol{\eta}) = 0 \text{ in } \omega\},\\
&\gamma_{\alpha\beta}(\boldsymbol{\eta}) := \frac{1}{2}(\partial_\beta \eta_\alpha + \partial_\alpha \eta_\beta) - \Gamma^\sigma_{\alpha\beta}\eta_\sigma - b_{\alpha\beta}\eta_3.
\end{aligned}$$

The associated **scaled two-dimensional variational problem $\mathcal{P}_F(\omega)$ of a linearly elastic flexural shell:** *Given $p^i \in L^2(\omega)$, find $\boldsymbol{\zeta} = (\zeta_i)$ that satisfies*

$$\begin{aligned}
&\boldsymbol{\zeta} \in \mathbf{V}_F(\omega),\\
&\frac{1}{3}\int_\omega a^{\alpha\beta\sigma\tau}\rho_{\sigma\tau}(\boldsymbol{\zeta})\rho_{\alpha\beta}(\boldsymbol{\eta})\sqrt{a}\,dy = \int_\omega p^i\eta_i\sqrt{a}\,dy\\
&\qquad\qquad \text{for all } \boldsymbol{\eta} = (\eta_i) \in \mathbf{V}_F(\omega),
\end{aligned}$$

where

$$\begin{aligned}
\rho_{\alpha\beta}(\boldsymbol{\eta}) &:= \partial_{\alpha\beta}\eta_3 - \Gamma^\sigma_{\alpha\beta}\partial_\sigma\eta_3 - b^\sigma_\alpha b_{\sigma\beta}\eta_3 + b^\sigma_\alpha(\partial_\beta\eta_\sigma - \Gamma^\tau_{\beta\sigma}\eta_\tau)\\
&\quad + b^\tau_\beta(\partial_\alpha\eta_\tau - \Gamma^\sigma_{\alpha\tau}\eta_\sigma) + (\partial_\alpha b^\tau_\beta + \Gamma^\tau_{\alpha\sigma}b^\sigma_\beta - \Gamma^\sigma_{\alpha\beta}b^\tau_\sigma)\eta_\tau,\\
a^{\alpha\beta\sigma\tau} &:= \frac{4\lambda\mu}{\lambda + 2\mu}a^{\alpha\beta}a^{\sigma\tau} + 2\mu\big(a^{\alpha\sigma}a^{\beta\tau} + a^{\alpha\tau}a^{\beta\sigma}\big),
\end{aligned}$$

has one and only one solution, which is also the unique solution of the minimization problem: Find $\boldsymbol{\zeta}$ such that

$$\begin{aligned}
&\boldsymbol{\zeta} \in \mathbf{V}_F(\omega) \text{ and } j_F(\boldsymbol{\zeta}) = \inf_{\boldsymbol{\eta}\in\mathbf{V}_F(\omega)} j_F(\boldsymbol{\eta}), \text{ where}\\
&j_F(\boldsymbol{\eta}) := \frac{1}{6}\int_\omega a^{\alpha\beta\sigma\tau}\rho_{\sigma\tau}(\boldsymbol{\eta})\rho_{\alpha\beta}(\boldsymbol{\eta})\sqrt{a}\,dy - \int_\omega p^i\eta_i\sqrt{a}\,dy.
\end{aligned}$$

Proof. The existence and uniqueness of a solution to the variational problem $\mathcal{P}_F(\omega)$, or to its equivalent minimization problem, is a consequence of the *inequality of Korn's type on a general surface* (Thm. 2.6-4), of the *uniform positive definiteness of the scaled two-dimensional elasticity tensor of the shell* (Thm. 3.3-2), of the inequalities $a(y) \geq a_0 > 0$ for all $y \in \overline{\omega}$ (Thm. 3.3-1), and of the *Lax-Milgram lemma.* ∎

The minimization problem encountered in Thm. 6.3-1 (or that in Thm. 6.3-2 below in its "de-scaled" formulation) provides an interesting example of a *minimization problem with "equality constraints"*, namely the relations "$\gamma_{\alpha\beta}(\boldsymbol{\eta}) = 0$ in ω" that the elements $\boldsymbol{\eta}$ of the space over which the functional is to be minimized must satisfy.

Normally, a *Lagrange multiplier* is then attached to such a problem (see, e.g., Ciarlet [1982, Thm. 7.2-2]), allowing in turn to write the *boundary value problem* that is (at least formally) equivalent to the minimization problem; this is the case for instance of the limit "multi-dimensional" equations found at the outcome of the asymptotic analysis of *elastic multi-structures* (see Vol. II, Chap. 2; especially Sect. 2.5). Finding the Lagrange multiplier in the present case seems more challenging. In this direction, see Sanchez-Hubert & Sanchez-Palencia [1997, Chap. 8], Brezzi, Gérard & Sanchez-Palencia [1998], Sanchez-Palencia [1999].

In order to get physically meaningful formulas, it remains to "de-scale" the unknowns ζ_i that satisfy the limit scaled problem $\mathcal{P}_F(\omega)$ found in Thm. 6.2-1. In view of the scalings

$$u_i(\varepsilon)(x) = u_i^\varepsilon(x^\varepsilon) \text{ for all } x^\varepsilon = \pi^\varepsilon x \in \overline{\Omega}^\varepsilon$$

made on the covariant components of the displacement field (Sect. 6.1), we are naturally led to defining for each $\varepsilon > 0$ the **covariant components** $\zeta_i^\varepsilon : \overline{\omega} \to \mathbb{R}$ of the **limit displacement field** $\tilde{\boldsymbol{\zeta}}^\varepsilon : \overline{\omega} \to \mathbb{R}^3$ *of the middle surface S of the shell* by letting (the vectors $\boldsymbol{a}^i$ form the contravariant basis at each point of S)

$$\boxed{\zeta_i^\varepsilon := \zeta_i \text{ and } \tilde{\boldsymbol{\zeta}}^\varepsilon := \zeta_i^\varepsilon \boldsymbol{a}^i.}$$

A word of caution. Like those found in the analysis of linearly elastic elliptic membrane shells (Sect. 4.5), *the fields $\boldsymbol{\zeta}^\varepsilon := (\zeta_i^\varepsilon)$ and $\tilde{\boldsymbol{\zeta}}^\varepsilon = \zeta_i^\varepsilon \boldsymbol{a}^i$ must be carefully distinguished*! The former is essentially a convenient *mathematical "intermediary"*, but only the latter has *physical significance.* ∎

Remark. Conceivably, the limit scaled displacement field *across the thickness of the shell* could also be de-scaled, resulting into a *limit displacement field* $\tilde{\boldsymbol{u}}^\varepsilon(0) : \overline{\Omega}^\varepsilon \to \mathbb{R}^3$ *inside the shell* defined by (the vectors $\boldsymbol{g}^{i,\varepsilon}$ form the contravariant basis at each point of the reference configuration $\boldsymbol{\Theta}(\overline{\Omega}^\varepsilon)$; cf. Sect. 3.1):

$$\tilde{\boldsymbol{u}}^\varepsilon(0) := \zeta_i^\varepsilon \boldsymbol{g}^{i,\varepsilon},$$

since the scaled limit $\boldsymbol{u} = (u_i)$ is independent of the transverse variable x_3 (Thm. 6.2-1). For the same reason, however, the de-scaled field does not inherit any remarkable structure as x_3^ε varies across the thickness of the shell.

By contrast, the limit displacement field across a *plate*, which is a linearly elastic flexural shell (Ex. 6.3), is a *Kirchhoff-Love* one (Vol. II, Thm. 1.7-1). However, it does inherit this richer structure only because *different* scalings can be made in this case on the "horizontal" and "vertical" components of the displacement field. ■

Recall that $f^{i,\varepsilon} \in L^2(\Omega^\varepsilon)$ and $h^{i,\varepsilon} \in L^2(\Gamma_+^\varepsilon \cup \Gamma_-^\varepsilon)$ represent the contravariant components of the applied body and surface forces *actually* acting on the shell and that λ^ε and μ^ε denote the *actual* Lamé constants of its constituting material. We then have the following immediate corollary to Thms. 6.2-1 and 6.3-1; naturally, the *existence* and *uniqueness* results of Thm. 6.3-1 apply *verbatim* to problem $\mathcal{P}_F^\varepsilon(\omega)$ (for this reason, they are not reproduced here):

Theorem 6.3-2. *Let the assumptions on the data and the definitions of the functions $\gamma_{\alpha\beta}(\boldsymbol{\eta})$ and $\rho_{\alpha\beta}(\boldsymbol{\eta})$ be as in* Thm. 6.2-1. *Then the vector field $\boldsymbol{\zeta}^\varepsilon := (\zeta_i^\varepsilon)$ formed by the covariant components of the limit displacement field $\zeta_i^\varepsilon \boldsymbol{a}^i$ of the middle surface S satisfies the following* **two-dimensional variational problem $\mathcal{P}_F^\varepsilon(\omega)$ of a linearly elastic flexural shell:**

$$\boldsymbol{\zeta}^\varepsilon \in \mathbf{V}_F(\omega) := \{\boldsymbol{\eta} = (\eta_i) \in H^1(\omega) \times H^1(\omega) \times H^2(\omega);$$
$$\eta_i = \partial_\nu \eta_3 = 0 \text{ on } \gamma_0,\ \gamma_{\alpha\beta}(\boldsymbol{\eta}) = 0 \text{ in } \omega\},$$
$$\frac{\varepsilon^3}{3} \int_\omega a^{\alpha\beta\sigma\tau,\varepsilon} \rho_{\sigma\tau}(\boldsymbol{\zeta}^\varepsilon) \rho_{\alpha\beta}(\boldsymbol{\eta}) \sqrt{a}\, dy = \int_\omega p^{i,\varepsilon} \eta_i \sqrt{a}\, dy$$
$$\text{for all } \boldsymbol{\eta} = (\eta_i) \in \mathbf{V}_F(\omega),$$

where

$$a^{\alpha\beta\sigma\tau,\varepsilon} := \frac{4\lambda^\varepsilon\mu^\varepsilon}{\lambda^\varepsilon + 2\mu^\varepsilon} a^{\alpha\beta}a^{\sigma\tau} + 2\mu^\varepsilon(a^{\alpha\sigma}a^{\beta\tau} + a^{\alpha\tau}a^{\beta\sigma}),$$
$$p^{i,\varepsilon} := \int_{-\varepsilon}^{\varepsilon} f^{i,\varepsilon}\, dx_3^\varepsilon + h_+^{i,\varepsilon} + h_-^{i,\varepsilon} \text{ and } h_\pm^{i,\varepsilon} := h^{i,\varepsilon}(\cdot, \pm\varepsilon).$$

Equivalently, the field $\boldsymbol{\zeta}^\varepsilon = (\zeta_i^\varepsilon)$ *satisfies the following minimization problem:*

$$\boldsymbol{\zeta}^\varepsilon \in \mathbf{V}_F(\omega) \text{ and } j_F^\varepsilon(\boldsymbol{\zeta}^\varepsilon) = \inf_{\boldsymbol{\eta}\in\mathbf{V}_F(\omega)} j_F^\varepsilon(\boldsymbol{\eta}), \text{ where}$$
$$j_F^\varepsilon(\boldsymbol{\eta}) := \frac{1}{6}\int_\omega \varepsilon^3 a^{\alpha\beta\sigma\tau,\varepsilon}\rho_{\sigma\tau}(\boldsymbol{\eta})\rho_{\alpha\beta}(\boldsymbol{\eta})\sqrt{a}\, dy - \int_\omega p^{i,\varepsilon}\eta_i\sqrt{a}\, dy.$$

■

Each one of the two formulations found in Thm. 6.3-2 constitutes the **two-dimensional equations of a linearly elastic flexural shell.**

We recall that the condition $\mathbf{V}_F(\omega) \neq \{\mathbf{0}\}$, which is the basis of the definition of a linearly elastic flexural shell, means that there exist *nonzero* **admissible linearized inextensional displacements** of the middle surface (Sect. 6.1), since the functions $\gamma_{\alpha\beta}(\boldsymbol{\eta})$ used in the definition of $\mathbf{V}_F(\omega)$ are the *covariant components of the* **linearized change of metric tensor** associated with a displacement field $\eta_i\boldsymbol{a}^i$ of the middle surface S.

In order to interpret the *boundary conditions* $\zeta_i^\varepsilon = \partial_\nu\zeta_3^\varepsilon$ on γ_0 satisfied by the field $\boldsymbol{\zeta}^\varepsilon = (\zeta_i^\varepsilon)$, let $\eta_i\boldsymbol{a}^i$ be a displacement field of the middle surface $S = \boldsymbol{\theta}(\overline{\omega})$ with smooth enough, but otherwise arbitrary, covariant components $\eta_i : \overline{\omega} \to \mathbb{R}$. The tangent plane at an arbitrary point $\boldsymbol{\theta}(y) + \eta_i(y)\boldsymbol{a}^i(y)$, $y \in \overline{\omega}$, of the deformed surface $(\boldsymbol{\theta} + \eta_i\boldsymbol{a}^i)(\overline{\omega})$ is thus spanned by the vectors

$$\partial_\alpha(\boldsymbol{\theta} + \eta_i\boldsymbol{a}^i)(y) = \boldsymbol{a}_\alpha(y) + \partial_\alpha\eta_i(y)\boldsymbol{a}^i(y) + \eta_i(y)\partial_\alpha\boldsymbol{a}^i(y),$$

if these are linearly independent. Since

$$\eta_i = \partial_\nu\eta_3 = 0 \text{ on } \gamma_0 \ \Rightarrow\ \eta_i = \partial_\alpha\eta_3 = 0 \text{ on } \gamma_0,$$

it follows that

$$\boldsymbol{\theta}(y) + \eta_i(y)\boldsymbol{a}^i(y) = \boldsymbol{\theta}(y) \text{ for all } y \in \gamma_0,$$
$$\partial_\alpha(\boldsymbol{\theta} + \eta_i\boldsymbol{a}^i)(y) = \boldsymbol{a}_\alpha(y) + \partial_\alpha\eta_\beta(y)\boldsymbol{a}^\beta(y) \text{ for all } y \in \gamma_0.$$

These relations thus show that the points of *the deformed and undeformed middle surfaces, and their tangent spaces at those points where the vectors* $\partial_\alpha(\boldsymbol{\theta}+\eta_i\boldsymbol{a}^i)$ *are linearly independent, coincide along the set* $\boldsymbol{\theta}(\gamma_0)$. Such **two-dimensional boundary conditions of clamping** are suggested in Figs. 6.1-1 to 6.1-3.

The functions $\rho_{\alpha\beta}(\boldsymbol{\eta})$ are the *covariant components* of the **linearized change of curvature tensor** associated with a displacement field $\eta_i\boldsymbol{a}^i$ of the middle surface S and the functions $a^{\alpha\beta\sigma\tau,\varepsilon}$ are the *contravariant components* of the **two-dimensional elasticity tensor of the shell**, already encountered in the two-dimensional equations of linearly elastic elliptic membrane and generalized membrane shells (Thms. 4.5-2 and 5.7-2).

Finally, the functional $j_F^\varepsilon : \mathbf{V}_F(\omega) \to \mathbb{R}$ is the **two-dimensional energy,** and the functional

$$\boxed{\boldsymbol{\eta} \in \mathbf{V}_F(\omega) \to \frac{1}{6}\int_\omega \varepsilon^3 a^{\alpha\beta\sigma\tau,\varepsilon}\rho_{\sigma\tau}(\boldsymbol{\eta})\rho_{\alpha\beta}(\boldsymbol{\eta})\sqrt{a}\,dy}$$

is the **two-dimensional strain energy, of a linearly elastic flexural shell.**

Note that the *same* two-dimensional problem $\mathcal{P}_F^\varepsilon(\omega)$ is evidently obtained if the scalings on the unknowns are the same as before, i.e.,

$$\boxed{u_i^\varepsilon(x^\varepsilon) = u_i(\varepsilon)(x) \text{ for all } x^\varepsilon = \pi^\varepsilon x \in \overline{\Omega}^\varepsilon,}$$

but the following *more general* **assumptions on the data** are made:

$$\boxed{\begin{aligned} \lambda^\varepsilon &= \varepsilon^t\lambda \quad &&\text{and} \quad \mu^\varepsilon = \varepsilon^t\mu, \\ f^{i,\varepsilon}(x^\varepsilon) &= \varepsilon^{2+t}f^i(x) \quad &&\text{for all} \quad x^\varepsilon = \pi^\varepsilon x \in \Omega^\varepsilon, \\ h^{i,\varepsilon}(x^\varepsilon) &= \varepsilon^{3+t}h^i(x) \quad &&\text{for all} \quad x^\varepsilon = \pi^\varepsilon x \in \Gamma_+^\varepsilon \cup \Gamma_-^\varepsilon, \end{aligned}}$$

where the constants $\lambda > 0$ and $\mu > 0$ and the functions $f^i \in L^2(\Omega)$ and $h^i \in L^2(\Gamma_+ \cup \Gamma_-)$ are *independent of* ε and t *is an arbitrary real number*. Besides, the analysis of Sect. 3.4 shows that *these assumptions on the data are the only ones possible for flexural shells.*

Remark. A *different* kind of generalization is possible. For definiteness, assume that the Lamé constants are independent of ε. Then it is easily verified that the same scaled limit problem $\mathcal{P}_F(\omega)$ is obtained under the assumptions that

$$f^{i,\varepsilon}(x^\varepsilon) = \varepsilon^2 f^i(\varepsilon; x) \text{ for all } x^\varepsilon = \pi^\varepsilon x \in \Omega^\varepsilon,$$
$$h^{i,\varepsilon}(x^\varepsilon) = \varepsilon^3 h^i(\varepsilon; x) \text{ for all } x^\varepsilon = \pi^\varepsilon x \in \Gamma_+^\varepsilon \cup \Gamma_-^\varepsilon,$$

and

$$f^i(\varepsilon; \cdot) \to f^i \text{ in } L^2(\Omega) \text{ as } \varepsilon \to 0,$$
$$h^i(\varepsilon; \cdot) \to h^i \text{ in } L^2(\Gamma_+ \cup \Gamma_-) \text{ as } \varepsilon \to 0.$$

The functions p_i^ε appearing in the right-hand sides of problem $\mathcal{P}_F^\varepsilon(\omega)$ are then defined by

$$p_i^\varepsilon := \varepsilon\left\{\int_{-1}^{1} f^i \, dx_3 + h_+^i + h_-^i\right\}.$$

■

6.4. JUSTIFICATION OF THE TWO-DIMENSIONAL EQUATIONS OF A LINEARLY ELASTIC FLEXURAL SHELL; COMMENTARY AND REFINEMENTS

It remains to convert in terms of de-scaled unknowns the fundamental convergence theorem established in Sect. 6.2. As the "original" unknowns u_i^ε are defined over a domain that *varies with* ε (the set Ω^ε), their *averages* $\frac{1}{2\varepsilon}\int_{-\varepsilon}^{\varepsilon} u_i^\varepsilon \, dx_3^\varepsilon$ are more appropriate for this purpose, since they are defined over a *fixed* domain (the set ω).

The convergences $u_i(\varepsilon) \to u_i$ in $H^1(\Omega)$ (Thm. 6.2-1), the scalings $u_i(\varepsilon)(x) = u_i^\varepsilon(x^\varepsilon)$ for all $x^\varepsilon = \pi^\varepsilon x \in \overline{\Omega}^\varepsilon$ (Sect. 6.1), the de-scalings $\zeta_i^\varepsilon = \zeta_i = \frac{1}{2}\int_{-1}^{1} u_i \, dx_3$ (Sect. 6.3) and Thm. 4.2-1 together yield the following convergences of the averages $\frac{1}{2\varepsilon}\int_{-\varepsilon}^{\varepsilon} u_i^\varepsilon \, dx_3^\varepsilon$ of the covariant components of the "original" three-dimensional displacement:

$$\frac{1}{2\varepsilon}\int_{-\varepsilon}^{\varepsilon} u_i^\varepsilon \, dx_3^\varepsilon \to \zeta_i^\varepsilon \text{ in } H^1(\omega).$$

However, these convergences can be further improved and given a more intrinsic character by considering instead the *averages* of the *tangential component* $u_\alpha^\varepsilon \boldsymbol{g}^{\alpha,\varepsilon}$ and of the *normal component* $u_3^\varepsilon \boldsymbol{g}^{3,\varepsilon}$ of the *three-dimensional displacement vector* itself; note in this respect that, along a given normal direction to the surface S, the vectors $\boldsymbol{g}^{\alpha,\varepsilon}$ and $\boldsymbol{g}^{3,\varepsilon}$ remain respectively parallel to the tangent plane and normal to S since $\boldsymbol{g}_\alpha^\varepsilon = \boldsymbol{a}_\alpha - x_3^\varepsilon b_\alpha^\sigma \boldsymbol{a}_\sigma$, $\boldsymbol{g}^{3,\varepsilon} = \boldsymbol{a}_3$, and $\boldsymbol{g}_i^\varepsilon \cdot \boldsymbol{g}^{i,\varepsilon} = \delta_i^j$. More specifically, the above convergences combined with the behavior as $\varepsilon \to 0$ of the vectors $\boldsymbol{g}^{i,\varepsilon}$ lead to the following result:

Theorem 6.4-1. *Assume that $\boldsymbol{\theta} \in \mathcal{C}^3(\overline{\omega}; \mathbb{R}^3)$. Consider a family of linearly elastic flexural shells with thickness 2ε approaching zero, with each having the same middle surface $S = \boldsymbol{\theta}(\overline{\omega})$ and with each subjected to a boundary condition of place along a portion of its lateral face having the same set $\boldsymbol{\theta}(\gamma_0)$ as its middle curve, and let the assumptions on the data be as in* Sect. 6.1.

Let $(u_i^\varepsilon) \in \mathbf{H}^1(\Omega^\varepsilon)$ and $(\zeta_i^\varepsilon) \in H^1(\omega) \times H^1(\omega) \times H^2(\omega)$ respectively denote for each $\varepsilon > 0$ the solutions to the three-dimensional and two-dimensional problems $\mathcal{P}(\Omega^\varepsilon)$ and $\mathcal{P}_F^\varepsilon(\omega)$ (Sect. 3.1 and Thm. 4.6-1).

Finally, let $(\zeta_i) \in H^1(\omega) \times H^1(\omega) \times H^2(\omega)$ denote the solution to problem $\mathcal{P}_F(\omega)$ (Thm. 6.3-1), *which is thus independent of ε. Then*

$$\boxed{\begin{aligned} \zeta_\alpha^\varepsilon = \zeta_\alpha \text{ and thus } \zeta_\alpha^\varepsilon \boldsymbol{a}^\alpha &= \zeta_\alpha \boldsymbol{a}^\alpha \text{ in } \mathbf{H}^1(\omega) \text{ for all } \varepsilon > 0, \\ \frac{1}{2\varepsilon}\int_{-\varepsilon}^{\varepsilon} u_\alpha^\varepsilon \boldsymbol{g}^{\alpha,\varepsilon}\, dx_3^\varepsilon &\to \zeta_\alpha \boldsymbol{a}^\alpha \text{ in } \mathbf{H}^1(\omega) \text{ as } \varepsilon \to 0, \end{aligned}}$$

and

$$\boxed{\begin{aligned} \zeta_3^\varepsilon = \zeta_3 \text{ and thus } \zeta_3^\varepsilon \boldsymbol{a}^3 &= \zeta_3 \boldsymbol{a}^3 \text{ in } \mathbf{H}^2(\omega) \text{ for all } \varepsilon > 0, \\ \frac{1}{2\varepsilon}\int_{-\varepsilon}^{\varepsilon} u_3^\varepsilon \boldsymbol{g}^{3,\varepsilon}\, dx_3^\varepsilon &\to \zeta_3 \boldsymbol{a}^3 \text{ in } \mathbf{H}^1(\omega) \text{ as } \varepsilon \to 0. \end{aligned}}$$

Proof. The proof is similar to that of Thm. 4.6-1. The assumption $\boldsymbol{\theta} \in \mathcal{C}^3(\overline{\omega}; \mathbb{R}^3)$ implies that the vector fields $\boldsymbol{g}^\alpha(\varepsilon) : \overline{\Omega} \to \mathbb{R}^3$ defined by $\boldsymbol{g}^\alpha(\varepsilon)(x) := \boldsymbol{g}^{\alpha,\varepsilon}(x^\varepsilon)$ for all $x^\varepsilon = \pi^\varepsilon x \in \overline{\Omega}^\varepsilon$ are such that

$\boldsymbol{g}^\alpha(\varepsilon) - \boldsymbol{a}^\alpha = O(\varepsilon)$ in $\mathcal{C}^1(\overline{\Omega})$. Since

$$\frac{1}{2\varepsilon}\int_{-\varepsilon}^{\varepsilon} u_\alpha^\varepsilon \boldsymbol{g}^{\alpha,\varepsilon}\, dx_3^\varepsilon - \zeta_\alpha^\varepsilon \boldsymbol{a}^\alpha = \frac{1}{2}\int_{-1}^{1} u_\alpha(\varepsilon)\,(\boldsymbol{g}^\alpha(\varepsilon) - \boldsymbol{a}^\alpha)\, dx_3 - (\overline{u_\alpha(\varepsilon)} - \zeta_\alpha)\boldsymbol{a}^\alpha,$$

the convergences $u_\alpha(\varepsilon) \to u_\alpha$ in $H^1(\Omega)$ and $\boldsymbol{g}^\alpha(\varepsilon) \to \boldsymbol{a}^\alpha$ in $\mathcal{C}^1(\overline{\Omega})$ imply that

$$u_\alpha(\varepsilon)\,(\boldsymbol{g}^\alpha(\varepsilon) - \boldsymbol{a}^\alpha) \to 0 \text{ in } \mathbf{H}^1(\Omega);$$

hence $\frac{1}{2}\int_{-1}^{1} u_\alpha(\varepsilon)(\boldsymbol{g}^\alpha(\varepsilon) - \boldsymbol{a}^\alpha)\, dx_3 \to 0$ in $\mathbf{H}^1(\omega)$ by Thm. 4.2-1. The same theorem also shows that $(\overline{u_\alpha(\varepsilon)} - \zeta_\alpha)\boldsymbol{a}^\alpha \to 0$ in $\mathbf{H}^1(\omega)$. Since

$$\frac{1}{2\varepsilon}\int_{-\varepsilon}^{\varepsilon} u_3^\varepsilon \boldsymbol{g}^{3,\varepsilon}\, dx_3^\varepsilon - \zeta_3^\varepsilon \boldsymbol{a}^3 = (\overline{u_3(\varepsilon)} - \zeta_3)\boldsymbol{a}^3,$$

the same theorem shows that $(\overline{u_3(\varepsilon)} - \zeta_3)\boldsymbol{a}^3 \to 0$ in $\mathbf{H}^1(\Omega)$. ■

The fields $\tilde{\boldsymbol{\zeta}}_T^\varepsilon : \overline{\omega} \to \mathbb{R}^3$ and $\tilde{\boldsymbol{\zeta}}_N^\varepsilon : \overline{\omega} \to \mathbb{R}^3$ defined by

$$\boxed{\tilde{\boldsymbol{\zeta}}_T^\varepsilon := \zeta_\alpha^\varepsilon \boldsymbol{a}^\alpha \text{ and } \tilde{\boldsymbol{\zeta}}_N^\varepsilon := \zeta_3^\varepsilon \boldsymbol{a}^3,}$$

which appear in Thm. 4.6-1, are respectively called the **limit tangential displacement field** and the **limit normal displacement field** *of the middle surface S of the shell.* Naturally, they are related to the *limit displacement field* $\tilde{\boldsymbol{\zeta}}^\varepsilon = \zeta_i^\varepsilon \boldsymbol{a}^i$ *of S* (Sect. 6.3) by

$$\boxed{\tilde{\boldsymbol{\zeta}}^\varepsilon = \tilde{\boldsymbol{\zeta}}_T^\varepsilon + \tilde{\boldsymbol{\zeta}}_N^\varepsilon.}$$

Under the essential assumptions that the space $\mathbf{V}_F(\omega)$ contains nonzero elements, we have therefore justified by a convergence result (Thm. 6.4-1) *the two-dimensional equations of a linearly elastic flexural shell.* In so doing, we have justified the formal asymptotic approach of Sanchez-Palencia [1990] (see also Miara & Sanchez-Palencia [1996] and Caillerie & Sanchez-Palencia [1995b]) when *"bending is not inhibited"*, according to the terminology of E. Sanchez-Palencia.

Credit should be given in this respect to Sanchez-Palencia [1989a] for recognizing the *central rôle played by the space $\mathbf{V}_F(\omega)$ in the classification of linearly elastic shells.*

A word of caution. In a *flexural* shell, body forces of order $O(\varepsilon^2)$ thus produce an $O(1)$ limit displacement field. By contrast, body forces of order $O(1)$ are required to also produce $O(1)$ limit displacement fields in an *elliptic membrane* shell.

Both types of shells thus exhibit strikingly different limit behaviors!

This conclusion of a *mechanical* nature simply reflects that *strikingly different three-dimensional inequalities of Korn's type* (compare Thm. 4.3-1 with Thm. 5.3-1) are needed for deriving in each case the fundamental *a priori* bounds in the convergence theorems (compare parts (i) of the proofs of Thms. 4.4-1 and 6.2-1). ■

After the original work of Ciarlet, Lods & Miara [1996] described in this chapter, the asymptotic analysis of linearly elastic flexural shells underwent several refinements and generalizations:

First, Genevey [1999] has shown that the convergence result of Thm. 6.2-1 can be also obtained by resorting to Γ-*convergence* theory.

Other useful extensions include the justification by an asymptotic analysis of *linearly elastic flexural shells with variable thickness* (Busse [1998]; see Ex. 6.7) or made with a *nonhomogeneous* and *anisotropic material* (Giroud [1998]), the *convergence of the* (scaled) *stresses* and the *explicit forms of the limit stresses* (Collard & Miara [1999]; see Ex. 6.8), an asymptotic analysis of the associated *eigenvalue problem* (Kesavan & Sabu [1999a]) and *time-dependent problem* (Xiao Li-ming [1999a]).

The surprising phenomena appearing when a linearly elastic flexural shell "*becomes a plate*" are investigated in Ciarlet [1992a, 1992b], Sanchez-Palencia [1994], and Lods [1996].

An asymptotic analysis similar in its principles can be applied to *linearly elastic curved rods*, whose limit behavior is always "flexural"; see Alvarez-Dios & Viaño [1997], Jamal [1998].

We conclude this commentary by examining *linearly elastic plates*, which constitute examples of linearly elastic flexural shells (as already observed; cf. Fig. 6.1-3), since

$$\mathbf{V}_F(\omega) = \{\boldsymbol{\eta} = (0, 0, \eta_3);\ \eta_3 \in H^2(\omega);\ \eta_3 = \partial_\nu \eta_3 = 0 \text{ on } \gamma_0\} \neq \{\mathbf{0}\}$$

in this case; naturally, we assume that the mapping $\boldsymbol{\theta}$ is of the form $\boldsymbol{\theta}(y_1, y_2) = (y_1, y_2, 0)$ for all $(y_1, y_2) \in \overline{\omega}$.

Assume that there exist constants $\lambda > 0$ and $\mu > 0$ and functions $f^i \in L^2(\Omega)$ and $h^i \in L^2(\Gamma_+ \cup \Gamma_-)$ *independent of* ε such that, for each $\varepsilon > 0$, the Lamé constants λ^ε and μ^ε of the material constituting the plates and the *Cartesian* components $f^{i,\varepsilon} : \Omega^\varepsilon \to \mathbb{R}$ and $h^{i,\varepsilon} : \Gamma_+^\varepsilon \cup \Gamma_-^\varepsilon \to \mathbb{R}$ of the applied body and surface forces (the set $\overline{\Omega}^\varepsilon$ is now the reference configuration) satisfy

$$\begin{aligned} \lambda^\varepsilon = \lambda \quad &\text{and} \quad \mu^\varepsilon = \mu, \\ f^{i,\varepsilon}(x^\varepsilon) = \varepsilon^2 f^i(x) \quad &\text{for all} \quad x^\varepsilon = \pi^\varepsilon x \in \Omega^\varepsilon, \\ h^{i,\varepsilon}(x^\varepsilon) = \varepsilon^3 h^i(x) \quad &\text{for all} \quad x^\varepsilon = \pi^\varepsilon x \in \Gamma_+^\varepsilon \cup \Gamma_-^\varepsilon, \end{aligned}$$

and let the scaled unknown $\boldsymbol{u}(\varepsilon) = (u_i(\varepsilon)) \in \mathbf{H}^1(\Omega)$ be defined by

$$u_i(\varepsilon)(x) := u_i^\varepsilon(x^\varepsilon) \text{ for all } x^\varepsilon = \pi^\varepsilon x \in \overline{\Omega}^\varepsilon.$$

where the functions $u_i^\varepsilon : \overline{\Omega}^\varepsilon \to \mathbb{R}$ are the *Cartesian* components of the displacement field of the plate.

Then Thm. 6.2-1 shows that, in this case, there exists a function $u_3 \in H^2(\Omega)$ vanishing on $\Gamma_0 = \gamma_0 \times [-1, 1]$ such that

$$\boldsymbol{u}(\varepsilon) = (u_i(\varepsilon)) \to (0, 0, u_3) \text{ in } \mathbf{H}^1(\Omega) \text{ as } \varepsilon \to 0,$$

$$u_3 \text{ is independent of the transverse variable } x_3,$$

$$\zeta_3 := \frac{1}{2}\int_{-1}^{1} u_3 \, dx_3 \in V_3(\omega),$$

$$\frac{1}{3}\int_\omega b^{\alpha\beta\sigma\tau}\partial_{\sigma\tau}\zeta_3\partial_{\alpha\beta}\eta_3 \, dy = \int_\omega p^3\eta_3 \, dy \text{ for all } \eta_3 \in V_3(\omega),$$

where

$$V_3(\omega) := \{\eta_3 \in H^2(\omega);\ \eta_3 = \partial_\nu\eta_3 = 0 \text{ on } \gamma_0\},$$

$$b^{\alpha\beta\sigma\tau} := \frac{4\lambda\mu}{\lambda + 2\mu}\delta^{\alpha\beta}\delta^{\sigma\tau} + 2\mu(\delta^{\alpha\sigma}\delta^{\beta\tau} + \delta^{\alpha\tau}\delta^{\beta\sigma}),$$

$$p^3 := \int_{-1}^{1} f^3 \, dx_3 + (h_+^3 + h_-^3) \text{ and } h_\pm^3 := h^3(\cdot, \pm 1).$$

However, this result is only a *special case* of the "classical" convergence theorem for a plate; see Vol. II, Thm. 1.4-1. For, when the asymptotic analysis is applied in *ibid.* to a *linearly elastic plate, more freedom is allowed, as regards both the choice of scalings of the displacement field and the choice of the asymptotic assumptions on the applied forces.* More specifically, *another* scaled unknown $\tilde{\boldsymbol{u}}(\varepsilon) = (\tilde{u}_i(\varepsilon))$ is defined in this case by letting

$$\tilde{u}_\alpha(\varepsilon)(x) = \varepsilon u_\alpha^\varepsilon(x^\varepsilon) \text{ and } \tilde{u}_3(\varepsilon)(x) = u_3^\varepsilon(x^\varepsilon) \text{ for all } x^\varepsilon \in \overline{\Omega}^\varepsilon,$$

and the applied forces are such that there exist functions $\tilde{f}^i \in L^2(\Omega)$ and $\tilde{h}^i \in L^2(\Gamma_+ \cup \Gamma_-)$ *independent of* ε such that

$$f^{\alpha,\varepsilon}(x^\varepsilon) = \varepsilon \tilde{f}^\alpha(x) \text{ and } f^{3,\varepsilon}(x^\varepsilon) = \varepsilon^2 \tilde{f}^3(x) \text{ for all } x \in \Omega,$$
$$h^{\alpha,\varepsilon}(x^\varepsilon) = \varepsilon^2 \tilde{h}^\alpha(x) \text{ and } h^{3,\varepsilon}(x^\varepsilon) = \varepsilon^3 \tilde{h}^3(x) \text{ for all } x \in \Gamma_+ \cup \Gamma_-.$$

The *cruder* scalings and assumptions made when the plate is considered as a special case of a shell have two consequences:

First, the horizontal components $f^{\alpha,\varepsilon}$ and $h^{\alpha,\varepsilon}$ of the applied forces *do not* contribute to the definition of the limit two-dimensional problem satisfied by the unknown ζ_3.

Second, the de-scaled components of the limit displacement field $(0, 0, \zeta_3)$ found in this fashion are *all of order zero* with respect to ε. Therefore, the *de-scaled* horizontal components of the limit displacement field that are of order *one* with respect to ε are necessarily "ignored" in this approach. These horizontal components, together with the transverse component of order *zero*, form a *Kirchhoff-Love displacement field* (Vol. II, Thm. 1.7-1) inside the plate.

Another major difference is that the asymptotic analysis of a *linearly elastic plate* yields a limit problem that *simultaneously* includes *flexural* and *membrane* equations. The reason why limit membrane equations are in addition obtained is again that horizontal components of order *one* of the limit displacement field can be recovered, thanks again to the refined scalings and assumptions that are allowed for a plate.

That *cruder scalings and assumptions are unavoidable for "genuine" shells* has been rigorously established in Sect. 3.4, after Miara & Sanchez-Palencia [1996].

EXERCISES

6.1. Let $\boldsymbol{f} = (f_\alpha) \in \mathcal{C}^2([0, 1]; \mathbb{R}^2)$ be an injective mapping such that $\boldsymbol{f}'(t) \neq \boldsymbol{0}$ for all $t \in [0, 1]$ and let $S = \boldsymbol{\theta}(\overline{\omega})$, where $\omega =]0, 1[\times]0, 1[$ and $\boldsymbol{\theta}(t, z) = f_\alpha(t)\boldsymbol{e}^\alpha + z\boldsymbol{e}^3$ for $(t, z) \in \overline{\omega}$. The surface S is thus a portion of a *cylinder* orthogonal to, and passing through, the planar curve $\boldsymbol{f}([0, 1])$; cf. Fig. 6.1-1.

(1) Assume that

$$\gamma_0 \subset \{(0, z) \in \mathbb{R}^2;\ 0 \leq z \leq 1\} \cup \{(1, z) \in \mathbb{R}^2;\ 0 \leq z \leq 1\}.$$

Show that the space

$$\mathbf{V}_F(\omega) = \{\boldsymbol{\eta} = (\eta_i) \in H^1(\omega) \times H^1(\omega) \times H^2(\omega); \\ \eta_i = \partial_\nu \eta_3 = 0 \text{ on } \gamma_0,\ \gamma_{\alpha\beta}(\boldsymbol{\eta}) = 0 \text{ in } \omega\}$$

contains nonzero functions.

(2) Is the space $\mathbf{V}_F(\omega)$ infinite-dimensional?

(3) Assume that the curve $\boldsymbol{f}([0, 1])$ contains a straight segment and that $\gamma_0 = \partial\omega$. Show that $\dim \mathbf{V}_F(\omega) = +\infty$.

(4) Assume that the curve $\boldsymbol{f}([0, 1])$ does not contain any straight segment and that $\gamma_0 = \partial\omega$. Does the space $\mathbf{V}_F(\omega)$ still contain nonzero functions?

6.2. Let $\boldsymbol{f} = (f_\alpha) \in \mathcal{C}^2([0, 1]; \mathbb{R}^2)$ be an injective mapping such that $\boldsymbol{f}'(t) \neq \mathbf{0}$ for all $t \in [0, 1]$ and let $S = \boldsymbol{\theta}(\overline{\omega})$, where

$$\omega =]0, 1[\times]s_0, 1[\text{ for some } 0 < s_0 < 1,$$
$$\boldsymbol{\theta}(t, s) = s f_\alpha(t) \boldsymbol{e}^\alpha + (1 - s)\boldsymbol{e}^3 \text{ for } (t, s) \in \overline{\omega}.$$

The surface S is thus a portion of a *cone* with vertex $\boldsymbol{e}^3$ and passing through the planar curve $\boldsymbol{f}([0, 1])$; cf. Fig. 6.1-2.

(1) Assume that $\gamma_0 = \{(0, s) \in \mathbb{R}^2;\ s_0 \leq s \leq 1\}$. Show that $\mathbf{V}_F(\omega) = \{\mathbf{0}\}$.

(2) Assume that

$$\gamma_0 = \{(0, s) \in \mathbb{R}^2;\ s_0 \leq s \leq 1\} \cup \{(1, s) \in \mathbb{R}^2;\ s_0 \leq s \leq 1\}.$$

Does the space $\mathbf{V}_F(\omega)$ still contain nonzero functions?

6.3. Show that a plate subjected to a boundary condition of place along any portion of its lateral face whose middle line γ_0 satisfies *length* $\gamma_0 > 0$ (Fig. 6.1-3) constitutes an instance of a linearly elastic flexural shell.

Hint: Show that the space $\mathbf{V}_F(\omega)$ consists in this case of vector fields of the form $\boldsymbol{\eta} = (0, 0, \eta_3)$ with $\eta_3 \in H^2(\omega)$ and $\eta_3 = \partial_\nu \eta_3 = 0$ on γ_0.

6.4. In this exercise, it is assumed that the injective mappings $\boldsymbol{\theta}$ are in $\mathcal{C}^2(\overline{\omega}; \mathbb{R}^3)$ and that $\gamma_0 = \gamma$.

(1) Show that, if the surface $S = \boldsymbol{\theta}(\omega)$ contains a planar portion, then $\mathbf{V}_F(\omega) \neq \{\mathbf{0}\}$, so that the associated shell is a linearly elastic flexural shell.

(2) Assume that the surface S does not contain any planar portion. Does the space $\mathbf{V}_F(\omega)$ still contain nonzero functions?

6.5. Do there exist instances where the space $\mathbf{V}_F(\omega)$ contains nonzero functions and is finite-dimensional?

6.6. The notations used below should be self-explanatory.

(1) Let a surface S be equipped with two systems of curvilinear coordinates $(y_\alpha) \in \overline{\omega}$ and $(\tilde{y}_\alpha) \in \{\tilde{\omega}\}^-$, i.e., $S = \boldsymbol{\theta}(\overline{\omega}) = \tilde{\boldsymbol{\theta}}(\{\tilde{\omega}\}^-)$, and assume that $\boldsymbol{\theta}(\gamma_0) = \tilde{\boldsymbol{\theta}}(\tilde{\gamma}_0)$. Show that $\mathbf{V}_F(\omega) \neq \{\mathbf{0}\}$ implies that $\widetilde{\mathbf{V}}_F(\tilde{\omega}) \neq \{\mathbf{0}\}$. In other words, the definition of a linearly elastic flexural shell (Sect. 6.1) is independent of the system of curvilinear coordinates employed for representing the surface S.

(2) Show that the dimensions (finite or infinite) of the spaces $\mathbf{V}_F(\omega)$ and $\widetilde{\mathbf{V}}_F(\tilde{\omega})$ are equal.

6.7. Consider a family of *linearly elastic flexural shells* (i.e., the space $\mathbf{V}_F(\omega)$ contains nonzero functions) *with variable thickness*, i.e., whose reference configurations are the sets $\boldsymbol{\Theta}^e(\overline{\Omega}^\varepsilon)$, where the mapping $\boldsymbol{\Theta}^e : \overline{\Omega}^\varepsilon \to \mathbb{R}^3$ is defined for each $\varepsilon > 0$ by

$$\boldsymbol{\Theta}^e(y, x_3^\varepsilon) := \boldsymbol{\theta}(y) + x_3^\varepsilon e(y)\boldsymbol{a}^3(y)$$

for all $(y, x_3^\varepsilon) \in \overline{\Omega}^\varepsilon = \overline{\omega} \times [-\varepsilon, \varepsilon]$, the given function $e \in W^{2,\infty}(\omega)$ satisfying $0 < e_0 \leq e(y)$ for all $y \in \overline{\omega}$.

For such a family of flexural shells, carry out an asymptotic analysis analogous to that of Thm. 6.2-1. Assuming for simplicity that there are no surface forces, show in particular that the limit scaled two-dimensional problem takes the form

$$\frac{1}{3}\int_\omega a^{\alpha\beta\sigma\tau}\rho^e_{\sigma\tau}(\boldsymbol{\zeta})\rho^e_{\alpha\beta}(\boldsymbol{\eta})e\sqrt{a}\,dy = \int_\omega \left\{\int_{-1}^{1} f^i\,dx_3\right\}\eta_i e\sqrt{a}\,dy$$

for all $\boldsymbol{\eta} \in \mathbf{V}_F(\omega)$, where the space $\mathbf{V}_F(\omega)$ is the same as in Thm. 6.2-1, but the covariant components of the linearized change of metric and change of curvature tensors are now replaced by the more general expressions:

$$\begin{aligned}\gamma^e_{\alpha\beta}(\boldsymbol{\eta}) &:= \frac{1}{2}(\partial_\beta\eta_\alpha + \partial_\alpha\eta_\beta) - \Gamma^\sigma_{\alpha\beta}\eta_\sigma - \frac{1}{e}b_{\alpha\beta}\eta_3,\\ \rho^e_{\alpha\beta}(\boldsymbol{\eta}) &:= \partial_{\alpha\beta}\eta_3 - \Gamma^\sigma_{\alpha\beta}\partial_\sigma\eta_3 - (b^\sigma_\alpha b_{\sigma\beta} + \frac{1}{e}\{\partial_{\alpha\beta}e - \Gamma^\sigma_{\alpha\beta}\partial_\sigma e\})\eta_3\\ &\quad + e\{b^\sigma_\alpha(\partial_\beta\eta_\sigma - \Gamma^\tau_{\beta\sigma}\eta_\tau) + b^\tau_\beta(\partial_\alpha\eta_\tau - \Gamma^\sigma_{\alpha\tau}\eta_\sigma)\\ &\quad + (\partial_\alpha b^\tau_\beta + \Gamma^\tau_{\alpha\sigma}b^\sigma_\beta - \Gamma^\sigma_{\alpha\beta}b^\tau_\sigma)\eta_\tau\}\\ &\quad + \frac{2}{e^2}\{\partial_\alpha e\partial_\beta e\}\eta_3 - \frac{1}{e}\{\partial_\alpha e\partial_\beta\eta_3 + \partial_\beta e\partial_\alpha\eta_3\}.\end{aligned}$$

Remark. This asymptotic analysis is treated in Busse [1998, Thm. 5.3].

6.8. We established in Thm. 6.2-1 that the scaled displacements converge and that their limits can be fully identified. The present problem shows that, under the same assumptions, the *convergence of the* (scaled) *stresses* can be also established and that the corresponding *limit stresses* can be likewise explicitly computed. These results are due to Collard & Miara [1999, Thm. 8].

Given a family of linearly elastic flexural shells that satisfies the assumptions of Thm. 6.2-1, let

$$\sigma^{ij,\varepsilon} = A^{ijkl,\varepsilon}e^\varepsilon_{k\|l}(\boldsymbol{u}^\varepsilon)$$

denote for each $\varepsilon > 0$ the contravariant components of the *linearized stress tensor* field (Thm. 1.6-1) inside each shell and define the *scaled stresses* $\sigma^{ij}(\varepsilon) : \overline{\Omega} \to \mathbb{R}$ by letting

$$\sigma^{ij,\varepsilon}(x^\varepsilon) =: \sigma^{ij}(\varepsilon)(x) \text{ for all } x^\varepsilon = \pi^\varepsilon x \in \overline{\Omega}.$$

Note that the scaled stresses then satisfy

$$\sigma^{ij}(\varepsilon) = A^{ijkl}(\varepsilon)e_{k\|l}(\varepsilon;\, \boldsymbol{u}(\varepsilon)).$$

Show that

$$\begin{aligned}&\frac{1}{\varepsilon}\sigma^{\alpha\beta}(\varepsilon) \to \sigma^{\alpha\beta,1} \text{ in } L^2(\Omega),\\ &\frac{1}{\varepsilon^2}\sigma^{i3}(\varepsilon) \to \sigma^{i3,2} \text{ in } H^1(-1,1;\, H^{-1}(\omega))\end{aligned}$$

as $\varepsilon \to 0$, where the limits are given by (covariant derivatives $n^{\alpha\beta}|_\sigma$ are defined in Thm. 4.5-1)

$$\begin{aligned}
\sigma^{\alpha\beta,1} &= -x_3 a^{\alpha\beta\sigma\tau}\rho_{\sigma\tau}(\overline{\boldsymbol{u}}),\\
\sigma^{\alpha 3,2} &= \frac{1}{2}(x_3^2-1)(a^{\alpha\beta\sigma\tau}\rho_{\sigma\tau}(\overline{\boldsymbol{u}}))|_\beta,\\
\sigma^{33,2} &= \frac{1}{2}(x_3^2-1)a^{\alpha\beta\sigma\tau}b_{\alpha\beta}\rho_{\sigma\tau}(\overline{\boldsymbol{u}}).
\end{aligned}$$

Remark. These results should be profitably compared with those established for linearly elastic membrane shells (Ex. 4.4) and plates (Vol. II, Sect. 1.6).

CHAPTER 7

KOITER'S EQUATIONS AND OTHER TWO-DIMENSIONAL LINEAR SHELL THEORIES

INTRODUCTION

Consider as in the previous chapters a linearly elastic shell with middle surface $S = \boldsymbol{\theta}(\overline{\omega})$ and thickness 2ε, subjected to a boundary condition of place along a portion of its lateral face with $\boldsymbol{\theta}(\gamma_0)$ as its middle curve, and subjected to applied forces.

This problem can thus be modeled either by the *equations of three-dimensional linearized elasticity* (Chap. 1) or by *two-dimensional equations* obtained by an *asymptotic analysis* of the three-dimensional equations. We showed that the form of these two-dimensional equations depends on, and only on, the geometry of the surface S and on the set γ_0 (Chaps. 4 to 6).

Founding his approach on *a priori assumptions* of a *geometrical* and *mechanical* nature about the solution of the three-dimensional equations (these assumptions are described in Sect. 7.1), W.T. Koiter has devised in the sixties yet another means of modeling the same problem by *two-dimensional equations.*

The resulting *two-dimensional Koiter equations for a linearly elastic shell* take the following form, when they are expressed as a *variational problem*: The unknown $\boldsymbol{\zeta}_K^\varepsilon = (\zeta_{K,i}^\varepsilon)$, whose components are the covariant components $\zeta_{K,i}^\varepsilon : \overline{\omega} \to \mathbb{R}$ of the displacement field $\zeta_{K,i}^\varepsilon \boldsymbol{a}^i$ of the points of the middle surface S, satisfies:

$$\boldsymbol{\zeta}_K^\varepsilon = (\zeta_{K,i}^\varepsilon) \in \mathbf{V}_K(\omega) = \{\boldsymbol{\eta} = (\eta_i) \in H^1(\omega) \times H^1(\omega) \times H^2(\omega);$$
$$\eta_i = \partial_\nu \eta_3 = 0 \text{ on } \gamma_0\},$$

$$\int_\omega \left\{ \varepsilon a^{\alpha\beta\sigma\tau,\varepsilon} \gamma_{\sigma\tau}(\boldsymbol{\zeta}^\varepsilon)\gamma_{\alpha\beta}(\boldsymbol{\eta}) + \frac{\varepsilon^3}{3} a^{\alpha\beta\sigma\tau,\varepsilon} \rho_{\sigma\tau}(\boldsymbol{\zeta}^\varepsilon)\rho_{\alpha\beta}(\boldsymbol{\eta}) \right\} \sqrt{a}\, dy$$
$$= \int_\omega p^{i,\varepsilon} \eta_i \sqrt{a}\, dy \text{ for all } \boldsymbol{\eta} = (\eta_i) \in \mathbf{V}_K(\omega),$$

where $a^{\alpha\beta\sigma\tau,\varepsilon}$ are the components of the two-dimensional elasticity tensor of the shell, $\gamma_{\alpha\beta}(\boldsymbol{\eta})$ and $\rho_{\alpha\beta}(\boldsymbol{\eta})$ are the components of the linearized change of metric and change of curvature tensors associated with a displacement field $\eta_i \boldsymbol{a}^i$ of S, and the given functions $p^{i,\varepsilon} \in L^2(\omega)$ account for the applied forces.

Equivalently, the unknown $\boldsymbol{\zeta}_K^\varepsilon$ satisfies the following *minimization problem*:

$$\boldsymbol{\zeta}_K^\varepsilon \in \mathbf{V}_K(\omega) \text{ and } j_K^\varepsilon(\boldsymbol{\zeta}_K^\varepsilon) = \inf_{\boldsymbol{\eta} \in \mathbf{V}_K(\omega)} j_K^\varepsilon(\boldsymbol{\eta}),$$

where the *two-dimensional Koiter energy* $j_K^\varepsilon : \mathbf{V}_K(\omega) \to \mathbb{R}$ is defined by

$$j_K^\varepsilon(\boldsymbol{\eta}) = \frac{1}{2}\int_\omega \Big\{ \varepsilon a^{\alpha\beta\sigma\tau,\varepsilon} \gamma_{\sigma\tau}(\boldsymbol{\eta})\gamma_{\alpha\beta}(\boldsymbol{\eta}) + \frac{\varepsilon^3}{3} a^{\alpha\beta\sigma\tau,\varepsilon} \rho_{\sigma\tau}(\boldsymbol{\eta})\rho_{\alpha\beta}(\boldsymbol{\eta}) \Big\} \sqrt{a}\, dy - \int_\omega p^{i,\varepsilon} \eta_i \sqrt{a}\, dy.$$

As already shown in Sect. 2.6, the *existence* and *uniqueness* of a solution to these equations essentially rely on an *inequality of Korn's type on a general surface*, itself a consequence of a crucial *lemma of J.L. Lions* and of an *infinitesimal rigid displacement lemma on a general surface*; this existence and uniqueness result is recalled in Sect. 7.1, together with its extension to shells whose middle surface has "little regularity" (Thms. 7.1-1 and 7.1-2). In addition, we identify the associated *boundary value problem* and we give sufficient conditions for the *regularity* of its solution (Thm. 7.1-3).

It is remarkable that *Koiter's equations can be fully justified for all types of shells*, even though it is clear that these equations cannot be recovered as the outcome of an asymptotic analysis of the three-dimensional equations, since Chaps. 4 to 6 exhaust all such possible outcomes!

More specifically, we show that, *for each category of linearly elastic shells* (elliptic membrane, generalized membrane, or flexural), *the fields* $\boldsymbol{\zeta}_K^\varepsilon$ *and* $\frac{1}{2\varepsilon}\int_{-\varepsilon}^{\varepsilon} \boldsymbol{u}^\varepsilon\, dx_3^\varepsilon$, where $\boldsymbol{u}^\varepsilon$ denotes the three-dimensional solution, *have the same asymptotic behavior in* ad hoc *functional spaces as* $\varepsilon \to 0$ (Thms. 7.2-1 to 7.2-3).

So, *even though Koiter's linear model is not a limit model, it is in a sense the "best" two-dimensional one for linearly elastic shells* !

One can thus only marvel at the insight that led W.T. Koiter to conceive the "right" equations, whose versatility is indeed remarkable (see the commentary in Sect. 7.3), out of purely mechanical and geometrical intuitions!

While Koiter's equations belong to the family of *Kirchhoff-Love theories* (Sect. 7.1), two-dimensional shell equations that rely on the notion of *one-director Cosserat surfaces* have also been proposed by P.M. Naghdi, again in the sixties. After describing the associated *two-dimensional Naghdi equations for a linearly elastic shell* (Sect. 7.4), we review the *existence* and *uniqueness* theory for these equations, which is akin to that for Koiter's equations (Thm. 7.4-1).

Finally, we briefly review in Sects. 7.5 and 7.6 other two-dimensional equations that have also been proposed for modeling *linearly elastic shells*, notably *"shallow" shells.*

7.1. THE TWO-DIMENSIONAL KOITER EQUATIONS FOR A LINEARLY ELASTIC SHELL: EXISTENCE, UNIQUENESS, AND REGULARITY OF SOLUTIONS; FORMULATION AS A BOUNDARY VALUE PROBLEM

Let ω be a domain in $\mathbb{R}^2$ with boundary γ, let $\boldsymbol{\theta} \in \mathcal{C}^3(\overline{\omega}; \mathbb{R}^3)$ be an injective mapping such that the two vectors $\boldsymbol{a}_\alpha = \partial_\alpha \boldsymbol{\theta}$ are linearly independent at all points of $\overline{\omega}$, and let γ_0 be a portion of γ that satisfies *length* $\gamma_0 > 0$.

Consider as in the previous chapters a *linearly elastic shell* with *middle surface* $S = \boldsymbol{\theta}(\overline{\omega})$ and *thickness* $2\varepsilon > 0$, i.e., a linearly elastic body whose reference configuration is the set $\boldsymbol{\Theta}(\overline{\Omega}^\varepsilon)$, where

$$\Omega^\varepsilon := \omega \times]-\varepsilon, \varepsilon[,$$
$$\boldsymbol{\Theta}(y, x_3^\varepsilon) := \boldsymbol{\theta}(y) + x_3^\varepsilon \boldsymbol{a}_3(y) \text{ for all } (y, x_3^\varepsilon) \in \overline{\Omega}^\varepsilon.$$

The material constituting the shell is *homogeneous* and *isotropic* and the reference configuration is a *natural state*, so that the material is characterized by its two *Lamé constants* $\lambda^\varepsilon > 0$ and $\mu^\varepsilon > 0$. The shell is subjected to a *boundary condition of place* along the portion $\boldsymbol{\Theta}(\Gamma_0^\varepsilon)$ of its lateral face, where $\Gamma_0^\varepsilon := \gamma_0 \times [-\varepsilon, \varepsilon]$, i.e., the three-dimensional displacement vanishes on $\boldsymbol{\Theta}(\Gamma_0^\varepsilon)$. Finally, the shell is subjected to *applied body forces* in its interior $\boldsymbol{\Theta}(\Omega^\varepsilon)$ and to *applied surface forces* on its "upper" and "lower" faces $\boldsymbol{\Theta}(\Gamma_+^\varepsilon)$ and $\boldsymbol{\Theta}(\Gamma_-^\varepsilon)$, their densities being given by their contravariant components $f^{i,\varepsilon} \in L^2(\Omega^\varepsilon)$ and $h^{i,\varepsilon} \in L^2(\Gamma_+^\varepsilon \cup \Gamma_-^\varepsilon)$, where $\Gamma_\pm^\varepsilon = \omega \times \{\pm\varepsilon\}$.

In a seminal work on shells, John [1965, 1971] has shown that, if the thickness is small enough, the state of stress is "approximately" planar and the stresses parallel to the middle surface vary "approximately linearly" across the thickness, at least "away from the lateral face". In *Koiter's approach* (Koiter [1960, 1966, 1970]), these approximations are taken as an *a priori* assumption of a *mechanical nature* (a precise statement of this assumption is found in Sect. 11.1) and combined with another *a priori* assumption of a *geometrical nature*, called the *Kirchhoff-Love assumption*: Any point on a normal to the middle surface remains on the normal to the deformed middle surface after the deformation has taken place and the distance between such a point and the middle surface remains constant (Ex. 7.1(1)). In fact, this assumption is required to hold only "to within the first order" in the linearized theory considered in this chapter, in which case it is called the *linearized Kirchhoff-Love assumption* (Ex. 7.1(2)).

Remark. In a pioneering contribution, Oden, Wempner & Kross [1968] have suggested to incorporate some discretized version of the linearized Kirchhoff-Love assumption into finite element codes, typically by requiring that it be only satisfied on a finite set of points. In myriad numerical simulations, this idea has since then proved to be highly judicious. ■

Taking these two *a priori* assumptions into account, W.T. Koiter then shows that the displacement field across the thickness of the shell can be completely determined from the sole knowledge of the displacement field of the points of the *middle surface* S, and he identifies the *two-dimensional problem*, i.e., posed over the two-dimensional set $\overline{\omega}$, that this displacement field should satisfy. As in the two-dimensional theories encountered so far, the *unknown* is a vector field, now denoted $\boldsymbol{\zeta}_K^\varepsilon = (\zeta_{i,K}^\varepsilon) : \overline{\omega} \to \mathbb{R}^3$, whose components $\zeta_{i,K}^\varepsilon : \overline{\omega} \to \mathbb{R}$ are the covariant components of the *displacement field of the middle surface* S. This means that $\zeta_{i,K}^\varepsilon(y)\boldsymbol{a}^i(y)$ is the displacement of the point $\boldsymbol{\theta}(y)$; cf. Fig. 7.1-1.

In their *linearized* version (the nonlinear one is given in Sect. 11.1), the equations found by W.T. Koiter consist in solving the following

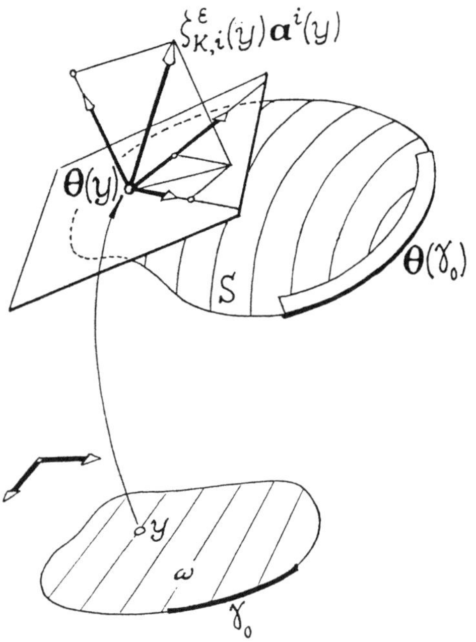

Fig. 7.1-1: *A linearly elastic shell modeled by Koiter's two-dimensional equations.* The three unknowns are the covariant components $\zeta^{\varepsilon}_{i,K} : \overline{\omega} \to \mathbb{R}$ of the displacement field of the middle surface S; this means that $\zeta^{\varepsilon}_{i,K}(y)\boldsymbol{a}^i(y)$ is the displacement of the point $\boldsymbol{\theta}(y)$ of the middle surface S. The " two-dimensional boundary conditions of clamping" $\zeta^{\varepsilon}_{i,K} = \partial_\nu \zeta^{\varepsilon}_{3,K} = 0$ on γ_0 mean that the points of, and the tangent spaces to, the deformed and undeformed middle surfaces coincide along the set $\boldsymbol{\theta}(\gamma_0)$.

variational problem $\mathcal{P}^{\varepsilon}_K(\omega)$ (the associated boundary value problem is given in Thm. 7.1-3): Find $\boldsymbol{\zeta}^{\varepsilon}_K = (\zeta^{\varepsilon}_{K,i})$ such that

$$\boldsymbol{\zeta}^{\varepsilon}_K \in \mathbf{V}_K(\omega) := \{\boldsymbol{\eta} = (\eta_i) \in H^1(\omega) \times H^1(\omega) \times H^2(\omega);\ \eta_i = \partial_\nu \eta_3 = 0 \text{ on } \gamma_0\},$$

$$\int_\omega \left\{ \varepsilon a^{\alpha\beta\sigma\tau,\varepsilon} \gamma_{\sigma\tau}(\boldsymbol{\zeta}^{\varepsilon}_K)\gamma_{\alpha\beta}(\boldsymbol{\eta}) + \frac{\varepsilon^3}{3} a^{\alpha\beta\sigma\tau,\varepsilon} \rho_{\sigma\tau}(\boldsymbol{\zeta}^{\varepsilon}_K)\rho_{\alpha\beta}(\boldsymbol{\eta}) \right\} \sqrt{a}\, dy$$

$$= \int_\omega p^{i,\varepsilon} \eta_i \sqrt{a}\, dy \text{ for all } \boldsymbol{\eta} = (\eta_i) \in \mathbf{V}_K(\omega),$$

where (the functions $a^{\alpha\beta}$, $b_{\alpha\beta}$, b_α^σ, $\Gamma_{\alpha\beta}^\sigma$, and a are defined as usual; see, e.g., Sect. 3.1):

$$a^{\alpha\beta\sigma\tau,\varepsilon} := \frac{4\lambda^\varepsilon\mu^\varepsilon}{\lambda^\varepsilon + 2\mu^\varepsilon} a^{\alpha\beta} a^{\sigma\tau} + 2\mu^\varepsilon (a^{\alpha\sigma} a^{\beta\tau} + a^{\alpha\tau} a^{\beta\sigma}),$$

$$\gamma_{\alpha\beta}(\boldsymbol{\eta}) := \frac{1}{2}(\partial_\beta \eta_\alpha + \partial_\alpha \eta_\beta) - \Gamma_{\alpha\beta}^\sigma \eta_\sigma - b_{\alpha\beta}\eta_3,$$

$$\begin{aligned}\rho_{\alpha\beta}(\boldsymbol{\eta}) := {} & \partial_{\alpha\beta}\eta_3 - \Gamma_{\alpha\beta}^\sigma \partial_\sigma \eta_3 - b_\alpha^\sigma b_{\sigma\beta}\eta_3 \\ & + b_\alpha^\sigma(\partial_\beta \eta_\sigma - \Gamma_{\beta\sigma}^\tau \eta_\tau) + b_\beta^\tau(\partial_\alpha \eta_\tau - \Gamma_{\alpha\tau}^\sigma \eta_\sigma) \\ & + (\partial_\alpha b_\beta^\tau + \Gamma_{\alpha\sigma}^\tau b_\beta^\sigma - \Gamma_{\alpha\beta}^\sigma b_\sigma^\tau)\eta_\tau,\end{aligned}$$

$$p^{i,\varepsilon} := \int_{-\varepsilon}^{\varepsilon} f^{i,\varepsilon}\, dx_3^\varepsilon + h_+^{i,\varepsilon} + h_-^{i,\varepsilon} \text{ and } h_\pm^{i,\varepsilon} := h^{i,\varepsilon}(\cdot, \pm\varepsilon).$$

The functions $\gamma_{\alpha\beta}(\boldsymbol{\eta})$ and $\rho_{\alpha\beta}(\boldsymbol{\eta})$ are the customary covariant components of the **linearized change of metric** and **change of curvature tensors** associated with a displacement field $\eta_i \boldsymbol{a}^i$ of the middle surface S and the functions $a^{\alpha\beta\sigma\tau,\varepsilon}$ are the customary contravariant components of the **two-dimensional elasticity tensor of the shell.**

Note that Destuynder [1985] has found an illuminating way of deriving the same linear Koiter equations from three-dimensional elasticity, which uses *a priori* assumptions only of a *geometrical* nature. ■

The *existence* and *uniqueness* of a solution to problem $\mathcal{P}_K^\varepsilon(\omega)$, which essentially follow from the $\mathbf{V}_K(\omega)$-ellipticity of the bilinear form, was first established by Bernadou & Ciarlet [1976]; a more natural proof was subsequently proposed by Ciarlet & Miara [1992b], then combined with the first one in Bernadou, Ciarlet & Miara [1994].

This proof, which has been already given and discussed at length in Sect. 2.6, is simply outlined in the next theorem. Observe that $\boldsymbol{\theta} \in \mathcal{C}^3(\overline{\omega}; \mathbb{R}^3)$ is precisely the assumption that will afford the justification of Koiter's equations in Sect. 7.2.

Theorem 7.1-1. *Let ω be a domain in $\mathbb{R}^2$, let γ_0 be a subset of $\gamma = \partial\omega$ with* length $\gamma_0 > 0$, *and let $\boldsymbol{\theta} \in \mathcal{C}^3(\overline{\omega}; \mathbb{R}^3)$ be an injective mapping such that the two vectors $\boldsymbol{a}_\alpha = \partial_\alpha \boldsymbol{\theta}$ are linearly independent at all points of $\overline{\omega}$.*

Then the variational problem $\mathcal{P}_K^\varepsilon(\omega)$ has one and only one solution, which is also the unique solution to the minimization problem: Find $\boldsymbol{\zeta}_K^\varepsilon = (\zeta_{K,i}^\varepsilon)$ such that

$$\boldsymbol{\zeta}_K^\varepsilon \in \mathbf{V}_K(\omega) \text{ and } j_K^\varepsilon(\boldsymbol{\zeta}_K^\varepsilon) = \inf_{\boldsymbol{\eta}\in\mathbf{V}_K(\omega)} j_K^\varepsilon(\boldsymbol{\eta}), \text{ where}$$

$$j_K^\varepsilon(\boldsymbol{\eta}) := \frac{1}{2}\int_\omega \Big\{\varepsilon a^{\alpha\beta\sigma\tau,\varepsilon}\gamma_{\sigma\tau}(\boldsymbol{\eta})\gamma_{\alpha\beta}(\boldsymbol{\eta}) + \frac{\varepsilon^3}{3}a^{\alpha\beta\sigma\tau,\varepsilon}\rho_{\sigma\tau}(\boldsymbol{\eta})\rho_{\alpha\beta}(\boldsymbol{\eta})\Big\}\sqrt{a}\,dy - \int_\omega p^{i,\varepsilon}\eta_i\sqrt{a}\,dy.$$

Proof. The assumptions $f^{i,\varepsilon} \in L^2(\Omega^\varepsilon)$ and $h^{i,\varepsilon} \in L^2(\Gamma_+^\varepsilon \cup \Gamma_-^\varepsilon)$ imply that $p^{i,\varepsilon} \in L^2(\omega)$. The existence and uniqueness of a solution to the variational problem $\mathcal{P}_K^\varepsilon(\omega)$, or to its equivalent minimization problem, are consequences of the *inequality of Korn's type on a general surface* (Thm. 2.6-4), of the *uniform positive definiteness of the two-dimensional elastic tensor of the shell* (Thm. 3.3-2), of the existence of a_0 such that $a(y) \geq a_0 > 0$ for all $y \in \overline{\omega}$ (Thm. 3.3-1), and of the *Lax-Milgram lemma.* ∎

In Sect. 2.6, we also described how Blouza & Le Dret [1994a, 1994b, 1999] showed that the introduction of new expressions, denoted $\tilde{\gamma}_{\alpha\beta}(\tilde{\boldsymbol{\eta}})$ and $\tilde{\rho}_{\alpha\beta}(\tilde{\boldsymbol{\eta}})$ (see *infra*), for the functions $\gamma_{\alpha\beta}(\boldsymbol{\eta})$ and $\rho_{\alpha\beta}(\boldsymbol{\eta})$ affords to consider more general situations, where the mapping $\boldsymbol{\theta}$ need only be in the space $W^{2,\infty}(\omega;\mathbb{R}^3)$. For a *linearly elastic shell, simply supported along its entire boundary* (boundary conditions of clamping along a portion of its boundary can be handled as well, provided they are first re-interpreted in an *ad hoc* manner), the associated **Koiter's equations for shells whose middle surface has little regularity** accordingly take the following form: The *unknown* $\tilde{\boldsymbol{\zeta}}_K^\varepsilon$, which is now the *displacement field of the middle surface*, satisfies the variational problem $\tilde{\mathcal{P}}_K^\varepsilon(\omega)$:

$$\tilde{\boldsymbol{\zeta}}_K^\varepsilon \in \widetilde{\mathbf{V}}_K^s(\omega) := \{\tilde{\boldsymbol{\eta}} \in \mathbf{H}_0^1(\omega);\ \partial_{\alpha\beta}\tilde{\boldsymbol{\eta}}\cdot\boldsymbol{a}_3 \in L^2(\omega)\},$$

$$\int_\omega \Big\{\varepsilon a^{\alpha\beta\sigma\tau,\varepsilon}\tilde{\gamma}_{\sigma\tau}(\tilde{\boldsymbol{\zeta}}_K^\varepsilon)\tilde{\gamma}_{\alpha\beta}(\tilde{\boldsymbol{\eta}}) + \frac{\varepsilon^3}{3}a^{\alpha\beta\sigma\tau,\varepsilon}\tilde{\rho}_{\sigma\tau}(\tilde{\boldsymbol{\zeta}}_K^\varepsilon)\tilde{\rho}_{\alpha\beta}(\tilde{\boldsymbol{\eta}})\Big\}\sqrt{a}\,dy = \int_\omega \tilde{\boldsymbol{p}}^\varepsilon\cdot\tilde{\boldsymbol{\eta}}\sqrt{a}\,dy \text{ for all } \tilde{\boldsymbol{\eta}} \in \widetilde{\mathbf{V}}_K^s(\omega),$$

where

$$\tilde{\gamma}_{\alpha\beta}(\tilde{\boldsymbol{\eta}}) := \frac{1}{2}(\partial_\beta \tilde{\boldsymbol{\eta}} \cdot \boldsymbol{a}_\alpha + \partial_\alpha \tilde{\boldsymbol{\eta}} \cdot \boldsymbol{a}_\beta),$$
$$\tilde{\rho}_{\alpha\beta}(\tilde{\boldsymbol{\eta}}) := (\partial_{\alpha\beta} \tilde{\boldsymbol{\eta}} - \Gamma^\sigma_{\alpha\beta} \partial_\sigma \tilde{\boldsymbol{\eta}}) \cdot \boldsymbol{a}_3,$$

the given function $\tilde{\boldsymbol{p}}^\varepsilon \in \mathbf{L}^2(\omega)$ account for the applied forces, and $a^{\alpha\beta\sigma\tau,\varepsilon}$ are the usual contravariant components of the two-dimensional elasticity tensor of the shell.

Recall that $\tilde{\gamma}_{\alpha\beta}(\tilde{\boldsymbol{\eta}}) = \gamma_{\alpha\beta}(\boldsymbol{\eta})$ and $\tilde{\boldsymbol{\rho}}_{\alpha\beta}(\tilde{\boldsymbol{\eta}}) = \rho_{\alpha\beta}(\boldsymbol{\eta})$ if $\tilde{\boldsymbol{\eta}} = \eta_i \boldsymbol{a}^i$ is such that $\boldsymbol{\eta} = (\eta_i) \in H^1(\omega) \times H^1(\omega) \times H^2(\omega)$.

A proof similar to that of Thm. 7.1-1, now based on the *inequality of Korn's type on a general surface with little regularity* (Thm. 2.6-6), then produces the following existence and uniqueness result:

Theorem 7.1-2. *Let there be given a domain ω in $\mathbb{R}^2$ and an injective mapping $\boldsymbol{\theta} \in W^{2,\infty}(\omega; \mathbb{R}^3)$ such that the two vectors $\boldsymbol{a}_\alpha = \partial_\alpha \boldsymbol{\theta}$ are linearly independent at all points of $\overline{\omega}$.*

Then the variational problem $\tilde{\mathcal{P}}^\varepsilon_K(\omega)$ has one and only one solution, which is also the unique solution to the minimization problem: Find $\tilde{\boldsymbol{\zeta}}^\varepsilon_K$ such that

$$\tilde{\boldsymbol{\zeta}}^\varepsilon_K \in \tilde{\mathbf{V}}^s_K(\omega) \text{ and } \tilde{j}^\varepsilon_K(\tilde{\boldsymbol{\zeta}}^\varepsilon_K) = \inf_{\tilde{\boldsymbol{\eta}} \in \tilde{\mathbf{V}}^s_K(\omega)} \tilde{j}^\varepsilon_K(\tilde{\boldsymbol{\eta}}), \text{ where}$$
$$\tilde{j}^\varepsilon_K(\tilde{\boldsymbol{\eta}}) := \frac{1}{2}\int_\omega \Big\{\varepsilon a^{\alpha\beta\sigma\tau,\varepsilon} \tilde{\gamma}_{\sigma\tau}(\tilde{\boldsymbol{\eta}}) \tilde{\gamma}_{\alpha\beta}(\tilde{\boldsymbol{\eta}})$$
$$+ \frac{\varepsilon^3}{3} a^{\alpha\beta\sigma\tau,\varepsilon} \tilde{\rho}_{\sigma\tau}(\tilde{\boldsymbol{\eta}}) \tilde{\rho}_{\alpha\beta}(\tilde{\boldsymbol{\eta}})\Big\} \sqrt{a}\, dy - \int_\omega \tilde{\boldsymbol{p}}^\varepsilon \cdot \tilde{\boldsymbol{\eta}} \sqrt{a}\, dy.$$

■

We emphasize that, in this approach, the unknown $\tilde{\boldsymbol{\zeta}}^\varepsilon_K$ and the fields $\tilde{\boldsymbol{\eta}}$ are *displacement fields of the middle surface*, no longer recovered as $\tilde{\boldsymbol{\zeta}}^\varepsilon_K = \zeta^\varepsilon_{K,i} \boldsymbol{a}^i$ or $\tilde{\boldsymbol{\eta}} = \eta_i \boldsymbol{a}^i$ by means of their covariant components $\zeta^\varepsilon_{K,i}$ or η_i.

We next derive the *boundary value problem* that is (at least formally) equivalent to Koiter's variational problem $\mathcal{P}^\varepsilon_K(\omega)$. We only consider here the case where $\gamma_0 = \gamma$; see Ex. 7.2 for the case where *length* $\gamma_0 <$ *length* γ.

We also state a *regularity* result that provides instances where the *weak solution* (the solution of the variational problem) becomes a *classical solution* (a solution of the boundary value problem).

Theorem 7.1-3. *Assume that* $\gamma_0 = \gamma$, *so that*

$$\mathbf{V}_K(\omega) = H_0^1(\omega) \times H_0^1(\omega) \times H_0^2(\omega).$$

(a) *If the solution* $\boldsymbol{\zeta}_K^\varepsilon$ *of the corresponding variational problem* $\mathcal{P}_K^\varepsilon(\omega)$ (Thm. 7.1-1) *is smooth enough, it also satisfies the boundary value problem:*

$$\boxed{\begin{aligned} m^{\alpha\beta,\varepsilon}|_{\alpha\beta} - b_\alpha^\sigma b_{\sigma\beta} m^{\alpha\beta,\varepsilon} - b_{\alpha\beta} n^{\alpha\beta,\varepsilon} &= p^{3,\varepsilon} \text{ in } \omega, \\ -(n^{\alpha\beta,\varepsilon} + b_\sigma^\alpha m^{\sigma\beta,\varepsilon})|_\beta - b_\sigma^\alpha (m^{\sigma\beta,\varepsilon}|_\beta) &= p^{\alpha,\varepsilon} \text{ in } \omega, \\ \zeta_{i,K}^\varepsilon = \partial_\nu \zeta_{3,K}^\varepsilon &= 0 \text{ on } \gamma, \end{aligned}}$$

where

$$\boxed{n^{\alpha\beta,\varepsilon} := \varepsilon a^{\alpha\beta\sigma\tau,\varepsilon} \gamma_{\sigma\tau}(\boldsymbol{\zeta}_K^\varepsilon), \quad m^{\alpha\beta,\varepsilon} := \frac{\varepsilon^3}{3} a^{\alpha\beta\sigma\tau,\varepsilon} \rho_{\sigma\tau}(\boldsymbol{\zeta}_K^\varepsilon),}$$

and, for an arbitrary tensor with twice differentiable covariant components $n^{\alpha\beta}$,

$$\boxed{\begin{aligned} n^{\alpha\beta}|_\beta &:= \partial_\beta n^{\alpha\beta} + \Gamma_{\beta\sigma}^\alpha n^{\beta\sigma} + \Gamma_{\beta\sigma}^\beta n^{\alpha\sigma}, \\ n^{\alpha\beta}|_{\alpha\beta} &:= \partial_\alpha (n^{\alpha\beta}|_\beta) + \Gamma_{\alpha\sigma}^\sigma (n^{\alpha\beta}|_\beta). \end{aligned}}$$

(b) *Assume that, for some integer* $m \geq 0$ *and some real number* $q > 1$, γ *is of class* $\mathcal{C}^{m+4}$, $\boldsymbol{\theta} \in \mathcal{C}^{m+4}(\overline{\omega}; \mathbb{R}^3)$, $p^{\alpha,\varepsilon} \in W^{m+1,q}(\omega)$, *and* $p^{3,\varepsilon} \in W^{m,q}(\omega)$. *Then*

$$\boxed{\boldsymbol{\zeta}_K^\varepsilon = (\zeta_i^\varepsilon) \in W^{m+3,q}(\omega) \times W^{m+3,q}(\omega) \times W^{m+4,q}(\omega).}$$

Proof. For notational convenience, we omit the exponents "ε" and the indices "K" throughout the proof, i.e., we let

$$\boldsymbol{\zeta} := \boldsymbol{\zeta}_K^\varepsilon,\ n^{\alpha\beta} := a^{\alpha\beta\sigma\tau} \gamma_{\sigma\tau}(\boldsymbol{\zeta}),\ m^{\alpha\beta} := \frac{1}{3} a^{\alpha\beta\sigma\tau} \rho_{\sigma\tau}(\boldsymbol{\zeta}),\ p^i := p^{i,\varepsilon}.$$

Assume that the solution $\boldsymbol{\zeta}$ is "smooth enough" in the sense that $n^{\alpha\beta} \in H^1(\omega)$ and $m^{\alpha\beta} \in H^2(\omega)$.

We already saw in the proof of Thm. 4.5-1 that

$$\int_\omega a^{\alpha\beta\sigma\tau}\gamma_{\sigma\tau}(\boldsymbol{\zeta})\gamma_{\alpha\beta}(\boldsymbol{\eta})\sqrt{a}\,dy = -\int_\omega \sqrt{a}\{(n^{\alpha\beta}|_\beta)\eta_\alpha + b_{\alpha\beta}n^{\alpha\beta}\eta_3\}\,dy$$

for all $\boldsymbol{\eta} = (\eta_i) \in H_0^1(\omega) \times H_0^1(\omega) \times L^2(\omega)$, hence *a fortiori* for all $\boldsymbol{\eta} \in H_0^1(\omega) \times H_0^1(\omega) \times H_0^2(\omega)$. It thus remains to transform the other integral appearing in the left-hand side of the variational equations, viz.,

$$\begin{aligned}
\frac{1}{3}\int_\omega a^{\alpha\beta\sigma\tau}\rho_{\sigma\tau}(\boldsymbol{\zeta})\rho_{\alpha\beta}(\boldsymbol{\eta})\sqrt{a}\,dy &= \int_\omega m^{\alpha\beta}\rho_{\alpha\beta}(\boldsymbol{\eta})\sqrt{a}\,dy \\
&= \int_\omega \sqrt{a}\,m^{\alpha\beta}\partial_{\alpha\beta}\eta_3\,dy \\
&+ \int_\omega \sqrt{a}\,m^{\alpha\beta}(2b_\alpha^\sigma\partial_\beta\eta_\sigma - \Gamma_{\alpha\beta}^\sigma\partial_\sigma\eta_3)\,dy \\
&+ \int_\omega \sqrt{a}\,m^{\alpha\beta}(-2b_\beta^\tau\Gamma_{\alpha\tau}^\sigma\eta_\sigma + b_\beta^\sigma|_\alpha\eta_\sigma - b_\alpha^\sigma b_{\sigma\beta}\eta_3)\,dy,
\end{aligned}$$

where $\boldsymbol{\eta} = (\eta_i) \in H_0^1(\omega) \times H_0^1(\omega) \times H_0^2(\omega)$. Using the symmetry $m^{\alpha\beta} = m^{\beta\alpha}$, the relation (already established in the proof of Thm. 4.5-1):

$$\partial_\beta\sqrt{a} = \sqrt{a}\,\Gamma_{\beta\sigma}^\sigma,$$

and the *Green formula in Sobolev spaces* (see, e.g., Vol. I, Thm. 6.1-9), we obtain

$$\begin{aligned}
\int_\omega m^{\alpha\beta}\rho_{\alpha\beta}(\boldsymbol{\eta})\sqrt{a}\,dy = &-\int_\omega \sqrt{a}(\partial_\beta m^{\alpha\beta} + \Gamma_{\beta\sigma}^\sigma m^{\alpha\beta} + \Gamma_{\sigma\beta}^\alpha m^{\sigma\beta})\partial_\alpha\eta_3\,dy \\
&+2\int_\omega \sqrt{a}\,m^{\alpha\beta}b_\alpha^\sigma\partial_\beta\eta_\sigma\,dy \\
&+\int_\omega \sqrt{a}\,m^{\alpha\beta}(-2b_\beta^\tau\Gamma_{\alpha\tau}^\sigma\eta_\sigma + b_\beta^\sigma|_\alpha\eta_\sigma - b_\alpha^\sigma b_{\sigma\beta}\eta_3)\,dy.
\end{aligned}$$

The same Green formula shows that

$$-\int_\omega \sqrt{a}(\partial_\beta m^{\alpha\beta} + \Gamma^\sigma_{\beta\sigma} m^{\alpha\beta} + \Gamma^\alpha_{\sigma\beta} m^{\sigma\beta})\partial_\alpha \eta_3 \, dy$$

$$= -\int_\omega \sqrt{a}(m^{\alpha\beta}|_\beta)\partial_\alpha \eta_3 \, dy = \int_\omega \partial_\alpha(\sqrt{a}\, m^{\alpha\beta}|_\beta)\eta_3 \, dy$$

$$= \int_\omega \sqrt{a}(m^{\alpha\beta}|_{\alpha\beta})\eta_3 \, dy,$$

$$2\int_\omega \sqrt{a}\, m^{\alpha\beta} b^\sigma_\alpha \partial_\beta \eta_\sigma \, dy = -2\int_\omega \sqrt{a}\{\partial_\beta(b^\sigma_\alpha m^{\alpha\beta}) + \Gamma^\tau_{\beta\tau} b^\sigma_\alpha m^{\alpha\beta}\}\eta_\sigma \, dy.$$

Consequently,

$$\int_\omega m^{\alpha\beta}\rho_{\alpha\beta}(\boldsymbol{\eta})\sqrt{a}\, dy = \int_\omega \sqrt{a}\{-2(b^\alpha_\sigma m^{\sigma\beta})|_\beta + (b^\alpha_\beta|_\sigma)m^{\sigma\beta}\}\eta_\alpha \, dy$$
$$+ \int_\omega \sqrt{a}\{m^{\alpha\beta}|_{\alpha\beta} - b^\sigma_\alpha b_{\sigma\beta} m^{\alpha\beta}\}\eta_3 \, dy.$$

Using in this relation the easily verified formula

$$(b^\alpha_\sigma m^{\sigma\beta})|_\beta = (b^\alpha_\beta|_\sigma)m^{\sigma\beta} + b^\alpha_\sigma(m^{\sigma\beta}|_\beta)$$

and the symmetry $b^\alpha_\beta|_\sigma = b^\alpha_\sigma|_\beta$ (Thm. 2.5-1), we finally obtain

$$\int_\omega m^{\alpha\beta}\rho_{\alpha\beta}(\boldsymbol{\eta})\sqrt{a}\, dy = -\int_\omega \sqrt{a}\{(b^\alpha_\sigma m^{\sigma\beta})|_\beta + b^\alpha_\sigma(m^{\sigma\beta}|_\beta)\}\eta_\alpha \, dy$$
$$- \int_\omega \sqrt{a}\{b^\sigma_\alpha b_{\sigma\beta} m^{\alpha\beta} - m^{\alpha\beta}|_{\alpha\beta}\}\eta_3 \, dy.$$

Hence the variational equations

$$\int_\omega \Big\{a^{\alpha\beta\sigma\tau}\gamma_{\sigma\tau}(\boldsymbol{\zeta})\gamma_{\alpha\beta}(\boldsymbol{\eta}) + \frac{1}{3}a^{\alpha\beta\sigma\tau}\rho_{\sigma\tau}(\boldsymbol{\zeta})\rho_{\alpha\beta}(\boldsymbol{\eta}) - p^i\eta_i\Big\}\sqrt{a}\, dy = 0$$

for all $\boldsymbol{\eta} = (\eta_i) \in \mathbf{V}_K(\omega)$ imply that

$$\int_\omega \sqrt{a}\{(n^{\alpha\beta} + b^\alpha_\sigma m^{\sigma\beta})|_\beta + b^\alpha_\sigma(m^{\sigma\beta}|_\beta) + p^\alpha\}\eta_\alpha \, dy$$
$$+ \int_\omega \sqrt{a}\{b_{\alpha\beta} n^{\alpha\beta} + b^\sigma_\alpha b_{\sigma\beta} m^{\alpha\beta} - m^{\alpha\beta}|_{\alpha\beta} + p^3\}\eta_3 \, dy = 0$$

for all $(\eta_i) \in \mathbf{V}_K(\omega)$. The announced partial differential equations are thus satisfied in ω.

The regularity result of part (b), which is due to Alexandrescu [1994], is left as a problem (Ex. 7.3). ■

Each one of the three formulations found in Thms. 7.1-1 and 7.1-3 constitutes the **two-dimensional Koiter equations for a linearly elastic shell.** We recall that the functions $\gamma_{\alpha\beta}(\boldsymbol{\eta})$ and $\rho_{\alpha\beta}(\boldsymbol{\eta})$ are the covariant components of the **linearized change of metric** and **change of curvature tensors** associated with a displacement field $\eta_i \boldsymbol{a}^i$ of the middle surface S, the functions $a^{\alpha\beta\sigma\tau,\varepsilon}$ are the contravariant components of the **two-dimensional elasticity tensor of the shell.** The functions $n^{\alpha\beta,\varepsilon}$ and $m^{\alpha\beta,\varepsilon}$ are the contravariant components of the **stress resultant** and **stress couple**, or **bending moment, tensor** fields.

As shown in Sect. 6.3, the **two-dimensional boundary conditions of clamping** $\zeta^\varepsilon_{K,i} = \partial_\nu \zeta^\varepsilon_{K,3} = 0$ on γ_0 express that *the points of, and the tangent spaces to, the deformed and undeformed middle surfaces coincide along the set* $\boldsymbol{\theta}(\gamma_0)$; cf. Fig. 7.1-1.

The functional $j^\varepsilon_K : \mathbf{V}_K(\omega) \to \mathbb{R}$ in Thm. 7.1-1 is the **two-dimensional Koiter energy of a linearly elastic shell.** *The associated* **Koiter strain energy:**

$$\boldsymbol{\eta} \in \mathbf{V}_K(\omega) \to \frac{1}{2}\int_\omega \Big\{\varepsilon a^{\alpha\beta\sigma\tau,\varepsilon}\gamma_{\sigma\tau}(\boldsymbol{\eta})\gamma_{\alpha\beta}(\boldsymbol{\eta}) + \frac{\varepsilon^3}{3} a^{\alpha\beta\sigma\tau,\varepsilon}\rho_{\sigma\tau}(\boldsymbol{\eta})\rho_{\alpha\beta}(\boldsymbol{\eta})\Big\}\sqrt{a}\,dy$$

is thus the sum of the strain energy of a linearly elastic elliptic membrane shell (Sect. 4.5) *and of the strain energy of a linearly elastic flexural shell* (Sect. 6.3)!

Showing that this uncanny and eerie "addition" is in fact of the utmost pertinence is the object of the next section.

Remarks. (1) In the original derivation of Koiter's equations, the functions $n^{\alpha\beta,\varepsilon}$ and $m^{\alpha\beta,\varepsilon}$ found in Thm. 7.1-3 are first defined by means of an *ad hoc* "two-dimensional variational principle" (Koiter [1970, Sect. 4]); they are then shown to be related to the components $\sigma^{\alpha\beta,\varepsilon}$ of the *linearized stress tensor* field (Thm. 1.6-1) inside the shell by the equations (Koiter [1970, eqs. (4.4) and (4.5)]):

$$n^{\alpha\beta,\varepsilon} = \int_{-\varepsilon}^{\varepsilon} \{\sigma^{\alpha\beta,\varepsilon} - (x_3^\varepsilon)^2 b^\alpha_\lambda b^\beta_\mu \sigma^{\lambda\mu,\varepsilon}\}\sqrt{\frac{g^\varepsilon}{a}}\,dx_3^\varepsilon,$$

$$m^{\alpha\beta,\varepsilon} = \int_{-\varepsilon}^{\varepsilon} \{-x_3^\varepsilon \sigma^{\alpha\beta,\varepsilon} + \frac{1}{2}(x_3^\varepsilon)^2 (b_\lambda^\alpha \sigma^{\lambda\beta,\varepsilon} + b_\mu^\beta \sigma^{\alpha\mu,\varepsilon})\} \sqrt{\frac{g^\varepsilon}{a}}\, dx_3^\varepsilon.$$

These formulas justify the terminology *"stress resultants"* and *"stress couples"* attached to the functions $n^{\alpha\beta,\varepsilon}$ and $m^{\alpha\beta,\varepsilon}$.

(2) Analogous relations can be rigorously derived from the two-dimensional equations of a linearly elastic *elliptic membrane* shell, the functions $n^{\alpha\beta,\varepsilon} := \varepsilon a^{\alpha\beta\sigma\tau,\varepsilon}\gamma_{\sigma\tau}(\boldsymbol{\zeta}^\varepsilon)$ being then related to the *limit stresses* $\sigma^{\alpha\beta,\varepsilon}(0)$ by the formulas (Ex. 4.4)

$$n^{\alpha\beta,\varepsilon} = \int_{-\varepsilon}^{\varepsilon} \sigma^{\alpha\beta,\varepsilon}(0)\, dx_3^\varepsilon.$$

(3) In the two-dimensional linear Kirchhoff-Love theory of *plates*, the "stress couples" are usually defined by the formulas (Vol. I, Thm. 1.7-3)

$$m^{\alpha\beta,\varepsilon} = \int_{-\varepsilon}^{\varepsilon} x_3^\varepsilon \sigma^{\alpha\beta,\varepsilon}(0)\, dx_3^\varepsilon.$$

Note the *change of sign*!

(4) The functions $m^{\alpha\beta,\varepsilon}|_{\alpha\beta}$ are instances of *second-order covariant derivatives* of a *second-order tensor field.*

(5) Koiter's equations can also accommodate loads and couples applied along the remaining portion $\boldsymbol{\theta}(\gamma - \gamma_0)$ of the boundary of S; see Koiter [1970, eqs. (3.8), (3.9), and (3.13) to (3.15)]. ■

7.2. JUSTIFICATION OF KOITER'S EQUATIONS FOR ALL TYPES OF LINEARLY ELASTIC SHELLS

Consider the same linearly elastic shell problem as in Sect. 7.1. Then the corresponding *three-dimensional variational problem* $\mathcal{P}(\Omega^\varepsilon)$ consists in finding $\boldsymbol{u}^\varepsilon = (u_i^\varepsilon)$ such that (all notations used *infra* are defined as in, e.g., Sect. 3.1):

$$\boxed{\begin{aligned}
&\boldsymbol{u}^\varepsilon \in \mathbf{V}(\Omega^\varepsilon) = \{\boldsymbol{v}^\varepsilon = (v_i^\varepsilon) \in \mathbf{H}^1(\Omega^\varepsilon);\ \boldsymbol{v}^\varepsilon = \mathbf{0} \text{ on } \Gamma_0^\varepsilon\},\\
&\int_{\Omega^\varepsilon} A^{ijkl,\varepsilon} e^\varepsilon_{k\|l}(\boldsymbol{u}^\varepsilon) e^\varepsilon_{i\|j}(\boldsymbol{v}^\varepsilon)\sqrt{g^\varepsilon}\, dx^\varepsilon\\
&\qquad = \int_{\Omega^\varepsilon} f^{i,\varepsilon} v_i^\varepsilon \sqrt{g^\varepsilon}\, dx^\varepsilon + \int_{\Gamma_+^\varepsilon \cup \Gamma_-^\varepsilon} h^{i,\varepsilon} v_i^\varepsilon \, d\Gamma^\varepsilon\\
&\qquad\qquad \text{for all } \boldsymbol{v}^\varepsilon \in \mathbf{V}(\Omega^\varepsilon),
\end{aligned}}$$

where

$$\boxed{\begin{aligned}
A^{ijkl,\varepsilon} &:= \lambda^\varepsilon g^{ij,\varepsilon} g^{kl,\varepsilon} + \mu^\varepsilon (g^{ik,\varepsilon} g^{jl,\varepsilon} + g^{il,\varepsilon} g^{jk,\varepsilon}),\\
e^\varepsilon_{i\|j}(\boldsymbol{v}^\varepsilon) &:= \frac{1}{2}(\partial_j^\varepsilon v_i^\varepsilon + \partial_i^\varepsilon v_j^\varepsilon) - \Gamma^{p,\varepsilon}_{ij}(\boldsymbol{v}^\varepsilon).
\end{aligned}}$$

The unknown functions u_i^ε in problem $\mathcal{P}(\Omega^\varepsilon)$ represent the covariant components of the *displacement field* $u_i^\varepsilon \boldsymbol{g}^{i,\varepsilon}$ of the points of the reference configuration $\boldsymbol{\Theta}(\overline{\Omega}^\varepsilon)$, the functions $A^{ijkl,\varepsilon}$ denote the contravariant components of the *three-dimensional elasticity tensor*, the functions $e^\varepsilon_{i\|j}(\boldsymbol{v}^\varepsilon)$ denote the covariant components of the *linearized strain tensor*, and $d\Gamma^\varepsilon$ denote the area element along the boundary of the set Ω^ε.

Consider now a *family* of such shells, with each having the *same* middle surface $S = \boldsymbol{\theta}(\overline{\omega})$ and with each subjected to a boundary condition of place along a portion of its lateral face having the *same* set $\boldsymbol{\theta}(\gamma_0)$ as its middle curve. All the linearly elastic shells in such a family are thus either *elliptic membrane*, or *generalized membrane*, or *flexural*, according to the definitions given in Chaps. 4, 5, and 6. We now show that, *in each case, the asymptotic behaviors as* $\varepsilon \to 0$ *of the average* $\frac{1}{2\varepsilon}\int_{-\varepsilon}^{\varepsilon} \boldsymbol{u}^\varepsilon \, dx_3^\varepsilon$ *of the solution to the three-dimensional variational problem* $\mathcal{P}(\Omega^\varepsilon)$ *and of the solution* $\boldsymbol{\zeta}_K^\varepsilon$ *to the two-dimensional Koiter equations formulated as the variational problem* $\mathcal{P}_K^\varepsilon(\omega)$ (Thm. 7.1-1) *are identical.* For *elliptic membrane* and *flexural* shells, we show in addition that the same property holds, separately for the *tangential* components $\frac{1}{2\varepsilon}\int_{-\varepsilon}^{\varepsilon} u_\alpha^\varepsilon \boldsymbol{g}^{\alpha,\varepsilon}\, dx_3^\varepsilon$ and $\zeta^\varepsilon_{K,\alpha}\boldsymbol{a}^\alpha$ and for the *normal* components $\frac{1}{2\varepsilon}\int_{-\varepsilon}^{\varepsilon} u_3^\varepsilon \boldsymbol{g}^{3,\varepsilon}\, dx_3^\varepsilon$ and $\zeta^\varepsilon_{K,3}\boldsymbol{a}^3$ of the associated *displacement fields.*

To this end, we use in an essential way the convergence theorems established in Chaps. 4, 5, and 6, together with slight improvements over former results of Destuynder [1985], Sanchez-Palencia [1989a, 1989b, 1992], and Caillerie & Sanchez-Palencia [1995a] (see also Caillerie [1996]) about the asymptotic behavior of the solution of Koiter's equation as ε approaches zero.

The forthcoming analyses have been recently extended by Xiao Liming [1998, 1999a, 1999b], who likewise justified the *time-dependent Koiter equations* for *elliptic membrane* and *flexural* shells.

To begin with, we consider *elliptic membrane shells.* We follow here Ciarlet & Lods [1996c, Thm. 2.1].

Theorem 7.2-1. *Assume that* $\boldsymbol{\theta} \in \mathcal{C}^3(\overline{\omega}; \mathbb{R}^3)$. *Consider a family of linearly elastic elliptic membrane shells, according to the definition given in* Sect. 4.1, *with thickness* 2ε *approaching zero and with each having the same elliptic middle surface* $S = \boldsymbol{\theta}(\overline{\omega})$, *and let the assumptions on the data be as in* Sect. 4.1; *in particular then,* $\gamma_0 = \gamma$.

For each $\varepsilon > 0$ *let*

$$(u_i^\varepsilon) \in \mathbf{H}^1(\Omega^\varepsilon) \text{ and } \boldsymbol{\zeta}_K^\varepsilon = (\zeta_{i,K}^\varepsilon) \in H_0^1(\omega) \times H_0^1(\omega) \times H_0^2(\omega)$$

respectively denote the solutions to the three-dimensional and two-dimensional variational problems $\mathcal{P}(\Omega^\varepsilon)$ *and* $\mathcal{P}_K^\varepsilon(\omega)$. *Let also*

$$\boldsymbol{\zeta} = (\zeta_i) \in H_0^1(\omega) \times H_0^1(\omega) \times L^2(\omega)$$

denote the solution to the two-dimensional scaled variational problem $\mathcal{P}_M(\omega)$ (Thm. 4.5-1), *solution which is thus independent of* ε. *Then*

$$\boxed{\begin{aligned} \frac{1}{2\varepsilon}\int_{-\varepsilon}^{\varepsilon} u_\alpha^\varepsilon \boldsymbol{g}^{\alpha,\varepsilon}\, dx_3^\varepsilon &\longrightarrow \zeta_\alpha \boldsymbol{a}^\alpha \text{ in } \mathbf{H}^1(\omega) \text{ as } \varepsilon \to 0,\\ \zeta_{K,\alpha}^\varepsilon \boldsymbol{a}^\alpha &\longrightarrow \zeta_\alpha \boldsymbol{a}^\alpha \text{ in } \mathbf{H}^1(\omega) \text{ as } \varepsilon \to 0,\end{aligned}}$$

and

$$\boxed{\begin{aligned} \frac{1}{2\varepsilon}\int_{-\varepsilon}^{\varepsilon} u_3^\varepsilon \boldsymbol{g}^{3,\varepsilon}\, dx_3^\varepsilon &\longrightarrow \zeta_3 \boldsymbol{a}^3 \text{ in } \mathbf{L}^2(\omega) \text{ as } \varepsilon \to 0,\\ \zeta_{3,K}^\varepsilon \boldsymbol{a}^3 &\longrightarrow \zeta_3 \boldsymbol{a}^3 \text{ in } \mathbf{L}^2(\omega) \text{ as } \varepsilon \to 0.\end{aligned}}$$

Proof. Under the assumptions that there exist constants $\lambda > 0$ and $\mu > 0$ and functions $f^i \in L^2(\Omega)$ and $h^i \in L^2(\Gamma_+ \cup \Gamma_-)$ independent of ε such that

$$\begin{aligned} \lambda^\varepsilon = \lambda \quad &\text{and} \quad \mu^\varepsilon = \mu, \\ f^{i,\varepsilon}(x^\varepsilon) = f^i(x) \quad &\text{for all} \quad x^\varepsilon = \pi^\varepsilon x \in \Omega^\varepsilon, \\ h^{i,\varepsilon}(x^\varepsilon) = \varepsilon h^i(x) \quad &\text{for all} \quad x^\varepsilon = \pi^\varepsilon x \in \Gamma_+^\varepsilon \cup \Gamma_-^\varepsilon \end{aligned}$$

(these are the assumptions on the data for a family of linearly elastic elliptic membrane shells; cf. Sect. 4.1) and that $\boldsymbol{\theta} \in \mathcal{C}^3(\overline{\omega}; \mathbb{R}^3)$, the convergences

$$\frac{1}{2\varepsilon} \int_{-\varepsilon}^{\varepsilon} u_\alpha^\varepsilon \boldsymbol{g}^{\alpha,\varepsilon} \, dx_3^\varepsilon \longrightarrow \zeta_\alpha \boldsymbol{a}^\alpha \text{ in } \mathbf{H}^1(\omega)$$

and

$$\frac{1}{2\varepsilon} \int_{-\varepsilon}^{\varepsilon} u_3^\varepsilon \boldsymbol{g}^{3,\varepsilon} \, dx_3^\varepsilon \longrightarrow \zeta_3 \boldsymbol{a}^3 \text{ in } \mathbf{L}^2(\omega)$$

as $\varepsilon \to 0$ have been established in Thm. 4.6-1, as corollaries to the fundamental convergence result of Thm. 4.4-1.

The convergence $\boldsymbol{\zeta}_K^\varepsilon \to \boldsymbol{\zeta}$ in $H^1(\omega) \times H^1(\omega) \times L^2(\omega)$ was first established by Destuynder [1985, Thm. 7.1]; it was also noted by Sanchez-Palencia [1989a, Thm. 4.1] (see also Caillerie & Sanchez-Palencia [1995a]), who observed that it is a consequence of general results in perturbation theory, as found for instance in Sanchez-Palencia [1980]). We give here a simple and "self-contained" proof. Let

$$a^{\alpha\beta\sigma\tau} := \frac{2\lambda\mu}{\lambda + 2\mu} a^{\alpha\beta} a^{\sigma\tau} + 2\mu(a^{\alpha\sigma} a^{\beta\tau} + a^{\alpha\tau} a^{\beta\sigma}),$$

$$B_M(\boldsymbol{\zeta}, \boldsymbol{\eta}) := \int_\omega a^{\alpha\beta\sigma\tau} \gamma_{\sigma\tau}(\boldsymbol{\zeta}) \gamma_{\alpha\beta}(\boldsymbol{\eta}) \sqrt{a} \, dy,$$

$$B_F(\boldsymbol{\zeta}, \boldsymbol{\eta}) := \frac{1}{3} \int_\omega a^{\alpha\beta\sigma\tau} \rho_{\sigma\tau}(\boldsymbol{\zeta}) \rho_{\alpha\beta}(\boldsymbol{\eta}) \sqrt{a} \, dy,$$

$$L(\boldsymbol{\eta}) := \int_\omega p^i \eta_i \sqrt{a} \, dy, \text{ where } p^i := \int_{-1}^{1} f^i \, dx_3 + h_+^i + h_-^i,$$

$$\mathbf{V}_M(\omega) := H_0^1(\omega) \times H_0^1(\omega) \times L^2(\omega),$$

$$\|\boldsymbol{\eta}\|_{\mathbf{V}_M(\omega)} := \Big\{ \sum_\alpha \|\eta_\alpha\|_{1,\omega}^2 + |\eta_3|_{0,\omega}^2 \Big\}^{1/2}.$$

Recall that there exists a constant $c_e > 0$ such that (Thm. 3.3-2)

$$\sum_{\alpha,\beta} |t_{\alpha\beta}|^2 \leq c_e a^{\alpha\beta\sigma\tau}(y) t_{\sigma\tau} t_{\alpha\beta}$$

for all $y \in \overline{\omega}$ and all symmetric matrices $(t_{\alpha\beta})$ and that, according to the *inequality of Korn's type on an elliptic surface* (Thm. 2.7-3), there exists a constant c_M such that

$$\|\boldsymbol{\eta}\|_{\mathbf{V}_M(\omega)} \leq c_M \Big\{ \sum_{\alpha,\beta} |\gamma_{\alpha\beta}(\boldsymbol{\eta})|_{0,\omega}^2 \Big\}^{1/2}$$

for all $\boldsymbol{\eta} = (\eta_i) \in \mathbf{V}_M(\omega)$. Finally, note that the space $\mathbf{V}_K(\omega)$, which here is simply $H_0^1(\omega) \times H_0^1(\omega) \times H_0^2(\omega)$, is contained in the space $\mathbf{V}_M(\omega)$.

By virtue of the assumptions on the applied forces, the solution $\boldsymbol{\zeta}_K^\varepsilon$ of the two-dimensional Koiter equations also satisfies the *scaled Koiter equations for an elliptic membrane shell*, viz.,

$$B_M(\boldsymbol{\zeta}_K^\varepsilon, \boldsymbol{\eta}) + \varepsilon^2 B_F(\boldsymbol{\zeta}_K^\varepsilon, \boldsymbol{\eta}) = L(\boldsymbol{\eta}) \text{ for all } \boldsymbol{\eta} \in \mathbf{V}_K(\omega).$$

Hence letting $\boldsymbol{\eta} = \boldsymbol{\zeta}_K^\varepsilon$ in these scaled equations yields the inequality

$$\frac{1}{c_M^2} \|\boldsymbol{\zeta}_K^\varepsilon\|_{\mathbf{V}_M(\omega)}^2 + \frac{1}{3} \sum_{\alpha,\beta} |\varepsilon \rho_{\alpha\beta}(\boldsymbol{\zeta}_K^\varepsilon)|_{0,\omega}^2 \leq c_e \|L\| \|\boldsymbol{\zeta}_K^\varepsilon\|_{\mathbf{V}_M(\omega)},$$

where

$$\|L\| := \Big\{ \sum_i |p^i|_{0,\omega}^2 \Big\}^{1/2}.$$

This inequality shows that the family $(\boldsymbol{\zeta}_K^\varepsilon)_{\varepsilon>0}$ is bounded in $\mathbf{V}_M(\omega)$ and the families $(\varepsilon \rho_{\alpha\beta}(\boldsymbol{\zeta}_K^\varepsilon))_{\varepsilon>0}$ are bounded in $L^2(\omega)$.

Consequently, there exists a subsequence, still denoted $(\boldsymbol{\zeta}_K^\varepsilon)_{\varepsilon>0}$ for convenience, and there exist a field $\boldsymbol{\zeta}^* \in \mathbf{V}_M(\omega)$ and functions $\rho_{\alpha\beta}^{-1} \in L^2(\omega)$ such that (as usual $\to$ and $\rightharpoonup$ respectively denote strong and weak convergences):

$$\boldsymbol{\zeta}_K^\varepsilon \rightharpoonup \boldsymbol{\zeta}^* \text{ in } \mathbf{V}_M(\omega) \text{ and } \varepsilon \rho_{\alpha\beta}(\boldsymbol{\zeta}_K^\varepsilon) \rightharpoonup \rho_{\alpha\beta}^{-1} \text{ in } L^2(\omega).$$

Fix $\boldsymbol{\eta} \in \mathbf{V}_K(\omega)$ in the scaled Koiter equations and let $\varepsilon \to 0$; then the above weak convergences yield $B_M(\boldsymbol{\zeta}^*, \boldsymbol{\eta}) = L(\boldsymbol{\eta})$. Since the space $\mathbf{V}_K(\omega)$ is dense in $\mathbf{V}_M(\omega)$, we conclude that $B_M(\boldsymbol{\zeta}^*, \boldsymbol{\eta}) = L(\boldsymbol{\eta})$ for all $\boldsymbol{\eta} \in \mathbf{V}_M(\omega)$. Hence

$$\boldsymbol{\zeta}^* = \boldsymbol{\zeta},$$

where $\boldsymbol{\zeta} \in \mathbf{V}_M(\omega)$ is the unique solution to the scaled problem $\mathcal{P}_M(\omega)$ (Thm. 4.5-1), and the weak convergence

$$\boldsymbol{\zeta}_K^\varepsilon \rightharpoonup \boldsymbol{\zeta} \text{ in } \mathbf{V}_M(\omega)$$

then holds for the whole family $(\boldsymbol{\zeta}_K^\varepsilon)_{\varepsilon>0}$.

By the inequality of Korn's type on an elliptic surface and by the positive definiteness of the scaled two-dimensional elasticity tensor of the shell, establishing the strong convergence $\boldsymbol{\zeta}_K^\varepsilon \to \boldsymbol{\zeta}$ in $\mathbf{V}_M(\omega)$ is equivalent to establishing the convergence

$$B_M(\boldsymbol{\zeta}_K^\varepsilon - \boldsymbol{\zeta}, \boldsymbol{\zeta}_K^\varepsilon - \boldsymbol{\zeta}) \to 0.$$

Letting $\boldsymbol{\eta} = \boldsymbol{\zeta}_K^\varepsilon$ in the scaled Koiter equations shows that

$$B_M(\boldsymbol{\zeta}_K^\varepsilon, \boldsymbol{\zeta}_K^\varepsilon) \le L(\boldsymbol{\zeta}_K^\varepsilon).$$

Hence

$$\begin{aligned} 0 &\le B_M(\boldsymbol{\zeta}_K^\varepsilon - \boldsymbol{\zeta}, \boldsymbol{\zeta}_K^\varepsilon - \boldsymbol{\zeta}) \\ &= B_M(\boldsymbol{\zeta}_K^\varepsilon, \boldsymbol{\zeta}_K^\varepsilon) - 2B_M(\boldsymbol{\zeta}_K^\varepsilon, \boldsymbol{\zeta}) + B_M(\boldsymbol{\zeta}, \boldsymbol{\zeta}) \\ &\le L(\boldsymbol{\zeta}_K^\varepsilon) - 2B_M(\boldsymbol{\zeta}_K^\varepsilon, \boldsymbol{\zeta}) + L(\boldsymbol{\zeta}), \end{aligned}$$

since $B_M(\boldsymbol{\zeta}, \boldsymbol{\zeta}) = L(\boldsymbol{\zeta})$. Noting that the weak convergence $\boldsymbol{\zeta}_K^\varepsilon \rightharpoonup \boldsymbol{\zeta}$ in $\mathbf{V}_M(\omega)$ implies that

$$B_M(\boldsymbol{\zeta}_K^\varepsilon, \boldsymbol{\zeta}) \to B_M(\boldsymbol{\zeta}, \boldsymbol{\zeta}) \text{ and } L(\boldsymbol{\zeta}_K^\varepsilon) \to L(\boldsymbol{\zeta}),$$

we have thus shown that $B_M(\boldsymbol{\zeta}_K^\varepsilon - \boldsymbol{\zeta}, \boldsymbol{\zeta}_K^\varepsilon - \boldsymbol{\zeta}) \to 0$, hence that

$$\boldsymbol{\zeta}_K^\varepsilon \to \boldsymbol{\zeta} \text{ in } \mathbf{V}_M(\omega).$$

This convergence clearly implies that

$$\zeta_{K,\alpha}^\varepsilon \boldsymbol{a}^\alpha \to \zeta_\alpha \boldsymbol{a}^\alpha \text{ in } \mathbf{H}^1(\omega) \text{ and } \zeta_{K,3}^\varepsilon \boldsymbol{a}^3 \to \zeta_3 \boldsymbol{a}^3 \text{ in } \mathbf{L}^2(\omega).$$

■

The convergence results of Thm. 7.2-1 have been improved by Mardare [1998b, Thm. 5.1], who showed that

$$\|\boldsymbol{\zeta}_K^\varepsilon - \boldsymbol{\zeta}\|_{H^1(\omega)\times H^1(\omega)\times L^2(\omega)} = O(\varepsilon^{1/5}),$$

and by Lods & Mardare [1998b, 1999c], who showed that

$$\|\frac{1}{2\varepsilon}\int_{-\varepsilon}^{\varepsilon} \boldsymbol{u}^{\varepsilon}\, dx_3^{\varepsilon} - \boldsymbol{\zeta}_K^{\varepsilon}\|_{H^1(\omega)\times H^1(\omega)\times L^2(\omega)} = O(\varepsilon^{1/5}).$$

Remark. Under the assumptions of Thm. 7.2-1, the function $\zeta_{3,K}^{\varepsilon}$ thus "looses its boundary condition" as ε approaches zero. This phenomenon, together with its accompanying *boundary layer*, can be already observed on a simple one-dimensional model of Koiter's equations (Ex. 7.4). As already noted (Sect. 4.6), a similar "loss of boundary condition" is shared by the average $\frac{1}{2\varepsilon}\int_{-\varepsilon}^{\varepsilon} u_3^{\varepsilon}\, dx_3^{\varepsilon}$ as ε approaches zero. ■

We next turn our attention to *generalized membrane shells of the first kind.* We follow here Ciarlet & Lods [1996d, Thms. 6.1 and 6.2].

In the same way that we required the applied forces to be "admissible" (Sect. 5.5) in order to carry out our asymptotic analysis of the solutions of the corresponding three-dimensional equations (Sect. 5.6), we need to assume that *the applied forces enter Koiter's equations* in such a way that the corresponding scaled linear forms are *continuous with respect to the norm* $|\cdot|_{\omega}^{M}$ *of the "limit" space* $\mathbf{V}_M^{\sharp}(\omega)$ and *uniformly so with respect to* ε.

More specifically, we set the following definition (notice the analogy with that given in Sect. 5.5): *Applied forces are* **admissible for the two-dimensional Koiter equations** if there exist functions $\varphi^{\alpha\beta} = \varphi^{\beta\alpha} \in L^2(\omega)$ such that, for each $\varepsilon > 0$, the right-hand side in Koiter's equations can also be written as

$$\boxed{\int_{\omega} p^{i,\varepsilon}\eta_i\sqrt{a}\, dy = \varepsilon\int_{\omega}\varphi^{\alpha\beta}\gamma_{\alpha\beta}(\boldsymbol{\eta})\sqrt{a}\, dy \text{ for all } \boldsymbol{\eta} = (\eta_i) \in \mathbf{V}_K(\omega).}$$

Remark. For simplicity, we assume here that the scaled linear forms (denoted L_M in the next proof) are *independent of* ε. But the conclusions of the next theorem would be unaltered under the more general assumption that, for each $\varepsilon > 0$,

$$\int_{\omega} p^{i,\varepsilon}\eta_i\sqrt{a}\, dy = \varepsilon\int_{\omega}\varphi^{\alpha\beta}(\varepsilon)\gamma_{\alpha\beta}(\boldsymbol{\eta})\sqrt{a}\, dy \text{ for all } \boldsymbol{\eta} = (\eta_i) \in \mathbf{V}_K(\omega),$$

with functions $\varphi^{\alpha\beta}(\varepsilon) \in L^2(\omega)$ converging strongly to $\varphi^{\alpha\beta}$ in $L^2(\omega)$ as ε approaches zero. ■

Theorem 7.2-2. *Assume that $\boldsymbol{\theta} \in \mathcal{C}^3(\overline{\omega}; \mathbb{R}^3)$. Consider a family of linearly elastic generalized membrane shells of the first kind, according to the definitions given in* Sect. 5.1 *and* 5.4, *with thickness 2ε approaching zero, with each having the same middle surface $S = \boldsymbol{\theta}(\overline{\omega})$, with each subjected to a boundary condition of place along a portion of its lateral face having the same set $\boldsymbol{\theta}(\gamma_0)$ as its middle curve, and subjected to applied forces that are admissible for both the three-dimensional equations* (Sect. 5.5) *and the two-dimensional Koiter equations, the functions $\varphi^{\alpha\beta} \in L^2(\omega)$ coinciding in addition with those found in* Thm. 5.6-1.

For each $\varepsilon > 0$, let

$$\boldsymbol{u}^\varepsilon \in \mathbf{H}^1(\Omega^\varepsilon) \text{ and } \boldsymbol{\zeta}_K^\varepsilon \in H^1(\omega) \times H^1(\omega) \times H^2(\omega)$$

respectively denote the solutions to the three-dimensional and two-dimensional variational problems $\mathcal{P}(\Omega^\varepsilon)$ and $\mathcal{P}_K^\varepsilon(\omega)$. Let also

$$\begin{aligned} \boldsymbol{\zeta} \in \mathbf{V}_M^\sharp(\omega) &:= \text{completion of } \mathbf{V}(\omega) \text{ with respect to } |\cdot|_\omega^M, \text{ where} \\ \mathbf{V}(\omega) &:= \{\boldsymbol{\eta} = (\eta_i) \in \mathbf{H}^1(\omega);\ \boldsymbol{\eta} = \mathbf{0} \text{ on } \gamma_0\}, \\ |\boldsymbol{\eta}|_\omega^M &:= \Big\{ \sum_{\alpha,\beta} |\gamma_{\alpha\beta}(\boldsymbol{\eta})|_{0,\omega}^2 \Big\}^{1/2}, \end{aligned}$$

denote the solution to the two-dimensional scaled variational problem $\mathcal{P}_M^\sharp(\omega)$ (Thm. 5.7-1), *solution which is thus independent of ε. Then*

$$\begin{aligned} \frac{1}{2\varepsilon}\int_{-\varepsilon}^{\varepsilon} \boldsymbol{u}^\varepsilon \, dx_3^\varepsilon &\longrightarrow \boldsymbol{\zeta} \text{ in } \mathbf{V}_M^\sharp(\omega) \text{ as } \varepsilon \to 0, \\ \boldsymbol{\zeta}_K^\varepsilon &\longrightarrow \boldsymbol{\zeta} \text{ in } \mathbf{V}_M^\sharp(\omega) \text{ as } \varepsilon \to 0. \end{aligned}$$

Proof. Under the assumptions that $\boldsymbol{\theta} \in \mathcal{C}^3(\overline{\omega}; \mathbb{R}^3)$ and that the applied forces are admissible in the sense of Sect. 5.5, the convergence

$$\frac{1}{2\varepsilon}\int_{-\varepsilon}^{\varepsilon} \boldsymbol{u}^\varepsilon \, dx_3^\varepsilon \longrightarrow \boldsymbol{\zeta} \text{ in } \mathbf{V}_M^\sharp(\omega)$$

as $\varepsilon \to 0$ has been established in Thm. 5.8-1, as a corollary to the fundamental convergence result of Thm. 5.6-1.

The rest of the proof is an elaboration over Caillerie & Sanchez-Palencia [1995a, Thm. 4.5], who established the *weak* convergence $\boldsymbol{\zeta}_K^\varepsilon \rightharpoonup \boldsymbol{\zeta}$ in $\mathbf{V}_M^\sharp(\omega)$ as $\varepsilon \to 0$. In particular, we establish here that this convergence is *strong*.

Since the space $\mathbf{V}_K(\omega)$ is dense in the space $\mathbf{V}(\omega)$ with respect to the norm $\|\cdot\|_{1,\omega}$ and there exists c such that $|\boldsymbol{\eta}|_\omega^M \leq c\|\boldsymbol{\eta}\|_{1,\omega}$ for all $\boldsymbol{\eta} \in \mathbf{V}(\omega)$, the space $\mathbf{V}_K(\omega)$ is dense in $\mathbf{V}(\omega)$ with respect to $|\cdot|_\omega^M$ and thus *the space* $\mathbf{V}_M^\sharp(\omega)$ *is also the completion of* $\mathbf{V}_K(\omega)$ *with respect to* $|\cdot|_\omega^M$.

Let $B_M^\sharp$ and $L_M^\sharp$ denote the unique continuous extensions from $\mathbf{V}(\omega)$ to $\mathbf{V}_M^\sharp(\omega)$ of the bilinear and linear forms B_M and L_M defined by

$$B_M(\boldsymbol{\zeta}, \boldsymbol{\eta}) := \int_\omega a^{\alpha\beta\sigma\tau}\gamma_{\sigma\tau}(\boldsymbol{\zeta})\gamma_{\alpha\beta}(\boldsymbol{\eta})\sqrt{a}\,dy,$$

$$L_M(\boldsymbol{\eta}) := \int_\omega \varphi^{\alpha\beta}\gamma_{\alpha\beta}(\boldsymbol{\eta})\sqrt{a}\,dy.$$

The applied forces being admissible for the two-dimensional Koiter equations, their solution $\boldsymbol{\zeta}_K^\varepsilon$ satisfies the *scaled Koiter equations for a generalized membrane shell*, viz.,

$$B_M(\boldsymbol{\zeta}_K^\varepsilon, \boldsymbol{\eta}) + \varepsilon^2 B_F(\boldsymbol{\zeta}_K^\varepsilon, \boldsymbol{\eta}) = L_M(\boldsymbol{\eta}) \text{ for all } \boldsymbol{\eta} \in \mathbf{V}_K(\omega),$$

where

$$B_F(\boldsymbol{\zeta}, \boldsymbol{\eta}) := \frac{1}{3}\int_\omega a^{\alpha\beta\sigma\tau}\rho_{\sigma\tau}(\boldsymbol{\zeta})\rho_{\alpha\beta}(\boldsymbol{\eta})\sqrt{a}\,dy.$$

Hence letting $\boldsymbol{\eta} = \boldsymbol{\zeta}_K^\varepsilon$ in these scaled equations yields the inequality (the constant c_e stems from the positive definiteness of the scaled two-dimensional elasticity tensor of the shell; cf. Thm. 3.3-2):

$$\left(|\boldsymbol{\zeta}_K^\varepsilon|_\omega^M\right)^2 + \frac{1}{3}\sum_{\alpha,\beta}|\varepsilon\rho_{\alpha\beta}(\boldsymbol{\zeta}_K^\varepsilon)|_{0,\omega}^2 \leq c_e\|L_M\||\boldsymbol{\zeta}_K^\varepsilon|_\omega^M,$$

where $\|L_M\| := \{\sum_{\alpha,\beta}|\varphi^{\alpha\beta}|_{0,\omega}^2\}^{1/2}$. This shows that the family $(\boldsymbol{\zeta}_K^\varepsilon)_{\varepsilon>0}$ is bounded in the space $\mathbf{V}_M^\sharp(\omega)$ and the families $(\varepsilon\rho_{\alpha\beta}(\boldsymbol{\zeta}_K^\varepsilon))_{\varepsilon>0}$ are bounded in $L^2(\omega)$.

Consequently, there exists a subsequence, still denoted $(\boldsymbol{\zeta}_K^\varepsilon)_{\varepsilon>0}$ for convenience, and there exist $\boldsymbol{\zeta}^* \in \mathbf{V}_M^\sharp(\omega)$ and $\rho_{\alpha\beta}^{-1} \in L^2(\omega)$ such that (as usual $\rightharpoonup$ denotes weak convergence):

$$\boldsymbol{\zeta}_K^\varepsilon \rightharpoonup \boldsymbol{\zeta}^* \text{ in } \mathbf{V}_M^\sharp(\omega) \text{ and } \varepsilon\rho_{\alpha\beta}(\boldsymbol{\zeta}_K^\varepsilon) \rightharpoonup \rho_{\alpha\beta}^{-1} \text{ in } L^2(\omega).$$

Fix $\boldsymbol{\eta} \in \mathbf{V}_K(\omega)$ in the scaled Koiter equations and let $\varepsilon \to 0$; then the above weak convergences yield $B_M^\sharp(\boldsymbol{\zeta}^*, \boldsymbol{\eta}) = L_M(\boldsymbol{\eta})$. Since $\mathbf{V}_K(\omega)$ is dense in $\mathbf{V}_M^\sharp(\omega)$, we conclude that $B_M^\sharp(\boldsymbol{\zeta}^*, \boldsymbol{\eta}) = L_M^\sharp(\boldsymbol{\eta})$ for all $\boldsymbol{\eta} \in \mathbf{V}_M^\sharp(\omega)$. Hence

$$\boldsymbol{\zeta}^* = \boldsymbol{\zeta},$$

where $\boldsymbol{\zeta} \in \mathbf{V}_M^\sharp(\omega)$ is the unique solution to the scaled problem $\mathcal{P}_M^\sharp(\omega)$ (Thm. 5.7-1), and the weak convergence

$$\boldsymbol{\zeta}_K^\varepsilon \rightharpoonup \boldsymbol{\zeta} \text{ in } \mathbf{V}_M^\sharp(\omega)$$

then holds for the whole family $(\boldsymbol{\zeta}_K^\varepsilon)_{\varepsilon>0}$.

By the positive definiteness of the scaled two-dimensional elasticity tensor of the shell and by the definitions of the norm $|\cdot|_\omega^M$ and of the bilinear form B_M and of its extension $B_M^\sharp$, establishing the strong convergence $\boldsymbol{\zeta}_K^\varepsilon \to \boldsymbol{\zeta}$ in $\mathbf{V}_M^\sharp(\omega)$ is equivalent to establishing the convergence

$$B_M^\sharp(\boldsymbol{\zeta}_K^\varepsilon - \boldsymbol{\zeta}, \boldsymbol{\zeta}_K^\varepsilon - \boldsymbol{\zeta}) \to 0.$$

Letting $\boldsymbol{\eta} = \boldsymbol{\zeta}_K^\varepsilon$ in the scaled Koiter equations shows that

$$B_M(\boldsymbol{\zeta}_K^\varepsilon, \boldsymbol{\zeta}_K^\varepsilon) \leq L_M(\boldsymbol{\zeta}_K^\varepsilon).$$

Hence

$$\begin{aligned}
0 &\leq B_M^\sharp(\boldsymbol{\zeta}_K^\varepsilon - \boldsymbol{\zeta}, \boldsymbol{\zeta}_K^\varepsilon - \boldsymbol{\zeta}) \\
&= B_M(\boldsymbol{\zeta}_K^\varepsilon, \boldsymbol{\zeta}_K^\varepsilon) - 2B_M^\sharp(\boldsymbol{\zeta}_K^\varepsilon, \boldsymbol{\zeta}) + B_M^\sharp(\boldsymbol{\zeta}, \boldsymbol{\zeta}) \\
&\leq L_M(\boldsymbol{\zeta}_K^\varepsilon) - 2B_M^\sharp(\boldsymbol{\zeta}_K^\varepsilon, \boldsymbol{\zeta}) + L_M^\sharp(\boldsymbol{\zeta}),
\end{aligned}$$

since $B_M^\sharp(\boldsymbol{\zeta}, \boldsymbol{\zeta}) = L_M^\sharp(\boldsymbol{\zeta})$. Noting that the weak convergence $\boldsymbol{\zeta}_K^\varepsilon \rightharpoonup \boldsymbol{\zeta}$ in $\mathbf{V}_M^\sharp(\omega)$ implies that

$$B_M^\sharp(\boldsymbol{\zeta}_K^\varepsilon, \boldsymbol{\zeta}) \to B_M^\sharp(\boldsymbol{\zeta}, \boldsymbol{\zeta}) \text{ and } L_M^\sharp(\boldsymbol{\zeta}_K^\varepsilon) \to L_M^\sharp(\boldsymbol{\zeta}),$$

we have thus shown that

$$\boldsymbol{\zeta}_K^\varepsilon \to \boldsymbol{\zeta} \text{ in } \mathbf{V}_M^\sharp(\omega),$$

and the proof is complete. ∎

Remark. The kind of erratic behavior that may be expected when the applied forces cease to be admissible is displayed by an ingenious one-dimensional example, due to Sanchez-Hubert & Sanchez-Palencia [1993, p. 55] and Leguillon, Sanchez-Hubert & Sanchez-Palencia [1999]; cf. Ex. 7.4. ■

Finally, we consider *flexural shells.* We follow here Ciarlet & Lods [1996c, Thm. 2.2].

Theorem 7.2-3. *Assume that* $\boldsymbol{\theta} \in \mathcal{C}^3(\overline{\omega}; \mathbb{R}^3)$. *Consider a family of linearly elastic flexural shells, according to the definition given in* Sect. 6.1, *with thickness* 2ε *approaching zero, with each having the same middle surface* $S = \boldsymbol{\theta}(\overline{\omega})$ *and with each subjected to a boundary condition of place along a portion of its lateral face having the same set* $\boldsymbol{\theta}(\gamma_0)$ *as its middle curve, and let the assumptions on the data be as in* Sect. 6.1.

For each $\varepsilon > 0$, *let*

$$(u_i^\varepsilon) \in \mathbf{H}^1(\Omega^\varepsilon) \text{ and } \boldsymbol{\zeta}_K^\varepsilon = (\zeta_{i,K}^\varepsilon) \in H^1(\omega) \times H^1(\omega) \times H^2(\omega)$$

respectively denote the solutions to the three-dimensional and two-dimensional variational problems $\mathcal{P}(\Omega^\varepsilon)$ *and* $\mathcal{P}_K^\varepsilon(\omega)$. *Let also*

$$\boldsymbol{\zeta} = (\zeta_i) \in H^1(\omega) \times H^1(\omega) \times H^2(\omega)$$

denote the solution to the two-dimensional scaled variational problem $\mathcal{P}_F(\omega)$ (Thm. 6.3-1), *solution which is thus independent of* ε. *Then*

$$\boxed{\begin{aligned} \frac{1}{2\varepsilon}\int_{-\varepsilon}^{\varepsilon} u_\alpha^\varepsilon \boldsymbol{g}^{\alpha,\varepsilon}\, dx_3^\varepsilon &\longrightarrow \zeta_\alpha \boldsymbol{a}^\alpha \text{ in } \mathbf{H}^1(\omega) \text{ as } \varepsilon \to 0, \\ \zeta_{\alpha,K}^\varepsilon \boldsymbol{a}^\alpha &\longrightarrow \zeta_\alpha \boldsymbol{a}^\alpha \text{ in } \mathbf{H}^1(\omega) \text{ as } \varepsilon \to 0, \end{aligned}}$$

and

$$\boxed{\begin{aligned} \frac{1}{2\varepsilon}\int_{-\varepsilon}^{\varepsilon} u_3^\varepsilon \boldsymbol{g}^{3,\alpha}\, dx_3^\varepsilon &\longrightarrow \zeta_3 \boldsymbol{a}^3 \text{ in } \mathbf{H}^1(\omega) \text{ as } \varepsilon \to 0, \\ \zeta_{3,K}^\varepsilon \boldsymbol{a}^3 &\longrightarrow \zeta_3 \boldsymbol{a}^3 \text{ in } \mathbf{H}^2(\omega) \text{ as } \varepsilon \to 0. \end{aligned}}$$

Proof. Under the assumptions that there exist constants $\lambda > 0$ and $\mu > 0$ and functions $f^i \in L^2(\Omega)$ and $h^i \in L^2(\Gamma_+ \cup \Gamma_-)$ independent of ε such that

$$\lambda^\varepsilon = \lambda \quad \text{and} \quad \mu^\varepsilon = \mu,$$
$$f^{i,\varepsilon}(x^\varepsilon) = \varepsilon^2 f^i(x) \text{ for all } x^\varepsilon = \pi^\varepsilon x \in \Omega^\varepsilon,$$
$$h^{i,\varepsilon}(x^\varepsilon) = \varepsilon^3 h^i(x) \text{ for all } x^\varepsilon = \pi^\varepsilon x \in \Gamma_+^\varepsilon \cup \Gamma_-^\varepsilon$$

(these are the assumptions on the data for a family of linearly elastic flexural shells; cf. Sect. 6.1) and that $\boldsymbol{\theta} \in \mathcal{C}^3(\overline{\omega}; \mathbb{R}^3)$, the convergences

$$\frac{1}{2\varepsilon}\int_{-\varepsilon}^{\varepsilon} u_\alpha^\varepsilon \boldsymbol{g}^{\alpha,\varepsilon}\, dx_3^\varepsilon \longrightarrow \zeta_\alpha \boldsymbol{a}^\alpha \text{ in } \mathbf{H}^1(\omega)$$

and

$$\frac{1}{2\varepsilon}\int_{-\varepsilon}^{\varepsilon} u_3^\varepsilon \boldsymbol{g}^{3,\varepsilon}\, dx_3^\varepsilon \longrightarrow \zeta_3 \boldsymbol{a}^3 \text{ in } \mathbf{H}^1(\omega)$$

as $\varepsilon \to 0$ have been established in Thm. 6.4-1, as corollaries to the fundamental convergence result of Thm. 6.2-1.

The *weak* convergence $\boldsymbol{\zeta}_K^\varepsilon \rightharpoonup \boldsymbol{\zeta}$ in $H^1(\omega)\times H^1(\omega)\times H^2(\omega)$ was first established by Sanchez-Palencia [1989a, Thm. 2.1], as a consequence of general results in perturbation theory. We show that, in fact, the *strong* convergence $\boldsymbol{\zeta}_K^\varepsilon \to \boldsymbol{\zeta}$ in $\mathbf{V}_K(\omega)$ holds. In addition to the notation introduced in the proof of Thm. 7.2-1, let

$$\mathbf{V}_F(\omega) := \{\boldsymbol{\eta} \in \mathbf{V}_K(\omega);\ \gamma_{\alpha\beta}(\boldsymbol{\eta}) = 0 \text{ in } \omega\} \subset \mathbf{V}_K(\omega),$$
$$\|\boldsymbol{\eta}\|_{\mathbf{V}_K(\omega)} := \Big\{\sum_\alpha \|\eta_\alpha\|_{1,\omega}^2 + \|\eta_3\|_{2,\omega}^2\Big\}.$$

Recall that, according to the *inequality of Korn's type on a general surface* (Thm. 2.6-4), there exists a constant c such that

$$\|\boldsymbol{\eta}\|_{\mathbf{V}_K(\omega)} \le c\Big\{\sum_{\alpha,\beta} |\gamma_{\alpha\beta}(\boldsymbol{\eta})|_{0,\omega}^2 + \sum_{\alpha,\beta} |\rho_{\alpha\beta}(\boldsymbol{\eta})|_{0,\omega}^2\Big\}^{1/2}$$

for all $\boldsymbol{\eta} \in \mathbf{V}_K(\omega)$ (by assumption, *length* $\gamma_0 > 0$).

By virtue of the assumptions on the applied forces, the solution $\boldsymbol{\zeta}_K^\varepsilon$ also satisfies the *scaled Koiter equations for a flexural shell* (not to be confused with those for a membrane shell introduced in the proof of Thm. 7.2-1!), viz.,

$$\frac{1}{\varepsilon^2}B_M(\boldsymbol{\zeta}_K^\varepsilon, \boldsymbol{\eta}) + B_F(\boldsymbol{\zeta}_K^\varepsilon, \boldsymbol{\eta}) = L(\boldsymbol{\eta}) \text{ for all } \boldsymbol{\eta} \in \mathbf{V}_K(\omega).$$

Hence letting $\boldsymbol{\eta} = \boldsymbol{\zeta}_K^\varepsilon$ in these scaled equations shows that (without loss of generality, we may assume $\varepsilon \leq 1$):

$$\frac{1}{3c^2}\|\boldsymbol{\zeta}_K^\varepsilon\|^2_{\mathbf{V}_K(\omega)} \leq \sum_{\alpha,\beta} |\frac{1}{\varepsilon}\gamma_{\alpha\beta}(\boldsymbol{\zeta}_K^\varepsilon)|^2_{0,\omega} + \frac{1}{3}\sum_{\alpha,\beta} |\rho_{\alpha\beta}(\boldsymbol{\zeta}_K^\varepsilon)|^2_{0,\omega}$$

$$\leq c_e\|L\|\|\boldsymbol{\zeta}_K^\varepsilon\|_{\mathbf{V}_K(\omega)}.$$

Consequently, there exists a subsequence, still denoted $(\boldsymbol{\zeta}_K^\varepsilon)_{\varepsilon>0}$ for convenience, and there exists a function $\boldsymbol{\zeta}^* \in \mathbf{V}_K(\omega)$ such that

$$\boldsymbol{\zeta}_K^\varepsilon \rightharpoonup \boldsymbol{\zeta}^* \text{ in } \mathbf{V}_K(\omega) \text{ and } \gamma_{\alpha\beta}(\boldsymbol{\zeta}_K^\varepsilon) \to 0 \text{ in } L^2(\omega).$$

The weak convergence $\boldsymbol{\zeta}_K^\varepsilon \rightharpoonup \boldsymbol{\zeta}^*$ in $\mathbf{V}_K(\omega)$ implies the weak convergences $\gamma_{\alpha\beta}(\boldsymbol{\zeta}^\varepsilon) \rightharpoonup \gamma_{\alpha\beta}(\boldsymbol{\zeta}^*)$ in $L^2(\omega)$; hence $\gamma_{\alpha\beta}(\boldsymbol{\zeta}^*) = 0$ and thus $\boldsymbol{\zeta}^* \in \mathbf{V}_F(\omega)$. Fix $\boldsymbol{\eta} \in \mathbf{V}_F(\omega)$ in the scaled Koiter equations and let $\varepsilon \to 0$; then the weak convergence $\boldsymbol{\zeta}_K^\varepsilon \rightharpoonup \boldsymbol{\zeta}^*$ in $\mathbf{V}_K(\omega)$ yields $B_F(\boldsymbol{\zeta}^*, \boldsymbol{\eta}) = L(\boldsymbol{\eta})$. Hence

$$\boldsymbol{\zeta}^* = \boldsymbol{\zeta},$$

where $\boldsymbol{\zeta} \in \mathbf{V}_F(\omega)$ is the unique solution to the scaled problem $\mathcal{P}_F(\omega)$ (Thm. 6.3-1) and the weak convergence then holds for the whole family $(\boldsymbol{\zeta}^\varepsilon)_{\varepsilon>0}$.

By the inequality of Korn's type on a general surface combined with the strong convergence $\gamma_{\alpha\beta}(\boldsymbol{\zeta}_K^\varepsilon) \to 0$ in $L^2(\omega)$ and the relations $\gamma_{\alpha\beta}(\boldsymbol{\zeta}) = 0$, and by the positive definiteness of the scaled two-dimensional elasticity tensor of the shell, establishing the strong convergence $\boldsymbol{\zeta}_K^\varepsilon \to \boldsymbol{\zeta}$ in $\mathbf{V}_K(\omega)$ is equivalent to establishing the convergence

$$B_F(\boldsymbol{\zeta}_K^\varepsilon - \boldsymbol{\zeta}, \boldsymbol{\zeta}_K^\varepsilon - \boldsymbol{\zeta}) \to 0.$$

Letting $\boldsymbol{\eta} = \boldsymbol{\zeta}_K^\varepsilon$ in the scaled Koiter equations shows that

$$B_F(\boldsymbol{\zeta}_K^\varepsilon, \boldsymbol{\zeta}_K^\varepsilon) \leq L(\boldsymbol{\zeta}_K^\varepsilon).$$

Hence

$$\begin{aligned} 0 &\leq B_F(\boldsymbol{\zeta}_K^\varepsilon - \boldsymbol{\zeta}, \boldsymbol{\zeta}_K^\varepsilon - \boldsymbol{\zeta}) \\ &= B_F(\boldsymbol{\zeta}_K^\varepsilon, \boldsymbol{\zeta}_K^\varepsilon) - 2B_F(\boldsymbol{\zeta}_K^\varepsilon, \boldsymbol{\zeta}) + B_F(\boldsymbol{\zeta}, \boldsymbol{\zeta}) \\ &\leq L(\boldsymbol{\zeta}_K^\varepsilon) - 2B_F(\boldsymbol{\zeta}_K^\varepsilon, \boldsymbol{\zeta}) + L(\boldsymbol{\zeta}), \end{aligned}$$

since $B_F(\boldsymbol{\zeta}, \boldsymbol{\zeta}) = L(\boldsymbol{\zeta})$. Noting that the weak convergence $\boldsymbol{\zeta}_K^\varepsilon \rightharpoonup \boldsymbol{\zeta}$ in $\mathbf{V}_K(\omega)$ implies that

$$B_F(\boldsymbol{\zeta}_K^\varepsilon, \boldsymbol{\zeta}) \to B_F(\boldsymbol{\zeta}, \boldsymbol{\zeta}) \text{ and } L(\boldsymbol{\zeta}_K^\varepsilon) \to L(\boldsymbol{\zeta}),$$

we have thus shown that $B_F(\boldsymbol{\zeta}_K^\varepsilon - \boldsymbol{\zeta}, \boldsymbol{\zeta}_K^\varepsilon - \boldsymbol{\zeta}) \to 0$, hence that

$$\boldsymbol{\zeta}_K^\varepsilon \to \boldsymbol{\zeta} \text{ in } \mathbf{V}_K(\omega).$$

This convergence clearly implies that

$$\zeta_{K,\alpha}^\varepsilon \boldsymbol{a}^\alpha \to \zeta_\alpha \boldsymbol{a}^\alpha \text{ in } \mathbf{H}^1(\omega) \text{ and } \zeta_{K,3}^\varepsilon \boldsymbol{a}^3 \to \zeta_3 \boldsymbol{a}^3 \text{ in } \mathbf{H}^2(\omega).$$

■

Remarks. (1) Lods [1997] has identified a specific class of applied forces for which the *error estimate* $\|\boldsymbol{\zeta}_K^\varepsilon - \boldsymbol{\zeta}\|_{\mathbf{V}_K(\omega)} = o(\varepsilon)$ upgrades the convergence $\boldsymbol{\zeta}_K^\varepsilon \to \boldsymbol{\zeta}$ in $\mathbf{V}_K(\omega)$ established in Thm. 7.2-3.

(2) For shells subjected to a boundary condition of place along their entire lateral face, Lods & Mardare [1998b, 1999c] have obtained *error estimates* between the scaled three-dimensional solution and the three-dimensional extension across the thickness, by means of the linearized Kirchhoff-Love assumption (Sect. 7.1), of the solution of Koiter's equations. ■

A *major conclusion* emerging from Thms. 7.2-1 to 7.2-3 is that *we have justified the two-dimensional linear Koiter equations* when they are applied either to an *elliptic membrane shell*, or to a *generalized membrane shell of the first kind* subjected to *admissible applied forces*, or to a *flexural shell.* For Thms. 7.2-1 to 7.2-3 show that the averages across the thickness of the three-dimensional solution and the solution of Koiter's equations have the *same principal part*, viz., in each case the solution $\boldsymbol{\zeta}$ to the corresponding two-dimensional scaled problem, *as the thickness approaches zero.*

Remarks. (1) Generalized membrane shells of the *second* kind are not covered by the present analysis. This restriction is compensated by the apparent lack of examples of such shells.

(2) Koiter's equations retain another property of the three-dimensional equations: Their solution converges to the solution of the two-dimensional equations of a *linearly elastic plate* when "the shell becomes a plate" (Ex. 3.5).

(3) The asymptotic analysis of the *eigenvalue problem for Koiter's equations* as ε approaches zero is extensively studied in Sanchez-Hubert & Sanchez-Palencia [1997, Chap. 10]. See also Sanchez-Palencia [1992], Sanchez-Palencia & Vassiliev [1992]. Special men-

tion must be also made in this direction of the pioneering works of Goldenveizer, Lidski & Tovstik [1979] and Vassiliev [1987, 1990]. ■

By virtue of the de-scalings, which are in each case of the form $\boldsymbol{\zeta}^\varepsilon = \boldsymbol{\zeta}$ (see Sects. 4.5, 5.7, and 6.3), our asymptotic analyses also show that *the solution* $\boldsymbol{\zeta}_K^\varepsilon$ *of Koiter's equations is asymptotically as good as the solution* $\boldsymbol{\zeta}^\varepsilon$ *obtained by solving either the two-dimensional problem* $\mathcal{P}_M^\varepsilon(\omega)$, *or the two-dimensional problem* $\mathcal{P}_M^{\sharp\varepsilon}(\omega)$, *or the two-dimensional problem* $\mathcal{P}_F^\varepsilon(\omega)$ (see Thms. 4.5-2, 5.7-2, and 6.3-2), *according to which category the shell falls into.*

Compared to these limit two-dimensional equations, Koiter's equations thus possess *two outstanding advantages: Not only does using Koiter's equations avoid a "preliminary" knowledge of the category in which a given shell problem falls into, but it also avoids the mathematical or numerical difficulties inherent to each such category,* briefly summarized below (see Chaps. 4 to 6 for more details).

If the shell is an *elliptic membrane* one, no boundary condition can be imposed on the normal component ζ_3^ε of the displacement field since ζ_3^ε is "only" in $L^2(\omega)$!

If the shell is a *generalized membrane* one, the solution $\boldsymbol{\zeta}^\varepsilon$ belongs to an "abstract" completion $\mathbf{V}_M^\sharp(\omega)$; the boundary conditions on $\boldsymbol{\zeta}^\varepsilon$ may thus be quite "exotic"!

If the shell is a *flexural* one, the unknown $\boldsymbol{\zeta}^\varepsilon$ is subjected to the constraints $\gamma_{\alpha\beta}(\boldsymbol{\zeta}^\varepsilon) = 0$ in ω, which certainly hinder its numerical approximation!

A word of caution. These conclusions could not be reached by an asymptotic analysis of Koiter's equations alone, for they definitely require an asymptotic analysis of the three-dimensional equations, which was the object of Chaps. 4 to 6. *Cela va sans dire certes, mais cela va encore mieux en le disant*! ■

These truly remarkable virtues of Koiter's equations certainly explain their "improbable success" in the manifold numerical simulations where they are so often blithely used.

7.3. KOITER'S EQUATIONS: ADDITIONAL COMMENTARY AND BIBLIOGRAPHICAL NOTES

As indicated in Sect. 7.1, W.T. Koiter based his derivation on two *a priori* assumptions, one of a *mechanical* nature stemming from delicate estimates of John [1965, 1971], and one of a *geometrical* nature, the *Kirchhoff-Love assumption.* The *linearized* version of this geometrical assumption (its explicit form is given in Ex. 7.1(2)) has been *a posteriori* justified for linearly elastic *elliptic membrane* shells by Lods & Mardare [1998b, 1999c]. Their justification relies on earlier estimates of Mardare [1998a, 1998b] and on an *error estimate* between the three-dimensional solution and the three-dimensional extension across the thickness, by means of the linearized Kirchhoff-Love assumption, of the displacement field obtained by solving Koiter's equations; this error estimate is valid for any shell subjected to a boundary condition of place along its entire lateral face.

As shown in Sect. 7.2, the solution of Koiter's equations possesses the virtue of mimicking the behavior of the three-dimensional solution as the thickness approaches zero. This virtue certainly explains why they are so often deemed adequate for modeling and approximating a variety of situations where three-dimensional equations would be otherwise untractable.

For instance, the modeling of linearly elastic shells with *periodically varying thickness* can be accomplished by an *homogenization of Koiter's equations* (cf. Telega & Lewiński [1998a, 1998b]; for illuminating introductions to homogenization theory in general, see the books by Sanchez-Hubert & Sanchez-Palencia [1992] and by Cioranescu & Donato [1999]); shells made of *anisotropic* and *nonhomogeneous* linearly elastic materials can be modeled by *ad hoc* extensions of Koiter's equations, where additional terms couple the linearized change of metric and change of curvature tensors (Caillerie & Sanchez-Palencia [1995a], Figueiredo & Leal [1998]); Koiter's equations are likewise well-suited for the mathematical modeling and numerical simulation of *junctions between linearly elastic shells* (see Bernadou & Cubier [1998a, 1998b]; an asymptotic analysis of the three-dimensional equations in the spirit of Ciarlet, Le Dret & Nzengwa [1989] and Le Dret [1989, 1990], or of Kozlov, Maz'ya & Movchan [1999], is yet to be carried out, however).

The *controllability* of a shell modeled by the *time-dependent Koiter equations*, in the form justified by Xiao Li-ming [1999b], has been established by Miara & Valente [1999], who showed that the *Hilbert uniqueness method* ("HUM") of Lions [1988a, 1988b, 1988c] could be applied if the middle surface meets *ad hoc* geometrical conditions (satisfied in particular if the shell is shallow; cf. Sect. 7.6). In this direction, see also Geymonat, Loreti & Valente [1993], Lasiecka, Triggiani & Valente [1996].

Koiter's equations are also well adapted for *shape optimization problems*. Typical problems consist in determining the "best" geometry of the shell (i.e., the surface S) or the "best" variation of its thickness, so as to minimize the total weight under *ad hoc* constraints (e.g., unilateral boundary conditions of place, or bounds on the thickness). In this direction, see Chenais [1987, 1994] (who considered the closely related Budiansky-Sanders equations; cf. Sect. 7.5), Bernadou, Palma & Rousselet [1991], Rao & Sokolowski [1996], Khludnev & Sokolowski [1997], Sokolowski [1997], Lewiński & Sokolowski [1998].

Engineers and experts in computational mechanics often base their *classification of shells* on the *relative orders of magnitudes of the "membrane" and "flexural" strain energies*, viz.,

$$\frac{\varepsilon}{2}\int_\omega a^{\alpha\beta\sigma\tau,\varepsilon}\gamma_{\sigma\tau}(\boldsymbol{\zeta}_K^\varepsilon)\gamma_{\alpha\beta}(\boldsymbol{\zeta}_K^\varepsilon)\sqrt{a}\,dy$$

and

$$\frac{\varepsilon^3}{6}\int_\omega a^{\alpha\beta\sigma\tau,\varepsilon}\rho_{\sigma\tau}(\boldsymbol{\zeta}_K^\varepsilon)\rho_{\alpha\beta}(\boldsymbol{\zeta}_K^\varepsilon)\sqrt{a}\,dy,$$

found in *Koiter's energy* j_K^ε (Sect. 7.1) evaluated at a *given* solution $\boldsymbol{\zeta}_K^\varepsilon$ (see, e.g., Leino & Pitkäranta [1994], Chapelle & Bathe [1998a]), rather than on an asymptotic analysis of the three-dimensional solution as here. This approach, in which the *applied forces* may thus also dictate either a *"membrane-dominated"*, or a *"flexural-dominated", behavior*, has been recently given a mathematical basis by Blouza, Brezzi & Lovadina [1999].

In the same spirit, Koiter's equations have been often applied to *benchmark problems* for testing the influence of the geometry of the middle surface and of the boundary conditions for determining the behavior of shells; see Kirmse [1993], Piila & Pitkäranta [1993a, 1993b, 1995], Piila [1994, 1996], Pitkäranta, Leino, Ovaskainen & Piila [1995]. Such benchmark problems are also of interest for identifying and approximating *boundary layers* in shells; see Hakula & Pitkäranta [1995], Hakula [1997], Gerdes, Matache & Schwab [1998].

By contrast with "boundary" layers, *"interior" layers*, i.e., "away from the lateral face", may appear inside shells with a *hyperbolic middle surface*. This challenging phenomenon seems to be again well modeled by Koiter's equations, as suggested by Sanchez-Palencia & Sanchez-Hubert [1998]. See also Karamian [1998a], Leguillon, Sanchez-Hubert & Sanchez-Palencia [1999], Pitkäranta, Matache & Schwab [1999].

Koiter's equations are often favored for the numerical simulation of shells with "small" thickness, Naghdi's equations (Sect. 7.4) being preferred for that of shells with "moderate" thickness. Accordingly, there exist myriad references on the *numerical analysis of shell problems* modeled by either kind of equations, the list given below constituting only a brief sample.

In particular, the book by Bernadou [1994] is highly recommended, as its treatment follows essentially the same lines as this chapter. For the same reason, the "numerical" chapters of the books by Destuynder [1990] and Sanchez-Hubert & Sanchez-Palencia [1997] and the articles by Bernadou, Mato-Eiroa & Trouvé [1994], Sanchez-Palencia [1995] and Chapelle & Bathe [1998a, 1998b, 1999] are likewise highly recommended. Of particular interest is also the approach of Le Tallec & Mani [1998] (see also Carrive [1995], Carrive, Le Tallec & Mouro [1995], Kerdid & Mato-Eiroa [1998]), where the *Cartesian* components of the displacement, rather than its covariant components, are approximated; in this case, the solution to be approximated is instead that of *Koiter's equations for shells whose middle surface has little regularity*, due to Blouza & Le Dret [1999] (Thm. 7.1-2).

Unless some care is exercised, the numerical approximation of shell problems is bound to be blurred by the *locking phenomenon*, first identified by Stolarski & Belytschko [1982], then defined by Babuška & Suri [1992] as the lack of uniformity of convergence with respect to ε as the discretization parameter approaches zero. While this phenomenon is thus inherent to the approximation of any physical problem with a "small" parameter, an even more serious phenomenon, called *membrane locking*, is specific to flexural shells: It reflects the inability of most finite element spaces to approximate "sufficiently well" the space $\mathbf{V}_F(\omega)$ of linearized inextensional displacements where the solution of the limit problem ultimately lies in this case (Thms. 6.4-1 and 7.2-3).

In view of their acute importance, these phenomena have generated a substantial literature; see in particular Kamoukalos [1991], Akian & Sanchez-Palencia [1992], Habbal & Chenais [1992], Pitkäranta [1992], Chenais & Zerner [1993], Leino & Pitkäranta [1994], Zerner [1994], Bathe [1996], Chapelle [1996], Hakula, Leino & Pitkäranta [1996], Ovaskainen & Pitkäranta [1996], Chapelle [1997], Sanchez-Hubert & Sanchez-Palencia [1997, Chap. 11], Chapelle & Bathe [1998a], Choi, Palma, Sanchez-Palencia & Vilariño [1998], Chapelle & Stenberg [1998]. Further references are provided at the end of the next section.

7.4. THE TWO-DIMENSIONAL NAGHDI EQUATIONS FOR A LINEARLY ELASTIC SHELL; EXISTENCE AND UNIQUENESS OF SOLUTIONS

Another two-dimensional linear shell theory, which appeals as much to the engineering community as Koiter's, is due to P.M. Naghdi. In this approach, which seems to be quite effective in numerical simulations of shells with "moderate" thicknesses, the shell is identified with a *one-director Cosserat surface*, i.e., a surface endowed with a *director* field (the field denoted $r_\alpha^\varepsilon \boldsymbol{a}^\alpha$ *infra*). In this sense, it is the analog for shells of the Reissner-Mindlin theory for plates (Vol. II, Sect. 1.9).

Conclusive attempts to establish the superiority of such theories over Koiter's theory for shells and Kirchhoff-Love's theory for plates are scarce. See, however, Lods & Mardare [1999a] for flexural shells and specific applied forces, Lods & Mardare [1999b] for shells subjected to a boundary condition of place along their entire lateral face, and Rössle [1999a] for plates.

Consider as in Sect. 7.1 a shell with middle surface $S = \boldsymbol{\theta}(\overline{\omega})$ and thickness $2\varepsilon > 0$, constituted by a homogeneous and isotropic linear elastic material with Lamé constants $\lambda^\varepsilon > 0$ and $\mu^\varepsilon > 0$, and subjected to applied body and surface forces with contravariant components $f^{i,\varepsilon} \in L^2(\Omega^\varepsilon)$ and $h^{i,\varepsilon} \in L^2(\Gamma_+^\varepsilon \cup \Gamma_-^\varepsilon)$.

In *Naghdi's approach* (Naghdi [1963, 1972]), the *a priori* assumption of a *mechanical* nature about the *stresses* inside the shell is the same as in Koiter's approach (Sect. 7.1), but the *a priori* assumption of a *geometrical* nature is different: The points situated on a line normal to S remain on a line and the lengths are unmodified along this line after the deformation has taken place, as in Koiter's

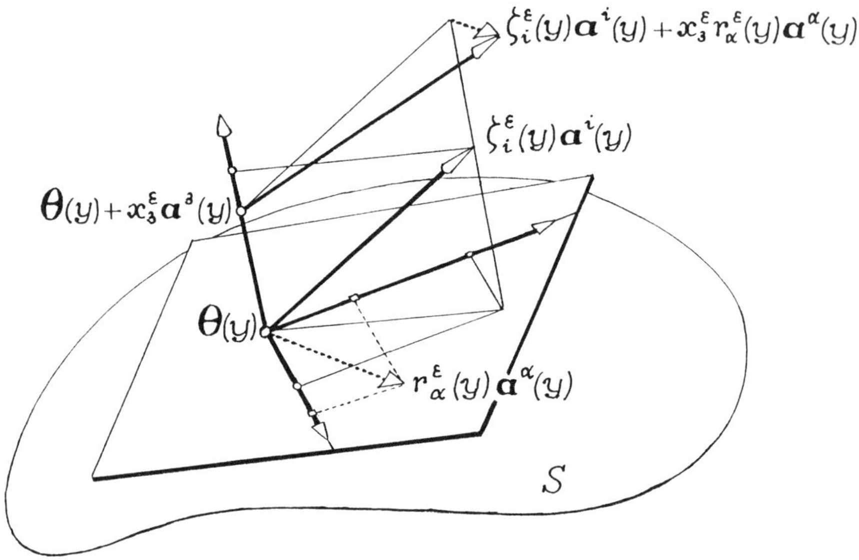

Fig. 7.4-1: *A linearly elastic shell modeled by Naghdi's two-dimensional equations.* The five unknowns are the three covariant components $\zeta_i^\varepsilon : \overline{\omega} \to \mathbb{R}$ of the displacement field of the middle surface S and the two covariant components $r_\alpha^\varepsilon : \overline{\omega} \to \mathbb{R}$ of the linearized rotation field of the unit normal vector along S; this means that $\zeta_i^\varepsilon(y)\boldsymbol{a}^i(y) + x_3^\varepsilon r_\alpha^\varepsilon(y)\boldsymbol{a}^\alpha(y)$ is the displacement of the point $(\boldsymbol{\theta}(y) + x_3^\varepsilon\boldsymbol{a}^3(y))$.

approach; however, *this line need no longer remain normal to the deformed middle surface.*

In the *linearized* version of this approach, there are *five* unknowns, the three covariant components $\zeta_i^\varepsilon : \overline{\omega} \to \mathbb{R}$ of the *displacement field* $\zeta_i^\varepsilon\boldsymbol{a}^i$ *of the middle surface* S and the two covariant components $r_\alpha^\varepsilon : \overline{\omega} \to \mathbb{R}$ of the *linearized rotation field* $r_\alpha^\varepsilon\boldsymbol{a}^\alpha$ *of the unit normal vector along* S (see Ex. 7.6 for the reason this field should be tangential in a linearized theory). This means that the displacement of the point $(\boldsymbol{\theta}(y) + x_3^\varepsilon\boldsymbol{a}^3(y))$ is the vector $(\zeta_i^\varepsilon(y)\boldsymbol{a}^i(y) + x_3^\varepsilon r_\alpha^\varepsilon(y)\boldsymbol{a}^\alpha(y))$; cf. Fig. 7.4-1.

In their weak formulation, **Naghdi's equations for a linearly**

elastic shell consist in solving the following *variational problem* $\mathcal{P}_N^\varepsilon(\omega)$:

Find $(\boldsymbol{\zeta}^\varepsilon, \boldsymbol{r}^\varepsilon) = ((\zeta_i^\varepsilon), (r_\alpha^\varepsilon))$ such that (the notation $\mathbf{H}^1(\Omega)$ stands for the space $(H^1(\omega))^5$ in the definition of the space $\mathbf{V}_N(\omega)$):

$$(\boldsymbol{\zeta}^\varepsilon, \boldsymbol{r}^\varepsilon) \in \mathbf{V}_N(\omega) := \{(\boldsymbol{\eta}, \boldsymbol{s}) = ((\eta_i), (s_\alpha)) \in \mathbf{H}^1(\omega);\ \eta_i = s_\alpha = 0 \text{ on } \gamma_0\},$$

$$\varepsilon \int_\omega \left\{ a^{\alpha\beta\sigma\tau,\varepsilon} \gamma_{\sigma\tau}(\boldsymbol{\zeta}^\varepsilon)\gamma_{\alpha\beta}(\boldsymbol{\eta}) + c\mu^\varepsilon a^{\alpha\beta}\gamma_{\alpha 3}(\boldsymbol{\zeta}^\varepsilon, \boldsymbol{r}^\varepsilon)\gamma_{\beta 3}(\boldsymbol{\eta}, \boldsymbol{s}) \right\} \sqrt{a}\, dy$$
$$+ \frac{\varepsilon^3}{3} \int_\omega a^{\alpha\beta\sigma\tau,\varepsilon} \rho_{\sigma\tau}^N(\boldsymbol{\zeta}^\varepsilon, \boldsymbol{r}^\varepsilon)\rho_{\alpha\beta}^N(\boldsymbol{\eta}, \boldsymbol{s})\sqrt{a}\, dy = \int_\omega p^{i,\varepsilon}\eta_i \sqrt{a}\, dy$$
$$\text{for all } (\boldsymbol{\eta}, \boldsymbol{s}) \in \mathbf{V}_N(\omega),$$

where (the functions $a^{\alpha\beta}$, $b_{\alpha\beta}$, b_α^σ, $\Gamma_{\alpha\beta}^\sigma$, and a are defined as usual; see, e.g., Sect. 3.1):

$$a^{\alpha\beta\sigma\tau,\varepsilon} := \frac{4\lambda^\varepsilon\mu^\varepsilon}{\lambda^\varepsilon + 2\mu^\varepsilon} a^{\alpha\beta}a^{\sigma\tau} + 2\mu^\varepsilon(a^{\alpha\sigma}a^{\beta\tau} + a^{\alpha\tau}a^{\beta\sigma}),$$
$$\gamma_{\alpha\beta}(\boldsymbol{\eta}) := \frac{1}{2}(\partial_\beta\eta_\alpha + \partial_\alpha\eta_\beta) - \Gamma_{\alpha\beta}^\sigma\eta_\sigma - b_{\alpha\beta}\eta_3,$$
$$\gamma_{\alpha 3}(\boldsymbol{\eta}, \boldsymbol{s}) := \frac{1}{2}(\partial_\alpha\eta_3 + b_\alpha^\sigma\eta_\sigma + s_\alpha),$$
$$\rho_{\alpha\beta}^N(\boldsymbol{\eta}, \boldsymbol{s}) := -\frac{1}{2}(\partial_\beta s_\alpha + \partial_\alpha s_\beta) + \Gamma_{\alpha\beta}^\sigma s_\sigma - b_\alpha^\sigma b_{\sigma\beta}\eta_3$$
$$+\frac{1}{2}b_\alpha^\sigma(\partial_\beta\eta_\sigma - \Gamma_{\beta\sigma}^\tau\eta_\tau) + \frac{1}{2}b_\beta^\tau(\partial_\alpha\eta_\tau - \Gamma_{\alpha\tau}^\sigma\eta_\sigma),$$
$$p^{i,\varepsilon} := \int_{-\varepsilon}^{\varepsilon} f^{i,\varepsilon}\, dx_3^\varepsilon + h_+^{i,\varepsilon} + h_-^{i,\varepsilon} \text{ and } h_\pm^{i,\varepsilon} := h^{i,\varepsilon}(\cdot, \pm\varepsilon),$$

and c is a strictly positive constant ($c = 8$ is commonly assumed; however, the "best" choice for this constant seems to be an unresolved issue).

The functions $a^{\alpha\beta\sigma\tau,\varepsilon}$ are the contravariant components of the **two-dimensional elasticity tensor of the shell** and the functions $\gamma_{\alpha\beta}(\boldsymbol{\eta})$ are the covariant components of the **linearized change of metric tensor** associated with a displacement field $\eta_i\boldsymbol{a}^i$ of the middle surface S, as before. The new functions $\gamma_{\alpha 3}(\boldsymbol{\eta}, \boldsymbol{s})$ and $\rho^N_{\alpha\beta}(\boldsymbol{\eta}, \boldsymbol{s})$ are the covariant components of the **linearized transverse shear strain tensor** and **Naghdi's linearized change of curvature tensor** associated with displacement and linearized rotation fields $\eta_i \boldsymbol{a}^i$ and $s_\alpha \boldsymbol{a}^\alpha$ of S; for a justification of these definitions, see, e.g., Bernadou [1994, Part I, Chap. 3].

The next *existence* and *uniqueness* result for the solution to the variational problem $\mathcal{P}^\varepsilon_N(\omega)$ is due to Bernadou, Ciarlet & Miara [1994, Thm. 3.1]. As its proof is akin to that of Thm. 7.1-1, it is left as a problem (Ex. 7.7).

Theorem 7.4-1. *Let ω be a domain in $\mathbb{R}^2$, let γ_0 be a subset of $\partial\omega$ with* length $\gamma_0 > 0$, *and let $\boldsymbol{\theta} \in \mathcal{C}^3(\overline{\omega}; \mathbb{R}^3)$ be an injective mapping such that the two vectors $\boldsymbol{a}_\alpha = \partial_\alpha \boldsymbol{\theta}$ are linearly independent at all points of $\overline{\omega}$.*

Then problem $\mathcal{P}^\varepsilon_N(\omega)$ has one and only one solution. ∎

The variational problem $\mathcal{P}^\varepsilon_N(\omega)$ is, at least formally, equivalent to a *boundary value problem*, which is again a *"uniformly"* and *"strongly elliptic"* system in the sense of Agmon, Douglis & Nirenberg [1964]. Accordingly, the *regularity* of its solution can be established as the regularity of the solution of Koiter's equations (Ex. 7.3); see Iosifescu [1999].

In the same manner that Blouza & Le Dret [1994a, 1994b, 1999] have generalized Thm. 7.1-1 to "Koiter's equations for shells whose middle surface has little regularity" (Thm. 7.1-2), Blouza [1997] has extended Thm. 7.4-1 to *Naghdi's equations for shells whose middle surface has little regularity* (the mapping $\boldsymbol{\theta}$ need only be in the space $W^{2,\infty}(\omega; \mathbb{R}^3)$).

Nonhomogeneous and *anisotropic* linearly elastic materials are likewise amenable to Naghdi's approach. The existence and uniqueness of the corresponding two-dimensional Naghdi's equations can be treated as in Thm. 7.3-1; see Figueiredo & Leal [1998]. The corresponding *infinitesimal rigid displacement lemma* is due to Coutris [1978].

Naghdi's equations are as much favored as Koiter's for analyzing the *locking phenomenon* and *membrane locking*, briefly described in Sect. 7.3; see Hughes & Franca [1988], Bathe, Brezzi & Fortin [1989], Pitkäranta [1992], Lyly, Stenberg & Vihinen [1993], Arnold & Brezzi [1997a, 1997b], Bramble & Sun [1997], Suri [1997], Gerdes, Matache & Schwab [1998], Chapelle & Bathe [1998a], Chapelle & Stenberg [1998].

7.5. OTHER LINEAR SHELL THEORIES

Linear "shallow" shell theories are treated separately (Sect. 7.6).

Budiansky-Sanders theory. Sanders [1959] and Koiter [1960] have proposed a linear shell theory akin to Koiter's, where the covariant components $\rho_{\alpha\beta}(\boldsymbol{\eta})$ of the linearized change of curvature tensor are replaced by the covariant components $\rho^{BS}_{\alpha\beta}(\boldsymbol{\eta})$ of the **Budiansky-Sanders linearized change of curvature tensor**, defined by

$$\rho^{BS}_{\alpha\beta}(\boldsymbol{\eta}) = \rho_{\alpha\beta}(\boldsymbol{\eta}) - \frac{1}{2}(b^{\sigma}_{\alpha}\gamma_{\sigma\beta}(\boldsymbol{\eta}) + b^{\tau}_{\beta}\gamma_{\tau\alpha}(\boldsymbol{\eta})).$$

The remaining terms in the equations are otherwise identical to those in Koiter's equations (Sect. 7.1).

The interest of using the modified functions $\rho^{BS}_{\alpha\beta}(\boldsymbol{\eta})$, rather than the "genuine" functions $\rho_{\alpha\beta}(\boldsymbol{\eta})$, has been discussed at length in Budiansky & Sanders [1967] and, for this reason, the resulting theory has become known as the *Budiansky-Sanders theory* (see also the commentary in Koiter [1970, Sect. 7.5]).

More recently, Destuynder [1985] has clearly shown how this theory can be derived from three-dimensional linearized elasticity, again on the basis of two *a priori* assumptions, *both* of a *geometrical* nature (one of them being the linearized Kirchoff-Love assumption; cf. Sect. 7.1), thus providing a mathematical justification of this theory.

Remarks. (1) A *nonlinear Budiansky-Sanders theory*, proposed by Destuynder [1982, 1985], is briefly presented in Sect. 11.2.

(2) The definition of the functions $\rho^{BS}_{\alpha\beta}(\boldsymbol{\eta})$ and the equivalence between the equations

$$\text{“}\gamma_{\alpha\beta}(\boldsymbol{\eta}) = \rho^{BS}_{\alpha\beta}(\boldsymbol{\eta}) = 0 \text{ in } \omega\text{” and “}\gamma_{\alpha\beta}(\boldsymbol{\eta}) = \rho_{\alpha\beta}(\boldsymbol{\eta}) = 0 \text{ in } \omega\text{”}$$

together imply that the *existence* and *uniqueness* theory for Koiter's equations (Sects. 2.6 and 7.1) extends almost *verbatim* to the Budiansky-Sanders equations. ■

Hierarchic shell theories. The underlying principle in a *hierarchic shell theory* consists in minimizing the three-dimensional energy of a linearly elastic shell, expressed in terms of its "natural" curvilinear coordinates $(y_1, y_2, x_3^\varepsilon)$, over a subspace of *displacement fields of a special form:* Each one of their covariant components is a finite sum of products of functions of (y_1, y_2) (to be determined through the minimization process) times linearly independent functions of x_3^ε (typically, *ad hoc* orthogonal polynomials). Increasing the number of linearly independent functions of the transverse variable produces a "hierarchy" of presumably increasingly accurate approximations. Due credit should be given in this respect to Vekua [1986, Chap. 1], for one of the earliest attempts to put such an approach for shells on a sound mathematical basis.

Hierarchic theories are especially advocated as efficient *numerical methods*: In particular, they have proved to be quite successful for the *numerical approximation* of *plate problems*; see Szabo & Sahrmann [1988], Babuška & Li [1991], Babuška, Szabo & Actis [1992], Babuška, d'Harcourt & Schwab [1993], Schwab [1994, 1996], Schwab & Wright [1995], Babuška & Schwab [1996] . They will certainly be increasingly used for the *numerical approximation of shell problems*, notably for their abilities to handle *multi-layered shells*, also called *composite* or *laminated* shells, *boundary layers*, or the *membrane locking* phenomenon (Sect. 7.3); see notably Oden & Cho [1996], Actis, Szabo & Schwab [1999].

Linear shell theories based on the method of internal constraints. The linearized Kirchhoff-Love assumption (Sect. 7.1; see also Ex. 7.1(2)) can be considered as an *internal constraint* imposed *ab initio* on the displacements that a shell can undergo.

In this approach, first advocated by Podio-Guidugli [1989] for linearly elastic *plates*, then by Podio-Guidugli [1990] for linearly elastic *shells*, special care must be exercised for the determination of the resulting two-dimensional constitutive equation, through a careful distinction between *"reactive"* and *"active"* stresses.

"Intrinsic" linear shell theory. An "intrinsic" approach to linear shell theory has been proposed by Delfour & Zolésio [1995] for modeling shells whose middle surface is the entire boundary Γ of a domain $\Omega \subset \mathbb{R}^3$. This approach makes an essential use of the "tangential differential calculus" developed in the book of Sokolowski

& Zolésio [1992] and of the "oriented distance function" $\Delta : \mathbb{R}^3 \to \mathbb{R}$, defined by $\Delta(x) = d(x, \Gamma)$ if $x \in \{\mathbb{R}^3 - \Omega\}$ and $\Delta(x) = -d(x, \Gamma)$ if $x \in \Omega$, also used in Delfour & Zolésio [1994, 1998] in a different context.

After integrating the three-dimensional energy across the thickness and making *ad hoc* simplifying assumptions, M.C. Delfour and J.P. Zolésio derive linear shell theories that bear strong resemblance with those of W.T. Koiter and P.M. Naghdi. An interesting feature of this approach is that local bases are not used.

This approach has since then undergone manifold extensions, which often parallel the asymptotic analyses of the previous chapters. See notably Delfour & Zolésio [1997a, 1997b], Delfour [1998, 1999].

7.6. LINEAR SHALLOW SHELL THEORIES

Linear shallow shell theories in *Cartesian* coordinates are treated at length in Vol. II, Chap. 3. Accordingly, the additional commentary and bibliographical notes found in this section apply mostly to linear shallow shell theories expressed in *curvilinear* coordinates.

According to the definition proposed by Ciarlet & Paumier [1986] in the nonlinear case, then justified by a convergence theorem as the thickness approaches zero by Ciarlet & Miara [1992a] in the linear case, a shell is **shallow** if the deviation of its middle surface S^ε from a plane is of the order of the thickness, i.e., if

$$\begin{aligned}
&S^\varepsilon = \boldsymbol{\theta}^\varepsilon(\overline{\omega}), \text{ where} \\
&\boldsymbol{\theta}^\varepsilon(y_1, y_2) = (y_1, y_2, \varepsilon\theta(y_1, y_2)) \text{ for all } (y_1, y_2) \in \overline{\omega},
\end{aligned}$$

and $\theta : \overline{\omega} \to \mathbb{R}$ is a smooth enough function that is *independent of* ε.

A word of caution. This specific "variation of the middle surface with ε" thus constitutes an additional *assumption on the data*, special to linear and nonlinear shallow shell theory. ■

As shown by Busse, Ciarlet & Miara [1997], who use the same definition of "shallowness", the *two-dimensional equations of a linearly elastic clamped shallow shell "in curvilinear coordinates"* can be given a rigorous justification by means of a *convergence theorem* as the thickness goes to zero. As the proof essentially resembles that given in *Cartesian coordinates* by Ciarlet & Miara [1992a] (see also Vol. II, Sect. 3.5), it is not reproduced in this volume. We simply list the limit equations that are found in this fashion, when they are expressed as a *minimization* problem: Let

$$b^{\alpha\beta\sigma\tau,\varepsilon} := \frac{4\lambda^\varepsilon\mu^\varepsilon}{\lambda^\varepsilon + 2\mu^\varepsilon}\delta^{\alpha\beta}\delta^{\sigma\tau} + 2\mu^\varepsilon(\delta^{\alpha\sigma}\delta^{\beta\tau} + \delta^{\alpha\tau}\delta^{\beta\sigma}),$$

$$p^{i,\varepsilon} := \int_{-\varepsilon}^{\varepsilon} f^{i,\varepsilon}\,dx_3^\varepsilon + h_+^{i,\varepsilon} + h_-^{i,\varepsilon},$$

$$q^{\alpha,\varepsilon} := \int_{-\varepsilon}^{\varepsilon} x_3^\varepsilon f^{\alpha,\varepsilon}\,dx_3^\varepsilon + \varepsilon(h_+^{\alpha,\varepsilon} - h_-^{\alpha,\varepsilon}),$$

$$e_{\alpha\beta}^{sh,\varepsilon}(\boldsymbol{\eta}) := \frac{1}{2}(\partial_\beta\eta_\alpha + \partial_\alpha\eta_\beta) - \varepsilon\eta_3\partial_{\alpha\beta}\theta,$$

and let $\boldsymbol{a}^{i,\varepsilon}$ designate the vectors of the contravariant bases along the middle surface S^ε. Then the *unknown* is the vector field $\boldsymbol{\zeta}^\varepsilon = (\zeta_i^\varepsilon)$, where the functions $\zeta_i^\varepsilon : \overline{\omega} \to \mathbb{R}$ are the covariant components of the displacement field $\zeta_i^\varepsilon \boldsymbol{a}^{i,\varepsilon}$ of S^ε, and $\boldsymbol{\zeta}^\varepsilon$ minimizes the energy $j^{sh,\varepsilon}$ defined by

$$j^{sh,\varepsilon}(\boldsymbol{\eta}) := \frac{1}{2}\int_\omega \Big\{\varepsilon b^{\alpha\beta\sigma\tau,\varepsilon} e_{\sigma\tau}^{sh,\varepsilon}(\boldsymbol{\eta}) e_{\alpha\beta}^{sh,\varepsilon}(\boldsymbol{\eta}) + \frac{\varepsilon^3}{3} b^{\alpha\beta\sigma\tau,\varepsilon}\partial_{\sigma\tau}\eta^3\partial_{\alpha\beta}\eta^3\Big\}\,dy$$
$$-\Big\{\int_\omega p^{i,\varepsilon}\eta_i\,dy - \int_\omega q^{\alpha,\varepsilon}\eta_\alpha\,dy\Big\},$$

over the space (the same as for Koiter's equations; cf. Sect. 7.1):

$$\mathbf{V}_K(\omega) := \{\boldsymbol{\eta} = (\eta_i) \in H^1(\omega)\times H^1(\omega)\times H^2(\omega);\ \eta_i = \partial_\nu\eta_3 = 0 \text{ on } \gamma_0\}.$$

An inspection of this minimization problem reveals that, even though it is expressed in curvilinear coordinates, it *"resembles more that of plate* (Vol. II, Chap. 1) *than that of shell"*! For the contravariant components of the metric tensor usually found in the two-dimensional elasticity tensor are now replaced by Kronecker deltas, the area element along the middle surface is replaced by dy, and finally, the components of the linearized change of metric and change of curvature tensors are replaced by the functions $e_{\alpha\beta}^{sh,\varepsilon}(\boldsymbol{\eta})$ and $\partial_{\alpha\beta}\eta^3$, where neither the Christoffel symbols nor any components of the curvature tensor of S^ε are to be found.

The equations found in this fashion constitute *Novozhilov's model of a shallow shell*, so named after Novozhilov [1959]. These equations, which were first analyzed by Shoiket [1974], were given a first justification by Destuynder [1980] for special geometries.

As shown by Andreoiu [1999a], it is a reassuring circumstance that the limit displacement fields found in either Cartesian or curvilinear coordinates, though not identical vector fields, are nevertheless "essentially the same", i.e., their components agree "to within their first orders", once they are expressed in a same basis.

The asymptotic analysis of the corresponding *eigenvalue problem* has been carried out in Cartesian coordinates by Kesavan & Sabu [1999b]; there is no doubt that it could be similarly carried out in curvilinear coordinates.

The exponential nature of the *boundary layers* that arise in linearly elastic shallow shells is analyzed in Pitkäranta, Matache & Schwab [1999].

Models of *multi-layered*, or *composite*, *linearly elastic shallow shells*, found in particular in hulls of sailboats, have been obtained by Kail [1994] by means of the method of formal asymptotic expansions.

The *control of vibrations* in linearly elastic shallow shells by means of piezo-ceramic actuators is studied in detail in Banks, Smith & Wang [1995, 1996].

Other definitions of "shallowness" have been proposed, which often make explicit reference to the *curvature* of the middle surface. For instance, Destuynder [1985, Sect. 1] considers that a shell is "shallow" if $\eta = \varepsilon^p$ for some $p \geq 2$, where the other "small" parameter η is the ratio of the thickness 2ε to the smallest absolute value of the radii of curvature along the middle surface, $p = 2$ corresponding to Novozhilov's model. In this direction, see also Vekua [1965], Green & Zerna [1968, p. 400], Gordeziani [1974], Dikmen [1982, p. 158].

EXERCISES

7.1. All the notations used in this problem should be self-explanatory. The *Kirchhoff-Love assumption for a shell* states that all points lying initially on a given normal to the middle surface remain on a single normal to the deformed middle surface and that the distance from any point of the shell to the middle surface remains unchanged.

Consider a shell with a reference configuration $\boldsymbol{\Theta}(\overline{\Omega}^\varepsilon)$ equipped with its "natural" curvilinear coordinates $(y_1, y_2, x_3^\varepsilon) \in \overline{\Omega}^\varepsilon$ and let $\boldsymbol{g}^{i,\varepsilon}$ denote the vectors of the corresponding contravariant bases in $\boldsymbol{\Theta}(\overline{\Omega}^\varepsilon)$, where $\Omega^\varepsilon = \omega \times]-\varepsilon, \varepsilon[$. Thus a displacement field $v_i^\varepsilon \boldsymbol{g}^{i,\varepsilon}$ of the set $\boldsymbol{\Theta}(\overline{\Omega}^\varepsilon)$ satisfies the Kirchhoff-Love assumption if and only if

$$v_i^\varepsilon \boldsymbol{g}^{i,\varepsilon} = \eta_i \boldsymbol{a}^i + x_3^\varepsilon \{\boldsymbol{a}_3(\boldsymbol{\eta}) - \boldsymbol{a}_3\},$$

where $\eta_i \boldsymbol{a}^i$ denotes the restriction to $x_3^\varepsilon = 0$ of the field $v_i^\varepsilon \boldsymbol{g}^{i,\varepsilon}$ and $\boldsymbol{a}_3(\boldsymbol{\eta})$ is the "usual" vector normal to the deformed middle surface $(\boldsymbol{\theta} + \eta_i \boldsymbol{a}^i)(\overline{\omega})$.

(1) Show that a displacement field $v_i^\varepsilon \boldsymbol{g}^{i,\varepsilon}$ of $\boldsymbol{\Theta}(\overline{\Omega}^\varepsilon)$ satisfies the Kirchhoff-Love assumption if and only if the covariant components $g_{ij}^\varepsilon(\boldsymbol{v}^\varepsilon)$ of the metric tensor of the associated deformed configuration, assumed to be equipped with the same curvilinear coordinates as those of $\boldsymbol{\Theta}(\overline{\Omega}^\varepsilon)$, are of the form

$$g_{\alpha\beta}^\varepsilon(\boldsymbol{v}^\varepsilon) = a_{\alpha\beta}(\boldsymbol{\eta}) - 2x_3^\varepsilon b_{\alpha\beta}(\boldsymbol{\eta}) + (x_3^\varepsilon)^2 b_\alpha^\sigma(\boldsymbol{\eta}) b_{\beta\sigma}(\boldsymbol{\eta}) \text{ and } g_{i3}^\varepsilon(\boldsymbol{v}^\varepsilon) = \delta_{i3},$$

where $a_{\alpha\beta}(\boldsymbol{\eta})$, $b_{\alpha\beta}(\boldsymbol{\eta})$, and $b_\alpha^\sigma(\boldsymbol{\eta})$ denote the covariant and mixed components of the metric and curvature tensors of the deformed middle surface $(\boldsymbol{\theta} + \eta_i \boldsymbol{a}^i)(\overline{\omega})$.

(2) A displacement field $v_i^\varepsilon \boldsymbol{g}^{i,\varepsilon}$ of $\boldsymbol{\Theta}(\overline{\Omega}^\varepsilon)$ satisfies the *linearized Kirchhoff-Love assumption* if only the linear terms with respect to $\boldsymbol{\eta}$ are retained in the difference $\{\boldsymbol{a}_3(\boldsymbol{\eta}) - \boldsymbol{a}_3\}$. Show that such a displacement is of the form (see, e.g., Bernadou & Boisserie [1982, eq. (1.3.19)]):

$$v_i^\varepsilon \boldsymbol{g}^{i,\varepsilon} = \eta_i \boldsymbol{a}^i - x_3^\varepsilon (\partial_\alpha \eta_3 + b_\alpha^\beta \eta_\beta) \boldsymbol{a}^\alpha.$$

7.2. Assume that the boundary γ of ω and the functions $p^{i,\varepsilon}$ are smooth enough and that *length* $\gamma_0 <$ *length* γ. Show that, if the solution $\boldsymbol{\zeta}_K^\varepsilon = (\zeta_{K,i}^\varepsilon)$ to the variational problem $\mathcal{P}_K^\varepsilon(\omega)$ (Thm. 7.1-1) is smooth enough, it is also a solution of the following boundary value problem:

$$m^{\alpha\beta,\varepsilon}|_{\alpha\beta} - b_\alpha^\sigma b_{\sigma\beta} m^{\alpha\beta,\varepsilon} - b_{\alpha\beta} n^{\alpha\beta,\varepsilon} = p^{3,\varepsilon} \text{ in } \omega,$$

$$-(n^{\alpha\beta,\varepsilon} + b_\sigma^\alpha m^{\sigma\beta,\varepsilon})|_\beta - b_\sigma^\alpha (m^{\sigma\beta,\varepsilon}|_\beta) = p^{\alpha,\varepsilon} \text{ in } \omega,$$

$$\zeta_{i,K}^\varepsilon = \partial_\nu \zeta_{3,K}^\varepsilon = 0 \text{ on } \gamma_0,$$

$$m^{\alpha\beta,\varepsilon} \nu_\alpha \nu_\beta = 0 \text{ on } \gamma_1,$$

$$(m^{\alpha\beta,\varepsilon}|_\alpha)\nu_\beta + \partial_\tau (m^{\alpha\beta,\varepsilon} \nu_\alpha \tau_\beta) = 0 \text{ on } \gamma_1,$$

$$(n^{\alpha\beta,\varepsilon} + 2b_\sigma^\alpha m^{\sigma\beta,\varepsilon})\nu_\beta = 0 \text{ on } \gamma_1,$$

where $\gamma_1 := \gamma - \gamma_0$, (ν_α) is the unit outer normal vector along γ, $\tau_1 := -\nu_2$, $\tau_2 := \nu_1$, and $\partial_\tau \theta := \tau_\alpha \partial_\alpha \theta$ denotes the tangential derivative of θ in the direction of the vector (τ_α).

Remark. It is instructive to compare these equations with those of a *linearly elastic plate* (Vol. II, Thms. 1.5-1, 1.5-2, and 1.7-2).

7.3. The following problem shows, after Alexandrescu [1994], how to establish the *regularity* (announced in Thm. 7.1-3 (b)) of the solution of the *two-dimensional Koiter equations for a linearly elastic shell.* It is assumed throughout this problem that $\gamma_0 = \gamma$, and in (1), (2), and (3) that the boundary γ is of class $\mathcal{C}^4$ and the mapping $\boldsymbol{\theta}$ is in the space $\mathcal{C}^4(\overline{\omega}; \mathbb{R}^3)$.

(1) Show that the linear system of partial differential equations found in Thm. 7.1-3, which is of the second order with respect to the unknowns $\zeta_{K,\alpha}^\varepsilon$ and of the fourth order with respect to the unknown $\zeta_{K,3}^\varepsilon$, is *"uniformly elliptic"* and satisfies the *"supplementing condition on L"* and the *"complementing boundary condition"* in the sense of Agmon, Douglis & Nirenberg [1964].

Remark. These properties are shared by the systems of partial differential equations found in Thm. 2.7-2 and in Ex. 4.2.

(2) Show that the same system is *"strongly elliptic"* in the sense of Nečas [1967, p. 185]. Using Nečas [1967, Lemma 3.2, p. 260], infer from this property that the solution $\boldsymbol{\zeta}_K^\varepsilon = (\zeta_{K,i}^\varepsilon)$ to the variational problem $\mathcal{P}_K^\varepsilon(\omega)$ (Thm. 7.1-1), which belongs to the space $H_0^1(\omega) \times H_0^1(\omega) \times H_0^2(\omega)$ since $\gamma_0 = \gamma$ by assumption, satisfies

$$\boldsymbol{\zeta}_K^\varepsilon = (\zeta_{K,i}^\varepsilon) \in H^3(\omega) \times H^3(\omega) \times H^4(\omega)$$

if $p^{\alpha,\varepsilon} \in H^1(\omega)$ and $p^{3,\varepsilon} \in L^2(\omega)$.

(3) Using Geymonat [1965, Thm. 3.5], shows that

$$\zeta_K^\varepsilon = (\zeta_{K,i}^\varepsilon) \in W^{3,q}(\omega) \times W^{3,q}(\omega) \times W^{4,q}(\omega)$$

if $p^{\alpha,\varepsilon} \in W^{1,q}(\omega)$ and $p^{3,\varepsilon} \in L^q(\omega)$ for some $q > 1$.

(4) Assume that, for some integer $m \geq 1$ and some real number $q > 1$, γ is of class $\mathcal{C}^{m+4}$, $\boldsymbol{\theta} \in \mathcal{C}^{m+4}(\overline{\omega}; \mathbb{R}^3)$, $p^{\alpha,\varepsilon} \in W^{m+1,q}(\omega)$, and $p^{3,\varepsilon} \in W^{m,q}(\omega)$. Show that

$$\zeta_K^\varepsilon = (\zeta_{K,i}^\varepsilon) \in W^{m+3,q}(\omega) \times W^{m+3,q}(\omega) \times W^{m+4,q}(\omega).$$

Hint: For (4), use Agmon, Douglis & Nirenberg [1964].

7.4. As shown by Sanchez-Hubert & Sanchez-Palencia [1993, p. 55] (see also Leguillon, Sanchez-Hubert & Sanchez-Palencia [1999]), the following simple, yet illuminating, two-point boundary value problem can be viewed as a convenient *one-dimensional model of Koiter's equations*, as regards in particular the behavior of their solution as ε approaches zero in the situations covered by Thms. 7.2-1 and 7.2-2. The space $H^{-m}(0, 1)$ denotes the dual space of $H_0^m(0, 1)$.

Given $f \in H^{-2}(0, 1)$, let $u^\varepsilon \in H_0^2(0, 1)$ denote for each $\varepsilon > 0$ the solution to the boundary value problem:

$$\begin{cases} \varepsilon^2 \dfrac{d^4 u^\varepsilon}{dx^4} - \dfrac{d^2 u^\varepsilon}{dx^2} = f \text{ in }]0, 1[, \\ u^\varepsilon(0) = (u^\varepsilon)'(0) = u^\varepsilon(1) = (u^\varepsilon)'(1) = 0. \end{cases}$$

(1) Assume that $f \in L^2(0, 1)$. Show that $u^\varepsilon \to u^0$ in $H_0^1(0, 1)$ as $\varepsilon \to 0$, where $u^0 \in H_0^1(0, 1)$ is the solution to the boundary value problem

$$\begin{cases} -\dfrac{d^2 u^0}{dx^2} = f \text{ in }]0, 1[, \\ u^0(0) = u^0(1) = 0. \end{cases}$$

(2) Compute explicitly and plot the solutions u^ε corresponding to $f = 1$. Show that *boundary layers* appear in the derivative $(u^\varepsilon)'$ as ε approaches zero.

Remark. Such boundary layers reflect that the boundary conditions $(u^\varepsilon)'(0) = (u^\varepsilon)'(1) = 0$ are "lost" at the limit.

(3) Let $f = \delta'_{1/2}$, where $\delta'_{1/2}$ denotes the derivative of the Dirac distribution at $x = \frac{1}{2}$. Verify first that $f \in H^{-2}(0, 1)$, but that $f \notin H^{-1}(0, 1)$; then show that

$$u^\varepsilon(x) \to x \quad \text{for each} \quad 0 \le x < \frac{1}{2} \quad \text{as } \varepsilon \to 0,$$
$$u^\varepsilon(x) \to 1 - x \quad \text{for each} \quad \frac{1}{2} < x \le 1 \quad \text{as } \varepsilon \to 0,$$

so that an *"interior" layer* around $x = \frac{1}{2}$ appears in the functions u^ε as ε approaches zero.

Remark. The choice $f = \delta'_{1/2}$ provides an example of a right-hand side that is not "admissible", in the sense that it is in the dual of the space $H_0^2(\omega)$ (to which u^ε belongs for each $\varepsilon > 0$), but *not* in the dual of the "limit" space $H_0^1(\omega)$ (to which the limit u^0 found in (1) belongs).

This one-dimensional model of Koiter's equations thus suggests why a boundary layer can even appear "away from the boundary of the middle surface" in a generalized membrane shell subjected to seemingly innocuous, but not admissible, applied forces.

7.5. Let $\boldsymbol{\theta} = (\theta_i) \in \mathcal{C}^3(\overline{\omega}; \mathbb{R}^3)$ be an injective mapping such that the two vectors $\partial_\alpha \boldsymbol{\theta}$ are linearly independent at all points of $\overline{\omega}$. Following Ciarlet [1992b], consider a one-parameter family of mappings $\boldsymbol{\theta}(t) = (\theta_i(t)) : \overline{\omega} \to \mathbb{R}^3$ defined for $0 \le t \le 1$ and $(y_1, y_2) \in \overline{\omega}$ by

$$\theta_\alpha(t)(y_1, y_2) = t\theta_\alpha(y_1, y_2) + (1 - t)y_\alpha,$$
$$\theta_3(t)(y_1, y_2) = t\theta_3(y_1, y_2).$$

Assuming that the functions $p^{i,\varepsilon} \in L^2(\omega)$ (which take into account the applied forces) are independent of t, let $\boldsymbol{\zeta}_K^\varepsilon(t) \in \mathbf{V}_K(\omega)$ denote for $0 \le t \le 1$ the solutions of the corresponding two-dimensional Koiter equations (Sect. 7.1), now parametrized by t.

Show that the mapping $t \in [0, 1] \to \boldsymbol{\zeta}_K^\varepsilon(t) \in \mathbf{V}_K(\omega)$ is continuous. In particular, $\boldsymbol{\zeta}_K^\varepsilon(t) \to \boldsymbol{\zeta}^\varepsilon$ in $H^1(\omega) \times H^1(\omega) \times H^2(\omega)$ when $t \to 0^+$, i.e., when *"the shell becomes a plate"*, where $\boldsymbol{\zeta}^\varepsilon = (\zeta_i^\varepsilon) \in \mathbf{V}_K(\omega)$ satisfies the *two-dimensional equations of a linearly elastic clamped plate* (Vol. II, Sect. 1.7):

$$\int_\omega \left\{ \varepsilon b^{\alpha\beta\sigma\tau,\varepsilon} e_{\sigma\tau}(\boldsymbol{\zeta}^\varepsilon) e_{\alpha\beta}(\boldsymbol{\eta}) + \frac{\varepsilon^3}{3} b^{\alpha\beta\sigma\tau,\varepsilon} \partial_{\sigma\tau}\zeta_3^\varepsilon \partial_{\alpha\beta}\eta_3 \right\} dy = \int_\omega p^{i,\varepsilon}\eta_i \, dy$$

for all $\boldsymbol{\eta} = (\eta_i) \in \mathbf{V}_K(\omega)$, where

$$b^{\alpha\beta\sigma\tau,\varepsilon} = \frac{4\lambda^\varepsilon \mu^\varepsilon}{\lambda^\varepsilon + 2\mu^\varepsilon}\delta^{\alpha\beta}\delta^{\sigma\tau} + 2\mu^\varepsilon(\delta^{\alpha\sigma}\delta^{\beta\tau} + \delta^{\alpha\tau}\delta^{\beta\sigma}).$$

Remark. This property of the solution of Koiter's equations is akin to a property of the three-dimensional solution (Ex. 3.5).

7.6. Show that the rotation vector of the unit normal vector associated with a displacement field of the middle surface is a *tangential* one "to within the first order". This explains why the "linearized" rotation vector found in Naghdi's equations (Sect. 7.4) has a zero component over the vector $\boldsymbol{a}^3$.

Remark. Part of the problem consists in specifying what "first order" precisely means.

7.7. This problem establishes the *existence* and *uniqueness* of a solution to *Naghdi's equations for a linearly elastic shell*, after Bernadou, Ciarlet & Miara [1994, Thm. 3.1]; the notations are those of Sect. 7.4 and the assumptions are those of Thm. 7.4-1. Let

$$|(\boldsymbol{\eta}, \boldsymbol{s})| := \Big\{\sum_{\alpha,\beta} |\gamma_{\alpha\beta}(\boldsymbol{\eta})|^2_{0,\omega} + \sum_{\alpha} |\gamma_{\alpha 3}(\boldsymbol{\eta}, \boldsymbol{s})|^2_{0,\omega} + \sum_{\alpha,\beta} |\rho^N_{\alpha\beta}(\boldsymbol{\eta}, \boldsymbol{s})|^2_{0,\omega}\Big\}^{1/2},$$

$$\|(\boldsymbol{\eta}, \boldsymbol{s})\| := \Big\{\sum_i |\eta_i|^2_{0,\omega} + \sum_\alpha |s_\alpha|^2_{0,\omega} + |(\boldsymbol{\eta}, \boldsymbol{s})|^2\Big\}^{1/2},$$

$$\|(\boldsymbol{\eta}, \boldsymbol{s})\|_{1,\omega} := \Big\{\|\boldsymbol{\eta}\|^2_{1,\omega} + \|\boldsymbol{s}\|^2_{1,\omega}\Big\}^{1/2},$$

where $\boldsymbol{\eta} = (\eta_i)$ and $\boldsymbol{s} = (s_\alpha)$.

(1) Show that there exists a constant c_0 such that

$$\|(\boldsymbol{\eta}, \boldsymbol{s})\|_{1,\omega} \le c_0 \|(\boldsymbol{\eta}, \boldsymbol{s})\|$$

for all $(\boldsymbol{\eta}, \boldsymbol{s}) \in \mathbf{H}^1(\omega) = \big(H^1(\omega)\big)^5$.

Hint: Use the *lemma of J.L. Lions* as in Thm. 2.6-1.

(2) Let $(\boldsymbol{\eta}, \boldsymbol{s}) \in \mathbf{H}^1(\omega)$ be such that

$$\gamma_{\alpha\beta}(\boldsymbol{\eta}) = \gamma_{\alpha 3}(\boldsymbol{\eta}, \boldsymbol{s}) = \rho^N_{\alpha\beta}(\boldsymbol{\eta}, \boldsymbol{s}) = 0 \text{ in } \omega.$$

Show that the vector field $\eta_i \boldsymbol{a}^i$ is an *infinitesimal rigid displacement* of the surface $S = \boldsymbol{\theta}(\overline{\omega})$, in the sense that there exist two vectors $\hat{\boldsymbol{c}}$, $\hat{\boldsymbol{d}} \in \mathbb{R}^3$ such that

$$\eta_i(y)\boldsymbol{a}^i(y) = \hat{\boldsymbol{c}} + \hat{\boldsymbol{d}} \wedge \boldsymbol{\theta}(y) \text{ for all } y \in \overline{\omega}.$$

Hint: Show that $\eta_3 \in H^2(\omega)$, then that $\rho^N_{\alpha\beta}(\boldsymbol{\eta}, \boldsymbol{s}) = \rho_{\alpha\beta}(\boldsymbol{\eta})$ for such fields $(\boldsymbol{\eta}, \boldsymbol{s})$. The conclusion then follows from Thm. 2.6-3.

(3) Let $(\boldsymbol{\eta}, \boldsymbol{s}) \in \mathbf{H}^1(\omega)$ be such that

$$\gamma_{\alpha\beta}(\boldsymbol{\eta}) = \gamma_{\alpha 3}(\boldsymbol{\eta}, \boldsymbol{s}) = \rho^N_{\alpha\beta}(\boldsymbol{\eta}, \boldsymbol{s}) = 0 \text{ in } \omega,$$

$$\eta_i = s_\alpha = 0 \text{ on } \gamma_0,$$

where $\gamma_0 \subset \gamma$ satisfies *length* $\gamma_0 > 0$. Show that $(\boldsymbol{\eta}, \boldsymbol{s}) = (\mathbf{0}, \mathbf{0})$.

Hint: Use Thm. 2.6-3 again.

(4) Show that there exists a constant c such that the following *inequality of Korn's type on a general surface* holds (compare with the inequality of Thm. 2.6-4):

$$\|(\boldsymbol{\eta}, \boldsymbol{s})\|_{1,\omega} \le c\|(\boldsymbol{\eta}, \boldsymbol{s})\|$$

for all $(\boldsymbol{\eta}, \boldsymbol{s}) \in \mathbf{V}_N(\omega)$, where

$$\mathbf{V}_N(\omega) := \{(\boldsymbol{\eta}, \boldsymbol{s}) = ((\eta_i)(s_\alpha)) \in \mathbf{H}^1(\omega);\ \eta_i = s_\alpha = 0 \text{ on } \gamma_0\}.$$

Hint: Argue by contradiction as in Thm. 2.6-4.

(5) Let $\varepsilon B_N : \mathbf{V}_N(\omega) \times \mathbf{V}_N(\omega) \to \mathbb{R}$ denote the bilinear form defined by the left-hand side of the variational equations in problem $\mathcal{P}^\varepsilon_N(\omega)$ (Sect. 7.4). Show that there exists a constant c_N such that

$$\|(\boldsymbol{\eta}, \boldsymbol{s})\|^2 \le c_N B_N((\boldsymbol{\eta}, \boldsymbol{s}), (\boldsymbol{\eta}, \boldsymbol{s}))$$

for all $(\boldsymbol{\eta}, \boldsymbol{s}) \in \mathbf{V}_N(\omega)$ and conclude that $\mathcal{P}^\varepsilon_N(\omega)$ has one and only one solution.

Remark. Together, the results of (2) and (3) constitute an *infinitesimal rigid displacement lemma on a general surface* (compare with that of Thm. 2.6-3), which is due to Coutris [1978] (see also Coutris [1973, 1976]).

PART B

NONLINEAR SHELL THEORY

CHAPTER 8

ASYMPTOTIC ANALYSIS OF NONLINEARLY ELASTIC SHELLS: PRELIMINARIES

INTRODUCTION

In the second part of this volume[1], we give a detailed account of recent justifications of *nonlinear shell theories* that are also based on an *asymptotic analysis* of the three-dimensional solution with the thickness as the "small" parameter.

A remarkable progress in the asymptotic analysis of *nonlinearly elastic shells* is due to B. Miara, then to B. Miara and V. Lods, who justified the two-dimensional equations of a *nonlinearly elastic "membrane" shell* and those of a *nonlinearly elastic "flexural" shell*, by means of the *method of formal asymptotic expansions* applied to the three-dimensional equations of a nonlinearly elastic shell modeled by a *St Venant-Kirchhoff material.*

Another remarkable progress is due to H. Le Dret and A. Raoult, who gave the first proof of *convergence* of the three-dimensional solutions to a "two-dimensional" one as the thickness approaches zero. Their approach, which also leads to a *nonlinear "membrane" shell theory*, is described in the next chapter (Sect. 9.5).

The purpose of this chapter is to lay the *preliminary grounds* for the formal approach.

As a preparation, we examine how the equations of *three-dimensional nonlinear elasticity* expressed in terms of *Cartesian coordinates* in a reference configuration $\{\hat{\Omega}\}^-$ (these equations are reviewed in Sect. 8.1) are transformed when they are expressed in terms of *curvilinear coordinates*, i.e., in terms of coordinates in a set of the form $\overline{\Omega} = \boldsymbol{\Theta}^{-1}(\{\hat{\Omega}\}^-)$, where Ω is a three-dimensional domain in $\mathbb{R}^3$ and $\boldsymbol{\Theta}$ is an injective mapping with other *ad hoc* properties. The vectors of the covariant and contravariant bases and the covariant and

[1]To a large extent, this second part can be read independently of the first one.

contravariant components of the metric tensor are then defined by the relations

$$\boldsymbol{g}_i = \partial_i \boldsymbol{\Theta}, \quad \boldsymbol{g}^i \cdot \boldsymbol{g}_j = \delta^i_j, \quad g_{ij} = \boldsymbol{g}_i \cdot \boldsymbol{g}_j, \quad g^{ij} = \boldsymbol{g}^i \cdot \boldsymbol{g}^j.$$

More specifically, we show, by means of a direct computation (Thm. 8.2-3), that the *variational*, or *weak, formulation* of these equations takes the form:

$$\boldsymbol{u} = (u_i) \in \mathbf{W}(\Omega) := \{\boldsymbol{v} = (v_i) \in \mathbf{W}^{1,4}(\Omega);\ \boldsymbol{v} = \mathbf{0} \text{ on } \Gamma_0\},$$

$$\int_\Omega A^{ijkl} E_{k\|l}(\boldsymbol{u}) F_{i\|j}(\boldsymbol{u}, \boldsymbol{v}) \sqrt{g}\, dx = \int_\Omega f^i v_i \sqrt{g}\, dx + \int_{\Gamma_1} h^i v_i \sqrt{g}\, d\Gamma$$

for all $\boldsymbol{v} = (v_i) \in \mathbf{W}(\Omega)$, where $u_i \boldsymbol{g}^i$ is the unknown displacement vector field inside the set $\{\hat{\Omega}\}^-$, Γ_0 is a subset of $\Gamma = \partial\Omega$ with *area* $\Gamma_0 > 0$,

$$A^{ijkl} = \lambda g^{ij} g^{kl} + \mu\left(g^{ik} g^{jl} + g^{il} g^{jk}\right),$$

λ and μ are the Lamé constants of the constituting St Venant-Kirchhoff material, $\Gamma_1 = \Gamma - \Gamma_0$, the functions f^i and h^i are the contravariant components of the applied body and surface force densities, $g = \det(g_{ij})$, the *strains in curvilinear coordinates* $E_{i\|j}(\boldsymbol{v}) \in L^2(\Omega)$ are defined for each $\boldsymbol{v} = (v_i) \in \mathbf{W}^{1,4}(\Omega)$ by

$$E_{i\|j}(\boldsymbol{v}) = \frac{1}{2}(v_{i\|j} + v_{j\|i} + g^{mn} v_{m\|i} v_{n\|j}),$$

where $v_{i\|j} = \partial_j v_i - \Gamma^p_{ij} v_p$ and $\Gamma^p_{ij} = \boldsymbol{g}^p \cdot \partial_i \boldsymbol{g}_j$, and, finally,

$$\begin{aligned} F_{i\|j}(\boldsymbol{u}, \boldsymbol{v}) &= \frac{1}{2}\left(v_{i\|j} + v_{j\|i} + g^{mn}\{u_{m\|i} v_{n\|j} + u_{n\|j} v_{m\|i}\}\right) \\ &= E'_{i\|j}(\boldsymbol{u})\boldsymbol{v}. \end{aligned}$$

This calculation introduces various fundamental notions, some familiar, such as the *covariant derivatives* $v_{i\|j}$ of a vector field $v_i \boldsymbol{g}^i$ (Chap. 1), and some new, such as the *change of metric tensor* associated with this vector field, found here by means of its covariant components $E_{i\|j}(\boldsymbol{v})$.

We then specialize the reference configuration in the usual way: Given a domain ω in $\mathbb{R}^2$ with boundary γ and given a smooth enough injective mapping $\boldsymbol{\theta} : \overline{\omega} \to \mathbb{R}^3$ such that the two vectors $\boldsymbol{a}_\alpha = \partial_\alpha \boldsymbol{\theta}$

are linearly independent at all points in $\overline{\omega}$, we consider a *shell* with middle surface $S = \boldsymbol{\theta}(\overline{\omega})$ and thickness $2\varepsilon > 0$, i.e., a body whose reference configuration is the set $\boldsymbol{\Theta}(\overline{\Omega}^\varepsilon)$, where $\Omega^\varepsilon = \omega \times]-\varepsilon, \varepsilon[$ and

$$\boldsymbol{\Theta}(x^\varepsilon) = \boldsymbol{\theta}(y) + x_3^\varepsilon \boldsymbol{a}_3(y) \text{ for all } x^\varepsilon = (y, x_3^\varepsilon) \in \overline{\Omega}^\varepsilon.$$

The shell is subjected to *applied body forces* with contravariant components $f^{i,\varepsilon} : \Omega^\varepsilon \to \mathbb{R}$ and is subjected to a *boundary condition of place* along a portion $\boldsymbol{\Theta}(\gamma_0 \times [-\varepsilon, \varepsilon])$ of its lateral face $\boldsymbol{\Theta}(\gamma \times [-\varepsilon, \varepsilon])$, where $\gamma_0 \subset \gamma$ and *length* $\gamma_0 > 0$. *Applied surface forces* acting on the upper and lower faces $\boldsymbol{\Theta}(\omega \times \{\varepsilon\})$ and $\boldsymbol{\Theta}(\omega \times \{-\varepsilon\})$ may be also considered; for simplicity, we assume in this introduction that they vanish. Let λ^ε and μ^ε denote the Lamé constants of the St Venant-Kirchhoff material constituting the shell.

It thus follows that the unknown $\boldsymbol{u}^\varepsilon = (u_i^\varepsilon)$, where $u_i^\varepsilon : \overline{\Omega}^\varepsilon \to \mathbb{R}$ are the covariant components of the displacement field of the points of the shell, satisfies the following *three-dimensional equations* in terms of the *curvilinear coordinates* x_i^ε of the reference configuration $\boldsymbol{\Theta}(\overline{\Omega}^\varepsilon)$ (Sect. 8.3):

$$\boldsymbol{u}^\varepsilon \in \mathbf{W}(\Omega^\varepsilon) = \{\boldsymbol{v}^\varepsilon = (v_i^\varepsilon) \in \mathbf{W}^{1,4}(\Omega^\varepsilon);\ \boldsymbol{v}^\varepsilon = \mathbf{0} \text{ on } \Gamma_0^\varepsilon\},$$

$$\int_{\Omega^\varepsilon} A^{ijkl,\varepsilon} E_{k\|l}^\varepsilon(\boldsymbol{u}^\varepsilon) F_{i\|j}^\varepsilon(\boldsymbol{u}^\varepsilon, \boldsymbol{v}^\varepsilon) \sqrt{g^\varepsilon}\, dx^\varepsilon = \int_{\Omega^\varepsilon} f^{i,\varepsilon} v_i^\varepsilon \sqrt{g^\varepsilon}\, dx^\varepsilon \text{ for all } \boldsymbol{v}^\varepsilon \in \mathbf{W}(\Omega^\varepsilon),$$

where

$$A^{ijkl,\varepsilon} = \lambda^\varepsilon g^{ij,\varepsilon} g^{kl,\varepsilon} + \mu^\varepsilon \left(g^{ik,\varepsilon} g^{jl,\varepsilon} + g^{il,\varepsilon} g^{jk,\varepsilon} \right),$$

$$E_{i\|j}^\varepsilon(\boldsymbol{v}^\varepsilon) = \frac{1}{2}(v_{i\|j}^\varepsilon + v_{j\|i}^\varepsilon + g^{mn,\varepsilon} v_{m\|i}^\varepsilon v_{n\|j}^\varepsilon),$$

$$F_{i\|j}^\varepsilon(\boldsymbol{u}^\varepsilon, \boldsymbol{v}^\varepsilon) = \frac{1}{2}(v_{i\|j}^\varepsilon + v_{j\|i}^\varepsilon + g^{mn,\varepsilon}\{u_{m\|i}^\varepsilon v_{n\|j}^\varepsilon + u_{n\|j}^\varepsilon v_{m\|i}^\varepsilon\}).$$

In Sect. 8.4, we transform this problem into an equivalent problem, but now posed over the set $\Omega = \omega \times]-1, 1[$, which is *independent of* ε. More specifically, we define the *scaled unknown* $\boldsymbol{u}(\varepsilon) = (u_i(\varepsilon))$ by letting

$$u_i^\varepsilon(x^\varepsilon) = u_i(\varepsilon)(x) \text{ for all } x^\varepsilon = \pi^\varepsilon x \in \overline{\Omega}^\varepsilon,$$

where $\pi^\varepsilon(x_1, x_2, x_3) = (x_1, x_2, \varepsilon x_3)$, we *assume* that there exist constants $\lambda > 0$ and $\mu > 0$ such that

$$\lambda^\varepsilon = \lambda \text{ and } \mu^\varepsilon = \mu,$$

and finally, we *define* functions $f^i(\varepsilon) : \Omega \to \mathbb{R}$ by letting

$$f^{i,\varepsilon}(x^\varepsilon) = f^i(\varepsilon)(x) \text{ for all } x^\varepsilon = \pi^\varepsilon x \in \Omega^\varepsilon.$$

Note that, *by contrast with the linear case*, it can no longer be *a priori* assumed that the scaled displacement is "of order 0 with respect to ε". The main objective of this preliminary chapter precisely consists in *proving* that this is indeed the case!

It is found in this fashion that the scaled unknown $\boldsymbol{u}(\varepsilon)$ satisfies a variational problem of the form (Thm. 8.4-1):

$$\boldsymbol{u}(\varepsilon) \in \mathbf{W}(\Omega) = \left\{ \boldsymbol{v} = (v_i) \in \mathbf{W}^{1,4}(\Omega);\ \boldsymbol{v} = \mathbf{0} \text{ on } \Gamma_0 \right\},$$

$$\int_\Omega A^{ijkl}(\varepsilon) E_{k\|l}(\varepsilon;\ \boldsymbol{u}(\varepsilon)) F_{i\|j}(\varepsilon;\ \boldsymbol{u}(\varepsilon),\ \boldsymbol{v}) \sqrt{g(\varepsilon)}\, dx$$
$$= \int_\Omega f^i(\varepsilon) v_i \sqrt{g(\varepsilon)}\, dx \text{ for all } \boldsymbol{v} \in \mathbf{W}(\Omega),$$

where the *scaled strains* $E_{i\|j}(\varepsilon;\ \boldsymbol{u}(\varepsilon)) \in L^2(\Omega)$ *in curvilinear coordinates* are defined by

$$E_{i\|j}(\varepsilon;\ \boldsymbol{u}(\varepsilon))(x) = E^\varepsilon_{i\|j}(\boldsymbol{u}^\varepsilon)(x^\varepsilon) \text{ for all } x^\varepsilon = \pi^\varepsilon x \in \overline{\Omega}^\varepsilon$$

and, for any $\boldsymbol{v} \in \mathbf{W}^{1,4}(\Omega)$, the functions $F_{i\|j}(\varepsilon;\ \boldsymbol{u}(\varepsilon),\ \boldsymbol{v}) \in L^2(\Omega)$ are defined by

$$F_{i\|j}(\varepsilon;\ \boldsymbol{u}(\varepsilon),\ \boldsymbol{v})(x) = F^\varepsilon_{i\|j}(\boldsymbol{u}^\varepsilon,\ \boldsymbol{v}^\varepsilon)(x^\varepsilon) \text{ for all } x^\varepsilon = \pi^\varepsilon x \in \overline{\Omega}^\varepsilon.$$

Once this variational problem is written *in extenso* (Thm. 8.4-1), its specific form suggests that we use the *method of formal asymptotic expansions*: First, we replace the scaled unknown $\boldsymbol{u}(\varepsilon)$ by a formal expansion:

$$\boldsymbol{u}(\varepsilon) = \frac{1}{\varepsilon^N}\boldsymbol{u}^{-N} + \frac{1}{\varepsilon^{N-1}}\boldsymbol{u}^{-N+1} + \cdots$$

in the variational equations; then we equate to zero the factors of the successive powers of ε found in the resulting equations until *the order* $-N$ *of the leading term is determined and* (once this order is known) *the leading term can be fully identified as the solution of an* ad hoc *variational problem, the "right" orders* (in terms of ε) *to be assumed on the applied forces being simultaneously determined.*

The "unreasonable efficiency" of this method (described in detail in Sect. 8.6) rests in particular on two essential *guiding principles*: *No restriction* should be put on the *applied forces* and the *linearization* of any *nonlinear* equation found in this process should produce an equation of the *linear* theory described in the first part of this volume.

After recapitulating in Sect. 8.5 various "geometrical" and "mechanical" preliminaries (in fact, the same as in the linear case), we show that these two guiding principles already yield *three preliminary conclusions of paramount importance:*

First, $N = 0$, i.e., the formal expansion of the scaled unknown is of the form (Thm. 8.7-1):

$$\boldsymbol{u}(\varepsilon) = \boldsymbol{u}^0 + \varepsilon \boldsymbol{u}^1 + \cdots .$$

Second, the applied body forces are "at least of order 0"; this means that there exist functions $f^{i,0} \in L^2(\Omega)$ *independent of* ε such that (Thm. 8.8-1):

$$f^{i,\varepsilon}(x^\varepsilon) = f^{i,0}(x) \text{ for all } x^\varepsilon = \pi^\varepsilon x \in \Omega^\varepsilon.$$

Third, the leading term $\boldsymbol{u}^0 : \overline{\Omega} \to \mathbb{R}^3$ is *independent of* x_3 and $\boldsymbol{\zeta}^0 = (\zeta_i^0) = \frac{1}{2}\int_{-1}^{1} \boldsymbol{u}^0\, dx_3$ satisfies the following *two-dimensional variational problem* (Thm. 8.8-1)

$$\boldsymbol{\zeta}^0 \in \mathbf{W}(\omega) := \{\boldsymbol{\eta} \in \mathbf{W}^{1,4}(\omega);\ \boldsymbol{\eta} = \mathbf{0} \text{ on } \gamma_0\},$$

$$\int_\omega a^{\alpha\beta\sigma\tau} E^0_{\sigma\|\tau} F^0_{\alpha\|\beta}(\boldsymbol{\eta})\sqrt{a}\, dy = \int_\omega p^{i,0}\eta_i \sqrt{a}\, dy$$

for all $\boldsymbol{\eta} = (\eta_i) \in \mathbf{W}(\omega)$, where

$$\begin{aligned}
E^0_{\alpha\|\beta} &= \frac{1}{2}(\zeta^0_{\alpha\|\beta} + \zeta^0_{\beta\|\alpha} + a^{mn}\zeta^0_{m\|\alpha}\zeta^0_{n\|\beta}), \quad a^{mn} = \boldsymbol{a}^m \cdot \boldsymbol{a}^n,\\
F^0_{\alpha\|\beta}(\boldsymbol{\eta}) &= \frac{1}{2}(\eta_{\alpha\|\beta} + \eta_{\beta\|\alpha} + a^{mn}\{\zeta^0_{m\|\alpha}\eta_{n\|\beta} + \zeta^0_{n\|\beta}\eta_{m\|\alpha}\}),\\
\eta_{\alpha\|\beta} &= \partial_\beta\eta_\alpha - \Gamma^\sigma_{\alpha\beta}\eta_\sigma - b_{\alpha\beta}\eta_3, \quad \eta_{3\|\beta} := \partial_\beta\eta_3 + b^\sigma_\beta\eta_\sigma,\\
a^{\alpha\beta\sigma\tau} &= \frac{4\lambda\mu}{\lambda + 2\mu}a^{\alpha\beta}a^{\sigma\tau} + 2\mu(a^{\alpha\sigma}a^{\beta\tau} + a^{\alpha\tau}a^{\beta\sigma}),\\
p^{i,0} &= \int_{-1}^{1} f^{i,0}\, dx_3.
\end{aligned}$$

As we shall see, there are then *two essentially distinct possibilities: Either* this problem constitutes the final step of the application of the method of formal asymptotic expansions; this case corresponds to the *nonlinearly elastic "membrane" shells* studied in Chap. 9. *Or* the induction must be continued (in which case the functions $f^{i,0}$ vanish); this case corresponds to the *nonlinearly elastic "flexural" shells* studied in Chap. 10.

8.1. THREE-DIMENSIONAL NONLINEAR ELASTICITY IN CARTESIAN COORDINATES

We briefly review in this section the equations of nonlinear elasticity, the constitutive equation being that of the "simplest" nonlinearly elastic material (more general elastic materials are considered in Sect. 9.5). Ample details about the various notions introduced may be found in Vol. I[1], Chaps. 1 to 3.

Let the "physical space" be endowed with a *Cartesian frame* and, for this reason, let it be henceforth identified with $\mathbb{R}^3$. We let $\hat{e}_i = \hat{e}^i$ denote the basis vectors of the Cartesian frame (these notions are discussed in more detail at the beginning of Sect. 1.1).

Let $\hat{\Omega}$ be a domain in $\mathbb{R}^3$ with boundary $\hat{\Gamma}$, let $d\hat{x}$ denote the *volume element* in $\hat{\Omega}$, let $d\hat{\Gamma}$ denote the *surface element* along $\hat{\Gamma}$, and let $\hat{\boldsymbol{n}} = \hat{n}_i \hat{\boldsymbol{e}}^i$ denote the *unit* ($|\hat{\boldsymbol{n}}| = 1$) *outer normal vector* along $\hat{\Gamma}$. Finally, let $\hat{\Gamma} = \hat{\Gamma}_0 \cup \hat{\Gamma}_1$ be a $d\hat{\Gamma}$-measurable partition ($\hat{\Gamma}_0 \cap \hat{\Gamma}_1 = \phi$) of the boundary that satisfies

$$\boxed{\textit{area}\ \hat{\Gamma}_0 > 0.}$$

The set $\{\hat{\Omega}\}^-$ is the **reference configuration** occupied by an **elastic body** in the absence of applied forces. The body is subjected to **applied body forces** in its interior, of *density* $(\hat{f}_i) : \hat{\Omega} \to \mathbb{R}^3$ per unit volume, and to **applied surface forces** on the portion $\hat{\Gamma}_1$ of its boundary, of *density* $(\hat{h}_i) : \hat{\Gamma}_1 \to \mathbb{R}^3$ per unit area. Since these densities do not depend on the unknown, the applied forces considered here are *dead loads* (Vol. I, Sect. 2.7). For definiteness, we assume that

$$\boxed{\hat{f}_i \in L^2(\hat{\Omega}) \text{ and } \hat{h}_i \in L^2(\hat{\Gamma}_1).}$$

[1]We recall that "Vol. I" stands for "Ciarlet, P.G. [1988]: *Mathematical Elasticity, Volume I: Three-Dimensional Elasticity*, North Holland, Amsterdam".

The *unknown* is the vector field $\hat{\boldsymbol{u}} = (\hat{u}_i) : \{\hat{\Omega}\}^- \to \mathbb{R}^3$, where the three functions $\hat{u}_i : \{\hat{\Omega}\}^- \to \mathbb{R}$ are the **Cartesian components** of the **displacement field** $\hat{u}_i\hat{\boldsymbol{e}}^i : \{\hat{\Omega}\}^- \to \mathbb{R}^3$ that the body undergoes when it is subjected to the applied forces. This means that $\hat{u}_i(\hat{x})\hat{\boldsymbol{e}}^i$ *is the displacement of the point* $\hat{x} \in \{\hat{\Omega}\}^-$ (Fig. 1.1-1). Notice that we identify $\hat{u}_i(\hat{x})\hat{\boldsymbol{e}}^i \in \mathbb{R}^3$ with the vector $\hat{\boldsymbol{u}}(\hat{x}) \in \mathbb{R}^3$ and write accordingly the displacement field as

$$\boxed{\hat{\boldsymbol{u}} = \hat{u}_i\hat{\boldsymbol{e}}^i.}$$

It is assumed that *the displacement vanishes on the set* $\hat{\Gamma}_0$, i.e., that it satisfies the **boundary condition of place**

$$\boxed{\hat{\boldsymbol{u}} = \boldsymbol{0} \text{ on } \hat{\Gamma}_0.}$$

Recall that a "boundary condition of place" is in this volume *always* meant to be *homogeneous.*

The set $\hat{\boldsymbol{\varphi}}(\{\hat{\Omega}\}^-)$, where

$$\boxed{\hat{\boldsymbol{\varphi}} := \boldsymbol{id}_{\mathbb{R}^3} + \hat{\boldsymbol{u}} : \hat{x} \in \{\hat{\Omega}\}^- \to (\hat{x} + \hat{\boldsymbol{u}}(\hat{x})) \in \mathbb{R}^3,}$$

is called a **deformed configuration** and the mapping $\hat{\boldsymbol{\varphi}}$ is called a **deformation** (of the reference configuration $\{\hat{\Omega}\}^-$). Since the approach in this section is essentially formal, we assume that the requirements that the deformation $\hat{\boldsymbol{\varphi}}$ should satisfy in order to be physically admissible (orientation-preserving character and injectivity; cf. Vol. I, Sect. 1.4) are satisfied.

We let $\hat{x} = (\hat{x}_i)$ denote a generic point in the set $\hat{\Omega}$ and we let $\hat{\partial}_i = \partial/\partial\hat{x}_i$.

The following **equations of equilibrium in Cartesian coordinates** (Vol. I, Sect. 2.6) are then satisfied in the reference configuration $\{\hat{\Omega}\}^-$:

$$\begin{aligned} -\hat{\partial}_j(\hat{\sigma}_{ij} + \hat{\sigma}_{kj}\hat{\partial}_k\hat{u}_i) &= \hat{f}_i \text{ in } \hat{\Omega}, \\ (\hat{\sigma}_{ij} + \hat{\sigma}_{kj}\hat{\partial}_k\hat{u}_i)\hat{n}_j &= \hat{h}_i \text{ on } \hat{\Gamma}_1, \end{aligned}$$

where $\hat{\boldsymbol{\Sigma}} = (\hat{\sigma}_{ij}) : \{\hat{\Omega}\}^- \to \mathbb{S}^3$ denotes the **second Piola-Kirchhoff stress tensor field** and its components $\hat{\sigma}_{ij} = \hat{\sigma}_{ji}$ are called the (second Piola-Kirchhoff) **stresses** (cf. Vol. I, Sect. 2.5; $\mathbb{S}^3$ denotes the space of all symmetric matrices of order three).

The functions

$$\hat{t}_{ij} := \hat{\sigma}_{ij} + \hat{\sigma}_{kj}\hat{\partial}_k\hat{u}_i$$

are the components of the **first Piola-Kirchhoff stress tensor field** $\hat{\mathbf{T}} = (\hat{t}_{ij}) : \{\hat{\Omega}\}^- \to \mathbb{M}^3$ (cf. again Vol. I, Sect. 2.5; $\mathbb{M}^3$ denotes the space of all matrices of order three).

The boundary conditions on the set $\hat{\Gamma}_1$ constitute a **boundary condition of traction.**

Let $\mathbf{W}(\hat{\Omega})$ denote a space of sufficiently smooth vector fields $\hat{\boldsymbol{v}} = (\hat{v}_i) : \{\hat{\Omega}\}^- \to \mathbb{R}^3$ that vanish on $\hat{\Gamma}_0$. Then, if the unknown displacement $\hat{\boldsymbol{u}}$ belongs to the space $\mathbf{W}(\hat{\Omega})$, it can be easily established (by integration by parts; cf. Vol. I, Thm. 2.6-2) that *the equations of equilibrium are formally equivalent to the* **principle of virtual work in Cartesian coordinates**, which states that:

$$\int_{\hat{\Omega}} (\hat{\sigma}_{ij} + \hat{\sigma}_{kj}\hat{\partial}_k\hat{u}_i)\hat{\partial}_j\hat{v}_i \, d\hat{x} = \int_{\hat{\Omega}} \hat{f}_i\hat{v}_i \, d\hat{x} + \int_{\hat{\Gamma}_1} \hat{h}_i\hat{v}_i \, d\hat{\Gamma}$$
$$\text{for all } (\hat{v}_i) \in \mathbf{W}(\hat{\Omega}).$$

Note that the principle of virtual work is nothing but the *weak*, or *variational*, form of the equations of equilibrium and that the functions $\hat{\boldsymbol{v}} \in \mathbf{W}(\hat{\Omega})$ that enter it are "variations" around the actual deformation $\hat{\boldsymbol{\varphi}} = \boldsymbol{id}_{\mathbb{R}^3} + \hat{\boldsymbol{u}}$ (Vol. I, Sect. 2.6).

We finally assume that the constituting material is **elastic, homogeneous**, and **isotropic** and that the reference configuration $\{\hat{\Omega}\}^-$ is "*stress-free*", i.e., is a **natural state** (these notions are defined in Vol. I, Chap. 3). We consider here the *simplest* **nonlinearly elastic material** that satisfies these assumptions.

More specifically, let the **strains**

$$\boxed{\hat{E}_{ij}(\hat{\boldsymbol{v}}) := \frac{1}{2}(\hat{\partial}_i\hat{v}_j + \hat{\partial}_j\hat{v}_i + \hat{\partial}_i\hat{v}_m\hat{\partial}_j\hat{v}_m)}$$

denote the **Cartesian components** of the **change of metric** or **Green-St Venant strain tensor** (Vol. I, Sect. 1.8):

$$\boxed{\hat{\mathbf{E}}(\hat{\boldsymbol{v}}) = (\hat{E}_{ij}(\hat{\boldsymbol{v}})) := \frac{1}{2}\left(\{\hat{\boldsymbol{\nabla}}\hat{\boldsymbol{v}}\}^T + \hat{\boldsymbol{\nabla}}\hat{\boldsymbol{v}} + \{\hat{\boldsymbol{\nabla}}\hat{\boldsymbol{v}}\}^T\hat{\boldsymbol{\nabla}}\hat{\boldsymbol{v}}\right)}$$

associated with an arbitrary displacement field $\hat{\boldsymbol{v}} = (\hat{v}_i) : \{\hat{\Omega}\}^- \to \mathbb{R}^3$ of the reference configuration $\{\hat{\Omega}\}^-$, where the matrix

$$\hat{\boldsymbol{\nabla}}\hat{\boldsymbol{v}} := \begin{pmatrix} \hat{\partial}_1\hat{v}_1 & \hat{\partial}_2\hat{v}_1 & \hat{\partial}_3\hat{v}_1 \\ \hat{\partial}_1\hat{v}_2 & \hat{\partial}_2\hat{v}_2 & \hat{\partial}_3\hat{v}_2 \\ \hat{\partial}_1\hat{v}_3 & \hat{\partial}_2\hat{v}_3 & \hat{\partial}_3\hat{v}_3 \end{pmatrix}$$

denotes the corresponding **displacement gradient.** Note in passing that the *linearized strain tensor* $\hat{\mathbf{e}}(\hat{\boldsymbol{v}}) = (\hat{e}_{ij}(\hat{\boldsymbol{v}}))$ introduced in Sect. 1.1 is indeed the linear part with respect to $\hat{\boldsymbol{v}}$ in the strain tensor $\hat{\mathbf{E}}(\hat{\boldsymbol{v}}) = (\hat{E}_{ij}(\hat{\boldsymbol{v}}))$.

We assume that the material constituting the elastic body is a **St Venant-Kirchhoff material.** This means (Vol. I, Sect. 3.9) that there exist two constants λ and μ such that the second Piola-Kirchhoff stress tensor is expressed in terms of the Green-St Venant strain tensor through a **constitutive equation** of the form:

$$\hat{\boldsymbol{\Sigma}} = \lambda(\mathrm{tr}\ \hat{\mathbf{E}}(\hat{\boldsymbol{u}}))\mathbf{I} + 2\mu\hat{\mathbf{E}}(\hat{\boldsymbol{u}}),$$

or, componentwise,

$$\boxed{\hat{\sigma}_{ij} = \lambda\hat{E}_{pp}(\hat{\boldsymbol{u}})\delta_{ij} + 2\mu\hat{E}_{ij}(\hat{\boldsymbol{u}}).}$$

The two constants λ and μ are the **Lamé constants** of the material; in accordance to experimental evidence, we assume that they satisfy

$$\boxed{\lambda > 0 \text{ and } \mu > 0.}$$

Remarks. (1) The constitutive equation may be also expressed in terms of the *Poisson ratio* and the *Young modulus* of the same material (Vol. I, Sect. 3.8).

(2) The *same nonlinearly elastic material* was used in Vol. II[1], Chaps. 4 and 5, for justifying, again by the method of formal asymptotic expansions, the two-dimensional equations of *nonlinearly elastic plates.* ■

We thus realize, simply by inspecting the equations, that *for a St Venant-Kirchhoff material* the minimum regularity needed on the

[1]We recall that "Vol. II" stands for "Ciarlet, P.G. [1997]: *Mathematical Elasticity, Volume II: Theory of Plates*, North Holland, Amsterdam".

components $\hat{v}_i$ of the fields $\hat{\boldsymbol{v}} = (v_i) \in \mathbf{W}(\hat{\Omega})$ in order that all integrals appearing in the left-hand sides of the principle of virtual work make sense is that they belong to the *Sobolev space*

$$W^{1,4}(\hat{\Omega}) := \{\hat{v} \in L^4(\hat{\Omega});\ \hat{\partial}_i \hat{v} \in L^4(\hat{\Omega})\}.$$

Hence the space $\mathbf{W}(\hat{\Omega})$ may be defined in the present case as:

$$\mathbf{W}(\hat{\Omega}) := \{\hat{\boldsymbol{v}} = (\hat{v}_i) \in \mathbf{W}^{1,4}(\hat{\Omega});\ \hat{\boldsymbol{v}} = \mathbf{0} \text{ on } \hat{\Gamma}_0\}.$$

If we *assume* that $\hat{\boldsymbol{u}} \in \mathbf{W}^{1,4}(\hat{\Omega})$, we also have

$$\lambda(\text{tr } \hat{\mathbf{E}}(\hat{\boldsymbol{u}}))\mathbf{I} + 2\mu\hat{\mathbf{E}}(\hat{\boldsymbol{u}}) = \hat{\boldsymbol{\Sigma}} \in \mathbf{L}^2_s(\hat{\Omega}),$$

where

$$\mathbf{L}^2_s(\hat{\Omega}) := \{(\hat{\tau}_{ij}) \in \mathbf{L}^2(\hat{\Omega});\ \hat{\tau}_{ij} = \hat{\tau}_{ji}\}.$$

To sum up, *the displacement field* $\hat{\boldsymbol{u}} = (\hat{u}_i)$ *satisfies the following nonlinear* **displacement-traction problem** ("displacement-traction" refers to the kind of boundary conditions considered here; cf. Vol. I, Sect. 5.1):

$$\begin{aligned} -\hat{\partial}_j(\hat{\sigma}_{ij} + \hat{\sigma}_{kj}\hat{\partial}_k\hat{u}_i) &= \hat{f}_i \text{ in } \hat{\Omega}, \\ \hat{u}_i &= 0 \text{ on } \hat{\Gamma}_0, \\ (\hat{\sigma}_{ij} + \hat{\sigma}_{kj}\hat{\partial}_k\hat{u}_i)\hat{n}_j &= \hat{h}_i \text{ on } \hat{\Gamma}_1, \end{aligned}$$

which is in turn equivalent to the *variational problem*:

$$\begin{aligned} &\hat{\boldsymbol{u}} \in \mathbf{W}(\hat{\Omega}) := \{\hat{\boldsymbol{v}} = (\hat{v}_i) \in \mathbf{W}^{1,4}(\hat{\Omega});\ \hat{\boldsymbol{v}} = \mathbf{0} \text{ on } \hat{\Gamma}_0\}, \\ &\int_{\hat{\Omega}} (\hat{\sigma}_{ij} + \hat{\sigma}_{kj}\hat{\partial}_k\hat{u}_i)\hat{\partial}_j\hat{v}_i \, d\hat{x} \\ &\qquad = \int_{\hat{\Omega}} \hat{f}_i\hat{v}_i \, d\hat{x} + \int_{\hat{\Gamma}_1} \hat{h}_i\hat{v}_i \, d\hat{\Gamma} \text{ for all } \hat{\boldsymbol{v}} \in \mathbf{W}(\hat{\Omega}), \end{aligned}$$

where

$$\begin{aligned} \hat{\sigma}_{ij} &:= \lambda\hat{E}_{pp}(\hat{\boldsymbol{u}})\delta_{ij} + 2\mu\hat{E}_{ij}(\hat{\boldsymbol{u}}), \\ \hat{E}_{ij}(\hat{\boldsymbol{u}}) &:= \frac{1}{2}(\hat{\partial}_i\hat{u}_j + \hat{\partial}_j\hat{u}_i + \hat{\partial}_i\hat{u}_m\hat{\partial}_j\hat{u}_m). \end{aligned}$$

Either formulation constitutes the **equations of three-dimensional nonlinear elasticity in Cartesian coordinates** for a *displacement-traction problem* associated with a *St Venant-Kirchhoff material.* Note in passing that the linearization of these equations yields precisely the linear equations considered in Sect. 1.1.

Because a St Venant-Kirchhoff material is **hyperelastic** (Vol. I, Thm. 4.4-3), solving the above variational problem is equivalent to finding the *stationary points* of an associated functional $\hat{J}$ (defined below), i.e., those points where the derivative of $\hat{J}$ vanishes (this equivalence is proved in the next section in the more general setting of nonlinear elasticity in curvilinear coordinates, which include the Cartesian ones as a special case; cf. Thm. 8.2-3).

Particular stationary points are thus obtained by solving a *minimization problem*, viz., find $\hat{\boldsymbol{u}}$ such that

$$\boxed{\hat{\boldsymbol{u}} \in \mathbf{W}(\hat{\Omega}) \text{ and } \hat{J}(\hat{\boldsymbol{u}}) = \inf_{\hat{\boldsymbol{v}} \in \mathbf{W}(\hat{\Omega})} \hat{J}(\hat{\boldsymbol{v}}),}$$

where the **three-dimensional energy** $\hat{J} : \mathbf{W}(\hat{\Omega}) \to \mathbb{R}$ **in Cartesian coordinates** associated with a *St Venant-Kirchhoff material* is defined for all $\hat{\boldsymbol{v}} = (\hat{v}_i) \in \mathbf{W}(\hat{\Omega})$ by

$$\boxed{\begin{aligned}\hat{J}(\hat{\boldsymbol{v}}) := \frac{1}{2} \int_{\hat{\Omega}} \left\{ \lambda (\text{tr } \hat{\mathbf{E}}(\hat{\boldsymbol{v}}))^2 + 2\mu \text{ tr } \hat{\mathbf{E}}^2(\hat{\boldsymbol{v}}) \right\} d\hat{x} \\ - \left\{ \int_{\hat{\Omega}} \hat{f}_i \hat{v}_i \, d\hat{x} + \int_{\hat{\Gamma}_1} \hat{h}_i \hat{v}_i \, d\hat{\Gamma} \right\}.\end{aligned}}$$

Let $\mathbb{S}^3$ denote the space of all symmetric matrices of order three. The mapping

$$\mathbf{E} \in \mathbb{S}^3 \to \left\{ \frac{\lambda}{2} (\text{tr } \mathbf{E})^2 + \mu \text{tr } \mathbf{E}^2 \right\} \in \mathbb{R}$$

that appears in a self-explanatory way in the first integrand of the energy $\hat{J}$ is called the **stored energy function** *of a St Venant-Kirchhoff material.*

A word of caution. *There is no available result guaranteeing the existence of a solution* to this *minimization problem* (all that can be proved is that $\hat{J}$ is coercive on $\mathbf{W}(\hat{\Omega})$; cf. Ex. 8.1). The only available *existence* result valid for a *St Venant-Kirchhoff material* is based on the *implicit function theorem* and is for this reason restricted to smooth boundaries, to "small enough" forces, and to special classes of boundary conditions, which *do not* include those considered here (see the discussion given in Vol. I, Sect. 6.7).

In a landmark paper, Ball [1977] has developed a powerful existence theory for *hyperelastic materials* whose stored energy is *polyconvex* and satisfies *ad hoc growth conditions* (a detailed account of this theory is given in Vol. I, Chap. 7). This theory accommodates non-smooth boundaries and boundary conditions of the type considered here and is not restricted to "small enough" forces. However, even within the class of elastic materials to which it applies, which does *not* include St Venant-Kirchhoff materials (their stored energy functions are not polyconvex, cf. Raoult [1986]), it neither provides the existence of a solution to the corresponding variational problem, because the energy is not differentiable in the spaces where the minimizers are found (Vol. I, Sect. 7.10). ■

Detailed expositions of the *modeling* of *three-dimensional nonlinear elasticity* are found in Truesdell & Noll [1965], Germain [1972], Wang & Truesdell [1973], Gurtin [1981a], Marsden & Hughes [1983, Chaps. 1-5], and Vol. I, Chaps. 1-5. Its *mathematical theory* is exposed in Ball [1977], Marsden & Hughes [1983], Valent [1988], and Vol. I, Chaps. 6 and 7.

8.2. THREE-DIMENSIONAL NONLINEAR ELASTICITY IN CURVILINEAR COORDINATES

To begin with, we slightly modify some of the notations of Sect. 8.1, so that the definitions and equations in Cartesian coordinates recalled there now obey the proper rules regarding covariant indices and contravariant exponents. Once so rewritten, they will more evidently appear as special cases of the more general ones in curvilinear coordinates found in this section.

We now denote by δ^{ij} the Kronecker symbol and by $\hat{f}^i := \hat{f}_i$ and $\hat{h}^i := \hat{h}_i$ the components of the applied forces. We also define the components $\hat{A}^{ijkl}$ of the **three-dimensional elasticity tensor in**

Cartesian coordinates by

$$\hat{A}^{ijkl} := \lambda\delta^{ij}\delta^{kl} + \mu(\delta^{ik}\delta^{jl} + \delta^{il}\delta^{jk}).$$

Remark. The *same* changes of notations were performed on the equations of three-dimensional *linearized* elasticity in Cartesian coordinates (Sect. 1.1). ■

With these notations, the minimization problem described at the end of Sect. 8.1 equivalently consists in finding $\hat{\boldsymbol{u}} = (\hat{u}_i)$ such that

$$\hat{\boldsymbol{u}} \in \mathbf{W}(\hat{\Omega}) := \{\hat{\boldsymbol{v}} = (\hat{v}_i) \in \mathbf{W}^{1,4}(\hat{\Omega});\ \hat{\boldsymbol{v}} = \mathbf{0} \text{ on } \hat{\Gamma}_0\} \text{ and}$$

$$\begin{aligned}
\hat{J}(\hat{\boldsymbol{u}}) &= \inf_{\hat{\boldsymbol{v}} \in \mathbf{W}(\hat{\Omega})} \hat{J}(\hat{\boldsymbol{v}}), \text{ where} \\
\hat{J}(\hat{\boldsymbol{v}}) &:= \frac{1}{2}\int_{\hat{\Omega}} \hat{A}^{ijkl}\hat{E}_{kl}(\hat{\boldsymbol{v}})\hat{E}_{ij}(\hat{\boldsymbol{v}})\,d\hat{x} - \left\{\int_{\hat{\Omega}} \hat{f}^i\hat{v}_i\,d\hat{x} + \int_{\hat{\Gamma}_1} \hat{h}^i\hat{v}_i\,d\hat{\Gamma}\right\}, \\
\hat{E}_{ij}(\hat{\boldsymbol{v}}) &:= \frac{1}{2}(\hat{\partial}_i\hat{v}_j + \hat{\partial}_j\hat{v}_i + \hat{\partial}_i\hat{v}_m\hat{\partial}_j\hat{v}_m).
\end{aligned}$$

This minimization problem is formulated in terms of the *Cartesian coordinates* $\hat{x}_i$ of the points $\hat{x} = (\hat{x}_i) \in \{\hat{\Omega}\}^-$ and of the *Cartesian components* $\hat{v}_i$, $\hat{f}^i$, $\hat{h}^i$ of the displacement and force densities.

Assume that we are also given a domain Ω in $\mathbb{R}^3$ and a smooth enough injective mapping $\boldsymbol{\Theta} : \overline{\Omega} \to \mathbb{R}^3$ such that $\boldsymbol{\Theta}(\overline{\Omega}) = \{\hat{\Omega}\}^-$ and the three vectors $\boldsymbol{g}_i(x) = \partial_i\boldsymbol{\Theta}(x)$ are linearly independent at all points $x \in \overline{\Omega}$.

Our *objective* consists in reformulating this problem in terms of the *curvilinear coordinates* x_i of the points $\hat{x} = \boldsymbol{\Theta}(x) \in \{\hat{\Omega}\}^-$, where $x = (x_i) \in \overline{\Omega}$. In other words, we wish to carry out a *change of variables*, from the "old" variables $\hat{x}_i$ to the "new" variables x_i, in each one of the integrals appearing in $\hat{J}(\hat{\boldsymbol{v}})$, integrals which we thus wish to write as

$$\int_{\hat{\Omega}} \cdots d\hat{x} = \int_{\Omega} \cdots dx \quad \text{and} \quad \int_{\hat{\Gamma}_1} \cdots d\hat{\Gamma} = \int_{\Gamma_1} \cdots d\Gamma,$$

where $\Gamma_1 \subset \Gamma$ and $\boldsymbol{\Theta}(\Gamma_1) = \hat{\Gamma}_1$. Thanks to the various formulas that were established in Chap. 1 for performing an analogous transformation on the equations of linearized elasticity, this change of variables becomes a simple task, which leads to the following result:

Theorem 8.2-1. *Let $\hat{\Omega}$ be a domain in $\mathbb{R}^3$, let $\hat{f}_i \in L^2(\hat{\Omega})$ and $\hat{h}_i \in L^2(\hat{\Gamma}_1)$ be given functions, and let $\hat{\boldsymbol{u}} = (\hat{u}_i) \in \mathbf{W}(\hat{\Omega})$ denote a minimizer of the energy $\hat{J}$ over the space $\mathbf{W}(\hat{\Omega})$.*

Let Ω be a domain in $\mathbb{R}^3$ and let $\boldsymbol{\Theta}$ be a $\mathcal{C}^2$-diffeomorphism of $\overline{\Omega}$ onto $\{\hat{\Omega}\}^- = \boldsymbol{\Theta}(\overline{\Omega})$, so that the three vectors $\boldsymbol{g}_i(x) = \partial_i \boldsymbol{\Theta}(x)$ are linearly independent at all points $x \in \overline{\Omega}$. Let the vectors $\boldsymbol{g}^i(x)$ be defined by the relations $\boldsymbol{g}^i(x) \cdot \boldsymbol{g}_j(x) = \delta^i_j$, let $g(x) = \det(\boldsymbol{g}_i(x) \cdot \boldsymbol{g}_j(x))$, and let $g^{ij}(x) = \boldsymbol{g}^i(x) \cdot \boldsymbol{g}^j(x)$, $x \in \overline{\Omega}$.

Then the vector field $\boldsymbol{u} = (u_i) : \overline{\Omega} \to \mathbb{R}^3$ defined by

$$\hat{u}_i(\hat{x})\hat{\mathbf{e}}^i := u_i(x)\boldsymbol{g}^i(x) \text{ for all } \hat{x} = \boldsymbol{\Theta}(x),\ x \in \overline{\Omega},$$

satisfies the following minimization problem:

$$\begin{aligned}
&\boldsymbol{u} \in \mathbf{W}(\Omega) := \{\boldsymbol{v} = (v_i) \in \mathbf{W}^{1,4}(\Omega);\ \boldsymbol{v} = \mathbf{0} \text{ on } \Gamma_0\},\\
&J(\boldsymbol{u}) = \inf_{\boldsymbol{v} \in \mathbf{W}(\Omega)} J(\boldsymbol{v}), \text{ where}\\
&J(\boldsymbol{v}) := \frac{1}{2}\int_\Omega A^{ijkl} E_{k\|l}(\boldsymbol{v}) E_{i\|j}(\boldsymbol{v}) \sqrt{g}\, dx\\
&\qquad\qquad - \left\{ \int_\Omega f^i v_i \sqrt{g}\, dx + \int_{\Gamma_1} h^i v_i \sqrt{g}\, d\Gamma \right\},
\end{aligned}$$

where $\Gamma_0 := \boldsymbol{\Theta}^{-1}(\hat{\Gamma}_0)$, $\Gamma_1 := \boldsymbol{\Theta}^{-1}(\hat{\Gamma}_1)$, the functions $f^i \in L^2(\Omega)$ and $h^i \in L^2(\Gamma_1)$ are defined by:

$$\begin{aligned}
\hat{f}^i(\hat{x})\hat{\mathbf{e}}_i\, d\hat{x} &=: \sqrt{g(x)} f^i(x) \boldsymbol{g}_i(x)\, dx,\ \hat{x} = \boldsymbol{\Theta}(x),\ x \in \Omega,\\
\hat{h}^i(\hat{x})\hat{\mathbf{e}}_i\, d\hat{\Gamma}(\hat{x}) &=: \sqrt{g(x)} h^i(x) \boldsymbol{g}_i(x)\, d\Gamma(x),\ \hat{x} = \boldsymbol{\Theta}(x),\ x \in \Gamma_1,
\end{aligned}$$

the functions $A^{ijkl} = A^{jikl} = A^{klij} \in \mathcal{C}^1(\overline{\Omega})$ are defined by:

$$A^{ijkl} := \lambda g^{ij} g^{kl} + \mu(g^{ik} g^{jl} + g^{il} g^{jk}),$$

and finally, the functions $E_{i\|j}(\boldsymbol{v}) = E_{j\|i}(\boldsymbol{v}) \in L^2(\Omega)$ are defined for all $\boldsymbol{v} = (v_i) \in \mathbf{W}^{1,4}(\Omega)$ by:

$$\begin{aligned}
E_{i\|j}(\boldsymbol{v}) &:= \frac{1}{2}(v_{i\|j} + v_{j\|i} + g^{mn} v_{m\|i} v_{n\|j}), \text{ where}\\
v_{i\|j} &:= \partial_j v_i - \Gamma^p_{ij} v_p \text{ and } \Gamma^p_{ij} := \boldsymbol{g}^p \cdot \partial_i \boldsymbol{g}_j = \Gamma^p_{ji} \in \mathcal{C}^0(\overline{\Omega}).
\end{aligned}$$

Proof. (i) Let

$$[\boldsymbol{g}_j(x)]^i := \boldsymbol{g}_j(x) \cdot \hat{\boldsymbol{e}}^i \text{ and } [\boldsymbol{g}^j(x)]_i := \boldsymbol{g}^j(x) \cdot \hat{\boldsymbol{e}}_i$$

denote for each $x \in \overline{\Omega}$ the i-th component of the vectors $\boldsymbol{g}_j(x)$ and $\boldsymbol{g}^j(x)$ over the basis $\{\hat{\boldsymbol{e}}^1, \hat{\boldsymbol{e}}^2, \hat{\boldsymbol{e}}^3\} = \{\hat{\boldsymbol{e}}_1, \hat{\boldsymbol{e}}_2, \hat{\boldsymbol{e}}_3\}$.

With any field $\hat{\boldsymbol{v}} = (\hat{v}_i) \in \mathbf{W}(\hat{\Omega})$, we associate as in Sect. 1.3 a field $\boldsymbol{v} = (v_i) : \overline{\Omega} \to \mathbb{R}^3$ by letting (see in particular Figs. 1.1-1 and 1.3-1, which illustrate the special case where $\hat{v}_i\hat{\boldsymbol{e}}^i$ is the displacement field $\hat{u}_i\hat{\boldsymbol{e}}^i$)

$$\hat{v}_i(\hat{x})\hat{\boldsymbol{e}}^i =: v_i(x)\boldsymbol{g}^i(x) \text{ for all } \hat{x} = \boldsymbol{\Theta}(x),\ x \in \overline{\Omega}.$$

Since both $\boldsymbol{\Theta}$ and its inverse mapping are Lipschitz-continuous by assumption, the functions $\hat{v}_j \circ \boldsymbol{\Theta}$ are in the space $W^{1,4}(\Omega)$ (see, e.g., Nečas [1967, Chap. 2, Sect. 3] or Adams [1975, Chap. 3]); consequently, the functions $v_i = (\hat{v}_j \circ \boldsymbol{\Theta})[\boldsymbol{g}_i]^j$ are also in $W^{1,4}(\Omega)$ since $[\boldsymbol{g}_i]^j \in \mathcal{C}^1(\overline{\Omega})$ by assumption. Hence $\hat{\boldsymbol{v}} \in \mathbf{W}(\hat{\Omega})$ implies that $\boldsymbol{v} \in \mathbf{W}(\Omega)$.

(ii) Again as in Sect. 1.3, we infer from the definitions of the functions $f^i \in L^2(\Omega)$ and $h^i \in L^2(\Gamma_1)$ that

$$\int_{\hat{\Omega}} \hat{f}^i \hat{v}_i \, d\hat{x} + \int_{\hat{\Gamma}_1} \hat{h}^i \hat{v}_i \, d\hat{\Gamma} = \int_{\Omega} f^i v_i \sqrt{g} \, dx + \int_{\Gamma_1} h^i v_i \sqrt{g} \, d\Gamma$$

for all $\hat{\boldsymbol{v}} = (\hat{v}_i) \in \mathbf{W}(\hat{\Omega})$ and $\boldsymbol{v} = (v_i) \in \mathbf{W}(\Omega)$ in the correspondence set in (i).

(iii) We established in part (iii) of the proof of Thm. 1.3-1 that

$$\hat{\partial}_j \hat{v}_i(\hat{x}) = \left(v_{k\|l}[\boldsymbol{g}^k]_i[\boldsymbol{g}^l]_j\right)(x) \text{ for all } \hat{x} = \boldsymbol{\Theta}(x),\ x \in \overline{\Omega},$$

for all $\hat{\boldsymbol{v}} = (\hat{v}_i) \in \mathbf{W}(\hat{\Omega})$ and $\boldsymbol{v} = (v_i) \in \mathbf{W}(\Omega)$ in the correspondence set in (i), where $v_{k\|l}(x) := \partial_l v_k(x) - \Gamma^p_{kl}(x) v_p(x)$. These relations in turn imply that

$$\hat{E}_{ij}(\hat{\boldsymbol{v}})(\hat{x}) = \frac{1}{2}(\hat{\partial}_j\hat{v}_i + \hat{\partial}_i\hat{v}_j + \hat{\partial}_i\hat{v}_m\hat{\partial}_j\hat{v}_m)(\hat{x})$$

$$= \frac{1}{2}\left((v_{k\|l} + v_{l\|k})[\boldsymbol{g}^k]_i[\boldsymbol{g}^l]_j + v_{r\|k}v_{s\|l}[\boldsymbol{g}^k]_i[\boldsymbol{g}^r]_m[\boldsymbol{g}^l]_j[\boldsymbol{g}^s]_m\right)(x)$$

$$= \frac{1}{2}\left(\{v_{k\|l} + v_{l\|k} + g^{rs}v_{r\|k}v_{s\|l}\}[\boldsymbol{g}^k]_i[\boldsymbol{g}^l]_j\right)(x)$$

since $[\boldsymbol{g}^r]_m[\boldsymbol{g}^s]_m = \boldsymbol{g}^r \cdot \boldsymbol{g}^s = g^{rs}$. Hence

$$\hat{E}_{ij}(\hat{\boldsymbol{v}})(\hat{x}) = \left(E_{k\|l}(\boldsymbol{v})[\boldsymbol{g}^k]_i[\boldsymbol{g}^l]_j\right)(x) \text{ for all } \hat{x} = \boldsymbol{\Theta}(x),\ x \in \overline{\Omega},$$

where the functions $E_{k\|l}(\boldsymbol{v}) \in L^2(\Omega)$ are those defined in the statement of the theorem.

(iv) Since the relations between the functions $\hat{E}_{ij}(\hat{\boldsymbol{v}}) \in L^2(\hat{\Omega})$ and $E_{i\|j}(\boldsymbol{v}) \in L^2(\Omega)$ are similar to those between the functions $\hat{e}_{ij}(\hat{\boldsymbol{v}}) \in L^2(\hat{\Omega})$ and $e_{i\|j}(\boldsymbol{v}) \in L^2(\Omega)$ established in part (iv) of the proof of Thm. 1.3-1, it likewise follows that

$$\left(\hat{A}^{ijkl}\hat{E}_{kl}(\hat{\boldsymbol{v}})\hat{E}_{ij}(\hat{\boldsymbol{v}})\right)(\hat{x}) = \left(A^{ijkl}E_{k\|l}(\boldsymbol{v})E_{i\|j}(\boldsymbol{v})\right)(x)$$
$$\text{for all } \hat{x} = \boldsymbol{\Theta}(x),\ x \in \overline{\Omega},$$

where the functions $A^{ijkl} \in \mathcal{C}^1(\overline{\Omega})$ are those defined in the statement of the theorem.

(v) *Conclusions:* Since $d\hat{x} = \sqrt{g(x)}\,dx$, part (iv) shows that

$$\int_{\hat{\Omega}} \hat{A}^{ijkl}\hat{E}_{kl}(\hat{\boldsymbol{v}})\hat{E}_{ij}(\hat{\boldsymbol{v}})\,d\hat{x} = \int_{\Omega} A^{ijkl}E_{k\|l}(\boldsymbol{v})E_{i\|j}(\boldsymbol{v})\sqrt{g}\,dx.$$

Together with that of part (ii), this relation shows that

$$\hat{J}(\hat{\boldsymbol{v}}) = J(\boldsymbol{v})$$

for all $\hat{\boldsymbol{v}} \in \mathbf{W}(\hat{\Omega})$ and $\boldsymbol{v} \in \mathbf{W}(\Omega)$ in the correspondence set in (i), and thus the proof is complete. ■

Naturally, if $\boldsymbol{\Theta} = \boldsymbol{id}_{\mathbb{R}^3}$, each vector $\boldsymbol{g}_i(x)$ is equal to $\hat{\boldsymbol{e}}_i$ and thus $\boldsymbol{g}^i(x) = \hat{\boldsymbol{e}}^i$, $g(x) = 1$, $g^{ij}(x) = \delta^{ij}$, and $v_{i\|j} = \partial_j v_i$; in addition, the fields (v_i) and $(\hat{v}_i)$, (f^i) and $(\hat{f}^i)$, (h^i) and $(\hat{h}^i)$, and (A^{ijkl}) and $(\hat{A}^{ijkl})$ coincide.

The functional $J : \mathbf{W}(\Omega) \to \mathbb{R}$ found in Thm. 8.2-1 is called the **three-dimensional energy in curvilinear coordinates** associated with a *St Venant-Kirchhoff material.*

The functions $A^{ijkl} : \overline{\Omega} \to \mathbb{R}$ are the **contravariant components** *of the* **three-dimensional elasticity tensor in curvilinear coordinates** (the same tensor appeared in linearized elasticity; cf. Sect. 1.3).

Let there be given a vector field $v_i\boldsymbol{g}^i$ on $\overline{\Omega}$ interpreted as an arbitrary **displacement field** of $\boldsymbol{\Theta}(\overline{\Omega})$ given by means of its **covariant components** $v_i : \overline{\Omega} \to \mathbb{R}$; this means that $v_i(x)\boldsymbol{g}^i(x)$ is the displacement of the point $\hat{x} = \boldsymbol{\Theta}(x)$. We now show that the matrix

$$(E_{i\|j}(\boldsymbol{v})) = \Big(\frac{1}{2}(v_{i\|j} + v_{j\|i} + g^{mn}v_{m\|i}v_{n\|j})\Big)$$

found in Thm. 8.2-1 generalizes to arbitrary *curvilinear coordinates* the *Green-St Venant strain tensor*

$$(\hat{E}_{ij}(\hat{\boldsymbol{v}})) = \Big(\frac{1}{2}(\hat{\partial}_j\hat{v}_i + \hat{\partial}_i\hat{v}_j + \hat{\partial}_i\hat{v}_m\hat{\partial}_j\hat{v}_m)\Big)$$

in Cartesian coordinates (Sect. 8.1).

More specifically, we show that *the matrix* $(E_{i\|j}(\boldsymbol{v}))$ *likewise measures* (half of) *the difference between the "new" metric in the deformed configuration* $(\boldsymbol{\Theta} + v_i\boldsymbol{g}^i)(\overline{\Omega})$ *and the "old" metric in the reference configuration* $\boldsymbol{\Theta}(\overline{\Omega})$, but now expressed by means of their covariant components in terms of the *curvilinear coordinates* x_i. We also record the relation between the functions $\hat{E}_{ij}(\hat{\boldsymbol{v}})$ and $E_{i\|j}(\boldsymbol{v})$ that was established in part (iii) of the proof of Thm. 8.2-1.

Theorem 8.2-2. *Let* Ω *be a domain in* $\mathbb{R}^3$, *let* $\boldsymbol{\Theta} : \overline{\Omega} \to \mathbb{R}^3$ *be a smooth enough injective mapping such that the three vectors* $\boldsymbol{g}_i = \partial_i\boldsymbol{\Theta}$ *are linearly independent at all points of* $\overline{\Omega}$ *and let the vectors* $\boldsymbol{g}^i$ *be defined by* $\boldsymbol{g}^i \cdot \boldsymbol{g}_j = \delta^i_j$. *Given a vector field* $v_i\boldsymbol{g}^i$ *on* $\overline{\Omega}$ *with smooth enough covariant components* $v_i : \overline{\Omega} \to \mathbb{R}$, *let the* **covariant components** *of the* **Green-St Venant strain tensor in curvilinear coordinates** *associated with this vector field be defined by*

$$\boxed{E_{i\|j}(\boldsymbol{v}) := \frac{1}{2}(v_{i\|j} + v_{j\|i} + g^{mn}v_{m\|i}v_{n\|j}),}$$

where $v_{i\|j} = \partial_j v_i - \Gamma^p_{ij}v_p$. *Then they also satisfy*

$$\boxed{E_{i\|j}(\boldsymbol{v}) = \frac{1}{2}(g_{ij}(\boldsymbol{v}) - g_{ij}),}$$

where g_{ij} *and* $g_{ij}(\boldsymbol{v})$ *denote the covariant components of the metric tensors respectively attached to the sets* $\boldsymbol{\Theta}(\overline{\Omega})$ *and* $(\boldsymbol{\Theta} + v_i\boldsymbol{g}^i)(\overline{\Omega})$.

For this reason, the functions $E_{i\|j}(\boldsymbol{v})$ are also called the **covariant components** *of the* **change of metric tensor in curvilinear coordinates,** *or simply the* **strains in curvilinear coordinates,** *associated with this vector field.*

The strains in curvilinear coordinates $E_{i\|j}(\boldsymbol{v})$ are related to the strains in Cartesian coordinates $\hat{E}_{ij}(\hat{\boldsymbol{v}})$ by:

$$\hat{E}_{ij}(\hat{\boldsymbol{v}})(\hat{x}) = \left(E_{k\|l}(\boldsymbol{v})[\boldsymbol{g}^k]_i[\boldsymbol{g}^l]_j\right)(x),\ \hat{x} = \boldsymbol{\Theta}(x).$$

Proof. The covariant components $g_{ij}(\boldsymbol{v})$ are defined in $\overline{\Omega}$ by

$$g_{ij}(\boldsymbol{v}) = \partial_i(\boldsymbol{\Theta} + v_m\boldsymbol{g}^m)\cdot\partial_j(\boldsymbol{\Theta} + v_n\boldsymbol{g}^n).$$

Using the relations

$$\begin{aligned}\partial_i(\boldsymbol{\Theta} + v_m\boldsymbol{g}^m) &= \partial_i\boldsymbol{\Theta} + \partial_i v_m\boldsymbol{g}^m + v_m\partial_i\boldsymbol{g}^m\\ &= \boldsymbol{g}_i + (\partial_i v_m - \Gamma^q_{im}v_q)\boldsymbol{g}^m \ = \ \boldsymbol{g}_i + v_{m\|i}\boldsymbol{g}^m\end{aligned}$$

(recall that $\partial_i\boldsymbol{g}^q = -\Gamma^q_{im}\boldsymbol{g}^m$; cf. Thm. 1.4-1), we obtain

$$\begin{aligned}g_{ij}(\boldsymbol{v}) &= (\boldsymbol{g}_i + v_{m\|i}\boldsymbol{g}^m)\cdot(\boldsymbol{g}_j + v_{n\|j}\boldsymbol{g}^n)\\ &= \boldsymbol{g}_i\cdot\boldsymbol{g}_j + v_{n\|j}\boldsymbol{g}_i\cdot\boldsymbol{g}^n + v_{m\|i}\boldsymbol{g}^m\cdot\boldsymbol{g}_j + v_{m\|i}v_{n\|j}\boldsymbol{g}^m\cdot\boldsymbol{g}^n\\ &= g_{ij} + v_{i\|j} + v_{j\|i} + g^{mn}v_{m\|i}v_{n\|j},\end{aligned}$$

as was to be proved. ∎

We next show that the minimization problem of Thm. 8.2-1 also admits a variational formulation, like its Cartesian counterpart (Sect. 8.1). We refer to, e.g., Vol. I, Sect. 1.2, for the definition and main properties of the Fréchet and Gâteaux derivatives.

Theorem 8.2-3. *As in* Thm. 8.2-1, *let the space $\mathbf{W}(\Omega)$ and the energy $J : \mathbf{W}(\Omega) \to \mathbb{R}$ be defined as*

$$\begin{aligned}\mathbf{W}(\Omega) &= \{\boldsymbol{v} = (v_i) \in \mathbf{W}^{1,4}(\Omega);\ \boldsymbol{v} = \mathbf{0} \text{ on } \Gamma_0\},\\ J(\boldsymbol{v}) &= \frac{1}{2}\int_\Omega A^{ijkl}E_{k\|l}(\boldsymbol{v})E_{i\|j}(\boldsymbol{v})\sqrt{g}\,dx\\ &\qquad - \left\{\int_\Omega f^i v_i\sqrt{g}\,dx + \int_{\Gamma_1} h^i v_i\sqrt{g}\,d\Gamma\right\},\end{aligned}$$

where the mappings $E_{i\|j} : \mathbf{W}^{1,4}(\Omega) \to L^2(\Omega)$ *are defined for each* $\boldsymbol{v} \in \mathbf{W}^{1,4}(\Omega)$ *by* (Thm. 8.2-2):

$$\boxed{\begin{aligned} E_{i\|j}(\boldsymbol{v}) &:= \frac{1}{2}(g_{ij}(\boldsymbol{v}) - g_{ij}) \\ &= \frac{1}{2}(v_{i\|j} + v_{j\|i} + g^{mn} v_{m\|i} v_{n\|j}). \end{aligned}}$$

Then the energy J *is differentiable over the space* $\mathbf{W}^{1,4}(\Omega)$, *hence over* $\mathbf{W}(\Omega)$, *and its Gâteaux derivatives* $J'(\boldsymbol{u})\boldsymbol{v}$ *are given by*

$$\boxed{\begin{aligned} J'(\boldsymbol{u})\boldsymbol{v} = &\int_\Omega A^{ijkl} E_{k\|l}(\boldsymbol{u}) F_{i\|j}(\boldsymbol{u}, \boldsymbol{v}) \sqrt{g}\, dx \\ &- \left\{ \int_\Omega f^i v_i \sqrt{g}\, dx + \int_{\Gamma_1} h^i v_i \sqrt{g}\, d\Gamma \right\}, \end{aligned}}$$

for all $\boldsymbol{u}, \boldsymbol{v} \in \mathbf{W}^{1,4}(\Omega)$, *where*

$$\boxed{\begin{aligned} F_{i\|j}(\boldsymbol{u}, \boldsymbol{v}) &:= E'_{i\|j}(\boldsymbol{u})\boldsymbol{v}, \\ &= \frac{1}{2}\left(v_{i\|j} + v_{j\|i} + g^{mn}\{u_{m\|i} v_{n\|j} + u_{n\|j} v_{m\|i}\}\right) \end{aligned}}$$

and $E'_{i\|j}(\boldsymbol{u}) \in \mathcal{L}(\mathbf{W}^{1,4}(\Omega); L^2(\Omega))$ *denotes the Fréchet derivative at any* $\boldsymbol{u} \in \mathbf{W}^{1,4}(\Omega)$ *of each mapping* $E_{i\|j} : \mathbf{W}^{1,4}(\Omega) \to L^2(\Omega)$. *Consequently,* $\boldsymbol{u} \in \mathbf{W}(\Omega)$ *is a stationary point of the energy* J, *i.e.,* $\boldsymbol{u}$ *satisfies* $J'(\boldsymbol{u}) = 0$, *if and only if it satisfies the variational problem:*

$$\boxed{\begin{aligned} &\boldsymbol{u} \in \mathbf{W}(\Omega) \text{ and, for all } \boldsymbol{v} \in \mathbf{W}(\Omega), \\ &\int_\Omega A^{ijkl} E_{k\|l}(\boldsymbol{u}) F_{i\|j}(\boldsymbol{u}, \boldsymbol{v}) \sqrt{g}\, dx = \int_\Omega f^i v_i \sqrt{g}\, dx + \int_{\Gamma_1} h^i v_i \sqrt{g}\, d\Gamma. \end{aligned}}$$

This is in particular the case if $\boldsymbol{u}$ *is a solution to the minimization problem of* Thm. 8.2-1.

Proof. The linear form L defined by

$$L(\boldsymbol{v}) = \int_\Omega f^i v_i \sqrt{g}\, dx + \int_{\Gamma_1} h^i v_i \sqrt{g}\, d\Gamma$$

is continuous and differentiable over $\mathbf{W}^{1,4}(\Omega)$, with

$$L'(\boldsymbol{u})\boldsymbol{v} = \int_\Omega f^i v_i \sqrt{g}\, dx + \int_{\Gamma_1} h^i v_i \sqrt{g}\, d\Gamma$$

for all $\boldsymbol{u}, \boldsymbol{v} \in \mathbf{W}^{1,4}(\Omega)$. For any $\boldsymbol{v} \in \mathbf{W}^{1,4}(\Omega)$, let

$$J_0(\boldsymbol{v}) = \frac{1}{2} \int_\Omega A^{ijkl} E_{k\|l}(\boldsymbol{v}) E_{i\|j}(\boldsymbol{v}) \sqrt{g}\, dx.$$

The mapping $J_0 : \mathbf{W}^{1,4}(\Omega) \to \mathbb{R}$ defined in this fashion can be written as $J_0 = B \circ \mathbf{E}$, where the mappings $\mathbf{E} : \mathbf{W}^{1,4}(\Omega) \to L^2(\Omega; \mathbb{S}^3)$ and $B : L^2(\Omega; \mathbb{S}^3) \to \mathbb{R}$ are defined by ($\mathbb{S}^3$ denotes the space of symmetric matrices of order three):

$$\mathbf{E}(\boldsymbol{v}) := (E_{i\|j}(\boldsymbol{v})) \in L^2(\Omega; \mathbb{S}^3) \text{ for all } \boldsymbol{v} = (v_i) \in \mathbf{W}^{1,4}(\Omega),$$
$$B(\mathbf{E}) := \frac{1}{2} \int_\Omega A^{ijkl} E_{k\|l} E_{i\|j} \sqrt{g}\, dx \text{ for all } \mathbf{E} = (E_{i\|j}) \in L^2(\Omega; \mathbb{S}^3).$$

As a sum of continuous linear and bilinear mappings, each mapping $E_{i\|j}$ is differentiable and its Gâteaux derivatives are given by (see, e.g., Vol. I, Sect. 1.2):

$$E'_{i\|j}(\boldsymbol{u})\boldsymbol{v} = \frac{1}{2}(v_{i\|j} + v_{j\|i} + g^{mn}\{u_{m\|i} v_{n\|j} + u_{n\|j} v_{m\|i}\}).$$

The mapping B is likewise differentiable and its Gâteaux derivatives are given for all $\mathbf{E} = (E_{i\|j})$ and $\mathbf{F} = (F_{i\|j})$ by

$$B'(\mathbf{E})\mathbf{F} = \int_\Omega A^{ijkl} E_{k\|l} F_{i\|j} \sqrt{g}\, dx.$$

By the chain rule (see, e.g., Vol. I, Thm. 1.2-1), we thus have

$$J_0'(\boldsymbol{u})\boldsymbol{v} = B'(\mathbf{E}(\boldsymbol{u}))\mathbf{E}'(\boldsymbol{u})\boldsymbol{v} = \int_\Omega A^{ijkl} E_{k\|l}(\boldsymbol{u}) E'_{i\|j}(\boldsymbol{u})\boldsymbol{v} \sqrt{g}\, dx$$

for all $\boldsymbol{u}, \boldsymbol{v} \in \mathbf{W}^{1,4}(\Omega)$. Hence the proof is complete (the other statements in the theorem are immediate consequences of the expression of $J'(\boldsymbol{u})\boldsymbol{v}$). ■

Remarks. (1) As a sum of continuous k-linear forms, $k = 1, 2, 3, 4$, the energy J is in fact *infinitely* differentiable over the space $\mathbf{W}^{1,4}(\Omega)$.

(2) Once it has been noticed that J is differentiable, a more direct way of computing $J'(\boldsymbol{u})\boldsymbol{v}$ simply consists in identifying the linear part with respect to $\boldsymbol{v}$ in the difference $J(\boldsymbol{u}+\boldsymbol{v}) - J(\boldsymbol{u})$. However, this procedure is less illuminating as it hinders the appearance of the derivatives $E'_{i\|j}(\boldsymbol{u})$.

(3) Naturally, the variational equations in curvilinear coordinates found in Thm. 8.2-3 by differentiating the energy J can be also directly recovered from the variational equations in Cartesian coordinates described in Sect. 8.1 (Ex. 8.2); however, the method followed in Thm. 8.2-3 is simpler. ■

To conclude this section, we derive the *boundary value problem* that is (at least formally) equivalent to the variational problem found in Thm. 8.2-3. We recall that $n_i(x)\boldsymbol{e}^i$ denotes the unit outer normal vector at $x \in \Gamma$.

Theorem 8.2-4. *Let the assumptions and notations be as in* Thms. 8.2-1 and 8.2-3. *If a solution* $\boldsymbol{u} = (u_i)$ *to the variational problem*

$$\boxed{\begin{aligned}&\boldsymbol{u} \in \mathbf{W}(\Omega) = \{\boldsymbol{v} = (v_i) \in \mathbf{W}^{1,4}(\Omega);\ \boldsymbol{v} = \mathbf{0} \text{ on } \Gamma_0\},\\ &\int_\Omega A^{ijkl} E_{k\|l}(\boldsymbol{u}) F_{i\|j}(\boldsymbol{u}, \boldsymbol{v})\sqrt{g}\, dx = \int_\Omega f^i v_i \sqrt{g}\, dx + \int_{\Gamma_1} h^i v_i \sqrt{g}\, d\Gamma\\ &\qquad\qquad\qquad\qquad\qquad\qquad\qquad\qquad \text{for all } \boldsymbol{v} \in \mathbf{W}(\Omega),\end{aligned}}$$

is smooth enough, it also satisfies the following **boundary value problem of three-dimensional nonlinear elasticity in curvilinear coordinates:**

$$\boxed{\begin{aligned} -(\sigma^{ij} + \sigma^{kj} g^{il} u_{l\|k})\|_j &= f_i \text{ in } \Omega,\\ u_i &= 0 \text{ on } \Gamma_0,\\ (\sigma^{ij} + \sigma^{kj} g^{il} u_{l\|k}) n_j &= h_i \text{ on } \Gamma_1,\end{aligned}}$$

where the functions

$$\boxed{\sigma^{ij} := A^{ijkl} E_{k\|l}(\boldsymbol{u})}$$

are the contravariant components of the **second Piola-Kirchhoff stress tensor field** *and*

$$\sigma^{ij}\|_k := \partial_k \sigma^{ij} + \Gamma^i_{pk}\sigma^{pj} + \Gamma^j_{kq}\sigma^{iq}$$

denote the **first-order covariant derivatives** *of any tensor field with differentiable contravariant components* $\sigma^{ij} : \overline{\Omega} \to \mathbb{R}$.

Proof. From the proof of Thm. 1.6-1, we infer that

$$\int_\Omega \sigma^{ij} e_{i\|j}(\boldsymbol{v})\sqrt{g}\,dx = -\int_\Omega \sqrt{g}(\sigma^{ij}\|_j)v_i\,dx + \int_\Gamma \sqrt{g}\sigma^{ij}n_j v_i\,d\Gamma,$$

$$\begin{aligned}\frac{1}{2}\int_\Omega \sigma^{ij} g^{mn}\{u_{m\|i}v_{n\|j} + u_{n\|j}v_{m\|i}\}\sqrt{g}\,dx \\ = \int_\Omega \sigma^{ij} g^{mn} u_{m\|i}v_{n\|j}\sqrt{g}\,dx \\ = -\int_\Omega \sqrt{g}\{(\sigma^{kj}g^{il}u_{l\|k})\|_j\}v_i\,dx + \int_\Gamma \sqrt{g}(\sigma^{kj}g^{il}u_{l\|k})n_j v_i\,d\Gamma\end{aligned}$$

for all $\boldsymbol{v} = (v_i) \in \mathbf{W}^{1,4}(\Omega)$. Hence the variational equations imply that

$$\int_\Omega \sqrt{g}\{(\sigma^{ij} + \sigma^{kj}g^{il}u_{l\|k})\|_j + f^i\}v_i\,dx = \int_{\Gamma_1} \sqrt{g}\{(\sigma^{ij} + \sigma^{kj}g^{il}u_{l\|k})n_j - h^i\}v_i\,d\Gamma$$

for all $(v_i) \in \mathbf{W}(\Omega)$. Letting (v_i) vary first in $(\mathcal{D}(\Omega))^3$, then in $\mathbf{W}(\Omega)$, yields the announced boundary value problem. ■

Naturally, the same boundary value problem in curvilinear coordinates can be derived from its Cartesian version (Sect. 8.1); cf. Ex. 8.3. It can also be directly derived in curvilinear coordinates from the axioms of continuum mechanics, as in Oden [1972, Chap. 1] or Washizu [1975, Chap. 4].

Either one of the formulations described in this section, i.e., as a minimization, or variational, or boundary value, problem constitutes the **equations of three-dimensional nonlinear elasticity in curvilinear coordinates** for a *displacement-traction problem* associated with a *St Venant-Kirchhoff material.* Naturally, it contains as a special case the equations in *Cartesian coordinates* considered in Sect. 8.1.

The variational equations

$$\int_{\Omega} \sigma^{ij} F_{i\|j}(\boldsymbol{u}, \boldsymbol{v})\sqrt{g}\, dx = \int_{\Omega} f^i v_i \sqrt{g}\, dx + \int_{\Gamma_1} h^i v_i \sqrt{g}\, d\Gamma$$
$$\text{for all } \boldsymbol{v} \in \mathbf{W}(\Omega)$$

constitute the **principle of virtual work in curvilinear coordinates.** By Thm. 8.2-4, this principle is (at least formally) equivalent to the **equations of equilibrium in curvilinear coordinates:**

$$-(\sigma^{ij} + \sigma^{kj} g^{il} u_{l\|k})\|_j = f^i \text{ in } \Omega,$$
$$(\sigma^{ij} + \sigma^{kj} g^{il} u_{l\|k}) n_j = h^i \text{ on } \Gamma_1.$$

Finally, the functions

$$t^{ij} := \sigma^{ij} + \sigma^{kj} g^{il} u_{l\|k}$$

are the **contravariant components** of the **first Piola-Kirchhoff stress tensor.**

Remark. As their Cartesian special cases, the principle of virtual work and the equations of equilibrium are valid for *any* continuum (Vol. I, Chap. 2). ■

8.3. THE THREE-DIMENSIONAL EQUATIONS OF A NONLINEARLY ELASTIC SHELL

To begin with, we reproduce those preliminaries from Sect. 3.1 that apply *verbatim* to nonlinearly elastic shells.

Let ω be a domain in $\mathbb{R}^2$ with boundary γ. Let $y = (y_\alpha)$ denote a generic point in the set $\overline{\omega}$ and let $\partial_\alpha := \partial/\partial y_\alpha$. Let $\boldsymbol{\theta} \in \mathcal{C}^2(\overline{\omega}; \mathbb{R}^3)$ be an *injective* mapping such that *the two vectors*

$$\boldsymbol{a}_\alpha(y) := \partial_\alpha \boldsymbol{\theta}(y)$$

are linearly independent at all points $y \in \overline{\omega}$. They form the *covariant basis* of the tangent plane to the *surface*

$$S := \boldsymbol{\theta}(\overline{\omega})$$

at the point $\boldsymbol{\theta}(y)$ and the two vectors $\boldsymbol{a}^\alpha(y)$ of the tangent plane at $\boldsymbol{\theta}(y)$ defined by the relations

$$\boldsymbol{a}^\alpha(y) \cdot \boldsymbol{a}_\beta(y) = \delta^\alpha_\beta$$

form its *contravariant basis*. Let also

$$\boldsymbol{a}_3(y) = \boldsymbol{a}^3(y) := \frac{\boldsymbol{a}_1(y) \wedge \boldsymbol{a}_2(y)}{|\boldsymbol{a}_1(y) \wedge \boldsymbol{a}_2(y)|}.$$

Then $|\boldsymbol{a}_3(y)| = 1$, the vector $\boldsymbol{a}_3(y)$ is normal to S at the point $\boldsymbol{\theta}(y)$, and the three vectors $\boldsymbol{a}^i(y)$ form the *contravariant basis at* $\boldsymbol{\theta}(y)$; cf. Fig. 2.3-1. Note that (y_1, y_2) constitutes a system of *curvilinear coordinates* (Sect. 2.1) for describing the surface S.

Let γ_0 denote a $d\gamma$-measurable subset of the boundary γ of ω satisfying

$$\boxed{\textit{length } \gamma_0 > 0.}$$

For each $\varepsilon > 0$, we define the sets

$$\boxed{\begin{gathered} \Omega^\varepsilon := \omega \times]-\varepsilon, \varepsilon[, \\ \Gamma^\varepsilon_+ := \omega \times \{\varepsilon\}, \quad \Gamma^\varepsilon_- := \omega \times \{-\varepsilon\}, \quad \Gamma^\varepsilon_0 := \gamma_0 \times [-\varepsilon, \varepsilon]. \end{gathered}}$$

Let $x^\varepsilon = (x^\varepsilon_i)$ denote a generic point in the set $\overline{\Omega}^\varepsilon$ and let $\partial^\varepsilon_i := \partial/\partial x^\varepsilon_i$; hence $x^\varepsilon_\alpha = y_\alpha$ and $\partial^\varepsilon_\alpha = \partial_\alpha$.

Consider an **elastic shell** with **middle surface** $S = \boldsymbol{\theta}(\overline{\omega})$ and **thickness** $2\varepsilon > 0$, i.e., an elastic body whose *reference configuration* consists of all points within a distance $\leq \varepsilon$ from S (Fig. 1.2-1). In other words, *the reference configuration of the shell is the image* $\boldsymbol{\Theta}(\overline{\Omega}^\varepsilon) \subset \mathbb{R}^3$ *of the set* $\overline{\Omega}^\varepsilon \subset \mathbb{R}^3$ *through a mapping* $\boldsymbol{\Theta} : \overline{\Omega}^\varepsilon \to \mathbb{R}^3$ *given by*

$$\boxed{\boldsymbol{\Theta}(x^\varepsilon) := \boldsymbol{\theta}(y) + x^\varepsilon_3 \boldsymbol{a}_3(y) \text{ for all } x^\varepsilon = (y, x^\varepsilon_3) = (y_1, y_2, x^\varepsilon_3) \in \overline{\Omega}^\varepsilon.}$$

Thm. 3.1-1 is essential for describing the "geometry" of such a shell: It shows that, if the mapping $\boldsymbol{\theta} : \overline{\omega} \to \mathbb{R}^3$ is injective and smooth enough, *the mapping* $\boldsymbol{\Theta} : \overline{\Omega}^\varepsilon \to \mathbb{R}^3$ *is also injective for* $\varepsilon > 0$ *small enough*; consequently, $(y_1, y_2, x^\varepsilon_3)$ constitutes in this case a *bona fide* system of *curvilinear coordinates* for describing the reference configuration $\boldsymbol{\Theta}(\overline{\Omega}^\varepsilon)$ of the shell, called the *"natural" curvilinear coordinates of the shell* (Sect. 1.2), and the physical problem is meaningful as *the set* $\boldsymbol{\Theta}(\overline{\Omega}^\varepsilon)$ *does not "interpenetrate itself"*. The curvilinear coordinate $x_3^\varepsilon \in [-\varepsilon, \varepsilon]$ is called the **transverse variable**.

The same theorem also shows that, again for $\varepsilon > 0$ small enough, the three vectors

$$\boldsymbol{g}_i^\varepsilon(x^\varepsilon) := \partial_i^\varepsilon \boldsymbol{\Theta}(x^\varepsilon)$$

form the *covariant basis* (of the tangent space, here $\mathbb{R}^3$, to the manifold $\boldsymbol{\Theta}(\overline{\Omega}^\varepsilon)$) at the point $\boldsymbol{\Theta}(x^\varepsilon)$, and the three vectors $\boldsymbol{g}^{i,\varepsilon}(x^\varepsilon)$ defined by

$$\boldsymbol{g}^{i,\varepsilon}(x^\varepsilon) \cdot \boldsymbol{g}_j^\varepsilon(x^\varepsilon) = \delta_j^i$$

form the *contravariant basis* at $\boldsymbol{\Theta}(x^\varepsilon)$ (Sect. 1.2).

The *covariant* and *contravariant* components g_{ij}^ε and $g^{ij,\varepsilon}$ of the *metric tensor* (Sect. 1.2) and the *Christoffel symbols* $\Gamma_{ij}^{p,\varepsilon}$ (Sect. 1.4) of the manifold $\boldsymbol{\Theta}(\overline{\Omega}^\varepsilon)$ are then defined by letting (the explicit dependence on x^ε is omitted):

$$g_{ij}^\varepsilon := \boldsymbol{g}_i^\varepsilon \cdot \boldsymbol{g}_j^\varepsilon, \quad g^{ij,\varepsilon} := \boldsymbol{g}^{i,\varepsilon} \cdot \boldsymbol{g}^{j,\varepsilon},$$
$$\Gamma_{ij}^{p,\varepsilon} := \boldsymbol{g}^{p,\varepsilon} \cdot \partial_i^\varepsilon \boldsymbol{g}_j^\varepsilon.$$

Note the symmetries:

$$g_{ij}^\varepsilon = g_{ji}^\varepsilon, \quad g^{ij,\varepsilon} = g^{ji,\varepsilon}, \quad \Gamma_{ij}^{p,\varepsilon} = \Gamma_{ji}^{p,\varepsilon}.$$

The *volume element* in the set $\boldsymbol{\Theta}(\overline{\Omega}^\varepsilon)$ is $\sqrt{g^\varepsilon}\, dx^\varepsilon$, where

$$g^\varepsilon := \det(g_{ij}^\varepsilon).$$

We assume that the material constituting the shell is a *St Venant-Kirchhoff material*, with *Lamé constants* $\lambda^\varepsilon > 0$ and $\mu^\varepsilon > 0$. The *unknown* of the problem is the vector field $\boldsymbol{u}^\varepsilon = (u_i^\varepsilon) : \overline{\Omega}^\varepsilon \to \mathbb{R}^3$, where the three functions $u_i^\varepsilon : \overline{\Omega}^\varepsilon \to \mathbb{R}$ are the *covariant components* of the displacement field $u_i^\varepsilon \boldsymbol{g}^{i,\varepsilon} : \overline{\Omega}^\varepsilon \to \mathbb{R}^3$ that the shell undergoes under the influence of applied forces; this means that $u_i^\varepsilon(x^\varepsilon)\boldsymbol{g}^{i,\varepsilon}(x^\varepsilon)$ is the displacement of the point $\boldsymbol{\Theta}(x^\varepsilon)$; see Fig. 3.1-1.

Finally, we assume that the shell is subjected to a *boundary condition of place* along the portion $\boldsymbol{\Theta}(\Gamma_0^\varepsilon)$ of its *lateral face* $\boldsymbol{\Theta}(\gamma \times [-\varepsilon, \varepsilon])$, in the sense that the displacement vanishes there.

It then follows from Thm. 8.2-3 that the unknown $\boldsymbol{u}^\varepsilon = (u_i^\varepsilon)$ satisfies the following **three-dimensional variational problem $\mathcal{P}(\Omega^\varepsilon)$ of a nonlinearly elastic shell in curvilinear coordinates**, i.e., expressed in terms of the "natural" curvilinear coordinates x_i^ε of the shell:

$$\boldsymbol{u}^\varepsilon \in \mathbf{W}(\Omega^\varepsilon) := \{\boldsymbol{v}^\varepsilon = (v_i^\varepsilon) \in \mathbf{W}^{1,4}(\Omega^\varepsilon);\ \boldsymbol{v}^\varepsilon = \mathbf{0} \text{ on } \Gamma_0^\varepsilon\},$$
$$\int_{\Omega^\varepsilon} A^{ijkl,\varepsilon} E_{k\|l}^\varepsilon(\boldsymbol{u}^\varepsilon) F_{i\|j}^\varepsilon(\boldsymbol{u}^\varepsilon, \boldsymbol{v}^\varepsilon)\sqrt{g^\varepsilon}\, dx^\varepsilon$$
$$= \int_{\Omega^\varepsilon} f^{i,\varepsilon} v_i^\varepsilon \sqrt{g^\varepsilon}\, dx^\varepsilon + \int_{\Gamma_+^\varepsilon \cup \Gamma_-^\varepsilon} h^{i,\varepsilon} v_i^\varepsilon \sqrt{g^\varepsilon}\, d\Gamma^\varepsilon$$
$$\text{for all } \boldsymbol{v}^\varepsilon \in \mathbf{W}(\Omega^\varepsilon),$$

where

$$A^{ijkl,\varepsilon} := \lambda^\varepsilon g^{ij,\varepsilon} g^{kl,\varepsilon} + \mu^\varepsilon (g^{ik,\varepsilon} g^{jl,\varepsilon} + g^{il,\varepsilon} g^{jk,\varepsilon})$$

designate the *contravariant* components of the *three-dimensional elasticity tensor*,

$$E_{i\|j}^\varepsilon(\boldsymbol{v}^\varepsilon) := \frac{1}{2}(v_{i\|j}^\varepsilon + v_{j\|i}^\varepsilon + g^{mn,\varepsilon} v_{m\|i}^\varepsilon v_{n\|j}^\varepsilon), \text{ where}$$
$$v_{i\|j}^\varepsilon := \partial_j^\varepsilon v_i^\varepsilon - \Gamma_{ij}^{p,\varepsilon} v_p^\varepsilon,$$

designate the *strains in curvilinear coordinates* associated with an arbitrary displacement field $v_i^\varepsilon \boldsymbol{g}^{i,\varepsilon}$ of the manifold $\boldsymbol{\Theta}(\overline{\Omega}^\varepsilon)$ (Thm. 8.2-2),

$$F_{i\|j}^\varepsilon(\boldsymbol{u}^\varepsilon, \boldsymbol{v}^\varepsilon) := (E_{i\|j}^\varepsilon)'(\boldsymbol{u}^\varepsilon)\boldsymbol{v}^\varepsilon$$
$$= \frac{1}{2}(v_{i\|j}^\varepsilon + v_{j\|i}^\varepsilon + g^{mn,\varepsilon}\{u_{m\|i}^\varepsilon v_{n\|j}^\varepsilon + u_{n\|j}^\varepsilon v_{m\|i}^\varepsilon\}),$$

and, finally, $f^{i,\varepsilon} \in L^2(\Omega^\varepsilon)$ and $h^{i,\varepsilon} \in L^2(\Gamma_+^\varepsilon \cup \Gamma_-^\varepsilon)$ denote the contravariant components of the *applied body* and *surface force densities*, respectively applied to the interior $\boldsymbol{\Theta}(\Omega^\varepsilon)$ of the shell and to its *"upper"* and *"lower" faces* $\boldsymbol{\Theta}(\Gamma_+^\varepsilon)$ and $\boldsymbol{\Theta}(\Gamma_-^\varepsilon)$, and $d\Gamma^\varepsilon$ designates the area element along $\partial\Omega^\varepsilon$. We thus assume that there are *no* surface forces applied to the portion $\boldsymbol{\Theta}((\gamma - \gamma_0) \times [-\varepsilon, \varepsilon])$ of the lateral face of the shell.

We record in passing the symmetries

$$A^{ijkl,\varepsilon} = A^{jikl,\varepsilon} = A^{klij,\varepsilon},$$

and the relations (satisfied because the mapping $\boldsymbol{\Theta}$ is of the special form given *supra*)

$$\Gamma^{3,\varepsilon}_{\alpha 3} = \Gamma^{p,\varepsilon}_{33} = 0 \text{ and } A^{\alpha\beta\sigma 3,\varepsilon} = A^{\alpha 333,\varepsilon} = 0 \text{ in } \overline{\Omega}^{\varepsilon}.$$

Our final objective consists in showing by means of the *method of formal asymptotic expansions* that, if the data are of an appropriate order with respect to ε as $\varepsilon \to 0$, the above three-dimensional problems are "asymptotically equivalent" to a *"two-dimensional problem posed over the middle surface of the shell"*. This means that *the new unknown should be* $\boldsymbol{\zeta}^{\varepsilon} = (\zeta_i^{\varepsilon})$, *where* ζ_i^{ε} *are the covariant components of the displacement* $\zeta_i^{\varepsilon}\boldsymbol{a}^i : \overline{\omega} \to \mathbb{R}^3$ *of the points of the middle surface* $S = \boldsymbol{\theta}(\overline{\omega})$. In other words, $\zeta_i^{\varepsilon}(y)\boldsymbol{a}^i(y)$ is the displacement of the point $\boldsymbol{\theta}(y) \in S$; see Fig. 3.1-2.

8.4. THE THREE-DIMENSIONAL EQUATIONS OVER A DOMAIN INDEPENDENT OF ε

We describe in this section the *basic preliminaries of the asymptotic analysis of a nonlinearly elastic shell.* Since these are essentially the same as in the *linear* case, we freely borrow those parts of Sect. 3.2 that apply *verbatim* to the nonlinear case.

"Asymptotic analysis" means that *our objective is to study the behavior of the displacement field* $u_i^{\varepsilon}\boldsymbol{g}^{i,\varepsilon} : \overline{\Omega}^{\varepsilon} \to \mathbb{R}^3$ *as* $\varepsilon \to 0$, an endeavor that will be achieved by studying the *behavior as* $\varepsilon \to 0$ *of the covariant components* $u_i^{\varepsilon} : \overline{\Omega}^{\varepsilon} \to \mathbb{R}$ *of the displacement field*, i.e, *the behavior of the unknown* $\boldsymbol{u}^{\varepsilon} = (u_i^{\varepsilon}) : \overline{\Omega}^{\varepsilon} \to \mathbb{R}^3$ of the three-dimensional shell problem $\mathcal{P}(\Omega^{\varepsilon})$ described in Sect. 8.3.

Since these fields are defined on sets $\overline{\Omega}^{\varepsilon}$ that themselves vary with ε, our first task naturally consists in *transforming the three-dimensional variational problems* $\mathcal{P}(\Omega^{\varepsilon})$ *into problems posed over a set that does not depend on* ε.

Let

$$\Omega := \omega \times]-1, 1[,$$
$$\Gamma_0 := \gamma_0 \times [-1, 1], \quad \Gamma_+ := \omega \times \{1\}, \quad \Gamma_- := \omega \times \{-1\}.$$

Let $x = (x_1, x_2, x_3)$ denote a generic point in the set $\overline{\Omega}$ and let $\partial_i := \partial/\partial x_i$; hence $x_\alpha = y_\alpha$ (a generic point in the set $\overline{\omega}$ is denoted $y = (y_1, y_2)$; cf. Sect. 8.3). The coordinate $x_3 \in [-1, 1]$ is called the **transverse variable**, like $x_3^\varepsilon \in [-\varepsilon, \varepsilon]$ in Sect. 8.3 ("scaled" transverse variable could thus be preferred; however, no confusion should arise).

With each point $x \in \overline{\Omega}$, we associate the point $x^\varepsilon \in \overline{\Omega}^\varepsilon$ through the bijection (Fig. 3.2-1)

$$\pi^\varepsilon : x = (x_1, x_2, x_3) \in \overline{\Omega} \longrightarrow x^\varepsilon = (x_i^\varepsilon) = (x_1, x_2, \varepsilon x_3) \in \overline{\Omega}^\varepsilon.$$

Consequently,

$$\partial_\alpha^\varepsilon = \partial_\alpha \text{ and } \partial_3^\varepsilon = \frac{1}{\varepsilon}\partial_3.$$

With the unknown $\boldsymbol{u}^\varepsilon = (u_i^\varepsilon) : \overline{\Omega}^\varepsilon \to \mathbb{R}^3$ and the vector fields $\boldsymbol{v}^\varepsilon = (v_i^\varepsilon) : \overline{\Omega}^\varepsilon \to \mathbb{R}^3$ appearing in the three-dimensional problem $\mathcal{P}(\Omega^\varepsilon)$, we associate the **scaled unknown** $\boldsymbol{u}(\varepsilon) = (u_i(\varepsilon)) : \overline{\Omega} \to \mathbb{R}^3$ and the **scaled vector fields** $\boldsymbol{v} = (v_i) : \overline{\Omega} \to \mathbb{R}^3$ defined by the **scalings**:

$$u_i^\varepsilon(x^\varepsilon) =: u_i(\varepsilon)(x) \text{ and } v_i^\varepsilon(x^\varepsilon) =: v_i(x) \text{ for all } x^\varepsilon = \pi^\varepsilon x \in \overline{\Omega}^\varepsilon,$$

and we call **scaled displacements** the three components $u_i(\varepsilon)$ of the scaled unknown $\boldsymbol{u}(\varepsilon)$.

We assume at the outset that *the Lamé constants* λ^ε *and* μ^ε *are independent of* ε, i.e., that there exist constants $\lambda > 0$ and $\mu > 0$ such that

$$\lambda^\varepsilon = \lambda \text{ and } \mu^\varepsilon = \mu \text{ for all } \varepsilon > 0.$$

Remark. A multiplication of both Lamé constants by a factor ε^t, $t \in \mathbb{R}$, is otherwise always possible, as we shall see in the next chapters (Sects. 9.1 and 10.2). The choice $t = 0$ is made here for definiteness. ∎

Then functions $f^i(\varepsilon) : \Omega \to \mathbb{R}$ and $h^i(\varepsilon) : \Gamma_+ \cup \Gamma_- \to \mathbb{R}$ are defined by letting

$$\boxed{\begin{aligned} f^{i,\varepsilon}(x^\varepsilon) &=: f^i(\varepsilon)(x) \text{ for all } x^\varepsilon = \pi^\varepsilon x \in \Omega^\varepsilon, \\ h^{i,\varepsilon}(x^\varepsilon) &=: h^i(\varepsilon)(x) \text{ for all } x^\varepsilon = \pi^\varepsilon x \in \Gamma_+^\varepsilon \cup \Gamma_-^\varepsilon. \end{aligned}}$$

A word of caution. At this stage, it *cannot* be *a priori* assumed as in Sect. 3.2 for the linear theory that the scaled displacements $u_i(\varepsilon)$ are "of order 0 with respect to ε". It so happens that this will be the case, in fact the *only* possible case, for the nonlinear theory, but proving this requires considerable efforts (Thm. 8.7-1)!

Likewise, the "right" orders on the contravariant components of the applied forces are *yet to be found* (as in the linear theory). Finding these (which depend on the kinds of shells that are considered) again requires considerable efforts (Thms. 8.8-1 and 10.1-2)! ∎

A simple computation then yields the variational problem that the scaled unknown $\boldsymbol{u}(\varepsilon)$ satisfies over the set Ω, thus over a domain that is *independent of* ε (the Christoffel symbols $\Gamma_{\alpha 3}^{3,\varepsilon}$ and $\Gamma_{33}^{p,\varepsilon}$ vanish in Ω^ε for the special class of mappings $\boldsymbol{\Theta}$ considered here; consequently, the functions $\Gamma_{\alpha 3}^3(\varepsilon)$ and $\Gamma_{33}^p(\varepsilon)$ defined below likewise vanish in Ω):

Theorem 8.4-1. *Let* ω *be a domain in* $\mathbb{R}^2$, *let* $\boldsymbol{\theta} \in \mathcal{C}^2(\overline{\omega}; \mathbb{R}^3)$ *be an injective mapping such that the two vectors* $\boldsymbol{a}_\alpha = \partial_\alpha \boldsymbol{\theta}$ *are linearly independent at all points of* $\overline{\omega}$, *and let* $\varepsilon_0 > 0$ *be chosen as in* Thm. 3.1-1. *With the functions* $g^{ij,\varepsilon}, \Gamma_{ij}^{p,\varepsilon}, g^\varepsilon, A^{ijkl,\varepsilon} : \overline{\Omega}^\varepsilon \to \mathbb{R}$ *appearing in problem* $\mathcal{P}(\Omega^\varepsilon)$ (Sect. 8.3), *we associate for each* $0 < \varepsilon \le \varepsilon_0$ *the "scaled" functions* $g^{ij}(\varepsilon), \Gamma_{ij}^p(\varepsilon), g(\varepsilon), A^{ijkl}(\varepsilon) : \overline{\Omega} \to \mathbb{R}$ *defined by*

$$\boxed{\begin{aligned} g^{ij}(\varepsilon)(x) &:= g^{ij,\varepsilon}(x^\varepsilon), \quad \Gamma_{ij}^p(\varepsilon)(x) := \Gamma_{ij}^{p,\varepsilon}(x^\varepsilon), \\ g(\varepsilon)(x) &:= g^\varepsilon(x^\varepsilon), \quad A^{ijkl}(\varepsilon)(x) := A^{ijkl,\varepsilon}(x^\varepsilon), \\ &\qquad \text{for all } x^\varepsilon = \pi^\varepsilon x \in \overline{\Omega}^\varepsilon. \end{aligned}}$$

Then for each $0 < \varepsilon \leq \varepsilon_0$, the scaled unknown $\boldsymbol{u}(\varepsilon)$ satisfies the following variational problem $\mathcal{P}(\varepsilon;\, \Omega)$, called the **scaled three-dimensional variational problem of a nonlinearly elastic shell:**

$$\boldsymbol{u}(\varepsilon) = (u_i(\varepsilon)) \in \mathbf{W}(\Omega) := \{\boldsymbol{v} = (v_i) \in \mathbf{W}^{1,4}(\Omega);\ \boldsymbol{v} = \mathbf{0} \text{ on } \Gamma_0\},$$

$$\int_\Omega A^{ijkl}(\varepsilon) E_{k\|l}(\varepsilon;\, \boldsymbol{u}(\varepsilon)) F_{i\|j}(\varepsilon;\, \boldsymbol{u}(\varepsilon), \boldsymbol{v}) \sqrt{g(\varepsilon)}\, dx$$

$$= \int_\Omega f^i(\varepsilon) v_i \sqrt{g(\varepsilon)}\, dx + \frac{1}{\varepsilon} \int_{\Gamma_+ \cup \Gamma_-} h^i(\varepsilon) v_i \sqrt{g(\varepsilon)}\, d\Gamma$$

$$\text{for all } \boldsymbol{v} \in \mathbf{W}(\Omega),$$

where

$$E_{i\|j}(\varepsilon;\, \boldsymbol{u}(\varepsilon)) := \frac{1}{2}\big(u_{i\|j}(\varepsilon) + u_{j\|i}(\varepsilon)\big) + \frac{1}{2} g^{mn}(\varepsilon) u_{m\|i}(\varepsilon) u_{n\|j}(\varepsilon),$$

$$F_{i\|j}(\varepsilon;\, \boldsymbol{u}(\varepsilon), \boldsymbol{v}) := \frac{1}{2}\big(v_{i\|j}(\varepsilon) + v_{j\|i}(\varepsilon)\big)$$

$$+ \frac{1}{2} g^{mn}(\varepsilon)\{u_{m\|i}(\varepsilon) v_{n\|j}(\varepsilon) + u_{n\|j}(\varepsilon) v_{m\|i}(\varepsilon)\},$$

and

$$u_{\beta\|\alpha}(\varepsilon) := \partial_\alpha u_\beta(\varepsilon) - \Gamma^p_{\alpha\beta}(\varepsilon) u_p(\varepsilon), \quad v_{\beta\|\alpha}(\varepsilon) := \partial_\alpha v_\beta - \Gamma^p_{\alpha\beta}(\varepsilon) v_p,$$

$$u_{3\|\alpha}(\varepsilon) := \partial_\alpha u_3(\varepsilon) - \Gamma^\sigma_{\alpha 3}(\varepsilon) u_\sigma(\varepsilon), \quad v_{3\|\alpha}(\varepsilon) := \partial_\alpha v_3 - \Gamma^\sigma_{\alpha 3}(\varepsilon) v_\sigma,$$

$$u_{\alpha\|3}(\varepsilon) := \frac{1}{\varepsilon}\partial_3 u_\alpha(\varepsilon) - \Gamma^\sigma_{\alpha 3}(\varepsilon) u_\sigma(\varepsilon), \quad v_{\alpha\|3}(\varepsilon) := \frac{1}{\varepsilon}\partial_3 v_\alpha - \Gamma^\sigma_{\alpha 3}(\varepsilon) v_\sigma,$$

$$u_{3\|3}(\varepsilon) := \frac{1}{\varepsilon}\partial_3 u_3(\varepsilon), \quad v_{3\|3}(\varepsilon) := \frac{1}{\varepsilon}\partial_3 v_3.$$

■

The functions $A^{ijkl}(\varepsilon)$ are called the **contravariant components of the scaled three-dimensional elasticity tensor of the shell.** The functions $E_{i\|j}(\varepsilon;\, \boldsymbol{u}(\varepsilon))$ are called the **scaled strains in curvilinear coordinates** because they satisfy

$$E_{i\|j}(\varepsilon;\, \boldsymbol{u}(\varepsilon))(x) = E^\varepsilon_{i\|j}(\boldsymbol{u}^\varepsilon)(x^\varepsilon) \text{ for all } x^\varepsilon = \pi^\varepsilon x \in \overline{\Omega}^\varepsilon.$$

Note that the above definitions likewise imply that

$$F_{i\|j}(\varepsilon;\, \boldsymbol{u}(\varepsilon),\, \boldsymbol{v})(x) = F^{\varepsilon}_{i\|j}(\boldsymbol{u}^{\varepsilon},\, \boldsymbol{v}^{\varepsilon})(x^{\varepsilon}) \text{ for all } x^{\varepsilon} = \pi^{\varepsilon}x \in \overline{\Omega}^{\varepsilon},$$
$$u_{i\|j}(\varepsilon)(x) = u^{\varepsilon}_{i\|j}(x^{\varepsilon}) \text{ and } v_{i\|j}(\varepsilon)(x) = v^{\varepsilon}_{i\|j}(x^{\varepsilon}) \quad \text{for all } x^{\varepsilon} = \pi^{\varepsilon}x \in \overline{\Omega}^{\varepsilon}.$$

8.5. GEOMETRICAL AND MECHANICAL PRELIMINARIES

Let ω be a domain in $\mathbb{R}^2$ and let $\boldsymbol{\theta} \in \mathcal{C}^3(\overline{\omega};\, \mathbb{R}^3)$ be an injective mapping such that the two vectors $\boldsymbol{a}_\alpha = \partial_\alpha \boldsymbol{\theta}$ are linearly independent at all points of $\overline{\omega}$. Let $\boldsymbol{a}_i$ and $\boldsymbol{a}^i$ denote the vectors of the covariant and contravariant bases along the surface $S = \boldsymbol{\theta}(\overline{\omega})$.

The *covariant* and *contravariant components* $a_{\alpha\beta}$ and $a^{\alpha\beta}$ of the *metric tensor* of the surface S (Sect. 2.1), the *covariant* and *mixed components* $b_{\alpha\beta}$ and b^β_α of the *curvature tensor* of S (Sect. 2.2), and the *Christoffel symbols* $\Gamma^\sigma_{\alpha\beta}$ of S (Sect. 2.3) are then defined by letting (whenever no confusion should arise, we henceforth drop the explicit dependence on the variable $y \in \overline{\omega}$):

$$a_{\alpha\beta} := \boldsymbol{a}_\alpha \cdot \boldsymbol{a}_\beta, \qquad a^{\alpha\beta} := \boldsymbol{a}^\alpha \cdot \boldsymbol{a}^\beta,$$
$$b_{\alpha\beta} := \boldsymbol{a}^3 \cdot \partial_\beta \boldsymbol{a}_\alpha, \qquad b^\beta_\alpha := a^{\beta\sigma} b_{\sigma\alpha},$$
$$\Gamma^\sigma_{\alpha\beta} := \boldsymbol{a}^\sigma \cdot \partial_\beta \boldsymbol{a}_\alpha.$$

Note the symmetries:

$$a_{\alpha\beta} = a_{\beta\alpha}, \quad a^{\alpha\beta} = a^{\beta\alpha}, \quad b_{\alpha\beta} = b_{\beta\alpha}, \quad \Gamma^\sigma_{\alpha\beta} = \Gamma^\sigma_{\beta\alpha}.$$

The *area element* along S is $\sqrt{a}\, dy$, where

$$a := \det(a_{\alpha\beta}).$$

The covariant derivatives $b^\sigma_\beta|_\alpha$ of the curvature tensor, defined in this case by means of its mixed components, are defined by (Thm. 2.5-1)

$$b^\sigma_\beta|_\alpha := \partial_\alpha b^\sigma_\beta + \Gamma^\sigma_{\alpha\tau} b^\tau_\beta - \Gamma^\tau_{\alpha\beta} b^\sigma_\tau.$$

All the functions $a_{\alpha\beta}$, $a^{\alpha\beta}$, $b_{\alpha\beta}$, b^β_α, $\Gamma^\sigma_{\alpha\beta}$, a, and $b^\sigma_\beta|_\alpha$ are thus at least continuous over the set $\overline{\omega}$.

Our first result, which is essentially Thm. 3.3-1 rewritten in a slightly different form, gathers all the *"geometrical" preliminaries* needed in the sequel for the asymptotic analysis of nonlinearly elastic shells, regarding *the behavior as* $\varepsilon \to 0$ *of the functions* $g^{ij}(\varepsilon)$, $\Gamma^p_{ij}(\varepsilon)$, *and* $g(\varepsilon)$ (defined in Thm. 8.4-1). Note that, while these are functions of $x = (y, x_3) \in \overline{\Omega} = \overline{\omega} \times [-1, 1]$, *their limits for* $\varepsilon = 0$ *are functions of* $y \in \overline{\omega}$ *only, i.e., the limits are independent of the transverse variable* x_3.

Theorem 8.5-1. *Let* ω *be a domain in* $\mathbb{R}^2$, *let* $\boldsymbol{\theta} \in \mathcal{C}^3(\overline{\omega}; \mathbb{R}^3)$ *be an injective mapping such that the two vectors* $\boldsymbol{a}_\alpha = \partial_\alpha \boldsymbol{\theta}$ *are linearly independent at all points of* $\overline{\omega}$ *and let* $\varepsilon_0 > 0$ *be chosen as in* Thm. 3.1-1. *Then*

$$\boxed{\begin{aligned} &g^{ij}(\varepsilon) = a^{ij} + \varepsilon x_3 g^{ij,1} + O(\varepsilon^2), \text{ where} \\ &a^{ij} := \boldsymbol{a}^i \cdot \boldsymbol{a}^j, \quad g^{\alpha\beta,1} := 2a^{\alpha\sigma} b^\beta_\sigma, \quad g^{i3,1} = 0, \end{aligned}}$$

$$\boxed{\begin{aligned} &\Gamma^p_{ij}(\varepsilon) = \Gamma^{p,0}_{ij} + \varepsilon x_3 \Gamma^{p,1}_{ij} + O(\varepsilon^2), \text{ where} \\ &\Gamma^{\sigma,0}_{\alpha\beta} := \Gamma^\sigma_{\alpha\beta}, \quad \Gamma^{3,0}_{\alpha\beta} := b_{\alpha\beta}, \quad \Gamma^{\sigma,0}_{\alpha 3} := -b^\sigma_\alpha, \quad \Gamma^{3,0}_{\alpha 3} = \Gamma^{p,0}_{33} := 0, \\ &\Gamma^{\sigma,1}_{\alpha\beta} := -b^\sigma_\beta|_\alpha, \; \Gamma^{3,1}_{\alpha\beta} := -b^\sigma_\alpha b_{\sigma\beta}, \; \Gamma^{\sigma,1}_{\alpha 3} := -b^\tau_\alpha b^\sigma_\tau, \; \Gamma^{3,1}_{\alpha 3} = \Gamma^{p,1}_{33} := 0, \end{aligned}}$$

and finally,

$$\boxed{g(\varepsilon) = a + O(\varepsilon),}$$

for all $0 < \varepsilon \le \varepsilon_0$, *where the functions* a^{ij}, $b_{\alpha\beta}$, b^σ_α, $\Gamma^\sigma_{\alpha\beta}$, $b^\sigma_\beta|_\alpha$, *and* a *are identified with functions in* $\mathcal{C}^0(\overline{\Omega})$ *and the order symbols* $O(\varepsilon)$ *and* $O(\varepsilon^2)$ *are meant with respect to the norm* $\|\cdot\|_{0,\infty,\overline{\Omega}}$ *defined by*

$$\|w\|_{0,\infty,\overline{\Omega}} := \sup\{|w(x)|;\, x \in \overline{\Omega}\}.$$

Proof. The asymptotic behavior of each function $g^{ij}(\varepsilon)$ immediately follows from the relations

$$\boldsymbol{g}^\alpha(\varepsilon) = \boldsymbol{a}^\alpha + \varepsilon x_3 b^\alpha_\sigma \boldsymbol{a}^\sigma + O(\varepsilon^2),$$

established in the course of the proof of Thm. 3.3-1. All the other relations were already proved in the same theorem. ■

Observe that the notation a^{ij} is consistent with that used for the contravariant components $a^{\alpha\beta} = \boldsymbol{a}^\alpha \cdot \boldsymbol{a}^\beta$ of the metric tensor of the surface S; note also that $a^{i3} = \delta^{i3}$.

Our second result, which is a restatement of Thm. 3.3-2 (b), gathers all the *"mechanical" preliminaries* needed in the sequel, in that it describes the *behavior as $\varepsilon \to 0$ of the scaled contravariant components $A^{ijkl}(\varepsilon)$ of the three-dimensional elasticity tensor* (defined in Thm. 8.4-1). Note that their limits for $\varepsilon = 0$ are again functions of $y \in \overline{\omega}$ only.

Theorem 8.5-2. *Let ω be a domain in $\mathbb{R}^2$, let $\boldsymbol{\theta} \in \mathcal{C}^3(\overline{\omega}; \mathbb{R}^3)$ be an injective mapping such that the two vectors $\boldsymbol{a}_\alpha$ are linearly independent at all points of $\overline{\omega}$, and let $\varepsilon_0 > 0$ be chosen as in* Thm. 3.1-1. *Then the contravariant components $A^{ijkl}(\varepsilon) = A^{jikl}(\varepsilon) = A^{klij}(\varepsilon)$ of the scaled three-dimensional elasticity tensor satisfy*

$$
\begin{aligned}
&A^{ijkl}(\varepsilon)\sqrt{g(\varepsilon)} = A^{ijkl}(0)\sqrt{a} + \varepsilon B^{ijkl,1} + \varepsilon^2 B^{ijkl,2} + o(\varepsilon^2), \text{ where}\\
&A^{\alpha\beta\sigma\tau}(0) := \lambda a^{\alpha\beta}a^{\sigma\tau} + \mu(a^{\alpha\sigma}a^{\beta\tau} + a^{\alpha\tau}a^{\beta\sigma}),\\
&A^{\alpha\beta 33}(0) := \lambda a^{\alpha\beta},\ A^{\alpha 3\sigma 3}(0) := \mu a^{\alpha\sigma},\ A^{3333}(0) := \lambda + 2\mu,\\
&A^{\alpha\beta\sigma 3}(0) = A^{\alpha 333}(0) := 0,
\end{aligned}
$$

for all $0 < \varepsilon \le \varepsilon_0$, where the functions $a \in \mathcal{C}^2(\overline{\omega})$ and $a^{\alpha\beta} \in \mathcal{C}^2(\overline{\omega})$ are identified with functions in $\mathcal{C}^2(\overline{\Omega})$, the functions $B^{ijkl,1}$ and $B^{ijkl,2}$ are continuous over $\overline{\Omega}$, and the order symbol $o(\varepsilon^2)$ is meant with respect to the norm $\|\cdot\|_{0,\infty,\overline{\Omega}}$ (defined in Thm. 8.5-1). ■

8.6. THE METHOD OF FORMAL ASYMPTOTIC EXPANSIONS

The specific dependence on ε of the equations of problem $\mathcal{P}(\varepsilon; \Omega)$ observed in Thm. 8.4-1 and the idea that ε is a "small" parameter naturally lead to apply the **method of formal asymptotic expansions**, briefly described *infra*. For general presentations, see Lions [1973] and Eckhaus [1979]. This method consists in using the following *basic* **Ansatz**:

(i) *Write* a priori $\boldsymbol{u}(\varepsilon)$ *as a* **formal asymptotic expansion**

$$\boldsymbol{u}(\varepsilon) = \frac{1}{\varepsilon^N}\boldsymbol{u}^{-N} + \frac{1}{\varepsilon^{N-1}}\boldsymbol{u}^{-N+1} + \cdots, \; N \in \mathbb{Z},$$

where the successive terms $\boldsymbol{u}^{-N}$, $\boldsymbol{u}^{-N+1}$, etc., in this expansion are *independent of* ε. The "first" *nonzero* term $\boldsymbol{u}^{-N}$ (nonzero if the applied forces are nonzero!) is called the **leading term**, and more generally $\boldsymbol{u}^q$, $q \geq -N$, is called the **term of order** q, of the formal expansion; the three dots "$\cdots$" account for the fact that the number of successive terms that will be eventually needed is left *unspecified* at this stage; the expansion is *formal* in that it is not required to prove that the terms $\boldsymbol{u}^q$, $q \geq -N+1$, do exist in the space $\mathbf{W}(\Omega)$, let alone that the above "series" converges!

Note that, while in the *linear* case there was no loss of generality in assuming *ab initio* that the leading term is of order 0 (Sect. 3.4), *this freedom is lost in the nonlinear case* (see Thm. 8.7-1)!

(ii) *Equate to zero the factors of the successive powers* ε^q, *arranged by increasing values of* q, *found in problem* $\mathcal{P}(\varepsilon;\Omega)$ *when* $\boldsymbol{u}(\varepsilon)$ *is replaced by its formal expansion, solve the resulting variational equations and, assuming* ad hoc *properties on whichever successive terms are needed, pursue this procedure until the order* $-N$ *of the leading term* $\boldsymbol{u}^{-N}$ *is determined.* In the present case, the "*ad hoc* properties" consist in assuming that each term needed in this process is in the space $\mathbf{W}^{1,4}(\Omega)$ and that it is required to satisfy the boundary condition on Γ_0 only if this is necessary for continuing the recursion.

(iii) Pursue likewise this procedure until the *leading term can be fully identified*, presumably as the solution of an *ad hoc* variational problem.

(iv) The following *two requirements* constantly guide the procedures described in (ii) and (iii). The first requirement was systematized by Miara [1994a] for linearly elastic plates (see also Vol. I, Sect. 1.10) and Miara & Sanchez-Palencia [1996] for linearly elastic shells (see also Sect. 3.4); the second was systematized by Miara [1994b, 1998] for nonlinearly elastic plates (see also Vol. I, Sect. 4.11) and shells (see Thms. 8.8-1 and 10.1-2).

The *first requirement*, valid in both the linear and nonlinear cases, asserts that *no restriction should be imposed on the applied forces entering the right-hand side of the equations used for determining the leading term*, since no such restriction is imposed on the "original" problem $\mathcal{P}(\varepsilon; \Omega)$.

The *second requirement*, valid in the nonlinear case only, asserts that, by retaining only the *linear* terms in any relation satisfied by terms of arbitrary order in the formal asymptotic expansion of the scaled unknown $\boldsymbol{u}(\varepsilon)$, a relation of the *linear* theory should be recovered. Loosely speaking, we thus wish that *the operations of taking "formal limits" as ε goes to 0 and of linearizing should commute.* For brevity, we shall call **"linearization trick"** this second requirement.

Applying the above Ansatz, we will first establish that $N = 0$, i.e., that *the leading term is of order* 0 (Thm. 8.7-1). We will then *identify* (first in Thm. 8.8-1, then in Thms. 9.2-1 and 10.3-4) *the two classes of variational problems that the leading term can satisfy*, according to *two classes of specific assumptions* bearing on the *geometry of the middle surface* of the shell and on the *boundary conditions*, together with the corresponding *assumptions* (in terms of powers of ε) *on the components of the applied forces.*

The proofs of these results, which are due to Miara [1998] and Lods & Miara [1998], are at times exceedingly delicate (they were already somewhat intricate in the linear case; cf. Sect. 3.4). They show the *highly improbable efficiency of the method of formal asymptotic expansions*: Seemingly endless and inextricable computations eventually lead to two-dimensional problems whose formulations exhibit a striking simplicity!

8.7. THE LEADING TERM IS OF ORDER ZERO

As shown by Miara [1998, Thm. 1], whom we follow in this section, a first virtue of the Ansatz described in Sect. 8.6 is to afford the specification of the order of the leading term in the expansion of the scaled unknown:

Theorem 8.7-1. *Assume that the scaled unknown satisfying problem* $\mathcal{P}(\varepsilon; \Omega)$ (Thm. 8.4-1) *admits for each* $0 < \varepsilon \leq \varepsilon_0$ *a formal* *asymptotic expansion of the form*

$$\boxed{\begin{gathered}\boldsymbol{u}(\varepsilon) = \frac{1}{\varepsilon^N}\boldsymbol{u}^{-N} + \frac{1}{\varepsilon^{N-1}}\boldsymbol{u}^{-N+1} + \cdots, \text{ with} \\ \boldsymbol{u}^{-N}, \boldsymbol{u}^{-N+1} \in \mathbf{W}(\Omega) = \{\boldsymbol{v} \in \mathbf{W}^{1,4}(\Omega),\ \boldsymbol{v} = \mathbf{0} \text{ on } \Gamma_0\},\ \boldsymbol{u}^{-N} \neq \mathbf{0},\end{gathered}}$$

for some integer $N \in \mathbb{Z}$. *Then* $N = 0$.

Proof. The proof is broken into seven parts. Before beginning the induction proper in (iv), we record several useful preliminaries:

(i) *Let the functions* $A^{ijkl}(0)$ *be defined as in* Thm. 8.5-2. *Then, for any symmetric matrices* (s_{kl}) *and* (t_{ij}),

$$\begin{aligned} A^{ijkl}(0)s_{kl}t_{ij} &= (\lambda a^{\alpha\beta}a^{\sigma\tau} + \mu\{a^{\alpha\sigma}a^{\beta\tau} + a^{\alpha\tau}a^{\beta\sigma}\})s_{\sigma\tau}t_{\alpha\beta} \\ &\quad + 4\mu a^{\alpha\sigma}s_{\alpha 3}t_{\sigma 3} + \lambda a^{\alpha\beta}s_{33}t_{\alpha\beta} \\ &\quad + \lambda a^{\sigma\tau}s_{\sigma\tau}t_{33} + (\lambda + 2\mu)s_{33}t_{33}. \end{aligned}$$

This formula, which immediately follows from the definitions, will be constantly put to use in the ensuing arguments.

(ii) *Let* $a^{ij} := \boldsymbol{a}^i \cdot \boldsymbol{a}^j$. *Then, for any* $y \in \overline{\omega}$ *and any matrix* (t_{ij}),

$$a^{ij}(y)a^{mn}(y)t_{im}t_{jn} \geq 0 \text{ and } a^{ij}(y)a^{mn}(y)t_{im}t_{jn} = 0 \Leftrightarrow t_{ij} = 0.$$

Given any $y \in \overline{\omega}$ and any matrix (t_{ij}), let $\boldsymbol{t}_i(y) := t_{im}\boldsymbol{a}^m(y)$ and let $[\boldsymbol{t}_i(y)]^q$ denote the q-th Cartesian component of the vector $\boldsymbol{t}_i(y)$. We thus have

$$\begin{aligned} a^{ij}(y)a^{mn}(y)t_{im}t_{jn} &= a^{ij}(y)\left\{(t_{im}\boldsymbol{a}^m(y)) \cdot (t_{jn}\boldsymbol{a}^n(y))\right\} \\ &= a^{ij}(y)\left\{\boldsymbol{t}_i(y) \cdot \boldsymbol{t}_j(y)\right\} = a^{ij}(y)[\boldsymbol{t}_i(y)]^p[\boldsymbol{t}_j(y)]^p \\ &= \left([\boldsymbol{t}_i(y)]^p\boldsymbol{a}^i(y)\right) \cdot \left([\boldsymbol{t}_j(y)]^p\boldsymbol{a}^j(y)\right) = \sum_p \left|[\boldsymbol{t}_i(y)]^p\boldsymbol{a}^i(y)\right|^2. \end{aligned}$$

Hence $a^{ij}(y)a^{mn}(y)t_{im}t_{jn} \geq 0$ and, since the three vectors $\boldsymbol{a}^i(y)$ are linearly independent,

$$\begin{aligned} a^{ij}(y)a^{mn}(y)t_{im}t_{jn} = 0 &\Rightarrow [\boldsymbol{t}_i(y)]^p\boldsymbol{a}^i(y) = \mathbf{0},\ 1 \leq p \leq 3, \\ &\Rightarrow \boldsymbol{t}_i(y) = t_{im}\boldsymbol{a}^m(y) = 0,\ 1 \leq i \leq 3, \\ &\Rightarrow t_{im} = 0,\ 1 \leq i, m \leq 3. \end{aligned}$$

(iii) Assume that the formal asymptotic expansion of the scaled unknown is of the form

$$\boldsymbol{u}(\varepsilon) = \frac{1}{\varepsilon^N}\boldsymbol{u}^{-N} + \frac{1}{\varepsilon^{N-1}}\boldsymbol{u}^{-N+1} + \cdots \text{ for some integer } N \geq 0,$$

with $\boldsymbol{u}^{-N} \in \mathbf{W}(\Omega)$ and $\boldsymbol{u}^{-N+1} \in \mathbf{W}(\Omega)$ (under these assumptions, it will be eventually shown that $N = 0$; thus the assumption $N < 0$ will be *de facto* ruled out).

Together with the asymptotic behavior of the functions $g^{ij}(\varepsilon)$ and $\Gamma^p_{ij}(\varepsilon)$ as $\varepsilon \to 0$ (Thm. 8.5-1), such an expansion induces specific formal asymptotic expansions of the various functions appearing in the formulation of problem $\mathcal{P}(\varepsilon;\, \Omega)$, viz.,

$$\begin{aligned}
u_{m\|\alpha}(\varepsilon) &= \frac{1}{\varepsilon^N} u^{-N}_{m\|\alpha} + \cdots, \\
u_{m\|3}(\varepsilon) &= \frac{1}{\varepsilon^{N+1}} u^{-N-1}_{m\|3} + \frac{1}{\varepsilon^N} u^{-N}_{m\|3} + \cdots, \\
E_{\alpha\|\beta}(\varepsilon;\, \boldsymbol{u}(\varepsilon)) &= \frac{1}{\varepsilon^{2N}} E^{-2N}_{\alpha\|\beta} + \cdots, \\
E_{\alpha\|3}(\varepsilon;\, \boldsymbol{u}(\varepsilon)) &= \frac{1}{\varepsilon^{2N+1}} E^{-2N-1}_{\alpha\|3} + \cdots, \\
E_{3\|3}(\varepsilon;\, \boldsymbol{u}(\varepsilon)) &= \frac{1}{\varepsilon^{2N+2}} E^{-2N-2}_{3\|3} + \cdots, \\
F_{\alpha\|\beta}(\varepsilon;\, \boldsymbol{u}(\varepsilon), \boldsymbol{v}) &= \frac{1}{\varepsilon^N} F^{-N}_{\alpha\|\beta}(\boldsymbol{v}) + \cdots, \\
F_{\alpha\|3}(\varepsilon;\, \boldsymbol{u}(\varepsilon), \boldsymbol{v}) &= \frac{1}{\varepsilon^{N+1}} F^{-N-1}_{\alpha\|3}(\boldsymbol{v}) + \cdots, \\
F_{3\|3}(\varepsilon;\, \boldsymbol{u}(\varepsilon), \boldsymbol{v}) &= \frac{1}{\varepsilon^{N+2}} F^{-N-2}_{3\|3}(\boldsymbol{v}) + \cdots,
\end{aligned}$$

where, *by definition*, $u^q_{i\|j}, E^q_{i\|j}$, and $F^q_{i\|j}(\boldsymbol{v})$ designate for each $q \in \mathbb{Z}$ the coefficient of ε^q in the induced expansions of $u_{i\|j}(\varepsilon)$, $E_{i\|j}(\varepsilon;\, \boldsymbol{u}(\varepsilon))$, and $F_{i\|j}(\varepsilon;\, \boldsymbol{u}(\varepsilon), \boldsymbol{v})$.

Note in passing that, while the functions factorizing the powers of ε are by definition *independent of* ε, they are *dependent* on *one* or *several* terms $\boldsymbol{u}^q$, $q \geq -N$. In this respect, *particular caution should be exercised as regards this dependence.* For instance,

$$u^{-N}_{m\|\alpha} = \partial_\alpha u^{-N}_m - \Gamma^{p,0}_{\alpha m} u^{-N}_p \text{ and } u^{-N}_{m\|3} = \partial_3 u^{-N+1}_m - \Gamma^{p,0}_{m3} u^{-N}_p,$$

i.e., the factor of $\frac{1}{\varepsilon^N}$ in $u_{m\|\alpha}(\varepsilon)$ depends only on $\boldsymbol{u}^{-N}$ but the one in $u_{m\|3}(\varepsilon)$ depends also on $\boldsymbol{u}^{-N+1}$ (a special notation will be introduced whenever such different dependences could lead to confusion; see, e.g., part (ii) of the proof of Thm. 8.8-1).

Likewise, it should be remembered that *the expression of some factor may differ according to which value of N is considered*; for instance,

$$\begin{aligned} E_{\alpha\|\beta}^{-2N} &= \frac{1}{2} a^{mn} u_{m\|\alpha}^{-N} u_{n\|\beta}^{-N} \quad \text{if } N \geq 1, \\ &= \frac{1}{2}\left(u_{\alpha\|\beta}^{0} + u_{\beta\|\alpha}^{0} + a^{mn} u_{m\|\alpha}^{0} u_{n\|\beta}^{0}\right) \text{ if } N = 0, \end{aligned}$$

$$\begin{aligned} F_{\alpha\|\beta}^{-N}(\boldsymbol{v}) &= \frac{1}{2} a^{mn}\left\{u_{m\|\alpha}^{-N} v_{n\|\beta} + u_{n\|\beta}^{-N} v_{m\|\alpha}\right\} \quad \text{if } N \geq 1, \\ &= \frac{1}{2}\left(v_{\alpha\|\beta} + v_{\beta\|\alpha} + a^{mn}\left\{u_{m\|\alpha}^{0} v_{n\|\beta} + u_{n\|\beta}^{0} v_{m\|\alpha}\right\}\right) \text{ if } N = 0, \end{aligned}$$

where

$$v_{m\|\alpha} := \partial_\alpha v_m - \Gamma_{\alpha m}^{p,0} v_p.$$

We are now in a position to start the cancellation of the factors of the successive powers of ε found in the variational equations of problem $\mathcal{P}(\varepsilon; \Omega)$ when $\boldsymbol{u}(\varepsilon)$ is replaced by its formal expansion. In what follows, L^r designates for any integer $r \geq -3N-4$ the linear form defined by

$$L^r(\boldsymbol{v}) := \int_\Omega f^{i,r} v_i \sqrt{a}\, dx + \int_{\Gamma_+ \cup \Gamma_-} h^{i,r+1} v_i \sqrt{a}\, d\Gamma,$$

and it is *always* understood that the functions $f^{i,r}$ and $h^{i,r+1}$ belong to the spaces $L^2(\Omega)$ and $L^2(\Gamma_+ \cup \Gamma_-)$ and are *independent of* ε.

(iv) *Assume that* $N \geq 0$. Since the lowest power of ε in the left-hand side is ε^{-3N-4}, we are naturally led to first try

$$f^i(\varepsilon) = \frac{1}{\varepsilon^{3N+4}} f^{i,-3N-4} \text{ and } h^i(\varepsilon) = \frac{1}{\varepsilon^{3N+3}} h^{i,-3N-3}.$$

Then the cancellation of the coefficient of ε^{-3N-4} leads to the equation (the functions $E_{3\|3}^{-2N-2}$ and $F_{3\|3}^{-N-2}(\boldsymbol{v})$ are defined in (iii)):

$$\int_\Omega (\lambda + 2\mu) E_{3\|3}^{-2N-2} F_{3\|3}^{-N-2}(\boldsymbol{v}) \sqrt{a}\, dx = L^{-3N-4}(\boldsymbol{v})$$

for all $\boldsymbol{v} \in \mathbf{W}(\Omega)$. Since

$$E_{3\|3}^{-2N-2} = \frac{1}{2} a^{mn} \partial_3 u_m^{-N} \partial_3 u_n^{-N} \text{ and } F_{3\|3}^{-N-2}(\boldsymbol{v}) = a^{mn} \partial_3 u_m^{-N} \partial_3 v_n,$$

we must have

$$L^{-3N-4}(\boldsymbol{v}) = \int_\Omega f^{i,-3N-4} v_i \sqrt{a}\, dx + \int_{\Gamma_+ \cup \Gamma_-} h^{i,-3N-3} v_i \sqrt{a}\, d\Gamma = 0$$

for all $\boldsymbol{v} \in \mathbf{W}(\Omega)$ that are *independent of* x_3. Consequently, the *first requirement* (that there be no restriction on the applied forces) implies that we must let $f^{i,-3N-4} = 0$ and $h^{i,-3N-3} = 0$.

Letting $\boldsymbol{v} = \boldsymbol{u}^{-N}$ in the resulting variational equations then shows that

$$\int_\Omega E_{3\|3}^{-2N-2} F_{3\|3}^{-N-2}(\boldsymbol{u}^{-N}) \sqrt{a}\, dx = \frac{1}{2} \int_\Omega (a^{mn} \partial_3 u_m^{-N} \partial_3 u_n^{-N})^2 \sqrt{a}\, dx = 0.$$

Since the symmetric matrix (a^{ij}) is positive definite, we conclude that

$$\partial_3 \boldsymbol{u}^{-N} = (\partial_3 u_m^{-N}) = \mathbf{0} \text{ in } \Omega,$$

i.e., that $\boldsymbol{u}^{-N}$ *is independent of* x_3. We thus infer that $E_{3\|3}^{-2N-2} = 0$ and $F_{3\|3}^{-N-2}(\boldsymbol{v}) = 0$ for all $\boldsymbol{v} \in \mathbf{W}(\Omega)$.

Cancelling the coefficient of ε^{-3N-4} thus yields the following relations (as usual, any function defined on $\overline{\Omega}$ that is independent of x_3 is identified with a function defined on $\overline{\omega}$):

$$\boldsymbol{u}^{-N} \in \mathbf{W}(\omega) := \{\boldsymbol{\eta} = (\eta_i) \in \mathbf{W}^{1,4}(\omega);\ \boldsymbol{\eta} = \mathbf{0} \text{ on } \gamma_0\},$$
$$E_{i\|j}^{-2N-2} = 0 \text{ in } \Omega \text{ and } F_{i\|j}^{-N-2}(\boldsymbol{v}) = 0 \text{ for all } \boldsymbol{v} \in \mathbf{W}(\Omega),$$

since $E_{3\|3}^{-2N-2} = 0$ and $F_{3\|3}^{-N-2}(\boldsymbol{v}) = 0$ and the leading terms in the expansions of $E_{\alpha\|j}(\varepsilon; \boldsymbol{u}(\varepsilon))$ and $F_{\alpha\|j}(\varepsilon; \boldsymbol{u}(\varepsilon), \boldsymbol{v})$ are of orders strictly higher than $(-2N-2)$ and $(-N-2)$, respectively.

Since $E_{\alpha\|\beta}^{-2N-1} = 0$ (the leading term in the formal expansion of $E_{\alpha\|\beta}(\varepsilon; \boldsymbol{u}(\varepsilon))$ is of order $-2N$) and $E_{i\|3}^{-2N-1} = 0$ (since $\partial_3 \boldsymbol{u}^{-N} = \mathbf{0}$, each factor of $\frac{1}{\varepsilon^{2N+1}}$ in the expansion of $E_{\alpha\|3}(\varepsilon; \boldsymbol{u}(\varepsilon))$ vanishes because it contains some derivative $\partial_3 u_m^{-N}$ and the leading term in the expansion of $E_{3\|3}(\varepsilon; \boldsymbol{u}(\varepsilon))$ is of order strictly higher than $(-2N-1)$), we also have

$$E_{i\|j}^{-2N-1} = 0 \text{ in } \Omega.$$

Our next try is thus

$$f^i(\varepsilon) = \frac{1}{\varepsilon^{3N+3}} f^{i,-3N-3} \text{ and } h^i(\varepsilon) = \frac{1}{\varepsilon^{3N+2}} h^{i,-3N-2}.$$

The cancellation of the coefficient of ε^{-3N-3} then yields the equations (the functions $A^{ijkl}(0)$ are defined in Thm. 8.5-2):

$$\int_\Omega A^{ijkl}(0) E^{-2N-1}_{k\|l} F^{-N-2}_{i\|j}(\boldsymbol{v}) \sqrt{a}\, dx = L^{-3N-3}(\boldsymbol{v})$$

for all $\boldsymbol{v} \in \mathbf{W}(\Omega)$. But since $E^{-2N-1}_{k\|l} = 0$, we must let $f^{i,-3N-3} = 0$ and $h^{i,-3N-2} = 0$ (first requirement) and accordingly try

$$f^i(\varepsilon) = \frac{1}{\varepsilon^{3N+2}} f^{i,-3N-2} \text{ and } h^i(\varepsilon) = \frac{1}{\varepsilon^{3N+1}} h^{i,-3N-1},$$

in which case the cancellation of the coefficient of ε^{-3N-2} yields the equations

$$\int_\Omega A^{ijkl}(0) E^{-2N}_{k\|l} F^{-N-2}_{i\|j}(\boldsymbol{v}) \sqrt{a}\, dx = L^{-3N-2}(\boldsymbol{v})$$

for all $\boldsymbol{v} \in \mathbf{W}(\Omega)$. But since $F^{-N-2}_{i\|j}(\boldsymbol{v}) = 0$ for all $\boldsymbol{v} \in \mathbf{W}(\Omega)$, we must let $f^{i,-3N-2} = 0$ and $h^{i,-3N-1} = 0$ (first requirement).

(v) *Assume that* $N \geq 1$. Our next try being thus

$$f^i(\varepsilon) = \frac{1}{\varepsilon^{3N+1}} f^{i,-3N-1} \text{ and } h^i(\varepsilon) = \frac{1}{\varepsilon^{3N}} h^{i,-3N},$$

the cancellation of the coefficient of ε^{-3N-1} in the variational equations of problem $\mathcal{P}(\varepsilon;\Omega)$ then yields the equations

$$\int_\Omega A^{ijkl}(0) E^{-2N}_{k\|l} F^{-N-1}_{i\|j}(\boldsymbol{v}) \sqrt{a}\, dx = L^{-3N-1}(\boldsymbol{v})$$

for all $\boldsymbol{v} \in \mathbf{W}(\Omega)$, where

$$\begin{aligned}
E^{-2N}_{\alpha\|\beta} &= \frac{1}{2} a^{mn} u^{-N}_{m\|\alpha} u^{-N}_{n\|\beta}, & F^{-N-1}_{\alpha\|\beta}(\boldsymbol{v}) &= 0,\\
E^{-2N}_{\alpha\|3} &= \frac{1}{2} a^{mn} u^{-N}_{m\|\alpha} u^{-N}_{n\|3}, & F^{-N-1}_{\alpha\|3}(\boldsymbol{v}) &= \frac{1}{2} a^{mn} u^{-N}_{m\|\alpha} \partial_3 v_n,\\
E^{-2N}_{3\|3} &= \frac{1}{2} a^{mn} u^{-N}_{m\|3} u^{-N}_{n\|3}, & F^{-N-1}_{3\|3}(\boldsymbol{v}) &= a^{mn} u^{-N}_{m\|3} \partial_3 v_n,
\end{aligned}$$

the functions $u_{m\|i}^{-N}$ being those defined in (iii); recall that $\partial_3 u_n^{-N} = 0$ by (iv) and that we assume $N \geq 1$ (otherwise some functions, such as $E_{\alpha\|\beta}^{-2N}$, have different expressions when $N = 0$; cf. (iii)).

Letting $\boldsymbol{v} \in \mathbf{W}(\Omega)$ be *independent of* x_3 then shows that we must let $f^{i,-3N-1} = 0$ and $h^{i,-3N} = 0$; hence

$$\int_\Omega A^{ijkl}(0) E_{k\|l}^{-2N} F_{i\|j}^{-N-1}(\boldsymbol{v})\sqrt{a}\, dx = 0$$

for all $\boldsymbol{v} \in \mathbf{W}(\Omega)$. Let the field $\boldsymbol{w}^N = (w_m^N)$ be defined for all $(y, x_3) \in \Omega$ by

$$w_m^N := u_m^{-N+1} - (1 + x_3)\Gamma_{m3}^{p,0} u_p^{-N}.$$

Then $\boldsymbol{w}^N \in \mathbf{W}(\Omega)$ because *both* $\boldsymbol{u}^{-N}$ and $\boldsymbol{u}^{-N+1}$ are assumed to be in the space $\mathbf{W}(\Omega)$ (this double assumption is thus crucial!).

Furthermore, $\partial_3 w_m^N = u_{m\|3}^{-N}$, so that

$$F_{\alpha\|3}^{-N-1}(\boldsymbol{w}^N) = E_{\alpha\|3}^{-2N} \text{ and } F_{3\|3}^{-N-1}(\boldsymbol{w}^N) = 2E_{3\|3}^{-2N}.$$

Letting $\boldsymbol{v} = \boldsymbol{w}^N$ in the last variational equations thus shows that

$$\int_\Omega A^{ijkl}(0) E_{k\|l}^{-2N} F_{i\|j}^{-N-1}(\boldsymbol{w}^N)\sqrt{a}\, dx = 0,$$

where

$$A^{ijkl}(0) E_{k\|l}^{-2N} F_{i\|j}^{-N-1}(\boldsymbol{w}^N) = 2\lambda a^{mn} E_{m\|n}^{-2N} E_{3\|3}^{-2N} + 4\mu a^{mn} E_{m\|3}^{-2N} E_{n\|3}^{-2N}.$$

Since

$$a^{mn} E_{m\|n}^{-2N} = \frac{1}{2} a^{ij} a^{mn} u_{i\|m}^{-N} u_{j\|n}^{-N} \geq 0 \text{ in } \Omega$$

(by (ii)) and

$$E_{3\|3}^{-2N} = \frac{1}{2} a^{mn} u_{m\|3}^{-N} u_{n\|3}^{-N} \geq 0 \text{ and } a^{mn} E_{m\|3}^{-2N} E_{n\|3}^{-2N} \geq 0 \text{ in } \Omega$$

(the matrix (a^{mn}) is positive definite), we conclude that

$$a^{mn} E_{m\|3}^{-2N} E_{n\|3}^{-2N} = 0 \text{ in } \Omega,$$

hence that

$$E_{m\|3}^{-2N} = 0 \text{ in } \Omega.$$

In particular then, $E_{3\|3}^{-2N} = \frac{1}{2}a^{mn}u_{m\|3}^{-N}u_{n\|3}^{-N} = 0$ and thus

$$u_{m\|3}^{-N} = 0.$$

(vi) *Assume that* $N \geq 2$ (the case $N = 1$ is considered separately; cf. (vii)). Our next try being thus

$$f^i(\varepsilon) = \frac{1}{\varepsilon^{3N}} f^{i,\,-3N} \text{ and } h^i(\varepsilon) = \frac{1}{\varepsilon^{3N-1}} h^{i,\,-3N+1},$$

the cancellation of the coefficient of ε^{-3N} in the variational equations of problem $\mathcal{P}(\varepsilon;\Omega)$ then yields the equations (note that two terms are needed here from the expansions of the functions $A^{ijkl}(\varepsilon)\sqrt{g(\varepsilon)}$; cf. Thm. 8.5-2)

$$\int_\Omega A^{ijkl}(0)\left\{E_{k\|l}^{-2N} F_{i\|j}^{-N}(\boldsymbol{v}) + E_{k\|l}^{-2N+1} F_{i\|j}^{-N-1}(\boldsymbol{v})\right\}\sqrt{a}\,dx$$
$$+ \int_\Omega B^{ijkl,1} E_{k\|l}^{-2N} F_{i\|j}^{-N-1}(\boldsymbol{v})dx = L^{-3N}(\boldsymbol{v})$$

for all $\boldsymbol{v} \in \mathbf{W}(\Omega)$, where

$$\begin{aligned}
E_{\alpha\|\beta}^{-2N} &= \frac{1}{2}a^{mn}u_{m\|\alpha}^{-N}u_{n\|\beta}^{-N},\\
E_{i\|3}^{-2N} &= 0 \text{ by (v)},\\
F_{\alpha\|\beta}^{-N-1}(\boldsymbol{v}) &= 0, \quad F_{\alpha\|3}^{-N-1}(\boldsymbol{v}) = \frac{1}{2}a^{mn}u_{m\|\alpha}^{-N}\partial_3 v_n,\\
F_{3\|3}^{-N-1}(\boldsymbol{v}) &= 0 \text{ by (v)},\\
F_{\alpha\|\beta}^{-N}(\boldsymbol{v}) &= \frac{1}{2}a^{mn}\left(u_{m\|\alpha}^{-N}v_{n\|\beta} + u_{n\|\beta}^{-N}v_{m\|\alpha}\right),\\
F_{3\|3}^{-N}(\boldsymbol{v}) &= a^{mn}u_{m\|3}^{-N+1}\partial_3 v_n,
\end{aligned}$$

the last expression of $F_{3\|3}^{-N}(\boldsymbol{v})$ *being valid only if* $N \geq 2$ (the expressions of $F_{\alpha\|3}^{-N}(\boldsymbol{v})$ are not needed since $A^{\alpha 3\sigma\tau}(0) = 0$ and $E_{\alpha\|3}^{-2N} = 0$).

Noting that $F_{\alpha\|3}^{-N-1}(\boldsymbol{v}) = F_{3\|3}^{-N}(\boldsymbol{v}) = 0$ if $\partial_3\boldsymbol{v} = \mathbf{0}$, we thus conclude that *the above variational equations reduce to*

$$\int_\Omega A^{\alpha\beta\sigma\tau}(0)E_{\sigma\|\tau}^{-2N}F_{\alpha\|\beta}^{-N}(\boldsymbol{v})\sqrt{a}\,dx = L^{-3N}(\boldsymbol{v})$$

for all $\boldsymbol{v} \in \mathbf{W}(\Omega)$ *that are independent of* x_3. Since each term in the sum $A^{\alpha\beta\sigma\tau}(0)E^{-2N}_{\sigma\|\tau}F^{-N}_{\alpha\|\beta}(\boldsymbol{v})$ is *cubic* with respect to the functions $u^{-N}_{m\|\alpha}$, the *linearization trick* (second requirement) implies that $L^{-3N}(\boldsymbol{v}) = 0$ for all $\boldsymbol{v} \in \mathbf{W}(\Omega)$ that are *independent of* x_3. Hence we must let $f^{i,-3N} = 0$ and $h^{i,-3N+1} = 0$.

Since $\boldsymbol{u}^{-N}$ is independent of x_3 by (iv), we may let $\boldsymbol{v} = \boldsymbol{u}^{-N}$ in the last variational equations. This gives

$$\int_\Omega A^{\alpha\beta\sigma\tau}(0)E^{-2N}_{\sigma\|\tau}E^{-2N}_{\alpha\|\beta}\sqrt{a}\,dx = 0,$$

since $F^{-N}_{\alpha\|\beta}(\boldsymbol{u}^{-N}) = 2E^{-2N}_{\alpha\|\beta}$. But

$$A^{\alpha\beta\sigma\tau}(0) = \lambda a^{\alpha\beta}a^{\sigma\tau} + \mu(a^{\alpha\sigma}a^{\beta\tau} + a^{\alpha\tau}a^{\beta\sigma})$$

(Thm. 8.5-2) and thus

$$E^{-2N}_{\alpha\|\beta} = \frac{1}{2}a^{mn}u^{-N}_{m\|\alpha}u^{-N}_{n\|\beta} = 0 \text{ in } \Omega$$

(to reach this conclusion, observe that $a^{\alpha\sigma}a^{\beta\tau}t_{\sigma\tau}t_{\alpha\beta} \geq 0$ and that $a^{\alpha\sigma}a^{\beta\tau}t_{\sigma\tau}t_{\alpha\beta} = 0$ only if $t_{\alpha\beta} = 0$ by (ii)); these relations in turn imply that

$$u^{-N}_{m\|\alpha} = 0.$$

By definition (cf. (iii) and Thm. 8.5-1),

$$\begin{aligned}
u^{-N}_{\beta\|\alpha} &= \partial_\alpha u^{-N}_\beta - \Gamma^{p,0}_{\alpha\beta}u^{-N}_p = \partial_\alpha u^{-N}_\beta - \Gamma^\sigma_{\alpha\beta}u^{-N}_\sigma - b_{\alpha\beta}u^{-N}_3,\\
u^{-N}_{3\|\alpha} &= \partial_\alpha u^{-N}_3 - \Gamma^{p,0}_{\alpha 3}u^{-N}_p = \partial_\alpha u^{-N}_3 + b^\sigma_\alpha u^{-N}_\sigma.
\end{aligned}$$

Let $\zeta_i := u^{-N}_i|_{x_3=0}$. Then $\zeta_i \in W^{1,4}(\omega)$ since $u^{-N}_i \in W^{1,4}(\Omega)$ and $\partial_3 u^{-N}_i = 0$ in Ω and $\zeta_i = 0$ on γ_0 since $u^{-N}_i = 0$ on Γ_0. The above relations combined with the Gauß and Weingarten formulas (Thm. 2.3-1) then imply that $\partial_\alpha(\zeta_i\boldsymbol{a}^i) = 0$ in ω, hence that $\zeta_i = 0$. We have thus shown that

$$\boldsymbol{u}^{-N} = \boldsymbol{0} \text{ for all } N \geq 2.$$

(vii) *Finally, assume that* $N = 1$. The *only* difference from (vi) is that now

$$F^{-1}_{3\|3}(\boldsymbol{v}) = \partial_3 v_3 + a^{mn}u^0_{m\|3}\partial_3 v_n.$$

But since the arguments that led in (vi) to the conclusion that $\boldsymbol{u}^{-N} = \mathbf{0}$ for $N \geq 2$ only required the consideration of functions $\boldsymbol{v} \in \mathbf{W}(\Omega)$ that are *independent of* x_3, in which case $F^{-1}_{3\|3}(\boldsymbol{v}) = 0$, they can be reproduced *verbatim* for $N = 1$, thus showing that

$$\boldsymbol{u}^{-1} = \mathbf{0},$$

and the proof is complete. ■

8.8. IDENTIFICATION OF A TWO-DIMENSIONAL VARIATIONAL PROBLEM SATISFIED BY THE LEADING TERM

It has just been shown (Thm. 8.7-1) that the asymptotic expansion of the scaled unknown is of the form $\boldsymbol{u}(\varepsilon) = \boldsymbol{u}^0 + \varepsilon \boldsymbol{u}^1 + \cdots$. According to the Ansatz of the method of formal asymptotic expansions (Sect. 8.6), there thus remains to continue the cancellation of the factors of the successive powers of ε in the variational equations of problem $\mathcal{P}(\varepsilon; \Omega)$ until the leading term $\boldsymbol{u}^0$ can be fully identified, as the solutions of an *ad hoc* variational problem. In this direction, the following result, due to Miara [1998, Thm. 2], constitutes the key step.

A word of caution. The next theorem only states (in the form of a variational problem) a *necessary* condition that the leading term should satisfy. While there are cases (studied in Chap. 9) where this condition is also sufficient, i.e., where this theorem also constitutes the final stage of the application of the method of formal asymptotic expansion, there are cases (studied in Chap. 10) where additional stages (which require considerable efforts!) are needed to complete the induction. ■

Theorem 8.8-1. *Assume that the scaled unknown* $\boldsymbol{u}(\varepsilon) = (u_i(\varepsilon))$ *satisfying problem* $\mathcal{P}(\varepsilon; \Omega)$ (Thm. 8.4-1) *admits a formal asymptotic expansion of the form*

$$\boxed{\begin{aligned}&\boldsymbol{u}(\varepsilon) = \boldsymbol{u}^0 + \varepsilon \boldsymbol{u}^1 + \varepsilon^2 \boldsymbol{u}^2 + \cdots, \text{ with}\\ &\boldsymbol{u}^0 \in \mathbf{W}(\Omega) \text{ and } \boldsymbol{u}^1, \boldsymbol{u}^2 \in \mathbf{W}^{1,4}(\Omega).\end{aligned}}$$

Then in order that no restriction be put on the applied forces and that the linearization trick be satisfied (Sect. 8.6), *the components of the applied forces must be of the form*

$$
\boxed{\begin{aligned}
f^{i,\varepsilon}(x^\varepsilon) &= f^{i,0}(x) \quad \text{for all} \quad x^\varepsilon = \pi^\varepsilon x \in \Omega^\varepsilon,\\
h^{i,\varepsilon}(x^\varepsilon) &= \varepsilon h^{i,1}(x) \quad \text{for all} \quad x^\varepsilon = \pi^\varepsilon x \in \Gamma_+^\varepsilon \cup \Gamma_-^\varepsilon,
\end{aligned}}
$$

where the functions $f^{i,0} \in L^2(\Omega)$ *and* $h^{i,1} \in L^2(\Gamma_+ \cup \Gamma_-)$ *are independent of* ε.

This being the case, the leading term $\boldsymbol{u}^0$ *is independent of the transverse variable* x_3 *and* $\boldsymbol{\zeta}^0 := (\zeta_i^0) = \frac{1}{2}\int_{-1}^{1} \boldsymbol{u}^0\, dx_3$ *satisfies the following two-dimensional variational problem:*

$$
\boxed{\begin{aligned}
&\boldsymbol{\zeta}^0 \in \mathbf{W}(\omega) := \{\boldsymbol{\eta} \in \mathbf{W}^{1,4}(\omega);\ \boldsymbol{\eta} = \mathbf{0} \text{ on } \gamma_0\},\\
&\int_\omega a^{\alpha\beta\sigma\tau} E^0_{\sigma\|\tau} F^0_{\alpha\|\beta}(\boldsymbol{\eta})\sqrt{a}\, dy = \int_\omega p^{i,0}\eta_i\sqrt{a}\, dy\\
&\qquad\qquad\qquad \text{for all } \boldsymbol{\eta} = (\eta_i) \in \mathbf{W}(\omega),
\end{aligned}}
$$

where (recall that $a^{mn} = \boldsymbol{a}^m \cdot \boldsymbol{a}^n$):

$$
\boxed{\begin{aligned}
E^0_{\alpha\|\beta} &:= \frac{1}{2}(\zeta^0_{\alpha\|\beta} + \zeta^0_{\beta\|\alpha} + a^{mn}\zeta^0_{m\|\alpha}\zeta^0_{n\|\beta}),\\
F^0_{\alpha\|\beta}(\boldsymbol{\eta}) &:= \frac{1}{2}(\eta_{\alpha\|\beta} + \eta_{\beta\|\alpha} + a^{mn}\{\zeta^0_{m\|\alpha}\eta_{n\|\beta} + \zeta^0_{n\|\beta}\eta_{m\|\alpha}\}),\\
\eta_{\alpha\|\beta} &:= \partial_\beta\eta_\alpha - \Gamma^\sigma_{\alpha\beta}\eta_\sigma - b_{\alpha\beta}\eta_3, \quad \eta_{3\|\beta} := \partial_\beta\eta_3 + b^\sigma_\beta\eta_\sigma,\\
a^{\alpha\beta\sigma\tau} &:= \frac{4\lambda\mu}{\lambda+2\mu}a^{\alpha\beta}a^{\sigma\tau} + 2\mu(a^{\alpha\sigma}a^{\beta\tau} + a^{\alpha\tau}a^{\beta\sigma}),\\
p^{i,0} &:= \int_{-1}^{1} f^{i,0}\, dx_3 + h^{i,1}_+ + h^{i,1}_- \text{ and } h^{i,1}_\pm := h^{i,1}(\cdot, \pm 1).
\end{aligned}}
$$

Proof. The proof comprises three parts.

(i) To begin with, we remark that the conclusions of part (iv) of the proof of Thm. 8.7-1 are valid for any $N \geq 0$ and *under the sole assumption that* $\boldsymbol{u}(\varepsilon)$ *can be expanded as* $\boldsymbol{u}(\varepsilon) = \boldsymbol{u}^{-N} + \cdots$, *with* $\boldsymbol{u}^{-N} \in \mathbf{W}(\Omega)$. Letting $N = 0$, we thus infer that $\partial_3 \boldsymbol{u}^0 = \mathbf{0}$ in Ω and that

$$\boldsymbol{\zeta}^0 := \frac{1}{2}\int_{-1}^{1} \boldsymbol{u}^0 \, dx_3 \in \mathbf{W}(\omega) := \{\boldsymbol{\eta} \in \mathbf{W}^{1,4}(\omega);\ \boldsymbol{\eta} = \mathbf{0} \text{ on } \gamma_0\},$$

and also that (recall that the functions $E^q_{i\|j}$ and $F^q_{i\|j}(\boldsymbol{v})$ are by definition the factors of ε^q in the formal expansions of the functions $E_{i\|j}(\varepsilon;\ \boldsymbol{u}(\varepsilon))$ and $F_{i\|j}(\varepsilon;\ \boldsymbol{u}(\varepsilon),\ \boldsymbol{v})$ defined in Thm. 8.4-1):

$$\begin{aligned} E^q_{i\|j} &= 0 \text{ for all integers } q \leq -1, \\ F^q_{i\|j}(\boldsymbol{v}) &= 0 \text{ for all integers } q \leq -2 \text{ and all } \boldsymbol{v} \in \mathbf{W}(\Omega), \end{aligned}$$

and, finally, that we must let $f^{i,-2} = 0$ and $h^{i,-1} = 0$.

(ii) Our next try is thus

$$f^i(\varepsilon) = \frac{1}{\varepsilon} f^{i,-1} \text{ and } h^i(\varepsilon) = h^{i,0},$$

where it is understood as in the proof of Thm. 8.7-1 that each function $f^{i,r} \in L^2(\Omega)$ and each function $h^{i,r+1} \in L^2(\Gamma_+ \cup \Gamma_-)$, $r \geq -1$, appearing here and subsequently is *independent of* ε; likewise, we again let

$$L^r(\boldsymbol{v}) := \int_\Omega f^{i,r} v_i \sqrt{a}\, dx + \int_{\Gamma_+ \cup \Gamma_-} h^{i,r+1} v_i \sqrt{a}\, d\Gamma,\ r \geq -1.$$

The cancellation of the coefficient of ε^{-1} in the variational equations of problem $\mathcal{P}(\varepsilon;\ \Omega)$ (Thm. 8.4-1) then yields the equations (the functions $A^{ijkl}(0)$ are defined in Thm. 8.5-2):

$$\int_\Omega A^{ijkl}(0) E^0_{k\|l} F^{-1}_{i\|j}(\boldsymbol{v}) \sqrt{a}\, dx = L^{-1}(\boldsymbol{v})$$

for all $\boldsymbol{v} \in \mathbf{W}(\Omega)$, where

$$
\begin{aligned}
E^0_{\alpha\|\beta} &= \frac{1}{2}(u^0_{\alpha\|\beta} + u^0_{\beta\|\alpha} + a^{mn} u^0_{m\|\alpha} u^0_{n\|\beta}),\\
E^0_{\alpha\|3} &= \frac{1}{2}(u^{(0)}_{\alpha\|3} + u^0_{3\|\alpha} + a^{mn} u^0_{m\|\alpha} u^{(0)}_{n\|3}),\\
E^0_{3\|3} &= u^{(0)}_{3\|3} + \frac{1}{2} a^{mn} u^{(0)}_{m\|3} u^{(0)}_{n\|3},\\
F^{-1}_{\alpha\|\beta}(\boldsymbol{v}) &= 0,\\
F^{-1}_{\alpha\|3}(\boldsymbol{v}) &= \frac{1}{2}(\partial_3 v_\alpha + a^{mn} u^0_{m\|\alpha} \partial_3 v_n),\\
F^{-1}_{3\|3}(\boldsymbol{v}) &= \partial_3 v_3 + a^{mn} u^{(0)}_{m\|3} \partial_3 v_n,
\end{aligned}
$$

and (the functions $\Gamma^{p,0}_{ij}$ are defined in Thm. 8.5-1)

$$
u^0_{m\|\alpha} := \partial_\alpha u^0_m - \Gamma^{p,0}_{\alpha m} u^0_p \text{ and } u^{(0)}_{m\|3} := \partial_3 u^1_m - \Gamma^{p,0}_{m3} u^0_p.
$$

The special notation $u^{(0)}_{m\|3}$ (which thus replaces the notation $u^0_{m\|3}$ used in the proof of Thm. 8.7-1) emphasizes that, by contrast with the functions $u^0_{m\|\alpha}$, which only depend on $\boldsymbol{u}^0 = (u^0_m)$, the functions $u^{(0)}_{m\|3}$ also depend on $\boldsymbol{u}^1 = (u^1_m)$ (recall that $u^{(0)}_{m\|3}$ is by definition the coefficient of ε^0 in the formal asymptotic expansion of $u_{m\|3}(\varepsilon)$).

For this reason, the occurrence of the functions $u^{(0)}_{m\|3}$ implies that the formal asymptotic expansion of $\boldsymbol{u}(\varepsilon)$ be "at least" of the form $\boldsymbol{u}(\varepsilon) = \boldsymbol{u}^0 + \varepsilon \boldsymbol{u}^1 + \cdots$.

The expressions of the functions $F^{-1}_{i\|j}(\boldsymbol{v})$ imply that $L^{-1}(\boldsymbol{v}) = 0$ for all $\boldsymbol{v} = \mathbf{W}(\Omega)$ that are *independent of* x_3. Hence we must let $f^{i,-1} = 0$ and $h^{i,0} = 0$ (first requirement), so that we are left with the equations

$$
\int_\Omega A^{ijkl}(0) E^0_{k\|l} F^{-1}_{i\|j}(\boldsymbol{v}) \sqrt{a}\, dx = 0
$$

for all $\boldsymbol{v} = (v_i) \in \mathbf{W}(\Omega)$. When the functions $F^{-1}_{i\|j}(\boldsymbol{v})$ are replaced by their expressions given *supra*, the integrand in the above integral takes the form $(w^\tau \partial_3 v_\tau + w^3 \partial_3 v_3)$ for *ad hoc* functions $w^i \in L^{4/3}(\Omega)$.

Then Thm. 3.4-1 shows that the functions w^τ and w^3 vanish in Ω, i.e., that

$$\begin{aligned}(\lambda a^{\alpha\beta} E^0_{\alpha\|\beta} + (\lambda+2\mu)E^0_{3\|3})a^{\sigma\tau}u^{(0)}_{\sigma\|3} \\ +2\mu E^0_{\alpha\|3}(a^{\alpha\tau} + a^{\alpha\sigma}a^{\beta\tau}u^0_{\beta\|\alpha}) = 0 \text{ in } \Omega,\end{aligned}$$

$$\begin{aligned}(\lambda a^{\alpha\beta} E^0_{\alpha\|\beta} + (\lambda+2\mu)E^0_{3\|3})(1 + u^{(0)}_{3\|3}) \\ +2\mu a^{\alpha\sigma} E^0_{\alpha\|3} u^0_{3\|\sigma} = 0 \text{ in } \Omega.\end{aligned}$$

One obvious solution to this system of three equations is

$$E^0_{\alpha\|3} = 0 \text{ and } \lambda a^{\alpha\beta}E^0_{\alpha\|\beta} + (\lambda+2\mu)E^0_{3\|3} = 0 \text{ in } \Omega,$$

but there might be other solutions to this nonlinear system. Denoting by $[\cdots]^{lin}$ the linear part with respect to (any component of) $\boldsymbol{u}^0$ or $\boldsymbol{u}^1$ in the expression $[\cdots]$ and using the notations of the linear case (Sect. 3.4), we have

$$\begin{aligned}[E^0_{\alpha\|3}]^{lin} &= e^0_{\alpha\|3},\\ [\lambda a^{\alpha\beta}E^0_{\alpha\|\beta} + (\lambda+2\mu)E^0_{3\|3}]^{lin} &= \lambda a^{\alpha\beta}e^0_{\alpha\|\beta} + (\lambda+2\mu)e^0_{3\|3},\end{aligned}$$

by definition of the functions $E^0_{i\|3}$ and $e^0_{i\|3}$ as the coefficients of ε^0 in the formal expansions of the functions $E_{i\|3}(\varepsilon;\, \boldsymbol{u}(\varepsilon))$ and $e_{i\|3}(\varepsilon;\, \boldsymbol{u}(\varepsilon))$, the latter being precisely the linear parts in the former!

Since it was found in the linear case (see step (ii) in Sect. 3.4) that

$$e^0_{\alpha\|3} = 0 \text{ and } e^0_{3\|3} = -\frac{\lambda}{\lambda+2\mu}a^{\alpha\beta}e^0_{\alpha\|\beta} \text{ in } \Omega,$$

the *linearization trick* (second requirement) suggests that we only retain the "obvious" solution found above (as shown by Collard [1999], this plausible argument can be made entirely rigorous; cf. Ex. 8.4).

(iii) Our next try is thus

$$f^i(\varepsilon) = f^{i,0} \text{ and } h^i(\varepsilon) = \varepsilon h^{i,1}.$$

The cancellation of the coefficient of ε^0 in the variational equations of problem $\mathcal{P}(\varepsilon;\,\Omega)$ then leads to the equations (two terms are needed here from the expansions of the functions $A^{ijkl}(\varepsilon)\sqrt{g(\varepsilon)}$; cf. Thm. 8.5-2):

$$\begin{aligned}&\int_\Omega A^{ijkl}(0)\{E^0_{k\|l}F^0_{i\|j}(\boldsymbol{v}) + E^1_{k\|l}F^{-1}_{i\|j}(\boldsymbol{v})\}\sqrt{a}\,dx\\ &+\int_\Omega B^{ijkl,1}E^0_{k\|l}F^{-1}_{i\|j}(\boldsymbol{v})\,dx = L^0(\boldsymbol{v})\end{aligned}$$

for all $\boldsymbol{v} \in \mathbf{W}(\Omega)$, where the functions $E^1_{i\|j}$ and $F^0_{i\|j}(\boldsymbol{v})$ are defined by means of the formal expansions

$$E_{i\|j}(\varepsilon;\, \boldsymbol{u}(\varepsilon)) = E^0_{i\|j} + \varepsilon E^1_{i\|j} + \cdots,$$

$$F_{i\|j}(\varepsilon;\, \boldsymbol{u}(\varepsilon),\, \boldsymbol{v}) = \frac{1}{\varepsilon} F^{-1}_{i\|j}(\boldsymbol{v}) + F^0_{i\|j}(\boldsymbol{v}) + \cdots.$$

Note that, while the functions $E^0_{i\|j}$, $F^{-1}_{i\|j}(\boldsymbol{v})$, and $F^0_{i\|j}(\boldsymbol{v})$ depend only on $\boldsymbol{u}^0$ and $\boldsymbol{u}^1$, the functions $E^1_{i\|j}$ depend also on $\boldsymbol{u}^2$ (but not on $\boldsymbol{u}^3$; each term involving $\boldsymbol{u}^3$ vanishes because it contains some derivative $\partial_3 u^0_m$ as a factor). For this reason the formal asymptotic expansion of $\boldsymbol{u}(\varepsilon)$ must be "at least" of the form

$$\boldsymbol{u}(\varepsilon) = \sum_{q=0}^{2} \varepsilon^q \boldsymbol{u}^q + \cdots.$$

In particular then, we must have

$$\int_\Omega A^{ijkl}(0) E^0_{k\|l} F^0_{i\|j}(\boldsymbol{v}) \sqrt{a}\, dx = L^0(\boldsymbol{v})$$

for all $\boldsymbol{v} \in \mathbf{W}(\Omega)$ that are *independent of* x_3 since $F^{-1}_{i\|j}(\boldsymbol{v}) = 0$ for such functions; equivalently, after performing the usual identification, we must have

$$\int_\Omega A^{ijkl}(0) E^0_{k\|l} F^0_{i\|j}(\boldsymbol{\eta}) \sqrt{a}\, dx = L^0(\boldsymbol{\eta})$$

for all $\boldsymbol{\eta} \in \mathbf{W}(\omega) = \{\boldsymbol{\eta} \in \mathbf{W}^{1,4}(\omega);\ \boldsymbol{\eta} = \mathbf{0} \text{ on } \gamma_0\}$.

Using the expressions of the functions $A^{ijkl}(0)$ and the relations satisfied by the functions $E^0_{i\|3}$ (see part (ii)), we are left with

$$\begin{aligned}
&\int_\Omega A^{ijkl}(0) E^0_{k\|l} F^0_{i\|j}(\boldsymbol{\eta}) \sqrt{a}\, dx \\
&= \int_\Omega \Big(\lambda a^{\alpha\beta} a^{\sigma\tau} + \mu(a^{\alpha\sigma} a^{\beta\tau} + a^{\alpha\tau} a^{\beta\sigma})\Big) E^0_{\sigma\|\tau} F^0_{\alpha\|\beta}(\boldsymbol{\eta}) \sqrt{a}\, dx \\
&\quad + \int_\Omega \Big(4\mu a^{\alpha\sigma} E^0_{\alpha\|3} F^0_{\sigma\|3}(\boldsymbol{\eta}) + \lambda a^{\alpha\beta} E^0_{3\|3} F^0_{\alpha\|\beta}(\boldsymbol{\eta})\Big) \sqrt{a}\, dx \\
&\quad + \int_\Omega \Big(\lambda a^{\sigma\tau} E^0_{\sigma\|\tau} + (\lambda + 2\mu) E^0_{3\|3}\Big) F^0_{3\|3}(\boldsymbol{\eta}) \sqrt{a}\, dx \\
&= \frac{1}{2} \int_\Omega a^{\alpha\beta\sigma\tau} E^0_{\sigma\|\tau} F^0_{\alpha\|\beta}(\boldsymbol{\eta}) \sqrt{a}\, dx = L^0(\boldsymbol{\eta})
\end{aligned}$$

for all $\boldsymbol{\eta} \in \mathbf{W}(\omega)$, where

$$a^{\alpha\beta\sigma\tau} = \frac{4\lambda\mu}{\lambda + 2\mu} a^{\alpha\beta} a^{\sigma\tau} + 2\mu(a^{\alpha\sigma} a^{\beta\tau} + a^{\alpha\tau} a^{\beta\sigma}).$$

Since $\boldsymbol{u}^0 \in \mathbf{W}(\Omega)$ is independent of x_3, it may be identified with a function $\boldsymbol{\zeta}^0 \in \mathbf{W}(\omega)$. Consequently, the functions

$$E^0_{\alpha\|\beta} = \frac{1}{2}(u^0_{\alpha\|\beta} + u^0_{\beta\|\alpha} + a^{mn} u^0_{m\|\alpha} u^0_{n\|\beta}) \in L^2(\Omega),$$

$$F^0_{\alpha\|\beta}(\boldsymbol{\eta}) = \frac{1}{2}(\eta_{\alpha\|\beta} + \eta_{\beta\|\alpha} + a^{mn}\{u^0_{m\|\alpha}\eta_{n\|\beta} + u^0_{n\|\beta}\eta_{m\|\alpha}\}) \in L^2(\Omega),$$

which are thus also independent of x_3, may be likewise identified with functions (denoted for convenience by the same symbols):

$$E^0_{\alpha\|\beta} = \frac{1}{2}(\zeta^0_{\alpha\|\beta} + \zeta^0_{\beta\|\alpha} + a^{mn} \zeta^0_{m\|\alpha} \zeta^0_{n\|\beta}) \in L^2(\omega),$$

$$F^0_{\alpha\|\beta}(\boldsymbol{\eta}) = \frac{1}{2}(\eta_{\alpha\|\beta} + \eta_{\beta\|\alpha} + a^{mn}\{\zeta^0_{m\|\alpha}\eta_{n\|\beta} + \zeta^0_{n\|\beta}\eta_{m\|\alpha}\}) \in L^2(\omega),$$

where

$$\eta_{\alpha\|\beta} = \partial_\beta \eta_\alpha - \Gamma^\sigma_{\alpha\beta}\eta_\sigma - b_{\alpha\beta}\eta_3 \text{ and } \eta_{3\|\beta} = \partial_\beta \eta_3 + b^\sigma_\beta \eta_\sigma$$

for all $\boldsymbol{\eta} = (\eta_i) \in \mathbf{W}(\omega)$. The last variational problem is thus indeed two-dimensional. ■

The functions $a^{\alpha\beta\sigma\tau}$ found in Thm. 8.8-1 are the **contravariant components** of the **scaled two-dimensional elasticity tensor of the shell**, already encountered in the *linear* theory (see, e.g., Sect. 3.4).

EXERCISES

8.1. The three-dimensional energy $\hat{J}$ in Cartesian coordinates associated with a St Venant-Kirchhoff material is defined by (the notations are those of Sects. 8.1 and 8.2):

$$\hat{J}(\boldsymbol{v}) = \frac{1}{2}\int_{\hat{\Omega}} \hat{A}^{ijkl}\hat{E}_{kl}(\hat{\boldsymbol{v}})\hat{E}_{ij}(\hat{\boldsymbol{v}})\, d\hat{x} - \left\{\int_{\hat{\Omega}} \hat{f}^i \hat{v}_i \, d\hat{x} + \int_{\hat{\Gamma}_1} \hat{h}^i \hat{v}_i \, d\hat{\Gamma}\right\}.$$

Show that there exist constants $\alpha > 0$ and $\beta \in \mathbb{R}$ such that

$$\hat{J}(\hat{\boldsymbol{v}}) \geq \alpha \|\hat{\boldsymbol{v}}\|^4_{\mathbf{W}^{1,4}(\Omega)} + \beta$$

for all $\hat{\boldsymbol{v}} \in \mathbf{W}(\hat{\Omega}) = \{\hat{\boldsymbol{v}} = (\hat{v}_i) \in \mathbf{W}^{1,4}(\hat{\Omega});\ \hat{\boldsymbol{v}} = \mathbf{0} \text{ on } \Gamma_0\}$.

Remark. The energy $\hat{J}$ is thus coercive on the space $\mathbf{W}(\hat{\Omega})$. It can be shown, however, that it is *not* weakly lower semi-continuous on $\mathbf{W}(\hat{\Omega})$; cf. Ball & Murat [1984], Raoult [1986], Dacorogna [1989], Le Dret & Raoult [1995a].

8.2. (1) Using the notations introduced at the beginning of Sect. 8.2, show that the variational equations in Cartesian coordinates of Sect. 8.1 may be equivalently rewritten as

$$\int_{\hat{\Omega}} \hat{A}^{ijkl} \hat{E}_{kl}(\hat{\boldsymbol{u}}) \hat{F}_{ij}(\hat{\boldsymbol{u}}, \hat{\boldsymbol{v}})\, d\hat{x} = \int_{\hat{\Omega}} \hat{f}^i \hat{v}_i \, d\hat{x} + \int_{\hat{\Gamma}_1} \hat{h}^i \hat{v}_i \, d\hat{\Gamma},$$

where

$$\hat{F}_{ij}(\hat{\boldsymbol{u}}, \hat{\boldsymbol{v}}) := \frac{1}{2}(\hat{\partial}_j \hat{v}_i + \hat{\partial}_i \hat{v}_j + \hat{\partial}_i \hat{u}_m \hat{\partial}_j \hat{v}_m + \hat{\partial}_j \hat{u}_m \hat{\partial}_i \hat{v}_m).$$

(2) Using the notations and definitions of Thms. 8.2-1 and 8.2-3, show directly, i.e., without resorting to the functionals $\hat{J}$ and J found in these theorems, that

$$\int_{\hat{\Omega}} \hat{A}^{ijkl} \hat{E}_{kl}(\hat{\boldsymbol{u}}) \hat{F}_{ij}(\hat{\boldsymbol{u}}, \hat{\boldsymbol{v}})\, d\hat{x} = \int_{\Omega} A^{ijkl} E_{k\|l}(\boldsymbol{u}) F_{i\|j}(\boldsymbol{u}, \boldsymbol{v}) \sqrt{g}\, dx.$$

8.3. Show that the boundary value problem of three-dimensional nonlinear elasticity in curvilinear coordinates (Thm. 8.2-4), viz.,

$$\begin{aligned} -(\sigma^{ij} + \sigma^{kj} g^{il} u_{l\|k})\|_j &= f_i \text{ in } \Omega, \\ u_i &= 0 \text{ on } \Gamma_0, \\ (\sigma^{ij} + \sigma^{kj} g^{il} u_{l\|k}) n_j &= h_i \text{ on } \Gamma_1, \end{aligned}$$

where $\sigma^{ij} = A^{ijkl} E_{k\|l}(\boldsymbol{u})$, can be directly derived from its Cartesian counterpart (Sect. 8.1), viz.,

$$\begin{aligned} -\hat{\partial}_j(\hat{\sigma}^{ij} + \hat{\sigma}^{kj} \hat{\partial}_k \hat{u}_i) &= \hat{f}_i \text{ in } \hat{\Omega}, \\ \hat{u}_i &= 0 \text{ on } \hat{\Gamma}_0, \\ (\hat{\sigma}^{ij} + \hat{\sigma}^{kj} \hat{\partial}_k \hat{u}_i) \hat{n}_j &= \hat{h}_i \text{ on } \hat{\Gamma}_1, \end{aligned}$$

where $\hat{\sigma^{ij}} = \hat{A}^{ijkl}\hat{E}_{kl}(\hat{\boldsymbol{u}})$ ("directly" means without recourse to the variational equations, as in Thm. 8.2-4).

8.4. It was observed in part (ii) of the proof of Thm. 8.8-1 that an obvious solution to the nonlinear system of equations:

$$\begin{aligned}
&\left(\lambda a^{\alpha\beta} E^0_{\alpha\|\beta} + (\lambda + 2\mu) E^0_{3\|3}\right) a^{\sigma\tau} u^{(0)}_{\sigma\|3} \\
&\qquad + 2\mu E^0_{\alpha\|3}\left(a^{\alpha\tau} + a^{\alpha\sigma} a^{\beta\tau} u^0_{\beta\|\alpha}\right) = 0 \text{ in } \Omega, \\
&\left(\lambda a^{\alpha\beta} E^0_{\alpha\|\beta} + (\lambda + 2\mu) E^0_{3\|3}\right)\left(1 + u^{(0)}_{3\|3}\right) \\
&\qquad + 2\mu a^{\alpha\sigma} E^0_{\alpha\|3} u^0_{3\|\sigma} = 0 \text{ in } \Omega,
\end{aligned}$$

is

$$E^0_{\alpha\|3} = 0 \text{ and } \lambda a^{\alpha\beta} E^0_{\alpha\|\beta} + (\lambda + 2\mu) E^0_{3\|3} = 0 \text{ in } \Omega.$$

The objective of this problem is to find *all* the other solutions.

(1) Show that another solution is

$$u^{(0)}_{\alpha\|3} = 0 \text{ and } u^{(0)}_{3\|3} = -1.$$

(2) Show that all the remaining solutions are of the form

$$u^{(0)}_{\alpha\|3} = (a_{\alpha\beta} + u^0_{\alpha\|\beta}) w^\beta \text{ and } u^{(0)}_{3\|3} = u^0_{3\|\alpha} w^\alpha - 1,$$

where the functions w^α depend only on the functions $u^0_{i\|\beta}$.

Remark. These results are due to Collard [1999].

CHAPTER 9

NONLINEARLY ELASTIC MEMBRANE SHELLS

INTRODUCTION

The purpose of this chapter is to *identify* and to *mathematically justify* the *two-dimensional equations of nonlinearly elastic "membrane" shells*. These equations are of two kinds, depending on whether they are obtained through the *method of formal asymptotic expansions* or by means of a *convergence theorem*.

We begin with the formal approach, which is due to B. Miara. Given a surface $S = \boldsymbol{\theta}(\omega)$ and a displacement field $\eta_i \boldsymbol{a}^i$ of S with smooth enough covariant components $\eta_i : \overline{\omega} \to \mathbb{R}$, let $a_{\alpha\beta}(\boldsymbol{\eta})$ denote the covariant components of the metric tensor of the associated deformed surface $(\boldsymbol{\theta} + \eta_i \boldsymbol{a}^i)(\overline{\omega})$.

A nonlinearly elastic shell with middle surface S subjected to a boundary condition of place along a portion of its lateral face with $\boldsymbol{\theta}(\gamma_0)$ as its middle curve, where $\gamma_0 \subset \gamma$, is a *nonlinearly elastic "membrane" shell* if *the manifold*

$$\boldsymbol{\mathcal{M}}_0(\omega) = \{\boldsymbol{\eta} = (\eta_i) \in \mathbf{W}^{1,4}(\omega);\ \boldsymbol{\eta} = \mathbf{0} \text{ on } \gamma_0,\ a_{\alpha\beta}(\boldsymbol{\eta}) - a_{\alpha\beta} = 0 \text{ in } \omega\}$$

reduces to $\{\mathbf{0}\}$ (Sect. 9.1). This means that $\mathbf{0}$ is the only displacement field $\eta_i \boldsymbol{a}^i$ of the surface S with covariant components η_i in $W^{1,4}(\omega)$ that is *admissible* and *inextensional*, i.e., that vanishes along the curve $\boldsymbol{\theta}(\gamma_0)$ and leaves invariant the metric of S.

This definition is motivated by the crucial observation that the induction begun in the previous chapter ends when $\boldsymbol{\mathcal{M}}_0(\omega) = \{\mathbf{0}\}$. The conclusions reached in Thm. 8.8-1 can then be reformulated as follows in this case (Thm. 9.2-1):

Assume that $\boldsymbol{\mathcal{M}}_0(\omega) = \{\mathbf{0}\}$. Then the contravariant components of the applied body forces must be of the form

$$f^{i,\varepsilon}(x^\varepsilon) = f^{i,0}(x) \text{ for all } x^\varepsilon = \pi^\varepsilon x \in \Omega^\varepsilon,$$

where the functions $f^{i,0}$ are *independent* of ε (for simplicity, we assume in this introduction that there are no surface forces), *the leading term* $\boldsymbol{u}^0 : \overline{\Omega} \to \mathbb{R}^3$ *of the formal asymptotic expansion of the scaled unknown is independent of the transverse variable, and* $\boldsymbol{\zeta}^0 = \frac{1}{2}\int_{-1}^{1} \boldsymbol{u}^0 \, dx_3$ *satisfies the following* (scaled) *two-dimensional variational problem of a nonlinearly elastic "membrane" shell:*

$$\boldsymbol{\zeta}^0 \in \mathbf{W}_M(\omega) = \{\boldsymbol{\eta} \in \mathbf{W}^{1,4}(\omega);\ \boldsymbol{\eta} = \mathbf{0} \text{ on } \gamma_0\},$$

$$\int_\omega a^{\alpha\beta\sigma\tau} G_{\sigma\tau}(\boldsymbol{\zeta}^0)\big(G'_{\alpha\beta}(\boldsymbol{\zeta}^0)\boldsymbol{\eta}\big)\sqrt{a}\, dy = \int_\omega p^{i,0}\eta_i \sqrt{a}\, dy$$

for all $\boldsymbol{\eta} = (\eta_i) \in \mathbf{W}_M(\omega)$, where, for any $\boldsymbol{\zeta}, \boldsymbol{\eta} \in \mathbf{W}^{1,4}(\omega)$,

$$G_{\alpha\beta}(\boldsymbol{\eta}) = \frac{1}{2}(a_{\alpha\beta}(\boldsymbol{\eta}) - a_{\alpha\beta}) = \frac{1}{2}(\eta_{\alpha\|\beta} + \eta_{\beta\|\alpha} + a^{mn}\eta_{m\|\alpha}\eta_{n\|\beta}),$$

$$G'_{\alpha\beta}(\boldsymbol{\zeta})\boldsymbol{\eta} = \frac{1}{2}(\eta_{\alpha\|\beta} + \eta_{\beta\|\alpha} + a^{mn}\{\zeta_{m\|\alpha}\eta_{n\|\beta} + \zeta_{n\|\beta}\eta_{m\|\alpha}\}),$$

$$a^{\alpha\beta\sigma\tau} = \frac{4\lambda\mu}{\lambda + 2\mu} a^{\alpha\beta}a^{\sigma\tau} + 2\mu(a^{\alpha\sigma}a^{\beta\tau} + a^{\alpha\tau}a^{\beta\sigma}),$$

$$p^{i,0} = \int_{-1}^{1} f^{i,0}\, dx_3.$$

The functions $G_{\alpha\beta}(\boldsymbol{\eta})$ and $a^{\alpha\beta\sigma\tau}$ are respectively the covariant components of the *change of metric tensor* associated with a displacement field $\eta_i \boldsymbol{a}^i$ of the middle surface S and the contravariant components of the (scaled) *two-dimensional elasticity tensor of the shell.*

These equations also express that $\boldsymbol{\zeta}^0$ *is a stationary point of a certain functional over the space* $\mathbf{W}_M(\omega)$, so that particular solutions may be obtained by solving a *minimization problem* (Thm. 9.3-1), written here directly in its "de-scaled" version (Thm. 9.4-1): The vector field $\boldsymbol{\zeta}^\varepsilon = (\zeta_i^\varepsilon)$ formed by the covariant components of the *limit displacement field* $\zeta_i^\varepsilon \boldsymbol{a}^i$ of the middle surface S satisfies:

$$\boldsymbol{\zeta}^\varepsilon \in \mathbf{W}_M(\omega) \text{ and } j_M^\varepsilon(\boldsymbol{\zeta}^\varepsilon) = \inf_{\boldsymbol{\eta} \in \mathbf{W}_M(\omega)} j_M^\varepsilon(\boldsymbol{\eta}),$$

where the *two-dimensional energy* j_M^ε *of a nonlinearly elastic membrane shell* is defined by

$$j_M^\varepsilon(\boldsymbol{\eta}) = \frac{\varepsilon}{8}\int_\omega a^{\alpha\beta\sigma\tau,\varepsilon}(a_{\sigma\tau}(\boldsymbol{\eta}) - a_{\sigma\tau})(a_{\alpha\beta}(\boldsymbol{\eta}) - a_{\alpha\beta})\sqrt{a}\, dy - \int_\omega p^{i,\varepsilon}\eta_i\sqrt{a}\, dy,$$

where

$$a^{\alpha\beta\sigma\tau,\varepsilon} = \frac{4\lambda^\varepsilon\mu^\varepsilon}{\lambda^\varepsilon + 2\mu^\varepsilon} a^{\alpha\beta}a^{\sigma\tau} + 2\mu^\varepsilon(a^{\alpha\sigma}a^{\beta\tau} + a^{\alpha\tau}a^{\beta\sigma}),$$

$$p^{i,\varepsilon} = \int_{-\varepsilon}^{\varepsilon} f^{i,\varepsilon}\, dx_3^\varepsilon.$$

The functions $a^{\alpha\beta\sigma\tau}$ are the contravariant components of the *two-dimensional elasticity tensor of the shell*, λ^ε and μ^ε are the Lamé constants of the material constituting the shell, and the functions $f^{i,\varepsilon}$ are the contravariant components of the body force density applied to the shell.

The *stored energy function* of a nonlinearly elastic membrane shell is thus remarkably simple: It is a quadratic and positive definite expression in terms of the *exact* difference between the *metric tensor* of the *deformed* middle surface and that of the *undeformed* one. Incidentally, this shows that the equations of the *linear* theory of "membrane" shells are immediately recovered (at least formally) under linearization of the nonlinear equations.

We also briefly discuss in Sect. 9.4 approaches that could lead to an existence result for the above *minimization problem* or for the associated *boundary value problem* (Thm. 9.4-2).

We conclude this chapter by reviewing in Sect. 9.5 a justification of yet another two-dimensional "membrane" shell theory, this time by means of a *convergence theorem*, the *first of its kind for shells.*

More specifically, *another* two-dimensional minimization problem for a *nonlinearly elastic shell* has been derived from nonlinear three-dimensional elasticity by H. Le Dret and A. Raoult for a realistic class of hyperelastic materials. Using Γ-*convergence theory*, they have shown that the minimizing deformations of the three-dimensional energies, once appropriately scaled over a fixed domain $\Omega \subset \mathbb{R}^3$, *weakly converge* in the space $\mathbf{W}^{1,p}(\Omega)$ as the thickness of the shell approaches zero (cf. Thm. 9.5-1; the exponent $p > 1$ is "governed" by the hyperelastic material considered). The limit deformation found in this fashion is *independent of the transverse variable* and *minimizes a limit energy obtained by computing the* Γ*-limit of the three-dimensional energies*; hence *the existence of a minimizer of this limit energy is* de facto *established.*

The limit stored energy function is again that of a *nonlinearly elastic "membrane" shell*, in the sense that it contains only first derivatives of the unknown deformation. For a St Venant-Kirchhoff material, it does *not* coincide, however, with the energy found *via* the method of formal asymptotic expansions, save when the singular values of the limit deformation gradients belong to a specific subset of the plane. For this reason, it does not always reduce under linearization to the linear theory of "membrane" shells.

The origin of such differences may lie in that the latter approach models shells made with *"soft"* elastic materials (like the sails of a sailboat or a balloon), while the former models shells made with *"rigid"* elastic materials (like the hull of a ship or the roof of the Hong Kong Convention and Exhibition Centre). But this affirmation is yet to be substantiated.

9.1. NONLINEARLY ELASTIC MEMBRANE SHELLS: DEFINITION, EXAMPLES, AND ASSUMPTIONS ON THE DATA

The analysis of Chap. 8 culminated in Thm. 8.8-1, where it was shown that the leading term $\boldsymbol{u}^0$ of the formal asymptotic expansion of the scaled unknown $\boldsymbol{u}(\varepsilon)$ is independent of x_3 and that $\boldsymbol{\zeta}^0 := \frac{1}{2}\int_{-1}^{1} \boldsymbol{u}^0 \, dx_3$ satisfies:

$$\boldsymbol{\zeta}^0 = (\zeta_i^0) \in \mathbf{W}(\omega) := \{\boldsymbol{\eta} \in \mathbf{W}^{1,4}(\omega); \boldsymbol{\eta} = \mathbf{0} \text{ on } \gamma_0\}$$

and

$$\int_\omega a^{\alpha\beta\sigma\tau} E^0_{\sigma\|\tau} F^0_{\alpha\|\beta}(\boldsymbol{\eta})\sqrt{a}\, dy = \int_\omega p^{i,0} \eta_i \sqrt{a}\, dy$$

for all $\boldsymbol{\eta} = (\eta_i) \in \mathbf{W}(\omega)$, where

$$E^0_{\alpha\|\beta} := \frac{1}{2}(\zeta^0_{\alpha\|\beta} + \zeta^0_{\beta\|\alpha} + a^{mn}\zeta^0_{m\|\alpha}\zeta^0_{n\|\beta}).$$

An inspection of the functions $E^0_{\alpha\|\beta}$ appearing in the left-hand side of the variational equations thus immediately reveals that *two fundamentally distinct situations* must be considered, depending on whether there are *non-zero* fields $\boldsymbol{\eta} = (\eta_i)$ in the space $\mathbf{W}(\omega)$ that satisfy

$$\eta_{\alpha\|\beta} + \eta_{\beta\|\alpha} + a^{mn}\eta_{m\|\alpha}\eta_{n\|\beta} = 0 \text{ in } \omega.$$

If there are no such fields, the induction is terminated (this situation is studied in this chapter), while the induction may be continued if there are such fields (see Sect. 10.1).

In this direction, the first task consists in recognizing in the functions $E^0_{\alpha\|\beta}$ (half of) the differences between the covariant components of the metric tensor of the "deformed" surface $(\boldsymbol{\theta} + \zeta_i^0 \boldsymbol{a}^i)(\overline{\omega})$ and those of the metric tensor of the "undeformed" surface $S = \boldsymbol{\theta}(\overline{\omega})$ and then in showing that the functions $F^0_{\alpha\|\beta}(\boldsymbol{\eta})$ are simply Gâteaux derivatives at $\boldsymbol{\zeta}^0$ of these differences. Together, these expressions will in turn provide an illuminating interpretation of the variational problem satisfied by $\boldsymbol{\zeta}^0$ (Thm. 9.2-1) (the definitions of the covariant and mixed components $b_{\alpha\beta}$ and b_β^α of the curvature tensor and the Christoffel symbols $\Gamma^\sigma_{\alpha\beta}$ of S appearing in the next theorem are recalled in Sect. 8.5).

Theorem 9.1-1. *Let ω be a domain in $\mathbb{R}^2$ and let $\boldsymbol{\theta} \in \mathcal{C}^2(\overline{\omega}; \mathbb{R}^3)$ be an injective mapping such that the two vectors $\boldsymbol{a}_\alpha = \partial_\alpha \boldsymbol{\theta}$ are linearly independent at all points of $\overline{\omega}$, let $\boldsymbol{a}_3 = \dfrac{\boldsymbol{a}_1 \wedge \boldsymbol{a}_2}{|\boldsymbol{a}_1 \wedge \boldsymbol{a}_2|}$, and let the vectors $\boldsymbol{a}^i$ be defined by $\boldsymbol{a}^i \cdot \boldsymbol{a}_j = \delta^i_j$. Given a displacement field $\eta_i \boldsymbol{a}^i$ of the surface $S = \boldsymbol{\theta}(\overline{\omega})$ with smooth enough covariant components $\eta_i : \overline{\omega} \to \mathbb{R}$, let the covariant components of the* **change of metric tensor** *associated with this displacement field be defined by*

$$\boxed{G_{\alpha\beta}(\boldsymbol{\eta}) := \frac{1}{2}\left(a_{\alpha\beta}(\boldsymbol{\eta}) - a_{\alpha\beta}\right),}$$

where $a_{\alpha\beta}$ and $a_{\alpha\beta}(\boldsymbol{\eta})$ are the covariant components of the metric tensors of the surfaces $\boldsymbol{\theta}(\overline{\omega})$ and $(\boldsymbol{\theta} + \eta_i \boldsymbol{a}^i)(\overline{\omega})$. Then

$$\boxed{\begin{aligned} G_{\alpha\beta}(\boldsymbol{\eta}) &= \frac{1}{2}(\eta_{\alpha\|\beta} + \eta_{\beta\|\alpha} + a^{mn}\eta_{m\|\alpha}\eta_{n\|\beta}), \text{ where} \\ \eta_{\alpha\|\beta} &:= \partial_\beta \eta_\alpha - \Gamma^\sigma_{\alpha\beta}\eta_\sigma - b_{\alpha\beta}\eta_3 \text{ and } \eta_{3\|\beta} := \partial_\beta \eta_3 + b^\sigma_\beta \eta_\sigma. \end{aligned}}$$

The Gâteaux derivatives of each function $G_{\alpha\beta} : \mathbf{W}^{1,4}(\omega) \to L^2(\omega)$ are given by

$$\boxed{G'_{\alpha\beta}(\boldsymbol{\zeta})\boldsymbol{\eta} = \frac{1}{2}(\eta_{\alpha\|\beta} + \eta_{\beta\|\alpha} + a^{mn}\{\zeta_{m\|\alpha}\eta_{n\|\beta} + \zeta_{n\|\beta}\eta_{m\|\alpha}\})}$$

for all $\boldsymbol{\zeta}, \boldsymbol{\eta} \in \mathbf{W}^{1,4}(\omega)$. *In particular then, the functions* $E^0_{\alpha\|\beta}$ *and* $F^0_{\alpha\|\beta}(\boldsymbol{\eta})$ *appearing in the variational problem found in* Thm. 8.8-1 *are also given by*

$$\boxed{E^0_{\alpha\|\beta} = G_{\alpha\beta}(\boldsymbol{\zeta}^0) \text{ and } F^0_{\alpha\|\beta}(\boldsymbol{\eta}) = G'_{\alpha\beta}(\boldsymbol{\zeta}^0)\boldsymbol{\eta}.}$$

Proof. Let $\tilde{\boldsymbol{\eta}} := \eta_i \boldsymbol{a}^i$. By definition,

$$\begin{aligned} a_{\alpha\beta}(\boldsymbol{\eta}) &= (\boldsymbol{a}_\alpha + \partial_\alpha \tilde{\boldsymbol{\eta}}) \cdot (\boldsymbol{a}_\beta + \partial_\beta \tilde{\boldsymbol{\eta}}) \\ &= a_{\alpha\beta} + \partial_\beta \tilde{\boldsymbol{\eta}} \cdot \boldsymbol{a}_\alpha + \partial_\alpha \tilde{\boldsymbol{\eta}} \cdot \boldsymbol{a}_\beta + \partial_\alpha \tilde{\boldsymbol{\eta}} \cdot \partial_\beta \tilde{\boldsymbol{\eta}}. \end{aligned}$$

By the formulas of Gauß and Weingarten (Thm. 2.3-1),

$$\partial_\alpha \tilde{\boldsymbol{\eta}} = \eta_{\beta\|\alpha} \boldsymbol{a}^\beta + \eta_{3\|\alpha} \boldsymbol{a}^3.$$

Since $\boldsymbol{a}^i \cdot \boldsymbol{a}_\alpha = \delta^i_\alpha$ and $\boldsymbol{a}^m \cdot \boldsymbol{a}^n = a^{mn}$, it thus follows that

$$a_{\alpha\beta}(\boldsymbol{\eta}) - a_{\alpha\beta} = \eta_{\alpha\|\beta} + \eta_{\beta\|\alpha} + a^{mn}\eta_{m\|\alpha}\eta_{n\|\beta}.$$

Hence $E^0_{\alpha\|\beta} = G_{\alpha\beta}(\boldsymbol{\zeta}^0)$. That $F^0_{\alpha\|\beta}(\boldsymbol{\eta}) = G'_{\alpha\beta}(\boldsymbol{\zeta}^0)\boldsymbol{\eta}$ simply results from the way derivatives of continuous linear and bilinear mappings are computed (see, e.g., Vol. I, Sect. 1.2). ∎

Remarks. (1) The functions $\eta_{i\|\beta}$ may be also expressed in terms of the "two-dimensional" covariant derivatives $\eta_{i|\beta}$ of the vector field $\tilde{\boldsymbol{\eta}} = \eta_i \boldsymbol{a}^i$ (Thm. 2.3-1) as

$$\eta_{\alpha\|\beta} = \eta_{\alpha|\beta} - b_{\alpha\beta}\eta_3 \text{ and } \eta_{3\|\beta} = \eta_{3|\beta} - b^\sigma_\beta \eta_\sigma.$$

The double bars used in the functions $\eta_{i\|\beta}$ remind that they are restrictions for $x_3 = 0$ of "three-dimensional" covariant derivatives (Sect. 1.4).

(2) The components $G_{\alpha\beta}(\boldsymbol{\eta})$ of the change of metric tensor are the restrictions for $x^\varepsilon_3 = 0$ of the "three-dimensional" strains $E^\varepsilon_{i\|j}(\boldsymbol{v}^\varepsilon)$ with $\boldsymbol{\eta} := \boldsymbol{v}^\varepsilon|_{x_3=0}$. This was to be expected in view of their respective significances (Thms. 8.2-2 and 9.1-1). ∎

In this chapter, we consider the case where the only field $\boldsymbol{\eta} = (\eta_i)$ *in the space* $\mathbf{W}(\omega) = \{\boldsymbol{\eta} \in \mathbf{W}^{1,4}(\omega);\ \boldsymbol{\eta} = \mathbf{0} \text{ in } \omega\}$ *that satisfies*

$$\eta_{\alpha\|\beta} + \eta_{\beta\|\alpha} + a^{mn}\eta_{m\|\alpha}\eta_{n\|\beta} = a_{\alpha\beta}(\boldsymbol{\eta}) - a_{\alpha\beta} = 0 \text{ in } \omega$$

is $\boldsymbol{\eta} = \mathbf{0}$. This assumption, combined with Thm. 8.8-1, then leads to the following fundamental definition:

Let ω be a domain in $\mathbb{R}^2$ with boundary γ and let $\boldsymbol{\theta} \in \mathcal{C}^2(\overline{\omega}; \mathbb{R}^3)$ be an injective mapping such that the two vectors $\partial_\alpha \boldsymbol{\theta}$ are linearly independent at all points of $\overline{\omega}$. A *nonlinearly elastic shell* with middle surface $S = \boldsymbol{\theta}(\overline{\omega})$ is called a **nonlinearly elastic membrane shell** if the following two conditions are simultaneously satisfied:

(i) The shell is subjected to a *boundary condition of place* along a portion of its lateral face with $\boldsymbol{\theta}(\gamma_0)$ as its middle curve, where the subset $\gamma_0 \subset \gamma$ satisfies

$$\boxed{\textit{length } \gamma_0 > 0.}$$

(ii) Define the *manifold*:

$$\boxed{\begin{aligned}\boldsymbol{\mathcal{M}}_0(\omega) := \big\{\boldsymbol{\eta} = (\eta_i) \in \mathbf{W}^{1,4}(\omega);\ \boldsymbol{\eta} = \mathbf{0} \text{ on } \gamma_0,\\ a_{\alpha\beta}(\boldsymbol{\eta}) - a_{\alpha\beta} = 0 \text{ in } \omega\big\}.\end{aligned}}$$

Then

$$\boxed{\boldsymbol{\mathcal{M}}_0(\omega) = \{\mathbf{0}\}.}$$

A displacement field $\eta_i \boldsymbol{a}^i$ of the middle surface S is an **inextensional displacement** of S if $a_{\alpha\beta}(\boldsymbol{\eta}) - a_{\alpha\beta} = 0$ in ω. These relations imply that the surfaces $S = \boldsymbol{\theta}(\overline{\omega})$ and $(\boldsymbol{\theta} + \eta_i \boldsymbol{a}^i)(\overline{\omega})$ are **isometric**, i.e., that their metrics are the same (in particular then, the lengths of curves are the same; cf. Sect. 2.1).

A displacement field $\eta_i \boldsymbol{a}^i$ of the middle surface S is **admissible** if it satisfies *ad hoc* boundary conditions along the curve $\boldsymbol{\theta}(\gamma_0)$, in the present case *"boundary conditions of simple support"*, expressed by means of the boundary conditions $\boldsymbol{\eta} = \mathbf{0}$ on γ_0 on the associated field $\boldsymbol{\eta} = (\eta_i)$ (these boundary conditions are interpreted later; cf. Sect. 9.4).

The assumption $\boldsymbol{\mathcal{M}}_0(\omega) = \{\mathbf{0}\}$ thus expresses that *the only admissible* (i.e., such that $\eta_i = 0$ on γ_0) *inextensional displacement* $\eta_i \boldsymbol{a}^i$ *of the middle surface* S *with covariant components* η_i *in the space* $W^{1,4}(\omega)$ *is zero.*

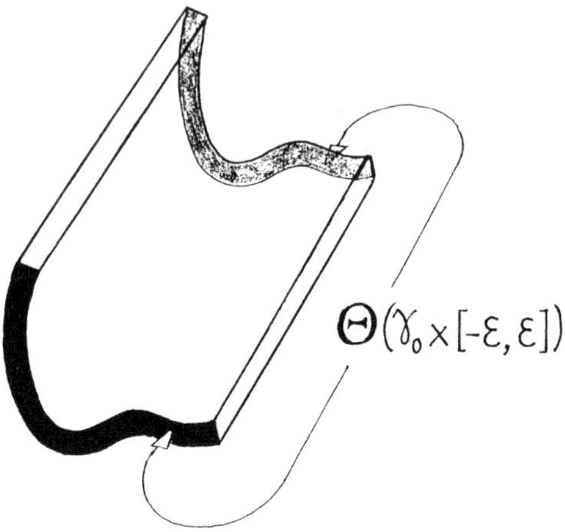

Fig. 9.1-1: *A nonlinearly elastic membrane shell.* A shell whose middle surface $S = \boldsymbol{\theta}(\overline{\omega})$ is a portion of a cylinder and which is subjected to a boundary condition of place (i.e., of vanishing displacement field) along a portion (darkened on the figure) of its lateral face whose middle curve $\boldsymbol{\theta}(\gamma_0)$ contains the two "end-curves" of S provides an instance of a nonlinearly elastic membrane shell, i.e., one for which the manifold

$$\boldsymbol{\mathcal{M}}_0(\omega) = \{\boldsymbol{\eta} = (\eta_i) \in \mathbf{W}^{1,4}(\omega);\ \boldsymbol{\eta} = \mathbf{0} \text{ on } \gamma_0,\ a_{\alpha\beta}(\boldsymbol{\eta}) - a_{\alpha\beta} = 0 \text{ in } \omega\}$$

reduces to $\{\mathbf{0}\}$; cf. Ex. 9.1.

Examples of nonlinearly elastic membrane shells are shown on Figs. 9.1-1 and 9.1-2 and analyzed in Exs. 9.1 and 9.2.

A word of caution. The definition of a nonlinearly elastic membrane shell depends only on the subset of the lateral face where the shell is subjected to a boundary condition of place (*via* the set γ_0) and on the geometry of the middle surface of the shell (*via* its two fundamental forms and Christoffel symbols, which appear in the functions $\eta_{i\|\beta}$); see Ex. 9.3. ■

Another word of caution. Although the above definition is strongly suggested by the asymptotic analysis, it is admittedly perfectible, as it leaves aside nonlinearly elastic shells that should also be deemed "membrane" ones.

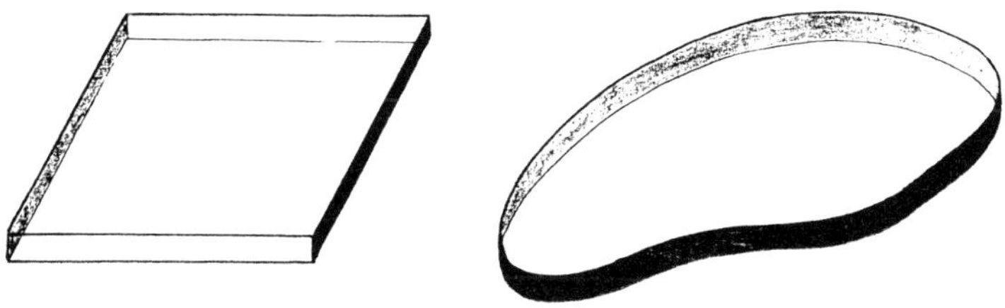

Fig. 9.1-2: *Another example of a nonlinearly elastic membrane shell.* Consider a plate subjected to a boundary condition of place on a portion of its lateral face whose middle line γ_0 has the following property: There exists (at least) one direction such that the intersection of $\overline{\omega}$ with any line parallel to this direction is a finite union of segments whose end-points belong to γ_0 (this is thus the case if $\gamma_0 = \gamma$). Then such a plate is a nonlinearly elastic membrane shell; cf. Ex. 9.2.

By contrast, such a plate is always modeled as a flexural shell (even if $\gamma_0 = \gamma$) when it is viewed as a linearly elastic body! See Fig. 6.1-3.

Consider for instance a shell whose middle surface is a half-sphere, subjected to a boundary condition of place along its entire lateral face (i.e., $\gamma_0 = \gamma$). Then $\mathcal{M}_0(\omega) \neq \{\mathbf{0}\}$, since symmetries with respect to planes orthogonal to the axis of revolution produce a *family of continuously varying isometric surfaces* $(\boldsymbol{\theta} + \eta_i \boldsymbol{a}^i)(\overline{\omega})$ with $\boldsymbol{\eta} = (\eta_i) \in \mathbf{W}^{1,\infty}(\omega)$ and $\boldsymbol{\eta} \neq \mathbf{0}$. Hence such a nonlinearly elastic shell is not a "membrane" one according to the above definition. Yet it is neither a "flexural" one according to the definition given in Sect. 10.2, as it seems plausible that the manifold $\mathcal{M}_F(\omega)$ introduced in *ibid.* contains only two elements (the covariant components of the admissible inextensional displacements in $\mathcal{M}_F(\omega)$ are then required to be in $W^{2,4}(\omega)$; see again Sect. 10.2). As hinted by this example, whether or not a manifold of inextensional displacements reduces to a finite set or not critically depends on the *smoothness* allowed on the admissible inextensional displacements.

This challenging question is related to that of the *rigidity of surfaces*, a classical problem in differential geometry: A compact surface "without boundary" is *rigid* if any other such surface in $\mathbb{R}^3$ with the same metric differs from it by a rigid deformation, i.e., a mapping of the form $x \in \mathbb{R}^3 \to \boldsymbol{a} + \mathbf{Q}\boldsymbol{ox}$, where $a \in \mathbb{R}^3$ and $\mathbf{Q}$ is an orthogonal matrix of order three; again, the allowed smoothness on the surfaces play a critical rôle. See, e.g., Pogorelov [1956], Nirenberg [1962], Klingenberg [1973, Thm. 6.2.8], Spivak [1975, Vol. V, Chap. 12], Berger & Gostiaux [1992, Thm. 11.14.1].

The same question can also be viewed as one about the *existence of everted states under vanishing applied forces.* For details about such *eversion problems for shells* (examples are given in Vol. I, Figs. 5.8-1 and 5.8-2), see in particular Srubshchik [1968, 1972, 1980], Vorovich [1969], Mel'nik & Srubshchik [1973], Antman [1979], Podio-Guidugli, Rosati, Schiaffino & Valente [1989], Szeri [1990], Geymonat & Léger [1994], Antman [1995, pp. 501-503], Antman & Srubshchik [1998]. ■

Since the formal asymptotic analysis ends with Thm. 8.8-1 when $\mathcal{M}_0(\omega) = \{\mathbf{0}\}$, we are naturally led to make the following **assumptions on the data** *for a family of nonlinearly elastic membrane shells*, with each having the *same* middle surface $S = \boldsymbol{\theta}(\overline{\omega})$ and with each subjected to a boundary condition of place along a portion of its lateral face having the *same* set $\boldsymbol{\theta}(\gamma_0)$ as its middle curve, as their thickness 2ε approaches zero:

We *require* that the *Lamé constants* and the *applied body and surface force densities* be such that

$$\begin{aligned} \lambda^\varepsilon = \lambda \quad &\text{and} \quad \mu^\varepsilon = \mu, \\ f^{i,\varepsilon}(x^\varepsilon) = f^{i,0}(x) \quad &\text{for all} \quad x^\varepsilon = \pi^\varepsilon x \in \Omega^\varepsilon, \\ h^{i,\varepsilon}(x^\varepsilon) = \varepsilon h^{i,1}(x) \quad &\text{for all} \quad x^\varepsilon = \pi^\varepsilon x \in \Gamma_+^\varepsilon \cup \Gamma_-^\varepsilon, \end{aligned}$$

where the constants $\lambda > 0$ *and* $\mu > 0$ *and the functions* $f^{i,0} \in L^2(\Omega)$ *and* $h^{i,1} \in L^2(\Gamma_+ \cup \Gamma_-)$ *are independent of* ε (Fig. 3.2-1 recapitulates the definitions of the sets Ω^ε, Ω, Γ_+^ε, Γ_+, Γ_-^ε, and Γ_-).

Note that the *same* limit two-dimensional equations are evidently obtained if the following *more general* **assumptions on the data** are made:

$$\begin{aligned} \lambda^\varepsilon = \varepsilon^t \lambda \quad &\text{and} \quad \mu^\varepsilon = \varepsilon^t \mu, \\ f^{i,\varepsilon}(x^\varepsilon) = \varepsilon^t f^{i,0}(x) \quad &\text{for all} \quad x^\varepsilon = \pi^\varepsilon x \in \Omega^\varepsilon, \\ h^{i,\varepsilon}(x^\varepsilon) = \varepsilon^{1+t} h^{i,1}(x) \quad &\text{for all} \quad x^\varepsilon = \pi^\varepsilon x \in \Gamma_+^\varepsilon \cup \Gamma_-^\varepsilon, \end{aligned}$$

where the constants $\lambda > 0$ and $\mu > 0$ and the functions $f^{i,0} \in L^2(\Omega)$ and $h^{i,1} \in L^2(\Gamma_+ \cup \Gamma_-)$ are again *independent of* ε and t *is an arbitrary real number*. Besides, the analysis of Chap. 8 (see in particular Thm. 8.8-1) shows that *these assumptions on the data are the only ones possible for nonlinearly elastic membrane shells.*

9.2. THE TWO-DIMENSIONAL EQUATIONS AS A VARIATIONAL PROBLEM

We know from Thm. 8.8-1 that the leading term $\boldsymbol{u}^0$ of the formal asymptotic expansion of the scaled unknown $\boldsymbol{u}(\varepsilon)$ is independent of the transverse variable x_3 and that, once identified with a function $\boldsymbol{\zeta}^0$ of two variables, it satisfies a *two-dimensional variational problem. When the manifold* $\boldsymbol{\mathcal{M}}_0(\omega)$ *reduces to* $\{\mathbf{0}\}$ (Sect. 9.1), no inconsistency arises in the formulation of this problem, and thus *the induction stops here.*

We also saw in Thm. 9.1-1 that the functions $E^0_{\alpha\|\beta}$ and $F^0_{\alpha\|\beta}(\boldsymbol{\eta})$ occurring in the formulation of this problem have more illuminating expressions than those found in Thm. 8.8-1; as a result, these new expressions shed a new light on the understanding of this variational problem, henceforth denoted $\mathcal{P}_M(\omega)$. The equivalent *boundary value problem* is given later; cf. Thm. 9.4-2.

The following result, which is an immediate corollary to Thms. 8.8-1 and 9.1-1, is due to Miara [1998]. Notice that a *new notation*, viz., $\mathbf{W}_M(\omega)$, is used *from now on in this chapter* for the space heretofore denoted $\mathbf{W}(\omega)$.

Theorem 9.2-1. *Consider a family of nonlinearly elastic membrane shells according to the definition of* Sect. 9.1, *with thickness* $2\varepsilon > 0$, *with each having the same middle surface* $S = \boldsymbol{\theta}(\overline{\omega})$ *and with each satisfying a boundary condition of place along a portion of its lateral face having the same set* $\boldsymbol{\theta}(\gamma_0)$ *as its middle curve, and let the assumptions on the data be as in* Sect. 9.1. *Finally, assume that* $\boldsymbol{\theta} \in \mathcal{C}^3(\overline{\omega}; \mathbb{R}^3)$.

Then the leading term $\boldsymbol{u}^0 : \overline{\Omega} \to \mathbb{R}^3$ *of the formal asymptotic expansion of the scaled unknown* $\boldsymbol{u}(\varepsilon)$ *is independent of the transverse variable and* $\boldsymbol{\zeta}^0 := \frac{1}{2}\int_{-1}^{1} \boldsymbol{u}^0\, dx_3$ *satisfies the following* **scaled two-dimensional variational problem $\mathcal{P}_M(\omega)$ of a nonlinearly elastic membrane shell:**

$$\boldsymbol{\zeta}^0 \in \mathbf{W}_M(\omega) := \{\boldsymbol{\eta} \in \mathbf{W}^{1,4}(\omega);\ \boldsymbol{\eta} = \mathbf{0} \text{ on } \gamma_0\},$$

$$\int_\omega a^{\alpha\beta\sigma\tau} G_{\sigma\tau}(\boldsymbol{\zeta}^0)\big(G'_{\alpha\beta}(\boldsymbol{\zeta}^0)\boldsymbol{\eta}\big)\sqrt{a}\, dy = \int_\omega p^{i,0}\eta_i \sqrt{a}\, dy$$

$$\text{for all } \boldsymbol{\eta} = (\eta_i) \in \mathbf{W}_M(\omega),$$

where, for any $\boldsymbol{\zeta}, \boldsymbol{\eta} \in \mathbf{W}^{1,4}(\omega)$ (the functions $a_{\alpha\beta}(\boldsymbol{\eta})$ are defined in Thm. 9.1-1),

$$G_{\alpha\beta}(\boldsymbol{\eta}) := \frac{1}{2}(a_{\alpha\beta}(\boldsymbol{\eta}) - a_{\alpha\beta})$$

$$= \frac{1}{2}(\eta_{\alpha\|\beta} + \eta_{\beta\|\alpha} + a^{mn}\eta_{m\|\alpha}\eta_{n\|\beta}),$$

$$G'_{\alpha\beta}(\boldsymbol{\zeta})\boldsymbol{\eta} = \frac{1}{2}(\eta_{\alpha\|\beta} + \eta_{\beta\|\alpha} + a^{mn}\{\zeta_{m\|\alpha}\eta_{n\|\beta} + \zeta_{n\|\beta}\eta_{m\|\alpha}\}),$$

$$\eta_{\alpha\|\beta} := \partial_\beta\eta_\alpha - \Gamma^\sigma_{\alpha\beta}\eta_\sigma - b_{\alpha\beta}\eta_3 \text{ and } \eta_{3\|\beta} := \partial_\beta\eta_3 + b^\sigma_\beta\eta_\sigma,$$

$$a^{\alpha\beta\sigma\tau} := \frac{4\lambda\mu}{\lambda + 2\mu}a^{\alpha\beta}a^{\sigma\tau} + 2\mu(a^{\alpha\sigma}a^{\beta\tau} + a^{\alpha\tau}a^{\beta\sigma}),$$

$$p^{i,0} := \int_{-1}^{1} f^{i,0}\, dx_3 + h^{i,1}_+ + h^{i,1}_- \text{ and } h^{i,1}_\pm := h^{i,1}(\cdot, \pm 1),$$

and $G'_{\alpha\beta}(\boldsymbol{\zeta}) \in \mathcal{L}(\mathbf{W}^{1,4}(\omega); L^2(\omega))$ *denotes the Fréchet derivative at* $\boldsymbol{\zeta} \in \mathbf{W}^{1,4}(\omega)$ *of each mapping* $G_{\alpha\beta} : \mathbf{W}^{1,4}(\omega) \to L^2(\omega)$. ■

Notice the analogy between the two-dimensional variational problem $\mathcal{P}_M(\omega)$ and the variational problem of three-dimensional nonlinear elasticity we started from (Thm. 8.2-3).

9.3. THE TWO-DIMENSIONAL EQUATIONS AS A MINIMIZATION PROBLEM

As noted by Miara [1998], the aforementioned analogy can be pursued further: First, the variational equations found in the two-dimensional and three-dimensional problems simply express in each case that the Gâteaux derivatives of an *ad hoc* functional vanish; second, the integrand in the functional is in each case (apart from its linear term accounting for the applied forces) a *quadratic and positive definite expression* (*via* an elasticity tensor) in terms of a *change of metric tensor.*

These crucial observations in turn allow to recast the two-dimensional variational problem $\mathcal{P}_M(\omega)$ as a *minimization problem*, which now exhibits a remarkable *simplicity*. In this problem lies the apex of the application of the method of formal asymptotic expansions to nonlinearly elastic membrane shells.

Theorem 9.3-1. *Given a nonlinearly elastic membrane shell according to the definition given in* Sect. 9.1, *let the space* $\mathbf{W}_M(\omega)$ *and the functional* $j_M : \mathbf{W}_M(\omega) \to \mathbb{R}$ *be defined by*

$$\begin{aligned} \mathbf{W}_M(\omega) &:= \{\boldsymbol{\eta} = (\eta_i) \in \mathbf{W}^{1,4}(\omega);\ \boldsymbol{\eta} = \mathbf{0} \text{ on } \gamma_0\}, \\ j_M(\boldsymbol{\eta}) &:= \frac{1}{8}\int_\omega a^{\alpha\beta\sigma\tau}(a_{\sigma\tau}(\boldsymbol{\eta}) - a_{\sigma\tau})(a_{\alpha\beta}(\boldsymbol{\eta}) - a_{\alpha\beta})\sqrt{a}\,dy \\ &\qquad - \int_\omega p^{i,0}\eta_i\sqrt{a}\,dy, \end{aligned}$$

where the functions $a_{\alpha\beta}(\boldsymbol{\eta})$ *are defined in* Thm. 9.1-1 *and the functions* $a^{\alpha\beta\sigma\tau}$ *and* $p^{i,0}$ *are defined in* Thm. 9.2-1.

Then the functional j_M *is differentiable over the space* $\mathbf{W}^{1,4}(\omega)$, *hence over* $\mathbf{W}_M(\omega)$, *and* $\boldsymbol{\zeta}^0 \in \mathbf{W}_M(\omega)$ *is a solution to the variational problem* $\mathcal{P}_M(\omega)$ *of* Thm. 9.2-1 *if and only if it is a stationary point of the functional* j_M *over the space* $\mathbf{W}_M(\omega)$, i.e., *it satisfies* $j'_M(\boldsymbol{\zeta}^0) = 0$. *Particular solutions to problem* $\mathcal{P}_M(\omega)$ *are thus obtained by solving the minimization problem: Find* $\boldsymbol{\zeta}$ *such that*

$$\boldsymbol{\zeta} \in \mathbf{W}_M(\omega) \text{ and } j_M(\boldsymbol{\zeta}) = \inf_{\boldsymbol{\eta}\in\mathbf{W}_M(\omega)} j_M(\boldsymbol{\eta}).$$

Proof. Since

$$j_M(\boldsymbol{\eta}) = \frac{1}{2}\int_\omega a^{\alpha\beta\sigma\tau} G_{\sigma\tau}(\boldsymbol{\eta}) G_{\alpha\beta}(\boldsymbol{\eta})\sqrt{a}\,dy - \int_\omega p^{i,0}\eta_i\sqrt{a}\,dy,$$

where

$$\begin{aligned} G_{\alpha\beta}(\boldsymbol{\eta}) &= \frac{1}{2}(a_{\alpha\beta}(\boldsymbol{\eta}) - a_{\alpha\beta}) \\ &= \frac{1}{2}(\eta_{\alpha\|\beta} + \eta_{\beta\|\alpha} + a^{mn}\eta_{m\|\alpha}\eta_{n\|\beta}), \end{aligned}$$

an argument similar to that used in the proof of Thm. 8.2-3 shows that j_M is differentiable over $\mathbf{W}^{1,4}(\omega)$ and that its Gâteaux derivatives $j'_M(\boldsymbol{\zeta})\boldsymbol{\eta}$ are given by

$$j'_M(\boldsymbol{\zeta})\boldsymbol{\eta} = \int_\omega a^{\alpha\beta\sigma\tau} G_{\sigma\tau}(\boldsymbol{\zeta})\big(G'_{\alpha\beta}(\boldsymbol{\zeta})\boldsymbol{\eta}\big)\sqrt{a}\,dy - \int_\omega p^{i,0}\eta_i\sqrt{a}\,dy$$

for all $\boldsymbol{\zeta}, \boldsymbol{\eta} \in \mathbf{W}^{1,4}(\omega)$, where

$$G'_{\alpha\beta}(\boldsymbol{\zeta})\boldsymbol{\eta} = \frac{1}{2}(\eta_{\alpha\|\beta} + \eta_{\beta\|\alpha} + a^{mn}\{\zeta_{m\|\alpha}\eta_{n\|\beta} + \zeta_{n\|\beta}\eta_{m\|\alpha}\}).$$

The variational equations satisfied by $\boldsymbol{\zeta}^0 \in \mathbf{W}_M(\omega)$ (Thm. 9.2-1) thus coincide with the equations

$$j'_M(\boldsymbol{\zeta}^0)\boldsymbol{\eta} = 0 \text{ for all } \boldsymbol{\eta} \in \mathbf{W}_M(\omega),$$

which are themselves equivalent to the equation $j'_M(\boldsymbol{\zeta}^0) = 0$.

That it makes sense to consider the above minimization problem stems from the observation that

$$\inf_{\boldsymbol{\eta}\in\mathbf{W}_M(\omega)} j_M(\boldsymbol{\eta}) > -\infty,$$

an inequality itself a consequence of the coerciveness of the functional j_M over the space $\mathbf{W}_M(\omega)$ (Ex. 9.4). The existence theory proper is otherwise fragmentary at the present time; see the discussion in the next section. ■

The functional $j_M : \mathbf{W}_M(\omega) \to \mathbb{R}$ is called the **scaled two-dimensional energy of a nonlinearly elastic membrane shell.**

9.4. THE TWO-DIMENSIONAL EQUATIONS OF A NONLINEARLY ELASTIC MEMBRANE SHELL DERIVED BY MEANS OF A FORMAL ASYMPTOTIC ANALYSIS; COMMENTARY

In order to get physically meaningful formulas, it remains to *de-scale* the components ζ_i^0 of the vector field $\boldsymbol{\zeta}^0$ that satisfies the scaled two-dimensional problems found in Thms. 9.2-1 and 9.3-1. In view of the scalings

$$u_i(\varepsilon)(x) = u_i^\varepsilon(x^\varepsilon) \text{ for all } x^\varepsilon = \pi^\varepsilon x \in \overline{\Omega}^\varepsilon$$

made on the covariant components of the displacement (Sect. 8.4), we are naturally led to defining for each $\varepsilon > 0$ the **covariant components** $\zeta_i^\varepsilon : \overline{\omega} \to \mathbb{R}$ of the **limit displacement field** $\tilde{\boldsymbol{\zeta}}^\varepsilon : \overline{\omega} \to \mathbb{R}^3$ *of the middle surface S of the shell* by letting (the vectors $\boldsymbol{a}^i$ form the contravariant basis at each point of S):

$$\boxed{\zeta_i^\varepsilon := \zeta_i^0 \text{ and } \tilde{\boldsymbol{\zeta}}^\varepsilon := \zeta_i^\varepsilon \boldsymbol{a}^i.}$$

Remark. I realize that, *for once*, my notations (ζ_i^ε vs. ζ_i^0) are somewhat misleading! ■

A word of caution. Naturally, *the fields* (ζ_i^ε) *and* $\tilde{\boldsymbol{\zeta}}^\varepsilon = \zeta_i^\varepsilon \boldsymbol{a}^i$ *must be carefully distinguished*! The former is essentially a convenient *mathematical "intermediary"*, but only the latter has *physical significance.* ■

Recall that $f^{i,\varepsilon} \in L^2(\Omega^\varepsilon)$ and $h^{i,\varepsilon} \in L^2(\Gamma_+^\varepsilon \cup \Gamma_-^\varepsilon)$ represent the contravariant components of the applied body and surface forces *actually* acting on the shell and that λ^ε and μ^ε denote the *actual* Lamé constants of its constituting material. We then have the following immediate corollary to Thms. 9.2-1 and 9.3-1.

Theorem 9.4-1. *Let the assumptions be as in* Thm. 9.2-1. *Then the vector field* $\boldsymbol{\zeta}^\varepsilon := (\zeta_i^\varepsilon)$ *formed by the covariant components of the limit displacement field* $\zeta_i^\varepsilon \boldsymbol{a}^i$ *of the middle surface S satisfies the following* **two-dimensional variational problem** $\mathcal{P}_M^\varepsilon(\omega)$ **of a**

nonlinearly elastic membrane shell:

$$\boxed{\begin{aligned}&\boldsymbol{\zeta}^\varepsilon \in \mathbf{W}_M(\omega) := \{\boldsymbol{\eta} \in \mathbf{W}^{1,4}(\omega);\ \boldsymbol{\eta} = \mathbf{0} \text{ on } \gamma_0\},\\ &\varepsilon \int_\omega a^{\alpha\beta\sigma\tau,\varepsilon} G_{\sigma\tau}(\boldsymbol{\zeta}^\varepsilon)\big(G'_{\alpha\beta}(\boldsymbol{\zeta}^\varepsilon)\boldsymbol{\eta}\big)\sqrt{a}\,dy = \int_\omega p^{i,\varepsilon}\eta_i\sqrt{a}\,dy\\ &\qquad\qquad \text{for all } \boldsymbol{\eta} = (\eta_i) \in \mathbf{W}_M(\omega),\end{aligned}}$$

where

$$\boxed{\begin{aligned}G_{\alpha\beta}(\boldsymbol{\eta}) &:= \frac{1}{2}(a_{\alpha\beta}(\boldsymbol{\eta}) - a_{\alpha\beta})\\ &= \frac{1}{2}(\eta_{\alpha\|\beta} + \eta_{\beta\|\alpha} + a^{mn}\eta_{m\|\alpha}\eta_{n\|\beta}),\\ G'_{\alpha\beta}(\boldsymbol{\zeta})\boldsymbol{\eta} &= \frac{1}{2}(\eta_{\alpha\|\beta} + \eta_{\beta\|\alpha} + a^{mn}\{\zeta_{m\|\alpha}\eta_{n\|\beta} + \zeta_{n\|\beta}\eta_{m\|\alpha}\}),\\ \eta_{\alpha\|\beta} &:= \partial_\beta\eta_\alpha - \Gamma^\sigma_{\alpha\beta}\eta_\sigma - b_{\alpha\beta}\eta_3 \text{ and } \eta_{3\|\beta} := \partial_\beta\eta_3 + b^\sigma_\beta\eta_\sigma,\\ a^{\alpha\beta\sigma\tau,\varepsilon} &:= \frac{4\lambda^\varepsilon\mu^\varepsilon}{\lambda^\varepsilon + 2\mu^\varepsilon}a^{\alpha\beta}a^{\sigma\tau} + 2\mu^\varepsilon(a^{\alpha\sigma}a^{\beta\tau} + a^{\alpha\tau}a^{\beta\sigma}),\\ p^{i,\varepsilon} &:= \int_{-\varepsilon}^{\varepsilon} f^{i,\varepsilon}\,dx_3^\varepsilon + h_+^{i,\varepsilon} + h_-^{i,\varepsilon} \text{ and } h_\pm^{i,\varepsilon} := h^{i,\varepsilon}(\cdot, \pm\varepsilon).\end{aligned}}$$

The field $\boldsymbol{\zeta}^\varepsilon = (\zeta_i^\varepsilon)$ is a solution to problem $\mathcal{P}_M^\varepsilon(\omega)$ if and only if it is a stationary point of the functional $j_M^\varepsilon : \mathbf{W}_M(\omega) \to \mathbb{R}$ defined by

$$\boxed{\begin{aligned}j_M^\varepsilon(\boldsymbol{\eta}) := \frac{\varepsilon}{8}\int_\omega a^{\alpha\beta\sigma\tau,\varepsilon}(a_{\sigma\tau}(\boldsymbol{\eta}) - a_{\sigma\tau})(a_{\alpha\beta}(\boldsymbol{\eta}) - a_{\alpha\beta})\sqrt{a}\,dy&\\ - \int_\omega p^{i,\varepsilon}\eta_i\sqrt{a}\,dy.&\end{aligned}}$$

Particular solutions to problem $\mathcal{P}_M^\varepsilon(\omega)$ may thus be obtained by solving the following minimization problem: Find $\boldsymbol{\zeta}^\varepsilon$ such that

$$\boxed{\boldsymbol{\zeta}^\varepsilon \in \mathbf{W}_M(\omega) \ \text{ and } \ j_M^\varepsilon(\boldsymbol{\zeta}^\varepsilon) = \inf_{\boldsymbol{\eta}\in\mathbf{W}_M(\omega)} j_M^\varepsilon(\boldsymbol{\eta}).}$$

■

We next derive the *boundary value problem* that is (at least formally) equivalent to the variational problem $\mathcal{P}_M^\varepsilon(\omega)$ found in Thm. 9.4-1. In the next theorem, (ν_α) denotes the unit outer normal vector along γ and $\gamma_1 := \gamma - \gamma_0$.

Theorem 9.4-2. *Let the assumptions and notations be as in* Thm. 9.4-1. *If a solution* $\boldsymbol{\zeta}^\varepsilon = (\zeta_i^\varepsilon)$ *to the variational problem*

$$
\boxed{\begin{aligned}
&\boldsymbol{\zeta}^\varepsilon \in \mathbf{W}_M(\omega) = \{\boldsymbol{\eta} \in \mathbf{W}^{1,4}(\omega);\ \boldsymbol{\eta} = \mathbf{0} \text{ on } \gamma_0\},\\
&\varepsilon \int_\omega a^{\alpha\beta\sigma\tau,\varepsilon} G_{\sigma\tau}(\boldsymbol{\zeta}^\varepsilon)\big(G'_{\alpha\beta}(\boldsymbol{\zeta}^\varepsilon)\boldsymbol{\eta}\big)\sqrt{a}\,dy = \int_\omega p^{i,\varepsilon}\eta_i\sqrt{a}\,dy\\
&\hspace{16em}\text{for all } \boldsymbol{\eta} = (\eta_i) \in \mathbf{W}_M(\omega)
\end{aligned}}
$$

is smooth enough, it also satisfies the following boundary value problem:

$$
\boxed{\begin{aligned}
-(n^{\alpha\beta,\varepsilon} + n^{\sigma\beta,\varepsilon}a^{\alpha\tau}\zeta^\varepsilon_{\tau\|\sigma})|_\beta + b^\alpha_\beta n^{\sigma\beta,\varepsilon}\zeta^\varepsilon_{3\|\sigma} &= p^{\alpha,\varepsilon} \text{ in } \omega,\\
-b_{\alpha\beta}(n^{\alpha\beta,\varepsilon} + n^{\sigma\beta,\varepsilon}a^{\alpha\tau}\zeta^\varepsilon_{\tau\|\sigma}) - (n^{\alpha\beta,\varepsilon}\zeta^\varepsilon_{3\|\alpha})|_\beta &= p^{3,\varepsilon} \text{ in } \omega,\\
\zeta_i^\varepsilon &= 0 \text{ on } \gamma_0,\\
(n^{\alpha\beta,\varepsilon} + n^{\sigma\beta,\varepsilon}a^{\alpha\tau}\zeta^\varepsilon_{\tau\|\sigma})\nu_\beta &= 0 \text{ on } \gamma_1,\\
n^{\alpha\beta}\zeta^\varepsilon_{3\|\alpha}\nu_\beta &= 0 \text{ on } \gamma_1,
\end{aligned}}
$$

where

$$
\boxed{n^{\alpha\beta,\varepsilon} := \varepsilon a^{\alpha\beta\sigma\tau,\varepsilon} G_{\sigma\tau}(\boldsymbol{\zeta}^\varepsilon)}
$$

the functions $G_{\alpha\beta}(\boldsymbol{\zeta}^\varepsilon)$, $a^{\alpha\beta\sigma\tau,\varepsilon}$, *and* $p^{i,\varepsilon}$ *are defined as in* Thm. 9.4-1 *and, for any vector field with differentiable contravariant components* $\eta^\alpha : \overline{\omega} \to \mathbb{R}$ *and any tensor field with differentiable contravariant components* $n^{\alpha\beta} : \overline{\omega} \to \mathbb{R}$,

$$
\boxed{\eta^\alpha|_\beta := \partial_\beta\eta^\alpha + \Gamma^\alpha_{\beta\sigma}\eta^\sigma \text{ and } n^{\alpha\beta}|_\sigma := \partial_\sigma n^{\alpha\beta} + \Gamma^\alpha_{\sigma\tau}n^{\beta\tau} + \Gamma^\beta_{\sigma\tau}n^{\alpha\tau}.}
$$

Proof. For notational brevity, the dependence on ε is omitted. Using Green's formula as in the proof of Thm. 4.5-1, we obtain

$$\frac{1}{2}\int_\omega \sqrt{a}n^{\alpha\beta}(\eta_{\alpha\|\beta}+\eta_{\beta\|\alpha})dy = -\int_\omega \sqrt{a}\{(n^{\alpha\beta}|_\beta)\eta_\alpha + b_{\alpha\beta}n^{\alpha\beta}\eta_3\}\,dy + \int_\gamma \sqrt{a}n^{\alpha\beta}\nu_\beta\eta_\alpha\,d\gamma,$$

$$\begin{aligned}\frac{1}{2}\int_\omega & \sqrt{a}n^{\alpha\beta}a^{mn}\{\zeta_{m\|\alpha}\eta_{n\|\beta}+\zeta_{n\|\beta}\eta_{m\|\alpha}\}\,dy \\ &= \int_\omega \sqrt{a}n^{\alpha\beta}a^{mn}\zeta_{m\|\alpha}\eta_{n\|\beta}\,dy \\ &= -\int_\omega \sqrt{a}\{(n^{\sigma\beta}a^{\alpha\tau}\zeta_{\tau\|\sigma})|_\beta - b^\alpha_\beta n^{\sigma\beta}\zeta_{3\|\sigma}\}\eta_\alpha\,dy \\ &\quad -\int_\omega \sqrt{a}\{b_{\alpha\beta}n^{\sigma\beta}a^{\alpha\tau}\zeta_{\tau\|\sigma} + (n^{\alpha\beta}\zeta_{3\|\alpha})|_\beta\}\eta_3\,dy \\ &\quad +\int_\gamma \sqrt{a}\{n^{\sigma\beta}a^{\alpha\tau}\zeta_{\tau\|\sigma}\nu_\beta\eta_\alpha + n^{\alpha\beta}\zeta_{3\|\alpha}\nu_\beta\eta_3\}\,d\gamma\end{aligned}$$

for all $(\eta_i) \in \mathbf{W}^{1,4}(\omega)$. Letting (η_i) vary first in $(\mathcal{D}(\omega))^3$, then in $\mathbf{W}_M(\omega)$, yields the announced boundary value problem. ■

Each one of the three formulations found in Thms. 9.4-1 and 9.4-2 constitutes one version of the **two-dimensional equations of a nonlinearly elastic membrane shell** (the specific meaning conveyed by "membrane" is given below).

The functions $G_{\alpha\beta}(\boldsymbol{\eta}) = \frac{1}{2}(a_{\alpha\beta}(\boldsymbol{\eta}) - a_{\alpha\beta})$ are the *covariant components* of the **change of metric tensor** associated with a displacement field $\eta_i\boldsymbol{a}^i$ of the middle surface S (Thm. 9.1-1) and the functions $a^{\alpha\beta\sigma\tau,\varepsilon}$ are the *contravariant components* of the **two-dimensional elasticity tensor of the shell.** The functional $j^\varepsilon_M : \mathbf{W}_M(\omega) \to \mathbb{R}$ is the **two-dimensional energy**, and the functional

$$\boxed{\boldsymbol{\eta} \in \mathbf{W}_M(\omega) \to \frac{\varepsilon}{8}\int_\omega a^{\alpha\beta\sigma\tau,\varepsilon}(a_{\sigma\tau}(\boldsymbol{\eta}) - a_{\sigma\tau})(a_{\alpha\beta}(\boldsymbol{\eta}) - a_{\alpha\beta})\sqrt{a}\,dy}$$

is the **two-dimensional strain energy**, of a **nonlinearly elastic membrane shell.**

Finally, the functions $n^{\alpha\beta,\varepsilon}$ are the *contravariant components* of the **stress resultant tensor** field. The functions $\eta^\alpha|_\beta$ are instances of *first-order covariant derivatives* of a vector field defined by means of its *contravariant* components η^α and the functions $n^{\alpha\beta}|_\sigma$ are *first-order covariant derivatives* of a tensor field defined by means of its contravariant components $n^{\alpha\beta}$ (such covariant derivatives have already been encountered in Thm. 4.5-1).

Considered as relations *with respect to the functions* $n^{\alpha\beta,\varepsilon}$, the partial differential relations in ω and the boundary conditions on γ_1 found in Thm. 9.4-2 constitute the **two-dimensional equations of equilibrium of a nonlinearly elastic membrane shell.**

The relations $n^{\alpha\beta,\varepsilon} = \varepsilon a^{\alpha\beta\sigma\tau,\varepsilon} G_{\alpha\beta}(\boldsymbol{\zeta}^\varepsilon)$ constitute the **two-dimensional constitutive equation of a nonlinearly elastic membrane shell.**

Finally, the boundary conditions $\eta_i = 0$ on γ_0 express that the points of the undeformed and deformed middle surfaces $\boldsymbol{\theta}(\overline{\omega})$ and $(\boldsymbol{\theta} + \eta_i \boldsymbol{a}^i)(\overline{\omega})$ coincide along the curve $\boldsymbol{\theta}(\gamma_0)$. For this reason, they are called **two-dimensional boundary conditions of simple support.**

A *major conclusion* is thus that, without any recourse to any *a priori* assumption of a geometrical or of a mechanical nature, *the method of formal asymptotic expansions provides a justification of the two-dimensional equations of a nonlinearly elastic membrane shell*, in the forms found in Thm. 9.4-1. This justification by Miara [1998] constitutes a generalization to *shells* of the formal analysis of Fox, Raoult & Simo [1993], who likewise identified and justified the two-dimensional equations of a *"nonlinearly elastic planar membrane"*, i.e., one for which the mapping $\boldsymbol{\theta}$ is of the form $\boldsymbol{\theta}(y_1, y_2) = (y_1, y_2, 0)$ for all $y = (y_1, y_2) \in \overline{\omega}$ (see also Vol. II, Sect. 4.12).

Further important conclusions and comments are in order about the present justification:

First and foremost, the resulting shell theory is a **nonlinear "membrane" theory** in the sense that the **stored energy function of a nonlinearly elastic membrane shell**, defined by

$$\boldsymbol{\eta} = (\eta_i) \in \mathbf{W}^{1,4}(\omega) \to \frac{\varepsilon}{8} a^{\alpha\beta\sigma\tau,\varepsilon}(a_{\sigma\tau}(\boldsymbol{\eta}) - a_{\sigma\tau})(a_{\alpha\beta}(\boldsymbol{\eta}) - a_{\alpha\beta}),$$

is a quadratic and positive definite expression (*via* the two-dimensional elasticity tensor of the shell) in terms of the *change of metric tensor*, i.e., of the *exact difference* between the *metric tensor* of the *deformed* middle surface and that of the *undeformed* one.

Note in passing the truly remarkable simplicity of the above stored energy function!

Second, the resulting theory is **frame-indifferent** in the sense that the value of the above stored energy function is unaltered if $\boldsymbol{\psi} := \boldsymbol{\theta} + \eta_i \boldsymbol{a}^i$ is replaced by $\mathbf{Q}\boldsymbol{\psi}$, where $\mathbf{Q}$ is any orthogonal matrix of order three (Ex. 9.5).

Third, the resulting theory is a **large displacement**, or equivalently a **large deformation, theory** in the sense that *the de-scaling produces a displacement field that is* $O(1)$ *with respect to* ε, at least in a formal sense.

Finally, a noteworthy characteristic of the equations found in Thm. 9.4-1 is that their *formal linearization* produces the equations of the *linear "membrane"* theory (Chaps. 4 and 5). This is perhaps best seen from the expression of the energy, in which the covariant components $G_{\alpha\beta}(\boldsymbol{\eta}) = \frac{1}{2}(a_{\alpha\beta}(\boldsymbol{\eta}) - a_{\alpha\beta})$ of the change of metric tensor become *by definition* the covariant components $\gamma_{\alpha\beta}(\boldsymbol{\eta})$ of the *linearized* change of metric tensor (Thm. 4.5-2). An inspection of the *partial differential equations* in the associated boundary value problem (Thm. 9.4-2) likewise shows that they reduce to those of the linear theory when $\gamma_0 = \gamma$ and the middle surface S is elliptic (Thm. 4.5-2).

Remarks. (1) Otherwise the *function spaces* and the *boundary conditions* have to be appropriately modified in the linearization process, so that the two-dimensional equations of the linear membrane theory become well-posed over the resulting spaces $\mathbf{V}_M(\omega)$ or $\mathbf{V}^\sharp_M(\omega)$ (Thms. 4.5-1 and 5.7-1).

(2) As observed by Miara [1998], another unexpected link with the *linearized* equations occur if higher-order forces are considered; cf. Ex. 9.6. ■

Note that Collard & Miara [1999] have shown that the formal analysis of Miara [1998] described *supra* also leads to the explicit computation of the *limit stresses* in a nonlinearly elastic membrane shell.

The *existence theory* for the *minimization problem* is challenging: The energy j^ε_M is *coercive* on the space $\mathbf{W}_M(\omega)$ (Ex. 9.4), but, as shown by Genevey [1997], j^ε_M is *not* sequentially weakly lower semi-continuous on $\mathbf{W}_M(\omega)$ (essential to her proof is the non-polyconvexity of the stored energy function of a St Venant-Kirchhoff material established by Raoult [1986]). Hence the fundamental theorem of the calculus of variations cannot be applied (by contrast, this theorem can be successfully applied to nonlinearly elastic "flexural" shells; cf. Sect. 10.6).

Another tool that is sometimes successful for proving existence theorems in nonlinear elasticity is the *implicit function theorem* applied to the associated system of partial differential equations, which in the present case is *quasilinear* (Thm. 9.4-2). But, even though the existence and regularity results for the *linearized* equations of a linearly elastic elliptic membrane shell are firmly established (Thm. 4.5-1), the implicit function theorem cannot be applied to the nonlinear membrane shell equations, expressed as a set of three nonlinear partial differential equations together with the boundary conditions $\zeta_i^\varepsilon = 0$ on γ, if only because the boundary condition $\zeta_3^\varepsilon = 0$ on γ is "lost" in the linearization. As a result, the derivative at $\boldsymbol{\zeta}^\varepsilon = \mathbf{0}$ of the associated nonlinear operator is not an isomorphism between the "right" spaces, i.e., those of the linearized boundary value problem; cf. Genevey [1997].

As shown by Coutand [1997b, 1999b, 1999c, 1999d, 1999e] in the *planar* case, the implicit function theorem can nevertheless be successfully used in *particular* situations, e.g., if the forces are themselves planar, or if the boundary conditions $\zeta_i = 0$ on γ are replaced by boundary conditions of "planar tension". Again in the *planar* case when $\overline{\omega}$ is a disk and $\gamma_0 = \gamma$, Genevey [1998] has likewise shown that the implicit function theorem provides the existence of radial solutions around an explicitly computable nonzero radial solution. Nonlinearly elastic membrane shells may thus be likewise amenable to this approach if the applied forces, or the boundary conditions, or the middle surface, are of special kinds.

9.5$^\flat$. THE TWO-DIMENSIONAL EQUATIONS OF A NONLINEARLY ELASTIC MEMBRANE SHELL DERIVED BY MEANS OF Γ-CONVERGENCE THEORY; COMMENTARY

A remarkable progress in the asymptotic analysis of *nonlinearly elastic shells* is due to Le Dret & Raoult [1996], who gave the *first* proof of *convergence as the thickness approaches zero.* In so doing, they extended to shells the analysis that they had successfully applied to *"nonlinearly elastic planar membranes"* in Le Dret & Raoult [1995a] (see also Vol. II, Sect. 4.13), where they had likewise given the *first* convergence result in the planar case. Note that their analysis has since then been extended by Ben Belgacem [1997] and El Bachari [1997, 1998] to stored energy functions such that $\hat{W}(\mathbf{F}) \to +\infty$ as $\det \mathbf{F} \to 0^+$ and to *junctions between shells* and *multi-layered shells.*

More specifically, H. Le Dret and A. Raoult showed that *a subsequence of deformations that minimize* (or rather "almost minimize" in a sense explained below) *the scaled three-dimensional energies weakly converges in* $\mathbf{W}^{1,p}(\Omega)$ *as* $\varepsilon \to 0$ (the number $p \in]1, \infty[$ is governed by the growth properties of the stored energy function). They showed in addition that the weak limit minimizes a *"membrane" energy* that is the Γ-*limit* of the scaled energies. We now give an abridged account of their analysis.

Let ω be a domain in $\mathbb{R}^2$ with boundary γ and let $\boldsymbol{\theta} \in \mathcal{C}^2(\overline{\omega}; \mathbb{R}^3)$ be an injective mapping such that the two vectors $\boldsymbol{a}_\alpha(y) = \partial_\alpha \boldsymbol{\theta}(y)$ are linearly independent at all points $y = (y_\alpha) \in \overline{\omega}$. For each $\varepsilon > 0$, we define the set

$$\Omega^\varepsilon := \omega \times]-\varepsilon, \varepsilon[,$$

we let $x^\varepsilon = (x_i^\varepsilon)$ denote a generic point in the set $\overline{\Omega}^\varepsilon$ and we let $\partial_i^\varepsilon := \partial/\partial x_i^\varepsilon$; hence $x_\alpha^\varepsilon = y_\alpha$ and $\partial_\alpha^\varepsilon = \partial_\alpha$.

Consider as in Sect. 8.3 a family of *elastic shells* with the same *middle surface* $S = \boldsymbol{\theta}(\overline{\omega})$ and whose *thickness* $2\varepsilon > 0$ approaches zero. The reference configuration of each shell is thus the image $\boldsymbol{\Theta}(\overline{\Omega}^\varepsilon) \subset \mathbb{R}^3$ of the set $\overline{\Omega}^\varepsilon \subset \mathbb{R}^3$ through a mapping $\boldsymbol{\Theta} : \overline{\Omega}^\varepsilon \to \mathbb{R}^3$ defined by

$$\boldsymbol{\Theta}(x^\varepsilon) := \boldsymbol{\theta}(y) + x_3^\varepsilon \boldsymbol{a}_3(y) \text{ for all } x^\varepsilon = (y, x_3^\varepsilon) = (y_1, y_2, x_3^\varepsilon) \in \overline{\Omega}^\varepsilon.$$

By Thm. 3.1-1, if the injective mapping $\boldsymbol{\theta} : \overline{\omega} \to \mathbb{R}^3$ is smooth enough, *the mapping* $\boldsymbol{\Theta} : \overline{\Omega}^\varepsilon \to \mathbb{R}^3$ *is also injective for* $\varepsilon > 0$ *small enough* and $y_1, y_2, x_3^\varepsilon$ then constitute the *"natural" curvilinear coordinates* for describing each reference configuration $\boldsymbol{\Theta}(\overline{\Omega}^\varepsilon)$.

Assume that *the shells are made for all* $\varepsilon > 0$ *of the same hyperelastic homogeneous material*, satisfying the following properties (hyperelastic materials are studied in detail in Vol. I, Chap. 4): Let $\mathbb{M}^3$ denote the space of all real square matrices of order three and let $|\cdot|$ denote any norm on $\mathbb{M}^3$; it is then assumed that the *stored energy function* $\hat{W} : \mathbb{M}^3 \to \mathbb{R}$ of the hyperelastic material satisfies the following assumptions: There exist constants $C > 0$, $\alpha > 0$, $\beta \in \mathbb{R}$, and $1 < p < \infty$ such that

$$\begin{aligned} |\hat{W}(\mathbf{F})| &\le C(1 + |\mathbf{F}|^p) \quad \text{for all } \mathbf{F} \in \mathbb{M}^3, \\ \hat{W}(\mathbf{F}) &\ge \alpha|\mathbf{F}|^p + \beta \quad \text{for all } \mathbf{F} \in \mathbb{M}^3, \\ |\hat{W}(\mathbf{F}) - \hat{W}(\mathbf{G})| &\le C(1 + |\mathbf{F}|^{p-1} + |\mathbf{G}|^{p-1})|\mathbf{F} - \mathbf{G}| \\ &\qquad\qquad \text{for all } \mathbf{F}, \mathbf{G} \in \mathbb{M}^3. \end{aligned}$$

It can be verified that the stored energy function of a *St Venant-Kirchhoff material*, which is given by

$$\hat{W}(\mathbf{F}) = \frac{\mu}{4} \operatorname{tr} (\mathbf{F}^T\mathbf{F} - \mathbf{I})^2 + \frac{\lambda}{8} \left\{\operatorname{tr} (\mathbf{F}^T\mathbf{F} - \mathbf{I})\right\}^2,$$

satisfies such inequalities with $p = 4$; cf. Ex. 9.7. By contrast, the stored energy function of a *linearly* elastic material, which is given by

$$\hat{W}(\mathbf{F}) = \frac{\mu}{4}\|\mathbf{F} + \mathbf{F}^T - 2\mathbf{I}\|^2 + \frac{\lambda}{8} \left\{\operatorname{tr} (\mathbf{F}^T + \mathbf{F} - 2\mathbf{I})\right\}^2,$$

where $\|\mathbf{F}\| := \{\operatorname{tr} \mathbf{F}^T\mathbf{F}\}^{1/2}$, satisfies the first inequality with $p = 2$, but *not* the second one.

It is further assumed that, for each $\varepsilon > 0$, the shells are subjected in their interior to *applied body forces* of density $\hat{\boldsymbol{f}}^\varepsilon = (\hat{f}_i^\varepsilon) : \hat{\Omega}^\varepsilon \to \mathbb{R}^3$ per unit volume, where $\hat{f}_i^\varepsilon \in L^q(\hat{\Omega}^\varepsilon)$ and $\frac{1}{p} + \frac{1}{q} = 1$. Since these densities do not depend on the unknown, the applied forces are *dead loads* (Vol. I, Sect. 2.7). Applied surface forces on the "upper" and "lower" faces of the shells could be likewise considered, but are omitted for simplicity; see in this respect Le Dret & Raoult [1996] who consider a *pressure load* (cf. Appendix in *ibid.*), an example of applied surface force that is a *live load* (Vol. I, Sect. 2.7).

Finally, it is assumed that each shell is subjected to a *boundary condition of place* along its entire lateral face $\boldsymbol{\Theta}(\gamma \times [-\varepsilon, \varepsilon])$, i.e., that the displacement vanishes there.

For each $\varepsilon > 0$, let

$$\hat{\Omega}^\varepsilon := \boldsymbol{\Theta}(\Omega^\varepsilon),$$

let $\hat{x}^\varepsilon = (\hat{x}_i^\varepsilon)$ denote a generic point in the reference configuration $\{\hat{\Omega}^\varepsilon\}^-$, let $\hat{\partial}_i^\varepsilon := \partial/\partial\hat{x}_i^\varepsilon$, and let the *deformation gradient* associated with any deformation $\hat{\boldsymbol{\psi}}^\varepsilon = (\hat{\psi}_i^\varepsilon) : \{\hat{\Omega}^\varepsilon\}^- \to \mathbb{R}^3$ of the reference configuration be the matrix field $\hat{\boldsymbol{\nabla}}^{\varepsilon}\hat{\boldsymbol{\psi}}^{\varepsilon} : \{\hat{\Omega}^{\varepsilon}\}^{-} \to \mathbb{M}^3$ defined by

$$\hat{\boldsymbol{\nabla}}^{\varepsilon}\hat{\boldsymbol{\psi}}^{\varepsilon} := \begin{pmatrix} \hat{\partial}_1^{\varepsilon}\hat{\psi}_1^{\varepsilon} & \hat{\partial}_2^{\varepsilon}\hat{\psi}_1^{\varepsilon} & \hat{\partial}_3^{\varepsilon}\hat{\psi}_1^{\varepsilon} \\ \hat{\partial}_1^{\varepsilon}\hat{\psi}_2^{\varepsilon} & \hat{\partial}_2^{\varepsilon}\hat{\psi}_2^{\varepsilon} & \hat{\partial}_3^{\varepsilon}\hat{\psi}_2^{\varepsilon} \\ \hat{\partial}_1^{\varepsilon}\hat{\psi}_3^{\varepsilon} & \hat{\partial}_2^{\varepsilon}\hat{\psi}_3^{\varepsilon} & \hat{\partial}_3^{\varepsilon}\hat{\psi}_3^{\varepsilon} \end{pmatrix}.$$

The three-dimensional problem is then posed as a *minimization problem* in terms of the *unknown deformation field*

$$\boxed{\hat{\boldsymbol{\varphi}}^{\varepsilon} := \boldsymbol{id}_{\mathbb{R}^3} + \hat{\boldsymbol{u}}^{\varepsilon}}$$

of the reference configuration, where $\hat{\boldsymbol{u}}^{\varepsilon} : \{\hat{\Omega}^{\varepsilon}\}^{-} \to \mathbb{R}^3$ is its displacement field: It consists in finding $\hat{\boldsymbol{\varphi}}^{\varepsilon}$ such that

$$\boxed{\begin{aligned} &\hat{\boldsymbol{\varphi}}^{\varepsilon} \in \boldsymbol{\Phi}(\hat{\Omega}^{\varepsilon}) \text{ and } \hat{I}^{\varepsilon}(\hat{\boldsymbol{\varphi}}^{\varepsilon}) = \inf_{\hat{\boldsymbol{\psi}}^{\varepsilon} \in \boldsymbol{\Phi}(\hat{\Omega}^{\varepsilon})} \hat{I}^{\varepsilon}(\hat{\boldsymbol{\psi}}^{\varepsilon}), \text{ where} \\ &\boldsymbol{\Phi}(\hat{\Omega}^{\varepsilon}) := \{\hat{\boldsymbol{\psi}}^{\varepsilon} \in \mathbf{W}^{1,p}(\hat{\Omega}^{\varepsilon});\ \hat{\boldsymbol{\psi}}^{\varepsilon} = \boldsymbol{id}_{\mathbb{R}^3} \text{ on } \boldsymbol{\Theta}(\gamma \times [-\varepsilon, \varepsilon])\}, \\ &\hat{I}^{\varepsilon}(\hat{\boldsymbol{\psi}}^{\varepsilon}) := \int_{\hat{\Omega}^{\varepsilon}} \hat{W}(\hat{\boldsymbol{\nabla}}^{\varepsilon}\hat{\boldsymbol{\psi}}^{\varepsilon})\, d\hat{x}^{\varepsilon} - \int_{\hat{\Omega}^{\varepsilon}} \hat{\boldsymbol{f}}^{\varepsilon} \cdot \hat{\boldsymbol{\psi}}^{\varepsilon}\, d\hat{x}^{\varepsilon}. \end{aligned}}$$

This minimization problem may have *no* solution; however, this is not a shortcoming as only the existence of a "diagonal infimizing family", whose existence is always guaranteed, is required in the ensuing analysis (Thm. 9.5-1).

For each $\varepsilon > 0$, this problem is first transformed into a *problem posed over the set* $\overline{\Omega}^{\varepsilon}$, i.e., expressed in terms of the "natural" curvilinear coordinates of the shell, the *unknown* $\boldsymbol{\varphi}^{\varepsilon} = (\varphi_i^{\varepsilon}) : \overline{\Omega}^{\varepsilon} \to \mathbb{R}^3$ of this new problem being defined by

$$\boxed{\boldsymbol{\varphi}^{\varepsilon}(x^{\varepsilon}) := \hat{\boldsymbol{\varphi}}^{\varepsilon}(\hat{x}^{\varepsilon}) \text{ for all } \hat{x}^{\varepsilon} = \boldsymbol{\Theta}(x^{\varepsilon}),\ x^{\varepsilon} \in \overline{\Omega}^{\varepsilon}.}$$

If $\varepsilon > 0$ is small enough, the mapping $\boldsymbol{\Theta}$ is a $\mathcal{C}^1$-diffeomorphism of $\overline{\Omega}^{\varepsilon}$ onto its image $\boldsymbol{\Theta}(\overline{\Omega}^{\varepsilon}) \subset \mathbb{R}^3$ and $\det \boldsymbol{\nabla}\boldsymbol{\Theta}^{\varepsilon} > 0$ in $\overline{\Omega}^{\varepsilon}$ (Thm. 3.1-1). The formula for changing variables in multiple integrals then shows that φ^ε satisfies the following *minimization problem*:

$$\begin{aligned}
&\boldsymbol{\varphi}^\varepsilon \in \boldsymbol{\Phi}(\Omega^\varepsilon) \text{ and } I^\varepsilon(\boldsymbol{\varphi}^\varepsilon) = \inf_{\boldsymbol{\psi}^\varepsilon \in \boldsymbol{\Phi}(\Omega^\varepsilon)} I^\varepsilon(\boldsymbol{\psi}^\varepsilon), \text{ where}\\
&\boldsymbol{\Phi}(\Omega^\varepsilon) := \{\boldsymbol{\psi}^\varepsilon \in \mathbf{W}^{1,p}(\Omega^\varepsilon);\ \boldsymbol{\psi}^\varepsilon = \boldsymbol{\Theta} \text{ on } \gamma \times [-\varepsilon, \varepsilon]\},\\
&I^\varepsilon(\boldsymbol{\psi}^\varepsilon) = \int_{\Omega^\varepsilon} \hat{W}\big(\boldsymbol{\nabla}^\varepsilon \boldsymbol{\psi}^\varepsilon (\boldsymbol{\nabla}^\varepsilon \boldsymbol{\Theta})^{-1}\big) \det \boldsymbol{\nabla}^\varepsilon \boldsymbol{\Theta}\, dx^\varepsilon\\
&\qquad\qquad - \int_{\Omega^\varepsilon} \tilde{\boldsymbol{f}}^\varepsilon \cdot \boldsymbol{\psi}^\varepsilon \det \boldsymbol{\nabla}^\varepsilon \boldsymbol{\Theta}\, dx^\varepsilon,
\end{aligned}$$

where the matrix field $\boldsymbol{\nabla}^\varepsilon \boldsymbol{\psi}^\varepsilon : \overline{\Omega}^\varepsilon \to \mathbb{M}^3$ is defined by (the matrix field $\boldsymbol{\nabla}^\varepsilon \boldsymbol{\Theta} : \overline{\Omega}^\varepsilon \to \mathbb{M}^3$ is analogously defined):

$$\boldsymbol{\nabla}^\varepsilon \boldsymbol{\psi}^\varepsilon := \begin{pmatrix} \partial_1^\varepsilon \psi_1^\varepsilon & \partial_2^\varepsilon \psi_1^\varepsilon & \partial_3^\varepsilon \psi_1^\varepsilon \\ \partial_1^\varepsilon \psi_2^\varepsilon & \partial_2^\varepsilon \psi_2^\varepsilon & \partial_3^\varepsilon \psi_2^\varepsilon \\ \partial_1^\varepsilon \psi_3^\varepsilon & \partial_2^\varepsilon \psi_3^\varepsilon & \partial_3^\varepsilon \psi_3^\varepsilon \end{pmatrix},$$

and the vector field $\tilde{\boldsymbol{f}}^\varepsilon : \overline{\Omega}^\varepsilon \to \mathbb{R}^3$ is defined by

$$\tilde{\boldsymbol{f}}^\varepsilon(x^\varepsilon) := \hat{\boldsymbol{f}}^\varepsilon(\hat{x}^\varepsilon) \text{ for all } \hat{x}^\varepsilon = \boldsymbol{\Theta}(x^\varepsilon),\ x^\varepsilon \in \Omega^\varepsilon.$$

A word of caution. The function $\det \boldsymbol{\nabla}^\varepsilon \boldsymbol{\Theta}$ is equal to the function $\sqrt{g^\varepsilon}$, where $g^\varepsilon = \det(g_{ij}^\varepsilon)$; cf. Sect. 8.3.

However, the resemblance with the change of coordinates, also from Cartesian to curvilinear ones, performed in Sect. 8.3 stops here! For in Sect. 8.3, the new unknowns are the *covariant* components $u_i^\varepsilon : \overline{\Omega}^\varepsilon \to \mathbb{R}$ of the displacement field $u_i^\varepsilon \boldsymbol{g}^{i,\varepsilon}$, while in the present case the new unknowns are still *Cartesian* components $\varphi_i^\varepsilon : \overline{\Omega}^\varepsilon \to \mathbb{R}$ of the deformation field $\varphi_i^\varepsilon \boldsymbol{e}^i$. ■

Like the variational problem in Sect. 8.4, this minimization problem is then transformed into an analogous one, but now posed over the *fixed domain* $\Omega := \omega\times]-1, 1[$. As in Sect. 8.4, we denote by $x = (x_1, x_2, x_3)$ a generic point in $\overline{\Omega}$, we let $\partial_i := \partial/\partial x_i$, and with each point $x \in \overline{\Omega}$, we associate the point $x^\varepsilon \in \overline{\Omega}^\varepsilon$ through the bijection

$$\pi^\varepsilon : x = (x_1, x_2, x_3) \in \overline{\Omega} \to x^\varepsilon = (x_i^\varepsilon) = (x_1, x_2, \varepsilon x_3) \in \overline{\Omega}^\varepsilon.$$

Then, the unknown is *scaled*, by letting

$$\boldsymbol{\varphi}^\varepsilon(x^\varepsilon) =: \boldsymbol{\varphi}(\varepsilon)(x) \text{ for all } x^\varepsilon = \pi^\varepsilon x,\ x \in \overline{\Omega},$$

and it is *assumed* that there exists a vector field $\tilde{\boldsymbol{f}} \in \mathbf{L}^2(\Omega)$ *independent of* ε such that

$$\tilde{\boldsymbol{f}}^\varepsilon(x^\varepsilon) = \tilde{\boldsymbol{f}}(x) \text{ for all } x^\varepsilon = \pi^\varepsilon x,\ x \in \overline{\Omega}.$$

Remarks. (1) Inasmuch as the outcome of the asymptotic analysis carried out in this section will also be a nonlinear theory of "membrane" shells, the above assumption that the applied body forces are of order $O(1)$ with respect to ε is consistent with the assumption made in Sect. 9.1 on the applied body forces.

(2) Should applied surface forces act on the upper and lower faces of the shells, they should be of order $O(\varepsilon)$ with respect to ε, again as in Sect. 9.1. ■

These scalings and assumptions then imply that the **scaled deformation** $\boldsymbol{\varphi}(\varepsilon)$ satisfies the following minimization problem:

$$\boldsymbol{\varphi}(\varepsilon) \in \boldsymbol{\Phi}(\varepsilon;\Omega) \text{ and } I(\varepsilon)(\boldsymbol{\varphi}(\varepsilon)) = \inf_{\boldsymbol{\psi}\in\boldsymbol{\Phi}(\varepsilon;\Omega)} I(\varepsilon)(\boldsymbol{\psi}), \text{ where}$$
$$\boldsymbol{\Phi}(\varepsilon;\Omega) := \{\boldsymbol{\psi} \in \mathbf{W}^{1,p}(\Omega);\ \boldsymbol{\psi} = \boldsymbol{\varphi}_0(\varepsilon) \text{ on } \gamma \times [-1,1]\},$$
$$I(\varepsilon)(\boldsymbol{\psi}) := \int_\Omega \hat{W}\Big((\partial_1\boldsymbol{\psi};\ \partial_2\boldsymbol{\psi};\ \frac{1}{\varepsilon}\partial_3\boldsymbol{\psi})(\mathbf{G}(\varepsilon))^{-1}\Big) \det \mathbf{G}(\varepsilon)\,dx - \int_\Omega \tilde{\boldsymbol{f}}\cdot\boldsymbol{\psi} \det \mathbf{G}(\varepsilon)\,dx,$$

where the vector field $\boldsymbol{\varphi}_0(\varepsilon) : \overline{\Omega} \to \mathbb{R}^3$ is defined for each $\varepsilon > 0$ by

$$\boldsymbol{\varphi}_0(\varepsilon)(x) := \boldsymbol{\Theta}(x^\varepsilon) \text{ for all } x^\varepsilon = \pi^\varepsilon x,\ x \in \overline{\Omega},$$

the matrix field $\mathbf{G}(\varepsilon) : \overline{\Omega} \to \mathbb{M}^3$ is defined by

$$\mathbf{G}(\varepsilon)(x) := \boldsymbol{\nabla}^\varepsilon\boldsymbol{\Theta}(x^\varepsilon) \text{ for all } x^\varepsilon = \pi^\varepsilon x,\ x \in \overline{\Omega},$$

and the notation $(\boldsymbol{b}_1;\ \boldsymbol{b}_2;\ \boldsymbol{b}_3)$ stands for the matrix in $\mathbb{M}^3$ whose three column vectors are $\boldsymbol{b}_1$, $\boldsymbol{b}_2$, $\boldsymbol{b}_3$ (in this order).

Remark. The function $\det \mathbf{G}(\varepsilon)$ is equal to the function $\sqrt{g(\varepsilon)}$, where $g(\varepsilon)$ is defined in Thm. 8.4-1. ■

The *scaled displacement*

$$\tilde{\boldsymbol{u}}(\varepsilon) := \boldsymbol{\varphi}(\varepsilon) - \boldsymbol{\varphi}_0(\varepsilon)$$

therefore satisfies the following minimization problem:

$$\tilde{\boldsymbol{u}}(\varepsilon) \in \tilde{\mathbf{W}}(\Omega) \text{ and } J(\varepsilon)(\tilde{\boldsymbol{u}}(\varepsilon)) = \inf_{\tilde{\boldsymbol{v}} \in \tilde{\mathbf{W}}(\Omega)} J(\varepsilon)(\tilde{\boldsymbol{v}}), \text{ where}$$

$$\tilde{\mathbf{W}}(\Omega) := \{\tilde{\boldsymbol{v}} \in \mathbf{W}^{1,p}(\Omega);\ \tilde{\boldsymbol{v}} = \mathbf{0} \text{ on } \gamma \times [-1, 1]\}$$

$$J(\varepsilon)(\tilde{\boldsymbol{v}}) := \int_\Omega \hat{W}\Big(\mathbf{I} + \big(\partial_1 \tilde{\boldsymbol{v}};\ \partial_2 \tilde{\boldsymbol{v}};\ \frac{1}{\varepsilon}\partial_3 \tilde{\boldsymbol{v}}\big)(\mathbf{G}(\varepsilon))^{-1}\Big) \det \mathbf{G}(\varepsilon)\, dx - \int_\Omega \tilde{\boldsymbol{f}} \cdot (\boldsymbol{\varphi}_0(\varepsilon) + \tilde{\boldsymbol{v}}) \det \mathbf{G}(\varepsilon)\, dx.$$

A word of caution. The notations $\tilde{\boldsymbol{u}}(\varepsilon)$ and $\tilde{\boldsymbol{v}}$ remind that, although this minimization problem is still expressed in terms of (scaled) *curvilinear* coordinates (the coordinates x_i of the points $x \in \overline{\Omega}$), the unknowns are *no longer* (scaled) *covariant* components of the displacement field. Instead, the unknowns are now (scaled) *Cartesian* components, i.e., over a *fixed* Cartesian frame, of the (scaled) displacement field $\tilde{\boldsymbol{u}}(\varepsilon) : \overline{\Omega} \to \mathbb{R}^3$. ■

Central to the subsequent analysis is the notion of *quasiconvexity*, due to Morrey [1952, 1966] (an account of its importance in the calculus of variations is provided in Dacorogna [1989, Chap. 5]): Let $\mathbb{M}^{m\times n}$ denote the space of all real matrices with m rows and n columns; a function $\hat{W} : \mathbb{M}^{m\times n} \to \mathbb{R}$ is *quasiconvex* if, for all bounded open subsets $U \subset \mathbb{R}^n$, all $\mathbf{F} \in \mathbb{M}^{m\times n}$, and all $\boldsymbol{\xi} = (\xi_i)_{i=1}^m \in \mathbf{W}_0^{1,\infty}(U)$,

$$\hat{W}(\mathbf{F}) \le \frac{1}{\text{meas } U} \int_U \hat{W}(\mathbf{F} + \boldsymbol{\nabla}\boldsymbol{\xi}(x))\, dx,$$

where $\boldsymbol{\nabla}\boldsymbol{\xi}$ denotes the matrix $(\partial_j \xi_i) \in \mathbb{M}^{m\times n}$. Given any function $\hat{W} : \mathbb{M}^{m\times n} \to \mathbb{R}$, its *quasiconvex envelope* $\mathcal{Q}\hat{W} : \mathbb{M}^{m\times n} \to \mathbb{R}$ is the function defined by

$$\mathcal{Q}\hat{W} := \sup\{\hat{X} : \mathbb{M}^{m\times n} \to \mathbb{R};\ \hat{X} \text{ is quasi-convex and } \hat{X} \le \hat{W}\}.$$

Remark. An illuminating instance of actual computation of a quasiconvex envelope is found in Le Dret & Raoult [1995b], who explicitly determine the quasiconvex envelope of the stored energy function of a *St Venant-Kirchhoff material.* ■

Also central to the ensuing analysis is the notion of Γ-**convergence**, a powerful theory initiated by De Giorgi [1975, 1977] (see also De Giorgi & Franzoni [1975]); an illuminating introduction is found in De Giorgi & Dal Maso [1983] and thorough treatments are given in the books of Attouch [1984] and Dal Maso [1993]. As shown in the pioneering works of Acerbi, Buttazzo & Percivale [1991] for *nonlinearly elastic strings*, then of Le Dret & Raoult [1995a, 1996] for *nonlinearly elastic planar membranes* and *membrane shells* (reviewed here), this approach has thus far provided the only known *convergence theorems* for justifying lower-dimensional nonlinear theories of elastic bodies.

We then recall a fundamental definition (various complements on Γ-convergence may be found in Vol. II, Sect. 1.11, where a detailed application of this theory is treated): Let V be a metric space and let $J(\varepsilon) : V \to \mathbb{R}$ be functionals defined for all $\varepsilon > 0$. *The family* $(J(\varepsilon))_{\varepsilon>0}$ is said to **Γ-converge** *as* $\varepsilon \to 0$ if there exists a functional $J : V \to \mathbb{R} \cup \{+\infty\}$, called the **$\Gamma$-limit** *of the functionals* $J(\varepsilon)$, such that

$$v(\varepsilon) \to v \text{ as } \varepsilon \to 0 \Rightarrow J(v) \leq \liminf_{\varepsilon\to 0} J(\varepsilon)(v(\varepsilon)),$$

on the one hand and, given any $v \in V$, there exist $v(\varepsilon) \in V$, $\varepsilon > 0$, such that

$$v(\varepsilon) \to v \text{ as } \varepsilon \to 0 \text{ and } J(v) = \lim_{\varepsilon\to 0} J(\varepsilon)(v(\varepsilon)),$$

on the other.

As a preparation to the application of Γ-convergence theory, the *scaled energies* $J(\varepsilon) : \tilde{\mathbf{W}}(\Omega) \to \mathbb{R}$ are first extended to the *larger* space $\mathbf{L}^p(\Omega)$ by letting

$$\tilde{J}(\varepsilon)(\tilde{\boldsymbol{v}}) = \begin{cases} J(\varepsilon)(\tilde{\boldsymbol{v}}) \text{ if } \tilde{\boldsymbol{v}} \in \tilde{\mathbf{W}}(\Omega), \\ +\infty \text{ if } \tilde{\boldsymbol{v}} \in \mathbf{L}^p(\Omega) \text{ but } \tilde{\boldsymbol{v}} \notin \tilde{\mathbf{W}}(\Omega). \end{cases}$$

Such an extension, customary in Γ-convergence theory, has *inter alia* the advantage of "incorporating" the boundary condition into the extended functional.

Le Dret & Raoult [1996] then establish that *the family* $(\tilde{J}(\varepsilon))_{\varepsilon>0}$ *of extended energies* Γ*-converges as* $\varepsilon \to 0$ *in* $\mathbf{L}^p(\Omega)$ *and that its* Γ*-limit can be computed by means of quasiconvex envelopes.*

More precisely, their analysis leads to the following remarkable *convergence theorem*, where the limit minimization problems are directly posed as two-dimensional problems (part c)); this is licit since the solutions of these limit problems do not depend on the transverse variable (part (b)). Note that, while minimizers of $J(\varepsilon)$ over $\tilde{\mathbf{W}}(\Omega)$ need not exist, the existence of a "diagonal infimizing family" in the sense understood below is *always* guaranteed because $\inf_{\boldsymbol{v}\in\tilde{\mathbf{W}}(\Omega)} J(\varepsilon)(\boldsymbol{v}) > -\infty$.

In what follows, the notation $(\boldsymbol{b}_1; \boldsymbol{b}_2)$ stands for the matrix in $\mathbb{M}^{3\times 2}$ with $\boldsymbol{b}_1$, $\boldsymbol{b}_2$ (in this order) as its column vectors and $\sqrt{a}\,dy$ denotes as usual the area element along the surface S.

Theorem 9.5-1. *Assume that there exist $C > 0$, $\alpha > 0$, $\beta \in \mathbb{R}$, and $1 < p < \infty$ such that the stored energy function $\hat{W} : \mathbb{M}^3 \to \mathbb{R}$ satisfies the following growth conditions:*

$$\begin{aligned} |\hat{W}(\mathbf{F})| &\le C(1 + |\mathbf{F}|^p) \quad \text{for all } \mathbf{F} \in \mathbb{M}^3, \\ \hat{W}(\mathbf{F}) &\ge \alpha|\mathbf{F}|^p + \beta \quad \text{for all } \mathbf{F} \in \mathbb{M}^3, \\ |\hat{W}(\mathbf{F}) - \hat{W}(\mathbf{G})| &\le C(1 + |\mathbf{F}|^{p-1} + |\mathbf{G}|^{p-1})|\mathbf{F} - \mathbf{G}| \\ &\qquad \text{for all } \mathbf{F},\, \mathbf{G} \in \mathbb{M}^3. \end{aligned}$$

Let the space $\tilde{\mathbf{W}}(\Omega)$ be defined by

$$\tilde{\mathbf{W}}(\Omega) := \{\tilde{\boldsymbol{v}} \in \mathbf{W}^{1,p}(\Omega);\ \tilde{\boldsymbol{v}} = \mathbf{0} \text{ on } \gamma \times [-1, 1]\},$$

and let $(\tilde{\boldsymbol{u}}(\varepsilon))_{\varepsilon>0}$ be a "diagonal infimizing family" of the scaled energies, i.e., a family that satisfies

$$\tilde{\boldsymbol{u}}(\varepsilon) \in \tilde{\mathbf{W}}(\Omega) \text{ and } J(\varepsilon)(\tilde{\boldsymbol{u}}(\varepsilon)) \le \inf_{\tilde{\boldsymbol{v}}\in\tilde{\mathbf{W}}(\Omega)} J(\varepsilon)(\tilde{\boldsymbol{v}}) + h(\varepsilon) \quad \text{for all } \varepsilon > 0,$$

where h is any positive function that satisfies $h(\varepsilon) \to 0$ as $\varepsilon \to 0$. Then:

(a) *The family $(\tilde{\boldsymbol{u}}(\varepsilon))_{\varepsilon>0}$ lies in a weakly compact subset of the space $\mathbf{W}^{1,p}(\Omega)$.*

(b) *The limit $\tilde{\boldsymbol{u}} \in \tilde{\mathbf{W}}(\Omega)$ as $\varepsilon \to 0$ of any weakly convergent subsequence of $(\tilde{\boldsymbol{u}}(\varepsilon))_{\varepsilon>0}$ satisfies $\partial_3\tilde{\boldsymbol{u}} = \mathbf{0}$ in Ω and is thus independent of the transverse variable.*

(c) *The vector field* $\tilde{\boldsymbol{\zeta}} := \frac{1}{2}\int_{-1}^{1} \tilde{\boldsymbol{u}}\, dx_3$ *satisfies the following minimization problem* ($\mathcal{Q}\hat{W}_0(y, \cdot)$ denotes for each $y \in \overline{\omega}$ the quasiconvex envelope of $\hat{W}_0(y, \cdot)$):

$$\tilde{\boldsymbol{\zeta}} \in \mathbf{W}_0^{1,p}(\omega) \text{ and } \tilde{j}_M(\tilde{\boldsymbol{\zeta}}) = \inf_{\tilde{\boldsymbol{\eta}} \in \mathbf{W}_0^{1,p}(\omega)} \tilde{j}_M(\tilde{\boldsymbol{\eta}}), \text{ where}$$

$$\tilde{j}_M(\tilde{\boldsymbol{\eta}}) = 2\int_\omega \mathcal{Q}\hat{W}_0(y, (\boldsymbol{a}_1 + \partial_1\tilde{\boldsymbol{\eta}}; \boldsymbol{a}_2 + \partial_2\tilde{\boldsymbol{\eta}}))\sqrt{a}\, dy - \int_\omega \left\{\int_{-1}^{1} \tilde{\boldsymbol{f}}\, dx_3\right\} \cdot \tilde{\boldsymbol{\eta}}\sqrt{a}\, dy,$$

$$\hat{W}_0(y, (\boldsymbol{b}_1; \boldsymbol{b}_2)) := \inf_{\boldsymbol{b}_3 \in \mathbb{R}^3} \hat{W}((\boldsymbol{b}_1; \boldsymbol{b}_2; \boldsymbol{b}_3)\mathbf{G}^{-1}(y)) \quad \text{for all } (y, (\boldsymbol{b}_1; \boldsymbol{b}_2)) \in \overline{\omega} \times \mathbb{M}^{3\times 2},$$

$$\mathbf{G}(y) = (\boldsymbol{a}_1(y), \boldsymbol{a}_2(y), \boldsymbol{a}_3(y)),$$

the vectors $\boldsymbol{a}_i(y)$ *forming for each* $y \in \overline{\omega}$ *the covariant basis at the point* $\boldsymbol{\theta}(y) \in S$. ∎

It remains to *de-scale* the vector field $\tilde{\boldsymbol{\zeta}}$. In view of the scalings performed on the deformations, we are naturally led to defining for each $\varepsilon > 0$ the **limit displacement field** $\tilde{\boldsymbol{\zeta}}^\varepsilon : \overline{\omega} \to \mathbb{R}^3$ *of the middle surface* S by

$$\tilde{\boldsymbol{\zeta}}^\varepsilon := \tilde{\boldsymbol{\zeta}}.$$

It is then immediately verified that $\tilde{\boldsymbol{\zeta}}^\varepsilon$ *satisfies the minimization problem* (the notations are those of Thm. 9.5-1):

$$\tilde{\boldsymbol{\zeta}}^\varepsilon \in \mathbf{W}_0^{1,p}(\omega) \text{ and } \tilde{j}_M^\varepsilon(\tilde{\boldsymbol{\zeta}}^\varepsilon) = \inf_{\tilde{\boldsymbol{\eta}} \in \mathbf{W}_0^{1,p}(\omega)} \tilde{j}_M^\varepsilon(\tilde{\boldsymbol{\eta}}), \text{ where}$$

$$\tilde{j}_M^\varepsilon(\tilde{\boldsymbol{\eta}}) = 2\varepsilon\int_\omega \mathcal{Q}\hat{W}_0(y, (\boldsymbol{a}_1 + \partial_1\tilde{\boldsymbol{\eta}}; \boldsymbol{a}_2 + \partial_2\tilde{\boldsymbol{\eta}}))\sqrt{a}\, dy - \int_\omega \left\{\int_{-\varepsilon}^{\varepsilon} \tilde{\boldsymbol{f}}^\varepsilon\, dx_3^\varepsilon\right\} \cdot \tilde{\boldsymbol{\eta}}\sqrt{a}\, dy.$$

A word of caution. As their three-dimensional counterparts $\tilde{\boldsymbol{u}}(\varepsilon)$ and $\tilde{\boldsymbol{v}}$, the notations $\tilde{\boldsymbol{\zeta}}^\varepsilon$ and $\tilde{\boldsymbol{\eta}}$ emphasize that, even though the limit minimization problem is expressed in terms of *curvilinear* coordinates (the coordinates y_α of the points $y \in \overline{\omega}$), the unknowns are *no longer* the *covariant* components of the limit displacement field. Instead, the unknowns are now the *Cartesian* components, i.e., over a *fixed* Cartesian frame, of the displacement field $\tilde{\boldsymbol{\zeta}}^\varepsilon : \overline{\omega} \to \mathbb{R}^3$ of the middle surface. Likewise, $\tilde{\boldsymbol{f}}^\varepsilon : \Omega^\varepsilon \to \mathbb{R}^3$ denotes the vector field formed by the *Cartesian* components of the applied body force density. ■

A natural question immediately arises: How does this de-scaled minimization problem compare with the de-scaled minimization problem derived in Sect. 9.4 by means of the method of formal asymptotic expansions? As we now explain, it presents *similarities* with, as well as *differences* from, the nonlinear membrane theory previously found by formal means.

First, it shares with the nonlinear membrane theory of Sect. 9.4 the property that the unknown $\tilde{\boldsymbol{\eta}}$ appears only by means of its *first-order partial derivatives* $\partial_\alpha \tilde{\boldsymbol{\eta}}$ in the **stored energy function**

$$\boxed{\tilde{\boldsymbol{\eta}} \in \mathbf{W}^{1,p}(\omega) \to \varepsilon \mathcal{Q}\hat{W}_0(\cdot, (\boldsymbol{a}_1 + \partial_1 \tilde{\boldsymbol{\eta}}; \boldsymbol{a}_2 + \partial_2 \tilde{\boldsymbol{\eta}}))}$$

found in the integrand of the **energy** $\tilde{j}^\varepsilon_M$. It is likewise a **large displacement**, or equivalently a **large deformation, theory** in the sense that *the de-scaling produces a displacement field that is* $O(1)$ *with respect to* ε.

Second, assume that the original stored energy function is *frame-indifferent* in the sense that

$$\hat{W}(\mathbf{RF}) = \hat{W}(\mathbf{F}) \text{ for all } \mathbf{R} \in \mathbb{O}^3_+, \ \mathbf{F} \in \mathbb{M}^3,$$

where $\mathbb{O}^3_+$ denotes the set of all real orthogonal matrices $\mathbf{R}$ of order three satisfying $\det \mathbf{R} = 1$. This relation is stronger than the usual one, which holds only for $\mathbf{F} \in \mathbb{M}^3$ with $\det \mathbf{F} > 0$ (Vol. I, Thm. 4.2-1); it is, however, verified by the kinds of stored energy functions to which the present analysis applies, e.g., that of a St Venant-Kirchhoff material. Under this stronger assumption, Le Dret & Raoult [1996, Thm. 10] establish the crucial properties that *the stored energy function found in* $\tilde{j}^\varepsilon_M$, *once expressed as a function of the points of* S, *is frame-indifferent* and that *it depends only on the metric of the deformed middle surface.* Hence this theory is also a **frame-indifferent, nonlinear "membrane" shell theory**, like the one derived in Sect. 9.4 by means of a formal method.

It is remarkable that the stored energy function found in $\tilde{j}_M^\varepsilon$ can be *explicitly computed* when the original three-dimensional stored energy function is that of a *St Venant-Kirchhoff material*; see Le Dret & Raoult [1996, Sect. 6].

It is no less remarkable that Genevey [1997, Thm. 2.2.1] has been able to show that, for such a material, *this stored energy function coincides with that found* by Miara [1998] (Thm. 9.4-1), *provided that the singular values of the limit deformation gradients belong to an* ad hoc *compact subset of* $\mathbb{R}^2$, which can be explicitly identified (Le Dret & Raoult [1995a, Prop. 16] had already made a similar observation in the planar case). As a result, the linearization of the present nonlinear membrane theory does not always produce the equations of the linear "membrane" theory, by contrast with the nonlinear membrane theory of Sect. 9.4 (see the commentary at the end of *ibid.*).

A word of caution. These two approaches thus provide an intriguing instance where the limit equations found by a *formal* asymptotic analysis *do not always* coincide with those found by a *convergence* theorem. This is all the more puzzling, since the original three-dimensional equations, the scalings, and the assumptions on the data are the same! ■

Le Dret & Raoult [1996, Sect. 6] have further shown that, if the stored energy function is frame-indifferent and satisfies $\hat{W}(\mathbf{F}) \geq \hat{W}(\mathbf{I})$ for all $\mathbf{F} \in \mathbb{M}^3$ (as does the stored energy function of a St Venant-Kirchhoff material), then *the corresponding shell energy is constant under compression.* This result has the striking consequence that *"nonlinear membrane shells offer no resistance to crumpling. This is an empirical fact, witnessed by anyone who ever played with a deflated balloon"* (to quote H. Le Dret and A. Raoult).

This is why the equations found by Le Dret & Raoult [1996] seem to be especially appropriate for membrane shells made with a *"soft" elastic material*, but this assertion is yet to be rigorously established. Indeed, the modeling and numerical simulation of shells made of "soft", or "rubberlike", elastic materials is of paramount importance as they are so often encountered in practice; think of, e.g., a trampoline, a sail, or a tire! In this direction, see Muttin [1991], Schieck, Pietraszkiewicz & Stumpf [1992], Basar & Itskov [1998], Antman & Schuricht [1999].

Another challenging open problem consists in determining whether Γ-convergence theory could produce a two-dimensional "flexural" theory of nonlinearly elastic shells, analogous to the theory (described in the next chapter) obtained through the formal approach.

EXERCISES

9.1. Let $\boldsymbol{f} = (f_\alpha) \in C^2([0, 1]; \mathbb{R}^2)$ be an injective mapping such that $\boldsymbol{f}'(t) \neq \mathbf{0}$ for all $t \in [0, 1]$ and let $S = \boldsymbol{\theta}(\overline{\omega})$, where $\omega =]0, 1[\times]0, 1[$ and $\boldsymbol{\theta}(t, z) = f_\alpha(t)\boldsymbol{e}^\alpha + z\boldsymbol{e}^3$ for $(t, z) \in \overline{\omega}$. The surface S is thus a portion of a *cylinder* orthogonal to, and passing through, the planar curve $\boldsymbol{f}([0, 1])$. Finally, let $\gamma_0 \subset \gamma$ be such that

$$\{(t, 0) \in \mathbb{R}^2;\ 0 \le t \le 1\} \cup \{(t, 1) \in \mathbb{R}^2;\ 0 \le t \le 1\} \subset \gamma_0.$$

Show that the manifold

$$\{\boldsymbol{\eta} = (\eta_i) \in \mathbf{H}^1(\omega);\ \boldsymbol{\eta} = \mathbf{0} \text{ on } \gamma_0,\ a_{\alpha\beta}(\boldsymbol{\eta}) - a_{\alpha\beta} = 0 \text{ in } \omega\},$$

which contains the manifold $\mathcal{M}_0(\omega)$ (Sect. 9.1), reduces to $\{\mathbf{0}\}$. This exercise (due to D. Coutand) thus provides an instance of a *nonlinearly elastic membrane shell* (Fig. 9.1-1).

9.2. Let ω be a domain in $\mathbb{R}^2$, let γ_0 be a subset of $\gamma = \partial\omega$, and let $\boldsymbol{\theta}(y_1, y_2) = (y_1, y_2, 0)$ for all $(y_1, y_2) \in \overline{\omega}$. Show that the manifold

$$\{\boldsymbol{\eta} = (\eta_i) \in \mathbf{H}^1(\omega);\ \boldsymbol{\eta} = \mathbf{0} \text{ on } \gamma_0,\ a_{\alpha\beta}(\boldsymbol{\eta}) - \delta_{\alpha\beta} = 0 \text{ in } \omega\}$$

reduces to $\{\mathbf{0}\}$ if $\gamma_0 = \gamma$ or if $\gamma_0 \subset \gamma$ has the following property: There exists (at least) one direction such that the intersection of $\overline{\omega}$ with any line parallel to this direction is a (finite, since ω is a domain) union of segments whose end-points belong to γ_0.

This exercise (due to C. Mardare) shows that a *plate* subjected to a boundary condition of place along a "sufficiently large" (in the above sense) portion of its lateral face constitutes an instance of a *nonlinearly elastic membrane shell* (Fig. 9.1-2).

9.3. The notations used in this exercise should be self-explanatory. Let a surface $S = \boldsymbol{\theta}(\overline{\omega}) = \tilde{\boldsymbol{\theta}}(\{\tilde{\omega}\}^-)$ be equipped with two systems of curvilinear coordinates $(y_\alpha) \in \overline{\omega}$ and $(\tilde{y}_\alpha) \in \{\tilde{\omega}\}^-$ and assume that $\boldsymbol{\theta}(y_0) = \tilde{\boldsymbol{\theta}}(\tilde{\gamma}_0)$.

Show that $\mathcal{M}_0(\omega) = \{\mathbf{0}\}$ implies that $\mathcal{M}_0(\tilde{\omega}) = \{\mathbf{0}\}$ (the manifold $\mathcal{M}_0(\omega)$ is defined in Sect. 9.1). This means that the definition of a nonlinearly elastic membrane shell (Sect. 9.1) is independent of the system of curvilinear coordinates employed for representing the surface S.

9.4. The scaled two-dimensional energy j_M of a nonlinearly elastic membrane shell is defined by (the notations are those of Thm. 9.3-1):

$$j_M(\boldsymbol{\eta}) = \frac{1}{8}\int_\omega a^{\alpha\beta\sigma\tau}(a_{\sigma\tau}(\boldsymbol{\eta}) - a_{\sigma\tau})(a_{\alpha\beta}(\boldsymbol{\eta}) - a_{\alpha\beta})\sqrt{a}\,dy - \int_\omega p^{i,0}\eta_i\sqrt{a}\,dy.$$

Show that there exist constants $\alpha > 0$ and $\beta \in \mathbb{R}$ such that

$$j_M(\boldsymbol{\eta}) \geq \alpha\|\boldsymbol{\eta}\|^4_{\mathbf{W}^{1,4}(\omega)} + \beta$$

for all $\boldsymbol{\eta} \in \mathbf{W}_M(\omega) = \{\boldsymbol{\eta} = (\eta_i) \in \mathbf{W}^{1,4}(\omega);\ \boldsymbol{\eta} = \mathbf{0} \text{ on } \gamma_0\}$.

Remark. The energy j_M is thus coercive on the space $\mathbf{W}_M(\omega)$. Genevey [1997, Thm. 1.4.3] has shown that it is *not* weakly lower semi-continuous on $\mathbf{W}_M(\omega)$, however (analogous properties hold in the three-dimensional case; cf. Ex. 8.1).

9.5. Show that the stored energy function of a nonlinearly elastic membrane shell, viz.,

$$\boldsymbol{\eta} = (\eta_i) \in \mathbf{W}^{1,4}(\omega) \to \frac{\varepsilon}{8} a^{\alpha\beta\sigma\tau,\varepsilon}(a_{\sigma\tau}(\boldsymbol{\eta}) - a_{\sigma\tau})(a_{\alpha\beta}(\boldsymbol{\eta}) - a_{\alpha\beta}),$$

is *frame-indifferent*, in the following sense: Given any $\boldsymbol{\eta} \in \mathbf{W}^{1,4}(\omega)$, its value is unaltered if $\boldsymbol{\psi} := \boldsymbol{\theta} + \eta_i \boldsymbol{a}^i$ is replaced by $\mathbf{Q}\boldsymbol{\psi}$, where Q is any orthogonal matrix of order three.

9.6. This exercise shows that, if *higher order forces* are considered, the leading term of the formal expansion of the scaled unknown (then also of a higher order) satisfies the equations of a *linearly* elastic "membrane" shell found in the linear theory (Thm. 3.4-2). This observation is due to Miara [1998, p. 352].

Let the assumptions be as in Thm. 9.2-1, save that the applied body and surface force densities are now such that

$$f^{i,\varepsilon}(x^\varepsilon) = \varepsilon^r f^{i,r}(x) \text{ for all } x^\varepsilon = \pi^\varepsilon x \in \Omega^\varepsilon,$$
$$h^{i,\varepsilon}(x^\varepsilon) = \varepsilon^{1+r} h^{i,1+r}(x) \text{ for all } x^\varepsilon = \pi^\varepsilon x \in \Gamma_+^\varepsilon \cup \Gamma_-^\varepsilon,$$

where r is an arbitrary integer ≥ 1 and the functions $f^{i,r} \in L^2(\Omega)$ and $h^{i,1+r} \in L^2(\Gamma_+ \cup \Gamma_-)$ are independent of ε.

(1) Show that the leading term of the formal asymptotic expansion of the scaled unknown $\boldsymbol{u}(\varepsilon)$ is $\boldsymbol{u}^r$ (in particular then, $\boldsymbol{u}^0$ vanishes).

(2) Show that $\boldsymbol{u}^r : \overline{\Omega} \to \mathbb{R}^3$ is independent of the transverse variable and that, once identified with a function $\boldsymbol{\zeta}^r : \overline{\omega} \to \mathbb{R}^3$, it satisfies the variational problem:

$$\boldsymbol{\zeta}^r \in \mathbf{W}_M(\omega) = \{\boldsymbol{\eta} \in \mathbf{W}^{1,4}(\omega);\ \boldsymbol{\eta} = \mathbf{0} \text{ on } \gamma_0\},$$
$$\int_\omega a^{\alpha\beta\sigma\tau}\gamma_{\sigma\tau}(\boldsymbol{\zeta}^r)\gamma_{\alpha\beta}(\boldsymbol{\eta})\sqrt{a}\,dy = \int_\omega p^{i,r}\eta_i\sqrt{a}\,dy$$

for all $\boldsymbol{\eta} = (\eta_i) \in \mathbf{W}_M(\omega)$, where the functions $\gamma_{\alpha\beta}(\boldsymbol{\eta})$ are the covariant components of the *linearized* change of metric tensor associated with a displacement field $\eta_i \boldsymbol{a}^i$ of S and

$$p^{i,r} = \int_{-1}^{1} f^{i,r}\,dx_3 + h_+^{i,1+r} + h_-^{i,1+r} \text{ and } h_\pm^{i,1+r} = h^{i,1+r}(\cdot, \pm 1).$$

9.7. Let $\mathbb{M}^3$ denote the set of all matrices of order three, let $|\cdot|$ denote any norm on $\mathbb{M}^3$, and let the mapping $\hat{W} : \mathbb{M}^3 \to \mathbb{R}$ defined by

$$\hat{W}(\mathbf{F}) = \frac{\mu}{4}\operatorname{tr}(\mathbf{F}^T\mathbf{F} - \mathbf{I})^2 + \frac{\lambda}{8}\{\operatorname{tr}(\mathbf{F}^T\mathbf{F} - \mathbf{I})\}^2 \text{ for all } \mathbf{F} \in \mathbb{M}^3$$

denote the stored energy function of a *St Venant-Kirchhoff material.* Show that there exist constants $C > 0$, $\alpha > 0$, and $\beta \in \mathbb{R}$ such that

$$|\hat{W}(\mathbf{F})| \leq C(1 + |\mathbf{F}|^4) \quad \text{for all } \mathbf{F} \in \mathbb{M}^3,$$
$$|\hat{W}(\mathbf{F})| \geq \alpha|\mathbf{F}|^4 + \beta \quad \text{for all } \mathbf{F} \in \mathbb{M}^3,$$
$$|\hat{W}(\mathbf{F}) - \hat{W}(\mathbf{G})| \leq C(1 + |\mathbf{F}|^3 + |\mathbf{G}|^3)|\mathbf{F} - \mathbf{G}| \text{ for all } \mathbf{F}, \mathbf{G} \in \mathbb{M}^3.$$

Remark. Such a stored energy function thus satisfies all the assumptions needed in Thm. 9.5-1, with $p = 4$.

CHAPTER 10

NONLINEARLY ELASTIC FLEXURAL SHELLS

INTRODUCTION

The purpose of this chapter is to *identify* and *mathematically justify* the *two-dimensional equations of a nonlinearly elastic "flexural" shell.* To this end, we follow the approach of V. Lods and B. Miara, who showed how these may be obtained through the *method of formal asymptotic expansions.*

Given a surface $S = \boldsymbol{\theta}(\overline{\omega})$ and a displacement field $\eta_i \boldsymbol{a}^i$ of S with smooth enough covariant components $\eta_i : \omega \to \mathbb{R}$, let $a_{\alpha\beta}(\boldsymbol{\eta})$ denote the covariant components of the metric tensor of the associated deformed surface $(\boldsymbol{\theta} + \eta_i \boldsymbol{a}^i)(\overline{\omega})$ and let

$$G_{\alpha\beta}(\boldsymbol{\eta}) = \frac{1}{2}(a_{\alpha\beta}(\boldsymbol{\eta}) - a_{\alpha\beta})$$

denote those of the associated *change of metric tensor.*

Consider a nonlinearly elastic shell with middle surface S, subjected to a boundary condition of place along a portion of its lateral face with $\boldsymbol{\theta}(\gamma_0)$ as its middle curve, where $\gamma_0 \subset \gamma$. Such a shell is a *nonlinearly elastic "flexural" shell* if *the manifold*

$$\boldsymbol{\mathcal{M}}_F(\omega) = \{\boldsymbol{\eta} = (\eta_i) \in \mathbf{W}^{2,4}(\omega);\ \boldsymbol{\eta} = \partial_\nu \boldsymbol{\eta} = \mathbf{0} \text{ on } \gamma_0,\\ a_{\alpha\beta}(\boldsymbol{\eta}) - a_{\alpha\beta} = 0 \text{ in } \omega\}$$

contains nonzero elements and possesses nonzero tangent vectors at each one of its points (Sect. 10.2). The condition $\boldsymbol{\mathcal{M}}_F(\omega) \neq \{\mathbf{0}\}$ means that there exist *nonzero* displacement fields $\eta_i \boldsymbol{a}^i$ of the middle surface S that are *admissible* and *inextensional*, i.e., that satisfy *ad hoc* boundary conditions along the curve $\boldsymbol{\theta}(\gamma_0)$ (these boundary conditions are interpreted in Sect. 10.5) and that leave invariant the metric of S.

This definition is motivated by a careful analysis (Thms. 10.1-1 and 10.1-2) of the remaining steps needed to complete the induction begun in Chap. 8 when the shell is not a "membrane" one according to the definition given in the previous chapter. More precisely, the formal asymptotic analysis begun in Chap. 8 is concluded as follows when the shell is a "flexural" one (Thm. 10.3-4):

Assume that the manifold $\boldsymbol{\mathcal{M}}_F(\omega)$ does not reduce to $\{\mathbf{0}\}$ and that it possesses nonzero tangent vectors at each one of its points. Then the contravariant components of the applied body forces must be of the form

$$f^{i,\varepsilon}(x^\varepsilon) = \varepsilon^2 f^{i,2}(x) \text{ for all } x^\varepsilon = \pi^\varepsilon x \in \Omega^\varepsilon,$$

where the functions $f^{i,2}$ are *independent of* ε (for simplicity, we assume in this introduction that there are no surface forces), *the leading term* $\boldsymbol{u}^0 : \overline{\Omega} \to \mathbb{R}^3$ *of the formal asymptotic expansion of the scaled unknown is independent of the transverse variable, and* $\boldsymbol{\zeta}^0 = \frac{1}{2}\int_{-1}^{1} \boldsymbol{u}^0 \, dx_3$ *satisfies the following* (scaled) *two-dimensional variational problem of a nonlinearly elastic "flexural" shell*:

$$\boldsymbol{\zeta}^0 \in \boldsymbol{\mathcal{M}}_F(\omega) = \{\boldsymbol{\eta} \in \mathbf{W}^{2,4}(\omega); \boldsymbol{\eta} = \partial_\nu \boldsymbol{\eta} = \mathbf{0} \text{ on } \gamma_0, \\ G_{\alpha\beta}(\boldsymbol{\eta}) = 0 \text{ in } \omega\},$$

$$\frac{1}{3}\int_\omega a^{\alpha\beta\sigma\tau} R^\flat_{\sigma\tau}(\boldsymbol{\zeta}^0)\big((R^\flat_{\alpha\beta})'(\boldsymbol{\zeta}^0)\boldsymbol{\eta}\big)\sqrt{a}\, dy = \int_\omega p^{i,2}\eta_i \sqrt{a}\, dy$$

for all $\boldsymbol{\eta} = (\eta_i) \in \mathbb{T}_{\boldsymbol{\zeta}^0}\boldsymbol{\mathcal{M}}_F(\omega)$, *where*

$$\mathbb{T}_{\boldsymbol{\zeta}^0}\boldsymbol{\mathcal{M}}_F(\omega) = \{\boldsymbol{\eta} \in \mathbf{W}^{2,4}(\omega);\ \boldsymbol{\eta} = \partial_\nu \boldsymbol{\eta} = \mathbf{0} \text{ on } \gamma_0, \\ G'_{\alpha\beta}(\boldsymbol{\zeta}^0)\boldsymbol{\eta} = 0 \text{ in } \omega\},$$

$$a^{\alpha\beta\sigma\tau} = \frac{4\lambda\mu}{\lambda + 2\mu} a^{\alpha\beta}a^{\sigma\tau} + 2\mu(a^{\alpha\sigma}a^{\beta\tau} + a^{\alpha\tau}a^{\beta\sigma}),$$

$$p^{i,2} = \int_{-1}^{1} f^{i,2}\, dx_3.$$

The functions $R^\flat_{\alpha\beta}(\boldsymbol{\eta})$, which are well-defined as functions in $L^2(\omega)$ for any field $\boldsymbol{\eta} \in \mathbf{W}^{2,4}(\omega)$, have the crucial property that they satisfy (Thm. 10.3-1):

$$R^\flat_{\alpha\beta}(\boldsymbol{\eta}) = b_{\alpha\beta}(\boldsymbol{\eta}) - b_{\alpha\beta} \text{ for all } \boldsymbol{\eta} \in \boldsymbol{\mathcal{M}}_F(\omega),$$

where $b_{\alpha\beta}(\boldsymbol{\eta})$ denote the covariant components of the curvature tensor of the deformed middle surface associated with any displacement field $\eta_i \boldsymbol{a}^i$ of S with smooth enough covariant components $\eta_i : \omega \to \mathbb{R}$ (in addition, the vectors $\partial_\alpha(\boldsymbol{\theta} + \eta_i \boldsymbol{a}^i)$ should be linearly independent a.e. in ω, to ensure that the normal vectors to the associated deformed surface are a.e. well defined).

The above variational equations also express that $\boldsymbol{\zeta}^0$ *is a stationary point of a certain functional over the manifold* $\boldsymbol{\mathcal{M}}_F(\omega)$ (Thm. 10.4-1). Particular solutions are thus obtained by solving a *minimization problem*, which, by virtue of the relations $R^\flat_{\alpha\beta}(\boldsymbol{\eta}) = b_{\alpha\beta}(\boldsymbol{\eta}) - b_{\alpha\beta}$ for all $\boldsymbol{\eta} \in \boldsymbol{\mathcal{M}}_F(\omega)$, takes the following remarkably simple form (directly written here in its "de-scaled" version): The vector field $\boldsymbol{\zeta}^\varepsilon = (\zeta_i^\varepsilon)$ formed by the covariant components of the *limit displacement field* $\zeta_i^\varepsilon \boldsymbol{a}^i$ of the middle surface S satisfies (Thm. 10.5-1):

$$\boldsymbol{\zeta}^\varepsilon \in \boldsymbol{\mathcal{M}}_F(\omega) \text{ and } j_F^\varepsilon(\boldsymbol{\zeta}^\varepsilon) = \inf_{\boldsymbol{\eta} \in \boldsymbol{\mathcal{M}}_F(\omega)} j_F^\varepsilon(\boldsymbol{\eta}),$$

where the *two-dimensional energy* $j_F^\varepsilon : \boldsymbol{\mathcal{M}}_F(\omega) \to \mathbb{R}$ *of a nonlinearly elastic flexural shell* is defined by

$$j_F^\varepsilon(\boldsymbol{\eta}) = \frac{\varepsilon^3}{6} \int_\omega a^{\alpha\beta\sigma\tau,\varepsilon}(b_{\sigma\tau}(\boldsymbol{\eta}) - b_{\sigma\tau})(b_{\alpha\beta}(\boldsymbol{\eta}) - b_{\alpha\beta})\sqrt{a}\,dy \\ - \int_\omega p^{i,\varepsilon}\eta_i \sqrt{a}\,dy,$$

where

$$a^{\alpha\beta\sigma\tau,\varepsilon} = \frac{4\lambda^\varepsilon \mu^\varepsilon}{\lambda^\varepsilon + 2\mu^\varepsilon} a^{\alpha\beta} a^{\sigma\tau} + 2\mu^\varepsilon (a^{\alpha\sigma} a^{\beta\tau} + a^{\alpha\tau} a^{\beta\sigma}),$$

$$p^{i,\varepsilon} = \int_{-\varepsilon}^{\varepsilon} f^{i,\varepsilon}\,dx_3^\varepsilon.$$

The functions $a^{\alpha\beta\sigma\tau,\varepsilon}$ are the contravariant components of the *two-dimensional elasticity tensor of the shell*, λ^ε and μ^ε are the Lamé constants of the material constituting the shell, and the functions $f^{i,\varepsilon}$ are the contravariant components of the body force density applied to the shell. The functions $(b_{\alpha\beta}(\boldsymbol{\eta}) - b_{\alpha\beta})$ are the covariant components of the *change of curvature tensor* associated with a displacement field $\eta_i \boldsymbol{a}^i$ of the middle surface S.

The *stored energy function* of a nonlinearly elastic flexural shell is thus remarkably simple: It is a quadratic and positive definite expression in terms of the *exact* difference between the *curvature tensor* of the *deformed* middle surface and that of the *undeformed* one, on the one hand. Since, on the other hand, the elements of the *manifold* $\mathcal{M}_F(\omega)$ over which the energy j_F^ε is to be minimized correspond to *displacements that leave invariant the metric of the surface* S, these two observations combined show that the equations of the *linear* theory of "flexural" shells are immediately recovered (at least formally) under linearization of the above nonlinear equations.

Even if the formal approach has its inherent drawback (precisely that of being only "formal"!), it seems to be the only way to justify a nonlinear "flexural" shell theory, up to now at least. If only for this reason, there should be at least one nonlinear theory that reduces to the linear flexural theory (justified in Chap. 6) under linearization!

We conclude this chapter by establishing the existence of a solution to the above minimization problem. The proof, which is based on the fundamental theorem of the calculus of variations, essentially hinges on a careful analysis of the properties of admissible inextensional displacements in $\mathbf{H}^2(\omega)$ (Thm. 10.6-1).

10.1. IDENTIFICATION OF A TWO-DIMENSIONAL VARIATIONAL PROBLEM SATISFIED BY THE LEADING TERM WHEN THERE ARE NONZERO ADMISSIBLE INEXTENSIONAL DISPLACEMENTS

We consider in this chapter the case where there exist *nonzero* admissible inextensional displacements, according to the definition given in Sect. 9.1.

It turns out that a *stronger* assumption is needed in order to pursue the analysis of this case, according to which the manifold $\mathcal{M}_0(\omega)$ introduced in Sect. 9.1 (its definition is recalled in the next theorem) has *nonzero tangent vectors* at each one of its points, i.e., *at each* $\boldsymbol{\zeta} \in \mathcal{M}_0(\omega)$, *the tangent space* $\mathbb{T}_{\boldsymbol{\zeta}}\mathcal{M}_0(\omega)$ *to* $\mathcal{M}_0(\omega)$ *contains nonzero functions.*

Note that, by introducing this additional assumption, I depart here from Lods & Miara [1998]; however, the conclusions eventually reached will be identical.

We begin the analysis of this case by showing that the *linearization trick* (the second requirement set in part (iv) of the Ansatz of the method of formal asymptotic expansions; cf. Sect. 8.6) leads to two important preliminary conclusions: First, *the induction is not com-*

plete (whereas it terminates with Thm. 8.8-1 for a membrane shell) and, second, *the leading term is itself an inextensional displacement.*

Words of caution. (1) It must be emphasized that $\mathcal{M}_0(\omega)$ *is a manifold, not a vector space*!

(2) The conclusions of Thm. 10.1-1 are perfectly compatible with those reached in Thm. 8.8-1. Simply, the variational equations found in this theorem reduce to "$0 = 0$ for all $\boldsymbol{\eta} \in \mathbf{W}(\omega)$" in the present case! ∎

Note that, when the shell is a nonlinearly elastic "membrane" shell (Chap. 9), and *only in this case*, the space $\mathbf{W}(\Omega)$ of the next theorem is denoted $\mathbf{W}_M(\Omega)$.

Theorem 10.1-1. *Assume that the scaled unknown* $\boldsymbol{u}(\varepsilon) = (u_i(\varepsilon))$ *appearing in problem* $\mathcal{P}(\varepsilon; \Omega)$ (Thm. 8.4-1) *admits a formal asymptotic expansion of the form*

$$\boxed{\begin{aligned}&\boldsymbol{u}(\varepsilon) = \boldsymbol{u}^0 + \varepsilon \boldsymbol{u}^1 + \varepsilon^2 \boldsymbol{u}^2 + \cdots, \text{ with}\\ &\boldsymbol{u}^0 \in \mathbf{W}(\Omega) := \{\boldsymbol{v} \in \mathbf{W}^{1,4}(\Omega);\ \boldsymbol{v} = \mathbf{0} \text{ on } \Gamma_0\},\\ &\boldsymbol{u}^1, \boldsymbol{u}^2 \in \mathbf{W}^{1,4}(\Omega).\end{aligned}}$$

Define the space $\mathbf{W}(\omega)$ *and the manifold* $\mathcal{M}_0(\omega) \subset \mathbf{W}(\omega)$ *by:*

$$\boxed{\begin{aligned}\mathbf{W}(\omega) &:= \{\boldsymbol{\eta} \in \mathbf{W}^{1,4}(\omega);\ \boldsymbol{\eta} = \mathbf{0} \text{ on } \gamma_0\},\\ \mathcal{M}_0(\omega) &:= \{\boldsymbol{\eta} = (\eta_i) \in \mathbf{W}(\omega);\ G_{\alpha\beta}(\boldsymbol{\eta}) = 0 \text{ in } \omega\},\end{aligned}}$$

where (Thm. 9.1-1)

$$\boxed{G_{\alpha\beta}(\boldsymbol{\eta}) := \frac{1}{2}(\eta_{\alpha\|\beta} + \eta_{\beta\|\alpha} + a^{mn}\eta_{m\|\alpha}\eta_{n\|\beta}) = \frac{1}{2}(a_{\alpha\beta}(\boldsymbol{\eta}) - a_{\alpha\beta}).}$$

Assume that

$$\boxed{\begin{aligned}&\mathcal{M}_0(\omega) \neq \{\mathbf{0}\} \text{ and, at each } \boldsymbol{\zeta} \in \mathcal{M}_0(\omega),\\ &\mathbb{T}_{\boldsymbol{\zeta}}\mathcal{M}_0(\omega) := \{\boldsymbol{\eta} \in \mathbf{W}(\omega);\ G'_{\alpha\beta}(\boldsymbol{\zeta})\boldsymbol{\eta} = 0 \text{ in } \omega\} \neq \{\mathbf{0}\},\end{aligned}}$$

where $G'_{\alpha\beta}(\boldsymbol{\zeta})\boldsymbol{\eta}$ *designate the Gâteaux derivatives of each function* $G_{\alpha\beta} : \mathbf{W}^{1,4}(\omega) \to L^2(\omega)$.

Then the functions $f^{i,0} \in L^2(\Omega)$ *and* $h^{i,1} \in L^2(\Gamma_+ \cup \Gamma_-)$ *found in* Thm. 8.8-1 *must vanish, i.e.,*

$$\boxed{f^{i,0} = 0 \text{ in } \Omega \text{ and } h^{i,1} = 0 \text{ on } \Gamma_+ \cup \Gamma_-,}$$

and the leading term $\boldsymbol{u}^0$, *which is independent of the transverse variable* x_3 (Thm. 8.8-1), *is such that its average* $\boldsymbol{\zeta}^0 := \frac{1}{2}\int_{-1}^{1} \boldsymbol{u}^0 \, dx_3$ *satisfies* $G_{\alpha\beta}(\boldsymbol{\zeta}^0) = 0$ *in* ω. *Consequently,*

$$\boxed{\boldsymbol{\zeta}^0 \in \boldsymbol{\mathcal{M}}_0(\omega).}$$

Proof. The assumption about the tangent spaces to the manifold $\boldsymbol{\mathcal{M}}_0(\omega)$ implies in particular that

$$\mathbb{T}_{\mathbf{0}}\boldsymbol{\mathcal{M}}_0(\omega) \neq \{\mathbf{0}\}.$$

But an inspection of the Gâteaux derivatives (Thm. 9.1-1)

$$G'_{\alpha\beta}(\boldsymbol{\zeta})\boldsymbol{\eta} = \frac{1}{2}(\eta_{\alpha\|\beta} + \eta_{\beta\|\alpha} + a^{mn}\{\zeta_{m\|\alpha}\eta_{n\|\beta} + \zeta_{n\|\beta}\eta_{m\|\alpha}\})$$

reveals that

$$G'_{\alpha\beta}(\mathbf{0})\boldsymbol{\eta} = \frac{1}{2}(\partial_\alpha\eta_\beta + \partial_\beta\eta_\alpha) - \Gamma^\sigma_{\alpha\beta}\eta_\sigma - b_{\alpha\beta}\eta_3,$$

i.e., that

$$G'_{\alpha\beta}(\mathbf{0})\boldsymbol{\eta} = \gamma_{\alpha\beta}(\boldsymbol{\eta}),$$

where the functions $\gamma_{\alpha\beta}(\boldsymbol{\eta})$ are the covariant components of the *linearized* change of metric tensor (Sect. 2.4). In other words,

$$\mathbb{T}_{\mathbf{0}}\boldsymbol{\mathcal{M}}_0(\omega) = \{\boldsymbol{\eta} \in \mathbf{W}(\Omega);\ \gamma_{\alpha\beta}(\boldsymbol{\eta}) = 0 \text{ in } \omega\} \neq \{\mathbf{0}\},$$

and thus the space

$$\mathbf{V}_0(\omega) = \{\boldsymbol{\eta} \in \mathbf{H}^1(\omega);\ \boldsymbol{\eta} = \mathbf{0} \text{ on } \gamma_0,\ \gamma_{\alpha\beta}(\boldsymbol{\eta}) = 0 \text{ in } \omega\}$$

of the *linear* theory (Sect. 3.4) *does not reduce to* $\{\mathbf{0}\}$.

According to the *linearization trick, we are thus compelled to let* $f^{i,0} = 0$ *and* $h^{i,1} = 0$ as in the *linear* case (see step (v) in Sect. 3.4).

These relations in turn imply that the leading term $\boldsymbol{u}^0 = \boldsymbol{\zeta}^0$ satisfies (Thm. 9.2-1)

$$\int_\omega a^{\alpha\beta\sigma\tau} G_{\sigma\tau}(\boldsymbol{\zeta}^0)\big(G'_{\alpha\beta}(\boldsymbol{\zeta}^0)\boldsymbol{\eta}\big)\sqrt{a}\,dy = 0 \text{ for all } \boldsymbol{\eta} \in \mathbf{W}(\omega),$$

or equivalently (Thm. 9.3-1), that $\boldsymbol{\zeta}^0$ *is a stationary point over the space* $\mathbf{W}(\omega)$ *of the functional* $j : \mathbf{W}(\omega) \to \mathbb{R}$ *defined by*

$$j(\boldsymbol{\eta}) = \frac{1}{2}\int_\omega a^{\alpha\beta\sigma\tau} G_{\sigma\tau}(\boldsymbol{\eta}) G_{\alpha\beta}(\boldsymbol{\eta})\sqrt{a}\,dy.$$

If $\boldsymbol{\zeta}^0 \in \mathbf{W}(\omega)$ is such that $G_{\alpha\beta}(\boldsymbol{\zeta}^0) = 0$ in ω, then $\boldsymbol{\zeta}^0$ is a stationary point of j over the space $\mathbf{W}(\omega)$ since then

$$j(\boldsymbol{\zeta}^0) = \inf_{\boldsymbol{\eta}\in\mathbf{W}(\omega)} j(\boldsymbol{\eta}) = 0$$

(recall that the scaled two-dimensional elasticity tensor of a shell is uniformly positive definite; cf. Thm. 3.3-2). Since

$$[G_{\alpha\beta}(\boldsymbol{\zeta}^0)]^{lin} = \gamma_{\alpha\beta}(\boldsymbol{\zeta}^0),$$

where $[\cdots]^{lin}$ denotes the linear part with respect to $\boldsymbol{\zeta}^0$ in the expression $[\cdots]$ and since $\gamma_{\alpha\beta}(\boldsymbol{\zeta}^0) = 0$ in ω in the *linear* case (step (v) in Sect. 3.4), the *linearization trick* leads us to only retain those $\boldsymbol{\zeta}^0 \in \mathbf{W}(\omega)$ that satisfy $G_{\alpha\beta}(\boldsymbol{\zeta}^0) = 0$ in ω. ∎

Equipped with the information contained in Thm. 10.1-1, about the applied forces and the leading term $\boldsymbol{\zeta}^0$, we are now in a position to continue the induction until $\boldsymbol{\zeta}^0$ can be fully identified as the solution of a variational problem when $\boldsymbol{\mathcal{M}}_0(\omega) \neq \{\mathbf{0}\}$ and $\mathbb{T}_{\boldsymbol{\zeta}}\boldsymbol{\mathcal{M}}_0(\omega) \neq \{\mathbf{0}\}$ at each $\boldsymbol{\zeta} \in \boldsymbol{\mathcal{M}}_0(\omega)$. The next theorem, which is due to Lods & Miara [1998, Thm. 1], constitutes the final stage of the induction in this case (more illuminating, and more concise, versions of this final stage will be presented in Sects. 10.3 and 10.4).

Not the least stunning consequence of the assumptions made on the manifold $\boldsymbol{\mathcal{M}}_0(\omega)$ is an "automatic refinement" of the regularity of the leading term $\boldsymbol{\zeta}^0$ and of the boundary conditions it should satisfy (a similar phenomenon was already encountered in the linear theory; cf. Thm. 3.4-3).

Theorem 10.1-2. *Assume that* (the notations are those of Thm. 10.1-1):

$$\mathcal{M}_0(\omega) := \{\boldsymbol{\eta} = (\eta_i) \in \mathbf{W}(\omega);\ G_{\alpha\beta}(\boldsymbol{\eta}) = 0 \text{ in } \omega\} \neq \{\mathbf{0}\},$$

and that, at each $\boldsymbol{\zeta} \in \mathcal{M}_0(\omega)$, the tangent space $\mathbb{T}_{\boldsymbol{\zeta}}\mathcal{M}_0(\omega)$ to $\mathcal{M}_0(\omega)$ at $\boldsymbol{\zeta}$ contains nonzero functions. Finally, assume that the scaled unknown $\boldsymbol{u}(\varepsilon) = (u_i(\varepsilon))$ appearing in problem $\mathcal{P}(\varepsilon;\Omega)$ (Thm. 8.4-1) *admits a formal asymptotic expansion of the form*

$$\begin{aligned}&\boldsymbol{u}(\varepsilon) = \boldsymbol{u}^0 + \varepsilon\boldsymbol{u}^1 + \varepsilon^2\boldsymbol{u}^2 + \varepsilon^3\boldsymbol{u}^3 + \varepsilon^4\boldsymbol{u}^4 + \cdots, \text{ with}\\ &\boldsymbol{u}^0,\ \boldsymbol{u}^1 \in \mathbf{W}(\Omega),\ \boldsymbol{u}^0 \neq \mathbf{0}, \text{ and } \boldsymbol{u}^q \in \mathbf{W}^{1,4}(\Omega),\ 2 \le q \le 4.\end{aligned}$$

Then in order that no restriction be put on the applied forces and that the linearization trick be satisfied (Sect. 8.6), *the components of the applied forces must be of the form*

$$\begin{aligned} f^{i,\varepsilon}(x^\varepsilon) &= \varepsilon^2 f^{i,2}(x) \quad \text{for all} \quad x^\varepsilon = \pi^\varepsilon x \in \Omega^\varepsilon,\\ h^{i,\varepsilon}(x^\varepsilon) &= \varepsilon^3 h^{i,3}(x) \quad \text{for all} \quad x^\varepsilon = \pi^\varepsilon x \in \Gamma_+^\varepsilon \cup \Gamma_-^\varepsilon,\end{aligned}$$

where the functions $f^{i,2} \in L^2(\Omega)$ and $h^{i,3} \in L^2(\Gamma_+ \cup \Gamma_-)$ are independent of ε.

This being the case, the leading term $\boldsymbol{u}^0$ is independent of the transverse variable x_3 and $\boldsymbol{\zeta}^0 := \frac{1}{2}\int_{-1}^{1} \boldsymbol{u}^0\, dx_3$ satisfies the following two-dimensional variational problem (∂_ν denotes the outer normal derivative operator along the boundary γ):

$$\begin{aligned}&\boldsymbol{\zeta}^0 \in \mathcal{M}_F(\omega) := \{\boldsymbol{\eta} = (\eta_i) \in \mathbf{W}^{2,4}(\omega);\ \boldsymbol{\eta} = \partial_\nu\boldsymbol{\eta} = \mathbf{0} \text{ on } \gamma_0,\\ &\qquad\qquad G_{\alpha\beta}(\boldsymbol{\eta}) = 0 \text{ in } \omega\},\\ &\frac{1}{3}\int_\omega a^{\alpha\beta\sigma\tau}\hat{E}^0_{\sigma\|\tau}\hat{F}^0_{\alpha\|\beta}(\boldsymbol{\eta})\sqrt{a}\,dy = \int_\omega p^{i,2}\eta_i\sqrt{a}\,dy\\ &\qquad\qquad \text{for all } \boldsymbol{\eta} = (\eta_i) \in \mathbb{T}_{\boldsymbol{\zeta}^0}\mathcal{M}_F(\omega),\end{aligned}$$

where

$$\begin{aligned}&\mathbb{T}_{\boldsymbol{\zeta}^0}\mathcal{M}_F(\omega) := \{\boldsymbol{\eta} \in \mathbf{W}^{2,4}(\omega);\ \boldsymbol{\eta} = \partial_\nu\boldsymbol{\eta} = \mathbf{0} \text{ on } \gamma_0,\\ &\qquad\qquad G'_{\alpha\beta}(\boldsymbol{\zeta}^0)\boldsymbol{\eta} = 0 \text{ in } \omega\}\end{aligned}$$

designates the tangent space to the manifold $\mathcal{M}_F(\omega)$ at $\boldsymbol{\zeta}^0$, the functions $\hat{E}^0_{\alpha\|\beta}$ depend on $\boldsymbol{\zeta}^0$, the functions $\hat{F}^0_{\alpha\|\beta}(\boldsymbol{\eta})$ depend on $\boldsymbol{\zeta}^0$ and linearly on $\boldsymbol{\eta}$ (the complete, but lengthy, definitions of these functions are given respectively in parts (iv) and (vi) of the subsequent proof and also repeated *in extenso* in Thms. 10.3-1 and 10.3-3), *and finally,*

$$a^{\alpha\beta\sigma\tau} := \frac{4\lambda\mu}{\lambda+2\mu} a^{\alpha\beta} a^{\sigma\tau} + 2\mu(a^{\alpha\sigma} a^{\beta\tau} + a^{\alpha\tau} a^{\beta\sigma}),$$

$$p^{i,2} := \int_{-1}^{1} f^{i,2}\, dx_3 + h^{i,3}_+ + h^{i,3}_- \text{ and } h^{i,3}_\pm := h^{i,3}(\cdot, \pm 1).$$

Proof. Let us first specify some notations frequently used below. By $E^q_{i\|j}$, $q \geq 0$, and $F^q_{i\|j}(\boldsymbol{v})$, $q \geq -1$, we denote the coefficient of ε^q in the formal asymptotic expansions

$$E_{i\|j}(\varepsilon; \boldsymbol{u}(\varepsilon)) = E^0_{i\|j} + \varepsilon E^1_{i\|j} + \varepsilon^2 E^2_{i\|j} + \varepsilon^3 E^3_{i\|j} + \cdots,$$

$$F_{i\|j}(\varepsilon; \boldsymbol{u}(\varepsilon), \boldsymbol{v}) = \frac{1}{\varepsilon} F^{-1}_{i\|j}(\boldsymbol{v}) + F^0_{i\|j}(\boldsymbol{v}) + \varepsilon F^1_{i\|j}(\boldsymbol{v}) + \cdots,$$

induced by that of the scaled unknown $\boldsymbol{u}(\varepsilon)$. We recall that the functions $E_{i\|j}(\varepsilon; \boldsymbol{u}(\varepsilon))$ and $F_{i\|j}(\varepsilon; \boldsymbol{u}(\varepsilon), \boldsymbol{v})$ are those appearing in the formulation of problem $\mathcal{P}(\varepsilon; \Omega)$ (Thm. 8.4-1); hence $E^q_{i\|j}$ and $F^q_{i\|j}(\boldsymbol{v})$ are themselves functions defined over the set $\overline{\Omega}$. For any integer $r \geq 0$, L^r designates the linear form defined by

$$L^r(\boldsymbol{v}) := \int_\Omega f^{i,r} v_i \sqrt{a}\, dx + \int_{\Gamma_+ \cup \Gamma_-} h^{i,r+1} v_i \sqrt{a}\, d\Gamma,$$

where it is understood that the functions $f^{i,r}$ and $h^{i,r+1}$ respectively belong to the spaces $L^2(\Omega)$ and $L^2(\Gamma_+ \cup \Gamma_-)$ and that they are *independent of* ε. Finally, recall that

$$\eta_{\alpha\|\beta} = \partial_\beta \eta_\alpha - \Gamma^\sigma_{\alpha\beta} \eta_\sigma - b_{\alpha\beta} \eta_3 \text{ and } \eta_{3\|\beta} = \partial_\beta \eta_3 + b^\sigma_\beta \eta_\sigma.$$

Our *point of departure* is Thm. 8.8-1, amended as indicated in Thm. 10.1-1. Before continuing the induction proper (part (v)), we take a closer look at the cancellation of the coefficients of $\frac{1}{\varepsilon}$ and ε^0 (parts (i) to (iv)):

(i) *When* $\mathbb{T}_{\boldsymbol{\zeta}}\boldsymbol{\mathcal{M}}_0(\omega) \neq \{\mathbf{0}\}$ *at each* $\boldsymbol{\zeta} \in \boldsymbol{\mathcal{M}}_0(\omega)$, *the cancellation of the coefficient of* $\frac{1}{\varepsilon}$ *leads to the following supplementary conclusions: First,*

$$E^0_{i\|3} = 0 \text{ in } \Omega.$$

Second, the coefficients $u^{(0)}_{m\|3}$ *of* ε^0 *in the formal expansions of the functions* $u_{m\|3}(\varepsilon)$ *are given by* (the functions $\zeta^0_{i\|3}$ are redefined *infra*):

$$\begin{aligned}
u^{(0)}_{1\|3} &= -(1 + a^{\alpha 2}\zeta^0_{\alpha\|2})\zeta^0_{3\|1} + a^{\alpha 2}\zeta^0_{\alpha\|1}\zeta^0_{3\|2},\\
u^{(0)}_{2\|3} &= -(1 + a^{\alpha 1}\zeta^0_{\alpha\|1})\zeta^0_{3\|2} + a^{\alpha 1}\zeta^0_{\alpha\|2}\zeta^0_{3\|1},\\
u^{(0)}_{3\|3} &= a^{\alpha\beta}\zeta^0_{\alpha\|\beta} + a^{\alpha 1}a^{\beta 2}(\zeta^0_{\alpha\|1}\zeta^0_{\beta\|2} - \zeta^0_{\alpha\|2}\zeta^0_{\beta\|1}).
\end{aligned}$$

Third, the term of order one in the formal expansion of $\boldsymbol{u}(\varepsilon)$ *is of the form*

$$\boldsymbol{u}^1 = \boldsymbol{\zeta}^1 - x_3\boldsymbol{\psi}^0 \text{ with } \boldsymbol{\zeta}^1 \in \mathbf{W}(\omega) \text{ and } \boldsymbol{\psi}^0 \in \mathbf{W}(\omega),$$

where $\boldsymbol{\psi}^0 = (\psi^0_i)$ *is given by*

$$\begin{aligned}
\psi^0_1 &:= b^\alpha_1\zeta^0_\alpha + (1 + a^{\alpha 2}\zeta^0_{\alpha\|2})\zeta^0_{3\|1} - a^{\alpha 2}\zeta^0_{\alpha\|1}\zeta^0_{3\|2},\\
\psi^0_2 &:= b^\alpha_2\zeta^0_\alpha + (1 + a^{\alpha 1}\zeta^0_{\alpha\|1})\zeta^0_{3\|2} - a^{\alpha 1}\zeta^0_{\alpha\|2}\zeta^0_{3\|1},\\
\psi^0_3 &:= -a^{\alpha\beta}\zeta^0_{\alpha\|\beta} - a^{\alpha 1}a^{\beta 2}(\zeta^0_{\alpha\|1}\zeta^0_{\beta\|2} - \zeta^0_{\alpha\|2}\zeta^0_{\beta\|1}).
\end{aligned}$$

Finally,

$$\begin{aligned}
\boldsymbol{\zeta}^0 \in \boldsymbol{\mathcal{M}}_F(\omega) := \{\boldsymbol{\eta} \in \mathbf{W}^{2,4}(\omega);\ \boldsymbol{\eta} = \partial_\nu\boldsymbol{\eta} = \mathbf{0} \text{ on } \gamma_0,\\
G_{\alpha\beta}(\boldsymbol{\eta}) = 0 \text{ in } \omega\}.
\end{aligned}$$

To prove that $E^0_{i\|3} = 0$ in Ω, simply recall that we showed in part(ii) of the proof of Thm. 8.8-1 and in Thm. 10.1-1 that

$$E^0_{\alpha\|3} = 0 \text{ and } \lambda a^{\alpha\beta}E^0_{\alpha\|\beta} + (\lambda + 2\mu)E^0_{3\|3} = 0 \text{ in } \Omega,$$

and that the assumptions $\boldsymbol{\mathcal{M}}_0(\omega) \neq \{\mathbf{0}\}$ and $\mathbb{T}_{\boldsymbol{\zeta}}\boldsymbol{\mathcal{M}}_0(\omega) \neq \{\mathbf{0}\}$ at all $\boldsymbol{\zeta} \in \boldsymbol{\mathcal{M}}_0(\omega)$ imply that

$$E^0_{\alpha\|\beta} = G_{\alpha\beta}(\boldsymbol{\zeta}^0) = 0 \text{ in } \omega, \text{ hence in } \Omega.$$

The equations $2E^0_{i\|3} = 0$ take the form of a *nonlinear system with respect to the functions* $u^{(0)}_{i\|3}$, viz.,

$$\begin{aligned} u^{(0)}_{1\|3} + \zeta^0_{3\|1} + a^{\alpha\beta}\zeta^0_{\alpha\|1}u^{(0)}_{\beta\|3} + \zeta^0_{3\|1}u^{(0)}_{3\|3} &= 0, \\ u^{(0)}_{2\|3} + \zeta^0_{3\|2} + a^{\alpha\beta}\zeta^0_{\alpha\|2}u^{(0)}_{\beta\|3} + \zeta^0_{3\|2}u^{(0)}_{3\|3} &= 0, \\ 2u^{(0)}_{3\|3} + a^{\alpha\beta}u^{(0)}_{\alpha\|3}u^{(0)}_{\beta\|3} + u^{(0)}_{3\|3}u^{(0)}_{3\|3} &= 0, \end{aligned}$$

where

$$\zeta^0_{\alpha\|\beta} = \partial_\beta\zeta^0_\alpha - \Gamma^\sigma_{\alpha\beta}\zeta^0_\sigma - b_{\alpha\beta}\zeta^0_3 \quad\text{and}\quad \zeta^0_{3\|\beta} = \partial_\beta\zeta^0_3 + b^\sigma_\alpha\zeta^0_\sigma,$$
$$u^{(0)}_{\alpha\|3} = \partial_3 u^1_\alpha + b^\sigma_\alpha\zeta^0_\sigma \quad\text{and}\quad u^{(0)}_{3\|3} = \partial_3 u^1_3,$$

and the relations $u^0_{i\|\beta} = \zeta^0_{i\|\beta}$, which follow from $\partial_3\boldsymbol{u}^0 = \mathbf{0}$ in Ω, have been used.

As we shall see in part (ii), *this nonlinear system can be explicitly solved*, thanks in particular to the relations $E^0_{\alpha\|\beta} = 0$. More specifically, one first shows that $u^{(0)}_{3\|3}$ *satisfies a quadratic equation*, viz.,

$$\left(1 + u^{(0)}_{3\|3}\right)^2 = \left(1 + a^{\alpha\beta}\zeta^0_{\alpha\|\beta} + a^{\alpha 1}a^{\beta 2}(\zeta^0_{\alpha\|1}\zeta^0_{\beta\|2} - \zeta^0_{\alpha\|2}\zeta^0_{\beta\|1})\right)^2.$$

Hence *either*

$$u^{(0)}_{3\|3} = a^{\alpha\beta}\zeta^0_{\alpha\|\beta} + a^{\alpha 1}a^{\beta 2}(\zeta^0_{\alpha\|1}\zeta^0_{\beta\|2} - \zeta^0_{\alpha\|2}\zeta^0_{\beta\|1}),$$

or

$$u^{(0)}_{3\|3} = -2 - (a^{\alpha\beta}\zeta^0_{\alpha\|\beta} + a^{\alpha 1}a^{\beta 2}(\zeta^0_{\alpha\|1}\zeta^0_{\beta\|2} - \zeta^0_{\alpha\|2}\zeta^0_{\beta\|1})).$$

Since $u^{(0)}_{3\|3} = \partial_3 u^1_3$ and $a^{\alpha\beta}\zeta^0_{\alpha\|\beta} = a^{\alpha\beta}\gamma_{\alpha\beta}(\boldsymbol{\zeta}^0)$ and since $\partial_3 u^1_3 = 0$ and $\gamma_{\alpha\beta}(\boldsymbol{\zeta}^0) = 0$ in the *linear* case (step (v) in Sect. 3.4), the *linearization trick* leads us to *only retain the first solution.*

One then solves the first two equations of the nonlinear system, which constitute a linear system with respect to $u^{(0)}_{1\|3}$ and $u^{(0)}_{2\|3}$. As shown in (ii), this gives

$$\begin{aligned} u^{(0)}_{1\|3} &= -(1 + a^{\alpha 2}\zeta^0_{\alpha\|2})\zeta^0_{3\|1} + a^{\alpha 2}\zeta^0_{\alpha\|1}\zeta^0_{3\|2}, \\ u^{(0)}_{2\|3} &= -(1 + a^{\alpha 1}\zeta^0_{\alpha\|1})\zeta^0_{3\|2} + a^{\alpha 1}\zeta^0_{\alpha\|2}\zeta^0_{3\|1}. \end{aligned}$$

Using these expressions for the functions $u^{(0)}_{i\|3}$, we then find that

$$\partial_3 u^1_\alpha = -b^\sigma_\alpha \zeta^0_\sigma + u^{(0)}_{\alpha\|3} = -\psi^0_\alpha \quad \text{and} \quad \partial_3 u^1_3 = u^{(0)}_{3\|3} = -\psi^0_3,$$

where

$$\begin{aligned}
\psi^0_1 &:= b^\alpha_1 \zeta^0_\alpha + (1 + a^{\alpha 2}\zeta^0_{\alpha\|2})\zeta^0_{3\|1} - a^{\alpha 2}\zeta^0_{\alpha\|1}\zeta^0_{3\|2},\\
\psi^0_2 &:= b^\alpha_2 \zeta^0_\alpha + (1 + a^{\alpha 1}\zeta^0_{\alpha\|1})\zeta^0_{3\|2} - a^{\alpha 1}\zeta^0_{\alpha\|2}\zeta^0_{3\|1},\\
\psi^0_3 &:= -a^{\alpha\beta}\zeta^0_{\alpha\|\beta} - a^{\alpha 1}a^{\beta 2}(\zeta^0_{\alpha\|1}\zeta^0_{\beta\|2} - \zeta^0_{\alpha\|2}\zeta^0_{\beta\|1}).
\end{aligned}$$

Since $\boldsymbol{u}^1 \in \mathbf{W}(\Omega)$ by assumption and the three functions ψ^0_i are independent of x_3, we conclude that $\boldsymbol{u}^1$ is of the form

$$\boldsymbol{u}^1 = \boldsymbol{\zeta}^1 - x_3\boldsymbol{\psi}^0 \text{ with } \boldsymbol{\zeta}^1 \in \mathbf{W}(\omega) \text{ and } \boldsymbol{\psi}^0 \in \mathbf{W}(\omega),$$

the last relation implying that the functions ζ^0_i should be in the space $W^{2,4}(\omega)$ in order that the products $\partial_\alpha\zeta^0_m\partial_\beta\zeta^0_n$ found in the functions ψ^0_i be themselves in the space $W^{1,4}(\omega)$ (recall that this space is an algebra; see, e.g., Vol. I, Thm. 6.1-4). We likewise infer that the derivatives $\partial_\alpha\zeta^0_i$ should vanish on γ_0 in order that the functions ψ^0_i vanish on γ_0 (we already know that $\zeta^0_i = 0$ on γ_0); hence $\partial_\nu\zeta^0_i = 0$ on γ_0. We have thus shown that

$$\begin{aligned}
\boldsymbol{\zeta}^0 \in \boldsymbol{\mathcal{M}}_F(\omega) = \{\boldsymbol{\eta} \in \mathbf{W}^{2,4}(\omega);\ \boldsymbol{\eta} = \partial_\nu\boldsymbol{\eta} = \mathbf{0} \text{ on } \gamma_0,\\
G_{\alpha\beta}(\boldsymbol{\eta}) = 0 \text{ in } \omega\}.
\end{aligned}$$

(ii) *Solution of the nonlinear system* (found in part (i)) *satisfied by the functions* $u^{(0)}_{i\|3}$, viz.,

$$\begin{aligned}
u^{(0)}_{1\|3} + \zeta^0_{3\|1} + a^{\alpha\beta}\zeta^0_{\alpha\|1}u^{(0)}_{\beta\|3} + \zeta^0_{3\|1}u^{(0)}_{3\|3} &= 0,\\
u^{(0)}_{2\|3} + \zeta^0_{3\|2} + a^{\alpha\beta}\zeta^0_{\alpha\|2}u^{(0)}_{\beta\|3} + \zeta^0_{3\|2}u^{(0)}_{3\|3} &= 0,\\
2u^{(0)}_{3\|3} + a^{\alpha\beta}u^{(0)}_{\alpha\|3}u^{(0)}_{\beta\|3} + u^{(0)}_{3\|3}u^{(0)}_{3\|3} &= 0.
\end{aligned}$$

Recall that the explicit solution of this system, formed by the equations $2E^0_{i\|3} = 0$, was crucially needed in part (i). Let

$$\boldsymbol{x} := \begin{pmatrix} u^{(0)}_{1\|3} \\ u^{(0)}_{2\|3} \end{pmatrix} \quad x_3 := u^{(0)}_{3\|3}, \quad \mathbf{A} := \begin{pmatrix} a^{11} & a^{12} \\ a^{21} & a^{22} \end{pmatrix},$$

$$\boldsymbol{z} := \begin{pmatrix} \zeta^0_{3\|1} \\ \zeta^0_{3\|2} \end{pmatrix}, \quad \mathbf{Z} := \begin{pmatrix} \zeta^0_{1\|1} & \zeta^0_{1\|2} \\ \zeta^0_{2\|1} & \zeta^0_{2\|2} \end{pmatrix},$$

$$\mathbf{M} := \begin{pmatrix} 1 + a^{\alpha 1}\zeta^0_{\alpha\|1} & a^{\alpha 2}\zeta^0_{\alpha\|1} \\ a^{\alpha 1}\zeta^0_{\alpha\|2} & 1 + a^{\alpha 2}\zeta^0_{\alpha\|2} \end{pmatrix}, \quad d := \det \mathbf{M},$$

and assume that the matrix $\mathbf{M}$ is invertible at all points in $\overline{\omega}$ (in fact, we shall see later that this assumption is not indispensable) as is the case if, e.g., the norms $\|\zeta^0_\alpha\|_{W^{2,4}(\omega)}$, hence the norms $\|\zeta^0_\alpha\|_{C^1(\overline{\omega})}$, are small enough. With these notations, the nonlinear system becomes

$$\begin{cases} (1 + x_3)\boldsymbol{z} + \mathbf{M}\boldsymbol{x} = \mathbf{0}, \\ (1 + x_3)^2 + \boldsymbol{x}^T\mathbf{A}\boldsymbol{x} = 1, \end{cases}$$

so that letting $\boldsymbol{x} = -(1 + x_3)\mathbf{M}^{-1}\boldsymbol{z}$ in the last equation gives

$$(1 + x_3)^2(1 + \boldsymbol{z}^T\mathbf{M}^{-T}\mathbf{A}\mathbf{M}^{-1}\boldsymbol{z}) = 1.$$

The rest of the argument essentially consists in computing the number $(1 + \boldsymbol{z}^T\mathbf{M}^{-T}\mathbf{A}\mathbf{M}^{-1}\boldsymbol{z})$, which turns out to be a simple expression in terms of $d = \det \mathbf{M}$. To this end, we first remark that

$$\mathbf{M}\mathbf{A}^{-1}\mathbf{M}^T = \mathbf{A}^{-1} + \mathbf{Z}^T + \mathbf{Z} + \mathbf{Z}^T\mathbf{A}\mathbf{Z}$$

(since $\mathbf{M} = \mathbf{I} + \mathbf{Z}^T\mathbf{A}$ and $\mathbf{A} = \mathbf{A}^T$) and that the relations

$$2E^0_{\alpha\|\beta} = \zeta^0_{\alpha\|\beta} + \zeta^0_{\beta\|\alpha} + a^{\sigma\tau}\zeta^0_{\sigma\|\alpha}\zeta^0_{\tau\|\beta} + \zeta^0_{3\|\alpha}\zeta^0_{3\|\beta} = 0$$

established in Thm. 10.1-1 can be equivalently written as

$$\mathbf{Z} + \mathbf{Z}^T + \mathbf{Z}^T\mathbf{A}\mathbf{Z} + \boldsymbol{z}\boldsymbol{z}^T = \mathbf{0},$$

so that we also have

$$\mathbf{M}\mathbf{A}^{-1}\mathbf{M}^T = \mathbf{A}^{-1} - \boldsymbol{z}\boldsymbol{z}^T = (\mathbf{I} - \boldsymbol{z}\boldsymbol{z}^T\mathbf{A})\mathbf{A}^{-1}.$$

Consequently,

$$\begin{aligned} d^2 &= \det \mathbf{M}\mathbf{M}^T = \det \mathbf{A} \det \mathbf{M}\mathbf{A}^{-1}\mathbf{M}^T \\ &= \det \mathbf{A} \det(\mathbf{I} - \boldsymbol{z}\boldsymbol{z}^T\mathbf{A})\mathbf{A}^{-1} = \det(\mathbf{I} - \boldsymbol{z}\boldsymbol{z}^T\mathbf{A}) \\ &= 1 - tr\, \boldsymbol{z}\boldsymbol{z}^T\mathbf{A} = 1 - \boldsymbol{z}^T\mathbf{A}\boldsymbol{z} \end{aligned}$$

(for any matrix of order two, $\det(\mathbf{I}+\mathbf{B}) = 1 + tr\,\mathbf{B} + \det \mathbf{B}$; besides, if $\mathbf{B} = \boldsymbol{z}\boldsymbol{z}^T\mathbf{A}$, then $\det \mathbf{B} = 0$ since the matrix $\boldsymbol{z}\boldsymbol{z}^T$ is singular); in particular then, $1 - \boldsymbol{z}^T\mathbf{A}\boldsymbol{z} \neq 0$ since $\mathbf{M}$ is assumed to be invertible. From the above expression of $\mathbf{M}\mathbf{A}^{-1}\mathbf{M}^T$, we also consecutively deduce that

$$\mathbf{M}\mathbf{A}^{-1}\mathbf{M}^T\mathbf{A}\boldsymbol{z} = (\mathbf{I} - \boldsymbol{z}\boldsymbol{z}^T\mathbf{A})\boldsymbol{z} = (1 - \boldsymbol{z}^T\mathbf{A}\boldsymbol{z})\boldsymbol{z},$$

$$A\boldsymbol{z} = (\mathbf{M}\mathbf{A}^{-1}\mathbf{M}^T)^{-1}\mathbf{M}\mathbf{A}^{-1}\mathbf{M}^T\mathbf{A}\boldsymbol{z} = (1 - \boldsymbol{z}^T\mathbf{A}\boldsymbol{z})\mathbf{M}^{-T}\mathbf{A}\mathbf{M}^{-1}\boldsymbol{z},$$

$$1 + \boldsymbol{z}^T\mathbf{M}^{-T}\mathbf{A}\mathbf{M}^{-1}\boldsymbol{z} = 1 + \frac{\boldsymbol{z}^T\mathbf{A}\boldsymbol{z}}{1 - \boldsymbol{z}^T\mathbf{A}\boldsymbol{z}} = \frac{1}{1 - \boldsymbol{z}^T\mathbf{A}\boldsymbol{z}} = \frac{1}{d^2}.$$

Hence the relation $(1 + x_3)^2(1 + \boldsymbol{z}^T\mathbf{M}^{-T}\mathbf{A}\mathbf{M}^{-1}\boldsymbol{z}) = 1$ simply becomes

$$d^2 = (1 + x_3)^2,$$

or, *in extenso*,

$$\left(1 + u^{(0)}_{3\|3}\right)^2 = \left(1 + a^{\alpha\beta}\zeta^0_{\alpha\|\beta} + a^{\alpha1}a^{\beta2}(\zeta^0_{\alpha\|1}\zeta^0_{\beta\|2} - \zeta^0_{\alpha\|2}\zeta^0_{\beta\|1})\right)^2.$$

As shown in part (i), the linearization trick leads to only retain the solution $(1 + x_3) = d$. We are thus left with

$$\boldsymbol{x} = -(1 + x_3)\mathbf{M}^{-1}\boldsymbol{z} = -(\det \mathbf{M})\mathbf{M}^{-1}\boldsymbol{z} = -(\mathbf{Cof}\ \mathbf{M}^T)\boldsymbol{z},$$

or, *in extenso*,

$$\begin{aligned} u^{(0)}_{1\|3} &= -(1 + a^{\alpha2}\zeta^0_{\alpha\|2})\zeta^0_{3\|1} + a^{\alpha2}\zeta^0_{\alpha\|1}\zeta^0_{3\|2}, \\ u^{(0)}_{2\|3} &= -(1 + a^{\alpha1}\zeta^0_{\alpha\|1})\zeta^0_{3\|2} + a^{\alpha1}\zeta^0_{\alpha\|2}\zeta^0_{3\|1}. \end{aligned}$$

It is then easily verified that the relations $d^2 = (1 + x_3)^2$ and $\boldsymbol{x} = -(\mathbf{Cof}\ \mathbf{M}^T)\boldsymbol{z}$ are still valid if $\mathbf{M}$ is singular.

(iii) *When $\mathbb{T}_{\boldsymbol{\zeta}}\boldsymbol{\mathcal{M}}_0(\omega) \neq \{\mathbf{0}\}$ at each $\boldsymbol{\zeta} \in \boldsymbol{\mathcal{M}}_0(\omega)$, the cancellation of the coefficient of ε^0 leads to the supplementary conclusion that the functions $E^1_{i\|3}$ satisfy*

$$E^1_{\alpha\|3} = 0 \text{ and } \lambda a^{\alpha\beta}E^1_{\alpha\|\beta} + (\lambda + 2\mu)E^1_{3\|3} = 0 \text{ in } \Omega.$$

Since all the functions $E^0_{i\|j}$ vanish in Ω (the relations $E^0_{\alpha\|\beta} = 0$ and $E^0_{i\|3} = 0$ have been respectively established in Thm. 10.1-1 and in part (i) of the present proof) and since $L^0 = 0$ (the functions $f^{i,0}$ and $h^{i,1}$ vanish; cf. Thm. 10.1-1), the cancellation of the coefficient of ε^0 in the variational equations of problem $\mathcal{P}(\varepsilon;\,\Omega)$ reduces to the equations

$$\int_\Omega A^{ijkl}(0) E^1_{k\|l} F^{-1}_{i\|j}(\boldsymbol{v}) \sqrt{a}\, dx = 0 \text{ for all } \boldsymbol{v} \in \mathbf{W}(\Omega).$$

Observing that the cancellation of the coefficient of ε^{-1} in the variational equations of problem $\mathcal{P}(\varepsilon;\,\Omega)$ led in part (ii) of the proof of Thm. 8.8-1 to analogous equations, only with $E^0_{k\|l}$ in lieu of $E^1_{k\|l}$, we conclude in the same manner that

$$\begin{aligned}
&\big(\lambda a^{\alpha\beta} E^1_{\alpha\|\beta} + (\lambda + 2\mu) E^1_{3\|3}\big) a^{\sigma\tau} u^{(0)}_{\sigma\|3} \\
&\qquad\qquad + 2\mu E^1_{\alpha\|3} \big(a^{\alpha\tau} + a^{\alpha\sigma} a^{\beta\tau} u^0_{\beta\|\alpha}\big) = 0 \text{ in } \Omega, \\
&\big(\lambda a^{\alpha\beta} E^1_{\alpha\|\beta} + (\lambda + 2\mu) E^1_{3\|3}\big) \big(1 + u^{(0)}_{3\|3}\big) \\
&\qquad\qquad + 2\mu a^{\alpha\sigma} E^1_{\alpha\|3} u^0_{3\|\sigma} = 0 \text{ in } \Omega.
\end{aligned}$$

One obvious solution to this system is

$$E^1_{\alpha\|3} = 0 \text{ and } \lambda a^{\alpha\beta} E^1_{\alpha\|\beta} + (\lambda + 2\mu) E^1_{3\|3} = 0 \text{ in } \Omega.$$

Denoting by $[\cdots]^{lin}$ the linear part with respect to (any component of) $\boldsymbol{u}^0$, $\boldsymbol{u}^1$, or $\boldsymbol{u}^2$ in the expression $[\cdots]$ and using the notations of the *linear* case (Sect. 3.4), we have

$$\begin{aligned}
[E^1_{\alpha\|3}]^{lin} &= e^1_{\alpha\|3}, \\
[\lambda a^{\alpha\beta} E^1_{\alpha\|\beta} + (\lambda + 2\mu) E^1_{3\|3}]^{lin} &= \lambda a^{\alpha\beta} e^1_{\alpha\|\beta} + (\lambda + 2\mu) e^1_{3\|3},
\end{aligned}$$

by definition of the functions $E^1_{i\|3}$ and $e^1_{i\|3}$. Since it was found in the linear case (see step (v) in Sect. 3.4) that

$$e^1_{\alpha\|3} = 0 \text{ and } e^1_{3\|3} = -\frac{\lambda}{\lambda + 2\mu} a^{\alpha\beta} e^1_{\alpha\|\beta} \text{ in } \Omega,$$

the *linearization trick* suggests that we only retain the "obvious" solution found above.

(iv) *Equipped with the information that the term of order one is of the form* $\boldsymbol{u}^1 = \boldsymbol{\zeta}^1 - x_3\boldsymbol{\psi}^0$ *with* $\boldsymbol{\zeta}^1 \in \mathbf{W}(\omega)$ *and* $\boldsymbol{\psi}^0 \in \mathbf{W}(\omega)$ (part (i)), *we now show that the functions* $E^1_{\alpha\|\beta}$ (the coefficients of ε in the formal expansion of $E_{\alpha\|\beta}(\varepsilon; \boldsymbol{u}(\varepsilon))$, whose explicit expressions will be needed in part (v) for pursuing the induction) *are given by* (the functions a^{ij}, $g^{\alpha\beta,1}$, and $\Gamma^{p,1}_{i\beta}$ are defined in Thm. 8.5-1):

$$E^1_{\alpha\|\beta} = F^0_{\alpha\|\beta}(\boldsymbol{\zeta}^1) - x_3\hat{E}^0_{\alpha\|\beta},$$

where

$$F^0_{\alpha\|\beta}(\boldsymbol{\eta}) := \frac{1}{2}(\eta_{\alpha\|\beta} + \eta_{\beta\|\alpha} + a^{mn}\{\zeta^0_{m\|\alpha}\eta_{n\|\beta} + \zeta^0_{n\|\beta}\eta_{m\|\alpha}\}),$$

$$\begin{aligned}\hat{E}^0_{\alpha\|\beta} &:= F^0_{\alpha\|\beta}(\boldsymbol{\psi}^0) - \frac{1}{2}g^{\sigma\tau,1}\zeta^0_{\sigma\|\alpha}\zeta^0_{\tau\|\beta} \\ &\quad + \big(\Gamma^{p,1}_{\alpha\beta} + \frac{1}{2}a^{mn}\{\Gamma^{p,1}_{m\alpha}\zeta^0_{n\|\beta} + \Gamma^{p,1}_{n\beta}\zeta^0_{m\|\alpha}\}\big)\zeta^0_p.\end{aligned}$$

An immediate computation, simply based on the definition of the functions $E^1_{\alpha\|\beta}$, shows that (recall that $g^{i3,1} = 0$):

$$2E^1_{\alpha\|\beta} = u^{(1)}_{\beta\|\alpha} + u^{(1)}_{\alpha\|\beta} + a^{mn}\{\zeta^0_{m\|\alpha}u^{(1)}_{n\|\beta} + \zeta^0_{n\|\beta}u^{(1)}_{m\|\alpha}\} + x_3 g^{\sigma\tau,1}\zeta^0_{\sigma\|\alpha}\zeta^0_{\tau\|\beta},$$

where (the functions $\Gamma^{p,0}_{m\alpha}$ are defined in Thm. 8.5-1):

$$u^{(1)}_{m\|\alpha} := \partial_\alpha u^1_m - \Gamma^{p,0}_{m\alpha}u^1_p - x_3\Gamma^{p,1}_{m\alpha}\zeta^0_p,$$

the notation $u^{(1)}_{m\|\alpha}$, which by definition designates the coefficient of ε in the formal expansion of $u_{m\|\alpha}(\varepsilon)$, emphasizing that the functions $u^{(1)}_{m\|\alpha}$ do not depend solely on $\boldsymbol{u}^1 = (u^1_p)$.

Since $\boldsymbol{u}^1 = \boldsymbol{\zeta}^1 - x_3\boldsymbol{\psi}^0$ by (i),

$$\begin{aligned}u^{(1)}_{m\|\alpha} &= \big(\partial_\alpha\zeta^1_m - \Gamma^{p,0}_{m\alpha}\zeta^1_p\big) - x_3\big(\partial_\alpha\psi^0_m - \Gamma^{p,0}_{m\alpha}\psi^0_p\big) - x_3\Gamma^{p,1}_{m\alpha}\zeta^0_p \\ &= \zeta^1_{m\|\alpha} - x_3\psi^0_{m\|\alpha} - x_3\Gamma^{p,1}_{m\alpha}\zeta^0_p,\end{aligned}$$

on the one hand; by definition,

$$2F^0_{\alpha\|\beta}(\boldsymbol{\eta}) = \eta_{\alpha\|\beta} + \eta_{\beta\|\alpha} + a^{mn}\{\zeta^0_{m\|\alpha}\eta_{n\|\beta} + \zeta^0_{n\|\beta}\eta_{m\|\alpha}\},$$

with $\eta_{m||\alpha} = \partial_\alpha \eta_m - \Gamma^{p,0}_{m\alpha}\eta_p$, on the other. Hence

$$\begin{aligned} E^1_{\alpha||\beta} &= F^0_{\alpha||\beta}(\boldsymbol{\zeta}^1 - x_3\boldsymbol{\psi}^0) + \frac{1}{2}x_3 g^{\sigma\tau,1}\zeta^0_{\sigma||\alpha}\zeta^0_{\tau||\beta} \\ &\quad - x_3\big(\Gamma^{p,1}_{\alpha\beta} + \frac{1}{2}a^{mn}\{\Gamma^{p,1}_{m\alpha}\zeta^0_{n||\beta} + \Gamma^{p,1}_{n\beta}\zeta^0_{m||\alpha}\}\big)\zeta^0_p. \end{aligned}$$

(v) *When* $\mathbb{T}_{\boldsymbol{\zeta}}\mathcal{M}_0(\omega) \neq \{\mathbf{0}\}$ *at each* $\boldsymbol{\zeta} \in \mathcal{M}_0(\omega)$, *the cancellation of the coefficient of* ε *leads to the following relations:*

$$f^{i,1} = 0 \text{ in } \Omega \text{ and } h^{i,2} = 0 \text{ on } \Gamma_+ \cup \Gamma_-,$$

$$\boldsymbol{\zeta}^1 \in \mathbb{T}_{\boldsymbol{\zeta}^0}\mathcal{M}_0(\omega) := \{\boldsymbol{\eta} \in \mathbf{W}(\omega);\ F^0_{\alpha||\beta}(\boldsymbol{\eta}) = 0 \text{ in } \omega\},$$

$$E^1_{\alpha||\beta} = -x_3 \hat{E}^0_{\alpha||\beta},$$

where the functions $F^0_{\alpha||\beta}(\boldsymbol{\eta})$ *and* $\hat{E}^0_{\alpha||\beta}$ *are those found in* (iv).

Since $f^{i,0} = 0$ and $h^{i,1} = 0$ when $\mathbb{T}_{\boldsymbol{\zeta}}\mathcal{M}_0(\omega) \neq \{\mathbf{0}\}$ at each $\boldsymbol{\zeta} \in \mathcal{M}_0(\omega)$ by Thm. 10.1-1, our next try is

$$f^i(\varepsilon) = \varepsilon f^{i,1} \text{ and } h^i(\varepsilon) = \varepsilon^2 h^{i,2}.$$

The cancellation of the coefficient of ε in the variational equations of problem $\mathcal{P}(\varepsilon; \Omega)$ then leads to the equations (two terms are needed here from the expansions of the functions $A^{ijkl}(\varepsilon)\sqrt{g(\varepsilon)}$; cf. Thm. 8.5-2):

$$\begin{aligned} &\int_\Omega A^{ijkl}(0)\{E^1_{k||l}F^0_{i||j}(\boldsymbol{v}) + E^2_{k||l}F^{-1}_{i||j}(\boldsymbol{v})\}\sqrt{a}\,dx \\ &+ \int_\Omega B^{ijkl,1}E^1_{k||l}F^{-1}_{i||j}(\boldsymbol{v})\,dx = L^1(\boldsymbol{v}) \end{aligned}$$

for all $\boldsymbol{v} \in \mathbf{W}(\Omega)$. In particular then, we must have

$$\int_\Omega A^{ijkl}(0)E^1_{k||l}F^0_{i||j}(\boldsymbol{v})\sqrt{a}\,dx = L^1(\boldsymbol{v})$$

for all $\boldsymbol{v} \in \mathbf{W}(\Omega)$ that are *independent of* x_3, since $F^{-1}_{i||j}(\boldsymbol{v}) = 0$ for such functions $\boldsymbol{v}$; or equivalently, after the usual identification has been performed,

$$\int_\Omega A^{ijkl}(0)E^1_{k||l}F^0_{i||j}(\boldsymbol{\eta})\sqrt{a}\,dx = L^1(\boldsymbol{\eta})$$

for all $\boldsymbol{\eta} \in \mathbf{W}(\omega) = \{\boldsymbol{\eta} \in \mathbf{W}^{1,4}(\omega);\ \boldsymbol{\eta} = \mathbf{0} \text{ on } \gamma_0\}$.

Using the expressions of the functions $A^{ijkl}(0)$ (Thm. 8.5-2) and the relations satisfied by the functions $E^1_{i\|3}$ (part (iii)), we are left with (an analogous computation is detailed in part (iii) of the proof of Thm. 8.8-1):

$$\int_\Omega A^{ijkl}(0)E^1_{k\|l}F^0_{i\|j}(\boldsymbol{\eta})\sqrt{a}\,dx$$

$$= \frac{1}{2}\int_\Omega a^{\alpha\beta\sigma\tau}E^1_{\sigma\|\tau}F^0_{\alpha\|\beta}(\boldsymbol{\eta})\sqrt{a}\,dx = L^1(\boldsymbol{\eta})$$

for all $\boldsymbol{\eta} \in \mathbf{W}(\omega)$, where

$$a^{\alpha\beta\sigma\tau} = \frac{4\lambda\mu}{\lambda+2\mu}a^{\alpha\beta}a^{\sigma\tau} + 2\mu(a^{\alpha\sigma}a^{\beta\tau} + a^{\alpha\tau}a^{\beta\sigma}).$$

Since

$$\boldsymbol{\zeta}^0 \in \boldsymbol{\mathcal{M}}_0(\omega) = \{\boldsymbol{\eta} \in \mathbf{W}(\omega);\, G_{\alpha\beta}(\boldsymbol{\eta}) = 0 \text{ in } \omega\}$$

(Thm. 10.1-1) and since the tangent space

$$\mathbb{T}_{\boldsymbol{\zeta}^0}\boldsymbol{\mathcal{M}}_0(\omega) = \{\boldsymbol{\eta} \in \mathbf{W}(\omega);\, F^0_{\alpha\|\beta}(\boldsymbol{\eta}) = 0 \text{ in } \omega\} \subset \mathbf{W}(\omega)$$

to the manifold $\boldsymbol{\mathcal{M}}_0(\omega)$ at $\boldsymbol{\zeta}^0$ (recall that $E^0_{\alpha\|\beta} = G_{\alpha\beta}(\boldsymbol{\zeta}^0)$ and that $F^0_{\alpha\|\beta}(\boldsymbol{\eta}) = G'_{\alpha\beta}(\boldsymbol{\zeta}^0)\boldsymbol{\eta}$; cf. Thm. 9.1-1) contains nonzero functions *by assumption*, using such functions in the last variational equations shows that we must let $f^{i,1} = 0$ and $h^{i,2} = 0$ in order to fulfill the *first requirement* (the linearization trick could be also called upon here since the same conclusion was reached in the linear case; cf. part (vi) in Sect. 3.4).

The functions $E^1_{\alpha\|\beta}$ being of the form (part (iv))

$$E^1_{\alpha\|\beta} = F^0_{\alpha\|\beta}(\boldsymbol{\zeta}^1) - x_3\hat{E}^0_{\alpha\|\beta},$$

where the functions $F^0_{\alpha\|\beta}(\boldsymbol{\zeta}^1)$ and $\hat{E}^0_{\alpha\|\beta}$ are independent of the transverse variable x_3, we are left with the equations

$$\int_\omega a^{\alpha\beta\sigma\tau}F^0_{\sigma\|\tau}(\boldsymbol{\zeta}^1)F^0_{\alpha\|\beta}(\boldsymbol{\eta})\sqrt{a}\,dx = 0$$

for all $\boldsymbol{\eta} \in \mathbf{W}(\omega)$. Letting $\boldsymbol{\eta} = \boldsymbol{\zeta}^1$ in these equations (this is licit since $\boldsymbol{\zeta}^1 \in \mathbf{W}(\omega)$ by (i)) shows that (the scaled two-dimensional elasticity tensor of a shell is positive definite; cf. Thm. 3.3-2):

$$F^0_{\alpha\|\beta}(\boldsymbol{\zeta}^1) = 0.$$

Hence $\boldsymbol{\zeta}^1 \in \mathbb{T}_{\boldsymbol{\zeta}^0}\boldsymbol{\mathcal{M}}_0(\omega)$ and $E^1_{\alpha\|\beta} = -x_3\hat{E}^0_{\alpha\|\beta}$.

(vi) *When $\mathbb{T}_{\boldsymbol{\zeta}}\mathcal{M}_0(\omega) \neq \{\mathbf{0}\}$ at each $\boldsymbol{\zeta} \in \mathcal{M}_0(\omega)$, the cancellation of the coefficient of ε^2 shows that the field $\boldsymbol{\zeta}^0$, which belongs to the space $\mathcal{M}_F(\omega)$ by* (i), *satisfies the following variational equations:*

$$\frac{1}{3}\int_\omega a^{\alpha\beta\sigma\tau}\hat{E}^0_{\sigma\|\tau}\hat{F}^0_{\alpha\|\beta}(\boldsymbol{\eta})\sqrt{a}\,dy = L^2(\boldsymbol{\eta})$$

for all $\boldsymbol{\eta} = (\eta_i) \in \mathbb{T}_{\boldsymbol{\zeta}^0}\mathcal{M}_F(\omega)$, where

$$\mathbb{T}_{\boldsymbol{\zeta}^0}\mathcal{M}_F(\omega) = \{\boldsymbol{\eta} \in \mathbf{W}^{2,4}(\omega);\ \boldsymbol{\eta} = \partial_\nu\boldsymbol{\eta} = \mathbf{0} \text{ on } \gamma_0;\\ F^0_{\alpha\|\beta}(\boldsymbol{\eta}) = 0 \text{ in } \omega\}.$$

The functions $\hat{E}^0_{\alpha\|\beta}$ are those found in (iv) *and the functions $\hat{F}^0_{\alpha\|\beta}(\boldsymbol{\eta})$ are given by* (see (iv) for some of the notations employed *infra*):

$$\begin{aligned}\hat{F}^0_{\alpha\|\beta}(\boldsymbol{\eta}) := {} & F^0_{\alpha\|\beta}(\boldsymbol{\varphi}^0(\boldsymbol{\eta})) - \frac{1}{2}g^{\sigma\tau,1}\{\zeta^0_{\sigma\|\alpha}\eta_{\tau\|\beta} + \zeta^0_{\tau\|\beta}\eta_{\sigma\|\alpha}\}\\ & + \Big(\Gamma^{p,1}_{\alpha\beta} + \frac{1}{2}a^{mn}\{\Gamma^{p,1}_{m\alpha}\zeta^0_{n\|\beta} + \Gamma^{p,1}_{n\beta}\zeta^0_{m\|\alpha}\}\Big)\eta_p\\ & + \frac{1}{2}a^{mn}\{(\psi^0_{m\|\alpha} + \Gamma^{p,1}_{m\alpha}\zeta^0_p)\eta_{n\|\beta} + (\psi^0_{n\|\beta} + \Gamma^{q,1}_{n\beta}\zeta^0_q)\eta_{m\|\alpha}\},\end{aligned}$$

where $\boldsymbol{\varphi}^0(\boldsymbol{\eta}) = (\varphi^0_i(\boldsymbol{\eta})) \in \mathbf{W}(\omega)$ is defined by

$$\begin{aligned}\varphi^0_1(\boldsymbol{\eta}) := {} & b^\alpha_1\eta_\alpha + \eta_{3\|1}\\ & + a^{\alpha 2}(\zeta^0_{\alpha\|2}\eta_{3\|1} + \zeta^0_{3\|1}\eta_{\alpha\|2} - \zeta^0_{3\|2}\eta_{\alpha\|1} - \zeta^0_{\alpha\|1}\eta_{3\|2}),\\ \varphi^0_2(\boldsymbol{\eta}) := {} & b^\alpha_2\eta_\alpha + \eta_{3\|2}\\ & + a^{\alpha 1}(\zeta^0_{\alpha\|1}\eta_{3\|2} + \zeta^0_{3\|2}\eta_{\alpha\|1} - \zeta^0_{3\|1}\eta_{\alpha\|2} - \zeta^0_{\alpha\|2}\eta_{3\|1}),\\ \varphi^0_3(\boldsymbol{\eta}) := {} & -a^{\alpha\beta}\eta_{\alpha\|\beta}\\ & - a^{\alpha 1}a^{\beta 2}(\zeta^0_{\alpha\|1}\eta_{\beta\|2} + \zeta^0_{\beta\|2}\eta_{\alpha\|1} - \zeta^0_{\alpha\|2}\eta_{\beta\|1} - \zeta^0_{\beta\|1}\eta_{\alpha\|2}).\end{aligned}$$

Since $f^{i,1} = 0$ and $h^{i,2} = 0$ when $\mathbb{T}_{\boldsymbol{\zeta}}\mathcal{M}_0(\omega) \neq \{\mathbf{0}\}$ at each $\boldsymbol{\zeta} \in \mathcal{M}_0(\omega)$ by (v), our next try (and final one as it will turn out) is

$$f^i(\varepsilon) = \varepsilon^2 f^{i,2} \text{ and } h^i(\varepsilon) = \varepsilon^3 h^{i,3}.$$

The cancellation of the coefficient of ε^2 in the variational equations of problem $\mathcal{P}(\varepsilon;\Omega)$ then leads to the equations (no less than three terms are needed here from the expansions of the functions $A^{ijkl}(\varepsilon)\sqrt{g(\varepsilon)}$; cf. Thm. 8.5-2):

$$\int_\Omega A^{ijkl}(0)\{E^1_{k\|l}F^1_{i\|j}(\boldsymbol{v}) + E^2_{k\|l}F^0_{i\|j}(\boldsymbol{v}) + E^3_{k\|l}F^{-1}_{i\|j}(\boldsymbol{v})\}\sqrt{a}\,dx$$
$$+ \int_\Omega B^{ijkl,1}\{E^1_{k\|l}F^0_{i\|j}(\boldsymbol{v}) + E^2_{k\|l}F^{-1}_{i\|j}(\boldsymbol{v})\}dx$$
$$+ \int_\Omega B^{ijkl,2}E^1_{k\|l}F^{-1}_{i\|j}(\boldsymbol{v})\,dx = L^2(\boldsymbol{v})$$

for all $\boldsymbol{v} \in \mathbf{W}(\Omega)$. Note that the highest order term in the formal asymptotic expansion of $\boldsymbol{u}(\varepsilon)$ appearing in the functions $E^3_{i\|j}$ is $\boldsymbol{u}^4$ (each term involving $\boldsymbol{u}^5$ vanishes because it contains some derivative $\partial_3 u^0_m$ as a factor); this is why this expansion must be "at least" of the form $\boldsymbol{u}(\varepsilon) = \sum_{q=0}^4 \varepsilon^q \boldsymbol{u}^q + \cdots$.

In particular then, these variational equations must be satisfied for all $\boldsymbol{v} \in \mathbf{W}(\Omega)$ that are *independent of* x_3 since $F^{-1}_{i\|j}(\boldsymbol{v}) = 0$ for such functions $\boldsymbol{v}$. Equivalently, after the usual identification has been performed, we must have

$$\int_\Omega A^{ijkl}(0)\{E^1_{k\|l}F^1_{i\|j}(\boldsymbol{\eta}) + E^2_{k\|l}F^0_{i\|j}(\boldsymbol{\eta})\}\sqrt{a}\,dx$$
$$+ \int_\Omega B^{ijkl,1}E^1_{k\|l}F^0_{i\|j}(\boldsymbol{\eta})\,dx = L^2(\boldsymbol{\eta})$$

for all $\boldsymbol{\eta} \in \mathbf{W}(\omega) = \{\boldsymbol{\eta} \in \mathbf{W}^{1,4}(\omega);\ \boldsymbol{\eta} = \mathbf{0} \text{ on } \gamma_0\}$, on the one hand. On the other, the cancellation of the coefficient of ε shows that we also have

$$\int_\Omega A^{ijkl}(0)\{E^1_{k\|l}F^0_{i\|j}(\boldsymbol{v}) + E^2_{k\|l}F^{-1}_{i\|j}(\boldsymbol{v})\}\sqrt{a}\,dx$$
$$+ \int_\Omega B^{ijkl,1}E^1_{k\|l}F^{-1}_{i\|j}(\boldsymbol{v})\,dx = 0$$

for all $\boldsymbol{v} \in \mathbf{W}(\Omega)$, since it was proved in (v) that $L^1 = 0$.

As we shall see later by means of an independent proof (Thm. 10.1-3), it is remarkable that, given any function $\boldsymbol{\eta}$ in the tangent space $\mathbb{T}_{\zeta^0}\boldsymbol{\mathcal{M}}_F(\omega)$ to the manifold $\boldsymbol{\mathcal{M}}_F(\omega)$, there exists a function $\boldsymbol{v}(\boldsymbol{\eta}) \in \mathbf{W}(\Omega)$ such that

$$F^0_{i\|j}(\boldsymbol{\eta}) = F^{-1}_{i\|j}(\boldsymbol{v}(\boldsymbol{\eta})).$$

Subtracting the last variational equations with $\boldsymbol{v} = \boldsymbol{v}(\boldsymbol{\eta})$ from the penultimate ones and noting that $\mathbb{T}_{\zeta^0}\mathcal{M}_F(\omega) \subset \mathbf{W}(\omega)$, we are thus left with the variational equations

$$\int_\Omega A^{ijkl}(0)E^1_{k\|l}\{F^1_{i\|j}(\boldsymbol{\eta}) - F^0_{i\|j}(\boldsymbol{v}(\boldsymbol{\eta}))\}\sqrt{a}\,dx = L^2(\boldsymbol{\eta})$$

for all $\boldsymbol{\eta} \in \mathbb{T}_{\zeta^0}\mathcal{M}_F(\omega)$, which in turn reduce to

$$\frac{1}{2}\int_\Omega a^{\alpha\beta\sigma\tau}E^1_{\sigma\|\tau}\{F^1_{\alpha\|\beta}(\boldsymbol{\eta}) - F^0_{\alpha\|\beta}(\boldsymbol{v}(\boldsymbol{\eta}))\}\sqrt{a}\,dx = L^2(\boldsymbol{\eta})$$

for all $\boldsymbol{\eta} \in \mathbb{T}_{\zeta^0}\mathcal{M}_F(\omega)$, in view of the expressions of the functions $A^{ijkl}(0)$ and of the relations satisfied by the functions $E^1_{i\|3}$ (see Thm. 8.5-2 and part (iii) of this proof).

As we shall likewise see later by means of yet another independent proof (Thm. 10.1-4), it is no less remarkable that, given any function $\boldsymbol{\eta} \in \mathbb{T}_{\zeta^0}\mathcal{M}_F(\omega)$, the differences $\{F^1_{\alpha\|\beta}(\boldsymbol{\eta}) - F^0_{\alpha\|\beta}(\boldsymbol{v}(\boldsymbol{\eta}))\}$ appearing in the last variational equations are of the form

$$F^1_{\alpha\|\beta}(\boldsymbol{\eta}) - F^0_{\alpha\|\beta}(\boldsymbol{v}(\boldsymbol{\eta})) = \hat{F}^1_{\alpha\|\beta}(\boldsymbol{\eta}) - x_3\hat{F}^0_{\alpha\|\beta}(\boldsymbol{\eta}),$$

where the functions $\hat{F}^1_{\alpha\|\beta}(\boldsymbol{\eta})$ are *independent of* x_3 and the functions $\hat{F}^0_{\alpha\|\beta}(\boldsymbol{\eta})$, which are defined as indicated *supra*, are also *independent of* x_3. Since

$$E^1_{\alpha\|\beta} = -x_3\hat{E}^0_{\alpha\|\beta},$$

where the functions $\hat{E}^0_{\alpha\|\beta}$ are again *independent of* x_3 (part (iv)), we are thus eventually left with the variational equations

$$\frac{1}{3}\int_\omega a^{\alpha\beta\sigma\tau}\hat{E}^0_{\sigma\|\tau}\hat{F}^0_{\alpha\|\beta}(\boldsymbol{\eta})\,\sqrt{a}\,dy = L^2(\boldsymbol{\eta}) \text{ for all } \boldsymbol{\eta} \in \mathbb{T}_{\zeta^0}\mathcal{M}_F(\omega),$$

as was to be proved. ■

It remains to establish the two properties stated without proof in part (vi) above. Their demonstrations, which are exceedingly delicate and lengthy, were for these reasons postponed in order not to disrupt the otherwise familiar course of the proof of Thm. 10.1-2.

The next two theorems are due to Lods & Miara [1998, Lemmas 4 and 5].

Theorem 10.1-3. *Let $F^q_{i\|j}(\boldsymbol{v})$, $q \geq -1$, denote the coefficient of ε^q in the formal asymptotic expansion*

$$F_{i\|j}(\varepsilon;\, \boldsymbol{u}(\varepsilon),\, \boldsymbol{v}) = \frac{1}{\varepsilon} F^{-1}_{i\|j}(\boldsymbol{v}) + F^0_{i\|j}(\boldsymbol{v}) + \varepsilon F^1_{i\|j}(\boldsymbol{v}) + \cdots .$$

Given any $\boldsymbol{\eta} = (\eta_i) \in \mathbb{T}_{\boldsymbol{\zeta}^0}\boldsymbol{\mathcal{M}}_F(\omega)$, where

$$\mathbb{T}_{\boldsymbol{\zeta}^0}\boldsymbol{\mathcal{M}}_F(\omega) = \{\boldsymbol{\eta} \in \mathbf{W}^{2,4}(\omega);\, \boldsymbol{\eta} = \partial_\nu \boldsymbol{\eta} = \mathbf{0} \text{ on } \gamma_0,\ F^0_{\alpha\|\beta}(\boldsymbol{\eta}) = 0 \text{ in } \omega\},$$

let $\boldsymbol{\varphi}^0(\boldsymbol{\eta}) = (\varphi^0_i(\boldsymbol{\eta})) \in \mathbf{W}(\omega) = \{\boldsymbol{\eta} \in \mathbf{W}^{1,4}(\omega);\, \boldsymbol{\eta} = \mathbf{0}$ on $\gamma_0\}$ be defined by

$$\begin{aligned}
\varphi^0_1(\boldsymbol{\eta}) &:= b^\alpha_1 \eta_\alpha + \eta_{3\|1} \\
&\quad + a^{\alpha 2}(\zeta^0_{\alpha\|2}\eta_{3\|1} + \zeta^0_{3\|1}\eta_{\alpha\|2} - \zeta^0_{3\|2}\eta_{\alpha\|1} - \zeta^0_{\alpha\|1}\eta_{3\|2}), \\
\varphi^0_2(\boldsymbol{\eta}) &:= b^\alpha_2 \eta_\alpha + \eta_{3\|2} \\
&\quad + a^{\alpha 1}(\zeta^0_{\alpha\|1}\eta_{3\|2} + \zeta^0_{3\|2}\eta_{\alpha\|1} - \zeta^0_{3\|1}\eta_{\alpha\|2} - \zeta^0_{\alpha\|2}\eta_{3\|1}), \\
\varphi^0_3(\boldsymbol{\eta}) &:= -a^{\alpha\beta}\eta_{\alpha\|\beta} \\
&\quad - a^{\alpha 1}a^{\beta 2}(\zeta^0_{\alpha\|1}\eta_{\beta\|2} + \zeta^0_{\beta\|2}\eta_{\alpha\|1} - \zeta^0_{\alpha\|2}\eta_{\beta\|1} - \zeta^0_{\beta\|1}\eta_{\alpha\|2}),
\end{aligned}$$

and let

$$\boldsymbol{v}(\boldsymbol{\eta}) := x_3 \boldsymbol{\varphi}^0(\boldsymbol{\eta}).$$

Then

$$\boxed{\begin{aligned}
&\boldsymbol{v}(\boldsymbol{\eta}) \in \mathbf{W}(\Omega) = \{\boldsymbol{\eta} \in \mathbf{W}^{1,4}(\Omega);\, \boldsymbol{v} = \mathbf{0} \text{ on } \Gamma_0\}, \\
&F^0_{i\|j}(\boldsymbol{\eta}) = F^{-1}_{i\|j}(\boldsymbol{v}(\boldsymbol{\eta})), \text{ and} \\
&\boldsymbol{\varphi}^0(\boldsymbol{\eta}) = \boldsymbol{\psi}'(\boldsymbol{\zeta}^0)\boldsymbol{\eta} \text{ for all } \boldsymbol{\eta} \in \mathbb{T}_{\boldsymbol{\zeta}^0}\boldsymbol{\mathcal{M}}_F(\omega),
\end{aligned}}$$

where $\boldsymbol{\psi}(\boldsymbol{\eta}) = (\psi_i(\boldsymbol{\eta})) \in \mathbf{W}^{1,4}(\omega)$ is defined for any $\boldsymbol{\eta} \in \mathbf{W}^{2,4}(\omega)$ by

$$\begin{aligned}
\psi_1(\boldsymbol{\eta}) &:= b^\alpha_1 \eta_\alpha + (1 + a^{\alpha 2}\eta_{\alpha\|2})\eta_{3\|1} - a^{\alpha 2}\eta_{\alpha\|1}\eta_{3\|2}, \\
\psi_2(\boldsymbol{\eta}) &:= b^\alpha_2 \eta_\alpha + (1 + a^{\alpha 1}\eta_{\alpha\|1})\eta_{3\|2} - a^{\alpha 1}\eta_{\alpha\|2}\eta_{3\|1}, \\
\psi_3(\boldsymbol{\eta}) &:= -a^{\alpha\beta}\eta_{\alpha\|\beta} - a^{\alpha 1}a^{\beta 2}(\eta_{\alpha\|1}\eta_{\beta\|2} - \eta_{\alpha\|2}\eta_{\beta\|1}).
\end{aligned}$$

Proof. The following notations will be used throughout the proof:

$$\eta_{\alpha\|\beta} := \partial_\beta \eta_\alpha - \Gamma^\sigma_{\alpha\beta}\eta_\sigma - b_{\alpha\beta}\eta_3, \quad \eta_{3\|\beta} := \partial_\beta \eta_3 + b^\sigma_\alpha \eta_\sigma,$$

$$\boldsymbol{z}(\boldsymbol{\eta}) := \begin{pmatrix} \eta_{3\|1} \\ \eta_{3\|2} \end{pmatrix}, \ \boldsymbol{z} := \begin{pmatrix} \zeta^0_{3\|1} \\ \zeta^0_{3\|2} \end{pmatrix} = \boldsymbol{z}(\boldsymbol{\zeta}^0), \ \mathbf{A} := \begin{pmatrix} a^{11} & a^{12} \\ a^{21} & a^{22} \end{pmatrix},$$

$$\mathbf{M}(\boldsymbol{\eta}) := \begin{pmatrix} 1 + a^{\alpha 1}\eta_{\alpha\|1} & a^{\alpha 2}\eta_{\alpha\|1} \\ a^{\alpha 1}\eta_{\alpha\|2} & 1 + a^{\alpha 2}\eta_{\alpha\|2} \end{pmatrix}, \ d(\boldsymbol{\eta}) := \det \mathbf{M}(\boldsymbol{\eta}),$$

$$\mathbf{M} := \begin{pmatrix} 1 + a^{\alpha 1}\zeta^0_{\alpha\|1} & a^{\alpha 2}\zeta^0_{\alpha\|1} \\ a^{\alpha 1}\zeta^0_{\alpha\|2} & 1 + a^{\alpha 2}\zeta^0_{\alpha\|2} \end{pmatrix} = \mathbf{M}(\zeta^0), \ d := \det \mathbf{M} = d(\boldsymbol{\zeta}^0),$$

$$\mathbf{Z}(\boldsymbol{\eta}) := \begin{pmatrix} \eta_{1\|1} & \eta_{1\|2} \\ \eta_{2\|1} & \eta_{2\|2} \end{pmatrix}, \ \mathbf{Z} := \begin{pmatrix} \zeta^0_{1\|1} & \zeta^0_{1\|2} \\ \zeta^0_{2\|1} & \zeta^0_{2\|2} \end{pmatrix} = \mathbf{Z}(\boldsymbol{\zeta}^0),$$

$$\hat{\mathbf{N}}(\boldsymbol{\eta}) := \mathbf{Z}(\boldsymbol{\eta})^T \mathbf{A} = \begin{pmatrix} a^{\alpha 1}\eta_{\alpha\|1} & a^{\alpha 2}\eta_{\alpha\|1} \\ a^{\alpha 1}\eta_{\alpha\|2} & a^{\alpha 2}\eta_{\alpha\|2} \end{pmatrix},$$

$$\mathbf{L} := \begin{pmatrix} \mathbf{M} & \boldsymbol{z} \\ -d\boldsymbol{z}^T \mathbf{M}^{-T} \mathbf{A} & d \end{pmatrix}, \ \mathbf{N}(\boldsymbol{\eta}) := \begin{pmatrix} \hat{\mathbf{N}}(\boldsymbol{\eta}) & \mathbf{0} \\ \mathbf{0} & 0 \end{pmatrix}.$$

Some of these notations were introduced in part (ii) of the proof of the preceding theorem; there, it was shown in particular that they satisfy

$$\begin{pmatrix} u^{(0)}_{1\|3} \\ u^{(0)}_{2\|3} \end{pmatrix} = -d\mathbf{M}^{-1}\boldsymbol{z}, \quad 1 + u^{(0)}_{3\|3} = d,$$

$$\mathbf{M}\mathbf{A}^{-1}\mathbf{M}^T\mathbf{A} = I - \boldsymbol{z}\boldsymbol{z}^T\mathbf{A}, \quad 1 + \boldsymbol{z}^T\mathbf{M}^{-T}\mathbf{A}\mathbf{M}^{-1}\boldsymbol{z} = \frac{1}{d^2},$$

$$\mathbf{M}^{-T}\mathbf{A}\mathbf{M}^{-1}\boldsymbol{z} = \frac{1}{d^2}\mathbf{A}\boldsymbol{z}, \quad \boldsymbol{z}^T\mathbf{A}\boldsymbol{z} = 1 - d^2.$$

These relations will be repeatedly used in the sequel.

Since $F^{-1}_{\alpha\|\beta}(\boldsymbol{v}) = 0$ for all $\boldsymbol{v} \in \mathbf{W}(\Omega)$ and $F^0_{\alpha\|\beta}(\boldsymbol{\eta}) = 0$ for all $\boldsymbol{\eta} \in \mathbb{T}_{\boldsymbol{\zeta}^0}\boldsymbol{\mathcal{M}}_F(\omega)$, it therefore suffices to show that the relations $F^{-1}_{i\|3}(\boldsymbol{v}(\boldsymbol{\eta})) = F^0_{i\|3}(\boldsymbol{\eta})$ are satisfied for all $\boldsymbol{\eta} \in \mathbb{T}_{\boldsymbol{\zeta}^0}\boldsymbol{\mathcal{M}}_F(\omega)$.

(i) *Let* $\boldsymbol{v} = (v_i) \in \mathbf{W}(\Omega)$ *and* $\boldsymbol{\eta} = (\eta_i) \in \mathbf{W}(\omega)$ *be given. A necessary and sufficient condition that*

$$F^{-1}_{i\|3}(\boldsymbol{v}) = F^{0}_{i\|3}(\boldsymbol{\eta})$$

is that $\boldsymbol{v}$ *be affine with respect to* x_3 *and that*

$$\mathbf{L}\begin{pmatrix}\partial_3 v_1\\ \partial_3 v_2\\ \partial_3 v_3\end{pmatrix} = \mathbf{L}\begin{pmatrix} b_1^\alpha \eta_\alpha \\ b_2^\alpha \eta_\alpha \\ 0\end{pmatrix} + \mathbf{N}(\boldsymbol{\eta})\begin{pmatrix} u^{(0)}_{1\|3}\\ u^{(0)}_{2\|3}\\ u^{(0)}_{3\|3}\end{pmatrix} + (1 + u^{(0)}_{3\|3})\begin{pmatrix}\eta_{3\|1}\\ \eta_{3\|2}\\ 0\end{pmatrix}.$$

To prove this relation, we express in matrix form the column vector fields with components $F^{-1}_{i\|3}(\boldsymbol{v}) \in L^2(\Omega)$ and $F^{0}_{i\|3}(\boldsymbol{\eta}) \in L^2(\omega)$. First, let $\boldsymbol{v} \in \mathbf{W}(\Omega)$ be given. Then the definition of the functions $F^{-1}_{i\|3}(\boldsymbol{v})$ shows that

$$\begin{aligned}
2F^{-1}_{1\|3}(\boldsymbol{v}) &= (1 + a^{\alpha 1}\zeta^0_{\alpha\|1})\partial_3 v_1 + a^{\alpha 2}\zeta^0_{\alpha\|1}\partial_3 v_2 + \zeta^0_{3\|1}\partial_3 v_3,\\
2F^{-1}_{2\|3}(\boldsymbol{v}) &= a^{\alpha 1}\zeta^0_{\alpha\|2}\partial_3 v_1 + (1 + a^{\alpha 2}\zeta^0_{\alpha\|2})\partial_3 v_2 + \zeta^0_{3\|2}\partial_3 v_3,\\
F^{-1}_{3\|3}(\boldsymbol{v}) &= a^{\alpha 1}u^{(0)}_{\alpha\|3}\partial_3 v_1 + a^{\alpha 2}u^{(0)}_{\alpha\|3}\partial_3 v_2 + (1 + u^{(0)}_{3\|3})\partial_3 v_3,
\end{aligned}$$

where the functions $u^{(0)}_{i\|3}$ are given by (cf. part (ii) of the proof of Thm. 10.1-2):

$$\begin{aligned}
u^{(0)}_{1\|3} &= -(1 + a^{\alpha 2}\zeta^0_{\alpha\|2})\zeta^0_{3\|1} + a^{\alpha 2}\zeta^0_{\alpha\|1}\zeta^0_{3\|2},\\
u^{(0)}_{2\|3} &= -(1 + a^{\alpha 1}\zeta^0_{\alpha\|1})\zeta^0_{3\|2} + a^{\alpha 1}\zeta^0_{\alpha\|2}\zeta^0_{3\|1},\\
u^{(0)}_{3\|3} &= d - 1.
\end{aligned}$$

Hence these relations together imply that

$$\begin{pmatrix} 2F^{-1}_{1\|3}(\boldsymbol{v})\\ 2F^{-1}_{2\|3}(\boldsymbol{v})\\ F^{-1}_{3\|3}(\boldsymbol{v})\end{pmatrix} = \mathbf{L}\begin{pmatrix}\partial_3 v_1\\ \partial_3 v_2\\ \partial_3 v_3\end{pmatrix}.$$

Next, let $\boldsymbol{\eta} \in \mathbf{W}(\omega)$ be given. Since $\boldsymbol{\eta}$ may be identified in the usual way with a field (still denoted) $\boldsymbol{\eta} \in \mathbf{W}(\Omega)$, the relation $\partial_3\boldsymbol{\eta} = \mathbf{0}$ in Ω shows that $F^{-1}_{i\|3}(\boldsymbol{\eta}) = 0$. The functions $F^{0}_{i\|3}(\boldsymbol{\eta})$, which in this case thus become the leading terms in the formal asymptotic expansions of the functions $F_{i\|j}(\varepsilon; \boldsymbol{u}(\varepsilon), \boldsymbol{\eta})$, are then given by

$$\begin{aligned} 2F^0_{\alpha\|3}(\boldsymbol{\eta}) &= \partial_\alpha\eta_3 + 2b^\sigma_\alpha\eta_\sigma + a^{\sigma\tau}(\partial_\alpha\zeta^0_\sigma - \Gamma^\beta_{\alpha\sigma}\zeta^0_\beta - b_{\alpha\sigma}\zeta^0_3)b^\lambda_\tau\eta_\lambda \\ &\quad + a^{\sigma\tau}(\partial_3 u^1_\tau + b^\sigma_\tau\zeta^0_\sigma)(\partial_\alpha\eta_\sigma - \Gamma^\beta_{\alpha\sigma}\eta_\beta - b_{\alpha\sigma}\eta_3) \\ &\quad + \partial_3 u^1_3(\partial_\alpha\eta_3 + b^\sigma_\alpha\eta_\sigma), \\ F^0_{3\|3}(\boldsymbol{\eta}) &= a^{\sigma\tau}(\partial_3 u^1_\sigma + b^\alpha_\sigma\zeta^0_\alpha)b^\beta_\tau\eta_\beta. \end{aligned}$$

Thanks to the relations

$$\zeta^0_{\alpha\|\beta} = \partial_\beta\zeta^0_\alpha - \Gamma^\sigma_{\alpha\beta}\zeta^0_\sigma - b_{\alpha\beta}\zeta^0_3 \quad \text{and} \quad \zeta^0_{3\|\beta} = \partial_\beta\zeta^0_3 + b^\sigma_\alpha\zeta^0_\sigma,$$
$$u^{(0)}_{\alpha\|3} = \partial_3 u^1_\alpha + b^\sigma_\alpha\zeta^0_\sigma \quad \text{and} \quad u^{(0)}_{3\|3} = \partial_3 u^1_3,$$

the relations giving the functions $F^0_{i\|3}(\boldsymbol{\eta})$ can be rewritten in vector form as

$$\begin{pmatrix} 2F^0_{1\|3}(\boldsymbol{\eta}) \\ 2F^0_{2\|3}(\boldsymbol{\eta}) \\ F^0_{3\|3}(\boldsymbol{\eta}) \end{pmatrix} = \mathbf{L}\begin{pmatrix} b^\alpha_1\eta_\alpha \\ b^\alpha_2\eta_\alpha \\ 0 \end{pmatrix} + \mathbf{N}(\boldsymbol{\eta})\begin{pmatrix} u^{(0)}_{1\|3} \\ u^{(0)}_{2\|3} \\ u^{(0)}_{3\|3} \end{pmatrix} + (1 + u^{(0)}_{3\|3})\begin{pmatrix} \eta_{3\|1} \\ \eta_{3\|2} \\ 0 \end{pmatrix}.$$

for all $\boldsymbol{\eta} \in \mathbf{W}(\omega)$. Hence the announced assertion is established.

(ii) *Given any field* $\boldsymbol{\eta} = (\eta_i) \in \mathbb{T}_{\boldsymbol{\zeta}^0}\boldsymbol{\mathcal{M}}_F(\omega) \subset \mathbf{W}(\omega)$, *let the field* $\boldsymbol{\tau}^0(\boldsymbol{\eta}) = (\tau^0_i(\boldsymbol{\eta})) \in \mathbf{W}(\omega)$ *be defined by*

$$\begin{aligned} \tau^0_1(\boldsymbol{\eta}) &:= b^\alpha_1\eta_\alpha + \frac{1}{d^2}\Big((1 + a^{\alpha 2}\zeta^0_{\alpha\|2})\eta_{3\|1} - a^{\alpha 2}\zeta^0_{\alpha\|1}\eta_{3\|2}\Big) \\ &\qquad + \frac{1}{d}a^{\alpha\beta}\zeta^0_{3\|\beta}\eta_{1\|\alpha}, \\ \tau^0_2(\boldsymbol{\eta}) &:= b^\alpha_2\eta_\alpha + \frac{1}{d^2}\Big(-a^{\alpha 1}\zeta^0_{\alpha\|2}\eta_{3\|1} + (1 + a^{\alpha 1}\zeta^0_{\alpha\|1})\eta_{3\|2}\Big) \\ &\qquad + \frac{1}{d}a^{\alpha\beta}\zeta^0_{3\|\beta}\eta_{2\|\alpha}, \\ \tau^0_3(\boldsymbol{\eta}) &:= \frac{1}{d}a^{\alpha\beta}\zeta^0_{3\|\alpha}\eta_{3\|\beta}. \end{aligned}$$

Then

$$\boldsymbol{v}(\boldsymbol{\eta}) := x_3\boldsymbol{\tau}^0(\boldsymbol{\eta}) \in \mathbf{W}(\Omega) \text{ and } F^{-1}_{i\|3}(\boldsymbol{v}(\boldsymbol{\eta})) = F^0_{i\|3}(\boldsymbol{\eta}).$$

It is easily verified that $\boldsymbol{\tau}^0(\boldsymbol{\eta}) \in \mathbf{W}(\omega)$ since $W^{1,4}(\omega)$ is an algebra; consequently, $x_3\boldsymbol{\tau}^0(\boldsymbol{\eta}) \in \mathbf{W}(\Omega)$.

A straightforward computation, based in particular on the relations

$$\mathbf{M}\mathbf{A}^{-1}\mathbf{M}^T\mathbf{A} = \mathbf{I} - \boldsymbol{z}\boldsymbol{z}^T\mathbf{A} \quad \text{and} \quad 1 + \boldsymbol{z}^T\mathbf{M}^{-T}\mathbf{A}\mathbf{M}^{-1}\boldsymbol{z} = \frac{1}{d^2},$$

shows that if the matrix $\mathbf{M}$ is invertible, then so is the matrix $\mathbf{L}$ with an inverse given by

$$\mathbf{L}^{-1} = \begin{pmatrix} \mathbf{M}^{-1}(\mathbf{I} - \boldsymbol{z}\boldsymbol{z}^T\mathbf{A}) & -d\mathbf{M}^{-1}\boldsymbol{z} \\ \boldsymbol{z}^T\mathbf{A} & d \end{pmatrix}.$$

To establish that $F^{-1}_{i\|3}(\boldsymbol{v}(\boldsymbol{\eta})) = F^0_{i\|3}(\boldsymbol{\eta})$, it suffices by (i) to establish that

$$\begin{pmatrix} \tau^0_1(\boldsymbol{\eta}) \\ \tau^0_2(\boldsymbol{\eta}) \\ \tau^0_3(\boldsymbol{\eta}) \end{pmatrix} = \begin{pmatrix} b^\alpha_1\eta_\alpha \\ b^\alpha_2\eta_\alpha \\ 0 \end{pmatrix} + \mathbf{L}^{-1}\mathbf{N}(\boldsymbol{\eta}) \begin{pmatrix} u^{(0)}_{1\|3} \\ u^{(0)}_{2\|3} \\ u^{(0)}_{3\|3} \end{pmatrix} + (1 + u^{(0)}_{3\|3})\mathbf{L}^{-1} \begin{pmatrix} \eta_{3\|1} \\ \eta_{3\|2} \\ 0 \end{pmatrix},$$

since $\partial_3\boldsymbol{v}(\boldsymbol{\eta}) = \boldsymbol{\tau}^0(\boldsymbol{\eta})$.

Thanks to the expressions of the matrices $\mathbf{L}^{-1}$ and $\mathbf{N}(\boldsymbol{\eta})$ and to the relations

$$\begin{pmatrix} u^{(0)}_{1\|3} \\ u^{(0)}_{2\|3} \end{pmatrix} = -d\mathbf{M}^{-1}\boldsymbol{z}, \quad 1 + u^{(0)}_{3\|3} = d, \quad \mathbf{M}\mathbf{A}^{-1}\mathbf{M}^T\mathbf{A} = \mathbf{I} - \boldsymbol{z}\boldsymbol{z}^T\mathbf{A},$$

it is first found that

$$\mathbf{L}^{-1}\mathbf{N}(\boldsymbol{\eta}) \begin{pmatrix} u^{(0)}_{1\|3} \\ u^{(0)}_{2\|3} \\ u^{(0)}_{3\|3} \end{pmatrix} + (1 + u^{(0)}_{3\|3})\mathbf{L}^{-1} \begin{pmatrix} \eta_{3\|1} \\ \eta_{3\|2} \\ 0 \end{pmatrix}$$

$$= d \begin{pmatrix} \mathbf{A}^{-1}\mathbf{M}^T\mathbf{A}\{-\mathbf{Z}(\boldsymbol{\eta})^T\mathbf{A}\mathbf{M}^{-1}\boldsymbol{z} + \boldsymbol{z}(\boldsymbol{\eta})\} \\ -\boldsymbol{z}^T\mathbf{A}\mathbf{Z}(\boldsymbol{\eta})^T\mathbf{A}\mathbf{M}^{-1}\boldsymbol{z} + \boldsymbol{z}^T\mathbf{A}\boldsymbol{z}(\boldsymbol{\eta}) \end{pmatrix}.$$

It thus remains to properly incorporate into the right-hand side of the above vector equation the relations

$$F^0_{\alpha\|\beta}(\boldsymbol{\eta}) = \frac{1}{2}(\eta_{\alpha\|\beta} + \eta_{\beta\|\alpha} + a^{mn}\{\zeta^0_{m\|\alpha}\eta_{n\|\beta} + \zeta^0_{n\|\beta}\eta_{m\|\alpha}\}) = 0$$

satisfied by the function $\boldsymbol{\eta} \in \mathbb{T}_{\zeta^0}\mathcal{M}_F(\omega)$. Another straightforward computation shows that these may be equivalently written in matrix form as

$$\mathbf{MZ}(\boldsymbol{\eta}) + \mathbf{Z}(\boldsymbol{\eta})^T\mathbf{M}^T + \boldsymbol{z}\boldsymbol{z}(\boldsymbol{\eta})^T + \boldsymbol{z}(\boldsymbol{\eta})\boldsymbol{z}^T = \mathbf{0}.$$

Together with the relations

$$\mathbf{M}^{-T}\mathbf{A}\mathbf{M}^{-1}\boldsymbol{z} = \frac{1}{d^2}\mathbf{A}\boldsymbol{z} \text{ and } \boldsymbol{z}^T\mathbf{A}\boldsymbol{z} = 1 - d^2,$$

this matrix equation shows that

$$\begin{aligned}
-\boldsymbol{z}^T\mathbf{A}\mathbf{Z}(\boldsymbol{\eta})^T\mathbf{A}\mathbf{M}^{-1}\boldsymbol{z} &= -\frac{1}{2d^2}\boldsymbol{z}^T\mathbf{A}\{\mathbf{MZ}(\boldsymbol{\eta}) + \mathbf{Z}(\boldsymbol{\eta})^T\mathbf{M}^T\}\mathbf{A}\boldsymbol{z} \\
&= \frac{1}{2d^2}\boldsymbol{z}^T\mathbf{A}\{\boldsymbol{z}(\boldsymbol{\eta})\boldsymbol{z}^T + \boldsymbol{z}\boldsymbol{z}(\boldsymbol{\eta})^T\}\mathbf{A}\boldsymbol{z} \\
&= \frac{1}{d^2}\boldsymbol{z}^T\mathbf{A}\boldsymbol{z}(\boldsymbol{\eta})\boldsymbol{z}^T\mathbf{A}\boldsymbol{z} = \left(\frac{1}{d^2} - 1\right)\boldsymbol{z}^T\mathbf{A}\boldsymbol{z}(\boldsymbol{\eta}),
\end{aligned}$$

hence that

$$d\Big(- \boldsymbol{z}^T\mathbf{A}\mathbf{Z}(\boldsymbol{\eta})^T\mathbf{A}\mathbf{M}^{-1}\boldsymbol{z} + \boldsymbol{z}^T\mathbf{A}\boldsymbol{z}(\boldsymbol{\eta})\Big) = \frac{1}{d}\boldsymbol{z}^T\mathbf{A}\boldsymbol{z}(\boldsymbol{\eta}),$$

which takes care of the third component in the vector equation.

The relation $\mathbf{M}\mathbf{A}^{-1}\mathbf{M}^T\mathbf{A} = \mathbf{I} - \boldsymbol{z}\boldsymbol{z}^T\mathbf{A}$ shows that the vector formed by the first two components in the same vector equation is of the form

$$\begin{aligned}
&d\Big(\mathbf{A}^{-1}\mathbf{M}^T\mathbf{A}\{-\mathbf{Z}(\boldsymbol{\eta})^T\mathbf{A}\mathbf{M}^{-1}\boldsymbol{z} + \boldsymbol{z}(\boldsymbol{\eta})\}\Big) \\
&= d\Big(\mathbf{M}^{-1}\boldsymbol{z}\boldsymbol{z}^T\mathbf{A}\mathbf{Z}(\boldsymbol{\eta})^T\mathbf{A}\mathbf{M}^{-1}\boldsymbol{z} - \mathbf{M}^{-1}\boldsymbol{z}\boldsymbol{z}^T\mathbf{A}\boldsymbol{z}(\boldsymbol{\eta})\Big) \\
&\qquad -d\Big(- \mathbf{M}^{-1}\mathbf{Z}(\boldsymbol{\eta})^T\mathbf{A}\mathbf{M}^{-1}\boldsymbol{z} + \mathbf{M}^{-1}\boldsymbol{z}(\boldsymbol{\eta})\Big).
\end{aligned}$$

Let us transform this expression. The relation (which has just been established)

$$\boldsymbol{z}^T\mathbf{A}\mathbf{Z}(\boldsymbol{\eta})^T\mathbf{A}\mathbf{M}^{-1}\boldsymbol{z} = \left(1 - \frac{1}{d^2}\right)\boldsymbol{z}^T\mathbf{A}\boldsymbol{z}(\boldsymbol{\eta})$$

implies that

$$\begin{aligned}
&\mathbf{M}^{-1}\boldsymbol{z}\boldsymbol{z}^T\mathbf{A}\mathbf{Z}(\boldsymbol{\eta})^T\mathbf{A}\mathbf{M}^{-1}\boldsymbol{z} - \mathbf{M}^{-1}\boldsymbol{z}\boldsymbol{z}^T\mathbf{A}\boldsymbol{z}(\boldsymbol{\eta}) \\
&\quad = -\frac{1}{d^2}\mathbf{M}^{-1}\boldsymbol{z}\boldsymbol{z}^T\mathbf{A}\boldsymbol{z}(\boldsymbol{\eta}) = -\frac{1}{d^2}\mathbf{M}^{-1}\boldsymbol{z}\boldsymbol{z}(\boldsymbol{\eta})^T\mathbf{A}\boldsymbol{z}
\end{aligned}$$

on the one hand; the matrix equation expressing that $F^0_{\alpha\|\beta}(\boldsymbol{\eta}) = 0$ together with the relations $\mathbf{M}^{-T}\mathbf{A}\mathbf{M}^{-1}\boldsymbol{z} = \frac{1}{d^2}\mathbf{A}\boldsymbol{z}$ and $\boldsymbol{z}^T\mathbf{A}\boldsymbol{z} = 1-d^2$ imply that

$$\begin{aligned}
-\mathbf{M}^{-1}\mathbf{Z}(\boldsymbol{\eta})^T\mathbf{A}\mathbf{M}^{-1}\boldsymbol{z} &= -\frac{1}{d^2}\mathbf{M}^{-1}\mathbf{Z}(\boldsymbol{\eta})^T\mathbf{M}^T\mathbf{A}\boldsymbol{z} \\
&= \frac{1}{d^2}\mathbf{Z}(\boldsymbol{\eta})\mathbf{A}\boldsymbol{z} + \frac{1}{d^2}\mathbf{M}^{-1}\boldsymbol{z}(\boldsymbol{\eta})\boldsymbol{z}^T\mathbf{A}\boldsymbol{z} + \frac{1}{d^2}\mathbf{M}^{-1}\boldsymbol{z}\boldsymbol{z}(\boldsymbol{\eta})^T\mathbf{A}\boldsymbol{z} \\
&= \frac{1}{d^2}\mathbf{Z}(\boldsymbol{\eta})\mathbf{A}\boldsymbol{z} + \left(\frac{1}{d^2} - 1\right)\mathbf{M}^{-1}\boldsymbol{z}(\boldsymbol{\eta}) + \frac{1}{d^2}\mathbf{M}^{-1}\boldsymbol{z}\boldsymbol{z}(\boldsymbol{\eta})^T\mathbf{A}\boldsymbol{z}
\end{aligned}$$

on the other. These relations thus imply that

$$\begin{aligned}
&d\Big(\mathbf{M}^{-1}\boldsymbol{z}\boldsymbol{z}^T\mathbf{A}\mathbf{Z}(\boldsymbol{\eta})^T\mathbf{A}\mathbf{M}^{-1}\boldsymbol{z} - \mathbf{M}^{-1}\boldsymbol{z}\boldsymbol{z}^T\mathbf{A}\boldsymbol{z}(\boldsymbol{\eta})\Big) \\
&\quad -d\Big(\mathbf{M}^{-1}\mathbf{Z}(\boldsymbol{\eta})^T\mathbf{A}\mathbf{M}^{-1}\boldsymbol{z} + \mathbf{M}^{-1}\boldsymbol{z}(\boldsymbol{\eta})\Big) = \frac{1}{d}\mathbf{Z}(\boldsymbol{\eta})\mathbf{A}\boldsymbol{z} + \frac{1}{d}\mathbf{M}^{-1}\boldsymbol{z}(\boldsymbol{\eta}).
\end{aligned}$$

In this fashion, we have established that

$$\begin{pmatrix} b_1^\alpha \eta_\alpha \\ b_2^\alpha \eta_\alpha \\ 0 \end{pmatrix} + \mathbf{L}^{-1}\mathbf{N}(\boldsymbol{\eta}) \begin{pmatrix} u^{(0)}_{1\|3} \\ u^{(0)}_{2\|3} \\ u^{(0)}_{3\|3} \end{pmatrix} + (1 + u^{(0)}_{3\|3})\mathbf{L}^{-1} \begin{pmatrix} \eta_{3\|1} \\ \eta_{3\|2} \\ 0 \end{pmatrix} = \begin{pmatrix} \tau_1^0(\boldsymbol{\eta}) \\ \tau_2^0(\boldsymbol{\eta}) \\ \tau_3^0(\boldsymbol{\eta}) \end{pmatrix},$$

where

$$\begin{aligned}
\begin{pmatrix} \tau_1^0(\boldsymbol{\eta}) \\ \tau_2^0(\boldsymbol{\eta}) \end{pmatrix} &= \begin{pmatrix} b_1^\alpha \eta_\alpha \\ b_2^\alpha \eta_\alpha \end{pmatrix} + \frac{1}{d}\mathbf{Z}(\boldsymbol{\eta})\mathbf{A}\boldsymbol{z} + \frac{1}{d}\mathbf{M}^{-1}\boldsymbol{z}(\boldsymbol{\eta}), \\
\tau_3(\boldsymbol{\eta}) &= \frac{1}{d}\boldsymbol{z}^T\mathbf{A}\boldsymbol{z}(\boldsymbol{\eta}).
\end{aligned}$$

It then suffices to combine the definitions of the vector fields $\boldsymbol{z}$ and $\boldsymbol{z}(\boldsymbol{\eta})$ and of the matrix fields $\mathbf{A}$ and $\mathbf{Z}(\boldsymbol{\eta})$ with the expressions

$$\mathbf{M}^{-1} = \frac{1}{d}\begin{pmatrix} 1 + a^{\alpha 2}\zeta^0_{\alpha\|2} & -a^{\alpha 2}\zeta^0_{\alpha\|1} \\ -a^{\alpha 1}\zeta^0_{\alpha\|2} & 1 + a^{\alpha 1}\zeta^0_{\alpha\|1} \end{pmatrix},$$

in order to obtain the announced expressions, viz.,

$$\tau_1^0(\boldsymbol{\eta}) = b_1^\alpha \eta_\alpha + \frac{1}{d^2}\Big((1 + a^{\alpha 2}\zeta^0_{\alpha\|2})\eta_{3\|1} - a^{\alpha 2}\zeta^0_{\alpha\|1}\eta_{3\|2}\Big) + \frac{1}{d}a^{\alpha\beta}\zeta^0_{3\|\beta}\eta_{1\|\alpha},$$

$$\tau_2^0(\boldsymbol{\eta}) = b_2^\alpha \eta_\alpha + \frac{1}{d^2}\Big(-a^{\alpha 1}\zeta^0_{\alpha\|2}\eta_{3\|1} + (1 + a^{\alpha 1}\zeta^0_{\alpha\|1})\eta_{3\|2}\Big) + \frac{1}{d}a^{\alpha\beta}\zeta^0_{3\|\beta}\eta_{2\|\alpha},$$

$$\tau_3^0(\boldsymbol{\eta}) = \frac{1}{d}a^{\alpha\beta}\zeta^0_{3\|\alpha}\eta_{3\|\beta}.$$

(iii) *Let the mapping* $\boldsymbol{\psi} = (\psi_i) : \mathbf{W}^{2,4}(\omega) \to \mathbf{W}^{1,4}(\omega)$ *be defined for all* $\boldsymbol{\eta} = (\eta_i) \in \mathbf{W}^{2,4}(\omega)$ *by*

$$\begin{aligned}
\psi_1(\boldsymbol{\eta}) &:= b_1^\alpha \eta_\alpha + (1 + a^{\alpha 2}\eta_{\alpha\|2})\eta_{3\|1} - a^{\alpha 2}\eta_{\alpha\|1}\eta_{3\|2},\\
\psi_2(\boldsymbol{\eta}) &:= b_2^\alpha \eta_\alpha + (1 + a^{\alpha 1}\eta_{\alpha\|1})\eta_{3\|2} - a^{\alpha 1}\eta_{\alpha\|2}\eta_{3\|1},\\
\psi_3(\boldsymbol{\eta}) &:= -a^{\alpha\beta}\eta_{\alpha\|\beta} - a^{\alpha 1}a^{\beta 2}(\eta_{\alpha\|1}\eta_{\beta\|2} - \eta_{\alpha\|2}\eta_{\beta\|1}),
\end{aligned}$$

and let the mapping $\boldsymbol{\varphi} = (\varphi_i) : \mathbf{W}^{2,4}(\omega) \times \mathbf{W}^{2,4}(\omega) \to \mathbf{W}^{1,4}(\omega)$ *be defined for all* $\boldsymbol{\zeta} = (\zeta_i) \in \mathbf{W}^{2,4}(\omega)$ *and* $\boldsymbol{\eta} = (\eta_i) \in \mathbf{W}^{2,4}(\omega)$ *by*

$$\begin{aligned}
\varphi_1(\boldsymbol{\zeta}, \boldsymbol{\eta}) &:= b_1^\alpha \eta_\alpha + \eta_{3\|1}\\
&\quad + a^{\alpha 2}(\zeta_{\alpha\|2}\eta_{3\|1} + \zeta_{3\|1}\eta_{\alpha\|2} - \zeta_{3\|2}\eta_{\alpha\|1} - \zeta_{\alpha\|1}\eta_{3\|2}),\\
\varphi_2(\boldsymbol{\zeta}, \boldsymbol{\eta}) &:= b_2^\alpha \eta_\alpha + \eta_{3\|2}\\
&\quad + a^{\alpha 1}(\zeta_{\alpha\|1}\eta_{3\|2} + \zeta_{3\|2}\eta_{\alpha\|1} - \zeta_{3\|1}\eta_{\alpha\|2} - \zeta_{\alpha\|2}\eta_{3\|1}),\\
\varphi_3(\boldsymbol{\zeta}, \boldsymbol{\eta}) &:= -a^{\alpha\beta}\eta_{\alpha\|\beta}\\
&\quad - a^{\alpha 1}a^{\beta 2}(\zeta_{\alpha\|1}\eta_{\beta\|2} + \zeta_{\beta\|2}\eta_{\alpha\|1} - \zeta_{\alpha\|2}\eta_{\beta\|1} - \zeta_{\beta\|1}\eta_{\alpha\|2}).
\end{aligned}$$

Then

$$\boldsymbol{\psi}'(\boldsymbol{\zeta})\boldsymbol{\eta} = \boldsymbol{\varphi}(\boldsymbol{\zeta}, \boldsymbol{\eta}) \text{ for all } \boldsymbol{\zeta} \in \boldsymbol{\mathcal{M}}_F(\omega),\ \boldsymbol{\eta} \in \mathbb{T}_{\boldsymbol{\zeta}}\boldsymbol{\mathcal{M}}_F(\omega).$$

In addition, the field $\boldsymbol{\tau}^0(\boldsymbol{\eta}) = (\tau_i^0(\boldsymbol{\eta})) \in \mathbf{W}(\omega)$ *defined in* (ii) *satisfies*

$$\boldsymbol{\tau}^0(\boldsymbol{\eta}) = \boldsymbol{\psi}'(\boldsymbol{\zeta}^0)\boldsymbol{\eta} \text{ for all } \boldsymbol{\eta} \in \mathbb{T}_{\boldsymbol{\zeta}^0}\boldsymbol{\mathcal{M}}_F(\omega).$$

Consequently,

$$\boldsymbol{\tau}^0(\boldsymbol{\eta}) = \boldsymbol{\varphi}^0(\boldsymbol{\eta}), \text{ where } \boldsymbol{\varphi}^0(\boldsymbol{\eta}) := \boldsymbol{\varphi}(\boldsymbol{\zeta}^0, \boldsymbol{\eta}).$$

The proof of the above assertions is given for $\boldsymbol{\zeta} = \boldsymbol{\zeta}^0$. It is easy to check that it likewise holds for any $\boldsymbol{\zeta} \in \boldsymbol{\mathcal{M}}_F(\omega)$.

The definition of the function $\psi_3(\boldsymbol{\eta})$ shows that

$$\psi_3(\boldsymbol{\eta}) = 1 - d(\boldsymbol{\eta}).$$

To compute the Gâteaux derivative $d'(\boldsymbol{\zeta}^0)\boldsymbol{\eta}$, we do not use the definition $d(\boldsymbol{\eta}) = \det \mathbf{M}(\boldsymbol{\eta})$; we use instead the relation

$$d(\boldsymbol{\eta})^2 = 1 - \boldsymbol{z}(\boldsymbol{\eta})^T \mathbf{A} \boldsymbol{z}(\boldsymbol{\eta}),$$

also satisfied by $d(\boldsymbol{\eta})$. This gives

$$d(\boldsymbol{\zeta}^0) d'(\boldsymbol{\zeta}^0)\boldsymbol{\eta} = -\boldsymbol{z}(\boldsymbol{\zeta}^0)^T \mathbf{A} \boldsymbol{z}(\boldsymbol{\eta});$$

hence

$$\psi_3'(\boldsymbol{\zeta}^0)\boldsymbol{\eta} = -d'(\boldsymbol{\zeta}^0)\boldsymbol{\eta} = \frac{1}{d}\boldsymbol{z}^T \mathbf{A} \boldsymbol{z}(\boldsymbol{\eta}) = \tau_3^0(\boldsymbol{\eta}).$$

While the above computation is valid for any $\boldsymbol{\eta} \in \mathbf{W}^{2,4}(\omega)$, the next one makes a crucial use of the relations $F^0_{\alpha\|\beta}(\boldsymbol{\eta}) = 0$ in ω satisfied by any $\boldsymbol{\eta} \in \mathbb{T}_{\boldsymbol{\zeta}^0}\boldsymbol{\mathcal{M}}_F(\omega) \subset \mathbf{W}^{2,4}(\omega)$; incidentally, this explains why the preceding computation is substantially simpler to carry out than the next one.

The definition of the functions $\psi_\alpha(\boldsymbol{\eta})$ shows that

$$\begin{pmatrix} \psi_1(\boldsymbol{\eta}) \\ \psi_2(\boldsymbol{\eta}) \end{pmatrix} = \begin{pmatrix} b_1^\alpha \eta_\alpha \\ b_2^\alpha \eta_\alpha \end{pmatrix} + \hat{\boldsymbol{\psi}}(\boldsymbol{\eta}),$$

where

$$\hat{\boldsymbol{\psi}}(\boldsymbol{\eta}) = (\hat{\psi}_\alpha(\boldsymbol{\eta})) := d(\boldsymbol{\eta}) \mathbf{M}^{-1}(\boldsymbol{\eta}) \boldsymbol{z}(\boldsymbol{\eta}).$$

To compute the Gâteaux derivative $\hat{\boldsymbol{\psi}}'(\boldsymbol{\zeta}^0)\boldsymbol{\eta}$, we first use the formula for the Gâteaux derivative of the inverse matrix mapping, viz. (see, e.g., Vol. I, Sect. 1.2),

$$\left(\mathbf{M}^{-1}(\boldsymbol{\zeta}^0)\right)' \boldsymbol{\eta} = -\mathbf{M}^{-1}(\boldsymbol{\zeta}^0) \left(\mathbf{M}'(\boldsymbol{\zeta}^0)\boldsymbol{\eta}\right) \mathbf{M}^{-1}(\boldsymbol{\zeta}^0),$$

where

$$\mathbf{M}'(\boldsymbol{\zeta}^0)\boldsymbol{\eta} = \begin{pmatrix} a^{\alpha 1}\eta_{\alpha\|1} & a^{\alpha 2}\eta_{\alpha\|1} \\ a^{\alpha 1}\eta_{\alpha\|2} & a^{\alpha 2}\eta_{\alpha\|2} \end{pmatrix} = \mathbf{Z}(\boldsymbol{\eta})^T \mathbf{A}.$$

Combining this formula with that giving the derivative of a trilinear mapping and with the expression found *supra* for $d'(\boldsymbol{\zeta}^0)\boldsymbol{\eta}$ (when computing $\psi_3'(\boldsymbol{\zeta}^0)\boldsymbol{\eta}$; this is why we began our computation with the third component), we obtain

$$\begin{aligned}\hat{\boldsymbol{\psi}}'(\boldsymbol{\zeta}^0)\boldsymbol{\eta} &= (d'(\boldsymbol{\zeta}^0)\boldsymbol{\eta})\mathbf{M}^{-1}(\boldsymbol{\zeta}^0)\boldsymbol{z}(\boldsymbol{\zeta}^0)\\ &\quad -d(\boldsymbol{\zeta}^0)\mathbf{M}^{-1}(\boldsymbol{\zeta}^0)\mathbf{Z}(\boldsymbol{\eta})^T\mathbf{A}\mathbf{M}^{-1}(\boldsymbol{\zeta}^0)\boldsymbol{z}(\boldsymbol{\zeta}^0)\\ &\quad +d(\boldsymbol{\zeta}^0)\mathbf{M}^{-1}(\boldsymbol{\zeta}^0)\boldsymbol{z}(\boldsymbol{\eta})\\ &= -\frac{1}{d}(\boldsymbol{z}(\boldsymbol{\eta})^T\mathbf{A}\boldsymbol{z})\mathbf{M}^{-1}\boldsymbol{z} - d\mathbf{M}^{-1}\mathbf{Z}(\boldsymbol{\eta})^T\mathbf{A}\mathbf{M}^{-1}\boldsymbol{z}\\ &\quad +d\mathbf{M}^{-1}\boldsymbol{z}(\boldsymbol{\eta}).\end{aligned}$$

Incorporating the relations $F^0_{\alpha\|\beta}(\boldsymbol{\eta}) = 0$ under their equivalent matrix formulation

$$\mathbf{Z}(\boldsymbol{\eta})^T = -\mathbf{M}\mathbf{Z}(\boldsymbol{\eta})\mathbf{M}^{-T} - \boldsymbol{z}(\boldsymbol{\eta})\boldsymbol{z}^T\mathbf{M}^{-T} - \boldsymbol{z}\boldsymbol{z}(\boldsymbol{\eta})^T\mathbf{M}^{-T},$$

and noting that

$$\mathbf{M}^{-T}\mathbf{A}\mathbf{M}^{-1}\boldsymbol{z} = \frac{1}{d^2}\mathbf{A}\boldsymbol{z} \quad \text{and} \quad 1 + \boldsymbol{z}^T\mathbf{M}^{-T}\mathbf{A}\mathbf{M}^{-1}\boldsymbol{z} = \frac{1}{d^2},$$

we infer that

$$\begin{aligned}-\mathbf{M}^{-1}\mathbf{Z}(\boldsymbol{\eta})^T\mathbf{A}\mathbf{M}^{-1}\boldsymbol{z} &= \frac{1}{d^2}\mathbf{Z}(\boldsymbol{\eta})\mathbf{A}\boldsymbol{z}\\ &\quad +\left(\frac{1}{d^2} - 1\right)\mathbf{M}^{-1}\boldsymbol{z}(\boldsymbol{\eta}) + \frac{1}{d^2}\mathbf{M}^{-1}\boldsymbol{z}\boldsymbol{z}(\boldsymbol{\eta})^T\mathbf{A}\boldsymbol{z}.\end{aligned}$$

Consequently,

$$\begin{aligned}\begin{pmatrix}\hat{\psi}_1'(\boldsymbol{\zeta}^0)\boldsymbol{\eta}\\ \hat{\psi}_2'(\boldsymbol{\zeta}^0)\boldsymbol{\eta}\end{pmatrix} = \hat{\boldsymbol{\psi}}'(\boldsymbol{\zeta}^0)\boldsymbol{\eta} &= \frac{1}{d}\mathbf{Z}(\boldsymbol{\eta})\mathbf{A}\boldsymbol{z} + \frac{1}{d}\mathbf{M}^{-1}\boldsymbol{z}(\boldsymbol{\eta})\\ &= \begin{pmatrix}\tau_1^0(\boldsymbol{\eta})\\ \tau_2^0(\boldsymbol{\eta})\end{pmatrix} - \begin{pmatrix}b_1^\alpha\eta_\alpha\\ b_2^\alpha\eta_\alpha\end{pmatrix}.\end{aligned}$$

This relation shows that $\psi_\alpha'(\boldsymbol{\zeta}^0)\boldsymbol{\eta} = \tau_\alpha^0(\boldsymbol{\eta})$. We have thus established that $\psi_i'(\boldsymbol{\zeta}^0)\boldsymbol{\eta} = \tau_i^0(\boldsymbol{\eta})$.

Let us next directly compute the Gâteaux derivatives $\psi_i'(\boldsymbol{\zeta}^0)\boldsymbol{\eta}$. The functions $\psi_i : W^{2,4}(\omega) \to W^{1,4}(\omega)$ being sums of linear and quadratic mappings, this easy task produces the announced relations $\psi_i'(\boldsymbol{\zeta}^0)\boldsymbol{\eta} = \varphi_i^0(\boldsymbol{\eta})$.

Finally, it must be verified that, when $\boldsymbol{\eta} \in \mathbb{T}_{\boldsymbol{\zeta}^0}\boldsymbol{\mathcal{M}}_F(\omega)$, the field $\boldsymbol{\varphi}^0(\boldsymbol{\eta})$ belongs to the space $\mathbf{W}(\omega)$. To this end, it suffices to note that the relations $\boldsymbol{\eta} \in \mathbf{W}^{2,4}(\omega)$ and $\boldsymbol{\eta} = \partial_\nu \boldsymbol{\eta} = \mathbf{0}$ on γ_0 imply that

$$\begin{aligned}\partial_\beta \eta_\alpha - \Gamma^\sigma_{\alpha\beta}\eta_\sigma - b_{\alpha\beta}\eta_3 &= \eta_{\alpha\|\beta} \in W^{1,4}(\omega) \text{ and } \eta_{\alpha\|\beta} = 0 \text{ on } \gamma_0,\\ \partial_\alpha \eta_3 + b^\sigma_\alpha \eta_\sigma &= \eta_{3\|\alpha} \in W^{1,4}(\omega) \text{ and } \eta_{3\|\alpha} = 0 \text{ on } \gamma_0,\end{aligned}$$

and that $W^{1,4}(\omega)$ is an algebra. Hence $\boldsymbol{\varphi}^0(\boldsymbol{\eta}) \in \mathbf{W}(\omega)$ and consequently $\boldsymbol{v}(\boldsymbol{\eta}) = x_3\boldsymbol{\varphi}^0(\boldsymbol{\eta}) \in \mathbf{W}(\Omega)$ when $\boldsymbol{\eta} \in \mathbb{T}_{\boldsymbol{\zeta}^0}\boldsymbol{\mathcal{M}}_F(\omega)$. ■

We conclude this analysis by establishing the second property stated without proof in part (vi) of the proof of Thm. 10.1-2. The expressions of the functions $\hat{F}^1_{\alpha\|\beta}(\boldsymbol{\eta})$ are not provided below simply because they are nowhere needed.

Theorem 10.1-4. *Let the notations be as in* Thm. 10.1-3. *Given any* $\boldsymbol{\eta} = (\eta_i) \in \mathbb{T}_{\boldsymbol{\zeta}^0}\boldsymbol{\mathcal{M}}_F$, *let the functions* $\boldsymbol{\psi}(\boldsymbol{\eta}) = (\psi_i(\boldsymbol{\eta})) \in \mathbf{W}(\omega)$ *and* $\boldsymbol{\varphi}^0(\boldsymbol{\eta}) = (\varphi^0_i(\boldsymbol{\eta})) \in \mathbf{W}(\omega)$ *be defined as in* ibid. *Then*

$$\boxed{F^1_{\alpha\|\beta}(\boldsymbol{\eta}) - F^0_{\alpha\|\beta}(\boldsymbol{v}(\boldsymbol{\eta})) = \hat{F}^1_{\alpha\|\beta}(\boldsymbol{\eta}) - x_3\hat{F}^0_{\alpha\|\beta}(\boldsymbol{\eta}),}$$

where the functions $\hat{F}^1_{\alpha\|\beta}(\boldsymbol{\eta}) \in L^2(\omega)$ *are independent of* x_3 *and the functions* $\hat{F}^0_{\alpha\|\beta}(\boldsymbol{\eta}) \in L^2(\omega)$, *which are also independent of* x_3, *are defined by*

$$\begin{aligned}\hat{F}^0_{\alpha\|\beta}(\boldsymbol{\eta}) := {}& F^0_{\alpha\|\beta}(\boldsymbol{\varphi}^0(\boldsymbol{\eta})) - \frac{1}{2}g^{\sigma\tau,1}\{\zeta^0_{\sigma\|\alpha}\eta_{\tau\|\beta} + \zeta^0_{\tau\|\beta}\eta_{\sigma\|\alpha}\}\\ &+ \big(\Gamma^{p,1}_{\alpha\beta} + \frac{1}{2}a^{mn}\{\Gamma^{p,1}_{m\alpha}\zeta^0_{n\|\beta} + \Gamma^{p,1}_{n\beta}\zeta^0_{m\|\alpha}\}\big)\eta_p\\ &+ \frac{1}{2}a^{mn}\{(\psi^0_{m\|\alpha} + \Gamma^{p,1}_{m\alpha}\zeta^0_p)\eta_{n\|\beta} + (\psi^0_{n\|\beta} + \Gamma^{q,1}_{n\beta}\zeta^0_q)\eta_{m\|\alpha}\},\end{aligned}$$

and the functions ψ^0_i *are given by*

$$\psi^0_i = \psi_i(\boldsymbol{\zeta}^0),$$

where $\boldsymbol{\psi}(\boldsymbol{\eta}) \in \mathbf{W}^{1,4}(\omega)$ *is defined for any* $\boldsymbol{\eta} \in \mathbf{W}^{2,4}(\omega)$ *by*

$$\begin{aligned}\psi_1(\boldsymbol{\eta}) &:= b^\alpha_1\eta_\alpha + (1 + a^{\alpha 2}\eta_{\alpha\|2})\eta_{3\|1} - a^{\alpha 2}\eta_{\alpha\|1}\eta_{3\|2},\\ \psi_2(\boldsymbol{\eta}) &:= b^\alpha_2\eta_\alpha + (1 + a^{\alpha 1}\eta_{\alpha\|1})\eta_{3\|2} - a^{\alpha 1}\eta_{\alpha\|2}\eta_{3\|1},\\ \psi_3(\boldsymbol{\eta}) &:= -a^{\alpha\beta}\eta_{\alpha\|\beta} - a^{\alpha 1}a^{\beta 2}(\eta_{\alpha\|1}\eta_{\beta\|2} - \eta_{\alpha\|2}\eta_{\beta\|1}).\end{aligned}$$

Proof. By definition, the functions $F^1_{\alpha\|\beta}(\boldsymbol{v})$ are for each $\boldsymbol{v} \in \mathbf{W}(\Omega)$ the coefficients of ε in the formal asymptotic expansions induced by that of the scaled unknown $\boldsymbol{u}(\varepsilon) = \boldsymbol{u}^0 + \varepsilon \boldsymbol{u}^1 + \cdots$ in the functions

$$\begin{aligned} F_{\alpha\|\beta}(\varepsilon;\, \boldsymbol{u}(\varepsilon),\, \boldsymbol{v}) &= \frac{1}{2}(v_{\alpha\|\beta}(\varepsilon) + v_{\beta\|\alpha}(\varepsilon)) \\ &+ \frac{1}{2} g^{mn}(\varepsilon)\{u_{m\|\alpha}(\varepsilon) v_{n\|\beta}(\varepsilon) + u_{n\|\beta}(\varepsilon) v_{m\|\alpha}(\varepsilon)\}, \end{aligned}$$

where (the functions $\Gamma^{p,0}_{m\alpha}$, $\Gamma^{p,1}_{m\alpha}$, a^{mn}, and $g^{mn,1}$ are defined in Thm. 8.5-1):

$$\begin{aligned} v_{m\|\alpha}(\varepsilon) &= v_{m\|\alpha} - \varepsilon x_3 \Gamma^{p,1}_{m\alpha} v_p + \cdots, \\ u_{m\|\alpha}(\varepsilon) &= u^0_{m\|\alpha} + \varepsilon(\partial_\alpha u^1_m - \Gamma^{p,0}_{m\alpha} u^1_p - x_3 \Gamma^{p,1}_{m\alpha} u^0_p) + \cdots, \\ v_{m\|\alpha} &= \partial_\alpha v_m - \Gamma^{p,0}_{m\alpha} v_p, \quad u^0_{m\|\alpha} = \partial_\alpha \zeta^0_m - \Gamma^{p,0}_{m\alpha} \zeta^0_p, \\ g^{mn}(\varepsilon) &= a^{mn} + \varepsilon x_3 g^{mn,1} + \cdots. \end{aligned}$$

A simple computation, which in particular takes into account that $\boldsymbol{u}^0 = \boldsymbol{\zeta}^0$ and $\boldsymbol{u}^1 = \boldsymbol{\zeta}^1 - x_3 \boldsymbol{\psi}^0$ with

$$\boldsymbol{\zeta}^0 = (\zeta^0_i) \in \boldsymbol{\mathcal{M}}_F(\omega),\ \boldsymbol{\zeta}^1 = (\zeta^1_i) \in \mathbf{W}(\omega),\ \text{and } \boldsymbol{\psi}^0 = (\psi^0_i) \in \mathbf{W}(\omega)$$

(see part (i) of the proof of Thm. 10.1-2), then shows that, for any field $\boldsymbol{\eta} \in \mathbb{T}_{\boldsymbol{\zeta}^0} \boldsymbol{\mathcal{M}}_F(\omega) \subset \mathbf{W}(\omega)$ identified with a field $\boldsymbol{\eta} \in \mathbf{W}(\Omega)$,

$$\begin{aligned} F^1_{\alpha\|\beta}(\boldsymbol{\eta}) &= \frac{1}{2} a^{mn}\{\zeta^1_{m\|\alpha}\eta_{n\|\beta} + \zeta^1_{n\|\beta}\eta_{m\|\alpha}\} \\ &+ \frac{x_3}{2} g^{\sigma\tau,1}\{\zeta^0_{\sigma\|\alpha}\eta_{\tau\|\beta} + \zeta^0_{\tau\|\beta}\eta_{\sigma\|\alpha}\} \\ &- x_3\left(\Gamma^{p,1}_{\alpha\beta} + \frac{1}{2} a^{mn}\{\Gamma^{p,1}_{m\alpha}\zeta^0_{n\|\beta} + \Gamma^{p,1}_{n\beta}\zeta^0_{m\|\alpha}\}\right)\eta_p \\ &- \frac{x_3}{2} a^{mn}\{(\psi^0_{m\|\alpha} + \Gamma^{p,1}_{m\alpha}\zeta^0_p)\eta_{n\|\beta} + (\psi^0_{n\|\beta} + \Gamma^{q,1}_{n\beta}\zeta^0_q)\eta_{m\|\alpha}\} \end{aligned}$$

on the one hand; the definition $\boldsymbol{v}(\boldsymbol{\eta}) = x_3 \boldsymbol{\varphi}^0(\boldsymbol{\eta})$ shows that

$$F^0_{\alpha\|\beta}(\boldsymbol{v}(\boldsymbol{\eta})) = x_3 F^0_{\alpha\|\beta}(\boldsymbol{\varphi}^0(\boldsymbol{\eta}))$$

on the other hand. Subtracting these expressions thus gives

$$F^1_{\alpha\|\beta}(\boldsymbol{\eta}) - F^0_{\alpha\|\beta}(\boldsymbol{v}(\boldsymbol{\eta})) = \hat{F}^1_{\alpha\|\beta}(\boldsymbol{\eta}) - x_3 \hat{F}^0_{\alpha\|\beta}(\boldsymbol{\eta}),$$

where the functions

$$\hat{F}^1_{\alpha\|\beta}(\boldsymbol{\eta}) := \frac{1}{2} a^{mn} \{\zeta^1_{m\|\alpha} \eta_{n\|\beta} + \zeta^1_{n\|\beta} \eta_{m\|\alpha}\} \in L^2(\omega)$$

are indeed independent of x_3 and the functions $\hat{F}^0_{\alpha\|\beta}(\boldsymbol{\eta}) \in L^2(\omega)$ have the announced expressions and are thus also independent of ε. ■

The proof of Thm. 10.1-2 is thus complete, but the analysis is far from being terminated: There remains to express the variational equations satisfied by the leading term in more "decent" terms (Sect. 10.3). But first, it is time to formulate an essential *definition.*

10.2. NONLINEARLY ELASTIC FLEXURAL SHELLS: DEFINITION, EXAMPLES, AND ASSUMPTIONS ON THE DATA

The way the variational problem found in Thm. 10.1-2 is formulated clearly shows that the "interesting" cases are those where, *first*, the manifold $\mathcal{M}_F(\omega)$ contains nonzero elements and, *second*, $\mathcal{M}_F(\omega)$ possesses nonzero tangent vectors at each one of its points. This observation is the basis for the following definition:

Let ω be a domain in $\mathbb{R}^2$ with boundary γ and let $\boldsymbol{\theta} \in \mathcal{C}^2(\overline{\omega}; \mathbb{R}^3)$ be an injective mapping such that the two vectors $\partial_\alpha \boldsymbol{\theta}$ are linearly independent at all points y of $\overline{\omega}$. A *nonlinearly elastic shell* with middle surface $S = \boldsymbol{\theta}(\overline{\omega})$ is called a **nonlinearly elastic flexural shell** if the following three conditions are simultaneously satisfied:

(i) The shell is subjected to a *boundary condition of place* along a portion of its lateral face with $\boldsymbol{\theta}(\gamma_0)$ as its middle curve, where the subset $\gamma_0 \subset \gamma$ satisfies

$$\boxed{\textit{length}\ \gamma_0 > 0.}$$

(ii) Define the *manifold*:

$$\boxed{\begin{aligned} \mathcal{M}_F(\omega) := \{\boldsymbol{\eta} = (\eta_i) \in \mathbf{W}^{2,4}(\omega);\ \boldsymbol{\eta} = \partial_\nu \boldsymbol{\eta} = \mathbf{0} \text{ on } \gamma_0, \\ a_{\alpha\beta}(\boldsymbol{\eta}) - a_{\alpha\beta} = 0 \text{ in } \omega\}, \end{aligned}}$$

where $a_{\alpha\beta}(\boldsymbol{\eta})$ denote the covariant components of the metric tensor of the surface $(\boldsymbol{\theta} + \eta_i \boldsymbol{a}^i)(\overline{\omega})$. *Then*

$$\boxed{\boldsymbol{\mathcal{M}}_F(\omega) \neq \{\mathbf{0}\}.}$$

In other words, there exist *nonzero* displacement fields $\eta_i \boldsymbol{a}^i$ of the middle surface S with covariant components η_i in the space $W^{2,4}(\omega)$ that are **inextensional displacements** of S, in the sense that they *preserve the metric of the surface* S (this is what the relations $a_{\alpha\beta}(\boldsymbol{\eta}) - a_{\alpha\beta} = 0$ in ω express), and that are also **admissible**, in the sense that they satisfy *boundary conditions "of strong clamping"* along the curve $\boldsymbol{\theta}(\gamma_0)$, expressed here by means of the boundary conditions $\boldsymbol{\eta} = \partial_\nu \boldsymbol{\eta} = \mathbf{0}$ on γ_0 on the associated field $\boldsymbol{\eta} = (\eta_i)$ (these boundary conditions are interpreted later; cf. Sect. 10.5).

(iii) At each $\boldsymbol{\zeta} \in \boldsymbol{\mathcal{M}}_F(\omega)$, the tangent space to the manifold $\boldsymbol{\mathcal{M}}_F(\omega)$ contains *nonzero* functions, i.e.,

$$\begin{aligned}
\mathbb{T}_{\boldsymbol{\zeta}} \boldsymbol{\mathcal{M}}_F(\omega) &\neq \{\mathbf{0}\} \text{ at each } \boldsymbol{\zeta} \in \boldsymbol{\mathcal{M}}_F(\omega), \text{ where} \\
\mathbb{T}_{\boldsymbol{\zeta}} \boldsymbol{\mathcal{M}}_F(\omega) &:= \{\boldsymbol{\eta} \in \mathbf{W}^{2,4}(\omega);\ \boldsymbol{\eta} = \partial_\nu \boldsymbol{\eta} = \mathbf{0} \text{ on } \gamma_0, \\
&\qquad\qquad G'_{\alpha\beta}(\boldsymbol{\zeta})\boldsymbol{\eta} = 0 \text{ in } \omega\},
\end{aligned}$$

and (Thm. 9.1-1)

$$\begin{aligned}
G_{\alpha\beta}(\boldsymbol{\eta}) &= \frac{1}{2}(a_{\alpha\beta}(\boldsymbol{\eta}) - a_{\alpha\beta}) \\
&= \frac{1}{2}(\eta_{\alpha\|\beta} + \eta_{\beta\|\alpha} + a^{mn}\eta_{m\|\alpha}\eta_{n\|\beta}), \\
G'_{\alpha\beta}(\boldsymbol{\zeta})\boldsymbol{\eta} &= \frac{1}{2}(\eta_{\alpha\|\beta} + \eta_{\beta\|\alpha} + a^{mn}\{\zeta_{m\|\alpha}\eta_{n\|\beta} + \zeta_{n\|\beta}\eta_{m\|\alpha}\}).
\end{aligned}$$

Examples of nonlinearly elastic flexural shells are given in Figs. 10.2-1 to 10.2-3 and justified in Exs. 10.1 to 10.3.

A word of caution. Like that of a nonlinearly elastic membrane shell, the definition of a nonlinearly elastic flexural shell depends only on the subset of the lateral face where a boundary condition of place is imposed (*via* the set γ_0) and on the geometry of the middle surface of the shell (*via* its second fundamental form and Christoffel symbols, which appear in the functions $\eta_{i\|\beta}$); cf. Ex. 10.4. ■

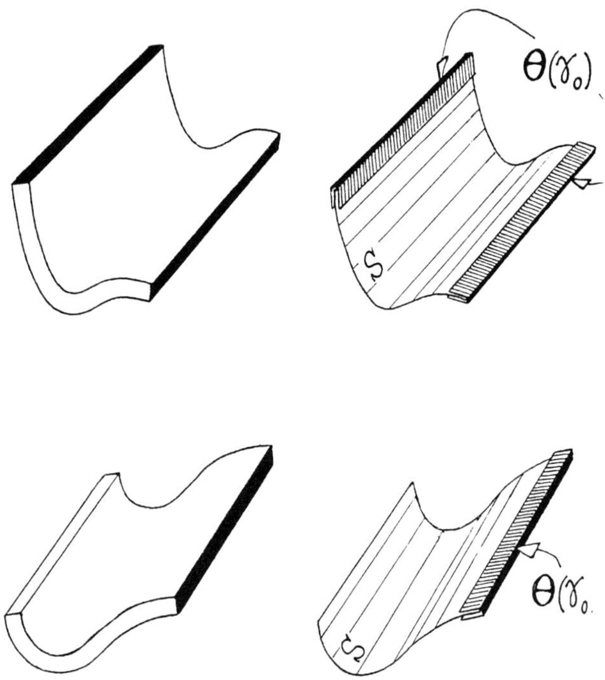

Fig. 10.2-1: *Nonlinearly elastic flexural shells.* A shell whose middle surface is a portion $S = \boldsymbol{\theta}(\overline{\omega})$ of a cylinder and which is subjected to a boundary condition of place (i.e., of vanishing displacement field) along a portion (darkened on the figure) of its lateral face whose middle curve $\boldsymbol{\theta}(\gamma_0)$ is contained in one or two generatrices of S provides an instance of a nonlinearly elastic flexural shell, i.e., one for which the manifold

$$\boldsymbol{\mathcal{M}}_F(\omega) = \{\boldsymbol{\eta} \in \mathbf{W}^{2,4}(\omega);\ \boldsymbol{\eta} = \partial_\nu \boldsymbol{\eta} = \mathbf{0} \text{ on } \gamma_0,\ a_{\alpha\beta}(\boldsymbol{\eta}) - a_{\alpha\beta} = 0 \text{ in } \omega\}$$

contains nonzero functions $\boldsymbol{\eta}$ and the tangent space at each one of its elements contains nonzero functions; cf. Ex. 10.1.

The "two-dimensional boundary conditions of strong clamping" $\boldsymbol{\eta} = \partial_\nu \boldsymbol{\eta} = \mathbf{0}$ on γ_0 are so named because they express that the points and the tangent vectors to the coordinate lines of the deformed and undeformed middle surfaces coincide along the set $\boldsymbol{\theta}(\gamma_0)$ (as suggested on the "two-dimensional" figures); cf. Sect. 10.5.

This example and that of Fig. 9.1-1 illustrate the crucial rôle played by the set $\boldsymbol{\theta}(\gamma_0)$ for determining the type of a shell, like in the linear theory (Figs. 5.8-2 and 6.1-1).

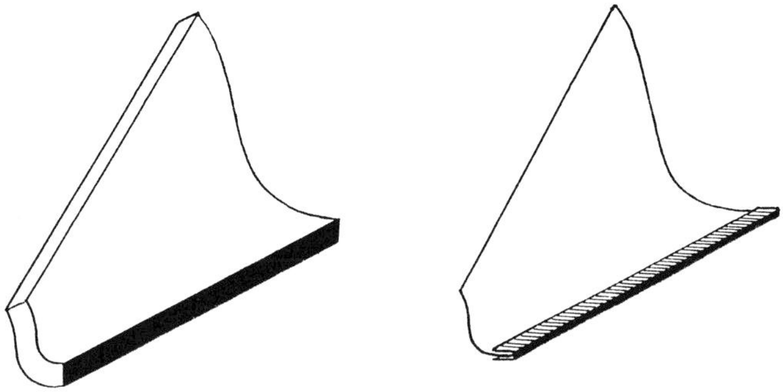

Fig. 10.2-2: *Another example of a nonlinearly elastic flexural shell.* A shell whose middle surface S is a portion of a cone (excluding its vertex) and which is subjected to a boundary condition of place along a portion (darkened on the figure) of its lateral face whose middle line is contained in one generatrix of S provides another example of a nonlinearly elastic flexural shell; cf. Ex. 10.2.

Another word of caution. Although the above definition is strongly suggested by the asymptotic analysis (Thm. 10.1-2), it is certainly perfectible, through an *ad hoc* weakening of the required regularity on the covariant components of the admissible inextensional displacements. Be that as it may, these should be "at least" in $W^{2,p}(\omega)$ for some $p \geq 1$, if only to insure that the traces $\partial_\nu \boldsymbol{\eta} = \mathbf{0}$ on γ_0 make sense.

An ideal classification of nonlinearly elastic shells would be one that leaves no gap between the "membrane" and "flexural" ones (in this respect, see a similar "another word of caution" in Sect. 9.1), like in the linear case! ∎

Remarks. (1) In the *linear* case (Sect. 6.1), only *one* condition was seemingly needed, viz., $\mathbf{V}_F(\omega) \neq \{\mathbf{0}\}$, where

$$\mathbf{V}_F(\omega) := \{\boldsymbol{\eta} = (\eta_i) \in H^1(\omega) \times H^1(\omega) \times H^2(\omega);\\ \eta_i = \partial_\nu \eta_3 = 0 \text{ on } \gamma_0,\ \gamma_{\alpha\beta}(\boldsymbol{\eta}) = 0 \text{ in } \omega\}.$$

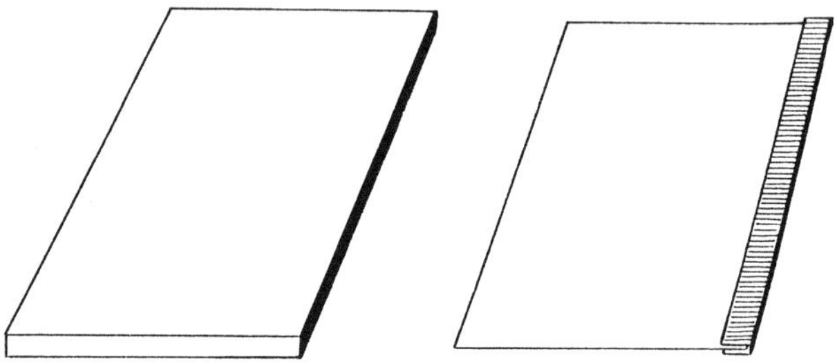

Fig. 10.2-3: *Another example of a nonlinearly elastic flexural shell.* A plate subjected to a boundary condition of place on a portion of its lateral face whose middle line γ_0 is a segment provides an instance of a nonlinearly elastic flexural shell; cf. Ex. 10.3.

By contrast, if γ_0 is the union of two parallel segments, the plate becomes a nonlinearly elastic membrane shell! cf. Fig. 9.1-2.

But this condition is likewise one (albeit "in disguise") on the tangent spaces to $\mathbf{V}_F(\omega)$, since $\mathbb{T}_{\boldsymbol{\zeta}}\mathbf{V}_F(\omega) = \mathbf{V}_F(\omega)$ at all $\boldsymbol{\zeta} \in \mathbf{V}_F(\omega)$ in the linear case.

(2) We shall see later (Sect. 10.6) that the proper framework for the *existence* theory rests on an *ad hoc* "enlargement" of the manifold $\mathcal{M}_F(\omega)$, to

$$\{\boldsymbol{\eta} \in \mathbf{H}^2(\omega);\ \boldsymbol{\eta} = \partial_\nu \boldsymbol{\eta} = \mathbf{0} \text{ on } \gamma_0,\ a_{\alpha\beta}(\boldsymbol{\eta}) - a_{\alpha\beta} = 0 \text{ in } \omega\},$$

a sign that the definition of nonlinearly elastic flexural shells proposed here is still perfectible (as hinted at above).

(3) By adding condition (iii) in the above definition, I depart here from the definition given in Lods & Miara [1998]. ■

The conclusions reached in Thm. 10.1-2 then naturally lead to the following **assumptions on the data** *for a family of nonlinearly elastic flexural shells*, with each having the *same* middle surface $S = \boldsymbol{\theta}(\overline{\omega})$ and with each subjected to a boundary condition of place along a portion of its lateral face having the *same* set $\boldsymbol{\theta}(\gamma_0)$ as its middle curve, as their thickness 2ε approaches zero:

We *require* that the *Lamé constants* and the *applied body and surface force densities* be such that

$$\boxed{\begin{array}{rll} \lambda^\varepsilon = \lambda & \text{and} & \mu^\varepsilon = \mu, \\ f^{i,\varepsilon}(x^\varepsilon) = \varepsilon^2 f^{i,2}(x) & \text{for all} & x^\varepsilon = \pi^\varepsilon x \in \Omega^\varepsilon, \\ h^{i,\varepsilon}(x^\varepsilon) = \varepsilon^3 h^{i,3}(x) & \text{for all} & x^\varepsilon = \pi^\varepsilon x \in \Gamma_+^\varepsilon \cup \Gamma_-^\varepsilon, \end{array}}$$

where the constants $\lambda > 0$ and $\mu > 0$ and the functions $f^{i,2} \in L^2(\Omega)$ and $h^{i,3} \in L^2(\Gamma_+ \cup \Gamma_-)$ are independent of ε (Fig. 3.2-1 recapitulates the definitions of the sets Ω^ε, Ω, Γ_+^ε, Γ_+, Γ_-^ε, and Γ_-).

Note that the *same* limit two-dimensional equations are evidently obtained if the following *more general* **assumptions on the data** are made:

$$\begin{array}{rll} \lambda^\varepsilon = \varepsilon^t \lambda & \text{and} & \mu^\varepsilon = \varepsilon^t \mu, \\ f^{i,\varepsilon}(x^\varepsilon) = \varepsilon^{2+t} f^{i,2}(x) & \text{for all} & x^\varepsilon = \pi^\varepsilon x \in \Omega^\varepsilon, \\ h^{i,\varepsilon}(x^\varepsilon) = \varepsilon^{3+t} h^{i,3}(x) & \text{for all} & x^\varepsilon = \pi^\varepsilon x \in \Gamma_+^\varepsilon \cup \Gamma_-^\varepsilon, \end{array}$$

where the constants $\lambda > 0$ and $\mu > 0$ and the functions $f^{i,2} \in L^2(\Omega)$ and $h^{i,3} \in L^2(\Gamma_+ \cup \Gamma_-)$ are again *independent of ε* and *t is an arbitrary real number.* Besides the analysis that led to Thm. 10.1-2 shows that *these assumptions on the data are the only ones possible for nonlinearly elastic flexural shells.*

10.3. THE TWO-DIMENSIONAL EQUATIONS AS A VARIATIONAL PROBLEM

It was shown in Thm. 10.1-2 that, when $\boldsymbol{\mathcal{M}}_0(\omega) \neq \{\mathbf{0}\}$ and $\mathbb{T}_{\boldsymbol{\zeta}} \boldsymbol{\mathcal{M}}_0(\omega)$ contains nonzero functions at each $\boldsymbol{\zeta} \in \boldsymbol{\mathcal{M}}_0(\omega)$, the leading term of the formal asymptotic expansion of the scaled unknown $\boldsymbol{u}(\varepsilon)$ is independent of the transverse variable x_3 and may thus be identified with a function $\boldsymbol{\zeta}^0$ of two variables; it was further shown in *ibid.* that $\boldsymbol{\zeta}^0$ belongs to the manifold $\boldsymbol{\mathcal{M}}_F(\omega) \subset \boldsymbol{\mathcal{M}}_0(\omega)$ and that $\boldsymbol{\zeta}^0$ satisfies *ad hoc* variational equations for all $\boldsymbol{\eta} \in \mathbb{T}_{\boldsymbol{\zeta}^0} \boldsymbol{\mathcal{M}}_F(\omega) \subset \mathbb{T}_{\boldsymbol{\zeta}^0} \boldsymbol{\mathcal{M}}_0(\omega)$. The definition of a nonlinearly elastic flexural shell given in the previous section precisely ensures that this variational problem makes sense for such a shell.

It remains, however, to shed a brighter light on the somewhat mysterious functions $\hat{E}^0_{\alpha\|\beta}$ and $\hat{F}^0_{\alpha\|\beta}(\boldsymbol{\eta})$ appearing in the variational equations of Thm. 10.1-2: In the same manner that the functions $E^0_{\alpha\|\beta}$ and $F^0_{\alpha\|\beta}(\boldsymbol{\eta})$ occurring in the variational equations of a nonlinearly elastic membrane shell are simply equal to $G_{\alpha\beta}(\boldsymbol{\zeta}^0)$ and $G'_{\alpha\beta}(\boldsymbol{\zeta}^0)\boldsymbol{\eta}$ (Thm. 9.1-1), where

$$G_{\alpha\beta}(\boldsymbol{\eta}) = \frac{1}{2}(a_{\alpha\beta}(\boldsymbol{\eta}) - a_{\alpha\beta})$$

are the covariant components of the *change of metric tensor* associated with a displacement field $\eta_i \boldsymbol{a}^i$ of the middle surface, the functions $\hat{E}^0_{\alpha\|\beta}$ and $\hat{F}^0_{\alpha\|\beta}(\boldsymbol{\eta})$ have definitely more illuminating expressions and interpretations, notably in terms of the covariant components

$$R_{\alpha\beta}(\boldsymbol{\eta}) = b_{\alpha\beta}(\boldsymbol{\eta}) - b_{\alpha\beta}$$

of the associated *change of curvature tensor* (see notably Thms. 10.3-1 and 10.4-1).

The following result takes care of the functions $\hat{E}^0_{\alpha\|\beta}$. As stated here, it extends Lods & Miara [1998, Thm. 2], in that I prove that it holds not only for the leading term $\boldsymbol{\zeta}^0$, but in fact for *any* element $\boldsymbol{\eta}$ in the manifold $\boldsymbol{\mathcal{M}}_F(\omega)$. This extension is particularly convenient for interpreting the two-dimensional equations as a minimization problem (Sect. 10.4).

Note that the intricate expressions given in the next theorem for the functions $R^\flat_{\alpha\beta}(\boldsymbol{\eta})$ can be considerably simplified; cf. Thm. 10.3-2.

Theorem 10.3-1. *Let ω be a domain in $\mathbb{R}^2$, and let $\boldsymbol{\theta} \in \mathcal{C}^3(\overline{\omega}; \mathbb{R}^3)$ be an injective mapping such that the two vectors $\boldsymbol{a}_\alpha = \partial_\alpha \boldsymbol{\theta}$ are linearly independent at all points of $\overline{\omega}$, let $\boldsymbol{a}_3 = \dfrac{\boldsymbol{a}_1 \wedge \boldsymbol{a}_2}{|\boldsymbol{a}_1 \wedge \boldsymbol{a}_2|}$, and let the vectors $\boldsymbol{a}^i$ be defined by $\boldsymbol{a}^i \cdot \boldsymbol{a}_j = \delta^i_j$. Given a displacement field $\eta_i \boldsymbol{a}^i$ of the surface $\boldsymbol{\theta}(\overline{\omega})$ with smooth enough covariant components $\eta_i : \overline{\omega} \to \mathbb{R}$ and such that the two vectors $\boldsymbol{a}_\alpha(\boldsymbol{\eta}) := \partial_\alpha(\boldsymbol{\theta} + \eta_i \boldsymbol{a}^i)$ are linearly independent at all points of ω, let the covariant components of the* **change of curvature tensor** *associated with this displacement field be defined by*

$$\boxed{R_{\alpha\beta}(\boldsymbol{\eta}) := b_{\alpha\beta}(\boldsymbol{\eta}) - b_{\alpha\beta},}$$

where $b_{\alpha\beta}$ and $b_{\alpha\beta}(\boldsymbol{\eta})$ are the covariant components of the curvature tensors of the surfaces $\boldsymbol{\theta}(\overline{\omega})$ and $(\boldsymbol{\theta} + \eta_i \boldsymbol{a}^i)(\overline{\omega})$ (Fig. 2.3-1).

Then the vectors $\boldsymbol{a}_\alpha(\boldsymbol{\eta})$ *are linearly independent, and consequently the functions* $b_{\alpha\beta}(\boldsymbol{\eta})$ *and* $R_{\alpha\beta}(\boldsymbol{\eta})$ *are well defined, for all* $\boldsymbol{\eta} \in \boldsymbol{\mathcal{M}}_F(\omega)$.

Let the functions $R^\flat_{\alpha\beta}(\boldsymbol{\eta}) \in L^2(\omega)$ *be defined for any* $\boldsymbol{\eta} \in \mathbf{W}^{2,4}(\omega)$ *by*

$$\boxed{\begin{aligned} R^\flat_{\alpha\beta}(\boldsymbol{\eta}) &:= F_{\alpha\|\beta}(\boldsymbol{\eta}, \boldsymbol{\psi}(\boldsymbol{\eta})) - \frac{1}{2} g^{\sigma\tau,1} \eta_{\sigma\|\alpha}\eta_{\tau\|\beta} \\ &\quad + \big(\Gamma^{p,1}_{\alpha\beta} + \frac{1}{2} a^{mn}\{\Gamma^{p,1}_{m\alpha}\eta_{n\|\beta} + \Gamma^{p,1}_{n\beta}\eta_{m\|\alpha}\}\big)\eta_p, \end{aligned}}$$

where $\eta_{\alpha\|\beta} := \partial_\beta\eta_\alpha - \Gamma^\sigma_{\alpha\beta}\eta_\sigma - b_{\alpha\beta}\eta_3$ *and* $\eta_{3\|\beta} := \partial_\beta\eta_3 + b^\sigma_\beta\eta_\sigma$, *the functions* $F_{\alpha\|\beta}(\boldsymbol{\eta}, \boldsymbol{\psi}) \in L^2(\omega)$ *are defined for any* $\boldsymbol{\eta} \in \mathbf{W}^{2,4}(\omega)$ *and any* $\boldsymbol{\psi} \in \mathbf{W}^{1,4}(\omega)$ *by*

$$\boxed{F_{\alpha\|\beta}(\boldsymbol{\eta}, \boldsymbol{\psi}) := \frac{1}{2}(\psi_{\alpha\|\beta} + \psi_{\beta\|\alpha} + a^{mn}\{\eta_{m\|\alpha}\psi_{n\|\beta} + \eta_{n\|\beta}\psi_{m\|\alpha}\})}$$

and finally, the field $\boldsymbol{\psi}(\boldsymbol{\eta}) = (\psi_i(\boldsymbol{\eta})) \in \mathbf{W}^{1,4}(\omega)$ *is defined for any* $\boldsymbol{\eta} \in \mathbf{W}^{2,4}(\omega)$ *by*

$$\boxed{\begin{aligned} \psi_1(\boldsymbol{\eta}) &:= b^\alpha_1\eta_\alpha + (1 + a^{\alpha 2}\eta_{\alpha\|2})\eta_{3\|1} - a^{\alpha 2}\eta_{\alpha\|1}\eta_{3\|2}, \\ \psi_2(\boldsymbol{\eta}) &:= b^\alpha_2\eta_\alpha + (1 + a^{\alpha 1}\eta_{\alpha\|1})\eta_{3\|2} - a^{\alpha 1}\eta_{\alpha\|2}\eta_{3\|1}, \\ \psi_3(\boldsymbol{\eta}) &:= -a^{\alpha\beta}\eta_{\alpha\|\beta} - a^{\alpha 1}a^{\beta 2}(\eta_{\alpha\|1}\eta_{\beta\|2} - \eta_{\alpha\|2}\eta_{\beta\|1}). \end{aligned}}$$

Then

$$\boxed{R^\flat_{\alpha\beta}(\boldsymbol{\eta}) = R_{\alpha\beta}(\boldsymbol{\eta}) \text{ for all } \boldsymbol{\eta} \in \boldsymbol{\mathcal{M}}_F(\omega).}$$

Let the functions $\hat{E}^0_{\alpha\|\beta} \in L^2(\omega)$ *be those appearing in the variational problem satisfied by the leading term* $\boldsymbol{\zeta}^0 \in \boldsymbol{\mathcal{M}}_F(\omega)$ (cf. Thm. 10.1-2; their definitions are given in part (iv) of its proof), *viz.*,

$$\boxed{\begin{aligned} \hat{E}^0_{\alpha\|\beta} &:= F^0_{\alpha\|\beta}(\boldsymbol{\psi}^0) - \frac{1}{2} g^{\sigma\tau,1}\zeta^0_{\sigma\|\alpha}\zeta^0_{\tau\|\beta} \\ &\quad + \big(\Gamma^{p,1}_{\alpha\beta} + \frac{1}{2} a^{mn}\{\Gamma^{p,1}_{m\alpha}\zeta^0_{n\|\beta} + \Gamma^{p,1}_{n\beta}\zeta^0_{m\|\alpha}\}\big)\zeta^0_p, \end{aligned}}$$

where

$$F^0_{\alpha\|\beta}(\boldsymbol{\eta}) := \frac{1}{2}(\eta_{\alpha\|\beta} + \eta_{\beta\|\alpha} + a^{mn}\{\zeta^0_{m\|\alpha}\eta_{n\|\beta} + \zeta^0_{n\|\beta}\eta_{m\|\alpha}\}),$$
$$\boldsymbol{\psi}^0 := (\psi^0_i) \text{ with } \psi^0_i := \psi_i(\boldsymbol{\zeta}^0).$$

Then

$$\hat{E}^0_{\alpha\|\beta} = R^\flat_{\alpha\beta}(\boldsymbol{\zeta}^0) = R_{\alpha\beta}(\boldsymbol{\zeta}^0).$$

Proof. That $\hat{E}^0_{\alpha\|\beta} = R^\flat_{\alpha\beta}(\boldsymbol{\zeta}^0)$ immediately follows from the definitions of the functions $R^\flat_{\alpha\beta}(\boldsymbol{\eta})$ and $\hat{E}^0_{\alpha\|\beta}$. It thus remains to show that $b_{\alpha\beta}(\boldsymbol{\eta})$ is well defined and that $R^\flat_{\alpha\beta}(\boldsymbol{\eta}) = b_{\alpha\beta}(\boldsymbol{\eta}) - b_{\alpha\beta}$ for all $\boldsymbol{\eta} \in \mathcal{M}_F(\omega)$.

(i) *For any $\boldsymbol{\eta} \in \mathcal{M}_F(\omega)$, the vectors $a_\alpha(\boldsymbol{\eta})$ are linearly independent at all points in $\overline{\omega}$.*

Since the set $\overline{\omega}$ is compact, the assumption that the vectors $\boldsymbol{a}_\alpha \in \mathcal{C}^2(\overline{\omega}; \mathbb{R}^3)$ are linearly independent implies that there exist constants c_1 and c_2 such that

$$0 \le c_1 < 1 \quad \text{and} \quad |\boldsymbol{a}_1 \cdot \boldsymbol{a}_2| \le c_1 |\boldsymbol{a}_1||\boldsymbol{a}_2| \text{ in } \overline{\omega},$$
$$0 < c_2 \le 1 \quad \text{and} \quad c_2 \le |\boldsymbol{a}_\alpha| \le c_2^{-1} \text{ in } \overline{\omega},$$

on the one hand. By definition of the manifold $\mathcal{M}_F(\omega)$,

$$|\boldsymbol{a}_1(\boldsymbol{\eta})| = \sqrt{a_{11}(\boldsymbol{\eta})} = \sqrt{a_{11}} = |\boldsymbol{a}_1| \text{ in } \overline{\omega},$$
$$|\boldsymbol{a}_2(\boldsymbol{\eta})| = \sqrt{a_{22}(\boldsymbol{\eta})} = \sqrt{a_{22}} = |\boldsymbol{a}_2| \text{ in } \overline{\omega},$$
$$\boldsymbol{a}_1(\boldsymbol{\eta}) \cdot \boldsymbol{a}_2(\boldsymbol{\eta}) = a_{12}(\boldsymbol{\eta}) = a_{12} \text{ in } \overline{\omega},$$

on the other. Hence

$$|\boldsymbol{a}_1(\boldsymbol{\eta}) \cdot \boldsymbol{a}_2(\boldsymbol{\eta})| \le c_1 |\boldsymbol{a}_1(\boldsymbol{\eta})||\boldsymbol{a}_2(\boldsymbol{\eta})| \text{ in } \overline{\omega},$$
$$c_2 \le |\boldsymbol{a}_\alpha(\boldsymbol{\eta})| \le c_2^{-1} \text{ in } \overline{\omega}.$$

Since $c_1 < 1$, these relations mean that the vectors $\boldsymbol{a}_1(\boldsymbol{\eta})$ and $\boldsymbol{a}_2(\boldsymbol{\eta})$ are likewise linearly independent in $\overline{\omega}$ (they are even "uniformly" linearly independent in $\overline{\omega}$, a crucial property of "isometric surfaces" that will be put to another essential use in the existence theory of Sect. 10.6).

(ii) *For any* $\boldsymbol{\eta} \in \boldsymbol{\mathcal{M}}_F(\omega)$, *the vector* $\boldsymbol{a}_3(\boldsymbol{\eta}) = \dfrac{\boldsymbol{a}_1(\boldsymbol{\eta}) \wedge \boldsymbol{a}_2(\boldsymbol{\eta})}{|\boldsymbol{a}_1(\boldsymbol{\eta}) \wedge \boldsymbol{a}_2(\boldsymbol{\eta})|}$, *which is well defined since the vectors* $\boldsymbol{a}_\alpha(\boldsymbol{\eta})$ *are then linearly independent by* (i), *is also given by*

$$\boldsymbol{a}_3(\boldsymbol{\eta}) = \chi_i(\boldsymbol{\eta})\boldsymbol{a}^i,$$

where

$$\begin{aligned}
\chi_1(\boldsymbol{\eta}) &:= -(1 + a^{\alpha 2}\eta_{\alpha\|2})\eta_{3\|1} + a^{\alpha 2}\eta_{\alpha\|1}\eta_{3\|2}, \\
\chi_2(\boldsymbol{\eta}) &:= a^{\alpha 1}\eta_{\alpha\|2}\eta_{3\|1} - (1 + a^{\alpha 1}\eta_{\alpha\|1})\eta_{3\|2}, \\
\chi_3(\boldsymbol{\eta}) &:= 1 + a^{\alpha\beta}\eta_{\alpha\|\beta} + a^{\alpha 1}a^{\beta 2}(\eta_{\alpha\|1}\eta_{\beta\|2} - \eta_{\alpha\|2}\eta_{\beta\|1}).
\end{aligned}$$

The formulas of Gauß and Weingarten (Thm. 2.3-1) imply that

$$\begin{aligned}
\boldsymbol{a}_1(\boldsymbol{\eta}) \wedge \boldsymbol{a}_2(\boldsymbol{\eta}) &= \big(\boldsymbol{a}_1 + \partial_1(\eta_i \boldsymbol{a}^i)\big) \wedge \big(\boldsymbol{a}_2 + \partial_2(\eta_j \boldsymbol{a}^j)\big) \\
&= \boldsymbol{a}_1 \wedge \boldsymbol{a}_2 + \eta_{\alpha\|1}\boldsymbol{a}^\alpha \wedge \boldsymbol{a}_2 + \eta_{\beta\|2}\boldsymbol{a}_1 \wedge \boldsymbol{a}^\beta \\
&\quad + \eta_{3\|1}\boldsymbol{a}^3 \wedge \boldsymbol{a}_2 + \eta_{3\|2}\boldsymbol{a}_1 \wedge \boldsymbol{a}^3 + \eta_{\alpha\|1}\eta_{\beta\|2}\boldsymbol{a}^\alpha \wedge \boldsymbol{a}^\beta \\
&\quad + \eta_{\alpha\|1}\eta_{3\|2}\boldsymbol{a}^\alpha \wedge \boldsymbol{a}^3 + \eta_{3\|1}\eta_{\beta\|2}\boldsymbol{a}^3 \wedge \boldsymbol{a}^\beta.
\end{aligned}$$

Incorporating the formulas (the first ones were established in parts (i) and (ii) of the proof of Thm. 2.5-1; the others are immediately proved):

$$\begin{aligned}
&\boldsymbol{a}_1 \wedge \boldsymbol{a}_2 = \sqrt{a}\,\boldsymbol{a}^3, \quad \boldsymbol{a}_1 \wedge \boldsymbol{a}^3 = -\sqrt{a}\boldsymbol{a}^2, \quad \boldsymbol{a}^3 \wedge \boldsymbol{a}_2 = -\sqrt{a}\,\boldsymbol{a}^1, \\
&\boldsymbol{a}^\alpha \wedge \boldsymbol{a}_2 = a^{\alpha\beta}\boldsymbol{a}_\beta \wedge \boldsymbol{a}_2 = \sqrt{a} \quad a^{\alpha 1}\boldsymbol{a}^3, \quad \boldsymbol{a}_1 \wedge \boldsymbol{a}^\beta = \sqrt{a}\,a^{\beta 2}\boldsymbol{a}^3, \\
&\boldsymbol{a}^\alpha \wedge \boldsymbol{a}^3 = a^{\alpha\beta}\boldsymbol{a}_\beta \wedge \boldsymbol{a}^3 = \sqrt{a}\,(a^{\alpha 2}\boldsymbol{a}^1 - a^{\alpha 1}\boldsymbol{a}^2), \\
&\boldsymbol{a}^\alpha \wedge \boldsymbol{a}^\beta = \sqrt{a}\,(a^{\alpha 1}a^{\beta 2} - a^{\alpha 2}a^{\beta 1})\boldsymbol{a}^3,
\end{aligned}$$

where $a = \det(a_{\alpha\beta})$, into the expression of $\boldsymbol{a}_1(\boldsymbol{\eta}) \wedge \boldsymbol{a}_2(\boldsymbol{\eta})$ then gives

$$\boldsymbol{a}_1(\boldsymbol{\eta}) \wedge \boldsymbol{a}_2(\boldsymbol{\eta}) = \sqrt{a}\,\chi_i(\boldsymbol{\eta})\boldsymbol{a}^i,$$

where the functions $\chi_i(\boldsymbol{\eta})$ are of the announced forms. It remains to observe that, by definition of the manifold $\boldsymbol{\mathcal{M}}_F(\omega)$,

$$|\boldsymbol{a}_1(\boldsymbol{\eta}) \wedge \boldsymbol{a}_2(\boldsymbol{\eta})| = \sqrt{\det a_{\alpha\beta}(\boldsymbol{\eta})} = \sqrt{\det a_{\alpha\beta}} = \sqrt{a}.$$

(iii) *For any* $\boldsymbol{\eta} \in \boldsymbol{\mathcal{M}}_F(\omega)$, *the functions* $b_{\alpha\beta}(\boldsymbol{\eta}) = \partial_\alpha \boldsymbol{a}_\beta(\boldsymbol{\eta}) \cdot \boldsymbol{a}_3(\boldsymbol{\eta})$, *which are well defined since the vector* $\boldsymbol{a}_3(\boldsymbol{\eta})$ *is then well defined by* (i), *are also given by*

$$b_{\alpha\beta}(\boldsymbol{\eta}) = \frac{1}{2}\big(\xi^p_{\alpha\beta}(\boldsymbol{\eta}) + \xi^p_{\beta\alpha}(\boldsymbol{\eta})\big)\chi_p(\boldsymbol{\eta}),$$

where the functions $\chi_p(\boldsymbol{\eta})$ *are defined in* (ii) *and*

$$\begin{aligned}
\xi^\sigma_{\alpha\beta}(\boldsymbol{\eta}) &:= \Gamma^\sigma_{\alpha\beta} + a^{\sigma\tau}(\partial_\alpha\{\eta_{\tau\|\beta}\} - \Gamma^\lambda_{\alpha\tau}\eta_{\lambda\|\beta}) - b^\sigma_\alpha \eta_{3\|\beta},\\
\xi^3_{\alpha\beta}(\boldsymbol{\eta}) &:= b_{\alpha\beta} + \partial_\alpha\{\eta_{3\|\beta}\} + b^\tau_\alpha \eta_{\tau\|\beta}.
\end{aligned}$$

The definition of the functions $b_{\alpha\beta}(\boldsymbol{\eta}) = b_{\beta\alpha}(\boldsymbol{\eta})$, the expression found in (ii) for the vector $\boldsymbol{a}_3(\boldsymbol{\eta})$, and the formulas of Gauß and Weingarten together give

$$\begin{aligned}
b_{\alpha\beta}(\boldsymbol{\eta}) &= \partial_\alpha \boldsymbol{a}_\beta(\boldsymbol{\eta}) \cdot \boldsymbol{a}_3(\boldsymbol{\eta})\\
&= \big(\partial_\alpha\{\boldsymbol{a}_\beta + \eta_{\sigma\|\beta}\boldsymbol{a}^\sigma + \eta_{3\|\beta}\boldsymbol{a}^3\}\big) \cdot (\chi_p(\boldsymbol{\eta})\boldsymbol{a}^p)\\
&= \xi^p_{\alpha\beta}(\boldsymbol{\eta})\chi_p(\boldsymbol{\eta}) = \frac{1}{2}\big(\xi^p_{\alpha\beta}(\boldsymbol{\eta}) + \xi^p_{\beta\alpha}(\boldsymbol{\eta})\big)\chi_p(\boldsymbol{\eta}),
\end{aligned}$$

where the functions $\xi^p_{\alpha\beta}(\boldsymbol{\eta})$ are of the announced forms.

(iv) *For any* $\boldsymbol{\eta} \in \boldsymbol{\mathcal{M}}_F(\omega)$, *the functions* $R^\flat_{\alpha\beta}(\boldsymbol{\eta})$ *as defined in the statement of the theorem satisfy*

$$R^\flat_{\alpha\beta}(\boldsymbol{\eta}) = -F_{\alpha\|\beta}(\boldsymbol{\eta}, \boldsymbol{\chi}(\boldsymbol{\eta})) - b_{\alpha\beta},$$

where $\boldsymbol{\chi}(\boldsymbol{\eta}) =: (\chi_i(\boldsymbol{\eta}))$ *and the functions* $\chi_i(\eta)$ *are defined in* (ii).

The definitions of the fields $\boldsymbol{\psi}(\boldsymbol{\eta})$ (see the statement of the theorem) and $\boldsymbol{\chi}(\boldsymbol{\eta})$ immediately reveal that

$$-\boldsymbol{\psi}(\boldsymbol{\eta}) = \boldsymbol{\chi}(\boldsymbol{\eta}) + \boldsymbol{\mathcal{K}}(\boldsymbol{\eta}),$$

where $\boldsymbol{\mathcal{K}}(\boldsymbol{\eta}) = (\mathcal{K}_i(\boldsymbol{\eta}))$, the functions $\mathcal{K}_i(\boldsymbol{\eta})$ being defined by

$$\mathcal{K}_\alpha(\boldsymbol{\eta}) := -b^\sigma_\alpha \eta_\sigma \text{ and } \mathcal{K}_3(\boldsymbol{\eta}) := -1.$$

Consequently, the functions $R^\flat_{\alpha\beta}(\boldsymbol{\eta})$ may be also written as

$$\begin{aligned} R^\flat_{\alpha\beta}(\boldsymbol{\eta}) &= -F_{\alpha\|\beta}(\boldsymbol{\eta}, \boldsymbol{\chi}(\boldsymbol{\eta})) - F_{\alpha\|\beta}(\boldsymbol{\eta}, \boldsymbol{\mathcal{K}}(\boldsymbol{\eta})) \\ &\quad -\frac{1}{2} g^{\sigma\tau,1} \eta_{\sigma\|\alpha}\eta_{\tau\|\beta} \\ &\quad +\left(\Gamma^{p,1}_{\alpha\beta} + \frac{1}{2} a^{mn}\{\Gamma^{p,1}_{m\alpha}\eta_{n\|\beta} + \Gamma^{p,1}_{n\beta}\eta_{m\|\alpha}\}\right)\eta_p. \end{aligned}$$

There thus remains to compute the functions

$$\begin{aligned} F_{\alpha\|\beta}(\boldsymbol{\eta}, \boldsymbol{\mathcal{K}}(\boldsymbol{\eta})) &= \frac{1}{2}\left(\{\mathcal{K}_\alpha(\boldsymbol{\eta})\}_{\|\beta} + \{\mathcal{K}_\beta(\boldsymbol{\eta})\}_{\|\alpha}\right) \\ &\quad + \frac{1}{2} a^{mn}\left(\eta_{m\|\alpha}\{\mathcal{K}_n(\boldsymbol{\eta})\}_{\|\beta} + \eta_{n\|\beta}\{\mathcal{K}_m(\boldsymbol{\eta})\}_{\|\alpha}\right). \end{aligned}$$

The definitions of the functions $\mathcal{K}_i(\boldsymbol{\eta})$, of the covariant derivatives $b^\sigma_\alpha|_\beta$ (Thm. 2.5-1), and of the functions $\Gamma^{p,1}_{\alpha\beta}$ (Thm. 8.5-1) together show that

$$\begin{aligned} \{\mathcal{K}_\alpha(\boldsymbol{\eta})\}_{\|\beta} &= \partial_\beta\{\mathcal{K}_\alpha(\boldsymbol{\eta})\} - \Gamma^\sigma_{\alpha\beta}\mathcal{K}_\sigma(\boldsymbol{\eta}) - b_{\alpha\beta}\mathcal{K}_3(\boldsymbol{\eta}) \\ &= -(\partial_\beta b^\sigma_\alpha)\eta_\sigma - b^\sigma_\alpha\partial_\beta\eta_\sigma + \Gamma^\sigma_{\alpha\beta}b^\tau_\sigma\eta_\tau + b_{\alpha\beta} \\ &= b_{\alpha\beta} - b^\sigma_{\alpha|\beta}\eta_\sigma - b^\sigma_\alpha\eta_{\sigma\|\beta} - b^\sigma_\alpha b_{\sigma\beta}\eta_3 \\ &= b_{\alpha\beta} + \Gamma^{p,1}_{\alpha\beta}\eta_p - b^\sigma_\alpha\eta_{\sigma\|\beta}, \\ \{\mathcal{K}_3(\boldsymbol{\eta})\}_{\|\beta} &= \partial_\beta\{\mathcal{K}_3(\boldsymbol{\eta})\} + b^\sigma_\beta\mathcal{K}_\sigma(\boldsymbol{\eta}) = -b^\sigma_\beta b^\tau_\sigma\eta_\tau = \Gamma^{\sigma,1}_{\beta 3}\eta_\sigma. \end{aligned}$$

Consequently,

$$\begin{aligned} F_{\alpha\|\beta}(\boldsymbol{\eta}, \boldsymbol{\mathcal{K}}(\boldsymbol{\eta})) &= b_{\alpha\beta} + \Gamma^{p,1}_{\alpha\beta}\eta_p - \frac{1}{2}(b^\sigma_\alpha\eta_{\sigma\|\beta} + b^\tau_\beta\eta_{\tau\|\alpha}) \\ &\quad +\frac{1}{2}a^{\sigma\tau}(b_{\tau\beta} + \Gamma^{p,1}_{\tau\beta}\eta_p - b^\lambda_\tau\eta_{\lambda\|\beta})\eta_{\sigma\|\alpha} \\ &\quad +\frac{1}{2}a^{\sigma\tau}(b_{\sigma\alpha} + \Gamma^{p,1}_{\sigma\alpha}\eta_p - b^\mu_\sigma\eta_{\mu\|\alpha})\eta_{\tau\|\beta} \\ &\quad +\frac{1}{2}\Gamma^{\sigma,1}_{\alpha 3}\eta_{3\|\beta}\eta_\sigma + \frac{1}{2}\Gamma^{\tau,1}_{\beta 3}\eta_{3\|\alpha}\eta_\tau \\ &= b_{\alpha\beta} - \frac{1}{2}g^{\sigma\tau,1}\eta_{\sigma\|\alpha}\eta_{\tau\|\beta} \\ &\quad +\left(\Gamma^{p,1}_{\alpha\beta} + \frac{1}{2}a^{mn}\{\Gamma^{p,1}_{m\alpha}\eta_{n\|\beta} + \Gamma^{p,1}_{n\beta}\eta_{n\|\alpha}\}\right)\eta_p, \end{aligned}$$

since $a^{\sigma\tau}b^\lambda_\tau = \frac{1}{2}g^{\sigma\lambda,1}$ (Thm. 8.5-1). Therefore,

$$R^\flat_{\alpha\beta}(\boldsymbol{\eta}) = -F_{\alpha\|\beta}(\boldsymbol{\eta}, \boldsymbol{\chi}(\boldsymbol{\eta})) - b_{\alpha\beta}.$$

(v) *For any* $\boldsymbol{\eta} \in \boldsymbol{\mathcal{M}}_F(\omega)$, *the functions* $F_{\alpha\|\beta}(\boldsymbol{\eta}, \boldsymbol{\chi}(\boldsymbol{\eta}))$ *satisfy*

$$F_{\alpha\|\beta}(\boldsymbol{\eta}, \boldsymbol{\chi}(\boldsymbol{\eta})) = -b_{\alpha\beta}(\boldsymbol{\eta}).$$

Consequently, for any $\boldsymbol{\eta} \in \boldsymbol{\mathcal{M}}_F(\omega)$,

$$R^\flat_{\alpha\beta}(\boldsymbol{\eta}) = -F_{\alpha\|\beta}(\boldsymbol{\eta}, \boldsymbol{\chi}(\boldsymbol{\eta})) - b_{\alpha\beta} = b_{\alpha\beta}(\boldsymbol{\eta}) - b_{\alpha\beta}.$$

A straightforward computation, based on the definitions of the functions $\chi_i(\boldsymbol{\eta})$, shows that these functions satisfy

$$\chi_\beta(\boldsymbol{\eta}) + a^{mn}\eta_{m\|\beta}\chi_n(\boldsymbol{\eta}) = 0 \text{ for all } \boldsymbol{\eta} \in \mathbf{W}^{1,4}(\omega).$$

Then the relations

$$\partial_\alpha\left(\chi_\beta(\boldsymbol{\eta}) + a^{mn}\eta_{m\|\beta}\chi_n(\boldsymbol{\eta})\right) = 0,$$

together with the relations $\partial_\alpha a^{m3} = 0$ (recall that $a^{m3} = \delta^{m3}$) and

$$\partial_\alpha a^{\sigma\tau} = \partial_\alpha \boldsymbol{a}^\sigma \cdot \boldsymbol{a}^\tau + \boldsymbol{a}^\sigma \cdot \partial_\alpha \boldsymbol{a}^\tau = -\Gamma^\sigma_{\alpha\beta}a^{\beta\tau} - \Gamma^\tau_{\alpha\beta}a^{\sigma\beta},$$

imply that

$$\begin{aligned} -a^{mn}\partial_\alpha\{\eta_{m\|\beta}\}\chi_n(\boldsymbol{\eta}) &= \partial_\alpha\{\chi_\beta(\boldsymbol{\eta})\} + a^{mn}\eta_{m\|\beta}\partial_\alpha\chi_n(\boldsymbol{\eta}) \\ &\quad -(a^{\sigma\lambda}\Gamma^\tau_{\alpha\lambda} + a^{\mu\tau}\Gamma^\sigma_{\alpha\mu})\eta_{\sigma\|\beta}\chi_\tau(\boldsymbol{\eta}). \end{aligned}$$

Incorporating these relations into the expressions of the functions $\xi^p_{\alpha\beta}(\boldsymbol{\eta})$ defined in (iii) then gives

$$\begin{aligned} \xi^p_{\alpha\beta}(\boldsymbol{\eta})\chi_p(\boldsymbol{\eta}) &= \Gamma^\sigma_{\alpha\beta}\chi_\sigma(\boldsymbol{\eta}) + b_{\alpha\beta}\chi_3(\boldsymbol{\eta}) + a^{mn}\partial_\alpha\{\eta_{m\|\beta}\}\chi_n(\boldsymbol{\eta}) \\ &\quad -(a^{\sigma\tau}\Gamma^\lambda_{\alpha\tau}\eta_{\lambda\|\beta} + b^\sigma_\alpha\eta_{3\|\beta})\chi_\sigma(\boldsymbol{\eta}) + b^\tau_\alpha\eta_{\tau\|\beta}\chi_3(\boldsymbol{\eta}) \\ &= -\{\chi_\beta(\boldsymbol{\eta})\}_{\|\alpha} - a^{mn}\{\chi_m(\boldsymbol{\eta})\}_{\|\alpha}\eta_{n\|\beta}, \end{aligned}$$

so that the expressions found in (iii) for the functions $b_{\alpha\beta}(\boldsymbol{\eta})$ when $\boldsymbol{\eta} \in \boldsymbol{\mathcal{M}}_F(\omega)$ become

$$\begin{aligned} b_{\alpha\beta}(\boldsymbol{\eta}) &= \frac{1}{2}\left(\xi^p_{\alpha\beta}(\boldsymbol{\eta}) + \xi^p_{\beta\alpha}(\boldsymbol{\eta})\right)\chi_p(\boldsymbol{\eta}) \\ &= -\frac{1}{2}\left(\{\chi_\beta(\boldsymbol{\eta})\}_{\|\alpha} + \{\chi_\alpha(\boldsymbol{\eta})\}_{\|\beta}\right) \\ &\quad -\frac{1}{2}a^{mn}\left(\eta_{m\|\alpha}\{\chi(\boldsymbol{\eta})\}_{\|\beta} + \eta_{n\|\beta}\{\chi(\boldsymbol{\eta})\}_{\|\alpha}\right) \\ &= -F_{\alpha\|\beta}(\boldsymbol{\eta}, \boldsymbol{\chi}(\boldsymbol{\eta})). \end{aligned}$$

The proof is thus complete. ∎

Remark. The proof shows that, in fact, $R^{\flat}_{\alpha\beta}(\boldsymbol{\eta}) = R_{\alpha\beta}(\boldsymbol{\eta})$ for all $\boldsymbol{\eta} \in \mathbf{W}^{2,4}(\omega)$ that satisfy $a_{\alpha\beta}(\boldsymbol{\eta}) - a_{\alpha\beta} = 0$ in ω, i.e., the boundary conditions satisfied by the fields $\boldsymbol{\eta}$ in the manifold $\boldsymbol{\mathcal{M}}_F(\omega)$ play no rôle here. ∎

We next show that the functions $R^{\flat}_{\alpha\beta}(\boldsymbol{\eta})$ have considerably simpler expressions in some realistic cases; this observation is due to Roquefort [2000]. ∎

Theorem 10.3-2. *Let ω be a domain in $\mathbb{R}^2$, let $\boldsymbol{\theta} \in \mathcal{C}^3(\overline{\omega}; \mathbb{R}^3)$ be an injective mapping such that the two vectors $\boldsymbol{a}_\alpha = \partial_\alpha \boldsymbol{\theta}$ are linearly independent at all points in ω, and let the functions $R^{\flat}_{\alpha\beta}(\boldsymbol{\eta}) \in L^2(\omega)$ be defined for each $\boldsymbol{\eta} \in \mathbf{W}^{2,4}(\omega)$ as in* Thm. 10.3-1. *Then one also has*

$$R_{\alpha\beta}(\boldsymbol{\eta}) = \frac{1}{\sqrt{a}}\partial_{\alpha\beta}(\boldsymbol{\theta} + \eta_i \boldsymbol{a}^i) \cdot \{\boldsymbol{a}_1(\boldsymbol{\eta}) \wedge \boldsymbol{a}_2(\boldsymbol{\eta})\} - b_{\alpha\beta} =: R^{\sharp}_{\alpha\beta}(\boldsymbol{\eta})$$
$$\text{for all } \boldsymbol{\eta} \in \mathbf{W}^{2,4}(\omega) \text{ such that } \det(a_{\alpha\beta}(\boldsymbol{\eta})) > 0 \text{ in } \overline{\omega},$$

where $a_\alpha(\boldsymbol{\eta}) := \partial_\alpha(\boldsymbol{\theta} + \eta_i \boldsymbol{a}^i)$. These relations imply in particular that

$$R^{\flat}_{\alpha\beta}(\boldsymbol{\eta}) = b_{\alpha\beta}(\boldsymbol{\eta}) - b_{\alpha\beta} = R_{\alpha\beta}(\boldsymbol{\eta}) \text{ for all } \boldsymbol{\eta} \in \boldsymbol{\mathcal{M}}_F(\omega),$$

as was already established in Thm. 10.3-1.

Proof. The proof follows the same lines as that of Thm. 10.3-1. If, instead of the relations $a_{\alpha\beta}(\boldsymbol{\eta}) - a_{\alpha\beta} = 0$ in ω, a vector field $\boldsymbol{\eta} \in \mathbf{W}^{2,4}(\omega)$ satisfies the less restrictive relation

$$a(\boldsymbol{\eta}) := \det(a_{\alpha\beta}(\boldsymbol{\eta})) > 0 \text{ in } \omega,$$

then the vector $\boldsymbol{a}_3(\boldsymbol{\eta})$ and the functions $b_{\alpha\beta}(\boldsymbol{\eta})$ are now given by (compare with parts (ii) and (iii) of the proof of Thm. 10.3-1, whose notations are employed here):

$$\boldsymbol{a}_3(\boldsymbol{\eta}) = \sqrt{\frac{a}{a(\boldsymbol{\eta})}}\chi_i(\boldsymbol{\eta})\boldsymbol{a}^i,$$

$$b_{\alpha\beta}(\boldsymbol{\eta}) = \frac{1}{2}\sqrt{\frac{a}{a(\boldsymbol{\eta})}}(\xi^p_{\alpha\beta}(\boldsymbol{\eta}) + \xi^p_{\beta\alpha}(\boldsymbol{\eta}))\chi_p(\boldsymbol{\eta}),$$

so that (compare with part (v) of the proof in *ibid.*)

$$b_{\alpha\beta}(\boldsymbol{\eta}) = -\sqrt{\frac{a}{a(\boldsymbol{\eta})}} F_{\alpha\|\beta}(\boldsymbol{\eta}, \boldsymbol{\chi}(\boldsymbol{\eta}))$$

for such a vector field, and the assertion is established. ∎

The next result, due to Lods & Miara [1998, Thm. 3], takes care of the functions $\hat{F}^0_{\alpha\|\beta}(\boldsymbol{\eta})$.

Theorem 10.3-3. *Let the functions* $R^\flat_{\alpha\beta}(\boldsymbol{\eta}) \in L^2(\omega)$ *be defined for any* $\boldsymbol{\eta} \in \mathbf{W}^{2,4}(\omega)$ *as in* Thm. 10.3-1 *and let the functions* $\hat{F}^0_{\alpha\|\beta}(\boldsymbol{\eta}) \in L^2(\omega)$ *be those appearing in the variational equations satisfied for all* $\boldsymbol{\eta} \in \mathbb{T}_{\boldsymbol{\zeta}^0}\boldsymbol{\mathcal{M}}_F(\omega)$ *by the leading term* $\boldsymbol{\zeta}^0 \in \boldsymbol{\mathcal{M}}_F(\omega)$ (cf. Thm. 10.1-2; their definitions are given in part (vi) of its proof), *viz.*,

$$\boxed{\begin{aligned}\hat{F}^0_{\alpha\|\beta}(\boldsymbol{\eta}) &= F^0_{\alpha\|\beta}(\boldsymbol{\varphi}^0(\boldsymbol{\eta})) - \frac{1}{2} g^{\sigma\tau,1}\{\zeta^0_{\sigma\|\alpha}\eta_{\tau\|\beta} + \zeta^0_{\tau\|\beta}\eta_{\sigma\|\alpha}\} \\ &+ \big(\Gamma^{p,1}_{\alpha\beta} + \frac{1}{2}a^{mn}\{\Gamma^{p,1}_{m\alpha}\zeta^0_{n\|\beta} + \Gamma^{p,1}_{n\beta}\zeta^0_{m\|\alpha}\}\big)\eta_p \\ &+ \frac{1}{2}a^{mn}\{(\psi^0_{m\|\alpha} + \Gamma^{p,1}_{m\alpha}\zeta^0_p)\eta_{n\|\beta} + (\psi^0_{n\|\beta} + \Gamma^{q,1}_{n\beta}\zeta^0_q)\eta_{m\|\alpha}\},\end{aligned}}$$

where

$$F^0_{\alpha\|\beta}(\boldsymbol{\eta}) := \frac{1}{2}(\eta_{\alpha\|\beta} + \eta_{\beta\|\alpha} + a^{mn}\{\zeta^0_{m\|\alpha}\eta_{n\|\beta} + \zeta^0_{n\|\beta}\eta_{m\|\alpha}\}),$$
$$\eta_{\alpha\|\beta} := \partial_\beta\eta_\alpha - \Gamma^\sigma_{\alpha\beta}\eta_\sigma - b_{\alpha\beta}\eta_3, \quad \eta_{3\|\beta} := \partial_\beta\eta_3 + b^\sigma_\beta\eta_\sigma,$$

and the fields $\boldsymbol{\varphi}^0(\boldsymbol{\eta}) = (\varphi^0_i(\boldsymbol{\eta})) \in \mathbf{W}(\omega)$ *and* $\boldsymbol{\psi}^0 = (\psi^0_i) \in \mathbf{W}(\omega)$ *are given by:*

$$\begin{aligned}\varphi^0_1(\boldsymbol{\eta}) &:= b^\alpha_1\eta_\alpha + \eta_{3\|1} \\ &\quad + a^{\alpha 2}(\zeta^0_{\alpha\|2}\eta_{3\|1} + \zeta^0_{3\|1}\eta_{\alpha\|2} - \zeta^0_{3\|2}\eta_{\alpha\|1} - \zeta^0_{\alpha\|1}\eta_{3\|2}), \\ \varphi^0_2(\boldsymbol{\eta}) &:= b^\alpha_2\eta_\alpha + \eta_{3\|2} \\ &\quad + a^{\alpha 1}(\zeta^0_{\alpha\|1}\eta_{3\|2} + \zeta^0_{3\|2}\eta_{\alpha\|1} - \zeta^0_{3\|1}\eta_{\alpha\|2} - \zeta^0_{\alpha\|2}\eta_{3\|1}), \\ \varphi^0_3(\boldsymbol{\eta}) &:= -a^{\alpha\beta}\eta_{\alpha\|\beta} \\ &\quad - a^{\alpha 1}a^{\beta 2}(\zeta^0_{\alpha\|1}\eta_{\beta\|2} + \zeta^0_{\beta\|2}\eta_{\alpha\|1} - \zeta^0_{\alpha\|2}\eta_{\beta\|1} - \zeta^0_{\beta\|1}\eta_{\alpha\|2}),\end{aligned}$$

$$\psi_1^0 := \psi_1(\boldsymbol{\zeta}^0) = b_1^\alpha \zeta_\alpha^0 + (1 + a^{\alpha 2}\zeta_{\alpha\|2}^0)\zeta_{3\|1}^0 - a^{\alpha 2}\zeta_{\alpha\|1}^0\zeta_{3\|2}^0,$$
$$\psi_2^0 := \psi_2(\boldsymbol{\zeta}^0) = b_2^\alpha \zeta_\alpha^0 + (1 + a^{\alpha 1}\zeta_{\alpha\|1}^0)\zeta_{3\|2}^0 - a^{\alpha 1}\zeta_{\alpha\|2}^0\zeta_{3\|1}^0,$$
$$\psi_3^0 := \psi_3(\boldsymbol{\zeta}^0) = -a^{\alpha\beta}\zeta_{\alpha\|\beta}^0 - a^{\alpha 1}a^{\beta 2}(\zeta_{\alpha\|1}^0\zeta_{\beta\|2}^0 - \zeta_{\alpha\|2}^0\zeta_{\beta\|1}^0).$$

Then

$$\boxed{\hat{F}_{\alpha\|\beta}^0(\boldsymbol{\eta}) = (R_{\alpha\beta}^\flat)'(\boldsymbol{\zeta}^0)\boldsymbol{\eta} \text{ for all } \boldsymbol{\eta} \in \mathbb{T}_{\boldsymbol{\zeta}^0}\boldsymbol{\mathcal{M}}_F(\omega),}$$

where $(R_{\alpha\beta}^\flat)'(\boldsymbol{\zeta}^0)$ *designates the Fréchet derivative at* $\boldsymbol{\zeta}^0 \in \boldsymbol{\mathcal{M}}_F(\omega)$ *of each mapping* $R_{\alpha\beta}^\flat : \mathbf{W}^{2,4}(\omega) \to L^2(\omega)$.

Proof. For details about the notions of differential calculus used *infra*, see, e.g., Vol. I, Sect. 1.2. The functions $R_{\alpha\beta}^\flat(\boldsymbol{\eta})$ are of the form

$$R_{\alpha\beta}^\flat(\boldsymbol{\eta}) = G_{\alpha\|\beta}(\boldsymbol{\eta}) + H_{\alpha\|\beta}(\boldsymbol{\eta}),$$

where

$$\begin{aligned} G_{\alpha\|\beta}(\boldsymbol{\eta}) &:= F_{\alpha\|\beta}(\boldsymbol{\eta}, \boldsymbol{\psi}(\boldsymbol{\eta})), \\ H_{\alpha\|\beta}(\boldsymbol{\eta}) &:= -\frac{1}{2}g^{\sigma\tau,1}\eta_{\sigma\|\alpha}\eta_{\tau\|\beta} \\ &\quad + \big(\Gamma_{\alpha\beta}^{p,1} + \frac{1}{2}a^{mn}\{\Gamma_{m\alpha}^{p,1}\eta_{n\|\beta} + \Gamma_{n\beta}^{p,1}\eta_{m\|\alpha}\}\big)\eta_p. \end{aligned}$$

The mappings $(\boldsymbol{\eta}, \boldsymbol{\psi}) : \mathbf{W}^{2,4}(\omega) \times \mathbf{W}^{1,4}(\omega) \to F_{\alpha\|\beta}(\boldsymbol{\eta}, \boldsymbol{\psi}) \in L^2(\omega)$ are affine with respect to $\boldsymbol{\eta}$, linear with respect to $\boldsymbol{\psi}$, and continuous; hence they are differentiable. As a sum of continuous linear and quadratic mappings, the mapping $\boldsymbol{\psi} : \mathbf{W}^{2,4}(\omega) \to \mathbf{W}^{1,4}(\omega)$ is likewise differentiable. The expressions of the Gâteaux derivatives for such functions together with the chain rule then give

$$\begin{aligned} G'_{\alpha\|\beta}(\boldsymbol{\zeta})\boldsymbol{\eta} &= F_{\alpha\|\beta}(\boldsymbol{\zeta}, \boldsymbol{\psi}'(\boldsymbol{\zeta})\boldsymbol{\eta}) \\ &\quad + \frac{1}{2}a^{mn}\left(\{\psi_m(\boldsymbol{\zeta})\}_{\|\alpha}\eta_{n\|\beta} + \{\psi_n(\boldsymbol{\zeta})\}_{\|\beta}\eta_{m\|\alpha}\right) \end{aligned}$$

for all $\boldsymbol{\zeta}, \boldsymbol{\eta} \in \mathbf{W}^{2,4}(\omega)$. Besides (Thm. 10.1-3),

$$\boldsymbol{\psi}'(\boldsymbol{\zeta}^0)\boldsymbol{\eta} = \boldsymbol{\varphi}^0(\boldsymbol{\eta}) \text{ for all } \boldsymbol{\eta} \in \mathbb{T}_{\boldsymbol{\zeta}^0}\boldsymbol{\mathcal{M}}_F(\omega) \subset \mathbf{W}^{2,4}(\omega),$$

so that

$$F_{\alpha\|\beta}(\boldsymbol{\zeta}^0, \boldsymbol{\psi}'(\boldsymbol{\zeta}^0)\boldsymbol{\eta}) = F_{\alpha\|\beta}(\boldsymbol{\zeta}^0, \boldsymbol{\varphi}^0(\boldsymbol{\eta})) = F_{\alpha\|\beta}^0(\boldsymbol{\varphi}^0(\boldsymbol{\eta}))$$

for all $\boldsymbol{\eta} \in \mathbb{T}_{\boldsymbol{\zeta}^0}\boldsymbol{\mathcal{M}}_F(\omega)$.

As sums of continuous linear and quadratic mappings, the mappings $H_{\alpha\|\beta} : \mathbf{W}^{2,4}(\omega) \to L^2(\omega)$ are differentiable and their Gâteaux derivatives are given for any $\boldsymbol{\zeta}, \boldsymbol{\eta} \in \mathbf{W}^{2,4}(\omega)$ by

$$\begin{aligned} H'_{\alpha\|\beta}(\boldsymbol{\zeta})\boldsymbol{\eta} &= -\frac{1}{2} g^{\sigma\tau,1}\{\zeta_{\sigma\|\alpha}\eta_{\tau\|\beta} + \zeta_{\tau\|\beta}\eta_{\sigma\|\alpha}\} \\ &\quad + \big(\Gamma^{p,1}_{\alpha\beta} + \frac{1}{2} a^{mn}\{\Gamma^{p,1}_{m\alpha}\zeta_{n\|\beta} + \Gamma^{p,1}_{n\beta}\zeta_{m\|\alpha}\}\big)\eta_p \\ &\quad + \frac{1}{2} a^{mn}\{\Gamma^{p,1}_{m\alpha}\eta_{n\|\beta} + \Gamma^{p,1}_{n\beta}\eta_{m\|\alpha}\}\zeta_p. \end{aligned}$$

Combining the above relations thus yields

$$R^\flat_{\alpha\beta}(\boldsymbol{\zeta}^0)\boldsymbol{\eta} = G'_{\alpha\|\beta}(\boldsymbol{\zeta}^0)\boldsymbol{\eta} + H'_{\alpha\|\beta}(\boldsymbol{\zeta}^0)\boldsymbol{\eta} = \hat{F}^0_{\alpha\|\beta}(\boldsymbol{\eta})$$

for all $\boldsymbol{\eta} \in \mathbb{T}_{\boldsymbol{\zeta}^0}\boldsymbol{\mathcal{M}}_F(\omega)$ and the proof is complete. ■

We are now in a position to recast the variational problem of Thm. 10.1-2 in a more suitable form, as an immediate corollary to Thms. 10.3-1 and 10.3-3:

Theorem 10.3-4. *Consider a family of nonlinearly elastic flexural shells according to the definition of* Sect. 10.2, *with thickness* $2\varepsilon > 0$, *with each having the same middle surface* $S = \boldsymbol{\theta}(\overline{\omega})$ *and with each satisfying a boundary condition of place along a portion of its lateral face having the same set* $\boldsymbol{\theta}(\gamma_0)$ *as its middle curve, and let the assumptions on the data be as in* Sect. 10.2. *Finally, assume that* $\boldsymbol{\theta} \in \mathcal{C}^3(\overline{\omega}; \mathbb{R}^3)$.

Then the leading term $\boldsymbol{u}^0$ *of the formal asymptotic expansion of the scaled unknown* $\boldsymbol{u}(\varepsilon)$ *is independent of the transverse variable* x_3 *and* $\boldsymbol{\zeta}^0 := \frac{1}{2}\int_{-1}^{1} \boldsymbol{u}^0\, dx_3$ *satisfies the following* **scaled two-dimensional variational problem** $\mathcal{P}_F(\omega)$ **of a nonlinearly elastic flexural shell:**

$$\boldsymbol{\zeta}^0 \in \boldsymbol{\mathcal{M}}_F(\omega) := \{\boldsymbol{\eta} \in \mathbf{W}^{2,4}(\omega);\ \boldsymbol{\eta} = \partial_\nu \boldsymbol{\eta} = \mathbf{0} \text{ on } \gamma_0,\ G_{\alpha\beta}(\boldsymbol{\eta}) = 0 \text{ in } \omega\},$$

$$\frac{1}{3}\int_\omega a^{\alpha\beta\sigma\tau} R^\flat_{\sigma\tau}(\boldsymbol{\zeta}^0)\big((R^\flat_{\alpha\beta})'(\boldsymbol{\zeta}^0)\boldsymbol{\eta}\big)\sqrt{a}\, dy = \int_\omega p^{i,2}\eta_i\sqrt{a}\, dy$$

$$\text{for all } \boldsymbol{\eta} = (\eta_i) \in \mathbb{T}_{\boldsymbol{\zeta}^0}\boldsymbol{\mathcal{M}}_F(\omega), \text{ where}$$

$$\mathbb{T}_{\boldsymbol{\zeta}^0}\boldsymbol{\mathcal{M}}_F(\omega) := \{\boldsymbol{\eta} \in \mathbf{W}^{2,4}(\omega);\ \boldsymbol{\eta} = \partial_\nu \boldsymbol{\eta} = \mathbf{0} \text{ on } \gamma_0,\ G'_{\alpha\beta}(\boldsymbol{\zeta}^0)\boldsymbol{\eta} = 0 \text{ in } \omega\},$$

where the functions $G_{\alpha\beta}(\boldsymbol{\eta}) = \frac{1}{2}(a_{\alpha\beta}(\boldsymbol{\eta}) - a_{\alpha\beta})$ *and* $R^\flat_{\alpha\beta}(\boldsymbol{\eta})$ *are defined in* Thms. 9.1-1 *and* 10.3-1, *and finally,*

$$a^{\alpha\beta\sigma\tau} := \frac{4\lambda\mu}{\lambda+2\mu} a^{\alpha\beta} a^{\sigma\tau} + 2\mu(a^{\alpha\sigma} a^{\beta\tau} + a^{\alpha\tau} a^{\beta\sigma}),$$

$$p^{i,2} := \int_{-1}^{1} f^{i,2}\, dx_3 + h_+^{i,3} + h_-^{i,3} \text{ and } h_\pm^{i,3} := h^{i,3}(\cdot, \pm 1).$$

■

10.4. THE TWO-DIMENSIONAL EQUATIONS AS A MINIMIZATION PROBLEM

A crucial property has yet to be exploited: *When a field* $\boldsymbol{\eta} = (\eta_i)$ *belongs to the manifold* $\mathcal{M}_F(\omega)$, *the functions* $R^\flat_{\alpha\beta}(\boldsymbol{\eta})$ *become equal to the covariant components of the change of curvature tensor associated with the displacement field* $\eta_i \boldsymbol{a}^i$ *of the middle surface* S (Thm. 10.3-1).

This goal is achieved in two stages: First, by recognizing as in Lods & Miara [1998, Thm. 3] that the variational equations of problem $\mathcal{P}_F(\omega)$ (Thm. 10.3-4) express that *the Gâteaux derivatives* $(j_F^\flat)'(\boldsymbol{\zeta}^0)\boldsymbol{\eta}$ *of an* ad hoc *functional* $j_F^\flat$ *vanish for all the fields* $\boldsymbol{\eta}$ *in the tangent space to the manifold* $\mathcal{M}_F(\omega)$ *at* $\boldsymbol{\zeta}^0$; second, by noting that, over this manifold, $j_F^\flat$ becomes equal to a functional j_F whose integrand is (apart from its linear terms accounting for the applied forces) a *quadratic and positive definite expression*, *via* the *scaled two-dimensional elasticity tensor of the shell*, in terms of the *change of curvature tensor* introduced in Thm. 10.3-1.

These decisive observations in turn allow to recast the two-dimensional variational problem $\mathcal{P}_F(\omega)$ as a *minimization problem*, which now exhibits a remarkable *simplicity*. In this minimization problem, which will be studied for its own sake in Sect. 10.6, lies the apex of the application of the method formal asymptotic expansions to nonlinearly elastic flexural shells.

Theorem 10.4-1. *Given a nonlinearly elastic flexural shell according to the definition given in* Sect. 10.2, *let the manifold* $\mathcal{M}_F(\omega)$

and the functional $j_F^\flat : \mathbf{W}^{2,4}(\omega) \to \mathbb{R}$ *be defined by*

$$\begin{aligned}\boldsymbol{\mathcal{M}}_F(\omega) &:= \{\boldsymbol{\eta} = (\eta_i) \in \mathbf{W}^{2,4}(\omega);\ \boldsymbol{\eta} = \partial_\nu \boldsymbol{\eta} = \mathbf{0} \text{ on } \gamma_0, \\ &\qquad a_{\alpha\beta}(\boldsymbol{\eta}) - a_{\alpha\beta} = 0 \text{ in } \omega\}, \\ j_F^\flat(\boldsymbol{\eta}) &:= \frac{1}{6}\int_\omega a^{\alpha\beta\sigma\tau} R_{\sigma\tau}^\flat(\boldsymbol{\eta}) R_{\alpha\beta}^\flat(\boldsymbol{\eta})\sqrt{a}\,dy - \int_\omega p^{i,2}\eta_i\sqrt{a}\,dy \\ &\qquad \text{for all } \boldsymbol{\eta} \in \mathbf{W}^{2,4}(\omega),\end{aligned}$$

where the functions $a_{\alpha\beta}(\boldsymbol{\eta})$ *and* $R_{\alpha\beta}^\flat(\boldsymbol{\eta})$ *are defined in* Thms. 9.1-1 *and* 10.3-1 *and the functions* $a^{\alpha\beta\sigma\tau}$ *and* $p^{i,2}$ *are defined in* Thm. 10.3-3.

Then the functional $j_F^\flat$ *is differentiable over the space* $\mathbf{W}^{2,4}(\omega)$ *and* $\boldsymbol{\zeta}^0 \in \boldsymbol{\mathcal{M}}_F(\omega)$ *is a solution to the variational problem* $\mathcal{P}_F(\omega)$ *of* Thm. 10.3-4 *if and only if it is a stationary point of the functional* $j_F^\flat$ *over the manifold* $\boldsymbol{\mathcal{M}}_F(\omega)$, i.e., *it satisfies* $(j_F^\flat)'(\boldsymbol{\zeta}^0)\boldsymbol{\eta} = 0$ *for all* $\boldsymbol{\eta}$ *in the tangent space* $\mathbb{T}_{\boldsymbol{\zeta}^0}\boldsymbol{\mathcal{M}}_F(\omega)$ *to the manifold* $\boldsymbol{\mathcal{M}}_F(\omega)$ *at* $\boldsymbol{\zeta}^0$.

Since the functions $b_{\alpha\beta}(\boldsymbol{\eta})$ *are well defined for all* $\boldsymbol{\eta} \in \boldsymbol{\mathcal{M}}_F(\omega)$ *and* $R_{\alpha\beta}^\flat(\boldsymbol{\eta}) = b_{\alpha\beta}(\boldsymbol{\eta}) - b_{\alpha\beta}$ *for all* $\boldsymbol{\eta} \in \boldsymbol{\mathcal{M}}_F(\omega)$ (Thm. 10.3-2), *particular solutions to problem* $\mathcal{P}_F(\omega)$ *are obtained by solving the minimization problem: Find* $\boldsymbol{\zeta}$ *such that*

$$\begin{aligned}&\boldsymbol{\zeta} \in \boldsymbol{\mathcal{M}}_F(\omega) \text{ and } j_F(\boldsymbol{\zeta}) = \inf_{\boldsymbol{\eta}\in\boldsymbol{\mathcal{M}}_F(\omega)} j_F(\boldsymbol{\eta}), \text{ where} \\ &j_F(\boldsymbol{\eta}) := \frac{1}{6}\int_\omega a^{\alpha\beta\sigma\tau}(b_{\sigma\tau}(\boldsymbol{\eta}) - b_{\sigma\tau})(b_{\alpha\beta}(\boldsymbol{\eta}) - b_{\alpha\beta})\sqrt{a}\,dy \\ &\qquad - \int_\omega p^{i,2}\eta_i\sqrt{a}\,dy.\end{aligned}$$

Proof. The differentiability of the functional $j_F^\flat$ over the space $\mathbf{W}^{2,4}(\omega)$ is a simple consequence of the definition of the functions $R_{\alpha\beta}^\flat(\boldsymbol{\eta})$ (Thm. 10.3-1). An argument similar to that found in the proof of Thm. 8.2-3 then shows that the Gâteaux derivatives $(j_F^\flat)'(\boldsymbol{\zeta})\boldsymbol{\eta}$ are given by

$$(j_F^\flat)'(\boldsymbol{\zeta})\boldsymbol{\eta} = \frac{1}{3}\int_\omega a^{\alpha\beta\sigma\tau} R_{\sigma\tau}^\flat(\boldsymbol{\zeta})\big((R_{\alpha\beta}^\flat)'(\boldsymbol{\zeta})\boldsymbol{\eta}\big)\sqrt{a}\,dy - \int_\omega p^{i,2}\eta_i\sqrt{a}\,dy$$

for all $\boldsymbol{\zeta}, \boldsymbol{\eta} \in \mathbf{W}^{2,4}(\omega)$. Hence an inspection of the variational equations satisfied by $\boldsymbol{\zeta}^0 \in \boldsymbol{\mathcal{M}}_F(\omega)$ for all $\boldsymbol{\eta} \in \mathbb{T}_{\boldsymbol{\zeta}^0}\boldsymbol{\mathcal{M}}_F(\omega)$ (Thm. 10.3-4) immediately reveals that they coincide with the equations

$$(j_F^\flat)'(\boldsymbol{\zeta}^0)\boldsymbol{\eta} = 0 \text{ for all } \boldsymbol{\eta} \in \mathbb{T}_{\boldsymbol{\zeta}^0}\mathcal{M}_F(\omega).$$

Conversely, if $\boldsymbol{\zeta}^0 \in \mathcal{M}_F(\omega)$ is a stationary point of the functional $j_F^\flat$ over the manifold $\mathcal{M}_F(\omega)$, it also satisfies the variational problem $\mathcal{P}_F(\omega)$ of Thm. 10.3-4, since the argument that led in Thm. 10.3-3 to the relations

$$(R_{\alpha\beta}^\flat)'(\boldsymbol{\zeta}^0)\boldsymbol{\eta} = \hat{F}^0_{\alpha\|\beta}(\boldsymbol{\eta}) \text{ for all } \boldsymbol{\eta} \in \mathbb{T}_{\boldsymbol{\zeta}^0}\mathcal{M}_F(\omega)$$

holds in fact if $\boldsymbol{\zeta}^0$ is replaced by *any* $\boldsymbol{\zeta}$ in the manifold $\mathcal{M}_F(\omega)$ (this observation relies in particular on part (iii) of the proof of Thm. 10.1-3).

Particular solutions to problem $\mathcal{P}_F(\omega)$ may thus be obtained by solving the minimization problem: Find $\boldsymbol{\zeta}$ such that

$$\boldsymbol{\zeta} \in \mathcal{M}_F(\omega) \text{ and } j_F^\flat(\boldsymbol{\zeta}) = \inf_{\boldsymbol{\eta} \in \mathcal{M}_F(\omega)} j_F^\flat(\boldsymbol{\eta}).$$

Since *the functionals* $j_F^\flat$ *and* j_F *coincide over the manifold* $\mathcal{M}_F(\omega)$, this minimization problem may be equivalently defined in terms of the functional j_F. ∎

The functional $j_F : \mathcal{M}_F(\omega) \to \mathbb{R}$ *is called the* **scaled two-dimensional energy of a nonlinearly elastic flexural shell.**

We shall see in Sect. 10.6 that the minimization problem in terms of the functional j_F is in fact well-posed over a manifold *larger* than $\mathcal{M}_F(\omega)$, in the definition of which the space $\mathbf{H}^2(\omega)$ replaces the space $\mathbf{W}^{2,4}(\omega)$.

10.5. THE TWO-DIMENSIONAL EQUATIONS OF A NONLINEARLY ELASTIC FLEXURAL SHELL DERIVED BY MEANS OF A FORMAL ASYMPTOTIC ANALYSIS; COMMENTARY

In order to get physically meaningful formulas, it remains to *descale* the components ζ_i^0 of the vector field $\boldsymbol{\zeta}^0$ that satisfies the scaled two-dimensional problems found in Thms. 10.3-4 and 10.4-1. In view of the scalings

$$u_i(\varepsilon)(x) = u_i^\varepsilon(x^\varepsilon) \text{ for all } x^\varepsilon = \pi^\varepsilon x \in \overline{\Omega}^\varepsilon$$

made on the covariant components of the displacement field (Sect. 8.4), we are naturally led to defining for each $\varepsilon > 0$ the **covariant components** $\zeta_i^\varepsilon : \overline{\omega} \to \mathbb{R}$ of the **limit displacement field** $\tilde{\boldsymbol{\zeta}}^\varepsilon : \overline{\omega} \to \mathbb{R}^3$ *of the middle surface S of the shell* by letting (the vectors $\boldsymbol{a}^i$ form the contravariant basis at each point of S):

$$\boxed{\zeta_i^\varepsilon := \zeta_i^0 \text{ and } \tilde{\boldsymbol{\zeta}}^\varepsilon := \zeta_i^\varepsilon \boldsymbol{a}^i.}$$

A word of caution. As always, *the fields $\boldsymbol{\zeta}^\varepsilon := (\zeta_i^\varepsilon)$ and $\tilde{\boldsymbol{\zeta}}^\varepsilon = \zeta_i^\varepsilon \boldsymbol{a}^i$ must be carefully distinguished*! The former is essentially a convenient *mathematical "intermediary"*, but only the latter has *physical significance.* ■

Recall that $f^{i,\varepsilon} \in L^2(\Omega^\varepsilon)$ and $h^{i,\varepsilon} \in L^2(\Gamma_+^\varepsilon \cup \Gamma_-^\varepsilon)$ represent the contravariant components of the applied body and surface forces *actually* acting on the shell and that λ^ε and μ^ε denote the *actual* Lamé constants of its constituting material. We then have the following immediate corollary to Thms. 10.3-4 and 10.4-1:

Theorem 10.5-1. *Let the assumptions and notations be as in* Thm. 10.3-4. *Then the vector field $\boldsymbol{\zeta}^\varepsilon := (\zeta_i^\varepsilon)$ formed by the covariant components of the limit displacement field $\zeta_i^\varepsilon \boldsymbol{a}^i$ of the middle surface S satisfies the following* **two-dimensional variational problem $\mathcal{P}_F^\varepsilon(\omega)$ of a nonlinearly elastic flexural shell:**

$$\boxed{\begin{aligned} &\boldsymbol{\zeta}^\varepsilon \in \boldsymbol{\mathcal{M}}_F(\omega) := \{\boldsymbol{\eta} = (\eta_i) \in \mathbf{W}^{2,4}(\omega);\ \boldsymbol{\eta} = \partial_\nu \boldsymbol{\eta} = \mathbf{0} \text{ on } \gamma_0, \\ &\qquad\qquad G_{\alpha\beta}(\boldsymbol{\eta}) = 0 \text{ in } \omega\}, \\ &\frac{\varepsilon^3}{3}\int_\omega a^{\alpha\beta\sigma\tau,\varepsilon} R_{\sigma\tau}^\flat(\boldsymbol{\zeta}^\varepsilon)\big((R_{\alpha\beta}^\flat)'(\boldsymbol{\zeta}^\varepsilon)\boldsymbol{\eta}\big)\sqrt{a}\,dy = \int_\omega p^{i,\varepsilon}\eta_i\sqrt{a}\,dy \\ &\text{for all } \boldsymbol{\eta} \in \mathbb{T}_{\boldsymbol{\zeta}^\varepsilon}\boldsymbol{\mathcal{M}}_F(\omega) := \{\boldsymbol{\eta} \in \mathbf{W}^{2,4}(\omega);\ \boldsymbol{\eta} = \partial_\nu \boldsymbol{\eta} = \mathbf{0} \text{ on } \gamma_0, \\ &\qquad\qquad G'_{\alpha\beta}(\boldsymbol{\zeta}^\varepsilon)\boldsymbol{\eta} = 0 \text{ in } \omega\}, \end{aligned}}$$

where

$$G_{\alpha\beta}(\boldsymbol{\eta}) := \frac{1}{2}(a_{\alpha\beta}(\boldsymbol{\eta}) - a_{\alpha\beta})$$
$$a^{\alpha\beta\sigma\tau,\varepsilon} := \frac{4\lambda^\varepsilon\mu^\varepsilon}{\lambda^\varepsilon + 2\mu^\varepsilon} a^{\alpha\beta}a^{\sigma\tau} + 2\mu^\varepsilon(a^{\alpha\sigma}a^{\beta\tau} + a^{\alpha\tau}a^{\beta\sigma}),$$
$$p^{i,\varepsilon} := \int_{-\varepsilon}^{\varepsilon} f^{i,\varepsilon}\, dx_3^\varepsilon + h_+^{i,\varepsilon} + h_-^{i,\varepsilon} \text{ and } h_\pm^{i,\varepsilon} := h^{i,\varepsilon}(\cdot, \pm\varepsilon),$$

and the functions $R^\flat_{\alpha\beta}(\boldsymbol{\eta})$ *are defined in* Thm. 10.3-1.

The field $\boldsymbol{\zeta}^\varepsilon = (\zeta_i^\varepsilon)$ *is a solution to problem* $\mathcal{P}_F^\varepsilon(\omega)$ *if and only if it is a stationary point over the manifold* $\boldsymbol{\mathcal{M}}_F(\omega)$ *of the functional* $j_F^{\flat\varepsilon}$ *defined by*

$$j_F^{\flat\varepsilon}(\boldsymbol{\eta}) := \frac{\varepsilon^3}{6}\int_\omega a^{\alpha\beta\sigma\tau,\varepsilon} R^\flat_{\sigma\tau}(\boldsymbol{\eta}) R^\flat_{\alpha\beta}(\boldsymbol{\eta})\sqrt{a}\, dy - \int_\omega p^{i,\varepsilon}\eta_i\sqrt{a}\, dy$$

for all $\boldsymbol{\eta} = (\eta_i) \in \mathbf{W}^{2,4}(\omega)$.

Particular solutions to problem $\mathcal{P}_F^\varepsilon(\omega)$ *are thus obtained by solving the minimization problem: Find* $\boldsymbol{\zeta}^\varepsilon$ *such that* (the functions $b_{\alpha\beta}(\boldsymbol{\eta})$ are the covariant components of the curvature tensor of the surface $(\boldsymbol{\theta} + \eta_i\boldsymbol{a}^i)(\overline{\omega})$; cf. Thm. 10.3-1)

$$\boldsymbol{\zeta}^\varepsilon \in \boldsymbol{\mathcal{M}}_F(\omega) \text{ and } j_F^\varepsilon(\boldsymbol{\zeta}^\varepsilon) = \inf_{\boldsymbol{\eta}\in\boldsymbol{\mathcal{M}}_F(\omega)} j_F^\varepsilon(\boldsymbol{\eta}), \text{ where}$$
$$j_F^\varepsilon(\boldsymbol{\eta}) := \frac{\varepsilon^3}{6}\int_\omega a^{\alpha\beta\sigma\tau,\varepsilon}(b_{\sigma\tau}(\boldsymbol{\eta}) - b_{\sigma\tau})(b_{\alpha\beta}(\boldsymbol{\eta}) - b_{\alpha\beta})\sqrt{a}\, dy - \int_\omega p^{i,\varepsilon}\eta_i\sqrt{a}\, dy.$$

■

Any one of these problems satisfied by $\boldsymbol{\zeta}^\varepsilon$ constitutes one version of the **two-dimensional equations of a nonlinearly elastic flexural shell** (the specific meaning conveyed by "flexural" is given below). The functions

$$G_{\alpha\beta}(\boldsymbol{\eta}) = \frac{1}{2}(a_{\alpha\beta}(\boldsymbol{\eta}) - a_{\alpha\beta}) \text{ and } R_{\alpha\beta}(\boldsymbol{\eta}) = (b_{\alpha\beta}(\boldsymbol{\eta}) - b_{\alpha\beta})$$

are respectively the *covariant components* of the **change of metric tensor** and **change of curvature tensor** associated with a displacement field $\eta_i \boldsymbol{a}^i$ of the middle surface S (Thms. 9.1-1 and 10.3-1) and the functions $a^{\alpha\beta\sigma\tau,\varepsilon}$ are the *contravariant components* of the **two-dimensional elasticity tensor of the shell.**

The functional $j_F^\varepsilon : \boldsymbol{\mathcal{M}}_F(\omega) \to \mathbb{R}$ is the **two-dimensional energy**, and the functional

$$\boldsymbol{\eta} \in \boldsymbol{\mathcal{M}}_F(\omega) \to \frac{\varepsilon^3}{6} \int_\omega a^{\alpha\beta\sigma\tau,\varepsilon}(b_{\sigma\tau}(\boldsymbol{\eta}) - b_{\sigma\tau})(b_{\alpha\beta}(\boldsymbol{\eta}) - b_{\alpha\beta})\sqrt{a}\, dy$$

is the **two-dimensional strain energy, of a nonlinearly elastic flexural shell.** Finally, the boundary conditions $\boldsymbol{\eta} = \partial_\nu \boldsymbol{\eta} = \mathbf{0}$ on γ_0 express that the shell is **strongly clamped** along γ_0 (the geometric interpretation of these boundary conditions is given *infra*).

A *major conclusion* is thus that, without any recourse to any *a priori* assumption of a geometrical or mechanical nature, *the method of formal asymptotic expansions provides a justification of the two-dimensional equations of a nonlinearly elastic flexural shell*, in the forms found in Thm. 10.5-1. This justification by Lods & Miara [1998] constitutes a generalization to *shells* of the formal analysis of Fox, Raoult & Simo [1993] in the *planar case*, i.e., when the mapping $\boldsymbol{\theta}$ is of the form $\boldsymbol{\theta}(y_1, y_2) = (y_1, y_2, 0)$ for all $y = (y_1, y_2) \in \overline{\omega}$ (see also Vol. II, Sect. 4.12).

Further important conclusions and comments are in order about the present justification:

First and foremost, the resulting shell theory is a **nonlinear flexural theory** in the sense that the **stored energy function of a nonlinearly elastic flexural shell**, defined by:

$$\boldsymbol{\eta} = (\eta_i) \in \boldsymbol{\mathcal{M}}_F(\omega) \to \frac{\varepsilon^3}{6} a^{\alpha\beta\sigma\tau,\varepsilon}(b_{\sigma\tau}(\boldsymbol{\eta}) - b_{\sigma\tau})(b_{\alpha\beta}(\boldsymbol{\eta}) - b_{\alpha\beta}),$$

is a quadratic and positive definite expression (*via* the two-dimensional elasticity tensor of the shell) in terms of the *change of curvature tensor*, i.e., of the *exact difference* between the *curvature tensor* of the *deformed* middle surface and that of the *undeformed* one, while the associated energy is to be minimized over displacement fields for which the *change of metric tensor*, i.e., the *exact difference* between the *metric tensor* of the *deformed* middle surface and that of the *undeformed* one, vanishes.

The *existence theory* for this minimization problem is treated in the next section, where further relevant comments are also to be found.

Note in passing the truly remarkable simplicity of this minimization problem!

Second, the resulting theory is **frame-indifferent** in the sense that the value of the above stored energy function is unaltered if, given any $\boldsymbol{\eta} = (\eta_i) \in \boldsymbol{\mathcal{M}}_F(\omega)$, $\boldsymbol{\psi} := \boldsymbol{\theta} + \eta_i \boldsymbol{a}^i$ is replaced by $\mathbf{Q}\boldsymbol{\psi}$, where $\mathbf{Q}$ is any orthogonal matrix of order three (Ex. 10.5). It is also a **large displacement**, or equivalently a **large deformation, theory**, in the sense that *the de-scaling produces a displacement field that is $O(1)$ with respect to ε.*

We now interpret the boundary conditions $\boldsymbol{\eta} = \partial_\nu \boldsymbol{\eta} = \mathbf{0}$ on γ_0. To this end, let $\eta_i \boldsymbol{a}^i$ be a displacement field of the middle surface $S = \boldsymbol{\theta}(\overline{\omega})$ with smooth enough, but otherwise arbitrary, covariant components $\eta_i : \overline{\omega} \to \mathbb{R}$. The tangent plane at an arbitrary point $\boldsymbol{\theta}(y) + \eta_i(y)\boldsymbol{a}^i(y)$, $y \in \overline{\omega}$, of the deformed surface $(\boldsymbol{\theta} + \eta_i \boldsymbol{a}^i)(\overline{\omega})$ is thus spanned by the vectors

$$\partial_\alpha(\boldsymbol{\theta} + \eta_i \boldsymbol{a}^i)(y) = \boldsymbol{a}_\alpha(y) + \partial_\alpha \eta_i(y)\boldsymbol{a}^i(y) + \eta_i(y)\partial_\alpha \boldsymbol{a}^i(y),$$

if these are linearly independent. Since

$$\boldsymbol{\eta} = \partial_\nu \boldsymbol{\eta} = \mathbf{0} \text{ on } \gamma_0 \ \Rightarrow\ \eta_i = \partial_\alpha \eta_i = 0 \text{ on } \gamma_0,$$

it follows that

$$\boldsymbol{\theta}(y) + \eta_i(y)\boldsymbol{a}^i(y) = \boldsymbol{\theta}(y) \text{ and } \partial_\alpha(\boldsymbol{\theta} + \eta_i \boldsymbol{a}^i)(y) = \boldsymbol{a}_\alpha(y) \text{ for all } y \in \gamma_0.$$

These relations thus show that *not only the points and the tangent spaces* (as when only the weaker "boundary conditions of clamping" $\eta_i = \partial_\nu \eta_3 = 0$ are imposed on γ_0; cf. Sect. 6.3), *but also the tangent vectors to the coordinate lines, of the deformed and undeformed middle surfaces coincide along the curve $\boldsymbol{\theta}(\gamma_0)$*. Such **two-dimensional boundary conditions of strong clamping** are suggested in Figs. 10.2-1 to 10.2.3.

The *formal linearization* of the equations found in Thm. 10.5-1 produces the equations of a *linearly elastic flexural shell* (Chap. 6). This is perhaps best seen on the minimization problem: In the linearization process, the covariant components $R_{\alpha\beta}(\boldsymbol{\eta}) = b_{\alpha\beta}(\boldsymbol{\eta}) - b_{\alpha\beta}$

of the change of curvature tensor entering the energy become *by definition* the covariant components $\rho_{\alpha\beta}(\boldsymbol{\eta})$ of the *linearized* change of curvature tensor (Sect. 2.5); likewise, the covariant components $G_{\alpha\beta}(\boldsymbol{\eta}) = \frac{1}{2}(a_{\alpha\beta}(\boldsymbol{\eta}) - a_{\alpha\beta})$ of the change of metric tensor entering the definition of the manifold $\mathcal{M}_F(\omega)$ become *by definition* the covariant components $\gamma_{\alpha\beta}(\boldsymbol{\eta})$ of the *linearized* change of metric tensor (Sect. 2.4), so that the *energy* becomes that of the *linear* flexural theory (Sect. 6.3) and the *manifold* $\mathcal{M}_F(\omega)$ likewise becomes the *vector space*

$$\mathbf{V}_F(\omega) := \{\boldsymbol{\eta} = (\eta_i) \in H^1(\omega) \times H^1(\omega) \times H^2(\omega); \\ \eta_i = \partial_\nu \eta_3 = 0 \text{ on } \gamma_0,\ \gamma_{\alpha\beta}(\boldsymbol{\eta}) = 0 \text{ in } \omega\}$$

of the *linear* flexural theory (Sect. 6.3)

Remarks. (1) The *function spaces* and the *boundary conditions*, which become in the linearization process those of the space $\mathbf{V}_F(\omega)$, have to be modified so that the two-dimensional equations of the linear flexural shell theory become well-posed (Thm. 6.3-1).

(2) As observed by Lods & Miara [1998], the equations of a *linearly elastic flexural shell* are also recovered by considering forces of higher order; cf. Ex. 10.6. ∎

Finally, note that Collard & Miara [1999] have shown that the formal analysis of Lods & Miara [1998] described *supra* also leads to the explicit computation of the *limit stresses* in a nonlinearly elastic flexural shell.

10.6. EXISTENCE OF SOLUTIONS TO THE MINIMIZATION PROBLEM

In this section, which is based on Ciarlet & Coutand [1998], we show that the minimization problem for a nonlinearly elastic flexural shell found in Thm. 10.5-1 has at least one solution, once the manifold $\mathcal{M}_F(\omega)$ is replaced by a *larger* manifold, where the space $\mathbf{H}^2(\omega)$ replaces the space $\mathbf{W}^{2,4}(\omega)$.

For notational convenience, we keep the *same* notation $\mathcal{M}_F(\omega)$ for the enlarged manifold and we suppress in this section all the *exponents* "ε" appearing in the formulation of this problem. Our *point of departure* is thus the following:

The *unknowns* are the three covariant components $\zeta_i : \overline{\omega} \to \mathbb{R}$ of the *limit displacement field* $\zeta_i \boldsymbol{a}^i$ of the middle surface S of the shell. This means that $\zeta_i(y)\boldsymbol{a}^i(y)$ is the *displacement* of the point $\boldsymbol{\theta}(y)$ of the middle surface for each $y \in \overline{\omega}$. The unknown vector field $\boldsymbol{\zeta} := (\zeta_i) : \overline{\omega} \to \mathbb{R}^3$ should then satisfy the following *minimization problem:*

$$\boldsymbol{\zeta} \in \boldsymbol{\mathcal{M}}_F(\omega) = \{\boldsymbol{\eta} = (\eta_i) \in \mathbf{H}^2(\omega);\ \boldsymbol{\eta} = \partial_\nu \boldsymbol{\eta} = \mathbf{0} \text{ on } \gamma_0,\ a_{\alpha\beta}(\boldsymbol{\eta}) - a_{\alpha\beta} = 0 \text{ in } \omega\},$$

$$j_F(\boldsymbol{\zeta}) = \inf_{\boldsymbol{\eta} \in \boldsymbol{\mathcal{M}}_F(\omega)} j_F(\boldsymbol{\eta}), \text{ where}$$

$$j_F(\boldsymbol{\eta}) = \frac{\varepsilon^3}{6} \int_\omega a^{\alpha\beta\sigma\tau}(b_{\sigma\tau}(\boldsymbol{\eta}) - b_{\sigma\tau})(b_{\alpha\beta}(\boldsymbol{\eta}) - b_{\alpha\beta})\sqrt{a}\, dy - \int_\omega p^i \eta_i \sqrt{a}\, dy,$$

where γ_0 is a portion of the boundary γ satisfying

$$\textit{length}\ \gamma_0 > 0,$$

∂_ν denotes the outer normal derivative operator along γ, the functions $a_{\alpha\beta}(\boldsymbol{\eta}) : \overline{\omega} \to \mathbb{R}$ and $b_{\alpha\beta}(\boldsymbol{\eta}) : \overline{\omega} \to \mathbb{R}$ are the covariant components of the metric and curvature tensors of the *deformed* surface $(\boldsymbol{\theta} + \eta_i \boldsymbol{a}^i)(\overline{\omega})$ corresponding to an arbitrary field $\boldsymbol{\eta} = (\eta_i) \in \boldsymbol{\mathcal{M}}_F(\omega)$,

$$a^{\alpha\beta\sigma\tau} := \frac{4\lambda\mu}{\lambda + 2\mu} a^{\alpha\beta} a^{\sigma\tau} + 2\mu(a^{\alpha\sigma} a^{\beta\tau} + a^{\alpha\tau} a^{\beta\sigma})$$

designate the contravariant components of the *two-dimensional elasticity tensor of the shell*, and finally, the given functions $p^i \in L^2(\omega)$ account for the applied forces.

Two immediate comments are in order about this minimization problem: *First*, it is clear that the "interesting" situations covered by the present theory are those where *the manifold* $\boldsymbol{\mathcal{M}}_F(\omega)$ *contains other fields* $\boldsymbol{\eta}$ *than* $\boldsymbol{\eta} = \mathbf{0}$. Observe in this respect that such an assumption, which in effect depends only upon the *geometry of* S and upon the *subset* γ_0 *of* γ, is one of the assumptions made in the *definition of a nonlinearly elastic flexural shell* (Sect. 10.2).

As already noted, this assumption means that there exist *nonzero* displacement fields $\eta_i \boldsymbol{a}^i$ of the middle surface S that are *inextensional*, in the sense that they preserve the metric of the surface S, and that are also *admissible*, in the sense that they satisfy the two-dimensional boundary conditions of strong clamping $\boldsymbol{\eta} = \partial_\nu \boldsymbol{\eta} = \mathbf{0}$ on γ_0. This is the case for instance if the middle surface S is either a portion of a cone (excluding its vertex) or a portion of a non-planar cylinder and the shell is subjected to a boundary condition of place along a portion of its lateral face whose middle curve $\boldsymbol{\theta}(\gamma_0)$ is a subset of one of its generatrices (if S is a non-planar cylinder, $\boldsymbol{\theta}(\gamma_0)$ may consist of subsets of two generatrices); cf. Figs. 10.1-1 and 10.1-2 and Exs. 10.1 and 10.2. These developable surfaces constitute canonical examples, but more "exotic" examples can be found; cf. Ex. 10.7.

Second, we must verify that *the functional j_F is well defined on the manifold $\boldsymbol{\mathcal{M}}_F(\omega)$*. To this end, we will show (part (ii) of the proof of Thm. 10.6-1) that the vectors $\partial_\alpha(\boldsymbol{\theta} + \eta_i \boldsymbol{a}^i)$ are linearly independent if $\boldsymbol{\eta} = (\eta_i) \in \boldsymbol{\mathcal{M}}_F(\omega)$ and that the functions $b_{\alpha\beta}(\boldsymbol{\eta})$, which are then well defined, belong to $L^2(\omega)$.

It turns out that the existence theory is substantially easier to carry out if, *instead of $\boldsymbol{\zeta} = (\zeta_i)$ and $\boldsymbol{\eta} = (\eta_i)$, the vector fields*

$$\boxed{\boldsymbol{\varphi} := \boldsymbol{\theta} + \zeta_i \boldsymbol{a}^i \text{ and } \boldsymbol{\psi} := \boldsymbol{\theta} + \eta_i \boldsymbol{a}^i}$$

are taken as the new unknown and new "trial functions". In other words, the new unknown is the **limit deformation field** of the middle surface of the shell: This means that $\boldsymbol{\varphi}(y) = \boldsymbol{\theta}(y) + \zeta_i(y)\boldsymbol{a}^i(y)$ is the *deformation* of the point $\boldsymbol{\theta}(y)$ of the middle surface for each $y \in \overline{\omega}$.

A word of caution. The minimization problem is thus still expressed in terms of *curvilinear coordinates* (the coordinates y_α of the points $y \in \overline{\omega}$), but the unknowns are *no longer covariant components* (the functions $\zeta_i : \overline{\omega} \to \mathbb{R}$) over the contravariant bases along the surface S. Instead, the unknowns are now the *Cartesian components*, i.e., over a *fixed* Cartesian frame, of the unknown deformation field $\boldsymbol{\varphi} : \overline{\omega} \to \mathbb{R}^3$. ∎

Using in particular the relations $\eta_j = (\eta_i \boldsymbol{a}^i) \cdot \boldsymbol{a}_j$, we remark that the equivalence

$$\boldsymbol{\eta} = (\eta_i) \in \mathbf{H}^2(\omega) \iff \boldsymbol{\theta} + \eta_i \boldsymbol{a}^i = \boldsymbol{\psi} \in \mathbf{H}^2(\omega)$$

holds, since the assumption $\boldsymbol{\theta} \in \mathcal{C}^3(\overline{\omega}; \mathbb{R}^3)$ made throughout this chapter (see, e.g., Thm. 10.3-4) implies that $\boldsymbol{a}_i \in \mathcal{C}^2(\overline{\omega}; \mathbb{R}^3)$ and $\boldsymbol{a}^i \in \mathcal{C}^2(\overline{\omega}; \mathbb{R}^3)$.

If $\boldsymbol{\psi} \in \mathbf{H}^2(\omega)$, we define a.e. in ω (here and subsequently, the abbreviation "a.e." stands for "almost everywhere") the vector fields

$$\boldsymbol{a}_\alpha(\boldsymbol{\psi}) := \partial_\alpha \boldsymbol{\psi}$$

and the functions

$$a_{\alpha\beta}(\boldsymbol{\psi}) := \boldsymbol{a}_\alpha(\boldsymbol{\psi}) \cdot \boldsymbol{a}_\beta(\boldsymbol{\psi}) = a_{\beta\alpha}(\boldsymbol{\psi}).$$

If $\boldsymbol{\psi} \in \mathbf{H}^2(\omega)$ is such that the two vectors $\boldsymbol{a}_\alpha(\boldsymbol{\psi})$ are linearly independent a.e. in ω, we also define a.e. in ω the vector field

$$\boldsymbol{a}_3(\boldsymbol{\psi}) := \frac{\boldsymbol{a}_1(\boldsymbol{\psi}) \wedge \boldsymbol{a}_2(\boldsymbol{\psi})}{|\boldsymbol{a}_1(\boldsymbol{\psi}) \wedge \boldsymbol{a}_2(\boldsymbol{\psi})|},$$

the functions

$$b_{\alpha\beta}(\boldsymbol{\psi}) := \partial_{\alpha\beta}\boldsymbol{\psi} \cdot \boldsymbol{a}_3(\boldsymbol{\psi}) = b_{\beta\alpha}(\boldsymbol{\psi}),$$

and finally, the vector fields $\boldsymbol{a}^i(\boldsymbol{\psi})$ by means of the relations

$$\boldsymbol{a}^i(\boldsymbol{\psi}) \cdot \boldsymbol{a}_j(\boldsymbol{\psi}) = \delta^i_j.$$

The vectors $\boldsymbol{a}_\alpha(\boldsymbol{\psi})(y)$ and $\boldsymbol{a}^\beta(\boldsymbol{\psi})(y)$ thus form respectively the covariant and contravariant bases of the tangent plane to the *deformed* surface $\boldsymbol{\psi}(\overline{\omega})$ at the point $\boldsymbol{\psi}(y)$, for almost all $y \in \omega$.

Remark. I hope, but am not entirely convinced, that I will be forgiven for the lack of consistency observed between, e.g., the notations $\boldsymbol{a}_i(\boldsymbol{\eta})$ used so far and $\boldsymbol{a}_i(\boldsymbol{\psi})$ used in this section. At least, these inconsistencies spared me the burden of introducing yet further notations! ■

We now establish an *existence result for the minimization problem of a nonlinearly elastic flexural shell*, reformulated in terms of the new unknown $\boldsymbol{\varphi}$ and trial functions $\boldsymbol{\psi}$. Note that, for simplicity, we impose only *one* boundary condition on γ_0, viz., $\boldsymbol{\varphi} = \boldsymbol{\varphi}_0$ on γ_0 (this boundary condition is thus more general than its special case $\boldsymbol{\varphi} = \boldsymbol{\theta}$ on γ_0 considered until now), as the following existence theory holds in this case of a shell that is **simply supported**, as well as in the case (considered so far) of a shell that is *strongly clamped.*

Note also that this existence result holds under a substantially *weaker regularity assumption on the mapping* $\boldsymbol{\theta}$ than that made so far, viz., $\boldsymbol{\theta} \in \mathcal{C}^3(\overline{\omega}; \mathbb{R}^3)$. This assumption was nevertheless essential for carrying out the asymptotic analysis to its term (Thm. 10.3-4).

The following *existence* result is due to Ciarlet & Coutand [1998]. It extends to shells the existence theorem established by Coutand [1997a] (see also Vol. II, Thm. 4.12-3) for the minimization problem justified by Fox, Raoult & Simo [1993] in the *planar case.* The proof is, however, substantially different.

The *uniqueness* or *multiplicity* of solutions to this problem are studied in Coutand [1999a, 1999f].

Theorem 10.6-1. *Let there be given a mapping* $\boldsymbol{\theta} \in W^{2,p}(\omega; \mathbb{R}^3)$ *with* $p > 2$, *such that the two vectors* $\boldsymbol{a}_\alpha = \partial_\alpha \boldsymbol{\theta}$ *are linearly independent at all points of* $\overline{\omega}$. *Let there be given a subset* $\gamma_0 \subset \gamma$ *such that* length $\gamma_0 > 0$ *and let there be given a mapping* $\boldsymbol{\varphi}_0 : \gamma_0 \to \mathbb{R}^3$ *such that the* **manifold of admissible inextensional deformations**

$$\boldsymbol{\Phi}_F(\omega) := \{\boldsymbol{\psi} \in \mathbf{H}^2(\omega); \boldsymbol{\psi} = \boldsymbol{\varphi}_0 \text{ on } \gamma_0, a_{\alpha\beta}(\boldsymbol{\psi}) - a_{\alpha\beta} = 0 \text{ in } \omega\}$$

is not empty. Then if $\boldsymbol{\psi} \in \boldsymbol{\Phi}_F(\omega)$, *the vectors* $\boldsymbol{a}_\alpha(\boldsymbol{\psi}) = \partial_\alpha \boldsymbol{\psi}$ *are linearly independent a.e. in* ω *and the functions* $b_{\alpha\beta}(\boldsymbol{\psi})$ *are in* $L^2(\omega)$.

Given a continuous linear form L *on* $\mathbf{H}^2(\omega)$, *define the* **two-dimensional energy** $I_F : \boldsymbol{\Phi}_F(\omega) \to \mathbb{R}$ **of a nonlinearly elastic flexural shell** *by*

$$I_F(\boldsymbol{\psi}) = \frac{\varepsilon^3}{6} \int_\omega a^{\alpha\beta\sigma\tau}(b_{\sigma\tau}(\boldsymbol{\psi}) - b_{\sigma\tau})(b_{\alpha\beta}(\boldsymbol{\psi}) - b_{\alpha\beta})\sqrt{a}\, dy - L(\boldsymbol{\psi})$$

for all $\boldsymbol{\psi} \in \boldsymbol{\Phi}_F(\omega)$. *Then there is at least one* $\boldsymbol{\varphi}$ *such that*

$$\boldsymbol{\varphi} \in \boldsymbol{\Phi}_F(\omega) \text{ and } I_F(\boldsymbol{\varphi}) = \inf_{\boldsymbol{\psi} \in \boldsymbol{\Phi}_F(\omega)} I_F(\boldsymbol{\psi}).$$

Proof. For the sake of clarity, the proof is broken into a series of seven parts, numbered (i) to (vii). In the first five parts, we establish various properties of the *manifold* $\mathbf{\Phi}_F(\omega)$, while in the remaining two, we establish properties of the *functional* I_F over this manifold, the combination of which eventually leads to the existence theorem. Note in passing that the properties established in parts (i) to (v) may also be viewed as *properties of isometric surfaces.*

The norms in the spaces $L^2(\omega)$ and $H^m(\omega)$, $m \geq 1$, are denoted as usual $|\cdot|_{0,\omega}$ and $\|\cdot\|_{m,\omega}$ and those in the spaces $L^\infty(\omega)$ and $W^{1,\infty}(\omega)$ are denoted $|\cdot|_{0,\infty,\omega}$ and $\|\cdot\|_{1,\infty,\omega}$. The same notations are used for the norms in the corresponding spaces of vector fields, such spaces being then denoted by boldface letters. Strong and weak convergences are denoted by $\to$ and $\rightharpoonup$, respectively (a review of all the properties relevant here about weak convergence and lower semi-continuity is found, e.g., in Vol. I, Sects. 7.1 and 7.2).

The proof follows a pattern familiar in the calculus of variations: After showing that *the manifold* $\mathbf{\Phi}_F(\omega)$ *is sequentially weakly closed* (part (i)), we establish that *the functional* I_F *is sequentially weakly lower semi-continuous and coercive over* $\mathbf{\Phi}_F(\omega)$ (parts (vi) and (vii)), all these properties holding with respect to the topology of the space $\mathbf{H}^2(\omega)$. The *existence of a minimizer of* I_F *over* $\mathbf{\Phi}_F(\omega)$ then classically follows from these properties; see, e.g., Dacorogna [1989, Chap. 1] or Struwe [1990, Chap. 1].

Note that the coerciveness of the functional hinges on the crucial property that *the manifold* $\mathbf{\Phi}_F(\omega)$ *lies in a bounded subset of* $\mathbf{W}^{1,\infty}(\omega)$ (part (iii)).

(i) *As a subset of* $\mathbf{H}^2(\omega)$, *the manifold* $\mathbf{\Phi}_F(\omega)$ *is sequentially weakly closed, i.e.,*

$$\boldsymbol{\psi}^l \in \mathbf{\Phi}_F(\omega),\ l \geq 1,\ \text{and}\ \boldsymbol{\psi}^l \rightharpoonup \boldsymbol{\psi}\ \text{in}\ \mathbf{H}^2(\omega) \Rightarrow \boldsymbol{\psi} \in \mathbf{\Phi}_F(\omega).$$

Let $\boldsymbol{\psi}^l \in \mathbf{\Phi}_F(\omega)$, $l \geq 1$, be such that $\boldsymbol{\psi}^l \rightharpoonup \boldsymbol{\psi}$ in $\mathbf{H}^2(\omega)$. Since the trace operator *tr* from $H^2(\omega)$ into $L^2(\gamma_0)$ is continuous with respect to the strong topologies of both spaces, it remains so with respect to the weak topologies of both spaces (see, e.g., Brezis [1983, Thm. III.9]). Hence tr $\boldsymbol{\psi}^l \rightharpoonup$ tr $\boldsymbol{\psi}$ in $L^2(\gamma_0)$ and thus tr $\boldsymbol{\psi} = \boldsymbol{\varphi}_0$ on γ_0 since tr $\boldsymbol{\psi}^l = \boldsymbol{\varphi}_0$ on γ_0 for all $l \geq 1$.

By the Rellich-Kondrašov imbedding theorem (see, e.g., Vol. I, Thm. 6.1-5), $\boldsymbol{\psi}^l \to \boldsymbol{\psi}$ in $\mathbf{H}^1(\omega)$; hence

$$a_{\alpha\beta}(\boldsymbol{\psi}^l) = \boldsymbol{a}_\alpha(\boldsymbol{\psi}^l) \cdot \boldsymbol{a}_\beta(\boldsymbol{\psi}^l) \to \boldsymbol{a}_\alpha(\boldsymbol{\psi}) \cdot \boldsymbol{a}_\beta(\boldsymbol{\psi}) = a_{\alpha\beta}(\boldsymbol{\psi}) \text{ in } L^1(\omega).$$

Since $a_{\alpha\beta}(\boldsymbol{\psi}^l) = a_{\alpha\beta}$ a.e. in ω for all l, we conclude that $a_{\alpha\beta}(\boldsymbol{\psi}) = a_{\alpha\beta}$ a.e. in ω; hence $\boldsymbol{\psi} \in \boldsymbol{\Phi}_F(\omega)$ as was to be proved.

Should the manifold $\boldsymbol{\Phi}_F(\boldsymbol{\omega})$ include a second boundary condition of the form $\partial_\nu \boldsymbol{\psi} = \boldsymbol{\varphi}_1$ on γ_0 (recall that the manifold $\boldsymbol{\mathcal{M}}_F(\omega)$ comprises a boundary condition of the form $\partial_\nu \boldsymbol{\eta} = \mathbf{0}$ on γ_0), a similar argument shows that such a manifold is again sequentially weakly closed.

(ii) *There exists* C_1 *such that, for all vector fields* $\boldsymbol{\psi} \in \mathbf{H}^2(\omega)$ *satisfying* $a_{\alpha\beta}(\boldsymbol{\psi}) = a_{\alpha\beta}$ *a.e. in* ω,

$$0 < C_1 \leq |\boldsymbol{a}_1(\boldsymbol{\psi}) \wedge \boldsymbol{a}_2(\boldsymbol{\psi})| \text{ a.e. in } \omega,$$
$$C_1^{-1} \leq |\boldsymbol{a}_\alpha(\boldsymbol{\psi})| \leq C_1 \text{ a.e. in } \omega.$$

Consequently, the vectors $\boldsymbol{a}_i(\boldsymbol{\psi})$ *and* $\boldsymbol{a}^i(\boldsymbol{\psi})$ *associated with such vector fields* $\boldsymbol{\psi}$ *are well defined and "uniformly" linearly independent a.e. in* ω*, the corresponding functions* $b_{\alpha\beta}(\boldsymbol{\psi})$ *are in* $L^2(\omega)$*, and the functional* I_F *is well defined over the manifold* $\boldsymbol{\Phi}_F(\omega)$ (that the functions $b_{\alpha\beta}(\boldsymbol{\psi})$ are indeed well-defined when $\boldsymbol{\psi}$ belongs to the manifold $\boldsymbol{\Phi}_F(\omega)$ was already observed in Thm. 10.3-1).

Since the set $\overline{\omega}$ is compact, the vectors $\boldsymbol{a}_\alpha = \partial_\alpha \boldsymbol{\theta}$ are "uniformly" linearly independent in $\overline{\omega}$ (they belong to the space $W^{1,p}(\omega; \mathbb{R}^3)$, which is continuously imbedded into the space $\mathcal{C}^0(\overline{\omega}; \mathbb{R}^3)$ since $p > 2$), in the sense that there exist c_1 and c_2 such that

$$0 \leq c_1 < 1 \text{ and } |\boldsymbol{a}_1 \cdot \boldsymbol{a}_2| \leq c_1 |\boldsymbol{a}_1||\boldsymbol{a}_2| \text{ in } \overline{\omega},$$
$$0 < c_2 \leq 1 \text{ and } c_2 \leq |\boldsymbol{a}_\alpha| \leq c_2^{-1} \text{ in } \overline{\omega}.$$

Furthermore,

$$\boldsymbol{a}_1(\boldsymbol{\psi}) \cdot \boldsymbol{a}_2(\boldsymbol{\psi}) = a_{12}(\boldsymbol{\psi}) = a_{12} = \boldsymbol{a}_1 \cdot \boldsymbol{a}_2 \text{ a.e. in } \omega,$$
$$|\boldsymbol{a}_1(\boldsymbol{\psi})|^2 = a_{11}(\boldsymbol{\psi}) = a_{11} = |\boldsymbol{a}_1|^2 \text{ a.e. in } \omega,$$
$$|\boldsymbol{a}_2(\boldsymbol{\psi})|^2 = a_{22}(\boldsymbol{\psi}) = a_{22} = |\boldsymbol{a}_2|^2 \text{ a.e. in } \omega;$$

consequently,

$$|\boldsymbol{a}_1(\boldsymbol{\psi}) \cdot \boldsymbol{a}_2(\boldsymbol{\psi})| \leq c_1 |\boldsymbol{a}_1(\boldsymbol{\psi})||\boldsymbol{a}_2(\boldsymbol{\psi})| \text{ a.e. in } \omega.$$

This inequality shows that the vectors $\boldsymbol{a}_1(\boldsymbol{\psi})$ and $\boldsymbol{a}_2(\boldsymbol{\psi})$ are likewise "uniformly" linearly independent a.e. in ω, since $c_1 < 1$. Hence there exists $c_3 > 0$ such that

$$c_3|\boldsymbol{a}_1(\boldsymbol{\psi}) \wedge \boldsymbol{a}_2(\boldsymbol{\psi})| \geq |\boldsymbol{a}_1(\boldsymbol{\psi})||\boldsymbol{a}_2(\boldsymbol{\psi})| = |\boldsymbol{a}_1||\boldsymbol{a}_2| \text{ a.e. in } \omega,$$

and thus there exists a constant C_1 such that the two announced inequalities hold. The vector

$$\boldsymbol{a}_3(\boldsymbol{\psi}) = \boldsymbol{a}^3(\boldsymbol{\psi}) = \frac{\boldsymbol{a}_1(\boldsymbol{\psi}) \wedge \boldsymbol{a}_2(\boldsymbol{\psi})}{|\boldsymbol{a}_1(\boldsymbol{\psi}) \wedge \boldsymbol{a}_2(\boldsymbol{\psi})|}$$

is thus well defined a.e. in ω. Consequently,

$$b_{\alpha\beta}(\boldsymbol{\psi}) = \partial_{\alpha\beta}\boldsymbol{\psi} \cdot \boldsymbol{a}_3(\boldsymbol{\psi}) \in L^2(\omega),$$

since $|\boldsymbol{a}_3(\boldsymbol{\psi})| = 1$ a.e. in ω. The vectors $\boldsymbol{a}^\alpha(\boldsymbol{\psi})$ are likewise well defined and "uniformly" linearly independent a.e. in ω.

(iii) *Let* $\boldsymbol{\psi} \in \mathbf{H}^2(\omega)$ *be such that* $a_{\alpha\beta}(\boldsymbol{\psi}) = a_{\alpha\beta}$ a.e. *in* ω. *Then* $\boldsymbol{\psi} \in \mathbf{W}^{1,\infty}(\omega)$ *and there exists* C_2 *such that*

$$|\partial_\alpha\boldsymbol{\psi}|_{0,\infty,\omega} \leq C_2 \text{ for all } \boldsymbol{\psi} \in \mathbf{H}^2(\omega).$$

In addition, there exists C_3 *such that*

$$\|\boldsymbol{\psi}\|_{1,\infty,\omega} \leq C_3 \text{ for all } \boldsymbol{\psi} \in \boldsymbol{\Phi}_F(\omega).$$

Let $\boldsymbol{\psi} = (\psi_i) \in \mathbf{H}^2(\omega)$ be such that $a_{\alpha\beta}(\boldsymbol{\psi}) = a_{\alpha\beta}$ a.e. in ω. We already noticed that there exists $c_2 > 0$ independent of such fields $\boldsymbol{\psi}$ such that, for almost all $y \in \omega$,

$$|\partial_\alpha\boldsymbol{\psi}(y)|^2 = |\boldsymbol{a}_\alpha(\boldsymbol{\psi})(y)|^2 = |\boldsymbol{a}_\alpha(y)|^2 \leq c_2^{-2}.$$

This shows that $\partial_\alpha\boldsymbol{\psi} \in \mathbf{L}^\infty(\omega)$, hence that $\boldsymbol{\psi} \in \mathbf{W}^{1,\infty}(\omega)$ since in addition $\boldsymbol{\psi} \in \mathbf{H}^2(\omega) \hookrightarrow \mathcal{C}^0(\overline{\omega}; \mathbb{R}^3)$; it also shows that there exists C_2 independent of such fields $\boldsymbol{\psi}$ such that $|\partial_\alpha\boldsymbol{\psi}|_{0,\infty,\omega} \leq C_2$.

Let $q > 2$ be fixed. By assumption, *length* $\gamma_0 > 0$; hence, by the *generalized Poincaré inequality* (see, e.g., Vol. I, Thm. 6.1-8), there exists $c_4 = c_4(q, \gamma_0)$ such that, for all $\boldsymbol{\psi} \in \mathbf{W}^{1,q}(\omega)$,

$$\int_\omega |\boldsymbol{\psi}|^q \, dy \leq c_4\Big\{\int_\omega \sum_\alpha |\partial_\alpha\boldsymbol{\psi}|^q \, dy + \Big|\int_{\gamma_0} \boldsymbol{\psi} d\gamma\Big|^q\Big\}.$$

Let $\boldsymbol{\psi} \in \boldsymbol{\Phi}_F(\omega)$. Then

$$\int_\omega \sum_\alpha |\partial_\alpha \boldsymbol{\psi}|^q \, dy \le 2C_2^q \int_\omega dy \text{ and } \left| \int_{\gamma_0} \boldsymbol{\psi} \, d\gamma \right|^q = \left| \int_{\gamma_0} \boldsymbol{\varphi}_0 \, d\gamma \right|^q.$$

Since the field $\boldsymbol{\varphi}_0 : \gamma_0 \to \mathbb{R}^3$ is continuous on γ_0 (as the trace on γ_0 of a function in $\mathbf{H}^2(\omega)$; the set $\boldsymbol{\Phi}_F(\omega)$ is not empty by assumption), there exists $c_5 = c_5(c_4, C_2, \boldsymbol{\varphi}_0)$ such that, for all $\boldsymbol{\psi} \in \boldsymbol{\Phi}_F(\omega)$,

$$\|\boldsymbol{\psi}\|^q_{\mathbf{W}^{1,q}(\omega)} = \int_\omega \left\{ |\boldsymbol{\psi}|^q + \sum_\alpha |\partial_\alpha \boldsymbol{\psi}|^q \right\} dy \le c_5.$$

The Sobolev imbedding $W^{1,q}(\omega) \hookrightarrow \mathcal{C}^0(\overline{\omega})$ then implies the existence of $c_6 = c_6(c_5)$ such that, for all $\boldsymbol{\psi} \in \boldsymbol{\Phi}_F(\omega)$,

$$|\boldsymbol{\psi}|_{0,\infty,\omega} \le c_6,$$

and the second assertion is proved.

If $\boldsymbol{\psi} \in \boldsymbol{\Phi}_F(\omega)$, the components $\partial_{\alpha\beta}\boldsymbol{\psi} \cdot \boldsymbol{a}_\sigma(\boldsymbol{\psi})$ of the vector fields $\partial_{\alpha\beta}\boldsymbol{\psi} = \partial_\alpha \boldsymbol{a}_\beta(\boldsymbol{\psi}) \in \mathbf{L}^2(\omega)$ over the vectors $\boldsymbol{a}^\sigma(\boldsymbol{\psi})$ of the contravariant basis of the tangent plane to the *deformed* surface $\boldsymbol{\psi}(\overline{\omega})$ are in $L^2(\omega)$, since $\boldsymbol{a}_\sigma(\boldsymbol{\psi}) = \partial_\sigma \boldsymbol{\psi} \in \mathbf{L}^\infty(\omega)$ by (iii). We next show that these components remain in a bounded subset of the space $L^2(\omega)$ when $\boldsymbol{\psi}$ varies in the set $\boldsymbol{\Phi}_F(\omega)$:

(iv) *There exists* C_4 *such that*

$$|\partial_{\alpha\beta}\boldsymbol{\psi} \cdot \boldsymbol{a}_\sigma(\boldsymbol{\psi})|_{0,\omega} \le C_4 \text{ for all } \boldsymbol{\psi} \in \boldsymbol{\Phi}_F(\omega).$$

By assumption, $\boldsymbol{\theta} \in W^{2,p}(\omega; \mathbb{R}^3)$ with $p > 2$; as a consequence, $\partial_{\alpha\beta}\boldsymbol{\theta} \cdot \partial_\sigma \boldsymbol{\theta} \in L^p(\omega) \subset L^2(\omega)$. Differentiating the relations

$$\partial_\alpha \boldsymbol{\psi} \cdot \partial_\beta \boldsymbol{\psi} = a_{\alpha\beta}(\boldsymbol{\psi}) = a_{\alpha\beta} = \partial_\alpha \boldsymbol{\theta} \cdot \partial_\beta \boldsymbol{\theta}$$

in the sense of distributions (which is licit, as is immediately verified) then shows that there exists c_7 such that, for all $\boldsymbol{\psi} \in \boldsymbol{\Phi}_F(\omega)$,

$$\begin{aligned}
|\partial_{11}\boldsymbol{\psi} \cdot \partial_1 \boldsymbol{\psi}|_{0,\omega} &\le c_7, \quad |\partial_{12}\boldsymbol{\psi} \cdot \partial_1 \boldsymbol{\psi}|_{0,\omega} \le c_7, \\
|\partial_{12}\boldsymbol{\psi} \cdot \partial_2 \boldsymbol{\psi}|_{0,\omega} &\le c_7, \quad |\partial_{22}\boldsymbol{\psi} \cdot \partial_2 \boldsymbol{\psi}|_{0,\omega} \le c_7, \\
|\partial_{11}\boldsymbol{\psi} \cdot \partial_2 \boldsymbol{\psi} + \partial_{12}\boldsymbol{\psi} \cdot \partial_1 \boldsymbol{\psi}|_{0,\omega} &\le c_7, \\
|\partial_{22}\boldsymbol{\psi} \cdot \partial_1 \boldsymbol{\psi} + \partial_{12}\boldsymbol{\psi} \cdot \partial_2 \boldsymbol{\psi}|_{0,\omega} &\le c_7.
\end{aligned}$$

The relations

$$\begin{aligned}\partial_{11}\boldsymbol{\psi}\cdot\partial_2\boldsymbol{\psi} &= (\partial_{11}\boldsymbol{\psi}\cdot\partial_2\boldsymbol{\psi}+\partial_{12}\boldsymbol{\psi}\cdot\partial_1\boldsymbol{\psi})-\partial_{12}\boldsymbol{\psi}\cdot\partial_1\boldsymbol{\psi},\\ \partial_{22}\boldsymbol{\psi}\cdot\partial_1\boldsymbol{\psi} &= (\partial_{22}\boldsymbol{\psi}\cdot\partial_1\boldsymbol{\psi}+\partial_{12}\boldsymbol{\psi}\cdot\partial_2\boldsymbol{\psi})-\partial_{12}\boldsymbol{\psi}\cdot\partial_2\boldsymbol{\psi},\end{aligned}$$

then imply that

$$|\partial_{11}\boldsymbol{\psi}\cdot\partial_2\boldsymbol{\psi}|_{0,\omega}\le 2c_7,\quad |\partial_{22}\boldsymbol{\psi}\cdot\partial_1\boldsymbol{\psi}|_{0,\omega}\le 2c_7.$$

Thanks to parts (ii) to (iv), a lower bound for the norms $|b_{\alpha\beta}(\boldsymbol{\psi})|_{0,\omega}$ when $\boldsymbol{\psi}\in\boldsymbol{\Phi}_F(\omega)$ can now be established. This lower bound will be essential for proving in part (vii) the coerciveness of the functional I_F over the manifold $\boldsymbol{\Phi}_F(\omega)$.

(v) *There exists* C_5 *such that*

$$\sum_{\alpha,\beta}|b_{\alpha\beta}(\boldsymbol{\psi})|_{0,\omega}^2\ge\|\boldsymbol{\psi}\|_{2,\omega}^2+C_5 \text{ for all } \boldsymbol{\psi}\in\boldsymbol{\Phi}_F(\omega).$$

Let $\boldsymbol{\psi}\in\boldsymbol{\Phi}_F(\omega)$. For almost all $y\in\omega$, the vectors $\boldsymbol{a}_i(\boldsymbol{\psi})(y)$ are linearly independent by (ii), so that the vectors $\boldsymbol{a}^i(\boldsymbol{\psi})(y)$ are well defined by the relations $\boldsymbol{a}^i(\boldsymbol{\psi})(y)\cdot\boldsymbol{a}_j(\boldsymbol{\psi})(y)=\delta^i_j$ for almost all $y\in\omega$. We can then expand $\partial_{\alpha\beta}\boldsymbol{\psi}$ as

$$\partial_{\alpha\beta}\boldsymbol{\psi}=\{\partial_{\alpha\beta}\boldsymbol{\psi}\cdot\boldsymbol{a}_\sigma(\boldsymbol{\psi})\}\boldsymbol{a}^\sigma(\boldsymbol{\psi})+\{\partial_{\alpha\beta}\boldsymbol{\psi}\cdot\boldsymbol{a}_3(\boldsymbol{\psi})\}\boldsymbol{a}^3(\boldsymbol{\psi}) \text{ a.e. in } \omega.$$

Since $b_{\alpha\beta}(\boldsymbol{\psi})=\partial_{\alpha\beta}\boldsymbol{\psi}\cdot\boldsymbol{a}_3(\boldsymbol{\psi})$, $|\boldsymbol{a}^3(\boldsymbol{\psi})|=1$, and $\boldsymbol{a}^3(\boldsymbol{\psi})\cdot\boldsymbol{a}^\sigma(\boldsymbol{\psi})=0$, we have, for almost all $y\in\omega$,

$$|\partial_{\alpha\beta}\boldsymbol{\psi}(y)|^2=|\{\partial_{\alpha\beta}\boldsymbol{\psi}(y)\cdot\boldsymbol{a}_\sigma(\boldsymbol{\psi})(y)\}\boldsymbol{a}^\sigma(\boldsymbol{\psi})(y)|^2+|b_{\alpha\beta}(\boldsymbol{\psi})(y)|^2.$$

Since the vector fields $\boldsymbol{a}^\sigma(\boldsymbol{\psi})$ lie in a bounded subset of $\mathbf{L}^\infty(\omega)$ when $\boldsymbol{\psi}$ varies in the set $\boldsymbol{\Phi}_F(\omega)$ (parts (ii) and (iii)) and since the functions $\partial_{\alpha\beta}\boldsymbol{\psi}\cdot\boldsymbol{a}_\sigma(\boldsymbol{\psi})$ lie in a bounded subset of $L^2(\omega)$ when $\boldsymbol{\psi}$ varies in $\boldsymbol{\Phi}_F(\omega)$ (part (iv)), there exists c_8 such that

$$|\{\partial_{\alpha\beta}\boldsymbol{\psi}\cdot\boldsymbol{a}_\sigma(\boldsymbol{\psi})\}\boldsymbol{a}^\sigma(\boldsymbol{\psi})|_{0,\omega}\le c_8 \text{ for all } \boldsymbol{\psi}\in\boldsymbol{\Phi}_F(\omega).$$

Consequently, there exists c_9 such that, for all $\boldsymbol{\psi}\in\boldsymbol{\Phi}_F(\omega)$,

$$0\le\sum_{\alpha,\beta}|\partial_{\alpha\beta}\boldsymbol{\psi}|_{0,\omega}^2-\sum_{\alpha,\beta}|b_{\alpha\beta}(\boldsymbol{\psi})|_{0,\omega}^2\le c_9.$$

Since there exists c_{10} such that $\|\boldsymbol{\psi}\|_{1,\omega} \leq c_{10}$ for all $\boldsymbol{\psi} \in \boldsymbol{\Phi}_F(\omega)$ (see part (iii)), we finally have

$$\begin{aligned}\sum_{\alpha,\beta} |b_{\alpha\beta}(\boldsymbol{\psi})|^2_{0,\omega} &\geq \sum_{\alpha,\beta} |\partial_{\alpha\beta}\boldsymbol{\psi}|^2_{0,\omega} + \|\boldsymbol{\psi}\|^2_{1,\omega} - c_9 - \|\boldsymbol{\psi}\|^2_{1,\omega} \\ &\geq \|\boldsymbol{\psi}\|^2_{2,\omega} - c_9 - c_{10}^2 \text{ for all } \boldsymbol{\psi} \in \boldsymbol{\Phi}_F(\omega).\end{aligned}$$

We now turn our attention to the functional I_F.

(vi) *The functional I_F is sequentially weakly lower semi-continuous over the manifold $\boldsymbol{\Phi}_F(\omega)$, i.e.,*

$$\boldsymbol{\psi}^l \in \boldsymbol{\Phi}_F(\omega),\ l \geq 1, \text{ and } \boldsymbol{\psi}^l \rightharpoonup \boldsymbol{\psi} \in \boldsymbol{\Phi}_F(\omega) \text{ in } \mathbf{H}^2(\omega)$$

implies that

$$I_F(\boldsymbol{\psi}) \leq \liminf_{l\to\infty} I_F(\boldsymbol{\psi}^l).$$

The weak convergence $\boldsymbol{\psi}^l \rightharpoonup \boldsymbol{\psi}$ in $\mathbf{H}^2(\omega)$ clearly implies that

$$\partial_{\alpha\beta}\boldsymbol{\psi}^l \rightharpoonup \partial_{\alpha\beta}\boldsymbol{\psi} \text{ in } \mathbf{L}^2(\omega) \text{ and } \boldsymbol{a}_\alpha(\boldsymbol{\psi}^l) \to \boldsymbol{a}_\alpha(\boldsymbol{\psi}) \text{ in } \mathbf{L}^2(\omega),$$

the last convergences being consequences of the Rellich-Kondrašov imbedding theorem. We first show that it also implies that

$$\boldsymbol{a}_3(\boldsymbol{\psi}^l) \to \boldsymbol{a}_3(\boldsymbol{\psi}) \text{ in } \mathbf{L}^2(\omega).$$

To this end, we observe that $|\boldsymbol{a}_3(\boldsymbol{\psi}^l)| = 1$ a.e. in ω and that there exists a subsequence $(\boldsymbol{\psi}^m)_{m=1}^\infty$ of $(\boldsymbol{\psi}^l)_{l=1}^\infty$ such that

$$\boldsymbol{a}_\alpha(\boldsymbol{\psi}^m)(y) \to \boldsymbol{a}_\alpha(\boldsymbol{\psi})(y) \text{ for almost all } y \in \omega,$$

since $\boldsymbol{a}_\alpha(\boldsymbol{\psi}^l) \to \boldsymbol{a}_\alpha(\boldsymbol{\psi})$ in $\mathbf{L}^2(\omega)$. The definition

$$\boldsymbol{a}_3(\boldsymbol{\psi}^m)(y) = \frac{\boldsymbol{a}_1(\boldsymbol{\psi}^m)(y) \wedge \boldsymbol{a}_2(\boldsymbol{\psi}^m)(y)}{|\boldsymbol{a}_1(\boldsymbol{\psi}^m)(y) \wedge \boldsymbol{a}_2(\boldsymbol{\psi}^m)(y)|}$$

thus shows that

$$\boldsymbol{a}_3(\boldsymbol{\psi}^m)(y) \to \frac{\boldsymbol{a}_1(\boldsymbol{\psi})(y) \wedge \boldsymbol{a}_2(\boldsymbol{\psi})(y)}{|\boldsymbol{a}_1(\boldsymbol{\psi})(y) \wedge \boldsymbol{a}_2(\boldsymbol{\psi})(y)|} = \boldsymbol{a}_3(\boldsymbol{\psi})(y) \text{ as } m \to \infty$$

for almost all $y \in \omega$ (by (ii), the vectors $\boldsymbol{a}_\alpha(\boldsymbol{\psi}^m)$, $m \geq 1$, and $\boldsymbol{a}_\alpha(\boldsymbol{\psi})$ are well defined and "uniformly" linearly independent a.e.

in ω). Therefore, $\boldsymbol{a}_3(\boldsymbol{\psi}^m) \to \boldsymbol{a}_3(\boldsymbol{\psi})$ in $\mathbf{L}^2(\omega)$ by Lebesgue's dominated convergence theorem. Since the limit $\boldsymbol{a}_3(\boldsymbol{\psi})$ is unique, the whole sequence $(\boldsymbol{a}_3(\boldsymbol{\psi}^l))_{l=1}^\infty$ strongly converges in $\mathbf{L}^2(\omega)$ to this limit.

Using these properties, we next show that

$$b_{\alpha\beta}(\boldsymbol{\psi}^l) = \partial_{\alpha\beta}\boldsymbol{\psi}^l \cdot \boldsymbol{a}_3(\boldsymbol{\psi}^l) \rightharpoonup \partial_{\alpha\beta}\boldsymbol{\psi} \cdot \boldsymbol{a}_3(\boldsymbol{\psi}) = b_{\alpha\beta}(\boldsymbol{\psi}) \text{ in } L^2(\omega).$$

To this end, fix α and β, let $f^l \in L^2(\omega)$ denote one component of $\partial_{\alpha\beta}\boldsymbol{\psi}^l$ (the same for all $l \geq 1$), let $g^l \in L^\infty(\omega)$ denote the same component of $\boldsymbol{a}_3(\boldsymbol{\psi}^l)$, and finally, let $f \in L^2(\omega)$ and $g \in L^\infty(\omega)$ likewise denote the corresponding components of $\partial_{\alpha\beta}\boldsymbol{\psi}$ and $\boldsymbol{a}_3(\boldsymbol{\psi})$. In this fashion, the two sequences $(f^l)_{l=1}^\infty$ and $(g^l)_{l=1}^\infty$ satisfy:

$$\begin{aligned} &f^l \rightharpoonup f \text{ in } L^2(\omega),\\ &g^l \to g \text{ in } L^2(\omega) \text{ and } |g^l|_{0,\infty,\omega} \leq 1 \text{ for all } l. \end{aligned}$$

It then follows that $fg \in L^2(\omega)$ and

$$f^l g^l \rightharpoonup fg \text{ in } L^2(\omega).$$

Although these implications are standard, we provide a proof for completeness. For any $\varphi \in \mathcal{D}(\omega)$, the bilinear form

$$(f, g) \in L^2(\omega) \times L^2(\omega) \to \int_\omega fg\varphi \, dy$$

is strongly continuous; hence

$$f^l \rightharpoonup f \text{ in } L^2(\omega) \text{ and } g^l \to g \text{ in } L^2(\omega) \Rightarrow \int_\omega f^l g^l \varphi \, dy \to \int_\omega fg\varphi \, dy.$$

Let $(f^m g^m)_{m=1}^\infty$ be an arbitrary subsequence of $(f^l g^l)_{l=1}^\infty$. Since $|f^m g^m|_{0,\omega} \leq |f^m|_{0,\omega}$ and the weakly convergent sequence $(f^m)_{m=1}^\infty$ is bounded in $L^2(\omega)$, there is a subsequence $(f^n g^n)_{n=1}^\infty$ of $(f^m g^m)_{m=1}^\infty$ that weakly converges in $L^2(\omega)$ to some $h \in L^2(\omega)$. Therefore,

$$\int_\omega f^n g^n \varphi \, dy \to \int_\omega h\varphi \, dy = \int_\omega fg\varphi \, dy \text{ for all } \varphi \in \mathcal{D}(\omega),$$

and thus $h = fg$. Since the limit fg of the subsequence $(f^n g^n)_{n=1}^\infty$ is unique, the whole sequence $(f^l g^l)_{l=1}^\infty$ weakly converges in $L^2(\omega)$ to this limit.

In particular then, we have established that $b_{\alpha\beta}(\boldsymbol{\psi}^l) \rightharpoonup b_{\alpha\beta}(\boldsymbol{\psi})$ in $L^2(\omega)$.

We are now in a position to establish the sequential weak lower semi-continuity of I_F over $\mathbf{\Phi}_F(\omega)$. Let $\mathbf{L}^2_s(\omega)$ denote the space of all fields of symmetric matrices of order two with components in $L^2(\omega)$. The symmetric bilinear form $B : \mathbf{L}^2_s(\omega) \times \mathbf{L}^2_s(\omega) \to \mathbb{R}$ defined by

$$B(\mathbf{S},\, \mathbf{T}) := \int_\omega a^{\alpha\beta\sigma\tau} s_{\sigma\tau} t_{\alpha\beta} \sqrt{a}\, dy$$

for all $(\mathbf{S},\, \mathbf{T}) = ((s_{\alpha\beta}),\, (t_{\alpha\beta})) \in \mathbf{L}^2_s(\omega) \times \mathbf{L}^2_s(\omega)$ is strongly continuous and positive definite since there exist constants c_{11} and c_{12} such that

$$0 < c_{11} \text{ and } a^{\alpha\beta\sigma\tau}(y) t_{\sigma\tau} t_{\alpha\beta} \geq c_{11} \sum_{\alpha,\beta} |t_{\alpha\beta}|^2$$

for all $y \in \overline{\omega}$ and all symmetric matrices $(t_{\alpha\beta})$ (Thm. 3.3-2) and

$$0 < c_{12} \leq 1 \text{ and } 0 < c_{12} \leq \sqrt{a(y)} \leq c_{12}^{-1}$$

for all $y \in \overline{\omega}$. Being strongly continuous and (strictly) convex, the mapping

$$\mathbf{S} \in \mathbf{L}^2_s(\omega) \to B(\mathbf{S},\, \mathbf{S})$$

is thus weakly lower semi-continuous. Let

$$s^l_{\alpha\beta} := b_{\alpha\beta}(\boldsymbol{\psi}^l) - b_{\alpha\beta} \text{ and } s_{\alpha\beta} := b_{\alpha\beta}(\boldsymbol{\psi}) - b_{\alpha\beta}.$$

Then

$$\mathbf{S}^l := (s^l_{\alpha\beta}) \rightharpoonup \mathbf{S} := (s_{\alpha\beta}) \text{ in } \mathbf{L}^2_s(\omega),$$

since $b_{\alpha\beta}(\boldsymbol{\psi}^l) \rightharpoonup b_{\alpha\beta}(\boldsymbol{\psi})$ in $L^2(\omega)$, and thus

$$B(\mathbf{S},\, \mathbf{S}) \leq \liminf_{l\to\infty} B(\mathbf{S}^l,\, \mathbf{S}^l).$$

This shows that the functional I_F, which is defined by

$$I_F(\boldsymbol{\psi}) = \frac{\varepsilon^3}{6} \int_\omega a^{\alpha\beta\sigma\tau} (b_{\sigma\tau}(\boldsymbol{\psi}) - b_{\sigma\tau})(b_{\alpha\beta}(\boldsymbol{\psi}) - b_{\alpha\beta}) \sqrt{a}\, dy - L(\boldsymbol{\psi})$$

for all $\boldsymbol{\psi} \in \mathbf{\Phi}_F(\omega)$ is sequentially weakly lower semi-continuous on $\mathbf{\Phi}_F(\omega)$ (recall that L is by assumption a continuous linear form on $\mathbf{H}^2(\omega)$).

(vii) *There exist constants C_6 and C_7 such that*

$$C_6 > 0 \text{ and } I_F(\boldsymbol{\psi}) \geq C_6 \|\boldsymbol{\psi}\|^2_{2,\omega} + C_7 \text{ for all } \boldsymbol{\psi} \in \mathbf{\Phi}_F(\omega).$$

Consequently, the functional I_F is coercive on the manifold $\mathbf{\Phi}_F(\omega)$.

By definition of the functional I_F, we have (the constants c_{11} and c_{12} appeared in the proof of part (vi))

$$I_F(\boldsymbol{\psi}) \geq \frac{\varepsilon^3}{6} c_{11} c_{12} \sum_{\alpha,\beta} |b_{\alpha\beta}(\boldsymbol{\psi}) - b_{\alpha\beta}|^2_{0,\omega} - c_{13} \|\boldsymbol{\psi}\|_{2,\omega},$$

where c_{13} denotes the norm of the continuous linear form L. Since

$$|b_{\alpha\beta}(\boldsymbol{\psi}) - b_{\alpha\beta}|^2_{0,\omega} \geq \frac{1}{2} |b_{\alpha\beta}(\boldsymbol{\psi})|^2_{0,\omega} - |b_{\alpha\beta}|^2_{0,\omega},$$

and since, by (v),

$$\sum_{\alpha,\beta} |b_{\alpha\beta}(\boldsymbol{\psi})|^2_{0,\omega} \geq \|\boldsymbol{\psi}\|^2_{2,\omega} + C_5 \text{ for all } \boldsymbol{\psi} \in \boldsymbol{\Phi}_F(\omega),$$

the assertion follows, and the existence of a minimizer of the functional I_F over the manifold $\boldsymbol{\Phi}_F(\omega)$ is thus established. ■

Remarks. (1) The manifold $\boldsymbol{\Phi}_F(\omega)$ is *not* convex (save if it contains only one element); cf. Ex. 10.8.

(2) A property analogous to that proved in (iii) also holds for *isometric three-dimensional manifolds* in $\mathbb{R}^3$; for a precise statement of this property, which is due to Luc Tartar, see, e.g., Vol. I, Ex. 1.16. ■

An interesting question consists in identifying the *Lagrange multiplier* associated with the "constraints" $a_{\alpha\beta}(\boldsymbol{\psi}) - a_{\alpha\beta} = 0$ in ω. In this direction, see the formal approach of Fox, Raoult & Simo [1993, Sect. 5.4] in the planar case.

EXERCISES

10.1. Let $\boldsymbol{f} = (f_\alpha) \in \mathcal{C}^3([0,\, 1]; \mathbb{R}^2)$ be an injective mapping such that $\boldsymbol{f}'(t) \neq \boldsymbol{0}$ for all $t \in [0,\, 1]$ and $\boldsymbol{f}([0,\, 1])$ is not a segment. Let $S = \boldsymbol{\theta}(\overline{\omega})$, where $\omega =]0,\, 1[\times]0,\, 1[$ and $\boldsymbol{\theta}(t,\, z) = f_\alpha(t)\boldsymbol{e}^\alpha + z\boldsymbol{e}^3$ for $(t,\, z) \in \overline{\omega}$. The surface S is thus a portion of a *cylinder* orthogonal to, and passing through, the planar curve $\boldsymbol{f}([0,\, 1])$; cf. Fig. 10.2-1.

(1) Assume that

$$\gamma_0 \subset \{(0,\, z) \in \mathbb{R}^2;\ 0 \leq z \leq 1\} \cup \{(1,\, z) \in \mathbb{R}^2;\ 0 \leq z \leq 1\}.$$

Show that the manifold

$$\boldsymbol{\mathcal{M}}_F(\omega) = \{\boldsymbol{\eta} = (\eta_i) \in \mathbf{W}^{2,4}(\omega);\ \boldsymbol{\eta} = \partial_\nu \boldsymbol{\eta} = \mathbf{0} \text{ on } \gamma_0;\\ a_{\alpha\beta}(\boldsymbol{\eta}) - a_{\alpha\beta} = 0 \text{ in } \omega\}$$

contains nonzero functions $\boldsymbol{\eta}$.

(2) Show that, at each $\boldsymbol{\zeta} \in \boldsymbol{\mathcal{M}}_F(\omega)$, the tangent space $\mathbb{T}_{\boldsymbol{\zeta}}\boldsymbol{\mathcal{M}}_F(\omega)$ to $\boldsymbol{\mathcal{M}}_F(\omega)$ contains nonzero functions (such a tangent space is defined, e.g., in Sect. 10.2).

10.2. Let $\boldsymbol{f} = (f_\alpha) \in \mathcal{C}^3([0, 1]; \mathbb{R}^2)$ be an injective mapping such that $\boldsymbol{f}'(t) \neq \mathbf{0}$ for all $t \in [0, 1]$.

Let $S = \boldsymbol{\theta}(\overline{\omega})$, where $\omega =]0, 1[\times]s_0, 1[$ for some $0 < s_0 < 1$ and $\boldsymbol{\theta}(t, s) = sf_\alpha(t)\boldsymbol{e}^\alpha + (1 - s)\boldsymbol{e}^3$ for $(t, s) \in \overline{\omega}$. The surface S is thus a portion of a *cone* with vertex $\boldsymbol{e}^3$ and passing through the planar curve $\boldsymbol{f}([0, 1])$; cf. Fig. 10.2-2.

(1) Assume that $\gamma_0 = \{(0, s) \in \mathbb{R}^2;\ s_0 \leq s \leq 1\}$. Show that the manifold

$$\boldsymbol{\mathcal{M}}_F(\omega) = \{\boldsymbol{\eta} = (\eta_i) \in \mathbf{W}^{2,4}(\omega);\ \boldsymbol{\eta} = \partial_\nu \boldsymbol{\eta} = \mathbf{0} \text{ on } \gamma_0;\\ a_{\alpha\beta}(\boldsymbol{\eta}) - a_{\alpha\beta} = 0 \text{ in } \omega\}$$

contains nonzero functions $\boldsymbol{\eta}$.

(2) Show that, at each $\boldsymbol{\zeta} \in \boldsymbol{\mathcal{M}}_F(\omega)$, the tangent space $\mathbb{T}_{\boldsymbol{\zeta}}\boldsymbol{\mathcal{M}}_F(\omega)$ to $\boldsymbol{\mathcal{M}}_F(\omega)$ contains nonzero functions.

(3) Assume that

$$\gamma_0 = \{(0, s) \in \mathbb{R}^2;\ s_0 \leq s \leq 1\} \cup \{(1, s) \in \mathbb{R}^2;\ s_0 \leq s \leq 1\}.$$

Are the conclusions of (1) and (2) still valid?

10.3. Let $\omega \subset \mathbb{R}^2$ be a domain and let $\gamma_0 \subset \partial\omega$ be a segment; cf. Fig. 10.2-3.

(1) Show that the manifold

$$\boldsymbol{\mathcal{M}}_F(\omega) = \{\boldsymbol{\eta} = (\eta_i) \in \mathbf{W}^{2,4}(\omega);\ \boldsymbol{\eta} = \partial_\nu \boldsymbol{\eta} = \mathbf{0} \text{ on } \gamma_0;\\ a_{\alpha\beta}(\boldsymbol{\eta}) - \delta_{\alpha\beta} = 0 \text{ in } \omega\}$$

contains nonzero functions $\boldsymbol{\eta}$.

(2) Show that, at each $\boldsymbol{\zeta} \in \boldsymbol{\mathcal{M}}_F(\omega)$, the tangent space $\mathbb{T}_{\boldsymbol{\zeta}}\boldsymbol{\mathcal{M}}_F(\omega)$ to $\boldsymbol{\mathcal{M}}_F(\omega)$ contains nonzero functions.

10.4. The notations used in this exercise should be self-explanatory. Let the manifold $\mathcal{M}_F(\omega)$ and its tangent spaces $\mathbb{T}_{\zeta}\mathcal{M}_F(\omega)$ be defined as in Sect. 10.2. Let a surface $S = \boldsymbol{\theta}(\overline{\omega}) = \tilde{\boldsymbol{\theta}}(\{\tilde{\omega}\}^-)$ be equipped with two systems of curvilinear coordinates $(y_\alpha) \in \overline{\omega}$ and $(\tilde{y}_\alpha) \in \{\tilde{\omega}\}^-$ and assume that $\boldsymbol{\theta}(\gamma_0) = \tilde{\boldsymbol{\theta}}(\tilde{\gamma}_0)$.

(1) Show that $\mathcal{M}_F(\omega) \neq \{\mathbf{0}\}$ implies that $\mathcal{M}_F(\tilde{\omega}) \neq \{\mathbf{0}\}$.

(2) Assume that, at each $\boldsymbol{\zeta} \in \mathcal{M}_F(\omega)$, the tangent space $\mathbb{T}_{\zeta}\mathcal{M}_F(\omega)$ to $\mathcal{M}_F(\omega)$ contains nonzero functions. Then show that, at each $\tilde{\boldsymbol{\zeta}} \in \mathcal{M}_F(\tilde{\omega})$, the tangent space $\mathbb{T}_{\tilde{\zeta}}\mathcal{M}_F(\tilde{\omega})$ contains nonzero functions.

In other words, the definition of a nonlinearly elastic flexural shell (Sect. 10.2) is independent of the system of curvilinear coordinates employed for representing the surface S.

10.5. Show that the stored energy function of a nonlinearly elastic flexural shell:

$$\boldsymbol{\eta} = (\eta_i) \in \mathcal{M}_F(\omega) \to \frac{\varepsilon^3}{6} a^{\alpha\beta\sigma\tau,\varepsilon}(b_{\sigma\tau}(\boldsymbol{\eta}) - b_{\sigma\tau})(b_{\alpha\beta}(\boldsymbol{\eta}) - b_{\alpha\beta})$$

is *frame-indifferent*, in the following sense: Given any $\boldsymbol{\eta} \in \mathcal{M}_F(\omega)$, its value is unaltered if $\boldsymbol{\psi} := \boldsymbol{\theta} + \eta_i \boldsymbol{a}^i$ is replaced by $\mathbf{Q}\boldsymbol{\psi}$, where $\mathbf{Q}$ is any orthogonal matrix of order three.

10.6. This exercise shows that, if *higher-order forces* are considered, the leading term of the formal expansion (then also of a higher order) satisfies the equations of a *linearly* elastic flexural shell, as found in the linear theory (Thm. 3.4-3). This observation is due to Lods & Miara [1998, p. 367].

Let the assumptions be as in Thm. 10.3-4, save that the applied body and surface forces are now such that

$$\begin{aligned} f^{i,\varepsilon}(x^\varepsilon) &= \varepsilon^{2+r} f^{i,2+r}(x) \quad \text{for all} \quad x^\varepsilon = \pi^\varepsilon x \in \Omega^\varepsilon, \\ h^{i,\varepsilon}(x^\varepsilon) &= \varepsilon^{3+r} h^{i,3+r}(x) \quad \text{for all} \quad x^\varepsilon = \pi^\varepsilon x \in \Gamma_+^\varepsilon \cup \Gamma_-^\varepsilon, \end{aligned}$$

where r is an arbitrary integer ≥ 1 and the functions $f^{i,2+r} \in L^2(\Omega)$ and $h^{i,3+r} \in L^2(\Gamma_+ \cup \Gamma_-)$ are independent of ε.

(1) Show that the leading term of the formal asymptotic expansion of the scaled unknown $\boldsymbol{u}(\varepsilon)$ is $\boldsymbol{u}^r$ (in particular then, $\boldsymbol{u}^0$ vanishes).

(2) Show that $\boldsymbol{u}^r : \overline{\Omega} \to \mathbb{R}^3$ is independent of the transverse variable and that, once identified with a function $\boldsymbol{\zeta}^r : \overline{\omega} \to \mathbb{R}^3$, it satisfies the variational problem

$$\boldsymbol{\zeta}^r \in \mathbf{W}_F(\omega) = \{\boldsymbol{\eta} = (\eta_i) \in W^{1,4}(\omega) \times W^{1,4}(\omega) \times W^{2,4}(\omega);\\ \eta_i = \partial_\nu \eta_3 = 0 \text{ on } \gamma_0,\ \gamma_{\alpha\beta}(\boldsymbol{\eta}) = 0 \text{ in } \omega\},$$

$$\frac{1}{3}\int_\omega a^{\alpha\beta\sigma\tau} \rho_{\sigma\tau}(\boldsymbol{\zeta}^r)\rho_{\alpha\beta}(\boldsymbol{\eta})\sqrt{a}\,dy = \int_\omega p^{i,2+r}\eta_i\sqrt{a}\,dy$$

for all $\boldsymbol{\eta} = (\eta_i) \in \mathbf{W}_F(\omega)$, where the functions $\gamma_{\alpha\beta}(\boldsymbol{\eta})$ and $\rho_{\alpha\beta}(\boldsymbol{\eta})$ are the covariant components of the *linearized* change of metric and change of curvature tensors and

$$p^{i,2+r} = \int_{-1}^{1} f^{i,2+r}\,dx_3 + h_+^{i,3+r} + h_-^{i,3+r} \text{ and } h_\pm^{i,3+r} = h^{i,3+r}(\cdot, \pm 1).$$

10.7. The following problem provides an "exotic" example of a *manifold* $\boldsymbol{\mathcal{M}}_F(\omega)$, as used in the definition of a nonlinearly elastic flexural shell (Sect. 10.2). Given any function $f \in \mathcal{C}^\infty([0, 1])$ that satisfies $f(t) = 1$ if $0 \le t \le \frac{1}{3}$ and $f(t) = 0$ if $\frac{2}{3} \le t \le 1$, let the mapping $\boldsymbol{\theta} \in \mathcal{C}^\infty(\overline{\omega}; \mathbb{R}^2)$ be defined by $\boldsymbol{\theta}(y_1, y_2) = (y_1, y_2, f(\sqrt{y_1^2 + y_2^2}))$ for all $(y_1, y_2) \in \overline{\omega}$, where $\omega = \{(y_1, y_2) \in \mathbb{R}^2;\ y_1^2 + y_2^2 < 1\}$.

(1) Show that the manifold

$$\boldsymbol{\mathcal{M}}_F(\omega) = \{\boldsymbol{\eta} = (\eta_i) \in \mathbf{W}^{2,4}(\omega);\ \boldsymbol{\eta} = \partial_\nu \boldsymbol{\eta} = \mathbf{0} \text{ on } \gamma,\\ a_{\alpha\beta}(\boldsymbol{\eta}) - a_{\alpha\beta} = 0 \text{ in } \omega\}$$

contains the two distinct elements $\boldsymbol{\eta} = \mathbf{0}$ and $\boldsymbol{\eta} = -2\boldsymbol{\theta}\cdot\boldsymbol{a}_3$, the second one thus corresponding to an *everted state.*

(2) Are there other elements in the manifold $\boldsymbol{\mathcal{M}}_F(\omega)$?

(3) Are the conclusions of (1) and (2) modified if the space $\mathbf{H}^2(\omega)$ replaces the space $\mathbf{W}^{2,4}(\omega)$ in the definition of the manifold $\boldsymbol{\mathcal{M}}_F(\omega)$?

Remark. This example can be generalized in various ways, for instance by letting $\boldsymbol{\theta}(y_1, y_2) = (y_1, y_2, f^n(\sqrt{y_1^2 + y_2^2}))$ for some $n \ge 1$, where the functions $f^n \in \mathcal{C}^\infty([0, 1])$, $n \ge 1$, are defined by

$$f^0 = f \text{ and } f^n(t) = \begin{cases} f^{n-1}(t) + f^{n-1}(3^n t) \text{ if } 0 \le t \le \dfrac{1}{3^n}, \\ f^{n-1}(t) \text{ if } \left(\dfrac{2}{3}\right)^n \le t \le 1,\ n \ge 1. \end{cases}$$

10.8. Let $\gamma_0 \subset \gamma$ be such that *length* $\gamma_0 > 0$, let the manifold $\mathbf{\Phi}_F(\omega)$ be defined as in Thm. 10.6-1, viz.,

$$\mathbf{\Phi}_F(\omega) = \{\boldsymbol{\psi} \in \mathbf{H}^2(\omega);\ \boldsymbol{\psi} = \boldsymbol{\varphi}_0 \text{ on } \gamma_0,\ a_{\alpha\beta}(\boldsymbol{\psi}) - a_{\alpha\beta} = 0 \text{ in } \omega\},$$

and let $\boldsymbol{\psi}_1$ and $\boldsymbol{\psi}_2$ be two distinct elements in $\mathbf{\Phi}_F(\omega)$.

(1) Show that there exist an index $i_0 \in \{1, 2, 3\}$ and a subset σ of ω with *area* $\sigma > 0$ such that

$$\partial_{i_0}\boldsymbol{\psi}_1(y) \neq \partial_{i_0}\boldsymbol{\psi}_2(y) \text{ for all } y \in \sigma.$$

(2) Let $\boldsymbol{\psi}(t) := t\boldsymbol{\psi}_1 + (1-t)\boldsymbol{\psi}_2$. Show that

$$\int_\omega |\partial_{i_0}\boldsymbol{\psi}(t)|^2\, dy < \int_\omega |\partial_{i_0}\boldsymbol{\theta}|^2\, dy \text{ for all } 0 < t < 1,$$

and conclude that *the manifold* $\mathbf{\Phi}_F(\omega)$ *is not convex.*

Remark. This elegant proof of the non-convexity of $\mathbf{\Phi}_F(\omega)$ is due to D. Coutand.

CHAPTER 11

KOITER'S EQUATIONS AND OTHER TWO-DIMENSIONAL NONLINEAR SHELL THEORIES

INTRODUCTION

The objective of this short chapter is to provide a succinct repertoire of, and *ad hoc* references to, "other" *two-dimensional nonlinear shell theories*, i.e., that are neither the membrane nor the flexural theories justified in the previous chapters. No exhaustiveness is to be expected, however, even within the few topics selected for this chapter.

Among these, *Koiter's nonlinear shell theory* (Sect. 11.1) is particularly appealing: Its corresponding strain energy is simply the sum of the strain energy of a nonlinearly elastic membrane shell and that of a nonlinearly elastic flexural shell, justified respectively in Chaps. 9 and 10.

Finally, we briefly review in Sects. 11.2 and 11.3 other two-dimensional equations that have been proposed for modeling *nonlinearly elastic shells*, notably *Naghdi's nonlinear shell theory*, which is based on the notion of *one-director Cosserat surface*, and those theories that are deemed appropriate for *nonlinearly elastic "shallow" shells.*

11.1. THE TWO-DIMENSIONAL KOITER EQUATIONS FOR A NONLINEARLY ELASTIC SHELL

In a landmark paper, Koiter [1966] has proposed nonlinear equations[1] which may well constitute, possibly after some amendments, an *"all-purpose nonlinear shell theory"*, in the same way that Koiter's linear equations constitute an "all-purpose linear shell theory" in the sense that they are valid for all kinds of geometries of the middle surface and boundary conditions (Sect. 7.2).

[1] An excerpt from this paper is reproduced on page lxi.

Another major advance in nonlinear shell theory is due to John [1965, 1971], who established (by means of exceedingly delicate computations) that, "away from the lateral face", the state of stress inside a shell is approximately planar and the stresses parallel to the middle surface vary approximately linearly across the thickness if the thickness is small enough. This means that, given any point $\hat{y} \in S = \boldsymbol{\theta}(\omega)$ sufficiently far from the "boundary" $\boldsymbol{\theta}(y)$ of S and a local *Cartesian frame* with origin $\hat{y}$ and basis vectors $\hat{\boldsymbol{e}}_\alpha$ in the tangent plane to S at $\hat{y}$, the components $\hat{\sigma}^{ij,\varepsilon}$ of the stress tensor are approximately of the form

$$\hat{\sigma}^{\alpha\beta,\varepsilon} = \hat{\sigma}^{\alpha\beta,0} + \hat{x}_3^\varepsilon \sigma^{\alpha\beta,1} \text{ and } \hat{\sigma}^{i3,\varepsilon} = 0$$

in a sufficiently small three-dimensional neighborhood of $\hat{y}$, where $\hat{x}_3^\varepsilon$ denotes the Cartesian coordinates along the basis vector $\hat{\boldsymbol{e}}_3$ (normal to S at $\hat{y}$) and the functions $\hat{\sigma}^{\alpha\beta,0}$ and $\hat{\sigma}^{\alpha\beta,1}$ are independent of $\hat{x}_3^\varepsilon$.

In Koiter's approach, these approximations constitute an *a priori* assumption of a *mechanical* nature, which is associated with another *a priori* assumption, of a *geometrical* nature. This second assumption is the **Kirchhoff-Love assumption**, so named after Kirchhoff [1876] and Love [1934]; it asserts that any point situated on a normal to the middle surface in the reference configuration remains on the normal to the deformed middle surface after the deformation has taken place and that, in addition, the distance between such a point and the middle surface remains constant (complements about this assumption are found in Ex. 7.1).

Combining these two *a priori* assumptions, W.T. Koiter concludes that the displacement field across the thickness of the shell can be completely determined from the sole knowledge of the displacement field of the middle surface and he identifies the two-dimensional problem (described *infra*) that this field should satisfy.

More specifically, let ω be a domain in $\mathbb{R}^2$ with boundary γ, let $\boldsymbol{\theta} \in \mathcal{C}^3(\overline{\omega}; \mathbb{R}^3)$ be an injective mapping such that the two vectors $\boldsymbol{a}_\alpha = \partial_\alpha \boldsymbol{\theta}$ are linearly independent at all points of $\overline{\omega}$, and let γ_0 be a portion of γ that satisfies *length* $\gamma_0 > 0$.

Consider a *nonlinearly elastic shell* with *middle surface* $S = \boldsymbol{\theta}(\overline{\omega})$ and *thickness* $2\varepsilon > 0$, i.e., a nonlinearly elastic body whose reference configuration is the set $\boldsymbol{\Theta}(\overline{\Omega}^\varepsilon)$, where

$$\Omega^\varepsilon := \omega \times]-\varepsilon, \varepsilon[,$$
$$\boldsymbol{\Theta}(y, x_3^\varepsilon) := \boldsymbol{\theta}(y) + x_3^\varepsilon \boldsymbol{a}_3(y) \text{ for all } (y, x_3^\varepsilon) \in \overline{\Omega}^\varepsilon.$$

The material constituting the shell is *homogeneous* and *isotropic* and the reference configuration is a *natural state*; hence the material is characterized by its two *Lamé constants* $\lambda^\varepsilon > 0$ and $\mu^\varepsilon > 0$. The shell is subjected to a *boundary condition of place* along a portion of its lateral face with $\boldsymbol{\theta}(\gamma_0)$ as its middle curve, i.e., the three-dimensional displacement vanishes on $\boldsymbol{\Theta}(\Gamma_0^\varepsilon)$, where $\Gamma_0^\varepsilon = \gamma_0 \times [-\varepsilon, \varepsilon]$.

Finally, the shell is subjected to *applied body forces* in its interior $\boldsymbol{\Theta}(\Omega^\varepsilon)$ and to *applied surface forces* on its "upper" and "lower" faces $\boldsymbol{\Theta}(\Gamma_+^\varepsilon)$ and $\boldsymbol{\Theta}(\Gamma_-^\varepsilon)$, given by the contravariant components $f^{i,\varepsilon} \in L^2(\Omega^\varepsilon)$ and $h^{i,\varepsilon} \in L^2(\Gamma_+^\varepsilon \cup \Gamma_-^\varepsilon)$ of their respective densities, where $\Gamma_\pm^\varepsilon = \omega \times \{\pm\varepsilon\}$.

Let

$$\begin{aligned} a^{\alpha\beta\sigma\tau,\varepsilon} &:= \frac{4\lambda^\varepsilon\mu^\varepsilon}{\lambda^\varepsilon + 2\mu^\varepsilon} a^{\alpha\beta} a^{\sigma\tau} + 2\mu^\varepsilon (a^{\alpha\sigma} a^{\beta\tau} + a^{\alpha\tau} a^{\beta\sigma}), \\ p^{i,\varepsilon} &:= \int_{-\varepsilon}^{\varepsilon} f^{i,\varepsilon}\, dx_3^\varepsilon + h_+^{i,\varepsilon} + h_-^{i,\varepsilon} \text{ and } h_\pm^{i,\varepsilon} := h^{i,\varepsilon}(\cdot, \pm\varepsilon), \end{aligned}$$

the functions $a^{\alpha\beta\sigma\tau,\varepsilon}$ thus denoting as usual the contravariant components of the *two-dimensional elasticity tensor of the shell.* Then the **two-dimensional Koiter equations for a nonlinearly elastic shell** take the following form when they are expressed as a *minimization problem* (Koiter [1966, eqs. (4.2) and (8.3)]): The unknown $\boldsymbol{\zeta}_K^\varepsilon = (\zeta_{K,i}^\varepsilon)$, where the functions $\zeta_{K,i}^\varepsilon : \overline{\omega} \to \mathbb{R}$ are the covariant components of the displacement field $\zeta_{K,i}^\varepsilon \boldsymbol{a}^i$ of the middle surface S (this means that $\zeta_{i,K}^\varepsilon(y)\boldsymbol{a}^i(y)$ is the displacement of the point $\boldsymbol{\theta}(y)$) should minimize the *energy* j_K^ε defined by

$$\begin{aligned} j_K^\varepsilon(\boldsymbol{\eta}) := \frac{\varepsilon}{8} \int_\omega & a^{\alpha\beta\sigma\tau,\varepsilon} (a_{\sigma\tau}(\boldsymbol{\eta}) - a_{\sigma\tau})(a_{\alpha\beta}(\boldsymbol{\eta}) - a_{\alpha\beta}) \sqrt{a}\, dy \\ + \frac{\varepsilon^3}{6} \int_\omega & a^{\alpha\beta\sigma\tau,\varepsilon} (b_{\sigma\tau}(\boldsymbol{\eta}) - b_{\sigma\tau})(b_{\alpha\beta}(\boldsymbol{\eta}) - b_{\alpha\beta}) \sqrt{a}\, dy \\ & - \int_\omega p^{i,\varepsilon} \eta_i \sqrt{a}\, dy \end{aligned}$$

over an appropriate manifold of vector fields $\boldsymbol{\eta} = (\eta_i)$ satisfying *ad hoc* boundary conditions, for instance the *boundary conditions of strong clamping* $\boldsymbol{\eta} = \partial_\nu \boldsymbol{\eta} = \mathbf{0}$ on γ_0 found in the two-dimensional equations of a nonlinearly elastic flexural shell (Sect. 10.5).

The functions $\frac{1}{2}(a_{\alpha\beta}(\boldsymbol{\eta}) - a_{\alpha\beta})$ and $(b_{\alpha\beta}(\boldsymbol{\eta}) - b_{\alpha\beta})$, already encountered and denoted $G_{\alpha\beta}(\boldsymbol{\eta})$ and $R_{\alpha\beta}(\boldsymbol{\eta})$ in Chaps. 9 and 10 (see, e.g., Thms. 9.4-1 and 10.5-1), are the covariant components of the *change of metric* and *change of curvature tensors* associated with a displacement field $\eta_i \boldsymbol{a}^i$ of the middle surface S. We recall that $a_{\alpha\beta}(\boldsymbol{\eta})$ and $b_{\alpha\beta}(\boldsymbol{\eta})$ respectively denote the covariant components of the metric and curvature tensors of the deformed surface $(\boldsymbol{\theta} + \eta_i \boldsymbol{a}^i)(\overline{\omega})$.

In other words, the **strain energy**

$$\boldsymbol{\eta} \to \frac{\varepsilon}{8} \int_\omega a^{\alpha\beta\sigma\tau,\varepsilon}(a_{\sigma\tau}(\boldsymbol{\eta}) - a_{\sigma\tau})(a_{\alpha\beta}(\boldsymbol{\eta}) - a_{\alpha\beta})\sqrt{a}\, dy$$

$$+ \frac{\varepsilon^3}{6} \int_\omega a^{\alpha\beta\sigma\tau,\varepsilon}(b_{\sigma\tau}(\boldsymbol{\eta}) - b_{\sigma\tau})(b_{\alpha\beta}(\boldsymbol{\eta}) - b_{\alpha\beta})\sqrt{a}\, dy$$

found in the energy j_K^ε *is the sum of the strain energy of a nonlinearly elastic membrane shell* (Sect. 9.4) *and of the strain energy of a nonlinearly elastic flexural shell* (Sect. 10.5).

Note in passing that this expression of the strain energy immediately shows that Koiter's equations for a *linearly* elastic shell (Sect. 7.1) are indeed obtained by linearizing the nonlinear ones.

But the functions $b_{\alpha\beta}(\boldsymbol{\eta})$ are *not* defined at those points of $\overline{\omega}$ where the two vectors $a_\alpha(\boldsymbol{\eta}) := \partial_\alpha(\boldsymbol{\theta} + \eta_i \boldsymbol{a}^i)$ are linearly dependent. For this reason, it is not clear to decide over which manifold the energy j_K^ε should be minimized.

One way to circumvent this difficulty may consist in replacing in the strain energy the functions $R_{\alpha\beta}(\boldsymbol{\eta}) = (b_{\alpha\beta}(\boldsymbol{\eta}) - b_{\alpha\beta})$ by the functions $R^\flat_{\alpha\beta}(\boldsymbol{\eta})$ introduced in Thm. 10.3-1, which are well defined for *all* smooth enough fields $\boldsymbol{\eta} = (\eta_i)$, irrespective of whether or not the vectors $a_\alpha(\boldsymbol{\eta})$ are linearly dependent in a subset of $\overline{\omega}$. We recall that these functions, which are also given by (Thm. 10.3-2)

$$R^\sharp_{\alpha\beta}(\boldsymbol{\eta}) := \frac{1}{\sqrt{a}} \partial_{\alpha\beta}(\boldsymbol{\theta} + \eta_i \boldsymbol{a}^i) \cdot \{\boldsymbol{a}_1(\boldsymbol{\eta}) \wedge \boldsymbol{a}_2(\boldsymbol{\eta})\} - b_{\alpha\beta}$$

when $\det a_{\alpha\beta}(\boldsymbol{\eta}) > 0$, coincide with the functions $(b_{\alpha\beta}(\boldsymbol{\eta}) - b_{\alpha\beta})$ when $\boldsymbol{\eta}$ corresponds to an *inextensional displacement* $\eta_i \boldsymbol{a}^i$ of the middle surface, i.e., when $a_{\alpha\beta}(\boldsymbol{\eta}) - a_{\alpha\beta} = 0$ in ω (Thm. 10.3-1).

It is interesting to note that the *same* functions $R^\sharp_{\alpha\beta}(\boldsymbol{\eta})$ as above have also been proposed by W.T. Koiter (cf. Koiter [1966, eq. (4.11)]), most likely out of different considerations!

11.2. OTHER NONLINEAR SHELL THEORIES

Nonlinear "shallow" shell theories are treated separately (see Sect. 11.3.

Naghdi's nonlinear shell theory with directors and extensions. While Koiter's theory relies in particular on the Kirchhoff-Love assumption, Naghdi [1963] has proposed another nonlinear shell theory, in which the *a priori* assumption on the stresses is the same as in Koiter's, but the *a priori* assumption of a geometrical nature affords more freedom on the displacements inside the shell; more specifically, the points situated on a line normal to the undeformed middle surface again stay on a line after the deformation has taken place and the distances are unmodified along this line, but this line need no longer remain normal to the deformed middle surface. Unlike Koiter's, Naghdi's theory may thus accommodate *shear* inside the shell.

In this approach, the displacement of any point $\boldsymbol{\theta}(y) + x_3^\varepsilon \boldsymbol{a}_3(y)$ inside the undeformed shell is of the form $\tilde{\boldsymbol{\zeta}}^\varepsilon(y) + x_3^\varepsilon \boldsymbol{d}^\varepsilon(y)$, where $\tilde{\boldsymbol{\zeta}}^\varepsilon(y) = \zeta_i^\varepsilon(y)\boldsymbol{a}^i(y)$ is the *unknown* displacement vector of the point $\boldsymbol{\theta}(y)$ of the middle surface $S = \boldsymbol{\theta}(\overline{\omega})$ and $\boldsymbol{d}^\varepsilon(y)$ is another *unknown* vector, called the *director* at $\boldsymbol{\theta}(y)$, that measures the "rotation" of the normal vector after the deformation has taken place. In this fashion, the shell is modeled as a *one-director Cosserat surface*, whose deformed configuration is specified not only by the displacement field $\zeta_i^\varepsilon \boldsymbol{a}^i : \overline{\omega} \to \mathbb{R}^3$ of the points of the middle surface, but also by a *director field* $\boldsymbol{d}^\varepsilon : \overline{\omega} \to \mathbb{R}^3$. This notion is due to Cosserat & Cosserat [1909].

Remark. An example of a one-director Cosserat surface is provided by *Naghdi's linear shell theory*, with $\boldsymbol{d}^\varepsilon = r_\alpha^\varepsilon \boldsymbol{a}^\alpha$ as the director field; cf. Sect. 7.4. A planar example is provided by the *Reissner-Mindlin theory* for linearly elastic plates (Vol. II, Sect. 1.9). ■

After its foundations were properly laid (see notably Naghdi [1963, 1972, 1982], Green, Naghdi & Wainwright [1965], Green & Naghdi [1974], and Reissner [1974]), Naghdi's theory has undergone significant developments, often under the appellations of *multi-director shell theories* or *geometrically exact shell theories.*

In these directions, see in particular the illuminating introduction to such theories given in Antman [1995, Chap. 14]. See also Basar [1987], Basar & Krätzig [1988], Antman [1989], Simo & Fox [1989], Simo, Fox & Rifai [1989, 1990a, 1990b], Fox & Simo [1992], Antman [1997], Valid [1995]; Ge, Kruse & Marsden [1996], Kirchgässner & Djurdjevic [1997], Djurdjevic [1999] for *time-dependent nonlinearly elastic shells*; and Basar, Ding & Schultz [1993], Krätzig [1993], Basar & Ding [1995] for the modeling of *multi-layered*, or *laminated*, nonlinearly elastic shells by multi-director theories.

"First-order" and "higher-order" shell theories. Shell theories, linear and nonlinear, have often been elaborated by assuming that the ratio of ε to the smallest absolute value of the radii of curvature of the middle surface is also a "small" parameter (for instance the linear Novozhilov and nonlinear Donnell-Mushtari-Vlasov models for shallow shells can be derived in this manner; cf. Sects. 7.6 and 11.3). A shell theory obtained in this fashion is deemed *"of the first order"* if all terms of order ≥ 2 with respect to η are neglected in the various formal asymptotic expansions considered and *"of a higher order"* otherwise.

This approach, which leads to theories that are often scarcely distinguishable from the one-director or multi-director theories mentioned *supra*, is frequently used for modeling *multi-layered*, or *laminated, shells*, whose constituting elastic material is thus *anisotropic.* See notably Librescu [1975], Krätzig [1974, 1976], Basar & Krätzig [1985], Ambartsumian [1991], Simmonds [1992], Pomp [1996].

Nonlinear shell theories based on the method of internal constraints. Another approach advocated by P.M. Naghdi (see notably Green, Laws & Naghdi [1968] and Green & Naghdi [1970]) for generating nonlinear shell theories consists in directly incorporating *ad hoc internal constraints* in the three-dimensional equations. For instance, the *Kirchhoff-Love assumption* can play the rôle of such an internal constraint, imposed *a priori* on the admissible displacements (in its linearized version, this is the approach of Podio-Guidugli [1990]; cf. Sect. 7.5).

Special care must be exercised, however, as serious inconsistencies arise if the *Lagrange multipliers* associated with such internal constraints (in the sense of optimization theory) are not properly interpreted. Such inconsistencies have been neatly clarified by Antman & Marlow [1991], who showed that the Lagrange multipliers are "reactive" stresses that maintain the constraints, while *ad hoc* "active" stresses must be added to them so as to make up the total stresses (which are no longer given by constitutive equations); see also Antman [1995, Chap. 14].

Time-dependent nonlinear shell theories can likewise be developed by the method of internal constraints; see Antman [1995, Chap. 14] and Lembo [1996].

A nonlinear Budiansky-Sanders theory. Destuynder [1982, 1983] has proposed a *nonlinear Budiansky-Sanders theory*, where the covariant components $\rho^{BS}_{\alpha\beta}(\boldsymbol{\eta})$ of the "modified linearized change of curvature tensor" are the same as in the linear Budiansky-Sanders theory (Sect. 7.5) and the covariant components

$$\begin{aligned}G_{\alpha\beta}(\boldsymbol{\eta}) &= \frac{1}{2}(a_{\alpha\beta}(\boldsymbol{\eta}) - a_{\alpha\beta})\\ &= \gamma_{\alpha\beta}(\boldsymbol{\eta}) + \frac{1}{2}(a^{\sigma\tau}\eta_{\sigma\|\alpha}\eta_{\tau\|\beta} + \eta_{3\|\alpha}\eta_{3\|\beta})\end{aligned}$$

of the change of metric tensor (see, e.g., Thm. 9.1-1) are replaced by the shorter functions

$$G^{BS}_{\alpha\beta}(\boldsymbol{\eta}) = \gamma_{\alpha\beta}(\boldsymbol{\eta}) + \frac{1}{2}\eta_{3\|\alpha}\eta_{3\|\beta}.$$

Nonlinear terms thus only appear in the "membrane" part of the strain energy.

P. Destuynder then establishes an existence theorem for the associated minimization problem, provided the Christoffel symbols of the middle surface are small enough or an *ad hoc* geometric assumption is satisfied. His proof makes an essential use of the *inequality of Korn's type on a general surface* (Thm. 2.6-4).

"Intrinsic" nonlinear shell theory. The "intrinsic" approach of Delfour & Zolésio [1995], which is based on a "tangential differential calculus" and on the "oriented distance function" (Sect. 7.4), has been recently extended to the modeling of nonlinearly elastic shells by Delfour & Zhao [1999].

11.3. NONLINEAR SHALLOW SHELL THEORIES

Nonlinear shallow shell theories in *Cartesian* coordinates are already treated in Vol. II, Sects. 4.14 and 5.12. Accordingly, the additional commentary and bibliographical notes found in this section concern mostly nonlinear shallow shell theories expressed in *curvilinear* coordinates.

We recall that, according to the definition proposed and justified by a formal asymptotic method by Ciarlet & Paumier [1986], an elastic shell is deemed *"shallow"* if, in its reference configuration, the deviation of its middle surface from a plane is of the order of the thickness. More specifically, there exists a smooth enough function $\theta : \overline{\omega} \to \mathbb{R}$ *independent of* ε such that

$$\boxed{\begin{aligned} &S^\varepsilon = \boldsymbol{\theta}^\varepsilon(\overline{\omega}), \text{ where} \\ &\boldsymbol{\theta}^\varepsilon(y_1, y_2) = (y_1, y_2, \varepsilon\theta(y_1, y_2)) \text{ for all } (y_1, y_2) \in \overline{\omega}. \end{aligned}}$$

A word of caution. As in the linear case, this specification of how the middle surface should "vary with ε" thus also constitutes an *assumption on the data*, specific to shallow shell theory. ■

The *two-dimensional equations of nonlinearly elastic shallow shells in curvilinear coordinates* have been justified by Busse [1995, 1997, 1999] and Andreoiu-Banica [1998, 1999], Andreoiu [1999b] by means of the method of formal asymptotic expansions. S. Busse has considered the case of a *clamped shallow shell*, while G. Andreoiu has identified *Marguerre-von Kármán equations in curvilinear coordinates* for a specific class of boundary conditions (introduced by Ciarlet [1980] in the plane case).

Since the asymptotic justification and the existence theory for such equations essentially resemble those for the equations of nonlinearly shallow shells in *Cartesian* coordinates (Vol. II, Sects. 4.14 and 5.12), they are not incorporated into this volume.

To give a flavor of these theories, we simply record the *two-dimensional equations of a nonlinearly elastic clamped shallow shell in curvilinear coordinates*, expressed here as a *minimization problem*:

Let

$$b^{\alpha\beta\sigma\tau,\varepsilon} := \frac{4\lambda^\varepsilon\mu^\varepsilon}{\lambda^\varepsilon + 2\mu^\varepsilon}\delta^{\alpha\beta}\delta^{\sigma\tau} + 2\mu^\varepsilon(\delta^{\alpha\sigma}\delta^{\beta\tau} + \delta^{\alpha\tau}\delta^{\beta\sigma}),$$

$$p^{i,\varepsilon} := \int_{-\varepsilon}^{\varepsilon} f^{i,\varepsilon}\, dx_3^\varepsilon + h_+^{i,\varepsilon} + h_-^{i,\varepsilon},$$

$$q^{\alpha,\varepsilon} := \int_{-\varepsilon}^{\varepsilon} x_3^\varepsilon f^{\alpha,\varepsilon}\, dx_3^\varepsilon + \varepsilon(h_+^{\alpha,\varepsilon} - h_-^{\alpha,\varepsilon}),$$

$$E_{\alpha\beta}^{sh,\varepsilon} := \frac{1}{2}(\partial_\beta\eta_\alpha + \partial_\alpha\eta_\beta) - \varepsilon\eta_3\partial_{\alpha\beta}\theta + \frac{1}{2}\partial_\alpha\eta_3\partial_\beta\eta_3,$$

and let $\boldsymbol{a}^{i,\varepsilon}$ designate the vectors of the contravariant bases along the middle surface S^ε.

Then the unknown $\boldsymbol{\zeta}^\varepsilon = (\zeta_i^\varepsilon)$, where the functions $\zeta_i^\varepsilon : \overline{\omega} \to \mathbb{R}$ are the covariant components of the displacement field $\zeta_i^\varepsilon \boldsymbol{a}^{i,\varepsilon}$ of the middle surface S^ε, minimizes the *energy* $j^{sh,\varepsilon}$ defined by

$$\begin{aligned} j^{sh,\varepsilon}(\boldsymbol{\eta}) := \frac{1}{2}\int_\omega \Big\{ & \varepsilon b^{\alpha\beta\sigma\tau,\varepsilon} E_{\sigma\tau}^{sh,\varepsilon}(\boldsymbol{\eta}) E_{\alpha\beta}^{sh,\varepsilon}(\boldsymbol{\eta}) \\ & + \frac{\varepsilon^3}{3} b^{\alpha\beta\sigma\tau,\varepsilon}\partial_{\sigma\tau}\eta^3\partial_{\alpha\beta}\eta^3 \Big\}\, dy \\ & - \Big\{ \int_\omega p^{i,\varepsilon}\eta_i\, dy - \int_\omega q^{\alpha,\varepsilon}\eta_\alpha\, dy \Big\}, \end{aligned}$$

over the space (the same space as for Koiter's linear equations; cf. Sect. 7.1):

$$\mathbf{V}_K(\omega) := \{\boldsymbol{\eta} = (\eta_i) \in H^1(\omega)\times H^1(\omega)\times H^2(\omega);\ \eta_i = \partial_\nu\eta_3 = 0 \text{ on } \gamma_0\}.$$

An inspection of this minimization problem immediately reveals a strong resemblance with the minimization problem found for the "same" nonlinearly elastic shallow shell, but where the unknowns are the *Cartesian* components of the unknown displacement field (Vol. II, Sect. 4.14). As in the linear case (Sect. 7.6), the resulting theory is more reminiscent of a plate theory than of a shell theory!

Using a different definition of *"shallowness"*, requiring in particular that the absolute value of the radii of curvature be everywhere on the middle surface "sufficiently large", Koiter [1966, eqs. (11.43) to (11.50)] also obtains a *nonlinear shallow shell theory in curvilinear coordinates*, which takes the following form when it is expressed as a minimization problem: The unknown $\boldsymbol{\zeta}^\varepsilon = (\zeta_i^\varepsilon)$, where the functions $\zeta_i^\varepsilon : \overline{\omega} \to \mathbb{R}$ are the covariant components of the displacement field of the middle surface, should minimize the *energy* $j_K^{sh,\varepsilon}$ defined by

$$j_K^{sh,\varepsilon}(\boldsymbol{\eta}) = \frac{\varepsilon}{8}\int_\omega a^{\alpha\beta\sigma\tau,\varepsilon} G_{\sigma\tau}^{sh}(\boldsymbol{\eta}) G_{\alpha\beta}^{sh}(\boldsymbol{\eta})\sqrt{a}\,dy$$
$$+\frac{\varepsilon^3}{6}\int_\omega a^{\alpha\beta\sigma\tau,\varepsilon}\eta_{3|\sigma\tau}\eta_{3|\alpha\beta}\sqrt{a}\,dy$$
$$-\int_\omega p^{i,\varepsilon}\eta_i\sqrt{a}\,dy,$$

where

$$G_{\alpha\beta}^{sh}(\boldsymbol{\eta}) := \gamma_{\alpha\beta}(\boldsymbol{\eta}) + \frac{1}{2}\partial_\alpha\eta_3\partial_\beta\eta_3 \text{ and } \eta_{3|\alpha\beta} := \partial_{\alpha\beta}\eta_3 - \Gamma_{\alpha\beta}^\sigma\partial_\sigma\eta_3,$$

over the space (again the same as for Koiter's linear equations):

$$\mathbf{V}_K(\omega) := \{\boldsymbol{\eta} = (\eta_i) \in H^1(\omega)\times H^1(\omega)\times H^2(\omega);\ \eta_i = \partial_\nu\eta_3 = 0 \text{ on } \gamma_0\}.$$

Using the theory of pseudo-monotone operators developed by Lions [1969], Bernadou & Oden [1981] have established the existence of a solution to the associated variational problem if the tangential components of the applied forces and the curvature of the middle surface are "sufficiently small". Their approach makes an essential use of the *inequality of Korn's type on a general surface* (Thm. 2.6-4).

Introducing an Airy stress function in curvilinear coordinates and using the same method as in Ciarlet & Rabier [1980] for establishing the existence of a solution to the von Kármán equations (see also Vol. II, Sect. 5.8; this method is itself an elaboration over Berger [1967, 1977]), Alexandrescu-Iosifescu [1995a] has also shown that the same nonlinear shallow shell model of W.T. Koiter has at least one solution when the applied tangential components of the applied forces vanish. In addition, Alexandrescu-Iosifescu [1995b] has studied the behavior of the solutions to this model when the shell "becomes a plate".

The analysis of the *eversion problem* for shallow shells modeled by the above Koiter equations has been carried out by Geymonat, Rosati & Valente [1989], Podio-Guidugli, Rosati, Schiaffino & Valente [1989], Geymonat & Léger [1994]. In this direction, see also Paumier [1978a, 1978b], Paumier & Rao [1989].

The same nonlinear shallow shell model is also called the *Donnell-Mushtari-Vlasov model*, so named by Sanders [1963] after Donnell [1933], Vlasov [1944], and Mushtari-Galimov [1961]. Under this name, it has been justified by Figueiredo [1990a] by means of a formal asymptotic method with two "small" parameters, the thickness 2ε and the ratio of ε to the smallest absolute value of the radii of curvature of the middle surface. Figueiredo [1990b] has also established the existence of a solution, by means of the implicit function theorem.

Other "classical" equations have been likewise deemed appropriate for modeling nonlinearly elastic shallow shells. A thorough treatment of their mathematical properties, together with a detailed discussion of the various *a priori* assumptions they are based upon, is found in the recent book by Vorovich [1999], which contains in addition an extensive list of references from the Russian literature on shell theory.

The modeling of *nonlinearly elastic shallow shells lying on an obstacle* leads to interesting *"unilateral eigenvalue problems"*, which can be solved by means of pseudo-monotone operator theory as shown by Gratie [1998, 1999a, 1999b]. In this direction, see also Goeleven, Nguyen & Théra [1993a, 1993b], Goeleven [1996], Gratie & Pascali [1997, 1999], Pauchard, Pomeau & Rica [1997], Le & Schmitt [1997], Goeleven & Motreanu [1998].

REFERENCES

ABRAHAM, R.; MARSDEN, J.E.; RATIU, T. [1983]: *Manifolds, Tensor Analysis, and Applications*, Global Analysis Series, No. 2, Addison-Wesley, Reading.

ACERBI, E.; BUTTAZZO, G.; PERCIVALE, D. [1988]: Thin inclusions in linear elasticity: a variational approach, *J. reine angew. Math.* **386**, 99-115.

ACERBI, E.; BUTTAZZO, G.; PERCIVALE, D. [1991]: A variational definition of the strain energy for an elastic string, *J. Elasticity* **25**, 137-148.

ACTIS, R.L.; SZABO, B.A.; SCHWAB, C. [1999]: Hierarchic models for laminated plates and shells, *Comput. Methods Appl. Mech. Engrg.* **172**, 79-107.

ADAMS, R. A. [1975]: *Sobolev Spaces*, Academic Press, New York.

AGANOVIĆ, I; JURAK, M.; MARUŠIĆ-PALOKA, E.; TUTEK, Z. [1998]: Moderately wrinkled plate, *Asymptotic Anal.* **16**, 273-297.

AGANOVIĆ, I.; MARUŠIĆ-PALOKA, E.; TUTEK, Z. [1995]: Slightly wrinkled plate, *Asymptotic Anal.* **13**, 1-29.

AGANOVIĆ, I.; TUTEK, Z. [1986]: A justification of the one-dimensional model of an elastic beam, *Math. Methods Appl. Sci.* **8**, 1-14.

AGMON, S.; DOUGLIS, A.; NIRENBERG, L. [1964]: Estimates near the boundary for solutions of elliptic partial differential equations satisfying general boundary conditions. II, *Comm. Pure Appl. Math.* **17**, 35-92.

AKIAN, J.L.; SANCHEZ-PALENCIA, E. [1992]: Approximation de coques élastiques minces par facettes planes. Phénomènes de blocage membranaire, *C.R. Acad. Sci. Paris, Sér. I*, **315**, 363-369.

ALESSANDRINI, S.M.; ARNOLD, D.N.; FALK, R.S.; MADUREIRA, A. [1999]: Derivation and justification of plate models by variational methods, in *Plates and Shells* (M. FORTIN, Editor), pp. 1-20, CRM Proceedings and Lecture Notes, Vol. 21, American Mathematical Society, Providence.

ALEXANDRESCU, O. [1994]: Théorème d'existence pour le modèle bidimensionnel de coque non linéaire de W.T. Koiter, *C.R. Acad. Sci. Paris, Sér. I*, **319**, 899-902.

ALEXANDRESCU-IOSIFESCU, O. [1995a]: Existence et régularité de la solution du modèle bidimensionnel non linéaire de coque faiblement courbée de W.T. Koiter, *C.R. Acad. Sci. Paris, Sér. I*, **321**, 1269-1274.

ALEXANDRESCU-IOSIFESCU, O. [1995b]: Comportement de la solution des modèles non linéaires bidimensionnels de coque "faiblement courbée" de W.T. Koiter et de Marguerre-von Kármán lorsque "la coque devient une plaque", *C.R. Acad. Sci. Paris, Sér. I*, **321**, 1389-1394.

ALVAREZ-DIOS, J.A.; VIAÑO, J.M. [1997]: An asymptotic bending model for a general planar curved rod, in *Shells: Mathematical Modelling and Scientific Computing* (M. BERNADOU, P.G. CIARLET & J.M. VIAÑO, Editors), pp. 11-15, Universidade de Santiago de Compostela.

AMBARTSUMIAN, S.A. [1991]: *Fragments of the Theory of Anisotropic Shells*, World Scientific, Singapore (translation from Russian).

AMROUCHE, C.; GIRAULT, V. [1994]: Decomposition of vector spaces and application to the Stokes problem in arbitrary dimension, *Czech. Math. J.* **44**, 109-140.

ANDREOIU, G.: *Alias* ANDREOIU-BANICA, G.; see ANDREOIU-BANICA, G. [1998] and ANDREOIU-BANICA [1999].

ANDREOIU, G. [1999a]: Comparaison entre modèles bidimensionnels de coques faiblement courbées, *C.R. Acad. Sci. Paris, Sér. I*, **329**, 339-342.

ANDREOIU, G. [1999b]: *Analyse des Coques Faiblement Courbées*, Doctoral Dissertation, Université Pierre et Marie Curie, Paris.

ANDREOIU-BANICA, G. [1998]: Les équations de Marguerre-von Kármán en coordonnées curvilignes, *C.R. Acad. Sci. Paris, Sér. I*, **327**, 319-322.

ANDREOIU-BANICA, G. [1999]: Justification of the Marguerre-von Kármán equations in curvilinear coordinates, *Asymptotic Anal.* **19**, 35-55.

ANTMAN, S.S. [1972]: The theory of rods, in *Handbuch der Physik VIa/2* (C. TRUESDELL, Editor), pp. 641-703, Springer-Verlag, Berlin.

ANTMAN, S.S. [1976]: Ordinary differential equations of one-dimensional nonlinear elasticity I: Foundations of the theories of nonlinearly elastic rods and shells, *Arch. Rational Mech. Anal.* **61**, 307-351.

ANTMAN, S.S. [1979]: The eversion of thick spherical shells, *Arch. Rational Mech. Anal.* **70**, 113-123.

ANTMAN, S.S. [1983]: The influence of elasticity on analysis: Modern developments, *Bull. Amer. Math. Soc.* **9**, 267-291.

ANTMAN, S.S. [1989]: Nonlinear problems of geometrically exact shell theories, in *Analytic and Computational Models for shells* (A.K. NOOR, T. BELYTSCHKO & J.C. SIMO, Editors), pp. 109-131, American Society of Mechanical Engineers, New York.

ANTMAN, S.S. [1995]: *Nonlinear Problems of Elasticity*, Springer-Verlag, Berlin.

ANTMAN, S.S. [1997]: Convergence properties of hierarchies of dynamical theories of rods and shells, *Z. angew. Math. Phys.* **48**, 874-884.

ANTMAN, S.S.; MARLOW, R.S. [1991]: Material constraints, Lagrange multipliers, and incompatibility, *Arch. Rational Mech. Anal.* **116**, 257-299.

ANTMAN, S.S.; SCHURICHT, F. [1999]: Incompressibility in rod and shell theories, *Math. Modelling Numer. Anal.* **33**, 289-304.

ANTMAN, S.S.; SRUBSHCHIK, L.S. [1998]: Asymptotic analysis of the eversion of nonlinearly elastic shells, *J. Elasticity* **50**, 129-179; Corrigendum, *J. Elasticity* **52** (1999), 293-294.

ANZELLOTTI, G.; BALDO, S.; PERCIVALE, D. [1994]: Dimension reduction in variational problems, asymptotic development in Γ-convergence and thin structures in elasticity, *Asymptotic Anal.* **9**, 61-100.

ARNOLD, D.N. [1981]: Discretization by finite elements of a model parameter dependent problem, *Numer. Math.* **37**, 405-421.

ARNOLD, D.N.; BREZZI, F. [1993]: Some new elements for the Reissner-Mindlin plate model, in *Boundary Value Problems for Partial Differential Equations and Applications* (J.L. LIONS & C. BAIOCCHI, Editors), pp. 287-292, Masson, Paris.

ARNOLD, D.N.; BREZZI, F. [1997a]: The partial selective reduced integration method and applications to shell problems, *Computers & Structures* **64**, 879-880.

ARNOLD, D.N.; BREZZI, F. [1997b]: Locking-free finite element methods for shells, *Math. Comp.* **66**, 1-14.

ARNOLD, D.N.; FALK, R.S. [1989]: A uniformly accurate finite element method for the Mindlin-Reissner plate, *SIAM J. Numer. Anal.* **26**, 1276-1290.

ARNOLD, D.N.; FALK, R.S. [1990]: The boundary layer for the Reissner-Mindlin plate model, *SIAM J. Math. Anal.* **21**, 281-312.

ARNOLD, D.N.; FALK, R.S. [1996]: Asymptotic analysis of the boundary layer for the Reissner-Mindlin plate model, *SIAM J. Math. Anal.* **27**, 486-514.

ARONSZAJN, N. [1956]: Sur l'unicité du prolongement des solutions des équations aux dérivées partielles elliptiques du second ordre, *Comptes-Rendus* **242**, 723-725.

ATTOUCH, H. [1984]: *Variational Convergence for Functions and Operators*, Pitman, Boston.

AXELRAD, E.L. [1987]: *Theory of Flexible Shells*, North-Holland, Amsterdam.

BABUŠKA, I.; D'HARCOURT, J.M; SCHWAB, C. [1993]: Optimal shear correction factors in hierarchical plate modelling, *Math. Modelling Scientific Computing* **1**, 1-30.

BABUŠKA, I.; LI, L. [1991]: Hierarchic modeling of plates, *Computers & Structures* **40**, 419-430.

BABUŠKA, I.; LI, L. [1992]: The problem of plate modeling - Theoretical and computational results, *Comput. Methods Appl. Mech. Engrg.* **100**, 249-273.

BABUŠKA, I.; SCHWAB, C. [1996]: A posteriori error estimation for hierarchic models of elliptic boundary value problems on thin domains, *SIAM J. Numer. Anal.* **33**, 221-246.

BABUŠKA, I.; SURI, M. [1992]: On locking and robustness in the finite element method, *SIAM J. Numer. Anal.* **29**, 1261-1293.

BABUŠKA, I.; SZABO, B.A.; ACTIS, R. [1992]: Hierarchic models for laminated composites, *Internat. J. Numer. Methods Engrg.* **33**, 503-536.

BALL, J.M. [1977]: Convexity conditions and existence theorems in nonlinear elasticity, *Arch. Rational Mech. Anal.* **63**, 337-403.

BALL, J.M.; MURAT, F. [1984]: $W^{1,p}$-quasiconvexity and variational problems for multiple integrals, *J. Functional Anal.* **58**, 225-253.

BANKS, H.T.; SMITH, R.C.; WANG, Y. [1995]: The modeling of piezoceramic patch interactions with shells, plates, and beams, *Quart. Appl. Math.* **53**, 353-381.

BANKS, H.T.; SMITH, R.C.; WANG, Y. [1996]: *Smart Material Structures: Modeling, Estimation and Control*, Masson, Paris & John Wiley, New York.

BASAR, Y [1987]: A consistent theory of geometrically non-linear shells with an independent rotation vector, *Internat. J. Solids Structures* **23**, 1401-1415.

BASAR, Y.; DING, Y. [1995]: Interlaminar stress analysis of composites: Layer-wise shell finite elements including transverse strains, *Composites Engineering* **5**, 485-499.

BASAR, Y.; DING, Y.; SCHULTZ, R. [1993]: Refined shear-deformation models for composite laminates with finite rotations, *Internat. J. Solids Structures* **30**, 2611-2638.

BASAR, Y.; ITSKOV, M. [1998]: Finite element formulation of the Ogden material model with application to rubber-like shells, *Intern. J. Numer. Methods Engrg.* **42**, 1279-1305.

BASAR, Y.; KRÄTZIG, W.B. [1985]: *Mechanik der Flächentragwerke*, Vieweg Verlag, Braunschweig.

BASAR, Y.; KRÄTZIG, W.B. [1988]: A consistent shell theory for finite deformations, *Acta Mechanica* **76**, 73-87.

BATHE, K.J. [1996]: *Finite Element Procedures*, Prentice Hall, Englewood Cliffs.

BATHE, K.J.; BREZZI, F. [1985]: On the convergence of a four-node plate bending element based on Mindlin-Reissner plate theory and a mixed interpolation, in *The Mathematics of Finite Elements and Applications V* (J.R. WHITEMAN, Editor), pp. 491-503, Academic Press, London.

BATHE, K.J.; BREZZI, F.; FORTIN, M. [1989]: Mixed-interpolated elements for Reissner-Mindlin plates, *Internat. J. Numer. Methods Engrg.* **28**, 1787-1801.

BEN AMAR, M.; POMEAU, Y. [1997]: Crumpled paper, *Proc. Royal Soc. London A*, **453**, 729-755.

BEN BELGACEM, H. [1997]: Une méthode de Γ-convergence pour un modèle de membrane non linéaire, *C.R. Acad. Sci. Paris, Sér. I*, **323**, 845-849.

BERGER, M.S. [1967]: On the von Kármán equations and the buckling of a thin elastic plate. I. The clamped plate, *Comm. Pure Appl. Math.* **20**, 687-719.

BERGER, M.S. [1977]: *Nonlinearity and Functional Analysis*, Academic Press, New York.

BERGER, M.; GOSTIAUX, B. [1992]: *Géométrie Différentielle: Variétés, Courbes et Surfaces, Second Edition*, Presses Universitaires de France, Paris.

BERMUDEZ, A.; VIAÑO, J.M. [1984]: Une justification des équations de la thermoélasticité des poutres à section variable par des méthodes asymptotiques, *RAIRO Analyse Numérique* **18**, 347-376.

BERNADOU, M. [1994]: *Méthodes d'Eléments Finis pour les Coques Minces*, Masson, Paris (English translation: *Finite Element Methods for Thin Shell Problems*, John Wiley, New York, 1995).

BERNADOU, M.; BOISSERIE, J.M. [1982]: *The Finite Element Method for Thin Shell Problems; Application to Arch Dam Problems*, Birkhäuser, Boston.

BERNADOU, M.; CIARLET, P.G. [1976]: Sur l'ellipticité du modèle linéaire de coques de W.T. Koiter, in *Computing Methods in Applied Sciences and Engineering* (R. GLOWINSKI & J.L. LIONS, Editors), pp. 89-136, Lecture Notes in Economics and Mathematical Systems, **134**, Springer-Verlag, Heidelberg.

BERNADOU, M.; CIARLET, P.G.; MIARA, B. [1994]: Existence theorems for two-dimensional linear shell theories, *J. Elasticity* **34**, 111-138.

BERNADOU, M.; CUBIER, A. [1998a]: Numerical analysis of junctions between thin shells, Part I: Continuous problems, *Comput. Methods Appl. Mech. Engrg.* **161**, 349-363.

BERNADOU, M.; CUBIER, A. [1998b]: Numerical analysis of junctions between thin shells, Part II: Approximations by finite element methods, *Comput. Methods Appl. Mech. Engrg.* **161**, 365-387.

BERNADOU, M.; MATO-EIROA, P.; TROUVÉ, P. [1994]: On the convergence of a discrete Kirchhoff triangle method valid for shells of arbitrary shape, *Comput. Methods Appl. Mech. Engrg.* **118**, 373-391.

BERNADOU, M.; ODEN, J.T. [1981]: An existence theorem for a class of nonlinear shallow shell problems, *J. Math. Pures Appl.* **60**, 285-308.

BERNADOU, M.; PALMA, F.J.; ROUSSELET, B. [1991]: Shape optimization of an elastic thin shell under various criteria, *Structural Optim.* **3**, 7-21.

BERS, L.; JOHN, F.; SCHECHTER, M. [1964]: *Partial Differential Equations*, Interscience Publishers, New York.

BLANCHARD, D.; FRANCFORT, G.A. [1987]: Asymptotic thermoelastic behavior of flat plates, *Quart. Appl. Math.* **45**, 645-667.

BLOUZA, A. [1997]: Existence et unicité pour le modèle de Naghdi pour une coque peu régulière, *C.R. Acad. Sci. Paris, Sér. I*, **324**, 839-844.

BLOUZA, A.; BREZZI, F.; LOVADINA, C. [1999]: Sur la classification des coques linéairement élastiques, *C.R. Acad. Sci. Paris, Sér. I*, **328**, 831-836.

BLOUZA, A.; LE DRET, H. [1994a]: Sur le lemme du mouvement rigide, *C.R. Acad. Sci. Paris, Sér. I*, **319**, 1015-1020.

BLOUZA, A.; LE DRET, H. [1994b]: Existence et unicité pour le modèle de Koiter pour une coque peu régulière, *C.R. Acad. Sci. Paris, Sér. I*, **319**, 1127-1132.

BLOUZA, A.; LE DRET, H. [1999]: Existence and uniqueness for the linear Koiter model for shells with little regularity, *Quart. Appl. Math.* **57**, 317-337.

BLUME, J.A. [1989]: Compatibility conditions for a left Cauchy-Green strain field, *J. Elasticity* **21**, 271-308.

BOLLEY, P.; CAMUS, J. [1976]: Régularité pour une classe de problèmes aux limites elliptiques dégénérés variationnels, *C.R. Acad. Sci. Paris, Sér. A*, **282**, 45-47.

BONNET, O. [1848]. Mémoire sur la théorie générale des surfaces, *Journal de l'Ecole Polytechnique* **19**, 1-146.

BOOTHBY, W.M. [1975]: *An Introduction to Differentiable Manifolds and Riemannian Geometry*, Academic Press, New York.

BORCHERS, W.; SOHR, H. [1990]: On the equations rot $v = g$ and div $u = f$ with zero boundary conditions, *Hokkaido Math. J.* **19**, 67-87.

BOURQUIN, F.; CIARLET, P.G.; GEYMONAT, G.; RAOULT, A. [1992]: Γ-convergence et analyse asymptotique des plaques minces, *C.R. Acad. Sci. Paris, Sér. I*, **315**, 1017-1024.

BRAMBLE, J.; SUN, T. [1997]: A locking free finite element method for Naghdi shells, *J. Computational Appl. Math.* **89**, 119-133.

BRENNER, S.C.; SCOTT, L.R. [1994]: *The Mathematical Theory of Finite Element Methods*, Springer-Verlag, Berlin.

BREZIS, H. [1983]: *Analyse Fonctionnelle, Théorie et applications*, Masson, Paris.

BREZZI F. [1994]: Personal Communication.

BREZZI, F.; FORTIN, M. [1986]: Numerical approximation of Mindlin-Reissner plates, *Math. Comp.* **47**, 151-158.

BREZZI, F.; FORTIN, M. [1991]: *Mixed and Hybrid Finite Element Methods*, Springer-Verlag, Berlin.

BREZZI, F.; FORTIN, M.; STENBERG, R. [1991]: Error analysis of mixed-interpolated elements for Reissner-Mindlin plates, *Math. Models Methods Appl. Sci.* **1**, 125-151.

BREZZI, F.; GÉRARD, P.; SANCHEZ-PALENCIA, E. [1998]: Remarks on ellipticity and regularity for non-inhibited shells and other constrained systems, *C.R. Acad. Sci. Paris, Sér. I*, **326**, 1155-1162.

BUDIANSKY, B.; SANDERS, J.L. [1967]: On the "best" first-order linear shell theory, in *Progress in Applied Mechanics, W. Prager Anniversary Volume*, pp. 129-140, MacMillan, New York.

BUSSE, S. [1995]: *Sur quelques Questions Liées à la Théorie Mathématique des Coques en Coordonnées Curvilignes*, Doctoral Dissertation, Université Pierre et Marie Curie, Paris.

BUSSE, S. [1997]: Modélisation des coques non linéairement élastiques faiblement courbées en coordonnées curvilignes, *C.R. Acad. Sci. Paris, Sér. I*, **324**, 1175-1179.

BUSSE, S. [1998]: Asymptotic analysis of linearly elastic shells with variable thickness, *Rev. Roumaine Math. Pures Appl.* **43**, 553-590.

BUSSE, S. [1999]: Modeling of nonlinearly elastic shallow shells in curvilinear coordinates, to appear.

BUSSE, S.; CIARLET, P.G.; MIARA, B. [1996]: Coques "faiblement courbées" en coordonnées curvilignes, *C.R. Acad. Sci. Paris, Sér. I*, **322**, 1093-1098.

BUSSE, S.; CIARLET, P.G.; MIARA, B. [1997]: Justification d'un modèle bidimensionnel de coques "faiblement courbées" en coordonnées curvilignes, *Math. Modelling Numer. Anal.* **31**, 409-434.

CAILLERIE, D. [1980]: The effect of a thin inclusion of high rigidity in an elastic body, *Math. Methods Appl. Sci.* **2**, 251-270.

CAILLERIE, D. [1996]: Etude générale d'un type de problèmes raides et de perturbation singulière, *C.R. Acad. Sci. Paris, Sér. I*, **323**, 835-840.

CAILLERIE, D.; SANCHEZ-PALENCIA, E. [1995a]: A new kind of singular-stiff problems and application to thin elastic shells, *Math. Models Methods Appl. Sci.* **5**, 47-66.

CAILLERIE, D.; SANCHEZ-PALENCIA, E. [1995b]: Elastic thin shells: asymptotic theory in the anisotropic and heterogeneous cases, *Math. Models Methods Appl. Sci.*, **5**, 473-496.

CALDERÓN, A.P. [1958]: Uniqueness in the Cauchy problem for partial differential equations, *Amer. J. Math.* **80**, 16-36.

CALLADINE, C.R. [1983]: *Theory of Shell Structures*, Cambridge University Press.

CARLEMAN, T. [1938]: Sur un problème d'unicité pour les systèmes d'équations aux dérivées partielles à deux variables indépendantes, *Arkiv für Mathematik* **26 B**, 1-9.

CARRIVE, M. [1995]: *Modélisation Intrinsèque et Analyse Numérique d'un Problème de Coque Mince en Grands Déplacements*, Doctoral Dissertation, Université Paris-Dauphine, Paris.

CARRIVE, M.; LE TALLEC, P.; MOURO, J. [1995]: Approximation par éléments finis d'un modèle de coques minces géométriquement exact, *Revue Européenne des Eléments Finis* **4**, 633-662.

DO CARMO, M.P. [1976]: *Differential Geometry of Curves and Surfaces*, Prentice-Hall, Englewood Cliffs.

DO CARMO, M.P. [1994]: *Differential Forms and Applications*, Universitext, Springer-Verlag, Berlin (English translation of: *Formas Diferenciais e Apliçöes*, Instituto da Matematica, Pura e Aplicada, Rio de Janeiro, 1971).

CHAPELLE, D. [1994]: Personal communication.

CHAPELLE, D. [1996]: *Etude des Phénomènes de Verrouillage Numérique pour les Problèmes de Coques Minces*, Doctoral Dissertation, Université Pierre et Marie Curie, Paris.

CHAPELLE, D. [1997]: A locking-free approximation of curved rods by straight beam elements, *Numer. Math.* **77**, 299-322.

CHAPELLE, D.; BATHE, K.J. [1998a]: Fundamental considerations for the finite element analysis of shell structures, *Computers & Structures* **66**, 19-36.

CHAPELLE, D.; BATHE, K.J. [1998b]: On general shell finite elements and mathematical shell models, in *Advances in Finite Element Procedures and Techniques* (B.H.V. TOPPING, Editor), pp. 25-30, Civil-Comp Press, Edinburgh.

CHAPELLE, D.; BATHE, K.J. [1999]: The mathematical shell model underlying general shell elements, to appear.

CHAPELLE, D.; STENBERG, R. [1998]: Stabilized finite element formulations for shells in a bending dominated state, *SIAM J. Numer. Anal.* **36**, 32-73.

CHENAIS, D. [1987]: Optimal design of midsurface of shells: Differentiability proof and sensitivity computations, *Appl. Math. Optim.* **16**, 93-133.

CHENAIS, D. [1994]: Shape optimization of shells, *Control and Cybernetics* **23**, 351-382.

CHENAIS, D.; PAUMIER, J.C. [1994]: On the locking phenomenon for a class of elliptic problems, *Numer. Math.* **67**, 427-440.

CHENAIS, D.; ZERNER, M. [1993]: Conditions nécessaires pour éviter le verrouillage numérique. Applications aux arches, *C.R. Acad. Sci. Paris, Sér. I*, **316**, 1097-1102.

CHOI, D. [1993]: Rigidité infinitésimale d'un type de pli d'une surface en déplacement inextensionnel, *C.R. Acad. Sci. Paris, Sér. I*, **317**, 323-327.

CHOI, D.; SANCHEZ-PALENCIA, E. [1993]: Rigidification de surfaces réglées par des plis. Application à la statique et à la dynamique des coques élastiques minces, *C.R. Acad. Sci. Paris, Sér. I*, **317**, 415-422.

CHOI, D.; PALMA, F.J.; SANCHEZ-PALENCIA, E.; VILARIÑO, M.A. [1998]: Membrane locking in the finite element computation of very thin elastic shells, *Math. Modelling Numer. Anal.* **32**, 131-152.

CHOQUET-BRUHAT, Y.; DEWITT-MORETTE, C.; DILLARD-BLEICK, M. [1977]: *Analysis, Manifolds and Physics*, North-Holland, Amsterdam.

CIARLET, P.G. [1978]: *The Finite Element Method for Elliptic Problems*, North-Holland, Amsterdam.

CIARLET, P.G. [1980]: A justification of the von Kármán equations, *Arch. Rational Mech. Anal.* **73**, 349-389.

CIARLET, P.G. [1982]: *Introduction à l'Analyse Numérique Matricielle et à l'Optimisation*, Masson, Paris (English translation: *Introduction to Numerical Linear Algebra and Optimisation*, Cambridge University Press, Cambridge, 1989).

CIARLET, P.G. [1988][1]: *Mathematical Elasticity, Volume I: Three-Dimensional Elasticity*, North-Holland, Amsterdam.

CIARLET, P.G. [1990]: *Platcs and Junctions in Elastic Multi-Structures: An Asymptotic Analysis*, Masson, Paris and Springer-Verlag, Berlin.

CIARLET, P.G. [1991]: Basic error estimates for elliptic problems, in *Handbook of Numerical Analysis, Vol. II* (P.G. CIARLET & J.L. LIONS, Editors), pp. 17-351, North-Holland, Amsterdam.

CIARLET, P.G. [1992a]: Echanges de limites en théorie asymptotique de coques. I. En premier lieu, "la coque devient une plaque", *C.R. Acad. Sci. Paris, Sér. I*, **315**, 107-111.

CIARLET, P.G. [1992b]: Echanges de limites en théorie asymptotique de coques. II. En premier lieu, "l'épaisseur tend vers zéro", *C.R. Acad. Sci. Paris, Sér. I*, **315**, 227-233.

CIARLET, P.G. [1993]: Modèles bi-dimensionnels de coques: Analyse asymptotique et théorèmes d'existence, in *Boundary Value Problems for Partial Differential Equations and Applications* (J.L. LIONS & C. BAIOCCHI, Editors), pp. 61-80, Masson, Paris.

CIARLET, P.G. [1997][2]: *Mathematical Elasticity, Volume II: Theory of Plates*, North-Holland, Amsterdam.

CIARLET, P.G. [1998a]: Un lemme de J.-L. Lions et les inégalités de Korn sur les surfaces, in *Equations aux Dérivées Partielles et Applications. Articles Dédiés à Jacques-Louis Lions*, pp. 357-382, Gauthier-Villars, Paris.

CIARLET, P.G. [1998b]: Inequalities of Korn's type on surfaces, in *Il Problema di de Saint-Venant: Aspetti Teorici e Applicativi*, pp. 105-134, *Atti dei Convegni Lincei* **140**, Accademia Nazionale dei Lincei, Roma.

CIARLET, P.G. [1998c]: *Introduction to Linear Shell Theory*, Gauthier-Villars, Paris and North-Holland, Amsterdam.

CIARLET, P.G.; COUTAND, D. [1998]: An existence theorem for nonlinearly elastic "flexural" shells, *J. Elasticity* **50**, 261-277.

CIARLET, P.G.; DESTUYNDER, P. [1979a]: A justification of the two-dimensional plate model, *J. Mécanique* **18**, 315-344.

CIARLET, P.G.; DESTUYNDER, P. [1979b]: A justification of a nonlinear model in plate theory, *Comput. Methods Appl. Mech. Engrg.* **17/18**, 227-258.

CIARLET, P.G.; KESAVAN, S. [1981]: Two-dimensional approximation of three-dimensional eigenvalue problems in plate theory, *Comput. Methods Appl. Mech. Engrg.* **26**, 149-172.

CIARLET, P.G.; LARSONNEUR, F. [2000]: Sur la détermination d'une surface dans $\mathbb{R}^3$ à partir de ses deux formes fondamentales, to appear.

CIARLET, P.G.; LE DRET, H.; NZENGWA, R. [1989]: Junctions between three-dimensional and two-dimensional linearly elastic structures, *J. Math. Pures Appl.* **68**, 261-295.

CIARLET, P.G.; LODS, V. [1996a]: On the ellipticity of linear membrane shell equations, *J. Math. Pures Appl.* **75**, 107-124.

CIARLET, P.G.; LODS, V. [1996b]: Asymptotic analysis of linearly elastic shells. I. Justification of membrane shell equations, *Arch. Rational Mech. Anal.* **136**, 119-161.

[1]Referred to in the text as "Vol. I".

[2]Referred to in the text as "Vol. II".

CIARLET, P.G.; LODS, V. [1996c]: Asymptotic analysis of linearly elastic shells. III. Justification of Koiter's shell equations, *Arch. Rational Mech. Anal.* **136**, 191-200.

CIARLET, P.G.; LODS, V. [1996d]: Asymptotic analysis of linearly elastic shells: "Generalized membrane shells", *J. Elasticity* **43**, 147-188.

CIARLET, P.G.; LODS, V.; MIARA, B. [1996]: Asymptotic analysis of linearly elastic shells. II. Justification of flexural shell equations, *Arch. Rational Mech. Anal.* **136**, 163-190.

CIARLET, P.G.; MIARA, B. [1992a]: Justification of the two-dimensional equations of a linearly elastic shallow shell, *Comm. Pure Appl. Math.* **45**, 327-360.

CIARLET, P.G.; MIARA, B. [1992b]: On the ellipticity of linear shell models, *Z. angew. Math. Phys.* **43**, 243-253.

CIARLET, P.G.; PAUMIER, J.C. [1986]: A justification of the Marguerre-von Kármán equations, *Computational Mech.* **1**, 177-202.

CIARLET, P.G.; RABIER, P. [1980]: *Les Equations de von Kármán*, Lecture Notes in Mathematics, Vol. 826, Springer Verlag, Berlin.

CIARLET, P.G.; SANCHEZ-PALENCIA, E. [1996]: An existence and uniqueness theorem for the two-dimensional linear membrane shell equations, *J. Math. Pures Appl.* **75**, 51-67.

CIMETIÈRE, A.; GEYMONAT, G.; LE DRET, H.; RAOULT, A.; TUTEK, Z. [1988]: Asymptotic theory and analysis for displacements and stress distribution in nonlinear elastic straight slender rods, *J. Elasticity* **19**, 111-161.

CIORANESCU, D.; DONATO, P. [1999]: *An Introduction to Homogenization*, Oxford Lecture Series in Mathematics and its Applications, Vol. 17, Oxford University Press.

CIORANESCU, D.; SAINT JEAN PAULIN, J. [1995]: Conductivity problems for thin tall structures depending on several small parameters, *Advances Math. Sci. Appl.* **5**, 287-320.

CIORANESCU, D.; SAINT JEAN PAULIN, J. [1999]: *Homogenization of Reticulated Structures*, Applied Mathematical Sciences, Vol. 136, Springer-Verlag, Berlin.

COLLARD, C. [1999]: A remark on the formal asymptotic analysis of nonlinearly elastic membrane shells, to appear.

COLLARD, C.; MIARA, B. [1999]: Asymptotic analysis of the stresses in thin elastic shells, *Arch. Rational Mech. Anal.* **148**, 233-264.

CONNÉTABLE, B. [1995]: *Applications du Calcul Formel à des Problèmes de Mécanique des Solides*, Doctoral Dissertation, Université Pierre et Marie Curie, Paris.

COURANT, R.; HILBERT, D. [1962]: *Methods of Mathematical Physics, Vol. II*, Interscience Publishers, New York.

COSSERAT, E.; COSSERAT, F. [1909]: *Théorie des Corps Déformables*. Hermann, Paris.

COUTAND, D. [1997a]: Existence d'un minimiseur pour le modèle "proprement invariant" de plaque "en flexion" non linéairement élastique, *C.R. Acad. Sci. Paris, Sér. I*, **324**, 245-248.

COUTAND, D. [1997b]: Théorèmes d'existence pour un modèle membranaire "proprement invariant" de plaque non linéairement élastique, *C.R. Acad. Sci. Paris, Sér. I*, **324**, 1181-1184.

COUTAND, D. [1999a]: Théorèmes d'unicité et de non-unicité pour une coque non linéairement élastique "en flexion", *C.R. Acad. Sci. Paris, Sér. I*, **328**, 627-630.

COUTAND, D. [1999b]: *Analyse Mathématique de Modèles de Plaques et de Coques en Elasticité Non Linéaire*, Doctoral Dissertation, Université Pierre et Marie Curie, Paris.

COUTAND, D. [1999c]: Existence of a solution for a nonlinearly elastic plane membrane subject to plane forces, *J. Elasticity* **53**, 147-159.

COUTAND, D. [1999d]: Existence of minimizing solutions for a nonlinearly elastic clamped plane membrane, *Chinese Annals Math.*, **20B**, 279-296.
COUTAND, D. [1999e]: Existence of a solution for a nonlinearly elastic membrane "under tension", *Math. Modelling Numer. Anal.* **33**, 1019-1032.
COUTAND, D. [1999f]: Uniqueness and non-uniqueness theorems for nonlinearly elastic "flexural shells", to appear.
COUTRIS, N. [1973]: Flexions élastique et élastoplastique d'une coque mince, *J. Mécanique* **12**, 463-475.
COUTRIS, N. [1976]: Théorème d'existence et d'unicité pour un problème de flexion élastique de coques dans le cadre de la modélisation de P.M. Naghdi, *C.R. Acad. Sci. Paris* **283**, 951-953.
COUTRIS, N. [1978]: Théorème d'existence et d'unicité pour un problème de coque élastique dans le cas d'un modèle linéaire de P.M. Naghdi, *RAIRO Analyse Numérique* **12**, 51-57.
DACOROGNA, B. [1989]: *Direct Methods in the Calculus of Variations*, Springer-Verlag, Berlin.
DAL MASO, G. [1993]: *An Introduction to Γ-Convergence*, Birkhäuser, Boston.
DAUGE, M.; DJURDJEVIC, I.; FAOU, E.; RÖSSLE, A. [1999]: Eigenmode asymptotics in thin elastic plates, *J. Math. Pures Appl.* **78**, 925-964.
DAUGE, M.; DJURDJEVIC, I.; RÖSSLE, A. [1998a]: Higher-order bending and membrane responses of thin linearly elastic plates, *C.R. Acad. Sci. Paris, Sér. I*, **326**, 519-524.
DAUGE, M.; DJURDJEVIC, I.; RÖSSLE, A. [1998b]: Full asymptotic expansions for thin elastic free plates, *C.R. Acad. Sci. Paris, Sér. I*, **326**, 1243-1248.
DAUGE, M.; GRUAIS, I. [1996]: Asymptotics of arbitrary order in thin elastic plates and optimal estimates for the Kirchhoff-Love model, *Asymptotic Anal.* **13**, 167-197.
DAUGE, M.; GRUAIS, I. [1998a]: Asymptotics of arbitrary order for a thin elastic clamped plate. II. Analysis of the boundary layer terms, *Asymptotic Anal.* **16**, 99-124.
DAUGE, M.; GRUAIS, I. [1998b]: Edge layers in thin elastic plates, *Comput. Methods Appl. Mech. Engrg.* **157**, 335-347.
DAUGE, M.; GRUAIS, I.; RÖSSLE, A. [1999]: The influence of lateral boundary conditions on the asymptotics in thin elastic plates, *SIAM J. Math. Anal.*, to appear.
DAUTRAY, R.; LIONS, J.L. [1984]: *Analyse Mathématique et Calcul Numérique pour les Sciences et les Techniques, Tome 1*, Masson, Paris.
DAVET, J.L. [1986]: Justification de modèles de plaques non linéaires pour des lois de comportement générales, *Modélisation Math. Anal. Numér.* **20**, 225-249.
DE GIORGI, E. [1975]: Sulla convergenza di alcune successioni di integrali del tipo dell' area, *Rend. Mat. Roma* **8**, 227-294.
DE GIORGI, E. [1977]: Γ-convergenza e G-convergenza, *Boll. Un. Mat. Ital.* **5**, 213-220.
DE GIORGI, E.; DAL MASO, G. [1983]: *Γ-Convergence and Calculus of Variations*, Lecture Notes in Mathematics, Vol. 979, Springer-Verlag, Berlin.
DE GIORGI, E.; FRANZONI, T. [1975]: Su un tipo di convergenza variazionale, *Atti Accad. Naz. Lincei Rend. Cl. Sci. Mat.* **58**, 842-850.
DELFOUR, M.C. [1998]: Intrinsic differential geometric methods in the asymptotic analysis of linear thin shells, in *Boundaries, Interfaces, and Transitions* (M.C. DELFOUR, Editor), pp. 19-90, American Mathematical Society, Providence.
DELFOUR, M.C. [1999]: Membrane shell equation: Characterization of the space of solutions, in *Control of Distributed Parameter and Stochastic Systems* (S. CHEH, X. LI, J. YONG & X.Y. ZHOU, Editors), pp. 21-29, Kluwer Academic Publishers, Boston.

DELFOUR, M.C.; ZHAO, J. [1999]: Intrinsic nonlinear models of shells, to appear.

DELFOUR, M.C.; ZOLÉSIO, J.P. [1994]: Shape analysis via distance functions, *J. Functional Anal.* **123**, 129-201.

DELFOUR, M.C.; ZOLÉSIO, J.P. [1995]: A boundary differential equation for thin shells, *J. Differential Equations* **119**, 426-449.

DELFOUR, M.C.; ZOLÉSIO, J.P. [1997a]: Convergence to the asymptotic model for linear thin shells, in *Optimization Methods in Partial Differential Equations* (S. COX & I. LASIECKA, Editors), pp. 75-93, Contemporary Mathematics, Vol. 209, American Mathematical Society, Providence.

DELFOUR, M.C.; ZOLÉSIO, J.P. [1997b]: Differential equations for linear shells: Comparison between intrinsic and classical models, in *Advances in the Mathematical Sciences: CRM's 25 Years* (L. VINET, Editor), pp. 42-124, CRM Proceedings and Lecture Notes, Vol. 11, American Mathematical Society, Providence.

DELFOUR, M.C.; ZOLÉSIO, J.P. [1998]: Shape analysis via distance functions: Local theory, in *Boundaries, Interfaces, and Transitions* (M.C. DELFOUR, Editor), pp. 91-123, American Mathematical Society, Providence.

DESTUYNDER, P. [1980]: *Sur une Justification des Modèles de Plaques et de Coques par les Méthodes Asymptotiques*, Doctoral Dissertation, Université Pierre et Marie Curie, Paris.

DESTUYNDER, P. [1981]: Comparaison entre les modèles tri-dimensionnels et bi-dimensionnels de plaques en élasticité, *RAIRO Analyse Numérique* **15**, 331-369.

DESTUYNDER, P. [1982]: On nonlinear membrane theory, *Comput. Methods Appl. Mech. Engrg.* **32**, 377-399.

DESTUYNDER, P. [1983]: An existence theorem for a nonlinear shell model in large displacement analysis, *Math. Methods Appl. Sci.* **5**, 68-83.

DESTUYNDER, P. [1985]: A classification of thin shell theories, *Acta Applicandae Mathematicae* **4**, 15-63.

DESTUYNDER, P. [1986]: *Une Théorie Asymptotique des Plaques Minces en Elasticité Linéaire*, Masson, Paris.

DESTUYNDER, P. [1990]: *Modélisation des Coques Minces Elastiques*, Masson, Paris.

DESTUYNDER, P.; GRUAIS, I. [1995]: Error estimation for the linear three-dimensional elastic plate model, in *Asymptotic Methods for Elastic Structures* (P.G. CIARLET, L. TRABUCHO & J.M. VIAÑO, Editors), pp. 75-88, de Gruyter, Berlin.

DESTUYNDER, P.; SALAUN, M. [1996]: *Mathematical Analysis of Thin Plate Models*, Mathématiques et Applications, Vol. 24, Springer-Verlag, Paris.

DIEUDONNÉ, J. [1968]: *Eléments d'Analyse, Tome 1: Fondements de l'Analyse Moderne*, Gauthier-Villars, Paris (translation from: *Foundations of Modern Analysis*, Academic Press, New York, 1960).

DIKMEN, M. [1982]: *Theory of Thin Elastic Shells*, Pitman, Boston.

DJURDJEVIC, I. [1999]: *Ein Beitrag zur Modellierung der Eigendynamik elastischer Schalen mit Hamiltonscher Strukturen*, Doctoral Dissertation, Mathematisches Institut A, Universität Stuttgart.

DONNELL, L.H. [1933]: Stability of thin-walled tubes under torsion, NACA Report TN 479.

DUVAUT, G.; LIONS, J.L. [1972]: *Les Inéquations en Mécanique et en Physique*, Dunod, Paris (English translation: *Inequalities in Mechanics and Physics*, Springer-Verlag, Berlin, 1976).

DYM, C.L. [1990]: *Introduction to the Theory of Shells*, Hemisphere Publishing Corporation, New York.

ECKHAUS, W. [1979]: *Asymptotic Analysis of Singular Perturbations*, North-Holland, Amsterdam.

EL BACHARI, E. [1997]: Modélisation d'une jonction (3d-2d) en élasticité non linéaire par Γ-convergence, *C.R. Acad. Sci. Paris, Sér. I*, **11**, 1241-1246.

EL BACHARI, E. [1998]: *Modélisation de Structures Minces et de Jonctions en Elasticité Non Linéaire*, Doctoral Dissertation, Université Pierre et Marie Curie, Paris.

ERINGEN, A.C. [1962]: *Nonlinear Theory of Continuous Media*, McGraw-Hill, New York.

FICHERA, G. [1972a]: Existence theorems in elasticity, in *Handbuch der Physik, Vol. VIa/2*, pp. 347-389, Springer-Verlag, Berlin.

FICHERA, G. [1972b]: Boundary value problems of elasticity with unilateral constraints, *Handbuch der Physik*, VIa/2, 391-424, Springer-Verlag, Berlin.

FIGUEIREDO, I.N. [1990a]: A justification of the Donnell-Mushtari-Vlasov model by the asymptotic expansion method, *Asymptotic Analysis* **36**, 221-234.

FIGUEIREDO, I.N. [1990b]: Local existence and regularity of the solutions of the nonlinear thin shell model of Donnell-Mushtari-Vlasov, *Applicable Anal.* **36**, 221-234.

FIGUEIREDO, I.N.; LEAL, C. [1998]: Ellipticity of Koiter's and Naghdi's models for nonhomogeneous anisotropic shells, *Applicable Anal.* **70**, 75-84.

FIGUEIREDO, I.N.; TRABUCHO, L. [1992]: A Galerkin approximation for linear elastic shallow shells, *Computational Mechanics* **10**, 107-119.

FIGUEIREDO, I.N.; TRABUCHO, L. [1993]: A Galerkin approximation for curved beams, *Comput. Methods Appl. Mech. Engrg.* **102**, 235-253.

FLÜGGE, W. [1934]: *Statik und Dynamik der Schalen*, Springer-Verlag, Berlin.

FLÜGGE, W. [1973]: *Stresses in Shells, Second Edition*, Springer-Verlag, Berlin.

FOX, D.D.; RAOULT, A.; SIMO, J.C. [1993]: A justification of nonlinear properly invariant plate theories, *Arch. Rational Mech. Anal.* **124**, 157-199.

FOX, D.D.; SIMO, J.C. [1992]: A drill rotation formulation for geometrically exact shells, *Comput. Methods Appl. Mech. Engrg.* **98**, 329-343.

FRIEDRICHS, K.O.; DRESSLER, R.F. [1961]: A boundary-layer theory for elastic plates, *Comm. Pure Appl. Math.* **14**, 1-33.

GAUSS, C.F. [1828]: Disquisitiones generales circas superficies curvas, *Commentationes societatis regiae scientiarum Gottingensis recentiores* **6**, Göttingen.

GE, Z.; KRUSE, H.P.; MARSDEN, J.E. [1996]: The limits of Hamiltonian structures in three-dimensional elasticity, shells, and rods, *J. Nonlinear Sci.* **6**, 19-57.

GENEVEY, K. [1996]: A regularity result for a linear membrane shell problem, *Math. Modelling Numer. Anal.* **30**, 467-488.

GENEVEY, K. [1997]: Remarks on nonlinear membrane shell problems, *Math. Mech. Solids* **2**, 215-237.

GENEVEY, K. [1998]: Existence d'une solution radiale pour un problème de membrane non linéaire, to appear.

GENEVEY, K. [1999]: Justification of two-dimensional linear shell models by the use of Γ-convergence theory, pp. 185-197, CRM Proceedings and Lecture Notes, Vol. 21, American Mathematical Society, Providence.

GÉRARD, P.; SANCHEZ-PALENCIA, E. [1999]: Sensitivity phenomena for certain thin elastic shells with edges, *Math. Models Methods Appl. Sci.*, to appear.

GERDES, K.; MATACHE, A.M.; SCHWAB, C. [1998]: Analysis of membrane locking in *hp*-FEM for a cylindrical shell, *Z. angew. Math. Mech.* **78**, 663-686.

GERMAIN, P. [1972]: *Mécanique des Milieux Continus, Tome I*, Masson, Paris.

GEYMONAT, G. [1965]: Sui problemi ai limiti per i sistemi lineari ellittici, *Ann. Mat. Pura Appl.* **69**, 207-284.

GEYMONAT, G.; GILARDI, G. [1998]: Contre-exemples à l'inégalité de Korn et au lemme de Lions dans des domaines irréguliers, in *Equations aux Dérivées Partielles et Applications. Articles Dédiés à Jacques-Louis Lions*, pp. 541-548, Gauthier-Villars, Paris.

GEYMONAT, G.; KRASUCKI, F.; MARIGO, J.J. [1987]: Stress distribution in anisotropic elastic composite beams, in *Applications of Multiple Scalings in Mechanics* (P.G. CIARLET & E. SANCHEZ-PALENCIA, Editors), pp. 118-133, Masson, Paris.

GEYMONAT, G.; LÉGER, A. [1994]: Effets de la courbure dans la limite coque sphérique non linéaire/plaque circulaire, *C.R. Acad. Sci. Paris, Sér. I*, **319**, 305-310.

GEYMONAT, G.; LORETI, P.; VALENTE, V. [1993]: Exact controllability of a thin elastic hemispherical shell *via* harmonic analysis, in *Boundary Value Problems for Partial Differential Equations and Applications* (J.L. LIONS & C. BAIOCCHI, Editors), pp. 379-385, Masson, Paris.

GEYMONAT, G.; ROSATI, M.; VALENTE, V. [1989]: Numerical analysis for eversion in elastic spherical caps equilibrium, *Comput. Methods Appl. Mech. Engrg.* **75**, 39-52.

GEYMONAT, G.; SANCHEZ-PALENCIA, E. [1991]: Remarques sur la rigidité infinitésimale de certaines surfaces elliptiques non régulières, non convexes et applications, *C.R. Acad. Sci. Paris, Sér. I*, **313**, 645-651.

GEYMONAT, G.; SANCHEZ-PALENCIA, E. [1995]: On the rigidity of certain surfaces with folds and applications to shell theory, *Arch. Rational Mech. Anal.* **129**, 11-45.

GEYMONAT, G.; SUQUET, P. [1986]: Functional spaces for Norton-Hoff materials, *Math. Methods Appl. Sci.* **8**, 206-222.

GIROUD, P. [1998]: Analyse asymptotique de coques inhomogènes en élasticité linéarisée anisotrope, *C.R. Acad. Sci. Paris, Sér. I*, **327**, 1011-1014.

GLOWINSKI, R. [1984]: *Numerical Methods for Nonlinear Variational Problems*, Springer-Verlag, Berlin.

GOELEVEN, D. [1996]: *Noncoercive Variational Problems and Related Results*, Pitman Research Notes in Mathematics Series, Vol. 357, Longman, Harlow.

GOELEVEN, D.; MOTREANU, D. [1998]: Asymptotic eigenvalues and spectral analysis of variational inequalities, *Comm. Appl. Anal.* **2**, 343-372.

GOELEVEN, D.; NGUYEN Van Hien; THÉRA, M. [1993a]: Méthode du degré topologique et branche de bifurcation pour les inéquations variationnelles de von Kármán, *C.R. Acad. Sci. Paris, Sér. I*, **317**, 631-635.

GOELEVEN, D.; NGUYEN Van Hien; THÉRA, M. [1993b]: Nonlinear eigenvalue problems governed by a variational inequality of von Kármán type: A degree theoretic approach, *Topological Methods in Nonlinear Analysis*, **2**, 253-276.

GOLDENVEIZER, A.L. [1961]: *Theory of Elastic Thin Shells*, Pergamon Press, Oxford.

GOLDENVEIZER, A.L. [1962]: Derivation of an approximate theory of bending of a plate by the method of asymptotic integration of the equations of the theory of elasticity, *Prikl. Mat. Mech.* **26**, 668-686 (English translation: *J. Appl. Math. Mech.* (1964), 1000-1025).

GOLDENVEIZER, A.L. [1963]: Derivation of an approximate theory of shells by means of asymptotic integration of the equations of the theory of elasticity, *Prikl. Mat. Mech.* **27**, 593-608.

GOLDENVEIZER, A.L. [1964]: The principles of reducing three-dimensional problems of elasticity to two-dimensional problems of the theory of plates and shells, in *Proceedings of the Eleventh International Congress of Theoretical and Applied Mechanics* (H. GÖRTLER, Editor), pp. 306-311, Springer-Verlag, Berlin.

GOLDENVEIZER, A.L. [1969]: Boundary layer and its interaction with the interior state of stress of an elastic thin shell, *Prikl. Math. Mech.* **33**, 971-1001.

GOLDENVEIZER, A.L.; LIDSKI, V.B.; TOVSTIK, P.E. [1979]: *Free Oscillations of Thin Elastic Shells*, Nauka, Moscow (in Russian).

GORDEZIANI, D.G. [1974]: On the solvability of some boundary value problems for a variant of the theory of thin shells, *Dokl. Akad. Nauk SSSR* **215** (English translation: *Soviet Math. Dokl.* **15** (1974), 677-680).

GOULD P.L. [1987]: *Analysis of Shells and Plates*, Springer-Verlag, Berlin.

GRATIE, L. [1998]: Unilateral eigenvalue problems for nonlinearly elastic plates and pseudo-monotone operators, *C.R. Acad. Sci. Paris, Sér. I*, **327**, 959-964.

GRATIE, L. [1999a]: Unilateral problems for nonlinearly elastic plates and shallow shells, *Math. Mech. Solids*, to appear.

GRATIE, L. [1999b]: Unilateral eigenvalue problems for nonlinearly elastic plates: An approach *via* pseudo-monotone operators, *Chinese Ann. Math.*, to appear.

GRATIE, L.; PASCALI, D. [1997]: Recession methods for noncoercive variational inequalities, *Libertas Math.* **17**, 155-164.

GRATIE, L.; PASCALI, D. [1999]: The index jump principle and bifurcation results for nonlinearly elastic plates, *Rev. Roumaine Math. Pures Appl.*, to appear.

GREEN, A.E. [1962]: On the linear theory of thin elastic shells, *Proc. Royal Soc. London* **A266**, 143-160.

GREEN, A.E.; ADKINS, J.E. [1970]: *Large Elastic Deformations, Second Edition*, Clarendon Press, Oxford.

GREEN, A.E.; LAWS, N.; NAGHDI, P.M. [1968]: Rods, plates and shells, *Proc. Cambridge Phil. Soc.* **64**, 895-913.

GREEN, A.E.; NAGHDI, P.M. [1965]: Some remarks on the linear theory of shells, *Quart. J. Mech. Appl. Math.* **18**, 257-276.

GREEN, A.E.; NAGHDI, P.M. [1970]: Non-isothermal theory of rods, plates, and shells, *Intern. J. Solids Structures* **6**, 209-244.

GREEN, A.E.; NAGHDI, P.M. [1974]: On the derivation of shell theories by direct approach, *J. Appl. Mech.* **41**, 173-176.

GREEN, A.E.; NAGHDI, P.M.; WAINWRIGHT, W.L. [1965]: A general theory of a Cosserat surface, *Arch. Rational Mech. Anal.* **20**, 287-308.

GREEN, A.E.; ZERNA, W. [1968]: *Theoretical Elasticity, Second Edition*, Oxford University Press.

GRUBB, G.; GEYMONAT, G. [1977]: The essential spectrum of elliptic systems of mixed order, *Math. Anal.* **227**, 247-276.

GURTIN, M.E. [1972]: The linear theory of elasticity, in *Handbuch der Physik, Vol. VIa/2* (S. FLÜGGE & C. TRUESDELL, Editors), pp. 1-295, Springer-Verlag, Berlin.

GURTIN, M.E. [1981a]: *Topics in Finite Elasticity*, CBMS-NSF Regional Conference Series in Applied Mathematics, SIAM, Philadelphia.

GURTIN, M.E. [1981b]: *An Introduction to Continuum Mechanics*, Academic Press, New York.

HABBAL, A.; CHENAIS, D. [1992]: Deterioration of a finite element method for arch structures when thickness goes to zero, *Numer. Math.* **62**, 321-341.

HAKULA, H. [1997]: *High-Order Finite Element Tools for Shell Problems*, Doctoral Dissertation, Helsinki University of Technology.

HAKULA, H.; LEINO, Y.; PITKÄRANTA, J. [1996]: Scale resolution, locking, and high-order finite element modelling of shells, *Comput. Methods Appl. Mech. Engrg.* **133**, 157-182.

HAKULA, H.; PITKÄRANTA, J. [1995]: Pinched shells of revolution: Experiments on high-order FEM, in *Proceedings of the Third International Conference on Spectral and Higher-Order Methods* (ICOSAHOM'95) (A.V. ILIN & R. SCOTT, Editors), pp. 193-201.

HANYGA, A. [1985]: *Mathematical Theory of Non-Linear Elasticity*, Polish Scientific Publishers, Warsaw and Ellis Horwood, Chichester.

HAMDACHE, K.; SANCHEZ-PALENCIA, E. [1997]: Homogénéisation d'un problème modèle en théorie des coques membranaires, *C.R. Acad. Sci. Paris, Sér. I*, **325**, 683-688.

HÖRMANDER, L. [1983]: Uniqueness theorems for second order elliptic differential equations, *Comm. Partial Differential Equations* **8**, 21-64.

HSIUNG, C.C. [1997]: *A First Course in Differential Geometry*, American Mathematical Society, Providence.

HUGHES, T.J.R. [1987]: *The Finite Element Method: Linear Static and Dynamic Finite Element Analysis*, Prentice-Hall, Englewood Cliffs.

HUGHES, T.J.R.; FRANCA, L.P. [1988]: A mixed finite element formulation for Reissner-Mindlin plate theory: Uniform convergence of all higher order spaces, *Comput. Methods Appl. Mech. Engrg.* **67**, 85-96.

IOSIFESCU, O.: *Alias* ALEXANDRESCU, O. or ALEXANDRESCU-IOSIFESCU, O.; see ALEXANDRESCU, O. [1994], ALEXANDRESCU-IOSIFESCU, O. [1995a], and ALEXANDRESCU-IOSIFESCU, O. [1995b].

IOSIFESCU, O. [1999]: Regularity for Naghdi's shell equations, *Math. Mech. Solids*, to appear.

JAMAL, R. [1998]: *Modélisation Asymptotique des Comportements Statistiques et Vibratoires des Tiges Courbées Elastiques*, Doctoral Dissertation, Université Pierre et Marie Curie, Paris.

JOHN, F. [1965]: Estimates for the derivatives of the stresses in a thin shell and interior shell equations, *Comm. Pure Appl. Math.* **18**, 235-267.

JOHN, F. [1971]: Refined interior equations for thin elastic shells, *Comm. Pure Appl. Math.* **24**, 583-615.

KAIL, R. [1994]: *Modélisation Asymptotique et Numérique de Plaques et Coques Stratifiées*, Doctoral Dissertation, Université Pierre et Marie Curie, Paris.

KAMOUKALOS, A. [1991]: A catenoidal patch test for the inextensional bending of thin shell finite elements, *Comput. Methods Appl. Mech. Engrg.* **92**, 1-32.

KARAMIAN, P. [1998a]: Réflexion des singularités dans les coques hyperboliques inhibées, *C.R. Acad. Sci. Paris, Sér. IIb*, **326**, 609-614.

KARAMIAN, P. [1998b]: Nouveaux résultats numériques concernant les coques minces hyperboliques inhibées: Cas du paraboloïde hyperbolique, *C.R. Acad. Sci. Paris, Sér. IIb*, **326**, 755-760.

VON KÁRMÁN, T. [1910]: Festigkeitsprobleme im Maschinenbau, in *Encyclopädie der Mathematischen Wissenschaften, Vol. IV/4*, pp. 311-385, Leipzig.

KARWOWSKI, A. [1990]: Asymptotic models for a long elastic cylinder, *J. Elasticity* **24**, 229-287.

KARWOWSKI, A. [1993]: Dynamical models for plates and membranes. An asymptotic approach, *J. Elasticity* **32**, 93-153.

KARWOWSKI, A. [1996a]: Shallow arch models. An asymptotic approach, *J. Elasticity* **43**, 1-29.

KARWOWSKI, A. [1996b]: Elastic arches with large elevation. Singularly perturbed problem of nonlinear elasticity, *J. Elasticity* **45**, 153-214.

KERDID, N.; MATO-EIROA, P. [1998]: Approximation par éléments finis conformes d'un modèle de coques peu régulières, *C.R. Acad. Sci. Paris, Sér. I*, **326**, 1335-1340.

KESAVAN, S.; SABU, N. [1999a]: Two-dimensional approximation of eigenvalue problems in shell theory: Flexural shells, *Chinese Ann. Math.*, to appear.

KESAVAN, S.; SABU, N. [1999b]: Two-dimensional approximation of eigenvalue problems in shallow shell theory, *Math. Mech. Solids*, to appear.

KHLUDNEV, A.M.; SOKOLOWSKI, J. [1997]: *Modelling and Control in Solid Mechanics*, Birkhäuser, Basel.

KLINGENBERG, W. [1973]: *Eine Vorlesung über Differentialgeometrie*, Springer-Verlag, Berlin (English translation: *A Course in Differential Geometry*, Springer-Verlag, Berlin, 1978).

KIRCHGÄSSNER, K.; DJURDJEVIC, I. [1997]: On the dynamics of a thin stress-free ring, in *Shells: Mathematical Modelling and Scientific Computing* (M. BERNADOU, P.G. CIARLET & J.M. VIAÑO, Editors), pp. 77-80, Universidade de Santiago de Compostela.

KIRCHHOFF, G. [1876]: *Vorlesungen über Mathematische Physik*, Mechanik, Leipzig.

KIRMSE, A. [1993]: Bending-dominated deformation of thin spherical shells: analysis and finite element approximation, *SIAM J. Numer. Anal.* **30**, 1015-1040.

KNOPS, R.J; PAYNE, L.E. [1971]: *Uniqueness Theorems in Linear Elasticity*, Springer Tracts in Natural Philosophy, Vol. 19, Springer-Verlag, Berlin.

KOHN, R.V.; VOGELIUS, M. [1984]: A new model for thin plates with rapidly varying thickness, I, *Internat. J. Engrg. Sci.* **20**, 333-350.

KOHN, R.V.; VOGELIUS, M. [1985]: A new model for thin plates with rapidly varying thickness, II. A convergence proof, *Quart. Appl. Math.* **43**, 1-22.

KOHN, R.V.; VOGELIUS, M. [1986]: A new model for thin plates with rapidly varying thickness, III. Comparison of different scalings, *Quart. Appl. Math.* **44**, 35-48.

KOITER, W.T. [1960]: A consistent first approximation in the general theory of thin elastic shells, in *Proceedings, IUTAM Symposium on the Theory of Thin Elastic Shells, Delft, August 1959*, pp. 12-33, Amsterdam.

KOITER, W.T. [1966]: On the nonlinear theory of thin elastic shells, *Proc. Kon. Ned. Akad. Wetensch.* **B69**, 1-54.

KOITER, W.T. [1970]: On the foundations of the linear theory of thin elastic shells, *Proc. Kon. Ned. Akad. Wetensch.* **B73**, 169-195.

KOITER, W.T.; SIMMONDS, J.C. [1973]: Foundations of shell theory, in *Applied Mechanics, Proceedings of the Thirteenth International Congress of Theoretical and Applied Mechanics, Moscow, August 1972* (E. BECKER & G.K. MIKHAILOV, Editors), pp. 150-176, Springer-Verlag, Berlin.

KOZLOV, V.; MAZ'YA, V.; MOVCHAN, A. [1999]: *Asymptotic Analysis of Fields in Multi-Structures*, Clarendon Press, Oxford.

KRÄTZIG, W.B. [1974]: Optimale Schalengrundgleichungen und deren Leistungsfähigkeit, *Z. angew. Math. Mech.* **54**, 265-276.

KRÄTZIG, W.B. [1976]: Die erste Approximation der Schalentheorie und deren Unschärfebereich, in *Konstruktiver Ingenieurbau in Forschung und Praxis*, Düsseldorf, Werner-Verlag.

KRÄTZIG, W.B. [1993]: Best transverse shearing and stretching shell theory for nonlinear finite element simulations, *Comput. Methods Appl. Mech. Engrg.* **103**, 135-160.

LADEVÈZE, P. [1976]: Justification de la théorie linéaire des coques élastiques, *J. Mécanique* **15**, 813-856.

LADEVÈZE, P. [1980]: On the validity of linear shell theories, in *Theory of Shells* (W.T. KOITER & G.K. MIKHAILOV, Editors), pp. 369-391, North-Holland, Amsterdam.

LANDAU, L.; LIFCHITZ, E. [1967]: *Théorie de l'Elasticité*, Mir, Moscou.

LASIECKA, I.; TRIGGIANI, R.; VALENTE, V. [1996]: Uniform stabilization of spherical shells by boundary dissipation, *Advances Differential Equations* **4**, 635-674.

LE, V.K.; SCHMITT, K. [1997]: *Global Bifurcation in Variational Inequalities - Applications to Obstacle and Unilateral Problems*, Springer-Verlag, Berlin.

LE DRET, H. [1989]: Folded plates revisited, *Computational Mech.* **5**, 345-365.

LE DRET, H. [1990]: Modeling of a folded plate, *Computational Mech.* **5**, 401-416.

LE DRET, H. [1991]: *Problèmes Variationnels dans les Multi-Domaines: Modélisation des Jonctions et Applications*, Masson, Paris.

LE DRET, H. [1995]: Convergence of displacements and stresses in linearly elastic slender rods as the thickness goes to zero, *Asymptotic Anal.* **10**, 367-402.

LE DRET, H.; RAOULT, A. [1995a]: The nonlinear membrane model as variational limit of nonlinear three-dimensional elasticity, *J. Math. Pures Appl.* **74**, 549-578.

LE DRET, H.; RAOULT, A. [1995b]: The quasiconvex envelope of the Saint Venant-Kirchhoff stored energy function, *Proc. Royal Soc. Edinburgh* **125A**, 1179-1192.

LE DRET, H.; RAOULT, A. [1996]: The membrane shell model in nonlinear elasticity: A variational asymptotic derivation, *J. Nonlinear Sci.* **6**, 59-84.

LEGUILLON, D.; SANCHEZ-HUBERT, J.; SANCHEZ-PALENCIA, E. [1999]: Model problem of singular perturbation without limit in the space of finite energy and its computation, *C.R. Acad. Sci. Paris, Sér. IIb*, **327**, 485-492.

LEINO, Y.; PITKÄRANTA, J. [1994]: On the membrane locking of *h-p* finite elements in a cylindrical shell problem, *Internat. J. Numer. Methods Engrg.* **37**, 1053-1070.

LEMBO, M. [1996]: On the dynamics of shells in the theory of Kirchhoff and Love, *European J. Mech. A/Solids* **15**, 647-666.

LE TALLEC, P. [1994]: Numerical methods for nonlinear three-dimensional elasticity, in *Handbook of Numerical Analysis, Vol. III* (P.G. CIARLET & J.L. LIONS, Editors), pp. 465-622, North-Holland, Amsterdam.

LE TALLEC, P.; MANI, S. [1998]: Analyse numérique d'un modèle de coques de Koiter discrétisé en base cartésienne par éléments finis DKT, *Modélisation Math. Anal. Numér.* **32**, 433-450.

LEWIŃSKI, T.; SOKOLOWSKI, J. [1998]: Optimal shells formed on a sphere. The topological derivative method, Research Report No. 3495, INRIA-Lorraine.

LIBAI, A.; SIMMONDS, J.G. [1998]: *The Nonlinear Theory of Elastic Shells, Second Edition*, Cambridge University Press.

LIBRESCU, L. [1975]: *Elastostatics and Kinetics of Anisotropic and Heterogeneous Shell-Type Structures*, Nordhoff, Leyden.

LIONS, J.L. [1961]: *Equations Différentielles Opérationnelles et Problèmes aux Limites*, Springer-Verlag, Berlin.

LIONS, J.L. [1969]: *Quelques Méthodes de Résolution des Problèmes aux Limites Non Linéaires*, Dunod, Paris.

LIONS, J.L. [1973]: *Perturbations Singulières dans les Problèmes aux Limites et en Contrôle Optimal*, Lecture Notes in Mathematics, Vol. 323, Springer-Verlag, Berlin.

LIONS, J.L. [1988a]: Exact controllability, stabilization and perturbations for distributed systems, John von Neumann Lecture, *SIAM Rev.* **30**, 1-68.

LIONS, J.L. [1988b]: *Contrôlabilité Exacte, Perturbations et Stabilisation de Systèmes Distribués, Tome 1: Contrôlabilité Exacte*, Masson, Paris.

LIONS, J.L. [1988c]: *Contrôlabilité Exacte, Perturbations et Stabilisation de Systèmes Distribués, Tome 2: Perturbations*, Masson, Paris.

LIONS, J.L.; MAGENES, E. [1968]: *Problèmes aux Limites Non Homogènes et Applications, Vol. 1*, Dunod, Paris (English translation: *Non-Homogeneous Boundary Value Problems and Applications, Vol. 1*, Springer-Verlag, Berlin, 1972).

LIONS, J.L.; SANCHEZ-PALENCIA, E. [1994]: Problèmes aux limites sensitifs, *C.R. Acad. Sci. Paris, Sér. I*, **319**, 1021-1026.

LIONS, J.L.; SANCHEZ-PALENCIA, E. [1996]: Problèmes sensitifs et coques élastiques minces, in *Partial Differential Equations and Functional Analysis - In Memory of Pierre Grisvard* (J. CÉA, D. CHENAIS, G. GEYMONAT & J.L. LIONS, Editors), pp. 207-220, Birkhäuser, Boston.

LIONS, J.L.; SANCHEZ-PALENCIA, E. [1997a]: Sur quelques espaces de la théorie des coques et la sensitivité, in *Homogenization and Applications to Material Sciences* (D. CIORANESCU, A. DAMLAMIAN & P. DONATO, Editors), pp. 271-278, Gakkotosho, Tokyo.

LIONS, J.L.; SANCHEZ-PALENCIA, E. [1997b]: Examples of sensitivity in shells with edges, in *Shells - Mathematical Modelling and Scientific Computing* (M. BERNADOU, P.G. CIARLET & J.M. VIAÑO, Editors), pp. 151-154, Universidade de Santiago de Compostela.

LIONS, J.L; SANCHEZ-PALENCIA, E. [1998]: Instabilities produced by edges in thin shells, in *Proceedings, IUTAM Symposium "Variation of Domains and Free-Boundary Problems"* (P. ARGOUL, M. FREMOND & NGUYEN Quoc Son, Editors), pp. 277-284, Kluwer Academic Publishers, Boston.

LIONS, J.L.; STAMPACCHIA, G. [1967]: Variational inequalities, *Comm. Pure Appl. Math.* **20**, 493-519.

LODS, V. [1996]: Limit behavior of a shell, when first the thickness approaches zero and then the shell becomes a plate, *Asymptotic Anal.* **12**, 187-211.

LODS, V. [1997]: Sur l'analyse asymptotique du modèle bidimensionnel de Koiter d'une coque en flexion linéairement élastique, *C.R. Acad. Sci. Paris, Sér. I*, **325**, 1337-1340.

LODS, V.; MARDARE, C. [1998a]: The space of inextensional displacements for a partially clamped linearly elastic shell with an elliptic middle surface, *J. Elasticity* **51**, 127-144.

LODS, V.; MARDARE, C. [1998b]: Justification asymptotique des hypothèses de Kirchhoff-Love pour une coque encastrée linéairement élastique, *C.R. Acad. Sci. Paris, Sér. I*, **326**, 909-912.

LODS, V.; MARDARE, C. [1999a]: Une justification du modèle de coques de Naghdi, *C.R. Acad. Sci. Paris, Sér. I*, **328**, 951-954.

LODS, V; MARDARE, C. [1999b]: Estimations d'erreur entre le problème tridimensionnel de coque linéairement élastique et le modèle de Naghdi, *C.R. Acad. Sci. Paris, Sér. I*, to appear.

LODS, V.; MARDARE, C. [1999c]: Asymptotic justification of the Kirchhoff-Love assumptions for a linearly elastic clamped shell, *J. Elasticity*, to appear.

LODS, V.; MIARA, B. [1998]: Nonlinearly elastic shell models. II. The flexural model, *Arch. Rational Mech. Anal.* **142**, 355-374.

LOVE, A.E.H. [1927]: *A Treatise on the Mathematical Theory of Elasticity*, Cambridge University Press, Cambridge (reprinted by Dover Publications, New York, 1944).

LOVE, A.E.H. [1934]: *The Mathematical Theory of Elasticity*, Cambridge University Press.

LUKASIEWICZ, S. [1979]: *Local Loads in Plates and Shells*, Sijthoff & Noordhoff, Warszawa.

LYLY, M.; STENBERG, R.; VIHINEN, T. [1993]: A stable bilinear element for Reissner-Mindlin plates, *Comput. Methods Appl. Mech. Engrg.* **110**, 343-357.

MADUREIRA, A. [1999]: *Asymptotics and Hierarchical Modeling of Thin Domains*, Doctoral Dissertation, Penn State University, University Park.

MAGENES, E.; STAMPACCHIA, G. [1958]: I problemi al contorno per le equazioni differenziali di tipo ellittico, *Ann. Scuola Norm. Sup. Pisa* **12**, 247-358.

MALLIAVIN, P. [1972]: *Géometrie Différentielle Intrinsèque*, Hermann, Paris.

MARDARE, C. [1998a]: Asymptotic analysis of linearly elastic shells: Error estimates in the membrane case, *Asymptotic Anal.* **17**, 31-51.

MARDARE, C. [1998b]: Two-dimensional models of linearly elastic shells: Error estimates between their solutions, *Math. Mech. Solids* **3**, 303-318.
MARDARE, C. [1998c]: The generalized membrane problem for linearly elastic shells with hyperbolic or parabolic middle surface, *J. Elasticity* **51**, 145-165.
MARSDEN, J.E.; HUGHES, T.J.R. [1983]: *Mathematical Foundations of Elasticity*, Prentice-Hall, Englewood Cliffs.
MASCARENHAS, M.L.; TRABUCHO, L. [1992]: Asymptotic, homogenisation and Galerkin methods in three-dimensional beam theory, in *Computational and Applied Mathematics II* (W.F. AMES & P.J. VAN DER HOUWEN, Editors), pp. 85-91, North-Holland, Amsterdam.
MEL'NIK, V.V.; SRUBSHCHIK, L.S. [1973]: Nonstiffness of nonshallow strictly convex shells of revolution with movable hinge supports, Izv. Severo-Kavkaskogo Nauchnogo Tsentr Vyssheĭ Shkoly, 52-55 (in Russian).
METSKHOVRISHVILI, S.S. [1957]: On infinitesimal bendings of a toroidal shell, *Soobshcheniya A.N. Gruz. S.S.R.* **18**, 521-527.
MIARA, B. [1989]: Optimal spectral approximation in linearized plate theory, *Applicable Anal.* **31**, 291-307.
MIARA, B. [1994a]: Justification of the asymptotic analysis of elastic plates. I. The linear case, *Asymptotic Anal.* **8**, 259-276.
MIARA, B. [1994b]: Justification of the asymptotic analysis of elastic plates. II. The nonlinear case, *Asymptotic Anal.* **9**, 119-134.
MIARA, B. [1998]: Nonlinearly elastic shell models. I. The membrane model, *Arch. Rational Mech. Anal.* **142**, 331-353.
MIARA, B.; SANCHEZ-PALENCIA, E. [1996]: Asymptotic analysis of linearly elastic shells, *Asymptotic Anal.* **12**, 41-54.
MIARA, B.; TRABUCHO, L. [1992]: A Galerkin spectral approximation in linearized beam theory, *Modélisation Math. Anal. Numér.* **26**, 425-446.
MIARA, B.; VALENTE, V. [1999]: Exact controllability of a Koiter shell by a boundary action, *J. Elasticity* **52**, 267-287.
MIELKE, A. [1988]: Saint Venant's problem and semi-inverse solutions in nonlinear elasticity, *Arch. Rational Mech. Anal.* **102**, 205-229; Corrigendum, *Arch. Rational Mech. Anal.* **110** (1990), 351-352.
MIELKE, A. [1990]: Normal hyperbolicity of center manifolds and Saint Venant's principle, *Arch. Rational Mech. Anal.* **110**, 353-372.
MIELKE, A. [1995]: On the justification of plate theories I: The linear case and exponential decay estimates, *J. Elasticity* **38**, 165-208.
MORGENSTERN, D. [1959]: Herleitung der Plattentheorie aus der dreidimensionalen Elastizitätstheorie, *Arch. Rational Mech. Anal.* **4**, 145-152.
MORGENSTERN, D.; SZABÓ, I. [1961]: *Vorlesungen über Theoretische Mechanik*, Springer-Verlag, Berlin.
MORREY, JR., C.B. [1952]: Quasi-convexity and the lower semicontinuity of multiple integrals, *Pacific J. Math.* **2**, 25-53.
MORREY, JR., C.B. [1966]: *Multiple Integrals in the Calculus of Variations*, Springer-Verlag, Berlin.
MORREY, C.B.; NIRENBERG, L. [1957]: On the analyticity of the solution of linear elliptic systems of partial differential equations, *Comm. Pure Appl. Math.* **10**, 271-290.
MURNAGHAN, F.D. [1951]: *Finite Deformations of an Elastic Solid*, John Wiley, New York.
MUSHTARI, K.M.; GALIMOV, K.Z. [1961]: *Non-Linear Theory of Thin Elastic Shells*, Israel Program for Scientific Translation, Jerusalem (English translation of *Nelineinaya Theriya Uprugikh Obolochek*, Tatknigoizdat, 1957).
MUTTIN, F. [1991]: Structural analysis of sails, *European J. Mech. A/Solids* **10**, 517-534.

NAGHDI, P.M. [1963]: Foundations of elastic shell theory, in *Progress in Solid Mechanics, Vol. 4* (I.N. SNEDDON & R. HILL, Editors), pp. 1-90, North-Holland, Amsterdam.

NAGHDI, P.M. [1972]: The theory of shells and plates, in *Handbuch der Physik, Vol. VIa/2* (S. FLÜGGE & C. TRUESDELL, Editors), pp. 425-640, Springer-Verlag, Berlin.

NAGHDI, P.M. [1982]: Finite deformation of elastic rods and shells, in *Finite Elasticity* (D.E. CARLSON & R.T. SHIELD, Editors), pp. 47-103, Martinus Nijhoff Publishers, The Hague.

NEČAS, J. [1967]: *Les Méthodes Directes en Théorie des Equations Elliptiques*, Masson, Paris.

NEČAS, J.; HLAVÁČEK, I. [1981]: *Mathematical Theory of Elastic and Elasto-Plastic Bodies: An Introduction*, Elsevier, Amsterdam.

NIORDSON, F.I. [1985]: *Shell Theory*, North-Holland, Amsterdam.

NIRENBERG, L. [1962]: Rigidity of a class of closed surfaces, in *Nonlinear Problems*, pp. 177-193, University of Wisconsin Press, Madison.

NOLL, W. [1959]: The Foundations of classical mechanics in the light of recent advances in continuum mechanics, in *The Axiomatic Method, with Special Reference to Geometry and Physics*, pp. 266-281, North-Holland, Amsterdam.

NOLL, W. [1966]: The foundations of mechanics, in *C.I.M.E., Non Linear Continuum Theories* (G. GRIOLI & C. TRUESDELL, Editors), pp. 159-200, Cremonese, Rome.

NOLL, W. [1972]: A new mathematical theory of simple materials, *Arch. Rational Mech. Anal.* **48**, 1-50.

NOLL, W. [1973]: Lectures on the foundation of continuum mechanics and thermodynamics, *Arch. Rational Mech. Anal.* **52**, 62-92.

NOLL, W. [1978]: A general framework for problems in the statics of finite elasticity, in *Contemporary Developments in Continuum Mechanics and Partial Differential Equations* (G.M. DE LA PENHA & L.A.J. MEDEIROS, Editors), pp. 363-387, North-Holland, Amsterdam.

NORDGREN, R.P. [1971]: A bound on the error in plate theory, *Quart. Appl. Math.* **28**, 587-595.

NORDGREN, R.P. [1972]: A bound on the error in Reissner's theory of plate, *Quart. Appl. Math.* **29**, 551-556.

NOVOZHILOV, V.V. [1953]: *Foundations of the Nonlinear Theory of Elasticity*, Graylock Press, Rochester.

NOVOZHILOV, V.V. [1959]: *Thin Shell Theory*, Noordhoff, Groningen.

NOVOZHILOV, V.V. [1961]: *Theory of Elasticity*, Pergamon Press, Oxford.

NOVOZHILOV, V.V. [1970]: *Thin Shell Theory, Second Edition*, Wolters-Noordhoff, Groningen.

ODEN, J.T. [1972]: *Finite Elements of Nonlinear Continua*, McGraw-Hill, New York.

ODEN, J.T. [1986]: *Qualitative Methods in Nonlinear Mechanics*, Prentice-Hall, Englewood Cliffs.

ODEN, J.T.; CHO, J.R. [1996]: Adaptive *hpq*-finite element methods of hierarchical models for plate- and shell-like structures, *Comput. Methods Appl. Mech. Engrg.* **136**, 317-345.

ODEN, J.T.; WEMPNER, G.A.; KROSS, D.A. [1968]: Finite element analysis of thin shells, *J. Engrg. Mech. Division, American Society of Civil Engineers*, **94**, 1273-1294.

OGDEN, R.W. [1984]: *Non-Linear Elastic Deformations*, Ellis Horwood, Chichester and John Wiley, New York.

OVASKAINEN, O.; PITKÄRANTA, J. [1996]: An *h-p-n* adaptive finite element scheme for shell problems, *Advances Engrg. Software* **26**, 201-208.

PAUCHARD, L.; POMEAU, Y.; RICA, S. [1997]: Déformation des coques élastiques, *C.R. Acad. Sci. Paris, Sér. IIb*, **324**, 411-418.

PAUMIER, J.C. [1978a]: Solutions stables d'un problème simplifié de coque élastique, *Rev. Française Automat. Informat. Recherche Opérationnelle, Sér. Rouge Anal. Numér.* **12**, 283-293.

PAUMIER, J.C. [1978b]: Stable solutions to a shell problem, *Comput. Methods Appl. Mech. Engrg.* **20**, 91-103.

PAUMIER, J.C. [1991]: Existence and convergence of the expansion in the asymptotic theory of elastic thin plates, *Math. Modelling Numer. Anal.* **25**, 371-391.

PAUMIER, J.C.; RAO, Bo-peng [1989]: Qualitative and quantitative analysis of buckling of shallow shells, *European J. Mech. A/Solids* **8**, 461-489.

PAUMIER, J.C.; RAOULT, A. [1997]: Asymptotic consistency of the polynomial approximation in the linearized plate theory - Application to the Reissner-Mindlin model, in *Elasticité, Viscoélasticité et Contrôle Optimal, ESAIM Proceedings* **2**, 203-213.

PIETRASZKIEWICZ, W. [1977]: *Introduction to the Non-Linear Theory of Shells*, Mitteilungen aus dem Institut für Mechanik, Ruhr-Universität, Bochum.

PIETRASZKIEWICZ, W. [1979]: *Finite Rotations and Lagrangean Description in the Non-Linear Theory of Shells*, Polish Scientific Publishers, Warszawa.

PIILA, J. [1994]: Characterization of the membrane theory of a clamped shell. The elliptic case, *Math. Models Methods Appl. Sci.* **4**, 147-177.

PIILA, J. [1996]: Characterization of the membrane theory of a clamped shell. The hyperbolic case, *Math. Models Methods Appl. Sci.* **6**, 169-194.

PIILA, J.; PITKÄRANTA, J. [1993a]: Energy estimates relating different linear elastic models of a thin cylindrical shell. I. The membrane-dominated case, *SIAM J. Math. Anal.* **24**, 1-22.

PIILA, J.; PITKÄRANTA, J. [1993b]: Characterization of the membrane theory of a clamped shell. The parabolic case, *Math. Models Methods Appl. Sci.* **3**, 417-442.

PIILA, J.; PITKÄRANTA, J. [1995]: Energy estimates relating different linear elastic models of a thin cylindrical shell. II. The case of a free boundary, *SIAM J. Math. Anal.* **26**, 820-849.

PITKÄRANTA, J. [1988]: Analysis of some low-order finite element schemes for Mindlin-Reissner and Kirchhoff plates, *Numer. Math.* **53**, 237-254.

PITKÄRANTA, J. [1992]: The problem of membrane locking in finite element analysis of cylindrical shells, *Numer. Math.* **61**, 523-542.

PITKÄRANTA, J.; LEINO, Y.; OVASKAINEN, O.; PIILA, J. [1995]: Shell deformation states and the finite element method: A benchmark study of cylindrical shells, *Comput. Methods Appl. Mech. Engrg.* **128**, 81-121.

PITKÄRANTA, J.; MATACHE, A.M.; SCHWAB, C. [1999]: Fourier mode analysis of layers in shallow shell deformations, Report 99-16, *Seminar für Angewandte Mathematik*, Eidgenössische Technische Hochschule, Zürich.

PITKÄRANTA, J.; SANCHEZ-PALENCIA, E. [1997]: On the asymptotic behavior of sensitive shells with small thickness, *C.R. Acad. Sci. Paris, Sér. IIb*, **325**, 127-134.

PITKÄRANTA, J.; SURI, M. [1996]: Design principles and error analysis for reduced-shear plate-bending finite elements, *Numer. Math.* **75**, 223-266.

PODIO-GUIDUGLI, P. [1989]: An exact derivation of the thin plate equation, *J. Elasticity* **22**, 121-133.

PODIO-GUIDUGLI, P. [1990]: Constraint and scaling methods to derive shell theory from three-dimensional elasticity, *Riv. Mat. Univ. Parma* **16**, 73-83.

PODIO-GUIDUGLI, P. [1991]: *Lezioni sulla Teoria Lineare dei Gusci Elastici Sottili*, Masson, Milano.

PODIO-GUIDUGLI, P.; ROSATI, M.; SCHIAFFINO, A.; VALENTE, V. [1989]: Equilibrium of an elastic spherical cap pulled at the rim, *SIAM J. Math. Anal.* **20**, 643-663.

POGORELOV, A.V. [1956]: *Die Eindeutige Bestimmung Allgemeiner Konvexer Flächen*, Akademie-Verlag, Berlin.

POGORELOV, A.V. [1966]: *Geometrical Theory of the Stability of Shells*, Nauka, Moscow (in Russian).

POGORELOV, A.V. [1967]: *Geometrical Methods in the Nonlinear Theory of Elastic Shells*, Nauka, Moscow (in Russian).

POGORELOV, A.V. [1988]: *Bending of Surfaces and Stability of Shells*, American Mathematical Society, Providence (translation from Russian edition, Nauka, 1986).

POMEAU, Y. [1995]: Papier froissé, *C.R. Acad. Sci. Paris, Sér. I*, **320**, 975-979.

POMP, A. [1996]: A linear, first-order shell model for transverse isotropic material under a uniform change of temperature, *Math. Methods Appl. Sci.* **19**, 1177-1197.

QUINTELA-ESTEVEZ, P. [1989]: A new model for nonlinear elastic plates with rapidly varying thickness, *Applicable Anal.* **32**, 107-127.

RAMOS, O. [1995]: *Applications du Calcul Différentiel Intrinsèque aux Modèles Bi-Dimensionels Linéaires de Coques*, Doctoral Dissertation, Université Pierre et Marie Curie, Paris.

RAO, Bo-peng [1994]: A justification of a nonlinear model of spherical shell, *Asymptotic Anal.* **9**, 47-60.

RAO, M.; SOKOLOWSKI, J. [1996]: Optimal shape design of plates and shells with obstacles, in *World Congress of Nonlinear Analysts'92* (V. LAKSHMIKANTHAM, Editor), de Gruyter, Berlin.

RAOULT, A. [1985]: Construction d'un modèle d'évolution de plaques avec termes d'inertie de rotation, *Annali di Matematica Pura ed Applicata* **139**, 361-400.

RAOULT, A. [1986]: Non-polyconvexity of the stored energy function of a Saint Venant-Kirchhoff material, *Aplikace Matematiky* **6**, 417-419.

RAOULT, A. [1988]: *Analyse Mathématique de Quelques Modèles de Plaques et de Poutres Elastiques ou Elasto-Plastiques*, Doctoral Dissertation, Université Pierre et Marie Curie, Paris.

RAPPAZ, J.; SANCHEZ-HUBERT, J.; SANCHEZ-PALENCIA, E.; VASSILIEV, D. [1997]: On spectral pollution in the finite element approximation of thin elastic "membrane" shells, *Numer. Math.* **75**, 473-500.

REISSNER, E. [1974]: Linear and nonlinear theory of shells, in *Thin Shell Structures: Theory, Experiment and Design* (Y.C. FUNG & E.E. SECHLER, Editors), pp. 29-44, Prentice-Hall, Englewood Cliffs.

RIGOLOT, A. [1972]: Sur une théorie asymptotique des poutres, *J. Mécanique* **11**, 673-703.

RIGOLOT, A. [1976]: *Sur une Théorie Asymptotique des Poutres Droites*, Doctoral Dissertation, Université Pierre et Marie Curie, Paris.

ROBERT, J.E.; THOMAS, J.M. [1991]: Mixed and hybrid methods, in *Handbook of Numerical Analysis, Vol. II* (P.G. CIARLET & J.L. LIONS, Editors), pp. 523-633, North-Holland, Amsterdam.

RÖSSLE, A. [1999a]: On the derivation of an asymptotically correct shear correction factor for the Reissner-Mindlin plate model, *C.R. Acad. Sci. Paris, Sér. I*, 269-274.

RÖSSLE, A. [1999b]: *Asymptotische Entwicklungen für dünne Platten im Rahmen der linearen Elastostatik*, Doctoral Dissertation, Mathematisches Institut A, Universität Stuttgart.

RÖSSLE, A.; BISCHOFF, M.; WENDLAND, W.; RAMM, E. [1999]: On the mathematical foundation of the (1, 1, 2)-plate model, *Internat. J. Solids Structures* **36**, 2143-2168.

ROQUEFORT, A. [2000]: *Sur Quelques Questions Liées au Modèle Non Linéaire de Coques de W.T. Koiter*, Doctoral Dissertation, Université Pierre et Marie Curie, Paris.

ROUGÉE, P. [1969]: *Equilibre des Coques Elastiques Minces Inhomogènes en Théorie Non Linéaire*, Doctoral Dissertation, Université Pierre et Marie Curie, Paris.

RUTTEN, H.S. [1973]: *Theory and Design of Shells on the Basis of Asymptotic Analysis*, Rutten & Kruisman Consulting Engineers, Holland.

SANCHEZ-HUBERT, J.; SANCHEZ-PALENCIA, E. [1992]: *Introduction aux Méthodes Asymptotiques et à l'Homogénéisation. Application à la Mécanique des Milieux Continus*, Masson, Paris.

SANCHEZ-HUBERT, J.; SANCHEZ-PALENCIA, E. [1993]: *Exercices sur les Méthodes Asymptotiques et l'Homogénéisation*, Masson, Paris.

SANCHEZ-HUBERT, J.; SANCHEZ-PALENCIA, E. [1997]: *Coques Elastiques Minces: Propriétés Asymptotiques*, Masson, Paris.

SANCHEZ-PALENCIA, E. [1980]: *Nonhomogenous Media and Vibration Theory*, Springer-Verlag, Berlin.

SANCHEZ-PALENCIA, E. [1989a]: Statique et dynamique des coques minces. I. Cas de flexion pure non inhibée, *C.R. Acad. Sci. Paris, Sér. I*, **309**, 411-417.

SANCHEZ-PALENCIA, E. [1989b]: Statique et dynamique des coques minces. II. Cas de flexion pure inhibée - Approximation membranaire, *C.R. Acad. Sci. Paris, Sér. I*, **309**, 531-537.

SANCHEZ-PALENCIA, E. [1990]: Passages à la limite de l'élasticité tri-dimensionnelle à la théorie asymptotique des coques minces, *C.R. Acad. Sci. Paris, Sér. II*, **311**, 909-916.

SANCHEZ-PALENCIA, E. [1992]: Asymptotic and spectral properties of a class of singular-stiff problems, *J. Math. Pures Appl.* **71**, 379-406.

SANCHEZ-PALENCIA, E. [1993]: On the membrane approximation for thin elastic shells in the hyperbolic case, *Revista Matematica de la Universidad Complutense de Madrid* **6**, 311-331.

SANCHEZ-PALENCIA, E. [1994]: Coques élastiques minces peu courbées. Transition continue entre les théories de plaques et de coques, *C.R. Acad. Sci. Paris, Sér. I*, **318**, 783-790.

SANCHEZ-PALENCIA, E. [1995]: Surfaces et coques élastiques minces: Problèmes et défis, *La Vie des Sciences* **12**, 239-258.

SANCHEZ-PALENCIA, E. [1999]: On sensitivity and related phenomena in thin shells which are not geometrically rigid, *Math. Models Methods Appl. Sci.* **9**, 139-160.

SANCHEZ-PALENCIA, E.; SANCHEZ-HUBERT, J. [1998]: Pathological phenomena in computation of thin elastic shells, *Trans. Canadian Soc. Mech. Engrg.* **22**, 435-446.

SANCHEZ-PALENCIA, E.; VASSILIEV, D.G. [1992]: Remarks on vibrations of thin elastic shells and their numerical computation, *C.R. Acad. Sci. Paris, Sér. II*, **314**, 445-452.

SANDERS, J.L. [1959]: An improved first-approximation theory for thin shells, NASA Report No. 24.

SANDERS, J.L. [1963]: Nonlinear theories for thin shells, *Quart. Appl. Math.* **21**, 21-36.

SCHIECK, B.; PIETRASZKIEWICZ, W.; STUMPF, H. [1992]: Theory and numerical analysis of shells undergoing large elastic strains, *Internat. J. Solids Structures* **29**, 689-709.

SCHWAB, C. [1994]: Boundary layer resolution in hierarchical models of laminated composites, *Modélisation Math. Numér.* **28**, 517-537.

SCHWAB, C. [1995]: *Hierarchic models of elliptic boundary value problems on thin domains - a-posteriori error estimation and Fourier analysis*, Habilitation Thesis, University of Stuttgart.

SCHWAB, C. [1996]: A-posteriori modeling error estimation for hierarchic plate models, *Numer. Math.* **74**, 221-259.

SCHWAB, C.; SURI, M. [1994]: Locking and boundary layer effects in the finite element approximation of the Reissner-Mindlin plate model, *Amer. Math. Soc. Proc. Symp. Appl. Math.* **48**, 367-371.

SCHWAB, C.; WRIGHT, S. [1995]: Boundary layers of hierarchical beam and plate models, *J. Elasticity* **38**, 1-40.

SCHWARTZ, L. [1966]: *Théorie des Distributions*, Hermann, Paris.

SCHWARTZ, L. [1992]: *Analyse II: Calcul Différentiel et Equations Différentielles*, Hermann, Paris.

SCHWARTZ, L. [1993]: *Analyse IV: Applications de la Théorie de la Mesure*, Hermann, Paris.

SHOIKHET, B.A. [1974]: On existence theorems in linear shell theory, *Prikl. Math. Mech.* **38**, 527-531.

SHOIKHET, B.A. [1976]: An energy identity in physically nonlinear elasticity and error estimates of the plate equations, *Prikl. Mat. Mech.* **40**, 317-326.

SIMMONDS, J.G. [1971a]: An improved estimate for the error in the classical linear theory of plate bending, *Quart. Appl. Math.* **29**, 439-447.

SIMMONDS, J.G. [1971b]: Extension of Koiter's L_2-error estimate to approximate shell solutions with no strain energy functional, *Z. angew. Math. Phys.* **22**, 339-345.

SIMMONDS, J.G. [1992]: An asymptotic analysis of end effects in the axisymmetric deformation of elastic tubes weak in shear: Higher-order shell theories are inadequate and unnecessary, *Intern. J. Solids Structures* **29**, 2441-2461.

SIMMONDS, J.G. [1994]: *A Brief on Tensor Analysis, Second Edition,* Springer-Verlag, Berlin.

SIMO, J.C.; FOX, D.D. [1989]: On a stress resultant geometrically exact shell model, Part I: Formulation and optimal parametrization, *Comput. Methods Appl. Mech. Engrg.* **72**, 267-304.

SIMO, J.C.; FOX, D.D.; RIFAI, M.S. [1989]: On a stress resultant geometrically exact shell model, Part II: The linear theory; computational aspects, *Comput. Methods Appl. Mech. Engrg.* **73**, 53-92.

SIMO, J.C.; FOX, D.D.; RIFAI, M.S. [1990a]: On a stress resultant geometrically exact model, Part III: Computational aspects of the nonlinear theory, *Comput. Methods Appl. Mech. Engrg.* **79**, 21-70

SIMO, J.C.; FOX, D.D.; RIFAI, M.S. [1990b]: On a stress resultant geometrically exact model, Part IV: Variable thickness with through-the-thickness stretching, *Comput. Methods Appl. Mech. Engrg.* **81**, 91-126.

SLICARU, S. [1997]: On the ellipticity of the middle surface of a shell and its application to the asymptotic analysis of "membrane shells", *J. Elasticity* **46**, 33-42.

SLICARU, S. [1998]: *Quelques Résultats dans la Théorie des Coques Linéairement Elastiques à Surface Moyenne Uniformément Elliptique ou Compacte Sans Bord*, Doctoral Dissertation, Université Pierre et Marie Curie, Paris.

SOKOLNIKOFF, I.S. [1956]: *Mathematical Theory of Elasticity*, McGraw-Hill.

SOKOLOWSKI, J. [1997]: Displacement derivatives in shape optimization of thin shells, *Contemporary Math.* **209**, 247-266.

SOKOLOWSKI, J.; ZOLÉSIO, J.P. [1992]: *Introduction to Shape Optimization: Shape Sensitivity Analysis*, Springer-Verlag, Heidelberg.

SPIVAK, M. [1975]: *A Comprehensive Introduction to Differential Geometry, Volumes I to V*, Publish or Perish, Boston.

SRUBSHCHIK, L.S. [1968]: Nonstiffness of a nonshallow spherical dome, *J. Appl. Math. Mech.* **32**, 435-445.

SRUBSHCHIK, L.S. [1972]: On the problem of non-stiffness in the nonlinear theory of shallow shells, *Izv. Akad. Nauk SSSR, Ser. Math.*, **36**, 890-909 (in Russian).

SRUBSHCHIK, L.S. [1980]: Precritical equilibrium of a thin shallow shell of revolution and its stability, *J. Appl. Math. Mech.* **44**, 229-235.

STEIN, E. [1970]: *Singular Integrals and Differentiability Properties of Functions*, Princeton University Press.

STOKER, J.J. [1968]: *Nonlinear Elasticity*, Gordon and Breach, New York.

STOKER, J.J. [1969]: *Differential Geometry*, John Wiley, New York.

STOLARSKI, H.; BELYTSCHKO, T. [1982]: Membrane locking and reduced integration for curved elements, *J. Appl. Mech.* **49**, 172-176.

STOLARSKI, H.; BELYTSCHKO, T.; LEE, S.H. [1995]: *A Review of Shell Finite Elements and Corotational Theories*, in Computational Mechanics Advances, Vol. 2, 125-212.

STRUIK, D.J. [1961]: *Lectures on Classical Differential Geometry, Second Edition*, Addison-Wesley, Reading.

STRUWE, M. [1990]: *Variational Methods*, Springer-Verlag, Berlin.

SURI, M. [1997]: A reduced constraint *hp* finite element method for shell problems, *Math. Comp.* **66**, 15-29.

SURI, M.; BABUŠKA, I.; SCHWAB, C. [1995]: Locking effects in the finite element approximation of plate models, *Math. Comp.* **64**, 461-482.

SZABO, B.A.; SAHRMANN, G.J. [1988]: Hierarchic plate and shell models based on p-extension, *Internat. J. Numer. Methods Engrg.* **26**, 1855-1881.

SZERI, A.J. [1990]: On the everted states of spherical and cylindrical shells, *Quart. Appl. Math.* **47**, 49-58.

TARTAR, L. [1978]: *Topics in Nonlinear Analysis*, Publications Mathématiques d'Orsay No. 78.13, Université de Paris-Sud, Orsay.

TELEGA, J.J.; LEWIŃSKI, T. [1998a]: Homogenization of linear elastic shells: Γ-convergence and duality. Part I. Formulation of the problem and the effective model, *Bull. Polish Acad. Sci., Technical Sci.*, **46**, 1-9.

TELEGA, J.J.; LEWIŃSKI, T. [1998b]: Homogenization of linear elastic shells: Γ-convergence and duality. Part II. Dual homogenization, *Bull. Polish Acad. Sci., Technical Sci.*, **46**, 11-21.

THOMAS, T.Y. [1934]: Systems of total differential equations defined over simply connected domains, *Annals Math.* **35**, 730-734.

TIMOSHENKO, S.P. [1951]: *Theory of Elasticity*, McGraw-Hill, New York.

TIMOSHENKO, S.P.; WOINOWSKY-KRIEGER, S. [1970]: *Theory of Plates and Shells*, McGraw-Hill, New York.

TRABUCHO, L.; VIAÑO, J.M. [1987]: Derivation of generalized models for linear elastic beams by asymptotic expansion methods, in *Applications of Multiple Scalings in Mechanics* (P.G. CIARLET & E. SANCHEZ-PALENCIA, Editors), pp. 302-315, Masson, Paris.

TRABUCHO, L.; VIAÑO, J.M. [1996]: Mathematical modelling of rods, in *Handbook of Numerical Analysis, Vol. IV* (P.G. CIARLET & J.L. LIONS, Editors), pp. 487-974, North Holland, Amsterdam.

TRUESDELL, C. [1978]: Comments on rational continuum mechanics, in *Contemporary Developments in Continuum Mechanics and Partial Differential Equations* (G.M. DE LAPENHA & L.A. MEDEIROS, Editors), pp. 495-603, North-Holland, Amsterdam.

TRUESDELL, C. [1983]: The influence of elasticity on analysis: The classic heritage, *Bull. Amer. Math. Soc.* **9**, 293-310.

TRUESDELL, C [1991]: *A First Course in Rational Continuum Mechanics, Vol. 1, Second Edition*, Academic Press, New York.

TRUESDELL, C.; NOLL, W. [1965]: The non-linear field theories of mechanics, in *Handbuch der Physik, Vol. III/3* (S. FLÜGGE, Editor), pp. 1-602, Springer-Verlag, Berlin.

TRUESDELL, C.; TOUPIN, R.A. [1960]: The classical field theories, in *Handbuch der Physik, Vol. III/1* (S. FLÜGGE, Editor), pp. 226-793, Springer-Verlag, Berlin.

VALENT, T. [1988]: *Boundary Value Problems of Finite Elasticity*, Springer Tracts in Natural Philosophy, Vol. 31, Springer-Verlag, Berlin.

VALENTE, V. [1997]: Relaxed exact spectral controllability of membrane shells, *J. Math. Pures Appl.* **76**, 551-562.

VALID, R. [1995]: *The Nonlinear Theory of Shells through Variational Principles: From Elementary Algebra to Differential Geometry*, John Wiley, New York.

VALIRON, G. [1950]: *Equations Fonctionnelles – Applications*, Masson, Paris.

VASSILIEV, D.G. [1987]: Asymptotics of the spectrum of a boundary value problem, *Trans. Moscow Math. Soc.* **49**, 173-245.

VASSILIEV, D.G. [1990]: Two-term asymptotics of the spectrum of natural frequencies in a thin elastic shell, *Soviet Mat. Doklady* **41**, 108-112.

VEIGA, M.F. [1995]: Asymptotic method applied to a beam with a variable cross section, in *Asymptotic Methods for Elastic Structures* (P.G. CIARLET, L. TRABUCHO & J.M. VIAÑO, Editors), pp. 237-254, de Gruyter, Berlin.

VEKUA, I.N. [1962]: *Generalized Analytic Functions*, Pergamon, New York.

VEKUA, I.N. [1965]: Theory of thin shallow shells of variable thickness, Acad. Nauk Gruzin. S.S.R. Trudy Tbilissi Mat. Inst. Razmadze **30**, 3-103.

VEKUA, I.N. [1986]: *Shell Theory: General Methods of Construction*, John Wiley, New York (English translation from Russian).

VILLAGGIO, P. [1977]: *Qualitative Methods in Elasticity*, Noorhoff, Leyden.

VILLAGGIO, P. [1997]: *Mathematical Models for Elastic Structures*, Cambridge University Press.

VINSON, J.P. [1974]: *Structural Mechanics: The Behavior of Plates and Shells*, Wiley, New York.

VLASOV, V.Z. [1944]: The basic differential equations in the general theory of elastic shells, *Prikl. Mat. Mek.* **8**, 109-140.

VLASOV, V.Z. [1958]: *Allgemeine Schalentheorie und ihre Anwendungen in der Technik*, Akademie Verlag, Berlin.

VOROVICH, I.I. [1969]: The unique solution of the boundary problems of the nonlinear theory of shells and the problem of stiffness of shells, *Proc. Inst. Appl. Math. Tbilisi Univ.* **2**, 49-55 (in Russian).

VOROVICH, I.I. [1999]: *Nonlinear Theory of Shallow Shells*, Springer-Verlag, Heidelberg (English translation of a revised and expanded manuscript of *Mathematical Problems in the Nonlinear Theory of Shallow Shells*, Published in Russian by Nauka, 1989).

WANG, C.C.; TRUESDELL, C. [1973]: *Introduction to Rational Elasticity*, Noordhoff, Groningen.

WASHIZU, K. [1975]: *Variational Methods in Elasticity and Plasticity*, Pergamon, Oxford.

XIAO, Li-ming [1998]: Asymptotic analysis of dynamic problems for linearly elastic shells – Justification of equations for dynamic membrane shells, *Asymptotic Anal.* **17**, 121-134.

XIAO, Li-ming [1999a]: Asymptotic analysis of dynamic problems for linearly elastic shells. – Justification of the dynamic flexural shell equations, to appear.

XIAO, Li-ming [1999b]: Asymptotic analysis of dynamic problems for linearly elastic shells. – Justification of the dynamic Koiter shell equations, to appear.

YOSIDA, K. [1966]: *Functional Analysis*, Springer-Verlag, Berlin.

ZERNER, M. [1994]: An asymptotically optimal finite element scheme for the arch problem, *Numer. Math.* **69**, 117-123.

INDEX

a priori assumption of a geometrical or mechanical nature : 336, 363, 546
admissible :
— applied force : 265, 292, 293, 298, 351
— displacement : 303
— inextensional deformation : 530
— inextensional displacement : 439, 503
— linearized inextensional displacement : 303
analytic functions: 121, 125
anisotropic elastic material : 162, 178, 234, 326, 360, 366, 550
Ansatz of the method of formal asymptotic expansions : 164, 413
applied body force : 7, 25, 386
assumptions on — s : 152, 162, 164, 179, 180, 199, 229, 232, 233, 247, 265, 305, 322, 326, 414, 442, 443, 458, 507
applied force (admissible —) : 265, 292, 293, 298, 351
applied surface force : 7, 26, 386
assumptions on — s : 152, 162, 164, 179, 180, 199, 229, 247, 265, 305, 322, 414, 442, 443, 507
area element : 19
area element on a surface : 70
assumptions on the data : See "applied body force", "applied surface force", "Lamé constants", "shallow shell"
justification of — : 162, 425, 443, 476, 507
asymptotic expansions (method of formal —) : 164, 413
asymptotic line : 83, 295
benchmark problem : 361
bending :
badly-inhibited — : 292
not inhibited — : 325
well-inhibited — : 232
bending in shells : See "flexural-dominated behavior", "flexural theory", "linearly elastic flexural shell", "nonlinearly elastic flexural shell"
bending moment : 344
body force : See "applied body force"
boundary condition :
— of place : 8, 32, 387
— of pressure : 455
— of traction : 388
complementing — : 123, 236, 237, 373
two-dimensional — s of clamping : 101, 116, 304, 305, 306, 322, 344
two-dimensional — s of simple support : 115, 134, 339, 439, 451, 530
two-dimensional — s of strong clamping : 503, 504, 524, 525, 530, 548
two-dimensional — s of weak simple support : 234

boundary layer : 351, 362, 368, 371, 374

BUDIANSKY-SANDERS linearized change of curvature tensor : 367

BUDIANSKY-SANDERS shell theory :
- linear — : 367
- nonlinear — : 551

canonical extension : 144

CARTESIAN components : 6
- — of the displacement field : 7, 387, 459, 463

CARTESIAN coordinates : 6, 67
- GREEN-ST VENANT strain tensor in — : 388
- KORN's inequality in — : See "Korn's inequality"
- linearized change of metric tensor in — : 9
- linearized elasticity in — : 10
- linearized strain tensor in — : 9
- linearized strains in — : 9, 36
- nonlinear elasticity in — : 391
- three-dimensional elasticity tensor in — : 11

CARTESIAN frame: 6

CAUCHY-GREEN strain tensor (right —) : 9

change of curvature tensor on a surface : 508, 519, 524, 548

change of curvature tensor on a surface (linearized —) : 93, 174, 181, 308, 322, 338, 344
- BUDIANSKY-SANDERS — : 367
- NAGHDI'S — : 366

change of metric tensor in a three-dimensional domain :
- — in CARTESIAN coordinates : 388
- — in curvilinear coordinates : 398

change of metric tensor in a three-dimensional domain (linearized —) :
- — in CARTESIAN coordinates : 9
- — in curvilinear coordinates : 36

change of metric tensor on a surface : 437, 450, 524, 548

change of metric tensor on a surface (linearized —) : 91, 170, 181, 210, 228, 308, 321, 338, 344, 366

CHRISTOFFEL symbols : 29, 33
- — of the first kind : 55, 60
- — of the second kind : 55, 60

CHRISTOFFEL symbols on a surface : 88, 136
- — of the first kind : 130
- — of the second kind : 130

clamping :
- two-dimensional boundary conditions of — : 101, 116, 304, 305, 306, 322, 344
- two-dimensional boundary conditions of strong — : 503, 504, 524, 525, 530, 548

CODAZZI-MAINARDI identities : 131, 133

coerciveness of a functional : 431, 452, 466, 538

complementing boundary condition : 123, 236, 237, 373

composite material (shell made with a —) : 368, 371

cone : 296, 305, 329, 505, 540

constitutive equation :
three-dimensional — : 389
two-dimensional — : 229, 451
continuation theorem (unique —) : 125, 126, 294
contravariant basis : 21
contravariant basis on a surface : 86
— of the tangent plane : 72
contravariant components :
— of a vector : 56
— of the applied body force density : 25
— of the applied surface force density : 26
— of the first PIOLA-KIRCHHOFF stress tensor : 403
— of the linearized stress tensor : 38
— of the metric tensor : 20, 57
— of the metric tensor of a surface : 72
— of the resultant stress tensor : 224
— of the second PIOLA-KIRCHHOFF stress tensor : 402
— of the three-dimensional elasticity tensor : 32
— of the two-dimensional elasticity tensor of a shell : 101
controllability of shells : 234, 361, 371
convergence :
— of displacements : 183, 209, 231, 233, 268, 291, 308, 324, 370, 461
— of stresses : 234, 237, 326, 331
Γ- — : 233, 326, 459
coordinate line : 17
— on a surface : 68
coordinates (see also "CARTESIAN coordinates" and "curvilinear coordinates") :
cylindrical — : 15
spherical — : 15, 67
stereographic — : 67
COSSERAT surface (one director —) : 363, 549
covariant basis : 17
— of the tangent plane : 69
— on a surface : 86
covariant components :
— of a vector : 25, 56
— of the change of metric tensor : 398
— of the curvature tensor : 81
— of the displacement field : 25
— of the displacement field of a surface : 86
— of the GREEN-ST VENANT strain tensor : 397
— of the linearized change of curvature tensor : 93
— of the linearized change of metric tensor : 36
— of the linearized change of metric tensor on a surface : 91
— of the linearized strain tensor : 36
— of the metric tensor : 18, 57
— of the metric tensor of a surface : 69
covariant derivative :
— of a tensor field : 38, 94, 133, 224, 402
— of a vector field : 30, 33, 451
— of a vector field on a surface : 88
second order — : 94, 345

curvature (see also "curvature tensor on a surface") :
center of — : 74, 77
— of a planar curve : 75, 76
GAUSSIAN — : 82, 83, 84, 121, 133
line of — : 82
mean — : 82
principal — : 82
principal radius of — : 82
radius of — : 74
total — : 82

curvature tensor on a surface :
covariant components of the — : 81
mixed components of the — : 81
RIEMANN — : 136

curvilinear coordinates : 15
GREEN-ST VENANT strain tensor in — : 397
KORN's inequality in — : 44, 48, 53, 59
linearized change of metric tensor in — : 36
linearized elasticity in — : 37
linearized strain tensor in — : 36
linearized strains in — : 36
natural — for a shell : 14, 145
nonlinear elasticity in — : 392
three-dimensional elasticity tensor in — : 32

curvilinear coordinates on a surface : 65

cylinder : 296, 328, 440, 465, 504, 539
circular — : 66, 191

cylindrical coordinates : 15

deformation: 9, 387, 456
admissible inextensional — : 530
large — nonlinear flexural shell theory : 525
large — nonlinear membrane shell theory : 452, 463
scaled — : 458

deformation gradient : 10

deformed configuration : 9, 287

director field on a surface : 363

developable surface : 83

développée : 75

director shell theory :
multi-— : 549
one-— : 363, 549

displacement :
admissible — : 303, 439, 503
CARTESIAN components of the — : 7, 387, 459
convergence of the — s : 183, 209, 231, 233, 268, 291, 308, 324, 370, 461
covariant components of the — : 25
covariant components of the — of a surface : 86
— field in three-dimensional elasticity : 7
— field of a surface : 86
error estimates for — s : 233, 350, 351, 358, 360
inextensional — : 439, 503

limit — s : 226, 232, 289, 319, 447, 462, 522
linearized inextensional — : 162
normal — of the middle surface : 232, 325
scaled — s : 152, 199, 246, 303, 408, 458
tangential — of the middle surface : 232, 325
displacement gradient : 10, 389
displacement-traction problem : 390
linearized — : 10
domain in $\mathbb{R}^n$: 6, 40
DONNELL-MUSHTARI-VLASOV nonlinear shallow shell theory : 555
edge : See "fold"
eigenvalue problems for shells : 234, 326, 358, 371
unilateral — : 555
elastic material :
anisotropic — : 162, 178, 234, 326, 360, 366
homogeneous — : 6, 388
isotropic — : 6, 388
linearly — : See "linearized elasticity"
nonhomogeneous — : 162, 178, 234, 326, 360, 366
nonlinearly — : 389, 392, 454
soft — : 464
ST VENANT-KIRCHHOFF — : 389, 392, 430, 455
elasticity tensor :
three-dimensional — in CARTESIAN coordinates : 11, 392
three-dimensional — in curvilinear coordinates : 32, 57
two-dimensional — of a shell : 101, 117, 228, 291, 322, 338, 344, 365, 450, 527
elastodynamics for a shell : See "time-dependent equations"
ellipsoid : 121, 197, 297
elliptic membrane shell (linearly elastic —) : See "linearly elastic elliptic membrane shell"
elliptic middle surface (shells with —) : 120, 196, 293
elliptic point : 83
elliptic surface : 120, 122, 126, 130, 196
elliptic system :
strongly — : 236, 373
uniformly — : 123, 236, 237, 373
energy : See also "three-dimensional energy", "two-dimensional energy", "two-dimensional strain energy"
existence of a minimum : 224, 228, 290, 321, 339, 340, 391, 462, 530, 551
minimizer of an — : See "minimization of a functional"
stored — function : 391, 392
strain — : See "two-dimensional strain energy"
equations of equilibrium :
— in CARTESIAN coordinates : 387
— in curvilinear coordinates : 403
linearized — in CARTESIAN coordinates : 12
linearized — in curvilinear coordinates : 40
two-dimensional — : 229, 451
error estimates for displacements : 233, 350, 351, 358, 360

EUCLIDEAN space :
three-dimensional affine — : 5
three-dimensional — : 5
EULER characteristic of a surface : 83
eversion of shells : 442, 542, 554
existence of solutions : 52, 101, 111, 115, 126, 223, 317, 338, 340, 366, 367, 376, 392, 452, 453, 529, 551, 554
face of a shell :
lateral — : 146
lower — : 148
upper — : 148
flexural-dominated behavior : 361
flexural shell (linearly elastic —) : See "linearly elastic flexural shell"
flexural shell (nonlinearly elastic —) : See "nonlinearly elastic flexural shell"
flexural theory : See "shell theory"
fold : 76, 295, 296, 297
force : See "applied body force", "applied force", "applied surface force"
formal asymptotic expansion : 164, 414
leading term in a — : 164, 414
method of — s : 164, 413
term of order q in a — : 141
frame-indifferent flexural theory : 525, 541
frame-indifferent membrane theory : 452, 463, 466
functional (see also "energy" and "minimization of a functional") :
coerciveness of a — : 431, 452, 466, 538
— as a Γ-limit : 460
stationary point of a — : 399, 445, 448, 520
weak lower semi-continuity of a — : 431, 452, 536
fundamental form of a surface :
first — : 69, 72
second — : 81
third — : 99
Γ-convergence : 233, 326, 459
Γ-limit : 460
GAUSS :
formula of — : 88
Theorema egregium of — : 131, 136
GAUSS-BONNET theorem : 83, 133
GAUSSIAN curvature : 82, 83, 84, 121, 133
generalized membrane shell (linearly elastic —) : See "linearly elastic generalized membrane shell"
genus of a surface : 83
geometrically exact shell theory : 549
GREEN-ST VENANT strain tensor :
— in CARTESIAN coordinates : 388
— in curvilinear coordinates : 397
hierarchic shell theory : 368
HILBERT uniqueness method (HUM) : 361

HOLMGREN's uniqueness theorem : 124

homogeneous elastic material : 6, 388

homogenization of shell equations : 234, 360

hyperbolic middle surface (shells with —) : 295, 362

hyperbolic paraboloid : 132

hyperbolic point : 83

hyperboloid of revolution : 294

hyperelastic material : 391, 392, 454

implicit function theorem : 144, 392, 453, 555

inextensional deformation : 530

inextensional displacement : 439, 503
    linearized — : 162, 303

infinitesimal rigid displacement : 46
    — on a surface : 109, 115

infinitesimal rigid displacement lemma : 49
    — on a general surface : 109, 366, 377
    — on a general surface with little regularity : 114
    — on an elliptic surface : 122, 295

infinitesimal rigidity of a surface : 297

infinitesimal rotation field : 133

interior layer : 362

internal constraints (method of —) : 368, 550

intrinsic shell theory :
    linear — : 368
    nonlinear — : 551

isometric surfaces : 85, 439, 441, 531

isotropic elastic material : 6, 388

junctions between shells : 360, 454

KIRCHHOFF-LOVE assumption : 336, 372, 546, 550
    linearized — : 336, 360, 368, 372

KIRCHHOFF-LOVE plate theory : 227, 320, 328

KOITER energy : 344, 548

KOITER strain energy : 344, 548

KOITER's linear shell equations : 100, 111, 115, 337, 339, 341, 344, 360, 373, 375
    admissible applied forces for — : 351
    eigenvalue problem for — : 358
    — for shells whose middle surface has little regularity : 115, 339, 362
    time-dependent — : 347, 361

KOITER's nonlinear shallow shell equations : 553

KOITER's nonlinear shell equations : 547

KORN's inequality :
    — in CARTESIAN coordinates : 49, 113, 207, 217, 253, 257
    — in curvilinear coordinates : 44, 48, 53, 59

KORN's type (inequality of —) :
— on a general surface : 103, 111, 118, 134, 318, 339, 356, 377, 551, 554
— on a general surface with little regularity : 115, 340
— on an elliptic surface : 126, 197, 208, 216, 225, 261, 349
three-dimensional — for a family of linearly elastic elliptic membrane shells : 205, 212, 220, 221
three-dimensional — for a family of linearly elastic shells : 259, 269, 271, 310

LAGRANGE multiplier : 319, 539, 550

LAMÉ constants : 7, 389
assumptions on the — : 152, 162, 199, 229, 247, 303, 322, 442, 443, 507

laminated shell : 368, 550

large deformation, or large displacement, flexural theory : 525

large deformation, or large displacement, membrane theory : 452, 463

lateral face of a shell : 146

LAX-MILGRAM lemma : 54, 101, 117, 192, 225, 278, 285, 288, 319, 339

layer :
boundary — : 351, 362, 368, 371, 374
interior — : 362

length element : 20

length element on a surface : 70

limit displacements : 226, 232, 289, 319, 447, 462, 522

limit stresses : 234, 237, 326, 331, 452, 546

line of curvature : 82

linear shell theory : See "shell theory"

linearization trick : 415

linearized change of curvature tensor : 93, 174, 181, 308, 322, 338, 344, 526
BUDIANSKY-SANDERS — : 367
NAGHDI'S — : 366

linearized change of metric tensor in a three-dimensional domain :
— in CARTESIAN coordinates : 9, 36
— in curvilinear coordinates : 36

linearized change of metric tensor on a surface : 91, 170, 181, 210, 228, 308, 321, 338, 344, 366, 452, 526

linearized elasticity (three-dimensional —) :
— in CARTESIAN coordinates : 8, 10, 455
— in curvilinear coordinates : 32, 38, 52

linearized inextensional displacement : 162, 303

linearized KIRCHHOFF-LOVE assumption : 336, 360, 368, 372

linearized rotation field : 364, 376

linearized strain (see also "change of metric") :
— in CARTESIAN coordinates : 9
— in curvilinear coordinates : 36
— tensor : 9
scaled — : 153

linearized stress
— in CARTESIAN coordinates : 11
— in curvilinear coordinates : 38
linearized transverse shear strain tensor : 366
linearly elastic elliptic membrane shell :
definition of a — : 196
example of a — : 197
KOITER's equations for a — : 347
two-dimensional constitutive equations for a — : 229
two-dimensional energy of a — : 224, 229
two-dimensional equations of a — : 117, 126, 210, 223, 227, 359
two-dimensional equilibrium equations for a — : 229
two-dimensional strain energy of a — : 229
linearly elastic flexural shell : 111, 541
definition of a — : 199, 302
example of a — : 303, 304, 305, 328, 329, 330
KOITER's equations for a — : 355
two-dimensional energy of a — : 322
two-dimensional equations of a — : 101, 309, 318, 320, 359
two-dimensional strain energy of a — : 322
linearly elastic generalized membrane shell :
admissible applied forces for a — : 265, 292, 293, 298
definition of a — : 245
example of a — : 293, 294, 295, 296, 297
— of the first kind : 262, 268, 293
— of the second kind : 262, 283, 293
KOITER's equations for a — : 351
two-dimensional energy of a — : 290
two-dimensional equations of a — : 268, 283, 359
two-dimensional strain energy of a — : 290
linearly elastic material : See "linearized elasticity"
linearly elastic membrane shell : 199, 246, 466
linearly elastic shallow shell :
definition of a — : 369, 371
two-dimensional energy of a — : 370
LIONS (lemma of J.L. —) : 42, 105, 119, 218, 256, 279, 376
locking:
— phenomenon: 362, 366
membrane — : 362, 366, 368
lower face of a shell : 148
manifold : 439, 472, 502, 520, 526, 529, 542, 543
tangent space to a — : 472, 503, 520
MARGUERRE-VON KÁRMÁN equations in curvilinear coordinates : 552
mean curvature : 82
membrane-dominated behavior : 361
membrane locking : 362, 366
membrane shell (linearly elastic —): See "linearly elastic elliptic membrane shell", "linearly elastic generalized membrane shell", "shell theory"
membrane shell (nonlinearly elastic —) : See "nonlinearly elastic membrane shell"
membrane theory : See "planar membrane theory", "shell theory"

metric tensor : 55
contravariant components of the — : 19
covariant components of the — : 18
mixed components of the — : 57

metric tensor on a surface :
contravariant components of the — : 72
covariant components of the — : 69

middle surface of a shell : 143
— with folds : 76
— with little regularity : 76, 114, 115, 339, 362, 366
— with no boundary : 143, 234, 292, 297

minimization of a functional : 224, 228, 290, 321, 339, 340, 391, 446, 448, 452, 456, 462, 520, 523, 529, 553

mixed component :
— of the curvature tensor on a surface : 81
— of the metric tensor : 57

multi-director shell theory : 549

multi-layered shell : 368, 371, 454, 550

multiplicity of solutions : 530

NAGHDI's linear shell equations : 365, 376
— for shells whose middle surface has little regularity : 366

NAGHDI's linearized change of curvature tensor : 366

NAGHDI's nonlinear shell theory : 549

natural curvilinear coordinates for a shell : 14, 145

natural state : 6, 388

nonhomogeneous elastic material (shell made with a —) : 162, 178, 234, 326, 360, 366

nonlinear elasticity (three-dimensional —) :
— in CARTESIAN coordinates : 391
— in curvilinear coordinates : 392

nonlinear planar membrane theory : 451, 453, 454, 460

nonlinear shell theory : See "shell theory"

nonlinearly elastic flexural shell :
definition of a — : 502
example of a — : 504, 505, 506, 539, 540
two-dimensional energy of a — : 524
two-dimensional equations of a — : 518, 522, 523
two-dimensional strain energy of a — : 524

nonlinearly elastic material : 389, 392, 454

nonlinearly elastic membrane shell (theory derived by formal analysis) :
definition of a — : 439
example of a — : 440, 441, 465
two-dimensional constitutive equations for a — : 451
two-dimensional energy of a — : 445, 446, 450
two-dimensional equations of a — : 444, 448, 450
two-dimensional equilibrium equations for a — : 451
two-dimensional strain energy of a — : 450

nonlinearly elastic membrane shell (theory derived by Γ-convergence) :
stored energy function of a — : 463
two-dimensional energy of a — : 463
nonlinearly elastic shallow shell:
definition of a — : 552, 553, 555
two-dimensional energy of a — : 553, 554
nonuniqueness of solutions : 530
NOVOZHILOV's shell theory : 371
numerical approximation of shell problems : 362, 363, 368
obstacle (shell lying on an —) : 555
one-director COSSERAT surface : 363, 549
one-director shell theory : 363, 549
parabolic middle surface (shells with —) : 296
parabolic point : 83
PIOLA-KIRCHHOFF stress tensor (first —): 388, 403
PIOLA-KIRCHHOFF stress tensor (second —) : 387
planar membrane theory (nonlinear —) : 451, 453, 454, 460
planar point : 82, 132
plate : 306, 329, 465, 506
shell becoming a — : 192, 234, 326, 358, 375
plate theory : 152, 326, 358, 370
KIRCHHOFF-LOVE — : 227, 320, 328
REISSNER-MINDLIN — : 363, 549
point on a surface :
elliptic — : 83
hyperbolic — : 83
parabolic — : 83
planar — : 82, 132
umbilical — : 82, 133
POISSON ratio : 389
polyconvex stored energy function : 392
pressure load : 455
principal curvature : 82
principal direction : 82
principal radius of curvature : 82
principle of virtual work :
linearized — in CARTESIAN coordinates : 11
linearized — in curvilinear coordinates : 40
— in CARTESIAN coordinates : 388
— in curvilinear coordinates : 403
quasiconvex :
— envelope : 459, 462
— function : 459
reference configuration : 6, 386
— of a shell : 12, 143
regularity of solutions : 60, 122, 235, 341, 373
REISSNER-MINDLIN plate theory : 363, 549
RIEMANN curvature tensor on a surface : 136

rigid displacement lemma : See "infinitesimal rigid displacement", "infinitesimal rigid displacement lemma"
rigidification of a surface : 297
rigidity of a surface : 442
    infinitesimal — : 297
rotation field :
    infinitesimal — : 133
    linearized — : 364, 376
scaled strain : See "strain"
scaled three-dimensional elasticity tensor of a shell : 154, 410
scaled three-dimensional equations :
    — for a linearly elastic shell : 154
    — for a nonlinearly elastic shell : 410
scaled two-dimensional elasticity tensor of a shell : 158, 170, 181, 210, 308, 430
scaled unknown : See "scalings of displacements"
scaled vector field : 152, 408
scalings of deformations : 458
scalings of displacements : 152, 199, 246, 303, 408, 458
    justification of — : 162, 415
scalings of stresses : 237
sensitive variational problem : 293, 295, 296
shallow shell :
    assumption on the middle surface of a — : 369, 371, 552
    linearly elastic — : 369
    nonlinearly elastic — : 552
shape optimization for shells : 361
shear :
    linearized transverse — tensor : 366
shell (see also "shallow shell") : 12, 143
    boundary layer in a — : 351, 362
    clamped — : 101, 116, 146, 339
    controllability of — s : 234, 361
    eigenvalue problem for — s : 234, 326, 358, 371, 555
    eversion of — s : 442, 542, 554
    homogenization of — equations : 234, 360
    junctions between — s : 360, 454
    laminated — : 368, 550
    lateral face of a — : 146
    linearly elastic elliptic membrane — : See "linearly elastic elliptic membrane shell"
    linearly elastic flexural — : See "linearly elastic flexural shell"
    linearly elastic generalized membrane — : See "linearly elastic generalized membrane shell"
    lower face of a — : 148
    middle surface of a — : 143
    multi-layered — : 368, 371, 454, 550
    natural curvilinear coordinates for a — : 14, 145
    nonlinearly elastic flexural — : See "nonlinearly elastic flexural shell"
    nonlinearly elastic membrane — : See "nonlinearly elastic membrane shell"

numerical approximation of — problems : 362, 363, 368
shape optimization for — s : 361
— becoming a plate : 192, 234, 326, 358, 375
— made with a composite material : 368, 371
— made with a nonhomogeneous material : 162, 234, 326, 360, 366
— made with a soft material : 464
— made with an anisotropic material : 162, 234, 326, 360, 366, 550
— whose middle surface has little regularity : 76, 114, 115, 339, 362, 366
— whose middle surface has no boundary : 143, 234, 292, 297
— with folds : 76, 295, 296, 297
— with variable thickness : 234, 238, 326, 330, 360
simply supported — : 115, 134, 339, 439, 451, 530
strongly clamped — : 503, 504, 525, 530, 548
thickness of a — : 143
upper face of a — : 148
weakly simply supported — : 234

shell theory :
COSSERAT — : 363, 549
DONNELL-MUSHTARI-VLASOV nonlinear shallow — : 555
first-order — : 550
frame-indifferent — : 452, 463, 466, 525, 541
geometrically exact — : 549
hierarchic — : 368
higher-order — : 550
intrinsic linear — : 368
intrinsic nonlinear — : 551
large deformation, or large displacement, — : 452, 463, 525
linear — based on the method of internal constraints : 368
linear BUDIANSKY-SANDERS — : 367
linear flexural — : 101, 162, 171, 180, 182, 199, 526
linear KOITER's — : 100, 115, 336, 346, 548
linear membrane — : 117, 126, 162, 170, 178, 182, 196, 199, 232, 246, 452, 466
linear NAGHDI's — : 363, 365
linear shallow — : 369
MARGUERRE-VON KÁRMÁN — : 552
multi-director — : 549
nonlinear BUDIANSKY-SANDERS — : 551
nonlinear flexural — : 502, 505, 525
nonlinear KOITER's shallow — : 553
nonlinear KOITER's — : 547
nonlinear membrane — : 439, 451, 463, 464, 505
nonlinear NAGHDI's — with directors : 549
nonlinear shallow — : 552
nonlinear — based on the method of internal constraints : 550
NOVOZHILOV — : 371
one-director — : 363, 549
— of higher-order : 550
— of the first order : 550
time-dependent — : 234, 326, 347, 550, 551

simple support :
two-dimensional boundary conditions of — : 115, 134, 339, 439, 451, 530
two-dimensional boundary conditions of weak — : 234

singular perturbation problem : 154

SOBOLEV spaces : 41

soft elastic material : 464

sphere : 67, 133, 135, 190, 441

spherical coordinates : 15, 67

ST VENANT-KIRCHHOFF material : 389, 392, 430
stored energy function of a — : 391, 455, 459, 464, 467

stationary point of a functional : 391, 448, 520, 523

stereographic coordinates : 67

stored energy function : 391, 392, 451, 455, 463

strain :
linearized — s in CARTESIAN coordinates : 9
linearized — s in curvilinear coordinates : 36
scaled linearized — : 153
scaled — : 410
— s in CARTESIAN coordinates : 388
— s in curvilinear coordinates : 398

strain energy : See "two-dimensional strain energy"

strain tensor :
CAUCHY-GREEN — : 9
linearized — in CARTESIAN coordinates : 9
GREEN-ST VENANT — in CARTESIAN coordinates : 388
GREEN-ST VENANT — in curvilinear coordinates : 397

stress :
convergence of — es : 234, 237, 326, 331
limit — es : 234, 237, 326, 331, 452, 546
linearized — es in CARTESIAN coordinates : 11
linearized — es in curvilinear coordinates : 38, 237
— es in CARTESIAN coordinates : 388
— es in curvilinear coordinates : 402

stress couple : 344, 345

stress resultant : 224, 229, 344, 345, 450

stress tensor :
first PIOLA-KIRCHHOFF — : 388
linearized — : 38
second PIOLA-KIRCHHOFF — : 387, 402

strongly elliptic system : 236

supplementary condition on L : 123, 236, 373

surface : 65
developable — : 83
elliptic — : 120, 293
hyperbolic — : 295
infinitesimal rigidity of a — : 297
isometric — s : 439, 441, 531
parabolic — : 296
rigidification of a — : 297
rigidity of a — : 442

surface force : See "applied surface force"

tensor :
BUDIANSKY-SANDERS linearized change of curvature — : 367
CAUCHY-GREEN strain — (right) : 9
change of curvature — : 508
change of metric — : 388, 398
change of metric — on a surface : 437
curvature — : 81
first PIOLA-KIRCHHOFF stress — : 388, 403
GREEN-ST VENANT strain — : 388
linearized change of curvature — : 93
linearized change of metric — : 9, 36
linearized change of metric — on a surface : 91
linearized strain — : 9, 36
linearized stress — : 38
linearized transverse shear strain — : 366
metric — : 18
metric — on a surface : 69
RIEMANN curvature — : 136
second PIOLA-KIRCHHOFF stress — : 387, 402
three-dimensional elasticity — : 11, 32, 392, 396
three-dimensional elasticity — of a shell : 148
two-dimensional elasticity — of a shell : 101, 117

Theorema egregium : 131, 136

thickness :
— shell with variable — : 234, 238, 326, 330, 360
— of a shell : 143

three-dimensional affine EUCLIDEAN space : 5

three-dimensional elasticity tensor :
scaled — : 154, 410
— in CARTESIAN coordinates : 11, 59, 392
— in curvilinear coordinates : 32, 50, 59, 182, 396

three-dimensional energy :
minimizer of a — : 392
— in CARTESIAN coordinates : 391
— in curvilinear coordinates : 396

three-dimensional equations for a linearly elastic shell : 146
scaled — : 154, 200, 247, 307

three-dimensional equations for a nonlinearly elastic shell : 406
scaled — : 410

three-dimensional EUCLIDEAN space : 5

three-dimensional linearized elasticity :
— in CARTESIAN coordinates : 8
— in curvilinear coordinates : 32

three-dimensional nonlinear elasticity :
— in CARTESIAN coordinates : 391
— in curvilinear coordinates : 401

time-dependent equations :
— for a linearly elastic shell : 234, 326, 347
— for a nonlinearly elastic shell : 550, 551

torus : 66, 133, 292

total curvature : 82

transverse variable : 145, 150
average with respect to the — : 201
two-dimensional boundary conditions :
— of clamping : 101, 116, 304, 305, 306, 322, 344
— of simple support : 115, 134, 439, 451, 530
— of strong clamping : 503, 504, 524, 525, 530, 548
— of weak simple support : 234
two-dimensional constitutive equation : 229, 451
two-dimensional elasticity tensor of a shell : 101, 117, 190, 228, 291, 322, 338, 344, 365, 450, 524
scaled — : 158, 170, 181, 210, 308, 430, 519
two-dimensional energy :
— of a linearly elastic elliptic membrane shell : 229
— of a linearly elastic flexural shell : 322
— of a linearly elastic generalized membrane shell : 290
— of a nonlinearly elastic flexural shell : 524, 530
— of a nonlinearly elastic membrane shell (theory derived by formal analysis) : 450
— of a nonlinearly elastic membrane shell (theory derived by Γ-convergence) : 463
two-dimensional equations :
— for a linearly elastic elliptic membrane shell : 117, 126, 210, 223, 227
— for a linearly elastic flexural shell : 101, 309, 318, 320, 321
— for a linearly elastic generalized membrane shell : 268, 283
— for a nonlinearly elastic flexural shell : 518, 530
— for a nonlinearly elastic membrane shell (theory derived by formal analysis) : 444, 450
— for a nonlinearly elastic membrane shell (theory derived by Γ-convergence) : 463
two-dimensional equations of equilibrium : 229, 451
two-dimensional KOITER energy :
— of a linearly elastic shell : 344
— of a nonlinearly elastic shell : 548
two-dimensional KOITER equations :
eigenvalue problem for the — : 358
time-dependent — : 347, 361
— for a linearly elastic shell : 100, 111, 115, 337, 339, 341, 344, 360, 373, 375
— for a nonlinearly elastic shell : 547
two-dimensional KOITER strain energy :
— of a linearly elastic shell : 344
— of a nonlinearly elastic shell : 548
two-dimensional strain energy (see also "two-dimensional KOITER strain energy") :
— of a linearly elastic elliptic membrane shell : 229, 344, 361
— of a linearly elastic flexural shell : 322, 344, 361
— of a linearly elastic generalized membrane shell : 290
— of a nonlinearly elastic flexural shell : 524, 548
— of a nonlinearly elastic membrane shell (theory derived by formal analysis) : 450, 548
— of a nonlinearly elastic membrane shell (theory derived by Γ-convergence) : 463

umbilical point : 82, 133
uniformly elliptic system : 123, 236, 237
uniqueness of solutions : 52, 101, 111, 115, 122, 124, 126, 223, 317, 338, 340, 366, 367, 376, 530
upper face of a shell : 148
volume element : 18
weak lower semi-continuity of a functional : 431, 452, 536
WEINGARTEN (formula of —) : 88
YOUNG modulus : 389